New and Future Developments in Microbial Biotechnology and Bioengineering

New and Future Developments in Microbial Biotechnology and Bioengineering

Microbes in Soil, Crop and Environmental Sustainability

Edited by

Jay Shankar Singh
Department of Environmental Microbiology, School for Environmental Sciences, Babasaheb Bhimrao Ambedkar (Central) University, Lucknow, India

ELSEVIER

Elsevier
Radarweg 29, PO Box 211, 1000 AE Amsterdam, Netherlands
The Boulevard, Langford Lane, Kidlington, Oxford OX5 1GB, United Kingdom
50 Hampshire Street, 5th Floor, Cambridge, MA 02139, United States

British Library Cataloguing-in-Publication Data
A catalogue record for this book is available from the British Library

Library of Congress Cataloging-in-Publication Data
A catalog record for this book is available from the Library of Congress

ISBN: 978-0-12-818258-1

For Information on all Elsevier publications
visit our website at https://www.elsevier.com/books-and-journals

Publisher: Susan Dennis
Acquisition Editor: Kostas Marinakis
Editorial Project Manager: Alexandra Romano
Production Project Manager: Nilesh Kumar Shah
Cover Designer: Greg Harris

Typeset by MPS Limited, Chennai, India

Contents

List of Contributors

Shimaila Ali, A&L Biologicals, Agroecological Research Services Centre, London, ON, Canada

Y.S. Amaresh, Department of Plant Pathology, College of Agriculture, UAS, Raichur, India

Anshu, CSIR-National Botanical Research Institute, Lucknow, India; Lucknow University, Lucknow, India

S. Avinash, Department of Plant Pathology, College of Agriculture, UAS, Raichur, India

Arpita Bhattacharya, Division of Plant Microbe Interactions, CSIR-National Botanical Research Institute, Lucknow, India

Utpal Bora, Department of Biosciences and Bioengineering, Indian Institute of Technology Guwahati, Guwahati, India; Centre for the Environment, Indian Institute of Technology Guwahati, Guwahati, India

Siddharth Boudh, Department of Environmental Microbiology, Babasaheb Bhimrao Ambedkar University, Lucknow, India

Siddhardha Busi, Department of Microbiology, School of Life Sciences, Pondicherry University, Kalapet, India

Avantika Chandra, Department of Environmental Science and Engineering, Indian Institute of Technology (ISM), Dhanbad, India

Antra Chatterjee, Center of Advanced Study in Botany, Institute of Science, Banaras Hindu University, Varanasi, India

Preeti Chaturvedi, Aquatic Toxicology Laboratory, Environmental Toxicology Division, CSIR—Indian Institute of Toxicology Research, Lucknow, India

Shruti Chaturvedi, Department of Microbial Technology and Nematology, CSIR—Central Institute of Medicinal and Aromatic Plants, Lucknow, India

Neha Chaurasia, Department of Biotechnology and Bioinformatics, North Eastern Hill University, Shillong, India

G. Chennappa, Department of Processing and Food Engineering, CAE, UAS, Raichur, India

R.L. Choudhary, ICAR-National Institute of Abiotic Stress Management, Baramati, India

N. Deepa, Department of Studies in Microbiology, University of Mysore, Mysore, India

Aline Meireles dos Santos, Department of Food Science and Technology, Federal University of Santa Maria, Santa Maria, Brazil

C.K. Dotaniya, College of Agriculture, SKRAU, Bikaner, India

M.L. Dotaniya, ICAR-Indian Institute of Soil Science, Bhopal, India

Jintu Dutta, Centre for the Environment, Indian Institute of Technology Guwahati, Guwahati, India

Touseef Fatima, Integral University, Lucknow, India; CSIR-National Botanical Research Institute, Lucknow, India

Avijit Ghosh, ICAR-Indian Grassland and Fodder Research Institute, Jhansi, India

Bernard R. Glick, Department of Biology, University of Waterloo, Waterloo, ON, Canada

Gobinath R, ICAR-Indian Institute of Rice Research, Rajendranagar, Hyderabad, India

Pratishtha Gupta, Laboratory of Applied Microbiology, Department of Environmental Science and Engineering, Indian Institute of Technology (ISM), Dhanbad, India

Rupali Gupta, Department of Microbial Technology and Nematology, CSIR—Central Institute of Medicinal and Aromatic Plants, Lucknow, India

Eduardo Jacob-Lopes, Department of Food Science and Technology, Federal University of Santa Maria, Santa Maria, Brazil

Ratna Katiyar, Lucknow University, Lucknow, India

Neha Khanna, Department of Agricultural Chemistry and Soil Science, Dr. B.R. Ambedkar University, Agra, India

Irina K. Kravchenko, Laboratory of Microbial Survival, Federal State Institution, Federal Research Centre Fundamentals of Biotechnology of the Russian Academy of Sciences, Moscow, Russian Federation

Rajesh Kumar, Rhizosphere Biology Laboratory, Department of Environmental Microbiology, Babasaheb Bhimrao Ambedkar University, Lucknow, India

Sanjeev Kumar, CSIR-National Botanical Research Institute, Lucknow, India; Academy of Scientific and Innovative Research (AcSIR), Ghaziabad, India

Vipin Kumar, Laboratory of Applied Microbiology, Department of Environmental Science and Engineering, Indian Institute of Technology (ISM), Dhanbad, India

Madan Kumar Mandal, Department of Biotechnology and Bioinformatics, North Eastern Hill University, Shillong, India

P.F. Mathad, Department of Processing and Food Engineering, CAE, UAS, Raichur, India

Himani Meena, Department of Microbiology, School of Life Sciences, Pondicherry University, Kalapet, India

M.D. Meena, ICAR-Directorate of Rapeseed-Mustard Research, Bharatpur, India

Shiv Singh Meena, Department of Soil Science, GBPUA&T, Pantnagar, India

V.S. Meena, ICAR-Vivekananda Parvatiya Krishi Anusandhan Sansthan, Almora, India

Ekaterina V. Menko, Laboratory of Microbial Survival, Federal State Institution, Federal Research Centre Fundamentals of Biotechnology of the Russian Academy of Sciences, Moscow, Russian Federation

Aradhana Mishra, Division of Plant Microbe Interactions, CSIR-National Botanical Research Institute, Lucknow, India

Nishtha Mishra, Division of Plant Microbe Interactions, CSIR-National Botanical Research Institute, Lucknow, India

M.K. Naik, Department of Plant Pathology, College of Agriculture, UAS, Raichur, India

Maitreyie Narayan, Department of Environmental Science, GBPUA&T, Pantnagar, India

Rakesh Pandey, Department of Microbial Technology and Nematology, CSIR—Central Institute of Medicinal and Aromatic Plants, Lucknow, India

Umesh Pankaj, Divion of Agronomy and Soil Science, CSIR-Central Institute of Medicinal and Aromatic Plants, Lucknow, India

Ranjan Paul, Division of Soil Resource Studies, ICAR-National Bureau of Soil Survey and Land Use Planning, Nagpur, India

Yongqiang Qin, Guangdong Province Key Laboratory of Microbial Signals and Disease Control, Guangdong Engineering Research Center for Litchi, Guangdong Engineering Research Center for Grass Science, College of Horticulture, South China Agricultural University, Guangzhou, P.R. China

Rupa Rani, Laboratory of Applied Microbiology, Department of Environmental Science and Engineering, Indian Institute of Technology (ISM), Dhanbad, India

Shatrupa Ray, Department of Mycology and Plant Pathology, Institute of Agricultural Sciences, Banaras Hindu University, Varanasi, India

Sonalika Sahoo, Division of Soil Resource Studies, ICAR-National Bureau of Soil Survey and Land Use Planning, Nagpur, India

Beenu Shastri, Rhizosphere Biology Laboratory, Department of Environmental Microbiology, Babasaheb Bhimrao Ambedkar University, Lucknow, India

Akanksha Singh, Department of Microbial Technology and Nematology, CSIR—Central Institute of Medicinal and Aromatic Plants, Lucknow, India

Geetu Singh, Divion of Agronomy and Soil Science, CSIR-Central Institute of Medicinal and Aromatic Plants, Lucknow, India

Harikesh Bahadur Singh, Department of Mycology and Plant Pathology, Institute of Agricultural Sciences, Banaras Hindu University, Varanasi, India

Jay Shankar Singh, Department of Environmental Microbiology, Babasaheb Bhimrao Ambedkar University, Lucknow, India

Jyoti Singh, Department of Mycology and Plant Pathology, Institute of Agricultural Sciences, Banaras Hindu University, Varanasi, India; Department of Botany, Institute of Sciences, Banaras Hindu University, Varanasi, India

Poonam C. Singh, CSIR-National Botanical Research Institute, Lucknow, India; Academy of Scientific and Innovative Research (AcSIR), Ghaziabad, India

Prachi Singh, Department of Mycology and Plant Pathology, Institute of Agricultural Sciences, Banaras Hindu University, Varanasi, India

Praveen Solanki, Department of Environmental Science, GBPUA&T, Pantnagar, India

M.Y. Sreenivasa, Department of Studies in Microbiology, University of Mysore, Mysore, India

R.K. Srivastava, Department of Environmental Science, GBPUA&T, Pantnagar, India

Sonal Srivastava, Academy of Scientific and Innovative Research (AcSIR), Ghaziabad, India; CSIR-National Botanical Research Institute, Lucknow, India

Subhi Srivastava, Lucknow University, Lucknow, India; CSIR-National Botanical Research Institute, Lucknow, India

Suchi Srivastava, CSIR-National Botanical Research Institute, Lucknow, India; Academy of Scientific and Innovative Research (AcSIR), Ghaziabad, India

M.P. Sudhakar, National Institute of Ocean Technology, Chennai, India

Ashmita Tandon, Academy of Scientific and Innovative Research (AcSIR), Ghaziabad, India; CSIR-National Botanical Research Institute, Lucknow, India

Ekaterina N. Tikhonova, Laboratory of Microbial Survival, Federal State Institution, Federal Research Centre Fundamentals of Biotechnology of the Russian Academy of Sciences, Moscow, Russian Federation

S. Udayakumar, Department of Agronomy, Horticulture, and Plant Science, South Dakota State University, Brookings, SD, United States

Nidoni Udaykumar, Department of Processing and Food Engineering, CAE, UAS, Raichur, India

Zeba Usmani, Laboratory of Applied Microbiology, Department of Environmental Science and Engineering, Indian Institute of Technology (ISM), Dhanbad, India

Rajesh Kumar Verma, Divion of Agronomy and Soil Science, CSIR-Central Institute of Medicinal and Aromatic Plants, Lucknow, India

M. Vidya, Centre for Nanotechnology, CAE, UAS, Raichur, India

Karem Rodrigues Vieira, Department of Food Science and Technology, Federal University of Santa Maria, Santa Maria, Brazil

Shashirekha Viswanaathan, Shri A.M.M. Murugappa Chettiar Research Centre, Chennai, India

Udit Yadav, CSIR-National Botanical Research Institute, Lucknow, India; Academy of Scientific and Innovative Research (AcSIR), Ghaziabad, India

Qing Yao, State Key Laboratory of Applied Microbiology Southern China, Guangdong Provincial Key Laboratory of Microbial Culture Collection and Application, Guangdong Institute of Microbiology, Guangzhou, P.R. China; Guangdong Province Key Laboratory of Microbial Signals and Disease Control, Guangdong Engineering Research Center for Litchi, Guangdong Engineering Research Center for Grass Science, College of Horticulture, South China Agricultural University, Guangzhou, P.R. China

Leila Queiroz Zepka, Department of Food Science and Technology, Federal University of Santa Maria, Santa Maria, Brazil

Yang Zhou, State Key Laboratory of Applied Microbiology Southern China, Guangdong Provincial Key Laboratory of Microbial Culture Collection and Application, Guangdong Institute of Microbiology, Guangzhou, P.R. China

Honghui Zhu, State Key Laboratory of Applied Microbiology Southern China, Guangdong Provincial Key Laboratory of Microbial Culture Collection and Application, Guangdong Institute of Microbiology, Guangzhou, P.R. China

Preface

The tremendous achievements in microbial technology during the past few decades necessitate a review of these achievements. In agriculture, plant–microbe associations are frequently manipulated to enhance the crop production, mainly via the pathways of increasing soil fertility, elevating the crop resistance to pathogens, and producing phytohormones. This book will highlight the developments and achievements of microbial resources and microbial biotechnology in the sustainable management of soil and agriculture productivity.

Soil microbial biodiversity enhances nutrient use efficiency of applied chemicals by P solubilization/mobilization/mineralization and decomposition of organic matter to release trapped nutrients for plant uptake. Microbial consortia are very much useful in degrading various species of xenobiotics and persistent organic pollutants. Waste recycling and management by efficient soil microbial communities have the largest economic benefit in terms of monetary value. This book reviews the exploitation of soil microbial agents and their services with respect to enhancing nutrient use efficiency, and improving and maintaining physical and chemical conditions of soil, which all together can provide economic and environmental benefit towards maintaining agricultural sustainability.

Major emphasis has been given on documenting and exploitation of soil microbial resources to improve the physicochemical and biological properties of soil. Various facets of microbial mediated soil health restoration strategies in sustainable management of degraded lands of different origin have been discussed in this very book. The editor and experts on microbiology consider that this is a favorable moment to strategically analyze the role of microbial services and resources for human wellbeing. The well acquainted experts in the subject were requested to contribute with the most appropriate and informative articles to solve the existing problems related to soil, agriculture, and environment. The various innovative expert opinions have been considered in this book entitled *New and Future Developments in Microbial Biotechnology and Bioengineering: Microbes in Soil, Crop and Environmental Sustainability*. I believe this book will be useful equally to students, teachers, and researchers in the disciplines of restoration ecology, agricultural and environmental microbiology, ecotoxicology, and environmental management.

Jay Shankar Singh
(Editor)

Chapter 1

Bacterial fertilizers for soybean plants: current status and future prospects

Ekaterina N. Tikhonova, Ekaterina V. Menko and Irina K. Kravchenko

Laboratory of Microbial Survival, Federal State Institution, Federal Research Centre Fundamentals of Biotechnology of the Russian Academy of Sciences, Moscow, Russian Federation

Chapter Outline

1.1 Introduction

Soy is one of the most important crops worldwide and soybean grains are important as protein meal and vegetable oil. The crop is grown on an estimated 6% of the world's arable land, and since the 1970s, the area in soybean production has had the highest increase compared with other major crops. Because of avoidance of environmental problems, problems with human health, and more crop integrated nutrient management, plant-beneficial living microbial cultures (biofertilizers) are supposed to be a safe alternative to chemical fertilizers (Singh et al., 2018; Tiwari et al., 2018; Singh and Gupta, 2018). A number of microorganisms are considered as challenging agents for agriculture to promote better nutrient uptake and availability for plant use (Singh, 2013; Singh, 2015; Kumar et al., 2018). Biofertilizers play an important role in increasing yield through the natural processes of nitrogen fixation, solubilizing phosphorus, and stimulating plant growth through the synthesis of growth-promoting substances, improving soil texture, pH, and other properties of soil (Tiwari and Singh, 2017; Vimal et al., 2018).

Macronutrients such as nitrogen, phosphorus, and potassium play a crucial role in plant growth and yield. Soybean nitrogen requirements are met in a complex manner, as this crop is capable of utilizing both soil N (mostly in the form of nitrate) and atmospheric N (through symbiotic nitrogen fixation). Some researchers suggested that N fertilization is not necessary for inoculated soybean (Simonis and Setatou, 1996), whereas others indicated that N fertilization is necessary to improve yield and quality of soybean depending on application rate (Shober and Taylor, 2014).

New and Future Developments in Microbial Biotechnology and Bioengineering. DOI: https://doi.org/10.1016/B978-0-12-818258-1.00001-7

Soybean depends on its symbionts *Bradyrhizobium japonicum* for effective growth and dry matter production. Phosphate solubilizing bacteria (PSB) improved nodulation, root and shoot biomass, straw and grain yield, and P and N uptake of the soya crop (Singh et al., 2010). It was shown that coinoculation of *Bradyrhizobium* and PSB significantly improved soybean growth and its yield components as compared with the sole application (Argaw, 2012). Significant increase in biomass, root and nodule number of soybean was also observed under single or binary inoculation with *Bradyrhizobium* and *Bacillus* (Atieno et al., 2012).

Application of a small amount of N at planting called as ''starter N'' was reported to be beneficial to improve early growth and yield of soybean in most cases (Gai et al., 2017). Plant–microorganism interactions are very complex and often have important consequences for the plants. Some bacteria that live in the rhizosphere (rhizobacteria) are able to influence nodule formation and biological nitrogen fixation of legumes; however, the mechanisms of these effects are not well understood. Usually discussed mechanisms include the release by bacteria of phytohormones (auxins, gibberellins, cytokinins, ethylene) (Egamberdieva et al., 2017) or flavonoids (Phillips and Tsai, 1992). We hypothesize that nitrogen-fixing rhizobacteria may apply the nitrogen to soybeans at the early stages of plant growth, when the symbiosis is developing and soil nitrogen supply is not high enough. To test this idea the diversity of diazotrophs in rhizosphere of soybean plant was investigated. This review is aimed at discussing factors that can modulate soya rhizosphere microbiome with focus on the contributions of N fixing bacteria toward sustainable agricultural development.

1.2 Current status of soybean study in Russia

Soybean (*Glycine max*) is one of the most important crops in the world. It is highly suitable for human and animal diets and is the source of 30% of the world's oil derived from processed crops. From 2010 to 2014 soya planted area has increased from 111 to 124 million ha, and its production quantity increased from 280 to 320 million tons. Currently the largest world producers are the United States (34%), Brazil (27%), and Argentina (17%) (FAO, 2017). Russia has abundant land resources and soybean is a regional crop that is distributed mainly in the Far East parts of the country (Belyshkina, 2013). Currently there are seven regions of the soybean cultivation in the Russian Federation, and three of them are located in the Far East District and four in the European part; mainly in North Caucasia and Central Chernozem Region, South European District (Fig. 1.1). The Amur Region of the Far East District has produced nearly 1/3 of the 3.34 million tons of soybean yield in 2016. Six major regions, including Amur Region, Primorye Territory, Belgorod Region, Krasnodar Territory, Kursk Region, and Jewish Autonomous Region, accounted for nearly 80% of the entire production of soybean (Sinegovskii et al., 2018).

FIGURE 1.1 Major soybean-producing regions in the Russian Federation (colored, gray in print version).
1, Amur Region; *2*, Primorye Territory; *3*, Belgorod Region; *4*, Krasnodar Territory; *5*, Kursk Region; *6*, Jewish Autonomous Region. *Data from courtesy www.ros-soya.su.*

The soybean cultivating area in Russia has increased significantly in recent years. In 2009 the planted area was 752 thousand ha; in 2015 it reached 2 million ha and grew by 8% in 2016. Although the total soybean production volumes in Russia are lower than that in Ukraine, the largest producer in Europe, Russia is the major globally producer of pure non-GM soybean globally. According to the Moscow branch of the USDA Russia will produce 3.87 million mt of soybeans in the 2018/19 marketing year.

According to the Web of Science database, from 1995 to 2016, more than 100 soybean-related articles were published from Russia. The top three institutions by article number were the Russian Academy of Sciences, Lomonosov Moscow State University, and Plekhanov Russian University of Economics. Most frequently articles discuss problems of biochemistry, molecular biology, chemistry, plant sciences, biotechnology, and applied microbiology. In the Far East regions, the major soybean research institutions are All-Russian Scientific Research Institute of Soybean, Primorskiy Institute of Agriculture, Far East Institute of Agriculture, and Far East State Agrarian University. The All-Russian Scientific Research Institute of Soybean, which is the center of the Russian soybean research, constitutes a strong soybean breeding system in Russia.

1.3 Microbial communities in soya rhizosphere

1.3.1 Rhizosphere, overview

The term *rhizosphere* was first proposed in 1904 by Lorentz Hiltner as the zone of soil surrounding the root, which is affected by the root (Sugiyama et al., 2014). It is rich in nutrient and microbe concentration and exhibits intense biological and chemical activities as compared with the bulk soil (Prashar et al., 2014). Rhizosphere is a complex system consisting of the microorganisms (soil inhabiting and endophytes), the soil, and the roots of plants. A number of narrow layers may be isolated in the rhizosphere with varying degrees of specificity compared with the bulk soil. Their thickness can range from a few millimeters to several centimeters; depending on soil conditions and the type of colonized plant that produces carbon-containing energy compounds, so-called rhizodeposits, including root exudates and root litter (Jones et al., 2009). Three zones have been described in the rhizosphere: the endorhizosphere, inside the apoplastic space between root cells; the rhizoplane, as the middle zone next to the root epidermal cells and mucilage; and the ectorhizosphere, which extends from the rhizoplane to the bulk soil (McNear, 2013). Generally regarded as a thin zone, rhizosphere holds a large volume that has no clear boundary line and varies with the plant, soil, and root structure (Hinsinger et al., 2005).

Most studies indicate an increased (10–100 times) number and activity of microorganisms in the rhizosphere zone compared with the bulk soil and this phenomenon is referred as the rhizosphere effect (Morgan and Whipps, 2001). It is calculated in terms of rhizosphere ratio, by dividing the total number of microorganisms in the rhizosphere by the corresponding number in the bulk soil (Aneja, 2003). According to most studies, the selection of microorganisms into the rhizosphere of soybeans and other plants is carried out by root exudates in the form of root litter and root exometabolites, reaching up to 20%–40% of the photosynthetic products of the plant (Lynch, 1990). This acts as a driving force for the setup of active and enhanced microbial populations in the root zone (Grayston et al., 1996). Recognition of partners of the plant–microbial interactions occurs in the form of the exchange of specific signal molecules such as enzymes, vitamins, alkaloids, hormones, glycosides, flavonoids, and lectins (Venturi and Keel, 2016). The variations in bacterial and fungal community structures in the rhizosphere are determined by plant-related factors such as the crop variety (Berg et al., 2006), plant growth developmental stages (Gomes et al., 2003), and soil characteristics (Nie et al., 2009).

Low–molecular weight organic compounds in root exudates play a key role in plant–microorganism interactions by influencing the structure and function of soil microbial communities. It was shown that root secretions affected not only the number of microorganisms in the rhizosphere, but also the structure of the microbial community (Grayston et al., 1998). In addition, they are involved in the phytohormonal regulation of plant growth and development and the regulation of allelopathic relationships, which, in turn, actively interferes with the microbial community of the soil (Sanon et al., 2009). The effects of plant root exudates on microbial populations demonstrated important beneficial (Ryu et al., 2004) and sometimes harmful interactions (Timmuck et al., 2005). It was shown that there was the development of a mutualistic symbiotic relationship between the plant and the microorganisms based on trophic, protective, and carrying out biological control. Rhizosphere microbiota inhibit a number of plant pathogens, such as *Fusarium oxysporum*, *Pythum*, *Phytophthora*, etc. (Whips, 1987). Soil type is the major factor in determining the bacterial community in the soybean rhizosphere. Researchers note the importance of such factors as soil structure, organic matter content, pH, and salinity (Xu et al., 2009; Wang et al., 2014).

Soybean selects a specific microbial community inhabiting the rhizosphere based on functional traits, which may be related to benefits to the plant, such as growth promotion and nutrition (Mendes et al., 2014). The rhizosphere community is collected from bulk soil microorganisms, but bacterial interactions were stronger and specific functional groups were more representative in the rhizosphere than in the bulk soil (Yan et al., 2017). It has been shown that the microbial community of the soybean rhizosphere is different from the community of the surrounding soil. It was suggested that soybean selects a specific microbial community based on functional traits, related to benefits of growth promotion and nutrition of the plants, such as membrane transport and nitrogen, phosphorous, potassium, and iron metabolism (Sugiyama et al., 2014). This selection follows largely the niche-based theory, indicating the selection power of the plant and other environmental variables in shaping the microbial community both at the taxonomic and functional level.

It therefore means that plant's root exudates can influence the diversity of resident microorganisms in the rhizosphere and these organisms can as well influence the plants by releasing regulatory substances. Hence, rhizosphere organisms are considered as a well-developed external functional environment for plants and they are regarded as a plant's second genome (Turner et al., 2013a,b). Since plants are regarded as metaorganisms (Vryzas, 2016), understanding the actual contributions of rhizosphere microbiome [particularly true for the nitrogen (N) fixing bacteria] towards plant health and productivity is very important. Further analysis is needed to better understand the mechanisms by which the plant selects the rhizospheric community, and the study of the role of rhizodeposits in microbial communities shift is of prime importance.

1.4 Methods to study the microbial communities in soya rhizosphere and bulk soil

1.4.1 Direct plating on the laboratory growth media

Enumeration of cultivable heterotrophic bacteria can be achieved using solid or liquid media, as well as the isolation of pure cultures of target organisms. In the case of solid culture media, a precise volume of sample (usually 50–100 μL) is spread on the agar surface of the medium. After incubation under optimal for investigated bacteria conditions of temperature, oxygenation, and duration, the colony-forming units (CFU) are counted to calculate the number of culturable bacteria present in the sample.

Ecological studies of members of the family Rhizobiaceae require the quantification and isolation of the indigenous and introduced inoculant strains. Despite the fact that the share of uncultured microorganisms in the rhizosphere is large and only 1% of the microbiome can be isolated in pure cultures (Lakshmanan et al., 2014), cultural methods are an integral part of the study of the soya rhizosphere. Traditional methods are used to isolate strains and their genetic characteristics.

Yeast Extract Mannitol Agar (YEM) with Congo Red is widely used for cultivation of rhizobia and for studying root nodulation (Rao, 1977). Congo Red inhibits penicillin-susceptible strains and colonies stand out as white, translucent, glistening, and elevated, with entire margins.

Some modifications done to the culture media can increase the cultivability. The selective agents used have included antibiotics, dyes, and metabolic inhibitors (Danso et al., 1973; Giller et al., 1989; Top et al., 1990). We have developed a new selective nutrient medium for estimating the number of viable bradyrhizobia based on YEM with Congo Red. The modified YEM-CV medium was amended by cyclohexemide (70 mg L^{-1}), growth inhibitor of the filamentous fungi, and vancomycin (1 mg L^{-1}), inhibitor of Gram-positive bacteria. The YEM-CV media was tested in experiments with mixed bacterial cultures and soil samples. Addition of antibiotics significantly reduced the number of nontarget organisms growing, and due to this the effectiveness of bradyrhizobia was increased (Fig. 1.2, Table 1.1).

Information on the diazotroph diversity in the rhizosphere of soybean for a long time was based mainly on the results of using traditional cultivation methods. Classical microbiological methods, implying, above all, cultivation, have significant limitations, which are repeatedly emphasized by the authors (Amann and Ludwig, 2000; Van Elsas and Bersma, 2011).

1.4.2 Quantitative immunofluorescence technique

The immunoassay techniques were widely used previously to quantify rhizobia in soils and inoculants (Schmidt et al., 1968). These techniques are based on the use of specific antibodies. There are monoclonal antibodies (recognizes only one type of epitope), which are very specific and reproducible, and polyclonal antibodies (mixture of antibodies recognizing different epitopes on the same antigen) that are less specific. Among the various immunoassays, immunofluorescence consists of chemically modifying an antibody by adding a fluorochrome without changing the specificity of the

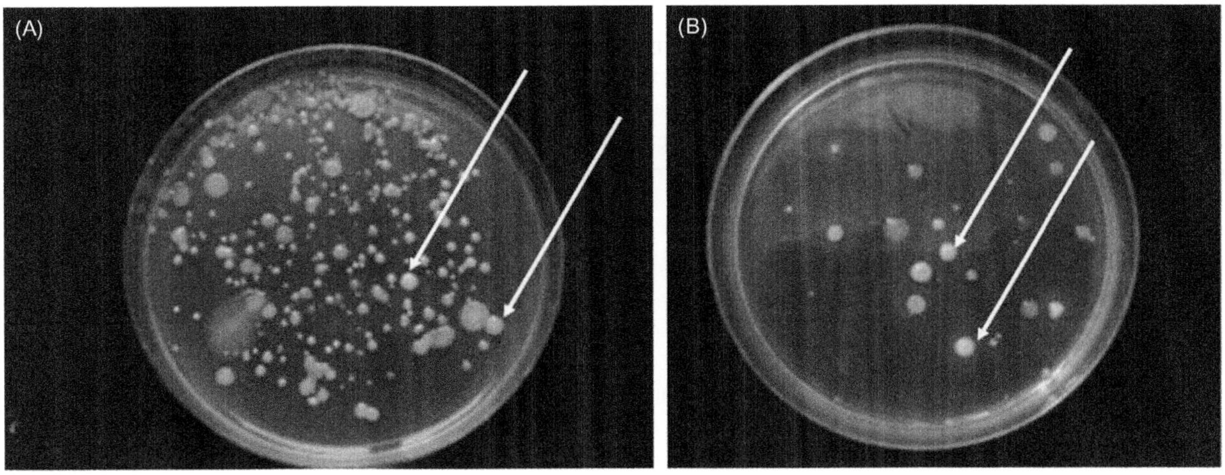

FIGURE 1.2 An example of the soil microorganisms growth on YEM (A) and YEM-media (B) (10^{-4} soil dilution, 2 weeks incubation). The arrows shows the target bradirizobia colonies.

TABLE 1.1 Dynamics of culturable bradyrhizobia (CFU) introduced into the soil (laboratory experiment).

	YEM medium, $\times 10^4$ CFU g^{-1} soil		YEM-CV medium, $\times 10^4$ CFU g^{-1} soil	
Variant of experiment	Total	*Bradyrhizobium*	Total	*Bradyrhizobium*
Control	115 ± 7	0	7 ± 2	0
Time after inoculation, weeks 0	277 ± 48	20 ± 5	60 ± 6	21 ± 4
1	253 ± 63	15 ± 3	18 ± 6	13 ± 3
2	252 ± 53	14 ± 2	18 ± 9	7 ± 2
3	189 ± 28	8 ± 3	29 ± 6	6 ± 4
4	112 ± 22	3 ± 2	23 ± 3	5 ± 3

antibody. This methodology, however, tends to overestimate population sizes and is subject to problems associated with overall sensitivity, the optimal dispersal, and flocculation of bacteria from different soil types (Postma et al., 1988).

1.4.3 In situ hybridization with fluorescent gene probes

In situ hybridization with fluorescent probes (fluorescent in situ hybridization; FISH) is a cytogenetic approach based on the principle of complementarity between nucleic acid chains and allows the specific pairing of a labeled oligonucleotide probe to an RNA or DNA sequence, ribosomes, or messengers. So it is an approach conceptually close to gene amplification by PCR followed by sequencing, with the advantage of a possible eye control by microscopy. The effectiveness of the FISH technique is proportional to the number of ribosomes and ultimately to the physiological state of the cell. The probes are oligonucleotide sequences of about 20 bases. It is necessary that they have a short size to allow their penetration into the cells and also to have a melting temperature close to room temperature to protect the cell structures. The undoubted advantages of the method are speed and that they allow detection and spatial distribution of more than one sample simultaneously (Moter and Göbel, 2000).

We have applied the ARB software (Ludwig et al., 2004) for design of specific gene probe for *Bradyrhizobium.* The sequences of the target group were aligned and analyzed using the ARB software, which allows designing an oligonucleotide sequence whose specificity can be checked in silico. The most effective nucleotide sequence for detection of *B. japonicum* was found to be 5′-TCGCTGCCCACTGTC-3′ (RHIZOB). Checking in Ribosomal Database Project (RDB) software revealed that the created gene probe does not have a high level of specificity. Except for the target microorganism related to Bradyrhizobiaceae family it had demonstrated nonspecific binding with Rhizobiaceae (*Rhizobium*, *Mesorhizobium*, and *Agrobacterium*).

Experimental verification of C3-labeled probe revealed that 20% formamide was optimal for hybridization for *Bradyrhizobium* in pure and mixed cultures. We have concluded that RHIZOB probe may be successfully applied for primary screening of *Bradyrhizobium* isolates, and estimation of the viable cell fraction in the preparation for inoculation of soybean. The results of using the probe for soil sample study were found to be unsatisfactory due to a weak fluorescence signal.

1.4.4 Quantitative PCR: real-time q-PCR

Nowadays, virtually all investigations involve molecular analyses, which are necessary for detailed characterization of species and investigation of microbial interactions with host plants in the rhizosphere. Significant progress in the study of advertising soil microorganisms was achieved in connection with the development and improvement of methods based on the analysis of soil nucleic acids via polymerase chain reaction (PCR). The most popular method for gene quantification is quantitative PCR (q-PCR) or real-time PCR.

Real-time quantitative PCR (q-PCR) is a technique that collects amplification data during PCR processing (Higuchi et al., 1993). Compared with other PCR-based techniques, q-PCR provides quantitative data on gene and transcript abundances, and does not require post-PCR handling, avoiding potential contamination of samples (Heid et al., 1996). Q-PCR is usually restricted to a relatively low number of sequences as it requires specific primers, although high-throughput q-PCR has also been developed (McGrath et al., 2005).

A system was developed for the detection of bradyrhizobia by the amplification of specific 16S rRNA and *nodZ* gene fragments with PCR. It was shown that *nodZ* gene can be successfully used as a genetic functional marker of *Bradyrizobium* (Moulin et al., 2004). As a result, two pairs of primers were selected, which, according to in silico analysis, are specific for *B. japonicum* and *B. elcanii*, which are most usually used for inoculation of soybeans (Table 1.2). Both primers amplified successfully for all of the test strains and the specificity of the amplified products was confirmed by subsequent sequencing. These results suggest the suitability of the method for the qualitative detection of bradyrhizobia bacteria in environmental samples.

This was shown by applying the primer combination developed in this study to total DNA preparations from soil habitats. The created system of primers has been successfully applied to quantify (the number of gene copies) bradyrhizobia in soil samples to study the dynamics of introduced *Bradyrhizobium* in laboratory experiments with soil samples and grown soybean plants by real-time PCR (Table 1.3).

The disadvantages include the fact that this method can be used only for targeting of known sequences; DNA impurities and artifacts may create false-positives or inhibit amplification. The advantages of q-PCR include that it is a quick, accurate, and highly sensitive method for sequence quantification that can also be used to quantify microbial groups; it is relatively cheap and easy to implement; and specific amplification can be confirmed by melting curve analysis. (Smith and Osborn, 2009).

TABLE 1.2 Novel primers developed for amplification of 16S rRNA and *nodZ* genes fragments of *Bradyrhizobium*.

Primer	Target gene	Primer sequence (5′–3′)
BradyF	16S rRNA	AMTKCCTTTGAKWYTKAAGATCTTG
BradyR	16S rRNA	GTCACATCTCTGCGACCGGTC
nodZ-AF	*nodZ*	GGTTTGGCGACTGTCTGTGGTC
nodZ-AR	*nodZ*	TTCCACCATGTTGGAAAGAATGGTCC

TABLE 1.3 Estimation of *Bradyrhizobium* in soil by q-PCR technique with novel primers systems.

	Bradyrhizobium, $\times 10^6$ gene copies g^{-1} soil		*Eubacteria*, $\times 10^8$ gene copies g^{-1} soil
Variant of the experiment	*nodZ*	16S rRNA	16S rRNA
Soil without plants			
Control	0.016	0.028	6.6
Inoculation (weeks)			
0	9.68	12.7	7.4
1	1.43	1.64	8.0
2	1.26	1.17	7.6
3	0.42	1.30	7.4
4	0.78	0.77	7.7
Soil with soybean plants			
8 weeks	1.92	1.94	12.36

1.4.5 Fingerprinting methods based on PCR analysis

Denaturing gradient gel electrophoresis (DGGE) was developed to separate RCR-amplified ribosomal DNA fragments of DNA with the same lengths, but with variation in nucleotide composition. This method was adapted to the analysis of the identified bacterial communities of the rhizosphere. DGGE is sensitive to variation in DNA sequences; bands can be excised, cloned, and sequenced for identification. However, this method is very time consuming; multiple bands for a single species can be generated due to microheterogeneity; it can be used only for short fragments; complex communities may appear smeared due to a large number of bands; and it is difficult to reproduce (gel to gel variation) (Kirk et al., 2004).

In all the communities, six to eight major bands were identified; four of them were found in all the samples and could be regarded as the widespread aboriginal organisms (Fig. 1.3). Sequence analysis of the typical bands has revealed a high degree of similarity with those of *Firmicutes*, *Acidobacteria*, and some environmental clones. Among the dominant communities of plant rhizosphere inoculated with bacterial preparations, introduced *B. japonicum* was not found. It was probably displaced with aboriginal microorganisms and only remained in the material of soybean root nodules.

The dinitrogenase reductase gene (*nif*H) is the most widely established molecular marker for the study of nitrogen-fixing prokaryotes in nature. Its analysis is widely used for identification of diazotroph species that occur in the microbiomes of crop plants and has shown them to be a subset of the soil diazotroph community (Sarita et al., 2008; Kizilova et al., 2012; Calderoli et al., 2017; Gaby et al., 2018). Comparative analysis of this gene sequences with the known nucleotide sequences may provide important information regarding the phylogenetic position of nitrogen-fixing bacteria in soils. Operational taxonomic unit (OTU)-based analyses showed that *nif*H phylotypes related to phylum *Firmicutes* are key microbial components of N_2 fixation in soybean rhizosphere. Most of the sequences formed a compact cluster together with the obligate and facultative anaerobic bacteria (*Clostridium, Paenibacillus*) (Fig. 1.4). Diazotrophs of the other dominant cluster demonstrated similarity to members of *Alphaproteobacteria* from *Leptospirillum, Derxia*, and *Azohydromonas* genera. These results are in accordance with earlier data. In the soybean rhizosphere in the Primorskii region, Russian Far East, numerous vegetative *Clostridium* cells were detected (Tilba and Golodyaev, 1966).

The domination of anaerobes in the soybean rhizosphere soil may be due to the specific composition of its organic compounds, mechanical properties of the soil, and specific features of microbial communities. The high representation of *nif*H phylotypes related to *Firmicutes* and *Alphaproteobacteria* provides solid bases to consider them as key players of free-nitrogen fixation in soybean rhizosphere.

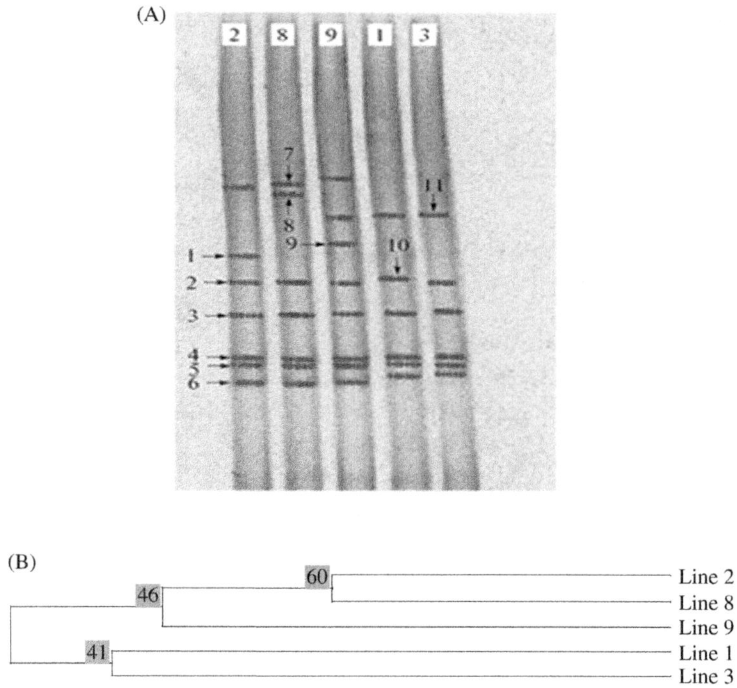

FIGURE 1.3 Example profiles of separation of the 16S rRNA gene fragments of rhizosphere soil from the field experiment with the soybean Gorlitsa variety by DGGE (A) and cluster analysis of the obtained fingerprints (B).
1, Control without inoculation; *2*, inoculation with *Bradyrhizobium japonicum* 71t; *3*, inoculation with *B. japonicum* 2c; *8*, inoculation with *B. japonicum* 71t and *Bacillus megaterium*; 9, inoculation with *B. japonicum* 71t sensibilized by genisteine. *Data from courtesy Kizilova, A.K., Titova, L.V., Kravchenko, I.K., Iutinskaya, G.A., 2012. Evaluation of the diversity of nitrogen fixing bacteria in soybean rhizosphere by nifH gene analysis. Microbiology 81 (5), 621–629 (Kizilova et al., 2012).*

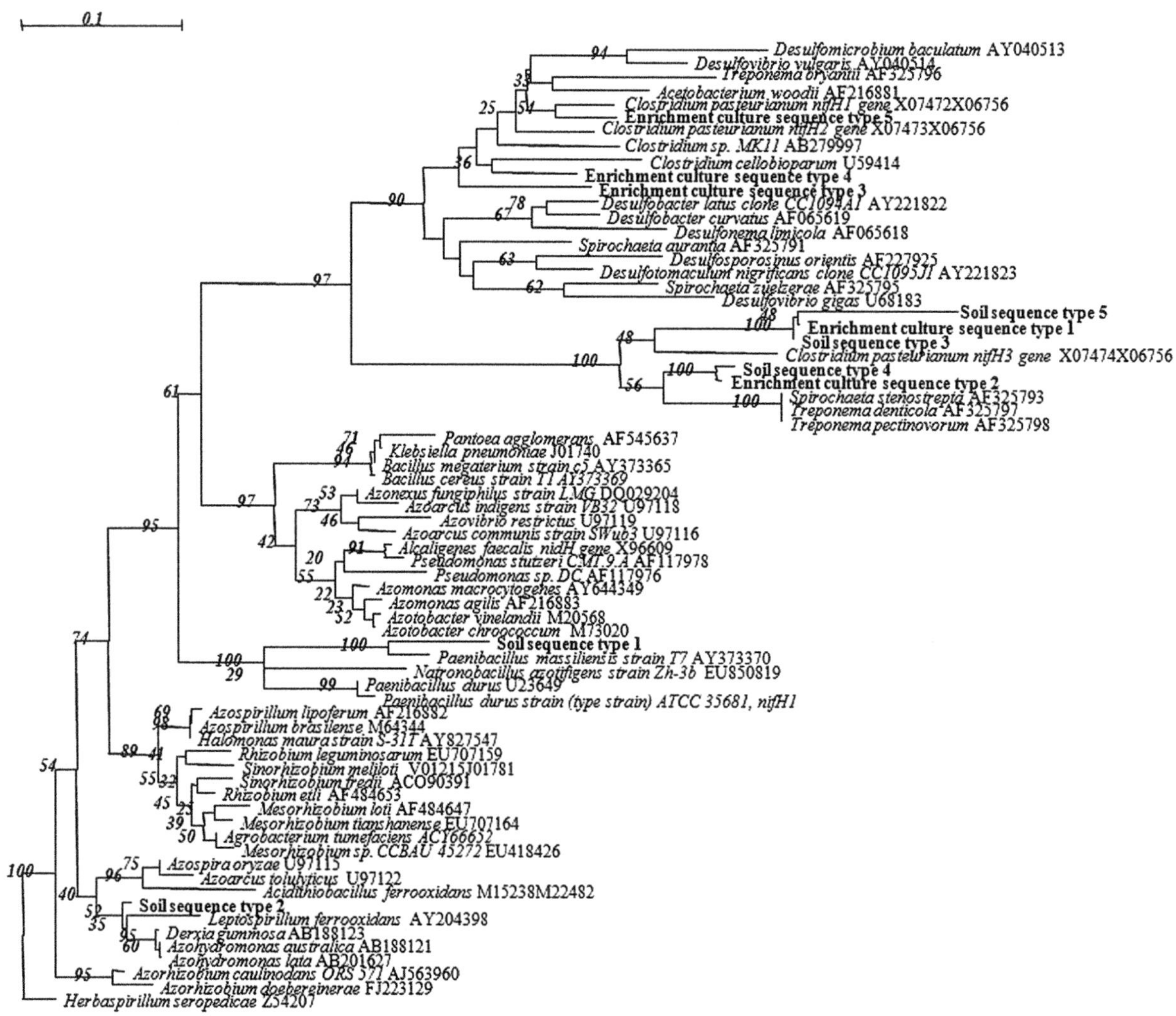

FIGURE 1.4 Phylogenetic based on the translated amino acid sequences of the NifH fragments, obtained by analysis of soybean rhizosphere soil samples. The sequences obtained in the present work are shown in boldface. Scale bar corresponds to 10 replacements per 100 amino acid residues (evolutionary distance). The numerals show the significance of the branching order determined using "bootstrap" analysis of 500 alternative trees (only bootstrap values over 50 are shown). *Data from courtesy of Kizilova, A.K., Titova, L.V., Kravchenko, I.K., Iutinskaya, G.A., 2012. Evaluation of the diversity of nitrogen fixing bacteria in soybean rhizosphere by nifH gene analysis. Microbiology 81 (5), 621–629 (Kizilova et al., 2012).*

1.4.6 High-throughput sequencing technologies

High-throughput sequencing approaches [also referred to as next-generation sequencing (NGS)] are increasingly being used for estimates of microbial diversity in various environments including the rhizosphere in a culture-independent manner. Due to advances in nanotechnology and bioinformatics, alternative technologies have been created to increase the throughput of DNA and RNA sequencing has emerged. Such technologies play a major role in metagenomic (DNA-based), and metatranscriptomic (RNA based) approaches, which provide a comprehensive picture of potential and active functions of microbial communities, respectively. The most widely used platforms for massive parallel sequencing for assessing soil microbial diversity are Roche 454 Genome Sequencer (Roche Diagnostics Corp., Branford, CT, USA) (Loman et al., 2012; Mardis 2008), HiSeq 2000 (Illumina Inc., San Diego, CA, USA) (Loman et al., 2012), and AB SOLiDTM System (Life Technologies Corp., Carlsbad, CA, USA) (Shokralla et al., 2012; Magi et al., 2010). On their basis, studies are conducted on the native communities composition (Sugiyama et al., 2014; Li et al., 2016), as well as on the biodiversity of the rhizosphere microbiome upon inoculation of bioformulation (Iutynska et al., 2017).

Currently there is a transition from the metagenomics approaches to metatranscriptomics techniques. Metatranscriptomics give information about the diversity and functional molecules of the microbial community unlike metagenomics, which only show diversity. It was recently noted that the functional diversity of microbiome is probably more dominant in ecological niches than genomic diversity (Turner et al., 2013a,b; Barret et al., 2011). It is also shown that the assembly of the microbial community in the rhizosphere is based on niche-based processes as a result of the selection power of the plant and other environmental factors (Mendes et al., 2014).

Among the currently used molecular techniques, NGS techniques have the greatest impact on DNA and RNA based analyses. NGS techniques help to find solutions to problems that could not be solved previously, due to financial and technical constraints.

1.5 Endosymbionts of soybean plants

Plant associations with beneficial microorganisms attract the attention of scientists not only as an object for the study of the fundamental interactions of various organisms, but also in terms of the possible their use in organic management practices.

1.5.1 Rhizobial endosymbionts

Soybeans are nodulated by fast- and slow-growing rhizobia, which are collectively called "soybean rhizobia." The genus *Bradyrhizobium* is a slow growing, Gram-negative kind of soil bacteria, and is a major symbiont of soybeans and under appropriate environmental conditions, soybean and *Bradyrhizobium* can initiate a symbiotic interaction, resulting in the development of nitrogen-fixing root nodules. Based on 16S rRNA gene sequences, the genus *Bradyrhizobium* was classified into a clade in the *Proteobacteria* along with oligotrophic soil, or aquatic bacteria such as *Rhodopseudomonas palustris*, *Rhodoplanes roseus*, *Nitrobacter winogradskyi, Blastobacter denitrificans*, and the pathogen *Afipia* spp. (Saito et al., 1998; Sawada et al., 2003; Van Berkum and Eardly, 2002; Willems et al., 2001). Currently there are 40 valid species in this genus, and *B. japonicum, B. elkanii, B. liaoningense, B. yuanmingense, B. betae*, and *B. canariense* most usually nodulate soybean plants. The genus *Bradyrhizobium* is known to display a large degree of antigenic heterogeneity, and can be categorized into several serological groups, based on differences in somatic antigens (Vincent, 1982).

1.5.2 Community composition of *Bradyrhizobium* in soybean plants nodules

Since soybean has been cropped for many centuries, it is reasonable to expect that soils are very rich in a high diversity of rhizobia. This idea is based on the hypothesis that coevolution between the two symbionts, the plant and the bacteria, has taken place over a long period of time, and, consequently, it should be possible to isolate different, or even new, soybean rhizobia species and also bacterial strains showing superior symbiotic performance with soybeans. Taxonomic studies have proved that this assumption is true, since new soybean rhizobia able to nodulate soybeans have been isolated from many distinct geographical areas.

Soils usually lack *Bradyrhizobium japonicum* strains unless soybean is grown on them for at least five or more years. It was reported (Hiltbold et al., 1985) that numbers of *B. japonicum* in 52 Iowa fields were correlated with whether soybeans had been grown at the site within the previous 13 years. It is therefore important to inoculate seeds with relevant strains of bacteria before sowing especially if the crop is to be grown for the first time. In any case, to get the maximum benefit out of inoculation there is a need to follow correct and careful inoculation procedures, and the inoculant should carry live and effective bacterial cells.

We have compared the genetic structure in population diversity of *Bradyrhizobium* in soybean nodules in recently inoculated plants (site A) and plants grown in the field without soybean cultivation for more than 10 years (site B). Phylogenetic relationships between bradyrhizobia associated with cropped soybeans and native legumes are shown in Fig. 1.5. To specify that our results were due to microevolutionary genetic changes or were changes in bacterial community composition, we investigated evolutionary relationships among clones and strains inoculated and noninoculated fields with phylogenetic analysis.

Nodulating bacteria from *Bradyrhizobium japonicum* 532C inoculated soybean plants exhibited the same genotype as inoculant strain (OTU 1). In contrast, all sequences of soybean-nodulating bacteria from noninoculated plants clustered into 8 OTUs. Phylogenetic analyses indicated that the *Bradyrhizobium* populations associated with soybeans at two field sites with contrasting histories of cultivation and inoculation were highly structured: analyses supported division of 16S rRNA sequences in five lineages corresponding either to *B. japonicum* group (OTUs 2 and 3), *B. elkanii* (OTU 4), *B. liaoningense* (OTU 6), and likely novel lineages (OUTs 5, 7, 8, 9) within the genus *Bradyrhizobium*. Previously only 20% *Bradyrhizobium* isolated from soil associated with root zone of the soybean plants from the recently inoculated site were attributed to inoculation sources (Tang et al., 2012). Very little information is available concerning the fate of bradyrhizobial populations during crop rotations typically used under agricultural conditions. When examined the populations of *Bradyrhizobium japonicum* in a cotton–corn–soybean rotation in Alabama (Hiltbold et al., 1985) the numbers of *B. japonicum* declined rapidly with either cotton or corn as the field crop (Table 1.4).

It is interesting to mention that the majority of the *Bradyrhizobium* clones, found in noninoculated soybean nodules, demonstrated the most similarity with *B. liaoningense*. These extra-slowly growing soybean rhizobia were isolated from root nodules in China (and have never been introduced into soil under investigation).

Taken together, our results indicate that field populations of unique *Bradyrhizobium* are present in soils with long-term (more than 10 years) delay in soybean cropping and their bacteria are very competitive for nodulation of currently

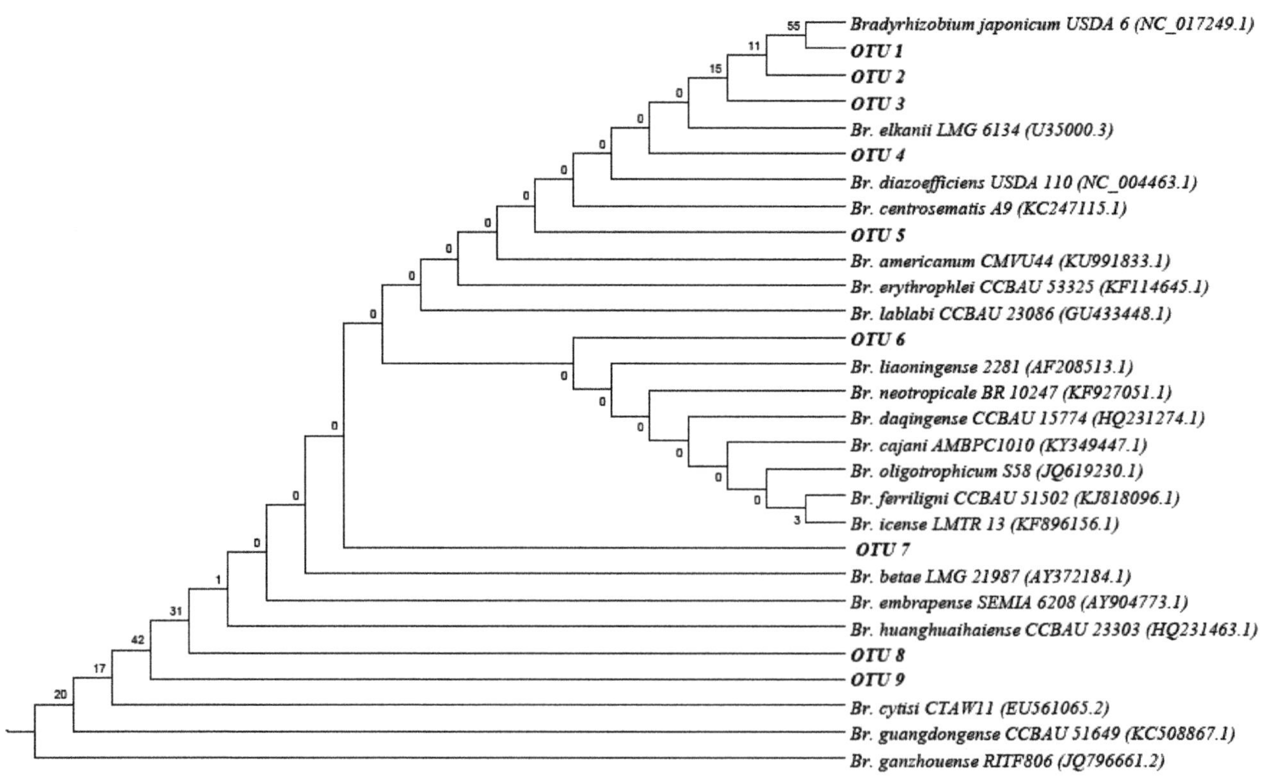

FIGURE 1.5 Phylogenetic based on the 16S rRNA sequences obtained by analysis of soybean nodules. The sequences obtained in the present work are shown in boldface.

TABLE 1.4 Relative abundance of 16S rRNA sequences of *Bradyrhizobium* from soybean nodules samples from two field sites.

#	Origin	Nearest relative	Proportion (% of total clone number)
OTU1	Site A	*Bradyrhizobium japonicum*	100
OTU2	Site B	*B. japonicum*	10
OTU3	Site B	*B. japonicum*	3
OTU4	Site B	*B. elcanii*	10
OTU5	Site B	*Bradyrhizobium* sp.	3
OTU6	Site B	*B. liaoningense*	63
OTU7	Site B	*Bradyrhizobium* sp.	3
OTU8	Site B	*Bradyrhizobium* sp.	6
OTU9	Site B	*Bradyrhizobium* sp.	3

soybean varieties. The data suggest that soybean-nodulating bacteria associated with such plants represent a novel source of ecologically adapted bacteria for soybean inoculation.

1.5.3 Nonrhizobial endosymbionts of soya

In various legumes, in addition to rhizobia that are responsible for nodulation and N_2 fixation, other endophytic bacteria, called nonrhizobial (NR) bacteria, are also found in the nodules. Some of these NR bacteria have proven beneficial to their legume hosts, as they enhance plant growth by producing plant hormones, fixing atmospheric N_2, and solubilizing phosphate (Peix et al., 2015).

Most studies have found that rhizobia makes up 80–90% of the nodule bacterial communities, however a number of studies have shown an equal ratio of rhizobial and NR endosymbionts (De Meyer et al., 2015; Zhang et al., 2018; Leite et al., 2017).

NR symbionts of the legumes were investigated in various regions of the world and different bacterial taxa dominated. NR symbionts were found in various parts of soybean, but mainly in the roots (Kuklinsky-Sobral et al., 2004). The most abundant phylum throughout soybean lines tested was *Proteobacteria* (58%–79%). Partial sequencing of the 16S rRNA gene revealed a large diversity of different taxa from the classes *Alphaproteobacteria, Betaproteobacteria, Gammaproteobacteria, Actinobacteria, Firmibacteria, Flavobacteria*, and *Sphingobacteria* in nodules. Many of the isolates belonged to the genera *Bacillus* and *Pseudomonas* (De Meyer et al., 2015; Leite et al., 2017; Zhao et al., 2018). In Brazil *γ-proteobacteria* subgroup was dominant, within which Enterobacteriaceae most frequently showed the highest diversity of cultivable epiphytic and endophytic bacteria associated with soybean cultivars (Kuklinsky-Sobral et al., 2004). In China *Firmicutes* and *Proteobacteria* were found to dominate NR bacteria in soybean nodules (Zhang et al., 2018).

It was demonstrated that *Gammaproteobacteria* were dominant (21%–72%) in Nod+ soybeans, and *Pseudomonas* were significantly abundant. On the contrary, high abundance of *Alphaproteobacteria* was observed in Nod− soybeans, which could be related to the increase in Rhizobiaceae and Sphingomonadaceae. A far greater abundance of *Firmicutes* was observed in Nod− and Nod++ mutant soybeans than in Nod+ soybeans (Okubo et al., 2009).

1.6 Factors influencing soybean rhizosphere microbial communities

Diverse rhizobia belonging to *Bradyrhizobium* and *Sinorhizobium* are associated with soybeans, and typically, biogeographic patterns have been found among the soybean rhizobia (Han et al., 2009). Various biotic and abiotic parameters modulate or influence the microbial diversity and composition in the region surrounding the root. The distribution, abundance, and nitrogen-fixing efficiency of soybean rhizobia are strongly related to genotypes or cultivars of soybeans, the latitude, and soil environmental conditions, like the organic carbon content, pH, available phosphorus, and other factors, climate change, and anthropogenic activities. The level to which microbial communities are influenced is not totally understood.

1.6.1 Geographic location and soil type

Moreover, factors such as soil type determine the composition of both endophytic and free-living rhizosphere microorganisms. Many studies have shown that soil type is a key factor that determines the bacterial community composition in the rhizosphere of plants (Buyer et al., 2002). Based on the molecular techniques it was shown that rhizospheric microbial communities were shaped mainly by soil properties from different geographical locations (Girvan et al., 2003). In a similar study, it was demonstrated that plants were stronger modulators of microbial richness than the soil type (Kowalchuk et al., 2002). Thus, although we know that both plant (crop) type and soil can direct the structure of microbial communities, the details of these interactions are incompletely elucidated and warrant further study.

Soil can differ in structure, organic matter, pH, texture, and nutrient status. These soil properties can select specific microorganisms by creating conducive environments. In particular, soil pH and availability of nutrients such as carbon have been observed to affect the diversity of crop pathogenic nematodes, bacteria, fungi, and beneficial microorganisms (Miransari, 2013). In some cases, soil properties may lead to soil type–specific composition of rhizosphere microbiome (Igwe et al., 2018). This was further confirmed by demonstration of the fact that bacterial community structures were alike in soils of the same type. This suggests that soil properties and soil type can determine the types of microorganisms that colonize the rhizosphere, and that different soil types can contain different microbial species.

1.6.2 Plant genotype as the determinant of the structure of rhizosphere microbial community

Plant cultivar/genotype influences the indigenous microorganisms present in the plant rhizosphere (Hartmann et al., 2009). While physicochemical characteristics of the soil can significantly influence the composition of soil microorganisms, plant root exudates are able to modify the rhizosphere environment that slowly alters the soil microorganisms to support the establishment of a rhizobiome. These root exudates together with plant root immune system would further select those microorganisms that have the ability to colonize root surface (rhizoplane) and inner root tissue (endosphere). Furthermore, certain metabolites liberated into the root region can elicit several responses in various soil microorganisms. In particular, flavonoids released from plants attract symbiotic microorganisms such as *Bradyrhizobium japonicum*, and disease-causing microorganisms, for example, *Phytophthora sojae*. The flavonoid naringenin produced by legumes activates germination of mycorrhizal spores and hyphal branching while the flavonoid catechin regulates quorum sensing (Graham, 1991). Similarly, some defense metabolites (e.g., pyrrolizidine alkaloids) can affect the rhizosphere microbial structure by enhancing the growth of microorganisms that are able to break down these metabolites.

Recent study shows that variations between plant cultivars in a single gene could significantly affect the rhizosphere microbiome. The release of a single exogenous glucosinolates changed the microbial population on transgenic *Arabidopsis* roots (Bressan et al,. 2009) in which fungi and *Alphaproteobacteria* were predominant. This result further suggests that plant cultivar can influence the accumulation of bacteria that protect the plant against pathogens.

1.6.3 Climate change and anthropogenic disturbance

Other factors like climate change and anthropogenic activities can also modulate the microbial communities in a specific plant host. Climate change has different effects, ranging from local cooling to global warming, shifting vegetation zone and augmenting extreme weather events, and all these effects have indirect impacts on rhizosphere microbiome.

Increase in carbon dioxide levels, a component that is alleged to be the key driver of climate change, could also affect rhizosphere interactions by changing root exudation patterns (Lam et al., 2012). Climate change can cause abiotic stresses and some of the stresses associated with it are drought, cold, and heat. It has been found that rhizobia can survive in dry environments, but their diversities were significantly lower in dry soil environments (Zahran, 1999). Moreover, drought does not only affect the ability of *Rhizobium* species to fix N but also the growth and development of plant legumes. Suitable rhizobial species that can survive under drought environments in symbiosis with leguminous crops are of utmost importance and therefore more research has to be focused in this area using modern state-of-the-art techniques.

Anthropogenic activity such as the use of N-based chemical fertilizer is associated with environmental destruction. The abundance of OTUs in the genera *Bradyrhizobium* and *Burkholderia* were higher in the rhizosphere microbiome from the field of the low-level N fertilizer than standard level N fertilizer. (Yan et al., 2014). These outcomes indicate that low-N fertilizer management is a crucial factor that modulates rhizosphere microbiome community structure. Coupled with other negative impacts of N-based fertilizers it demonstrates the need for a more ecofriendly means of enhancing N level in agricultural land, which can be achieved by application of N-fixing endophytic and free-living rhizobacteria.

Herbicides are chemical compounds commonly used to control weeds in agriculture and many of them cause environmental contamination and human health problems due to their toxicity and persistence in the soil. The rhizobacteria community can be affected by the presence of herbicides such as glyphosate, frequently used where genetically modified herbicide-tolerant crops are grown. Numerous studies have investigated the impacts of glyphosate on soil microbial properties methods such as microbial biomass, enzyme activity, and respiration (Bünemann et al., 2006; Newman et al., 2016). Typically the results of these studies have shown no or transitory effects but the effect may be masked by changes in composition in specific microbial communities mediating the key functions (Imfeld and Vuilleumier, 2012). Alterations to soil microbial community composition and changes in microbial diversity could have pronounced long-term effects on soil quality as well as plant health and crop production (Bending et al., 2007).

Herbicides do not only affect microbial diversity, they also affect microbial functions (e.g., N fixation) in the soil. N fixation in legumes such as soybean can be triggered by symbiotic interaction between rhizobia and soybean roots and there is an indication that the presence of sublethal amounts of glyphosate in growth media results in the accumulation of shikimate and decrease in *Bradyrhizobium japonicum* growth (Zablotowisz and Reddy, 2004). But, different *B. japonicum* strains exhibit different sensitivities to the herbicide. Some species in the family Rhizobiaceae might break down glyphosate and utilize it as the main source of phosphorus in the presence of aromatic amino acids (Liu et al., 1991).

1.7 Biofertilizers for soybean plants

1.7.1 Overview of biofertilizers for soybean crops

The use of beneficial microorganisms has been proven to be an environmentally sound option to increase crop yields. Plant–microbe interactions in the rhizosphere are the determinants of plant health, productivity, and soil fertility. Plant growth-promoting bacteria (PGPB) are bacteria that can enhance plant growth and protect plants from disease and abiotic stresses through a wide variety of mechanisms; those that establish close associations with plants, such as the endophytes, could be more successful in plant growth promotion (PGP). Several important bacterial characteristics, such as biological nitrogen fixation, phosphate solubilization, ACC deaminase activity, and production of siderophores and phytohormones, can be assessed as PGP traits.

Biofertilizer is a substance containing living microorganisms that, when applied to seed, plant surfaces, or soil, colonizes the rhizosphere or the interior of the plant and promotes growth by increasing the supply or availability of primary nutrients to the host plant (Vessey, 2003). This definition distinguishes biofertilizer from organic fertilizer. The latter contains organic compounds that directly, or by decay, increase soil fertility. Likewise, the term biofertilizer should not be used interchangeably with terms such as green manure or manure, intercrop, or organic-supplemented chemical fertilizer. Not all plant growth-promoting rhizobacteria (PGPR) can be considered as biofertilizers. Bacteria that promote plant growth by control of deleterious organisms are biopesticides. Similarly, such bacteria can enhance plant growth by producing phytohormones and are regarded as bioenhancers, not biofertilizer. Interestingly, some PGPR can promote growth by acting as both biofertilizer and biopesticide or bioenhancer.

There are several ways in which PGPR can stimulate soybean growth. They may fix atmospheric nitrogen and supply it to plants, although this is usually a minor component of the benefit; they can synthesize siderophores, which can sequester iron from the soil and provide it to plant cells; they can synthesize phytohormones such as auxins, cytokinins, and gibberellins; they can convert nonplant available forms of soil nutrients into plant available forms (phosphorus and potassium solubilizing bacteria); and they can compete with plant pathogenic fungi and harmful bacteria (Bashan and de-Bashan, 2010). Another point is that soybean sometimes has to grow in stressful environments and identification and manipulation of their relationships with PGPR may be an effective strategy of modern agriculture (Boiero et al., 2006; Esitken et al., 2006).

The microsymbiont of soybean *B. japonicum* is a highly efficient diazotroph forming symbiotic association with soybean. Some host genotypes are superior to others in the ability to fix N_2, and, in turn, some rhizobial strains have similar superior capability. Amounts of N_2-fixed (kg ha^{-1}) by soybean have been up to 450 kg N ha^{-1} (Unkovich and Pate, 2000). Thus soybean depends on its symbionts for a large part of its N requirements for effective growth and dry matter production. Additionally, PSB have the capacity to stimulate soybean growth. Coinoculation of *Bradyrhizobium* and PGPR microorganisms significantly improved soybean growth and its yield components as compared with the sole application of *Bradyrhizobium* (Dubey 1996; Wasule et al. 2007). It was shown that inoculation with PSB improved nodulation, root and shoot biomass, straw and grain yield, and P and N uptake by the crop (Linu et al., 2009). This approach is in agreement with modern demands of agricultural, economic, social, and environmental sustainability (Chaparro et al., 2012).

To be efficient in PGP, the PGPB should remain active under a large range of conditions, such as fluctuating pH, temperature, and concentration of different ions. These requirements are not easy to be fulfilled, which explains why several commercial inoculant products are not successful. In addition, to express beneficial features, inoculated strains should also be able to compete successfully with other organisms for nutrients from the root and for niches on the root as well as to escape in sufficient numbers from predators (Jousset et al., 2006). The increase of our understanding about the mechanisms of PGP and on the selection procedures of beneficial bacteria will improve the development of PGPB-based inoculants (Lugtenberg and Kamilova, 2009).

1.7.2 Mechanisms of soybean crops growth promotion by biofertilizers

Soil inoculants/biostimulants based on complex microbial formulas represent a next step in the agricultural industry. The effectiveness of microbial products based on single species is strongly tied to the conditions within the soil to which it is applied and reduced or eliminated outside of optimal environmental conditions. A second key limitation is the nutrient needs of the microbial species. Many or most metabolic processes require multiple input factors. For example, in the case of nitrogen fixation, the primary inputs are atmospheric N and the ATP molecules, a source of cellular energy for an energy-intensive process. Therefore, if a single strain microbial formula is used and the conditions are not correct for the ATP generation, microbial products often fail to perform the task for which they are intended.

Complex microbial products include the essential microbial ecosystem needed to perform designed product capabilities and to provide commercially important processes in a wide variety of soil conditions. The effect of individual inoculation and coinoculation with both strains of *Bradyrhizobium japonicum* and phosphorus-solubilizing *Bacillus* sp. on soybean seedling growth was evaluated. Fig. 1.6 shows the effect of inoculation on the structural characteristics of the soybean leaves.

Our results indicate that inoculated soybean produced maximum seed yield under coinoculation with *B. japonicum* and *Bacillus* sp. (2600 kg ha^{-1}) and minimum in plots without inoculation (1380 kg ha^{-1}) (Fig. 1.7).

The important question is if any selected taxa of soil bacteria could be used as the indicator of the agronomic management. Based on molecular analysis of *nif*H gene we have found that use of the abundance of two bacterial phyla could potentially fulfill this task. It was found that composition of soil diazotrophic communities was different when control and inoculated variants were compared (Fig. 1.8). It was shown that *Alphaproteobacteria* related to

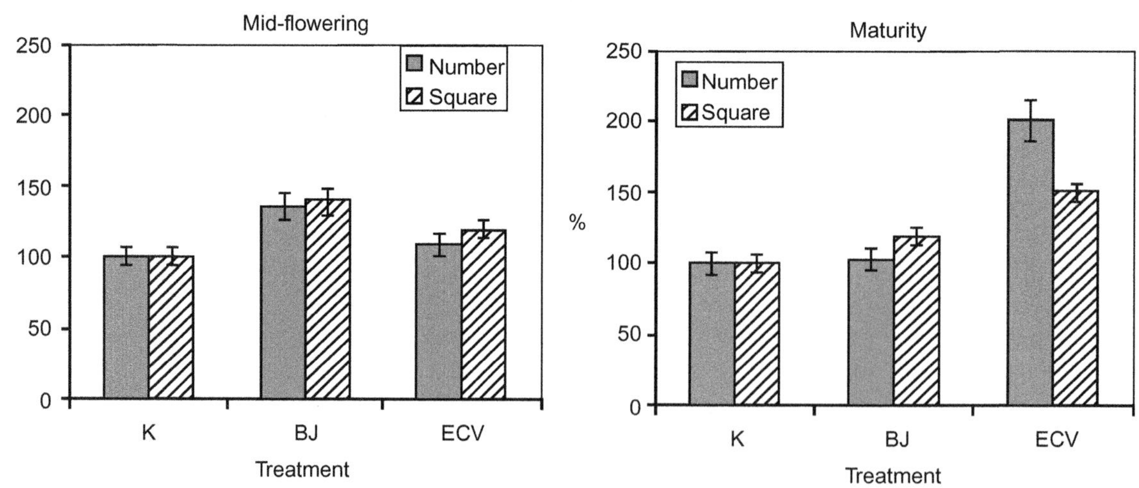

FIGURE 1.6 Effect of bacterial fertilization on structural characteristics of soybean green leaves (number and square for one plant) at midflowering and mature phases in field experiments with Romantica biovar in comparison with noninoculated seeds; % Inoculation of seeds by *Bradyrhizobium* alone (BJ) had no significant ($P < .01$) influence on leaf number and average area as compared with noninoculated seeds (K). Coinoculation with *Bacillus* sp. (ECV) was the most effective especially in maturity stage. Complex microbial inoculation enhanced the chlorophyll content up to 36%, but the effect was insignificant for control and monoinoculation. These data are in accordance with earlier findings (Zarei et al., 2011).

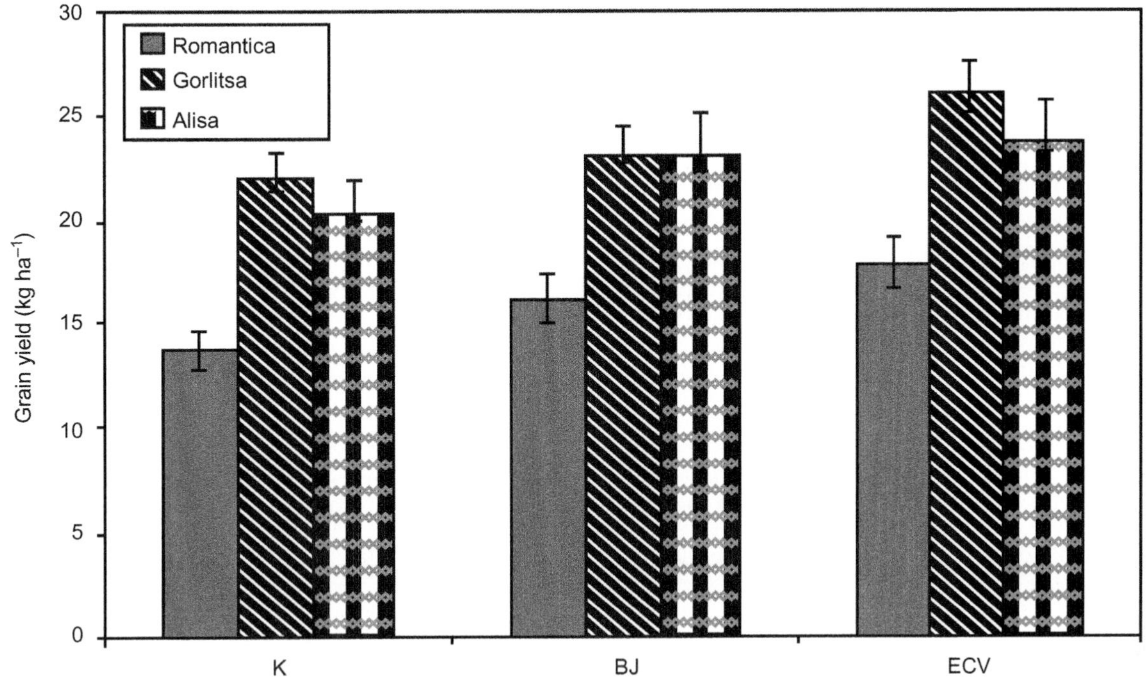

FIGURE 1.7 Effect of bacterial fertilization on grain yield of different soybean biovars.
Abbreviations: K, control without inoculation; BJ, inoculation with *B. japonicum*; ECV, complex inoculation with *B. japonicum* and *Bacillus* sp. *Data from courtesy of Tytova, L.V., Brovko, I.S., Kizilova, A.K., Kravchenko, I.K., Iutynska, G.A., 2013. Effect of complex microbial inoculants on the number and diversity of rhizospheric microorganisms and the yield of soybean. Int. J. Microbiol. Res. 4 (3), 267–274. doi: 10.5829/idosi.ijmr.2013.4.3.1110 (Tytova et al., 2013).*

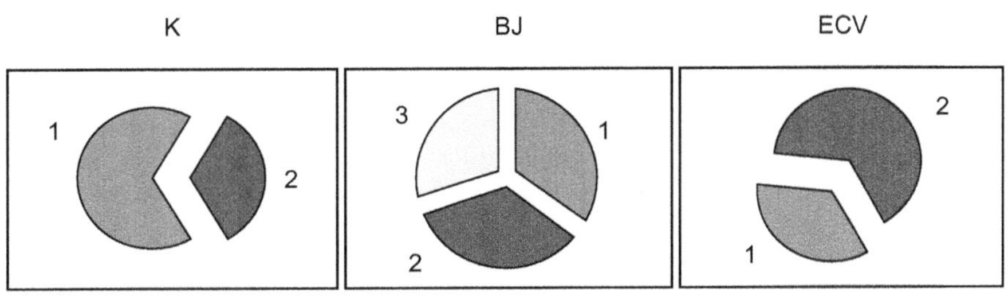

FIGURE 1.8 Effect of bacterial fertilization on *nif*H gene diversity in rhizosphere of Gorlitsa biovar at midflowering stage.
1, Alphaproteobacteria; 2, Clostridium; 3, Paenibacillus. Data from courtesy of Tytova, L.V., Brovko, I.S., Kizilova, A.K., Kravchenko, I.K., Iutynska, G.A., 2013. Effect of complex microbial inoculants on the number and diversity of rhizospheric microorganisms and the yield of soybean. Int. J. Microbiol. Res. 4 (3), 267–274. doi: 10.5829/idosi.ijmr.2013.4.3.1110.

Leptospirillum, Derxia, Azohydromonas were the most abundant in control variants. Both mono- and coinoculation significantly changed the community structure. Predominated diazotrophs were clustered with obligate (*Clostridium*) and facultative (*Paenibacillus*) anaerobes. So, we concluded that the ratio between nitrogen-fixing *Alphaproteobacteria* and *Firmicutes* may be the indicator for effectiveness of bacterial inoculation of soybean.

1.8 Future directions and perspectives

A major focus in the coming decades should be on safe and ecofriendly methods by exploiting the beneficial microorganisms in sustainable crop production and development of low-cost and long shelf-life complex liquid biofertilizers to be used in an innovative and high productivity cultivation system. Future research should focus on managing plant–microbe interactions, particularly with respect to their mode of action and adaptability to conditions under extreme environments. Furthermore, certain issues need to be addressed, such as how to improve the efficacy of biofertilizers, what should be an ideal and universal delivery system, how to stabilize these microbes in the soil systems, and how to control/facilitate nutritional and root exudation aspects to get maximum benefits from biofertilizer application.

The main directions of research in the near future on soybean biofertilizers as the frontiers in soil fertility and fertilizer technology can be formulated as follows:

1. Ecological-friendly soybean production with respect to environmental principles and standards has the highest priority.
2. The observed rise in the N-gap in high-yielding conditions suggests the need to have an additional source for supplying N to the crops. Development of strategies of low-rate N supply, especially during the seed-filling period, which reduce the negative impact of nitrate formation, seems to be a likely solution.
3. Combined use of biofertilizers with organic (citric, malic, humic acids) and mineral, including micronutrient (Zn, Mn, Mo) fertilizers may be the most effective solution to maximize yield and economic profitability of soybean production.
4. Application of modified and new biological technologies in crop production tends to the production of safe food, which would affect human health and environmental quality in a positive way. Yield in organic production is lower compared with conventional production, but looking at the unit price, ecological soybean production is economically justified.
5. In-depth study of microorganisms selected by the plant under certain conditions will allow modeling the rhizosphere for "difficult" habitats of soybean.

1.9 Conclusions

Use of chemical fertilizers has received preference over biological fertilizers as they can be used irrespective of the environment, are easy to store and transport, and have wide-ranging applications. The use of biological fertilizers has gained attention mainly due to the (1) environment-friendly nature of bioinoculants and (2) hazards associated with the chemical fertilizers. The practical use of biological fertilizers is well below its full potential, mainly due to nonavailability of suitable inoculants. In view of the ecological specificity associated with the naturally occurring microorganisms, concerted efforts from various research laboratories are required for developing microbial inoculants for a specific set of climatic conditions. The chance of obtaining a successful inoculant in the end will be greatly enhanced when ecological principles are applied throughout the procedure of development of microbial inoculants. However, before PGPR can contribute the desired benefits, scientists need to learn more and explore ways and means for their better utilization.

Biotechnological and molecular approaches should help in the development of more understanding about the mode of PGPR action leading to successful plant–microbe interaction. Furthermore, proper guidelines for the production and commercialization of biofertilizers should be framed to popularize the use of such bioagents for maintaining the sustainability of agro-ecosystems across the globe, taking due care of required safety measures associated with the use of live cultures.

References

Amann, R., Ludwig, W., 2000. Ribosomal RNA-targeted nucleic acid probes for studies in microbial ecology. FEMS Microbiol. Rev. 24, 555–565.

Aneja, K.R., 2003. Experiments in Microbiology, Plant Pathology and Biotechnology. Age International (P) Limited Publishers, New Delhi.

Argaw, A., 2012. Evaluation of co-inoculation of *Bradyrhizobium japonicum* and phosphate solubilizing *Pseudomonas* spp. Effect on soybean (*Glycine max* L. (Merr.)) in Assossa area. J. Agric. Sci. Technol. 14, 213–224.

Atieno, M., Herrmann, L., Okalebo, R., Lesueur, D., 2012. Efficiency of different formulations of *Bradyrhizobium japonicum* and effect of co-inoculation of *Bacillus subtilis* with two different strains of *Bradyrhizobium japonicum*. World J. Microbiol. Biotechnol. 28 (7), 2541–2550.

Barret, M., Morrissey, J.P., O'Gara, F., 2011. Functional genomics analysis of plant growth-promoting rhizobacterial traits involved in rhizosphere competence. Biol. Fertil. Soils. 47, 729.

Bashan, Y., de-Bashan, L.E., 2010. How the plant growth-promoting bacterium *Azospirillum* promotes plant growth – a critical assessment. Adv. Agron. 108, 77–136.

Belyshkina, M.E., 2013. Analysis and prospective for soybean production in Russia and the world. Kormoproizvodstvo. 7, 3–7. In Russian.

Bending, G.D., Rodriguez-Cruz, M.S., Lincoln, S.D., 2007. Fungicide impacts on microbial communities in soils with contrasting management histories. Chemosphere 69, 82–88.

Berg, G., Opelt, K., Zachow, C., Lottmann, J., Gotz, M., Costa, R., et al., 2006. The rhizosphere effect on bacteria antagonistic towards the pathogenic fungus *Verticillium* differs depending on plant species and site. FEMS Microbiol. Ecol. 56, 250–261.

Boiero, L., Perrig, D., Masciarelli, O., Penna, C., Cassan, F., Luna, V., 2006. Phytohormone production by three strains of *Bradyrhizobium japonicum*, and possible physiological and technological implications. Appl. Microbiol. Biotechnol. 74, 874–880.

Bressan, M., Roncato, M.-A., Bellvert, F., Comte, G., Haichar, F.Z., Achouak, W., et al., 2009. Exogenous glucosinolate produced by *Arabidopsis thaliana* has an impact on microbes in the rhizosphere and plant roots. ISME J. 3, 1243–1257.

Bünemann, E., Schwenke, G., Van Zwieten, L., 2006. Impact of agricultural inputs on soil organisms—a review. Soil Res. 44, 379–406.

Buyer, J.S., Roberts, D.P., Russek-Cohen, E., 2002. Soil and plant effects on microbial community structure. Can. J. Microbiol. 48, 955.

Calderoli, P.A., Mónica, M., Collavino, F., Héctor, B.K., Morrás, J.M., Aguila, O.M., 2017. Analysis of *nif*H-RNA reveals phylotypes related to *Geobacter* and *Cyanobacteria* as important functional components of the N_2-fixing community depending on depth and agricultural use of soil. Microbiol. Open 6, e502. Available from: https://doi.org/10.1002/mbo3.502.

Chaparro, J.M., Sheflin, A.M., Manter, D.K., Vivanco, J.M., 2012. Manipulating the soil microbiometo increase soil health and plant fertility. Biol. Fertil. Soils 48, 489–499.

Danso, S.K., Habte, A.M., Alexander, M., 1973. Estimating the density of individual bacterial populations introduced into natural ecosystems. Can. J. Microbiol. 19, 1145–1151.

De Meyer, S.E., De Beuf, K., Vekemann, B., Willems, A., 2015. A large diversity of nonrhizobial endophytes found in legume root nodules in Flanders (Belgium). Soil Biol. Biochem. 1–11. Available from: https://doi.org/10.1016/j.soilbio.2015.01.002.

Dubey, S.K., 1996. Combined effect of *Bradyrhizobium japonicum* and phosphate-solubilizing *Pseudomonas striata* on nodulation, yield attributes and yield of rainfed soybean (*Glycine max*) under different sources of phosphorus in Vertisols. Ind. J. Microbiol. 33, 61–65.

Egamberdieva, D., Wirth, S.J., Abdulaziz, A., Alqarawi, E.F., Allah, A., Hashem, A., 2017. Phytohormones and beneficial microbes: essential components for plants to balance stress and fitness. Front. Microbiol. 8, 2104.

Esitken, A., Pirlak, L., Turan, M., Sahin, F., 2006. Effects of floral and foliar application of plant growth promoting rhizobacteria (PGPR) on yield, growth and nutrition of sweet cherry. Sci. Hortic. 110, 324–327.

FAO, 2017 http://www.fao.org/news/archive/news-by-date/2017/en/.

Gaby, J.C., Rishishwar, L., Valderrama-Aguirre, L.C., Green, S.J., Valderrama-Aguirre, A., Jordan, I.K., et al., 2018. Diazotroph community characterization via a highthroughput *nif*H amplicon sequencing and analysis pipeline. Appl. Environ. Microbiol. 84, e01512–e01517. Available from: https://doi.org/10.1128/AEM.01512-17.

Gai, Z., Zhang, J., Li, C., 2017. Effects of starter nitrogen fertilizer on soybean root activity, leaf photosynthesis and grain yield. PLoS One. 12 (4), e0174841.

Giller, K.E., McGrath, S.P., Hirsch, P.R., 1989. Absence of nitrogen fixation in clover grown on soil subject to long-term contamination with heavy metals is due to survival of only ineffective *Rhizobium*. Soil. Biol. Biochem. 21, 841–848.

Girvan, M.S., Bullimore, J., Pretty, J.N., Osborn, A.M., Ball, A.S., 2003. Soil type is the primary determinant of the composition of the total and active bacterial communities in arable soils. Appl. Environ. Microbiol. 69, 1800.

Gomes, N.C.M., Fagbola, O., Costa, R., Rumjanek, N.G., Buchner, A., Mendona-Hagler, L., et al., 2003. Dynamics of fungal communities in bulk and maize rhizosphere soil in the tropics. Appl. Environ. Microbiol. 69, 3758–3766.

Graham, T.L., 1991. Flavonoid and isoflavonoid distribution in developing soybean seedling tissues and in seed and root exudates. Plant Physiol. 95, 594–603.

Grayston, S.J., Vaughan, D., Jones, D., 1996. Rhizosphere carbon flow in trees, in comparison with annual plants: the importance of root exudation and its impact on microbial activity and nutrient availability. Appl. Soil. Ecol. 5, 29–56.

Grayston, S.J., Wang, S.Q., Campbell, C.D., Edwards, A.C., 1998. Selective influence of plant species on microbial diversity in the rhizosphere. Soil Biol. Biochem. 30, 369–378.

Han, L.L., Wang, E.T., Han, T.X., Liu, J., Sui, X.H., Chen, W.F., et al., 2009. Unique community structure and biogeography of soybean rhizobia in the saline-alkaline soils of Xinjiang, China. Plant Soil 324, 291–305.

Hartmann, A., Schmid, M., van Tuinen, D., Berg, G., 2009. Plant-driven selection of microbes. Plant Soil 321 (1–2), 235–257.

Heid, C.A., Stevens, J., Livak, K.J., Williams, P.M., 1996. Real-time quantitative PCR. Genome Res. 6, 986–994.

Higuchi, R., Fockler, C., Dollinger, G., Watson, R., 1993. Kinetic PCR analysis: real-time monitoring of DNA amplification reactions. Biotechnology 11, 1026–1030.

Hiltbold, A.E., Patterson, R.M., Reed, R.B., 1985. Soil populations of *Rhizobium japonicum* in a cotton corn-soybean rotation. Soil Sci. Soc. Am. J. 49, 343–348.

Hinsinger, P., Gobran, G.R., Gregory, P.J., Wenzel, W.W., 2005. Rhizosphere geometry and heterogeneity arising from root-mediated physical and chemical processes. New Phytol. 168, 293–303. Available from: https://doi.org/10.1111/j.1469-8137.2005.01512.x.

Igwev A., Vannette R.L., 2018. Plant species and soil type influence rhizosphere bacterial composition and seedling establishment on serpentine soils. bioRxiv 489344. Available from: https://doi.org/10.1101/489344.

Imfeld, G., Vuilleumier, S., 2012. Measuring the effects of pesticides on bacterial communities in soil: a critical review. Eur. J. Soil Biol. 49, 22–30.

Iutynska, G.O., Tytova, L.V., Pinaev, O.G., Andronow, E.E., Vozniuk, S.V., 2017. Microbiome biodiversity of soybean rhizosphere under application of fungicides and inoculation by microbial bioformulation Ecovital. Microbiol. Biotechnol. 1, 23–35. in Russian.

Jones, J.D.L., Nguyen, C., Finlay, R.D., 2009. Carbon flow in the rhizosphere: carbon trading at the soil–root interface. Plant Soil. 321, 5–33. Available from: https://doi.org/10.1007/s11104-009-9925-0.

Jousset, A., Lara, E., Wall, L.G., Valverde, C., 2006. Secondary metabolites help biocontrol strain *Pseudomonas fluorescens* CHA0 to escape protozoan grazing. Appl. Environ. Microbiol. 72, 7083–7090.

Kirk, J.L., Beaudette, L.A., Hart, M., Moutoglis, P., Klironomos, J.N., Lee, H., et al., 2004. Methods of studying soil microbial diversity. J. Microbiol. Methods 58, 169–188.

Kizilova, A.K., Titova, L.V., Kravchenko, I.K., Iutinskaya, G.A., 2012. Evaluation of the diversity of nitrogen fixing bacteria in soybean rhizosphere by *nif*H gene analysis. Microbiology 81 (5), 621–629.

Kowalchuk, G.A., Buma, D.S., De Boer, W., Klinkhamer, P.G.L., van Veen, J.A., 2002. Effects of above-ground plant species composition and diversity on the diversity of soil-borne microorganisms. Antonie Van Leeuwenhoek 81, 509–520.

Kuklinsky-Sobral, J., Araújo, W.L., Mendes, R., Geraldi, I.O., Pizzirani-Kleiner, A.A., Azevedo, J.L., 2004. Isolation and characterization of soybean-associated bacteria and their potential for plant growth promotion. Environ. Microbiol. 6 (12), 1244–1251. Available from: https://doi.org/10.1111/j.1462-2920.2004.00658.x.

Kumar, A., Kaushal, S., Saraf, S.A., Singh, J.S., 2018. Microbial bio-fuels: a solution to carbon emissions and energy crisis. Front. Biosci. (Landmark ED) 1 (23), 1789–1802.

Lakshmanan, V., Selvaraj, G., Bais, H.P., 2014. Functional soil microbiome: Belowground solutions to an aboveground problem. Plant Physiol. 166, 689–700.

Lam, S.K., Hao, X., Lin, E., Han, X., Norton, R., Mosier, A.R., et al., 2012. Effect of elevated carbon dioxide on growth and nitrogen fixation of two soybean cultivars in northern China. Biol. Fertil. Soils 48 (5), 603–606. Available from: https://doi.org/10.1007/s00374-011-0648-z.

Leite, J., Fischer, D., Rouws, L.F.M., Fernandes-Júnior, P.I., Hofmann, A., Kublik, S., et al., 2017. Nodules harbor Nonrhizobial bacterial communities that are shaped by soil type rather than plant genotype. Front. Plant Sci. 7, 2064. Available from: https://doi.org/10.3389/fpls.2016.02064.

Li, X., Sun, M., Zhang, H., Xu, N., Sun, G., 2016. Use of mulberry-soybean intercropping in salt-alkali soil impacts the diversity of the soil bacterial community. Microb. Biotechnol. 9 (3), 293–304.

Linu, M.S., Stephen, J., Jisha, M.S., 2009. Phosphate solubilizing *Gluconacetobacter* sp. (MTCC 8368) and *Burkholderia* sp. (MTCC 8369) and their potential interaction with cowpea (*Vigna unguiluiculata* L. walp). Int. J. Agric. Res. 4 (2), 79–87. Available from: https://doi.org/10.3923/ijar.2009.79.87.

Liu, C.M., McLean, P.A., Sookdeo, C.C., Cannon, F.C., 1991. Degradation of the herbicide glyphosate by members of the family *Rhizobiaceae*. Appl. Environ. Microbiol. 57 (6), 1799–1804.

Loman, N.J., Constantinidou, C., Chan, J.Z., Halachev, M., Sergeant, M., Penn, C.W., et al., 2012. High-throughput bacterial genome sequencing: an embarrassment of choice, a world of opportunity. Nat. Rev. Microbiol. 10, 599-206.

Ludwig, W., Strunk, O., Westram, R., Richter, L., Meier, H., Yadhukumar, et al., 2004. ARB: a software environment for sequence data. Nucleic Acids Res. 32, 1363–1371.

Lugtenberg, B., Kamilova, F., 2009. Plant-growth-promoting rhizobacteria. Annu. Rev. Microbiol. 63, 541–556.

Lynch, J.M., 1990. Some consequences of microbial rhizosphere competence for plant and soil. In: Lynch, J.M. (Ed.), The Rhizosphere. Wiley Interscience, New York, pp. 1–10.

Magi, A., Benelli, M., Gozzini, A., Girolami, F., Torricelli, F., Brandi, M.L., 2010. Bioinformatics for next generation sequencing data. Genes 1, 294–307.

Mardis, E.R., 2008. Next-generation DNA sequencing methods. Ann. Rev. Genomics Hum. Genet. 9, 387–402.

McGrath, K.C., Dombrecht, B., Manners, J.M., Schenk, P.M., Edgar, C.I., Maclean, D.J., et al., 2005. Repressor- and activator-type ethylene response factors functioning in jasmonate signaling and disease resistance identified via a genome-wide screen of *Arabidopsis* transcription factor gene expression. Plant Physiol. 139, 949–959.

McNear, J.D.H., 2013. The Rhizosphere - roots, soil and everything in between. Nat. Educ. Knowl. 4 (3), 1.

Mendes, L.W., Kuramae, E.E., Navarrete, A.A., van Veen, J.A., Tsai, S.M., 2014. Taxonomical and functional microbial community selection in soybean rhizosphere. ISME J. 8 (8), 1577–1587.

Miransari, M., 2013. Soil microbes and the availability of soil nutrients. Acta. Physiol. Plant. 35 (11), 3075–3084. Available from: https://doi.org/10.1007/s11738-013-1338-2.

Morgan, J.A.W., Whipps, J.M., 2001. Methodological approaches to the study of rhizosphere carbon flow and microbial population dynamics. In: Pinton, A., Varanini, Z., Nannipieri, P. (Eds.), The Rhizosphere: Biochemistry and Organic Substances at the Soil-Plant Interface. Marcel Dekker, New York, pp. 373–409.

Moter, A., Göbel, U.B., 2000. Fluorescence in situ hybridization (FISH) for direct visualization of microorganisms. J. Microbiol. Methods 41, 85–112.

Moulin, L., Béna, G., Boivin-Masson, C., Stepkowski, T., 2004. Phylogenetic analyses of symbiotic nodulation genes support vertical and lateral gene co-transfer within the *Bradyrhizobium* genus. Mol. Phylogenet. Evol. 30, 720–732.

Newman, M.M., Hoilett, N., Lorenz, N., Dick, R.P., Liles, M.R., Ramsier, C., et al., 2016. Glyphosate effects on soil rhizosphere-associated bacterial communities. Sci. Total Environ. 543, 155–160.

Nie, M., Zhang, X.D., Wang, J.Q., et al., 2009. Rhizosphere effects on soil bacterial abundance and diversity in the Yellow river deltaic ecosystem as influenced by petroleum contamination and soil salinization. Soil Biol. Biochem. 41, 2535–2542.

Okubo, T., Ikeda, S., Kaneko, T., Eda, S., Mitsui, H., Sato, S., et al., 2009. Nodulation-dependent communities of culturable bacterial endophytes from stems of field-grown soybeans. Microbes Environ. 24, 253–258.

Peix, A., Ramírez-Bahena, M.H., Velázquez, E., Bedmar, E.J., 2015. Bacterial associations with legumes. Crit. Rev. Plant Sci. 34, 17–42. Available from: https://doi.org/10.1080/07352689.2014.897899.

Phillips, D.A., Tsai, S.M., 1992. Flavonoids as plant signals to rhizosphere microbes. Mycorrhiza 1 (2), 55–58.

Postma, J., van Elsas, J.D., Govaert, J.M., van Veen, J.A., 1988. The dynamics of the *Rhizobium leguminosarum* biovar trifolii introduced as determined by immunofluorescence and selective plating technique. FEMS Microbiol. Ecol. 53, 251–260.

Prashar, P., Kapoor, N., Sachdeva, S., 2014. Rhizosphere: its structure, bacterial diversity and significance. Rev. Environ. Sci. Biotechnol. 13 (1), 63–77. Available from: https://doi.org/10.1007/s11157-013-9317-z.

Rao, N.S., 1977. Soil Microorganisms and Plant Growth. Oxford and IBH Publishing Co.

Ryu, C.M., Farag, M.A., Hu, C.H., Reddy, M.S., Kloepper, J.W., Pare, P.W., 2004. Bacterial volatiles induce systemic resistance in Arabidopsis. Plant Physiol. 134, 1017–1026.

Saito, A., Mitsui, H., Hattori, R., Minamisawa, K., Hattori, T., 1998. Slow-growing and oligotrophic soil bacteria phylogenetically close to *Bradyrhizobium japonicum*. FEMS Microb. Ecol. 25, 277–286.

Sanon, A., Andrianjaka, Z.N., Prin, Y., 2009. Comparison of soil bacterial communities in rhizosphere of three plant species and the interspaces in an Arid Grassland. Plant Soil 321 (1–2), 259–278.

Sarita, S., Priefer, U.B., Prell, J., Sharma, P.K., 2008. Diversity of *nif*H gene amplified from rhizosphere soil DNA. Curr. Sci. 94 (1), 109–115.

Sawada, H., Kuykendall, L.D., Young, J.M., 2003. Changing concepts in the systematics of bacterial nitrogen-fixing legume symbionts. J. Gen. Appl. Microbiol. 49, 155–179.

Schmidt, E.L., Bankole, R., Bohlool, B.B., 1968. Fluorescent-antibody approach to study the rhizobia in soil, *J. Bacteriol.* 51, 1212–1215.

Shober, A.L., Taylor, R.W., 2014. Nitrogen Management for Soybean. University of Delaware, Newark, DE. Available from: http://extension.udel.edu/factsheet/nitrogen-management-for-soybean.

Shokralla, S., Spall, J.L., Gibson, J.F., Hajibabaei, M., 2012. Next-generation sequencing technologies for environmental DNA research. Mol. Ecol. 21, 1794–1805.

Simonis, A.D., Setatou, H.B., 1996. Effect of nitrogen fertilization and plant population on the yield of soybeans. In: Rodriguez-Barrueco, C. (Ed.), Fertilizers and Environment. Developments in Plant and Soil Sciences, 66. Springer, Dordrecht, pp. 241–244.

Sinegovskii, M., Yuan, S., Sinegovskaya, V., Han, T., 2018. Current status of the Soybean Industry and Research in the Russian Federation. Soybean Sci. 37 (1), 1–7. Available from: https://doi.org/10.11861/j.issn.1000-9841.2018.01.0001.

Singh, J.S., 2013. Plant growth promoting rhizobacteria: potential microbes for sustainable agriculture. Resonance 18 (3), 275–281.

Singh, J.S., 2015. Microbes: the chief ecological engineers in reinstating equilibrium in degraded ecosystems. Agric. Ecosyst. Environ. 203, 80–82.

Singh, J.S., Gupta, V.K., 2018. Soil microbial biomass: a key soil driver in management of ecosystem functioning. Sci. Total Environ. 634, 497–500.

Singh, A.V., Shan, S., Prasad, B., 2010. Effect of phosphate solubilizing bacteria on plant growth promotion and nodulation in soybean (*Glycine max* (L.) Merr.). J. Hill Agric. 1 (1), 35–39.

Singh, C., Tiwari, S., Gupta, V.K., Singh, J.S., 2018. The effect of rice husk biochar on soil nutrient status, microbial biomass and paddy productivity of nutrient poor agriculture soils. Catena 171, 485–493.

Smith, C.J., Osborn, A.M., 2009. Advantages and limitations of quantitative PCR (qPCR)-based approaches in microbial ecology. FEMS Microbiol. Ecol. 67, 6–20.

Sugiyama, A., Ueda, Y., Zushi, T., Takase, H., Yazaki, K., 2014. Changes in the bacterial community of soybean rhizospheres during growth in the field. PLoS One 9 (6), e100709. Available from: https://doi.org/10.1371/journal.pone.0100709.

Tang, J.E., Bromfield, S.P., Rodrigue, N., Cloutier, S., Tambong, J.T., 2012. Microevolution of symbiotic *Bradyrhizobium* populations associated with soybeans in east North America. Ecol. Evol. 2 (12), 2943–2961. Available from: https://doi.org/10.1002/ece3.404.

Tilba, V.A., Golodyaev, G.P., 1966. Nitrogen fixing bacteria of Primorskii krai arable soils. In: Velikov, I.F. (Ed.), Problemy biologii na Dal'nem Vostoke (Problems of Biology in the Far East). DVF SO ANSSSR, Vladivostok, pp. 152–170. (in Russian).

Timmuck, S., Grantcharova, N., Wagner, E.G.H., 2005. *Paenibacillus polymyxa* invades plant roots and forms biofilm. Appl. Environ. Microbiol. 71 (11), 7292–7300.

Tiwari, P., Singh, J.S., 2017. A plant growth promoting rhizospheric *Pseudomonas aeruginosa* strain inhibits seed germination in *Triticum aestivum* (L) and *Zea mays* (L). Microbiol. Res. 8, 1–7.

Tiwari, S., Singh, C., Singh, J.S., 2018. Land use changes: a key ecological driver regulating methanotrophs abundance in upland soils. Energy Ecol. Environ. 3, 355–371.

Top, E., Mergeay, M., Springael, D., Vestrate, W., 1990. Gene escape model: transfer of heavy metal resistance genes from *Escherichia coli* to *Alcaligenes eutrophus* on agar plates and soil samples. Appl. Environ. Microbiol. 56, 2471–2478.

Turner, T.R., James, E.K., Poole, P.S., 2013a. The plant microbiome. Genome Biol. 14, 209–218.

Turner, T.R., Ramakrishnan, K., Walshaw, J., Heavens, D., Alston, M., Swarbreck, D., et al., 2013b. Comparative metatranscriptomics reveals kingdom level changes in the rhizosphere microbiome of plants. ISME J. 7, 2248–2258.

Tytova, L.V., Brovko, I.S., Kizilova, A.K., Kravchenko, I.K., Iutynska, G.A., 2013. Effect of complex microbial inoculants on the number and diversity of rhizospheric microorganisms and the yield of soybean. Int. J. Microbiol. Res. 4 (3), 267–274. Available from: https://doi.org/10.5829/idosi.ijmr.2013.4.3.1110.

Unkovich, M.J., Pate, J.S., 2000. An appraisal of recent field measurements of symbiotic N_2 fixation by annual legumes. Field Crop. Res. 65, 211–228.

Van Berkum, P., Eardly, B.D., 2002. The aquatic budding bacterium *Blastobacter denitrificans* is a nitrogen-fixing symbiont of *Aeschynomene indica*. Appl. Environ. Microbiol. 68, 1132–1136.

Van Elsas, J.D., Bersma, F.G.H., 2011. A review of molecular methods to study the microbiota of soil and the mycosphere. Eur. J. Soil. Biol. 47, 77–87.

Venturi, V., Keel, C., 2016. Signaling in the rhizosphere. Trends Plant Sci. 21 (3), 187–198. Available from: https://doi.org/10.1016/j.tplants.2016.01.005.

Vessey, J.K., 2003. Plant growth promoting rhizobacteria as biofertilizers. Plant Soil 255, 571–586.

Vimal, S.R., Patel, V.K., Singh, J.S., 2018. Plant growth promoting *Curtobacterium albidum* strain SRV4: an agriculturally important microbe to alleviate salinity stress in paddy plants. Ecol. Indic (Article in Press).

Vincent, J.M., 1982. The basic serology of rhizobia, p. 13–26. In: Vincent, J.M. (Ed.), Nitrogen Fixation in Legumes. Academic Press, Inc, New York.

Vryzas, Z., 2016. The Plant as metaorganism and research on next-generation systemic pesticides – prospects and challenges. Front. Microbiol. 7, 1968. Available from: https://doi.org/10.3389/fmicb.2016.01968.

Wang, H., Wang, S.D., Jiang, Y., Zhao, S.J., Chen, W.X., 2014. Diversity of rhizosphere bacteria associated with different soybean cultivars in two soil conditions. Soil Sci. Plant Nutr. 60 (5), 630–639. Available from: https://doi.org/10.1080/00380768.2014.942212.

Wasule, D., Wadyalkar, S., Buldeo, A., 2007. Effect of phosphate solubilizing bacteria on role of Rhizobium on nodulation by soybean. In: Velazquez, E, Rodriguez-Barrueco, C. (Eds.), First International Meeting on Microbial Phosphate Solubilization, Springer, pp. 139–142.

Whips, J.M., 1987. Effect of media on growth and interactions between a range of soil-borne glasshouse pathogens and antagonistic fungi. New Phytol. 107, 127–142.

Willems, A., Coopman, R., Gillis, M., 2001. Phylogenetic and DNA-DNA hybridization analyses of *Bradyrhizobium* species. Int. J. Syst. Evol. Microbiol. 51, 111–117.

Xu, Y., Wang, G., Jin, J., Liu, J., Zhang, Q., Liu, X., 2009. Bacterial communities in soybean rhizosphere in response to soil type, soybean genotype, and their growth stage. Soil Biol. Biochem. 41, 919–925.

Yan, J., Han, X.Z., Ji, Z.J., Li, Y., Wang, E.T., Xie, Z.H., et al., 2014. Abundance and diversity of soybean-nodulating *Rhizobia* in Black soil are impacted by land use and crop management. Appl. Environ. Microbiol. 80 (17), 5394–5402. Available from: https://doi.org/10.1128/AEM.01135-14.

Yan, Y., Kuramae, E.E., de Hollander, M., Klinkhamer, P.G.L., van Veen, J.A., 2017. Functional traits dominate the diversity-related selection of bacterial communities in the rhizosphere. ISME J. 11, 56–66. Available from: https://doi.org/10.1038/ismej.2016.108.

Zablotowisz, M., Reddy, K.N., 2004. Impact of glyphosate on the *Bradyrhizobium japonicum* symbiosis with glyphosate-resistant transgenic soybean: a minireview. J. Environ. Qual. 33 (3), 825–831.

Zahran, H.H., 1999. Rhizobium-legume symbiosis and nitrogen fixation under severe conditions and in an arid climate. Microbiol. Mol. Biol. Rev. 63 (4), 968–989.

Zarei, I., Khah, E.M., Mohammadi, G., Petropoulos, S., 2011. Assessment of growth and yield components following the application of different biological fertilizers on soybean (*Glycine max* L.) cultivation. Aust. J. Crop Sci. 5 (13), 1776–1782.

Zhang, K.R., Song, C.H., Zhang, Y.L., Dang, H.S., Cheng, X.L., Zhang, Q.F., 2018. Global-scale patterns of nutrient density and partitioning in forests in relation to climate. Glob. Change Biol. 24, 536. Available from: https://doi.org/10.1111/gcb.13860.

Zhao, T., Zhang, Y., Di, H.W., Wang, Y.C., Mei-Jun, Z.L., Ma, Z., 2018. Extracellular aminopeptidase modulates biofilm development of *Pseudomonas aeruginosa* by affecting matrix exopolysaccharide and bacterial cell death. Environ. Microbiol. Rep. https://doi.org/10.1111/1758-2229.12682.

Further reading

Ros-soya. su., 2018. Analytics of the Russian Soy Union. Map of the current state of soybean production in Russia regions. Available from: http://www.ros-soya.su/public.aspx?DB47E393.

Chapter 2

Plant–bacterial interactions in management of plant growth under abiotic stresses

Shimaila Ali[1] and Bernard R. Glick[2]

[1]*A&L Biologicals, Agroecological Research Services Centre, London, ON, Canada,* [2]*Department of Biology, University of Waterloo, Waterloo, ON, Canada*

Chapter Outline

2.1 Introduction to plant growth-promoting bacteria

Soil serves as a rich habitat for many life forms including bacteria. One gram of soil typically contains 10^7–10^9 bacterial colony-forming units (CFU). These bacteria may interact with plants and can influence them harmfully, neutrally, or beneficially. Soil bacteria that support plants growth and development are termed plant growth-promoting bacteria (PGPB) (Bashan and Holguin, 1998). These bacteria can inhabit a variety of plant tissues including the plant interior and the plant surface. Plant–microbe interactions have been shown to be key to the adaptation and survival of both the plants and their microbial partners in many ecosystems.

2.1.1 Rhizospheric and endophytic bacteria

Soil bacteria that reside in the area immediately surrounding a plant's root system (rhizosphere) are termed rhizospheric bacteria. Plants continuously exude a portion of their photosynthesized food through their roots to their surroundings; this high concentration of plant nutrients around plant roots subsequently attracts large numbers of soil microorganisms (Glick, 2015). Another group of PGPB is one that goes one step further in interacting with plants and enters and colonizes the interior of plant tissues it enjoys a secure habitat and a more continuous food supply. These plant tissue colonizers are generally harmless to their hosts and are called endophytes. In fact, endophytic bacteria offer more or less the same benefits to their host plants as their rhizospheric counterpart. Fig. 2.1 illustrates the interactions and different PGPB niches with the plants. Bacterial endophytes can be subdivided into two main categories, one that are nonspecific plant colonizers and the other category includes those that are plant and tissue specific.

New and Future Developments in Microbial Biotechnology and Bioengineering. DOI: https://doi.org/10.1016/B978-0-12-818258-1.00002-9

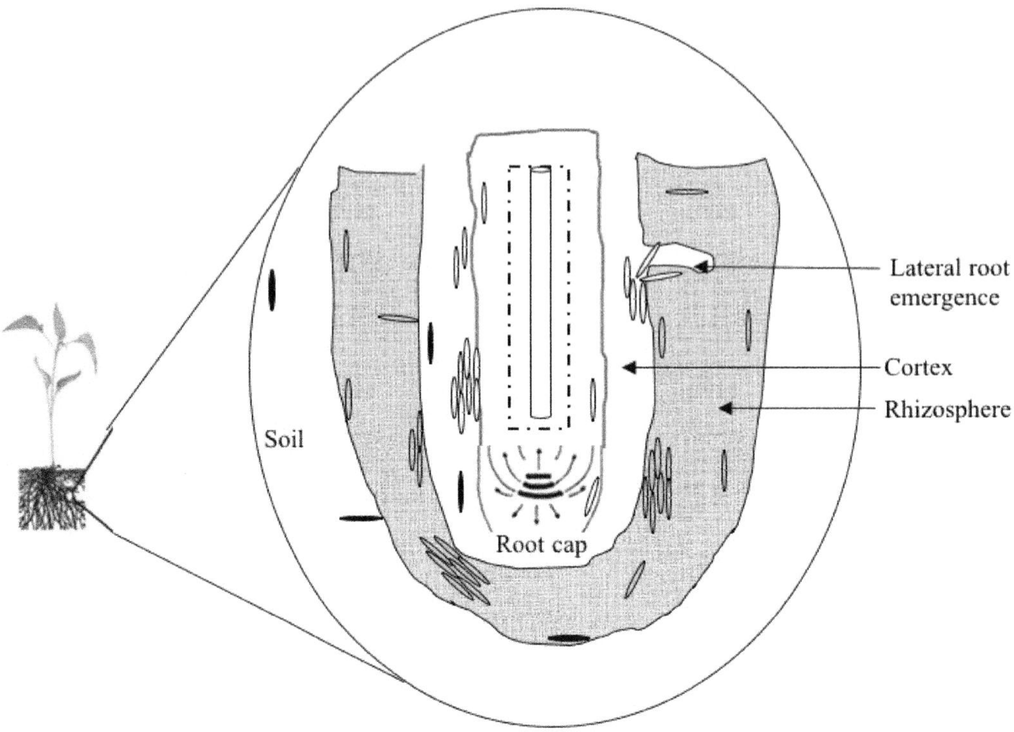

FIGURE 2.1 Schematic representation of plant associated bacteria.
Color key: green rods represent bacterial endophytes, blue rods represent rhizospheric bacteria, and red rods represent plant pathogen.

2.2 Benefits that plants get from PGPB: direct and indirect mechanisms

Whichever environmental niche PGPB colonizes (i.e., the rhizosphere or the plant endosphere) the bacteria always receive benefits from the plants (Singh, 2015a,b, 2016; Singh et al., 2016). In return, PGPBs offer a variety of benefits to plants, which are mainly dependent on the environmental conditions but generally include plant nutrient acquisition, plant hormone production and modulation, and production of bioweapons that the host plant can use in helping the plant to fight the inhibitory effects of phytopathogens that are also found in soil environments. PGPB utilize various mechanisms to benefit plants. The mechanisms where PGPB are actively involved in promoting plant growth are called direct mechanisms, however, PGPB also employ mechanisms that protect plants by relieving some of the growth inhibition caused by numerous phytopathogens; these are called indirect mechanisms of plant growth promotion (Singh and Gupta, 2018; Vimal et al., 2018). With direct plant growth promotion, bacteria either supply or modulate essential phytohormone levels, including auxin, cytokinin, ethylene, and gibberellin, or provide plants with nutrients such as phosphate, fixed nitrogen, potassium, zinc, and iron. On the other hand, the indirect mechanisms include production of lytic enzymes, antibiotics, siderophores, induction of systemic resistance, alteration in ethylene levels, and direct competition with phytopathogens (Glick, 2012).

2.3 Plant colonization

Plant–microbe interactions are initiated as soon as a seed has germinated its very first root is in the soil. The PGPB tend to migrate toward the root, attach onto the root surface, and multiply there to form microcolonies. Soil is probably the most diverse and the most complex environmental microbial niche. Soil typically contains 0.8%–2.0% total organic matter, and for this reason, soil bacteria continuously suffer from limited nutrition supply (Timmusk et al., 2011). Under environmental conditions where the nutrient concentration is insufficient for their continued growth, bacterial cells tend to enter into the stationary phase of growth. Moreover, continuous nutritional depletion compels bacteria to remain in stationary phase (Kolter et al., 1993). With stationary phase bacteria in the same ecological niche as plants that not only make their food but continuously exudes small molecules such as amino acids, sugars, and other organic acids through their roots into the surrounding area (rhizosphere), a number of these starving bacteria (and other microorganisms) are attracted to the rhizosphere to take advantage of this nutrient-rich niche. The root exudation process boosts associated microbial growth and helps to sustain their relationship with plants. Nevertheless, the relationships that are mutualistic in nature have the greatest chance to persist (PGPB–plant) and that those are harmful for the plants (pathogen–plant) end up with the eventual demise of the relationship. On the same note, some bacteria have evolved to possess special apparatus that helps them in initiating and establishing this beneficial interaction. For example, next generation sequence analysis of 16S rRNA genes from the sugarcane rhizosphere and from bulk soils has shown that the microbial population [operational taxonomic unit (OTU)] retrieved from rhizospheric soil exhibited a much higher abundance of the genes involved in biofilm formation, polygalacturonase activity, and with evidence for previous horizontal gene transfer (HGT) processes than the bulk soil microbial communities (Lopes et al., 2016). In a similar study, it was found that *Pseudomonas fluorescens* F113, which produces two flagellar systems, has shown increased motility and better root colonization in alfalfa in root–tip competitive colonization assay (Barahona et al., 2016).

Some PGPB, those that enter and colonize plant tissues (endophytes), utilize mechanisms similar to phytopathogens. In fact, endophytes are sometimes considered disarmed pathogens that have lost their pathogenicity and have evolved into more plant (host) friendly ways. PGPB endophytes may enter the plant through different points; these entry points include tissue wounds (Agarwhal and Shende, 1987), stomata, lenticels (Scot et al., 1996), root cracks (Sørensen and Sessitsch, 2006), and germinating radicles (Gagn et al., 1987). Entry through root cracks is recognized as the main portal of entry for bacterial colonization. Other mechanisms may include penetration of bacteria via root hair cells (Huang, 1986) and through the production of cell wall degradative enzymes (Quadt-Hallmann et al., 1997).

Upon reaching the inside of the plant, an endophyte may localize itself at the point of entry or it may spread throughout the plant (Hallmann et al., 1997). If endophytes are systemically spread, they can reside in intracellular spaces (Patriquin and Döbereiner, 1978), the vascular system (Bell et al., 1995), and/or within the cells (Jacobs et al., 1985). Specialized cellular environments are selected by certain bacterial species because of their tendency to become established in those environments (van Overbeek et al., 2006). For example, the plant xylem was demonstrated to be a selective environment for nitrogen-fixing endophytic bacteria such as *Acetobacter diazotrophicus* (Dong et al., 1994), *Bacillus pumilus* (Benhamou et al., 1996; Benhamou et al., 1998), *Gluconacetobacter diazotrophicus* (Cocking, 2003), *Herbaspirillum seropedicae* (James et al., 2002), *Klebsiella pneumoniae* (Dong et al., 2003), and *Serratia marcescens* (Gyaneshwar et al., 2001). On the other hand, sieve tubes (phloem) are a habitat for certain pathogenic phytoplasma (Bové and Garnier, 2003).

2.4 Microbiome

The term *microbiome* refers to the entire collection of microorganisms present in a specific environmental niche at a given time. Microbiomes vary from environment to environment and are generally considered to be tissue specific. A plant starts establishing its microbiome from the time that it is a seed, which is referred to as the seed-borne microbiome. The most crucial time of a plant's life cycle is the time when a seed is underground and interacts with the surrounding soil, which includes microbes that are potentially endophytic, epiphytic, and rhizospheric and most likely these interactions are imprinted on the subsequent plant health, ecological behavior, environmental fitness, and productivity (Nelson, 2018). A plant may have significantly different microbiome across its tissues, that is, different sets of microbes colonize the rhizosphere, roots, and shoots (leaves and stem). The rhizosphere is typically the first place where most of the plant–microbe interactions occur. Plant roots are the hotspots and host various microbiota at the soil–root interface, rhizoplane, and endorhizosphere, and select them to colonize the plant (Reinhold-Hurek et al., 2015). Moreover, the plant microbiome can change significantly over time. In one study carried out on the microbiome of various tissues of corn over time (Ali et al., unpublished data), it was shown that the microbial population of corn rhizospheric soil (soil close to the roots) is significantly different than the root microbiome and also than the sap microbiome (Fig. 2.2A and B). These differences persisted over the time from 30 days after planting (DAP) to 60–90 DAP. Similarly, a large-scale field study on the corn rhizospheric microbiome demonstrated that various corn lines harbor specific microbiota that can be heritable, and their relative abundance is influenced by plant age, soil fertility, corn genotype, and weather (Walters et al., 2018). It is to be noted that a microbiome is a mixed population of different taxa, where each of the members is somehow adapted and synchronized with the other members of the community. Due to the complex nature of the mechanisms these microbial communities utilize to get assembled, it is quite challenging to study these interactions in vivo. However, a study of an artificial microbiome consisting of seven common and abundant genera found in corn root was conducted (Niu et al., 2017). In this study, corn seedlings were grown and maintained axenically and inoculated with this simple synthetic bacterial community by using a selective culture-dependent method that can track down the population of each of the strain. The bacteria that were used in this study included *Enterobacter cloacae*, *Stenotrophomonas maltophilia*, *Ochrobactrum pituitosum*, *Herbaspirillum frisingense*,

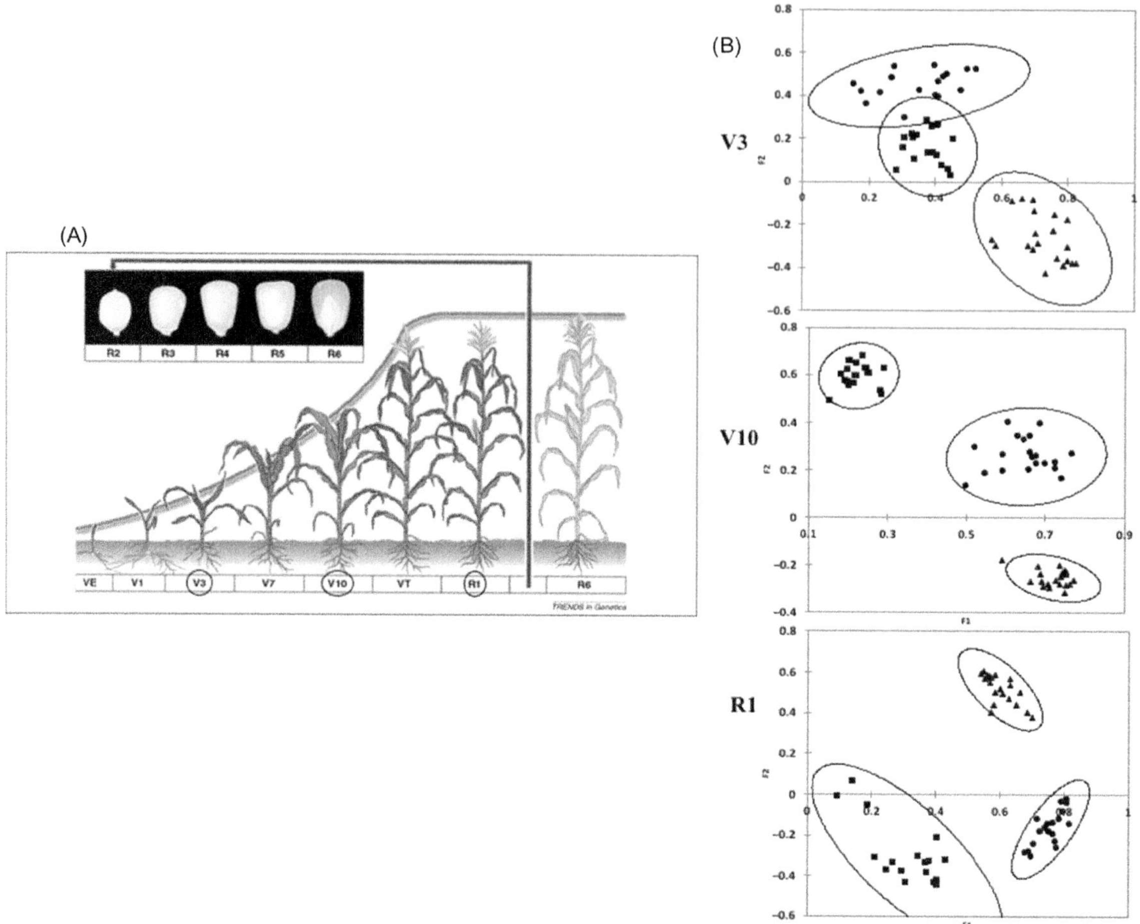

FIGURE 2.2 (A) Growth stages of corn. (B) Terminal restriction fragment length polymorphism (TRFLP) analysis of bacterial populations in various corn tissues sampled at three different growth stages. The TRFLP data was analyzed by principal component analysis (PCA); each dot represents one plant (20 replications). Color key: green rods represent corn sap microbiome, blue rods represent rhizosphere microbiome, and red rods represent plant root microbiome.

Pseudomonas putida, *Curtobacterium pusillum*, and *Chryseobacterium indologenes* (Niu et al., 2017). Furthermore, it was found that when *E. cloacae* was removed from this bacterial consortium, the complete structure of the community was lost and in return, *C. pusillum* took over (Niu et al., 2017). PGPB (both rhizospheric and endophytic) have advantage over the other soil bacteria in that they establish a mutualistic plant–bacterial interaction where both partners receive benefits, since the selection and colonization of a microbiome are hugely influenced by a plant's genotype and by abiotic factors that may create both negative and positive selection pressures (Gaiero et al., 2013). For instance, a bloom in the population of PGP-*Pseudomonas* was observed in corn plants (in the field study mentioned above) at week 8 and onwards, where the population was increased from 2.7% to 44.5% (Walters et al., 2018).

Interestingly, when the soil microbiome of an organic farm was compared with the microbiome of a conventional farm by using 16S rRNA gene next generation sequences, it was found that the organic farm contained higher microbial diversity than the conventional farm (Lupatini et al., 2017). Similarly, the type of cropping systems practiced in specific farming have been found to be significant in selecting and enriching specific microbiota in their rhizosphere and root systems that ultimately resulted in different crop yields. It was documented that intercropping millet with woody shrub significantly changed the root zone microbiome (both bacteria and fungi) and improved the yields of millet (Debenport et al., 2015). The natural soil microbiome has been described to affect flowering timing in field mustard (*Brassica rapa*), where soil microbiota accelerated the flowering under drought conditions compared with the wet-adapted control plants grown with the same microbial community (Lau and Lennon, 2012); similar impacts of the soil microbial community have been described for *Boechera stricta* (a wild relative of *Arabidopsis*), which was grown in four natural soils, and the timing to flowers was monitored and compared with a control grown in sterilized soil (Wagner et al., 2014).

The microbiome of the soils was quantified, and genera including *Proteobacteria*, *Acidobacteria*, *Verrucomicrobia*, *Actinobacteria*, *Gemmatimonadetes*, *Chloroflexi*, *Planctomycetes*, *Bacteroidetes*, and *Crenarchaeota* were suggested to play an important role in changing flowering time in *B. stricta* (Wagner et al., 2014).

2.5 Plants' response to abiotic stress

Like other living beings, plants require water and energy to drive their growth and development. Plants are unique in that they prepare their food through photosynthesis using sunlight, water, and carbon dioxide. Unlike other life forms, plants are sessile and have to face and respond to the surrounding environmental conditions. The abiotic conditions that are not favorable for plant growth and development include temperature extremes, water stress (flooding and drought), and the presence of growth inhibitors (heavy metals, organic phytotoxic compounds, salt, etc.). Plants promptly sense the changes in the environment and respond accordingly. The physiological symptoms that a typical plant shows under stress include reduction in leaf size, leaf curling or flatting, prolonged stomata opening, reduction in seed number, size and viability, delay in flowering and fruiting, stunted growth, and limited productivity (Kamlesh et al., 2017). Changes at cellular levels that plants undergo during stress vary depending on the nature of the stress stimulus; for instance, prolonged light exposure induces the production of high reactive oxygen species (ROS) that interacts with the biomolecules and enzymes, which are ultimately inhibitory to plant health. Similarly, pH variation in the surrounding soil affects nutrient availability, and at extreme pHs plants face growth-limiting nutrient deficiencies (Rorison, 1986). Under prolonged salinity stress, plants suffer from osmotic and ionic imbalances within cells that results in stunted plant growth (Munns and Tester, 2008). Some plants have evolved mechanisms to combat salinity stress and have become tolerant up to certain levels of salts found in the rhizosphere (Kamlesh et al., 2017). The most common mechanisms that plants employ to resist and to protect plant tissues from salt stress is to transport and exude ions into the root zone to make it neutral and suitable for growth and to create a compartment between ions (salt) and the cytoplasm of physiologically active cells (Silva et al., 2010). Plants initiate instant responses upon sensing stress signals from their surroundings. These responses primarily are stress specific that are started in an effective and coordinated manner and may include accumulation of phenolic and flavonoid compounds, increased production of ROS, secretion of osmolytes and molecules that can serve as antioxidants, activation of transcription regulatory factors, stress related gene activation, and production of plant hormones (mainly abscisic acid, jasmonic acid, salicylic acid, and ethylene) (Spoel and Dong, 2008; Qin et al., 2011; Tiwari et al., 2011; Todaka et al., 2012; Kamlesh et al., 2017).

2.6 Overview of PGPB direct mechanisms of facilitating plant growth

2.6.1 Resource acquisition

2.6.1.1 Nitrogen fixation

Nitrogen is a key macronutrient that plants require for their growth and development. Plants cannot utilize atmospheric nitrogen directly; the nitrogen must first be reduced to be utilized by plants (Lichtfouse, 2012). Many plant associated microflorae can reduce nitrogen and provide it to plants; that is, the process of biological nitrogen fixation (BNF). The BNF offers an environmentally friendly alternative to the use of synthetically produced N-fertilizers, which can have a major disruptive impact on the nitrogen cycle (Döbereiner and Urquiaga, 1992; Singh et al., 2011; Maheshwari et al., 2013). Since endophytic diazotrophs (nitrogen-fixing bacteria) colonize the interior portion of plants where the partial oxygen pressure is lower than in the atmosphere, these organisms can provide greater amounts of fixed nitrogen to their host plants compared with many rhizospheric bacteria (James and Olivares, 1998; Maheshwari et al., 2013). The nitrogen fixation ability of a bacterium is characterized by one or more of the following methods: (1) total N difference method, (2) acetylene reduction assay, (3) comparison of nitrogen compounds in plant tissues, mainly xylem, (4) ^{15}N isotope labeling, and (5) direct detection of *nif* gene(s) (Gupta et al., 2012). The first four methods can quantify the amount of nitrogen fixed by the bacteria whereas the fifth method only provides information about the nitrogen fixation potential of a particular bacterium (qualitative). Molecular screening for nitrogenase protein can also be analyzed by (1) in situ hybridization with fluorophore-labeled probes and (2) immune blot and/or western blot (Gupta et al., 2012).

Nitrogen fixation by PGPB is a widespread as well as an important trait. Table 2.1 summarizes some of the commonly known diazotrophic PGPB. Members from various bacteria genera namely *Acetobacter*, *Azoarcus*, *Azospirillum*, *Azotobacter*, *Burkholderia*, *Enterobacter*, *Herbasprillum*, *Glucenobacter*, and *Pseudomonas* have been classified as diazotrophic PGPB (Baldani et al., 1997; Cocking, 2003; Ali et al., 2018). In one study, a diazotrophic endophyte from surface sterilized stems of wild cottonwood was isolated and subsequently identified as *Burkholderia vietnamiensis*

TABLE 2.1 Bacterial diazotrophs from different host plants.

Diazotrophic PGPB	Plant (s)	Reference
Acinetobacter sp., *Burkholderia* sp. *Herbaspirillum* sp., *Pseudomonas* sp. *Rahnella* sp., *Sphingomonas* sp. *Stenotrophomonas* sp.	Poplar and willow	Doty (2008)
Azospirillum sp., *Brevundimonas* sp. *Enterobacter* sp., *Herbaspirillum* sp. *Pantoea* sp., *Pseudomonas* sp. *Rhanella* sp., *Rhizobium* sp., *Agrobacterium* sp.	Corn	Montañez et al. (2012)
Azospirillum sp. *Azotobacter* sp., *Bacillus* sp., *Beijerinckia* sp., *Derxia* sp., *Enterobacter* sp., *Erwinia* sp. *Gluconacetobacter* sp.	Sugarcane	Maheshwari et al. (2013), James and Olivares (1998), Maheshwari et al. (2013)
Herbaspirillum sp.	Rice, elephant grass, maiden grass, Amur silver grass, prairie cord grass, maize, sorghum, banana, African palm, and pineapple	Maheshwari et al. (2013)
Acinetobacter sp. *Enterobacter* sp., *Ralstonia* sp. *Burkholderia* sp., *Pseudomonas* sp.	Soybean	Kuklinsky-Sobral et al. (2004), Maheshwari et al. (2013)
Azoarcus sp., *Gluconacetobacter* sp., *Herbaspirillum* sp.	Grass	Reinhold-Hurek and Hurek (1998)
Gluconobacter sp.	Pineapple, sorghum, African millet, and coffee	Maheshwari et al. (2013)
Burkholderia sp.	Sugarcane, maize, and teosinte plants Maize and teosinte plants	Govindarajan et al. (2006), Luvizotto et al. (2010), Reis et al. (2004)
Azospirillum sp.	Graminaceous and nonleguminous plants	Baldani et al. (1997)
Burkholderia sp.	Corn	Estrada et al. (2002)
B. cepacia	Rice	Singh et al. (2011)
B. tropica subspecies *nov.*	Corn	Reis et al. (2004)
B. vietnamiensis	Wild cottonwood	Xin et al. (2009)
Herbaspirillum sp.	Sorghum	James et al. (1997)
Herbaspirillum sp.	Palm tree	Reis et al. (2000)
H. seropedicae	Corn	Baldani et al. (1992), Dobereiner et al. (1993)
Klebsiella sp.	Sugarcane	Govindarajan et al. (2007)
K. pneumoniae	Corn	Chelius and Triplett (2000)
Rhizobium etli	Corn, common bean	Martínez-Romero (2003)
S. maltophilia	Ginseng, various crops	Park et al. (2005)
Rhizobium sp.	Ginseng	Maheshwari et al. (2013)
Azospirillum sp., *Enterobacter* sp. *Rhizobium* sp.	Potato	Naqqash et al. (2016)

(Xin et al., 2009). The molecular identification of this bacterium was done by analysis of *recA* and 16S rRNA genes, and its nitrogen fixation ability was tested by the amplification of the *nifHDK* gene cluster and then confirmed by ^{15}N isotope incorporation and acetylene reduction assays (Xin et al., 2009). This diazotrophic endophyte promoted the growth of Kentucky bluegrass where bacterially treated plants exhibited a significant increase in dry weight and nitrogen contents compared with untreated control plants (Xin et al., 2009). In another study, a group of 17 bacteria was isolated from surface sterilized stems of poplar and willow plants and grown on nitrogen-limited medium (Doty et al., 2009). The diazotrophic nature of these bacteria was tested by *nifH* gene amplification and acetylene reduction assay. Seven of the isolates tested positive for the presence of the *nifH* gene, five of the isolates reduced acetylene, and almost all were able to grow on nitrogen-free medium (Doty et al., 2009). It was therefore suggested that these isolates, which include *Burkholderia*, *Rahnella*, *Sphingomonas*, *Acinetobacter*, *Stenotrophomonas*, *Herbaspirillum*, and *Pseudomonas*, might be utilized as biofertilizers (Doty et al., 2009).

A study by Montañez et al. (2012) described the isolation and identification of diazotrophic bacteria in a variety of commercial corn cultivars by first testing the bacteria by the ^{15}N isotope dilution method. The nitrogen fixation capabilities of the isolated bacterial strains were then checked by the acetylene reduction assay and finally verified by amplifying the *nifH* gene. Only 6.2% of the isolates (11 of 178) were found to have this activity and hence considered to be corn diazotrophs. The isolated diazotrophs belonged to various genera including *Pantoea*, *Pseudomonas*, *Rhanella*, *Herbaspirillum*, *Azospirillum*, *Rhizobium* (*Agrobacterium*), *Enterobacter*, and *Brevundimonas* (Montañez et al., 2012).

In addition, a group of endophytic *Burkholderia* sp. was reported to have nitrogen fixation ability in corn cultivated in Mexico (Estrada et al., 2002), this group was confirmed to possess this activity by the acetylene reduction assay, PCR based amplification of the *nifH* gene, and a corn seed bacterial inoculation assay under laboratory conditions. *K. pneumoniae* was recovered from surface sterilized tissues of corn (Palus et al., 1996) and found to be capable of nitrogen fixation. This strain effectively colonizes corn roots (zone of root hair formation) and stems (intracellular space of stem cortex) (Chelius and Triplett, 2000). *Herbasprillum seropedicae* was documented to be a corn root diazotrophic endophyte (Baldani et al., 1992; Dobereiner et al., 1993) but was found to be pathogenic to other plants such as sorghum and *pennisetum* (Dobereiner et al., 1993). One study (Singh et al., 2011) described the potential of isolate RREM25, which was one of the six bacteria isolated from rice variety HUR-36 grown under limited nitrogen conditions. The isolate RREM25 was identified as *Burkholderia cepacia* by 16S rRNA gene amplification analysis, and the diazotrophic characterization was confirmed by the amplification of the *nifH* gene and the presence of dinitrogenase reductase in Western blot analysis (Maheshwari et al., 2013). *B. cepacia* strain RREM25 significantly increased nitrogen contents and nitrogenase activity, and hence promoted rice plant growth (Singh et al., 2011). Some endophytic bacteria from soybean have also been documented as diazotrophs (Kuklinsky-Sobral et al., 2004). In this study, a group of 75 endophytic potential diazotrophs were isolated and characterized. The primary screening was performed on nitrogen-free media where more than half (60%) of bacterial endophytes were able to grow whereas only a small fraction of the bacterial endophytes were confirmed as soybean diazotrophs by amplification of the *nifH* gene. Bacterial genera associated with soybean included *Acinetobacter*, *Burkholderia*, *Pseudomonas*, *Ralstonia*, and *Enterobacter* (Kuklinsky-Sobral et al., 2004; Maheshwari et al., 2013).

2.6.1.2 *Iron acquisition*

Siderophores are small, low molecular weight iron (III) chelating compounds that bind iron reversibly (Rashid et al., 2012) and are responsible for the solubilization and transport of iron (III) into bacterial cells (Neilands, 1981; Hider and Kong, 2010). Siderophores produced by bacteria and fungi are known as the strongest iron (III) binding agents in nature (Hider and Kong, 2010). Notwithstanding being an abundant metal, the bioavailability of iron in soil is very limited. Therefore, the siderophore producing plant associated microflora can provide soluble iron to their host plants for their growth requirements. Moreover, some siderophores have higher affinities toward iron than other siderophores, so that they can sequester iron from siderophores that are produced by other microorganisms with lower iron affinities and consequently outcompete them (Castignetti and Smarrelli, 1986; O'Sullivan and O'Gara, 1992; Whipps, 2001; Lodewyckx et al., 2002; Rajkumar et al., 2010). Siderophore producing PGPB can also serve as biocontrol agents, particularly against fungal phytopathogens. That is because fungal-produced siderophores generally have lower affinities for iron (III) than bacterial siderophores. A biocontrol agent with stronger siderophore capacities will leave less bioavailable iron behind for plant pathogenic fungi, ultimately making the pathogenic fungi less able to proliferate and infect plants (Kloepper et al., 1980; O'Sullivan and O'Gara, 1992; Loper and Henkels, 1999).

There is a limited amount of literature that specifically deals with siderophore production by PGPB. Nevertheless, there are some studies of siderophore production by PGPB dealing with antagonistic effects of these bacteria on plant pathogenic fungi (Lacuna and Azevedo, 2013). For example, one study (Li et al., 2009) described the biocontrol potential of a strain of *Pantoea* spp. (isolated from cotton roots) due to secreted siderophores. Similarly, rice endophytes of *Pseudomonas*, *Burkholderia*, and *Pantoea* exhibited siderophore production with strong pathogen antagonistic activities (Yang et al., 2008). Another group of three endophytic strains of *Streptomyces* spp. that were isolated from the roots of a Meliaceae plant (in the Mahogany family) has been documented to act as a biocontrol agent and exhibited significant growth promotion of tomato plants in the presence of early blight disease caused by *Alternaria alternate*; notably all three strains produced siderophores and IAA (Verma et al., 2011). Likewise, a siderophore producing rhizospheric PGPB *Azospirillum brasilense* has been demonstrated to promote cucumber plant growth by altering root structure, improving mineral uptake, and increasing bioavailability of nutrients under soil which was nutrients deprived (Mimmo et al., 2014). Another study (Vendan et al., 2010) described the capacity of seven bacteria isolated from ginseng to produce siderophores. These isolates were identified as *Bacillus cereus*, *B. flexus*, *B. megaterium*, *Lysinibacillus fusiformis*, *L. sphaericus*, *Microbacterium phyllosphaerae*, and *Micrococcus luteus* (Vendan et al., 2010).

A group of 131 endophytic actinomycete strains were isolated from banana roots and screened for their ability to inhibit *Fusarium* wilt disease of banana. A total of 24 strains (mainly *Streptomyces* spp.) inhibited the growth of pathogenic *Fusarium oxysporum* f. sp. cubense on banana tissue extract medium (Cao et al., 2005). However, the antagonism exhibited by *Streptomyces* was diminished when the medium was supplemented with sufficient iron (III) suggesting that the fungal inhibition is mainly from the iron competition by siderophores produced by the endophytes (Cao et al., 2005). In an in vivo biocontrol assay with one of the above-mentioned strains, *Streptomyces griseorubiginosus* strain S96 treated plants displayed increased fresh weight and decreased disease severity index compared with nontreated plants (Cao et al., 2005).

The ability to produce siderophores can help some bacteria to outcompete other microorganisms within the same ecological niche and thereby colonize inside plant tissues (Loaces et al., 2011). When several bacterial strains were isolated from various tissues of rice plants, it was found that rice tissues harbor different bacterial genera during different stages of plant growth. For instance, *Sphingomonas*, *Pseudomonas*, *Burkholderia*, and *Enterobacter* alternated during different growth stages of rice plants but on the other hand, *Pantoea ananatis* was found throughout the plant's life cycle and was the most prominent endophyte in roots and leaves at tillering and subsequent stages (Loaces et al., 2011). All of the isolates were first screened for siderophore production, and subsequently it was described that isolates from the genus *Burkholderia* showed antagonistic effects toward pathogenic fungi (Loaces et al., 2011) in an in vitro inhibition assay. However, it has been suggested that some phytopathogens may utilize the siderophores that were produced by endophytes or other PGPB in the same niche (Lacava et al., 2008). It was hypothesized that siderophores produced by an endophytic strain of *Methylobacterium* could be taken up and used by the bacterial pathogen *Xylella fastidiosa*, which shares the same niche (i.e., xylem vessels) with *Methylobacterium* in the host plant and causes citrus variegated chlorosis. The growth of *X. fastidiosa* was enhanced when the cell-free supernatant of siderophore producing endophytic strains of *Methylobacterium mesophilicum* was present in the medium (Lacava et al., 2008), which suggested that *X. fastidio* could take up the siderophores produced by the endophytic bacterium *M. mesophilicum* and utilize them in its growth (Lacava et al., 2008).

Furthermore, siderophore producing bacteria are generally considered as a useful tool in heavy metal phytoremediation processes (Burd et al., 2000; Rajkumar et al., 2010). Although siderophores show higher affinity toward Fe (III) ion they can form complexes with numerous divalent heavy metals as well (Evers et al., 1989), in this way siderophore production may be stimulated by the presence of heavy metals in the surrounding environment and confer heavy metal tolerance to their host plants (Rajkumar et al., 2009). In one report, a group of 83 siderophore producing bacterial endophytes, isolated from the roots, stems, and leaves of *Alyssum bertolonii* Desv. (Brassicaceae) promoted plant growth in the presence of nickel (Barzanti et al., 2007).

Recently, a new application of siderophore producing bacteria has been discussed, in which *Burkholderia phytofirmans* PsJN, which can secrete siderophores, was used to treat *Arabidopsis thaliana* plants (Zhao et al., 2016). The treated plants showed improved iron uptake, which resulted in an enlarged stem pith, optimal essential metal uptake, higher overall biomass, and most importantly, treated plants released higher levels of glucose and xylose that subsequently helped in cosaccharification (i.e., breakdown of complex polysaccharides into monosaccharides) in postharvest treatments of plants. This approach may offer a new direction in biofuels production from renewable biomass resources (Zhao et al., 2016).

There is one important caveat that needs to be kept in mind when discussing the role of siderophores from PGPB in plant growth promotion. That is, many of the reported studies with siderophore producing bacteria do not fully characterize all of the other plant growth-promoting activities of the bacterium being studied. Many of these bacteria contain a range of other activities that can promote plant growth. Thus, it is often difficult to assign all of the changes in a plant that are observed following treatment of the plant with a particular siderophore producing bacterium to siderophores per se. In this regard, the use of siderophore knockout mutants as well as siderophore overproducing strains would go a long way to clarifying the role of siderophores in various situations.

2.6.1.3 Phosphate solubilization

Phosphate is another essential nutrient for proper plant growth particularly in root system development, early flowering, cell division and development, and in other key elements of metabolism such as protein synthesis, signal transduction, photosynthesis and respiration, and energy transportation (Lacava et al., 2008; Khan et al., 2014). The concentration of available phosphorus in soil ranges from 0.001 to 1 mg/L (Khan et al., 2014). Phosphorus in the soil mainly occurs as minerals (inorganic) and organic compounds and much of it is inaccessible to plants. The most abundant form of organic phosphorus in the soil is known as phytin (a calcium–magnesium salt of phytic acid, also called phytate). The other forms of organic phosphorus found in soil, typically in small amounts, include sugar phosphates, nucleotides, phosphoprotein, phosphonates, and phospholipids (Yadav and Verma, 2012). On the other hand, inorganic phosphorus mainly contains minerals of phosphorus with calcium, iron, manganese, and aluminum (Khan et al., 2014). However, plants can only directly utilize ions of primary and secondary orthophosphate as nutrients. In addition, the phosphates that are applied to soil as fertilizer are often rapidly immobilized to complex insoluble forms that remain inaccessible to plants (Rodríguez and Fraga, 1999) and emerging plants that lack sufficient utilizable phosphate often suffer from stunted growth, dark leaves, root system undergrowth, and flowering inhibition (Khan et al., 2014).

The process of releasing phosphate from organic sources of phosphate is considered as phosphate mineralization and is primarily carried out by microbial enzymes including phosphatases, phytases, C–P lyases, and phosphonatases (Khan et al., 2014). On the other hand, the release of phosphates in soluble form from inorganic complex compounds of phosphorus (sometimes called rock phosphate) is generally referred to as phosphate solubilization, which is primarily driven by the acidification of minerals that subsequently releases soluble phosphate in the vicinity (Khan et al., 2014). Microorganisms (bacteria and fungi) can produce a variety of low molecular weight organic acids and secrete them in the neighboring environment of the rock phosphate, which in turn protonates insoluble sources of phosphate and makes them soluble and available to plants (Khan et al., 2014). A majority of soil microorganisms have the ability to solubilize and mineralize complex phosphates and make them available to plants (Lacava et al., 2008). As a consequence of their localization inside of plant tissues, one might speculate that endophytes are not especially good at solubilizing either inorganic or organic phosphates compared with rhizospheric bacteria. However, it is certainly possible for endophytes living within plant tissues to produce and secrete organic acids (to solubilize inorganic phosphates) or enzymes (to mineralize organic phosphates). However, once these compounds are secreted outside of the bacteria, to be efficient phosphate solubilizers, they need to be subsequently secreted by the plant root to come in contact with their (often insoluble) substrate.

A large number of reports have indicated the potential of PGPB in phosphate solubilization in a variety of plants such as corn (Ali et al., 2018) tomatoes (Rashid et al., 2012), cactus (Puente et al., 2009), sunflowers (Forchetti et al., 2007), peanuts and other legumes (Palaniappan et al., 2010), soybeans (Kuklinsky-Sobral et al., 2004), ginseng (Vendan et al., 2010), and strawberry (Dias et al., 2009). In most cases, phosphate solubilization activity is screened and measured using a plate assay, and upon characterization of these bacteria it has become clear that phosphate-solubilizing bacteria often carry one or more other plant growth promotion traits. For example, in one study, bacteria isolated from strawberry mainly represented *Bacillus* spp. that could also produce IAA along with the ability to solubilize phosphate (Dias et al., 2009). Similarly, 60% of the PGPB from soybean representing the genera *Burkholderiaceae*, *Enterobacteriaceae*, and *Pseudomonaceae* exhibited the abilities to fix nitrogen, produce IAA, and solubilize phosphate (Kuklinsky-Sobral et al., 2004). In a recent study (Joe et al., 2016), two bacterial endophytes namely *Acinetobacter* sp. and *Bacillus* sp. were isolated from *Phyllanthus amarus* Schum and Thonn (a medicinal herb) on a medium that contains insoluble complex phosphates and a high salt (160 mM) concentration. These two endophytes can promote plant growth as evidenced by increased biomass, higher vigor index, increased emergence count, higher phenolic contents, and higher phosphate contents. Moreover, these endophytes were able to produce hydrolytic enzymes and exhibited antioxidative activity that was involved in the promotion of plant growth under salt stress (Joe et al., 2016). Similarly, four representative bacteria from *Aloe vera* plants that were identified as *Pseudomonas synxantha*, *Burkholderia gladioli*, *Enterobacter hormaechei*, and *S. marcescens* were found to be able to solubilize 25–340 $\mu g\ mL^{-1}$ phosphate of tricalcium phosphate in addition to producing siderophores and IAA in vitro (Gupta et al., 2012). Upon inoculation of *Aloe barbadensis* Miller with these bacterial endophytes, individually or in combination, the resulting plants exhibited higher phenolic contents due to increased plant biomass (i.e., from 108% to 673%), higher phosphate intake, and increased levels of soluble phosphates found in adjacent soil compared with uninoculated control plants (Gupta et al., 2012).

For example, the cassava root endophyte, *Pantoea dispersa* was found to solubilize inorganic phosphate by production of organic acids including succinic acid, oxalic acid, citric acid, salicylic acid, and benzeneacetic acid. The process of solubilization also elicited the ability of the native soil microbial community to enhance this activity (Chen et al., 2014). Likewise, various species of *Pseudomonas* have been found to exhibit phosphate solubilization activity by producing gluconic acid (Oteino et al., 2015). Thus, phosphate-solubilizing strains of *Pseudomonas* promoted pea plant growth with the production of moderate to high levels of gluconic acid when grown in phosphate limiting soils (Oteino et al., 2015). The bottom line for these PGPB is that these microorganisms are apparently able to solubilize various forms of phosphate and provide it to plants thereby augmenting their value as biofertilizers to improve crop health and productivity.

2.6.2 Hormone level modulation

2.6.2.1 Auxin

Auxins are plant growth regulatory hormones that play a crucial role in regulating a myriad of processes in plant growth and development (Grossmann, 2010). Within auxins, indole-3-acetic acid (IAA) is the most common naturally occurring auxin that is known to affect many plant cellular and physiological processes including cell differentiation, lateral and adventitious roots initiation, cell division stimulation, stem and root elongation, and maintaining the root and shoot growth in response to light and gravity (Ali et al., 2017; Glick, 2015). Other than in plants, IAA has been described to be synthesized by bacteria and fungi; by at least three different tryptophan-dependent pathways (Duca et al., 2014). Since plants actively synthesize IAA during their vegetative growth, microbial production of IAA in the close proximity of, or within, a plant may cause either stimulatory or adverse effects for the plant host thereby upsetting plant homeostasis and altering plant growth and development (Mellor et al., 2016). Thus, not all IAA producing bacteria are PGPB; some are also phytopathogens. Furthermore, it has been suggested that phytopathogens often synthesize IAA using the indole-3-acetamide (IAM) pathway, while IAA production through the indole-3-pyruvate (IPyA) pathway is often considered to be the pathway of choice for many plant-beneficial bacteria (Ali et al., 2017; Hardoim et al., 2008). Numerous bacterial genera have been demonstrated to promote plant growth by producing IAA and include *Alcaligenes*, *Acetobacter*, *Azospirillum*, *Azotobacter*, *Bacillus*, *Burkholderia*, *Enterobacter*, *Herbaspirillum*, *Klebsiella*, *Pantoea*, *Pseudomonas*, *Rhizobium*, *Rhodococcus*, and *Streptomyces* (Ali et al., 2018; Duca et al., 2014). Some of the recent examples where PGPB considerably facilitated host plant growth by producing just the right amount of IAA include, but are not limited to, the application of two PGPB (*P. putida* and *B. megaterium*) to clover plants under water stress where these bacteria increased plant biomass and water content compared with untreated control plants (Marulanda et al., 2009). Similarly, IAA producing PGPB *Phyllobacterium brassicacearum* strain STM196 protected *Arabidopsis* plants under drought stress that exhibited increased root growth with a modified root structure that supported higher tolerance to the drought (Contesto et al., 2010). In this case, the plant morphological changes (root structure and increased biomass) were directly correlated with the physiological functioning of the PGPB (i.e., production of IAA in the treated plants) suggesting that the enhanced drought tolerance primarily was a consequence of the bacterial treatment (Contesto et al., 2010). The IAA producing soybean seed endophyte, *Enterobacter* sp. has also demonstrated a significant increase in the root dry biomass when soybean seeds were pretreated with this bacterial strain (Assumpção et al., 2009; Kuklinsky-Sobral et al., 2004) investigated the population dynamics of IAA producing PGPB in soybean plants. In this study, it was found that the endophytic IAA producing PGPB generated higher levels of IAA (34%) than their epiphytic PGPB counterpart (i.e., 21%) (Kuklinsky-Sobral et al., 2004).

2.6.2.2 Cytokinin

Cytokinins are a group of similar compounds that chemically resemble adenine (one of the DNA nitrogenous base) and were originally named because of their important role in plant cell division and cytokinesis (Glick, 2015). Based on their chemical precursor, cytokinin can be classified into two main groups, that is, adenine cytokinin and the diphenylurea cytokinin (García de Salamone et al., 2001). There are about 200 compounds that have been categorized as cytokinins, either synthetic or natural. It has been suggested that the cytokinins are primarily produced in the meristem region of plant roots and then transported to the other parts of the plant via xylem vessels (Ali et al., 2017) cytokinins can be found in roots, stems, leaves, flowers, fruits, and seeds of plants (Garcia De Salamone et al., 1996). A number of bacteria including *Azospirillum* spp., *Azotobacter* spp., *Rhizobium* spp., *P. agglomerans*, *Rhodosprillum rubrum*, *P. fluorescens*, *Bacillus subtilis*, and *Paenibacillus polymyxa* have been reported to contain cytokinins in their growth culture filtrate (Ali et al., 2017). Studies of the biosynthesis of cytokinins in PGPB and other soil bacteria, and their direct role in plant growth promotion and development, have been relatively limited. In fact, much of what is known about cytokinin promotion of plant growth comes from studies using purified plant cytokinins. Considerable additional research is required to elucidate details of cytokinin biosynthesis and regulation by PGPB and other bacteria. However, some soil bacteria have been described to produce various types of cytokinin. For example, *Corynebacterium fascians* could produce seven types of cytokinin namely *cis*-zeatin (c-Z), *trans*-zeatin (t-Z), zeatin riboside (ZR), metyltio-*cis*-zeatin (cmsZ), isopentenyladenine (2iP), isopentenyladeniosine (2iPA), and methyloaminopurine (meZ) (Donderski and Gluchowska, 2000). In addition to soil bacteria, water dwelling bacteria have also been documented to produce at least three types of cytokinins including isopentenyladenine (2iP), zeatin (Z), and zeatin riboside (ZR), and belong to *Vibrio* spp., *Bacillus* spp., *Aeromonas* spp., *Achromobacter* spp., and the *Enterobacteriaceae* family (Donderski and Gluchowska, 2000).

2.6.2.3 Gibberellin

Gibberellins (GAs) are plant growth regulators that play important roles in plant physiological and metabolic functions. The GAs mainly participates in seed germination, leaf and stem growth, flowering, and fruit formation and senescence (Ali et al., 2017). Chemically, GAs are *ent*-kaurene-derived diterpenoid phytohormones and are named after the fungus *Giberella fujikuroi*, the original source of this phytohormone. Generally different types of GAs are named numerically by the order of their discovery (Khan et al., 2015). A detailed gibberellin biosynthesis pathway by bacteria has recently been described and can be obtained on http://www.nature.com/nchembio/journal/v13/n1/full/nchembio.2232.html (Nett et al., 2017). Bacteria that have been suggested to produce various types of GAs include *A. diazotrophicus*, *H. seropedicae*, *Azospirillum lipoferum* strain op 33, *Rhizobium phaseoli*, *Enterococcus faecium* LKE12, and *Sphingomonas* sp. LK11 (Ali et al., 2017). As is the case with cytokinins, studies on the interactions and relationships between plant and microbes that are focused on GAs are quite limited. Most of the time, GA producing PGPB also exhibit other plant growth-promoting mechanisms so that the consequence of treating plants with a GA producing PGPB is generally a combination of several bacterially encoded mechanisms. For example, the above-mentioned bacteria, *A. diazotrophicus* and *H. seropedicae*, that produce gibberellins (GA1 and GA3) and IAA, have been documented to colonize graminaceae species and promote plant growth and overall yield (Bastián et al., 1998). Similarly, the GA4, GA9, and GA20 producing PGPB *Sphingomonas* sp. LK11 has been effectively used to promote tomato plant growth by increasing plant shoot length, chlorophyll contents, and shoot and root dry weights compared with control plants (Khan et al., 2014). In a separate study, a GA-deficient rice dwarf mutant (waito-C) and a normal GA producing rice cultivar (Hwayongbyeo) were treated with the GA and IAA producing PGPB, *E. faecium* LKE12. The bacterial treatment significantly increased the shoot length and biomass of both wild-type and mutant plants; thus, plant growth promotion in this case was suggested to be due to the bacterial production of both GAs and IAA (Li et al., 2016).

2.6.2.4 Ethylene

Ethylene is another important plant growth regulator, which was discovered in the late 19th century. Ethylene is chemically the simplest gaseous phytohormone, and is produced by all higher plants. The physiological roles of ethylene include involvement in seed germination, tissue differentiation, root development, lateral bud development, flowering, anthocyanin synthesis, fruit ripening, aroma production (by regulating VOCs production within fruits), leaf senescence, leaf and fruit abscission, and maintenance of plant–microbe interaction in Rhizobia nodule formation and mycorrhizae–plant interactions (Glick, 2015). Ethylene is also a plant stress hormone; the intracellular levels of ethylene change according to the growth stage of the plant and, more importantly, as a consequence of various environmental stresses. Ethylene is generally required in very low concentrations for normal plant growth and development (except during fruit ripening where ethylene levels are very high). The high levels of ethylene that plants produce when they are subject to either abiotic or biotic stresses are deleterious to plant growth and development (Ali et al., 2017; Glick, 2015). Since many of the plant senescence physiological responses are triggered by elevated levels of ethylene, a vast number of studies have been focused on demonstrating the ability of the microbial enzyme 1-aminocyclopropane-1-carboxylic acid (ACC) deaminase to ameliorate the harmful effects of stress ethylene by degrading ethylene immediate precursor, ACC, into ammonia and α-ketobutyrate (Glick et al., 2007; Glick, 2015; Ali et al., 2017) thus preventing the formation of ethylene. Functional ACC deaminase has been found and reported in α, β, and, γ Proteobacteria, Actinobacteria, Firmicutes, Bacteroidetes, archaea, various fungi, and yeast (Fujino et al., 2004; Hontzeas et al., 2004; Marques et al., 2010; Minami et al., 1998; Nascimento et al., 2014; Singh et al., 2015; Ekimova et al., 2018). Moreover, *Acidovorax*, *Bordetella*, *Brenneria*, *Burkholderia*, *Collimonas*, *Cupriavidus*, *Curvibacter*, *Dickeya*, *Herbaspirillum*, *Halomonas*, *Lonsdalea*, *Methylibium*, *Pantoea*, *Phytophthora*, *Polaromonas*, *Pseudomonas*, *Ralstonia*, *Serratia*, *Tatumella*, *Variovorax*, and *Xenophilus*, within the domain Archaea, strains *Archaeoglobus fulgidus*, *Pyrococcus abyssi*, *Pyrococcus furiosus*, and *Thermococcus nautili*, were found to contain ACC deaminase genes based on the genomic and metagenomic database search studies where *acd*S gene locus of *P. putida* UW4 was used as the search criterion; likewise, the phyla Ascomycota and Basidiomycota for fungi, and soybean, potato, maize, and castor oil for the kingdom Plantae were found to contain ACC deaminase genes (Singh et al., 2015). PGPB expressing the enzyme ACC deaminase have been shown to help to protect plants from a number of different biotic and abiotic stresses and also to promote the growth of plants in the absence of stressful conditions.

2.6.2.5 Abscisic acid

Initially abscisic acid (ABA) was believed to be involved in leaf abscission, which is now known to be true for only a limited number of plants. The ABA is an isoprenoid based plant growth regulator and is mainly produced in response to different environmental changes. Although ABA is synthesized in all parts of the plant, it is largely produced in leaves and is involved in many plant physiological functions. In leaves, ABA initiates formation of primordia in terminal buds in adaptation to cold weather. Within seeds, ABA helps seed to mature and establish dormancy, however, ABA inhibits seed germination. Roots can synthesize their own ABA, or it can be transported from leaves through the xylem system to the other parts of the plants. Under abiotic stress, ABA is accumulated at higher levels and causes stomatal closure by altering the turgor pressure on the guard cells (Sukweenadhi et al., 2015). During desiccation conditions, ABA promotes root elongation and inhibits shoot growth (Setter and Parra, 2010). One study on the isolation and utilization of ABA-metabolizing rhizobacteria suggested the use of these bacteria as a possible remedy to overcome the abiotic stress caused due to the elevated levels of ABA (Hasegawa et al., 1984). In one study, rhizobacteria were isolated on growth medium supplemented with ABA as a sole source of carbon and identified as *Corynebacterium* sp.; this bacterium converted ABA to dehydrovomifoliol via the action of the enzyme vomifoliol dehydrogenase thereby reducing ABA levels (Hasegawa et al., 1984). In another study, two PGPB were isolated using ABA-based selective medium from the rice rhizosphere and identified as *Rhodococcus* sp. P1Y and *Novosphingobium* sp. P6W. Treating rice seedlings with these two bacteria had no effect on the ABA concentration in their roots, however, the ABA concentration in shoots was considerably decreased in the seedlings treated with strain P6W (14%) and with strain P1Y (22%) (Belimov et al., 2014). When the same ABA-metabolizing PGPB were tested on wild-type tomato plants and ABA deficient mutant tomato plants, the bacterial treatment by strain P6W lowered the root ABA concentrations of both wild-type and mutant plants thus inhibiting ABA induced root elongation, and reducing shoot ABA content that consequently increased leaf biomass (Belimov et al., 2014). To rule out the possible role of ethylene induced effects in these experiments, the wild-type tomato plants were also treated with silver ions (an inhibitor of ethylene synthesis). In this case, both of the ABA-metabolizing strains significantly decreased root ABA levels, and strain P6W also reduced leaf ABA levels, suggesting that bacteria-mediated changes in ABA concentration did not come from ethylene (Belimov et al., 2014). Several other studies suggest that the PGPB that are able to increase ABA concentrations can promote plant growth by enhancing their tolerance to drought. In one experiment, lettuce plants treated with the PGPB *Bacillus* sp. displayed higher levels of ABA and higher drought tolerance than the uninoculated control (Arkhipova et al., 2007). Similarly, *Arabidopsis* plants treated with the PGPB *A. brasilense* Sp245 attained both higher drought tolerance and ABA levels compared with untreated control plants (Cohen et al., 2008). The increased drought tolerance from the elevated levels of ABA may be due to the up regulation of aquaporins (Zhou et al., 2012) and/or regulation of the transpiration rate and root hydraulic conductivity (Aroca et al., 2006). Notwithstanding a number of suggestive experiments, more research is required to elucidate the precise role of ABA mediated plant growth promotion under abiotic stress conditions.

2.6.2.6 Volatile organic compounds

Volatile organic compounds (VOCs) are secondary metabolites that a large number of microbes produce during their stationary phase of growth. VOCs are thought to regulate a number of physiological functions in plants (Liu and Zhang, 2015). The PGPB *Pseudomonas simiae* has been shown to produce VOCs that have the ability to stimulate induced salinity tolerance in soybean plants under salt stress by up regulating various proteins including γ-glutamyl hydrolase, the large subunit of ribulose bisphosphate carboxylase, and vegetative storage protein (Vaishnav et al., 2015). Moreover, the PGPB *Alcaligenes faecalis* strain JBCS1294 has demonstrated the ability to confer treated plants with induced salinity tolerance by producing hexanedioic acid and butanoic acid, VOCs that stimulate the auxin and gibberellin pathways (Bhattacharyya et al., 2015). Unfortunately, the detailed mechanism of how VOCs work is not as yet fully understood (Qin et al., 2016).

There is very little literature that specifically deals with the role of bacterial VOCs in protecting plants from abiotic stresses; most of the work on VOCs addresses the effects and potential of bacterial VOCs in fighting phytopathogens

TABLE 2.2 Volatile compounds production by PGPB and their effects on host plants under various abiotic stresses.

Bacterium	Chemical nature	Plant	Abiotic stress	Outcome	Mechanism	Reference
Bacillus amyloliquefaciens GB03	2,3-Butanediol	Arabidopsis	Salt stress	Lower Na accumulation, increased biomass	*hkt1* gene upregulated in shoots and downregulated in roots	Mayak et al., (2004), Liu and Zhang (2015)
P. simiae AU	NA	Soybean	Salt stress	Lower Na accumulation, higher proline contents	IST[a]	Vaishnav et al. (2015)
A. faecalis strain JBCS1294	Hexanedioic acid and butanoic acid	Arabidopsis	Salt stress	Enhanced plant biomass	Reprogram auxins and gibberellin interactions	Bhattacharyya et al. (2015)
B. amyloliquefaciens GB03	2,3-Butanediol	Arabidopsis	Water loss	Drought tolerance	Accumulation of osmoprotectant, choline and glycine betain, up regulation of *PEAMT*	Zhang et al. (2010)
Pseudomonas chlororaphis O6	2,3-Butanediol	Arabidopsis	Water loss	Increased stomatal closure	Improved cross talk between phytohormones, SA,[a] JA,[a] and ET[a]	Song et al. (2008)
Bacillus sp. B55	Dimethyl disulfide, DMDS	Tobacco	Sulfur deprivation	Improved plant growth	Sulfur acquisition	Meldau et al. (2013)
B. amyloliquefaciens GB03	NA	Arabidopsis	Leaf chlorosis	Improved uptake of iron	Induction of genes *IRT1* and *FIT1*	Zhang et al. (2009)

[a]*IST, induced salinity tolerance;* JA, *jasmonic acid;* SA, *salicylic acid;* ET, *ethylene.*

and facilitating plant growth under biotic stress conditions. Nevertheless, Table 2.2 illustrates some PGPB and the VOCs that they produce that have demonstrated significant interaction with plants in overcoming various abiotic stresses (Liu and Zhang, 2015).

2.7 Examples of PGPB overcoming abiotic stresses

2.7.1 High salt

Increasing salinity levels in agriculture lands has become a major problem worldwide. PGPB have been documented to protect and promote the growth of a wide range of plants under salt stress (Table 2.3). For instance, a group of seven rhizobacteria has been reported to enhance the growth of tomato seedlings under high salt stress (150 mM NaCl). These bacteria, mainly belonging to genera *Arthrobacter* and *Bacilli,* were initially isolated from wild plants grown in high salinity lands and it was suggested that they help plants by having capacities to hydrolyze starch, enhance phosphorus uptake, produce IAA, and regulate various plant growth promotion-related genes (Fan et al., 2016). A similar study has documented the effects of *Chryseobacterium gleum* sp. SUK on the growth of wheat. In this study, the bacterial inoculum was prepared using poultry feathers as a substrate and then the hydrolyzed material with the bacteria was used to treat wheat plants grown under 100 mM salt stress. *C. gleum* sp. SUK was able to produce ACC deaminase, hydrogen cyanide, siderophores, protease, cellulase, and IAA in vitro and facilitated plant growth under salinity stress conditions (Bhise et al., 2017). Plant growth-promoting bacterial endophytes isolated from the halophyte plant *Salicornia europaea*

TABLE 2.3 List of PGPB that have shown to ameliorate salinity stress in plant.

Bacteria	Source of isolation	Plant	Salinity (mM)	Mode of action	Activity (μM)	Growth enhancement (%)	Reference
Enterobacter sp. SBP-6	Sorghum	Wheat	150, 175, 200	ACC deaminase	2.84	10–40	Singh et al. (2016)
C. gleum sp. SUK	Soil	Wheat	100	ACC deaminase	10 ± 0.41	NA	Bhise et al. (2017)
Arthrobacter sp.	Wild plants	Tomato	150	Starch hydrolysis, P-solubilization, IAA, regulation of plant growth promotion-related genes	NA	NA	Fan et al. (2016)
Bacillus megaterium	Wild plants	Tomato	150	Starch hydrolysis, P-solubilization, IAA, regulation of plant growth promotion-related genes	NA	NA	Fan et al. (2016)
B. endophyticus	Glasswort	Glasswort	50, 250, 500	IAA, P-solubilization, ACC deaminase	1.7 ± 0.3	21–54	Zhao et al. (2016)
B. tequilensis	Glasswort	Glasswort	50, 250, 500	IAA, P-solubilization, ACC deaminase	–	21–54	Zhao et al. (2016)
Arthrobacter agilis	Glasswort	Glasswort	50, 250, 500	IAA, P-solubilization, ACC deaminase	1.3 ± 0.5	21–54	Zhao et al. (2016)
Planococcus rifietoensis	Glasswort	Glasswort	50, 250, 500	IAA, P-solubilization, ACC deaminase	2.6 ± 0.7	21–54	Zhao et al. (2016)
Variovorax paradoxus	Glasswort	Glasswort	50, 250, 500	IAA, P-solubilization, ACC deaminase	3.2 ± 0.3	21–54	Zhao et al. (2016)
P. putida H-2–3	Soil	Soybean	120	Gibberellin production	–	8.60	Kang et al. (2014)
Agrobacierium tumefaciens MBE01	Wild	Peanuts	100	ACCD, P-solubilization, IAA, N-fixation	0.12	19–31	Sharma et al. (2016)
Klebsiella sp. MBE02	Wild	Peanuts	100	ACCD, P-solubilization, IAA, N-fixation	0.17	19–31	Sharma et al. (2016)
Ochrobactrum anthropi MBE03	Wild	Peanuts	100	ACCD, P-solubilization, IAA, N-fixation	0.1	19–31	Sharma et al. (2016)
Pseudomonas stutzeri MBE04	Wild	Peanuts	100	ACCD, P-solubilization, IAA, N-fixation	0.31	19–31	Sharma et al. (2016)
Pseudomonas sp. MBE05	Wild	Peanuts	100	ACCD, P-solubilization, IAA, N-fixation	0.19	19–31	Sharma et al. (2016)
P. fluorescens	Soil	Cucumber	100	Exopolysaccharide production, ACC deaminase, P-solubilization, siderophore	170.2 ± 10.0	67	Nadeem et al. (2016)
B. megaterium	Soil	Cucumber	100	Exopolysaccharide production, ACC deaminase, P-solubilization, siderophore	18.9 ± 02.3	26	Nadeem et al. (2016)
V. paradoxus	Soil	Cucumber	100	ACC deaminase, siderophore	114.5 ± 9.2	46	Nadeem et al. (2016)

E. cloacae strain KBPD	Soil	Mung bean	150	Exopolysaccharide production, ACC deaminase, P-solubilization, siderophore	19.41	29	Bhise et al. (2017)
Mesorhizobium sp. KR48	Chickpea	Chickpea	80	ACC deaminase, N-fixation	NA	38	Chaudhary and Sindhu (2015)
Mesorhizobium sp. MHD2	Chickpea	Chickpea	80	ACC deaminase, N-fixation	NA	32	Chaudhary and Sindhu (2015)
Mesorhizobium sp. MBD26	Chickpea	Chickpea	80	ACC deaminase, N-fixation	NA	42	Chaudhary and Sindhu (2015)
P. frederiksbergensis OS261	Soil	Pepper	150	ACC deaminase	NA	67	Chatterjee et al. (2017)
Bacillus sp. *EN1*	Soil	Wheat	200	IAA, N-fixation	NA	15.58	Orhan (2016)
Zhihengliuella halotolerans EN3	Soil	Wheat	200	ACC deaminase	NA	10.27	Orhan (2016)
Staphylococcus succinus EN4	Soil	Wheat	200	P-solubilization	NA	6.50	Orhan (2016)
B. gibsonii EN6	Soil	Wheat	200	ACC deaminase, IAA, N-fixation	NA	30.65	Orhan (2016)
Oceanobacillus oncorhynchi EN8	Soil	Wheat	200	IAA	NA	20.89	Orhan (2016)
Halomonas sp. *IA*	Soil	Wheat	200	ACC deaminase	NA	25.85	Orhan (2016)
Thalassobacillus sp. *ID*	Soil	Wheat	200	ACC deaminase	NA	33.73	Orhan (2016)
S. marcescens CDP-13	Karira	Wheat	175	ACC deaminase, IAA, P-solubilization	0.172	22.33	Singh and Jha (2016a,b)
P. putida R4	Soil	Cotton	100	IAA	NA	55	Egamberdieva et al. (2015)
P. chlororaphis R5	Soil	Cotton	100	IAA	NA	44	Egamberdieva et al. (2015)
Brevibacterium epidermidis RS15	Soil	Canola	175	ACC deaminase	2.37	43.37	Siddikee et al. (2015)
B. aryabhattai RS341	Soil	Canola	175	ACC deaminase	1.11	39.20	Siddikee et al. (2015)
P. migulae 8R6	Tomato	Tomato	185	ACC deaminase, IAA, P-solubilization, siderophore	10.9	NA	Ali et al. (2014)
P. fluorescens YsS6	Tomato	Tomato	185	ACC deaminase, IAA, P-solubilization, siderophore	12.5	NA	Ali et al. (2014)

were characterized for their potential PGP abilities, and subsequently tested with *S. europaea* grown under various salt levels (50–500 mM of NaCl). Five of the bacterial endophytes were found to be highly effective at protecting and promoting plant health under high salinity levels (250 and 500 mM), four of them were ACC deaminase positive, and the highest plant protection against salinity inhibition came from the bacterial endophyte that had the highest level of ACC deaminase (Zhao et al., 2016). Another study demonstrated the amelioration of biotic stress caused by high levels of salt (100 mM) with rhizospheric PGPB that reduced the levels of ROS and maintained ion homeostasis in peanut plants compared with untreated control plants (Sharma et al., 2016). In this study, when five PGPB were screened for their potential plant growth-promoting traits, it was concluded that the plant growth promotion under salt stress was an outcome of several mechanisms including nitrogen fixation, IAA biosynthesis, phosphate solubilization, ACC deaminase production, and the ability to maintain a low Na^+/K^+ ratio within plant cells (Sharma et al., 2016). Similarly, two bacterial endophytes *P. fluorescens* YsS6 and *Pseudomonas migulae* 8R6, isolated from tomato plants, can provide protection and growth promotion to tomato plants treated with a high level (185 mM) of salt (Ali et al., 2014). These bacterial endophytes were both ACC deaminase positive, can produce IAA and siderophores, and were able to solubilize phosphate (Rashid et al., 2012). The plant growth inhibition under salinity is mainly from the disruption of several crucial physiological processes (Forni et al., 2017). High levels of salt stress have been reported to cause delays in seed germination as a consequence of a reduction in hydrolytic enzyme activities and mobilization of seed metabolites (Smith and Cobb, 1991). In addition, PGPB that contain ACC deaminase have shown increased amylase, invertase, and protease activities with simultaneous reduction in stress ethylene levels and significantly improved canola seed germination compared with untreated controls (Siddikee et al., 2015). Abiotic stress also stimulates the accumulation of ROS, which eventually leads to oxidative damages to the plant cell membrane and other cellular structures. The PGPB that contain ACC deaminase have been reported to increase the level of antioxidant enzymes and significantly facilitate pepper plant (Chatterjee et al., 2017) and potato growth (Gururani et al., 2013). Similarly, wheat plants that were inoculated with the PGPB *S. marcescens* CDP-13 exhibited enhanced salinity tolerance (Singh and Jha, 2016a,b). In this case, the bacterially treated plants producing significantly higher levels of antioxidant enzymes superoxide dismutase, SOD (34%), catalase, CAT (40.8%), and ascorbate peroxidase, POX (49%). The bacterially treated plants protected plants from membrane lipid peroxidation compared with uninoculated control plants, which exhibited higher levels of ROS resulting in loss of membrane stability and higher electrolytic leakage (increases of 39%–62%) in the presence of 150 mM salt (Singh and Jha, 2016a,b). High levels of salt can eventually result in cell desiccation and injury. PGPB may protect plants in a saline environment by stimulating the production of osmolytes, such as organic solutes, proline, and total soluble sugars in wheat (Singh and Jha, 2016a,b), tomato, and *Arabidopsis* (Kim et al., 2014). Proline and soluble sugars have been reported to protect enzymes from stress-mediated denaturation and function in osmoprotection and radical scavenging (Singh and Jha, 2016a,b). PGPB treatment of salt stressed plants has been reported to regulate the expression of genes related to ion homeostasis maintenance across the cell membrane (Sukweenadhi et al., 2015).

2.7.2 Drought

Worldwide, climate change has exacerbated the problem of drought and its effect on agricultural productivity. One means of addressing some of the negative effects of drought on crop plants is the use of PGPB. For example, the PGPB *P. putida* H-2-3 significantly enhanced the growth of soybean seedlings in the laboratory under 15% polyethylene glycol mediated drought stress (Kang et al., 2014). The PGPB-treated soybeans demonstrated a reduced stress response (lower ABA and salicylic acid, and an increased level of jasmonic acid) compared with untreated control plants (Kang et al., 2014). The expression levels of two genes, namely *AtERD15* and *AtRAB18* in *A. thaliana*, were found to be significantly higher under drought stress. These genes are induced under various stresses (mainly drought) and have been reported to be associated with ABA sensitivity (Sukweenadhi et al., 2015). Furthermore, RNAi silencing of gene *AtERD15* has been reported to reprogram the ABA signaling resulting in increased tolerance to drought in *A. thaliana* (Kariola et al., 2006). Interestingly, PGPB *Paenibacillus yonginensis* DCY84T treated *A. thaliana* plants exhibited a higher level of induction of gene *AtERD15*, which should have resulted in higher sensitivity to ABA and therefore drought stress, however, despite the higher level of gene induction, bacterial treatment significantly facilitated *A. thaliana* growth under drought stress (Sukweenadhi et al., 2015). Unfortunately, the precise mechanism of action of this PGPB is still unknown.

Different PGPB utilize different mechanisms to help their host plants in overcoming drought stress. For example, *Achromobacter piechaudii* ARV8 (Mayak et al., 2004) and *Pseudomonas* sp. (Arshad et al., 2008) have been found to utilize their ability to synthesize ACC deaminase and reduce the levels of deleterious stress ethylene, which was a consequence of drought stress in tomatoes and peas, respectively. Two bacterial isolates *Pseudomonas oryzihabitans* Ep4 and *Achromobacter xylosoxidans* Cm4 were found to facilitate potato growth in pot and field conditions (40% yield increase) where they were exposed to drought (Belimov et al., 2015). These PGPB were also shown to have ACC deaminase activity and the decreased concentration of ACC in the rhizosphere of treated plants compared with the untreated control indicated that the plant growth promotion under drought mediated stress came from the action of ACC deaminase produced by the applied bacteria (Belimov et al., 2015). In addition, the PGPB *P. putida* P45 has been reported to confer drought resistance in sunflower by rhizospheric soil aggregation by extracellular polysaccharide production (Sandhya et al., 2009) for *P. agglomerans* in wheat (Amellal et al., 1998). Moreover, the endophytic PGPB *B. phytofirmans* PsJN significantly enhanced wheat yield when the plants were under water stress at two different time points, that is, tilling and flowering, in the field (Naveed et al., 2014). The bacterial endophyte *B. phytofirmans* PsJN helped plants in improving plant physiology, water retention capacity, and antioxidant activity, all of which subsequently improved the final plant yield by 23% compared with untreated control (Naveed et al., 2014). In a separate study, it was shown that when *B. phytofirmans* PsJN colonized a plant that undergoes drought stress, the bacterial transcriptome changes and the genes responsible for oxidative phosphorylation, transcription regulation, cellular homeostasis, and detoxification of ROS are significantly upregulated whereas six different gene transcripts that are involved in extracytoplasmatic function (ECF) group IV sigma factors were modulated. Among those six sigma factors, the expression of one ECF group sigma factor was upregulated whereas the other resulted in downregulation of the genes involved. It is also interesting to note that the level of expression differs among different potato cultivars (Sheibani-Tezerji et al., 2015). Cytokinin producing PGPB have also been demonstrated to ameliorate desiccation stress in plants. *P. stutzeri* CHP413A treated corn plants show a significant increase (37%) in plant dry weight over untreated control plants under drought conditions (Raza et al., 2015). Likewise, *Bacillus* sp. strain IB-22 treated lettuce plants demonstrate a lower root/shoot ratio when they experience water stress (Arkhipova et al., 2007).

2.7.3 Flooding and hypoxic stress

Flooding stress mainly prevents gaseous exchange between plant cells due to restricted gas (oxygen and carbon dioxide) diffusion that in turns leads to oxygen deficiency in plant cells (Glick, 2015). Under flooding stress, plants undergo two key adaptive changes, that is, morphological changes and metabolic reprogramming. Depleted levels of oxygen induce the expression of the plant enzyme ACC synthase that increases the levels of ACC and subsequently the plant stress hormone ethylene (Glick, 2015; Mustroph, 2018). As the flooding response is mediated primarily by stress ethylene, plants show morphological changes that include adventitious root formation, stunted stem growth, wilting, chlorosis, epinasty, and leaf abscission (Glick, 2015). The PGPB, especially those that express ACC deaminase, can be utilized for their potential to lower the stress ethylene levels and to decrease the plant growth inhibitory effects. In one set of experiments, PGPB *P.* sp. UW4, *E. cloacae* CAL2, *P. putida* (ATCC17399/pRKACC), or *P. putida* (ATCC17399/pRK415) were utilized to treat tomato seeds, and the emerging plants subsequently underwent flooding stress for nine consequential days. Tomato plants inoculated with PGPB that expressed ACC deaminase established significant tolerance to flooding stress (Grichko and Glick, 2001). In other experiments, Li et al. (2013) not only showed that PGPB that expressed ACC deaminase could protect cucumber plants against hypoxic stress (caused by lack of oxygen), but they documented, using proteomics, the specific changes that occurred within the plant roots. Finally, it is important to note that the use of ACC deaminase-containing PGPB in protecting plants against flooding has been shown to be effective both in the laboratory and in the field (Glick, 2015).

2.7.4 Contaminating metals and organic compounds

Heavy metal contamination of soils is a major agriculture problem. A significant amount of work has been published describing the potential of PGPB to help plants to clean their surroundings of metal contaminants (i.e., phytoremediation). In one example, two PGPB (*Acinetobacter* sp. RG30 and *Acinetobacter* sp. RG40) that exhibit copper (Cu) tolerance were used in a greenhouse experiment where soil contaminated with 200 ppm of Cu (II) was used to evaluate the bacterial inoculation effects on the soil remediation and Cu accumulation within corn tissues (Rojas-Tapias et al., 2014). The results from this study indicated that the coinoculated bacterial treatment significantly increased the dry biomass of corn (19% dry shoots and 25% dry root biomass) over untreated control plants under the metal stress. Additionally, the accumulated Cu contents within corn tissues were significantly higher ~25% and 100% in treated corn shoot and root tissues, respectively (Rojas-Tapias et al., 2014). Soil communities play a crucial role in helping host plants to better adapt to the environment, and if the microbial community is disturbed this possibly could lead a physiological stress stimulus to plant. Thus, a natural soil rich in zinc (Zn) and cadmium (Cd) was severely damaged by gamma radiation (which has no effects on the plant growth and on the acid extractable levels of Zn and Cd) and used to grow the plant *Arabidopsis*. Upon harvest, plants that emerged from natural undisturbed soil contain 100% more cadmium and 15% more zinc accumulation than plants from the gamma-irradiated soil (Muehe et al., 2015). The rhizospheric microbiome composition, determined by 16S rRNA gene pyrosequencing, revealed that the natural and undisturbed soil had a higher abundance of the microbial taxa including *Lysobacter*, *Streptomyces*, *Agromyces*, *Nitrospira*, *Candidatus*, and *Chloracidobacterium* with the suggestion that these taxa play an important role in phytoremediation of Zn and Cd (Muehe et al., 2015).

Similarly, a group of 300 Mn-resistant endophytic bacteria were isolated from pokeweed (*Phytolacca americana*) and characterized for their ability to produce siderophores, IAA, and ACC deaminase, under Mn stress and the absence of stress (Zhang et al., 2015). It was found that under Mn stress, the ability of the endophytes to produce siderophores and ACC deaminase was significantly increased compared with nonstress conditions. Moreover, plants treated with one representative endophyte (*B. megaterium* 1Y31) exhibited increased biomass and 64% greater accumulation of Mn in plant shoots over the untreated control (Zhang et al., 2015). In another study, a number of bacterial endophytes isolated from copper tolerant plants were initially screened for ACC deaminase activity and then the ACC deaminase positive strains were further characterized for IAA production, siderophore production, and phosphate solubilization (Zhang et al., 2011a). Based on that characterization, strains that are able to produce high levels of ACC deaminase three separate strains, namely *Ralstonia* sp., *P. agglomerans*, and *Pseudomonas thivervalensis*, were further selected to test their ability to help canola plants grown in the presence of copper contaminants. Interestingly, the higher ACC deaminase producers also exhibited higher IAA and siderophore production (Zhang et al., 2011a). All three bacterial endophytes provided protection for canola cultivated in copper contaminated substrates with increased biomass and increased copper uptake in the above ground plant tissues (Zhang et al., 2011a). A similar study was carried out by the same researchers with the isolation of ACC deaminase-containing bacterial endophytes from lead tolerant plants (Zhang et al., 2011b). In another study, Idris et al. (2004) demonstrated that bacterial endophytes isolated from shoots of the nickel tolerant plant *Thlaspi goesingenese* provide higher plant tolerance to Ni than the bacteria isolated from the rhizosphere of the same plant and argued that endophytes may be better partners of plants than rhizosphere bacteria for heavy metal phytoremediation (Idris et al., 2004; Doty, 2008). The possible use of genetically manipulated endophytes in phytoremediation of heavy metals is another approach that has been discussed (Doty, 2008). The mechanisms that endophytes can use to help phytoremediation may include extracellular metal precipitation (Babu et al., 2015), metal sequestration and accumulation inside the cell (Shin et al., 2012), bioconversion of metal ions to less- or nontoxic forms (Zhu et al., 2014), and adsorption of heavy metal ions (Guo et al., 2010; Ma et al., 2016). In addition to bacteria, arbuscular mycorrhizas (AMs) are also able to help plants in detoxification of heavy metals (Gamalero et al., 2009) where AM fungi may be considered to be a type of fungal endophyte (Doty, 2008). In one study, the capacity of AM fungi to help the metal hyperaccumulating fern *Pteris vittata* by taking up arsenic was described (Trotta et al., 2006). The plants treated with the AM fungus *Glomus mosseae* were found to accumulate significantly greater amounts of arsenic than the untreated control plants (Trotta et al., 2006). PGPB that are actively involved in phytoremediation also generally demonstrate siderophore production, and phosphate solubilization activity (Burd et al., 2000; Trotta et al., 2006). In addition, some metals (e.g., arsenic) utilize the same mechanism of transportation to a plants' interior, as does phosphate, so that phosphate solubilization and transportation systems from endophytes may help plants in heavy metal detoxification (Doty, 2008; Wang et al., 2002).

Organic contaminants include, but are not limited to, petroleum, solvents, explosives, and polycyclic aromatic hydrocarbons (PAHs) (Doty, 2008). Most plants undergo stress in the presence of high levels of organic contaminants (Glick, 2014). A number of endophytes have been tested to assess their ability to help plants in the phytoremediation of organic pollutants (Doty, 2008). For example, endophytic strains of *Pseudomonas*, isolated from hybrid cottonwood that was grown in polluted site, have shown potential in phytoremediation of 2,4-dichlorophenoxyacetic acid (2,4-D), toluene, and naphthalene (Germaine et al., 2004). Another bacterial endophyte, *Methylobacterium* sp. strain BJ001 was isolated from hybrid poplar trees and was able to mineralize organic explosives trinitrotoluene (TNT) and Royal Demolition Explosive (RDX; hexahydro-1,3,5-trinitro-1,3,5-triazine) to carbon dioxide (Van Aken et al., 2004). Similarly, another endophyte of poplar, *Rhizobium tropici* bv. *populus* strain PTD1, was able to degrade RDX in solution (Doty et al., 2009). In addition, endophytes of hybrid cottonwood that were selected from different plant tissues demonstrated variable abilities in the phytoremediation of trichloroethylene (TCE) and BTEX (benzene, toluene, ethylbenzene, and xylene) (Moore et al., 2006). It is believed that plant compartmentalization plays an important role in selecting desirable endophytic communities (Doty, 2008). In a similar study PGPB were isolated and characterized from the rhizosphere and from the roots of hairy-fruited broom (*Cytisus striatus*) growing in a lindane-contaminated soil. In a subsequent experiment, bacterially treated plants were grown in soil spiked with gasoline with an unspiked control, which provided promising results in breaking down the diesel into its components and promoted plant growth and development in the presence of this stress (Balseiro-Romero et al., 2016).

The use of PGPB in plants other than their natural host is also a possible way of effecting phytoremediation. A native endophyte of poplar, *P. putida*, which can effectively degrade 2,4-dichlorophenoxyacetic acid (2,4-D) (Germaine et al., 2006), was used to inoculate pea plants and the colonization efficiency was confirmed by the expression of green fluorescent protein that was present on a mini-Tn*5* within the bacterium (Doty, 2008). The pea plants inoculated with the poplar PGPB *P. putida* show increased biomass over uninoculated control plants in the presence of the herbicide 2,4-D (Doty, 2008). Moreover, the PGPB, with its capacity to degrade 2,4-D, facilitated phytoremediation of this herbicide, leaving the soil completely free of 2,4-D (Doty, 2008).

2.8 Conclusions and future prospects

Plants are hugely affected by environmental changes in their surroundings and have adopted many ways to overcome such variation over time. Nevertheless, many environmental changes are deleterious to plant health and growth. PGPB, plant natural flora, have been shown to have beneficial effects on the growth and development of their host plants. A number of studies have documented plant–microbial interactions that facilitate plant growth in the presence of abiotic stress stimuli. The bacterial enzyme ACC deaminase plays a central role in helping plants to overcome abiotic stress by lowering the harmful levels of ethylene. Research on the precise role of other plant growth mediators (such as cytokinin and ABA) is still at a much more rudimentary stage. However, microbial nitrogen fixation and phosphate solubilization have been studies more intensely, although studies on the microbial uptake and supply of other nutrients (i.e., Zn, K, and Fe) that plants require for their growth and development are needed. Since PGPB mediated plant growth promotion is a combination of various microbial activities, it is still worthwhile to isolate and screen new bacteria from naturally stressed environmental niches where the microbial community faces environmental selection pressure and has adapted to survive under such stressful conditions; and upon thorough characterization, can be utilized as biofertilizers, biocontrol agents, and/or, biostimulants.

References

Agarwhal, S., Shende, S.T., 1987. Tetrazolium reducing microbes inside the roots of *brassica* species. Curr. Sci. 56, 187–188.

Ali, S., Charles, T.C., Glick, B.R., 2014. Amelioration of high salinity stress damage by plant growth-promoting bacterial endophytes that contain ACC deaminase. Plant Physiol. Biochem. 80, 160–167.

Ali, S., Charles, T.C., Glick, B.R., 2017. Endophytic phytohormones and their role in plant growth promotion. In: Doty, S.L. (Ed.), Functional Importance of the Plant Microbiome. Springer, Cham, pp. 89–105.

Ali, S., Isaacson, J., Kroner, Y., Lazarovits, G., 2018. Corn sap bacterial endophytes and their potential in plant growth-promotion. Environ. Sustain. 1–15.

Amellal, N., Burtin, G., Bartoli, F., Heulin, T., 1998. Colonization of wheat roots by an exopolysaccharide-producing *Pantoea agglomerans* strain and its effect on rhizosphere soil aggregation. Appl. Environ. Microbiol. 64 (10), 3740–3747.

Arkhipova, T.N., Prinsen, E., Veselov, S.U., Martinenko, E.V., Melentiev, A.I., Kudoyarova, G.R., 2007. Cytokinin producing bacteria enhance plant growth in drying soil. Plant Soil 292 (1–2), 305–315.

Aroca, R., Ferrante, A., Vernieri, P., Chrispeels, M.J., 2006. Drought, abscisic acid and transpiration rate effects on the regulation of PIP aquaporin gene expression and abundance in *Phaseolus vulgaris* plants. Ann. Bot. 98 (6), 1301–1310.

Arshad, M., Shahroona, B., Mahmood, T., 2008. Inoculation with *Pseudomonas* spp. containing ACC-deaminase partially eliminates the effects of drought stress on growth, yield, and ripening of pea (*Pisum sativum* L.). Pedosphere 18 (5), 611–620.

Assumpção, L.D.C., Lacava, P.T., Dias, A.C., de Azevedo, J.L., Menten, J.O.M., 2009. Diversity and biotechnological potential of endophytic bacterial community of soybean seeds. Pesqui. Agropecu. Bras. 44 (5), 503–510.

Babu, A.G., Shea, P.J., Sudhakar, D., Jung, I., Oh, B., 2015. Potential use of *Pseudomonas koreensis* AGB-1 in association with *Miscanthus sinensis* to remediate heavy metal (lead)-contaminated mining site soil. J. Environ. Manage. 151, 160–166.

Baldani, V.L.D., Baldani, J.I., Olivares, F., Dobereiner, J., 1992. Identification and ecology of *Herbaspirillum seropedicae* and the closely related *Pseudomonas rubrisubalbicans*. Symbiosis 13, 65–73.

Baldani, J.I., Caruso, L., Baldani, V.L.D., Goi, S.R., Dobereiner, J., 1997. Recent advances in BNF with nonlegume plants. Soil Biol. Biochem. 29 (5–6), 911–922.

Balseiro-Romero, M., Gkorezis, P., Kidd, P.S., Vangronsveld, J., Monterroso, C., 2016. Enhanced degradation of diesel in the rhizosphere of *Lupinus luteus* after inoculation with diesel-degrading and plant growth-promoting bacterial strains. J. Environ. Qual. 45 (3), 924–932.

Barahona, E., Navazo, A., Garrido-Sanz, D., Muriel, C., Martínez-Granero, F., Redondo-Nieto, M., et al., 2016. *Pseudomonas fluorescens* F113 can produce a second flagellar apparatus, which is important for plant root colonization. Front. Microbiol. 7 (SEP), 1471.

Barzanti, R., Ozino, F., Bazzicalupo, M., Gabbrielli, R., Galardi, F., Gonnelli, C., et al., 2007. Isolation and characterization of endophytic bacteria from the nickel hyperaccumulator plant *Alyssum bertolonii*. Microb. Ecol. 53 (2), 306–316.

Bashan, Y., Holguin, G., 1998. Proposal for the division of plant growth-promoting rhizobacteria into two classifications: biocontrol-PGPB (plant growth-promoting bacteria) and PGPB. Soil Biol. Biochem. 30 (8–9), 1225–1228.

Bastián, F., Cohen, A., Piccoli, P., Luna, V., Baraldi, R., Bottini, R., 1998. Production of indole-3-acetic acid and gibberellins A1 and A3 by *Acetobacter diazotrophicus* and *Herbaspirillum seropedicae* in chemically-defined culture media. Plant Growth Regul. 24 (1), 7–11.

Belimov, A.A., Dodd, I.C., Safronova, V.I., Dumova, V.A., Shaposhnikov, A.I., Ladatko, A.G., et al., 2014. Abscisic acid metabolizing rhizobacteria decrease ABA concentrations in planta and alter plant growth. Plant Physiol. Biochem. 74, 84–91.

Belimov, A.A., Dodd, I.C., Safronova, V.I., Shaposhnikov, A.I., Azarova, T.S., Makarova, N.M., et al., 2015. Rhizobacteria that produce auxins and contain 1-amino-cyclopropane-1-carboxylic acid deaminase decrease amino acid concentrations in the rhizosphere and improve growth and yield of well-watered and water-limited potato (*Solanum tuberosum*). Ann. Appl. Biol. 167 (1), 11–25.

Bell, C.R., Dickie, G.A., Harvey, W.L.G., Chan, J.W.Y.F., 1995. Endophytic bacteria in grapevine. Can. J. Microbiol. 41, 46–53.

Benhamou, N., Kloepper, J.W., Quadt-Hallman, A., Tuzun, S., 1996. Induction of defense-related ultrastructural modifications in pea root tissues inoculated with endophytic bacteria. Plant Physiol. 112 (3), 919–929.

Benhamou, N., Kloepper, J.W., Tuzun, S., 1998. Induction of resistance against *fusarium* wilt of tomato by combination of chitosan with an endophytic bacterial strain: ultrastructure and cytochemistry of the host response. Planta 204, 153–168.

Bhattacharyya, D., Yu, S.M., Lee, Y.H., 2015. Volatile compounds from *Alcaligenes faecalis* JBCS1294 confer salt tolerance in *Arabidopsis thaliana* through the auxin and gibberellin pathways and differential modulation of gene expression in root and shoot tissues. Plant Growth Regul. 75, 297.

Bhise, K.K., Bhagwat, P.K., Dandge, P.B., 2017. Synergistic effect of *Chryseobacterium gleum* sp. SUK with ACC deaminase activity in alleviation of salt stress and plant growth promotion in *Triticum aestivum* L. Biotech. 7 (2), 105.

Bové, J.M., Garnier, M., 2003. Phloem-and xylem-restricted plant pathogenic bacteria. Plant Sci. 164, 421–438.

Burd, G.I., Dixon, D.G., Glick, B.R., 2000. Plant growth-promoting bacteria that decrease heavy metal toxicity in plants. Can. J. Microbiol. 46 (3), 237–245.

Cao, L., Qiu, Z., You, J., Tan, H., Zhou, S., 2005. Isolation and characterization of endophytic streptomycete antagonists of *Fusarium* wilt pathogen from surface-sterilized banana roots. FEMS Microbiol. Lett. 247 (2), 147–152.

Castignetti, D., Smarrelli Jr., J., 1986. Siderophores, the iron nutrition of plants, and nitrate reductase. FEBS Lett. 209 (2), 147–151.

Chatterjee, P., Samaddar, S., Anandham, R., Kang, Y., Kim, K., Selvakumar, G., et al., 2017. Beneficial soil bacterium *Pseudomonas frederiksbergensis* Os261 augments salt tolerance and promotes red pepper plant growth. Front. Plant Sci. 8.

Chaudhary, D., Sindhu, S.S., 2015. Inducing salinity tolerance in chickpea (*Cicer arietinum* L.) by inoculation of 1-aminocyclopropane-1-carboxylic acid deaminase-containing *Mesorhizobium* strains. Afr. J. Microbiol. Res. 9 (2), 117–124.

Chelius, M.K., Triplett, E.W., 2000. Immunolocalization of dinitrogenase reductase produced by *Klebsiella pneumoniae* in association with *Zea mays* L. Appl. Environ. Microbiol. 66 (2), 783–787.

Chen, Y., Fan, J., Du, L., Xu, H., Zhang, Q., He, Y., 2014. The application of phosphate solubilizing endophyte *Pantoea dispersa* triggers the microbial community in red acidic soil. Appl. Soil Ecol. 84, 235–244.

Cocking, E.C., 2003. Endophytic colonisation of plant roots by nitrogen-fixing bacteria. Plant Soil 252, 169–175.

Cohen, A.C., Bottini, R., Piccoli, P.N., 2008. *Azospirillum brasilense* sp. 245 produces ABA in chemically-defined culture medium and increases ABA content in Arabidopsis plants. Plant Growth Regul. 54 (2), 97–103.

Contesto, C., Milesi, S., Mantelin, S., Zancarini, A., Desbrosses, G., Varoquaux, F., et al., 2010. The auxin-signaling pathway is required for the lateral root response of Arabidopsis to the rhizobacterium *Phyllobacterium brassicacearum*. Planta 232 (6), 1455–1470.

Debenport, S.J., Assigbetse, K., Bayala, R., Chapuis-Lardy, L., Dick, R.P., McSpadden Gardener, B.B., 2015. Association of shifting populations in the root zone microbiome of millet with enhanced crop productivity in the Sahel region (Africa). Appl. Environ. Microbiol. 81 (8), 2841–2851.

Dias, A.C.F., Costa, F.E.C., Andreote, F.D., Lacava, P.T., Teixeira, M.A., Assumpção, L.C., et al., 2009. Isolation of micropropagated strawberry endophytic bacteria and assessment of their potential for plant growth promotion. World J. Microbiol. Biotechnol. 25 (2), 189–195.

Döbereiner, J., Urquiaga, S., 1992. Soil biology and sustainable agriculture. An. Acad. Bras. Sci. 84, 127–133.

Dobereiner, J., Reis, V.M., Paula, M.A., Olivares, F., 1993. Endophytic diazotrophs in sugar cane, cereals and tuber plants. In: Palacios, R. (Ed.), New Horizons in Nitrogen Fixation. Kluwer Academic Publishers, Dordrecht, the Netherlands, pp. 671–676.

Donderski, W., Gluchowska, M., 2000. Production of cytokinin-like substances by planktonic bacteria isolated from lake Jeziorak. Pol. J. Environ. Stud. 9 (5), 369–376.

Dong, Z.M., Canny, M.J., McCully, M.E., Roboredo, M.R., Cabadilla, C.F., Ortega, E., et al., 1994. A nitrogen-fixing endophyte of sugarcane stems—a new role for the apoplast. Plant Physiol. 105, 1139–1147.

Dong, Y.M., Iniguez, A.L., Ahmer, B.M.M., Triplett, E.W., 2003. Kinetics and strain specificity of rhizosphere and endophytic colonisation by enteric bacteria on seedlings of *Medicago sativa* and *Medicago truncatula*. Appl. Environ. Microbiol. 69, 1783–1790.

Doty, S.L., 2008. Enhancing phytoremediation through the use of transgenics and endophytes. New Phytol. 179 (2), 318–333.

Doty, S.L., Oakley, B., Xin, G., Kang, J.W., Singleton, G., Khan, Z., et al., 2009. Diazotrophic endophytes of native black cottonwood and willow. Symbiosis 47 (1), 23–33.

Duca, D., Lorv, J., Patten, C.L., Rose, D., Glick, B.R., 2014. Indole-3-acetic acid in plant-microbe interactions. Antonie Leeuwenhoek, Int. J. Gen. Mol. Microbiol. 106 (1), 85–125.

Egamberdieva, D., Jabborova, D., Hashem, A., 2015. *Pseudomonas* induces salinity tolerance in cotton (*Gossypium hirsutum*) and resistance to fusarium root rot through the modulation of indole-3-acetic acid. Saudi J. Biol. Sci. 22 (6), 773–779.

Ekimova, G.A., Fedorov, D.N., Tani, A., Doronina, N.V., Trotsenko, Y.A., 2018. Distribution of 1-aminocyclopropane-1-carboxylate deaminase and d-cysteine desulfhydrase genes among type species of the genus *Methylobacterium*. Antonie Leeuwenhoek, Int. J. Gen. Mol. Microbiol. 111 (10), 1723–1734.

Estrada, P., Mavingui, P., Cournoyer, B., Fontaine, F., Balandreau, J., Caballero-Mellado, J., 2002. A N_2-fixing endophytic *Burkholderia* sp. associated with maize plants cultivated in Mexico. Can. J. Microbiol. 48 (4), 285–294.

Evers, A., Hancock, R.D., Martell, A.E., Motekaitis, R.J., 1989. Metal ion recognition in ligands with negatively charged oxygen donor groups. Complexation of Fe (III), Ga(III), In(III), Al(III), and other highly charged metal ions. Inorg. Chem. 28 (11), 2189–2195.

Fan, P., Chen, D., He, Y., Zhou, Q., Tian, Y., Gao, L., 2016. Alleviating salt stress in tomato seedlings using *Arthrobacter* and *Bacillus megaterium* isolated from the rhizosphere of wild plants grown on saline–alkaline lands. Int. J. Phytorem. 18 (11), 1113–1121.

Forchetti, G., Masciarelli, O., Alemano, S., Alvarez, D., Abdala, G., 2007. Endophytic bacteria in sunflower (*Helianthus annuus* L.): isolation, characterization, and production of jasmonates and abscisic acid in culture medium. Appl. Microbiol. Biotechnol. 76 (5), 1145–1152.

Forni, C., Duca, D., Glick, B.R., 2017. Mechanisms of plant response to salt and drought stress and their alteration by rhizobacteria. Plant Soil 410 (1–2), 335–356.

Fujino, A., Ose, T., Yao, M., Tokiwano, T., Honma, M., Watanabe, N., et al., 2004. Structural and enzymatic properties of 1-aminocyclopropane-1-carboxylate deaminase homologue from *Pyrococcus horikoshii*. J. Mol. Biol. 341 (4), 999–1013.

Gagn, S., Richard, C., Rousseau, H., Antoun, H., 1987. Xylem-residing bacteria in alfalfa. Can. J. Microbiol. 33, 996–1005.

Gaiero, J.R., McCall, C.A., Thompson, K.A., Day, N.J., Best, A.S., Dunfield, K.E., 2013. Inside the root microbiome: bacterial root endophytes and plant growth promotion. Am. J. Bot. 100 (9), 1738–1750.

Gamalero, E., Lingua, G., Berta, G., Glick, B.R., 2009. Beneficial role of plant growth promoting bacteria and arbuscular mycorrhizal fungi on plant responses to heavy metal stress. Can. J. Microbiol. 55 (5), 501–514.

Garcia De Salamone, I.E., Döbereiner, J., Urquiaga, S., Boddey, R.M., 1996. Biological nitrogen fixation in *Azospirillum* strain-maize genotype associations as evaluated by the ^{15}N isotope dilution technique. Biol. Fert. Soils 23 (3), 249–256.

García de Salamone, I.E., Hynes, R.K., Nelson, L.M., 2001. Cytokinin production by plant growth promoting rhizobacteria and selected mutants. Can. J. Microbiol. 47 (5), 404–411.

Germaine, K., Keogh, E., Garcia-Cabellos, G., Borremans, B., Van Der Lelie, D., Barac, T., et al., 2004. Colonisation of poplar trees by gfp expressing bacterial endophytes. FEMS Microbiol. Ecol. 48 (1), 109–118.

Germaine, K.J., Liu, X., Cabellos, G.G., Hogan, J.P., Ryan, D., Dowling, D.N., 2006. Bacterial endophyte-enhanced phytoremediation of the organochlorine herbicide 2,4-dichlorophenoxyacetic acid. FEMS Microbiol. Ecol. 57 (2), 302–310.

Glick, B.R., 2012. Plant growth-promoting bacteria: mechanisms and applications. Scientifica 2012.

Glick, B.R., 2014. Bacteria with ACC deaminase can promote plant growth and help to feed the world. Microbiol. Res. 169 (1), 30–39.

Glick, B.R., 2015. Beneficial Plant–Bacterial Interactions. Springer, pp. 1–243.

Glick, B.R., Cheng, Z., Czarny, J., Duan, J., 2007. Promotion of plant growth by ACC deaminase-producing soil bacteria. Eur. J. Plant Pathol. 119 (3), 329–339.

Govindarajan, M., Balandreau, J., Muthukumarasamy, R., Revathi, G., Lakshminarasimhan, C., 2006. Improved yield of micropropagated sugarcane following inoculation by endophytic *Burkholderia vietnamiensis*. Plant Soil 280 (1–2), 239–252.

Govindarajan, M., Kwon, S.W., Weon, H.Y., 2007. Isolation, molecular characterization and growth-promoting activities of endophytic sugarcane diazotroph *Klebsiella* sp. GR9. World J. Microbiol. Biotechnol. 23, 997–1006.

Grichko, V.P., Glick, B.R., 2001. Amelioration of flooding stress by ACC deaminase-containing plant growth-promoting bacteria. Plant Physiol. Biochem. 39 (1), 11–17.

Grossmann, K., 2010. Auxin herbicides: current status of mechanism and mode of action. Pest Manage. Sci. 66 (2), 113–120.

Guo, H., Luo, S., Chen, L., Xiao, X., Xi, Q., Wei, W., et al., 2010. Bioremediation of heavy metals by growing hyperaccumulator endophytic bacterium *Bacillus* sp. L14. Bioresour. Technol. 101 (22), 8599–8605.

Gupta, G., Panwar, J., Akhtar, M.S., Jha, P.N., 2012. Endophytic nitrogen-fixing bacteria as bio-fertilizer. In: Lichtfouse, E. (Ed.), Sustainable Agriculture Reviews. Springer, Dordrecht, pp. 183–221.

Gururani, M.A., Upadhyaya, C.P., Baskar, V., Venkatesh, J., Nookaraju, A., Park, S.W., 2013. Plant growth-promoting rhizobacteria enhance abiotic stress tolerance in *Solanum tuberosum* through inducing changes in the expression of ROS-scavenging enzymes and improved photosynthetic performance. J. Plant Growth Regul. 32 (2), 245–258.

Gyaneshwar, P., James, E.K., Mathan, N., Reddy, P.M., Reinhold-Hurek, B., Ladha, J.K., 2001. Endophytic colonisation of rice by a diazotrophic strain of *Serratia marcescens*. J. Bacteriol. 183, 2634–2645.

Hallmann, J., Quadt-Hallmann, A., Mahaffee, W.F., Kloepper, J.W., 1997. Bacterial endophytes in agricultural crops. Can. J. Microbiol. 43 (10), 895–914.

Hardoim, P.R., van Overbeek, L.S., Elsas, J.D., 2008. Properties of bacterial endophytes and their proposed role in plant growth. Trends Microbiol. 16 (10), 463–471.

Hasegawa, S., Poling, S.M., Maier, V.J., Bennett, R.D., 1984. Metabolism of abscisic acid: bacterial conversion to dehydrovomifoliol and vomifoliol dehydrogenase activity. Phytochemistry 23 (12), 2769–2771.

Hider, R.C., Kong, X., 2010. Chemistry and biology of siderophores. Nat. Prod. Rep. 27 (5), 637–657.

Hontzeas, N., Zoidakis, J., Glick, B.R., Abu-Omar, M.M., 2004. Expression and characterization of 1-aminocyclopropane-1-carboxylate deaminase from the rhizobacterium *Pseudomonas putida* UW4: a key enzyme in bacterial plant growth promotion. Biochem. Biophys. Acta 1703 (1), 11–19.

Huang, J., 1986. Ultrastructure of bacterial penetration in plants. Annu. Rev. Phytopathol. 24, 141–157.

Idris, R., Trifonova, R., Puschenreiter, M., Wenzel, W.W., Sessitsch, A., 2004. Bacterial communities associated with flowering plant of Ni hyperaccumulator *Thlaspi goesingense*. Appl. Environ. Microbiol. 70 (5), 2667–2677.

Jacobs, J., Bugbee, W.M., Gabrielson, G.A., 1985. Enumeration, location, and characterization of endophytic bacteria within sugar beet roots. Can. J. Bot. 63, 1262–1265.

James, E.K., Olivares, F.L., 1998. Infection and colonization of sugar cane and other graminaceous plants by endophytic diazotrophs. Crit. Rev. Plant Sci. 17 (1), 77–119.

James, E.K., Olivares, F.L., Baldani, J.I., Döbereiner, J., 1997. *Herbaspirillum*, an endophytic diazotroph colonizing vascular tissue in leaves of sorghum bicolor *L. moench*. J. Exp. Bot. 48 (308), 785–797.

James, E.K., Gyaneshwar, P., Mathan, N., Barraquio, W.L., Reddy, P.M., Iannetta, P.P.M., et al., 2002. Infection and colonisation of rice seedlings by the plant growth promoting bacterium *Herbaspirillum seropedicae* Z67. Mol. Plant Microb. Interact. 15, 894–906.

Joe, M.M., Devaraj, S., Benson, A., Sa, T., 2016. Isolation of phosphate solubilizing endophytic bacteria from *Phyllanthus amarus* Schum & Thonn: evaluation of plant growth promotion and antioxidant activity under salt stress. J. Appl. Res. Med. Aromat. Plants 3 (2), 71–77.

Kamlesh, M.K., Sorty, A.M., Bitla, U.M., Choudhary, K., Gupta, P., Pareek, A., et al., 2017. Abiotic stress responses and mediated mitigation in plants: the omics strategies. Front. Plant Sci. 8 (172).

Kang, S., Radhakrishnan, R., Khan, A.L., Kim, M., Park, J., Kim, B., et al., 2014. Gibberellin secreting rhizobacterium, *Pseudomonas putida* H-2-3 modulates the hormonal and stress physiology of soybean to improve the plant growth under saline and drought conditions. Plant Physiol. Biochem. 84, 115–124.

Kariola, T., Brader, G., Helenius, E., Li, J., Heino, P., Palva, E.T., 2006. Early responsive to dehydration 15, a negative regulator of abscisic acid responses in Arabidopsis. Plant Physiol. 142 (4), 1559–1573.

Khan, A.L., Waqas, M., Kang, S., Al-Harrasi, A., Hussain, J., Al-Rawahi, A., et al., 2014. Bacterial endophyte *Sphingomonas* sp. LK11 produces gibberellins and IAA and promotes tomato plant growth. J. Microbiol. 52 (8), 689–695.

Khan, A.L., Hussain, J., Al-Harrasi, A., Al-Rawahi, A., Lee, I., 2015. Endophytic fungi: resource for gibberellins and crop abiotic stress resistance. Crit. Rev. Biotechnol. 35 (1), 62–74.

Kim, K., Jang, Y.J., Lee, S.M., Oh, B.T., Chae, J.C., Lee, K.J., 2014. Alleviation of salt stress by *Enterobacter* sp. EJ01 in tomato and Arabidopsis is accompanied by up-regulation of conserved salinity responsive factors in plants. Mol. Cells 37, 109–117.

Kloepper, J.W., Leong, J., Teintze, M., Schroth, M.N., 1980. Enhanced plant growth by siderophores produced by plant growth-promoting rhizobacteria. Nature 286 (5776), 885–886.

Kolter, R., Siegele, D.A., Tormo, A., 1993. The stationary phase of the bacterial life cycle. Ann. Rev. Microbiol. 47, 855–874.

Kuklinsky-Sobral, J., Araújo, W.L., Mendes, R., Geraldi, I.O., Pizzirani-Kleiner, A.A., Azevedo, J.L., 2004. Isolation and characterization of soybean-associated bacteria and their potential for plant growth promotion. Environ. Microbiol. 6 (12), 1244–1251.

Lacava, P.T., Silva-Stenico, M.E., Araújo, W.L., Simionato, A.V.C., Carrilho, E., Siu, M.T., et al., 2008. Detection of siderophores in endophytic bacteria *Methylobacterium* spp. associated with *Xylella fastidiosa* subsp. pauca. Pesqui. Agropecu. Bras. 43 (4), 521–528.

Lacuna, P.T., Azevedo, J.L., 2013. Endophytic Bacteria: A Biotechnological Potential in Agrobiology System. Springer, Berlin.

Lau, J.A., Lennon, J.T., 2012. Rapid responses of soil microorganisms improve plant fitness in novel environments. Proc. Natl. Acad. Sci. 109 (35), 14058–14062.

Li, C.H., Zhao, M.W., Tang, C.M., Li, S.P., 2009. Population dynamics and identification of endophytic bacteria antagonistic toward plant-pathogenic fungi in cotton root. Microb. Ecol. 59, 344–356.

Li, J., McConkey, B.J., Cheng, Z., Guo, S., Glick, B.R., 2013. Identification of plant growth-promoting bacteria-responsive proteins in cucumber roots under hypoxic stress using a proteomic approach. J. Proteom. 84, 119–131.

Li, W., Yamaguchi, S., Khan, M.A., An, P., Liu, X., Tran, L.S., 2016. Roles of gibberellins and abscisic acid in regulating germination of *Suaeda salsa* dimorphic seeds under salt stress. Front. Plant Sci. 6 (JAN), 1235.

Lichtfouse, E. (Ed.), 2012. Sustainable Agriculture Reviews 1. Springer, Dordrecht.

Liu, X., Zhang, H., 2015. The effects of bacterial volatile emissions on plant abiotic stress tolerance. Front. Plant Sci. 6 (SEP), 774.

Loaces, I., Ferrando, L., Scavino, A.F., 2011. Dynamics, diversity and function of endophytic siderophore-producing bacteria in rice. Microb. Ecol. 61 (3), 606–618.

Lodewyckx, C., Vangronsveld, J., Porteous, F., Moore, E.R.B., Taghavi, S., Mezgeay, M., et al., 2002. Endophytic bacteria and their potential applications. Crit. Rev. Plant Sci. 21 (6), 583–606.

Loper, J.E., Henkels, M.D., 1999. Utilization of heterologous siderophores enhances levels of iron available to *Pseudomonas putida* in the rhizosphere. Appl. Environ. Microbiol. 65 (12), 5357–5363.

Lopes, L.D., Pereira e Silva, M.C., Andreote, F.D., 2016. Bacterial abilities and adaptation toward the rhizosphere colonization. Front. Microbiol. 7 (AUG), 1341.

Lupatini, M., Korthals, G.W., de Hollander, M., Janssens, T.K.S., Kuramae, E.E., 2017. Soil microbiome is more heterogeneous in organic than in conventional farming system. Front. Microbiol. 7 (JAN), 2064.

Luvizotto, D.M., Marcon, J., Andreote, F.D., Dini-Andreote, F., Neves, A.A.C., Araújo, W.L., et al., 2010. Genetic diversity and plant-growth related features of *Burkholderia* spp. from sugarcane roots. World J. Microbiol. Biotechnol. 26 (10), 1829–1836.

Ma, Y., Rajkumar, M., Zhang, C., Freitas, H., 2016. Beneficial role of bacterial endophytes in heavy metal phytoremediation. J. Environ. Manage. 174, 14–25.

Maheshwari, D.K., Aeron, A., Saraf, M., 2013. Bacteria in Agrobiology: Crop Productivity, 1–507.

Marques, A.P.G.C., Pires, C., Moreira, H., Rangel, A.O.S.S., Castro, P.M.L., 2010. Assessment of the plant growth promotion abilities of six bacterial isolates using *Zea mays* as indicator plant. Soil Biol. Biochem. 42 (8), 1229–1235.

Martínez-Romero, E., 2003. Diversity of *rhizobium-phaseolus vulgaris* symbiosis: overview and perspectives. Plant Soil 252 (1), 11–23.

Marulanda, A., Barea, J., Azcón, R., 2009. Stimulation of plant growth and drought tolerance by native microorganisms (AM fungi and bacteria) from dry environments: mechanisms related to bacterial effectiveness. J. Plant Growth Regul. 28 (2), 115–124.

Mayak, S., Tirosh, T., Glick, B.R., 2004. Plant growth-promoting bacteria that confer resistance to water stress in tomatoes and peppers. Plant Sci. 166 (2), 525–530.

Meldau, D.G., Meldau, S., Hoang, L.H., Underberg, S., Wünsche, H., Baldwin, I.T., 2013. Dimethyl disulfide produced by the naturally associated bacterium *Bacillus* sp B55 promotes *Nicotiana attenuata* growth by enhancing sulfur nutrition. Plant Cell 25 (7), 2731–2747.

Mellor, N., Band, L.R., Pением, A., Novák, O., Rashed, A., Holman, T., et al., 2016. Dynamic regulation of auxin oxidase and conjugating enzymes AtDAO1 and GH3 modulates auxin homeostasis. Proc. Natl. Acad. Sci. USA. 113 (39), 11022–11027.

Mimmo, T., Del Buono, D., Terzano, R., Tomasi, N., Vigani, G., Crecchio, C., et al., 2014. Rhizospheric organic compounds in the soil-microorganism-plant system: their role in iron availability. Eur. J. Soil Sci. 65 (5), 629–642.

Minami, R., Uchiyama, K., Murakami, T., Kawai, J., Mikami, K., Yamada, T., et al., 1998. Properties, sequence, and synthesis in *Escherichia coli* of 1-aminocyclopropane-1-carboxylate deaminase from *Hansenula saturnus*. J. Biochem. 123 (6), 1112–1118.

Montañez, A., Blanco, A.R., Barlocco, C., Beracochea, M., Sicardi, M., 2012. Characterization of cultivable putative endophytic plant growth promoting bacteria associated with maize cultivars (*Zea mays* L.) and their inoculation effects in vitro. Appl. Soil Ecol. 58, 21–28.

Moore, F.P., Barac, T., Borremans, B., Oeyen, L., Vangronsveld, J., Van Der Lelie, D., et al., 2006. Endophytic bacterial diversity in poplar trees growing on a BTEX-contaminated site: the characterization of isolates with potential to enhance phytoremediation. Syst. Appl. Microbiol. 29, 539–556.

Muehe, E.M., Weigold, P., Adaktylou, I.J., Planer-Friedrich, B., Kraemer, U., Kappler, A., et al., 2015. Rhizosphere microbial community composition affects cadmium and zinc uptake by the metal-hyperaccumulating plant *Arabidopsis halleri*. Appl. Environ. Microbiol. 81 (6), 2173–2181.

Munns, R., Tester, M., 2008. Mechanisms of salinity tolerance. Ann. Rev. Plant Biol. 59, 651–681.

Mustroph, A., 2018. Flooding Stress in Plants. Wiley, Chichester, In eLS (Ed.).

Nadeem, S.M., Ahmad, M., Naveed, M., Imran, M., Zahir, Z.A., Crowley, D.E., 2016. Relationship between in vitro characterization and comparative efficacy of plant growth-promoting rhizobacteria for improving cucumber salt tolerance. Archives of Microbiology 198 (4), 379–387.

Naqqash, T., Hameed, S., Imran, A., Hanif, M.K., Majeed, A., van Elsas, J.D., 2016. Differential response of potato toward inoculation with taxonomically diverse plant growth promoting rhizobacteria. Front. Plant Sci. 7.

Nascimento, F.X., Rossi, M.J., Soares, C.R.F.S., McConkey, B.J., Glick, B.R., 2014. New insights into 1-aminocyclopropane-1-carboxylate (ACC) deaminase phylogeny, evolution and ecological significance. PLoS One 9 (6).

Naveed, M., Hussain, M.B., Zahir, Z.A., Mitter, B., Sessitsch, A., 2014. Drought stress amelioration in wheat through inoculation with *Burkholderia phytofirmans* strain PsJN. Plant Growth Regul. 73 (2), 121–131.

Neilands, J.B., 1981. Iron absorption and transport in microorganisms. Annu. Rev. Nutr. 1, 27–46.

Nelson, E.B., 2018. The seed microbiome: origins, interactions, and impacts. Plant Soil 422 (1–2), 7–34.

Nett, R.S., Montanares, M., Marcassa, A., Lu, X., Nagel, R., Charles, T.C., et al., 2017. Elucidation of gibberellin biosynthesis in bacteria reveals convergent evolution. Nat. Chem. Biol. 13 (1), 69–74.

Niu, B., Paulson, J.N., Zheng, X., Kolter, R., 2017. Simplified and representative bacterial community of maize roots. Proc. Natl. Acad. Sci. USA. 114 (12), E2450–E2459.

Orhan, F., 2016. Alleviation of salt stress by halotolerant and halophilic plant growth-promoting bacteria in wheat (*Triticum aestivum*). Braz. J. Microbiol. 47 (3), 621–627.

O'Sullivan, D.J., O'Gara, F., 1992. Traits of fluorescent *Pseudomonas* spp. involved in suppression of plant root pathogens. Microbiol. Rev. 56 (4), 662–676.

Oteino, N., Lally, R.D., Kiwanuka, S., Lloyd, A., Ryan, D., Germaine, K.J., et al., 2015. Plant growth promotion induced by phosphate solubilizing endophytic *Pseudomonas* isolates. Front. Microbiol. 6.

Palaniappan, A., Goh, W.H., Tey, J.N., Wijaya, I.P.M., Moochhala, S.M., Liedberg, B., et al., 2010. Aligned carbon nanotubes on quartz substrate for liquid gated biosensing. Biosens. Bioelectron. 25 (8), 1989–1993.

Palus, J.A., Borneman, J., Ludden, P.W., Triplett, E.W., 1996. A diazotrophic bacterial endophyte isolated from stems of *Zea mays* l. and *Zea luxurians* iltis and doebley. Plant Soil 186 (1), 135–142.

Park, M., Kim, C., Yang, J., Lee, H., Shin, W., Kim, S., et al., 2005. Isolation and characterization of diazotrophic growth promoting bacteria from rhizosphere of agricultural crops of Korea. Microbiol. Res. 160 (2), 127–133.

Patriquin, D.G., Döbereiner, J., 1978. Light microscopy observations of tetrazolium-reducing bacteria in the endorhizosphere of maize and other grasses in Brazil. Can. J. Microbiol. 24 (6), 734–742.

Puente, M.E., Li, C.Y., Bashan, Y., 2009. Rock-degrading endophytic bacteria in cacti. Environ. Exp. Bot. 66 (3), 389–401.

Qin, F., Shinozaki, K., Yamaguchi-Shinozaki, K., 2011. Achievements and challenges in understanding plant abiotic stress responses and tolerance. Plant Cell Physiol. 52 (9), 1569–1582.

Qin, Y., Druzhinina, I.S., Pan, X., Yuan, Z., 2016. Microbially mediated plant salt tolerance and microbiome-based solutions for saline agriculture. Biotechnol. Adv. 34 (7), 1245–1259.

Quadt-Hallmann, A., Benhamou, N., Kloepper, J.W., 1997. Bacterial endophytes in cotton: mechanisms of entering the plant. Can. J. Microbiol. 43, 557–582.

Rajkumar, M., Ae, N., Freitas, H., 2009. Endophytic bacteria and their potential to enhance heavy metal phytoextraction. Chemosphere 77 (2), 153–160.

Rajkumar, M., Ae, N., Prasad, M.N.V., Freitas, H., 2010. Potential of siderophore-producing bacteria for improving heavy metal phytoextraction. Trends Biotechnol. 28 (3), 142–149.

Rashid, S., Charles, T.C., Glick, B.R., 2012. Isolation and characterization of new plant growth-promoting bacterial endophytes. Appl. Soil Ecol. 61, 217–224.

Raza, F.A., Amin, A., Faisal, M., 2015. Desiccation-tolerant rhizobacteria from Cholistan desert, Pakistan, and their impact on *Zea mays* L. Pol. J. Environ. Stud. 24 (4), 1773–1781.

Reinhold-Hurek, B., Hurek, T., 1998. Interactions of gramineous plants with *Azoarcus* spp. and other diazotrophs: identification, localization, and perspectives to study their function. Crit. Rev. Plant Sci. 17 (1), 29–54.

Reinhold-Hurek, B., Bünger, W., Burbano, C.S., Sabale, M., Hurek, T., 2015. Roots shaping their microbiome: global hotspots for microbial activity. Annu. Rev. Phytopathol. 53, 403–424.

Reis, V.M., Baldani, J.I., Baldani, V.L.D., Dobereiner, J., 2000. Biological dinitrogen fixation in *Gramineae* and palm trees. Crit. Rev. Plant Sci. 19 (3), 227–247.

Reis, V.M., Estrada-de los Santos, P., Tenorio-Salgado, S., Vogel, J., Stoffels, M., Guyon, S., et al., 2004. *Burkholderia tropica* sp. nov., a novel nitrogen-fixing, plant-associated bacterium. Int. J. Syst. Evol. Microbiol. 54 (6), 2155–2162.

Rodríguez, H., Fraga, R., 1999. Phosphate solubilizing bacteria and their role in plant growth promotion. Biotechnol. Adv. 17 (4–5), 319–339.

Rojas-Tapias, D.F., Bonilla, R., Dussán, J., 2014. Effect of inoculation and co-inoculation of *Acinetobacter* sp. RG30 and *Pseudomonas putida* GN04 on growth, fitness, and copper accumulation of maize (*Zea mays*). Water Air Soil Pollut. 225 (12).

Rorison, I.H., 1986. The response of plants to acid soils. Experientia 42 (4), 357–362.

Sandhya, V., Ali, S.Z., Grover, M., Reddy, G., Venkateswarlu, B., 2009. Alleviation of drought stress effects in sunflower seedlings by the exopolysaccharides producing *Pseudomonas putida* strain GAP-p45. Biol. Fert. Soils 46 (1), 17–26.

Scot, R.I., Chard, J.M., Hocart, M.J., Lennard, J.H., Graham, D.C., 1996. Penetration of potato tuber lenticels by bacteria in relation to biological control of blackleg disease. Potato Res. 39, 333–334.

Setter, T.L., Parra, R., 2010. Relationship of carbohydrate and abscisic acid levels to kernel set in maize under post pollination water deficit. Crop Sci. 50 (3), 980–988.

Sharma, S., Kulkarni, J., Jha, B., 2016. Halotolerant rhizobacteria promote growth and enhance salinity tolerance in peanut. Front. Microbiol. 7.

Sheibani-Tezerji, R., Rattei, T., Sessitsch, A., Trognitz, F., Mitter, B., 2015. Transcriptome profiling of the endophyte *Burkholderia phytofirmans* PsJN indicates sensing of the plant environment and drought stress. mBio 6 (5), 1–11.

Shin, M., Shim, J., You, Y., Myung, H., Bang, K., Cho, M., et al., 2012. Characterization of lead resistant endophytic *Bacillus* sp. MN3-4 and its potential for promoting lead accumulation in metal hyperaccumulator *Alnus firma*. J. Hazard. Mater. 199–200, 314–320.

Siddikee, M.A., Sundaram, S., Chandrasekaran, M., Kim, K., Selvakumar, G., Sa, T., 2015. Halotolerant bacteria with ACC deaminase activity alleviate salt stress effect in canola seed germination. J. Korean Soc. Appl. Biol. Chem. 58 (2), 237–241.

Silva, E.N., Ribeiro, R.V., Ferreira-Silva, S.L., Viégas, R.A., Silveira, J.A.G., 2010. Comparative effects of salinity and water stress on photosynthesis, water relations and growth of *Jatropha curcas* plants. J. Arid Environ. 74 (10), 1130–1137.

Singh, J.S., 2015a. Plant-microbe interactions: a viable tool for agricultural sustainability. Appl. Soil Ecol. 92, 45–46.

Singh, J.S., 2015b. Microbes: the chief ecological engineers in reinstating equilibrium in degraded ecosystems. Agric. Ecosyst. Environ. 203, 80–82.

Singh, J.S., 2016. Microbes play major roles in ecosystem services. Clim. Change Environ. Sustain. 3, 163–167.

Singh, J.S., Gupta, V.K., 2018. Soil microbial biomass: a key soil driver in management of ecosystem functioning. Sci. Tot. Environ. 634, 497–500.

Singh, R.P., Jha, P.N., 2016a. Mitigation of salt stress in wheat plant (*Triticum aestivum*) by ACC deaminase bacterium *Enterobacter* sp. SBP-6 isolated from *Sorghum bicolor*. Acta Physiol. Plant. 38 (5).

Singh, R.P., Jha, P.N., 2016b. The multifarious PGPR *Serratia marcescens* CDP-13 augments induced systemic resistance and enhanced salinity tolerance of wheat (*Triticum aestivum* L.). PLoS One 11 (16).

Singh, K., Agrawal, K.N., Bora, G.C., 2011. Advanced techniques for weed and crop identification for site specific weed management. Biosyst. Eng. 109 (1), 52–64.

Singh, R.P., Shelke, G.M., Kumar, A., Jha, P.N., 2015. Biochemistry and genetics of ACC deaminase: a weapon to "stress ethylene" produced in plants. Front. Microbiol. 6.

Singh, J.S., Abhilash, P.C., Gupta, V.K., 2016. Agriculturally important microbes in sustainable food production. Trends Biotechnol. 34, 773–775.

Smith, P.T., Cobb, B.G., 1991. Physiological and enzymatic activity of pepper seeds (*Capsicum annuum*) during priming. Physiol. Plant. 82 (3), 433–439.

Song, M.C., Beom, R.K., Song, H.H., Anderson, A.J., Park, J., Lee, Y., et al., 2008. 2R, 3R-butanediol, a bacterial volatile produced by *Pseudomonas chlororaphis* O6, is involved in induction of systemic tolerance to drought in *Arabidopsis thaliana*. Mol. Plant Microbe Interact. 21 (8), 1067–1075.

Sørensen, J., Sessitsch, A., 2006. Plant-associated bacteria lifestyle and molecular interactions. In: Van Elsas, J.D., Jansson, J.K., Trevors, J.T. (Eds.), Modern Soil Microbiology, second ed CRC Press, pp. 211–236.

Spoel, S.H., Dong, X., 2008. Making sense of hormone crosstalk during plant immune responses. Cell Host Microbe. 3 (6), 348–351.

Sukweenadhi, J., Kim, Y., Choi, E., Koh, S., Lee, S., Kim, Y., et al., 2015. *Paenibacillus yonginensis* DCY84T induces changes in *Arabidopsis thaliana* gene expression against aluminum, drought, and salt stress. Microbiol. Res. 172, 7–15.

Timmusk, S., Paalme, V., Pavlicek, T., Bergquist, J., Vangala, A., Danilas, T., et al., 2011. Bacterial distribution in the rhizosphere of wild barley under contrasting microclimates. PLoS One 6 (3).

Tiwari, S., Singh, P., Tiwari, R., Meena, K.K., Yandigeri, M., Singh, D.P., et al., 2011. Salt-tolerant rhizobacteria-mediated induced tolerance in wheat (*Triticum aestivum*) and chemical diversity in rhizosphere enhance plant growth. Biol. Fert. Soils 47 (8), 907–916.

Todaka, D., Nakashima, K., Maruyama, K., Kidokoro, S., Osakabe, Y., Ito, Y., et al., 2012. Rice phytochrome-interacting factor-like protein OsPIL1 functions as a key regulator of internode elongation and induces a morphological response to drought stress. Proc. Natl. Acad. Sci. USA. 109 (39), 15947–15952.

Trotta, A., Falaschi, P., Cornara, L., Minganti, V., Fusconi, A., Drava, G., et al., 2006. Arbuscular mycorrhizae increase the arsenic translocation factor in the As hyperaccumulating fern *Pteris vittata* L. Chemosphere 65 (1), 74–81.

Vaishnav, A., Kumari, S., Jain, S., Varma, A., Choudhary, D.K., 2015. Putative bacterial volatile-mediated growth in soybean (*Glycine max* L. merrill) and expression of induced proteins under salt stress. J. Appl. Microbiol. 119 (2), 539–551.

Van Aken, B., Yoon, J., Schnoor, J.L., 2004. Using populus species to degrade explosives for sustainable range management. Proceedings of the First Conference on Sustainable Range Management. pp. 1678–1720.

van Overbeek, L.S., van Vuurde, J., Van Elsas, J.D., 2006. Application of molecular fingerprinting techniques to explore the diversity of bacterial endophytic communities. Soil Biol. 9, 337–354.

Vendan, R.T., Yu, Y.J., Lee, S.H., Rhee, Y.H., 2010. Diversity of endophytic bacteria in ginseng and their potential for plant growth promotion. J. Microbiol. 48 (5), 559–565.

Verma, V.C., Singh, S.K., Prakash, S., 2011. Bio-control and plant growth promotion potential of siderophore producing endophytic *Streptomyces* from *Azadirachta indica* A. Juss. J. Basic Microbiol. 51 (5), 550–556.

Vimal, S.R., Patel, V.K., Singh, J.S., 2018. Plant growth promoting *Curtobacterium albidum* strain SRV4: an agriculturally important microbe to alleviate salinity stress in paddy plants. Ecol. Indic (Article in press).

Wagner, M.R., Lundberg, D.S., Coleman-Derr, D., Tringe, S.G., Dangl, J.L., Mitchell-Olds, T., 2014. Natural soil microbes alter flowering phenology and the intensity of selection on flowering time in a wild Arabidopsis relative. Ecol. Lett. 17 (6), 717–726.

Walters, W.A., Jin, Z., Youngblut, N., Wallace, J.G., Sutter, J., Zhang, W., et al., 2018. Large-scale replicated field study of maize rhizosphere identifies heritable microbes. Pro. Natl. Acad. Sci. USA. 115 (28), 7368–7373.

Wang, J., Zhao, F., Meharg, A.A., Raab, A., Feldmann, J., McGrath, S.P., 2002. Mechanisms of arsenic hyperaccumulation in *Pteris vittata*. Uptake kinetics, interactions with phosphate, and arsenic speciation. Plant Physiol. 130 (3), 1552–1561.

Whipps, J.M., 2001. Microbial interactions and biocontrol in the rhizosphere. J. Exp. Bot. 52, 487–511.

Xin, G., Zhang, G., Kang, J.W., Staley, J.T., Doty, S.L., 2009. A diazotrophic, indole-3-acetic acid-producing endophyte from wild cottonwood. Biol. Fert. Soils 45 (6), 669–674.

Yadav, B.K., Verma, A., 2012. Phosphate solubilization and mobilization in soil through soil microorganisms under arid ecosystems. In: Ali, M. (Ed.), The Functioning of Ecosystems. InTech.

Yang, J., Liu, H., Zhu, G., Pan, Y., Xu, L., Guo, J., 2008. Diversity analysis of antagonists from rice-associated bacteria and their application in biocontrol of rice diseases. J. Appl. Microbiol. 104 (1), 91–104.

Zhang, H., Sun, Y., Xie, X., Kim, M., Dowd, S.E., Paré, P.W., 2009. A soil bacterium regulates plant acquisition of iron via deficiency-inducible mechanisms. Plant J. 58 (4), 568–577.

Zhang, H., Murzello, C., Sun, Y., Kim, M., Xie, X., Jeter, R.M., et al., 2010. Choline and osmotic-stress tolerance induced in Arabidopsis by the soil microbe *Bacillus subtilis* (GB03). Mol. Plant Microbe Interact. 23 (8), 1097–1104.

Zhang, Y., He, L., Chen, Z., Wang, Q., Qian, M., Sheng, X., 2011a. Characterization of ACC deaminase-producing endophytic bacteria isolated from copper-tolerant plants and their potential in promoting the growth and copper accumulation of *Brassica napus*. Chemosphere 83 (1), 57–62.

Zhang, Y., He, L., Chen, Z., Zhang, W., Wang, Q., Qian, M., et al., 2011b. Characterization of lead-resistant and ACC deaminase-producing endophytic bacteria and their potential in promoting lead accumulation of rape. J. Hazard. Mater. 186 (2–3), 1720–1725.

Zhang, W., Chen, W., He, L., Wang, Q., Sheng, X., 2015. Characterization of Mn-resistant endophytic bacteria from Mn-hyperaccumulator *Phytolacca americana* and their impact on Mn accumulation of hybrid *Penisetum*. Ecotoxicol. Environ. Saf. 120, 369–376.

Zhao, S., Zhou, N., Zhao, Z., Zhang, K., Wu, G., Tian, C., 2016. Isolation of endophytic plant growth-promoting bacteria associated with the halophyte *Salicornia europaea* and evaluation of their promoting activity under salt stress. Curr. Microbiol. 73 (4), 574–581.

Zhou, S., Hu, W., Deng, X., Ma, Z., Chen, L., Huang, C., et al., 2012. Overexpression of the wheat aquaporin gene, TaAQP7, enhances drought tolerance in transgenic tobacco. PLoS One 7 (12).

Zhu, L., Guan, D., Luo, J., Rathinasabapathi, B., Ma, L.Q., 2014. Characterization of arsenic-resistant endophytic bacteria from hyperaccumulators *Pteris vittata* and *Pteris multifida*. Chemosphere 113, 9–16.

Chapter 3

Environmental applications of microalgae/cyanobacteria

Aline Meireles dos Santos, Karem Rodrigues Vieira, Leila Queiroz Zepka and Eduardo Jacob-Lopes

Department of Food Science and Technology, Federal University of Santa Maria, Santa Maria, Brazil

Chapter Outline

3.1 Introduction

Experts estimate that the world population will reach about 9.5 billion by 2050 (FAO, 2011). This growth is aggravated by the impacts of climate change, leading to unsustainable pressures on existing resources. The current linear "take make dispose" economic model needs to be exchanged for a more circular one, to relieve the escalating pressures on resources such as energy, materials, food, and water. The circular economy is receiving increasing attention worldwide as more industries have shown some commitment to it. The main objective of the circular economy is to overcome the current production and consumption model based on continuous growth and increasing resource throughput and to eliminate waste systematically, throughout the life cycles and uses of products and their components (UNFCCC, 2015; Flores et al., 2018).

In their Nationally Determined Contributions (NDCs) for the Paris Agreement, a number of countries included waste treatment as a priority sector given the growing demand for sanitation and the large electricity costs and greenhouse gas (GHG) emissions associated with water supply and waste treatment (UNFCCC, 2015). In a world where demands for freshwater are ever-growing, and where limited water resources are increasingly stressed by overabstraction, pollution, and climate change, the safe management of wastes towards a circular economy is indispensable (WWP, UN, 2017).

Microalgae/cyanobacteria can play a significant role in waste management and contribute to a circular economy with the potential to recover nutrients from wastes and form a variety of biobased chemicals that can be applied extensively as intermediate inputs and final products in different industrial sectors (Santos et al., 2016a). However, a number of technoeconomic barriers need to be overcome for the deployment of microalgae/cyanobacteria-based processes on a commercial scale. To understand these processes' performance, comprehensive analyses addressing environmental, economic and social indicators should be carried out following a life-cycle perspective (González-González et al., 2018).

Thus the objective of this chapter is to compile environmental applications of microalgae/cyanobacteria summarizing a range of useful technoeconomic opportunities to be applied to waste management to develop a circular economy.

3.2 Status of legislation on management of waste

Industrial wastes, by definition, are the set of fractions of raw materials that were not incorporated into the final product. In this way, the ideality of a technological process of the industry could be achieved in processes whose inputs are

New and Future Developments in Microbial Biotechnology and Bioengineering. DOI: https://doi.org/10.1016/B978-0-12-818258-1.00003-0

fully incorporated in the final products. As technological processes and process engineering are not aligned with this profile, what is verified in practice is the massive generation of liquid, solid, and gaseous waste proportionally to process efficiency (Garcia-Serna et al., 2007).

Regarding forms of wastes, the current legislation is ineffective as far as gaseous wastes are concerned, with concrete initiatives toward particulate pollution and in the medium to long-term perspective for GHGs (UNFCCC, 2015). Solid wastes, in turn, have been effectively transformed into coproducts associated with the main proceedings, where reuse as animal feed, fertilizer, and energy is already consolidated (USEPA, 1993). On the other hand, the rigor of the current legislation as regards management of liquid wastes requires greater efforts in the development and establishment of technologies (Council Directive 91/271/EEC, 1991; WWP, UN, 2017).

Sustaining economic development and maintaining ecosystems requires more effective management of water resources. This has become more urgent as we witness increasing pressure on water resources and competition for their use among different economic activities (Arundel and Sawaya, 2009). The setting up of adequate legislation for the protection of the quality of water resources is an essential point in the environmental and public health development of all countries (Von Sperling and Fattal, 2001).

National standards management of wastewater, defined by each country, have legal status and are based on the specific conditions of the country itself. Depending on the political structure of the country, regional standards may also be developed. Adequate consideration of the guidelines prior to the adoption of standards may be an invaluable tool in the health and environmental development of a country, whereas inadequate consideration may lead to discrediting, frustration, unnecessary monetary expenditure, and unsustainable systems (Von Sperling and Fattal, 2001; Von Sperling and Chernicharo, 2002).

The most common form of industrial pollution is abundant in organic matter and nutrients (nitrogen and phosphorus) of nature suspended and dissolved, beyond biological contaminants, normally expressed by thermotolerant coliforms and specific inorganic and organic pollutants, which are a function of the raw material and technological route taken. In this sense, large industries are required to have waste with lower concentrations of pollutants when compared with small-sized industries (CONAMA, 2005; Council Directive 91/271/EEC, 1991).

In addition, one must consider the possibility of attendance to fixed minimum efficiency in cases that do not reach the emission standard, expressed as concentration (mg/L), and regulatory agencies allow the disposal of effluent above the fixed value provided that there is proven removal efficiency (%), typically 75% for total nitrogen and total phosphorus and 95% for thermotolerant coliforms (CONAMA, 2005; Council Directive 91/271/EEC, 1991). This additional possibility stems from the currently existing lower technological level to mitigate these polluting forms.

The implementation of national water quality standards is intimately linked to the adoption of adequate technologies for the treatment of water and wastewater and depends largely on the political will and, even when this is present, financial constraints are the final barriers to undermine the necessary steps towards environmental restoration and public health maintenance (Von Sperling and Chernicharo, 2002).

Turning waste into a resource is an essential part of increasing resource efficiency and moving towards a more circular economy (EU, 2015). Industrial waste flows contain a number of useful materials, such as nutrients, metals, and organic material that, much like the water itself, can be extracted and used for other productive purposes. As such, industrial waste constitutes a valuable resource that, if sustainably managed, is set to become a central pillar of the circular economy. The upside to improving the way we manage waste is huge, with potential cobenefits to societies and the environment (WWP, UN, 2017).

The necessary transition to a circular and sustainable economy will not happen overnight and may never happen without supportive governmental policies. Sustainability is a global imperative and demands urgent international action (OECD, 2009).

3.3 Environmental applications of microalgae/cyanobacteria in the nutrient cycling

Nutrient cycling by microalgae/cyanobacteria emerges as a promising technology for environmental applications because it balances sustainable vectors by reuse of pollutants, like carbon, nitrogen, and phosphorus, present in wastes generated by the industry thus forming a microalgal/cyanobacterial sludge where one can extract a variety of biobased products with high added value (Santos et al., 2016a). Microalgae/cyanobacteria play critical roles in marine energy fluxes and nutrient cycles by incorporating and redistributing dissolved organic matter and inorganic nutrients in the oceans (Arrigo, 2005). These microorganisms have various forms of nutrient cycling, as shown in Table 3.1.

Carbon, in the form of carbon dioxide, may be fixed from the atmosphere and industrial exhaust gases through the photosynthetic activity of autotrophic microalgae/cyanobacteria. Photosynthesis can be defined as the synthesis of

TABLE 3.1 Mechanisms involved in nutrient cycling by microalgae/cyanobacteria.

Nutrients	Mechanisms
Carbon	
CO_2	Integration in the Calvin cycle
Organic carbon	Integration in the respiration, metabolism
Nitrogen	
N_2-N	Fixation by prokaryotic microalgae/cyanobacteria into ammonia, followed by conversion into amino acids
NH_4-N	Direct conversion into amino acids Stripping due to volatilization (high pH values and temperature)
Phosphorus	
PO_4-P	Phosphorylation Chemical precipitation (high pH values and dissolved oxygen concentrations)

Source: Adapted from Gonçalves, A.L., Pires, J.C.M., Simões, M., 2017. A review on the use of microalgal consortia for wastewater treatment. Algal Res. 24, 403–415.

organic compounds through the assimilation of CO_2 with the use of light as an energy source. CO_2 is incorporated into a 5-carbon acceptor, ribulose-1,5-bisphosphate (RuBP) in an energy-requiring reaction catalyzed by the primary carboxylating enzyme, RuBP carboxylase. The product splits into two molecules of a 3-carbon compound, phosphoglyceric acid (PGA), and the reduction of PGA, mediated by the electrons carrier nicotinamide adenine dinucleotide phosphate (NADPH) leads to the formation of a series of sugar phosphate intermediates and, finally, glucose. During this sequence of metabolic transformations, known as the Calvin cycle, the acceptor RuBP is regenerated, ready to accept another CO_2 molecule (Schlesinger, 1997).

Although there is some understanding of nutrient dynamics in photosynthetic conditions, it must be borne in mind that microalgae/cyanobacteria can exhibit many types of metabolism (e.g., autotrophic, heterotrophic, mixotrophic, photoheterotrophic) and are capable of a metabolic shift in response to changes in environmental conditions (Delgadillo-Mirquez et al., 2016).

Some microalgae/cyanobacteria display heterotrophic behavior, using organic forms of carbon present, for example, in wastewater, where it is metabolized through respiration to generate ATP. Heterotrophic growth is an aerobic process where assimilation of organic substrates generates energy through oxidative phosphorylation accompanied by oxygen consumption as the final electron acceptor (Queiroz et al., 2011; Perez-Garcia and Bashan, 2015). To use these organic compounds, transport occurs through the membrane. This substrate will be converted into glucose 6-phosphate so it can start the route through the oxidative pentose phosphate pathway (Fay, 1983; Santos et al., 2016b). The heterotrophic microalgae/cyanobacteria metabolism has as its characteristic the simultaneous conversion of the pollutants presents in wastewater in a single step, however, this results in slow growth rate compared with autotrophic mode (Queiroz et al., 2007; Perez-Garcia et al., 2011; Santos et al., 2015).

Carbon, nitrogen, and phosphorus are elemental constituents with a very important contribution in from microalgal/cyanobacterial cells; they are assimilated and found in a variety of biobased products (Santos et al., 2016c, 2017a). The most commonly used methods for nitrogen and phosphorus removal in wastewaters include biological processes, such as anaerobic digestion followed by nitrification and denitrification. These methodologies tend to be complex, expensive, and energy demanding (Queiroz et al., 2007). Microalgae/cyanobacteria offer a win–win alternative for tertiary wastewater treatment due to their ability to simultaneously remove inorganic nitrogen and phosphorus to meet the effluent standards and at the same time produce valuable biomass (González-Camejo et al., 2018).

Nitrogen in the wastewater is mainly available in the form of inorganic nitrogen, and the microalgae/cyanobacteria play an important role in both nitrogen fixation and assimilation. These species are able to fix nitrogen, incorporating into amino acids, proteins, and chlorophyll or excreting to the environment (Cuellar-Bermudez et al., 2017; Gonçalves et al., 2017).

Phosphorus is another macronutrient that supports the growth of microalgae/cyanobacteria that are presented in nucleic acids, phospholipids, lipopolysaccharides, and various cytoplasmic solutes (Vieira et al., 2012; Peccia et al., 2013). This

TABLE 3.2 Removal efficiency nitrogen and phosphorus from different wastewater by microalgae/cyanobacteria.

Microalgae/cyanobacteria	Wastewater	N (%)	P (%)	References
Chlamydomonas reinhardtii	Domestic	83	14.45	Kong et al. (2010)
Chlorella ellipsoidea	Municipal	99.3	95.2	Yang et al. (2011)
Brotryococcus braunii	Domestic	79.63	100	Sydney et al. (2011)
Chlorella kessleri	Municipal	83.94	51.20	Li et al. (2011a)
Chlorella sp.	Municipal	93.9	80.9	Li et al. (2011b)
Scenedesmus sp.	Municipal	90	90	McGinn et al. (2012)
Chlorella zofingiensis	Dairy	51.7	97.5	Huo et al. (2012)
Chlamydomonas	Industrial	100	33	Wu et al. (2012)
Chlorella vulgaris	Domestic	96.3	94.7	Li et al. (2013)
Phormidium autumnale	Industrial	87.5	100	Santos et al. (2017a)
Spirulina platensis	Industrial	99	99	Cuellar-Bermudez et al. (2017)

nutrient gets in microalgae/cyanobacteria cells through active transport at the plasma membrane (Gonçalves et al., 2017). Several studies reported that microalgae/cyanobacteria were widely used for effective removal of nitrogen and phosphorous and are presented in Table 3.2.

The factors that more influence nutrient cycling by microalgae/cyanobacteria are pH, light, and temperature. The pH is an important factor, as this parameter determines CO_2 solubility in the culture medium. Furthermore, high pH values are responsible for ammonium stripping and phosphorus precipitation (Vieira et al., 2012).

The light and temperature also play an important role in nutrient cycling by microalgae/cyanobacteria (Gonçalves et al., 2017). Microalgae/cyanobacteria growth is a direct function of the light intensity supplied as long as it remains under the optimum light intensity (González-Camejo et al., 2018). In general, an increase in light supply results in increased nutrient removal (Gonçalves et al., 2017). Temperature is another parameter that must be taken into account in microalgal growth. Many species of microalgae/cyanobacteria are able to grow in a wide range of temperature, but care should always be taken regarding optimum temperature of growth (Salama et al., 2017; González-Camejo et al., 2018).

From these macronutrients, the microalgae/cyanobacteria as renewable natural biomass possess molecular mechanisms that allow them to discriminate nonessential heavy metals from essential ones, for their growth, particularly metals, such as Cd, Co, Cr, Cu, Fe, Hg, Ni, Pb, and Zn (Kumar et al., 2015; Gonçalves et al., 2017). Although there is a negative effect due to heavy metals present in the industrial wastewater to the microalgae/cyanobacteria, this can be overcome by using growth simulators, or be subjected to dilution or mixing with organic compounds (Abinandan and Shanthakumar, 2015).

Microalgal/cyanobacterial affinities for polyvalent metals help establish their potential application in cleansing wastewater containing dissolved metallic ions (Bashan and Bashan, 2010). Conventional methods for the removal of heavy metal ions involve adsorption of metals on activated carbon, chemical precipitation, electrochemical treatment, evaporation, filtration, ion exchange, and membrane technologies (Paliwal et al., 2018). However, these techniques produce large amounts of sludge that require additional treatment.

Microalgae/cyanobacteria feature constitutive mechanisms for the removal of free metal ions in wastewater (Kumar et al., 2015). Microalgae/cyanobacteria can sequester up to 10% of their biomass as metals (Kumar et al., 2018). In addition to essential trace metals required for metabolism, microalgae/cyanobacteria can efficiently sequester toxic heavy metals (Paliwal et al., 2018).

Metals uptake by microalgal/cyanobacterial species can be both passive and active. The passive uptake mainly depends on the binding of metal ions to functional the surface ligands of the cell wall structure. Microalgae/cyanobacteria possess a high metal binding capacity (Paliwal et al., 2018). This property was attributed to the presence of polysaccharides, proteins, or lipids on the surface of their cell walls, which in turn contained some functional groups like amino, hydroxyl, carboxyl, and sulfate, which can act for the heavy metal ions' sequestration (Priya et al., 2014; Paliwal et al., 2018).

The active mechanism is based on the processes of bioaccumulation, in which living organisms remove metals metabolically, being called biosorbents (Peña-Castro et al., 2004). Biosorption is a method of removing heavy metals from wastewater, as it uses naturally abundant microalgae/cyanobacteria (Priya et al., 2014). Table 3.3 describes the heavy metal removal potential of various microalgae/cyanobacteria.

TABLE 3.3 Microalgae/cyanobacteria used for removal of heavy metals.

Metal	Microalgae/cyanobacteria	Metal uptake (mg/g)
Cd	*Chlorella sorokiniana*	33.5
	Chlorella vulgaris	86.6
	Phormidium spp.	9.6
	Scenedesmus obliquus	60.8
	Spirulina platensis	357
Co	*C. reinhardtii*	1.3
	Oscillatoria angustissima	15.32
	Spirogyra hyalina	12.82
	Spirulina spp.	0.01
Cr	*Spirulina* spp.	167
	C. vulgaris	23.6
	Phormidium bohneri	8.55
	S. obliquus	15.6
	Dunaliella sp.	58.3
Cu	*Anabaena cylindrica*	12.62
	C. reinhardtii	6.42
	C. vulgaris	37.6
	Phormidium spp.	10.1
	S. obliquus	20
Fe	*C. vulgaris*	24.52
	Microcystis sp.	0.03
Hg	*C. vulgaris*	18
	Scenedesmus acutus	20
	Spirulina spp.	1.34
	S. hyaline	35.71
	Porphyridium purpureum	0.51
Ni	*C. vulgaris*	59.29
	Phormidium spp.	5.7
	Spirulina spp.	515
	Spirogyra insignis	17.5
	Pithophora oedogonia	11.81
Pb	*Anabaena flosaquae*	70
	C. vulgaris	131.36
	Microcystis novacekii	80
	Phormidium spp.	13.6
	S. platensis	16.97
Zn	*C. vulgaris*	43.41
	Isochrysis galbana	0.3
	Phormidium spp.	9.4
	S. obliquus	22.3
	S. platensis	7.36

Source: Adapted from Kumar, K.S., Hans-Uwe, D., Eun-Ji, W., Jae-Seong, L., Kyung-Hoon, S., 2015. Microalgae/cyanobacteria—a promising tool for heavy metal remediation. Ecotoxicol. Environ. Saf. 113, 329–352.

Although the capability of microalgae/cyanobacteria-based wastewater treatment systems to remove nutrients and heavy metals has been studied the longest, few studies have focused on the removal of emerging organic contaminants (EOCs) (Meffe and Bustamante, 2014; Matamoros et al., 2015). EOCs include a wide range of compounds belonging to different chemical classes, such as pharmaceuticals, personal care products, plasticizers, flame retardants, surfactants, and certain pesticides, among others, the ecotoxicological effects of which are relatively unknown (Meffe and Bustamante, 2014).

The use of adsorption for the treatment of the EOCs is efficient and they do not imply the transformation these compounds into different products, which may be more dangerous than the original compounds (Escapa et al., 2017). Activated carbon, which is the most largely used adsorbent for industrial wastewater treatment, however, has a quite high price for its implementation in the treatment of wastewater (Meffe and Bustamante 2014).

TABLE 3.4 Removal and absorption emerging organic contaminants from wastewater by microalgae/cyanobacteria.

Microalgae/ cyanobacteria	Contaminant and removal efficiency (%)	Wastewater category	References
Chlorella sorokiniana	Diclofenac (60) Ibuprofen (100) Paracetamol (100) Metoprolol (100) Carbamazepine (30) Trimethoprim (40)	Urine, anaerobically treated black water and synthetic urine	Tran et al. (2013)
Chlorella sp.	Caffeine (99) Ibuprofen (99) Carbamazepine (20)	Urban or synthetic wastewater	Zhang and Zhou (2008)
Selenastrum capricornutum	Estradiol (100) 17α-ethynylestradiol (95)	Wastewater digestate and growth medium	Zhao et al. (2008)
Scenedesmus sp.	Caffeine (99) Ibuprofen (99) Carbamazepine (20)	Urban or synthetic wastewater	Zhang and Zhou (2008)

Sorption by microalgae/cyanobacteria has been pointed out as a key process in the removal of antibiotics in high-rate microalgal/cyanobacterial ponds (Escapa et al., 2017). On the other hand, Matamoros et al. (2015), when studying the removal of 28 different EOCs in high-rate microalgal/cyanobacterial ponds, found that sorption was relevant for the removal of hydrophobic compounds, while other EOCs, such as paracetamol, were mostly removed by biodegradation. Wilt et al. (2016) studied the removal of pharmaceuticals, by the microalgae *Chlorella sorokiniana* in batch experiments, and observed that sorption to microalgal biomass accounted for less than 20% of the removal. Shi et al. (2010) showed that natural and synthetic estrogens can be effectively removed in microalgae/cyanobacteria-based wastewater treatment systems.

Biotransformation is the dominant mechanism responsible for the removal of the EOCs by microalgae/cyanobacteria. In addition to biotransformation, photolysis appears to be a major removal mechanism in microalgal/cyanobacterial treatment systems (Wilt et al., 2016). Table 3.4 shows the ability of microalgae/cyanobacteria to remove and absorb EOCs from wastewater.

Biological CO_2 mitigation through microalgae/cyanobacteria is an emerging technology to solve the problem of GHG emission. Microalgae/cyanobacteria fix CO_2 for growth during photosynthesis (Jacob-Lopes et al., 2010; González-Camejo et al., 2018). These microorganisms have the ability to assimilate ambient CO_2, the CO_2 from the flue and flaring gas, and chemically fixed CO_2 in the form of soluble carbonates (Keffer and Kleinheinz, 2002). An advantage of microalgae/cyanobacteria is that they have the ability to survive when CO_2 is supplied from industrial flue gas, which contains other toxic gases, including sulfur dioxide (SO_2), nitrogen oxide, and volatile organic compounds (VOCs) (Borowitzka, 1998; Severo et al., 2018).

Another challenge in waste management is odor emissions abatement in the wastewater treatment plant (WWTP), which causes nuisance to populations living nearby. Odor emissions are important sources of air pollution from the anthropogenic origin are usually emitted from industrial and agricultural activities, including WWTPs, in different industrial plants (Rappert and Müller, 2005; Brancher et al., 2017).

Odors are emitted from wastewater collection, treatment, and storage systems through volatilization of organic compounds at the liquid surface (Capelli et al., 2011; Lebrero et al., 2014). The volatile profile form of WWTPs includes characteristic groups of odorants such as alcohols, volatile fatty acids, aldehydes, ketones, amines compounds, indole, and skatole. Thus the chronic presence of those pollutants in outdoor environments requires effective emissions control strategies on large emission sources (Lebrero et al., 2011, 2014).

Among conventional biotechnologies, biofiltration and biotrickling filtration are the more implemented technologies for odor abatement likely due to their ease of operation and the extensive design and operation experience. However, these biotechnologies are claimed to support low removal efficiencies for the hydrophobic fraction of malodorous emissions, whose elimination is mandatory for an efficient odor abatement (Lebrero et al., 2014).

As a cost-effective, environmentally friendly alternative, as well as the advantage of metabolic plasticity, converting polar and nonpolar molecules of the wastewater, the use of the microalgae/cyanobacteria-based processes can be an innovative technology for wastewater deodorization. Microalgae/cyanobacteria are capable of producing a variety of VOCs (Santos et al., 2016c; Hosoglu, 2018). The biosynthesis of these compounds will depend on the availability of building blocks, such as carbon, nitrogen, and energy supply from the primary metabolism. Therefore the availability of these compounds has a great impact on the concentration of VOCs (Santos et al., 2016b).

The emission of volatile compounds in microalgae/cyanobacteria depends upon several biotic and abiotic factors such as growth phase, species type, stresses (temperature, light intensity, pH, salinity), nutrients, gases (CO_2, O_3), aeration (mixing/turbulence) or static culture (Achyuthan et al., 2017). The different VOCs produced through the metabolism of microalgae/cyanobacteria may belong to organic classes such as aldehydes, alcohols, terpenes, sulfurized compounds, esters, ketones, and hydrocarbons (Hosoglu, 2018).

Several biosynthetic pathways are involved in the synthesis of VOCs released by microalgae/cyanobacteria. The main routes of formation of these compounds may be enzymatically or degradation reaction. From this principle, it is believed who the use of the microalgae/cyanobacteria-based processes for abatement or removal total of the compounds malodors is possible (Santos et al., 2016b; Achyuthan et al., 2017).

3.4 Dewatering and drying methods for microalgal/cyanobacterial sludge

Once the environmental applications of microalgae/cyanobacteria for nutrient cycling and sludge obtention are finished, the upstream processing ends and gives way to the downstream processing (Fig. 3.1) (Wang et al., 2008). The choice of the downstream processes used greatly influences the viability of the microalgae/cyanobacteria-based processes (Richardson et al., 2014). Dewatering and drying are stages that require a high amount of energy. These aspects lead not only to high operating costs (Gouveia et al., 2017) but also to unfavorable environmental impacts in terms of resource depletion and carbon footprint (Zamagni, 2012; Valente et al., 2019).

Dewatering of microalgal/cyanobacterial sludge is considered an important contributor to the total costs of biobased products obtention. Several studies report the dewatering costs at 20%–30% of the total production costs (Brennan and Owende, 2010; Santos et al., 2017b). The high capital expenditure and energy consumption result from the dilute microalgae/cyanobacteria solutions, due to the low biomass concentration in microalgae/cyanobacteria broth (0.6 g/L) (Shelef et al., 1984), the small size of cells ($<20\,\mu m$), nearly neutral buoyancy of cells (1.08–1.13 g/mL of density), and strongly negative surface charge (Park et al., 2011; Chen et al., 2015).

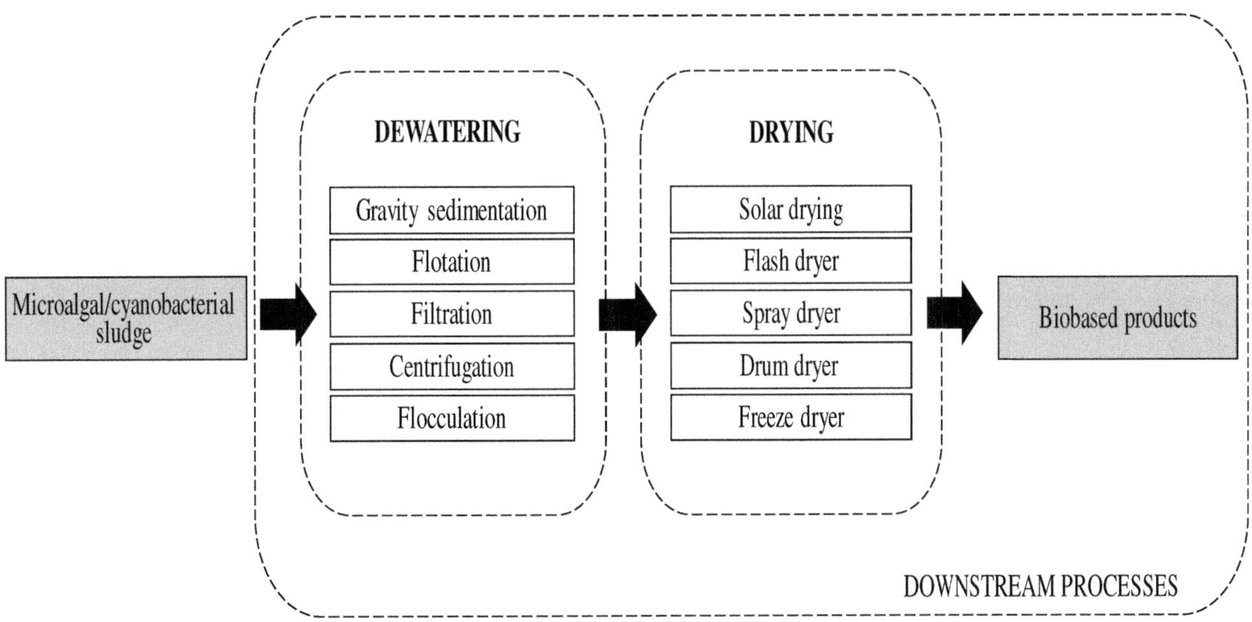

FIGURE 3.1 Diagram of downstream processes.

TABLE 3.5 Advantages and disadvantages of different dewatering methods applied to microalgal/cyanobacterial sludge.

Dewatering method	Advantages	Disadvantages	Dry solids (%)
Gravity sedimentation	• Simple and the inexpensive method.	• Time-consuming; • Possibility of sludge deterioration; • Low concentration of the microalgal/cyanobacterial sludge.	0.5–3
Flotation	• Feasible for large scale applications; • Low cost method; • Low space requirements; • Short operation times.	• Generally, requires the use of chemical flocculants; • Unfeasible for marine microalgae/cyanobacteria harvesting.	7
Filtration	• High recovery efficiencies; • Allows the separation of shear sensitive species.	• The possibility of fouling/clogging increases operational costs; • Membranes should be regularly cleaned; • Membrane replacement and pumping represent the major associated costs.	2–27
Centrifugation	• Fast method; • High recovery efficiencies; • Suitable for almost all microalgae/cyanobacteria species.	• Expensive method; • High energy requirements; • Suitable only for the recovery of high-valued products; • Possibility of cell damage due to high shear forces.	10–22
Chemical flocculation	• Simple and fast method; • Low energy requirements.	• Chemical flocculants may be expensive and toxic to microalgal/cyanobacterial sludge; • Recycling of culture medium is limited.	3–8

Source: Adapted from Milledge, J.J., Heaven, S., 2013. A review of the harvesting of micro-algae for biofuel production. Rev. Environ. Sci. Biotechnol. 12, 165–178 and Santos, A.M., Santos, A.M., Zepka, L.Q., Jacob-Lopes, E., 2017b. Downstream recovery of microalgal bioproducts with highlights on biorefineries. In: Biorefineries: Concepts, Advancements, and Research. Nova Science Publishers, Inc., New York.

The most common dewatering technologies used are gravity sedimentation, flotation, filtration, centrifugation, and flocculation (Table 3.5). However, there is no all-purpose dewatering method that is able to treat all microalgae/cyanobacteria suspensions under consideration of both cost and energy consumption (Uduman et al., 2010).

Although dewatering can result in a 50- to 200-fold concentration of microalgal/cyanobacterial sludge, the moisture in the dewatered sludge is still high enough to spoil the biomass slurry within a few hours at room temperature (Milledge and Heaven, 2013). In addition, the water content remaining could cause problems in downstream processing and biobased product isolation or conversion. A drying process is thus often needed to convert the sludge into a stable, storable form (Chen et al., 2015). Methods that have been used include solar drying (Prakash et al., 1997), flash dryer, spray dryer (Desmorieux and Decaen, 2005), drum dryer (Prakash et al., 1997), and freeze dryer (Table 3.6) (Santos et al., 2017b). Some research suggests that biobased products can be recovered from the humid biomass. However, is difficult to discuss processing since it is highly specific and strongly depends on the biobased products desired (Patil et al., 2011; Sathish and Sims, 2012; Su et al., 2017).

Over the years, the processes involving microalgae/cyanobacteria have undergone several critiques about their productivity and industrial scalability. One of the bottlenecks for commercialization that still persists in these processes is the high-energy consumption for the production and recovery of the microalgae/cyanobacteria biomass (Uduman et al., 2010). However, in an effort to neutralize the high energy expenditure and capital investment requirements, many researchers have suggested the exploitation of microalgae/cyanobacteria biobased products, since this microorganism is the only biomass source in nature that offers a high application potential, since they have a rich and versatile chemical composition, and thus could become an effective feedstock to produce many high value added products (Deprá et al., 2018).

3.5 Biobased products obtention from microalgal/cyanobacterial sludge

Microalgae/cyanobacteria have the potential to produce a wide range of biobased chemicals through environmental applications that turn waste into resources, thus relieving the negative effects associated with human activities, such as pollution of water bodies, GHG emissions, and scarcity of mineral resources (Abdel-Raouf et al., 2012; Singh et al., 2016).

In this context, microalgae/cyanobacteria are considered to be one of the most promising next-generation biobased feedstocks with biomass rich in lipids (Singh, 2015) carbohydrates, proteins, and other valuable compounds (Table 3.7). Lipids can be used as a source for biofuels, as building blocks in the chemical industry, as edible oils for the food industry, and as antioxidants, glycolipids, phospholipids, and polyunsaturated fatty acids (PUFAs) for the cosmetics and

TABLE 3.6 Comparison of drying methods.

Drying method	Electric power per kg of dry algae in kWh	Thermal energy per kg of algae in kWh	Advantages	Disadvantages
Solar drying	–	–	• Technically simplest and most economical method.	• Highly weather dependent with problems of overheating and unreliable operation; • Slow drying process, thereby affecting the product quality; • Not recommended for microalgal/ cyanobacterial products intended for human consumption.
Flash dryer	0.41	2.74	• Rapid removal of moisture.	• The cost of drying and the final microalgal/cyanobacterial product quality are greatly influenced by the hot gas source.
Spray dryer	0.43	3.23	• Rapid and efficient drying method; • Appropriate for production of microalgae/ cyanobacteria for human consumption.	• Degradation of the product quality due to high-pressure atomization; • High operating cost; • Low digestibility of dried microalgae/ cyanobacteria.
Drum dryer	0.09	4.32	• Dual advantage of sterilizing and disruption.	• High energy cost to run dryer.
Freeze dryer	2.28	3.28	• The most gentle of the drying methods.	• Process in the large scale is too costly.

Source: Adapted from Pandey, A., Höfer, R., Taherzadeh, M., Nampoothiri, M., Larroche, C. (Eds.), 2015. Industrial Biorefineries & White Biotechnology. Elsevier and Santos, A.M., Santos, A.M., Zepka, L.Q., Jacob-Lopes, E., 2017b. Downstream recovery of microalgal bioproducts with highlights on biorefineries. In: Biorefineries: Concepts, Advancements, and Research. Nova Science Publishers, Inc., New York.

TABLE 3.7 Comparison of the centesimal composition of microalgae and cyanobacteria.

Centesimal composition	*Chlorella vulgaris*	*Arthrospira platensis*
Protein	51–58	60–71
Lipids	14–22	6–7
Carbohydrates	12–17	13–16
Rhamnose (%)	45	30.1
Glucose (%)	4	30.9
Galactose (%)	26	21.4
Mannose (%)	7	1.42
Xylose (%)	19	2.23
Fructose (%)	–	1.21
Amino acids		
Aspartic acid (g)	10.94	5.80
Threonine	6.09	2.97
Serine	7.77	3.0
Glutamic acid	9.08	8.38
Glycine	8.60	3.10
Alanine	10.9	–
Cysteine	0.19	0.66
Valine	3.09	3.51

(*Continued*)

TABLE 3.7 (Continued)

Centesimal composition	*Chlorella vulgaris*	*Arthrospira platensis*
Methionine	0.65	1.15
Isoleucine	0.09	3.21
Leucine	7.49	4.95
Tyrosine	8.44	2.58
Phenylanine	5.81	2.77
Histidine	1.25	1.08
Lysine	6.83	3.02
Arginine	7.38	4.15
Tryptophan	2.21	0.93
Ornithine	0.13	2.38
Proline	2.97	–
Minerals		
Calcium (mg)	0.27	1.20
Iron (mg)	0.68	0.28
Magnesium (mg)	0.44	1.95
Phosphorus (mg)	0.96	1.18
Potassium (mg)	2.15	13.63
Sodium (mg)	1.35	10.48
Zinc (mg)	0.55	0.02
Copper (mg)	0.19	0.06
Manganese (mg)	0.40	0.01
Selenium (μg)	–	0.07
Vitamins		
C (ascorbic acid) (mg)	15.6	10.1
B6 (pyridoxine) (mg)	1.7	0.36
A (retinol) (IU)	13.2	570.0
E (tocopherol) (mg)	20.0	5.0
K (μg)	–	25.5
B1 (thiamine) (mg)	2.4	2.4
B2 (riboflavin) (mg)	4.8	3.7
B3 (niacin) (mg)	23.8	12.8
B5 (pantothenic acid) (mg)	1.3	3.5
B9 (folic acid) (μg)	26.9	94.0
Fatty acids		
Myristic acid (%)	–	7.5
Palmitic acid (%)	4.36	25.0
Stearic acid (%)	1.20	7.7
Palmitolytic acid (%)	23.47	32.8
Oleic acid (%)	21.81	34.7
Linoleic acid (%)	6.26	1,2.5
α-Linolenic acid (%)	20.0	17.84
Pigments		
β-Carotene (μg/g)	12	211.5
Astaxanthin (μg/g)	550	–
Cantaxanthin (μg/g)	362	–
Lutein (μg/g)	3830	–
Chlorophyll-a (μg/g)	9630	200.8
Chlorophyll-b (μg/g)	5770	
Violoxanthin (μg/g)	37	
Phycoerythrin (μg/g)	–	31.46
Phycocyanin (μg/g)	–	43.75

Source: Adapted from Deprá, M.C., Santos, A.M., Severo, I.A., Santos, A.B., Zepka, L.Q., Jacob-Lopes, E., 2018. Microalgal biorefineries for bioenergy production: can we move from concept to industrial reality? Bioenergy Res. 11, 1–21.

health market (Draaisma et al., 2013; Leu and Boussiba, 2014; Cuellar-Bermudez et al., 2015; Voort et al., 2015). The excreted carbohydrates such as complex polysaccharides might have health beneficial properties (Laurienzo, 2010; Wijesekara et al., 2011; Raposo and Morais, 2013) and be used as for the production of pharmaceuticals and cosmetics, whereas the carbohydrates in the cell such as starch and hemicellulose can be transitioned into bioethanol through fermentation or into CH_4 via anaerobic digestion (Zhu, 2014). Purified proteins can be used in the food, feed, and bulk chemical markets (Vanthoor-Koopmans et al., 2013; Zhu and Hiltunen, 2016).

Biofuels based on microalgae/cyanobacteria are considered to be the most promising alternative to replacing fossil fuels and obtain renewable energy (Demirbas, 2011). They are characterized as an environmentally friendly alternative to petroleum-based diesel, producing lower levels of CO, SO_2, and unburned hydrocarbons than petroleum-based diesel fuel, and are biodegradable, renewable, and nontoxic because they are obtained from fatty acids produced by the microalgae. In this context, microalgae/cyanobacteria are a potential feedstock for the production of different types of biofuels, such as biodiesel, bioethanol, biohydrogen, biomethane, syngas, bio-oil, and, more recently, VOCs (Baicha et al., 2016; Severo et al., 2018).

Since the beginning of this century, microalgae/cyanobacteria have received an increasing amount of attention for being a feedstock for next-generation biofuels (Pulz and Gross, 2004; Chisti, 2007; Schenk et al., 2008; Priyararshani and Rath, 2012; Rawat et al., 2013; Ho et al., 2014). However, for biofuels only, microalgae/cyanobacteria production appears to be too costly (Wijffels and Barbosa, 2010; Chisti, 2013; Torres and Rios, 2013) because one of the main bottlenecks in production is the amount of energy used in the whole cultivation process combined with the investment costs (Chisti, 2013; Norsker et al., 2011; Demirbas, 2011). Even though initially the lipid fraction was the main focus most microalgae/cyanobacteria ingredients currently sold commercially are high-value compounds such as PUFAs and pigments (Spolaore et al., 2016).

The lipid content in microalgae/cyanobacteria can reach up to 40% in dry biomass. The interest in microalgal/cyanobacterial lipids arises primarily from the fact that these organisms can synthesize considerable quantities of PUFAs that either can be used as food supplements or animal feed production. In addition, microalgae/cyanobacteria have been attracting interest as sources of fatty acids, especially the long-chain PUFAs such as γ-linolenic acid, arachidonic acid, eicosapentaenoic acid (EPA), and docosahexaenoic acid (DHA) (Deprá et al., 2018; Chauton et al., 2015). The EPA and DHA yield from microalgae/cyanobacteria are dependent on the biomass density that can be achieved per volume or area of culture, and the number of relevant lipids that can be obtained per weight unit of microalgal/cyanobacterial biomass (Griffiths and Harrison, 2009; Xu and Boeing, 2014).

Other high-value compounds are the pigments. The global market for carotenoids was estimated to US$ 1.4 billion in 2018, with projections of US$ 1.53 billion in 2021, with the bulk of the carotenoids being produced by chemical synthesis (D'Alessandro and Antoniosi Filho, 2016). β-Carotene from *Dunaliella salina* and astaxanthin from *Haematococcus pluvialis* (Cuellar-Bermudez et al., 2015; Voort et al., 2015; Ben-Amotz, 1995) do, however, represent a major part of the "natural" production of these carotenoids. The total market for β-carotene is estimated at about US$ 270 million, with natural β-carotene estimated at about 20%–30% of the market (Borowitzka, 2013).

The development of the "natural" astaxanthin market was more difficult than for β-carotene. The principal existing market for astaxanthin was for use as a feed additive for farmed salmon and trout as a pigment for the fish flesh. The annual worldwide aquaculture market of this pigment is estimated at US$ 200 million with an average price of US$ 2500/kg at a production cost of microalgae of US$ 5 to US$ 20/kg dry weight (Li et al., 2015; Pulz and Gross, 2004). However, Li et al. (2011c) surveyed the costs and concluded that China can sell astaxanthin at a lower price (US$ 718/kg) using low-cost photobioreactors, two phases in raceways (ponds), and low cost of labor and reagents.

The majority of microalgal/cyanobacterial biomass is currently sold a dried supplement for human health applications (Pulz and Gross, 2004; Spolaore et al., 2016). However, human consumption of microalgal/cyanobacterial biomass is still restricted to a very few taxa, for example, *Spirulina*, *Chlorella*, *Dunaliella*, *Tetraselmis*, *Nostoc*, and *Aphanizomenon* (Spolaore et al., 2016; Gouveia, 2008; Kovac et al., 2013). The incorporation of whole dried microalgal/cyanobacterial biomass into commodity food products has not happened yet because microalgae/cyanobacteria species tend to have a strong green color, fishy taste and odor, and a powdery consistency (Chacon-lee and Gonzalez-Marino, 2010). Therefore only minor amounts of microalgae/cyanobacteria powder (<1%) are currently blended with existing food products (Kovac et al., 2013; Eppink et al., 2017).

3.6 Life-cycle analysis of microalgae/cyanobacteria-based processes

In the past years, most efforts to improve WWTPs' performance have been focused on obtaining a good effluent quality (Bolong et al., 2009; Zanetti et al., 2012; Zhang et al., 2016). However, nowadays, new challenges are under consideration, oriented to ensure the sustainability of WWTPs in terms of their environmental impact and economic feasibility. GHG emissions and energy consumption are among the aspects that have become key factors concerning the overall performance of the WWTPs (Yerushalmi et al., 2013; Campos et al., 2016).

In this new scenario, microalgae/cyanobacteria-based wastewater treatment systems, mainly high-rate algal ponds are recognized as a cost-effective and efficient upgrade to conventional wastewater ponds for the treatment of municipal, agricultural, and industrial wastewater (Craggs et al., 2014). Compared with conventional ponds, high-rate algal ponds offer improved nutrient removal and natural disinfection, with the added benefit of resource recovery, in the form of microalgal/cyanobacterial biomass, for beneficial reuse as fertilizer, feed, or biofuel (Benemann, 2008; Rawat et al., 2011; Craggs et al., 2012).

Due to the avoidance of mechanical aeration, high-rate algal ponds would require significantly less energy consumption than conventional activated sludge systems: 0.02 kWh/m^3 versus 1 kWh/m^3 of water, respectively (Garfí et al., 2017; Passos et al., 2017). However, high-rate algal ponds occupy large land areas, which limit their use to rural areas and water evaporation losses could have impacts during summer in arid locations (Campos et al., 2016).

These benefits and disadvantages should then be weighed against the potential benefits offered by biomass recovery and valorization. Selecting locations where microalgal/cyanobacterial cultivation using wastewater indeed provides environmental and economic gains will, therefore, require careful assessment and optimization (Béchet et al., 2017).

Although the feasibility of high-rate algal ponds systems for wastewater treatment has already been demonstrated in terms of process performance (Garfí et al., 2017; Maga, 2016), more studies in the field of systems analysis are needed to overcome the technical and environmental difficulties in the expansion of this technology for industrial applications (Marín et al., 2018; Posadas et al., 2017).

Life-cycle assessments are usually performed using similar methodologies, however, there are variabilities in tool application, in terms of choice of functional unit, allocation, system boundary, and procedures followed to interpret the results. Therefore it is necessary to develop guidelines for the field of wastewater treatment capable of thoroughly assessing the environmental consequence of implementing a given wastewater treatment process and of proposing methods to reduce negative consequence (Luterbacher et al., 2009).

3.7 Conclusions

The race for the transformation of waste into resources is underway and microalgae/cyanobacteria-based processes play a central role in this issue, as these microorganisms are capable of reusing pollutants present in the waste generated by the industry, forming a microalgae/cyanobacteria sludge with the potential for use in several industrial sectors.

However, although the feasibility of microalgae/cyanobacteria-based processes has been studied for more than 50 years, there are still no prospects for applications of microalgae/cyanobacterial-based process for large scale of waste management due to operating and capital costs. For this reason, it is necessary to perform a life-cycle assessment before the pilot or large-scale microalgal/cyanobacterial productions to foreseen the pros and cons of the considered microalgal/cyanobacterial system. Life-cycle modeling is a powerful tool to guide research priorities and evaluate the environmental impact and sustainability of microalgae/cyanobacteria-based processes.

References

Abdel-Raouf, N., Al-Homaidan, A.A., Ibraheem, I.B.M., 2012. Microalgae and wastewater treatment. Saudi J. Biol. Sci. 19 (3), 257–275.

Abinandan, S., Shanthakumar, S., 2015. Challenges and opportunities in application of microalgae (Chlorophyta) for wastewater treatment: a review. Renew. Sustain. Energ. Rev. 52, 123–132.

Achyuthan, K.E., Harper, J.C., Manginell, R.P., Moorman, M.W., 2017. Volatile metabolites emission by in vivo microalgae/cyanobacteria—an overlooked opportunity? Metabolites 7, 39–85.

Arrigo, K.R., 2005. Marine microorganisms and global nutrient cycles. Nature 437, 349–355.

Arundel, A., Sawaya, D., 2009. The Bioeconomy to 2030: Designing a Policy Agenda. OECD, Paris.

Baicha, Z., Salar-García, M.J., Ortiz-Martínez, V.M., Hernández-Fernández, F.J., de los Ríos, A.P., Labjar, N., et al., 2016. A critical review on microalgae as an alternative source for bioenergy production: a promising low cost substrate for microbial fuel cells. Fuel Process. Technol. 154, 104–116.

Bashan, L.E., Bashan, Y., 2010. Immobilized microalgae for removing pollutants: review of practical aspects. Bioresour. Technol. 101, 1611–1627.

Béchet, Q., Plouviez, M., Chambonnière, P., Guieysse, B., 2017. Environmental impacts of full-scale algae cultivation. In: Muñoz, R., Gonzalez-Fernandez, C. (Eds.), Microalgae-Based Biofuels and Bioproducts. Woodhead Publishing, Madrid, Spain, pp. 505–525.

Ben-Amotz, A., 1995. New mode of *Dunaliella* biotechnology: two-phase growth for β-carotene production. J. Appl. Phycol. 7, 65–68.

Benemann, J.R., 2008. Opportunities and Challenges in Algae Biofuels Production. Algae World, Singapore, pp. 15, 17–18.

Bolong, N., Ismail, A.F., Salim, M.R., Matsuura, T., 2009. A review of the effects of emerging contaminants in wastewater and options for their removal. Desalination 238 (1–3), 229–246.

Borowitzka, M.A., 1998. Limits to growth. In: Wong, Y.S., Tam, N.E.Y. (Eds.), Wastewater Treatment With Algae. Elsevier, Amsterdam, pp. 203–226.

Borowitzka, M.A., 2013. High-value products from microalgae—their development and commercialisation. J. Appl. Psychol. 25 (3), 743–756.

Brancher, M., Griffiths, K.D., Franco, D., Lisboa, H.M., 2017. A review of odour impact criteria in selected countries around the world. Chemosphere 168, 1531–1570.

Brennan, L., Owende, P., 2010. Biofuels from microalgae—a review of technologies for production, processing, and extractions of biofuels and co-products. Renew. Sustain. Energ. Rev. 14, 557–577.

Campos, J.L., Valenzuela-Heredia, D., Pedrouso, A., Val del Rio, A., Belmonte, M., Mosquera-Corral, A., 2016. Greenhouse gases emissions from wastewater treatment plants: minimization, treatment, and prevention. J. Chem. 2016, 1–12.

Capelli, L., Sironi, S., Rosso, R.D., Céntola, P., Rossi, A., Austeri, C., 2011. Odour impact assessment in urban areas: case study of the city of Terni. Proc. Environ. Sci. 4, 151–157.

Chacon-lee, T.L., Gonzalez-Marino, G.E., 2010. Microalgae for "healthy" foods possibilities and challenges. Comp. Rev. Food Sci. Food Saf. 9, 655–675.

Chauton, M.S., Reitan, K.I., Norsker, N.H., Tveterås, R., Kleivdal, H.T., 2015. A techno-economic analysis of industrial production of marine microalgae as a source of EPA and DHA-rich raw material for aquafeed: research challenges and possibilities. Aquaculture 436, 95–103.

Chen, C.L., Chang, J.S., Lee, D., 2015. Dewatering and drying methods for microalgae. Dry. Technol. 33, 443–454.

Chisti, Y., 2007. Biodiesel from microalgae. Biotechnol. Adv. 25, 294–306.

Chisti, Y., 2013. Constraints to commercialization of algal fuels. J. Biotechnol. 167, 201–214.

CONAMA, Resolução n° 357/2005, 2005. Dispõe sobre a classificação dos corpos de água e diretrizes ambientais para o seu enquadramento, bem como estabelece as condições e padrões de lançamento de efluentes. In: Conselho Nacional do Meio Ambiente (Ed.), Brasília, Brasil.

Council Directive, 1991. Council Directive 91/271/EEC of 21 May 1991 concerning urban waste-water treatment. Off. J. L 135, 40–52.

Craggs, R., Sutherland, D., Campbell, H., 2012. Hectare-scale demonstration of high rate algal ponds for enhanced wastewater treatment and biofuel production. J. Appl. Phycol. 24, 329–337.

Craggs, R., Park, J., Heubeck, S., Sutherland, D., 2014. High rate algal pond systems for low-energy wastewater treatment, nutrient recovery and energy production. New Zeal. J. Bot. 52, 60–73.

Cuellar-Bermudez, S.P., Aguilar-Hernandez, I., Cardenas-Chavez, D.L., Ornelas-Soto, N., Romero-Ogawa, M.A., Parra-Saldivar, R., 2015. Extraction and purification of high-value metabolites from microalgae: essential lipids, astaxanthin and phycobiliproteins. Microb. Biotechnol. 8, 190–209.

Cuellar-Bermudez, S.P., Aleman-Nava, G.S., Chandra, R., Garcia-Perez, J.S., Contreras-Ângulo, J.R., Markou, G., et al., 2017. Nutrients utilization and contaminants removal. A review of two approaches of algae and cyanobacteria in wastewater. Algal Res. 24, 438–449.

D'Alessandro, E.B., Antoniosi Filho, N.R., 2016. Concepts and studies on lipid and pigments of microalgae: a review. Renew. Sustain. Energ. Rev. 58, 832–841.

Delgadillo-Mirquez, L., Lopes, F., Taidi, B., Pareau, D., 2016. Nitrogen and phosphate removal from wastewater with a mixed microalgae and bacteria culture. Biotechnol. Rep. 11, 18–26.

Demirbas, M.F., 2011. Biofuels from algae for sustainable development. Appl. Energy 88, 3473–3480.

Deprá, M.C., Santos, A.M., Severo, I.A., Santos, A.B., Zepka, L.Q., Jacob-Lopes, E., 2018. Microalgal biorefineries for bioenergy production: can we move from concept to industrial reality? Bioenergy Res. 11, 1–21.

Desmorieux, H., Decaen, N., 2005. Convective drying of spirulina in thin layer. J. Food Eng. 66, 497–503.

Draaisma, R.B., Wijffels, R.H., Slegers, P.M., Brentner, L.B., Roy, A., Barbosa, M.J., 2013. Food commodities from microalgae. Curr. Opin. Biotechnol. 24, 169–177.

Eppink, M.H.M., Olivieri, G., Reith, H., van den Berg, C., Barbosa, M.J., Wijffels, R.H., 2017. From current algae products to future biorefinery practices: a review. Adv. Biochem. Eng. Biotechnol. 166, 99–123.

Escapa, C., Coimbra, R.N., Nuevo, C., Vega, S., Paniagua, S., García, A.I., et al., 2017. Valorization of microalgae biomass by its use for the removal of paracetamol from contaminated water. Water 312, 1–9.

European Commission, 2015. Closing the Loop—An EU Action Plan for the Circular Economy, Com. 614 Communication from the Commission to the European Parliament, the Council, the European Economic and Social Committee and the Committee of the Regions. European Commission, Brussels.

FAO, F, 2011. The State of the World's Land and Water Resources for Food and Agriculture (SOLAW)—Managing Systems at Risk. Food and Agriculture Organization of the United Nations.

Fay, P., 1983. The blue-greens (Cyanophyta-Cyanobacteria)(Studies in biology/Institute of Biology; no. 160), Edward Arnold (Publishers) Ltd., London.

Flores, C.C., Bressers, H., Gutierrez, C., Boer, C., 2018. Towards circular economy—a wastewater treatment perspective, the Presa Guadalupe case. Manag. Res. Rev. 41 (5), 554–571.

Garcia-Serna, J., Perez-Barrigon, L., Cocero, M.J., 2007. New trends for design towards sustainability in chemical engineering: green engineering. Chem. Eng. J. 133, 7–30.

Garfí, M., Flores, L., Ferrer, I., 2017. Life cycle assessment of wastewater treatment systems for small communities: activated sludge, constructed wetlands and high rate algal ponds. J. Clean. Prod. 161, 211–219.

Gonçalves, A.L., Pires, J.C.M., Simões, M., 2017. A review on the use of microalgal consortia for wastewater treatment. Algal Res. 24, 403–415.

González-Camejo, J., Barat, R., Pachés, M., Murgui, M., Seco, A., Ferrer, J., 2018. Wastewater nutrient removal in a mixed microalgae-bacteria culture: effect of light and temperature on the microalgae-bacteria competition. Environ. Technol. 39, 503–515.

González-González, L.M., Correa, D.F., Ryan, S., Jensen, P.D., Pratt, S., Schenk, P.M., 2018. Integrated biodiesel and biogas production from microalgae: towards a sustainable closed loop through nutrient recycling. Renew. Sustain. Energ. Rev. 82, 1137–1148.

Gouveia, L., 2008. Microalgae in novel food products. In: Papa-Doupoulos, K. (Ed.), Food Chemistry Research Developments. Nova Science Publishers, New York, pp. 75–112.

Gouveia, L., Oliveira, A.C., Congestri, R., Bruno, L., Soares, A.T., Menezes, R.S., et al., 2017. Biodiesel from microalgae. In: González-Fernández, C., Muñoz, R. (Eds.), Microalgae-Based Biofuels and Bioproducts—From Feedstock Cultivation to End-Products. Woodhead Publishing, Duxford, pp. 235–258.

Griffiths, M.J., Harrison, S.T.L., 2009. Lipid productivity as a key characteristic for choosing algal species for biodiesel production. J. Appl. Psychol. 21, 493–507.

Ho, S.H., Ye, X., Hasunuma, T., Chang, J.S., Kondo, A., 2014. Perspectives on engineering strategies for improving biofuel production from microalgae—a critical review. Biotech. Adv. 32, 1448–1459.

Hosoglu, M.I., 2018. Aroma characterization of five microalgae/cyanobacteria species using solid-phase microextraction and gas chromatography-mass spectrometry/olfactometric. Food Chem. 240, 1210–1218.

Huo, S., Wang, Z., Zhu, S., Zhou, W., Dong, R., Yuan, Z., 2012. Cultivation of *Chlorella zofingiensis* in bench-scale outdoor ponds by regulation of pH using dairy wastewater in winter, South China. Bioresour. Technol. 121, 76–82.

Jacob-Lopes, E., Scoparo, C.H.G., Queiroz, M.I., Franco, T.T., 2010. Biotransformations of carbon dioxide in photobioreactors. Energy Convers. Manag. 51, 894–900.

Keffer, J.E., Kleinheinz, G.T., 2002. Use of *Chlorella vulgaris* for CO_2 mitigation in a photobioreactor. J. Ind. Microbiol. Biotechnol. 29, 275–280.

Kong, Q., Li, L., Martinez, B., Chen, P., Ruan, R., 2010. Culture of microalgae/cyanobacteria *Chlamydomonas reinhardtii* in wastewater for biomass feed stock production. Appl. Biochem. Biotechnol. 160, 9–18.

Kovac, D.J., Simeunovic, J.B., Babic, O.B., Misan, A.C., Milovanovic, I.L., 2013. Algae in food and feed. Food Feed Res. 40, 21–31.

Kumar, K.S., Hans-Uwe, D., Eun-Ji, W., Jae-Seong, L., Kyung-Hoon, S., 2015. Microalgae/cyanobacteria—a promising tool for heavy metal remediation. Ecotoxicol. Environ. Saf. 113, 329–352.

Kumar, A., Kaushal, S., Saraf, S.A., Singh, J.S., 2018. Microbial bio-fuels: a solution to carbon emissions and energy crisis. Front. Biosci. (Landmark ED) 1 (23), 1789–1802.

Laurienzo, P., 2010. Marine polysaccharides in pharmaceutical applications: an overview. Mar. Drugs 8, 2435–2465.

Lebrero, R., Bouchy, L., Stuetz, R., Muñoz, R., 2011. Odor assessment and management in wastewater treatment plants: a review. Crit. Rev. Environ. Sci. Technol. 41, 915–950.

Lebrero, R., Gondim, A.C., Pérez, R., García-Encina, P.A., Muñoz, R., 2014. Comparative assessment of a biofilter, a biotrickling filter and a hollow fiber membrane bioreactor for odor treatment in wastewater treatment plants. Water Res. 49, 339–350.

Leu, S., Boussiba, S., 2014. Advances in the production of high-value products by microalgae. Ind. Biotechnol. 10, 169–183.

Li, Y., Zhou, W., Hu, B., Min, M., Chen, P., Ruan, R., 2011a. Integration of algae cultivation as biodiesel production feed stock with municipal wastewater treatment: strains screening and significance evaluation of environmental factors. Bioresour. Technol. 102, 10861–10867.

Li, Y., Chen, Y.F., Chen, P., Min, M., Zhou, W., Martinez, B., et al., 2011b. Characterization of a microalga *Chlorella* sp. well adapted to highly concentrated municipal wastewater for nutrient removal and biodiesel production. Bioresour. Technol. 102, 5138–5144.

Li, J., Zhu, D., Niu, J., Shen, S., Wang, G., 2011c. An economic assessment of astaxanthin production by large scale cultivation of *Haematococcus pluvialis*. Biotechnol. Adv. 29, 568–574.

Li, C., Yang, H., Xia, X., Li, Y., Chen, L., Zhang, M., et al., 2013. High efficient treatment of citric acid effluent by *Chlorella vulgaris* and potential biomass utilization. Bioresour. Technol. 127, 248–255.

Li, J., Liu, Y., Cheng, J.J., Mos, M., Daroch, M., 2015. Biological potential of microalgae in China for biorefinery-based production of biofuels and high value compounds. New Biotechnol. 32, 588–596.

Luterbacher, J.S., Fröling, M., Vogel, F., Maréchal, F., Tester, J.W., 2009. Hydrothermal gasification of waste biomass: process design and life cycle assessment. Environ. Sci. Technol. 43 (5), 1578-–11583.

Maga, D., 2016. Life cycle assessment of biomethane produced from microalgae grown in municipal waste water. Biomass. Conv. Bioref. 7, 1–10.

Marín, D., Posadas, E., Cano, P., Pérez, V., Blanco, S., Lebrero, R., et al., 2018. Seasonal variation of biogas upgrading coupled with digestate treatment in an outdoors pilot scale algal-bacterial photobioreactor. Bioresour. Technol. 263, 58–66.

Matamoros, V., Gutiérrez, R., Ferrer, I., García, J., Bayona, J.M., 2015. Capability of microalgae/cyanobacteria-based wastewater treatment systems to remove emerging organic contaminants: a pilot-scale study. J. Hazard. Mater. 288, 34–42.

McGinn, P.J., Dickinson, K.E., Park, K.C., Whitney, C.G., MacQuarrie, S.P., Black, F.J., et al., 2012. Assessment of the bioenergy and bioremediation potentials of the microalga *Scenedesmus* sp. AMDD cultivated in municipal wastewater effluent in batch and continuous mode. Algal Res. 1, 155–165.

Meffe, R., Bustamante, I., 2014. Emerging organic contaminants in surface water and groundwater: a first overview of the situation in Italy. Sci. Tot. Environ. 481, 280–295.

Milledge, J.J., Heaven, S., 2013. A review of the harvesting of micro-algae for biofuel production. Rev. Environ. Sci. Biotechnol. 12, 165–178.

Norsker, N.H., Barbosa, M.J., Vermue, M.H., Wijffels, R.H., 2011. Microalgal production—a close look at the economics. Biotechnol. Adv. 29, 24–27.

Organization for Economic Cooperation and Development, 2009. The Bioeconomy to 2030: Designing a Policy Agenda. OECD Publishing, Paris.
Paliwal, R., Gusain, P., Kaphaliya, B., Uniyal, S., 2018. Algal potential for inorganic and organic pollutants decontamination. In: Khanna, D.R., Bhutiani, R. (Eds.), Water Biology. Discovery Publishing House Pvt. Ltd, New Delhi, India, pp. 309–317.
Pandey, A., Höfer, R., Taherzadeh, M., Nampoothiri, M., Larroche, C. (Eds.), 2015. Industrial Biorefineries & White Biotechnology. Elsevier, pp. 341–368.
Park, J.B.K., Craggs, R.J., Shilton, A.N., 2011. Wastewater treatment high rate algal ponds for biofuel production. Bioresour. Technol. 102, 35–42.
Passos, F., Gutiérrez, R., Uggetti, E., Garfí, M., García, J., Ferrer, I., 2017. Towards energy neutral microalgae-based wastewater treatment plants. Algal Res. 28, 235–243.
Patil, P.D., Gude, V.G., Mannarswamy, A., Deng, S., Cooke, P., Munson-McGee, S., et al., 2011. Optimization of direct conversion of wet algae to biodiesel under supercritical methanol conditions. Bioresour. Technol. 102, 118–122.
Peccia, J., Haznedaroglu, B., Gutierrez, J., Zimmerman, J.B., 2013. Nitrogen supply is an important driver of sustainable microalgae/cyanobacteria biofuel production. Trends Biotechnol. 31, 3.
Peña-Castro, J.M., Martinez-Jeronimo, F., Esparza-Garcia, F., Canizares-Villanueva, R.O., 2004. Phenotypic plasticity in *Scenedesmus incrassatulus* (Chlorophyceae) in response to heavy metals stress. Chemosphere 57, 1629–1636.
Perez-Garcia, O., Bashan, Y., 2015. Microalgal heterotrophic and mixotrophic culturing for bio-refining: from metabolic routes to techno-economics. In: 1st ed. Prokop, A., et al., (Eds.), Algal Biorefineries: Algal Growth, Products and Optimization, 2. Springer International Publishing, Switzerland, pp. 61–131. (Eds.).
Perez-Garcia, O., Bashan, Y., Puente, M.E., 2011. Organic carbon supplementation of sterilized municipal wastewater is essential for heterotrophic growth and removing ammonium by the microalga *Chlorella Vulgaris*. J. Phycol. 47, 190–199.
Posadas, E., Marín, D., Blanco, S., Lebrero, R., Muñoz, R., 2017. Simultaneous biogas upgrading and centrate treatment in an outdoors pilot scale high rate algal pond. Bioresour. Technol. 232, 133–141.
Prakash, J., Pushparaj, B., Carlozzi, P., Torzillo, G., Montaini, E., Materassi, R., 1997. Microalgal biomass drying by a simple solar device. Int. J. Sol. Energy 18, 303–311.
Priya, M., Gurung, N., Mukherjee, K., Bose, S., 2014. Microalgae/cyanobacteria in removal of heavy metal and organic pollutants from soil. In: Surajit, D. (Ed.), Microbial Biodegradation and Bioremediation. Elsevier, Rourkela, India, pp. 519–537.
Priyararshani, I., Rath, B., 2012. Commercial and industrial applications of micro algae—a review. J. Algal Biomass. Utln. 3, 89–100.
Pulz, O., Gross, W., 2004. Valuable products from biotechnology of microalgae. Appl. Microbiol. Biotechnol. 65, 635–648.
Queiroz, M.I., Jacob-Lopes, E., Zepka, L.Q., Bastos, R.G., Goldbeck, R., 2007. The kinetics of the removal of nitrogen and organic matter from parboiled rice effluent by cyanobacteria in a stirred batch reactor. Bioresour. Technol. 98, 2163–2169.
Queiroz, M.I., Hornes, M.O., Silva-Menetti, A.G., Jacob-Lopes, E., 2011. Single-cell oil production by cyanobacterium *Aphanothece microscopica Nägeli* cultivated heterotrophically in fish processing wastewater. Appl. Energy 88, 3438–3443.
Raposo, M.F., Morais, R.M.S.C., 2013. Bioactivity and applications of sulphated polysaccharides from marine microalgae. Mar. Drugs 11, 233–252.
Rappert, S., Müller, R., 2005. Microbial degradation of selected odorous substances. Waste Manag. 25, 940–954.
Rawat, I., Kumar, R.R., Mutanda, T., Bux, F., 2011. Dual role of microalgae: phycoremediation of domestic wastewater and biomass production for sustainable biofuels production. Appl. Energy 88, 3411–3424.
Rawat, I., Ranjith, R., Mutanda, T., Bux, F., 2013. Biodiesel from microalgae: a critical evaluation from laboratory to large scale production. Appl. Energy 103, 444–467.
Richardson, J.W., Johnson, M.D., Lacey, R., Oyler, J., Capareda, S., 2014. Harvesting and extraction technology contributions to algae biofuels economic viability. Algal Res. 5, 70–78.
Salama, E.S., Kurade, M.B., Abou-Shanab, R.A.I., El-Dalatony, M.M., Yang, S., Min, B., et al., 2017. Recent progress in microalgal biomass production coupled with wastewater treatment for biofuel generation. Renew. Sustain. Energ. Rev. 79, 1189–1211.
Santos, A.M., Deprá, A.C., Santos, A.M., Zepka, L.Q., Jacob-Lopes, E., 2015. Aeration energy requirements in microalgal heterotrophic bioreactors applied to agroindustrial wastewater treatment. Curr. Biotechnol. 4, 249–254.
Santos, A.M., Santos, A.M., Severo, I.A., Queiroz, M.I., Zepka, L.Q., Jacob-Lopes, E., 2016a. Nutrient cycling in wastewater treatment plants by microalgae-based processes. In: Barton, S.N. (Ed.), Industrial Waste: Management, Assessment and Environmental Issues. Nova Science Publishers, Inc, New York, pp. 41–63.
Santos, A.B., Vieira, K.R., Nogara, G.P., Wagner, R., Jacob-Lopes, E., Zepka, L.Q., 2016b. Biogeneration of volatile organic compounds by microalgae/cyanobacteria: occurrence, behavior, ecological implications and industrial applications. In: Moore, J.P. (Eds) (Ed.), Volatile Organic Compounds: Occurrence, Behavior and Ecological Implications, 1st ed. Nova Science Publishers, New York, pp. 1–18.
Santos, A.B., Fernandes, A.F., Wagner, R., Jacob-Lopes, E., Zepka, L.Q., 2016c. Biogeneration of volatile organic compounds produced by *Phormidium autumnale* in heterotrophic bioreactor. J. Appl. Phycol. 1, 1–10.
Santos, A.M., Santos, A.M., Sartori, R.B., Queiroz, L.Z., Barin, J.S., Jacob-Lopes, E., 2017a. Nutrient cycling in meat processing industry by microalgae/cyanobacteria-based processes. Desalin. Water Treat. 100, 91–99.
Santos, A.M., Santos, A.M., Zepka, L.Q., Jacob-Lopes, E., 2017b. Downstream recovery of microalgal bioproducts with highlights on biorefineries. Biorefineries: Concepts, Advancements, and Research. Nova Science Publishers, Inc, New York, pp. 1–38.
Sathish, A., Sims, R.C., 2012. Biodiesel from mixed culture algae via a wet lipid extraction procedure. Bioresour. Technol. 118, 643–647.
Schenk, P.M., Thomas-Hall, R.S., Stephens, E., Marx, U.C., Mussgnung, J.H., Posten, C., et al., 2008. Second generation biofuels: high-efficiency microalgae for biodiesel production. Bioenergy Res. 1, 20–43.
Schlesinger, W.H., 1997. Biogeochemistry: An Analysis of Global Change, 2nd ed. Academic, San Diego, CA, USA.

Severo, I.A., Deprá, M.C., Barin, J.S., Wagner, R., Menezes, C.R.D., Zepka, L.Q., et al., 2018. Bio-combustion of petroleum coke: the process integration with photobioreactors. Chem. Eng. Sci. 177, 422–430.

Shelef, G., Sukenik, A., Green, M., 1984. Microalgae harvesting and processing: a literature review. Report, Solar Energy Research Institute, Golden Colorado, SERI/STR-231-2396.

Shi, W., Wang, L., Rousseau, D.P.L., Lens, P.N.L., 2010. Removal of estrone, 17α-ethinylestradiol, and 17β-estradiol in algae and duckweed-based wastewater treatment systems. Environ. Sci. Pollut. Res. Int. 17, 824–833.

Singh, J.S., 2015. Microbes: the chief ecological engineers in reinstating equilibrium in degraded ecosystems. Agric. Ecosyst. Environ. 203, 80–82.

Singh, J.S., Kumar, A., Rai, A.N., Singh, D.P., 2016. Cyanobacteria: a precious bio-resource in agriculture, ecosystem, and environmental sustainability. Front. Microbiol. 7, 529.

Spolaore, P., Joannis-Cassan, C., Duran, E., Isambert, A., 2016. Commercial applications of microalgae. J. Biosci. Bioeng. 101, 87–96.

Su, Y., Song, K., Zhang, P., Su, Y., Cheng, J., Chen, X., 2017. Progress of microalgae biofuel's commercialization. Renew. Sustain. Energ. Rev. 74, 402–411.

Sydney, E., Da Silva, T., Tokarski, A., Novak, A., De Carvalho, J., Woiciecohwski, A., et al., 2011. Screening of microalgae with potential for biodiesel production and nutrient removal from treated domestic sewage. Appl. Energy 88, 3291–3294.

Torres, C.M., Rios, S.D., 2013. Microalgae-based biodiesel: a multicriteria analysis of the production process using realistic scenarios. Bioresour. Technol. 147, 7–16.

Tran, N.H., Urase, T., Ngo, H.H., Hu, J., Ong, S.L., 2013. Insight into metabolic and cometabolic activities of autotrophic and heterotrophic microorganisms in the biodegradation of emerging trace organic contaminants. Bioresour. Technol. 146, 721–731.

U.S. Environmental Protection Agency-USEPA, 1993. A Guide to the Biosolids Risk Assessments for the EPA Part 503 Rule. EPA/832-B-93-005. Office of Wastewater Management, Washington, DC, p. 144.

Uduman, N., Qi, Y., Danquah, M.K., Forde, G.M., Hoadley, A., 2010. Dewatering of microalgal cultures: a major bottleneck to algae-based fuels. J. Renew. Sustain. Energ. Rev. 2, 12701–12715.

UNFCCC, 2015. Intended Nationally Determined Contributions (INDCs). United Nations Framework Convention on Climate Change.

Valente, A., Iribarren, D., Dufour, J., 2019. How do methodological choices affect the carbon footprint of microalgal biodiesel? A harmonised life cycle assessment. J. Clean. Prod. 207, 560–568.

Vanthoor-Koopmans, M., Wijffels, R.H., Barbosa, M.J., Eppink, M.H.M., 2013. Biorefinery of microalgae for food and fuel. Bioresour. Technol. 135, 142–149.

Vieira, J.G., Manetti, A.G., Jacob-Lopes, E., Queiroz, M.I., 2012. Uptake of phosphorus from dairy wastewater by heterotrophic cultures of cyanobacteria. Desalin. Water Treat. 40, 224–230.

Von Sperling, M., Chernicharo, C.A.L., 2002. Urban wastewater treatment technologies and the implementation of discharge standards in developing countries. Urban Water 4 (1), 105–114.

Von Sperling, M., Fattal, B., 2001. Implementation of guidelines: some practical aspects. Water Quality: Guidelines, Standards and Health. Assessment of Risk and Risk Management for Water-Related Infectious Disease. IWA Publishing, London, pp. 361–376.

Voort, M.P.J., van der, Vulsteke, E., Visser, C.L.M. de. 2015. Macro-economics of algae products, Public Output Report of the EnAlgae Project, Swansea, 47.

Wang, B., Li, Y., Wu, N., Lan, C.Q., 2008. CO_2 bio-mitigation using microalgae. Appl. Microbiol. Biotechnol. 79, 707–718.

Wijesekara, I., Pangestuti, R., Kim, S.K., 2011. Biological activities and potential health benefits of sulfated polysaccharides derived from marine algae. Carbohydr. Polym. 84, 14–21.

Wijffels, R.H., Barbosa, M.J., 2010. An outlook on microalgal biofuels. Science 329, 796–799.

Wilt, A., Butkovskyi, A., Tuantet, K., Leal, L.H., Fernandes, T.V., Langenhoff, A., et al., 2016. Micropollutant removal in an algal treatment system fed with source separated wastewater streams. J. Hazard. Mater. 304, 84–92.

Wu, L.F., Chen, P.C., Huang, A.P., Lee, C.M., 2012. The feasibility of biodiesel production by microalgae/cyanobacteria using industrial wastewater. Bioresour. Technol. 113, 14–18.

WWP, UN. 2017. Wastewater: The untapped resource. The United Nations World Water Development Report, 2017.

Xu, Y.Y., Boeing, W.J., 2014. Modeling maximum lipid productivity of microalgae: review and next step. Renew. Sustain. Energ. Rev. 32, 29–39.

Yang, J., Li, X., Hua, H., Zhang, X., Yu, Y., Chen, Y., 2011. Growth and lipid accumulation properties of a freshwater microalga, *Chlorella ellipsoidea* YJ1, in domestic secondary effluents. Appl. Energy 88, 3295–3299.

Yerushalmi, L., Ashrafi, O., Haghighat, F., 2013. Reductions in greenhouse gas (GHG) generation and energy consumption in wastewater treatment plants. Water Sci. Technol. 67 (5), 1159–1164.

Zamagni, A., 2012. Life cycle sustainability assessment. Int. J. Life Cycle Assess. 17, 373–533.

Zanetti, L., Frison, N., Nota, E., Tomizioli, M., Bolzonella, D., Fatone, F., 2012. Progress in real-time control applied to biological nitrogen removal from wastewater. A short review. Desalination 286, 1–7.

Zhang, Y., Zhou, J.L., 2008. Occurrence and removal of endocrine disrupting chemicals in wastewater. Chemosphere 73, 848–853.

Zhang, Q., Hu, J., Lee, D.J., 2016. Aerobic granular processes: current research trends. Bioresour. Technol. 210, 74–80.

Zhao, J., Li, Y., Zhang, C., Zeng, Q., Zhou, Q., 2008. Sorption and degradation of bisphenol a by aerobic activated sludge. J. Hazard. Mater. 155, 305–311.

Zhu, L., 2014. The combined production of ethanol and biogas from microalgal residuals to sustain microalgal biodiesel: a theoretical evaluation. Biofuels Bioprod. Biorefin. 8, 7–15.

Zhu, L.D., Hiltunen, E., 2016. Application of livestock waste compost to cultivate microalgae for bioproducts production: a feasible framework. Renew. Sustain. Energ. Rev. 54, 1285–1290.

Chapter 4

Plant–microbe associations for enhancement of agricultural productivity

Honghui Zhu[1], Yang Zhou[1], Yongqiang Qin[2] and Qing Yao[1,2]

[1]State Key Laboratory of Applied Microbiology Southern China, Guangdong Provincial Key Laboratory of Microbial Culture Collection and Application, Guangdong Institute of Microbiology, Guangzhou, P.R. China, [2]Guangdong Province Key Laboratory of Microbial Signals and Disease Control, Guangdong Engineering Research Center for Litchi, Guangdong Engineering Research Center for Grass Science, College of Horticulture, South China Agricultural University, Guangzhou, P.R. China

Chapter Outline

4.1 Introduction

Soil microbes are phylogenetically diverse and functionally multifaceted. They can form different associations with plants, from beneficial symbiosis or mutualism to detrimental parasitism or pathogenesis, with some others neutral to plants in an independent status. "Plant microbiota" refers to those microbes in the term of phylogenetics, which depend on plants and reversely regulate plant performance (Wang et al., 2008; Müller et al., 2016). Recently, the term "plant microbiome" has been used to describe not only the phylogenetic but also the functional profiling of the plant microbiota (Lebeis et al., 2012; Klaus and Bulgarelli, 2015). In agricultural ecosystems, the beneficial plant–microbe associations are of special interest for potential application for better crop growth. These beneficial associations are frequently manipulated to enhance the crop production, mainly via the pathways of increasing soil fertility, elevating the crop resistance to pathogens, and producing plant hormones (Van Wees et al., 2008; Hayat et al., 2010; Chaparro et al., 2012). These microbes can be regarded as a subset of plant microbiome, which provides beneficial effects to the associated plants. In most agricultural ecosystems, nitrogen (N) and phosphorus (P) are the main nutrient constraints to crop growth and agricultural production without chemical fertilizer input, and thus rhizobia or AM fungi can be employed to increase soil fertility in an environment friendly way (Singh, 2013; Tiwari et al., 2018; Singh and Gupta, 2018; Vimal et al., 2018). From the view of field application, several groups are successfully applied in practice, or regarded as potential candidates, including arbuscular mycorrhizal (AM) fungi, rhizobia, and plant growth promoting rhizobacteria (PGPRs). In this context, it is valuable to review and discuss the recent advances in this field, covering

New and Future Developments in Microbial Biotechnology and Bioengineering. DOI: https://doi.org/10.1016/B978-0-12-818258-1.00004-2

the associations between plants and AM fungi, rhizobia, and PGPRs. Moreover, we focus special emphasis on the interactions of these beneficial microbes, which potentially provide synergistic effects on plant growth, for example, rhizobia and AM fungi, AM fungi and P solubilizing bacteria (PSBs, a group of PGPRs). This tripartite association of plants, AM fungi, and PSBs has recently been attracting tremendous attention (Taktek et al., 2015; Zhang et al., 2016, 2018), and potentially multiplies the soil P efficiency for better crop production. The objectives of this chapter are to review our overall knowledge on plant–microbe association, with special focus on AM fungi, rhizobia, and PGPRs; to demonstrate the recent advances in the application of these beneficial microbes under field conditions; and to discuss the mechanisms underlying these beneficial effects (Singh, 2015; Singh et al., 2018). Particularly, dual inoculation with these microbes is emphasized.

4.2 Plants and their associated microbes

Microbes are ubiquitous, representing indispensable components in all ecosystems. In agricultural ecosystems, soil microbes play crucial roles in diverse ecological processes, such as nutrient cycling, pathogen suppression, and organic degradation. To a certain degree, it is not surprising to claim that these microbes support the good functioning of the agricultural ecosystem (Kennedy and Smith, 1995). Particularly, some microbes associate closely with plants, forming an intimate relationship with them. Recently, the term "plant microbiome" has been proposed, focusing on the entirety of these microbes (Turner et al., 2013; Berg et al., 2014). These associated microbes establish mutualism with their hosts, transferring benefits to plants and acquiring carbon (C) from plants in return (Sasse et al., 2018). They colonize the rhizosphere niches, readily utilizing the root exudates (e.g., PGPRs), or enter the root cells, developing a symbiotic structure (e.g., AM fungi and rhizobia). These intimate relationships guarantee the mutualism is well executed.

4.2.1 Plants shape microbial community structure

The precondition of mutualism between plants and their associated microbes is the selection of microbes by hosts. It is well acknowledged that plants can shape soil microbial community structure (Cline and Zak, 2015; Ridl et al., 2016; Wagner et al., 2016; Leff et al., 2019). This regulation mainly depends on root exudation and nutrient requirement characteristics (Zhou et al., 2017a; Sasse et al., 2018). Root exudates vary greatly both in quantity and in quality (composition) among plant species and even genotype. Plants with C_4 photosynthetic pathway normally secrete more root exudate than plants with C_3 photosynthetic pathway, probably due to their higher photosynthetic efficiency than C_3 plants. Moreover, the root exudates of C_4 plants comprise higher level of amino acids and organic acids, while those of C_3 plants comprise more carbohydrate (Sivaram et al., 2018). Rhizosphere microbes highly depend on these organic compounds as C source and energy. On the other hand, plants have distinct nutrient requirement characteristics, which result in different soil chemical profiles (Badri and Vivanco, 2009). This difference can further differentiate microbial community composition and even function (Zhou et al., 2017b).

Many important field crops and economic crops are either legumes or grasses, which regulate the microbial community composition and function in different patterns. Legumes can enrich fungi in their rhizosphere soil compared with grasses (Zhou et al., 2017a, 2017b). For bacteria, legumes increase the abundance of *Verrucomicrobia*, while grasses decrease it (Zhou et al., 2017a). This selection can be attributed, at least partly, to soil extracts, as influenced by both root exudates and microbial decomposition (Zhou et al., 2017b). More interestingly, the difference in microbial community composition can further lead to difference in microbial community function, for example, P cycling. Zhou et al. (2018) demonstrated that bacterial community in legume rhizosphere tended to hydrolyze organic P while bacterial community in grass rhizosphere tended to solubilize sparingly soluble inorganic P. This study highlights the advantages of intercropping legume crops and cereal crops in fully mobilizing soil P for sustainable agriculture. In general, the regulation of microbial community by plants should be reevaluated in a coadaptation perspective, and plants tend to recruit those microbes capable of promoting their growth performance.

4.2.2 Plant–AM fungi association

AM fungi, phylogenetically belonging to *Glomeromycota*, are symbiotic soil fungi in almost all terrestrial ecosystems (Smith and Read, 2008). They establish symbiosis with more than 80% of land plants, including most agricultural crop species, and exert beneficial effects on hosts. Since Mosse (1957) demonstrated for the first time that AM fungi could drastically increase the uptake of readily immobilized nutrients (e.g., P, Zn, Cu) by apple roots, the underlying mechanisms have been intensively investigated, mainly including enlarged soil volume for nutrient acquisition, secretion of phosphatase, soil acidification, high efficiency in P transport from extraradical phase to intraradical phase, high compatibility of fungal P transporter (Bolan, 1991; George et al., 1995; Smith et al., 2011). The mutualism between AM fungi and hosts can be manifested via the exchange of nutrients and carbohydrate at the interface of the periarbuscule space (Ferrol et al., 2019). Once AM fungi penetrate root cortex cells, hyphal tips dichotomously branch and thus produce arbuscules, the key structure indicative of symbiotic effect (Luginbuehl and Oldroyd, 2017).

In most cases, AM fungi can promote host nutrient uptake and plant growth, especially under stressed conditions. Plant root systems often do not develop well for enough nutrient uptake under stressed conditions, and AM fungi can help take up more nutrients. This point explains, to the maximum extent, why the inoculation effect of AM fungi under stressed conditions is sometimes greater than that under nonstressed conditions (Ruiz-Lozano et al., 2016; Oliveira et al., 2017; Li et al., 2018). Therefore, in this context, it is necessary to evaluate the tolerance of AM fungi to stressed conditions. In acidic soil, low pH can be detrimental to both roots and AMF. With axenic culture of hairy roots, Wang et al. (2017) demonstrated that *Rhizophagus irregularis* DAO197198 was more tolerant to low pH than roots, although low pH greatly reduced the fungal colonization, with arbuscule formation the most affected. In general, AM fungi can drastically improve plant growth performance of most field crops, especially under stressed conditions. This feature guarantees the application of AM fungal inocula in agriculture systems, where the adverse conditions (e.g., heat stress, drought stress) become increasingly frequent due to the global climate change (Wiebe et al., 2015).

4.2.3 Legume–rhizobia association

Legume crops comprise of over 20,000 species, with many species as important crops for human society, for example, soybean, common bean, and chickpea. Legumes can establish symbiosis with rhizobia, and acquire atmosphere N_2 as N nutrient via biological N_2-fixation (BNF) by rhizobia. According to Herridge et al. (2008), the annual N input by BNF in agricultural systems is 21.45 Tg, with soybean–rhizobia symbiosis contributing up to 16.4 Tg. However, BNF only accounts for 13% of the total fertilization on arable land, indicating the potential of BNF in agricultural systems (Anglade et al., 2015). The year 2016 was declared as the International Year of Pulses (grain legumes) by the United Nations, in that legume crops provide nutritional seeds to human society in a sustainable manner (Foyer et al., 2016).

In the nodules, BNF is conducted with nitrogenase, a complex enzyme consisting of two metalloprotein components. To better understand its assembly, polyprotein strategy was successfully employed to reveal the stoichiometric assembly of nitrogenase components in views of synthetic biology (Yang et al., 2018). So far, much effort has been made to associate rhizobia with cereal crops, for example, electron-transport chain engineering (Yang et al., 2017), it is still far from practicability. Symbiosis genes can be horizontally transferred between bacteria, allowing the bacteria adapted to local soil conditions to become rhizobial symbionts of previously incompatible legumes growing in these soils (Andrews et al., 2018). Consequently, the ecological significance of this phenomenon can be great for isolation of appropriate rhizobial strains and agricultural practices.

4.2.4 Plant and PGPRs interactions

PGPR is an operationally defined term, referring to those microorganisms that are able to promote plant growth (Zahir et al., 2004). PGPRs are phylogenetically diverse and functionally multifaceted, with *Bacillus*, *Pseudomonas*, and *Burkholderia* frequently isolated from soils as PGPRs (Vessey, 2003; Tabassum et al., 2017). The roles of PGPR in plant growth performance have been recognized for decades, primarily including facilitating nutrient uptake, suppression of diseases, and production of plant hormones (Tabassum et al., 2017). Due to these diverse functions, the action modes employed by PGPRs also vary greatly, depending on microbial species, function types, and environmental cues. P solubilizing bacteria in the rhizosphere can solubilize the sparingly soluble inorganic phosphate mainly by secreting gluconic acid and 2-ketogluconic acid, with the latter more efficient (Khan et al., 2010), although they also produce and secrete other organic acids, such as citric acid, oxalic acid, and malic acid (Israr et al., 2016). Promoted resistance to pathogens by PGPRs can be attributed to production of siderophores limiting iron availability to pathogens, production of antibiotics killing pathogens, and induction of systemic resistance in plants (Siddiqui, 2005). Plant hormones are essential to the growth and development of plants, regulating almost all biological processes. PGPRs can synthesize and secrete auxin, or inhibit the synthesis of ethylene (Goswami et al., 2015; Tsukanova et al., 2017; Saleem et al., 2018), by which they exert strong influences on crop growth and production, especially under stressed conditions (Etesami and Maheshwari, 2018).

Sometimes, PGPRs promote plant growth not in a direct way but in an indirect way. They may help promote the survival of other inoculated beneficial microbes in rhizosphere niches. Biofilm is the key structure for microbial community to protect from stressed conditions. With well-established biofilm, microbes can survive, propagate, and function well in response to drought and salinity (Barthès et al., 2015). Biofilm also facilitate the cooperation among microbes in biofilm community. Swarnalakshmi et al. (2013) reported a novel photoautotropic cyanobacterial strain *Anabaena torulosa*, capable of forming matrix for biofilm development. *A. torulosa* increased acetylene-reducing activity of rhizobia in vitro condition, and also enhanced P uptake of wheat inoculated with PSB in pot experiment. This cooperation was also demonstrated with chickpea under field conditions (Bidyarani et al., 2016). This indirect interaction between plants and *A. torulosa* highlights a complex microbial cooperation in rhizosphere niches.

4.3 Enhancement of plant production via plant–microbe associations

Owing to the aforementioned manifest growth promoting effects of plant associated microbes, transition of the potential from laboratory to fields attracts tremendous interest. During the past few decades, a great deal of experiments have been carried out under field conditions or greenhouse conditions to test the effects of applying these microbes to crops (Tiwari and Singh, 2017; Kumar et al., 2018).

4.3.1 AM fungi

Although AM fungi are ubiquitous in terrestrial ecosystems, they have not been successfully pure cultured on medium without host plants so far (Smith and Read, 2008). Presently, AM fungal inocula can be produced only in symbiosis with plants, with the methods of pot culture with soil substrate, massive culture with light substrate such as perlite and vermiculite, or axenic culture on agar associated with transformed hairy roots (Gianinazzi and Vosátka, 2004; Berruti et al., 2016). These propagation techniques require larger area and longer period of cultivation than the conventional propagation of culturable microbes, making AM fungal inocula very expensive. Therefore, data on the application of AM fungi under field conditions is relatively limited.

Horticultural crops represent relatively high value crops compared with cereal crops, and thus most field studies investigate the practical effect of AM fungi with horticultural crops as host plants. Bona et al. (2017) reported that inoculation with a mixed AM fungal inocula under field conditions did not increase plant growth of tomato in terms of fruit fresh weight and marketable fruit number, but significantly improved fruit quality in terms of sweetness index to titratable acidity ratio.

Although exogenously inoculated AM fungi can effectively promote crop growth and production under field conditions, in some cases the native AM fungal community is also effective (Panja and Chaudhuri, 2004; Affokpon et al., 2011). In a field experiment with nematode infestation, native AM fungal inoculation increased tomato yields by up to 26% and carrot yields by up to 300% (Affokpon et al., 2011). This means that the cost on AM fungal inocula can be saved if the native AM fungi are managed properly. Proper management of native AM fungi involve several aspects, including soil properties, agrochemical inputs, arrangement of precrops. Carballo-Hernández et al. (2018) collected native AM fungal consortia from agricultural soils with low-, moderate-, and intense management, respectively, as inocula. They demonstrated that AM fungal consortium from soils with intense management had the lowest efficiency in promoting plant growth of pepper. Soil available P contents were averaged as 22.28, 36.55, and 60.49 mg/kg for low-, moderate-, and intense management, respectively, indicating the importance of soil chemical properties in regulating native AM fungal efficiency to a certain degree. More interestingly, planting seven crops (including mycrorrhizal and nonmycorrhizal species) as precrops resulted in much different plant growth of citrus, and citrus grew the best with maize and millet as precrops but the worst with mustard and ginger as precrops (Panja and Chaudhuri, 2004). This difference was mainly due to the influences of precrops on AM fungal propagule intensity in soils, which provides an alternative strategy for the utilization of native AM fungi. Cover cropping in orchards represents another example utilizing the native AM fungi with this strategy, where growing bahiagrass in citrus orchard significantly increased the spore number and hyphal length intensity in rhizosphere, and AM colonization in citrus roots (Wang et al., 2012).

4.3.2 Rhizobia

Rhizobia are the most widely utilized plant associated microbes so far, although they form symbiosis with only legumes. This is because some legumes are widely cultivated important crops, providing a huge amount of proteins to human society, for example, soybean. Due to the intimate relationship between rhizobia and legume hosts, the symbiotic efficiency is proved to be high in many agricultural systems (Herridge et al., 2008; Anglade et al., 2015).

Thilakarathna and Raizada (2017) conducted a meta-analysis of the effectiveness of rhizobia inoculants on soybean under field conditions. They concluded that a diversity of *Bradyrhizobium* and *Sinorhizobium* species/isolates are effective inoculants, even including some indigenous strains, particularly with the indigenous strain absent or in extremely low population. The indigenous strains can be well adapted to local field (stressed) conditions, and thus hold the potential for local commercialization. Moreover, the effectiveness of these rhizobia inoculants mainly depends on legume genotypes, legume and rhizobia interactions, inoculant amount and formulation, and application method (Denton et al., 2017; Thilakarathna and Raizada, 2017). Additionally, Mulas et al. (2015) reported that tillage system also affected the effectiveness of indigenous rhizobia strains for common bean, with no effect in no-tillage system compared with conventional system and cover cropping system. Nowadays, enhancing legume resistance to stressed conditions with rhizobia has increasingly become a research hot spot, primarily due to the limited soil N supply in most agricultural systems with stressed conditions. Some tolerant rhizobial strains isolated from saline or drought stressed soils were efficient in promoting plant growth and yield of common bean in the Egypt Nile delta, increasing plant weight in pot culture from 2.718 to 3.314 g and seed yield in saline or drought stressed fields by 9.3%–67.7% (Yanni et al., 2016).

When inoculating rhizobia to legume crops, the coupling of N and P nutrition should be borne in mind. In most cases, P application can enhance the rhizobial effect particularly in P deficient soils. Along with the increasing emphasis on agriculture technology in Africa, many field experiments have been carried out recently to test the effectiveness of rhizobia inoculation in combination with P application on legume crops (Table 4.1), which can curb the malnutrition of a large amount of population there. In northern Nigeria, rhizobial inoculation and P (single super phosphate) application in Luvisols increased the soybean yield by 447 and 452 kg/ha, respectively, while combination of rhizobial inoculation and P application increased it by 777 kg/ha (Ronner et al., 2016). Similar results were obtained in Ghana (Ulzen et al., 2018) and in Rwanda (Rurangwa et al., 2018). Rurangwa et al. (2018) indicated that the combination of rhizobia with application of P and manure greatly increased the yield of common bean and soybean by up to approximately twofold and threefold, respectively. More importantly, they demonstrated the residual effect on the subsequent crop maize, increasing the maize yield by up to approximately 6-fold and 2.5-fold following common bean and soybean, respectively. This study highlights the necessity to combine rhizobia inoculation with other technology, for example, P application and organic amendment in soils with poor fertility, and also suggests a systematic evaluation of rhizobia inoculation including the subsequent crop performance. In Zimbabwe, however, rhizobia inoculation did not seem to be

TABLE 4.1 Some experiments carried out under field conditions in Africa to examine the effects of rhizobia inoculation in combination with P application.

Legume–Rhizobia	Inoculation effect (I)	P application effect (P)	I × P interaction	Country	References
Soybean–*Bradyrhizobium japonicum* USDA 532c	Positive	Positive	Positive	Nigeria	Ronner et al. (2016)
Soybean–*Bradyrhizobium*	Positive	Positive	Positive	Ghana	Ulzen et al. (2018)
Common bean–*Rhizobium tropici* CIAT 899	Zero	Zero	Positive	Rwanda	Rurangwa et al. (2018)
Soybean–*B. japonicum* USDA 110	Zero	Zero	Positive	Rwanda	Rurangwa et al. (2018)
Chickpea–*Mesorhizobium* strains CP-41 and CP-029	Positive	Positive	Positive	Ethiopia	Wolde-meskel et al. (2018)
Soybean–*B. japonicum* SEMIA 5079	Positive	–	–	Mozambique	Chibeba et al. (2018)
Climbing bean–*Rhizobium* spp.	Positive	Zero	Zero	Tanzania	Mmbaga et al. (2015)
Common bean–*Rhizobium leguminosarum* biovar *phaseoli* HB-429 or *phaseoli* R.l.bv. GT-9	Positive	Positive	Positive	Ethiopia	Samago et al. (2018)
Soybean–*B. japonicum* SB6B1	Positive	Positive	Positive	Ethiopia	Fituma et al. (2018)
Soybean–*B. japonicum* FA3 and STM3043	Positive	Positive	Positive	Benin	Zoundji et al. (2015)
Common bean–*Rhizobium* spp.	Positive	Zero	Positive	Tanzania	Ndakidemi et al. (2006)
Soybean–*Rhizobium* spp.	Positive	Zero	Positive	Tanzania	Ndakidemi et al. (2006)

effective to increase common bean yield, compared with the application of N or P, either in degraded soil or nondegraded soil (Chekanai et al., 2018). Although the explanation for this ineffectiveness was not provided and needs further investigation, the interaction between plant genotype and rhizobia strain may be involved.

4.3.3 PGPRs

PGPRs can promote crop growth and production in various patterns, for example, promoting nutrient uptake, suppressing soil-borne diseases, and producing plant hormones. PSBs are a typical group of PGPRs that solubilize sparingly available P to increase soil P availability. Inoculation of six PSB strains (3 *Bacillus*, 1 *Serratia*, 1 *Arthrobacter*, 1 *Pantoea*) to rapeseed plants grown under field conditions enhanced the final yield by 21%–44% even without P application (Valetti et al., 2018). More frequently, some PGPRs possess multiple functions and thus promote crop production with more than one action mode. *Bacillus* M-3, *Burkholderia* OSU-7, and *Pseudomonas* BA-8 were capable of solubilizing P and producing indole acetic acid (IAA) and cytokinin (Aslantaş et al., 2007). Their inoculation to young apple trees by root dipping at planting increased the fruit yield by 73.7%, 88.2%, and 137.5%, respectively. This promotion can be attributed to both improved nutrient status and balanced plant hormones. *Pseudomonas fluorescens* N21.4 was able to release siderophores and chitinases in vitro and trigger defensive metabolism in plants (Garcia-Seco et al., 2015). Its inoculation to blackberry roots not only increased the plant growth and fruit number, but also promoted some secondary metabolites (total phenolics, flavonols, and epicatechins/catechins), which offer blackberry potential to human health.

Due to the high diversity of PGPRs and their practical application in agricultural systems, research on PGPRs has been accumulating progressively. Their functioning is also shifted to a broader scope, not only limited to increasing crop production, but also extended to enhancing the product quality and promoting functional secondary compounds. This shift can be of great significance in the research field of PGPRs, which meets more diverse demands on agricultural products from consumers.

4.3.4 Coinoculation of AM fungi, rhizobia, and PGPRs

Field conditions are complicated and very different from laboratory conditions and greenhouse conditions. It is common that several constraints limit crop growth, production, and product quality in agricultural systems. Since plant associated microbes are not always multifunctional, it is reasonable to coinoculate more than one group with different functions in practical application. According to the reported studies, it seems that dual inoculation of AM fungi, rhizobia, and PGPRs can be effective, and produces synergistic effects, as summarized in Fig. 4.1. In contrast, experiments with triple

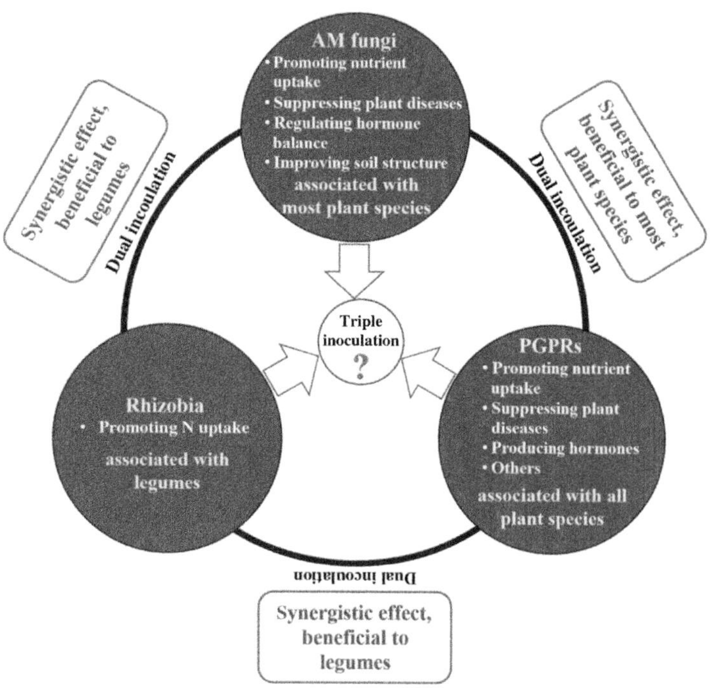

FIGURE 4.1 The main functions of AM fungi, rhizobia, and PGPRs, and possible synergistic effects of their dual inoculation.

inoculation are scarce (Requena et al., 1997), and whether it is necessary or more effective is yet to be investigated. If necessary, the interactions between three partners in the rhizosphere need in-depth investigation, and the fine-tuning of their respective functioning can contribute to the best application of this complex.

Rhizobia and PGPRs can provide additive benefits to legume crops even under field conditions. Hungria et al. (2015) reported that, in a field experiment (Londrina) in Brazil, dual inoculation with *Bradyrhizobium* spp. and *Azospirillum brasilense* significantly increased the soybean yield from 3360 to 3755 kg/ha, while inoculation with *Bradyrhizobium* spp. alone increased it to 3512 kg/ha. In this study, soybean plants of both dual inoculation and single inoculation received no chemical N fertilizer, however, the crop yield of dual inoculation was equivalent to that of chemical N fertilizer application (3760 kg/ha). In the other three field experiments (Ponta Grossa, Rio Verde, Cachoeira Dourada), they got similar results, indicating the potential to apply dual inoculation to sustainable agriculture. Pot experiment in net house revealed the potential application of cyanobacterium *Anabaena laxa* and *Trichoderma* sp. with rhizobia to produce synergistic effects on pea, chickpea, and lentil (Babu et al., 2015). Dual inoculation of *Schizolobium parahyba* var. *amazonicum* (Huber ex Ducke), a nonnodulation woody legume species, with AM fungi and PGPRs increased wood yield by about 20% (Cely et al., 2016). However, these dual inoculation strategies have been proven to be more efficient than single inoculation with a prerequisite, namely selection of appropriate combination for a given crop species (or a given crop variety).

4.4 Mechanisms involved in enhanced plant production

In agricultural systems, several biotic and abiotic environmental constraints represent common limiting factors for crop growth and production, such as soil fertility, pests, especially under stressed conditions. To cope with these stresses, plants recruit an appropriate subset of soil microbiome, establishing association with soil microbes in proper manner (Soussi et al., 2016; Vimal et al., 2017; Kwak et al., 2018), and in turn, these associated microbial groups promote host plants' performance via some common mechanisms.

4.4.1 Increased soil fertility via plant–microbe association

The most prominent feature of AM fungi is to greatly promote the uptake of nutrients by roots, especially those nutrients (e.g., P, Zn, Cu) readily immobilized in soils (Smith and Read, 2008). In other words, AM fungi increase soil fertility due to the enhanced bioavailability of these nutrients. P is one of the most important macronutrients in agricultural soils, and easily subjected to immobilization by Ca, Fe, Al, and thus much efforts have been focused on revealing how AM fungi increase P bioavailability (Bolan, 1991; Smith et al., 2011; Ferrol et al., 2019). The external hyphae of AM fungi function as root hairs and efficiently take up soil P via fungal P transporter with high affinity (Harrison and van Buuren, 1995; Olsson et al., 2006). This facilitation results in increased P uptake due to not only the different P transporter but also the enlarged soil volume for providing P, because external hyphae can extend far beyond the range of root hairs (Rhodes and Gerdemann, 1975). Moreover, AM fungi also produce and secrete phosphatase into the surrounding environments and promote the hydrolysis of soil organic P (Joner and Johansen, 2000; Sato et al., 2015). In the stressed conditions, root growth (e.g., root elongation) can be often severely inhibited, which is detrimental to nutrient uptake, however, external hyphae seems more tolerant than roots to these stresses, such as low pH (Wang et al., 2017), which highlights the advantages of AM fungi in nutrient uptake over roots.

Rhizobia in symbiosis with legumes enable plants to fix atmospheric N_2 to satisfy the N demand of symbionts, which is critical for sustainable agricultural systems. Rhizobia synthesize nitrogenase complex consisting of an Fe-protein (dinitrogenase reductase) and a MoFe-protein (dinitrogenase) (Schneider et al., 2019). Genes encoding nitrogenase are expressed only in the presence of host plants, namely, nodulation is required for the function of rhizoba (Spaink et al., 1987). Nodulation is strictly controlled by host plants for the best cost-to-benefit ratio, because nodulation represents a huge carbohydrate sink, consuming lots of photosynthates (Sachs et al., 2018). Subsequently, N_2-fixation activity is regulated by a series of environmental cues, such as N, P, drought, salinity, acidity, and other stressed conditions (Gordon et al., 1997; Hungria and Vargas, 2000; Isidra-Arellano et al., 2018).

The PSBs and potassium solubilizing bacteria are the typical PGPRs functioning via increasing soil fertility, although the latter have been less investigated than the former. PSBs solubilize sparingly soluble phosphate primarily via secreting organic acids, for example, gluconic acid. The action mode of solubilizing P by gluconic acid is the release of a large amount of protons rather than the chelation of metal cations (Illmer and Schinner, 1995; Lin et al., 2006). For example, *Burkholderia cepacia* CC-Al74, a high efficient PSB strain, increased soluble P from 0 to 200 μg/mL in parallel with a decrease in pH from 8.0 to 3.0 within 52 hours of incubation with tricalcium phosphate as sole

P source. The final concentration of gluconic acid reached 16.3 mM at pH 3.0, solubilizing 376 μg/mL P with the final proton concentration of 13.5 mM, in contrast to the solubilization of 35 μg/mL P by deionized water at pH 3.0 with the final proton concentration of 13.5 mM (Lin et al., 2006). Soil organic P represents a huge P pool for plants. However, it is available to plants only if it is hydrolyzed by phosphatase. Soil microbes can synthesize acidic and alkaline phosphatase, while plants can synthesize only acidic phosphatase (Spohn and Kuzyakov, 2013). Therefore, those PSBs with high activity of phosphatase can be applied to increase soil P availability (Malboobi et al., 2009). Potassium (K) is another macronutrient required by crops in large amount. Recently, K solubilizing bacteria has attracted much attention (Parmar and Sindhu, 2018; Sattar et al., 2018), and the mechanisms responsible for enhanced K availability are similar to those for P (Sattar et al., 2018), probably indicating the dual functions in solubilizing P and K for some PGPR strains.

4.4.2 Stimulated resistance to pathogens via plant–microbe associations

AM fungi have been proposed as biopesticide due to their facilitation in plant resistance to pathogens (Baum et al., 2015; Jacott et al., 2017). Several mechanisms are involved in the enhanced disease resistance by AM fungi. Zhu and Yao (2004) demonstrated that AM fungal inoculation significantly increased the contents of both free and cell wall bound phenolic substances, which contributed to resistance to tomato *Fusarium* wilt disease. This increase induced by AM fungi was both in systemic pattern and in local pattern signaled by H_2O_2, NO, and salicylic acid (Zhang et al., 2013; Zhu et al., 2015). For soil-borne diseases, the rhizosphere microbial community is a key factor controlling the pathogenesis. Ren et al. (2010) demonstrated that AM fungi inhibited *Fusarium* wilt in tomato via modifying rhizosphere microbial community, and thus enzyme activity. Moreover, improved P nutrition may function as a common trait employed by AM fungi to increase resistance to pathogens (Spagnoletti et al., 2018). Although most related experiments focused on the resistance to soil-borne diseases, a few studies targeted the resistance to airborne diseases, with intriguing results. Song et al. (2015) found that AM fungi significantly reduced early blight disease caused by *Alternaria solani* Sorauerthe in tomato. This reduction by AM fungi occurred only when plants were challenged with pathogens, along with the increased activities of some pathogenesis related protein, for example, β-1,3-glucanase, chitinase, phenylalanine ammonia-lyase, and lipoxygenase. However, contrasting results were reported in other experiments. Liu et al. (2018) indicated that AM fungi increased the susceptibility of standing milkvetch to powdery mildew even though plant growth was promoted.

Similar to AM fungi, PGPRs function to inhibit plant diseases with diverse mechanisms (Ramamoorthy et al., 2001; Beneduzi et al., 2012). PGPRs can produce and secrete hydrolytic enzymes to lyse pathogen cells. Naing et al. (2014) reported that a PGPR strain, *Paenibacillus ehimensis* KWN38, showed high antifungal activity against six tested fungal pathogens belonging to various taxonomic groups. Meanwhile this strain had high activities of chitinase, cellulase, glucanase, and protease, with the hyphal morphologies of fungal pathogens with abnormal bending, tip curling, and irregular branching. Some PGPRs with high activity of producing siderophores have a competitive advantage over pathogens, and thus exclude the potential pathogens from rhizosphere niches (Duijff et al., 1993). Antibiotics are among the other alternative tools employed by PGPRs to suppress pathogens. Prabhukarthikeyan et al. (2018) isolated a *P. fluorescens* strain FP7, harboring genes encoding for several antibiotic biosynthetic, phenazine, 2,4-diacetylphloroglucinol, pyrrolnitrin, pyoluteorin, and hydrogen cyanide. Field study demonstrated that application of this strain increased plant growth and yield in turmeric plants by reducing the incidence of rhizome rot disease. Other mechanisms such as induced resistance (both induced systemic resistance and systemic acquired resistance) have also been proposed (Beneduzi et al., 2012). In general, a particular PGPR strain can increase plant resistance via one or more mechanisms depending on environmental context.

4.4.3 Balanced plant hormones via plant–microbe associations

Plant hormones are essential to plant growth and development, which is mainly reflected in the balance of hormones. However, stressed conditions can destroy the balance and result in adverse effects on plant growth and development. Fortunately, most PGPRs are capable of producing or regulating hormones in varying degree, with the production of auxins the most investigated (Sarwar et al., 1992; Dodd et al., 2010). Moreover, these hormones function in not only controlling the growth and development of host plants but also in regulating the host–microbe interactions (Adolfsson et al., 2017).

It is well acknowledged that AM fungi can regulate the hormone levels in many plant species (Shaul-Keinan et al., 2002; Yao et al., 2005; Adolfsson et al., 2017). Inoculation of tobacco with AM fungi induced the accumulation of specific zeatin riboside (ZR)-like and isopentenyl adenosine (iPA)-like compounds in both roots and shoot, and gibberellin acids (GAs) biosynthetic pathway was significantly more abundant in roots, while IAA concentration was not altered (Shaul-Keinan et al., 2002). Yao et al. (2005) reported that AM fungi significantly increased the levels of IAA and iPAs in litchi, which was responsible for root morphological alteration induced by AM fungi. The dynamic of strigolactones in plants, emerging as novel hormone recently, are also regulated by AM fungi (Lanfranco et al., 2018). So far, there is no direct evidence that AM fungi can synthesize plant hormones, however, the synthesis of IAA-like compounds by ectomycorrhizal fungi has been confirmed (Strzelczyk et al., 1997). Even though AM fungi do not synthesize hormone, the growth promotion of plants can be achieved via the balanced hormone by AM fungi, which is normally referred to as "nonnutritional effect" of AM fungi (Delavaux et al., 2017). It is interesting that plant hormones can regulate AM fungal symbiosis (Liao et al., 2018). For example, auxins are key factor in promoting colonization and arbuscule formation (Etemadi et al., 2014; Ludwig-Müller and Güther, 2007), and abscisic acid (ABA) was also demonstrated to increase AM colonization and arbuscule formation with reduced arbuscule abundance in the ABA mutant (*Sitiens*) of tomato (Herrera-Medina et al., 2007). Some rhizobia are capable of producing IAA, which functions as symbiotic signaling as well as to promote plant growth (Alikhani et al., 2010; Puente et al., 2018; Buhian and Bensmihen, 2018). Garg et al. (2015) isolated 40 rhizobia strains from the root nodules of *Trifolium alexandrinum*, and all of them were capable of producing IAA. More importantly, overproduction of IAA in rhizobia can induce more efficient nodulation, and thus leading to higher efficient symbiotic effect (Camerini et al., 2008). Synthesizing auxins is one of the main mechanisms by which PGPRs promote plant growth (Tabassum et al., 2017; Etesami and Maheshwari, 2018). Plant roots can exudate L-tryptophan into the rhizosphere soil (Moe, 2013), while PGPRs subsequently transform this IAA precursor to IAA, and thus promote plant growth especially under stressed conditions (Goswami et al., 2015; Hassan and Bano, 2015; Yasmina et al., 2017).

4.5 Perspectives

4.5.1 Application of beneficial microbes in sustainable agriculture

Soil beneficial microbes are a ubiquitous biological resource, with enormous potential to promote crop growth and productivity. In conventional agricultural systems, crop production is characterized by high input and high production. This intensified agricultural pattern is not sustainable and easily subjected to pest outbreak, resulting in huge economic loss. According to the evidence aforementioned, soil beneficial microbes can be relied on to increase the soil fertility, enhance the pest resistance, promote the plant growth, or elevate the resistance to other biotic and abiotic stresses. In contrast to conventional agricultural systems, sustainable agricultural systems seek low input (mainly agrochemicals) and moderate production, where stressed conditions exist to a certain degree. Under these stressed conditions, PGPRs can play a crucial role in promoting crop production (Etesami and Maheshwari, 2018). Similarly for AMF, they can drastically promote crop growth under diverse stressed conditions, thus known as biofertilizers and biopesticides (Baum et al., 2015). On the other hand, high level of agrochemicals (e.g., phosphate, fungicide) can be inhibitory to AMF colonization and functioning. It is established that nodulation and N_2-fixation are normally reduced with high inorganic N (particularly nitrate) input (Peoples et al., 1995; Reid et al., 2011), and the rhizobia strains isolated from soil receiving high N level are not effective compared with those isolated from soil receiving low N level. Therefore, application of beneficial microbes and sustainable agriculture are highly compatible. Beneficial microbial technology is indispensable for sustainable agriculture, and in turn sustainable agriculture provides suitable soil environments for beneficial microbe functioning. In general, beneficial microbes are promising biological agents both in organic farming systems in developed countries and in low input agricultural systems in developing countries, particularly in Africa.

4.5.2 New insights into P exploration via plant–AM fungi–PSBs tripartite association

PSBs have been attracting much attention as an important PGPR group. When most studies concentrate on screening PSBs from rhizosphere niches, some researchers have recently shown great interest in those PSBs associated with AM fungi (Zhang et al., 2016, 2018). It is proposed that PSBs adhering to the external hyphae of AM fungi contribute much to the increased P mobilization by AM fungi. Zhang et al. (2016, 2018) revealed that AM fungi and PSBs establish a firm cooperation relationship in mobilizing P via C (fructose) and P exchange between AM fungi and PSBs, where fructose exudated from AM fungi functioned not only as C source but also as a signal triggering the expression of phosphatase gene and the secretion of this enzyme (Zhang et al., 2018). Since both AM fungi and PSBs are ubiquitous in agricultural soils, it is reasonable to expect their cooperative effects on P uptake by crops. Therefore, not only the

PSBs in rhizosphere but also the PSBs on AM fungal hyphae should be taken into consideration when isolating PSBs from environments, especially in the sustainable agricultural systems with AM technology. The tripartite association of plant–AM fungi–PSB can be expected in terms of dual inoculation.

Acknowledgments

The authors conceived this chapter with support from the Natural Science Foundation of China (31570395); Guangdong Academy of Science (2017GDASCX-0402, 2019GDASYL-0103012); and Guangdong Science and Technology Program (2018B020205001, 2016A020210071).

References

Adolfsson, L., Nziengui, H., Abreu, I.N., Šimura, J., Beebo, A., Herdean, A., et al., 2017. Enhanced secondary- and hormone metabolism in leaves of arbuscular mycorrhizal *Medicago truncatula*. Plant Physiol. 175, 392–411.

Affokpon, A., Coyne, D.L., Lawouin, L., Tossou, C., Agbèdè, R.D., Coosemans, J., 2011. Effectiveness of native West African arbuscular mycorrhizal fungi in protecting vegetable crops against root-knot nematodes. Biol. Fert. Soils 47 (2), 207–217.

Alikhani, H.A., Etesami, H., Mohammadi, L., 2010. Effect of superior indole-3-acetic acid producing rhizobia and the combination with Ag and L-tryptophan on wheat growth indices under in vitro conditions. J. Food Agric. Environ. 8 (3-4), 949–954.

Andrews, M., De Meyer, S., James, E., Stępkowski, T., Hodge, S., Simon, M., et al., 2018. Horizontal transfer of symbiosis genes within and between rhizobial genera: occurrence and importance. Genes 9 (7), 321.

Anglade, J., Billen, G., Garnier, J., 2015. Relationships for estimating N2 fixation in legumes: incidence for N balance of legume-based cropping systems in Europe. Ecosphere 6 (3), 1–24.

Aslantaş, R., Cakmakçi, R., Şahin, F., 2007. Effect of plant growth promoting rhizobacteria on young apple tree growth and fruit yield under orchard conditions. Sci. Hortic. 111 (4), 371–377.

Babu, S., Prasanna, R., Bidyarani, N., Nain, L., Shivay, Y.S., 2015. Synergistic action of PGP agents and *Rhizobium* spp. for improved plant growth, nutrient mobilization and yields in different leguminous crops. Biocatal. Agric. Biotechnol. 4 (4), 456–464.

Badri, D.V., Vivanco, J.M., 2009. Regulation and function of root exudates. Plant Cell Environ. 32, 666–681.

Barthès, A., Ten-Hage, L., Lamy, A., Rols, J.L., Leflaive, J., 2015. Resilience of aggregated microbial communities subjected to drought: small-scale studies. Microb. Ecol. 70 (1), 9–20.

Baum, C., El-Tohamy, W., Gruda, N., 2015. Increasing the productivity and product quality of vegetable crops using arbuscular mycorrhizal fungi: a review. Sci. Hortic. 187, 131–141.

Beneduzi, A., Ambrosini, A., Passaglia, L.M., 2012. Plant growth-promoting rhizobacteria (PGPR): their potential as antagonists and biocontrol agents. Genet. Mol. Biol. 35 (4), 1044–1051.

Berg, G., Grube, M., Schloter, M., Smalla, K., 2014. Unraveling the plant microbiome: looking back and future perspectives. Front. Microbiol. 5, 148.

Berruti, A., Lumini, E., Balestrini, R., Bianciotto, V., 2016. Arbuscular mycorrhizal fungi as natural biofertilizers: let's benefit from past successes. Front. Microbiol. 6, 1559.

Bidyarani, N., Prasanna, R., Babu, S., Hossain, F., Saxena, A.K., 2016. Enhancement of plant growth and yields in chickpea (*Cicer arietinum* L.) through novel cyanobacterial and biofilmed inoculants. Microbiol. Res. 188, 97–105.

Bolan, N.S., 1991. A critical review on the role of mycorrhizal fungi in the uptake of phosphorus by plants. Plant Soil 134 (2), 189–207.

Bona, E., Cantamessa, S., Massa, N., Manassero, P., Marsano, F., Copetta, A., et al., 2017. Arbuscular mycorrhizal fungi and plant growth-promoting pseudomonads improve yield, quality and nutritional value of tomato: a field study. Mycorrhiza 27 (1), 1–11.

Buhian, W.P., Bensmihen, S., 2018. Nod factor regulation of phytohormone signaling and homeostasis during rhizobia-legume symbiosis. Front. Plant Sci. 9, 1247.

Camerini, S., Senatore, B., Lonardo, E., Imperlini, E., Bianco, C., Moschetti, G., et al., 2008. Introduction of a novel pathway for IAA biosynthesis to rhizobia alters vetch root nodule development. Arch. Microbiol. 190, 67–77.

Carballar-Hernández, S., Hernández-Cuevas, L.V., Montaño, N.M., Ferrera-Cerrato, R., Alarcón, A., 2018. Species composition of native arbuscular mycorrhizal fungal consortia influences growth and nutrition of poblano pepper plants (*Capsicum annuum* L.). Appl. Soil Ecol. 130, 50–58.

Cely, M.V., Siviero, M.A., Emiliano, J., Spago, F.R., Freitas, V.F., Barazetti, A.R., et al., 2016. Inoculation of *Schizolobium parahyba* with mycorrhizal fungi and plant growth-promoting rhizobacteria increases wood yield under field conditions. Front. Plant Sci. 7, 1708.

Chaparro, J.M., Sheflin, A.M., Manter, D.K., Vivanco, J.M., 2012. Manipulating the soil microbiome to increase soil health and plant fertility. Biol. Fert. Soils 48 (5), 489–499.

Chekanai, V., Chikowo, R., Vanlauwe, B., 2018. Response of common bean (*Phaseolus vulgaris* L.) to nitrogen, phosphorus and rhizobia inoculation across variable soils in Zimbabwe. Agr. Ecosyst. Environ. 266, 167–173.

Chibeba, A.M., Kyei-Boahen, S., de Fátima Guimarães, M., Nogueira, M.A., Hungria, M., 2018. Feasibility of transference of inoculation-related technologies: a case study of evaluation of soybean rhizobial strains under the agro-climatic conditions of Brazil and Mozambique. Agr. Ecosyst. Environ. 261, 230–240.

Cline, L.C., Zak, D.R., 2015. Soil microbial communities are shaped by plant-driven changes in resource availability during secondary succession. Ecology 96 (12), 3374–3385.

Delavaux, C.S., Smith-Ramesh, L.M., Kuebbing, S.E., 2017. Beyond nutrients: a meta-analysis of the diverse effects of arbuscular mycorrhizal fungi on plants and soils. Ecology 98, 2111–2119.

Denton, M.D., Phillips, L.A., Peoples, M.B., Pearce, D.J., Swan, A.D., Mele, P.M., et al., 2017. Legume inoculant application methods: effects on nodulation patterns, nitrogen fixation, crop growth and yield in narrow-leaf lupin and faba bean. Plant Soil 419 (1–2), 25–39.

Dodd, I.C., Zinovkina, N.Y., Safronova, V.I., Belimov, A.A., 2010. Rhizobacterial mediation of plant hormone status. Ann. Appl. Biol. 157 (3), 361–379.

Duijff, B.J., Meijer, J.W., Bakker, P.A., Schippers, B., 1993. Siderophore-mediated competition for iron and induced resistance in the suppression of *Fusarium* wilt of carnation by fluorescent *Pseudomonas* spp. Neth. J. Plant Pathol. 99 (5–6), 277–289.

Etemadi, M., Gutjahr, C., Couzigou, J.M., Zouine, M., Lauressergues, D., Timmers, A., et al., 2014. Auxin perception is required for arbuscule development in arbuscular mycorrhizal symbiosis. Plant Physiol. 166, 281–292.

Etesami, H., Maheshwari, D.K., 2018. Use of plant growth promoting rhizobacteria (PGPRs) with multiple plant growth promoting traits in stress agriculture: action mechanisms and future prospects. Ecotox. Environ. Saf. 156, 225–246.

Ferrol, N., Azcón-Aguilar, C., Pérez-Tienda, J., 2019. Arbuscular mycorrhizas as key players in sustainable plant phosphorus acquisition: an overview on the mechanisms involved. Plant Sci. Available from: https://doi.org/10.1016/j.plantsci.2018.11.011.

Fituma, T., Tana, T., Alemneh, A.A., 2018. Response of soybean to *Bradyrhizobium japonicum* inoculation and phosphorus application under intercropping in the Central Rift Valley of Ethiopia. S. Afr. J. Plant Soil 35, 33–40.

Foyer, C.H., Lam, H.M., Nguyen, H.T., Siddique, K.H., Varshney, R.K., Colmer, T.D., et al., 2016. Neglecting legumes has compromised human health and sustainable food production. Nat. Plants 2 (8), 16112.

Garcia-Seco, D., Zhang, Y., Gutierrez-Mañero, F.J., Martin, C., Ramos-Solano, B., 2015. Application of *Pseudomonas fluorescens* to blackberry under field conditions improves fruit quality by modifying flavonoid metabolism. PLoS One 10 (11), e0142639.

Garg, V., Kukreja, K., Gera, R., Singla, A., 2015. Production of indole-3-acetic acid by berseem (*Trifolium alexandrinum* L.) rhizobia isolated from Haryana, India. Agric. Sci. Dig. 35, 229–232.

George, E., Marschner, H., Jakobsen, I., 1995. Role of arbuscular mycorrhizal fungi in uptake of phosphorus and nitrogen from soil. Crit. Rev. Biotechnol. 15 (3-4), 257–270.

Gianinazzi, S., Vosátka, M., 2004. Inoculum of arbuscular mycorrhizal fungi for production systems: science meets business. Can. J. Bot. 82 (8), 1264–1271.

Gordon, A.J., Minchin, F.R., Skot, L., James, C.L., 1997. Stress-induced declines in soybean N_2 fixation are related to nodule sucrose synthase activity. Plant Physiol. 114 (3), 937–946.

Goswami, D., Thakker, J.N., Dhandhukia, P.C., 2015. Simultaneous detection and quantification of indole-3-acetic acid (IAA) and indole-3-butyric acid (IBA) produced by rhizobacteria from l-tryptophan (Trp) using HPTLC. J. Microbiol. Methods 110, 7–14.

Harrison, M.J., van Buuren, M.L., 1995. A phosphate transporter from the mycorrhizal fungus *Glomus versiforme*. Nature 378 (6557), 626.

Hassan, T.U., Bano, A., 2015. The stimulatory effects of L-tryptophan and plant growth promoting rhizobacteria (PGPR) on soil health and physiology of wheat. J. Soil Sci. Plant Nutr. 15 (1), 190–201.

Hayat, R., Ali, S., Amara, U., Khalid, R., Ahmed, I., 2010. Soil beneficial bacteria and their role in plant growth promotion: a review. Ann. Microbiol. 60 (4), 579–598.

Herrera-Medina, M.J., Steinkellner, S., Vierheilig, H., Ocampo Bote, J.A., García Garrido, J.M., 2007. Abscisic acid determines arbuscule development and functionality in the tomato arbuscular mycorrhiza. New Phytol. 175, 554–564.

Herridge, D.F., Peoples, M.B., Boddey, R.M., 2008. Global inputs of biological nitrogen fixation in agricultural systems. Plant Soil 311 (1–2), 1–18.

Hungria, M., Vargas, M.A., 2000. Environmental factors affecting N_2 fixation in grain legumes in the tropics, with an emphasis on Brazil. Field Crop. Res. 65 (2–3), 151–164.

Hungria, M., Nogueira, M.A., Araujo, R.S., 2015. Soybean seed coinoculation with *Bradyrhizobium* spp. and *Azospirillum brasilense*: a new biotechnological tool to improve yield and sustainability. AJPS. 6 (06), 811.

Illmer, P., Schinner, F., 1995. Solubilization of inorganic calcium phosphates: solubilization mechanisms. Soil Biol. Biochem. 27 (3), 257–263.

Isidra-Arellano, M., Reyero-Saavedra, M., Sánchez-Correa, M., Pingault, L., Sen, S., Joshi, T., et al., 2018. Phosphate deficiency negatively affects early steps of the symbiosis between common bean and rhizobia. Genes 9 (10), 498.

Israr, D., Mustafa, G., Khan, K.S., Shahzad, M., Ahmad, N., Masood, S., 2016. Interactive effects of phosphorus and *Pseudomonas putida* on chickpea (*Cicer arietinum* L.) growth, nutrient uptake, antioxidant enzymes and organic acids exudation. Plant Physiol. Biochem. 108, 304–312.

Jacott, C., Murray, J., Ridout, C., 2017. Trade-offs in arbuscular mycorrhizal symbiosis: disease resistance, growth responses and perspectives for crop breeding. Agronomy 7 (4), 75.

Joner, E.J., Johansen, A., 2000. Phosphatase activity of external hyphae of two arbuscular mycorrhizal fungi. Mycol. Res. 104 (1), 81–86.

Kennedy, A.C., Smith, K.L., 1995. Soil microbial diversity and the sustainability of agricultural soils. Plant Soil 170 (1), 75–86.

Khan, M.S., Zaidi, A., Ahemad, M., Oves, M., Wani, P.A., 2010. Plant growth promotion by phosphate solubilizing fungi: current perspective. Arch. Agron. Soil Sci. 56 (1), 73–98.

Klaus, S., Bulgarelli, D., 2015. The plant microbiome at work. Mol. Plant Microbe Interact. 28 (3), 212–217.

Kumar, A., Kaushal, S., Saraf, S.A., Singh, J.S., 2018. Microbial bio-fuels: a solution to carbon emissions and energy crisis. Front. Biosci. (Landmark Ed) 1 (23), 1789–1802.

Kwak, M.J., Kong, H.G., Choi, K., Kwon, S.K., Song, J.Y., Lee, J., et al., 2018. Rhizosphere microbiome structure alters to enable wilt resistance in tomato. Nat. Biotechnol. 36 (11), 1100.

Lanfranco, L., Fiorilli, V., Venice, F., Bonfante, P., 2018. Strigolactones cross the kindoms: plants, fungi, and bacteria in the arbuscular mycorrhizal symbiosis. J. Exp. Bot. 69, 2175–2188.

Lebeis, S.L., Rott, M., Dangl, J.L., Schulze-Lefert, P., 2012. Culturing a plant microbiome community at the cross-rhodes. New Phytol. 196 (2), 341–344.

Leff, J.W., Bardgett, R.D., Wilkinson, A., Jackson, B.G., Pritchard, W.J., Long, J.R., et al., 2019. Predicting the structure of soil communities from plant community taxonomy, phylogeny, and traits. ISME J. Available from: https://doi.org/10.1038/s41396-018-0089-x.

Li, J., Sun, Y., Jiang, X., Chen, B., Zhang, X., 2018. Arbuscular mycorrhizal fungi alleviate arsenic toxicity to *Medicago sativa* by influencing arsenic speciation and partitioning. Ecotox. Environ. Saf. 157, 235–243.

Liao, D., Wang, S., Cui, M., Liu, J., Chen, A., Xu, G., 2018. Phytohormones regulate the development of arbuscular mycorrhizal symbiosis. Int. J. Mol. Sci. 19 (10), 3146.

Lin, T.F., Huang, H.I., Shen, F.T., Young, C.C., 2006. The protons of gluconic acid are the major factor responsible for the dissolution of tricalcium phosphate by *Burkholderia cepacia* CC-Al74. Bioresour. Technol. 97 (7), 957–960.

Liu, Y., Feng, X., Gao, P., Li, Y., Christensen, M.J., Duan, T., 2018. Arbuscular mycorrhiza fungi increased the susceptibility of *Astragalus adsurgens* to powdery mildew caused by *Erysiphe pisi*. Mycology 1–10.

Ludwig-Müller, J., Güther, M., 2007. Auxins as signals in arbuscular mycorrhiza formation. Plant Signal. Behav. 2, 194–196.

Luginbuehl, L.H., Oldroyd, G.E., 2017. Understanding the arbuscule at the heart of endomycorrhizal symbioses in plants. Curr. Biol. 27 (17), R952–R963.

Malboobi, M.A., Owlia, P., Behbahani, M., Sarokhani, E., Moradi, S., Yakhchali, B., et al., 2009. Solubilization of organic and inorganic phosphates by three highly efficient soil bacterial isolates. World J. Microbiol. Biotechnol. 25 (8), 1471–1477.

Mmbaga, G.W., Mtei, K.M., Ndakidemi, P.A., 2015. Yield and fiscal benefits of rhizobium inoculation supplemented with phosphorus (P) and potassium (K) in climbing beans (*Phaseolus vulgaris* L.) grown in northern Tanzania. Agri. Sci. 6 (08), 783.

Moe, L.A., 2013. Amino acids in the rhizosphere: from plants to microbes. Am. J. Bot. 100 (9), 1692–1705.

Mosse, B., 1957. Growth and chemical composition of mycorrhizal and non-mycorrhizal apples. Nature 179, 922–924.

Mulas, D., Seco, V., Casquero, P.A., Velázquez, E., González-Andrés, F., 2015. Inoculation with indigenous *Rhizobium* strains increases yields of common bean (*Phaseolus vulgaris* L.) in northern Spain, although its efficiency is affected by the tillage system. Symbiosis 67 (1–3), 113–124.

Müller, D.B., Vogel, C., Bai, Y., Vorholt, J.A., 2016. The plant microbiota: systems-level insights and perspectives. Annu. Rev. Genet. 50, 211–234.

Naing, K.W., Anees, M., Kim, S.J., Nam, Y., Kim, Y.C., Kim, K.Y., 2014. Characterization of antifungal activity of *Paenibacillus ehimensis* KWN38 against soilborne phytopathogenic fungi belonging to various taxonomic groups. Ann. Microbiol. 64 (1), 55–63.

Ndakidemi, P.A., Dakora, F.D., Nkonya, E.M., Ringo, D., Mansoor, H., 2006. Yield and economic benefits of common bean (*Phaseolus vulgaris*) and soybean (*Glycine max*) innoculation in northern Tanzania. Aust. J. Exp. Agric. 46, 571–577.

Oliveira, D.F.B.D., Endres, L., Barros, J.M.T.D.M., Santos, A.F.D.S., Gonzaga, E.P., 2017. Association and mycorrhizal dependency in *Jatropha curcas* L. seedlings under salt stress. Rev. Ceres 64 (6), 592–599.

Olsson, P.A., Hansson, M.C., Burleigh, S.H., 2006. Effect of P availability on temporal dynamics of carbon allocation and *Glomus intraradices* high-affinity P transporter gene induction in arbuscular mycorrhiza. Appl. Environ. Microbiol. 72 (6), 4115–4120.

Panja, B.N., Chaudhuri, S., 2004. Exploitation of soil arbuscular mycorrhizal potential for AM-dependent mandarin orange plants by pre-cropping with mycotrophic crops. Appl. Soil Ecol. 26 (3), 249–255.

Parmar, P., Sindhu, S.S., 2018. The novel and efficient method for isolating potassium solubilizing bacteria from rhizosphere soil. Geomicrobiol. J. 1–7.

Peoples, M.B., Herridge, D.F., Ladha, J.K., 1995. Biological nitrogen-fixation: an efficient source of nitrogen for sustainable agricultural production. Plant Soil 174, 3–28.

Prabhukarthikeyan, S.R., Keerthana, U., Raguchander, T., 2018. Antibiotic-producing *Pseudomonas fluorescens* mediates rhizome rot disease resistance and promotes plant growth in turmeric plants. Microbiol. Res. 210, 65–73.

Puente, M.L., Gualpa, J.L., Lopez, G.A., Molina, R.M., Carletti, S.M., Cassán, F.D., 2018. The benefits of foliar inoculation with *Azospirillum brasilense* in soybean are explained by an auxin signaling model. Symbiosis 76 (1), 41–49.

Ramamoorthy, V., Viswanathan, R., Raguchander, T., Prakasam, V., Samiyappan, R., 2001. Induction of systemic resistance by plant growth promoting rhizobacteria in crop plants against pests and diseases. Crop Prot. 20 (1), 1–11.

Reid, D.E., Ferguson, B.J., Gresshoff, P.M., 2011. Inoculation- and nitrate-induced CLE peptides of soybean control NARK-dependent nodule formation. Mol. Plant Microbe Interact. 24, 606–618.

Ren, L., Lou, Y., Sakamoto, K., Inubushi, K., Amemiya, Y., Shen, Q., et al., 2010. Effects of arbuscular mycorrhizal colonization on microbial community in rhizosphere soil and *Fusarium* wilt disease in tomato. Commun. Soil Sci. Plant Anal. 41 (11), 1399–1410.

Requena, N., Jimenez, I., Toro, M., Barea, J.M., 1997. Interactions between plant-growth-promoting rhizobacteria (PGPR), arbuscular mycorrhizal fungi and *Rhizobium* spp. in the rhizosphere of *Anthyllis cytisoides*, a model legume for revegetation in mediterranean semi-arid ecosystems. New Phytol. 136 (4), 667–677.

Rhodes, L.H., Gerdemann, J.W., 1975. Phosphate uptake zones of mycorrhizal and nonmycorrhizal onions. New Phytol. 75 (3), 555–561.

Ridl, J., Kolar, M., Strejcek, M., Strnad, H., Stursa, P., Paces, J., et al., 2016. Plants rather than mineral fertilization shape microbial community structure and functional potential in legacy contaminated soil. Front. Microbiol. 7, 995.

Ronner, E., Franke, A.C., Vanlauwe, B., Dianda, M., Edeh, E., Ukem, B., et al., 2016. Understanding variability in soybean yield and response to P-fertilizer and rhizobium inoculants on farmers' fields in northern Nigeria. Field Crop. Res. 186, 133–145.

Ruiz-Lozano, J.M., Aroca, R., Zamarreño, Á.M., Molina, S., Andreo-Jiménez, B., Porcel, R., et al., 2016. Arbuscular mycorrhizal symbiosis induces strigolactone biosynthesis under drought and improves drought tolerance in lettuce and tomato. Plant Cell Environ. 39 (2), 441–452.

Rurangwa, E., Vanlauwe, B., Giller, K.E., 2018. Benefits of inoculation, P fertilizer and manure on yields of common bean and soybean also increase yield of subsequent maize. Agr. Ecosyst. Environ. 261, 219–229.

Sachs, J.L., Quides, K.W., Wendlandt, C.E., 2018. Legumes versus rhizobia: a model for ongoing conflict in symbiosis. New Phytol. 219 (4), 1199–1206.

Saleem, A.R., Brunetti, C., Khalid, A., Della Rocca, G., Raio, A., Emiliani, G., et al., 2018. Drought response of *Mucuna pruriens* (L.) DC. inoculated with ACC deaminase and IAA producing rhizobacteria. PLoS One 13 (2), e0191218.

Samago, T.Y., Anniye, E.W., Dakora, F.D., 2018. Grain yield of common bean (*Phaseolus vulgaris* L.) varieties is markedly increased by rhizobial inoculation and phosphorus application in Ethiopia. Symbiosis 75, 245–255.

Sarwar, M., Arshad, M., Martens, D.A., Frankenberger, W.T., 1992. Tryptophan-dependent biosynthesis of auxins in soil. Plant Soil 147 (2), 207–215.

Sasse, J., Martinoia, E., Northen, T., 2018. Feed your friends: do plant exudates shape the root microbiome? Trends Plant Sci. 23 (1), 25–41.

Sato, T., Ezawa, T., Cheng, W., Tawaraya, K., 2015. Release of acid phosphatase from extraradical hyphae of arbuscular mycorrhizal fungus *Rhizophagus clarus*. Soil Sci. Plant Nutr. 61 (2), 269–274.

Sattar, A., Naveed, M., Ali, M., Zahir, Z.A., Nadeem, S.M., Yaseen, M., et al., 2018. Perspectives of potassium solubilizing microbes in sustainable food production system: a review. Appl. Soil Ecol. 133, 146–159.

Schneider, S., Schintlmeister, A., Becana, M., Wagner, M., Woebken, D., Wienkoop, S., 2019. Sulfate is transported at significant rates through the symbiosome membrane and is crucial for nitrogenase biosynthesis. Plant Cell Environ. Available from: https://doi.org/10.1111/pce.13481.

Shaul-Keinan, O., Gadkar, V., Ginzberg, I., Grünzweig, J.M., Chet, I., Elad, Y., et al., 2002. Hormone concentrations in tobacco roots change during arbuscular mycorrhizal colonization with *Glomus intraradices*. New Phytol. 154, 501–507.

Siddiqui, Z.A., 2005. PGPR: prospective biocontrol agents of plant pathogens. In: Z.A. Siddiqui (Ed.), PGPR: Biocontrol and Biofertilization. Springer International Publishing, Dordrecht, pp. 111–142.

Singh, J.S., 2013. Plant growth promoting rhizobacteria: potential microbes for sustainable agriculture. Resonance 18 (3), 275–281.

Singh, J.S., 2015. Microbes: the chief ecological engineers in reinstating equilibrium in degraded ecosystems. Agric. Ecosyst. Environ. 203, 80–82.

Singh, J.S., Gupta, V.K., 2018. Soil microbial biomass: a key soil driver in management of ecosystem functioning. Sci. Total Environ. 634, 497–500.

Singh, C., Tiwari, S., Gupta, V.K., Singh, J.S., 2018. The effect of rice husk biochar on soil nutrient status, microbial biomass and paddy productivity of nutrient poor agriculture soils. Catena 171, 485–493.

Sivaram, A.K., Logeshwaran, P., Subashchandrabose, S.R., Lockington, R., Naidu, R., Megharaj, M., 2018. Comparison of plants with C_3 and C_4 carbon fixation pathways for remediation of polycyclic aromatic hydrocarbon contaminated soils. Sci. Rep. 8 (1), 2100.

Smith, S.E., Read, D., 2008. Mycorrhizal Symbiosis, third ed. Academic Press, San Diego, CA.

Smith, S.E., Jakobsen, I., Grønlund, M., Smith, F.A., 2011. Roles of arbuscular mycorrhizas in plant phosphorus (P) nutrition: interactions between pathways of P uptake in arbuscular mycorrhizal (AM) roots have important implications for understanding and manipulating plant P acquisition. Plant Physiol. 156, 1050–1057.

Song, Y., Chen, D., Lu, K., Sun, Z., Zeng, R., 2015. Enhanced tomato disease resistance primed by arbuscular mycorrhizal fungus. Front. Plant Sci. 6, 786.

Soussi, A., Ferjani, R., Marasco, R., Guesmi, A., Cherif, H., Rolli, E., et al., 2016. Plant-associated microbiomes in arid lands: diversity, ecology and biotechnological potential. Plant Soil 405 (1–2), 357–370.

Spagnoletti, F.N., Leiva, M., Chiocchio, V., Lavado, R.S., 2018. Phosphorus fertilization reduces the severity of charcoal rot (*Macrophomina phaseolina*) and the arbuscular mycorrhizal protection in soybean. J. Plant Nutr. Soil Sci. 181 (6), 855–860.

Spaink, H.P., Wijffelman, C.A., Pees, E., Okker, R.J., Lugtenberg, B.J.J., 1987. *Rhizobium* nodulation gene *nodD* as a determinant of host specificity. Nature 328 (6128), 337.

Spohn, M., Kuzyakov, Y., 2013. Distribution of microbial-and root-derived phosphatase activities in the rhizosphere depending on P availability and C allocation: coupling soil zymography with ^{14}C imaging. Soil Biol. Biochem. 67, 106–113.

Strzelczyk, E., Dahm, H., Pachlewski, R., Rozycki, H., 1997. Production of indole compounds by the ectomycorrhizal fungus *Catharellus cibarius* Fr. Pedobiologia 41, 402–411.

Swarnalakshmi, K., Prasanna, R., Kumar, A., Pattnaik, S., Chakravarty, K., Shivay, Y.S., et al., 2013. Evaluating the influence of novel cyanobacterial biofilmed biofertilizers on soil fertility and plant nutrition in wheat. Eur. J. Soil Biol. 55, 107–116.

Tabassum, B., Khan, A., Tariq, M., Ramzan, M., Khan, M.S.I., Shahid, N., et al., 2017. Bottlenecks in commercialisation and future prospects of PGPR. Appl. Soil Ecol. 121, 102–117.

Taktek, S., Trépanier, M., Servin, P.M., St-Arnaud, M., Piché, Y., Fortin, J.A., et al., 2015. Trapping of phosphate solubilizing bacteria on hyphae of the arbuscular mycorrhizal fungus *Rhizophagus irregularis* DAOM 197198. Soil Biol. Biochem. 90, 1–9.

Thilakarathna, M.S., Raizada, M.N., 2017. A meta-analysis of the effectiveness of diverse rhizobia inoculants on soybean traits under field conditions. Soil Biol. Biochem. 105, 177–196.

Tiwari, P., Singh, J.S., 2017. A plant growth promoting rhizospheric *Pseudomonas aeruginosa* strain inhibits seed germination in *Triticum aestivum* (L) and *Zea mays* (L). Microbiol. Res. 8, 1–7.

Tiwari, S., Singh, C., Singh, J.S., 2018. Land use changes: a key ecological driver regulating methanotrophs abundance in upland soils. Energy Ecol. Environ. 3, 355–371.

Tsukanova, K.A., Meyer, J.J.M., Bibikova, T.N., 2017. Effect of plant growth-promoting rhizobacteria on plant hormone homeostasis. S. Afr. J. Bot. 113, 91–102.

Turner, T.R., James, E.K., Poole, P.S., 2013. The plant microbiome. Genome Biol. 14 (6), 209.

Ulzen, J., Abaidoo, R.C., Ewusi-Mensah, N., Masso, C., 2018. On-farm evaluation and determination of sources of variability of soybean response to *Bradyrhizobium* inoculation and phosphorus fertilizer in northern Ghana. Agric. Ecosyst. Environ. 267, 23–32.

Valetti, L., Iriarte, L., Fabra, A., 2018. Growth promotion of rapeseed (*Brassica napus*) associated with the inoculation of phosphate solubilizing bacteria. Appl. Soil Ecol. 132, 1–10.

Van Wees, S.C., Van der Ent, S., Pieterse, C.M., 2008. Plant immune responses triggered by beneficial microbes. Curr. Opin. Plant Biol. 11 (4), 443–448.

Vessey, J.K., 2003. Plant growth promoting rhizobacteria as biofertilizers. Plant Soil 255 (2), 571–586.

Vimal, S.R., Singh, J.S., Arora, N.K., Singh, S., 2017. Soil-plant–microbe interactions in stressed agriculture management: a review. Pedosphere 27 (2), 177–192.

Vimal, S.R., Patel, V.K., Singh, J.S., 2018. Plant growth promoting *Curtobacterium albidum* strain SRV4: an agriculturally important microbe to alleviate salinity stress in paddy plants. Ecol. Indic. (Article in press).

Wagner, M.R., Lundberg, D.S., Tijana, G., Tringe, S.G., Dangl, J.L., Mitchell-Olds, T., 2016. Host genotype and age shape the leaf and root microbiomes of a wild perennial plant. Nat. Commun. 7, 12151.

Wang, H.X., Geng, Z.L., Zeng, Y., Shen, Y.M., 2008. Enriching plant microbiota for a metagenomic library construction. Environ. Microbiol. 10 (10), 2684–2691.

Wang, P., Zhang, J.J., Shu, B., Xia, R.X., 2012. Arbuscular mycorrhizal fungi associated with citrus orchards under different types of soil management, southern China. Plant Soil Environ. 58 (7), 302–308.

Wang, N., Feng, Z., Zhou, Y., Zhu, H., Yao, Q., 2017. External hyphae of *Rhizophagus irregularis* DAOM 197198 are less sensitive to low pH than roots in arbuscular mycorrhizae: evidence from axenic culture system. Environ. Microbiol. Rep. 9 (5), 649–657.

Wiebe, K., Lotze-Campen, H., Sands, R., Tabeau, A., van der Mensbrugghe, D., Biewald, A., et al., 2015. Climate change impacts on agriculture in 2050 under a range of plausible socioeconomic and emissions scenarios. Environ. Res. Lett. 10 (8), 085010.

Wolde-meskel, E., van Heerwaarden, J., Abdulkadir, B., Kassa, S., Aliyi, I., Degefu, T., et al., 2018. Additive yield response of chickpea (*Cicer arietinum* L.) to rhizobium inoculation and phosphorus fertilizer across smallholder farms in Ethiopia. Agric. Ecosyst. Environ. 261, 144–152.

Yang, J., Xie, X., Yang, M., Dixon, R., Wang, Y.P., 2017. Modular electron-transport chains from eukaryotic organelles function to support nitrogenase activity. Proc. Natl. Acad. Sci. USA. 114 (12), E2460–E2465.

Yang, J., Xie, X., Xiang, N., Tian, Z.X., Dixon, R., Wang, Y.P., 2018. Polyprotein strategy for stoichiometric assembly of nitrogen fixation components for synthetic biology. Proc. Natl. Acad. Sci. USA. 115 (36), E8509–E8517.

Yanni, Y., Zidan, M., Dazzo, F., Rizk, R., Mehesen, A., Abdelfattah, F., et al., 2016. Enhanced symbiotic performance and productivity of drought stressed common bean after inoculation with tolerant native rhizobia in extensive fields. Agric. Ecosyst. Environ. 232, 119–128.

Yao, Q., Zhu, H.H., Chen, J.Z., 2005. Growth responses and endogenous IAA and iPAs changes of litchi (*Litchi chinensis* Sonn.) seedlings induced by arbuscular mycorrhizal fungal inoculation. Sci. Hortic. 105, 145–151.

Yasmina, H., Nosheena, A., Naza, R., Banob, A., Keyani, R., 2017. L-tryptophan-assisted PGPR-mediated induction of drought tolerance in maize (*Zea mays* L.). J. Plant Interact. 12, 567–578.

Zahir, Z.A., Arshad, M., Frankenberger, W.T., 2004. Plant growth promoting rhizobacteria: applications and perspectives in agriculture. Adv. Agron. 81, 98–169.

Zhang, R.Q., Zhu, H.H., Zhao, H.Q., Yao, Q., 2013. Arbuscular mycorrhizal fungal inoculation increases phenolic synthesis in clover roots via hydrogen peroxide, salicylic acid and nitric oxide signaling pathways. J. Plant Physiol. 170 (1), 74–79.

Zhang, L., Xu, M., Liu, Y., Zhang, F., Hodge, A., Feng, G., 2016. Carbon and phosphorus exchange may enable cooperation between an arbuscular mycorrhizal fungus and a phosphate-solubilizing bacterium. New Phytol. 210 (3), 1022–1032.

Zhang, L., Feng, G., Declerck, S., 2018. Signal beyond nutrient, fructose, exuded by an arbuscular mycorrhizal fungus triggers phytate mineralization by a phosphate solubilizing bacterium. ISME J. 12, 2339–2351.

Zhou, Y., Zhu, H., Fu, S., Yao, Q., 2017a. Variation in soil microbial community structure associated with different legume species is greater than that associated with different grass species. Front. Microbiol. 8, 1007.

Zhou, Y., Zhu, H., Fu, S., Yao, Q., 2017b. Metagenomic evidence of stronger effect of stylo (legume) than bahiagrass (grass) on taxonomic and functional profiles of the soil microbial community. Sci. Rep. 7, 10195.

Zhou, Y., Zhu, H., Yao, Q., 2018. Contrasting P acquisition strategies of the bacterial communities associated with legume and grass in subtropical orchard soil. Microbiol. Rep. 10 (3), 310–319.

Zhu, H., Yao, Q., 2004. Localized and systemic increase of phenols in tomato roots induced by *Glomus versiforme* inhibits *Ralstonia solanacearum*. J. Phytopathol. 152 (10), 537–542.

Zhu, H., Zhang, R., Chen, W., Gu, Z., Xie, X., Zhao, H., et al., 2015. The possible involvement of salicylic acid and hydrogen peroxide in the systemic promotion of phenolic biosynthesis in clover roots colonized by arbuscular mycorrhizal fungus. J. Plant Physiol. 178, 27–34.

Zoundji, C.C., Houngnandan, P., Amidou, M.H., Kouelo, F.A., Toukourou, F., 2015. Inoculation and phosphorus application effects on soybean [*Glycine max* (L.)Merrill] productivity grown in farmers' fields of Benin. J. Anim. Plant Sci. 25, 1384–1392.

Chapter 5

Microbial resources in management of C sequestration, greenhouse gases, and bioremediation processes

M.L. Dotaniya[1], C.K. Dotaniya[2], Praveen Solanki[3], V.S. Meena[4], M.D. Meena[5] and R.L. Choudhary[6]

[1]ICAR-Indian Institute of Soil Science, Bhopal, India, [2]College of Agriculture, SKRAU, Bikaner, India, [3]Department of Environmental Science, GBPUA&T, Pantnagar, India, [4]ICAR-Vivekananda Parvatiya Krishi Anusandhan Sansthan, Almora, India, [5]ICAR-Directorate of Rapeseed-Mustard Research, Bharatpur, India, [6]ICAR-National Institute of Abiotic Stress Management, Baramati, India

Chapter Outline

5.1 Introduction

Increasing food grain production from limited land with poor agricultural input is not a sustainable way to feed the huge, growing population in developing countries. Most of the natural resources are exploited in the name of higher production without conservation strategies of natural resources. Population pressure is growing at an exponential rate and millions of tonnes are needed to feed the hungry people. Overuse of land and water resources led to reaching their natural capacity of production. Most of the soils showed deficiency of multiple nutrients. Across the globe, soil health is much affected by the intensive cropping system without adding balance nutrient over a period. Decreasing partial factor productivity, nutrient use efficiencies, and crop quality and quantity are the common discussion points in scientific gatherings and research meetings during policy formulation with respect to agriculture. By 2020, India needs to feed its growing population with 280 million tonnes (MT) of food materials against the current production of 233 MT in the year 2016–17 (Singh et al., 2016). In one way, agricultural fields and fertilizer manufacturing units have

New and Future Developments in Microbial Biotechnology and Bioengineering. DOI: https://doi.org/10.1016/B978-0-12-818258-1.00005-4

exhausted their maximum capacity; on the other side of the coin, population pressure is increasing by the second. This has reduced the potential capacity and functions of ecosystems, ultimately reducing the quality life of humans on the Earth. Food webs and chains are disturbed due to increasing contamination by pollutants and the loss of biodiversity observed in every corner of the earth (Chary et al., 2008; Dotaniya et al., 2014g). Increasing global temperature has also affected the agricultural production potential and degradation rate of natural resources. Poor management promotes the land degradation and water contamination rate at a higher level, and much attention is needed to conserve them (Feigin et al., 1991; Sreeramulu, 2010). It is time to think in a sustainable way to enhance food grain production and aquatic food, decreasing postharvest losses without compromising the health of ecosystems.

Fresh water availability is decreasing per year and the crisis in terms of drinking water and water for agricultural use has been widely covered by the mass media. Overexploitation of ground water has created a chronic water shortage in almost every state of India especially in Rajasthan, Maharashtra, and Madhya Pradesh. Farmers are using poor quality water or agricultural inputs for agricultural uses without knowledge of its harmful effect on soil and human health (Debosz et al., 2002; Lenka et al., 2016; Saha et al., 2017e). Apart from these poor management practices, landfill leakage in water and soil, wet and dry deposition in mining areas, industrial emissions, sludge and sewage application, poor segregations, low grade agricultural amendments, household discharge, municipal solid waste application, and metal contaminated compost are also playing an important role in polluting natural resources worldwide (Ghosh et al., 2012; Saha et al., 2017d). Non–point source pollution also contributes significantly to natural ecosystems (Coumar et al., 2016a). In recent years, huge volumes of industrial waste, sewage, and contaminated water have been used for the cultivation of agricultural crops mostly in periurban areas. These are the easily available potential water resources and they contribute a significant amount of organic matter (OM) and plant nutrients to crop plants. Application of sewage improved soil OM as well as plant nutrients in crop field and enhanced the crop yield (Dotaniya et al., 2016a,b), but heavy metals (HMs) and pathogen load limit the crop yield mainly in periurban areas (Saha et al., 2010, 2013, 2017a; Meena et al., 2015). HM chemistry in soil greatly affects the soil pH and OM (Rajendiran et al., 2015). It is precipitate with oxide, sulfate, and carbonates, and takes part in oxidation reduction mechanisms, adsorption or desorption on soil components, vaporization in atmosphere, and uptake by plants (Gupta et al., 2016). Organic and inorganic pollutants have different types of environmental risk due to huge interrelation process with soil and microbial populations (Saha et al., 2017d). These processes directly and indirectly affect the living organisms both intracellularly and extracellularly. The interactive effect of metals on soil fertility or soil microbes can be computed as a single or multimetal toxicity with the help of advanced tools and techniques (McLaughlin et al., 2000). The food chain contamination of HM causes higher loss of biodiversity in an ecosystem (Meers et al., 2007). Most of the developing countries have HMs standards regarding use of poor natural resources in agricultural crop production systems. Environmental protection laws and their enforcement are very strong to control the metal concentration from input to soil and soil to human via food chain contamination. Several countries have soil protection policies to maintain the soil health and sustainability crop production to feed the growing population (Saha et al., 2017m). In this chapter the authors describe the role of microbial resources in C sequestration, HMs remediation, and mitigating the effect of climate change from ecosystems.

5.2 Microbial resources in agro-ecosystems

Soil is a complex, dynamic, nonrenewable resource formed after years of physical, chemical, and biological weathering processes on rock (Saha et al., 2017b). It shelters a huge microbial population and is a working platform for many important biogeochemical processes related to soil fertility and ecosystem services (Saha et al., 2017c). Microbial population consists of flora and fauna and further these are classified into macro (can be observed by the naked eye) and micro (cannot be observed by the naked eye) on the basis of size (Bharti et al., 2017). Most of the larger insects, annelids, beetles, grub, ants, millipedes, centipedes, and thicker plant roots are known as macroorganisms. Whereas, smaller organisms such as fungi, bacterial, virus, actinomycetes, and fine roots. are known as microorganisms (Fig. 5.1). Microorganisms play a valuable role in soil fertility and mineralization of plant nutrients. They also secrete plant growth promoting substances and antibiotics, and act as a controller of pathogens. These organisms are very important for ecosystem services. The population and diversity of an organism in soil is greatly affected by the type of soil, OM, plant nutrient, crop rotation, climatic conditions, and crop management options during the period. Higher amount of OM in soil provides food material to microorganisms and boosts the microorganism population after the application of organic inputs in soil. Upper surface of soil is more fertile and has a higher population and diversity of microorganisms in comparison to lower profile of soil. One gram of soil contains up to 10 billion microorganisms and thousands of various types of species (Saha et al., 2017c). Application of poor quality natural resources reduces the population and diversity of microorganisms in soil. Application of Cr more than $20\ mg\ kg^{-1}$ in black cotton soil

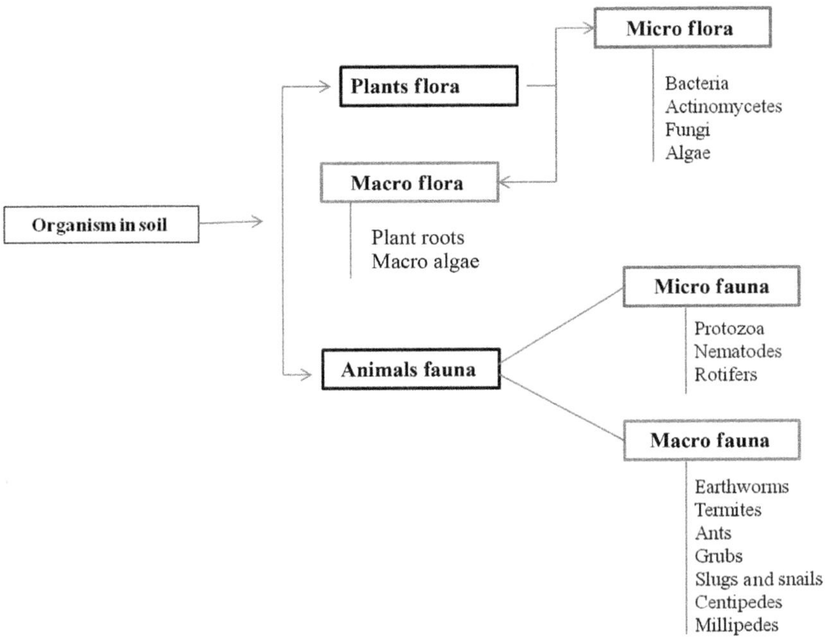

FIGURE 5.1 Soil microbial populations.

reduced the soil biomass C and acid and alkaline phosphatases (Dotaniya et al., 2017c). Higher amount of OM in sewage enhances the population of bacteria, fungi, and actinomycetes. For the remediation of problematic soils, microbial consortium is used in combination with OM sources. Intercropping or mixed cropping, having higher microbial biomass, reflects and multiplies microbial biodiversity compared with monocropping. Most microorganisms are nonculturable. This limits getting a clear picture of soil biodiversity. Recent tools and techniques (metagenomics tools, polymerase chain reaction [PCR]) have succeeded in identifying different types of genetic material from microorganisms (Dotaniya et al., 2018b). Cropping patterns and crop varieties also affect the rhizospheric microbial population due to secretion of greater amounts of root exudates (Kumar et al., 2018; Kushwah et al., 2014; Singh et al., 2017). Legume crops improve soil biological health more than cereal crops. Incorporation of succulent crop biomass at flowering stage enhances the microbial population in soil and positively affects the nutrient mineralization process in soil (Meena et al., 2018).

5.3 Soil enzymatic activities and their role

Soil is a biological active and a shelter for billions of microorganisms, and is a storehouse of plant nutrients and the basis of human civilization. It produces various types of metabolites by soil organisms. Bioindicators are indicators for monitoring environment qualities, and measurement of ecological changes and efficient resource management. These may be an organism or its part, or a biological process as well as a product mediated by organisms to find information regarding a system (Killham, 2002). Soil quality parameters are also assessed through bioindicators including nutrient cycling, OM mineralization, soil biomass C, microorganisms and diversity, soil enzymatic activities, fauna and flora community structure, etc. (Bucket and Dick, 1998). These parameters are very useful and directly and indirectly play a crucial role in crop production and maintaining soil fertility. Advanced tools and techniques are in use to assess the impact of bioindicators on ecosystem services. These techniques should be easy to measure, respond to external change, be more accurate and sensitive to system stress, be easy to understand, cost effective, and able to measure spatial and temporal variation, viable with response, etc. (Dale et al., 2008). Soil enzymes are biological crucial bioindicators for soil health and have the potential to affect the nutrient dynamics and crop quality (Das and Verma, 2011; Dindar et al., 2017). They are also known as biological fingerprints and are mediated by soil tillage and management strategies (Bandick and Dick, 1999). They are used for the computation of biological equilibrium (Frankenberger and Tabatabai, 1991), nutrient availability and soil fertility (Antonious, 2003), soil and ecosystem quality (Bucket and Dick, 1998), and biological parameters modified due to anthropogenic activities (Trasar-Cepeda et al., 2000). Soil enzymatic activities are the important factor for deciding plant nutrient chemistry in soil and nutrient uptake by crops (Dotaniya et al., 2013d; Dotaniya et al., 2018e). Soil enzymes mobilize the plant nutrients from immobilized sites of the soil matrix. They mediate and catalyze most of the soil functions like mineralization of plant nutrients, decomposition of OM

(Jat et al., 2018), release of inorganic nutrients, nitrogen fixation and phosphorus release, decomposition of pollutants, and also play a role in different plant nutrient cycles. In agricultural crop production systems, soil enzymes play a key role, that is, urease in N, acid and alkaline phosphatases in P, sulfatase in sulfur, and β-glucoside as well as β-galactosidase in C cycle during decomposition (Karaca et al., 2011). The level of soil enzymes is greatly affected by the natural composition of soil and its human management over a period (Utobo and Tewari, 2014). Most important human management factors may enhance or decrease soil quality. Use of poor natural resources reduces the soil quality whereas, addition of crop residues improves the soil fertility and crop productivity (Gianfreda and Bollag, 1996; Dotaniya, 2013; Dotaniya et al., 2013c; Dotaniya and Datta, 2014).

5.4 Global climate change

Global climate change is a burning issue among policy makers and research communities to combat the adverse effect on agricultural crop production systems. Increasing concentration of greenhouse gases (GHGs) in the atmosphere elevates the temperature, and increasing temperature causes imbalance to ecological services and ecosystem health (Dotaniya, et al. 2017c). This effect is known as climate change. It affects the different soil forming processes and crop yield (Kundu et al., 2013). In tropical regions crop yield little will reduce due to increase in respiration rate in most of the crops. The photosynthesis conversion from source to sink rate is slow and adversely affect the crop yield productivity (Dotaniya, 2015). Increasing the Earth's atmospheric temperature affects the soil microbial population and plant nutrient dynamics in soil (Mandal et al., 2013).

5.5 Composition of greenhouse gases

Solar energy comes from sun in the form of short waves and after it reaches Earth's atmosphere approximately 30% of this energy reflects back to space, which is known as the albedo. In this mechanism, aerosol, clouds gases, and Earth's surface play a vital role in converting short waves into long waves and enhancing the global temperature. As for the rest of the incoming energy, 20% is absorbed by the atmosphere, including clouds, and 80% is absorbed by Earth's surface. Clouds are the potential absorber of GHGs and emit infrared radiation and affect the global energy cycle in the short and long term. GHGs greatly affect the temperature of Earth and soil processes. Some of the important GHGs are as follows (Lenka et al., 2013; Table 5.1):

5.5.1 CO_2

Most of the C in the atmosphere is in the form of CO_2. By its chemical nature it is an inert gas and maintains Earth's atmospheric temperature (Meena and Dotaniya, 2017). It is easily mixed with other gases in the atmosphere. Its concentration is highly varied with less concentration in the green canopy.

5.5.2 CH_4

This is found in meager amounts in Earth's atmosphere. Chemically, it is active in nature and is mainly contributed by natural gas sources as well as the breakdown of OM by anaerobic process. Most marshlands contribute a significant amount of CH_4 to Earth's atmosphere. In the atmosphere it is converted into CO_2.

$$CH_4 + 2O_2 \rightarrow CO_2 + 2H_2$$

TABLE 5.1 Composition of major greenhouse gases.

Gas	Formula	Contribution (%)
Water vapor	H_2O	36–72
Carbon dioxide	CO_2	9–26
Methane	CH_4	4–9
Ozone	O_3	3–7

5.6 Why is global climate change a major concern?

Nowadays, increasing climate change disturbs the soil processes, causes climatic events, and affects crop productivity and wildlife across the globe. Short duration and erratic pattern of rainfall shortens crop duration and decreases crop productivity. Increasing the temperature enhances plant respiration rate and reduces the source to sink conversion rate in most of the crops. In this regard, global climate change is a major concern across the world and how the combine affords is fruitful to reduce or minimize the climate change effect on ecosystem services (Dotaniya and Meena, 2017).

5.7 Effect of climatic factors on crop productivity

The production of food grain during photosynthesis is mostly governed by soil and climatic factors. Plant leaves synthesize the carbohydrates in the presence of water, CO_2, temperature, and the chlorophyll content in the leaves as per the Eq. (5.1).

$$CO_2 + 2H_2O + \text{photons} \rightarrow (CH_2O) + O_2 + H_2O \tag{5.1}$$

Carbon dioxide + water + light energy → carbohydrate + oxygen + water

In which light energy is converted to chemical energy in plants or smaller microorganisms.

5.8 Carbon sequestration potential

Increasing concentration of GHGs in atmosphere mediates the atmospheric temperature and the soil–plant process. Increasing the temperature enhances the respiration rate of crop plants and reduces the conversion of photosynthates from the source to sink. Increasing the root exudates in the rhizosphere enhances the microbial count and diversity and mediates the soil and nutrient processes in the soil (Dotaniya and Kushwah, 2013; Dotaniya et al., 2014f). The CO_2 concentration in the atmosphere plays a crucial role in maintaining and governing the physical process on Earth. Soils act as a source and sink for the CO_2 during the C cycle. To reduce the global climate change effect on crop production, CO_2 concentration has to be minimized or reduced in the atmosphere. In this regard, capturing of C from the atmosphere and fixed into long lived pools in different ecosystems known as C sequestration process. Green plants act as a source and sink for C and play a crucial role in sequestration in terrestrial ecosystems (Dotaniya and Meena, 2013; Dotaniya et al., 2013a). Storing of C in soil is as long lived pools for mitigating the climate change adverse effect by screening or modifying the efficient crop cultivars. Phytoliths occluded C is resistant to easily mineralize with soil microbial degradation and retain the C for longer duration in soil (Prajapati et al., 2016).

5.9 Remediation of greenhouse gases

GHGs are the one of the key components in climate change phenomena across the globe. Increasing concentration of GHGs has mediated the atmospheric temperature exponentially as compared with the preindustrial era. This limits the crop yield productivity potential and enhances the rate depreciation of natural resources. Decreasing the forest cover and increasing the industrial growth with time are needed to minimize the GHGs adverse effect on ecosystem services. Some of the viable strategies to minimize GHGs emission and reduce their concentration in the atmosphere are as follows:

5.9.1 Reduction in greenhouse gases

Greenhouse gas emission is the key factor in global climate change phenomena and it has modified atmospheric events, for example, amount of rainfall, temperature, wind direction, atmospheric pressure, etc. (Dotaniya et al., 2018a). These are all the result of mass industrialization and faulty practices for minimization of environmental pollution. Most countries do not have a clear cut plan to reduce GHGs emission by minimizing industrial and household activities, especially in the developing countries. We should more focus on minimum emission of GHGs at smaller (household) to larger (industrial) scale. This will help to cut resources used for minimizing the GHGs' adverse effect on ecosystem processes.

5.9.2 Afforestation

Increasing the forest cover purifies the surroundings by liberating the oxygen. Green plants consume atmospheric CO_2 during photosynthesis and liberate O_2 in the process of photolysis. In ancient days, most land was covered with forest and the industrial growth rate was minimal. There is a prime need to enhance the forest cover by utilizing barren land, fallow land, and multicropping crop production systems. This also helps to minimize the soil degradation and improve ecosystem health (Dotaniya et al., 2016a,b).

5.9.3 Improved technology

Industrial growth is necessary for the development of a country at global scale. Most of the developing or developed countries are bound to enhance their industrial growth and generation of employment. However, most of the developed countries are focused on the refinement of technology that is highly fuel efficient with less emission of GHGs mostly in vulnerable areas (Meena et al., 2017). The GHGs and their residence time in the atmosphere are different with respect to a particular gas (Wallace and Hobbs, 2006). Research and policy wings of these countries are more powerful and functional than developing countries. Environmental protection norms and rules are strictly followed during development of a technology with respect to regional needs. In developing countries lack of funds and overcrowded situations are more dominant than environmental protection laws. People have been using outdated technologies for many years and have shifted from mega cities to local areas. The planning of highly efficient technologies that are economically viable and that emit less carbon is the need of today and tomorrow for the well-being of humankind across the globe.

5.10 Heavy metals

HMs are toxic in nature, having carcinogenic effect on humans, as well as reducing growth and yield in plants and causing different malfunctions in animals. They are highly toxic at lower concentration and reduce the health of an ecosystem. Metals or metalloids having atomic number more than calcium (Ca) or atomic density greater than 5 g cc-1 are known as HMs. HMs belong to the d-orbitals, having completely and partially filled outer orbitals. Among the HMs, a few are more common in nature such as arsenic (As), cadmium (Cd), chromium (Cr), nickel (Ni), selenium (Se), lead (Pb), zinc (Zn), copper (Cu), mercury (Hg), and aluminum (Al). (Saha et al., 1999). These metals originate geogenically, that is, by volcanic eruption or as a part of rock origin. HM concentration varies spatially and is greatly affected by the formation of parent materials (Rajendiran et al., 2018). Some HM-contaminated places include As contamination in west Bengal, Bangladesh and contaminated through utilization of As contaminated ground water as well as soil formed from As contaminated rocks enhances the As contamination in food chain. There is also mercury contamination in coal mine areas, Se contamination in some parts of Punjab (Dhillon and Dhillon, 1991), aluminum contamination in Al mining areas, Zn contamination in Zn smelter or mining area at Udaipur, Rajasthan, Cu contamination at Malajkhand, MP. The HM contamination is greater due to anthropogenic activities (manmade activities); it has more potential to spread the metal toxicity to natural resources. Technological advancement and poor disposal of HM-contaminated waste has created metal pollution in noncontaminated areas.

5.11 World scenario of metal production

The main source of HMs is geogenic due to the formation of land, volcanic eruption, etc. After the extraction and beneficiation of different metals for industrial purposes, HM concentration spreads into noncontaminated areas; this is known as an anthropogenic source of contamination (Saha et al., 2017h). Over time metal production and its consumption increased greatly across the global market (Nriagu, 1996). Metals reach regions where they are not generally available due to international trade and liberalization policies. These are the toxic metals having high electrical conductivity, luster, and malleability, and that cause ill effect at lower concentration in human, animal and plant systems. It adversely affects the ecological balance, evolutionary advancement, nutrient kinetics, and environmental health (Saha and Sharma, 2006; Nagajyoti et al., 2010; Saha et al., 2017i,j). Different sources, that is, soil degradation by erosion and natural weathering, use of poor quality agricultural inputs, use of metal contaminated industrial effluent, sewage sludge, use of insecticide, pesticides in crop production are enhanced the HM in different ecosystems. The periodic monitoring of contaminated sites and analysis with respect to bioavailable metal concentration in soil is the key factor for HM assessment in a particular ecosystem. For collection of soil samples, the following are needed: pollutant history, type of pollutant, duration of contamination, sampling strategies, site specific remediation procedure, spatial and temporal variation, information about soil physicochemical properties, climatic condition, and the crop history of the region (Saha et al., 2017f).

5.12 Heavy metal availability in different ecosystems

The availability of different metals is also affected by the properties of rocks and their chemical properties. The basic igneous rocks contain higher amounts of Cr, Mn, Co, and Ni and these reach soils via the soil forming process. Whereas, sedimentary rocks contribute mainly Cr, Co, Ni, Zn, and Pb (Levinson 1974; Alloway 1990). The release kinetics of HMs from soil to soil solution is mediated by the soil physicochemical properties (i.e., texture, CEC, OC, pH, soluble salt, Eh, soil moisture) and biochemical process (mineralization kinetics) in soil. Majority of soils contain Zn, Mn, Fe, and Cu in higher amount than other HMs, for example, Cd, Cr, Ni, and Pb. Industrial hubs also contribute a meaningful amount of heavy meals in surrounding agricultural fields as dry and wet depositions. The steel and iron industrial belt of Orissa contributes a significant amount of Fe and toxicity symptoms have appeared on crop plants. Zinc mining at Udaipur, Rajasthan contaminated the surrounding field of agriculture by adding waste mining material and industrial effluent, and through dry depositions. Kanpur tannery regions are contaminated by the application of Cr contaminated tannery effluent for the cultivation of crops, mainly vegetables in periurban belts. Long-term application of Cr wastewater accumulated a significant amount of Cr in the soil, which reached humans via food chain contamination (Devkota and Schmidt, 2000). Dotaniya et al. (2017b) analyzed soil data from a tannery contaminated area as well as tube well irrigated (fresh water) field of agriculture and found that Cr had highly contaminated the soil categorized under class VI as per the geoaccumulation index of Muller (1969). Application of Cr in soil reduced the germination and root–shoot growth of wheat (Dotaniya et al., 2014a) and pigeon pea (Dotaniya et al., 2014c) due to reduction of microbial biomass and soil enzymatic activities (Dotaniya et al., 2017c). HM contamination in soil can be assessed with the help of geoaccumulation index, enrichment factor, enrichment ratios, contamination index, integrated contamination index, Nemerow's pollution index, contamination degree, contamination factor pollution load index, etc. (Saha et al., 2017g). However, HM contamination level in plant systems are assessed with the help of bioconcentration factor, transfer efficiency, transfer factor. These factors are also used for the identification of phytoaccumulative plants for remediation of pollutant sites (Pipalde and Dotaniya, 2018).

5.13 Heavy metal contamination in ecosystems

Increasing global population reduces the per capita availability of natural resources and forces people to use poor quality resources. These situations are very dire in developing or underdeveloped countries. Most of the low-income countries are using contaminated/degraded land and polluted water for their necessary activities. These situations are particularly bad due to rudimentary advancement of science and technology in the field of natural resources. Use of low grade available natural resources for growing food is a necessity of today and tomorrow across the world. Use of proper tools and techniques for remediation of toxic metals' effect on soil health and crop quality is a key area of research in pollution sectors. HM toxicity can be categorized on the basis of mode of action into two groups. First, acute exposure means contact to hazardous material caused toxicity mostly in place of work and industrial workers. This type of exposure creates problems like loss of appetite, skin irritation, headache, hypertension, stomach pain, eye irritation, restlessness, vomiting, etc. (Singh, 2002). The second type of toxicity is chronic exposure, in which longer contact with metal contamination can cause paralysis, kidney problems, stomach damage, birth deformities, and loss of intelligence in children (Table 5.2).

5.14 Major heavy metals and their chemistry

5.14.1 Chromium

It is one of the toxic metals and causes health hazards in humans via food chain contamination. The International Agency of Research on Cancer categorizes Cr as a mutagenic element under rank 1. The major sources of Cr are leather industries, electroplating, wooden industries, paper and pulp units, paint and varnishing, and these contribute a significant amount of Cr in effluent (Nath et al., 2005). It has two major forms, Cr(III) and Cr(VI), on the basis of toxicity. Cr (III) is nontoxic to plants and immobile in soil, insoluble in water, and mostly exists in reduced form in the environment (Shanker et al., 2005). In contrast to Cr(III), Cr(VI) is toxic to plants, highly soluble in water, and mobile in soil solution and predominantly found in oxidation states (Dotaniya et al., 2014d). During the quantification of Cr concentration

TABLE 5.2 Heavy metal toxicity effect on humans.

Heavy metal	Effects
Cd	Interfere with Ca metabolism, mutagenic, anemia, failure of kidney, cancer
Cr	Hair loss, skin irritation, metabolic disturbance, vomiting
As	Interfere with cell metabolism, structure similar to phosphorus, skin diseases
Pb	Stunted growth, renal infection, cardiovascular disease, hair loss, toxicity (mainly in children)
Hg	Tiredness, memory loss, lung and kidney cancer, depression, hair loss
Se	Hair loss, fatigue, skin irritability, and foul breath odor (known as "garlic breath")

FIGURE 5.2 Chromium toxicity in spinach.

in soil and plants need to be analysised of Cr speciation with respect to their toxicity on plant and soil enzymatic activities (Gürkan et al., 2012; Dotaniya et al., 2018c). Application of OM and other electron donating substrates present in soil system converted hexavalent Cr into trivalent Cr in soil. In a similar pattern, higher oxygen concentration and electron donor subtracts enhance the conversion rate of Cr(III) to Cr(VI) in soil (Cervantes et al., 2001). Chromium toxicity affects the root growth drastically and reduces the uptake pattern of mineral nutrition, showing chlorosis and necrosis symptoms in plants (Fig. 5.2). Hexavalent Cr is isostructural with SO_4^{2-} and HPO_4^{2-} anions in soil and is easily uptaken by the plant root system. During the uptake of hexavalent Cr from soil system to plant root cell it easily passes via the cell membrane and affects the catalase, peroxidase, and cytochrome oxidate with Fe and affects the plant defense mechanisms. After the reach of the Cr(VI) in cells and again intracellular reduction mechanism converts it to activated Cr form. It mediates the reaction rate of biological reaction during reactions between hexavalent Cr and thiols and ascorbate biological reductants and produces reactive oxygen species (ROS), that is, superoxide ions, H_2O_2, OH-radicals, and creates extreme oxidative stress in cells, and damage to cell DNA was observed (Stohs and Bagchi, 1995).

5.14.2 Arsenic

This is a toxic metal that disrupts the ecological system and decreases biodiversity. It is the 20th most plentiful element in the Earth's crust, and is a geogenic source of contamination mostly in West Bengal and Bangladesh. Its two most known forms in nature are arsenite (AsO_3^{3-}) and arsenate (AsO_4^{3-}). Sometimes arsenic is taken deliberately during suicide attempts or by mistake, causing acute toxicity of As in humans (Saha et al., 1999; Jaishankar et al., 2014). In the human body, biotransformation of As converts the inorganic As into monomethylarsonic acid (MMA) to dimethylarsinic acid (DMA) and finally it is excreted by the excretory system in the form of urine.

$$\text{iAs (V)} \rightarrow \text{iAs (III)} \rightarrow \text{MMA (V)} \rightarrow \text{MMA (III)} \rightarrow \text{DMA (V)}$$

Biomethylation is a detoxification process in which MMA (V) and DMA (V) both are excreted through urine, whereas, MMA (III) is toxic in nature and not excreted during urination.

5.14.3 Lead

It is also one of the HMs and is extremely toxic for plant growth and development (Najeeb et al., 2014). Its toxicity clearly shows in children and adults as classical disease that mostly affects the central nervous system (Saha et al., 2017k). As per the guidelines of the Environmental Protection Agency (EPA) it is categorized it as a carcinogenic element for human health. Losses of chlorophyll and photosynthesis rate in cells through producing the ROS are the common indicator symptoms of Pb toxicity in plants. Yongsheng et al. (2011) reported that toxicity of Pb reduced the uptake pattern of essential plant nutrient and growth in tea plant. Increasing the concentration of Pb in soil reduces the uptake pattern of Ca, Mg, Fe, and Na in plants and disturbs the biological metabolic process of cells (Flora et al., 2008; Dotaniya and Pipalde, 2018). Lead poisoning in man causes loss of intracellular second messenger system and modifies the central nervous system (Zhuang et al., 2009), in which the plasma membrane is shifted into interstitial spaces of the brain, known as edema (Teo et al., 1997).

5.14.4 Mercury

Mercury is used in various scientific instruments like thermometer, barometers, Hg torch, and also agricultural inputs like insecticide–pesticide formulations, dental formulations, catalyst in industrial uses, batteries, paper and pulp, etc. It is odorless and colorless in nature and has higher threatening potential in aquatic animals. It also damages the tertiary and quaternary structure of protein in plants. It also reduces the cell structural growth and restricts the cell division and elongation in crop plants. Higher concentration in water bodies bioaccumulation in fish causes Hg toxicity in humans via consumption. Anthropogenic activities such as agricultural use, sewage sludge application, mining, and use of industrial effluent for crop productions are the key source of Hg in the environment (Chen et al., 2012). Element concentration of Hg in aquatic systems is converted in methylated form by microorganisms and causes human exposure by the consumption of contaminated aquatic animals (Trasande et al., 2005). Mercury poisoning is known as acrodynia or pink disease in humans (Morais et al., 2012). In ecosystems, Hg absorption rate by different sources will vary, for example, elemental Hg, methyl Hg, inorganic Hg are 75%–85% of vapor, 95%–100% in intestinal tract, and 7%–15% ingested dose absorbed, respectively.

5.14.5 Cadmium

It is a carcinogenic metal contributed through batteries industries, alloys production units, application of Cd contaminated effluents in agricultural crop production system and also through tobacco consumption. It is also contributed by P fertilizer industrial units (Lenka et al., 2016). It is also categorized as a carcinogen per the International Agency for Research on Cancer (Friberg et al., 1971). It accelerates bone mineralization causing fragile bones, which is known as osteoporosis, and disturbs the Ca metabolism in humans and animals. Imbalance of Ca in body enhances the Ca ions' concentration in urine (known as hypercalciuria) and deposition of Ca in body parts (Henson and Chedrese, 2004). Han et al. (2009) observed that increasing concentration of Cd enhances Cd uptake predominantly in fruits and vegetables. The elevated level of Cd accumulates more in roots than in shoots (Coumar et al., 2016b; Dotaniya et al., 2017a) and adversely affects the plant enzymatic and oxidative mechanism, causing nutrient deficiency (Irfan et al., 2013). Cadmium reduces the growth of the crops in Cd contaminated agricultural fields (Andresen and Küpper, 2013). Higher accumulation of Cd through industrial effluents causes deficiency of Zn and Fe in most crops (Flora et al. 2008).

5.15 Kinetics of heavy metal in soil

HM uptake by plants is directly proportional to metal concentration in soil solution and affected by the genetic makeup of crop plants. Most of the monovalent and divalent cations of HMs are governed by the active transport across the plasma membrane in higher plants. The chemistry of monovalent HM cations follows the relationship path of the Michaelis–Menten equation, which is mostly used in the study of enzymatic activities in soil.

$$\text{U} = \frac{CV_{max}}{C+K_s} \tag{5.2}$$

where U is the ion uptake velocity of a particular cation, C is the concentration of ions in external solution, V_{max} is the maximum rate of cation uptake during the all available carrier sites loaded or maximum transport rate, K_s is the Michaelis constant.

5.16 Heavy metal remediation mediated by genetic potential of plant

HM concentrations in soil solution are affected by the soil and plant characteristics (Ferguson, 1990). External application of metals through sewage sludge, agricultural inputs, dumping of waste materials on agricultural fields, dry and wet deposition, natural contamination in soil or ground water, etc. increase the natural concentration of HMs in soil solution. Different phytoremediation methods are available, that is, phytoextraction, phytovolatilization, rhizofiltration, and rhizodegradation to remediate the contaminated sites (Saha et al., 2017l). Plant genetic makeup plays a crucial role during the uptake of HM from soil (Fig. 5.3). Most crops perform three types of activities during the uptake, as follows:

5.16.1 Excluders

In such conditions plants restrict the HM uptake from soil solution to the shoot.

5.16.2 Index plants

Plants uptake the metal in direct proportion to the availability of metal ions in soil solution. This reflects the metal concentration in plant parts.

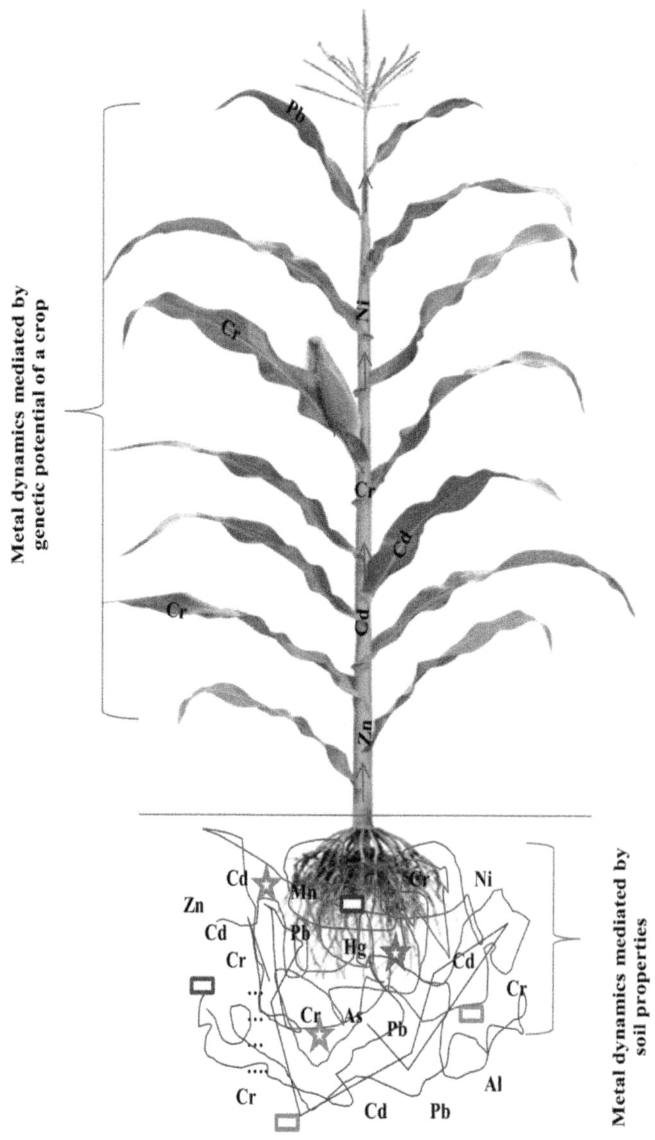

FIGURE 5.3 Heavy metal uptake dynamics in crop plant.

5.16.3 Accumulators

Plants uptake the significant amount of metal in parts without affecting the physiological and yield parameters up to a level of concentration. These types of plants accumulate a particular metal from contaminated land.

5.17 Defense mechanism of plants

HMs are a key threat and are associated with many health hazards for humankind. These are not essential for healthy growth of the body, but few have a pseudo role and interface with the metabolic functions and process. Metal toxicity is affected by the type, mode, and absorption kinetics governing soil or the human body, route of exposure, soil properties, duration of exposure to metal, and targeted organisms. Most of the HMs are imposing oxidative stress on plants by generating free radical formation and cause different disorder in crop plants (Jaishankar et al., 2014). During stress conditions plants generate different types of antioxidants to counter the adverse effect of ROS and free radicals (Stohs and Bagchi, 1995). Antioxidants like glutathione counter the H_2O_2 and save the cell from Pb toxicity (Fig. 5.4). Production of ROS and antioxidants both are antagonistic in nature in the plant system. Glutathione on the basis of oxidation and reduction reaction having into glutathione (GSH) in reduce and oxidized glutathione (GSSG) in oxidized state. In normal conditions, GSSG is having 10% whereas GSH in 90% account in plants. Increasing the oxidative stress conditions enhances the concentration of GSSG (Flora et al., 2008). Increasing the Pb concentration in soil adversely affects the nutrient dynamics of C mineralization through harmful effect on soil microbial count (Flora et al., 2008).

5.18 Remediation of heavy metals contamination by microbial resources

Remediation of metal contaminated soils is the need of today's agriculture to counter the needs of the growing population without deterioration of soil and environmental quality. It is a great challenge for agricultural research system to grow more healthy food with limited natural resources. The anthropogenic activities are generating more wastewater and this creates a problem for its safe discharge on soil and water bodies. All the methods of HM remediation should be more efficient, cheaper, have less technical requirement, and be without adverse effect on other natural ecosystems (Diels et al., 2010). Addition of crop residue enhances the microbial population and acts as a biosorption mechanism and reduces the metal toxicity. Coumar et al. (2016a) reported that addition of pigeon pea biochar as a soil amendment reduces the Cu toxicity and improves the soil health. Many HM remediation methods are in practice for the

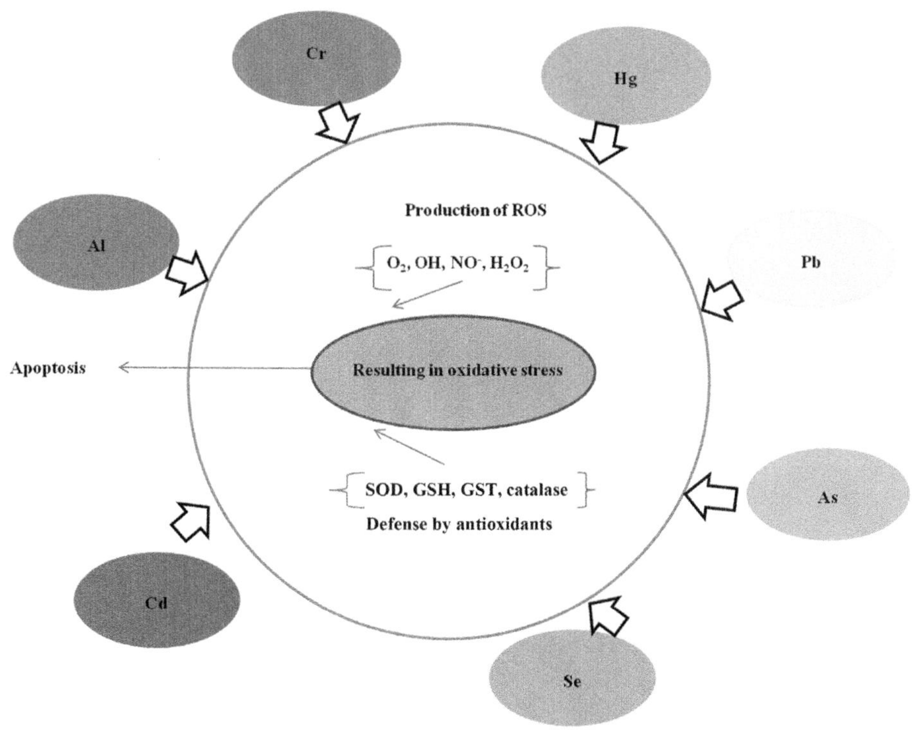

FIGURE 5.4 Defense mechanism of plant under heavy metals stress condition.

decontamination of organic and inorganic loads from different ecosystems. Soil microbial population is playing a crucial role in remediation of HMs (Bishnoi and Garima 2005; Mandal et al., 2017). Paranthaman and Karthikeyan (2015) isolated the *Pseudomonas* spp. from paper mill effluent and showed its positive effect on degradation of pollutants. Application of *Bacillus safensis* strains (RSA-4) decontaminated the Cd contamination from the marina ecosystem (Priyalaxmi et al., 2014). Use of organic and inorganic amendments reduces the availability of HMs in soil solution and improves the crop yield. Application of pressmud and FYM reduce the Cr toxicity in wheat crop, and FYM better performed than pressmud (Dotaniya et al., 2018d). After decomposition of crop residue produced various type of organic acids (Dotaniya et al., 2013b, 2014b,e, 2015, 2016a,b), which act as a bioabsorber of HMs (Dobson and Burgess, 2007).

5.19 Conclusion

Microbial population and diversity are the key components of a healthy soil. These are the bioindicators for measuring the health of ecosystems. Soil fertility is maintained by modifying OM decomposition and mineralization rate affects the plant nutrient supplying capacity of a soil. Higher population growth sequesters huge amounts of microbial biomass in soil and affects the plant nutrient dynamics in the shorter and longer term. Increasing global temperature also affects the microbial biomass in soil and these organisms are playing a crucial role in mitigation the climate change effect. HMs are highly toxic and carcinogenic in nature with respect to living systems. These also minimize the toxic effect of pollutants; they convert toxic substances into nontoxic and minimize the ecosystem losses. Use and identification of suitable microorganisms are helpful to enhance the soil organic C, minimize the climate change effect, and remediate HM toxicity in agro-ecosystems. Use of different public and private resources affords for control, prevention, and remediation of metal toxicity from the environment with respect to microbial tools and techniques.

References

Alloway, B.J., 1990. Heavy Metals in Soils, second ed. Blackie Academic & Professional, 368 p.

Andresen, E., Küpper, H., 2013. Cadmium toxicity in plants. Met. Ions Life Sci. 11, 395–413.

Antonious, G.F., 2003. Impact of soil management and two botanical insecticides on urease and invertase activity. J. Environ. Sci. Health B 38, 479–488.

Bandick, A.K., Dick, R.P., 1999. Field management effects on soil enzyme activities. Soil Biol. Biochem. 31, 1471–1479.

Bharti, V.S., Dotaniya, M.L., Shukla, S.P., Yadav, V.K., 2017. Managing soil fertility through microbes: prospects, challenges and future strategies. In: Singh, J.S., Seneviratne, G. (Eds.), Agro-environmental Sustainability. Springer, pp. 81–111.

Bishnoi, N.R., Garima, 2005. Fungus-an alternative for bioremediation of heavy metal containing wastewater: a review. J. Sci. Ind. Res. 64, 93–100.

Bucket, J.Z., Dick, R.P., 1998. Microbial and soil parameters in relation to n mineralization in soils of diverse genesis under differing management systems. Biol. Fert. Soils 27, 430–438.

Cervantes, C., Campos-García, J., Devars, S., Gutiérrez-Corona, F., Loza-Tavera, H., Torres-Guzmán, J.C., et al., 2001. Interactions of chromium with microorganisms and plants. FEMS Microbiol. Rev. 25 (3), 335–347.

Chary, N.S., Kamala, C.T., Raj, D.S., 2008. Assessing risk of heavy metals from consuming food grown on sewage irrigated soils and food chain transfer. Ecotoxicol. Environ. Saf. 69, 513–524.

Chen, C.W., Chen, C.F., Dong, C.D., 2012. Distribution and accumulation of mercury in sediments of Kaohsiung river mouth, Taiwan. APCBEE Procedia. 1, 153–158.

Coumar, M.V., Parihar, R.S., Dwivedi, A.K., Saha, J.K., Lakaria, B.L., Biswas, A.K., et al., 2016a. Pigeon pea biochar as a soil amendment to repress copper mobility in soil and its uptake by spinach. BioResources 11 (1), 1585–1595.

Coumar, M.V., Parihar, R.S., Dwivedi, A.K., Saha, J.K., Rajendiran, S., Dotaniya, M.L., et al., 2016b. Impact of pigeon pea biochar on cadmium mobility in soil and transfer rate to leafy vegetable spinach. Environ. Monit. Assess. 188, 31.

Dale, V.H., Peacock, A.D., Garten, C.T., Sobek, F., Walfe, A.K., 2008. Selecting indicators of soil, microbial and plant conditions to understand ecological changes in Georgia pine forests. Ecol. Indic. 8, 818–827.

Das, S.K., Verma, A., 2011. Role of soil enzymes in maintaining soil health. In: Shukla, G., Verma, A. (Eds.), Soil Enzymology. Springer International. Available from: https://doi.org/10.1007/978-3-642-14225-3_2.

Debosz, K., Petersen, S.O., Kure, L.K., Ambus, P., 2002. Evaluating effects of sewage sludge and household compost on soil physical, chemical and microbiological properties. Appl. Soil. Ecol. 19 (3), 237–248.

Dobson, R.S., Burgess, J.E., 2007. Biological treatment of precious metal refinery wastewater: A review. Miner. Eng. 20, 519–532.

Devkota, B., Schmidt, G.H., 2000. Accumulation of heavy metals in food plants and grasshoppers from the Taigetos Mountains, Greece. Agric. Ecosyst. Environ. 78, 85–91.

Dhillon, K.S., Dhillon, S.K., 1991. Selenium toxicity in soil, plant and animals in some parts of Punjab, India. Int. J. Environ. Stud. 37, 15–24.

Diels, L., Geets, J., Dejonghe, W., Van Roy, S., Vanbroekhoven, K., Szewczyk, A., et al., 2010. Heavy metal immobilization in groundwater by in situ bioprecipitation: comments and questions about efficiency and sustainability of the process. In: Proceedings of the Annual International Conference on Soils, Sediments, Water and Energy, vol. 11, no. 1. Chicago, p. 7.

Dindar, E., Topac, F.O., Baskaya, H.S., Kaya, T., 2017. Effect of wastewater sludge application on enzyme activities in soil contaminated with crude oil. J. Soil Sci. Plant Nutr. 17 (1), 180–193.

Dotaniya, M.L., 2013. Impact of various crop residue management practices on nutrient uptake by rice-wheat cropping system. Curr. Adv. Agric. Sci. 5 (2), 269–271.

Dotaniya, M.L., 2015. Impact of rising atmospheric CO_2 concentration on plant and soil process. In: Mohanty, M., Sinha, N.K., Hati, K.M., Chaudhary, R.S., Patra, A.K. (Eds.), Crop Growth Simulation Modelling and Climate Change. Scientific Publisher, Jodhpur, pp. 69–86.

Dotaniya, M.L., Datta, S.C., 2014. Impact of bagasse and press mud on availability and fixation capacity of phosphorus in an Inceptisol of north India. Sugar Tech. 16 (1), 109–112.

Dotaniya, M.L., Kushwah, S.K., 2013. Nutrients uptake ability of various rainy season crops grown in a Vertisol of central India. Afr. J. Agric. Res. 8 (44), 5592–5598.

Dotaniya, M.L., Meena, V.D., 2013. Rhizosphere effect on nutrient availability in soil and its uptake by plants -a review. Proc. Natl. Acad. Sci., India Sec. B: Biol. Sci. 85 (1), 1–12.

Dotaniya, M.L., Meena, B.P., 2017. Rhizodeposition by plants: A boon to soil health. In: Elanchezhian, R., Biswas, A.K., Ramesh, K., Patra, A.K. (Eds.), Advances in Nutrient Dynamics in Soil Plant System for Improving Nutrient Use Efficiency. New India Publishing Agency, New Delhi, pp. 207–224.

Dotaniya, M.L., Pipalde, J.S., 2018. Soil enzymatic activities as influenced by lead and nickel concentrations in a Vertisol of Central India. Bull. Environ. Contam. Toxicol. Available from: https://doi.org/10.1007/s00128-018-2402-9.

Dotaniya, M.L., Meena, H.M., Lata, M., Kumar, K., 2013a. Role of phytosiderophores in iron uptake by plants. Agric. Sci. Dig. 33 (1), 73–76.

Dotaniya, M.L., Prasad, D., Meena, H.M., Jajoria, D.K., Narolia, G.P., Pingoliya, K.K., et al., 2013b. Influence of phytosiderophore on iron and zinc uptake and rhizospheric microbial activity. Afr. J. Microbiol. Res. 7 (51), 5781–5788.

Dotaniya, M.L., Sharma, M.M., Kumar, K., Singh, P.P., 2013c. Impact of crop residue management on nutrient balance in rice-wheat cropping system in an Aquic hapludoll. J. Rural Agric. Res. 13 (1), 122–123.

Dotaniya, M.L., Datta, S.C., Biswas, D.R., Meena, B.P., 2013d. Effect of solution phosphorus concentration on the exudation of oxalate ions by wheat (*Triticum aestivum* L.). Proc. Natl. Acad. Sci., India Sec. B: Biol. Sci. 83 (3), 305–309.

Dotaniya, M.L., Das, H., Meena, V.D., 2014a. Assessment of chromium efficacy on germination, root elongation, and coleoptile growth of wheat (*Triticum aestivum* L.) at different growth periods. Environ. Monit. Assess. 186, 2957–2963.

Dotaniya, M.L., Datta, S.C., Biswas, D.R., Meena, H.M., Kumar, K., 2014b. Production of oxalic acid as influenced by the application of organic residue and its effect on phosphorus uptake by wheat (*Triticum aestivum* L.) in an Inceptisol of north India. Natl. Acad. Sci. Lett. 37 (5), 401–405.

Dotaniya, M.L., Meena, V.D., Das, H., 2014c. Chromium toxicity on seed germination, root elongation and coleoptile growth of pigeon pea (*Cajanus cajan*). Legume Res. 37 (2), 225–227.

Dotaniya, M.L., Thakur, J.K., Meena, V.D., Jajoria, D.K., Rathor, G., 2014d. Chromium pollution: a threat to environment. Agric. Rev. 35 (2), 153–157.

Dotaniya, M.L., Datta, S.C., Biswas, D.R., Kumar, K., 2014e. Effect of organic sources on phosphorus fractions and available phosphorus in Typic Haplustept. J. Ind. Soc. Soil Sci. 62 (1), 80–83.

Dotaniya, M.L., Kushwah, S.K., Rajendiran, S., Coumar, M.V., Kundu, S., Rao, A.S., 2014f. Rhizosphere effect of *kharif* crops on phosphatases and dehydrogenase activities in a Typic Haplustert. Natl. Acad. Sci. Lett. 37 (2), 103–106.

Dotaniya, M.L., Saha, J.K., Meena, V.D., Rajendiran, S., Coumar, M.V., Kundu, S., et al., 2014g. Impact of tannery effluent irrigation on heavy metal build up in soil and groundwater in Kanpur. Agrotechnol 2 (4), 77.

Dotaniya, M.L., Datta, S.C., Biswas, D.R., Meena, H.M., Rajendiran, S., Meena, A.L., 2015. Phosphorus dynamics mediated by bagasse, press mud and rice straw in Inceptisol of north India. Agrochimica 59 (4), 358–369.

Dotaniya, M.L., Meena, V.D., Kumar, K., Meena, B.P., Jat, S.L., Lata, M., et al., 2016a. Impact of Biosolids on Agriculture and Biodiversity. Today and Tomorrow's Printer and Publisher, New Delhi, pp. 11–20.

Dotaniya, M.L., Datta, S.C., Biswas, D.R., Dotaniya, C.K., Meena, B.L., Rajendiran, S., et al., 2016b. Use of sugarcane industrial byproducts for improving sugarcane productivity and soil health-a review. Int. J. Recycl. Org. Waste Agric. 5 (3), 185–194.

Dotaniya, M.L., Rajendiran, S., Coumar, M.V., Meena, V.D., Saha, J.K., Kundu, S., et al., 2017a. Interactive effect of cadmium and zinc on chromium uptake in spinach grown on Vertisol of Central India. Int. J. Environ. Sci. Technol. 15 (2), 441–448.

Dotaniya, M.L., Meena, V.D., Rajendiran, S., Coumar, M.V., Saha, J.K., Kundu, S., et al., 2017b. Geo-accumulation indices of heavy metals in soil and groundwater of Kanpur, India under long term irrigation of tannery effluent. Bull. Environ. Contam. Toxicol. 98 (5), 706–711.

Dotaniya, M.L., Rajendiran, S., Meena, V.D., Saha, J.K., Coumar, M.V., Kundu, S., et al., 2017c. Influence of chromium contamination on carbon mineralization and enzymatic activities in Vertisol. Agric. Res. 6 (1), 91–96.

Dotaniya, M.L., Dotaniya, C.K., Sanwal, R.C., Meena, H.M., 2018a. CO_2 sequestration and transformation potential of agricultural system. In: Martínez, L., Kharissova, O., Kharisov, B. (Eds.), Handbook of Ecomaterials. Springer. Available from: https://doi.org/10.1007/978-3-319-48281-1_87-1.

Dotaniya, M.L., Meena, V.D., Saha, J.K., Rajendiran, S., Patra, A.K., Dotaniya, C.K., et al., 2018b. Environmental impact measurements: tool and techniques. In: Martínez, L., Kharissova, O., Kharisov, B. (Eds.), Handbook of Ecomaterials. Springer, Cham. Available from: https://doi.org/10.1007/978-3-319-48281-1_87-1.

Dotaniya, M.L., Panwar, N.R., Meena, V.D., Dotaniya, C.K., Regar, K.L., Lata, M., et al., 2018c. Bioremediation of metal contaminated soils for sustainable crop production. In: Meena, V.S. (Ed.), Role of Rhizospheric Microbes in Soil. Springer, India. Available from: https://doi.org/978-981-10-8401-0,460132_1.

Dotaniya, M.L., Rajendiran, S., Coumar, M.V., Meena, V.D., Kundu, S., Saha, J.K., et al., 2018d. Remediation of chromium toxicity in wheat by use of FYM and pressmud. In: Oral Presentation in the National Conference "Organic waste management for food and environmental security" at ICAR-IISS, Bhopal During 08–10 February, 2018.

Dotaniya, M.L., Aparna, K., Dotaniya, C.K., Singh, M., Regar, K.L., 2018e. Role of soil enzymes in sustainable crop production. In: Khudus, et al., (Eds.), Enzymes in Food Biotechnology. Springer international.

Feigin, A., Ravina, I., Shalhevet, J., 1991. Irrigation With Treated Sewage Effluent: Management for Environmental Protection. Springer, Berlin, p. 224.

Ferguson, J.E., 1990. The Heavy Elements: Chemistry, Environmental Impact and Health Effects. Pergamon Press, New York.

Flora, S.J.S., Mittal, M., Mehta, A., 2008. Heavy metal induced oxidative stress & its possible reversal by chelation therapy. Indian J. Med. Res. 128, 501–523.

Frankenberger, W.T., Tabatabai, M.A., 1991. 2-Glutaminase activity of soils. Soil Biol. Biochem. 23, 869–874.

Friberg, L., Piscator, M., Nordberg, G.F., Kjellstrom, T., 1971. Cadmium in the Environment. CRC Press, Cleveland, p. 166.

Ghosh, A.K., Bhatt, M.A., Agrawal, H.P., 2012. Effect of long-term application of treated sewage water on heavy metal accumulation in vegetables grown in northern India. Environ. Monit. Assess. 184, 1025–1036.

Gianfreda, L., Bollag, M.J., 1996. Influence of natural and anthropogenic factors on enzyme activity in soil. In: Stotzky, G., Bollag, J.M. (Eds.), Soil Biochemistry. Marcel Dekker, New York.

Gürkan, R., Ulusoy, H.I., Akçay, M., 2012. Simultaneous determination of dissolved inorganic chromium species in wastewater/natural waters by surfactant sensitized catalytic kinetic spectrophotometry. Arabian J. Chem. 10 (1), S450–S460.

Gupta, A., Dixit, S.K., Senthil-Kumar, M., 2016. Drought stress predominantly endures *Arabidopsis thaliana* to *Pseudomonas syringae* infection. Front. Plant Sci. 7, 808.

Han, J.X., Shang, Q., Du, Y., 2009. Effect of environmental cadmium pollution on human health. Health 1 (3), 159–166.

Henson, M.C., Chedrese, P.J., 2004. Endocrine disruption by cadmium, a common environmental toxicant with paradoxical effects on reproduction. Exp. Biol. Med. (Maywood) 229 (5), 383–392.

Irfan, M., Hayat, S., Ahmad, A., Alyemeni, M.N., 2013. Soil cadmium enrichment: allocation and plant physiological manifestations. Saudi J. Biol. Sci. 20 (1), 1–10.

Jaishankar, M., Tseten, T., Anbalagan, N., Mathew, B.B., Beeregowda, K.N., 2014. Toxicity, mechanism and health effects of some heavy metals. Interdiscip. Toxicol. 7 (2), 60–72.

Jat, R.L., Jha, P., Dotaniya, M.L., Lakaria, B.L., Rashmi, I., Meena, B.P., et al., 2018. Carbon and nitrogen mineralization in Vertisol as mediated by type and placement method of residue. Environ. Monit. Assess. 190, 439.

Karaca, A., Cema, C.C., Turgay, O.C., Kizilkaya, R., 2011. Soil enzymes as indicator of soil quality. In: Shukla, G., Varma, A. (Eds.), Soil Enzymology. Springer-Verlag, Berlin Heidelberg, USA.

Killham, K., 2002. Bioindicators and sensors of soil health and the application of geostatistics. In: Burns, R.G., Dick, R.P. (Eds.), Enzymes in the Environment: Activity, Ecology and Applications. Marcel Dekker, Inc, USA.

Kumar, S., Meena, R.S., Lal, R., Yadav, G.S., Mitran, T., Meena, B.L., et al., 2018. Role of legumes in soil carbon sequestration. In: Meena, R., Das, A., Yadav, G., Lal, R. (Eds.), Legumes for Soil Health and Sustainable Management. Springer, Singapore, pp. 109–138.

Kundu, S., Dotaniya, M.L., Lenka, S., 2013. Carbon sequestration in Indian agriculture. In: Lenka, S., Lenka, N.K., Kundu, S., Rao, A.S. (Eds.), Climate Change and Natural Resources Management. New India Publishing Agency, pp. 269–289.

Kushwah, S.K., Dotaniya, M.L., Upadhyay, A.K., Rajendiran, S., Coumar, M.V., Kundu, S., et al., 2014. Assessing carbon and nitrogen partition *in kharif* crops for their carbon sequestration potential. Natl. Acad. Sci. Lett. 37 (3), 213–217.

Lenka, S., Lenka, N.K., Kundu, S., Rao, A.S., 2013. Climate change and natural resources management, first ed. New India Publishing Agency, New Delhi, India.

Lenka, S., Rajendiran, S., Coumar, M.V., Dotaniya M.L., Saha J.K., 2016. Impacts of fertilizers use on environmental quality. In: National Seminar on Environmental Concern for Fertilizer Use in Future at Bidhan Chandra Krishi Viswavidyalaya, Kalyani on February 26, 2016.

Levinson, A.A., 1974. Introduction to Exploration Geochemistry. Applied Publishing Ltd, Wilmette, Illinois, U.S.A, 924 pp.

Mandal, A., Radha, T.K., Neenu, S., 2013. Impact of climate change on rhizosphere microbial activity and nutrient cycling. In: Lenka, S., Lenka, N. K., Kundu, S., Rao, A.S. (Eds.), Climate Change and Natural Resource Management. New India Publishing Agency, New Delhi, pp. 93–116.

Mandal, A., Thakur, J.K., Sahu, A., Bhattacharjya, S., Manna, M.C., Patra, A.K., 2017. Plant–microbe interaction for the removal of heavy metal from contaminated site. In: Choudhary, D., Varma, A., Tuteja, N. (Eds.), Plant-Microbe Interaction: An Approach to Sustainable Agriculture. Springer, Singapore.

McLaughlin, M.J., Zarcinas, B.A., Stevens, D.P., Cook, N., 2000. Soil testing for heavy metals. Commun. Soil Sci. Plant Anal. 31, 1661–1700.

Meena, V.D., Dotaniya, M.L., 2017. Climate change, water scarcity and sustainable agriculture for food security. In: Kumar, P.S., Kanwat, M., Meena, P.D., Kumar, V., Alone, R.A. (Eds.), Climate Change & Sustainable Agriculture. New India Publishing Agency, New Delhi, pp. 123–142.

Meena, V.D., Dotaniya, M.L., Saha, J.K., Patra, A.K., 2015. Antibiotics and antibiotic resistant bacteria in wastewater: impact on environment, soil microbial activity and human health. Afr. J. Microbiol. Res. 9 (14), 965–978.

Meena, B.L., Meena, R.L., Kanwat, M., Kumar, A., Dotaniya, M.L., 2017. Impact of climate change under coastal ecosystem & adoption strategies. In: Kumar, P.S., Kanwat, M., Meena, P.D., Kumar, V., Alone, R.A. (Eds.), Climate Change & Sustainable Agriculture. New India Publishing Agency, New Delhi, pp. 55–66.

Meena, B.L., Fagodiya, R.K., Prajapat, K., Dotaniya, M.L., Kaledhonkar, M.J., Sharma, P.C., et al., 2018. Legume green manuring: an option for soil sustainability. In: Meena, R., Das, A., Yadav, G., Lal, R. (Eds.), Legumes for Soil Health and Sustainable Management. Springer, Singapore, pp. 387–408.

Meers, E., Samson, R., Tack, F.M.G., et al., 2007. Phytoavailability assessment of heavy metals in soils by single extractions and accumulation in *Phaseolus vulgaris*. Environ. Exp. Bot. 60, 385–396.

Morais, S., Costa, F.G., Pereira, M.L., 2012. Heavy metals and human health. In: Oosthuizen, J. (Ed.), Environmental Health – Emerging Issues and Practice. InTech, pp. 227–246.

Muller, G., 1969. Index of geo-accumulation in sediments of the Rhine river. Geo. J. 2, 109–118.

Nagajyoti, P.C., Lee, K.D., Sreekanth, T.V.M., 2010. Heavy metals, occurrence and toxicity for plants: a review. Environ. Chem. Lett. 8 (3), 199–216.

Najeeb, U., Ahmad, W., Zia, M.H., Malik, Z., Zhou, W., 2014. Enhancing the lead phytostabilization in wetland plant *Juncus effusus* L. through somaclonal manipulation and EDTA enrichment. Arabian J. Chem. 10, S3310–S3317.

Nath, K., Saini, S., Sharma, Y.K., 2005. Chromium in tannery industry effluent and its effect on plant metabolism and growth. J. Environ. Biol. 26, 197–204.

Nriagu, J.O., 1996. History of global metal pollution. Science 272, 223–224.

Paranthaman, S.R., Karthikeyan, B., 2015. Bioremediation of heavy metal in paper mill effluent using *Pseudomonas spp*. Intl. J. Microbiol. 1, 1–5.

Pipalde, J.S., Dotaniya, M.L., 2018. Interactive effects of lead and nickel contamination on nickel mobility dynamics in spinach. Int. J. Environ. Res. Available from: https://doi.org/10.1007/s41742-018-0107-x.

Prajapati, K., Rajendiran, S., Coumar, M.V., Dotaniya, M.L., Ajay, Kundu, S., et al., 2016. Carbon occlusion potential of rice phytoliths: implications for global carbon cycle and climate change mitigation. Appl. Ecol. Environ. Res. 14 (2), 265–281.

Priyalaxmi, R., Murugan, A., Raja, P., Raj, K.D., 2014. Bioremediation of cadmium by *Bacillus safensis* (JX126862), a marine bacterium isolated from mangrove sediments. Intl. J. Curr. Microbiol. Appl. Sci. 3, 326–335.

Rajendiran, S., Dotaniya, M.L., Coumar, M.V., Panwar, N.R., Saha, J.K., 2015. Heavy metal polluted soils in India: status and countermeasures. JNKVV Res. J. 49 (3), 320–337.

Rajendiran, S., Singh, T.B., Saha, J.K., Coumar, J.K., Dotaniya, M.L., Kundu, S., et al., 2018. Spatial distribution and baseline concentration of heavy metals in swell–shrink soils of Madhya Pradesh, India. In: Singh, V., Yadav, S., Yadava, R. (Eds.), Environmental Pollution. Water Sci. Tech. Lib. Springer, Singapore, pp. 135–145.

Saha, J.K., Sharma, A.K., 2006. Impact of the Use of Polluted Irrigation Water on Soil Quality and Crop Productivity Near Ratlam and Nagda Industrial Area. IISS Bulletin, Indian Institute of Soil Science, Bhopal, p. 26.

Saha, J.C., Dikshit, A.K., Bandyopadhyay, M., Saha, K.C., 1999. A review of arsenic poisoning and its effects on human health. Crit. Rev. Environ. Sci. Technol. 29 (3), 281–313.

Saha, J.K., Panwar, N., Srivastava, A., Biswas, A.K., Kundu, S., Rao, A.S., 2010. Chemical, biochemical, and biological impact of untreated domestic sewage water use on Vertisol and its consequences on wheat (*Triticum aestivum*) productivity. Environ. Monit. Assess. 161, 403–412.

Saha, J.K., Panwar, N., et al., 2013. Risk assessment of heavy metals in soil of a susceptible agroecological system amended with municipal solid waste compost. J. Indian Soc. Soil Sci. 61, 15–22.

Saha, J.K., Rajendiran, S., Coumar, M.V., Dotaniya, M.L., Kundu, S., Patra, A.K., 2017a. Agriculture, soil and environment. In: Saha, J.K., et al., (Eds.), Soil Pollution - An Emerging Threat to Agriculture. Springer, Singapore, pp. 1–9. Available from: http://doi.org/10.1007/978-981-10-4274-4_1.

Saha, J.K., Rajendiran, S., Coumar, M.V., Dotaniya, M.L., Kundu, S., Patra, A.K., 2017b. Soil and its role in the ecosystem. In: Saha, et al., (Eds.), Soil Pollution - An Emerging Threat to Agriculture. Springer, Singapore, pp. 11–36. Available from: http://doi.org/10.1007/978-981-10-4274-4_2.

Saha, J.K., Rajendiran, S., Coumar, M.V., Dotaniya, M.L., Kundu, S., Patra, A.K., 2017c. Impacts of soil pollution and their assessment. In: Saha, J. K., et al., (Eds.), Soil Pollution - An Emerging Threat to Agriculture. Springer, Singapore, pp. 37–73. Available from: http://doi.org/10.1007/978-981-10-4274-4_3.

Saha, J.K., Rajendiran, S., Coumar, M.V., Dotaniya, M.L., Kundu, S., Patra, A.K., 2017d. Major inorganic pollutants affecting soil and crop quality. In: Saha, J.K., et al., (Eds.), Soil Pollution - An Emerging Threat to Agriculture. Springer, Singapore, pp. 75–104. Available from: http://doi.org/10.1007/978-981-10-4274-4.

Saha, J.K., Rajendiran, S., Coumar, M.V., Dotaniya, M.L., Kundu, S., Patra, A.K., 2017e. Organic pollutants. In: Saha, J.K., et al., (Eds.), Soil Pollution - An Emerging Threat to Agriculture. Springer, Singapore, pp. 105–135. Available from: http://doi.org/10.1007/978-981-10-4274-5.

Saha, J.K., Rajendiran, S., Coumar, M.V., Dotaniya, M.L., Kundu, S., Patra, A.K., 2017f. Collection and processing of polluted soil for analysis. In: Saha, J.K., et al., (Eds.), Soil Pollution - An Emerging Threat to Agriculture. Springer, Singapore, pp. 137–153. Available from: http://doi.org/10.1007/978-981-10-4274-6.

Saha, J.K., Rajendiran, S., Coumar, M.V., Dotaniya, M.L., Kundu, S., Patra, A.K., 2017g. Assessment of heavy metals contamination in soil. In: Saha, J.K., et al., (Eds.), Soil Pollution - An Emerging Threat to Agriculture. Springer, Singapore, pp. 155–191. Available from: http://doi.org/10.1007/978-981-10-4274-7.

Saha, J.K., Rajendiran, S., Coumar, M.V., Dotaniya, M.L., Kundu, S., Patra, A.K., 2017h. Urban activities in India leading to soil pollution. In: Saha, J.K., et al., (Eds.), Soil Pollution - An Emerging Threat to Agriculture. Springer, Singapore, pp. 193–228. Available from: http://doi.org/10.1007/978-981-10-4274-8.

Saha, J.K., Rajendiran, S., Coumar, M.V., Dotaniya, M.L., Kundu, S., Patra, A.K., 2017i. Industrial activities in India and their impact on agroecosystem. In: Saha, J.K., et al., (Eds.), Soil Pollution - An Emerging Threat to Agriculture. Springer, Singapore, pp. 229–249. Available from: http://doi.org/10.1007/978-981-10-4274-9.

Saha, J.K., Rajendiran, S., Coumar, M.V., Dotaniya, M.L., Kundu, S., Patra, A.K., 2017j. Impact of different developmental projects on soil fertility. In: Saha, J.K., et al., (Eds.), Soil Pollution - An Emerging Threat to Agriculture. Springer, Singapore, pp. 251–269. Available from: http://doi.org/10.1007/978-981-10-4274-10.

Saha, J.K., Rajendiran, S., Coumar, M.V., Dotaniya, M.L., Kundu, S., Patra, A.K., 2017k. Status of soil pollution in India. In: Saha, J.K., et al., (Eds.), Soil Pollution - An Emerging Threat to Agriculture. Springer, Singapore, pp. 271–315. Available from: http://doi.org/10.1007/978-981-10-4274-11.

Saha, J.K., Rajendiran, S., Coumar, M.V., Dotaniya, M.L., Kundu, S., Patra, A.K., 2017l. Remediation and management of polluted sites. In: Saha, J.K., et al., (Eds.), Soil Pollution - An Emerging Threat to Agriculture. Springer, Singapore, pp. 317–372. Available from: http://doi.org/10.1007/978-981-10-4274-12.

Saha, J.K., Rajendiran, S., Coumar, M.V., Dotaniya, M.L., Kundu, S., Patra, A.K., 2017m. Soil protection policy. In: Saha, J.K., et al., (Eds.), Soil Pollution - An Emerging Threat to Agriculture. Springer, Singapore, pp. 373–382. Available from: http://doi.org/10.1007/978-981-10-4274-13.

Shanker, A.K., Cervantes, C., Loza-Tavera, H., Avudainayagam, S., 2005. Chromium toxicity in plants. Environ. Int. 31, 739–753.

Singh, B., 2002. Soil pollution and its control. In: Sekhon, G.S., Chhonkar, P.K., Das, D.K., Goswami, N.N., Narayanaswamy, G., Poonia, S.R., et al., Fundamental of Soil Science. Ind. Soc. Soil Sci, New Delhi, pp. 499–514.

Singh, M., Dotaniya, M.L., Mishra, A., Dotaniya, C.K., Regar, K.L., 2016. Role of biofertilizers in conservation agriculture. In: Bisht, J.K., Meena, V.S., Mishra, P.K., Pattanayak, A. (Eds.), Conservation Agriculture: An Approach to Combat Climate Change in Indian Himalaya. Springer, pp. 113–134.

Singh, V.S., Meena, S.K., Verma, J.P., Kumar, A., Aeron, A., Mishra, P.K., et al., 2017. Plant beneficial rhizospheric microorganism (PBRM) strategies to improve nutrients use efficiency: a review. Ecol. Eng. 107, 8–32.

Sreeramulu, U.S., 2010. Soil pollution: Indian scenario. In: Kundu, S., Jha, P., Hati, K.M., Mohanty, S.R., Adhikari, T. (Eds.), Souvenir of 75th Annual Convention. Indian Soc. Soil Sci, Bhopal, pp. 105–110. , chapter.

Stohs, S.J., Bagchi, D., 1995. Oxidative mechanisms in the toxicity of metal ions. Free Radic. Biol. Med. 18 (2), 321–336.

Teo, J., Goh, K., Ahuja, A., Ng, H., Poon, W., 1997. Intracranial vascular calcifications, glioblastoma multiforme, and lead poisoning. AJNR Am J Neuroradiol 18, 576–579.

Trasande, L., Landrigan, P.J., Schechter, C., 2005. Public health and economic consequences of methyl mercury toxicity to the developing brain. Environ. Health Perspect. 113 (5), 590–596.

Trasar-Cepeda, C., Leiros, M.C., Seoane, S., Gil-Sotres, F., 2000. Limitation of soil enzymes as indicators of soil pollution. Soil Biol. Biochem 32, 1867–1875.

Utobo, E.B., Tewari, L., 2014. Soil enzymes as bioindicators of soil ecosystem status. Appl. Ecol. Environ. Res 13 (1), 147–169.

Wallace, J.M., Hobbs, P.V., 2006. The earth system. In: Wallace, J.M., Hobbs, P.V. (Eds.), Atmospheric Science an Introductory Survey. Academic press.

Yongsheng, W., Qihui, L., Qian, T., 2011. Effect of Pb on growth, accumulation and quality component of tea plant. Proc. Eng. 18, 214–219.

Zhuang, P., Zou, H., Shu, W., 2009. Biotransfer of heavy metals along a soil-plant-insect-chicken food chain: field study. J. Environ. Sci. 21, 849–853.

Further reading

Coumar, M.V., Kundu, S., Saha, J.K., Rajendiran, S., Dotaniya, M.L., Meena, V.D., et al., 2018. Relative contribution of phosphorus from various sources to the upper lake, Bhopal. In: Singh, V., Yadav, S., Yadava, R. (Eds.), Environmental Pollution. Water Science and Technology Library. Springer, Singapore, pp. 459–467.

Dotaniya, M.L., Rajendiran, S., Meena, B.P., Meena, A.L., Meena, B.L., Jat, R.L., et al., 2016a. Elevated carbon dioxide (CO_2) and temperature vis-a-vis carbon sequestration potential of global terrestrial ecosystem. In: Bisht, J.K., Meena, V.S., Mishra, P.K., Pattanayak, A. (Eds.), Conservation Agriculture: An Approach to Combat Climate Change in Indian Himalaya. Springer, pp. 225–256.

Dotaniya, M.L., Meena, V.D., Basak, B.B., Meena, R.S., 2016b. Potassium uptake by crops as well as microorganisms. In: Meena, V.S., Maurya, B.R., Verma, J.P., Meena, R.S. (Eds.), Potassium Solubilizing Microorganisms for Sustainable Agriculture. Springer, India, pp. 267–280.

Dotaniya, M.L., Meena, V.D., Lata, M., Meena, B.L., 2017. Climate change impact on agriculture: adaptation strategies. In: Kumar, P.S., Kanwat, M., Meena, P.D., Kumar, V., Alone, R.A. (Eds.), Climate Change & Sustainable Agriculture. New India Publishing Agency, New Delhi, pp. 27–38.

Kabata-Pendias, A., 2000. Trace Element in Soils and Plants, third ed. CRC Press, Baton Raton, p. 432.

Seo, J.S., Keum, Y.S., Li, Q.X., 2009. Bacterial degradation of aromatic compounds. Intl. J. Environ. Res. Public Health 6 (1), 278–309.

Singh, A., Agrawal, M., 2010. Effects of municipal waste water irrigation on availability of heavy metals and morpho-physiological characteristics of *Beta vulgaris* L. J. Environ. Biol. 31, 727–736.

Chapter 6

Microbial secondary metabolites and plant–microbe communications in the rhizosphere

Beenu Shastri and Rajesh Kumar

Rhizosphere Biology Laboratory, Department of Environmental Microbiology, Babasaheb Bhimrao Ambedkar University, Lucknow, India

Chapter Outline

6.1 Introduction

Soil in general may be referred as the "reserve of microorganisms" where a vast array of microbes colonize and interact. Within the soil, complex and dynamic communication occurs between plant roots, soil microorganisms, and physicochemical environment of the soil. The microbial community associated with the plant fluctuates from phyllosphere to rhizospheric zone. However, the rhizospheric zone (zone of soil surrounding the plant's root) is known for its largest biological diversity in terms of microbial richness due to the favorable environment provided by the plant root for the recruitment of microorganisms. Campbell and Greaves (1990) stated that the density of the microbial population in the plant root region is 100-fold larger than the bulk soil due to the secretion of root exudates, which attract, select, and set up microorganisms in the rhizosphere. While microorganisms also absorb certain compounds in the plant root exudates and release certain additional compounds to establish an intimate communication with the plants (Singh, 2013, 2015; Vimal et al., 2018). Plants, being stationary in nature, adapted and coevolved with microorganisms recruited in their niche and acquired a number of mechanisms that influence their interactions either in positive direction (mutualistic and symbiotic) or in negative direction (competitive, antagonistic, pathogenic, and parasitic). Such mechanisms induce various morphological, physiological, and behavioral changes in the microorganisms when exposed to interact with plants. Chemical signals are the main factor that drives the interaction between plants and soil organisms at the root proximity. These signals may be viewed as secondary metabolites and other compounds that constitute a complex chemical system through which plant and soil microbe communicate at the interface of biotic and abiotic environment. Secondary metabolites are proactive compounds that are found to play an important role in the underground

New and Future Developments in Microbial Biotechnology and Bioengineering. DOI: https://doi.org/10.1016/B978-0-12-818258-1.00006-6

communication (below rhizosphere) in the soil. Soil microorganisms release secondary metabolites that act as signaling molecules to the plant, affecting their metabolism by providing the plant host with microbial derived compounds. In turn, plants make available root exudates that act as substrate and signaling molecule to the soil microbe (Bais et al., 2006). Thus, for cooperative existence, stable and balance response in the soil is required, which can be achieved by these signaling compounds or secondary metabolites, which function as messengers for communication amongst the diverse group of organisms found in the soil ecosystems (Seneviratne, 2015).

Plant–microbial interaction has always been an area of research as such interactions are vital for successful establishment and maintenance of soil ecosystems by contributing efforts in maintaining ecological processes such as nutrient availability and uptake by plants, increasing soil productivity, sustaining the healthy structure of soil, and increased immunity against biotic and abiotic stress (Tiwari and Singh, 2017; Singh et al., 2018; Singh and Gupta, 2018). Thus, via different mechanisms these interactions influence the plant health and growth. Hence, understanding how this bidirectional plant–microbe association maintains a correct balance is important for sustenance of soil ecosystems. At present, there is incomplete understanding of plant–microbial signaling compounds and the mechanisms underlying plant–microbe interaction in both symbiotic and defense associations. This chapter focuses on various metabolites or compounds influencing the plant–microbe communication in the rhizosphere, and also on how these interactions are responsible for the maintenance of soil fertility, nutrient availability, plant productivity, and biocontrol of phytopathogens by the signaling compounds.

6.2 Rhizosphere: root microbiome and root exudates

Lorentz Hiltner (1904) defined the "rhizosphere" as the zone of soil surrounding the plant root, which is influenced and inhabited by the microorganisms. Within the rhizosphere, plant roots have direct influence on the composition and density of soil microbial community due to increased level of nutrients and highest respiration rate of root. The microbial community in the rhizosphere is influenced by the plant species and development stage. Rhizosphere has mainly two components, namely root microbiome and root exudates. The root microbiome is collectively defined as group of all microorganisms residing in the root or rhizosphere (Lundberg et al., 2012). Biotic and abiotic factors mainly control the composition of the root microbiome. Root microbiome found in the rhizosphere includes bacteria, fungi, oomycetes, nematodes, protozoa, algae, viruses, archaea, and arthropods. Therefore, plant-associated complex microbial communities are referred to as the "second genome of the plants" (Berendsen et al., 2012). Root exudates are generally a variety of small molecules excreted or synthesized continuously by plant roots that get accumulated in the soil (Bais et al., 2006). Root exudate includes soluble sugar, amino acid, organic acid, protein, lipids, mucilage, or secondary metabolite (Badri and Vivanco, 2009) and these substances are used by the microbes living in the rhizosphere. Root exudates mediate the interaction between plant root and microbe in the rhizosphere. These interactions affect the microbial population in the rhizosphere by altering the physicochemical properties of soil (Badri et al., 2009).

However, this interaction between root microbiome and root exudates is bidirectional as both the components produce many metabolites for serving their community to achieve better fitness in dynamically changing soil conditions. For example, root exudates release carbon (C) content in the soil, which is then utilized by the rhizospheric microorganisms, which in turn mobilize nitrogen (N) and phosphorus (P) to make it available to the plants. Rhizospheric microorganisms including nitrogen fixing bacteria, mycorrhizal fungi, and plant growth promoting rhizobacteria (PGPR), biocontrol microorganisms, mycoparasitic fungi, and protozoa have been well studied for their constructive effects on plant growth and health since their discovery. The term "PGPR" was first used by Kloepper and Schroth (1978) to describe those bacteria that colonize the plant roots and help in promoting the growth of plants.

6.3 Rhizospheric microorganisms and their multifaceted role in plant–microbe communication

Plant–microbe interactions have been a fundamental topic of interest in the field of rhizospheric research. These interactions between plants and microorganisms can be beneficial, harmful, and neutral based on exchanges of complex molecular signals. The signaling compounds play diverse roles in ensuring benefits to both parties of the plant–microbe interaction. Plant root releases a wide range of potentially active compounds into the rhizosphere, which play a key role in the chemical, physical, and biological interaction between roots of the plants and the rhizosphere. The broad range of signaling molecules are synthesized by plants including primary metabolites (carbohydrates, proteins, organic acids, etc.) and secondary metabolites (flavonoids, phenol, phytohormones, etc.), which play an important role as chemical signals to attract the microorganisms in the rhizosphere (Singh et al., 2016a,b). While on the other hand,

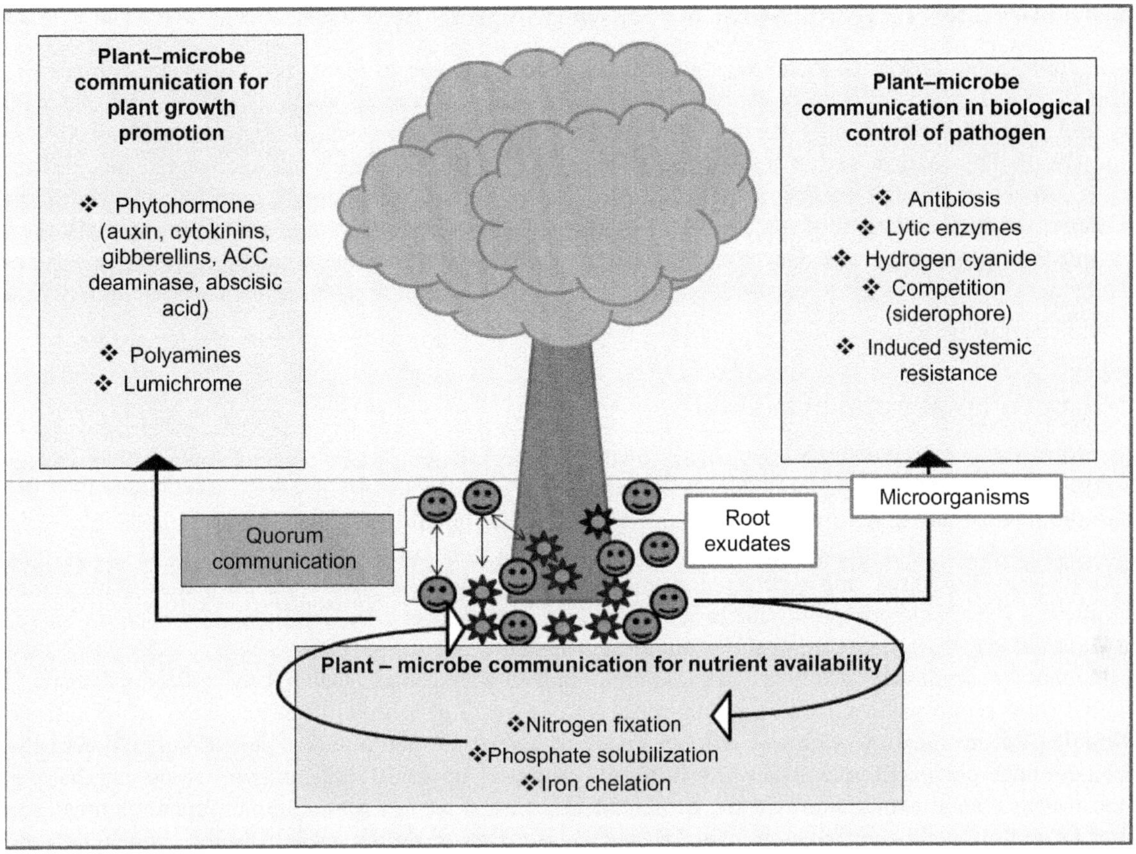

FIGURE 6.1 A diagrammatic representation of different plant–microbe communication.

microorganisms in the rhizosphere also releases wide variety of small molecules or volatile compounds that includes, phytohormones, enzymes, siderophores, hydrogen cyanide (HCN), *N*-acyl-l-homoserine lactones, etc. (reviewed in Ortíz-Castro et al., 2009), which are recognized by plants to adjust their defense and growth responses against the microorganisms exposed to it.

Long-term close interactions between different biological species, such as symbiosis and pathogenesis, are common between plants and soil microorganisms. Amongst them, number of such mutualistic associations like the legume–rhizobia nitrogen-fixing symbiosis of plant–bacterial nature and that of mycorrhizae with plant roots (plant–fungal nature) are well documented. There are a number of ways by which both plants and microbes communicate for the promotion of plant growth including synthesis of compound for acquisition of nutrient uptake by plants and disease prevention (Fig. 6.1).

Rhizospheric interactions are important for application in agriculture. Plant beneficial interactions are roughly divided into three categories. Firstly, the interaction of microorganisms, those in association with plants are highly responsible for improving the nutritional status (i.e., microorganisms that can increase the supply of mineral nutrients to the plant). In this case, while most may not directly interact with the plant, their effects on soil biotic and abiotic parameters certainly have an impact on plant growth. The second category involves interactions of those microorganisms with the plant that are responsible for direct growth promotion of the plant by releasing various plant growth regulating hormones, for example, by production of various phytohormones. The third category involves the communication where a group of microorganisms stimulate plant growth indirectly by preventing the growth or activity of phytopathogens by process of antibiosis, competition, and inducing systemic resistance (ISR). Such microorganisms are referred to as biocontrol agents, and they have been well documented. Thus, plant–microbe interaction in the rhizosphere plays a central role in maintaining health and sustainability of soil.

6.4 Plant–microbe communication for nutrient availability

Nutrient availability is an important source for the growth and development of plants and microorganisms both. Various physical and chemical reactions occurring in the rhizosphere zone influence microbial interactions that mediate the availability and acquisition of nutrients (Agrios, 2005). Rhizospheric soil represents a large dynamic source and sink of all nutrients (N, P, Fe, K, and S) but due to its unavailable nature to plants, soil microorganisms utilize various approaches in soil ecosystems for nutrient acquisition, cycling, and ensuring availability to the plants. Multiple microorganisms are involved in functioning of soil fertility, soil structure, soil properties, nutrient cycling through the complex interaction, influencing plant growth and development. Direct functional attributes including nitrogen fixation, phosphate solubilization, iron acquisition, zinc solubilization, and potassium mobilization are necessary for the uptake of insoluble nutrients by plants.

6.4.1 Nitrogen uptake and availability

Nitrogen is one of the essential elements for plant growth and development, and also contributes to the soil fertility by nitrogen cycle (Jetten, 2008). Nitrogen is abundantly present in gaseous form but due to this nature it is not readily accessible to the plant (Baas et al., 2014). Molecular nitrogen cannot be directly assimilated by plants, but with the process of biological nitrogen fixation assisted by microorganisms, it gets converted to ammonia, which becomes easily available to plants. Biological nitrogen fixation mainly occurs through symbiotic or nonsymbiotic association. Symbiotic nitrogen fixing microorganisms includes *Rhizobium, Frankia*, and the free-living soil bacteria including cyanobacteria, *Azotobacter*, *Azospirillum*, *Klebsiella*, and *Clostridium.*

In the mutualistic symbiotic Rhizobium–legume communication, the plant benefits the bacteria with fixed carbon (organic acids) from photosynthesis, which the bacterium requires for growth while it continues to fix nitrogen in the nodules. Nodules are the sites for symbiotic nitrogen fixation that are formed after a series of interactions between rhizobia and leguminous plants. Root exudates containing flavonoid compounds serve as specific signals that get synthesized by the roots of leguminous plants in the symbiotic association of *Rhizobium*, which enhance their growth and attract rhizobial cells toward the plant, through the induction of transcription of *nod* genes. *Nod* gene induction is dependent on flavonoid concentration in root exudates involved in the synthesis of lipo-chitooligosaccharide signals, called Nod factors (Kumar and Chauhan, 2014). Besides flavonoids, other compounds in exudates also play an important role in the determination of microbial community structure in the plant rhizosphere.

Flavonoids also act as signaling compounds in mycorrhizae symbiosis (Steinkellner et al., 2007). The arbuscular mycorrhiza fungi (AMF) symbiotically communicate with more than 80% of plant species (Parniske, 2008). Strigolactone is another component of root exudates exuded from the roots and considered to be the plant signal molecule for AMF symbiosis (Akiyama et al., 2005). AMF establishes its contact with the root of plants by forming hyphae branches and branching factor with the aid of strigolactone (Akiyama et al., 2005). It is hypothesized that AMF may use the ancestral receptor to perceive signals of strigolactones in the rhizosphere (Parniske, 2005). Hence, the identification of the receptors for strigolactones in fungi might be a fascinating topic for future research.

6.4.2 Phosphate solubilization

Phosphorus (P) is the second major limiting macronutrient after nitrogen, necessary for plant growth and development (Kumar and Shastri, 2017). Like nitrogen, it is also present in the immobilized form as organic and inorganic phosphates in the soil. Phosphorus is also not readily available to plants due to its high reactivity with some metal complexes such as Fe, Al, and Ca leading to precipitation in the soil (Kumar and Shastri, 2017).

Phosphate-solubilizing microorganisms mainly solubilize insoluble phosphate by the following mechanism:

First, by releasing mineral dissolving compounds such as organic acid, siderophores, protons, hydroxyl ions (H +), and CO_2 (Rodríguez and Fraga, 1999), which results in lowering the pH (Sharma et al., 2017). Gluconic acid and 2-ketogluconic acid are the main organic acids released by the phosphate solubilizing microorganisms, resulting in acidification of the microbial cells and their surroundings, hence, phosphate ions are released. Thus, the pH of the rhizosphere and phosphorous availability are inversely connected.

Second, by liberation of extracellular enzymes such as phosphatases, phytases, phosphonatases, and C–P lyases (Sharma et al., 2013).

At last, phosphate is released during substrate degradation (biological phosphate mineralization) (McGill and Cole, 1981). Some forms of phosphorus in soil exist as high molecular weight compounds (inositol phosphates, phosphoester, phosphodiester, and phosphotriesters), which are first converted into inorganic or low molecular weight organic acids before they can be easily assimilated by the plants. Therefore, microorganisms play an important role as a component of the soil in the phosphorous cycling.

Phosphate-solubilizing microorganisms (PSM) include several groups of fungi and bacteria, which assist the plants in the mobilization of insoluble forms of phosphate. Among the bacteria, *Bacillus* is the most powerful phosphate solubilizer. Besides this, *Azotobacter, Azospirillum, Rhizobium, Achromobacter*, *Agrobacterium*, *Enterobacter*, *Erwinia*, *Escherichia*, *Flavobacterium*, *Mycobacterium*, *Pseudomonas*, and *Serratia* are also highly efficient in solubilizing unavailable complexed phosphate into available inorganic phosphate ion (Saharan and Nehra, 2011). Amongst fungi, *Penicillium, Trichoderma, Rhizoctonia solani*, and *Aspergillus* are the most powerful phosphate solubilizers (Whitelaw, 1999; Jacobs et al., 2002). A nematofungus *Arthrobotrys oligospora* also has the ability to solubilize the phosphate rocks (Duponnois et al., 2006).

Mycorrhizal fungi are the other major mutualist group in phosphorus cycle that are the major phosphorus contributors to the plants in exchange for reduced carbon. AMF have been found to be an essential component of sustainable soil–plant systems (Schreiner et al., 1997). AMF enhances the phosphate nutrition of plants by scavenging the available phosphorus due to the large surface area of their hyphae, and by their high-affinity phosphate uptake mechanisms (Hayman, 1983). PSM improves the solubilization of fixed soil phosphate, resulting in higher crop yields, and therefore are used as biofertilizers.

6.4.3 Iron uptake

Iron (Fe) belongs to the transition group of elements and is an essential micronutrient for almost all microorganisms, including bacteria, fungi, and plants. It facilitates the processes of respiration, photosynthesis, and nitrogen fixation but because of its immobilized form as insoluble iron oxides or ferric hydroxide in soil, microbes find difficulty in obtaining enough iron to support their growth as well as plant growth. Iron acts as a limiting factor for plant growth even in iron-rich soils because its availability to the organism is very limited due to the rapid oxidation of ferrous (Fe^{2+}) to insoluble ferric (Fe^{3+}) state (Neilands, 1995).

Under iron deficit conditions, plants have evolved two mechanisms to increase iron solubility in the rhizosphere. Both strategies convert inorganic ferric (Fe^{3+}) into ferrous (Fe^{2+}), which can be easily accessible to plants. The first strategy involves a plasma membrane bound reductase enzyme to convert Fe^{3+} to accessible Fe^{2+} form. The second strategy is mediated by the secretion of organic Fe^{3+} chelating molecules known as phytosiderophores (Römheld, 1987). Phytosiderophores have been widely reported from the roots of different species of the grass family (including oat, barley, wheat, and rice) (Dotaniya et al., 2013).

Rhizospheric microorganisms also have evolved particular mechanisms for the assimilation of iron, by the production of low molecular weight iron-chelating compounds known as siderophores, which transport iron into their cells through membrane receptors and also to the plants (Loper and Buyer, 1991; Neilands, 1995). Siderophores have been implicated for both direct and indirect enhancement of plant growth (Neilands, 1981). Plants benefit from this increased Fe availability, mediated by the siderophore and through their signaling with root exudate compounds, select the beneficial microbes with siderophore producing ability to colonize their root to enhance iron availability (Hartmann et al., 2009).

Siderophores and phytosiderophores have equal importance as some researchers reported that bacterial siderophores are usually poor sources of Fe for both monocot and dicot plants (Bar-Ness et al., 1992; Crowley et al., 1992) while some reported microbial siderophores have alleviated Fe deficiency-induced chlorosis in dicots (Yehuda et al., 2000). On the other hand, plant derived Fe-Phytosiderophore complexes appear to be a good Fe source for bacteria (Marschner and Crowley, 1998). Thus, Fe-chelating compounds (microbial or plant origin) are highly soluble and are stable over a wide range of pH and equally important as they help in the transportation of iron as nutrient in soil and to plants.

6.5 Plant–microbe communication for plant growth promotion

Microbes produce a wide range of signaling molecules like phytohormones (auxin, cytokinin, and gibberellins) in plant–microbe communication that are involved in the direct control of plant growth and their development (Vacheron et al., 2013) as illustrated in Table 6.1. Such rhizospheric microorganisms are designated as PGPR as it facilitates the direct growth promotion of plants (Kloepper and Schroth, 1978). Another mechanism of plant growth stimulation by PGPR is the production of 1-aminocyclopropane-1-carboxylate (ACC) deaminase and various other compounds. The

TABLE 6.1 Role of secondary metabolites produced from microorganisms involved in plant–microbe communication for plant growth promotion.

S. no.	Secondary metabolites	Microorganisms	Role
1.	Auxin	*Azospirillum, Aeromonas, Alcaligenes, Bradyrhizobium, Enterobacter, Comamonas, Pseudomonas* and *Rhizobium*	Stimulation of root system, development and growth promotion of plant via increasing the density and length of root hairs; expansion of uptake mechanisms of water and mineral nutrients from a large volume of soil
2.	Cytokinins	*Bacillus subtilis, Sinorhizobium meliloti* and *Pseudomonas*	Promotion of the cell division, root development, and root hair formation; alleviation of drought stress
3.	Gibberellins	*Agrobacterium, Azotobacter, Arthrobacter, Azospirillum, Bacillus, Burkholderia, Clostridium, Flavobacterium, Micrococcus, Pseudomonas, Xanthomonas, Achromobacter xylosoxidans, Gluconobacter diazotrophicus, Acinetobacter calcoaceticus, Rhizobia* and *Herbaspirillum seropedicae*	Induce cell elongation within the subapical meristem resulting in stem elongation; increased N uptake
4.	ACC deaminase	*Acinetobacter, Achromobacter, Agrobacterium, Alcaligenes, Azospirillum, Bacillus, Burkholderia, Enterobacter, Pseudomonas, Ralstonia, Serratia* and *Rhizobium*	Reduction of pathogen; protecting plant from stress conditions; increase nodulation and mycorrhizal colonization
5.	Abscisic acid	*Azospirillum brasilense, A. xylosoxidans, Bacillus pumilus, Brevibacterium halotolerans, Bacillus subtilis, Bacillus licheniformis, B. pumilus, Lysinibacillus fusiformis, A. xylosoxidans* and *Pseudomonas putida*	Stomatal closure in water- stressed environment; regulation of the uptake by and transport in plant found in arid and semiarid climates

Source: Adapted from Olanrewaju, O.S., Glick, B.R., Babalola, O.O., 2017. Mechanisms of action of plant growth promoting bacteria. World J. Microbiol. Biotechnol. 33, 197.

following section includes certain metabolites or signaling compounds synthesized by PGPR, which uses certain mechanisms to interact in the rhizospheric zone and enhance plant growth.

6.5.1 Phytohormones

Phytohormones are compounds synthesized by plants and are involved in the developmental activities of plants like cell division, tissue differentiation, cell elongation, nutrient movement, apical dominance, ripening, and abscission. In addition to plants, phytohormones are also synthesized by rhizospheric microorganisms within the root zone to stimulate the plant growth by increasing the density and length of root hairs. This increment in root hair expands the root surface area and ultimately advances the plant uptake potential of water and mineral nutrients from a large volume of soil (Volkmar and Bremer, 1998). Microbe produces various phytohormones such as auxins, gibberellins, cytokinins, and abscisic acid (ABA). To cope with various stresses, plants also synthesize phytohormones (salicylic acid, jasmonic acid, and ethylene) that induce resistance against various pathogens. These phytohormones are involved in providing protection against both biotic and abiotic stresses (Fujita et al., 2006).

6.5.1.1 Auxin

Rhizospheric microorganisms are known for synthesizing the phytohormone auxin and are regarded as secondary metabolites or signaling molecules (Patten and Glick, 1996). This signaling property of microbial IAA/auxin affects gene expression in some microorganisms that interfere with the many plant developmental processes such as alteration in the root exudate secretion and plant auxin pools, thereby increasing root length and surface area (Ali et al., 2010). The auxin biosynthesis in bacteria can occur via multiple pathways as has been observed in plants. Most auxin/IAA is synthesized from the amino acid tryptophan present in plant root exudates at varying low concentrations based on the plant's genotype. Genus *Azospirillum, Aeromonas, Alcaligenes, Bradyrhizobium, Enterobacter, Comamonas,*

Pseudomonas, and *Rhizobium* are well known for production of IAA, which is responsible for the stimulation of root system development and growth promotion (Tailor and Joshi, 2014). IAA produced from fungus plays a crucial role in plant–fungal interaction in a bidirectional way such that fungal part uses IAA to colonize with the plants leading to plant growth stimulation and modification of basal plant defense mechanisms. *Trichoderma* inoculation affects root system architecture in *Zea mays* and *Arabidopsis thaliana*, which results in increased lateral root formation and root hair growth thereby enhancing the yield of plants by signaling mechanisms (Contreras-Cornejo et al., 2009).

6.5.1.2 Cytokinins

Soil microbes have the potential to produce cytokinins. However, relatively little information is available on the role of microbe produced cytokinins (Olanrewaju et al., 2017). Various studies showed that cytokinin gene is expressed in many plant growth promoting bacteria (PGPB). When such cytokinin producing bacteria is inoculated to the plant it then changes the composition of the plant's phytohormone. This can be observed in a study in which acceleration in the growth of lettuce plant and cytokinin content was noticed after inoculation of *Bacillus subtilis* (Arkhipova et al., 2005). Also, the alfalfa plant is protected against senescence arising from drought stress by a genetically engineered strain of *Sinorhizobium meliloti* that overproduces cytokinin approximately five times greater than the wild strains (Xu et al., 2012).

6.5.1.3 Gibberellins

Gibberellins are another member of phytohormone family that play a central role in processes such as stem elongation, seed germination, flowering, fruit setting, enhanced photosynthesis rate, and chlorophyll content (Zaidi et al., 2015). Gibberellins alter the hormonal balance by interacting with other phytohormones thereby affecting plant growth by regulating the root hair (Bottini et al., 2004). Reduction of the lateral root number and length is observed in the deficiency of GAs (Dodd et al., 2010). Bacterial derived gibberellin-like substances were first described in *Azospirillum brasilense* (Tien et al., 1979) and *Rhizobium* (Tewari and Arora, 2013). Many PGPB including *Agrobacterium, Azotobacter, Arthrobacter, Azospirillum, Bacillus, Burkholderia, Clostridium, Flavobacterium, Micrococcus, Pseudomonas, Xanthomonas, Achromobacter xylosoxidans, Gluconobacter diazotrophicus, Acinetobacter calcoaceticus, Rhizobium,* and *Herbaspirillum seropedicae* are known producers of gibberellins (Dodd et al., 2010; Deka et al., 2015). Gibberellins synthesized by fluorescence pseudomonads induce cell elongation within the subapical meristem resulting in stem elongation (Dobbelaere et al., 2003). Gibberellin production by *Azospirillum* and *Bacillus* has been implicated in the increased N uptake (Bottini et al., 2004).

6.5.1.4 1-Aminocyclopropane-1-carboxylate deaminase

Ethylene is an essential hormone produced by all the higher plants with the trait for influencing cell development and plant growth, and it plays a key role in combating various stresses exposed to plants (Abeles et al., 1992). Ethylene affects almost all the tissues (roots, stem, leaves, flower, and fruits) and developmental stages of plant. The condition called "stress ethylene" arises under salinity, drought, water logging, and heavy metals and pathogenicity stress conditions in which ethylene production gets increased by the action of ACC oxidase and adversely affects the plant tissue by hampering its growth, thus reducing the crop yield (Saleem et al., 2007; Bhattacharyya and Jha, 2012).

PGPR acquire the quality to alleviate such stress conditions by possessing ACC deaminase enzyme, which assists plant growth and development by decreasing ethylene levels, inducing salt tolerance, and reducing drought stress in plants (Nadeem et al., 2007). ACC deaminase alleviates the stress of plants caused by pathogen after lowering the ethylene concentration synthesized by the plants in response to pathogenicity (Toklikishvili et al., 2010; Glick, 2012). ACC deaminase, an enzyme capable of degrading the ethylene precursor ACC into α-ketobutyrate and ammonia was isolated from *Pseudomonas* in 1978 (Honma and Shimomura, 1978). Further, this enzyme was detected in wide range of soil microorganisms including the fungus *Penicillium citrinum* (Honma, 1993) and various bacteria. *Acinetobacter, Agrobacterium, Azospirillum, Achromobacter, Alcaligenes, Bacillus, Burkholderia, Enterobacter, Pseudomonas, Ralstonia, Rhizobium, Serratia,* etc. possess ACC deaminase activity (Nadeem et al., 2007; Kang et al., 2010). Such rhizobacteria take up the ethylene precursor ACC and convert it into α-Ketobutyrate and NH_3 with the aid of ACC deaminase (Arshad et al., 2007)

Duan et al. (2009) reported that 12% of the isolated *Rhizobium* possessed this enzyme, which helps in protecting plants from stress conditions as in nodulation and other symbiotic association (mycorrhizal) increases the ethylene content. In this context, ACC deaminase- producing bacteria, lowering the ethylene content in the plants, can increase both nodulation and mycorrhizal colonization, respectively (Ma et al., 2003).

6.5.1.5 Abscisic acid

ABA is acknowledged as a natural plant growth regulator that is involved in different physiological and developmental processes of plants. Stomatal closure is considered as one of the prime roles of ABA in water-stressed environments as it regulate the uptake by and transport in plants found in arid and semiarid climates (Frankenberger and Arshad, 1995).

In one study, it was speculated that inoculation of *A. brasilense* (PGPR) into *A. thaliana* plants resulted in twofold increase of ABA contents in plant. Forchetti et al. (2007) stated that ABA synthesized by PGPR defends the plant under stressful conditions and accelerates the plant growth. Presence of a significant amount of ABA production was also observed from the bacteria *A. xylosoxidans* and *Bacillus pumilus* isolated from the roots *of Helianthus annuus* in chemically defined medium. Sgroy et al. (2009) highlighted the production of ABA in chemically defined media for *Brevibacterium halotolerans, B. subtilis, Bacillus licheniformis, B. pumilus, Lysinibacillus fusiformis, A. xylosoxidans*, and *Pseudomonas putida*, which suggested role of ABA in enhancing plant growth promoting characters. Furthermore, supernatants of *Azospirillum* and *Rhizobium* cultures detected the presence of ABA by radioimmunoassay or TLC in culture (Dobbelaere et al., 2003).

6.6 Plant–microbe communication in biological control

Rhizospheric microorganisms provide an environment-friendly approach by offering a cutting edge defense response against the attack of phytopathogen by a variety of mechanisms and thus they are widely used as biocontrol agents (Lugtenberg and Kamilova, 2009; Glick, 2012). Mechanisms include production of antifungal metabolites *viz.* antibiotics, several volatile compounds like hydrogen cyanide, iron chelating siderophores (thus depriving pathogens of this important element), inducing systemic resistance, compete for nutrients with pathogen, and niche exclusion are the principal modes of biocontrol activity in PGPR's (Lugtenberg and Kamilova, 2009) as described in Table 6.2. Fungi are the most important plant pathogens followed by viruses and bacteria. Plant–microbe communication in controlling various diseases is done with the aid of elicitor signals sent by microbes, which are then recognized by a plant receptor with extremely high sensitivity and specificity (Felix et al., 1999). Subsequently, transduction of these signals results in a fortifying response by which the plant may restrain the attack by pathogens. Many genera including *Bacillus, Escherichia, Pseudomonas, Rhizobium,* and *Agrobacterium* produce such elicitors (Felix et al., 1999).

6.6.1 Antibiosis

Microorganisms perform an active role in the suppression of pathogenic microorganisms by secreting one or more antibiotics (Raaijmakers and Mazzola, 2012). These antibiotics are low molecular weight organic compounds that at even low concentration are detrimental to the growth and metabolism of pathogenic microorganisms. The first well known developed biocontrol agent *Agrobacterium radiobacter* strainK84 produces an antibiotic called agrocin 84 that contributes to its ability to control the pathogen *Agrobacterium tumefaciens. Pseudomonas* is an important genera that contributes multiple antibiotics namely phenazines (PHZ), pyoluteorin (PLT), 2,4-diacetylphloroglucinol (DAPG), phloroglucinol, cyclic lipopeptides (cLPs), pyrrolnitrin (PRN), and HCN. Some of the antibiotics possess broad

TABLE 6.2 Mechanisms and bioactive molecules involved in biocontrol of pathogen.

Mechanisms involved	Bioactive molecule	Example
	Antibiosis	Phloroglucinol (2,4-diacetylphloroglucinol), phenazines, pyrrolnitrin, pyoluteorin, cyclic lipopeptides, bacteriocins, zwittermicin a
	Lytic enzymes	Chitinase, glucanases, proteases, pectinase, cellulase, xylanse
	Volatile compounds	Hydrogen cyanide
Indirect antagonism	Competition	Siderophores,
	Induced systemic resistance	Jasmonic acid (JA), and ethylene
	Systemic acquired resistance	Salicylic acid (SA)

Source: Adapted from Adapted from Olanrewaju, O.S., Glick, B.R., Babalola, O.O., 2017. Mechanisms of action of plant growth promoting bacteria. World J. Microbiol. Biotechnol. 33, 197.

spectrum activity and inhibit the growth of different groups of macro- and microorganisms. For example, DAPG exhibits antibacterial, antifungal, and antihelminthic activities. Recently, antifungal compound have been isolated from *Bacillus* and *Streptomyces*. In case of *Bacillus*, they are either derived from the ribosome or the nonribosomal peptide synthetase (NRPS) and/or polyketide synthase (PKS) (NRPSs/PKS).

6.6.1.1 Phenazines

PHZ are a diverse group of heterocyclic ring compounds containing nitrogen identified in bacterial phyla Actinobacteria, Proteobacteria, and the archaeal phylum Euryarcheota (Mavrodi et al., 2001) with their antibiotic and virulent properties (Pierson and Pierson, 2010). To date, more than 180 PHZ based products are known with antibiotic, antifungal, insecticidal, antitumor, anticancer, antiprotozoan, and other antipathogenic activities (Guttenberger et al., 2017). Cell signaling, regulating gene expression, biofilm formation, acquisition of iron, and survival of bacteria are the diverse roles played by phenanzines. All these functions help bacteria in competition and colonization of the rhizosphere, thus providing advantage to their producers and keeping away the phytopathogens (Pierson and Pierson, 2010).

Different substituents attached to nitrogen containing heterocyclic pigments in PHZ give rise to compounds with different colors (Leisinger and Margraff, 1979; Budzikiewicz, 1993). For example pyocyanin (PYO) with blue color, phenazine-1-carboxylic acid (PCA) with yellow, phenazine-1-carboxamide (PCN) and 1-hydroxyphenazine with orange are the widely known PHZ isolated from *Pseudomonas* spp. (Briard et al., 2015). PYO is lethal to a wide variety of fungi including *Septoria tritici* and bacteria predominantly produced by *P. aeruginosa* (Demange et al., 1987; Flaishman et al., 1990). It has been proclaimed that PCA isolated from *P. fluorescens* (Gurusiddaiah et al., 1986), from *P. chlororaphis* (Pierson and Thomashow, 1992), from *P. aeruginosa* (Anjaiah et al., 1998), and from *P. putida* (Pathma et al., 2010) inhibit a series of fungal pathogens such as *R. solani*, *Gaeumannomyces graminis* var. *tritici*, *Pythium* sp., *Polyporus* sp., *Sarocladium oryzae, Macrophomina phaseolina, Pestalotia theae, Colletotrichum*, etc. and bacterial pathogens *Actinomyces viscosus*, *B. subtilis, Erwinia amylovora*, etc. (Thomashow et al., 1990; Ayyadurai et al., 2007; Pathma et al., 2010). PCN possessing broad-spectrum antifungal activity against *Pythium, Fusarium oxysporum* f. sp. *radiciopersici, S. oryzae* and *R. solani* has been documented (Chin–A–Woeng et al., 1998; Sunish Kumar et al., 2005).

6.6.1.2 Pyrrolnitrin

Arima et al. (1964) reported another antifungal broad spectrum secondary metabolite derived from L-tryptophan, i.e., PRN (3-chloro-4-(2′-nitro-3′-chlorophenyl) pyrrole). PRN is mainly produced by *P. fluorescens* (Kirner et al., 1998) and *P. aureofaciens* (Elander et al., 1968) and is found to be active against a wide range of fungi belonging to deuteromycota, ascomycota, and basidiomycota, and thus widely used as a fungicide in agriculture. Ligon et al. (2000) reported that *P. fluorescens* BL915 acts as a biocontrol agent against *R. solani* in cotton plant. *P. cepacia* and *Burkholderia cepacia* 5.5B also showed antifungal activity towards *R. solani* by suppressing stem rot of poinsettia (Cartwright et al., 1995; Hwang et al., 2002). Variants of PRN viz., isopyrrolnitrin, oxypyrrolnitrin from *Pseudomonas* spp. (Hashimoto and Hattori, 1966a,b) and monodechloropyrrolnitrin from *P. pyrrolnitrica* have also been reported; but in a comparative study it has been found that PRN has higher antifungal activity than these other variants (Hashimoto and Hattori, 1968).

6.6.1.3 Pyoluteorin

PLT composed of resorcinol ring attached to bichlorinated pyrrole moiety was reported for bactericidal, herbicidal, and fungicidal activities Takeda (1958) first isolated it from *P. aeruginosa* and later from other fluorescent pseudomonads (Nowak-Thompson et al., 1997). A combination of 10 genes (plt) is required for biosynthesis of PLT (secondary metabolite) identified in *Pseudomonas fluorescence* Pf-5 (Nowak-Thompson et al., 1999). The PLT contributes to the biological control of many soil-borne plant diseases by some strains of *Pseudomonas fluorescens*, including Pf-5 and CHA0 as it is found to be more effective against oomycetes *P. ultimum* causing damping-off disease (Nowak-Thompson et al., 1994; Kraus and Loper, 1995; Maurhofer et al., 1992). PLT is also reported to function as an autoinducer and as an intercellular signal between distinct populations of bacterial cells coinhabiting the rhizosphere (Brodhagen et al., 2004). The mode of action is by the selective inhibition of bacterial isoleucyl-tRNA synthetase (Bennett et al., 1999).

6.6.1.4 Phloroglucinol

Phloroglucinols are phenolic secondary metabolites exhibiting broad spectrum antiviral, antibacterial, antifungal, antihelminthic, and phytotoxic properties (Loper et al., 2012). Their production has been reported from various plants, microorganisms, and marine organisms. However, amongst the microbes, ability to produce phloroglucinol and its derivatives is highly conserved in fluorescent pseudomonads (Weller, 2007). DAPG is a well-known form of phloroglucinol derivative involved in biocontrol of phytopathogens (Sonnleitner and Haas, 2011) produced by fluorescent pseudomonads strains Pf-5, CHA0, Q2-87, F113, Q8r1-96. (Pathma et al., 2011;Troppens et al., 2013). Due to its multispectrum antimicrobial activity, DAPG has drawn great interest in agriculture to check the growth of phytopathogens (Duffy et al., 2004). DAPG is found to be effective against black root rot of tobacco, root rot of tomato, pythium damping-off of cucumber and sugar beet, cyst nematode and soft rot of potato, and take-all of wheat (Vincent et al., 1991; Fenton et al., 1992; Pierson and Weller, 1994; Rosales et al., 1995; Cronin et al., 1997; Raaijmakers and Weller, 1998; Pathma et al., 2011).

6.6.1.5 Cyclic lipopeptides

cLPs are short cyclic oligopeptides linked with lipid tail biosynthesized by NRPSs. They are reported from diverse genera of bacteria such as *Streptomyces, Pseudomonas,* and *Bacillus* (Finking and Marahiel, 2004) as well as from fungi (Raaijmakers et al., 2006). cLPs exhibit manifold functions in the ecosystem of the producing strains and particularly in interactions with coexisting organisms, including bacteria, fungi, oomycetes, protozoan predators, and plants (Raaijmakers et al., 2010). Due the presence of its activity to disturb the biological membrane, cLPs are considered to be an effective biocontrol metabolite, majorly produced by fluorescent pseudomonads (Raaijmakers et al., 2006). Currently, cLPs' role has been cited as biosurfactants and represents antimicrobial activities against a wide range of pathogenic microorganisms (enveloped viruses, mycoplasmas, and Gram-positive bacteria) (Raaijmakers et al., 2006; Raju et al., 2016). Viscosin, amphisin, tolaasin, and syringomycin are the most considered groups of cLPs (Nybroe and Sørensen, 2004). Tolaasin and syringomycin groups of CLPs function as cellular toxins by the formation of pores or tunnels (Nielsen et al., 2005). Viscosinamide, a cyclic lipopeptide produced by *P. fluorescens* DR54 (Nielsen et al., 1999) shows outstanding antifungal and biosurfactant properties (Nielsen et al., 2000; Nielsen et al., 2002) and is highly effective against *R. solani* (Thrane et al., 2001). Tensin, a cyclic lipodecapeptide, produced by *P. fluorescens* 96.578 (Nielsen et al., 2000) successfully inhibited *R. solani* in sugar beet (Nielsen et al., 2000). *P. fluorescence* strain SS101 also produces another cyclic lipopeptide antibiotic, the massetolides A, which is found to be effective in biocontrol of tomato late blight. The *Bacillus* genera are also producers of biologically active lipopeptide molecules. Surfactins, iturins, and fengycins are the three main families of Bacillus cLPs that are used as versatile weapons against a wide range of phytopathogens including bacteria, fungi, and oomycetes (Ongena and Jacques, 2008).

6.6.1.6 Bacteriocins

Bacteria surviving in a competitive environment are able to produce toxins known as bacteriocins (either ribosomally/nonribosomally synthesized), which are proteinaceous in nature (Mojgani, 2017) and show bacteriostatic or bactericidal/fungicidal activity against other closely related and unrelated microorganisms (Parret et al., 2005). To date, approximately 500 bacteriocins have been identified and characterized, of which the majority are produced by rhizospheric bacteria to control pathogenic microorganisms (Riley and Wertz, 2002). Bacillus is a major producer of bacteriocin with high antimicrobial activity as this can be well explored in a study conducted by Hammami et al. (2009) where strain *B. subtilis* 14B, isolated from the rhizosphere of almond plants, is capable of producing bacteriocins (Bac 14B) with high antimicrobial activity against *A. tumefaciens* (causal agent of crown gall disease). *Halobacteriaceae* a family extremely halophilic *Archaea* has also been reported for Bacteriocin production (Li et al., 2003). The importance of bacteriocins is well recognized in the agriculture industry for their role in reducing the use of fertilizers and chemical inputs such as fungicides and insecticides (Mojgani, 2017).

6.6.1.7 Zwittermicin A

Zwittermicin A is a promising aminopolyol antibiotic produced by *Bacillus* sp. with a broad spectrum of activity against various microbes (Elizabeth et al., 1999). Zwittermicin A is produced by *B. cereus* and *B. thuringiensis* and shows significant antifungal activity against oomycetes and other pathogenic fungi causing disease to various plants and is also responsible for the insecticidal activity (Emmert et al., 2004). It has ability to suppress damping-off disease driven by *Phytophthora medicaginis* in alfalfa (Pathma et al., 2011). Emmert et al., 2004 speculated that Zwittermicin A is synthesized by a mixed NRPS/PKS pathway due to the presence of homologs of NRPS and PKS. Kevany et al. (2009) characterized the complete zwittermicin A biosynthesis gene cluster from *B. cereus*.

6.6.2 Lytic enzyme

Production of cell wall degrading lytic enzyme by rhizospheric microorganism in response to attack caused by pathogenic fungi evolved as an ecological approach of biocontrol activity to protect the plant against phytopathogens. Chitinases, which degrade chitin, an important constituent of phytopathogenic fungi (Husson et al., 2017); proteases (degrades cell wall proteins); pectinases (degrades pectins); cellulases (cellulose lysis); xylanase (xylans); β-1, 3-glucanases (lyse β-1, 3 glucan, another cell wall carbohydrate of fungi) (Vaddepalli et al., 2017); etc. are some of the reported lytic enzymes that inhibit the growth and lyse the activities of pathogens (Gortari and Hours, 2008; Friedrich et al., 2012). Chitinase enzymes have been detected in bacteria (*Flavobacterium, Bacillus, Cytophaga, and Pseudomonas*) and fungus *(Aspergillus, Mucor, and Mortierella*) and are able to degrade chitin (Schlegel, 2006). Fluorescent pseudomonads produce cell wall degrading endochitinase enzyme, which inhibits *R. solani* in sugar beet (Nielsen et al., 1998). Extracellular proteases from *P. fluorescens* CHA0 act as an important biocontrol agent against the root-knot nematode disease caused by *Meloidogyne incognita* (Siddiqui and Shaukat, 2005). Chitinase and β-1, 3-glucanase can work synergistically against various fungal pathogens for better results (Mauch et al., 1988). Fluorescence pseudomonads strain, *Pseudomonas* GRC 2, produces chitinases that cause destruction and deformities of hypha, mycelia, and sclerotia of *M. phaseolina* and *Sclerotinia sclerotiorum* (Gupta et al., 2001).

6.6.3 Hydrogen cyanide

HCN is a volatile secondary metabolite produced by many bacterial genera such as *Rhizobium, Pseudomonas, Alcaligenes, Bacillus, Aeromonas Chromobacterium, Burkholderia,* and *Cyanobacteria* (Blumer and Haas, 2000; Ahemad and Kibret, 2014) that works as biocontrol agent against plant pathogens (Rodríguez and Fraga, 1999; Siddiqui et al., 2006). HCN is produced from HCN synthase enzyme and is encoded by three different genes: *hcn* A, *hcn* B, and *hcn* C (Ramette et al., 2003). HCN inhibits the cytochrome c oxidase as well as other important metalloenzyme components of electron transport systems of pathogenic organisms, which results in loss of energy and ultimately causes death of the organisms (Voisard et al., 1989; Nandi et al., 2017). HCN has been investigated for its suppression of *Meloidogyne javanica* (causing root knot disease in tomato) (Siddiqui et al., 2006), inhibiting fungus *Thielaviopsis basicola* (causing black root rot of tobacco) (Voisard et al., 1989), and quelling take-all disease of wheat caused by *G. graminis* var. *tritici*. Beside biocontrol of phytopathogens, HCNs are more likely involved in chelation of metals thereby increasing the availability of phosphate to the plant as reported by Rijavec and Lapanje (2016).

6.6.4 Competition

The PGPRs present in the vicinity of plant root exhibit competition with other detrimental microorganisms and pathogens for limited nutrients available in the rhizospheric zone and thus restrict their colonization around the plant root surface. Apart from root colonization, PGPR outcompetes the population of pathogens by limiting the nutrients available for them. Such competition can act to limit the binding of the phytopathogen to the plant thereby making it difficult for it to proliferate. Siderophores are low molecular iron-chelating compound produced by PGPR and they starve the pathogen for iron nutrient by sequestering most of the available Fe^{3+} in the rhizosphere, ultimately leading to suppression of the pathogen. Siderophore production has been cited in a large number of bacterial genera including *Rhizobium, Aeromonas, Bacillus, Pseudomonas*, and *Serratia*. Siderophores have been isolated from *Pseudomonas* demonstrate antifungal activity against the phytopathogenic fungi, including *Aspergillus niger, A. flavus, A. oryzae, F. oxysporum, M. phaseolina*, and *Sclerotium rolfsii* (Taguchi et al., 2010).

6.6.5 Induced systemic resistance

Plant–microbe interaction is regulated through signal transduction pathways allowing plants to prioritize defense response following stress conditions. The plant–microbe association involves molecular recognition between the two partners through a signaling network mediated by the plant hormones salicylic acid (SA), jasmonic acid (JA), and ethylene. JA and ethylene have been described as signal transduction molecules for ISR due to the effect induced by beneficiary microbes, and the signal transduction pathway through SA accumulation is found in the systemic acquired resistance (SAR) induced by the attack of pathogens (van Loon et al., 1998; Park and Kloepper, 2000).

ISR is a process in which beneficial microbes (PGPR) lessen the detrimental effects of phytopathogens by initiating a resistance mechanism in the plants (van Loon et al., 1998). It can be local or systemic in action and provides defense against phytopathogens (Jamsen, 2000). The fluorescent pseudomonads and other PGPRs have been reported to induce systemic resistance in the plants against bacterial, fungal, and viral diseases. ISR confers a high level of protection, which is controlled by a network of coordinated signaling pathways that are dominated and majorly regulated by plant hormones sharing signaling components (Walters et al., 2013; Olanrewaju et al., 2017). Bacterial components such as siderophores, cLPs, 2,4-diacetylphloroglucinol, and volatiles like acetoin and 2, 3-butanediol also induce ISR mechanisms in plants (Lugtenberg and Kamilova, 2009). Volatiles produced by *B. subtilis* and *Bacillus amyloliquefaciens* are able to elicit an ISR pathway in *A. thaliana* (Ryu et al., 2004). Priming of plants with an appropriate PGPR can react faster against pathogen attack by inducing plant defense mechanisms as can be observed in *P. putida* 89 B−27 stain, which showed resistance against *Colletotrichum orbiculare* (Wei et al., 1991). Elicitation of ISR has also been reported in growth inhibition of *F. oxysporum* f. sp. *dianthi* by action of *Pseudomonas fluorescens* strain WCS417r causing vascular wilts in *Arabidopsis* (Pieterse et al., 1996), and suppressing *Alternaria brassicicola* and *Pseudomonas syringae* pv. tomato causing necrotic lesions in radish (Hoffland et al., 1996).

Accumulation of pathogenesis-related (PR) proteins and induction of defense related compounds of the phenylpropanoid pathway have been linked to ISR-mediated enhanced resistance by PGPR. Phenylalanine ammonia lyase (PAL) is the first enzyme involved in the phenylpropanoid pathway and plays a key role in the biosynthesis of phenolics and phytoalexins. ISR mechanisms get induced by accumulation of PAL an important PR protein in tomato and pepper plants when treated with *P. fluorescence* in response to infection by *F. oxysporum* f. sp. *lycopersici* and *Colletotrichum capsici* (Ramamoorthy et al., 2001). Elicitation of ISR in sugar beet by *B. mycoides* and *B. pumilus* is associated with enhanced peroxidase activity (Kloepper et al., 2004).

Salicylic acid is another signal molecule that gives rise to "systemic acquired resistance" or SAR mechanisms in plants similar to ISR (Chen et al., 1999). SAR is usually induced by pathogenic microorganisms, and resistance offered by SAR is effective against a wide range of pathogens (Zhang et al., 2002). Furthermore, SAR is associated with the activation of PR genes that encode PR proteins and accumulation of salicylic acid. One of the best characterized PR genes is PR-1, which is mostly used as a biomarker for SAR (van Loon et al., 2006). PGPB-mediated ISR and SA-dependent SAR are coordinated by different signaling pathways, which are supported by the observations that both PGPB-mediated ISR and pathogen-induced SAR are effective control mechanisms, but their degree of effectiveness is slightly different (Ton et al., 2002; Olanrewaju et al., 2017).

6.7 Quorum sensing

Rhizospheric communication among microbes (intra- and interspecies) results in the eventual establishment of the microbial population. These communications between microbes occur through various signal molecules, which assist bacteria to sense their surroundings. Quorum sensing (QS) is the process that is known for bacterial cell−cell communication. This process involves the production and detection of signaling molecules/autoinducers, which allows bacterial communities to express genes collectively (Hawver et al., 2016). Diverse kinds of quorum-sensing networks are found in different bacterial species with varying molecular structure in signaling molecules or autoinducers. Acyl homoserine lactone (AHL) is a signal molecule that is widely studied in bacteria. These metabolites are released into the cellular environment to sense the quality of the ecological niche in terms of diffusion space and the density and distribution of their own population. *N*-acyl-homoserinelactone (AHL) autoinducers are common in Gram-negative bacteria, which facilitates several traits linked with plant−microbe interaction, while cyclic peptides as QS-signals are only to be found in Gram-positive bacteria. *Rhizobium, Sinorhizobium, Pseudomonas, Agrobacterium, Pantoea, Erwinia*, and *Xanthomonas* are the AHL producer plant-associated bacteria (Cha et al., 1998). Different bacterial species can produce the same AHLs or AHLs with similar structures and properties suggesting that crosstalk between populations occurs and it is evident that QS via AHLs is more common among plant-associated bacteria than the general population of soil bacteria (Elasri et al., 2001; d'Angelo-Picard et al., 2005).

Several ecological and interdependent key characteristics of bacteria, like antibiotic, siderophore, or enzyme secretion; virulence factors of phytopathogens; as well as plant−microbe communications, are also coordinated through QS (Altaf et al., 2017). Plants and microbes are dependent on chemical signals for multiple communications. Microbes get attracted to plants after sensing the release of phytohormones by plants and plant gets aware of microbial population via direction/synthesis of quorum-sensing compounds (Rowe et al., 2018). Plants have evolved various mechanisms to respond to these QS compounds in specific ways as it has been thought that QS compounds are early signals indicating that pathogens are in the surrounding environment to gather themselves for attack or that mutualists are about to interact with roots. The first expression of specific response so far of plant to bacterial AHLs was demonstrated for the legumes *Phaseolus vulgaris* (Joseph and Phillips, 2003) and *Medicago truncatula* (Mathesius et al., 2003).

Plants also secrete substances that imitate bacterial AHLs (Teplitski et al., 2000) or get involved in other activities that manipulate QS of bacteria associated with the plant (Gao et al., 2003; Bauer and Mathesius, 2004). In the symbiotic association of rhizobium and legume plant, flavonoids secretion increases the expression of AHLs synthesis genes in rhizobia (Pérez-Montaño et al., 2011). Similarly, in *Gypsophila* plant synthesis of IAA and cytokinin influences QS, type III secretion system, and gall formation activity by *Pantoea plantarum* (Chalupowicz et al., 2009). Fray et al., 1999 stated that tobacco has been genetically modified to secrete AHLs, which influence the bacterial population surrounding their root. In addition, transgenic tomato plants have also been engineered to produce AHL compounds that alter the activity of PGPR and result in increased salt tolerance (Barriuso et al., 2008). Thus, it is speculated that QS is a plant–microbe association system that should be regarded as bidirectional, that is, influenced from both plant and microbial partner.

6.8 Conclusion and future challenges

This current chapter highlights the complex interaction of plant-associated microorganisms in the rhizospheric zone that can enhance plant growth through different mechanisms. Chemical signals contribute to play a major role in establishing strong associations between plant and PGPRs. Several metabolites from the plant root (root exudates) as well as from microorganisms (secondary metabolite) are the only messengers that unravel the interaction in the rhizosphere. Root exudates and PGPR metabolites help in establishing beneficial or detrimental relationships. Such interaction facilitates nutrient transformation (by nitrogen fixation, mobilization, and immobilization of macro- and micronutrients), improving the plant growth factor, thwarting the phytopathogens and cross talk among beneficial and pathogenic microorganisms that ultimately improve plant growth, crop yields, and soil fertility. Significance of QS and signaling molecules has also drawn attention towards microbe–microbe and microbe–plant interaction in the rhizosphere with special reference to PGPRs. Although researchers have identified a number of factors responsible for initiating such a communication in the rhizosphere, still numerous other features are yet to be explored to understand plant–microbe interactions at the ecological level for the better benefit of agricultural usage. This can be achieved by the usage of molecular, bioinformatics, and modeling tools. Since the rhizospheric zone has always been of interest among researchers for their novel interactions, the application of recent “omics” technologies (proteomics, metabolomics, transcriptomics, and secretomics) will definitely strengthen the knowledge for multitrophic interaction study. This would also lead to novel ideas for enhancing the benefit for sustainable agricultural development with their application usage due to better understanding of plant–microbe and microbe–microbe interactions, eventually providing a supportive outlook in stabilizing the ecosystem by increasing the productivity of plants and minimizing the level of disease index against phytopathogens.

References

Abeles, F.B., Morgan, P.W., Saltveit, M.E., 1992. Ethylene in Plant Biology, second ed. Academic Press, New York.

Agrios, G.N., 2005. Plant Pathology., fifth ed. Elsevier Academic Press, San Diego, p. 922.

Ahemad, M., Kibret, M., 2014. Mechanisms and application s of plant growth promoting rhizobacteria: current perspective. J. King Saud Univ. Sci. 26, 1–20.

Akiyama, K., Matsuzaki, K.I., Hayashi, H., 2005. Plant sesquiterpenes induce hyphal branching in arbuscular mycorrhizal fungi. Nature 435, 824–827.

Ali, B., Sabri, A.N., Hasnain, S., 2010. Rhizobacterial potential to alter auxin content and growth of Vigna radiata (L.). World J. Microbiol. Biotechnol. 26, 1379–1384.

Anjaiah, V., Koedam, N., Nowak-Thompson, B., Loper, J.E., Hofte, M., Tambong, J.T., et al., 1998. Involvement of phenazines and anthranilate in the antagonism of *Pseudomonas aeruginosa* PNA1 and Tn5 derivatives toward *Fusarium* spp. and *Pythium* spp. Mol Plant Microbe Interact. 11, 847–854.

Arima, K., Imanaka, H., Kousaka, M., Fukuta, A., Tamura, G., 1964. Pyrrolnitrin, a new antibiotic substance, produced by Pseudomonas. Agri Biol Chem. 28, 575–576.

Arkhipova, T.N., Veselov, S.U., Melentiev, A.I., Martynenko, E.V., Kudoyarova, G.R., 2005. Ability of bacterium *Bacillus subtilis* to produce cytokinins and to influence the growth and endogenous hormone content of lettuce plants. Plant Soil 272, 201–209.

Arshad, M., Saleem, M., Hussain, S., 2007. Perspectives of bacterial ACC deaminase in phytoremediation. Trends Biotechnol. 25, 356–362.

Altaf, M.M., Khan, M.S.A., Abulreesh, H.H., Ahmad, I., 2017. Quorum sensing in plant growth-promoting rhizobacteria and its impact on plant–microbe interaction. In: Singh, D., Singh, H., Prabha, R. (Eds.), Plant–Microbe Interactions in Agro-Ecological Perspectives. Springer, Singapore, pp. 311–331.

Ayyadurai, N., Ravindra Naik, P., Sakthivel, N., 2007. Functional characterization of antagonistic fluorescent pseudomonads associated with rhizospheric soil of rice (*Oryza sativa* L.). J. Microbiol. Biotechnol. 17, 919–927.

Baas, P., Mohan, J.E., Markewitz, D., Knoepp, J.D., 2014. Assessing heterogeneity in soil nitrogen cycling: a plot-scale approach. Soil Sci. Soc. Am. J. 78 (S1), S237–S247.

Badri, D.V., Vivanco, J.M., 2009. Regulation and function of root exudates. Plant Cell Environ. 32 (6), 666–681.

Badri, D.V., Weir, T.L., van der Lelie, D., Vivanco, J.M., 2009. Rhizosphere chemical dialogues: plant–microbe interactions. Curr. Opin. Biotechnol. 20, 642–650.

Bais, H.P., Weir, T.L., Perry, L.G., Gilroy, S., Vivanco, J.M., 2006. The role of root exudates in rhizosphere interactions with plants and other organisms. Annu. Rev. Plant Biol. 57, 233–266.

Bar-Ness, E., Hadar, Y., Chen, Y., Römheld, V., Marschner, H., 1992. Short-term effects of rhizosphere microorganisms on Fe uptake from microbial siderophores by maize and oat. Plant Physiol. 100, 451–456.

Barriuso, J., Solano, B.R., Fray, R.G., Cámara, M., Hartmann, A., Gutierrez Manero, F.J., 2008. Transgenic tomato plants alter quorum sensing in plant growth-promoting rhizobacteria. Plant Biotech. J. 6, 442–452.

Bauer, W.D., Mathesius, U., 2004. Plant responses to bacterial quorum sensing signals. Curr. Opin. Plant Biol. 7, 429–433.

Bennett, I., Broom, N.J.P., Cassels, R., Elder, J.S., Masson, N.D., O'Hanlon, P.J., 1999. Synthesis and antibacterial properties of â-diketone acrylate bioisosteres of pseudomonic acid A. Bioorg. Med. Chem. Lett. 9, 1847–1852.

Berendsen, R.L., Pieterse, C.M., Bakker, P.A., 2012. The rhizosphere microbiome and plant health. Trends Plant Sci. 17, 478–486.

Bhattacharyya, P.N., Jha, D.K., 2012. Plant growth-promoting rhizobacteria (PGPR): emergence in agriculture. World J. Microbiol. Biotechnol. 28, 1327–1350.

Blumer, C., Haas, D., 2000. Mechanism, regulation, and ecological role of bacterial cyanide biosynthesis. Arch. Microbiol. 173, 170–177.

Bottini, R., Cassán, F., Piccoli, P., 2004. Gibberellin production by bacteria and its involvement in plant growth promotion and yield increase. Appl. Microbiol. Biotechnol. 65, 497–503.

Briard, B., Bomme, P., Lechner, B.E., Mislin, G.L.A., Lair, V., Prévost, M.-C., et al., 2015. *Pseudomonas aeruginosa* manipulates redox and iron homeostasis of its microbiota partner *Aspergillus fumigatus* via phenazines. Sci. Rep. 5, 8220.

Brodhagen, M., Henkels, M.D., Loper, J.E., 2004. Positive autoregulation and signaling properties of pyoluteorin, an antibiotic produced by the biological control organism *Pseudomonas fluorescens* Pf-5. Appl. Environ. Microbiol. 70, 1758–1766.

Budzikiewicz, H., 1993. Secondary metabolites from fluorescent pseudomonads. FEMS Microbiol. Ecol. 104, 209–228.

Campbell, R., Greaves, M.P., 1990. Anatomy and community structure of the rhizosphere. The Rhizosphere 11–34.

Cartwright, D.K., Chilton, W.S., Benson, D.M., 1995. Pyrrolnitrin and phenazine production by *Pseudomonas cepacia*, strain 5.5B, a biocontrol agent of *Rhizoctonia solani*. Appl. Microbiol. Biotechnol. 43, 211–216.

Cha, C., Gao, P., Chen, Y.C., Shaw, P.D., Farrand, S.K., 1998. Production of acyl-homoserine lactone quorum-sensing signals by Gram-negative plant-associated bacteria. Mol. Plant Microbe Interact. 11, 1119–1129.

Chalupowicz, L., Barash, I., Panijel, M., Sessa, G., Manulis-Sasson, S., 2009. Regulatory interaction between quorum-sensing, auxin, cytokinin, and the Hrp regulon in relation to gall formation and epiphytic fitness of *Pantoea agglomerans* pv. Gypsophilae. Mol. Plant Microbe Interact 22 (7), 849–856.

Chen, C., Bélanger, R.R., Benhamou, N., Paulitz, T.C., 1999. Role of salicylic acid in systemic resistance induced by *Pseudomonas* spp. against *Pythium aphanidermatum* in cucumber roots. Eur. J. Plant. Pathol. 105, 477–486.

Chin–A–Woeng, T.F.C., Bloemberg, G.V., Van der Bij, A.J., Van der Drift, K.M.G.M., Schripse-ma, J., Kroon, B., et al., 1998. Biocontrol by phenazine-1-carboxamide producing *Pseudomonas chlororaphis* PCL1391 of tomato root rot caused by *Fusarium oxysporum* f. sp. radicis-lycopersici. Mol. Plant Microbe Interact. 11, 1069–1077.

Contreras-Cornejo, H.A., Macías-Rodríguez, L., Cortés-Penagos, C., López-Bucio, J., 2009. *Trichoderma virens*, a plant beneficial fungus, enhances biomass production and promotes lateral root growth through an auxin-dependent mechanism in Arabidopsis. Plant Physiol. 149, 1579–1592.

Cronin, D., Moenne-Loccoz, Y., Fenton, A., Dunne, C., Dowling, D.N., O'Gara, F., 1997. Role of 2,4-diacetylphloroglucinol in the interaction of the biocontrol pseudomonad strain F113 with the potato cyst nematode *Globodera rostochiensis*. Appl. Environ. Microbiol. 63, 1357–1361.

Crowley, D.E., Römheld, V., Marschner, H., Szaniszlo, P.J., 1992. Root-microbial effects on plant iron uptake from siderophores and phytosiderophores. Plant Soil. 142, 1–7.

d'Angelo-Picard, C., Faure, D., Penot, I., Dessaux, Y., 2005. Diversity of *N*-acyl homoserine lactone-producing and -degrading bacteria in soil and tobacco rhizosphere. Environ. Microbiol. 7, 1796–1808.

Deka, H., Deka, S., Baruah, C.K., 2015. Plant growth promoting rhizobacteria for value addition: mechanism of action. Plant-Growth-Promoting Rhizobacteria (PGPR) and Medicinal Plants. Springer, New York, pp. 305–321.

Demange, P., Wendenbaum, S., Linget, C., Bateman, A., MacLeod, J., Dell, A., et al., 1987. Bacterial siderophores: structures and physicochemical properties of pyoverdines and related compounds. In: Winkelmann, G., Weinheim, V.C.H. (Eds.), Iron Transport in Microbes, Plants and Animals. VCH Press, Weinheim, Germany, pp. 189–205.

Dobbelaere, S., Vanderleyden, J., Okon, Y., 2003. Plant growth-promoting effects of diazotrophs in the rhizosphere. Crit. Rev. Plant Sci. 22, 107–149.

Dodd, I.C., Zinovkina, N.Y., Safronova, V.I., Belimov, A.A., 2010. Rhizobacterial mediation of plant hormone status. Ann. Appl. Biol. 157, 361–379.

Dotaniya, M.L., Meena, H.M., Lata, M., Kumar, K., 2013. Role of phytosiderophores in iron uptake by plants. Agric. Sci. Digest 33, 73–76.

Duan, J., Müller, K.M., Charles, T.C., Vesely, S., Glick, B.R., 2009. 1-Aminocyclopropane-1-carboxylate (ACC) deaminase genes in rhizobia from southern Saskatchewan. Microb. Ecol. 57, 423–436.

Duffy, B.K., Keel, C., Defago, G., 2004. Potential role of pathogen signaling in multitrophic plant–microbe interactions involved in disease protection. Appl. Environ. Microbiol. 70, 1836–1842.

Duponnois, R., Kisa, M., Plenchette, C., 2006. Phosphate-solubilizing potential of the nematophagous fungus *Arthrobotrys oligospora*. J. Plant Nutr. Soil Sci. 169, 280–282.

Elander, R.P., Mabe, J.A., Hamill, R.H., Gorman, M., 1968. Metabolism of Tryptophans by *Pseudomonas aureofaciens*: VI. Production of Pyrrolnitrin by Selected Pseudomonas Species. Appl Environ Microbiol. 16, 753–758.

Elasri, M., Delorme, S., Lemanceau, P., Stewart, G., Laue, B., Glickmann, E., et al., 2001. Acyl-homoserine lactone production is more common among plant-associated *Pseudomonas* spp. than among soilborne *Pseudomonas* spp. Appl. Environ. Microbiol. 67, 1198–1209.

Elizabeth, A.S., Milner, J.L., Handelsman, J., 1999. Zwittermicin A biosynthetic cluster. Gene 237, 403–411.

Emmert, B.A.E., Klimowicz, K.A., Thomas, G.M., Handelsman, J., 2004. Genetics of zwittermicin A production by *Bacillus cereus*. Appl. Environ. Microbiol. 70, 104–113.

Felix, G., Duran, J.D., Volko, S., Boller, T., 1999. Plants have a sensitive perception system for the most conserved domain of bacterial flagellin. The Plant J. 18, 265–276.

Fenton, A.M., Stephens, P.M., Crowley, J., O Callaghen, M., O'Gara, F., 1992. Exploitation of gene(s) involved in 2,4-diacetylphloroglucinol biosynthesis to confer a new biocontrol capability to a *Pseudomonas* strain. Appl. Environ. Microbiol. 58, 3873–3878.

Finking, R., Marahiel, M.A., 2004. Biosynthesis of nonribosomal peptides. Annu. Rev. Microbiol. 58, 453–488.

Flaishman, M., Eyal, Z., Voisard, C., Haas, D., 1990. Suppression of *Septoria tritici* by phenazine or siderophore deficient mutants of *Pseudomonas*. Curr. Microbiol. 20, 121–124.

Forchetti, G., Masciarelli, O., Alemano, S., Alvarez, D., Abdala, G., 2007. Endophytic bacteria in sunflower (*Helianthus annuus* L.): isolation, characterization, and production of jasmonates and abscisic acid in culture medium. Appl. Microbiol. Biotechnol. 76, 1145–1152.

Frankenberger, J.W.T., Arshad, M., 1995. Microbial synthesis of auxins. In: Frankenberger, W.T., Arshad, M. (Eds.), Phytohormones in Soils. Marcel Dekker, New York, pp. 35–71.

Fray, R.G., Throup, J.P., Daykin, M., Wallace, A., Williams, P., Stewart, G.S., Grierson, D., 1999. Plants genetically modified to produce N-acylhomoserine lactones communicate with bacteria. Nature Biotechnol. 17, 1017.

Friedrich, N., Hagedorn, M., Soldati-Favre, D., Soldati, T., 2012. Prison break: pathogens' strategies to egress from host cells. Microbiol. Mol. Biol. Rev. 76, 707–720.

Fujita, M., Fujita, Y., Noutoshi, Y., Takahashi, F., Narusaka, Y., Yamaguchi-Shinozaki, K., et al., 2006. Crosstalk between abiotic and biotic stress responses: a current view from the points of convergence in the stress signaling networks. Curr. Opin. Plant Biol. 9, 436–442.

Gao, M., Teplitski, M., Robinson, J.B., Bauer, W.D., 2003. Production of substances by *Medicago truncatul*a that affect bacterial quorum sensing. Mol. Plant Microbe Interact. 16, 827–834.

Glick, B.R., 2012. Plant Growth-Promoting Bacteria: Mechanisms and Applications. Hindawi Publishing Corporation, Scientifica.

Gortari, M.C., Hours, R.A., 2008. Fungal chitinases and their biological role in the antagonism onto nematode eggs: a review. Mycol. Prog. 7, 221–238.

Gupta, C.P., Dubey, R.C., Kang, S.C., Maheshwari, D.K., 2001. Antibiosis-mediated necrotrophic effect of *Pseudomonas* GRC 2 against two fungal plant pathogens. Curr. Sci. 91–94.

Gurusiddaiah, S., Weller, D.M., Sarkar, A., Cook, R.J., 1986. Characterization of an antibiotic produced by a strain of *Pseudomonas fluoroscens* inhibitory to *Gaeumannomyces graminis* var. tritici and *Pythium* spp. Antimicrob. Agents Chemotherapy 29, 488–495.

Guttenberger, N., Blankenfeldt, W., Breinbauer, R., 2017. Recent developments in the isolation, biological function, biosynthesis, and synthesis of phenazine natural products. Bioorg. Med. Chem. 25, 6149–6166.

Hammami, I., Rhouma, A., Jaouadi, B., Rebai, A., Nesme, X., 2009. Optimization and biochemical characterization of a bacteriocin from a newly isolated *Bacillus subtilis* strain 14B for biocontrol of *Agrobacterium* spp. strains. Lett. Appl. Microbiol. 48, 253–260.

Hartmann, A., Schmid, M., Van Tuinen, D., Berg, G., 2009. Plant-driven selection of microbes. Plant Soil 321, 235–257.

Hashimoto, M., Hattori, K., 1966a. Isopyrrolnitrin: a metabolite from *Pseudomonas*. Bull. Chem. Soc. Jpn. 39, 410.

Hashimoto, M., Hattori, K., 1966b. Oxypyrrolnitrin: a metabolite of *Pseudomonas*. Chem. Pharm. Bull. 14, 1314–1316.

Hashimoto, M., Hattori, K., 1968. A new metabolite from *Pseudomonas pyrrolnitrica*. Chem. Pharm. Bull. 16, 1144.

Hawver, L.A., Jung, S.A., Ng, W.L., 2016. Specificity and complexity inbacterial quorun-sensing systems. FEMS Microbiol. Rev. 40, 738–752.

Hayman, D.S., 1983. The physiology of vesicular-arbuscular endomycorrhizal symbiosis. Can. J. Bot. 61, 944–963.

Hiltner, L.T., 1904. U neuere Erfahrungen und Probleme auf dem Gebiet der Bodenbakteriologie und unter besonderer Berucksichtigung der Grundungung und Brache. Arb. Dtsch. Landwirt. Ges. 98, 59–78.

Hoffland, E., Hakulinen, J., Van Pelt, J.A., 1996. Comparison of systemic resistance induced by avirulent and nonpathogenic *Pseudomonas* species. Phytopathology 86, 757–762.

Honma, M., 1993. Stereospecific reaction of 1-aminocyclopropane-1-carboxylate deaminase. In: Pech, J.C., Latché, A., Balagué, C. (Eds.), Cellular and Molecular Aspects of the Plant Hormone Ethylene. Springer, Dordrecht, The Netherlands, pp. 111–116.

Honma, M., Shimomura, T., 1978. Metabolism of 1-aminocyclopropane-1-carboxylic acid. Agric. Biol. Chem. 42, 1825–1831.

Husson, E., Hadad, C., Huet, G., Laclef, S., Lesur, D., Lambertyn, V., et al., 2017. The effect of room temperature ionic liquids on the selective biocatalytic hydrolysis of chitin via sequential or simultaneous strategies. Green Chem. 19, 4122–4131.

Hwang, J., Chilton, W.S., Benson, D.M., 2002. Pyrrolnitrin production by *Burkholderia cepacia* and biocontrol of *Rhizoctonia* stem rot of poinsettia. Biol. Control 25, 56–63.

Jacobs, H., Boswell, G.P., Ritz, K., Davidson, F.A., Gadd, G.M., 2002. Solubilization of calcium phosphate as a consequence of carbon translocation by *Rhizoctonia solani*. FEMS Microbiol. Ecol. 40, 65–71.

Jamsen, P., 2000. Auxins and cytokinins in plant pathogen interactions—an overview. Plant Growth Regul. 32, 369–380.

Jetten, M.S., 2008. The microbial nitrogen cycle. Environ. Microbiol. 10, 2903–2909.

Joseph, C.M., Phillips, D.A., 2003. Metabolites from soil bacteria affect plant water relations. Plant Physiol. Biochem. 41, 189–192.

Kang, B.G., Kim, W.T., Yun, H.S., Chang, S.C., 2010. Use of plant growth-promoting rhizobacteria to control stress responses of plant roots. Plant Biotechnol. Rep. 4, 179–183.

Kevany, B.M., Rasko, D.A., Thomas, M.G., 2009. Characterization of the complete Zwittermicin. A biosynthesis gene cluster from *Bacillus cereus*. Appl. Environ. Microbiol. 75, 1144–1155.

Kirner, S., Hammer, P.E., Hill, D.S., Altmann, A., Fischer, I., Weislo, L.J., et al., 1998. Functions encoded by pyrrolnitrin biosynthetic genes from *Pseudomonas fluoroscens*. J. Bacteriol. 180, 1939–1943.

Kloepper, J.W., Schroth, M.N., 1978. Plant growth-promoting rhizobacteria on radishes. In: Proceedings of the 4th International Conference on Plant Pathogenic Bacteria, 2, pp. 879–882.

Kloepper, J.W., Ryu, C.-M., Zhang, S., 2004. Induced systemic resistance and promotion of plant growth by *Bacillus* spp. Phytopathology 94, 1259–1266.

Kraus, J., Loper, J.E., 1995. Characterization of a genomic region required for production of the antibiotic pyoluteorin by the biological control agent *Pseudomonas fluoroscens* Pf-5. Appl. Environ. Microbiol. 61, 849–854.

Kumar, R., Chauhan, S., 2014. Molecular biology of nodule development and nitrogen fixation in *Rhizobium*-legume symbiosis. In: Chandra, R., Raverkar, K.P. (Eds.), Bioresources for Sustainable Plant Nutrient Management. Satish Serial Publishing House, pp. 435–479.

Kumar, R., Shastri, B., 2017. Role of phosphate-solubilizing microorganisms in sustainable agricultural development. In: Singh, J.S., Sneiviratne, G. (Eds.), Agro Environmental Sustainability – Managing Crop Health (vol. I). Springer, Cham, pp. 271–303.

Leisinger, T., Margraff, R., 1979. Secondary metabolites of fluorescent pseudomonads. Microb. Rev. 43, 422–442.

Li, Y., Xiang, H., Liu, J., Zhou, M., Tan, H., 2003. Purification and biological characterization of halocin C8, a novel peptide antibiotic from Halobacterium strain AS7092. Extremophiles 7, 401–407.

Ligon, J.M., Hill, D.S., Hammer, P.E., Torkewitz, N.R., Hofmann, D., Kempf, H.J., et al., 2000. Natural products with antifungal activity from *Pseudomonas* biocontrol bacteria. Pest Manag. Sci. 56, 688–695.

Loper, J.E., Buyer, J.S., 1991. Siderophores in microbial interactions of plant surfaces. Mol. Plant. Microbe Interact. 4, 5–13.

Loper, J.E., Hassan, K.A., Mavrodi, D.V., Davis II, E.W., Lim, C.K., Shaffer, B.T., et al., 2012. Comparative genomics of plant-associated *Pseudomonas* spp.: insights into diversity and inheritance of traits involved in multitrophic interactions. PLoS Genet. 8, e.1002784.

Lugtenberg, B., Kamilova, F., 2009. Plant growth-promoting rhizobacteria. Annu. Rev. Microbiol. 63, 541–556.

Lundberg, D.S., Lebeis, S.L., Paredes, S.H., Yourstone, S., Gehring, J., Malfatti, S., et al., 2012. Defining the core *Arabidopsis thaliana* root microbiome. Nature 488, 86.

Ma, W., Guinel, F.C., Glick, B.R., 2003. The *Rhizobium leguminosarum* bv. viciae ACC deami-nase protein promotes the nodulation of pea plants. Appl. Environ. Microbiol. 69, 4396–4402.

Marschner, P., Crowley, D.E., 1998. Phytosiderophores decrease iron stress and pyoverdine production of *Pseudomonas fluoroscens* Pf-5 (pvd-ina Z). Soil Biol. Biochem. 30, 1275–1280.

Mathesius, U., Mulders, S., Gao, M., Teplitski, M., Caetano-Anollés, G., Rolfe, B.G., et al., 2003. Extensive and specific responses of a eukaryote to bacterial quorum-sensing signals. Proc. Natl. Acad. Sci. U.S.A. 100, 1444–1449.

Mauch, F., Mauch-Mani, B., Boller, T., 1988. Antifungal hydrolases in pea tissue: II. Inhibition of fungal growth by combinations of chitinase and β-1, 3-glucanase. Plant Physiol. 88, 936–942.

Maurhofer, M., Keel, C., Schnider, U., Viosard, C., Haas, D., Defago, G., 1992. Influence of enhanced antibiotic production in *Pseudomonas fluoroscens* strain CHA0 on its disease suppressive capacity. Phytopathol. 82, 190–195.

Mavrodi, D.V., Bonsall, R.F., Delaney, S.M., Soule, M.J., Phillips, G., Thomashow, L.S., 2001. Functional analysis of genes for biosynthesis of pyocyanin and phenazine-1-carboxamide from *Pseudomonas aeruginosa* PAO1. J. Bacteriol. 183, 6454–6465.

McGill, W.B., Cole, C.V., 1981. Comparative aspects of cycling of organic C, N, S and P through soil organic matter. Geoderma 26, 267–286.

Mojgani, N., 2017. Bacteriocin-producing rhizosphere bacteria and their potential as a biocontrol agent. In: Mehnaz, S. (Ed.), Rhizotrophs: Plant Growth Promotion to Bioremediation. Microorganisms for Sustainability, vol. 2. Springer, Singapore.

Nadeem, S.M., Zahir, Z.A., Naveed, M., Arshad, M., 2007. Preliminary investigations on inducing salt tolerance in maize through inoculation with rhizobacteria containing ACC deaminase activity. Can. J. Microbiol. 53, 1141–1149.

Nandi, M., Selin, C., Brawerman, G., Fernando, W.G.D., de Kievit, T., 2017. Hydrogen cyanide, which contributes to *Pseudomonas chlororaphis* strain PA23 biocontrol, is upregulated in the presence of glycine. Biol. Control. 108, 47–54.

Neilands, J.B., 1981. Microbial iron compounds. Annu. Rev. Biochem. 50, 715–731.

Neilands, J.B., 1995. Siderophores: structure and function of microbial iron transport compounds. J. Biol. Chem. 270, 26723–26726.

Nielsen, M.N., Sørensen, J., Fels, J., Pedersen, H.C., 1998. Secondary metabolite-and endochitinase-dependent antagonism toward plant-pathogenic microfungi of *Pseudomonas fluoroscens* isolates from sugar beet rhizosphere. Appl. Environ. Microbiol. 64, 3563–3569.

Nielsen, T.H., Christophersen, C., Anthoni, U., Sorensen, J., 1999. Viscosinamide, a new cyclic depsipeptide with surfactant and antifungal properties produced by *Pseudomonas fluoroscens* DR54. J. Appl. Microbiol. 86, 80–90.

Nielsen, T.H., Thrane, C., Christophersen, C., Anthoni, U., Sørensen, J., 2000. Structure, production characteristics and fungal antagonism of tensin – a new antifungal cyclic lipopeptide from *Pseudomonas fluoroscens* strain 96.578. J. Appl. Microbiol. 89, 992–1001.

Nielsen, T.H., Sorensen, D., Tobiasen, C., Andersen, J.B., Christophersen, C., Givskov, M., et al., 2002. Antibiotic and biosurfactant properties of cyclic lipopeptides produced by fluorescent *Pseudomonas* spp. from the sugar beet rhizosphere. Appl. Environ. Microbiol. 68, 3416–3423.

Nielsen, T.H., Nybroe, O., Koch, B., Hansen, M., Sørensen, J., 2005. Genes involved in cyclic lipopeptide production are important for seed and straw colonization by *Pseudomonas* sp. strain DSS73. Appl. Environ. Microbiol. 71, 4112–4116.

Nowak-Thompson, B., Gould, S.J., Kraus, J., Loper, J.E., 1994. Production of 2,4-diacetylphloroglucinol by the biocontrol agent *Pseudomonas fluoroscens* Pf 5. Can. J. Microbiol. 40, 1064–1066.

Nowak-Thompson, B., Gould, S.J., Loper, J.E., 1997. Identification and sequence analysis of the genes encoding a polyketide synthase required for pyoluteorin biosynthesis in *Pseudomonas fluoroscens* Pf-5. Gene 204, 17–24.

Nowak-Thompson, B., Chaney, N., Wing, J.S., Gould, S.J., Loper, J.E., 1999. Characterization of the pyoluteorin biosynthetic gene cluster of *Pseudomonas fluorescens* Pf-5. J. Bacteriol. 181 (7), 2166–2174.

Nybroe, O., Sørensen, J., 2004. Production of cyclic lipopeptides by fluorescent pseudomonads. Pseudomonas. Springer, Boston, MA, pp. 147–172.

Olanrewaju, O.S., Glick, B.R., Babalola, O.O., 2017. Mechanisms of action of plant growth promoting bacteria. World J. Microbiol. Biotechnol. 33, 197.

Ongena, M., Jacques, P., 2008. Bacillus lipopeptides: versatile weapons for plant disease biocontrol. Trends Microbiol. 16, 115–125.

Ortíz-Castro, R., Contreras-Cornejo, H.A., Macías-Rodríguez, L., López-Bucio, J., 2009. The role of microbial signals in plant growth and development. Plant Signal. Behav. 4, 701–712.

Park, K.S., Kloepper, J.W., 2000. Activation of PR-1a promotes by rhizobacteria which induce systemic resistance in tobacco against *P. syringae* pv. tabaci. Biol. Control 18, 2–9.

Parniske, M., 2005. Plant–fungal associations: cue for the branching connection. Nature 435, 750–751.

Parniske, M., 2008. Arbuscular mycorrhiza: the mother of plant root endosymbioses. Nat. Rev. Microbiol. 6, 763–775.

Parret, A.H.A., Temmerman, K., De Mot, R., 2005. Novel lectin-like bacteriocins of biocontrol strain *Pseudomonas fluoroscens* Pf-5. Appl. Environ. Microbiol. 71, 5197–5207.

Pathma, J., Ayyadurai, N., Sakthivel, N., 2010. Assessment of genetic and functional relationship of antagonistic fluorescent pseudomonads of rice rhizosphere by repetitive sequence, protein coding sequence and functional gene analyses. J. Microbiol. 48, 715–727.

Pathma, J., Rahul, G.R., Kamaraj, K.R., Subashri, R., Sakthivel, N., 2011. Secondary metabolite production by bacterial antagonists. J. Biol. Control. 25, 165–181.

Patten, C.L., Glick, B.R., 1996. Bacterial biosynthesis of indole-3-acetic acid. Can. J. Microbiol. 42, 207–220.

Pérez-Montaño, F., Guasch-Vidal, B., González-Barroso, S., López-Baena, F.J., Cubo, T., Ollero, F.J., et al., 2011. Nodulation-gene-inducing flavonoids increase overall production of autoinducers and expression of N-acyl homoserine lactone synthesis genes in rhizobia. Res. Microbiol. 162, 715–723.

Pierson, L.S., Thomashow, L.S., 1992. Cloning of heterologous expression of phenazine biosynthesis locus from *Pseudomonas aureofaciens* 30–84. Mol. Plant Microbe Interact. 53, 330–339.

Pierson, E.A., Weller, D.M., 1994. Use of mixtures of fluorescent pseudomonads to suppress take-all and improve the growth of wheat. Phytopathology 84, 940–947.

Pierson, L.S., Pierson, E.A., 2010. Metabolism and function of phenazines in bacteria in environment and biotechnological processes. Appl. Microbiol. Biotechnol. 86, 1659–1670.

Pieterse, C.M., Van Wees, S.C., Hoffland, E., Van Pelt, J.A., Van Loon, L.C., 1996. Systemic resistance in *Arabidopsis* induced by biocontrol bacteria is independent of salicylic acid accumulation and pathogenesis-related gene expression. The Plant Cell 8, 1225–1237.

Raaijmakers, J.M., Weller, D.M., 1998. Natural plant protection by 2,4-diacetylphloroglucinol-producing *Pseudomonas* spp. in take-all decline soils. Mol. Plant–microbe Interact. 11, 144–152.

Raaijmakers, J.M., Mazzola, M., 2012. Diversity and natural functions of antibiotics produced by beneficial and plant pathogenic bacteria. Annu. Rev. Phytopathol. 50, 403–424.

Raaijmakers, J.M., de Bruijn, I., de Kock, M.J.D., 2006. Cyclic lipopeptide production by plant-associated *Pseudomonas* spp. Diversity, activity, biosynthesis, and regulation. Mol. Plant Microbe Interact. 19, 699–710.

Raaijmakers, J.M., De Bruijn, I., Nybroe, O., Ongena, M., 2010. Natural functions of lipopeptides from *Bacillus* and *Pseudomonas*: more than surfactants and antibiotics. FEMS Microbiol. Rev. 34, 1037–1062.

Raju, R., Kandhasamy, S., Nalliappan, G.K., Natarajan, K.V., Gandhi, K., Chandrasekaran, B., 2016. Cyclic depsipeptide producing fluorescent pseudomonads exerts antifungal activity against fungal pathogens of maize (*Zea mays*). Afric. J. Microbiol. Res. 10, 1767–1774.

Ramamoorthy, V., Viswanathan, R., Raguchander, T., Prakasam, V., Samiyappan, R., 2001. Induction of systemic resistance by plant growth promoting rhizobacteria in crop plants against pests and diseases. Crop Prot. 20, 1–11.

Ramette, A., Frapolli, M., Defago, G., Moenne-Loccoz, Y., 2003. Phylogeny of HCN synthase encoding hcnBC genes in biocontrol fluorescent pseudomonads and its relationship with host plant species and HCN synthesis ability. Mol. Plant Microbe Interact. 16, 525–535.

Rijavec, T., Lapanje, A., 2016. Hydrogen cyanide in the rhizosphere: not suppressing plant pathogens, but rather regulating availability of phosphate. Front. Microbiol. 7, 1785.

Riley, M.A., Wertz, J.E., 2002. Bacteriocins: evolution, ecology, and application. Annu. Rev. Microbiol. 56, 117–137.

Rodríguez, H., Fraga, R., 1999. Phosphate solubilizing bacteria and their role in plant growth promotion. Biotechnol. Adv. 17, 319–339.

Römheld, V., 1987. Different strategies for iron acquisition in higher plants. Physiol. Plant. 70, 231–234.

Rosales, A.M., Thomashow, L.S., Cook, R.J., Mew, T.W., 1995. Isolation and identification of antifungal metabolite produced by rice associated antagonistic *Pseudomonas* spp. Phytopathology 85, 1028–1032.

Rowe, S.L., Norman, J.S., Friesen, M.L., 2018. Coercion in the evolution of plant–microbe communication: a perspective. Mol. Plant Microbe Interact. 31 (8), 789–794.

Ryu, C.M., Farag, M.A., Hu, C.H., Reddy, M.S., Kloepper, J.W., Paré, P.W., 2004. Bacterial volatiles induce systemic resistance in Arabidopsis. Plant Physiol. 134, 1017–1026.

Saharan, B.S., Nehra, V., 2011. Plant growth promoting rhizobacteria: a critical review. Life Sci. Med. Res. 21, 30.

Saleem, M., Arshad, M., Hussain, S., Bhatti, A.S., 2007. Perspective of plant growth promoting rhizobacteria (PGPR) containing ACC deaminase in stress agriculture. J. Ind. Microbiol. Biotechnol. 34, 635–648.

Schlegel, H.G., 2006. General Microbiology. Polish Scientific Publishers, Warsaw, Poland (in polish).

Schreiner, R.P., Mihara, K.L., McDaniel, H., Bethlenfalvay, G.J., 1997. Mycorrhizal fungi influence plant and soil functions and interactions. Plant Soil 188, 199–209.

Seneviratne, G., 2015. Signal transduction in edaphic ecosystems governs sustainability. Agric. Ecosyst. Environ. 210, 47–49.

Sgroy, V., Cassán, F., Masciarelli, O., Del Papa, M.F., Lagares, A., Luna, V., 2009. Isolation and characterization of endophytic plant growth-promoting (PGPB) or stress homeostasis-regulating (PSHB) bacteria associated to the halophyte *Prosopis strombulifera*. Appl. Microbiol. Biotechnol. 85, 371–381.

Sharma, S.B., Sayyed, R.Z., Trivedi, M.H., Gobi, T.A., 2013. Phosphate solubilizing microbes: sustainable approach for managing phosphorus deficiency in agricultural soils. Springer Plus 2, 587.

Sharma, S., Kumar, V., Tripathi, R.B., 2017. Isolation of phosphate solubilizing microorganism (PSMs) from soil. J. Microbiol. Biotechnol. Res. 1, 90–95.

Siddiqui, I.A., Shaukat, S., 2005. P*seudomonas* a*eruginosa* mediated induction of systemic resistance in tomato against root knot nematode. J. Phytopathol. 4, 21–25.

Siddiqui, I.A., Shaukat, S.S., Sheikh, I.H., Khan, A., 2006. Role of cyanide production by *Pseudomonas fluoroscens* CHA0 in the suppression of root-knot nematode, *Meloidogyne javanica* in tomato. World J. Microbiol. Biotechnol. 22, 641–650.

Singh, J.S., 2013. Plant growth promoting rhizobacteria: potential microbes for sustainable agriculture. Resonance 18 (3), 275–281.

Singh, J.S., 2015. Microbes: the chief ecological engineers in reinstating equilibrium in degraded ecosystems. Agric. Ecosyst. Environ. 203, 80–82.

Singh, J.S., Gupta, V.K., 2018. Soil microbial biomass: a key soil driver in management of ecosystem functioning. Sci. Total Environ. 634, 497–500.

Singh, J.S., Abhilash, P.C., Gupta, V.K., 2016a. Agriculturally important microbes in sustainable food production. Trend Biotechnol. 34, 773–775.

Singh, J.S., Kumar, A., Rai, A.N., Singh, D.P., 2016b. Cyanobacteria: a precious bio-resource in agriculture, ecosystem, and environmental sustainability. Front. Microbiol. 7, 1–19.

Singh, C., Tiwari, S., Gupta, V.K., Singh, J.S., 2018. The effect of rice husk biochar on soil nutrient status, microbial biomass and paddy productivity of nutrient poor agriculture soils. Catena 171, 485–493.

Sonnleitner, E., Haas, D., 2011. Small RNAs as regulators of primary and secondary metabolism in *Pseudomonas* species. Appl. Microbiol. Biotechnol. 91, 63–79.

Steinkellner, S., Lendzemo, V., Langer, I., Schweiger, P., Khaosaad, T., Toussaint, J.P., et al., 2007. Flavonoids and strigolactones in root exudates as signals in symbiotic and pathogenic plant-fungus interactions. Molecules 12, 1290–1306.

Sunish Kumar, R., Ayyadurai, N., Pandiaraja, P., Reddy, A.V., Venkateswarlu, Y., Prakash, O., et al., 2005. Characterization of antifungal metabolite produced by a new strain *Pseudomonas aeruginosa* PUPa3 that exhibits broad-spectrum antifungal activity and biofertilizing traits. J. Appl. Microbiol. 98, 145–154.

Taguchi, F., Suzuki, T., Inagaki, Y., Toyoda, K., Shiraishi, T., Ichinose, Y., 2010. The siderophore pyoverdine of *Pseudomonas syringae* pv. tabaci 6605 is an intrinsic virulence factor in host tobacco infection. J. Bacteriol. 192, 117–126.

Tailor, A.J., Joshi, B.H., 2014. Harnessing plant growth promoting rhizobacteria beyond nature. Rev. J. Plant Nutr. 37, 1534–1571.

Takeda, R., 1958. *Pseudomonas* pigments. III. Derivatives of pyoluteorin. J. Agric. Chem. Soc. Jpn. 23, 126–130.

Teplitski, M., Robinson, J.B., Bauer, W.D., 2000. Plants secrete substances that mimic bacterial N-acyl homoserine lactone signal activities and affect population density-dependent behaviors in associated bacteria. Mol. Plant Microbe Interact. 13, 637–648.

Tewari, S., Arora, N.K., 2013. Transactions among microorganisms and plant in the composite rhizosphere habitat. Plant Microbe Symbiosis: Fundamentals and Advances. Springer, New Delhi, pp. 1–50.

Thomashow, L.S., Weller, D.M., Bonsall, R.F., Pierson, L.S., 1990. Production of the antibiotic phenazine 1-carboxylic acid by fluorescent *Pseudomonas* species in the rhizosphere of wheat. Appl. Environ. Microbiol. 56, 908–912.

Thrane, C., Nielsen, M.N., Sorensen, J., Olsson, S., 2001. *Pseudomonas fluoroscens* DR54 reduces sclerotia formation, biomass development, and disease incidence of *Rhizoctonia solani* causing damping-off in sugar beet. Microb. Ecol. 42, 438–445.

Tien, T.M., Gaskins, M.H., Hubbell, D.H., 1979. Plant growth substances produced by *Azospirillum brasilense* and their effect on the growth of pearl millet (*Pennisetum americanum* L.). Appl. Environ. Microbiol. 37, 1016–1024.

Tiwari, P., Singh, J.S., 2017. A plant growth promoting rhizospheric *Pseudomonas aeruginosa* strain inhibits seed germination in *Triticum aestivum* (L) and *Zea mays* (L). Microbiol. Res. 8, 1–7.

Toklikishvili, N., Dandurishvili, N., Vainstein, A., Tediashvili, M., Giorgobiani, N., Lurie, S., et al., 2010. Inhibitory effect of ACC deaminase-producing bacteria on crown gall formation in tomato plants infected by *Agrobacterium tumefaciens* or *A. vitis*. Plant Pathol 59, 1023–1030.

Ton, J., Van Pelt, J.A., Van Loon, L.C., Pieterse, C.M., 2002. Differential effectiveness of salicylate-dependent and jasmonate/ethylene-dependent induced resistance in *Arabidopsis*. Mol. Plant Microbe Interact. 15, 27–34.

Troppens, D.M., Moynihan, J.A., Barret, M., O'Gara, F., Morrissey, J.P., 2013. Genetics and evolution of 2, 4-diacetylphloroglucinol synthesis in *Pseudomonas fluoroscens*. In: de Bruijn, F.J. (Ed.), Molecular Microbial Ecology of the Rhizosphere, vols. 1 and 2. John Wiley and Sons, Inc, New Jersey, USA, pp. 593–605.

Vacheron, J., Desbrosses, G., Bouffaud, M.L., Touraine, B., Moënne-Loccoz, Y., Muller, D., et al., 2013. Plant growth-promoting rhizobacteria and root system functioning. Front. Plant Sci. 4, 356.

Vaddepalli, P., Fulton, L., Wieland, J., Wassmer, K., Schaeffer, M., Ranf, S., et al., 2017. The cell wall-localized atypical β-1, 3 glucanase ZERZAUST controls tissue morphogenesis in *Arabidopsis thaliana*. Development 144, 2259–2269.

van Loon, L.C., Bakker, P.A.H.M., Pieterse, C.M.J., 1998. Systemic resistance induced by rhizosphere bacteria. Annu. Rev. Phytopathol. 36, 453–483.

van Loon, L.C., Rep, M., Pieterse, C.M., 2006. Significance of inducible defense-related proteins in infected plants. Annu. Rev. Phytopathol. 44, 135–162.

Vimal, S.R., Patel, V.K., Singh, J.S., 2018. Plant growth promoting *Curtobacterium albidum* strain SRV4: an agriculturally important microbe to alleviate salinity stress in paddy plants. Ecol Indic. (Article in Press).

Vincent, M.N., Harrison, L.A., Brackin, J.M., Kovacevich, P.A., Mukerji, P., Weller, D.M., et al., 1991. Genetic analysis of the antifungal activity of a soil borne *Pseudomonas aureofaciens* strain. Appl. Environ. Microbiol. 57, 2928–2934.

Voisard, C., Keel, C., Haas, D., Defago, G., 1989. Cyanide production by *Pseudomonas fluoroscens* helps suppress black root rot of tobacco under gnotobiotic conditions. EMBO J. 8, 351–358.

Volkmar, K.M., Bremer, E., 1998. Effects of seed inoculation with a strain of *Pseudomonas fluoroscens* on root growth and activity of wheat in well-watered and drought-stressed glass-fronted rhizotrons. Can. J. Plant Sci. 78, 545–551.

Walters, D.R., Ratsep, J., Havis, N.D., 2013. Controlling crop diseases using induced resistance: challenges for the future. J. Exp. Bot. 64, 1263–1280.

Wei, G., Kloepper, J.W., Tuzun, S., 1991. Induction of systemic resistance of cucumber to *Colletotrichum orbiculare* by select strains of plant growth-promoting rhizobacteria. Phytopathology 81, 1508–1512.

Weller, D.M., 2007. Pseudomonas biocontrol agents of soilborne pathogens: looking back over 30 years. Phytopathology 97, 250–256.

Whitelaw, M.A., 1999. Growth promotion of plants inoculated with phosphate-solubilizing fungi, Adv. Agron., 69. Academic Press, pp. 99–151.

Xu, J., Li, X.-L., Luo, L., 2012. Effects of engineered *Sinorhizobium meliloti* on cytokinin synthesis and tolerance of Alfalfa to extreme drought stress, Appl. Environ. Microbiol., 78. pp. 8056–8061.

Yehuda, Z., Shenker, M., Hadar, Y., Chen, Y., 2000. Remedy of chlorosis induced by iron deficiency in plants with the fungal siderophore rhizoferrin. J. Plant Nutr. 23, 1991–2006.

Zaidi, A., Ahmad, E., Khan, M.S., Saif, S., Rizvi, A., 2015. Role of plant growth promoting rhizobacteria in sustainable production of vegetables. Curr. Perspect. Sci. Hortic. 193, 231–239.

Zhang, S., Moyne, A.L., Reddy, M.S., Kloepper, J.W., 2002. The role of salicylic acid in induced systemic resistance elicited by plant growth-promoting rhizobacteria against blue mold of tobacco. Biol. Control 25, 288–296.

Further reading

Arguelles–Arias, A., Ongena, M., Halimi, B., Lara, Y., Brans, A., Joris, B., et al., 2009. *Bacillus amyloliquefaciens* GA1 as a source of potent antibiotics and other secondary metabolites for biocontrol of plant pathogens. Microb. Cell Factories. 8, 63–74.

Bulgarelli, D., Rott, M., Schlaeppi, K., van Themaat, E.V.L., Ahmadinejad, N., Assenza, F., et al., 2012. Revealing structure and assembly cues for Arabidopsis root-inhabiting bacterial microbiota. Nature 488, 91–95.

Bulgarelli, D., Garrido-Oter, R., Münch, P.C., Weiman, A., Dröge, J., Pan, Y., et al., 2015. Structure and function of the bacterial root microbiota in wild and domesticated barley. Cell Host Microbe 17, 392–403.

Clifford, J.C., Buchanan, A., Vining, O., Kidarsa, T.A., Chang, J.H., McPhail, K.L., et al., 2016. Phloroglucinol functions as an intracellular and intercellular chemical messenger influencing gene expression in *Pseudomonas protegens*. Environ. Microbiol. 18, 3296–3308.

Kloepper, J.W., Zehnder, G.W., Tuzum, S., Murphy, J.F., Wei, G., Yao, C., et al., 1996. Proceedings of the International Workshop on Biological Control of Plant Diseases. China Agricultural University Press, Beijing, pp. 165–174.

Kondo, Y., Shomura, T., Ogawa, Y., Tsuruoka, Elander, R.P., Mabe, J.A., et al., 1968. Metabolism of tryptophans by *Pseudomonas aureofaciens*. VI. Production of pyrrolnitrin by selected *Pseudomonas* species. Appl. Microbiol. 116, 753–758.

Phillips, D.A., Joseph, C.M., Yang, G.P., Martínez-Romero, E., Sanborn, J.R., Volpin, H., 1999. Identification of lumichrome as a *Sinorhizobium* enhancer of alfalfa root respiration and shoot growth. Proc. Natl. Acad. Sci. 96, 12275–12280.

Chapter 7

Mycorrhizal symbiosis: an effective tool for metal bioremediation

Aradhana Mishra, Arpita Bhattacharya and Nishtha Mishra

Division of Plant Microbe Interactions, CSIR-National Botanical Research Institute, Lucknow, India

Chapter Outline

7.1 Introduction

There is an ever increasing global concern regarding the countless adverse health issues arising due to heavy metal exposures leading to a myriad of death and disabilities worldwide. The varied range of sources of these pollutants including arsenic, chromium, and cadmium pose an alarming concern and needs immediate measures to overcome this challenge. The uncontrolled anthropogenic industrial and agricultural activities have further contributed to the heavy metal contamination that exists through various natural sources like volcanic eruptions, geogenic sources, etc. The traditional remediation approaches to deal with the persisting problem have shown limited results. Numerous studies on finding unconventional, eco-friendly approaches to overcome the problem have led us to the discovery that various heavy metal tolerant plant and microbial species not only have the potential to tolerate the these highly toxic metals ions but also grow and thrive in these contaminated areas (Kumar et al., 2018). This novel approach to use the biological material like plants and microbes has led to a new and emerging field of study and research called bioremediation. The term *bioremediation* brings under its umbrella all the biological approaches applied for the remediation of the contaminated environment, which includes using plants (phytoremediation), microbes (microbial bioremediation), and the interaction of the root of plants and microbes for a synergistic approach towards the remediation process (rhizoremediation) (Singh, 2015).

The rising awareness among for the conservation and protection of the ecosystem has led to further exploration into research for economical, efficient, and long lasting solutions to the rising concerns of the soil contamination. The use of metals has been part of human civilization for time immemorial. Metals have been part of daily life in one form or the other, either as items of utility, ornaments, or for making tools. Some of these metals when accumulated in a concentration higher than their threshold level lead to severe toxicity in humans, plants, and animals. Metals like arsenic (As),

New and Future Developments in Microbial Biotechnology and Bioengineering. DOI: https://doi.org/10.1016/B978-0-12-818258-1.00007-8

lead (Pb), chromium (Cr), tin (Sn), and mercury (Hg) have been found to be the major cause behind several chronic and acute disorders (Singh and Singh, 2017). These metals are generally termed as heavy metals.

The heavy metals are not only a threat to the humans, they also cause severe damage to the environment and biodiversity. The various anthropogenic activities like mining, atmospheric deposition, and agrochemicals such as pesticides and fertilizer have served as means of entry for these heavy metals into the ecosystem (Wuana and Okieimen, 2011). The famous cases of heavy metal toxicity in the recent past include Minimata disease due to heavy mercury poisoning of water bodies in Japan, Itai–Itai disease due to cadmium toxicity in the coal mines of Japan, and prevalent arsenic toxicity in several districts of West Bengal, India. In spite of all these recent cases, the accurate definition of heavy metals remains obscure in relation to life sciences. Adriano (1986) defined heavy metals as the elements having specific density higher than 5 $gm\ cm^{-3}$ (Holleman et al., 1985); though this definition is sufficient from a chemistry point of view, it doesn't fulfill the need of the biological sciences. In relevance to biological sciences, the term *heavy metal* brings with it the view of having some toxic effect on the biological system after crossing a threshold level of its concentration.

In recent times, various industrial and waste disposal activities leading to seepage from the landfills, direct discharge of industrial wastes into the soil, percolation of heavy metal contaminated water into the soil, and extraction from ores by smelting have significantly contributed to increased level of heavy metals (HMs) in soil thus leading to soil pollution. The agriculture sector has also significantly contributed to the growing soil pollution through the excessive application of chemical pesticides, herbicides or fertilizers, rupture of underground storage tanks, etc. (Tchounwou et al., 2012). The elevated levels of HMs in soil have also been found to be toxic to the plants growing in it and result in impaired metabolism and decrease in plant growth. This also adversely affects the soil microbial population. Several HMs like Cd, Zn, Ni, Cu, etc. usually act as a micronutrient in both plants and microbes but when accumulated in higher concentrations, they adversely affect the diversity and the activities of both plants as well as the microbial communities thriving in the soil (Ahemad, 2014).

Various new cleanup strategies of the contaminated wastelands extensively polluted with heavy metals are being explored. These heavy metals cannot be completely degraded; they only transform from one oxidation state or organic complex to another oxidation state or to simpler forms (Garbisu and Alkorta, 2001). For the remediation of heavy metal contaminated soil three approaches are generally employed, that is, through chemical, physical, or biological techniques (Yao et al., 2012). Out of the three, chemical and physical remediation measures are known to produce other pollutants as by-products, whose treatment and disposal is inconvenient at a large scale.

The available techniques are grouped into two categories based on the site of treatment: (1) ex situ techniques, which require removal of the contaminated soil for treatment on- or off-site; and (2) in situ methods, which remediate without excavation of contaminated soil. The ex situ measures involving excavation and landfill, thermal treatment, acid leaching, electroreclamation etc. are not economically feasible and practical because of their high cost, large level destruction of the soil structure and fertility, and high dependence on the contaminants of concern, soil properties, site conditions, etc. The classical remediation method involves two major methods: first, extraction, that is, the removal of contaminated material by physical means; and second, stabilization of the metals in the contaminated sites. The physical means of cleanup of contaminated sites involves the potential human exposure to contaminants during their transport and in the stabilization of the contaminants in the sites leads to their slow remediation; besides, the region remains polluted and long-term extensive monitoring is required. Keeping in mind all drawbacks new eco-friendly measures are being looked into, thus the bioremediation strategies have come forward as a novel means to overcome the problem.

7.2 Bioremediation: the way forward

Bioremediation chiefly involves the use of plants and microorganisms like bacteria, fungi, and actinomycetes and their biosynthesized nanoparticles (Pandey et al., 2015; Kumari et al., 2017) for accelerated degradation or immobilization of the contaminants and waste materials. Soil biota consists of the microorganisms (bacteria, fungi, actinomycetes), soil fauna (protozoa, nematodes, mites, springtails, spiders, insects, and earthworms) and flora living all or part of their lives in or on the soil or pedosphere (Singh and Seneviratne, 2017a,b). Within the soil, millions of soil organisms exist but only a fraction of an immense number of species have been cultured and identified through the available culturable and nonculturable techniques (Fortuna, 2012). These soil microbes are responsible for the soil organic matter turnover (Condron et al., 2010). Though it is not a new concept, during the last few decades, its application has gradually increased, as it is a relatively economic, environment friendly remediation measure. According to the Environment Protection Authority (EPA), three stages of soil bioremediation projects have been described, laboratory based study: on-site or pilot field trial and the full-scale bioremediation conducted on-site or at an approved EPA-licensed facility.

7.3 Why bioremediation?

The bioremediation is process is in situ, that is, no hazardous contaminated soil needs to relocate to treatment sites unlike the mechanical and chemical treatment plants. The plant based bioremediation measure is solar driven, and does not require any additional input for the treatment. The treatments lead to minimum by-product emission and it has been estimated that the cost of biological treatment of the contaminated soil is about 10%–20% cheaper than the mechanical and chemical treatments. Bioremediation measures like biostimulation and bioaugmentation of contaminants using varied organisms are widely and popularly used, and are efficient in ameliorating the problems of heavy metal accumulation. In spite of being widely accepted and an efficient method for remediation, they suffer from some limitations. They are restricted to the shallow soil level, groundwater, and streams. Contaminants and by-products of degradation can mobilize to groundwater sources or may enter the food chain through animal consumption of the plant parts consisting of accumulated heavy metals. Another major limitation involved is that the hyperaccumulator plants are generally slow-growing with low biomass and shallow root systems and the concern of proper disposal of the harvested plants. Phytoextraction studies carried out in hydroponic systems where the solution contains contaminants do not mirror the actual scenario of the affected field conditions. Phytoextraction coefficients measured under field conditions are likely to be less than those determined at the laboratory level.

7.4 Bioremediation of heavy metals

Heavy metals present in the disturbed sites, owing to their toxicity, pose a serious threat to the environment (Boudh and Singh, 2019). Initial colonizers of such heavily disturbed lands or polluted sites are often plant species along with fungi having high tolerance to extreme conditions and having the ability to accumulate high concentration of heavy metals in their shoot and hyphal body (Phieler et al., 2013; Shuab et al., 2017). The process to mitigate these heavily disturbed sites containing environmental pollutants such as metals or any other organic contaminants by the interference of plants is popularly known as phytoremediation. If not removed from the soil the heavy metals are likely to enter the food chain and disturb the entire food web. The factors that influence the accumulation of the heavy metal ions in the food web include the sources of these ions, their concentration in soil, the degree of uptake by the plants, and their absorbance by the animal system (Bolan et al., 2014). Various strategies are available under the umbrella of phytoremediation, that is phytovolatilization, phytoextraction, and phytostabilization (Ghasemi-Fasaei, 2012; Jiang et al., 2010; Rahimi et al., 2013). Different types of phytoremediation measures use different plant properties to ameliorate the problem of heavy metals present in the soil (Pilon-Smits, 2000).

The heavy metals extracted from the contaminated soil are translocated from soil to the plants and concentrate in shoots within the vacuole or are released through the stomata present in the leaves. As it is already known that heavy metals cannot be chemically degraded and need to be physically removed or be immobilized, the phytoremediation provides a low cost and effective means to mitigate the problem. The mechanism used by plants for the remediation through metal uptake involves either volatilization or compartmentation or sequestration within the vacuoles (Fig. 7.1).

The process of bioremediation is not limited to the use of plants, and it also includes the rich diverse microbial flora available at our disposal (Boudh et al., 2017; Kaushal and Singh, 2017). The microbial flora for the implementation of bioremediation process include both bacteria and fungi. Heavy metal tolerant microbes screened and isolated from the contaminated sites can be mass-produced and reintroduced as the prime agents for the bioremediation. The availability of the metal cations to the plant depends upon the ability of the native microflora for mobilization and immobilization of the ions (Kushwaha et al., 2018). The main criterion for selecting a microbe or microbial consortia for the task depends on the chemical nature of the contaminant, the tolerance level of the microbe to the pollutant, and its ability to degrade it (Poi et al., 2017). The "one size fits all" approach is not suitable while selecting the microbes for bioremediation, so it is important to consider that the different microbes show different tolerance towards different heavy metals ions present in the contaminated soil. The microbe selected for remediation of a particular heavy metal via exogenous introduction should be able to tolerate high concentration and it should be compatible with beneficial native microflora (Ojuederie and Babalola, 2017). The uptake of metal ions by microbes depends on several factors like the properties of the medium, the root zone, environmental conditions, chelating agents added, chemical properties of the contaminant, and the bioavailability of the metals (Gaur et al., 2014). Several cyanobacterial (Kumar and Singh, 2017) and bacterial species like *Rhodococcus* sp. *Enterobacter* sp., *Strenotrophomonas* sp., *Providencia* sp., *Chryseobacterium* sp., *Ochrobacterium* sp., and *Delftia* sp. as consortia have been found to remove heavy metals (elevated concentrations) and reduce the organic load of wastewater effluent (Bestawy et al., 2013; Shinwari et al., 2015; Šrut et al., 2018). Fungal species like *Penicillium chrysogenum, Aspergillus nidulans, Rhizopus arrhizus, Microsporum cookei, Chysosporium*

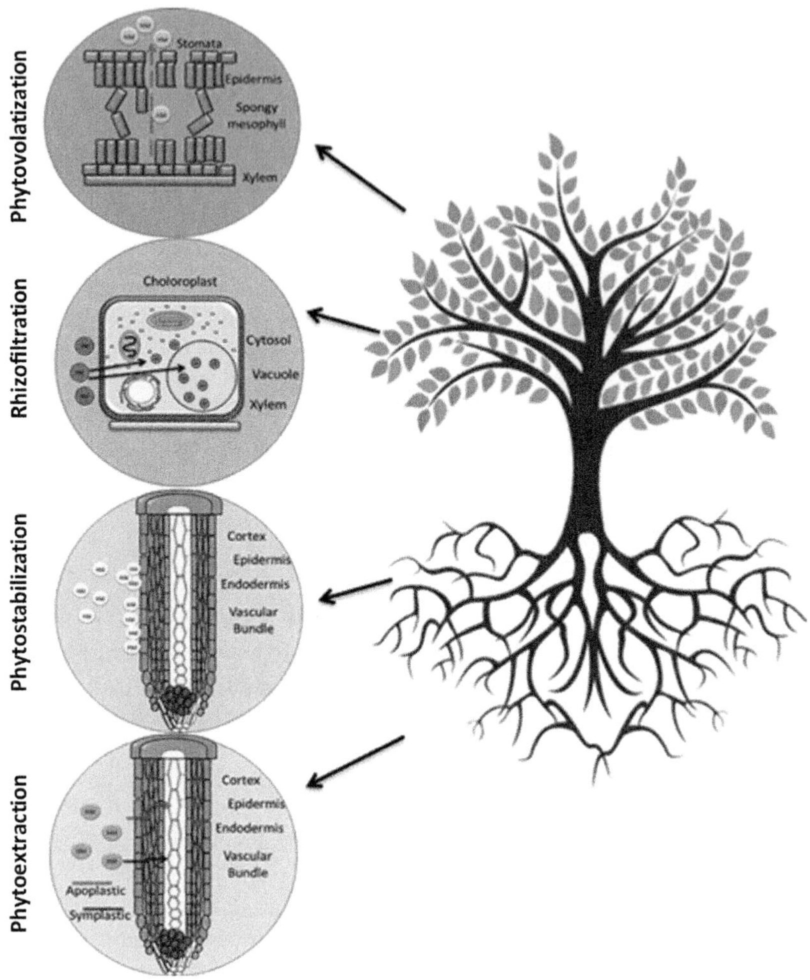

FIGURE 7.1 The mechanism used by plants for the remediation through metal uptake involves either volatilization, compartmentation, or sequestration within the vacuoles.

spp., *Trichosporon asteroids*, *Fomitopsis* cf. *meliae*, and *Ganoderma aff. steyaertanum* have been reported to have ability heavy metal tolerance and bioremediation activity (Kumar et al., 2014; Devaraj and Leeraj, 2015; Kaewdoung et al., 2016).

The fungus *Trichoderma*, widely known for its biocontrol activity, is also tolerant to a wide range of pollutants including heavy metals, pesticides, and polyaromatic hydrocarbons (Tripathi et al., 2012, 2013a,b). Plants growing in heavy metal contaminated sites have been reported to form a symbiotic relationship with the mycorrhizae (Karimi et al., 2011; Pichardo et al., 2012) suggesting that these fungi have coevolved with the plant with heavy metal tolerance strategies and may have a role in enhancing phytoremediation of the contaminated site. The application of mycorrhiza forming a symbiotic association between a fungus and the root of higher plants in the remediation of heavy metal contamination is a growing field of research owing to its economical, efficient, and eco-friendly nature (Singh et al., 2016).

7.5 Mycorrhiza: the step forward

Mycorrhizae form a symbiotic association where the fungal partner assists the host plant in the absorption of nutrient (phosphorus), minerals, and water from the soil and in turn provides the plant protection against different fungal phytopathogen and invading nematodes, while the plant provides the fungus nutrition in form of carbohydrates. Based on their localization the mycorrhiza are broadly classified into two groups in the host plant, that is, ectomycorrhiza and endomycorrhiza. Ectomycorrhiza generally surround the cells of the host plant. They do not penetrate within the host cells and forms a sheath around the root tips of the plant. They are associated with the roots of woody angiosperms and the coniferous trees (Anderson and Cairney, 2007). The symbiotic relation of ectomycorrhiza has shown to play a crucial role in forest resilience and in maintaining the ecosystem balance in the woodlands (Pickles and Simard, 2017).

The endomycorrhizal hyphae not only grow inside the host roots but also penetrate within the cells of the root cortex and form a sheath over the cell membrane, thus forming an invasive symbiotic relationship with the host plant (Bonfante and Genre, 2010). According to the classification by Peterson and Seligman (2004), the endomycorrhiza are divided further into five major groups: arbuscular, ericoid, arbutoid, monotropoid, and orchid mycorrhizae.

7.6 Mycorrhiza: role in heavy metal remediation

Field studies have highlighted the significance of the mutualistic relationship between the fungi and the roots of higher plants, which leads to enhanced plant survival in the polluted areas by improving water relations and enhancing the nutrient uptake from the surroundings, leading to possible efficient bioremediation. However, in spite of all these advantages, there has been limited application of mycorrhiza in bioremediation processes. Two possible reasons may exist behind this: first, the use of nonmycorrhizal plant families like *Brassicaceae* and *Caryophyllaceae* for the phytoremediation processes; and second, earlier workers did not consider the mycorrhiza as a potent tool for remediation processes (Khan, 2006). Recent findings suggest that the application of mycorrhiza has immense potential in the bioremediation process. This is because fungi have the ability to accumulate significant amounts of heavy metals within them varying from a few percent to 20% of the dry mass, suggesting that mobility of metals in the soil systems are affected by the microbial biomass and processes (Siddiquee et al., 2015). Mycorrhiza increases the root absorption area up to 47-fold providing the plants appropriate below ground ecosystems for better uptake of water and nutrients (Hakeem et al., 2014). Studies have also highlighted the role of the fungi involved in mycorrhizal association in detoxification of heavy metals and establishment of vegetation at the contaminated sites (Reddy et al., 2016). It has shown to reduce the metal toxicity in plants and trees by varied mechanisms like in Cd toxicity, suppressing metal uptake in *Lonicera japonica*, enhancing tolerance to Cu and Cd in *Eucalyptus tereticornis*; and reducing Mn and Zn root-to-shoot translocation in *Betula pubescens* (Jiang et al., 2016; Reddy et al., 2016; Fernández-Fuego et al., 2017). The arbuscular mycorrhiza take up the heavy metal ions through their hyphal surface absorption. The members of phylum Glomeromycota of arbuscular mycorrhizal fungi (AMF) widely found in the soil of most ecosystems have shown to produce a protein, glomulin, that is efficient in sequestering Cu, Cd, PD, Zn, and Mn (Gonzalez-Chavez et al., 2004).

According to the calculations by Berg and Söderström (1979), the surface of interaction between fungi and soil is up to 0.14 m^2 per gram of soil. The metals are removed from the wastes by different microorganisms through both metabolism dependent (bioaccumulation) and independent (biosorption) processes (Lin et al., 2017). Earlier, it was believed that pollution inhibits mycorrhizae development (Perrin and Estival et al., 1990) but the recent findings suggest otherwise. Previous studies have suggested that heavy metal stress may also lead to decrease in fruiting body abundance for some species and increase for others that are sometimes rare in nonpolluted ecosystems (Rühling et al., 1984). However, field studies have shown that the ectomycorrhizal fungi can accumulate a high level of metal contents in their fruiting bodies and this metal accumulation varied between different species (Melgar et al, 2016). The studies show that the mycorrhiza not only form a physical barrier between the toxic heavy metals and the host plants but also actively help the host through accumulation of phosphorus thus providing higher tolerance to the plant to survive in the contaminated soil (Soares and Siqueira, 2008).

The role of mycorrhiza in amelioration of the toxic metal ions is still a developing field of research. With the advances made in the field of research technologies, various mechanisms through which mycorrhiza protect the host plant from the toxic effect of the heavy metal ions are unraveling. The different mechanisms include (1) immobilization of the heavy metal ions and their retention with the fungal structures; (2) precipitation, chelation, and complex formations in rhizosphere to prevent the uptake; (3) sequestration of the heavy metals within the vacuole or arbuscule of the fungal cells; (4) use of metallothione, organic acids, and compound-specific chaperones for the cytosolic storage; (5) activation of antioxidation mechanism; (6) passive (across concentration gradient) and active (using ATP) movement of the heavy metal ions through different transporters or pore across the plasma membrane; and (7) dissemination of the accumulated heavy metal ions through specific and nonspecific pathways (Fig. 7.2 and Table 7.1).

7.6.1 Immobilization of heavy metal ions and their retention with fungal structures

The fungus, through secretion of various compounds, converts the available heavy metal ions in the soil to unavailable forms for uptake by the plants. In some cases, it has been observed that the fungal mycelia bind the available heavy metal ions with their cell wall components making them unavailable for uptake. Immobilization of the metals in the

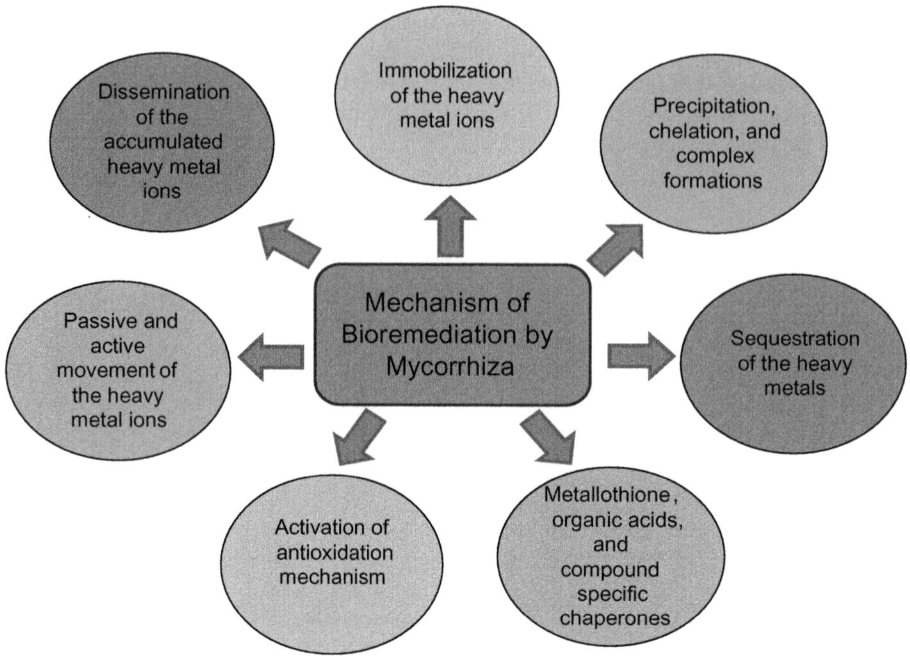

FIGURE 7.2 Different mechanism of mycorrhizal remediation.

TABLE 7.1 Different mycorrhizal association for bioremediation of heavy metal toxicity.

Mycorrhizal partner used	Plant partner	Heavy metal	Mechanism/effect on the host plant	References
Glomus mosseae	*Pteris vittata* L.;	As	Elevation in plant P concentration	Liu et al. (2005)
Rhizophagus irregularis	*Medicago truncatula* L.	As	Improves the plant P nutrition and influences As accumulation and speciation	Zhang et al. (2015)
R. irregularis	*Oryza sativa*	As	Increased the proportion of DMA, and aerobic cultivation decreased As accumulation	Zhang et al. (2016)
Rhizoglomus intraradices and Glomus etunicatum	*Triticum aestivum*	As	Strengthening antioxidant defense system and thiol metabolism	Sharma et al. (2017)
Funneliformis mosseae, Rhizophagus intraradices, and Claroideoglomus etunicatum	*Solanum lycopersicum* L.	Cd	Induction of acquired systemic tolerance	Hashem et al. (2016a,b)
Arbascular Mycorrhizal fungi	*Helianthus annus* L.	Cd	Higher activities of acid and alkaline phosphatases, which were reduced by cadmium stress	Abeer and Alqarawi (2015)
Funneliformis mosseae, Rhizophagus intraradices, and Claroideoglomus etunicatum	*Cassia italica* Mill	Cd	Reduced lipid peroxidation	Hashem et al. (2016a,b)
Funelliformis mosseae and Piriformospora indica	*T. aestivum*	Cd	Increased plant's dry mass and activities of catalase, ascorbate peroxidase, and glutathione S transferase activity	Shahabivand et al. (2016)

(Continued)

TABLE 7.1 (Continued)

Mycorrhizal partner used	Plant partner	Heavy metal	Mechanism/effect on the host plant	References
R. irregularis	*Phragmites australis*	Cd	Impacted Mn and P uptake and accumulation while coping with Cd toxicity	Huang et al. (2017)
Mycorrhizal fungi	*Linum usitatissimum*	Cd	Concentration of unsaturated fatty acids at 15 ppm Cd	Kaplan et al. (2015)
Glomus intraradices	*Zea mays*	Cd	Higher activities of superoxide dismutase (SOD), peroxidase (POD), catalase (CAT), and lower Cd concentrations in roots, stems, leaves, and ears	Liu et al. (2018)
Arbascular mycorrhizal fungi	*Elsholtzia splendens*	Cu	Increase in germination rate and the germination index of the seeds as well as the fresh weights of hypocotyl and radicle	Li et al. (2017)
R. irregularis	*Z. mays*	Cu	Increased accumulation of total phytochelatin content in shoots	Merlos et al. (2016)
Funneliformis mosseae; R. intraradices	*Capsicum annuum* L.	Cu	Higher total dry weight and the leaf area	Ruscitti et al. (2017)
F. mosseae	*Festuca arundinacea*	Ni	Affects the nickel translocation and expression of ABC transporter and metallothionein genes	Shabani and Sabzalian (2016)
Claroideoglomus claroideum	*Oenothera picensis*	Cu	Protect plant from metal toxicity, enhance both plant establishment and nutrition	Meier et al. (2015)
Glomus versiforme; G. caledonium	*Z. mays*	Zn	Increases root activity and antioxidant activities, help to acquire nutrients	Wang et al. (2016)
R. irregularis	*Medicago truncatula*	Zn	Positive plant biomass, increase in AMF-induced phosphate transporter gene *MtPT4* responses	Watts-Williams et al. (2017)
G. intraradices	*Trigonella foenum-gracum*	Zn	Diminished accumulation of ZnONPs; secretion of glycoprotein called glomalin	Sinani et al. (2007)
Gigaspora rosea	*Glycine max* (L.) Merr.	Fe	Improved shoot Zn and systemically modulated Mn and Fe	Ibiang et al. (2017)
R. irregularis; R. irregularis, Funneliformis mosseae, R. aggregates; Claroideoglomus etunicatum	*Lolium perenne* L.	Fe-CN	Reduction of total CN and free CN content in the mycorrhizal roots	Sut et al. (2016)

fungal biomass is an important mechanism employed by the mycorrhizal fungi. The fungi efficiently immobilize and act as biosorbants for a variety of heavy metals like Cd, Zn, Cu, Pb, and Ni (Gadd, 2009, 2010). In tobacco plant *Glomus intraradices* mycorrhizal symbiosis, the extraradial mycorrhiza immobilize the Cd in the soil, thus leading to lower accumulation of the heavy metal ion in the biomass than tobacco roots. It was found that accumulation in the extraradial mycorrhiza is 10–20 times more Cd per unit of biomass than the nonmycorrhizal plants (Janoušková et al., 2006). The mycorrhizal association with roots of rice plant shows higher accumulation of Cu in the roots but significantly lower accumulation within the shoots. The mycorrhiza produce organic substances to bind the Cu ions to the cell wall, changing the cell wall composition, immobilizing the ions for preventing further movement to the symplast of the host roots (Zhang et al., 2009). *Glomus mosseae* upon inoculation to *Zea mays* decreases the concentration of Cd toxicity within the host plant by accumulation of the excess Cd with the fungal biomass (Liu et al., 2014). Through the production of glomulin, a metal-sorbing glycoprotein, the AMF binds the Pb and Cd forms a metal bound complex. It was observed that although the total content of glomulin declined over a period the amount of metal bound to glomulin increases (Wu et al., 2014).

7.6.2 Precipitation, chelation, and complex formation in rhizosphere to prevent uptake

The heavy metals bind to the cell walls of ecto- or endo-mycorrhiza by the components like chitin, melanin, and other cellulosic derivatives (Meharg, 2003). In *Picea abeis, Laccaria laccata* forms the ectomycorrhiza, where it was found to associate Cd to its cell wall (Khan, 2006). In *Pisolithus tinctorius* for the amelioration of the heavy metal toxicity the fungi secretes extrahyphal slime and forms polyphosphate linkage of Zn and Cu (Meharg, 2003). Al and Cd have been found to accumulate in the fungal mantle and their concentration gradually decreases along the Hartig net towards the interior in *Rhizopogaon roseolus* (Agerer, 2012).

Chelation of free iron ions by siderophore production is another mechanism for complex formation for binding of Fe^{3+} ions. Although in *Laccaria bicolor*, an ectomycorrhizal fungi, the gene (Sid1/SidA) encoding the enzyme involved in the first step of fungal siderophore production is not yet reported, according to a recent report, a different set of hydroxamate production has been found (Haselwandter et al., 2013). Under iron deficient conditions ectomycorrhizal fungi (EMF), *Scleroderma verrucosum, Suillus luteus,* and *Rhizopogon luteolus* isolated from the fruiting bodies of *Pinus radiate* produce catecholates and hydroxamate siderophores for the chelation of the excess toxic iron ions (Machuca et al., 2007).

Some mycorrhiza in spite of being unable to produce siderophores themselves are able to take up the siderophore complexes produced by other organisms having bacterial or plant origins (Haas et al., 2008; Saha et al., 2013). Such mycorrhiza fungi comprise the siderophore-iron transporter (SIT), which helps them to translocated the chelate complex from soil to inside the fungal cell. These transporter SIT subfamilies are exclusively found within the fungal kingdom. Structurally they are predicted to consist of 12–14 transmembrane domains, which possibly function as the proton symporters driven by the plasma membrane potential (Fig. 7.3) (Haas et al., 2008; Philpott and Protchenko, 2008).

7.6.3 Sequestration of the heavy metals within the vacuole or arbuscule of the fungal and plant cells

Mycorrhizal structures like arbuscules, vacuoles, and vesicles are actively involved in the storage of toxic metal ions, preventing their translocation to the plants. Cu accumulation has been observed within the vacuoles of spores *Glomus intraradices* (Ferrol et al., 2009). Mycorrhizal ecotypes indigenous to the heavy metal polluted soil are found to be

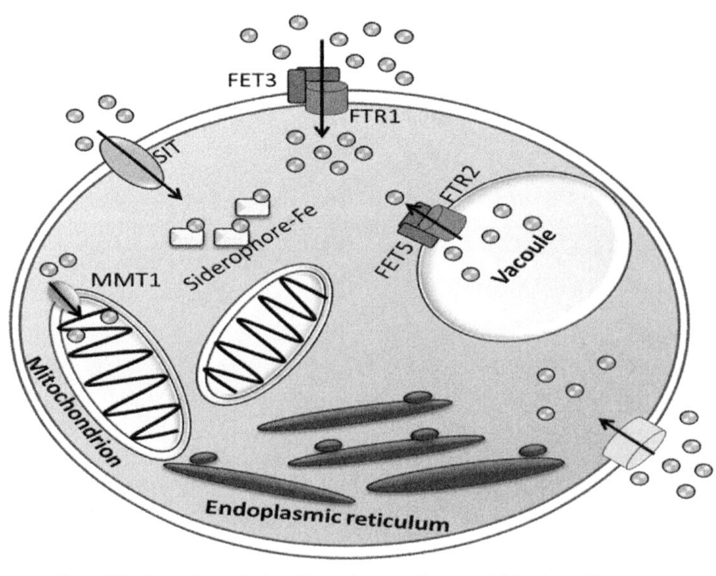

FIGURE 7.3 Mechanism of transportation of Fe ions through the siderophore pathway within the cell.

better adapted to the heavy metal toxicity than those that inhabit the unpolluted sites by the storage of heavy metal ions as inactive forms within the fungal cellular inclusion bodies. Vijver et al. (2004) divided these distinct inclusion bodies or the heat stable proteins formed by binding with the heavy metals into three types of granules: type I, containing acid phosphatases for the accumulation of heavy metals like Cd, Cu, Hg, and Ag; type II, for the storage of excess iron; and type III are the amorphous calcium phosphates for heavy metals like zinc (Vijver et al., 2004). Thus, for the enhanced heavy metal tolerance by such mycorrhizal fungi, several approaches have been suggested by various investigators, which include adsorption or accumulation of metals by ion exchange, the formation of complexes, precipitation, or crystallization (Hu et al., 2013). The mechanism involves the attachment of the heavy metals to the fungal cell wall thus leading to their subsequent accumulation within the vacuoles.

7.6.4 Metallothione, organic acids, and compound-specific chaperones for the cytosolic storage

Metallothione are low molecular weight and cysteine-rich metal-binding proteins having about 70–80 amino acids. These compounds are involved in the coordination, sequestration, and regulation of the concentration of heavy metal ions within the fungal cytosol. These compounds are synthesized by the fungi upon exposure to toxic concentrations of the heavy metal ions like Zn, Cu, and Cd (Folli-Pereira et al., 2012). The expression of metallothione or the phytochelatins varies within the host plant cell in response to activation by the mycorrhiza. Within the mycorrhiza or the plant cell, the heavy metals may form complexes with the metallothione or the phytochelatins (Merlos et al., 2016). During the heavy metal toxicity the expression level of genes encoding metallothioneins, phytochelatins, and heat shock proteins increases (Fuentes et al., 2016). In *Populus alba* grown in heavy metal contaminated soil containing high concentration of Cu and Zn, upon inoculating with the arbuscular mycorrhiza *Glomus mosseae* or *G. intraradices*, there was a upregulation of expression of foliar metallothionein and polyamine biosynthetic gene (Cicatelli et al., 2010).

The metallothione plays a significant role in regulating the mechanisms of Copper toxicity tolerance through maintaining the cytosolic Cu ion concentrations (González-Guerrero and Argüello, 2008). Upon experiencing high concentration of the Cu ions in the surroundings the fungi maintains the concentration through low molecular weight and cysteine-rich metal binding proteins metallothioniens (Tamayo et al., 2014). The dissolution of heavy metals by the fungi is mediated through the ligand promoted or proton promoted mechanisms, and the organic acids act as the source of protons and metal-chelating anions for the formation of complex compounds with the heavy metal cations (Finlay, 2008).

7.6.5 Activation of antioxidation mechanism

The antioxidant defense mechanisms include both the enzymes like catalases, superoxide dismutases, and glutathione S-transferases (Mayer et al., 2017), while nonenzymatic measures include glutathione (GSH) and vitamins C, E, and B6 (Cabral et al., 2015; Calvo et al., 2017). In sunflower plants (*Helianthus annuus* L.) mycorrhization leads to Cd induced guaiacol peroxidase activity in roots being less elevated in mycorrhizal plants than the nonmycorrhizal roots (de Andrade et al., 2008). The interaction of a saprophytic fungal consortium (*Bjerkandera adusta* and *Mortierella* sp.) and mycorrhizal fungus *Rhizophagus irregularis* alleviate oxidative stress in plants grown in heavy metal contaminated soil by increasing the ascorbate and guaicol peroxidise activity and decreasing the lipid peroxidation resulting in lower amount of malondialdehyde content (Fuentes et al., 2016). The mycorrhizal partner seems to play a key role in regulating the guiacol peroxidation and lipid peroxidation of the host plant for minimizing the oxidative stress induced by the heavy metal toxicity. *Solanum melongena* L. Pb contaminated soils amended with the arbuscular mycorrhizal fungus have shown higher antioxidative defense response of plants against heavy metal stress (Chaturvedi et al., 2018).

7.6.6 Active (using ATP) and passive (across concentration gradient) movement of the heavy metal ions through different transporters or pore across the plasma membrane

In arbuscular mycorrhizal fungus different transmembrane transporter proteins have also been reported. In AMF *Rhizophagus irregularis* genes encoding two families of copper transporters (CTR) and copper-transporting P-type ATPase family have been identified based on the complete study of its genome sequence. (Kim et al., 2013; Tamayo et al., 2014). These genes are responsible for the synthesis of Cu^+ transport specific transporter channels (Sinani et al., 2007)

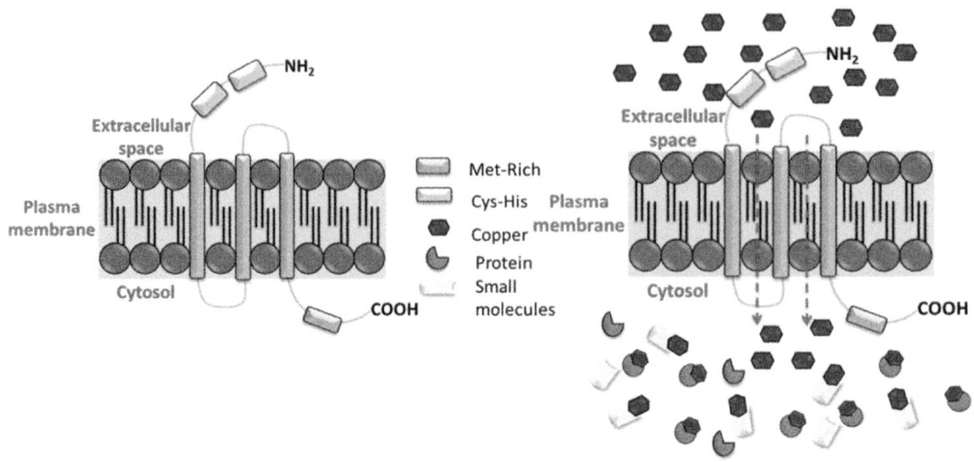

FIGURE 7.4 Membrane structure of copper transporters (CTR) family protein and transportation of Cu ions through the CTR membrane protein channel.

CTR protein and copper-transporting P-type ATPase families are highly conserved protein transporters among all the fungal species for mediating the transport of copper ions across the plasma membrane along the concentration gradient within the cell's cytoplasm. Structurally the CTR protein composed of a single integral protein with three transmembrane domains. The N-terminal lying in the extracellular region comprises of Cystein-Histidine (Cys-His) rich motif, while the C-terminal facing the cytosol has methionine (Met) rich motif. These specific stretches of amino acid residues are the key players in the copper transport across the plasma membrane. The methionine residues present in the C-terminal help in coordinating the metal transport across membranes. (Yuan et al., 2011). Three members of the CTR family have been reported in fungi to be high-affinity transporters of copper ions across the membranes. CTR1 and CTR3 are the Cu transporters found across the plasma membrane to acquire the environmental copper (Marjorette et al., 2000), while CRT2 is localized at the tonoplast, the vacuolar membrane (Pócsi, 2011) from where it pumps out Cu ions from the vacuole to the cytosol (Fig. 7.4) (Puig and Thiele, 2002).

Copper-transporting P-type ATPases belong to the family of heavy metal P-type ATPase transporter (HMA family) protein family. Structurally they comprise of eight transmembrane domains, ATP binding and phosphorylation site, long cytoplasmic loop and like most members of HMA family one conserved CPX motif. The Cu ions entering the cytosol through the CTR1/3 transmembrane protein are bound by the ATX1 chaperone. The CCC2 take the Cu ion from the ATX1 and pass it on to the post-Golgi network. The process of transfer between the CCC2 and the post-Golgi network involves the release of energy through the dephosphorylation of ATP on the CCC2 molecule. Upon entering the post-Golgi network the copper ion is allocated to enzymes like multicopper oxidase and other copper-dependent proteins. (Tamayo et al., 2014).

Zn homeostasis within the mycorrhiza is regulated mainly by two families of transporter proteins, that is, the ZIP (zinc-iron permease or ZRT-IRT-like Protein) and the Cation Diffusion Facilitator (CDF) families (Amich and Calera, 2014). Zinc acquisition is a key aspect in *Aspergillus fumigates* virulence. Mycopathologia, 1–7). Of these two zinc transporter families CDF is chiefly responsible for protecting the fungal cell against the zinc toxicity. So far only three Zn transporters belonging to the CDF family have been identified and characterized among the mycorrhizal fungi. The first was reported in *R. irregularis*, second in *Hebeloma cylindrosporum*, and finally the third was reported recently in *Oidiodendron maius* (González-Guerrero et al., 2005; Blaudez and Chalot, 2011; Khouja et al., 2013; Tamayo et al., 2014).

The ZIP protein comprises of eight transmembrane domains with both N- and C-terminal lying in the extracellular space. The ZIP protein transporters are responsible for transport of Zn or other ions across the membranes whether it be plasma membrane or tonoplast membrane to the cytosol. For selectivity of Zn ion in the ZIP protein third and fourth histidine-rich region are responsible (Nishida et al., 2008). Studies on expression analysis of the mycorrhizal fungus *R. irregularis* showed that the ZRT1 gene responsible for coding the ZIP protein was highly upregulated in mycorrhizal roots in comparison to the extradial mycelium and the germinated spores (Figs. 7.5 and 7.6) (Tamayo et al., 2014).

The cation diffusion facilitator (CDF) family consists of probably six transmembrane domains with N- and C-terminal lie facing the cytoplasm. It also has a large cytoplasmic loop facing the cytoplasm, lying between the fourth

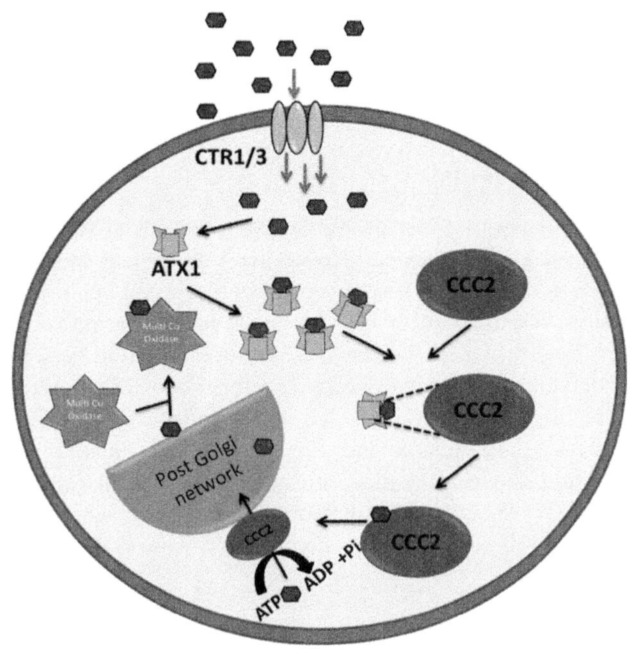

FIGURE 7.5 Mechanism of transportation of Cu ions through the copper transporters (CTR) membrane protein channel within the cell.

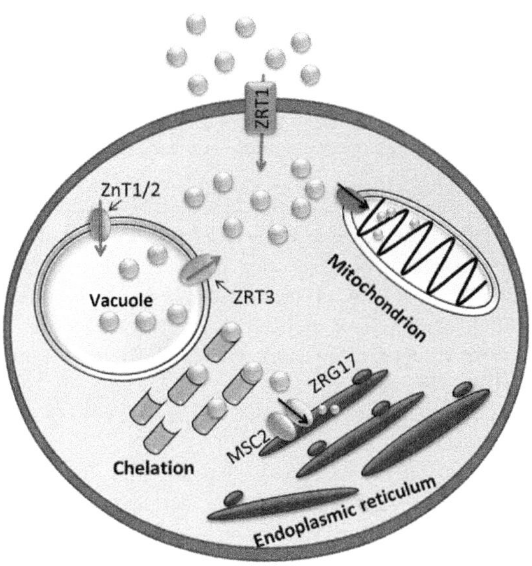

FIGURE 7.6 Mechanism of transportation of Zn ions through the ZRT1 membrane protein channel within the cell.

and fifth transmembrane domain. The function of CDF protein is relatively opposite to that of the ZIP proteins as it transports Zn and other ions into the organelles or secretes outside the cell from the cytoplasm. ZRT1 mediates entry of the Zn ions into the cytosol from the surroundings; it enters the vacuole via the ZnT1 transporter, which is responsible for the Zn tolerance in the mycorrhiza (González-Guerrero et al., 2005). MSC2, another member of the CDF family transports Zn ions into the endoplasmic reticulum (Ellis et al., 2004).

7.6.7 Dissemination of the accumulated heavy metal ions through specific and nonspecific pathways

The accumulated heavy metal ions after reaching a threshold level within the mycorrhizal cells are excreted out of the cells. The fate of the Cu ions within the cytosol depends on their cytosolic concentrations. Depending on the cell's homeostasis, the Cu ions that entered via the CTR proteins bind either to the small proteins or enzymes or with a chaperone molecule, which further guides them through the cytosol. The heavy metal ions are then extruded from the cytosol by specific heavy metal transporters located on plant membranes (Merlos et al., 2016).

7.7 Conclusions and future prospects

Bioremediation using mycorrhiza is an emerging low-cost alternative employed for cleaning up of contaminated heavy metal–containing soils. The technology is being tested at pilot level as well as at large scale and its future prospects are being analyzed. Some constraints or barriers in the way of large-scale application of mycorrhiza are its absorption capacities, its host plant translocation, stabilization of heavy metals, identification of genes responsible for the uptake and/or degradation of contaminant, length of time needed for phytoremediation to work, disposal of biomass or produce, and protecting wildlife from feeding on the host plants used for mycorrhizal remediation measures. The AMF are universal and ubiquitous rhizosphere microflora forming a symbiosis with plant roots and acting as biofertilizer, bioprotectants, and biodegraders. As the heavy metal uptake and tolerance also depends on the host plant and the soil factors including soil microflora, the interaction between the plant and the mycorrhiza plays an important role in the successful survival and growth of plants in the heavy metal contaminated soils. Further research and investigation is needed in the mechanism involved in the amelioration strategies of the mycorrhizal remediation for devising site and contaminant specific treatment methods.

References

Abeer, H., Alqarawi, 2015. Alleviation of adverse impact of cadmium stress in sunflower (*Helianthus annuus* L.) by arbuscular mycorrhizal fungi. Pak. J. Bot. 47, 785–795.

Adriano, D.C., 1986. Trace Elements in the Terrestrial Environment. Spring-Verlag. New York Inc, New York.

Agerer, R., 2012. Ectotrophic and ectendotrophic mycorrhizae, Genetics cell biology and physiology systematics and comparative morphology ecology and vegetation science. In: Progress in Botany, vol. 60. Springer, p. 471.

Ahemad, M., 2014. Remediation of metalliferous soils through the heavy metal resistant plant growth promoting bacteria: paradigms and prospects. Arab. J. Chem. In press.

Amich, J., Calera, J.A., 2014. Zinc acquisition: a key aspect in *Aspergillus fumigates* virulence. Mycopathologia. 178, 379–385.

Anderson, I.C., Cairney, J.W., 2007. Ectomycorrhizal fungi: exploring the mycelial frontier. FEMS Microbiol. Rev. 31, 388–406.

Berg, B., Söderström, B., 1979. Fungal biomass and nitrogen in decomposing Scots pine needle litter. Soil Biol. Biochem. 11, 339–341.

Bestawy, E.E., Helmy, S., Hussien, H., Fahmy, M., Amer, R., 2013. Bioremediation of heavy metal-contaminated effluent using optimized activated sludge bacteria. Appl. Water Sci. 3, 181–192.

Blaudez, D., Chalot, M., 2011. Characterization of the ER-located zinc transporter ZnT1 and identification of a vesicular zinc storage compartment in *Hebeloma cylindrosporum*. Fungal Genet Biol. 48, 496–503.

Bolan, N., Kunhikrishnan, A., Thangarajan, R., Kumpiene, J., Park, J., Makino, T., et al., 2014. Remediation of heavy metal (loid)s contaminated soils–to mobilize or to immobilize? J. Hazard. Mater. 266, 141–166.

Bonfante, P., Genre, A., 2010. Mechanisms underlying beneficial plant–fungus interactions in mycorrhizal symbiosis. Nat. Commun. 1, 48.

Boudh, S., Singh, J.S., 2019. Pesticide contamination: environmental problems and remediation strategies. In: Bharagava, R.N., Chowdhary, P. (Eds.), Emerging and Eco-Friendly Approaches for Waste Management. Springer International Publishing, Singapore, pp. 245–269.

Boudh, S., Tiwari, S., Singh, J.S., 2017. Microbial mediated lindane bioremediation. In: Singh, J.S., Seneviratne, G. (Eds.), Agro-Environmental Sustainability: Volume 2: Managing Environmental Pollution. Springer International Publishing, Dordrecht, Netherlands, pp. 213–233.

Cabral, L., Soares, C.R.F.S., Giachini, A.J., Siqueira, J.O., 2015. Arbuscular mycorrhizal fungi in phytoremediation of contaminated areas by trace elements: mechanisms and major benefits of their applications. World J. Microbiol. Biotechnol. 31, 1655–1664.

Calvo, J., Jung, H., Meloni, G., 2017. Copper metallothioneins. IUBMB Life 69, 236–245.

Chaturvedi, R., Favas, P., Pratas, J., Varun, M., Paul, M.S., 2018. Assessment of edibility and effect of arbuscular mycorrhizal fungi on *Solanum melongena* L. grown under heavy metal (loid) contaminated soil. Ecotoxicol. Environ. Saf. 148, 318–326.

Cicatelli, A., Lingua, G., Todeschini, V., Biondi, S., Torrigiani, P., Castiglione, S., 2010. Arbuscular mycorrhizal fungi restore normal growth in a white poplar clone grown on heavy metal-contaminated soil, and this is associated with upregulation of foliar metallothionein and polyamine biosynthetic gene expression. Ann. Bot. 106, 791–802.

Condron, L., Stark, C., O'Callaghan, M., Clinton, P., Huang, Z., 2010. The role of microbial communities in the formation and decomposition of soil organic matter. Soil Microbiology and Sustainable Crop Production. Springer, Dordrecht, pp. 81–118.

de Andrade, S.A.L., da Silveira, A.P.D., Jorge, R.A., de Abreu, M.F., 2008. Cadmium accumulation in sunflower plants influenced by arbuscular mycorrhiza. Int. J. Phytorem. 10, 1–13.

Devaraj, S., Leeraj, C., 2015. Screening of Potential Fungi from the Polluted Soil for Copper Bioremediation Doctoral dissertation). INTI International University.

Ellis, C.D., Wang, F., MacDiarmid, C.W., Clark, S., Lyons, T., Eide, D.J., 2004. Zinc and the Msc2 zinc transporter protein are required for endoplasmic reticulum function. J. Cell Biol. 166, 325–335.

Fernández-Fuego, D., Bertrand, A., González, A., 2017. Metal accumulation and detoxification mechanisms in mycorrhizal *Betula pubescens*. Environ. Pollut. 231, 1153–1162.

Ferrol, N., González-Guerrero, M., Valderas, A., Benabdellah, K., Azcón-Aguilar, C., 2009. Survival strategies of arbuscular mycorrhizal fungi in Cu-polluted environments. Phytochem. Rev. 8, 551.

Finlay, R.D., 2008. Ecological aspects of mycorrhizal symbiosis: with special emphasis on the functional diversity of interactions involving the extraradical mycelium. J. Exp. Bot. 59, 1115–1126.

Folli-Pereira, M.D.S., Meira-Haddad, L.S.A., Bazzolli, D.M.S., Kasuya, M.C.M., 2012. Arbuscular mycorrhiza and plant tolerance to stress. Rev. Bras. Ciênc. Solo. 36, 1663–1679.

Fortuna, A., 2012. The soil biota. Nat. Educ. Knowl. 3, 1.

Fuentes, A., Almonacid, L., Ocampo, J.A., Arriagada, C., 2016. Synergistic interactions between a saprophytic fungal consortium and *Rhizophagus irregularis* alleviate oxidative stress in plants grown in heavy metal contaminated soil. Plant Soil 407, 355–366.

Gadd, G.M., 2009. Biosorption: critical review of scientific rationale, environmental importance and significance for pollution treatment. J. Chem. Technol. Biotechnol. 84, 13–28.

Gadd, G.M., 2010. Metals, minerals and microbes: geomicrobiology and bioremediation. Microbiology 156, 609–643.

Garbisu, C., Alkorta, I., 2001. Phytoextraction: a cost-effective plant-based technology for the removal of metals from the environment. Bioresour. Technol. 77, 229–236.

Gaur, N., Flora, G., Yadav, M., Tiwari, A., 2014. A review with recent advancements on bioremediation-based abolition of heavy metals. Environ. Sci. Process. 16, 180–193.

Ghasemi-Fasaei, R., 2012. Malic acid and phosphorus influences on nickel phytoremediation efficiency and metal nutrients relationships in a Ni-polluted calcareous soil. Int. Res. J. Basic Appl. 3, 2805–2808.

Gonzalez-Chavez, M.C., Carrillo-Gonzalez, R., Wright, S.F., Nichols, K.A., 2004. The role of glomalin, a protein produced by arbuscular mycorrhizal fungi, in sequestering potentially toxic elements. Environ. Pollut. 130 (3), 317–323.

González-Guerrero, M., Argüello, J.M., 2008. Mechanism of Cu + -transporting ATPases: soluble Cu + chaperones directly transfer Cu + to transmembrane transport sites. Proc. Natl. Acad. Sci. 105, 5992–5997.

González-Guerrero, M., Azcón-Aguilar, C., Mooney, M., Valderas, A., MacDiarmid, C.W., Eide, D.J., et al., 2005. Characterization of a *Glomus intraradices* gene encoding a putative Zn transporter of the cation diffusion facilitator family. Fungal Genet. Biol. 42, 130–140.

Haas, H., Eisendle, M., Turgeon, B.G., 2008. Siderophores in fungal physiology and virulence. Annu. Rev. Phytopathol. 46, 149–187.

Hakeem, K., Sabir, M., Ozturk, M., Mermut, A.R. (Eds.), 2014. Soil Remediation and Plants: Prospects and Challenges. Academic Press.

Haselwandter, K., Häninger, G., Ganzera, M., Haas, H., Nicholson, G., Winkelmann, G., 2013. Linear fusigen as the major hydroxamate siderophore of the ectomycorrhizal Basidiomycota *Laccaria laccata* and *Laccaria bicolor*. Biometals 26, 969–979.

Hashem, A., Abd-Allah, E.F., Alqarawi, A.A., Egamberdieva, D., 2016a. Bioremediation of adverse impact of cadmium toxicity on *Cassia italica* Mill by arbuscular mycorrhizal fungi. Saudi J. Biol. Sci. 23, 39–47.

Hashem, A., Abd-Allah, E.F., Alqarawi, A.A., Al Huqail, A.A., Egamberdieva, D., Wirth, S., 2016b. Alleviation of cadmium stress in *Solanum lycopersicum* L. by arbuscular mycorrhizal fungi via induction of acquired systemic tolerance. Saudi J. Biol. Sci. 23, 272–281.

Holleman, A.F., Wiberg, E., Wiberg, N., 1985. Lehrbuch der anorganischen Chemie [Textbook of inorganic chemistry]. Available from: http://moef.gov.in/wp-content/uploads/2017/09/AR-11-12-En.pdf

Hu, J., Wu, S., Wu, F., Leung, H.M., Lin, X., Wong, M.H., 2013. Arbuscular mycorrhizal fungi enhance both absorption and stabilization of Cd by Alfred stonecrop (*Sedum alfredii* Hance) and perennial ryegrass (*Lolium perenne* L.) in a Cd-contaminated acidic soil. Chemosphere 93, 1359–1365.

Huang, X., Ho, S.H., Zhu, S., Ma, F., Wu, J., Yang, J., et al., 2017. Adaptive response of arbuscular mycorrhizal symbiosis to accumulation of elements and translocation in *Phragmites australis* affected by cadmium stress. J. Environ. Manage. 197, 448–455.

Ibiang, Y.B., Mitsumoto, H., Sakamoto, K., 2017. Bradyrhizobia and arbuscular mycorrhizal fungi modulate manganese, iron, phosphorus, and polyphenols in soybean (*Glycine max* (L.) Merr.) under excess zinc. Environ. Exp. Bot. 137, 1–13.

Janoušková, M., Pavlíková, D., Vosátka, M., 2006. Potential contribution of arbuscular mycorrhiza to cadmium immobilisation in soil. Chemosphere 65, 1959–1965.

Jiang, J., Wu, L., Li, N., Luo, Y., Liu, L., Zhao, Q., et al., 2010. Effects of multiple heavy metal contamination and repeated phytoextraction by *Sedum plumbizincicola* on soil microbial properties. Eur. J. Soil Biol. 46, 18–26.

Jiang, Q.Y., Zhuo, F., Long, S.H., Zhao, H.D., Yang, D.J., Ye, Z.H., et al., 2016. Can arbuscular mycorrhizal fungi reduce Cd uptake and alleviate Cd toxicity of *Lonicera japonica* grown in Cd-added soils? Sci. Rep. 6, 21805.

Kaewdoung, B., Sutjaritvorakul, T., Gadd, G.M., Whalley, A.J., Sihanonth, P., 2016. Heavy metal tolerance and biotransformation of toxic metal compounds by new isolates of wood-rotting fungi from Thailand. Geomicrobiol. J. 33, 283–288.

Kaplan, M.E., Simmons, E.R., Hawkins, J.C., Ruane, L.G., Carney, J.M., 2015. Influence of cadmium and mycorrhizal fungi on the fatty acid profile of flax (*Linum usitatissimum*) seeds. J. Sci. Food Agric. 95, 2528–2532.

Karimi, A., Khodaverdiloo, H., Sepehri, M., Sadaghiani, M.R., 2011. Arbuscular mycorrhizal fungi and heavy metal contaminated soils. Afr. J. Microbiol. Res. 5, 1571–1576.

Kaushal, S., Singh, J.S., 2017. Wastewater impact on human health and microorganism-mediated remediation and treatment through technologies. In: Singh, J.S., Seneviratne, G. (Eds.), Agro-Environmental Sustainability: Volume 2: Managing Environmental Pollution. Springer International Publishing, Dordrecht, Netherlands, pp. 235–250.

Khan, A.G., 2006. Mycorrhizoremediation—an enhanced form of phytoremediation. J. Zhejiang Univ. Sci. B 7, 503–514.

Khouja, H.R., Abbà, S., Lacercat-Didier, L., Daghino, S., Doillon, D., Richaud, P., et al., 2013. OmZnT1 and OmFET, two metal transporters from the metal-tolerant strain Zn of the ericoid mycorrhizal fungus *Oidiodendron maius*, confer zinc tolerance in yeast. Fungal Genet. Biol. 52, 53–64.

Kim, H., Wu, X., Lee, J., 2013. SLC31 (CTR) family of copper transporters in health and disease. Mol. Aspects Med. 34, 561–570.

Kumar, A., Singh, J.S., 2017. Cyanoremediation: a green-clean tool for decontamination of synthetic pesticides from agro-and aquatic ecosystems. In: Singh, J.S., Seneviratne, G. (Eds.), Agro-Environmental Sustainability: Volume 2: Managing Environmental Pollution. Springer International Publishing, Dordrecht, Netherlands, pp. 59–83.

Kumar, R., Sharma, A.K., Singh, P., Dhir, B., Mehta, D., 2014. Potential of some fungal and bacterial species in bioremediation of heavy metals. JNPMSRA 1, 213–223.

Kumar, A., Kaushal, S., Saraf, S.A., Singh, J.S., 2018. Screening of chlorpyrifos (CPF) tolerant cyanobacteria from paddy field soil of Lucknow. India Int. J. Appl. Adv. Sci. Res. 3, 100–105.

Kumari, R., Singh, J.S., Singh, D.P., 2017. Biogenic synthesis and spatial distribution of silver nanoparticles in the legume mungbean plant (*Vigna radiata* L.). Plant Physiol. Biochem. 110, 158–166.

Kushwaha, A., Hans, N., Kumar, S., Rani, R., 2018. A critical review on speciation, mobilization and toxicity of lead in soil-microbe-plant system and bioremediation strategies. Ecotoxicol. Environ. Saf. 147, 1035–1045.

Li, J., Liang, H., Yan, M., Chen, L., Zhang, H., Liu, J., et al., 2017. Arbuscular mycorrhiza fungi facilitate rapid adaptation of *Elsholtzia splendens* to copper. Sci. Total Environ. 599, 1462–1468.

Lin, D., Ji, R., Wang, D., Xiao, M., Zhao, J., Zou, J., et al., 2017. The research progress in mechanism and influence of biosorption between lactic acid bacteria and Pb (II): a review. Crit. Rev. Food Sci. Nutr. 1–16.

Liu, Y., Zhu, Y.G., Chen, B.D., Christie, P., Li, X.L., 2005. Influence of the arbuscular mycorrhizal fungus *Glomus mosseae* on uptake of arsenate by the As hyperaccumulator fern *Pteris vittata* L. Mycorrhiza 15, 187–192.

Liu, L., Gong, Z., Zhang, Y., Li, P., 2014. Growth, cadmium uptake and accumulation of maize (*Zea mays* L.) under the effects of arbuscular mycorrhizal fungi. Ecotoxicology 23, 1979–1986.

Liu, L., Li, J., Yue, F., Yan, X., Wang, F., Bloszies, S., et al., 2018. Effects of arbuscular mycorrhizal inoculation and biochar amendment on maize growth, cadmium uptake and soil cadmium speciation in Cd-contaminated soil. Chemosphere 194, 495–503.

Machuca, A., Pereira, G., Aguiar, A., Milagres, A.M.F., 2007. Metalchelating compounds produced by ectomycorrhizal fungi collected from pine plantations. Lett. Appl. Microbiol. 44, 7–12.

Marjorette, M., Peña, O., Puig, S., Thiele, D.J., 2000. Characterization of the *Saccharomyces cerevisiae* high affinity copper transporter Ctr3. J. Biol. Chem. 275, 33244–33251. Available from: https://doi.org/10.1074/jbc.M005392200.

Mayer, Z., Duc, N.H., Sasvári, Z., Posta, K., 2017. How arbuscular mycorrhizal fungi influence the defense system of sunflower during different abiotic stresses. Acta. Biol. Hung. 68, 376–387.

Meharg, A.A., 2003. The mechanistic basis of interactions between mycorrhizal associations and toxic metal cations. Mycol. Res. 107, 1253–1265.

Meier, S., Cornejo, P., Cartes, P., Borie, F., Medina, J., Azcón, R., 2015. Interactive effect between Cu adapted arbuscular mycorrhizal fungi and biotreated agrowaste residue to improve the nutritional status of *Oenothera picensis* growing in Cu polluted soils. J. Plant Nutr. Soil Sci. 178, 126–135.

Melgar, M.J., Alonso, J., García, M.A., 2016. Cadmium in edible mushrooms from NW Spain: bioconcentration factors and consumer health implications. Food Chem. Toxicol. 88, 13–20.

Merlos, M.A., Zitka, O., Vojtech, A., Azcón-Aguilar, C., Ferrol, N., 2016. The arbuscular mycorrhizal fungus *Rhizophagus irregularis* differentially regulates the copper response of two maize cultivars differing in copper tolerance. Plant Sci. 253, 68–76.

Nishida, S., Mizuno, T., Obata, H., 2008. Involvement of histidine-rich domain of ZIP family transporter TjZNT1 in metal ion specificity. Plant Physiol. Biochem. 46, 601–606.

Ojuederie, O.B., Babalola, O.O., 2017. Microbial and plant-assisted bioremediation of heavy metal polluted environments: a review. Int. J. Environ. Res. Public Health 14, 1504.

Pandey, S., Kumari, M., Singh, S.P., Bhattacharya, A., Mishra, S., Chauhan, P.S., et al., 2015. Bioremediation via nanoparticles: an innovative microbial approach. Handbook of Research on Uncovering New Methods for Ecosystem Management Through Bioremediation. IGI Global, pp. 491–515.

Perrin, R., Estivalet, D., 1990. Mycorrhizal, association and forest decline (yellowing of spruce). Agric. Ecosyst. Environ. 28, 381–387.

Peterson, C., Seligman, M.E., 2004. Character Strengths and Virtues: A Handbook and Classification, Vol. 1. Oxford University Press.

Phieler, R., Voit, A., Kothe, E., 2013. Microbially supported phytoremediation of heavy metal contaminated soils: strategies and applications. Geobiotechnology I. Springer, Berlin, Heidelberg, pp. 211–235.

Philpott, C.C., Protchenko, O., 2008. Response to iron deprivation in *Saccharomyces cerevisiae*. Eukaryotic Cell 7, 20–27.

Pichardo, S.T., Su, Y., Han, F.X., 2012. The potential effects of arbuscular mycorrhizae (AM) on the uptake of heavy metals by plants from contaminated soils. J. Bioremed. Biodegrad. 3, 124.

Pickles, B.J., Simard, S.W., 2017. Mycorrhizal networks and forest resilience to drought. Mycorrhizal Mediation of Soil. Elsevier, pp. 319–339.

Pilon-Smits, E., 2000. Phytoremediation. Annu. Rev. Plant Biol. 56, 15–39.

Pócsi, I., 2011. Toxic metal/metalloid tolerance in fungi—a biotechnology-oriented approach. Cellular Effects of Heavy Metals. Springer, Dordrecht, pp. 31–58.

Poi, G., Aburto-Medina, A., Mok, P.C., Ball, A.S., Shahsavari, E., 2017. Bioremediation of phenol-contaminated industrial wastewater using a bacterial consortium—from laboratory to field. Water Air Soil Pollut. 228, 89.

Puig, S., Thiele, D.J., 2002. Molecular mechanisms of copper uptake and distribution. Curr. Opin. Chem. Biol. 6, 171–180.

Rahimi, M., Farhadi, R., Mehdizadeh, R., 2013. Phytoremediation: using plants to clean up contaminated soils with heavy metals. Int. J. Agric. Res. 3, 148.

Reddy, M.S., Kour, M., Aggarwal, S., Ahuja, S., Marmeisse, R., FraissinetTachet, L., 2016. Metal induction of a *Pisolithus albus* metallothionein and its potential involvement in heavy metal tolerance during mycorrhizal symbiosis. Environ. Microbiol. 18, 2446–2454.

Rühling, Å., Bååth, E., Nordgren, A., Söderström, B., 1984. Fungi in metal-contaminated soil near the Gusum brass mill, Sweden. Ambio 34–36.

Ruscitti, M., Arango, M., Beltrano, J., 2017. Improvement of copper stress tolerance in pepper plants (*Capsicum annuum* L.) by inoculation with arbuscular mycorrhizal fungi. Theor. Exp. Plant Physiol. 29, 37–49.

Saha, R., Saha, N., Donofrio, R.S., Bestervelt, L.L., 2013. Microbial siderophores: a mini review. J. Basic Microbiol. 53, 303–317.

Shabani, L., Sabzalian, M.R., 2016. Arbuscular mycorrhiza affects nickel translocation and expression of ABC transporter and metallothionein genes in *Festuca arundinacea*. Mycorrhiza 26, 67–76.

Shahabivand, S., Maivan, H.Z., Mahmoudi, E., Soltani, B.M., Sharifi, M., Aliloo, A.A., 2016. Antioxidant activity and gene expression associated with cadmium toxicity in wheat affected by mycorrhizal fungus. Zemdirbyste-agriculture 103, 53–60.

Sharma, S., Anand, G., Singh, N., Kapoor, R., 2017. Arbuscular mycorrhiza augments arsenic tolerance in wheat (*Triticum aestivum* L.) by strengthening antioxidant defense system and thiol metabolism. Front. Plant Sci. 8, 906.

Shinwari, K.I., Shah, A.U., Afridi, M.I., Zeeshan, M., Hussain, H., Hussain, J., et al., 2015. Application of plant growth promoting rhizobacteria in bioremediation of heavy metal polluted soil. Asian J. Multidiscip. Stud. 3, 4.

Shuab, R., Lone, R., Ahmad, J., Reshi, Z.A., 2017. Arbuscular mycorrhizal fungi: a potential tool for restoration of degraded land. Mycorrhiza-Nutrient Uptake, Biocontrol, Ecorestoration. Springer, Cham, pp. 415–434.

Siddiquee, S., Rovina, K., Azad, S.A., Naher, L., Suryani, S., Chaikaew, P., 2015. Heavy metal contaminants removal from wastewater using the potential filamentous fungi biomass: a review. J. Microb. Biochem. Technol. 7, 384–393.

Sinani, D., Adle, D.J., Kim, H., Lee, J., 2007. Distinct mechanisms for Ctr1-mediated copper and cisplatin transport. J. Biochem. 282, 26775–26785.

Singh, J.S., 2015. Microbes play major roles in the ecosystem services. Clim. Change Environ. Sustain. 3, 163–167.

Singh, J.S., Seneviratne, G., 2017a. Agro-Environmental Sustainability: Volume 2: Managing Environmental Pollution, 2. Springer International Publishing, Dordrecht, Netherlands, p. 229.

Singh, J.S., Seneviratne, G., 2017b. Agro-Environmental Sustainability: Volume 1: Managing Crop Health, vol. 1. Springer International Publishing, Dordrecht, Netherlands, p. 316.

Singh, J.S., Singh, D.P., 2017. Methanotrophs: an emerging bioremediation tool with unique broad spectrum methane monooxygenase (MMO) enzyme. In: Singh, J.S., Seneviratne, G. (Eds.), Agro-Environmental Sustainability: Volume 2: Managing Environmental Pollution. Springer International Publishing, Dordrecht, Netherlands, pp. 1–18.

Singh, J.S., Abhilash, P.C., Gupta, V.K., 2016. Agriculturally important microbes in sustainable food production. Trends Biotechnol. 34, 773–775.

Soares, C.R., Siqueira, J.O., 2008. Mycorrhiza and phosphate protection of tropical grass species against heavy metal toxicity in multi-contaminated soil. Biol. Fertil. Soils 44, 833–841.

Šrut, M., Menke, S., Höckner, M., Sommer, S., 2018. Earthworms and cadmium-heavy metal resistant gut bacteria as indicators for heavy metal pollution in soils? BioRxiv 1, 295444.

Sut, M., Boldt-Burisch, K., Raab, T., 2016. Possible evidence for contribution of arbuscular mycorrhizal fungi (AMF) in phytoremediation of iron–cyanide (Fe–CN) complexes. Ecotoxicology 25, 1260–1269.

Tamayo, E., Gómez-Gallego, T., Azcón-Aguilar, C., Ferrol, N., 2014. Genome-wide analysis of copper, iron and zinc transporters in the arbuscular mycorrhizal fungus *Rhizophagus irregularis*. Front. Plant Sci. 5, 547.

Tchounwou, P.B., Yedjou, C.G., Patlolla, A.K., Sutton, D.J., 2012. Heavy metal toxicity and the environment. Molecular, Clinical and Environmental Toxicology. Springer, Basel, pp. 133–164.

Tripathi, P., Singh, P.C., Mishra, A., Chauhan, P.S., Dwivedi, S., Bais, R.T., et al., 2012. *Trichoderma*: A potential bioremediator for environmental cleanup. Clean Technol. Envir. Policy 15 (4), 541–550.

Tripathi, P., Singh, P.C., Mishra, A., Chaudhry, V., Mishra, S., Tripathi, R.D., et al., 2013a. Trichoderma inoculation ameliorates arsenic induced phytotoxic changes in gene expression and stem anatomy of chickpea (*Cicer arietinum*). Ecotoxicol. Environ. Saf. 89, 8–14.

Tripathi, P., Singh, P.C., Mishra, A., Chauhan, P.S., Dwivedi, S., Bais, R.T., et al., 2013b. *Trichoderma*: a potential bioremediator for environmental cleanup. Clean Technol. Environ. 15, 541–550.

Vijver, M.G., Van Gestel, C.A., Lanno, R.P., Van Straalen, N.M., Peijnenburg, W.J., 2004. Internal metal sequestration and its ecotoxicological relevance: a review. Environ. Sci. Technol. 38, 4705–4712.

Wang, F., Liu, X., Shi, Z., Tong, R., Adams, C.A., Shi, X., 2016. Arbuscular mycorrhizae alleviate negative effects of zinc oxide nanoparticle and zinc accumulation in maize plants–a soil microcosm experiment. Chemosphere 147, 88–97.

Watts-Williams, S.J., Tyerman, S.D., Cavagnaro, T.R., 2017. The dual benefit of arbuscular mycorrhizal fungi under soil zinc deficiency and toxicity: linking plant physiology and gene expression. Plant Soil 420, 375–388.

Wu, Z., McGrouther, K., Huang, J., Wu, P., Wu, W., Wang, H., 2014. Decomposition and the contribution of glomalin-related soil protein (GRSP) in heavy metal sequestration: field experiment. Soil Biol. Biochem. 68, 283–290.

Wuana, R.A., Okieimen, F.E., 2011. Heavy metals in contaminated soils: a review of sources, chemistry, risks and best available strategies for remediation. ISRN Ecol. Artilce ID 402647.

Yao, Z., Li, J., Xie, H., Yu, C., 2012. Review on remediation technologies of soil contaminated by heavy metals. Procedia Environ. Sci. 16, 722–729.

Yuan, M., Li, X., Xiao, J., Wang, S., 2011. Molecular and functional analyses of COPT/Ctr-type copper transporter-like gene family in rice. BMC Plant Biol. 11, 69.

Zhang, X.H., Lin, A.J., Gao, Y.L., Reid, R.J., Wong, M.H., Zhu, Y.G., 2009. Arbuscular mycorrhizal colonisation increases copper binding capacity of root cell walls of *Oryza sativa* L. and reduces copper uptake. Soil Biol. Biochem. 41, 930–935.

Zhang, X., Ren, B.H., Wu, S.L., Sun, Y.Q., Lin, G., Chen, B.D., 2015. Arbuscular mycorrhizal symbiosis influences arsenic accumulation and speciation in *Medicago truncatula* L. in arsenic-contaminated soil. Chemosphere 119, 224–230.

Zhang, X., Wu, S., Ren, B., Chen, B., 2016. Water management, rice varieties and mycorrhizal inoculation influence arsenic concentration and speciation in rice grains. Mycorrhiza 26, 299–309.

Further reading

Brundrett, M., 2004. Diversity and classification of mycorrhizal associations. Biol. Rev. 79, 473–495.

Citterio, S., Prato, N., Fumagalli, P., Aina, R., Massa, N., Santagostino, A., et al., 2005. The arbuscular mycorrhizal fungus *Glomus mosseae* induces growth and metal accumulation changes in *Cannabis sativa* L. Chemosphere 59, 21–29.

EPA, 2000. Introduction to Phytoremediation. Environmental Protection Agency, Cincinnati, 70.

Gaur, A., Adholeya, A., 2004. Prospects of arbuscular mycorrhizal fungi in phytoremediation of heavy metal contaminated soils. Curr. Sci. 528–534.

Gonzalez-Chavez, C., D'haen, J., Vangronsveld, J., Dodd, J.C., 2002. Copper sorption and accumulation by the extraradical mycelium of different *Glomus* spp. (arbuscular mycorrhizal fungi) isolated from the same polluted soil. Plant Soil 240, 287–297.

Machuca, Á., 2011. Metal-chelating agents from ectomycorrhizal fungi and their biotechnological potential. Diversity and Biotechnology of Ectomycorrhizae. Springer, Berlin, Heidelberg, pp. 347–369.

Peña, M.M.O., Puig, S., Thiele, D.J., 2000. Characterization of the *Saccharomyces cerevisiae* high affinity copper transporter Ctr3. J. Biol. Chem. 275, 33244–33251.

Prasad, M.N.V., 2011. A state-of-the-art report on bioremediation, its applications to contaminated sites in India. Ministry of Environment & Forests, Government of India.

Chapter 8

Microalgae: potential agents for CO_2 mitigation and bioremediation of wastewaters

Shashirekha Viswanaathan[1] and M.P. Sudhakar[2]

[1]Shri A.M.M. Murugappa Chettiar Research Centre, Chennai, India, [2]National Institute of Ocean Technology, Chennai, India

Chapter Outline

8.1 Introduction

The primitive Earth consisted of an atmosphere filled with carbon dioxide that could not support any life forms. The advent of cyanobacteria and other autotrophic algae that consumed CO_2 through photosynthesis, in turn releasing O_2 led to rapid decline in atmospheric CO_2 levels paving the way for other lives possible on Earth. Autotrophic microalgae require just sunlight, CO_2, and water for their growth, through the photosynthesis process. The conversion of solar energy to chemical energy is their characteristic feature; however it varies for each organism depending on its conversion efficiency. The harnessed energy is stored in the cells as proteins, carbohydrates, and lipids. Algae are the known to be most photosynthetically efficient organisms on the face of the earth and due to their high affinity for CO_2, they are considered to be prospective organisms for various applications including CO_2 mitigation. They are, therefore, undoubtedly the primary producers in the aquatic ecosystem and contribute to approximately half of the global net primary productivity. Algae are ubiquitous inhabiting damp terrestrial surroundings and also freshwater and marine environments. They are diverse in nature found in several different forms and structures varying from unicellular to filamentous, complex and differentiated levels (Field et al., 1998; Demirbaş, 2009a). They are versatile photoautotrophic

New and Future Developments in Microbial Biotechnology and Bioengineering. DOI: https://doi.org/10.1016/B978-0-12-818258-1.00008-X

entities with the inherent ability to adapt effortlessly to diverse ecological habitats, however challenging/adverse, ranging from freshwater, brackish water or seawater and also in moist soil because they are well equipped to thrive in various extreme temperatures and pH conditions (Falkowski et al., 2004; Diniz et al., 2017). Some water bodies with high organic and inorganic contaminants act as ideal substrates promoting algal growth; however, rapid uncontrolled increase in algal population could lead to algal blooms, a condition referred to as eutrophication, depriving other forms of life to coexist and disturbing a balanced ecosystem (Demirbaş, 2009b). Microalgae were one of the earliest life forms on Earth and they are the fastest growing plants in the world (Falkowski et al., 2004). They are much more efficient than terrestrial plants in capturing sunlight, utilizing toxic pollutants, requiring minimal resources and they are noncompetitive to food crops for resources (Singh et al., 2016; Kumar and Singh, 2017; Kumar et al., 2018). There are over 50,000 microalgal species, of which approximately 30,000 species have been explored (Mata et al., 2009). Depending upon the species and the conditions in which they survive, their growth rate varies from 24 hours, some species exhibiting the doubling time of just few hours of exponential phase, to several days, much less than that of terrestrial plants, producing a large quantity of biomass in a shorter duration and lesser area, with an average range somewhere between 10 and 50 g m^{-2} day^{-1}.

These organisms represent exciting possibilities as potential resources of a vast collection of biomolecules including carbohydrates, proteins, lipids and triglycerides, and metabolites of enormous medical and industrial importance (Spolaore et al., 2006). Their ability to acclimatize to extreme conditions and ecologies and grow rapidly make them apt models for understanding the evolutionary and metabolic processes, as well as miniature factories of useful products.

8.2 Valuable products from microalgae

A wide array of metabolites with various bioactive properties such as polysaccharides, fatty acids, essential amino acids, polypeptides, pigments, vitamins and minerals are being biosynthesized by them (Cardozo et al., 2007). In the last decade, microalgae have been explored for a diverse range of commercial uses in medicine, pharmaceutical, nutraceutical, and cosmeceutical industries and in agriculture as biofertilizers (Gupta et al., 2015). Selection of microalgal strain is a vital step in algal biotechnology as it determines the usefulness of that particular microalga for a specific application. For example, the species selection could depend upon high lipid accumulation and tolerance to flue gases when they find application in biodiesel production. In many studies related to CO_2 biomitigation and biodiesel production, *Chlorella* sp. and *Botryococcus braunii* have found to be ideal candidates. In general, strain selection is dictated by its high productivity, easy to harvest, production of secondary metabolites or desirable coproduct. The selection process can also be determined by the alga's ability to be mass cultured, ruggedness to the extreme environmental conditions, and its high growth rate with significant biomass yield (Borowitzka and Borowitzka, 1988; Becker, 1994).

8.2.1 Biopigments

Chlorophyll-*a* is the primary pigment responsible for photosynthesis in all microalgae. Depending on the species, other pigments such as carotenoids (β-carotene, lutein, fucoxanthin, zeaxanthin, etc.) and phycobiliproteins (phycoerythrin and phycocyanin) also participate in harvesting photons from light of different wavelengths. Cyanobacteria are rich sources of biopigments such as phycocyanin, phycoerythrin, allo-phycocyanin, and carotenoids having good antioxidant properties. Among them, phycocyanin and phycoerythrin have vast commercial applications in food and agriculture, cosmetics, biotechnology, diagnostics, and pharmaceuticals (Das, 2015).

8.2.2 Biopolymers

Algae-based polymers such as polyvinylalcohol, poly-3-hydroxybutyrate, and polysaccharides like agar, agarose, and carrageenans obtained mainly from macroalgae find many applications in the manufacture of many products like disposable plastic containers finding use in agriculture and horticulture for planting and algal bioplastics in packaging industries due to the advantage of its complete biodegradability properties (Sharma et al., 2011). From the cell wall of green microalgae that is, *Scenedesmus communis* and *B. braunii*, a resistant biopolymer called algaenans has been extracted (Allard et al., 1998). An extracellular matrix containing very thin fibers extracted from *Spirulina* biomass has been shown to have potential applications in culturing stem cells and for the treatment of injuries (de Morais et al., 2010).

8.2.3 Biofertilizers

Blue-green algae or cyanobacteria have immense agricultural importance due to their capacity to metabolize molecular nitrogen, secrete growth promoters as secondary metabolites, solubilize the insoluble phosphates, and add organic matter to the soil to improve its physical and chemical nature. Both the heterocystous and nonheterocystous forms of cyanobacteria can fix nitrogen, which may become available to plants (e.g., rice) either as extracellular products or as mineralized product of the intracellular contents by microbial decomposition (Srinivasan, 1978). The nitrogen fixation by cyanobacteria has a "switch on" mechanism that gets triggered when the combined nitrogen level drops below a threshold limit of ~40 ppm. This has been practically demonstrated by the addition of chemical fertilizers at recommended or lower dosages that was found to stimulate diazotrophic cyanobacterial growth and nitrogenase activity in a rice field. However, higher fertilizer levels were demonstrated to be inhibitory for the crop (Jha et al., 2001). In addition to nitrogen fixation, cyanobacteria metabolize atmospheric CO_2 during photosynthesis.

8.2.4 Biodiesel

The algal biomass is also widely used for biofuel production, which includes biodiesel, bioethanol, biobutanol, biohydrogen, and biomethane. Lipids produced by microalgae generally include neutral lipids, polar lipids, wax esters, sterols, and hydrocarbons, besides prenyl derivatives (tocopherols, carotenoids, terpenes, quinines, etc.) and pyrrole derivatives (chlorophylls). Microalgal lipids can be grouped into two categories, that is, structural or polar lipids and storage or nonpolar lipids. Structural lipids generally contain higher levels of polyunsaturated fatty acids (PUFAs), the essential nutrients for humans. Storage lipids are mainly in the form of triacylglycerides (TAGs) predominately composed of saturated fatty acids and some unsaturated fatty acids. TAGs act as energy reserve in microalgae, which when extracted can easily be converted into biodiesel through transesterification process (Bharanidharan et al., 2016). The productivity of oil per acre is the highest in microalgae when compared with any other crop-based oil and it can surpass the latter for biodiesel production. Additionally, arable land is not required for their growth and they can be sustained on wastewater (Demirbaş, 2009a). Table 8.1 summarizes the potential microalgal species with significant lipid content explored for biodiesel production (Sudhakar et al., 2011).

8.2.5 Bioethanol

Some microalgae have high carbohydrate content and hence act as highly potential source for bioethanol production (Hankamer et al., 2007; Schenk et al., 2008). Bioethanol from these organisms can be produced via two different

TABLE 8.1 Algae species and their typical oil content.

Microalgae	Lipid content (% dry weight)
Botryococcus braunii	25–75
Chlorella sp.	28–32
Crypthecodinium cohnii	20
Cylindrotheca sp.	16–37
Dunaliella primolecta	23
Isochrysis sp.	25–33
Monallanthus salina	>20
Nannochloris sp.	20–35
Nannochloropsis sp.	31–68
Neochloris oleoabundans	35–54
Nitzschina sp.	45–47
Schizochytrium sp.	50–77
Tetraselmis suecica	15–23

processes: dark fermentation or yeast fermentation. It has been estimated that approximately 47–140 m^3 of ethanol yield ha^{-1} $year^{-1}$ can be produced from microalgal biomass, which is several folds higher than the yields obtained from other feedstock materials (Cheryl, 2010). Matsumoto et al. (2003) have identified about 76 strains of marine microalgae possessing high carbohydrate content (40%–53%). Some of the other major advantages in the bioethanol production through fermentation of microalgal biomass include (1) usage of leftover biomass (after oil extraction) from other processes or intact biomass can be used, (2) no energy needs to be spent for drying the biomass since fermentation is a wet process, and (3) the cell disruption techniques using chemical can simultaneously breakdown complex sugars necessary for yeast fermentation, which is a well established technology in industry. Several authors have experimented with carbohydrate-rich microalgal species such as *Chlorella*, *Dunaliella*, *Chlamydomonas*, *Scenedesmus*, *Spirulina*, *Chlorococcum*, etc. and reported to have >50% ethanol yield (Hirano et al., 1997; Chen et al., 2009; Harun et al., 2010).

8.2.6 Biohydrogen

Cyanobacteria and green algae are the only organisms reported to be capable of carrying out both oxygenic photosynthesis and biohydrogen production. Hydrogen production by direct photolysis occurs in photosynthetic microorganisms like green algae, which convert sunlight into useful, stored chemical energy by the following general reaction: $2H_2O/2H_2 \rightarrow O_2$. Indirect photolysis occurs in cyanobacteria that can synthesize and evolve hydrogen through photosynthesis through a two phase process: light-dependent reaction $12H_2O \rightarrow 6CO_2/C_6H_{12}O_6 \rightarrow 6O_2$ catalyzed by nitrogenase followed by dark anaerobic condition $C_6H_{12}O_6 \rightarrow 12H_2O/12H_2 \rightarrow 6CO_2$ catalyzed by hydrogenase (Seshadri and Shashirekha, 2012).

8.2.7 Biomethane

Microalgae have been studied as substrates for methane production by anaerobic digestion (Sialve et al., 2009). This process can recycle the nutrient remains (such as nitrogen and phosphorus) available in the spent algal biomass after lipid extraction and use it for algae growth. This greatly minimizes the fertilizer use (Tornabene et al., 1983). Methane can be used as fuel and helps to offset the cost and energy in the microalgae-to-biofuel process. Oligonutrients such as iron, cobalt, and zinc present in algae along with the presence of lipid, starch, and protein and absence of lignin have also been found to activate methanogenesis, along with carbon, nitrogen, and phosphorus (Speece, 1996). For the fermentation of microalgal biomass, proteins, fats, and carbohydrates as the chemical components are of special

TABLE 8.2 Methane and biogas production of different algae species.

Substrate	Temperature (°C)	Methane production (L kg^{-1} DM)	Biogas production (L kg^{-1} DM)	Methane content (% by vol.)	Literature
Scenedesmus sp., *Chlorella* sp.	50	–	500	–	Golueke et al. (1957)
Scenedesmus sp., *Chlorella* sp.	35	–	405	–	Oswald and Goluecke (1960)
Scenedesmus sp., *Chlorella* sp.	45	–	611	–	Golueke and Oswald (1959)
Graclaria sp.	35	280–400	–	–	Bird et al. (1990)
Sargassum sp.	35	120–190	–	–	Bird et al. (1990)
Brown seaweeds	–	200–250	–	–	Horn (2000)
Laminaria japonica	35	250	–	–	Yokoyama et al. (2007)
Nannochloropsis sp.	38	–	388	80.5	Schmack (2008)
Spirulina sp.	38	–	556	76.3	Schmack (2008)
Green algae	38	–	420	73.9	Schmack (2008)

importance. Due to the high biogas potential of fats (compared with proteins and carbohydrates), the fat content is particularly important for high methane yields. From the fats, up to 1400 L biogas kg^{-1} oDM can be obtained according to the Buswell and Boyle formula and from both the proteins and carbohydrates approximately 900 L of biogas kg^{-1} oDM (Table 8.2).

8.3 The CO_2 emissions and their mitigation

As humankind tries to improve its standard of life and socioeconomic status through the development of industries, vehicles, machineries etc., every new advent has serious impact on the environment at a magnified level. Anthropogenic CO_2 emission has increased to an unimaginable level and hence, it has become the major contributor to global warming and climate change. Due to the world's insatiable demand for energy and realizing that fossil fuels are still inevitably the major energy source for several decades into the future, research and technologies will propel CO_2 management (capture, storage, and sequestration) to the center stage.

8.3.1 The CO_2 emissions and source points

Combustion of fossil fuels generates huge amounts of CO_2 into the atmosphere, a major source of greenhouse gases (GHGs), contributing to global warming. Burning 1 ton of carbon in fossil fuels releases more than 3.5 tons of CO_2. Total emissions were less than 25 Gt 20 years ago, and are projected to rise to over 50 Gt 20 years from now (Chisti, 2007). Human activity generates about 37 Gt of CO_2 emissions each year, with over 24 gigatons coming from direct combustion of carbon for energy production (Lewis and Nocera, 2006). In the total world emission the United States contributes 27% followed by Russia (24%). Both China and the European Union are emitting CO_2 accounting to 15% of the world emission. India emits three times less than China. France is emitting 2.2 times less CO_2 than Germany and 1.4 times less than the UK (Bataille and Birraux, 2006). Statistics show that fossil fuel based power generation contributes to nearly one-third of the total CO_2 emissions from fuel combustion (Hossain and de Lasa, 2007). As a result, the atmospheric CO_2 concentration has risen from 280 ppm to 389 ppm by volume over the last century leading to global warming, climate change, and resultant biological extinctions (Sayer, 2010).

The International Energy Agency (IEA) has estimated that by 2035, many countries across the world will have access to relatively cheap sources of fossil fuels, therefore, it is quite likely that CO_2 emissions will further increase dramatically without the application of carbon capture and storage (CCS) protocol (IEA, 2012). A number of international agreements such as the Kyoto Protocol promoted by the United Nations in 1997 have laid the platform for global efforts to reduce CO_2 emissions. More than 170 countries have ratified this protocol with the goal of reducing GHGs by 5.2% based on the emission statistics in 1990. Since then, numerous R&D studies are being carried out around the world to meet the objective/agreement on CO_2 mitigation (Razzak et al., 2013).

8.3.2 The CO_2 mitigation strategies

Attempts towards CO_2 emission reduction are in progress with a multipronged approach:

1. Efficient use of carbon-based energy sources
2. Use of alternative, carbon-free energy sources such as nuclear, solar, wind, biomass, etc.
3. Use of a posttreatment using CCS technology.

While considerable advances have been achieved on all three fronts, CCS technology has gained prominence. Ideally, the captured CO_2 could also be reused by conversion into useful fuels and chemicals thereby controlling global warming. However, the technologies developed should also be applied to other related issues such as the combustion of biomass-based fuels and in different industrial processes, such as the production of hydrogen, ammonia, iron and steel, or cement or thermal power plant (IPCC, 2001).

CCS technology could generally be through various methods such as the physical adsorption (using different adsorbents, membrane and cryogenic separation methods), scrubbing, geological sequestration (depleted oil and gas reservoirs), oceanographic sequestration (deep aquifers), chemical reaction-based approaches such as chemical reduction by metals at high temperatures, thermochemical conversion, radiochemical method, photochemical conversion, biochemical conversion, electrochemical conversion, biophotochemical conversion, electrophotochemical conversion, and CO_2 fixation by biological methods (IPCC, 2001; Wang et al., 2018; Khairul, 2009).

8.4 The CO_2 sequestration by microalgae

The CO_2 sequestration is either the prevention of release into the atmosphere or capture of CO_2 from the atmosphere resulting in the storage and conversion of CO_2 in a solid material through biological or physical processes. Biological methods mainly involve the photosynthesis process taking advantage of the photosynthetic machinery of the aquatic species like phytoplanktons in marine environment, terrestrial plants, algae, besides, the use of carbonic anhydrase (CA) enzyme and some nonphotosynthetic bacteria that use CO_2 for their metabolism (Kumar and Das, 2014). Algal biotechnology can provide a practical solution to many problems like CO_2 sequestration, bioremediation of wastewater, and generation of value added products (Demirbaş, 2009a).

Microalgae are the most promising production facilities, providing an effective solution for carbon capture and recycle in a sustained manner. They have the advantage of being the potential leaders in environment cleaning, owing to their CO_2 fixation capability, which is multifold higher CO_2 per unit area than agri-crops/trees (Singh et al., 2012). Thus, CCS using microalgae is considered a promising alternative to many existing CO_2 abatement strategies mainly due to its advantage of generating carbon credits (Hall et al., 1991; Perez-Garcia and Bashan, 2015). Time has come yet again that these tiny creatures salvage the Earth from the hazardous effect of global warming.

8.4.1 Photosynthesis

Photosynthesis is the process where plants and algae convert water, CO_2 and sunlight with the help of the pigment chlorophyll into carbohydrates. It comprises of two steps, that is, (1) light reactions occurring only when the cells are provided with illumination wherein cells transform light energy into chemical energy and (2) dark reactions, also known as carbon-fixation reactions, occurring both in the presence and absence of light, where the stored high-energy compounds are used. CO_2 is converted into carbohydrates in the algal cells during the light independent phase through the Calvin cycle, which takes place in the stroma of the chloroplasts. Carbohydrates such as sucrose are transported into the cytosol where they are stored as glycogen in cyanobacteria and starch in green algae. Calvin cycle is an energy-intensive process among well known pathways of CO_2 capture, requiring 7 ATP molecules and 4 NAD(P)H molecules to transform $2CO_2$ molecules into acetyl−CoA (Fast and Papoutsakis, 2012; Kumar and Das, 2014).

Biofixation of CO_2 takes place during algal growth through numerous processes such as biomass formation (photosynthesis), mineralization (conversion of gaseous CO_2 into chemical species such as HCO_3^-, CO_3^{2-}), and production of extracellular products such as polysaccharides, volatile organic compounds, hormones, etc. Among them, photosynthesis is the major factor that accounts for 70%−88% of total CO_2 fixation during algal cultivation. CO_2 fixed during other processes either go unutilized or lost to the environment. Therefore, CO_2 fixation taking only biomass formation through photosynthesis is generally taken into account. All of this carbon is typically derived from CO_2. The CO_2 fixation ability of microalgae and cyanobacteria varies (Sydney et al., 2014). It is estimated that about 1.83 tons of CO_2 gets converted to 1 ton of dry algal biomass under ideal conditions including continuous feeding of CO_2 during daylight hours. Microalgal biomass contains approximately 50% carbon on dry weight basis (Chisti, 2007).

8.4.2 The CO_2 uptake by microalgae

Generally, photosynthesis is considered a promising method for capturing CO_2 in the form of algal biomass and there are several viewpoints that substantiate the approach to use microalgae for CO_2 mitigation (Benemann, 1997):

1. Microalgal cultivation is inherently faster than other biomass systems, due to their very rapid growth rates.
2. Microalgae-based systems are essentially generic; thus, can be easily translated to new sites.
3. Microalgae R&D can produce "spin-offs" from CO_2 sequestration and wastewater treatment to chemicals or compounds of high commercial value.
4. Microalgae R&D can be easily extrapolated from small scales to larger systems.
5. The use and recycling of CO_2 on site is inherently preferable to the disposal options (such as geological sequestration).
6. Due to the carbon neutral nature of microalgae-based CO_2 mitigation process, the costs for CO_2 capture and sequestration are less when compared with the other methods.
7. In the case of fixation of CO_2 generate in the form of flue gas from industrial operations, high purity of the gas is not needed for the algae culture. Flue gas having ~2%−5% CO_2 can directly be fed into the algal cultivation system.

8. The capability of microalgae to use NO_x or SO_x, some combustion products, could help to reduce an additional step of flue gas scrubbing.
9. The commercial profit from the algal biomass produced could possibly offset the overall operational costs for CO_2 capture.

Microalgae can fix CO_2 from different sources, which can be categorized as (1) CO_2 from the atmosphere, (2) CO_2 from industrial exhaust gases like flue or flaring gas, and (3) fixed CO_2 in the form of soluble carbonates for example, $NaHCO_3$ and Na_2CO_3 (Wang et al., 2018). In this context, possibly the transfer of CO_2 from the atmosphere to microalgae through photosynthesis is the fundamental route for CO_2 capture. However, high concentration of CO_2 of the order of 380 ppm in the atmospheric air makes that economically nonfeasible (Fulke et al., 2010; Chakrabarthi et al., 2014). In contrast, direct utilization of power plant flue gas or from other emission sources is being largely considered for CO_2 sequestration systems photoautotrophic microalgal species not only due to the advantages stated above but also because of the cost reduction in separating the CO_2 gas (Gupta et al., 2015). Normally the concentration of CO_2 in power plant flue gas is high and hence, identifying CO_2 tolerant species is an essential step. Depending on the flue gas sources, the CO_2 concentrations vary; however, an average of 15%–20% (v/v) is typically found. To achieve effective and practical biological CO_2 fixation systems, selection of ideal microalgal species is very vital, which depends on specific strategies employed for CO_2 sequestration. On these lines, over the past four decades many researchers have tested the growth of several microalgal species including *B. braunii*, *Chlorella kessleri*, *Chlorococcum littorale*, *Chlamydomona reinhardtii*, *Chlorella vulgaris*, *Dunaliella* sp., *Euglena gracilis*, *Nannochloris* sp., *Synechococcus elongates*, *Spirulina* sp., *Tetraselmis* sp., and *Scenedesmus* sp. at CO_2 concentrations of over 15%. *C. littorale* has been demonstrated to grow up to 60% CO_2 by stage-wise adaptation of the alga (Kodama et al., 1993). Although *E. gracilis* showed the best growth with 5% CO_2 concentration, it is another high CO_2 tolerant species that survives up to 45% CO_2 concentration (Nakano et al., 1996). While Hirata et al. (1996a,b) described successful cultivation of *Chlorella* sp. UK001 at 10% CO_2 conditions, Hanagata et al. (1992) demonstrated the growth of *Chlorella* sp. under CO_2 concentrations as high as 40%. Interestingly, *Cyanidium caldarium* was found to grow in pure CO_2 (Seckbach et al., 1971). Furthermore, Maeda et al. (1995) found a strain of *Chlorella* sp. T-1 that could survive at 100% CO_2, although the maximum growth rate was observed only at 10% CO_2 levels. Similarly, *Scenedesmus* sp. was reported to grow at 80% CO_2 concentration; however, the maximum cell mass was recorded only in 10%–20% CO_2 concentrations (Hanagata et al., 1992). This trend of surviving at high concentrations of CO_2 (up to 50%), yet its better utilization and therefore, higher biomass yield at concentrations between 10% and 20% was reported in *Spirulina* sp. (MCRC-A0003) also (Sivakumar et al., 2014). The effect of elevated levels of CO_2 on the growth rate of three microalgal species, that is, *C. reinhardtii*, *Chlorella pyrenoidosa*, and *Scenedesmus obliquus* enhanced considerably with maximum specific growth rates at 30, 100, and 60 μM CO_2, respectively. Such an enhanced growth rate with enriched CO_2 was established at different levels of inorganic N and P. The highest rates of net photosynthesis, photosynthetic efficiency, and light-saturating point were observed in cells grown with high CO_2. The authors concluded that elevated CO_2 levels with low pH could affect the growth rate and photosynthetic physiology of all three microalgae (Yang and Gao, 2003). In a similar study using *S. obliquus*, Shashirekha et al. (2016) reported that the carbon uptake was better and faster at low carbon and high nitrogen and phosphorus concentrations than higher dosages of carbon. Furthermore, they indicated that the alga could adapt quite well to varying organic carbon concentrations demonstrating its versatility to switch from phototrophic to heterotrophic mode of growth.

In a more recent study carried out by Patil and Kaliwal (2017), 15% CO_2 was proved to be optimal for the growth of *Scenedesmus bajacalifornicus* BBKLP-07 showing the maximum biomass productivity of 0.061 ± 0.0007 g L^{-1} day^{-1}, protein content of 32.89% of dry cell weight, chlorophyll accounting to 25.07 μg mL^{-1} and carotenoid levels of 6.15 μg mL^{-1} and the CO_2 fixation rate was 0.12 ± 0.002 g L^{-1} day^{-1}. However, the carbohydrate and lipid contents were maximum at 25% CO_2 with 26.19% and 25.81% dry cell weight respectively. Way back in 1960, Witt and Borchardt demonstrated the acclimatization of algal culture to bicarbonate as a carbon source, when the CO_2 source was shunted, the cells start to deplete the reserve before their growth ceases at higher pH values. Many studies have been carried out by various researchers on the effect of CO_2 on the lipid production in microalgae. Cultivating microalgae *C. vulgaris* under 8% (v/v) CO_2, Montoya et al. (2014) achieved the maximum amount of saturated fatty acids and lipid productivity of 29.5 mg L^{-1} day^{-1}. The microalgae *Chlamydomonas* sp. JSC4 strain produced maximum lipid content (65.3%) and productivity (169.1 mg L^{-1} day^{-1}) at 4% (v/v) CO_2 (Nakanishi et al., 2014). Maximum lipid content (34% wt.) was achieved, when microalgae *C. littorale* was cultivated with 5% CO_2 concentration (Masaki et al., 2009). In another study, Ho et al. (2012) grew green microalgae *S. armatus* at CO_2 concentration of 2% and obtained the highest lipid content of 22.4% (Table 8.3).

Most of the studies on CO_2 mitigation using microalgae provide little information on the performance of ecotypes and their physiological response when grown under elevated CO_2 and increased temperature conditions.

TABLE 8.3 The CO_2 tolerance and utilization by various microalgal species.

Microalgae species	Tolerance to maximum CO_2 concentration (%)	CO_2 fixation rate ($g\ L^{-1}\ day^{-1}$)	Reference
Chlamydomonas sp.	15	–	Miura et al (1993)
Chlorella sp.	40	–	Hanagata et al. (1992)
	0.03	1.62	Fulke et al. (2013)
	15	0.46	Jin et al. (2006)
	5	0.7	Ryu et al. (2009)
	–	1.38	Zhao et al. (2011)
	15	–	Fulke et al. (2010)
Chlorella kessleri	18	0.16	de Morais and Costa (2007)
Chlorella vulgaris	10	0.25	Sydney et al. (2010)
	2	0.43	Yeh and Chang (2011)
	18	–	de Morais and Costa (2007)
Chlorococcum littorale	60	–	Kodama et al. (1993)
Chroococcus cohaerens	0.03	0.78	Fulke et al. (2013)
Cyanidium caldarium	100	–	Seckbach et al. (1971)
Dunaliella sp.	3	0.31	Kishimoto et al. (1994)
Dunaliella tertiolecta	10	0.27	Sydney et al. (2010)
	15	–	Nagase et al. (1998)
Eudorina spp.	20	–	Hanagata et al. (1992)
Euglena gracilis	45	–	Nakano et al. (1996)
Haematococcus pluvialis	34	0.14	Huntley and Redalje (2007)
Nannochloris sp.	15	–	Yoshihara et al. (1996)
Scenedesmus sp.	80	–	Hanagata et al. (1992)
	15	0.61	Jin et al. (2006)
Scenedesmus dimorphus	0.03	1.27	Fulke et al. (2013)
Scenedesmus incrassatulus	0.03	1.50	Fulke et al. (2013)
Scenedesmus obliquus	10	0.55	Ho et al. (2010)
	10	0.29	Tang et al. (2011)
	2.5	1.19	Ho et al. (2013)
	18	–	de Morais and Costa (2007)
Synechococcus elongatus	60	–	Miyairi (1995)
Spirulina sp.	18	–	de Morais and Costa (2007)
	20	0.14	Sivakumar et al. (2014)
Spirulina platensis	15	0.92	Kumar et al. (2010)
Tetraselmis sp.	14	–	Matsumoto et al. (1995)

To develop an algal-based system to sequester CO_2, either atmospheric or from industrial processes, it is important to have knowledge on the biochemical response of algae such as its tolerance limit, assimilation ability, and associated biochemical changes etc. (Sivakumar et al., 2014).

8.5 Factors influencing CO_2 uptake by microalgae

8.5.1 Mode of growth and nutrient sufficiency and deficiency

During photosynthesis, the major nutrients such as nitrogen, phosphorus, and potassium (NPK) in addition to several other minor nutrients such as magnesium and iron required for algal culture, depending on the specific need of each alga, are taken up along with CO_2 (Becker, 1994; Dalrymple et al., 2013). Therefore, NPK including common agricultural fertilizers can be used to sustain steady growth rates (Weissman and Tillett, 1992; Dalrymple et al., 2013). Algal cells grown under nutrient depletion conditions have shown to exhibit significant biochemical responses depending on the type and extent of nutrient depletions (Shashirekha et al., 2016). This has been encountered more frequently as one of the principal difficulties when the supply of carbon is limited during microalgal mass cultivation (Oswald, 1988). The primary consideration with respect to the CO_2 budget is that the gas concentration must neither be higher leading to inhibition nor fall below the minimal required quantum that stunts the growth (Rados et al., 1975).

The microalgae follow three different growth patterns. While some species are obligate photoautotrophs, most of them are obligate heterotrophs, and a few others are mixotrophs, which follow both phototrophy and heterotrophy to use inorganic (CO_2) as well as organic carbon (Gladue and Maxey, 1994; Chojnacka and Marquez-Rocha, 2004). The facultative heterotrophs choose to use organic carbon, whenever available, than fixing CO_2 (Shi et al., 2002; Kandimalla et al., 2015). Microalgae utilize organic or inorganic carbon, carbonates through heterotrophic nutrition mode that accounts for 25%–50% of their carbon (Borowitzka and Borowitzka, 1988). For example, in the case of *Spirulina*, carbon is the principal nutrient required by them and in alkaline lakes this organism is the dominant species because of the presence of high concentrations of sodium carbonate. Carbon source is not likely to become limiting in laboratory cultures as light and temperature are both kept at low levels. In open-air conditions, however, carbon becomes often limiting if sufficient CO_2 supply is not ensured. Although the nutrient medium is high in bicarbonates, as the alga grows, pH increases and bicarbonate gets converted to carbonate thus making it unavailable. CO_2 supply restores the balance and pH level of medium (Jeeji Bai et al., 2018). Carbon assimilation in algae such as *Chlorella* sp. and *Scenedesmus* sp. has been found to switch from autotrophy to heterotrophy whenever the carbon source changes from CO_2 to organic carbon, or when there is limitation or alternations in their concentration during their cultivation (Becker, 1994; Nirbhay and Dolly, 2011).

8.5.2 Sources of inorganic carbon

Air sparging is the most practiced method to provide homogenous mixture of nutrients (e.g., carbon, nitrogen, and phosphate) as well as equal light penetration for the microalgal cells around their surface. It is broadly accepted that this method supplies CO_2, the inorganic carbon, to the algal cultures (Rodríguez-Maroto et al., 2005). In an experiment conducted to compare the difference in CO_2 utilization efficiencies by direct bubbling of CO_2 to the algal system versus adding as carbonated water, the former method showed 96% ± 11% utilization efficiencies while the latter showed only 81% ± 11% efficiencies. On the other hand, Jiménez et al. (1995) reported that the contribution of air sparging of CO_2 to enhance the concentration of total inorganic carbon (TIC) in *Dunaliella viridis* was nil and that only the turbulence was responsible for considerable differences in the growth when bubbled and nonbubbled cultures were provided with equal quantities of TIC. They also demonstrated the loss of CO_2 from the air-bubbled HCO_3^- enriched medium in the form of air bubbles that escaped from the culture container, and there was no transference of CO_2 from the air bubbles to the medium.

According to Rodríguez-Maroto et al. (2005) bicarbonate-enriched aqueous solutions normally used for microalgal cultivation are far from equilibrium, and tend to lose inorganic carbon (as CO_2), with a simultaneous increase in the pH of the substrate. A fraction of the inorganic carbon supplied as $NaHCO_3$ is not utilized by the alga but escaped from the substrate as CO_2. Air sparging accelerates this process, due to the transfer of CO_2 from the solution to the air bubbles. An appropriate strategy for standardizing the use of inorganic carbon in bicarbonate enriched substrate would be to avoid CO_2 transfer from the aqueous substrate to the air. This can be achieved only when the CO_2 concentration in the substrate is in equilibrium with that in the incoming air.

8.5.3 Composition of other gases in flue gas

Flue gas is typically composed of carbon dioxide, water vapor, nitrogen, and small amounts of carbon monoxide, nitrogen oxides, and sulfur oxides. Recent studies have shown that direct injection of flue gases into microalgal cultivation ponds increases biomass productivity by 30% compared with direct injection of an equivalent concentration of pure CO_2 (Douskova et al., 2009). The flue gas stimulant effect may be due to the presence of supplemental nutrients (sulfur and nitrate) present in flue gases. Direct use of flue gas could reduce the cost of pretreatment to a large extent but it could impose extremely severe conditions on microalgae such as high CO_2 concentrations and other inhibitory compounds like NO_x and SO_x. Some researchers considered that it is important to study the effect of trace acid gases such as NO_x and SO_x on CO_2 uptake by microalgae. Most strains are known to be critically inhibited by air containing only 50 ppm SO_x (Kurano et al., 1995). Therefore, direct CO_2 fixation by culturing microalgae may only be possible with flue gas containing SO_x <50 ppm. To avoid the growth inhibition, physicochemical methods like absorption, adsorption, membrane separation, etc. can be used to scrub off the toxic compounds from flue gas before CO_2 fixation by culturing microalgae.

Yoshihara et al. (1996) reported that *Nannochloris* sp. could survive less than 100 ppm of nitric oxide (NO). It is interesting to note that the other flue gas such as nitric oxide even at very high concentrations (1000 ppm) was found to be removed by *Dunaliella tertiolecta* to the order of 51%–96% when 15% CO_2 concentration was available in the flue gas (Nagase et al., 1998). *Tetraselmis* sp. was shown to grow utilizing the flue gas containing 14.1% CO_2 besides 185 ppm of SO_x and 125 ppm of NO_x (Matsumoto et al., 1995). Maeda et al. (1995) investigated the tolerance of a strain of *Chlorella* and found that it could grow under various combinations and concentrations of trace elements. Kumar et al. (2014) studied the effect of oil producing industrial flue gas CO_2 on the growth of *Chlorella sorokiniana* by passing the gas through serially connected airlift reactors. The microalga removed the CO_2 content significantly from 15.65% (v/v) to 4.1% (v/v).

8.5.4 Temperature and light

Temperature strongly influences the cellular chemical composition, uptake of nutrient and CO_2, and the growth rates for every species of algae (Singh and Singh, 2015). Temperature also greatly influences the fatty acid production in algae. The optimum temperature for their growth ranges between 18°C and 24°C. The algal productivity increases with increasing temperature of the culture medium and beyond the optimum temperature, increasing algal respiration and photorespiration reduces the overall productivity. The optimal temperature for maximum algal growth rate under sufficient nutrient and light conditions varies from species to species (Table 8.4).

The temperature of exhaust gas from thermal power plants is around 120°C. Therefore, many studies are being conducted for the potential use of thermophilic or high temperature tolerant species that can grow in temperature ranging between 42°C and 100°C (Bayless et al., 2001). An apparent benefit of the use of thermophiles for CO_2 sequestration is the lower costs involved towards the cooling system. Furthermore, secondary metabolites produced by some thermophiles could possibly reduce the overall costs for CO_2 sequestration (Edwards, 1990). However, the great disadvantage is the loss of water on account of evaporation. *C. caldarium* is a thermophilic species that can grow under pure CO_2 (Seckbach et al., 1971). Miyairi (1995) studied the growth characteristics of *Synechococcus elongatus* under high CO_2 concentrations and at high temperature. 60% CO_2 at 60°C was found to be ideal for their growth. Some mesophiles have been shown to be productive under high temperature due to their tolerance during direct injection of flue gas, although they are not as effective as thermophiles (Edwards, 1990). Light is yet another important limiting factor for algal growth. The maximum algal growth rate has been shown to reach at the light saturation point, beyond which the algal growth is inhibited due to photoinhibition.

TABLE 8.4 Optimum and tolerance conditions for algal growth.

Conditions	Tolerable range	Optimum range
Temperature (°C)	16–27	18–24
Salinity (g L^{-1})	12–40	20–24
Light intensity (lux)	1000–10,000	2500–5000
Photoperiod–light: dark (h)	24:0 (max.)	16:8 (min.)
pH	7–9	8.2–8.7

Source: Sudhakar, K., Suresh, S., Premalatha, M., 2011. An overview of CO_2 mitigation using algae cultivation technology. Int. J. Chem. Res. 3 (3), 110–117.

8.5.5 The CO_2 sparging and mixing system

Adequate mixing of the algal culture is one of the key measures to keep the culture system as well as nutrients in a homogeneous distribution, to augment light capture efficiency and gas exchange and transfer (CO_2 supply and O_2 removal). To keep CO_2 available to cells, it needs to be dissolved in water. CO_2 largely exists as CO_2 gas at pH <6, HCO_3^- between pH 6 and 10 and CO_3^{2-} at pH >10. In the presence of light, the absorbed CO_2 is converted to glucose during the rate limiting step of photosynthesis using Rubisco enzyme. As Rubisco has higher affinity for O_2 than CO_2, algae have, thus, evolved carbon concentrating mechanisms using the enzyme CA inside the cell to support photosynthesis. Despite this evolution-selected mechanism, photorespiration or increased dissolved oxygen levels can inhibit formation of biomass to a large extent (Giordano et al., 2005). In this context, the effect of mixing has been demonstration to have significant influence on the algal productivities. As the saturation concentration of dissolved O_2 is only 10 ppm, removal of excess O_2 is very important and is possible through proper mixing and aeration. In this context, an efficient sparging and mixing pattern is needed to increase the dissolved CO_2 level in the medium and improve its uptake by algal cells. On the other hand, increased CO_2 concentration reduces pH of the culture medium, which is also detrimental to the algal growth (Kumar et al., 2011). Therefore, it is very important to maintain the balance between O_2 and CO_2 in the medium, which could be achieved by intermittent sparging of CO_2 and aeration.

8.5.6 Method of cultivation

In the debate about development of efficient microalgal cultivation systems, photobioreactor (PBR) has always scored over the raceway ponds, even in terms of efficient CO_2 utilization. While the outdoor culture systems have certain limitations they are not easy to control, showing low productivity due to variability in the (1) environmental temperatures, (2) system circulation, and (3) light utilization. However, PBR have some advantages (Jeeji Bai et al., 2018):

1. Efficient collectors of solar radiation due to their geometry (tubular channels with relatively narrow gauge).
2. High culture densities achieved, thus, resulting in higher areal yields.
3. Being closed systems, sterility can be maintained and contamination can be minimized to a large extent.
4. Good mixing and CO_2 conversion efficiency ensure good photosynthetic production.
5. Monitoring and regulation of parameters are easy.
6. The microalgae cultured in PBR can act as a bioscrubber for waste gas treatment in an energy-efficient and economical manner

The gas–liquid mass transfer in PBRs is important for two major aspects, that is, CO_2 fixation and O_2 release by microalgae, as a result of photosynthesis. The gas–liquid mass transfer in PBRs could happen either through the dissolution of gaseous CO_2 into the aqueous medium through separate gas exchange systems like horizontal tubular PBRs or by supplying gaseous CO_2 via internal sparging systems, e.g., bubble column and airlift reactors (Olivieri et al., 2014). To improve the mass transfer rate, some of the strategies that can be considered are (1) maintenance of sufficient liquid CO_2 concentration in the medium to avoid carbon limitation or excess, (2) increase the retention time of CO_2 along the reactor, (3) efficient CO_2 sparging systems, (4) implementation of multistage sequential rector, etc. De Morais and Costa (2007) demonstrated this strategy using a three-stage serial PBR, wherein 53.29% and 28.08% of maximum daily CO_2 fixation was recorded for *Spirulina* sp. and *Scenedesmus* sp., respectively. In yet another investigation by Cheng et al. (2013) using 9-stage and 14-stage sequential reactors the average CO_2 fixation efficiency of *Chlorella* PY-ZU1 was improved from 7.6% to 70.48% resulting in the biomass concentration of 5.42 gL and productivity of 0.95 g L^{-1} day^{-1}.

In the case of open pond or raceway pond algal systems, where the gaseous CO_2 when sparged directly into the system tends to escape leaving very short contact period with the algal cells, different sparging strategy needs to be followed. In this context, the use of carbonation column for CO_2 mass transfer into microalgal culture ponds not only aids improved efficiency of gas transfer but also meets the CO_2 demand of high-rate algae outdoor ponds. According to Putt et al. (2011), the mass-transfer efficiency was best observed in low alkaline water (pH 8–9). The transfer rate efficiency of carbonation column was twice when compared with direct sparging, and thus, the former offers the plausible solution to reduce the cost of algae pond carbonation significantly. The ease to fabricate the carbonation column and ease to use with various CO_2 rich air streams make it an economical device for carbon recycling.

8.6 Microalgae in wastewater treatment

There are multiple benefits of microalgae with wide range of purposes such as potential agents for CO_2 sequestration, miniature factories for the production of valuable metabolites that find applications in agriculture, medicine, pharmaceutical, nutraceutical, and cosmeceutical industries. Despite these benefits, there are certain key challenges to generate algal biomass at large scale; one among them is availability of huge volumes of water. The use of seawater is another option to reduce the freshwater usage; however, it cannot totally negate the freshwater requirement.

Algae have a huge water footprint in terms of energy returned on water invested (0.5–1.1 and 0.05–1.0 MJ L^{-1} for PBRs and raceway pond system, respectively). For cultivation in open ponds, algae require as high as 11–13 million liters of water per hectare in addition to macronutrients (N and P), which accounts for 10%–20% of the total cost for algal production. Given the huge requirement for energy, freshwater, and nutrients, algal cultivation would catalyze the water woes and energy crisis of the world. The rising cost of NPK fertilizers also reflects in the increasing cost of algal biomass production. For economic viability, the algal production cost is required to be reduced by 20–25 times. Therefore, it is imperative to identify alternative sources of water and cost-effective nutrients for mass cultivation of algae to produce biofuels and other value-added products. It has been suggested that wastewaters rich in organic and inorganic components can be used as substitutes for freshwater and fertilizers (Bhatnagar et al., 2011). The use of wastewater streams can reduce the requirement for freshwater by over 90%. Organic substances such as carbon, nitrogen, and phosphorus and inorganic salts available in the effluents act as nutritive substrate for microalgal growth (Morales-Amaral et al., 2015). Due to the nutrient removal efficiency, microalgae provide an additional advantage of wastewater treatment by reducing the requirement for huge amounts of toxic chemicals normally used in the conventional processes.

Conventional wastewater treatment techniques are expensive as they involve high energy mechanical aeration process to supply O_2 to aerobic bacteria that consume the organic components in the wastewater. In contrast, microalgae offer an efficient method to consume nutrients and provide the aerobic bacteria with the required O_2 through the photosynthesis process (Oswald et al., 1953). It is estimated that 1 kWh of electricity is required for aeration to remove approximately 1 kg of BOD in an activated sludge process. This generates about 1 kg of fossil CO_2 from the power consumption during operation. On the other hand, 1 kg of BOD removed by photosynthetic oxygenation not only requires any energy inputs but also produces sufficient microalgal biomass to produce methane, which can produce 1 kWh of electric power (Oswald, 2003).

8.6.1 Role of microalgae in bioremediation

Microalgae-based wastewater treatment involves photosynthetic conversion of solar energy into valuable biomass utilizing major nutrients such as nitrogen and phosphorus along with organic carbon that are available in wastewater. This wastewater without prior treatment, if discharged into the water bodies, would lead to eutrophication, which is a serious environmental problem and has become more widespread in the recent past leading to the reduction in biodiversity and replacement of dominant species, increased water toxicity, increased turbidity of the water, and decreased life-span of the lakes. Microalgae are used as the major biocatalysts in oxidation ponds, raceway ponds, and high rate algal ponds because they remove nutrients and are especially used for polishing purposes (Venkata Mohan et al., 2015). Furthermore, microalgae provide free aeration by release of oxygen through photosynthesis resulting in reduction of biochemical oxygen demand (BOD) and chemical oxygen demand (Cantrell et al., 2008; Sekabira et al., 2010).

Many microalgal species exhibit mixotrophic or heterotrophic mode of growth utilizing the organic and inorganic carbon sources in wastewaters and convert them to valuable biomass. In addition to the ability to grow heterotrophically in nutrient-rich organic environment, algae are also capable of removing toxic substances and heavy metals from wastewaters (Perez-Garcia and Bashan, 2015, Venkata Mohan et al., 2015). However, considering the dynamic nature of wastewaters, normally inconsistent due to variations in the quantum of unutilized portion of chemicals used at different stages, it is imperative to identify the right wastewater with appropriate nutrients and suitable microalgal species particularly for the biotreatment process development. Besides the nutrient removal, other advantages of using microalgae as an eco-friendly technology include: (1) lower cost due to utilization of naturally available sunlight, (2) biofixation of CO_2 through photosynthesis, (3) lesser external input of organic carbon (in contrast to biological nitrification–denitrification processes), (4) discharge of stabilized oxygenated effluents into water bodies, (5) decreased sludge problems, and (6) economic value addition to the harvested algal biomass at spinoff industries. Microalgal biomass is also regarded as a reasonable feedstock for producing bioenergy (e.g., biohydrogen, biodiesel, and bioethanol) and thus, a potential future alternative for fossil fuels (Wijffels and Barbosa, 2010).

8.6.2 Microalgal cultivation in wastewaters

To effectively couple advanced wastewater treatment and bioenergy production, selection of appropriate as well as robust microalgae strains that can grow consistently in wastewaters with wide range of variations in their physicochemical characteristics would be foremost key point of research in this field. The most extensively studied microalgae species for bioremediation of wastewater with special reference to nutrient removal are *Scenedesmus* sp., *Chlorella* sp., *Spirulina* sp. and *B. braunii.* Use of microalgae in wastewater treatment was pioneered by Oswald (1988) with promotion of naturally occurring algae (*Chlorella*) by providing suitable conditions. The earliest studies on growing *Spirulina* in wastewater are those of Chung et al. (1978) who grew *S. platensis* in swine wastewater and Oron et al. (1979) who grew *S. maxima* on cow manure wastes. The microalga *S. obliquus* has shown an extremely good adaptation to various kinds of wastewaters (e.g., domestic and livestock wastewaters). This freshwater green alga tolerates a wide range of nutrient concentrations, temperature, and pH, making it versatile for wastewater purification.

Many studies in the recent past have demonstrated the cultivation of various microalgae in different wastewaters such as municipal sewage, dairy, poultry, swine, aquaculture, paper and pulp, palm oil, agricultural runoff, etc. by assimilating carbon, nitrogen, and phosphorus and other chemical compounds as summarized in Table 8.5.

TABLE 8.5 Microalgae used in wastewater treatment.

Microalga	Nature of effluent	Reference
Chlorella minutissima	Municipal wastewaters	Bhatnagar and Chinnasamy (2010)
Chlorella sorokiniana	Food waste and municipal wastewater	Chi et al. (2011)
Chlorella sp.	Municipal wastewaters	Li et al. (2011)
Auxenochlorella protothecoides UMN280	Municipal wastewater	Zhou et al. (2012)
Scenedesmus sp. AMDD	Municipal wastewater	McGinn et al. (2012)
Rhizoclonium hieroglyphicum; Microspora willeana; Ulothrix ozonata; Odeogonium sp.	Dairy manure effluent	Mulbry et al. (2008)
Chlorella saccharophila UTEX 2911; *Chlamydomonas pseudococcum*, UTEX 214, *Scenedesmus* sp. UTEX 1185,	Dairy farm wastewater	Hena et al. (2015)
Spirulina platensis	Biogas plant effluent	Mahadevaswamy and Venkataraman (1986)
C. minutissima, C. sorokiniana; Scenedesmus bijuga	Poultry litter anaerobic digestion effluent	Singh et al. (2011)
Scenedesmus sp.	Fermented swine wastewater	Kim et al. (2007)
Chlorella sp.	Pretreated swine manure	Hu et al. (2012)
Chlorella vulgaris; Chlamydomona reinhardtii; Chlamydomona debaryana	Swine wastewater	Hasan et al. (2014)
Chlorella sp.	Piggery wastewater digestate	Michelon et al. (2016)
C. vulgaris; Scenedesmus obliquus	Aquaculture wastewater	Gao et al. (2016)
Chaetoceros calcitrans, Nannochloris maculate, Tetraselmis chuii	Aquaculture wastewater	Khatoon et al. (2016)
Oscillatoria chlorine; Scenedesmus *quadricauda*	Pulp and paper mill effluents	Saikia et al. (2011)
Scenedesmus sp.	Pulp and paper mill effluent	Usha et al. (2016)
S. obliquus	Sweetmeat factory waste media	Toyub et al. (2008)
C. sorokiniana	Palm oil mill effluent	Putri et al. (2011)
Chlorella sp.	Palm oil mill effluent	Hadiyanto and Nur (2012)
	Agricultural run-off	Anand (2010)

(*Continued*)

TABLE 8.5 (Continued)

Microalga	Nature of effluent	Reference
C. vulgaris	Agroindustrial wastewater	Mitra et al. (2012)
	Agroindustrial wastewater	Posadas et al. (2014)
S. platensis	Digested distillery effluent	Kaushik et al. (2006)
Spirulina sp., *Oscillatoria* sp., *Synechocystis* sp.	Tannery chrome effluent	Shashirekha et al. (2011)
Spirulina sp., *Nannochloropsis* sp.	Tannery soak liquor	Abinandan et al. (2014)
S. obliquus	Sugar mill effluent	Shashirekha et al. (2016)

Some reduction in medium costs can also be achieved by using industrial grade chemicals wherever possible and by substituting cheaper sources of alternatives (e.g., urea for nitrate). Some algal species and applications also allow for the use of wastewaters as the basis of the medium. A chance observation of *Spirulina* growing in tapioca starch factory waste stabilization pond by Tanticharoen et al. (1993) led to its mass production in effluent from secondary treated starch wastewater with suitable enrichment with $NaHCO_3$ and nitrogen and phosphorus fertilizers and the biomass is used as animal feed—grade *Spirulina* (Tanticharoen et al., 1993). Michelon et al. (2016) studied the phycoremediation of swine wastewaters using *Chlorella* sp. and found that the nitrogen and phosphorus contents in the effluent was completely removed in 3 and 11 days by the alga. In an experiment conducted with three mixotrophic microalgal species *Chlorella minutissima*, *C. sorokiniana*, and *Scenedesmus bijuga* to utilize poultry litter anaerobic digester effluent as nutritional growth medium, the algae were not only found to be efficient in removing 60% of total nitrogen and 80% of total phosphorus from the effluent, but also exhibited good biomass productivity (76 mg L^{-1} day^{-1}) under booster dosage conditions. An efficient continuous microalgae cultivation process for biomass production and nutrients removal from aquaculture wastewater was developed using a membrane photobioreactor (MPBR). *C. vulgaris* and *S. obliquus* were batch cultured in aquaculture wastewater. *C. vulgaris* showed better performance when continuously cultivated in MPBR with average biomass productivity of 42.6 mg L^{-1} day^{-1}, which is around fivefold higher than the batch cultivation method. The average reduction in total nitrogen and total phosphorus is 86.1% and 82.7% respectively, after stabilization (Gao et al., 2016).

8.7 Economic advantages of algal technology

8.7.1 Biofuel perspectives

The costs of producing biofuels from algae using the PBR technology are high at present, but this is likely to become a viable source in the near future due rising cost of fuels and stringent regulations being imposed on industries and corporate houses to reduce their energy intensity in production processes. Furthermore, the incentives for this technology are (Sudhakar et al., 2011):

- The generation of marketable biomass based renewable energy; reducing consumption and saving the cost of fossil fuels.
- Recycling CO_2 generated from industrial processes for culturing algae to produce combustible biomass for in-house consumption and sale to other industries.
- Building a corporate brand image fulfilling commitments to sustainable development goals.
- Creating value in CO_2 abatement credits.
- Extending the service life of existing power, cement, and other fossil fuels using industrial assets, by making them more profitable and environmentally acceptable and compliant.

8.7.2 Carbon capture and storage perspective

While chemical and physical methods to capture CO_2 from smoke stack emissions are available, the cost of implementing these technologies would result in a remarkable escalation in the cost of power. For example, the cost of stripping

CO_2 from a conventional coal-fired power station is estimated to range from \$35 to \$264 per ton of CO_2 (Olaizola, 2003); however, the sustainable nature of conventional methods is still not entirely clear, especially from technological and societal point of view. Given the current scenario, the cost of employing algae-based CO_2 capture technology will be the main driving force behind the much required change as this process is more sustainable, their products that can be monetized, and the algae must be price competitively with relatively conventional, nonbiological forms of carbon capture. Algae-based CO_2 capture might even prove to be a profitable business scheme in future as the price of oil keeps escalating. To make it a commercially successful by solving the biological, technical, and economic challenges, the factors to be taken into consideration include:

- Energy efficient technologies
- Easy scalability
- Maximum utilization of the algal biomass as coproducts
- Using the most economical sources/concepts at every stage

For nonbiological, conventional methods of CCS, projected calculations show the costs, in terms of energy penalties, can be approximately 13.5% for separation of CO_2 and an additional 9% for compression. However, the transportation and storage costs are relatively small. In economic terms, values of \$52 per ton of CO_2 for capture and an additional \$10 per ton of CO_2 for transport and storage have been calculated. In comparison, some studies on costs of algae production using CO_2 from power plants indicate that it could potentially be \$250 per ton of CO_2 in a PBR system and \$55 per ton of CO_2 in a raceway pond system. However the energy penalty associated with algae as a CCS technology would likely be zero or negative due to the production of large amounts of biomass that can be used as fuel (see http://www.powerplantccs.com/ccs/cap/fut/alg/algae_co2_capture_costs.html).

8.8 Conclusion

In general, algae are considered as tremendous creatures that have great potential finding use in various applications ranging from environmental cleanup to generation of value added products. Coupling of processes such as CO_2 sequestration and bioremediation of wastewater, agricultural runoffs etc. has shown to have serious impact on the environment. The sustainability of algal biotechnology in environmental, social, and economic aspects will depend largely on the integrated approach for the development of viable solutions. The use of gaseous waste (CO_2) and the liquid waste (effluent) from industries for algal biomass generation would not only offset the energy and operational cost for its production but also contribute extensively to the reduction in the usage of fossil fuels for energy purposes. This sustainable approach needs further development and refining particularly in the existing technologies for cost effective microalgal production.

Acknowledgments

The authors acknowledge Shri A.M.M. Murugappa Chettiar Research Centre (MCRC), Chennai for the facilities provided and the Technical Committee of the Centre for their valuable suggestions. The authors also wish to thank Dr. N. Jeeji Bai, Research Advisor, MCRC and Mr. M.R. Sridharan, Senior Principal Scientist, CSIR-CLRI, Chennai for their inputs and collection of reprints that aided the successful completion of this book chapter.

References

Abinandan, S., Shashirekha, V., Vedaraman, N., Khambhaty, Y., Seshadri, S., Sridharan, M.R., 2014. Feasibility of microalgal cultivation in soak liquors from leather processing unit. Indian Hydrobiol. 16 (2), 145–152.

Allard, B., Templier, J., Largeau, C., 1998. An improved method for the isolation of artifact-free algaenans from microalgae. Org. Geochem. 28 (9–10), 543–548.

Anand, M.N., 2010. Microalgal bioremediation of nutrients in wastewater and carbon-dioxide in flue gas. Masters Thesis, Environmental Engineering. Missouri University of Science and Technology.

Bataille, M.C., Birraux, C., 2006. 14 New Energy Technologies and CO_2 Sequestration. Parliamentary Office for Science and Technology Assessment, France.

Bayless, D.J., Kremer, G.G., Prudich, M.E., Stuart, B.J., Vis-Chiasson, M.L., Cooksey, K., et al., 2001. Enhanced practical photosynthetic CO_2 mitigation. In: Proceedings of the First National Conference on Carbon Sequestration, 5A4, pp. 1–14.

Becker, E.W., 1994. Microalgae: Biotechnology & Microbiology. Cambridge University Press.

Benemann, J.R., 1997. CO_2 mitigation with microalgae systems. Energy Convers. Manage. 38, S475–S479.

Bharanidharan, M., Mydeen, M.A., Sridharan, M.R., 2016. Algae as an alternative fuel: present and future challenges – a review. J. Bioenergy Biofuels 2 (1), 1–15.

Bhatnagar, M., Chinnasamy, S., 2010. *Chlorella minutissima* – a promising fuel alga for cultivation in municipal wastewaters. Appl. Biochem. Biotechnol. 161, 523–536.

Bhatnagar, A., Chinnasamy, S., Singh, M., Das, K.C., 2011. Renewable biomass production by mixotrophic algae in the presence of various carbon sources and wastewaters. Appl. Energy 88, 3425-343.

Bird, K.T., Chynoweth, D.P., Jerger, D.E., 1990. Effects of marine algal proximate composition on methane yields. J. Appl. Phycol. 2, 207–213.

Borowitzka, M.A., Borowitzka, I.J., 1988. Dunaliella. In: Borowitzka, M.A., Borowitzka, I.J. (Eds.), Microalgae Biotechnology. Cambridge University Press, Cambridge, pp. 27–58.

Cantrell, K.B., Ducey, T., Ro, K.S., Hunt, P.G., 2008. Livestock waste to bioenergy generation opportunities. Bioresour. Technol. 99, 7941–7953.

Cardozo, K.H.M., Guaratini, T., Barros, M.P., Vanessa, R.F., Tonon, A.P., Lopes, N.P., et al., 2007. Metabolites from algae with economical impact. Comp. Biochem. Physiol. C: Toxicol. Pharmacol. 146, 60–78.

Chakrabarthi, T., Krishnamurthi, K., Devi, S.S., Fulke, A.B., 2014. CO_2 sequestration by microalgae: advances and perspectives. In: Liu, Jin, Sun, Zheng, Gerken, Henri (Eds.), Recent Adv. in Microalgal Biotechnol. Publ: Omics Group eBooks, CA, USA, pp. 3–7.

Chen, P., Min, M., Chen, Y., Wang, L., Li, Y., Chen, Q., et al., 2009. Review of the biological and engineering aspects of algae to biofuels approach. Int. J. Agric. Biol. Eng. 2 (4), 1–24.

Cheng, J., Huang, Y., Feng, J., Sun, J., Zhou, J., Cen, K., 2013. Improving CO_2 fixation efficiency by optimizing *Chlorella* PY-ZU1 culture conditions in sequential bioreactors. Bioresour. Technol. 144, 321–327.

Cheryl, 2010. Algae becoming the new biofuel of choice. Available online 2008: http://duelingfuels.com/biofuels/non-food-biofuels/algae-biofuel.php#more-115N.

Chi, Z., Zheng, Y., Jiang, A., Chen, S., 2011. Lipid production and culturing oleaginous yeast and algae with food waste and municipal wastewater in an integrated process. Appl. Biochem. Biotechnol. 165, 442–453.

Chisti, Y., 2007. Biodiesel from microalgae. Biotechnol. Adv. 25, 294–306.

Chojnacka, K., Marquez-Rocha, F.J., 2004. Kinetic and stoichiometric relationships of the energy and carbon metabolism in the culture of microalgae. Biotechnol. 3 (1), 21–34.

Chung, P., Pond, L.W.G., Kingsbury, J.M., Walker, E.F., Krook, L., 1978. Production and nutritive value of *Arthrospira platensis*, a spiral blue–green alga grown on swine wastes. J. Anim. Sei. 47, 319–330.

Dalrymple, O.K., Halfhide, T., Udom, I., Gilles, B., Wolan, J., Zhang, Q., et al., 2013. Wastewater use in algae production for generation of renewable resources: a review and preliminary results. Aquat. Biosyst. 9, 2.

Das, D., 2015. Introduction. In: Das, D. (Ed.), Algal Biorefinery: An Integrated Approach. Springer International Publishing, Switzerland co-published with Capital Publishing Company, New Delhi, pp. 1–34.

de Morais, M.G., Costa, J.A.V., 2007. Carbon-dioxide fixation by *Chlorella kessleri*, *C. vulgaris*, *Scenedesmus obliquus* and *Spirulina* sp. cultivated in flasks and vertical tubular photobioreactors. Biotechnol. Lett. 29, 1349–1352.

de Morais, M.G., Stillings, C., Dersch, R., Rudisile, M., Pranke, P., Costa, J.A.V., et al., 2010. Preparation of nanofibers containing the microalga *Spirulina* (*Arthrospira*). Bioresour. Technol. 101 (8), 2872–2876.

Demirbaş, A., 2009a. Inexpensive oil and fats feedstocks for production of biodiesel. Energy Edu. Sci. Technol. A 23, 1–13.

Demirbaş, A., 2009b. Production of biodiesel from algae oils. Energy Sources A 31, 163–168.

Diniz, G.S., Silva, A.F., Araújo, O.Q.F., Chaloub, M.R., 2017. The potential of microalgal biomass production for biotechnological purposes using wastewater resources. J. Appl. Phycol. 29, 821–832.

Douskova, I., Doucha, J., Livansky, K., Machat, J., Novak, P., Umysova, D., et al., 2009. Simultaneous flue gas bioremediation and reduction of microalgal biomass production costs. Appl. Microbiol. Biotechnol. 82, 179–185.

Edwards, C., 1990. In: Edwards, C. (Ed.), Thermophiles. Microbiology of Extreme Environments. Open University Press, Milton Keynes, Great Britain, p. 218.

Falkowski, P.G., Katz, M.E., Knoll, A.H., Quigg, A., Raven, J.A., Schofield, O., et al., 2004. The evolution of modern eukaryotic phytoplankton. Science 305, 354–360.

Fast, A.G., Papoutsakis, E.T., 2012. Stoichiometric and energetic analyses of non-photosynthetic CO_2-fixation pathways to support synthetic biology strategies for production of fuels and chemicals. Curr. Opin. Chem. Eng. 1, 1–16.

Field, C.B., Behrenfeld, M.J., Randerson, J.T., Falkowski, P., 1998. Primary production of the biosphere: integrating terrestrial and oceanic components. Science 281 (5374), 237–240.

Fulke, A.B., Mudliar, S.N., Yadav, R., Shekh, A., Srinivasan, N., 2010. Bio-mitigation of CO_2, calcite formation and simultaneous biodiesel precursors production using *Chlorella* sp. Bioresour. Technol. 101, 8473–8476.

Fulke, A.B., Chambhare, K., Giripunje, M.D., Sangolkar, L., Krishnamurthi, K., 2013. Potential of wastewater grown algae for biodiesel production and CO_2 sequestration. Afr. J. Biotechnol. 12, 2939–2948.

Gao, F., Li, C., Yang, Z.H., Zeng, G.M., Feng, L.J., Liu, J.Z., et al., 2016. Continuous microalgae cultivation in aquaculture wastewater by a membrane photobioreactor for biomass production and nutrients removal. Ecol. Eng. 92, 55–61.

Giordano, M., Beardall, J., Raven, J.A., 2005. Mechanisms in algae: mechanisms, environmental modulation and evolution. Annu. Rev. Plant Biol. 56, 99–131.

Gladue, R.M., Maxey, J.E., 1994. Microalgal feeds for aquaculture. J Appl. Phycol. 6 (2), 131–141.

Golueke, C.G., Oswald, W., 1959. Biological conversion of light energy to the chemical energy of methane. Appl. Environ. Microbiol. 7 (4), 219–227.

Golueke, C., Oswald, W., Gotaas, H., 1957. Anaerobic digestion of algae. Appl. Environ. Microbiol. 5 (1), 47–55.

Gupta, P.L., Lee, S.M., Choi, H.J., 2015. A mini review: photobioreactors for large scale algal cultivation. World J. Microbiol. Biotechnol. 31 (9), 1409–1417.

Hadiyanto, H., Nur, M.M.A., 2012. Potential of palm oil mill effluent (POME) as medium growth of *Chlorella* sp. for bioenergy production. Int. J. Environ. Bioenergy 3 (2), 67–74.

Hall, D.O., Mynick, H.E., Williams, R.H., 1991. Cooling the greenhouse with bioenergy. Nature 353, 11.

Hanagata, N., Takeuchi, T., Fukuju, Y., Barnes, D.J., Karube, I., 1992. Tolerance of microalgae to high CO_2 and high temperature. Phytochemistry 31 (10), 3345–3348.

Hankamer, B., Lehr, F., Rupprecht, J., Mussgnug, J.H., Posten, C., Kruse, O., 2007. Photosynthetic biomass and H_2 production by green algae: from bioengineering to bioreactor scale-up. Physiol. Plant 131, 10–21.

Harun, R., Danquah, M.K., Forde, G.M., 2010. Microbial biomass as a fermentation feedstock for bioethanol production. J Chem. Technol. Biotechnol. 85, 199–203.

Hasan, R., Zhang, B., Wang, L.J., Shahbazi, A., 2014. Bioremediation of swine wastewater and biofuel potential by using *Chlorella vulgaris*, *Chlamydomonas reinhardtii* and *Chlamydomonas debaryana*. J. Pet. Environ. Biotechnol. 5 (3), 175.

Hena, S., Fatimah, S., Tabassum, S., 2015. Cultivation of algae consortium in a dairy farm wastewater for biodiesel production. Water Resour. Ind. 10, 1–14.

Hirano, A., Ryohei, U., Shin, H., Yasuyuki, O., 1997. CO_2 fixation and ethanol production with microalgal photosynthesis and intracellular anaerobic fermentation. Energy 22, 137–142.

Hirata, S., Hayashitani, M., Taya, M., Tone, S., 1996a. Carbon-dioxide fixation in batch culture of *Chlorella* sp. using a photobioreactior with a sunlight-collection device. J. Ferment. Bioeng. 81 (5), 470–472.

Hirata, S., Taya, M., Tone, S., 1996b. Characterization of *Chlorella* cell cultures in batch and continuous operations under a photoautotrophic condition. J. Chem. Eng. Jpn. 29 (6), 953–959.

Ho, S.H., Chen, C.Y., Yeh, K.L., Chen, W.M., Lin, C.Y., 2010. Characterization of photosynthetic carbon-dioxide fixation ability of indigenous *Scenedesmus obliquus* isolates. Biochem. Eng. J. 53, 57–62.

Ho, S.H., Chen, C.Y., Chang, J.S., 2012. Effect of light intensity and nitrogen starvation on CO_2 fixation and lipid/carbohydrate production of an indigenous microalga *Scenedesmus obliquus* CNW-N. Bioresour. Technol. 113, 244–252.

Ho, S.H., Li, P.J., Liu, C.C., Chang, J.S., 2013. Bioprocess development on microalgae-based CO_2 fixation and bioethanol production using *Scenedesmus obliquus* CNW-N. Bioresour. Technol. 145, 142–149.

Horn, S.J., 2000. Bioenergy From Brown Seaweeds. Norwegian University of Science and Technology NTNU, Department of Biotechnology, Trondheim.

Hossain, M.M., de Lasa, H.I., 2007. Reactivity and stability of Co–Ni/Al_2O_3 oxygen carrier in a multicycle CLC. AIChE J. 53 (6), 1817–1829.

Hu, B., Min, M., Zhou, W., Du, Z., Mohr, M., Chen, P., et al., 2012. Enhanced mixotrophic growth of microalga *Chlorella* sp. on pretreated swine manure for simultaneous biofuel feedstock production and nutrient removal. Bioresour. Technol.: Adv. Biol. Waste Treat. Bioconvers. Technol. 126, 71–79.

Huntley, M., Redalje, D., 2007. CO_2 mitigation and renewable oil from photosynthetic microbes: a new appraisal. Mitigation Adapt. Strat. Global Change 12, 573–608.

IEA, 2012. Energy Technology Perspectives: Pathways to a Clean Energy System. OECD/IEA.

IPCC, 2001. Climate change 2001 – mitigation. In: Metz, B., Davidson, O., Swart, R., Pan, J. (Eds.), The Third Assessment Report of the Intergovernmental Panel on Climate Change. Cambridge University Press, Cambridge.

Jeeji Bai, N., Nair, B.B., Shashirekha, V., 2018. *Spirulina* (*Arthrospira*): Taxonomy, Biology and Applications. Pub: Bishen Singh Mahendra Pal Singh, Dehra Dun, p. 112.

Jha, M.N., Prasad, A.N., Sharma, S.G., Bharti, R.C., 2001. Effects of fertilization rate and crop rotation on diazotrophic cyanobacteria. World J. Microbiol. Biotechnol. 17, 463.

Jiménez, C., Mercado, J., Aguilera, J., Rodríguez-Maroto, J.M., Niell, F.X., 1995. Effect of turbulence and inorganic carbon supply on growth of *Dunaliella viridis* Teodoresco. Int. J. Salt Lake Res. 4, 223–232.

Jin, H.F., Lim, B.R., Lee, K., 2006. Influence of nitrate feeding on carbon dioxide fixation by microalgae. J Environ. Sci. Health A: Tox Hazard. Subst. Environ. Eng. 41, 2813–2824.

Kandimalla, P., Desi, S., Vurimindi, H., 2015. Mixotrophic cultivation of microalgae using industrial flue gases for biodiesel production. Environ. Sci. Pollut. Res. 23 (10), 1–10.

Kaushik, R., Prasanna, R., Joshi, H.C., 2006. Utilization of anaerobically digested distillery effluent for the production of *Spirulina platensis* (ARM 730). J. Sci. Ind. Res. 65, 521–525.

Khairul, S.N.K., 2009. Synthesis and Modification of Micro and Mesoporous Materials as CO_2 Adsorbents. http://eprints.utm.my/id/eprint/9749/1/78207.pdf.

Khatoon, H., Banerjee Md., S., Syahiran, S., Noordin, N.B.M., Bolong, A.M.A., Endut, A., 2016. Re-use of aquaculture wastewater in cultivating microalgae as live feed for aquaculture organisms. Desalin. Water Treat. Available from: http://dx.doi.org/10.1080/19443994.2016.1156030.

Kim, M.K., Park, J.W., Park, C.S., Kim, S.J., Jeune, K.H., Chang, M.U., et al., 2007. Enhanced production of *Scenedesmus* spp. (green microalgae) using a new medium containing fermented swine wastewater. Bioresour. Technol. 98, 2220–2228.

Kishimoto, M., Okakura, T., Nagashima, H., Minowa, T., Yokoyama, S.Y., 1994. CO_2 fixation and oil production using microalgae. J. Ferment. Bioeng. 78, 479–482.

Kodama, M., Ikemoto, H., Miyachi, S., 1993. A new species of highly CO_2-tolreant fast-growing marine microalga suitable for high-density culture. J. Mar. Biotechnol. 1, 21–25.

Kumar, K., Das, D., 2014. Carbon-dioxide sequestration by biological processes. In: Bhanage, B.M., Arai, M. (Eds.), Transformation and Utilization of Carbon-dioxide. Springer, pp. 303–334.

Kumar, A., Singh, J.S., 2017. Cyanoremediation: a green-clean tool for decontamination of synthetic pesticides from agro- and aquatic ecosystems. In: Singh, J.S., Seneviratne, G. (Eds.), Agro-Environmental Sustainability, vol. 2: Managing Environmental Pollution. Springer, International Publishing AG, pp. 59–83.

Kumar, A., Yuan, X., Sahu, A.K., Erga, S.J., Langenhove, V., 2010. A hollow fiber membrane photo-bioreactor for CO_2 sequestration from combustion gas coupled with wastewater treatment: a process engineering approach. J. Chem. Technol. Biotechnol. 85, 387–394.

Kumar, K., Dasgupta, C.N., Nayak, B.K., Lindblad, P., Das, D., 2011. Development of suitable photobioreactors for CO_2 sequestration addressing global warming using green algae and cyanobacteria. Bioresour. Technol. 102 (8), 4945–4953.

Kumar, K., Banerjee, D., Das, D., 2014. Carbon-dioxide sequestration from industrial flue gas by *Chlorella sorokiniana*. Bioresour. Technol. 152, 225–233.

Kumar, A., Kaushal, S., Saraf, S.A., Singh, J.S., 2018. Microbial bio-fuels: a solution to carbon emissions and energy crisis. Front. Biosci. (Landmark ED) 1 (23), 1789–1802.

Kurano, N., Ikemoto, H., Miyashita, H., Hasegawa, H., Hata, T., Miyachi, S., 1995. Fixation and utilization of carbon dioxide by microalgal photosynthesis. Energy Convers. Manage. 36 (6–9), 689–692.

Lewis, N.S., Nocera, D.G., 2006. Powering the planet: chemical challenges in solar energy utilization. Proc. Natl. Acad. Sci. 103, 15729–15735.

Li, Y., Chen, Y.F., Chen, P., Min, M., Zhou, W., Martinez, B., et al., 2011. Characterization of a microalga *Chlorella* sp. well adapted to highly concentrated municipal wastewater for nutrient removal and biodiesel production. Bioresour. Technol. 102, 5138–5144.

Maeda, K., Owada, M., Kimura, N., Omata, K., Karube, I., 1995. CO_2 fixation from the flue gas on coal-fired thermal power plant by microalgae. Energy Convers. Manage. 36, 717–720.

Mahadevaswamy, M., Venkataraman, L.V., 1986. Bioconversion of poultry droppings for biogas and algal production. Agric. Wastes 18 (2), 93–101.

Masaki, O., Yoshitaka, K., Watanabe, H., Watanabe, M., Sato, Y., Richard, L., et al., 2009. Inomata, fatty acid production from a highly CO_2 tolerant alga, *Chlorocuccum littorale*, in the presence of inorganic carbon and nitrate. Bioresour. Technol. 100, 5237–5242.

Mata, T.M., Martins, A.A., Caetano, N.S., 2009. Microalgae for biodiesel production and other applications: a review. Renew. Sustain. Energy Rev. 14, 217–232.

Matsumoto, H., Shioji, N., Hamasaki, A., Ikuta, Y., Fukuda, Y., Sato, M., et al., 1995. Carbon dioxide fixation by microalgae photosynthesis using actual flue gas discharged from a boiler. Appl. Biochem. Biotechnol. 51/52, 681–692.

Matsumoto, M., Hiroko, Y., Nobukazu, S., Hiroshi, O., Tadashi, M., 2003. Saccharification of marine microalgae using marine bacteria for ethanol production. Appl. Biochem. Biotechnol. 105, 247–254.

McGinn, P.J., Dickinson, K.E., Park, K.C., Whitney, C.G., Mac Quarrie, S.P., Black, F.J., et al., 2012. Assessment of the bioenergy and bioremediation potentials of the microalga *Scenedesmus* sp. AMDD cultivated in municipal wastewater effluent in batch and continuous mode. Algal Res. 1, 155–165.

Michelon, W., Da Silva, M.L.B., Mezzari, M.P., Pirolli, M., Prandini, J.M., Soares, H.M., 2016. Effects of nitrogen and phosphorus on biochemical composition of microalgae polyculture harvested from phycoremediation of piggery wastewater digestate. Appl. Biochem. Biotechnol. 178 (7), 1407–1419.

Mitra, D., van Leeuwen, J.H., Lamsal, B., 2012. Heterotrophic/mixotrophic cultivation of oleaginous *Chlorella vulgaris* on industrial co-products. Algal Res. 1 (1), 40–48.

Miura, Y., Yamada, W., Hirata, K., Miyamoto, K., Kiyohara, M., 1993. Stimulation of hydrogen production in algal cells grown under high CO_2 concentration and low temperature. Appl. Biochem. Biotechnol. 39/40, 753–761.

Miyairi, S., 1995. CO_2 assimilation in a thermophilic cyanobacterium. Energy Convers. Manage. 36 (6–9), 763–766.

Montoya, E.Y.O., Casazza, A.A., Aliakbarian, B., Perego, P., Converti, A., de Carvalho, J.C.M., 2014. Production of *Chlorella vulgaris* as a source of essential fatty acids in a tubular photobioreactor continuously fed with air enriched with CO_2 at different concentrations. Biotechnol. Prog. 30, 916–922.

Morales-Amaral, M.M., Gómez-Serrano, C., Acién, F.G., Fernández-Sevilla, J.M., Molina-Grima, E., 2015. Production of microalgae using centrate from anaerobic digestion as the nutrient source. Algal Res. 9, 297–305.

Mulbry, W., Kondrad, S., Pizarro, C., Kebede-Westhead, E., 2008. Treatment of dairy manure effluent using freshwater algae: algal productivity and recovery of manure nutrients using pilot-scale algal turf scrubbers. Bioresour. Technol. 99, 8137–8142.

Nagase, H., Eguchi, K., Yoshihara, K., Hirata, K., Miyamoto, K., 1998. Improvement of microalgal NO_x removal in bubble column and airlift reactors. J. Ferment. Bioeng. 86 (4), 421–423.

Nakanishi, A., Aikawa, S., Ho, S.H., Chen, C.Y., Chang, J.S., Hasunuma, T., et al., 2014. Development of lipid productivities under different CO_2 conditions of marine microalgae *Chlamydomonas* sp. JSC4. Bioresour. Technol. 152, 247–252.

Nakano, Y., Miyatake, K., Okuno, H., Hamazaki, K., Takenaka, S., Honami, N., et al., 1996. Growth of photosynthetic algae *Euglena* in high CO_2 conditions and its photosynthetic characteristics. Acta Hortic. 440, 49–54.

Nirbhay, K.S., Dolly, W.D., 2011. Microalgae as second generation biofuel. A review. Agron. Sustain. Dev. 31, 605–629.

Olaizola, M., 2003. Commercial development of microalgal biotechnology: from test tube to marketplace. Biomol. Eng. 20, 459–466.

Olivieri, G., Salatinoa, P., Marzocchellaa, A., 2014. Advances in photobioreactors for intensive microalgal production: configurations, operating strategies and applications. J. Chem. Technol. Biotechnol. 89, 178–195.

Oron, G., Shelef, G., Levi, A., 1979. Growth of Spirulina maximaon cow manure waste. Biotechnol. Bioeng. 20, 2165–2173.

Oswald, J.A., 1988. Large-scale algal culture systems (engineering aspects). In: Borowitzka, L.J., Borowitzka, M.A. (Eds.), Microalgal Biotechnology. Cambridge University Press, Cambridge, pp. 357–395.

Oswald, W.J., 2003. My sixty years in applied algology. J. Appl. Phycol. 15, 99–106.

Oswald, W.J., Goluecke, C.G., 1960. Biological transformation of solar energy. Adv. Appl. Microbiol. 2, 223–262.

Oswald, W.J., Gotaas, H.B., Ludwig, H.F., Lynch, V., 1953. Algae symbiosis in oxidation ponds: photosynthetic oxygenation. Sew. Ind. Wastes 25 (6), 692–705.

Patil, L., Kaliwal, B., 2017. Effect of CO_2 concentration on growth and biochemical composition of newly isolated indigenous microalga *Scenedesmus bajacalifornicus* BBKLP-07. Appl. Biochem. Biotechnol. 182, 335. Available from: https://doi.org/10.1007/s12010-016-2330-2.

Perez-Garcia, O., Bashan, Y., 2015. Microalgal heterotrophic and mixotrophic culturing for bio-refining: from metabolic routes to techno-economics. In: Prokop, A., Bajpai, R.K., Zappi, M.E. (Eds.), Algal Biorefineries. Springer, Berlin, pp. 61–131.

Posadas, E., Bochon, S., Coca, M., García-González, M.C., García-Encina, P.A., Muñoz, R., 2014. Microalgae-based agro-industrial wastewater treatment: a preliminary screening of biodegradability. J. Appl. Phycol. 26 (6), 2335–2345.

Putri, E.V., Din, M.F.M., Ahmed, Z., Jamaluddin, H., Chelliapan, S., 2011. Investigation of microalgae for high lipid content using palm oil mill effluent (POME) as carbon source. In: Intl. Conf. on Environ. & Industrial Innovation, IACSIT Press, Singapore, vol. 12, pp. 85–89.

Putt, R., Singh, M., Chinnasamy, S., Das, K.C., 2011. An efficient system for carbonation of high-rate algal pond water to enhance CO_2 mass transfer. Bioresour. Technol. 102 (3), 3240–3245.

Rados, S., Vaclav, B., Frantisek, D., 1975. CO_2 Balance in industrial cultivation of algae. Arch. Hydrobiol. 46 (12), 297–310.

Razzak, S.A., Hossain, M.M., Lucky, R.A., Bassi, A.S., de Lasa, H., 2013. Integrated CO_2 capture, wastewater treatment and biofuel production by microalgae culturing – A review. Renew. Sustain. Energ. Rev. 27, 622–653.

Rodríguez-Maroto, J.M., Jiménez, C., Aguilera, J., Niell, F.X., 2005. Air bubbling results in carbon loss. Aquacult. Eng. 32, 493–508.

Ryu, H.J., Oh, K.K., Kim, Y.S., 2009. Optimization of the influential factors for the improvement of CO_2 utilization efficiency and CO_2 mass transfer rate. J Indus. Eng. Chem. 15, 471–475.

Saikia, M.K., Kalita, S., Sarma, G.C., 2011. An experimental investigation on growth stimulation (+) and inhibition (-) of algae (*Oscillatoria chlorina* and *Scenedesmus quadricauda*) treated with pulp and paper mill effluents. Int. J. Appl. Biol. Pharm. Technol. 2 (4), 87–94.

Sayer, R., 2010. Microalgae: the potential for carbon capture. Bioscience 60 (9), 722–727.

Schenk, P.M., Thomas-Hall, S.R., Stephens, E., Marx, U.C., Mussgnug, J.H., Posten, C., et al., 2008. Second generation biofuels: high-efficiency microalgae for biodiesel production. Bioenergy Res. 1, 20–43.

Schmack, B.A., 2008. Effizienssteigerung der biogasnutzung durch solarenergie – BSIE - Techni-kumsphase. Schmack Biogas AG.

Seckbach, J., Gross, H., Nathan, M.B., 1971. Growth and photosynthesis of *Cyanidium caldarium* cultured under pure CO_2. Isr. J. Bot. 20, 84–90.

Sekabira, K., Origa, H.O., Basamba, T.A., Mutumba, G., Kakudidi, E., 2010. Application of algae in biomonitoring and phytoextraction of heavy metals contamination in urban stream water. Int. J. Environ. Sci. Technol. 8, 115–128.

Seshadri, S., Shashirekha, V., 2012. Perspectives on Biohydrogen Production, vol. 5. Akshay Urja, RE Feature, MNRE Publications, pp. 18–20.

Sharma, S., Hunt, R., Cao, J. and Zeller, M.A., 2011. Bioplastics from algae biomass (Conference Paper) Society of Plastics Engineers—Global Plastics Environmental Conference.

Shashirekha, V., Sridharan, M.R., Swamy, M., 2011. Bioremediation of tannery effluents using a consortium of blue-green algal species. Clean Soil Air Water 39 (9), 863–873.

Shashirekha, V., Sivakumar, M., Seshadri, S., 2016. Effective C-N-P ratio for growth and nutrient removal efficiency of *Scenedesmus obliquus* in sugar mill effluent. Energy Ecol. Environ. 1 (5), 283–295.

Shi, X.M., Zhang, X.W., Chen, F., 2002. Heterotrophic production of biomass and lutein by *Chlorella protothecoides* in heterotrophic fed-batch culture. Biotechnol. Prog. 18, 723–727.

Sialve, B., Bernet, N., Bernard, O., 2009. Anaerobic digestion of microalgae as a necessary step to make microalgal biodiesel sustainable. Biotechnol. Adv. 27, 409–416.

Singh, S.P., Singh, P., 2015. Effect of temperature and light on the growth of algae species: a review. Renew. Sustain. Energy Rev. 50, 431–444.

Singh, M., Reynolds, D., Das, K.C., 2011. Microalgal system for treatment of effluent from poultry litter anaerobic digestion. Bioresour. Technol. 102 (23), 10841–10848.

Singh, A., Pant, D., Olsen, S.I., Nigam, P.S., 2012. Key issues to consider in microalgae based biodiesel production. Energy Educ. Sci. Technol. A Energy Sci. Res. 29, 563–576.

Singh, J.S., Kumar, A., Rai, A.N., Singh, D.P., 2016. Cyanobacteria: a precious bio-resource in agriculture, ecosystem, and environmental sustainability. Front. Microbiol. 7, 529.

Sivakumar, M., Ranjithkumar, R., Shashirekha, V., Seshadri, S., 2014. Influence of carbon-dioxide on the growth of *Spirulina* sp. (MCRC-A0003) isolated from Muttukadu backwaters, South India. World J. Microbiol. Biotechnol. 30 (10), 2775–2781.

Speece, R.E., 1996. Anaerobic Biotechnology for Industrial Wastewaters. Archae Press, Nashville.

Spolaore, P., Cassan, C.J., Duran, E., Isambert, A., 2006. Commercial applications of microalgae. J. Biosci. Bioeng. 101, 87–96.

Srinivasan, S., 1978. Algae multiplication fertilizer practices. Int. Rice Res. Newslett. 11 (4), 35–36.

Sudhakar, K., Suresh, S., Premalatha, M., 2011. An overview of CO_2 mitigation using algae cultivation technology. Int. J. Chem. Res. 3 (3), 110–117.

Sydney, E.B., Sturm, W., de Carvalho, J.C., Thomaz-Soccol, V., Larroche, C., 2010. Potential carbon dioxide fixation by industrially important microalgae. Bioresour. Technol. 101, 5892–5896.

Sydney, E.B., Novak, A.C., de Carcalho, J.C., Soccol, C.R., 2014. Respirometric balance and carbon fixation of industrially important algae. In: Paney, A., Lee, D.J., Chisti, Y., Soccol, C.R. (Eds.), Biofuels From Algae. Elsevier, MA, pp. 67–84.

Tang, D., Han, W., Li, P., Miao, X., Zhong, J., 2011. CO_2 biofixation and fatty acid composition of *Scenedesmus obliquus* and *Chlorella pyrenoidosa* in response to different CO_2 levels. Bioresour. Technol. 102, 3071–3076.

Tanticharoen, M., Bunnag, B., Vonshak, A., 1993. Cultivation of *Spirulina* using secondary treated starch waste water. Aust. Biotechnol. 3, 223–226.

Tornabene, T.G., Holzer, G., Lien, S., Burris, N., 1983. Lipid composition of the nitrogen starved green alga *Neochloris oleabundans*. Enzyme Microb. Technol. 5, 435–440.

Toyub, M.A., Miah, M.I., Habib, M.A.B., Rahman, M.M., 2008. Growth performance and nutritional value of *Scenedesmus obliquus* cultured in different concentrations of sweetmeat factory waste media. Bangladesh J. Anim. Sci. 37 (1), 86–93.

Usha, M.T., Chandra, T.S., Sarada, R., Chauhan, V.S., 2016. Removal of nutrients and organic pollution load from pulp and paper mill effluent by microalgae in outdoor open pond. Bioresour. Technol. 214, 856–860.

Venkata Mohan, S., Rohit, M.V., Chiranjeevi, P., Hariprasad, R., Arora, S., 2015. Integrating microalgae cultivation with wastewater treatment for biodiesel production. In: Das, D. (Ed.), Algal Biorefinery: An Integrated Approach. Springer International Publishing, Switzerland co-published with Capital Publishing Company, New Delhi, pp. 321–337.

Wang, B., Li, Y., Wu, N., Lan, C.Q., 2018. CO_2 bio-mitigation using microalgae. Appl. Microbiol. Biotechnol. 79, 707–718.

Weissman, J.C., Tillett, D.M., 1992. Design and operation of an outdoor microalgae test facility: large-scale system results. In: Brown, L.M., Sprague, S. (Eds.), Aquatic Species Project Report (FY 1989–90). National Renewable Energy Laboratory, Golden, CO, pp. 32–56. , NREL/TP-232-4174.

Wijffels, R.H., Barbosa, M.J., 2010. An outlook on microalgal biofuels. Science 329, 796–799.

Witt, V., Borchardt, J.A., 1960. The removal of nitrogen and phosphorus from sewage effluents through the use of algal culture. J. Biochem. Microbiol. Technol. Eng. 11 (2), 187–203.

Yang, Y., Gao, K., 2003. Effects of CO_2 concentrations on the freshwater microalgae, *Chlamydomonas reinhardtii*, *Chlorella pyrenoidosa* and *Scenedesmus obliquus* (Chlorophyta). J. Appl. Phycol. 1–11.

Yeh, K.L., Chang, J.S., 2011. Nitrogen starvation strategies and photobioreactor design for enhancing lipid content and lipid production of a newly isolated microalga *Chlorella vulgaris* ESP-31: implications for biofuels. Biotechnol. J. 6, 1358–1366.

Yokoyama, S., Jonouchi, K., Imou, K., 2007. Energy production from marine biomass: fuel cell power generation driven by methane produced from seaweed. In: Proc. World Acad. Sci. Eng. Technol. vol. 22, pp. 320–323.

Yoshihara, K., Nagase, H., Eguchi, K., Hirata, K., Miyamoto, K., 1996. Biological elimination of nitric oxide and carbon dioxide from flue gas by marine microalga NOA-113 cultivation in a long tubular photobioreactor. J. Ferment. Bioeng. 82 (4), 351–354.

Zhao, B., Zhang, Y., Xiong, K., Zhang, Z., Hao, X., 2011. Effect of cultivation mode on microalgal growth and CO_2 fixation. Chem. Eng. Res. Design. 89, 1758–1762.

Zhou, W., Li, Y., Min, M., Hu, B., Zhang, H., Ma, X., et al., 2012. Growing wastewater-borne microalga *Auxenochlorella protothecoides* UMN280 on concentrated municipal wastewater for simultaneous nutrient removal and energy feedstock production. Appl. Energy 98, 433–440.

Further reading

Beneman, J.R., Tillet, D.M., Weissman, J.C., 1987. Micro-algae biotechnology. Trends Biotechnol. 5 (2), 47–53.

Chapter 9

Sustainable approaches for biological control of mycotoxigenic fungi and mycotoxins in cereals

N. Deepa and M.Y. Sreenivasa
Department of Studies in Microbiology, University of Mysore, Mysore, India

Chapter Outline

9.1 Introduction

Cereals are a basic staple food that provides energy and protein for many populations. Cereals contains a wide range of micronutrients, vitamins B and E, along with sodium, magnesium, and zinc. Cereals are grown in greater quantities as they provide more food energy worldwide than any other type of crop, with 2534 MT including 245.5 MT in India, 1104 MT consumed as food by humans, and 876 MT as feed for animals, which finally will be consumed as meat by humans. In some developing nations, grain in the form of rice, wheat, or maize constitutes a majority of daily substance. In developed nations, cereal consumption is more moderate and varied but still substantial in the form of cereal-based food products like corn flakes, bread, biscuits, flour, oats, poultry and animal feeds, etc., and also used in industries for production of beverages, medicines, brown rice, etc., for both humans and animals.

Due to poor agricultural practices, incessant rainfall, and unfavorable environmental conditions during pre- and postharvest phases, fungi have been reported as an important concern for worldwide agriculture, global economy, and public health that has become an unavoidable, a worldwide problem causing yield and quality losses of crops (Nguyen et al., 2017). A number of fungi contaminate cereals making them unfit for human consumption due, especially those produced by species of *Aspergillus*, *Fusarium*, *Penicillium*, and *Alternaria* due to its phytopathological and mycotoxicological risks at preharvest and postharvest stages and processed food products (Castoria et al., 2008; Dass et al., 2007). The susceptibility of these cereals to various fungi has been well documented (Munkvold and Desjardins, 1997).

New and Future Developments in Microbial Biotechnology and Bioengineering. DOI: https://doi.org/10.1016/B978-0-12-818258-1.00009-1

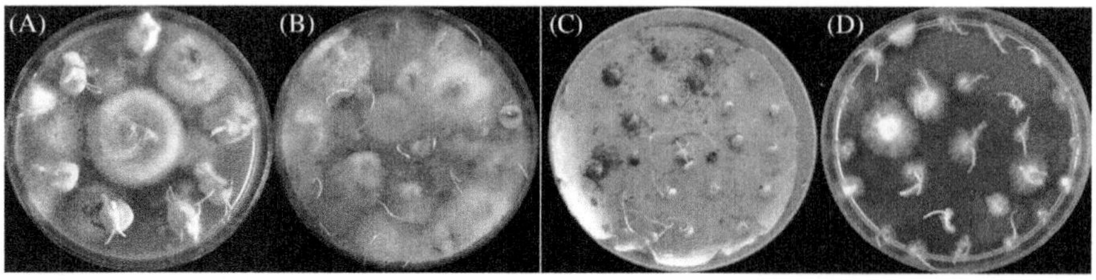

FIGURE 9.1 Mycotoxigenic fungi in association with cereals. (A) Maize; (B) paddy; (C) sorghum; (D) pearl millet.

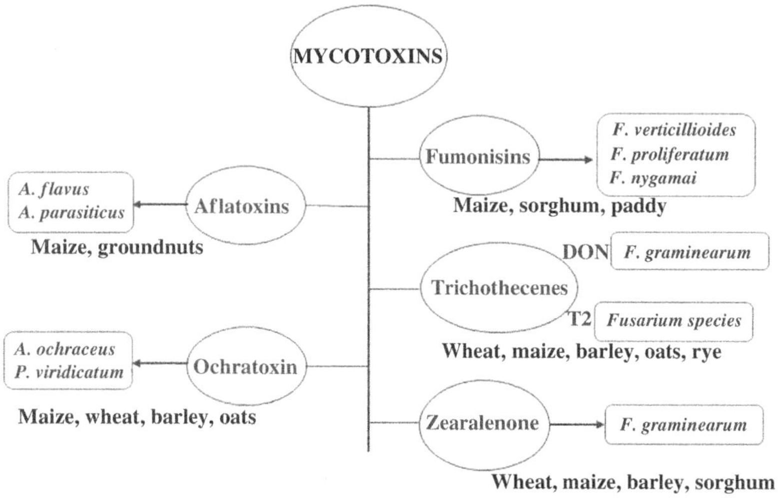

FIGURE 9.2 Representation of mycotoxigenic fungi and their major mycotoxins on cereals.

9.2 Mycotoxigenic fungi and its mycotoxins

Species of *Fusarium*, *Alternaria*, *Curvularia*, and *Cladosporium* are important field fungi as they require high moisture content for their growth and mycotoxin synthesis while *Penicillium* and *Aspergillus* are important storage fungi that can also grow well at lower moisture contents (Tsitsigiannis et al., 2012). During preharvest level *Fusarium* and *Alternaria* pose a high mycotoxicological risk whereas *Aspergillus* and *Penicillium* represent higher risk in stored food and feed products (Logrieco et al., 2003; Sreenivasa et al., 2013). These fungi, when growing on stored grains, can reduce the germination along with loss of carbohydrate, protein, and oil content; increase the moisture and free fatty acid content; and also bring about several other biochemical changes (Wilson and Abramson, 1992) (Fig. 9.1).

Mycotoxins produced by mycotoxigenic fungi receive the most attention as they are a potential carcinogen of global concern since they are the common contaminants of cereals and cereal-based foods (González et al., 1997; Navi, 2005; Nagaraja et al., 2016a,b). Approximately, 400 secondary metabolites with toxigenic potential are produced by more than 100 fungi and the Food and Agriculture Organization (FAO) has estimated that 25% of the world's agricultural commodities have been contaminated with mycotoxins, leading to significant economic losses. Fumonisins, aflatoxins, and ochratoxin receive highest attention among the mycotoxins produced by *Fusarium*, *Aspergillus*, and *Penicillium* species. Cereal samples collected from different districts of Karnataka were associated with 103 *Fusarium* isolates with the majority of 64 isolates infected with fumonisin producing *Fusarium verticillioides* (Sreenivasa et al., 2008). Consumption of such mycotoxin contaminated cereals mainly affects or targets liver, lung, and kidney in animals and causes skin lesions, wounds, and life-threatening cancer to humans (Deepa and Sreenivasa, 2017b; Sreenivasa, 2012). The International Agency for Research on Cancer (IARC) evaluated the toxin as a human carcinogen (Fandohan et al., 2003) (Fig. 9.2). However, isolated mycotoxigenic fungi and its mycotoxins have been confirmed using polymerase chain reaction and also by many recent methods such as nested polymerase chain reaction (PCR) and multiplex PCR for the confirmation of mycotoxins associated in cereals and plant parts (Sreenivasa et al., 2006; Dass et al., 2009; Deepa et al., 2016b,c). Further confirmation of produced mycotoxin was carried out using chromatographic methods for evading the misleading of mycotoxin synthesized from toxigenic strain but not of atoxigenic strains (Sreenivasa et al., 2012; Deepa et al., 2018).

9.3 Control of mycotoxigenic fungi

Plant pests seem to be an important biotic agent causing solemn losses to agricultural products and need to be meticulous to ensure cereal and cereal-based food products production quantitatively and qualitatively (Deepa and Sreenivasa, 2018). Earlier many physical methods like gamma irradiation methods and application of fungicides were used for the management of mycotoxigenic fungi and their toxins (Sreenivasa et al., 2009). Currently various strategies have been employed for the management of plant pests (Heydari, 2007). Usually plant diseases are controlled by the use of chemical pesticides. Excessive use of agrochemicals has led to environmental pollution leading to undesirable effects on nontarget organisms and possible carcinogenicity of chemicals. Initiatives have been in progress to change people's attitudes towards the use of pesticides in agriculture and strict regulations on chemical pesticide usage, and political pressure has been applied to eliminate the hazardous chemicals from the market (Heydari and Pessarakli, 2010).

Currently, the idea of controlling fungal pathogens with chemical fungicides or pesticides is being challenged by the biological control approach, which plays an important role in sustainable agriculture (Huwig et al., 2001). Biological control of plant diseases has become a viable substitute method for the management of plant diseases by inhibiting the growth or infection or reproduction of one organism using another organism (Baker, 1987). Biocontrol strategies has become environmentally safe and also is the only possible way left available for the effective management of mycotoxigenic fungi to protect plants, cereals, and cereal-based products (Prasad et al., 2010).

9.4 Strategy for biocontrol of *Fusarium* species

Among the mycotoxigenic fungi, *Fusarium* are the most common fungi associated with cereals all over the world (Munkvold and Desjardins, 1997). A total of 44 sorghum samples were analyzed for postharvest phytopathogenic fungi and reported majority of *Fusarium* (93.2%) and *Aspergillus* (88.6%) species (Sreenivasa et al., 2010). Marasas et al. (1984) estimated that approximately 20 species of *Fusarium* can occur on cereal grains. Nearly 64 mycotoxigenic *Fusarium* species were found to be associated with sorghum samples collected from different districts of Karnataka (Sreenivasa et al., 2007). Among the *Fusarium* species, *F. verticillioides* is the most frequently isolated species from cereals (Da Silva et al., 2004). *F. verticillioides* is primarily a pathogen mainly associated with maize and other crops like sorghum and has been largely responsible for economic losses worldwide (Deepa and Sreenivasa, 2017a).

The *Fusarium* species causes various diseases among the crops; Fusarium head blight and *Fusarium* ear rot are the major diseases affecting maize and wheat throughout the world. Species of *Fusarium* produces variety of mycotoxins namely fumonisins (FB1, FB2, FB3), trichothecenes (deoxynivalenol, T-2 toxin, nivalenol), and zearalenone (F-2 toxin). Rocha et al. (2011) collected maize grains from four different regions of Brazil reporting high frequency (96%) of *F. verticillioides*. Many researchers have been reported high incidence of tricothecene and fumonisin producing *Fusarium* spp. associated with maize and sorghum and its control of contamination (Sreenivasa et al., 2010, 2011b, 2013). Among 135 cereal samples from different districts of Karnataka, 69 samples were associated with phytopathogenic *Fusarium* infection in which 51 samples were associated with *F. verticillioides* (Deepa et al., 2016a). Biocontrol of *Fusarium* infection, growth, and its mycotoxins production plays a vital role and several trials have been conducted to effectively control it among cereal crops with different strategies using other bacteria and fungi as biocontrol agents, probiotics, endophytes, plant growth promoting rhizobacteria (PGPR), nontoxigenic strains, and plant products.

9.4.1 Plant growth promoting rhizobacteria and bacterial endophytes

These microbes serve as an efficient biocontrol agent by not only inhibiting the plant pathogen but also by promoting the growth of host plants (Lodewyckx et al., 2002; Rai et al., 2007; Hasegawa et al., 2006). *Bacillus amyloliquefaciens*, *Microbacterium oleovarus*, and *Enterobacter hormacchei* reduced the growth of *F. verticillioides* by acting as antagonists and its fumonisin production by improving the maize grain quality (Pereira et al., 2010). *Paenibacillus polymyxa* and *Citrobacter* detoxified deoxynivalenol (DON) produced by *Fusarium* species in vitro (Mousa et al., 2015). *Bacillus subtilis* reduced 50% of FB1 accumulation through vertical transmission and occupies the same ecological niche as *F. verticillioides* (Bacon et al., 2001). Antifungal efficacy of *Azotobacter nigricans* against trichothecene producing *Fusarium* species such as *F. sporotrichioides*, *F. graminearum*, *F. poae*, and *F. equiseti* showed a significant reduction (50%) of *Fusarium* infection incidence in all the treated cereals (Nagaraja et al., 2016a,b). Significant increase in plant growth, germination percentage, and mass of the maize seedling was observed on application of *Azotobacter* and *Azospirillum* (Biari et al., 2008). A total of 51 *Azotobacter* isolates were tested against *F. oxysporum* and *Rhizoctonia* spp. on cereal samples (Chauhan et al., 2012) (Fig. 9.3C).

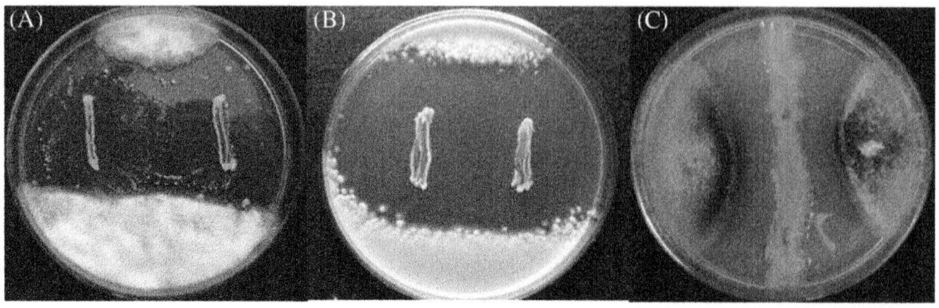

FIGURE 9.3 (A) *Enterococcus faecium* inhibiting *Fusarium verticillioides*. (B) *Lactobacillus plantarum* MYS6 inhibiting *Fusarium proliferatum*. (C) *Azotobacter* species inhibiting *F. verticillioides* and *Fusarium graminearum* by overlay and dual culture method, respectively.

The *Enterobacter cloacae* were reported to be the best biocontrol agents against root colonization against *F. verticillioides* with maize, whereas bacterial mixtures of *E. cloacae*, *Microbacterium oleovorans*, *Pseudomonas solanacearum*, and *B. subtilis*, respectively, exhibited reduction and prevention of *F. verticillioides* root colonization in maize seed inoculants (Cavaglieri et al., 2005b; Hinton and Bacon, 1995). *B. subtilis* proves to be a potent biocontrol agent in reducing the colonization of *F. verticillioides* at root level during greenhouse trials (Cavaglieri et al., 2005a). Application of *Bacillus* and *Pseudomonas* species on lentil plants controlled the disease Fusarium wilt (Akhtar and Siddiqui, 2008). *Arthrobacter globiformis*, *P. solanacearum*, and *Azotobacter armeniacus* exhibited antifungal activity on reduction of *F. verticillioides* growth and FB1 production in vitro (Cavaglieri et al., 2004, 2005b).

9.4.2 Bacterial and fungal species other than plant growth promoting rhizobacteria

Bacterial and fungal species act as the best biocontrol strategy for inhibition of mycotoxigenic *Fusarium* species. *Bacillus amyloliquifaciens* BA-S13, *M. oleovorans* DMS 16091, *Enterobacter hormomaechei* EM-562T, and *Kluyyeromyces* spp. L14 and L16 reduced the growth of toxigenic *F. verticillioides* associated on maize ears (Etcheverry et al., 2009). The *B. amyloliquifaciens* inhibited the infection of *F. verticillioides* association in maize root seedlings without changing antioxidant response in seedlings (Pereira et al., 2011). *F. graminearum* was inhibited by the bacterial isolates *B. subtilis* H-08-02, *B. cereus* L-07-01 and *B. mycoides* S-07-01 by 60%, 52%, and 55%, respectively, on wheat (Fernando et al., 2002). *Pseudomonas fluorescens* strains MKB 158/MKB 249, *Pseudomonas* sp. strain AS64.4, and *Pseudomonas frederiksbergensis* strain 202 reduced the *Fusarium* head blight disease symptoms caused by *F. culmorum* in wheat and barley reporting in reduction of 74%–78% of DON mycotoxin content (Khan and Doohan, 2009; Schisler et al., 2006). The *F. graminearum*, *F. culmorum*, and *F. avenaceum* are mainly associated with *Fusarium* head blight disease associated with wheat and other cereals in the Mediterranean region causing significant losses in reduction to yield and quality of grains (Logrieco et al., 2003).

The *Trichoderma* species determines the promising results in postharvest control of the growth and toxin accumulation from *F. moniliforme* in corn during storage (Yates et al., 1999b). *Trichoderma viridae* is mainly used in biofertilizers as a biocontrol agent for *F. verticillioides* on maize crops (Bacon et al., 2001). *T. gamsii* 6085 is found to be an effective antagonist against *F. culmorum* and *F. graminearum* and reduced its DON production by 92% (Matarese et al., 2012). Complete reduction of DON produced by *F. graminearum* was reported on application of *B. subtilis* RC218 and *Brevibacillus* spp. RC 263 (Palazzini et al., 2016). Two strains of *P. fluorescens* and one strain of *P. frederiksbergensis* on application against *F. culmorum* and DON production was reduced by 12% and 21%, respectively, on wheat and barley (Khan and Doohan, 2009). Two isolates of *F. graminearum*, *F. culmorum* and production of their zearalenone was reduced by 96% on application of *Trichoderma* species and *Clonostachys* in vitro (Gromadzka et al., 2009).

9.4.3 Probiotic bacteria

Being beneficial organisms they act as an antifungal biocontrol strategy for *Fusarium* species and its mycotoxin production. *Pediococcus pentosaceus* as biocontrol agent and its supernatant were tested for antifungal activity with pH dependence observed at the end phase of growth inhibiting *F. verticillioides* and *Fusarium proliferatum* growth (Dalie et al., 2010). *Saccharomyces cerevisiae* acts as an efficient biocontrol agent on controlling the growth of *F. moniliforme* as well as FB1 production in seeds (Fareid, 2011). *Lactobacillus rhamnous* and *S. cerevisiae* inhibited the growth of *F. moniliforme* in vitro and eliminated FB1 from the body of mature rats in vivo by orally administering the doses of biocontrol agents (Al-Masri et al., 2011). The antifungal activity of efficient probiotic *L. plantarum* MYS6 against *F. proliferatum* MYS9 inhibited its growth and fumonisin production by in vitro assays and its continuation study on

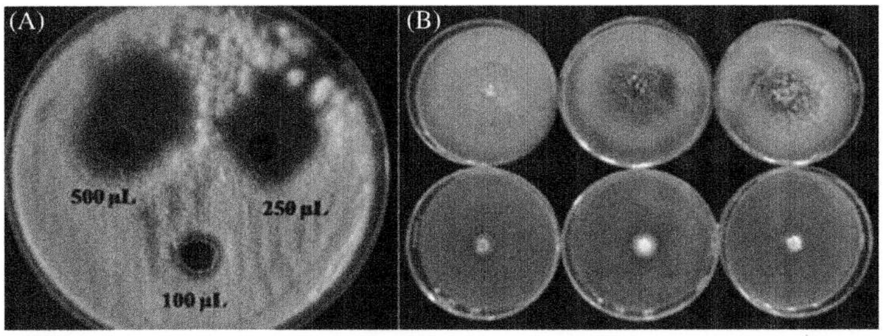

FIGURE 9.4 Antifungal activity exhibiting inhibition of *Fusarium* species. (A) Increase in inhibition of *Fusarium proliferatum* growth on increase in concentration of *Lactobacillus* species cell free supernatant (CFS). (B) Complete inhibition of three strains of *Fusarium verticillioides* on treating with methanol extract of *Prosopis juliflora*.

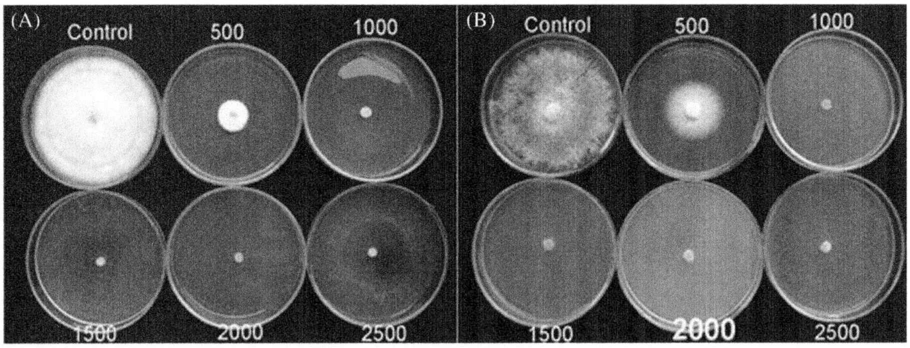

FIGURE 9.5 Antifungal activity of essential oils on mycelial growth of *Fusarium* species. (A) Effect of citronella oil on the growth of *Fusarium graminearum*. (B) Effect of lemon grass oil on the growth of *Fusarium anthophilum*.

in vivo reported the ameliorating effects of *L. plantarum* MYS6 on FB1 induced toxicity in vital organs and oxidative stress in broilers caused by *F. proliferatum* (Deepthi et al., 2016, 2017). Three strains of yeast *Cryptococcus* sp. on application to several field experiments reduced the DON level by 50–60 on wheat produced by *F. graminearum* (Schisler et al., 2014) (Fig. 9.3A and B and Fig. 9.4A).

9.4.4 Plant extracts and plant-based products

Like essential oils, phenolic compounds serve as an efficient biocontrol strategy for mycotoxigenic fungi. *F. verticillioides* M7075 were assessed and fumonisin production was inhibited by constituents of essential oils like *Origanum vulgare*, *Aloysia triphylla*, *Aloysia polystachya*, and *Mentha piperita* isolated from aromatic plants (Soliman and Badeea, 2002; Sokmen et al., 2004; Velluti et al., 2004b). Aqueous extract, methanol extract, and alkaloid extract of *Prosopis juliflora* plant inhibited more than 50% and 65% (400 $\mu g\ mL^{-1}$) and complete inhibition (300 $\mu g\ mL^{-1}$) of *F. verticillioides* strains, respectively (Deepa et al., 2012). Growth of *F. verticillioides*, *F. proliferatum*, and *F. graminearum* were used as a test fungus to check their inhibitory activity on screening for 37 essential oils among which cinnamon, clove, lemon grass, palmarosa, and oregano exhibited the best antifungal activity (Velluti et al., 2004a). Essential oils of clove, cedarwood, *Cymbopogon* species, peppermint, eucalyptus, and neem were tested for their efficacy against nine *Fusarium* species associated with maize and sorghum among which the oil *Cymbopogon nardus* (citronella) inhibited all the tested *Fusarium* species growth at 500 ppm (Sreenivasa et al., 2011c) (Fig. 9.4B; Fig. 9.5).

Recently, *Piriformo sporaindica* proved its ability to reduce disease caused by *F. graminearum* and its DON production in wheat by 70%–80% (Rabiey and Shaw, 2016). Butylated hydroxyanisole (BHA) and polypropylene (PP) has a good potential activity in minimizing the growth of *F. verticillioides* and *F. proliferatum* and its fumonisin production at various water activity and incubation period in vitro (Etccheverry et al., 2002). Combined effect of antioxidants as BHA/PP (500 $\mu g\ g^{-1}$), aw (0.95) of 10 to 100-fold treated for 28 days against *F. verticillioides* observed 77% reduction in log colony forming units (CFU) (Farnochi et al., 2005). Phenolic compounds vanillic acid and caffeic acid exhibited decrease in growth of *F. verticillioides* and *F. proliferatum* and FB1 production with increase in phenolic compound concentration at different water activities in maize (Samapundo et al., 2007) (Table 9.1).

9.4.5 Atoxigenic fungal strains

Atoxigenic strains are the strains of mycotoxigenic fungi without the production of toxins in it. In recent days nonpathogenic species are applied for the biocontrol of pathogenic strains (Liu et al., 2013). Two atoxigenic strains of *F. equiseti*

TABLE 9.1 Strategies on reducing the growth of *Fusarium* species and its toxin synthesis.

S. no.	Biocontrol strategies	Biocontrol of *Fusarium* species on application of	References
1.	PGPR	*Azotobacter nigricans, Azospirillum, Azotobacter* species, *Bacillus subtilis, P. solanaseacum, Arthrobacter globiformis, A. armeniacum*	Nagaraja et al. (2016a); Biari et al. (2008); Chauhan et al. (2012); Cavaglieri et al. (2004, 2005a,b); Hinton and Bacon (1995); Akhtar and Siddiqui (2008)
2.	Endophytes	*Enterobacter cloacae, M. oleovoraus, B. amyloliquifaciens, E. hormacchei, Paenibacillus polymyxa, Citrobacter* species	Pereira et al. (2010); Mousa et al. (2015); Bacon et al. (2001)
3.	Bacteria, fungi, and algae	*Baciilus amyloliquifaciens, B. subtilis, M. oleovorans, E. hormomochaie, B. mycoides, B. cereus, P. flourescens, P. frederiksbergensis, Brevi bacillus, T. viridae, T. gamsii* 6085, *Clonastachys, Trichodrma* sp.	Etcheverry et al. (2009); Pereira et al. (2011); Fernando et al. (2002); Khan and Doohan (2009); Schisler et al. (2006); Palazzini et al. (2016); Yates et al. (1999b); Bacon et al. (2001); Gromadzka et al. (2009); Matarese et al. (2012)
4.	Plant extracts and plant products	Essential oils: *Origanum vulgare, Alopiatri plulla, Aloysia polytachya, Mentha piperita*, clove, cinnamon, lemon grass, palmarosa, cedarwood, cynbopogo, peppermint, eucalyptus, neem, methanol, aqueous and alkaloid extract from *Prosopis juliflora, Piriformopora indica* plant, BHA and PP phenolic compounds, vanillic acid, caffeic acid	Soliman and Badeea (2002); Sokmen et al. (2004); Velluti et al. (2004a,b); Deepa et al. (2012); Sreenivasa et al. (2011c); Rabiey and Shaw (2016); Etcchevery et al. (2002); Farnochi et al. (2005); Samapundo et al. (2007)
5.	Atoxigenic strains	*F. equiseti, F. verticillioides, Fusarium* species	Dawson et al. (2004); Luongo et al. (2005); Liu et al. (2013)
6.	Probiotics	*P. pentosaceus, S. cerevisiae, L. rhamnous, L. plantarum* MYS6, *Cryptococcus* species	Dalie et al. (2010); Fareid (2011); Al-Masri et al. (2011); Deepthi et al. (2016, 2017); Schisler et al. (2014)

reported the reduced production of DON level by 70%–94% produced by *F. culmorum* and *F. graminearum* (Dawson et al., 2004). Luongo et al. (2005) suppressed the toxigenic *F. verticillioides* and *F. proliferatum* strains on application of nonpathogenic *Fusarium* strains (Table 9.1).

9.5 Strategies for biocontrol of *Aspergillus* species

Species of *Aspergillus* such as *Aspergillus flavus* and *A. parasiticus* infect both cereals and grains such as groundnuts, maize, cottonseed, soybean, and nuts in the field or during storage along with the production of secondary toxic metabolites such as aflatoxins (Horn and Dorner, 2009, Bayman et al., 2003). *A. flavus* is found to be highly invasive and more widespread when compared with *A. parasiticus* (Okwu et al., 2011). Peanuts, maize, and pistachios being the major crops of the Mediterranean area have been contaminated with *A. flavus*. In 2003 an *A. flavus* outbreak led to contamination of milk among dairy cattle that consumed contaminated maize (Logrieco and Moretti, 2008). Among a total of 130 maize and sorghum samples 10 diversified species of *Fusarium* and *Aspergillus* were isolated and similarly among 80 maize samples collected throughout Karnataka screened high frequency of (96.5%) *Fusarium* and (41.7%) *Aspergillus* species including species of *Penicillium, Drechslera, Nigrospora, Curvularia, Alternaria, Chaetomium,* and *Phomopsis* were isolated (Sreenivasa et al., 2011a,b). Mycotoxigenic *A. flavus* strains are reported to produce aflatoxin (AFB1, AFB2) whereas *A. parasiticus* produces AFB1, AFB2, AFG1, and AFG2 (Perrone et al., 2014). Reduction of AFs among different cereals has been widely performed applying different strategies such as nontoxigenic strains, probiotics, plant and plant products, different bacteria and fungi.

9.5.1 Atoxigenic fungal strains

These microbes have been extensively used as a biocontrol strategy. Atoxigenic strains of *A. flavus* or *A. parasiticus* reduced the aflatoxin contamination in peanuts (Dorner et al., 2003), paddy, and maize (Dorner et al., 1999). Brown et al. (1991) reported a nonaflatoxigenic strain of *A. flavus* that reduced aflatoxin contamination by 80%−95% in maize. Two atoxigenic *A. flavus* strains, AF36 and NRRL 21882, reduced the aflatoxin contamination in peanut (Chang and Hua, 2007). In the United States nontoxigenic strains CT3 and K49 of *A. flavus* were reported as efficient strains in reduction of aflatoxin contamination among corn (Abbas et al., 2006). Nontoxigenic strain BN30 in Africa served to be as effective aflatoxin reducing strain in maize (Cardwell and Henry, 2004). In Nigeria, two atoxigenic strains La3279 and La303 were applied successfully in maize field trials for the control of AFB1 and AFB2 up to 99.9% (Atehnkeng et al., 2008) (Table 9.2).

9.5.2 Plant products

These products served as antifungals for the biocontrol of mycotoxigenic fungi (Soliman and Badeea, 2002). Seven essential oils, namely cinnamon, peppermint, basil, origanum, epazote, clove, and thyme, were screened against

TABLE 9.2 Strategies on reducing the growth of *Aspergillus* species and its toxin synthesis.

Biocontrol strategies	Inhibition of *Aspergillus* species	References
Bacillus subtilis RCB 90	*Aspergillus* species AFB1	Bacon et al. (2001); Nesci et al. (2005); Line et al. (1994)
Bacillus amyloliquifaciens	*Aspergillus flavus*	Etcheverry et al. (2009)
Microbacterium oleovarans		
Enterobacter hormacchei		
Lactobacillus species	*A. fumigatus*, AFB1	Al-Masri et al. (2011); Ilavenil et al. (2015); Karlovosky (1999)
Lactobacillus casei	*Aspergillus parasiticus*	El-Gendy and Marth (1981)
Streptococcus lactis	*A. flavus*	Coallier-Ascah and Idziak (1985)
Lactobacillus sanfrancisco	*Aspergillus* spp.	Kabak et al. (2006)
Flavobacterium aurentiacum	*A. flavus*, AFB1	Line et al. (1994); D'Souza and Brackett (2000)
Lactobacillus plantarum MYS 44	*A. parasiticus*	Poornachandra Rao et al. (2017)
Trichoderma viridae	*A. flavus*	Calistru et al. (1997); Choudhary (1992); Bacon et al. (2001); Larkin and Fravel (1998); Yates et al. (1999b)
Trichoderma harzianum	*A. flavus*	
Pseudomonas species *Burkholderia*	*Aspergillus* species, *Aflatoxin*	Logrieco et al. (2003)
Bacillus pumilis	*Aspergillus* species	Munimbazi and Lloyd (1998)
Shewanella algae	Aflatoxin synthesis inhibition	Gong et al. (2015)
Candida parapilosis IP1698, *C. krusei*, *Pichia anomoda*	*A. flavus* and aflatoxin	Niknejad et al. (2012); Masoud and Kaltoft (2006)
Phomopsis, *Rhizopus*, *Alternaria*	AFB1	Shantha (1999)
Essential oils: cinnamon, peppermint, basil, origanum, epazote, clove, thyme; medicinal plants	*A. flavus*, *A. parasiticus*, *A. ochraceus*, *A. niger*	Montes-Belmont and Carvajal (1998); Yoousef et al. (2016); Viuda-Martos et al. (2007); Soliman and Badeea (2002)
A. flavus AF36, NRRL 21882, CT3, K49, BRC 30, La3279, La303, *A. parasiticus*	*A. flavus*, *A. parasiticus*, AFB1, AFB2	Dorner et al. (1999, 2003); Brown et al. (1991); Chang and Hua (2007); Abbas et al. (2006); Cardwell and Henry (2004); Atehnkeng et al. (2008)

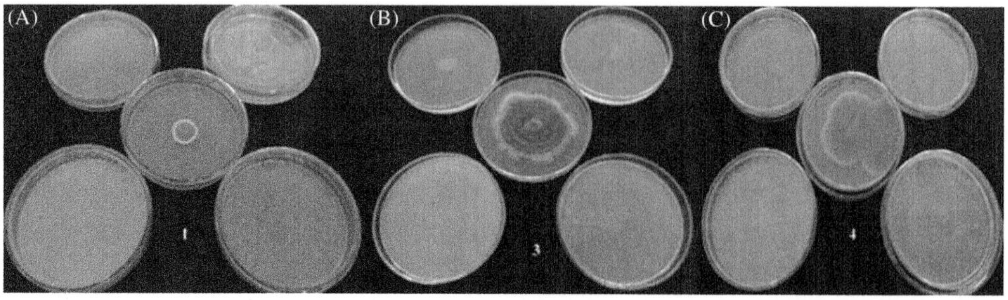

FIGURE 9.6 Antifungal activity of essential oil against test pathogens. (A) *Penicillium digitatum*, (B) *Aspergillus fumigates*, (C) *Aspergillus parasiticus*. Note: Increasing concentration in clockwise direction 0.25, 0.5, 0.75, 1 $\mu L\ mL^{-1}$ and control at the center.

A. flavus infection on maize kernels and the total inhibition of fungal development was found (Montes-Belmont and Carvaial, 1998). Among 17 essential oils screened for the antifungal activity on *Aspergillus parasiticus* by agar dilution method, the pathogen was found sensitive to cinnamon (Yooussef et al., 2016). A total of 12 medicinal plant essential oils were tested for inhibitory activity against *Aspergillus* species and among them thyme and cinnamon found complete inhibition of test fungi *A. flavus*, *A. parasiticus*, *A. ochraceous*, and *F. moniliforme* at a dose of 500 ppm (Solima and Badeea, 2002). *A. flavus* was more sensitive to thyme oil than *A. niger* when tested for the antifungal potential of essential oils of oregano, thyme, and clove against *Aspergillus* species (Viuda-Martos et al., 2007) (Fig. 9.6B and C).

9.5.3 Other microorganisms such as bacteria, fungi, and algae

These play a vital role in reduction of growth and aflatoxin production. Volatile compounds such as dimethyl trisulfide and 2,4-bis, phenol isolated from *Shewanella* algae YM8 reported complete inhibition on aflatoxin synthesis in maize and peanuts on storage at different water activities (Gong et al., 2015). Earlier *B. subtilis* RCB 90 was reported to have complete inhibition of AFB1 (Nesci et al., 2005). *Candida parapsilosis* IP1698 was found to inhibit 90%–99% of aflatoxin production at different pH and temperatures (Niknejad et al., 2012). *Candida krusei* and *Pichia anomala* served as a biocontrol agent against *A. flavus* inhibiting its growth and aflatoxin production (Masoud and Kaltoft, 2006).

Bacterial species *B. subtilis*, *Lactobacilli* spp., *Pseudomonas* spp., *Ralstonia* spp., and *Burkholderia* spp. reported its capability to inhibit fungal growth and aflatoxins production by *Aspergillus* spp. (Logrieco et al., 2003). A total of six *B. pumilus* strains reduced the growth of aflatoxin producing molds and its production through antagonism assays (Munimbazi and Lloyd, 1998). *B. subtilis* and *Flavobacterium aurantiacum* from groundnuts were found to inhibit the growth of *A. flavus* in groundnuts and removed the aflatoxin production from various foods (Line et al., 1994). Culture filtrates of *Trichoderma harzianum* and *T. viride* reported the capability of inhibiting the growth and its aflatoxin production of *A. flavus* (Calistru et al., 1997). Another strain of *T. viride* inhibits the growth of *A. flavus* and production of AFB1 by 73.5% and AFG1 by 100% (Choudhary, 1992). Shantha (1999) reported the degradation of Aflatoxin B1 by 65%–99% using species of *Trichoderma*, *Phoma*, *Rhizopus*, *Sporotrichium*, and *Alternaria* (Table 9.2).

9.5.4 Lactic acid bacteria

The antifungal properties of lactic acid bacteria were tested against *Aspergillus* species. Aflatoxin production by *A. flavus* was reduced by *Streptococcus lactis* and *Leuconostoc mesenteroides* (Coallier-Ascah and Idziak, 1985). Gourama and Bullerman (1997) reported *Lactobacillus casei*, *L. pseudoplantarum* inhibiting the production of AFB1 by 80% and AFG1 by 92%. *Lactobacillus sanfrancisco* CBI inhibited the spoilage molds of the genus *Maonilia*, *Aspergillus*, *Penicillium*, and *Fusarium* (Kabak et al., 2006). *L. casei* inhibited the growth and aflatoxin production in *A. parasiticus* (El-Gendy and Marth, 1981). Antifungal properties of *lactobacillus* strains against *A. fumigatus*, *Penicillium chrysogenum*, *Penicillium roqueforti*, *Botrytis elliptica*, and *Fusarium oxysporum* were reported (Ilavenil et al., 2015). Presence of Aflatoxin B1 in milk was degraded by using some strains of *Latobacillus*, *Streptococcus*, and *Bifidobacterium* (Karlovosky, 1999). Cell-free supernatant of *L. plantarum* strain MYS44 was tested for in vitro assays against the growth and aflatoxin production by *A. parasiticus* MTCC 411 (Poornachandra Rao et al., 2017). *F. aurantiacum* has been significantly eliminated AFB1 from corn, peanuts, milk, and other food products (D'Souza and Brackett, 2000).

9.6 Strategy for biocontrol of *Penicillium* and *Alternaria* species

Penicillium produces mycotoxins ochratoxin, patulin, citrinin, penicillic acid, and PR-toxin, and mainly on association with vegetables, fruits, spices, grains rather than on cereals. *Penicillium* strains *P. martensii*, *P. cyclopium*, and *P. frequentans* were mainly associated with wheat flour samples (Doolotkeldieva, 2010). Yeast *Pichia anomola* J121 reduced the infection of mold *P. roqueforti* on airtight storage of wheat (Druvefors et al., 2005). In many parts of the world, high frequency of *Alternaria* species was reported in association with wheat, barley, oats, and rye (Li and Yoshizawa, 2000; Logrieco et al., 1990). PGPR especially *P. fluorescens* and *Bacillus* species play a vital role as potent biocontrol plant pathogens as they improve plant growth by colonizing in its root system (Schroth and Haneock, 1982). Among six potential isolates screened from wheat fields, B28 strain was found to be best in improving the growth of wheat on *Alternaria triticina* inoculated plants (Siddiqui, 2007). *T. harzianum* was reported to be a potential biocontrol agent against *A. alternata* under different environmental conditions such as water activity and temperature (Sempere and Santamarina, 2007) (Fig. 9.6A).

9.7 Conclusions

Even though a variety of chemical, physical, and biological methods have been developed to minimize the growth of mycotoxigenic fungi and production of mycotoxins at pre- or postharvest stages, complete eradication of mycotoxin contaminated foods and feeds is not at all convincingly attainable. In addition to biocontrol methods, management tools and procedures have to be followed that facilitate the effective preservation of stored commodities with minimum loss in quality and quantity. Currently, research studies need to practice control measures to develop substantial strategies that respond to the effects of mycotoxigenic fungi and prevent their mycotoxin production in human and animal food chains. Recent developments in the use of biocontrol strategies have led to registration of commercial products with increased practical application for the benefit of growers in several countries.

References

Abbas, H.K., Zablotowicz, R.M., Bruns, H.A., Abel, C.A., 2006. Biocontrol of aflatoxin in corn by inoculation with nonaflatoxigenic *Aspergillus flavus* isolates. Biocontrol Sci. Technol. 16, 437–449.

Akhtar, M.S., Siddiqui, Z.A., 2008. Biocontrol of root rot disease complex of chicken pea by *Glomus intradadices*, *Rhizobium sp.* and *Pseudomonas straita*. Crop Prot. 410–417.

Al-Masri, S.A., El- Safty, S.M.S., Nada, S.A., Amra, H.A., 2011. *Saccharomyces cerevisiae* and probiotic bacteria potentially inhibit fumonisin B1 production in vitro and in vivo. J. Am. Sci. 7, 198–205.

Atehnkeng, J., Ojiambo, P.S., Ikotun, T., Sikora, R.A., Cotty, P.J., Bandyopadhyay, R., 2008. Evaluation of atoxigenic isolates of *Aspergillus flavus* as potential biocontrol agents for aflatoxin in maize. Food Add. Contamin. A: Chem. Anal. Control Expo Risk Assess 25, 1264–1271.

Bacon, C.W., Yates, I.E., Hinton, D.M., Meredith, F., 2001. Biological control of *Fusarium moniliforme* in maize. Environ. Health Perspect. 109, 325–332.

Baker, K.F., 1987. Evolving concepts of biological control of plant pathogens. Annu. Rev. Phytopathol. 25, 67–85.

Bayman, P., Baker, J.L., Mahoney, N.E., 2003. *Aspergillus* on tree nuts: incidence and associations. Mycopathologia 155, 161–169.

Biari, A., Gholami, A., Rahmani, H.A., 2008. Growth promotion and enhanced nutrient uptake of maize (*Zea mays* L.) by application of plant growth promoting rhizobacteria in arid region of Iran. J. Biol. Sci. 8, 1015–1020.

Brown, R.L., Cotty, P.J., Cleveland, T.E., 1991. Reduction in aflatoxin content of maize by atoxigenic strains of *Aspergillus flavus*. J. Food Prot. 54, 623–626.

Calistru, C., McLean, M., Berjak, P., 1997. In vitro studies on the potential for biological control of *Aspergillus flavus* and *Fusarium moniliforme* by *Trichoderma* species: a study of the production of extracellular metabolites by *Trichoderma* species. Mycopathologia 137, 115–121.

Cardwell, K.F., Henry, S.H., 2004. Risk of exposure to and mitigation of effect of aflatoxin on human health: a West African example. J. Toxicol. Toxin Rev. 23, 17–247.

Castoria, R., Wright, S.A.I., Droby, S., 2008. Biological control of mycotoxigenic fungi in fruits. In: Barkai-Golan, R., Paster, N. (Ed.), Mycotoxins in Fruits and Vegetables, Academic Press, pp. 311–333. ISBN 9780123741264, https://doi.org/10.1016/B978-0-12-374126-4.00016-4.

Cavaglieri, L., Passone, A., Etcheverry, M., 2004. Screening procedures for selecting rhizobacteria with biocontrol effects upon *F. vericillioides* growth and fumonisin B1 production. Res. Microbiol. 155, 747–754.

Cavaglieri, L., Orlando, J., Etcheverry, M., 2005b. In vitro influence of bacterial mixtures of *F. verticillioides* growth and fumonisin B1 production. Effect of seed treatment on maize root colonization. Lett. Appl. Microbiol. 41, 390–396.

Cavaglieri, L., Orlando, J., Rodríguez, M.I., 2005a. Biocontrol of *Bacillus subtilis* against *Fusarium verticillioides* in vitro and at the maize root level. Res. Microbiol. 156, 748–754.

Chang, P.K., Hua, S.S., 2007. Nonaflatoxigenic *Aspergillus flavus* TX9-8 competitively prevents aflatoxin accumulation by *A. flavus* isolates of large and small sclerotial morphotypes. Int. J. Food Microbiol. 114, 275–279.

Chauhan, S., Wadhwa, K., Vasudeva, M., Narula, N., 2012. Potential of *Azotobacter* spp. as biocontrol agents against *Rhizoctoniasolani* and *Fusarium oxysporum* in cotton (*Gossypium hirsutum*), guar (*Cyamopsistetragonoloba*) and tomato (*Lycopersicumesculentum*). Arch. Agron. Soil Sci. 58, 1365–1385.

Choudhary, A.K., 1992. Influence of microbial co-inhabitants on aflatoxin synthesis of *Aspergillus flavus* on miaze kernels. Lett. Appl. Microbiol. 14, 143–147.

Coallier-Ascah, E.S., Idziak, J., 1985. Interaction between *Streptococcus lactis* and *Aspergillus flavus* on production of aflatoxin. Appl. Environ. Microbiol 49, 163–167.

D'Souza, D.H., Brackett, R.E., 2000. The influence of divalent cations and chelators on aflatoxin B1 degradation by *Flavobacterium aurantiacum*. J. Food Prot. 63, 102–105.

Da Silva, J.B., Dilkin, P., Fonseca, H., Correa, B., 2004. Production of aflatoxin by *Aspergillus flavus* and fumonisin by *Fusarium* species isolated from Brazilian sorghum. Braz. J. Microbiol. 35, 182–186.

Dalie, D.K., Deschamps, A.M., Atanasova-Penichon, V., Richard-Forget, F., 2010. Potential of *Pediococcuspentosaceus* (L006) isolated from maize leaf to suppress fumonisin-producing fungal growth. J. Food Prot. 73, 1129–1137.

Dass, R.S., Sreenivasa, M.Y., Janardhana, G.R., 2007. High incidence of *Fusarium verticillioides* in animal and poultry feed mixtures produced in Karnataka, India. Plant Pathol. J. 6 (2), 174–178.

Dass, R.S., Sreenivasa, M.Y., Charith Raj, A.P., Janardhana, G.R., 2009. PCR-based assay for the rapid detection of Fumonisin-producing *Fusarium* species in Maize-based animal and poultry feeds in Karnataka, India. Arch. Phytopathol. Plant Prot. 42 (8), 796–804.

Dawson, W.A.A.J.M., Jestoi, M., Rizzo, A., Nicholson, P., Bateman, G.L., 2004. Field evaluation of fungal competitors of *F. culmorum* and *F. graminearum* casual agents of ear blight of winter wheat for the control of mycotoxin production in grain. Biocontrol Sci. Technol. 14, 783–799.

Deepa, N., Sreenivasa, M.Y., 2018. Biocontrol strategies for effective management of phytopathogenic fungi associated with cereals. In: Singh, J.S., Singh, D.P. (Eds.), Microbial Biotechnology in Agro-Environmental Sustainability. Elsevier Publishers.

Deepa, N., Nagaraja, H., Sreenivasa, M.Y., 2016a. Prevalence of fumonisin producing *Fusarium verticillioides* associated with cereals grown in Karnataka (India). FSHW 5, 156–162.

Deepa, N., Charith Raj, A.P., Sreenivasa, M.Y., 2016b. Multiplex PCR method for the early detection of fumonisin producing *Fusarium verticillioides*. Food Biosci. 13, 84–88.

Deepa, N., Charith Raj, A.P., Sreenivasa, M.Y., 2016c. Nested PCR method for the early detection of fumonisin producing *Fusarium verticillioides* in pure cultures, cereal samples and plant parts. Food Biotechnol. 30 (1), 18–29.

Deepa, N., Sreenivasa, M.Y., 2017a. *Fusarium verticillioides*, a globally important pathogen of Agriculture and Livestock: A review. J. Vet. Med. Res. 4, 1084.

Deepa, N., Sreenivasa, M.Y., 2017b. Fumonisins: a review on its global occurrence, epidemiology, toxicity and detection. J. Vet. Med. Res. 4, 1093.

Deepa, N., Rakesh, S., Sreenivasa, M.Y., 2018. Morphological, pathological and mycotoxicological variations among *Fusarium verticillioides* isolated from cereals. 3 Biotech 8, 105.

Deepa, N., Nayaka, C.S., Udaya Shankar, A.C., Krishna Kumar, V., Niranjana, S.R., Prakash, H.S., et al., 2012. Detection and management of seed-borne toxigenic *Fusarium verticillioides* by plant alkaloids. J. Mycol. Plant Pathol. 42, 161–166.

Deepthi, B.V., Poornachandra Rao, K., Chennapa, G., Naik, M.K., Chandrashekara, K.T., Sreenivasa, M.Y., 2016. Antifungal Attributes of *Lactobacillus plantarum* MYS6 against fumonisin producing *Fusarium proliferatum* associated with poultry feeds. PLoS One 11, e0155122.

Deepthi, B.V., Rakesh, S., Poornachandra Rao, K., Deepa, N., Dharanesha, N.K., Girish, K.S., et al., 2017. *Lactobacillus plantarum* MYS6 ameliorates fumonisin B1-induced hepatorenal damage in broilers. Front. Microbiol. 8, 2317.

Doolotkeldieva, T.D., 2010. Microbiological control of flour manufacturers. Dissemination of mycotoxins producing fungi in cereal products. Microbiol. Insights 3.

Dorner, J.W., Cole, R.J., Wicklow, D.T., 1999. Aflatoxin reduction in corn through field application of competitive fungi. J. Food Prot. 62, 650–656.

Dorner, J.W., Cole, R.J., Connick, W.J., Daigle, D.J., McGuire, M.R., Shasha, B.S., 2003. Evaluation of biological control formulations to reduce aflatoxin contamination in peanuts. Biol. Control 26, 318–324.

Druvefors, U.A., Passoth, V., Schnurer, J., 2005. Nutrient effects on biocontrol of *P. roqueforti* by *Pichia anomala* J121 during airtight storage of wheat. Appl. Environ. Microbiol. 71 (4), 1865–1869.

El-Gendy, S.M., Marth, E.H., 1981. Growth of aflatoxin production by *Aspergillus parasiticus* in the presence of *Lactobacillus casei*. J. Food Prot. 44, 211–212.

Etccheverry, M., Torres, A., Ramirez, M.L., Chulze, S., Magan, N., 2002. In vitro control of growth and fumonisin production by *Fusarium verticillioides* and *F. proliferatum* using antioxidants under different water availability and temperature regimes. J. Appl. Microbiol. 92, 624–632.

Etcheverry, M.G., Scandolara, A., Nesci, A., Boas Ribeiro, M.S.V., Pereira, P., Battilani, P., 2009. Biological interactions to select biocontrol agents against toxigenic strains of *Aspergillus flavus* and *Fusarium verticillioides* from maize. Mycopathologia 167, 287–295.

Fandohan, P., Hell, K., Marasas, W.F.O., Wingfield, M.J., 2003. Infection of maize by *Fusarium* species and contamination with fumonisins in Africa. Afr. J. Biotechnol. 2, 570–579.

Fareid, M.A., 2011. Biocontrol of *Fusarium* moulds and fumonisin B1 production. World Rural Observ. 3.

Farnochi, M.C., Torres, A.M., Magan, N., Chulze, S.N., 2005. Effect of antioxidants and competing mycoflora on *F. verticillioides* and *F. proliferatum* populations and fumonisin production on maize grain. J. Stored Prod. Res. 41, 211–219.

Fernando, W.G.D., Chen Y., Parks, P., 2002. Effect of three *Bacillus* sp. from wheat on FHB reduction. In: National Fusarium Head Blight Proceedings. U.S. Wheat and Barley Scab Initiative, East Lansing, MI, pp. 73–76.

Gong, A.D., Li, H.P., Shen, L., Zhang, J.B., Wu, A.B., He, W.J., et al., 2015. The *Shewanella* algae strain YM8 produces volatiles with strong inhibition activity against *Aspergillus* pathogens and aflatoxins. Front. Microbiol. 6, 1–12.

González, H.H.L., Martýnez, E.J., Resnik, S.L., 1997. Fungi associated with sorghum grain from Argentina. Mycopathologia 139, 35–41.

Gourama, H., Bullerman, L.B., 1997. Anti-aflatoxigenic activity of *Lactobacillus casei* subsp. *pseudoplantarum*. Int. J. Food Microbiol. 34, 131–143.

Gromadzka, K., Chelkowski, J., Popiel, D., Kachliki, P., Kostecki, M., Golinski, P., 2009. Solid substrate bioassay to evaluate the effect of *Trichoderma* and *Clonostachys* on the production of zearalenone by *Fusarium* species. World Mycotoxin J. 2, 45–52.

Hasegawa, S., Meguro, A., Shimizu, M., Nishimura, T., Kunoh, H., 2006. Endophytic actinomycetes and their interactions with host plants. Actinomycetologica 20, 72–81.

Heydari, A., 2007. Biological control of Turfgrass fungal diseases. In: Pessarakli, M. (Ed.), Turfgrass Management and Physiology. CRC Press, Boca Raton, FL.

Heydari, A., Pessarakli, M., 2010. A review on biological control of fungal plant pathogens using microbial antagonists. J. Biol. Sci. 10 (4), 273–290.

Hinton, D.M., Bacon, C.W., 1995. *Enterobacter cloacae* is an endophytic symbiont of corn. Mycopathologia 129, 117–125.

Horn, B.W., Dorner, J.W., 2009. Effect of nontoxigenic *Aspergillus flavus* and *A. parasiticus* on aflatoxin contamination of wounded peanut seeds inoculated with agricultural soil containing natural fungal populations. Biocontrol Sci. Technol. 19, 249–262.

Huwig, A.K., Freimund, S., Kappeli, O., 2001. Mycotoxin detoxification of animal feed by different adsorbents. Toxicol. Lett. 122, 179–188.

Ilavenil, S., Park, H.S., Vijayakumar, M., Valan Arasu, M., Kim, D.H., Ravikumar, S., et al., 2015. Probiotic potential of *Lactobacillus strains* with antifungal activity isolated from animal manure. Sci. World J. Available from: https://doi.org/10.1155/2015/802570.

Kabak, B., Dobson, A.D., Var, I.şl, 2006. Strategies to prevent mycotoxin contamination of food and animal feed: a review. Crit. Rev. Food Sci. Nut. 593–618.

Karlovosky, P., 1999. Biological detoxification of fungal toxins and it use in plant breeding, feed and food production. Nat. Tox. 7, 1–23.

Khan, M.R., Doohan, F.M., 2009. Bacterium-mediated control of *Fusarium* head blight disease of wheat and barley and associated mycotoxin contamination of grain. Biol. Control 48, 42–47.

Larkin, R.P., Fravel, R., 1998. Efficacy of various fungal and bacterial biocontrol organisms for control of *Fusarium* wilt of tomato. Plant Dis. 82, 1022–1028.

Li, F.Q., Yoshizawa, T., 2000. Alternaria mycotoxins in weathered wheat from china. J. Agric. Food Chem. 48 (7), 2920–2924.

Line, J.E., Brackett, R.E., Wilkison, R.E., 1994. Evidence for degradation of aflatoxin B1 by *Flavobacterium aurantiacum*. J. Food Prot. 57, 788–791.

Liu, J., Sui, Y., Wisniewski, M., Droby, S., Liu, Y., 2013. Review: utilization of antagonistic yeasts to manage postharvest fungal diseases of fruit. Int. J. Food Microbiol. 167, 153–160.

Lodewyckx, C., Vangronsveld, J., Porteous, F., Moore, E.R.B., Taghavi, S., Mezgeay, M., 2002. Endophytic bacteria and their potential applications. Crit. Rev. Plant Sci. 21, 583–606.

Logrieco, A.F., Moretti, A., 2008. Between emerging and historical problems: An overview of the main toxigenic fungi and mycotoxin concerns in Europe. In: Leslie, J.F., Bandyopadhyay, R., Visconti, A. (Eds.), Mycotoxins: Detection Methods, Management, Public Health, and Agricultural Trade. C.A.B. International, Wallingford, UK, pp. 139–154.

Logrieco, A., Bottalico, A., Solfizzo, M., Mule, G., 1990. Incidence of *Alternaria* species in grain from Mediterranean countries and their ability to produce mycotoxin. J. Mycol. 82 (4), 501–505.

Logrieco, A., Bottalico, A., Mule, G., Moretti, A., Perrone, G., 2003. Epidemiology of toxigenic fungi and their associated mycotoxins for some Mediterranean crops. Eur. J. Plant Pathol. 109 (7), 645–667.

Luongo, L., Galli, M., Corazza, L., Meekes, E., Hass, L., Plas, L.C., 2005. Potential of fungal antagonists for biocontrol of Fusarium species in wheat and maize through competition in crop debris. Biocontrol Sci. Technol. 15, 229–242.

Marasas, W.F.O., Kriek, N.P.J., Fincham, J.E., Van Rensburg, S.J., 1984. Primary liver cancer and oesophageal basal cell hyperplasia in rats caused by *Fusarium moniliforme*. Int. J. Cancer 34, 383–387.

Masoud, W., Kaltoft, C.H., 2006. The effects of yeasts involved in the fermentation of coffea arabica in East Africa on growth and ochratoxin A (OTA) production by *Aspergillus ochraceus*. Int. J. Food Microbiol. 106, 229–234.

Matarese, F., Sarrocco, S., Gruber, S., 2012. Biocontrol of Fusarium head blight: Interactions between *Trichoderma* and mycotoxigenic *Fusarium*. Microbiology 158, 98–106.

Montes-Belmont, R., Carvaial, M., 1998. Control of *Aspergillus flavus* in maize with plant essentials oils and their components. J. Food Prot. 61, 616–619.

Mousa, W.K., Shearer, C.R., Limay-Rios, V., 2015. Bacterial endophytes from wild maize suppress *Fusarium graminearum* in modern maize and inhibit mycotoxin accumulation. Front. Plant Sci. 6, 805.

Munimbazi, C., Lloyd, B.B., 1998. Inhibition of aflatoxin production of *Aspergillus parasiticus* NRRL 2999 by *Bacillus pumilus*. Mycopathology 140, 163–169.

Munkvold, G.P., Desjardins, A.E., 1997. Fumonisins in maize, can we reduce their occurrence? Plant Dis. 81, 556–565.

Nagaraja, H., Chennappa, G., Rakesh, S., Naik, M.K., Amaresh, Y.S., Sreenivasa, M.Y., 2016a. Antifungal activity of *Azotobacter nigricans* against trichothecene-producing *Fusarium* species associated with cereals. Food Sci. Biotechnol. 25 (4), 1197–1204.

Nagaraja, H., Chennappa, G., Poorna Chandra Rao, K., Mahadev prasad, G., Sreenivasa, M.Y., 2016b. Diversity of toxic and phytopathogenic Fusarium species occurring on cereals grown in Karnataka state, India. 3 Biotech. 6, 57.

Prasad, M.N.N., Shankara Bhat, S., Sreenivasa, M.Y., 2010. Antifungal activity of essential oils against *Phomopsis azadirachtae*—the causative agent of die-back disease of neem. J. Agric. Technol. 6 (1), 127–133.

Navi, S.S., 2005. Fungi associated with sorghum grain in rural Indian storages. J. New Seed 7, 51–68.

Nesci, A.V., Bluma, R.V., Etcheverry, M.G., 2005. In vitro selection of maize rhizobacteria to study potential biological control of *Aspergillus* section Flavi and aflatoxin production. Eur. J. Plant Pathol. 113, 159–171.

Nguyen, P., Strub, C., Durand, N., Alter, P., Fontana, A., Schorr-Galindo, S., 2017. Biocontrol of *Fusarium verticillioides* using organic amendments and their actinomycete isolates. Biol. Control 118, 55–66.

Niknejad, F., Zaini, F., Faramarzi, M., Amini, M., Kordbacheh, P., Mahmoudi, M., et al., 2012. *Candida parapsilosis* as a potent biocontrol agent against growth and aflatoxin production by *Aspergillus* species. Iran J. Public Health 41, 72–80.

Okwu, G.I., Achar, P.N., Ikenebomeh, M.J., Sreenivasa, M.Y., 2011. Studies of food thickeners used in Nigeria for contamination by aflatoxigenic forms of *Aspergillus* and their detection by PCR. Afr. J. Biotechnol. 10 (43), 8641–8646.

Palazzini, J.M., Alberione, E., Torres, A., Donat, C., Kohl, J., Chulze, S., 2016. Biological control of *Fusarium graminearum* sensustricto, causal agent of *Fusarium* head blight of wheat, using formulated antagonists under field conditions in Argentina. Biol. Control 94, 56–61.

Pereira, P., Nesci, A., Castillo, C., Etcheverry, M., 2010. Impact of bacterial biological control agents on fumonisin B 1 content and *Fusarium verticillioides* infection of field-grown maize. Biol. Control 53, 258–266.

Pereira, P., Sabrina, G.I., Agostini, E., Etcheverry, M., 2011. Effects of maize inoculation with *Fusarium verticillioides* and with 2 bacterial biocontrol agents on seedlings growth and antioxidative enzymatic activities. Appl. Soil Ecol. 51, 52–59.

Perrone, G., Gallo, A., Logrieco, A.F., 2014. Biodiversity of *Aspergillus* section Flavi in Europe in relation to the management of aflatoxin risk. Front. Microbiol. 5, 377.

Poornachandra Rao, K., Deepthi, B.V., Rakesh, S., Ganesh, T., Achar, P., Sreenivasa, M.Y., 2017. Antiaflatoxigenic potential of cell-free supernatant from *Lactobacillus plantarum* MYS44 against *Aspergillus parasiticus*. Probiotics Antimicrob. Proteins. Available from: https://doi.org/10.1007/s12602-017-9338-y.

Rabiey, M., Shaw, M.W., 2016. *Piriformo sporaindica* reduces *Fusarium* head blight disease severity and mycotoxin DON contamination in wheat under UK weather conditions. Plant Pathol. 65, 940–952.

Rai, R., Dash, P.K., Prasanna, B.M., Singh, A., 2007. Endophytic bacterial flora in the stem tissue of a tropical maize (Zea mays L) genotype: isolation, identification and enumeration. World J. Microb. Biotechnol. 23, 853–858.

Rocha, L.O., Reis, G.M., Silva, V.N., Braghini, R., et al., 2011. Molecular characterization and fumonisin production by *F. verticillioides* isolated from corn grains of different geographic origins in Brazil. Int. J. Food Microbiol. 145, 9–12.

Samapundo, S., De Meulenaer, B., Osei-Nimoh, D., Lamboni, Y., Debevere, J., Devlieghere, F., 2007. Can phenolic compounds be used for the protection of corn from fungal invasion and mycotoxin contamination during storage? Food Microbiol. 24, 465–473. Available from: https://doi.org/10.1016/j.fm.2006.10.003.

Schisler, D.A., Khan, N.I., Boehm, M.J., Lipps, P.E., Slininger, P.J., Zhang, S., 2006. Selection and evaluation of the potential of choline-metabolising microbial strains to reduce *Fusarium* head blight. Biol. Control 39, 497–506.

Schisler, D.A., Core, A.B., Boehm, M.J., Horst, L., Krause, C., Dunlap, C.A., et al., 2014. Population dynamics of the *Fusarium* head blight biocontrol agent *Cryptococcus flavescens* OH 182.9 on wheat anthers and heads. Biol. Control 70, 17–27.

Schroth, M.N., Hancock, J.G., 1982. Disease suppressive soil and root colonizing bacteria. Science 216, 1376–1381.

Sempere, F., Santamarina, M.P., 2007. *In vitro* biocontrol analysis of *Alternaria alternate* (Fr.) Keissler under different environmental conditions. Mycopathologia 163, 183–190.

Shantha, T., 1999. Fungal degradation of aflatoxin B1. Nat. Tox. 7, 175–178.

Siddiqui, Z.A., 2007. Biocontrol of *A. triticina* by plant growth promoting rhizobacteria on wheat. Arch. Phytopathol. Plant Prot. 40 (4), 301–308.

Sokmen, A., Gulluce, M., Akpulat, H.A., Daferera, D., Tepe, B., Polissiou, M., et al., 2004. The *in vitro* antimicrobial and antioxidant activities of the essential oils and methanol extracts of endemic *Thymus spathulifolius*. Food Control 15, 627–634.

Solimana, K.M., Badeaab, R.I., 2002. Effect of oil extracted from some medicinal plants on different mycotoxigenic fungi. Food Chem. Toxicol. 1669–1675.

Sreenivasa, M.Y., Dass, R.S., Charithraj, A.P., Janardhana, G.R., 2006. Molecular detection of fumonisin producing *Fusarium* species of freshly harvested maize kernels using polymerase chain reaction (PCR). Taiwania 51 (4), 251–257.

Sreenivasa, M.Y., Dass, R.S., Charithraj, A.P., Janardhana, G.R., 2007. PCR method for the detection of genus *Fusarium* and fuonisin producing isolates from freshly harvested sorghum grains grown in Karnataka, India. J. Food Saf. 28, 236–247.

Sreenivasa, M.Y., Gonzalez Jaen, M.T., Dass, R.S., Charithraj, A.P., Janrdhana, G.R., 2008. A PCR based assay for the detection and differentiation of potential fumonisin producing *Fusarium verticillioides* isolated from maize kernels. Food Biotechnol. 22, 160–170.

Sreenivasa, M.Y., Maheshwar, P.K., Sanjay, S.R., Diwakar, B.T., Naidu, K.A., Janardhana, G.R., 2009. Effect of gamma irradiation on the incidence and fumonisins production by *Fusarium* species occurring on maize and sorghum grains. J. Agric. Technol. 5 (2), 325–335.

Sreenivasa, M.Y., Dass, R.S., Janardhana, G.R., 2010. Survey of postharvest fungi associated with sorghum grain produced in Karnataka (India). J. Plant Prot. Res. 50, 335–339.

Sreenivasa, M.Y., Charithraj, A.P., Janardhana, G.R., 2011a. Diversity of *Fusarium* species associated with maize and sorghum grains in Karnataka, India. Fungal Sci. 26, 111–123.

Sreenivasa, M.Y., Dass, R.S., Charithraj, A.P., Janardhana, G.R., 2011b. Mycological evaluation of maize grains produced in Karnataka (India) for the postharvest fungal contamination. World Appl. Sci. J. 13, 688–692.

Sreenivasa, M.Y., Dass, R.S., Charithraj, A.P., Premila, N.A., Janardhana, G.R., 2011c. Assessment of the growth inhibiting effect of some plant essential growth inhibiting effect of some plant essential oils on different *Fusarium* species isolated from sorghum and maize grains. J. Plant Dis. Prot. 118, 208–213.

Sreenivasa, M.Y., 2012. Fumonisin—a potential carcinogen is of global concern. Res. J. Biotechnol. 7 (4), 1–2.

Sreenivasa, M.Y., Diwakar, B.T., Charith Raj, A.P., Dass, Regina Sharmila, Naidu, K.A., Janardhana, G.R., 2012. Determination of toxigenic potential of *Fusarium* species occurring on sorghum and maize grains produced in Karnataka, India by using thin layer chromatography. Int. J. Life Sci. 6 (1), 31–36.

Sreenivasa, M.Y., Diwakar, B.T., Charithraj, A.P., Dass, R.S., Naidu, K.A., Janardhana, G.R., 2013. Toxigenic Fusarium species and Fumonisin B1 and B2 associated with freshly harvested sorghum and maize grains produced in Karnataka, India. Ann. Food Sci. Technol. 14 (1), 100–107.

Tsitsigiannis, D.I., Dimakopoulou, M., Antoniou, P.P., Tjamos, E.C., 2012. Biological control strategies of mycotoxigenic fungi and associated mycotoxins in Mediterranean basin crops. Phytopathol. Mediterr.

Velluti, A., Marin, S., Gonzalez, P., Ramos, A.J., Sanchis, V., 2004a. Initial screening for inhibitory activity of essential oils on growth of *Fusarium verticillioides*, *F. proliferatum* and *F. graminearum*on maize-based agar media. Food Microbiol. 21, 649–656.

Velluti, A., Sanchis, V., Ramos, A.J., Marin, S., 2004b. Effect of essential oils of cinnamon, clove, lemon grass, oregano and palmarosa on growth and fumonisin B1 production by *Fusarium verticillioides* in maize. J. Sci. Food Agric. 84, 1141–1146.

Viuda-Martos, M., Ruiz-Navajas, Y., Fernandez-Lopez, J., Perez-Alvarez, J.A., 2007. Antifungal activities of thyme, clove and oregano essential oils. J. Food Saf. 27, 91–101.

Wilson, D.M. and Abramson, D., 1992. Mycotoxins. In: Sauer, D.B. (Ed.), Storage of Cereal Grains and Their Products, second ed. American Association of Cereal Chemists, St. Paul, MN, USA. pp. 341–391.

Yates, I.E., Meredith, F., Smart, W., Bacon, C.W., Jaworski, A.J., 1999b. *Trichoderma viride* suppresses fumonisin B1 production by *Fusarium moniliforme*. J. Food Prot. 62, 1326–1332.

Yooussef, M.M., Pham, Quyen, Achar, P.N., Sreenivasa, M.Y., 2016. Antifungal activity of essential oils on *Aspergillus parasiticus* isolated from peanuts. J. Plant Prot. Res. 56, 139–142.

Further reading

IARC, 1993. IARC monographs on the evaluation of carcinogenic risks to humans. In: Some Naturally Occurring Substances: Food Items and Constituents, Heterocyclic Aromatic Amines and Mycotoxins. International Agency for Research on Cancer, vol. 56, Lyon, France, pp. 397–488.

Chapter 10

Plant growth promoting microbes: a future trend for environmental sustainability

G. Chennappa[1], M.K. Naik[2], Nidoni Udaykumar[1], M. Vidya[3], M.Y. Sreenivasa[4], Y.S. Amaresh[2] and P.F. Mathad[1]

[1]*Department of Processing and Food Engineering, CAE, UAS, Raichur, India,* [2]*Department of Plant Pathology, College of Agriculture, UAS, Raichur, India,* [3]*Centre for Nanotechnology, CAE, UAS, Raichur, India,* [4]*Department of Studies in Microbiology, University of Mysore, Mysore, India*

Chapter Outline

10.1 Introduction

In agriculture, crop protection against phytopathogens relies heavily on agrochemicals. Different approaches are being used to prevent, mitigate, or control plant diseases. Beyond good agronomic practices, growers often rely heavily on chemical fertilizers and pesticides but also pesticides pose adverse effects on human health, environment, and soil biodiversity (Singh, 2013, 2015; Tiwari and Singh, 2017; Singh et al., 2018). Plant diseases result in severe losses of agricultural and horticultural crops every year and thus need to be controlled to maintain the quality and abundance of food, feed, and fiber. These losses can bring about reduced food supplies, poor quality products, and economic hardship for growers and processors. Due to unscientific methods and their extensive use, pesticides are largely distributed and contaminate soil and ground water and get deposited as sediments (Castillo et al., 2011). On the other hand, the demand for agricultural crops is increasing day by day due to the rapidly growing industrialization along with increasing population. Pesticides and chemical fertilizers reaching the soil in significant quantities have direct effect on soil microbiological aspects, environmental pollution, and health hazards (Martin et al., 2011; Chennappa et al., 2014). This to leading to alterations in ecological balance of the soil microflora, adversely affecting soil fertility and crop productivity, inhibiting nitrogen (N_2) fixing bacterial activity, suppressing nitrifying bacteria and soil microbe interaction, altering nitrogen balance of the soil, interfering with ammonification, and hampering mycorrhizal symbiosis or nodulation in plants, as well as plant growth, soil structure, organic matter decomposition, and biogeochemical cycling of elements

New and Future Developments in Microbial Biotechnology and Bioengineering. DOI: https://doi.org/10.1016/B978-0-12-818258-1.00010-8

(Aleem et al., 2003; Reinhardt et al., 2008; Sachin, 2009; Chennappa et al., 2013, 2014; Chennappa, 2015). Increasing awareness and long-term effects of chemical residues in the soil have resulted in the need for alternative agricultural practices that are less destructive to the environment (Tiwari et al., 2018; Singh and Gupta, 2018; Vimal et al., 2018). Understanding the application of microbes for plant growth and their role in disease management is very essential. An ideal antagonistic microorganism must have the potential to compete and colonize successively in the rhizosphere to exist in a nutritionally limited environment.

10.2 Plant growth promoting microbes

The group of microbes that have a positive influence on plant growth and health are referred to as plant growth promoting microorganisms (PGPMs). The group PGPMs includes *Azotobacter*, *Azospirillum*, *Pseudomonas*, *Acetobacter*, *Burkholderia*, *Bacillus*, *Paenibacillus*, and members of the fungi, *viz.*, *Trichoderma*, *Gleocladium*, and vascular arbuscular mycorrhiza. The beneficial effects of these PGPMs have been broadly attributed to their ability to produce various compounds such as phytohormones, organic acids, siderophores, atmospheric nitrogen fixation, phosphate solubilization, and antibiotics and induced systemic resistance to control plant diseases (Glick, 1995). These microbial inoculants are also used as biofertilizers, biocontrol agents, phytostimulators, and bioremediators as well (Lugtenberg et al., 2001; Chennappa, 2015; Chennappa et al., 2016a,b, 2017a,b). PGPMs produce secondary metabolites and play a crucial roles in microbe–host interactions; these metabolites are important for combating the plant pathogens by altering the environment and improving their ability to compete with pathogens by inhibiting the activity of pathogens or by triggering host defenses. Biological control involves the use of beneficial microorganisms, such as specialized fungi and bacteria control plant pathogens (Monte, 2001; Raaijmakers et al., 2002). Utilization of these PGPMs against plant pathogens in agricultural crops has been proposed as an alternative to chemical pesticides (Pal and McSpadden-Gardener, 2006). Isolation and exploitation of indigenous strains with more than one antimicrobial compound having multiple and broad spectrum action will be beneficial to the farming community, which will go a long way in ensuring sustainable crops and the environment. Biological control of soil-borne diseases by PGPMs is a well-established phenomenon and has been shown to play a major role in suppression of several plant pathogens (Handelsman and Stabb, 1996). Among PGPMs, plant growth promoting rhizobacteria (PGPR) are rhizosphere competent bacteria that aggressively colonize plant roots and have ability to multiply and colonize across the ecological niches found on the roots at all stages of plant growth in the presence of competing microflora (Antoun and Kloepper, 2001).

10.3 Plant growth promoting rhizobacteria

The term PGPR was first described by Kloepper and Schroth (1980). The PGPR are a group of bacteria that actively colonize plant roots and promote plant growth and increase yield of agricultural produce. Rhizobacterial strains are found to increase plant growth after inoculation of seeds and therefore are called PGPR. Free living, root-colonizing bacteria have been studied for the past century as possible inoculants for increasing plant productivity (Kloepper, 1992). PGPR are able to increase plant nutrient uptake by introducing nitrogen fixing bacteria associated with roots (*Azotobacter*, *Azospirillum*) for nitrogen uptake, iron uptake from siderophores producing bacteria (*Pseudomonas*), sulfur uptake from sulfur-oxidizing bacteria (*Thiobacillus*), phosphorus uptake from phosphate mineral solubilizing bacteria (*Bacillus*, *Pseudomonas*), and potassium uptake from potassium solubilizing bacteria (*Bacillus*) respectively (Myresiotis et al., 2012). *Azospirillum* is also one of the best-studied growth hormone producers and other bacteria belonging to the genera are *Aeromonas*, *Burkholderia*, *Azotobacter*, *Bacillus*, *Enterobacter*, *Pseudomonas*, and *Rhizobium* (Ahmad et al., 2008; Ghosh et al., 2010; Chennappa et al., 2013, 2014).

Enhanced supply of other plant nutrients by PGPR such as phytochrome production leads to increases in shoot and root length as well as seed germination of several agricultural crops (Ahmad et al., 2005). The production of plant growth regulators is one of the major mechanisms through which PGPR influence the plant growth and development (Javed et al., 2009). The ability to synthesize phytohormone is widely distributed among plant associated bacteria and 80% of the bacteria isolated from the plant rhizosphere are able to produce plant growth promoting substances. PGPR are characterized by a number of functions, which include improvement of plant establishment, increased plant nutrients, enhancement of nutrient uptake, improvement of soil structure, and protection against diseases (Glick, 1995). They survive in soil, multiply in the rhizosphere in response to the exudates rich in carbohydrates and amino acids, and attach to root surfaces and become endophytic by colonizing in the root cortex region (Kloepper, 1992). Plant growth promotion is also mediated by these rhizobacteria through different mechanisms, which include production of lytic enzymes

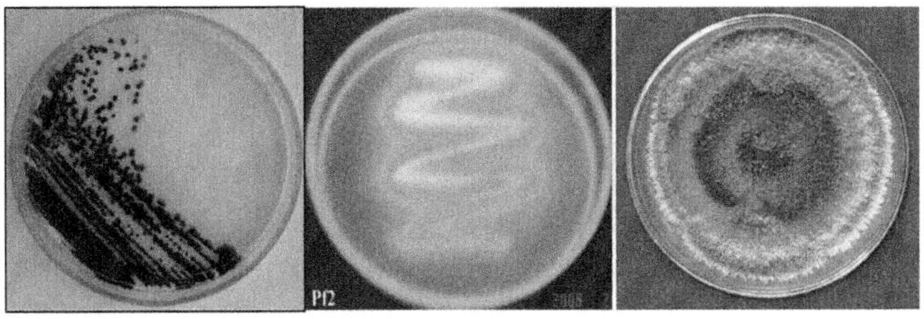

FIGURE 10.1 Pure cultures of *Azotobacter*, *Pseudomonas*, and *Trichoderma* species.

such as chitinase and β-1, 3 glucanase, phenylalanine ammonia lyase, and polyphenol oxidases (Anand et al., 2010). PGPR bacteria having these properties are important for crop health management. Among the PGPR group, *Azotobacter* and *Pseudomonas* are known for their beneficial activities.

10.4 *Azotobacter* species

Among the entire PGPR group, *Azotobacter* are ubiquitous, and free living in soil, water, and sediments. Being the major group of soil-borne bacteria, *Azotobacter* play different beneficial roles and are known to produce varieties of plant growth promoting substances in the rhizosphere soil (Fig. 10.1). These growth promoting substances have direct influence on shoot and root length as well as seed germination of several agriculture crops. *Azotobacter* species are efficient in fixation, production of indole acetic acid, gibberellic acid, and phosphate solubilization (Chennappa et al., 2013, 2016a,b, 2017a,b). The species of *Azotobacter* can grow and survive at extreme environmental conditions; they are tolerant to higher salt concentration, pH values, and they can survive even at maximum temperature. *Azotobacter* species are known to produce antimicrobial compounds that are able to inhibit the growth of common phytopathogens. The species of *Azotobacter* are known to tolerate and survive pesticide concentration and also have the ability to degrade heavy metals and pesticides (Chennappa et al., 2014, 2018a,b; Chennappa, 2015). *Azotobacter* strains are potential bioagents for sustainable agriculture and control the ecological imbalance in the environment.

10.4.1 Growth promoters (vitamins, amino acids, indole acetic acid, gibberellic acid)

Vitamins, amino acids, and plant growth hormones [indole acetic acid (IAA), gibberellic acid (GA)] are essential for the growth and development of the plant and are produced by certain species of PGPR bacteria. These growth promoters are produced under congenial conditions and *Azotobacter* species (*A. vinelandii* and *A. chroococcum*) produce different types of vitamins such as niacin, pantothenic acid, riboflavin, and biotin, which belong to the B-group vitamins. These vitamins produced by *Azotobacter* are used to maintain metabolic processes of living beings and several factors that are responsible for the production of vitamins include growth conditions, pH, temperature, and availability of nutrient sources (Almon, 1958; Revillas et al., 2000). Amino acids are the building blocks of life and amino acids are biologically important organic compounds that combine to form proteins composed of amine (NH_2) and carboxylic acid (−COOH). Species of *Azotobacter* produce different types of amino acids such as aspartic acid, serine, glutamic acid, glycine, histidine, threonine, arginine, alanine, proline, cysteine, tyrosine, valine, methionine, lysine, isoleucine, leucine, tryptophan, glutamic acid, and phenylalanine under diazotrophic conditions (Lopez et al., 1981; Revillas et al., 2000).

Different bacteria genera are known to produce IAA, which is common to inhabitants of all soil ecosystems (Barazani and Friedman, 1999). Some of the bacterial species that can produce the highest amount of IAA belong to the PGPR group. IAA producing bacteria grow in more saline soil and in saline soil 75% of the bacterial isolates are more active in IAA production. Several species of *Azotobacter* including *A. salinestris*, *A. tropicalis*, *A. vinelandii*, *A. armenicaus*, *A. nigricans*, and *A. chroococcum* are found to produce IAA in the range of $2.09-33.28\ \mu g\ mL^{-1}$ (Chennappa et al., 2013; Chennappa, 2015). IAA producing PGPR strains are known to increase plant vigor (root and shoot length), seed germination, and root biomass, resulting in greater root surface area, which enables plants to access more nutrients from the soil. IAA is responsible for the division, expansion, and differentiation of plant cells and tissues and stimulates root elongation (Patten and Glick, 2002; Spaepen et al., 2007). Another important growth promoter is GA, which is produced naturally within the plant rhizosphere by bacteria and fungi. GA was first discovered by Japanese scientist Eiichi Kurosawa from a fungus called *Gibberellafuji kuroi* in rice plants. Gibberellins are important in seed germination and enzyme production, which mobilizes growth of new cells. GA promotes flowering, cellular division, and seed growth after germination (Upadhyay et al., 2009). Similarly, different species of *Azotobacter* isolates (*A. salinestris*, *A. tropicalis*, *A. vinelandii A. armenicaus*, *A. nigricans*, and *A. chroococcum*) produced GA in the range of 10.81–17.55 μg 25 mL^{-1} (Chennappa et al., 2013, 2016a,b; Chennappa, 2015).

10.4.2 Hydrogen cyanide

Hydrogen cyanide (HCN) is a volatile, secondary metabolite that suppresses the growth or development of microorganisms and that also influences the growth and development of plants. HCN is one of the significant inhibitors of cytochrome C oxidases and other metalloenzymes, which are involved in the metabolic activities of phytopathogens. HCN is mainly synthesized by an enzyme, HCN synthetase, which is known to be associated with plasma membrane of certain rhizobacteria. Presently many bacterial species have shown to be capable of producing HCN, including species of *Azotobacter*, *Alcaligenes*, *Aeromonas*, *Bacillus*, *Pseudomonas*, and *Rhizobium* (Ahmad et al., 2008; Naik et al., 2013). As per the Lorck assay method *Azotobacter* species produced light orange to red color HCN metabolites (Chennappa, 2015). Some studies showed that about 50% of the *Pseudomonas* isolates obtained and characterized from wheat and potato rhizospheric soil are able to produce HCN. Various studies reported that HCN can inhibit the growth of phytopathogenic nematodes such as *Meloidogyne javanica* and *Thielaviopsis basicola*, which are known to cause root knot, and blackrot of tomato and tobacco in the rhizospheric soil. The subterranean termite *Odontotermes obesus*, an important pest in agricultural and forestry crops in India, is also controlled by HCN produced by rhizobacteria (Kannapiran and Ramkumar, 2011).

10.4.3 Siderophores

Siderophores are iron-chelating agents that are produced and utilized by a number of bacteria and fungi. These low molecular weight compounds are produced in soil rhizosphere under neutral. to alkaline pH condition. Iron is one of the essential components for the cellular growth and metabolic activities of the bacteria. The bacteria utilize iron through siderophore production, which plays an essential role in determining the ability of bacteria to colonize plant roots and also in competing for iron with other microbes (Johri et al., 2003; Naik et al., 2013). *A. vinelandii* produces siderophores under limited iron conditions. Recently, it has been found that *A. vinelandii* produces different types of siderophores that are antibiotic in nature such as 2,3-dihydroxybenzoic acid, aminochelin, azotochelin, protochelin, and azotobactin (Page and Von Tigerstrom, 1988; Kraepiel et al., 2009; Barrera and Soto, 2010). The main biotechnological applications of siderophores are as drug delivery agents, antimicrobial agents, and for soil remediation. The PGPR with the siderophores producing ability can prevent the root infection by phytopathogens in the rhizospheric region. In certain conditions, plants can also use microbial siderophores as iron sources where there is a lack of sufficient iron concentration in soil (Mollmann et al., 2009; Naik et al., 2013).

10.4.4 Biodegradation of pesticides

Several physical, chemical, and biological forces act upon the pesticides when they reach the soil. However, biological forces, particularly microbes, play a more significant role in degrading the pesticides than the physical and chemical forces. Several biodegradation studies have been carried out to minimize the pesticide residues in food and feed. Many soil microorganisms have the ability to act upon pesticides and convert them into simpler nontoxic compounds. Bacterial genera like *Pseudomonas*, *Rhodococcus*, *Arthrobacter* spp., *Burkholderia* spp., *Azotobacter*, *Clostridium*, *Bacillus*, *Thiobacillus*, *Achromobacter*, etc., and fungal genera like *Trichoderma*, *Penicillium*, *Aspergillus*, *Rhizopus*, and *Fusarium* play an important role in the degradation of the toxic chemicals or pesticides in soil. Species of *Azotobacter viz*., *A. vinelandii* (RCR-4), *A. tropicalis* (KOP-11), *A. armeniacus* (GVT-11), *A. salinestris* (GVT-1), and *A. chroococcum* (SND-4-2) biodegraded three prominent pesticides such as pendimethalin, chloropyrifos, and carbendazim (up to 97%). Among them, *A. tropicalis* (KOP-11), *A. vinelandii* (RCR-4), and *A. salinestris* (GVT-1) isolates degraded the pendimethalin by 100% over the control (Chennappa, 2015; Chennappa et al., 2016a,b, 2018a,b). Similarly, Phorate, an insecticide, has been degraded by *Rhizobium*, *Pseudomonas* species isolated from paddy cultivated fields (Bano and Musarrat, 2003). The biodegradation abilities of bacterial species including *Azotobacter* and *Pseudomonas fluorescens* were tested for glyphosate (herbicide) and it was found that all species degraded glyphosate efficiently (Moneke et al., 2010). *Azotobacter* facilitates the mobility of heavy metals in the soil and thus enhances bioremediation of soil from heavy metals, such as cadmium, mercury, and lead. Reports are also revealed that *A. chroococcum* has the ability to degrade hydrocarbon and can be used as a biosurfactant in the treatment of marine oil contamination (Helmy et al., 2010). Degradation of pesticides is often considered to be by mineralization process and the degradation pathways are catechol, protocatechuic acid, gentisic acid, ferulic acid, resorcinol, and 2,4-dichloro phenoxyacetic acid (Mrkovacki et al., 2002).

10.4.5 Biocontrol

The antibiotics produced by PGPR are shown to be effective against phytopathogens and can provide protection against soil-borne fungi such as *Aspergillus*, *Fusarium*, *Curvularia*, *Alternaria*, and *Helminthosporium* (Khan et al., 2008; Mali and Bodhankar, 2009). The species of *Azotobacter* produce different types of secondary metabolites, which include 2,3-dihydroxybenzoic acid, aminochelin, azotochelin, protochelin, and azotobactin (Kraepiel et al., 2009) and act as biocontrol agents by inhibiting the pathogen infections. *Azotobacter* species act as biocontrol agents for many plant pathogens (Agarwal and Singh, 2002; Mali and Bodhankar, 2009; Nagaraja et al., 2016). The species of *Azotobacter* have inhibited root colonization of different species of *Fusarium viz., F. verticillioides, F. poae, F. sporotrichiodes, F. culmorum*, etc., which are causal agents of infection in cereals. The metabolites produced by *Azotobacter* have suppressed fumonisin B-1 production by *A. nigricans* (Nagaraja et al., 2016; Chennappa et al., 2016a,b, 2017a,b). Antifungal activity of *A. vinelandii* showed maximum zone of inhibition (40 mm) against Fusarium *oxysporum*, which is commonly known to cause several diseases in agriculture crops such as chili and pigeon pea (Cavaglieri et al., 2005; Bhosale et al., 2013).

10.5 *Pseudomonas* species

Pseudomonas is a Gram-negative bacteria and it includes fluorescent as well as nonfluorescent species such as *P. fluorescens*, Pseudomonas *putida*, *Pseudomonas chlororaphis*, *Pseudomonas aureofaciens*, and *Pseudomonas aeruginosa* (Anand et al., 2010). Among the various PGPRs, fluorescent *Pseudomonas* is considered as the most important as they have both plant growth promotion activity, induced systemic resistance (ISR), and production of secondary metabolites (Fig. 10.1). In recent years, several plant root-colonizing *Pseudomonas* species have shown to be potent biocontrol agents in various plant–pathogen systems (Thomashow and Weller, 1996). The primary mechanism of biocontrol by fluorescent *Pseudomonas* involves production of antibiotics such as 2,4-diacetylphloroglucinol, pyoluteorin, pyrrolnitrin, phenazine-1-carboxyclic acid, 2-hydroxy phenazines, phenazine-1-carboxamide, rhamnolipids, oomycin A, cepaciamide A, ecomycins, viscosinamide, butyrolactones, N-butylbenzene, sulfonamide, pyocyanin, pseudomonic acid, azomycin, cepafungins, and karalicin. These antibiotics are known to possess antiviral, antimicrobial, insecticidal, antihelminthic, antioxidant, cytotoxic, and antitumor activities (Fernando et al., 2005). Among the various extracellular metabolites produced by *Pseudomonas* species 2,4-DAPG (2,4-di-acetyl-phloro-glucinol), HCN, phenazine, pyoverdine, and pyoluteorine have prime importance in plant protection because genes involved in the production have ubiquitous distribution among fluorescent pseudomonads diversity, different mode of action, and broad spectrum activity against plant diseases (Naik et al., 2016; Vinay et al., 2016).

10.5.1 DAPG

The 2,4-DAPG exhibits antibacterial, antifungal, antiviral, antihelminthic (Bowden et al., 1965), and herbicidal properties and has played a significant role in the biological control of tobacco, wheat, tomato, rice, and sugar beet diseases (Kataryan and Torgashova, 1976; Velusamy et al., 2006; Carine et al., 2012). DAPG is the major determinant secreted by *P. fluorescens* in the biological control of plant pathogens (Bangera and Thomashow, 1996). Fluorescent bacterial strains positive for 2,4-DAPG obtained from rice rhizosphere showed antibiosis and suppressed growth of *Xanthomonas oryzae* pv. *oryzae* (up to 64%), the causal agent of bacterial blight of rice disease (Vasudevan et al., 2002). These results suggest that the production of 2,4-DAPG in crop rhizosphere is yet another important mechanism that reduces the severity of bacterial blight in rice (Naik et al., 2014; Vinay et al., 2016). Furthermore, detection of 2,4-DAPG gene in the endophytic region was first reported and isolated from sorghum, chickpea, pigeon pea, and green gram. Indian soils are considered to be poor in 2,4-DAPG gene possessing isolates but few isolates were obtained having 2,4-DAPG gene indicating the possibility of exploiting the rhizospheric soils for obtaining wide ranging antibiotic and other useful traits for emerging pests and diseases (Rajalaxmi et al., 2012; Naik et al., 2014). DAPG positive *Pseudomonas* species evaluated against plant-parasitic nematodes such as *Helicotylenchus indicus*, *Xiphinema americanum*, and *Meloidogyne incognita* and were found effective against plant-parasitic nematodes (Yuan et al., 1998; Khan et al., 2012). Similarly, Pseudomonas *brassicacearum*, a 2,4-DAPG positive strain, was found to be effective against *Ralstonia solanacearum* and results showed strong antibacterial activity (Zhou et al., 2012; Vinay et al., 2016).

10.5.2 Pyoluteorine

Pyoluteorine is a aromatic chlorinated polyketide with resorcinol ring that is produced by fluorescent Pseudomonads and resorcinol ring is linked to a bichlorinated pyrrole moiety (Kitten et al., 1998; Nowak-Thompson et al., 1999). It was first isolated from *P. aeruginosa* followed by *P. fluorescens* Pf-5 (Bender et al., 1999). Pyoluteorin has bactericidal, herbicidal, and fungicidal properties. Application of Pyoluteorin to cotton seeds suppressed cotton damping-off disease. Pyoluteorin is found to be more effective especially against *Pythium ultimum*, which causes damping-off disease (Howell and Stipanovic, 1980; Maurhofer et al., 1992, 1994).

10.5.3 Siderophores

Siderophores are green fluorescent, low molecular weight compounds that are produced under iron limiting conditions, chelate the ferric ion (Fe + 3) with a high specific activity, and serve as vehicles for the transport of Fe + 3 into a microbial cell (Neilands, 1981), to play a vital role in controlling several plant diseases. The role of siderophores produced by fluorescent *Pseudomonas* in plant growth promotion was first reported by Kloepper and Schroth in 1981 (Bakker et al., 1986; Becker and Cook, 1988). Several species of fluorescent Pseudomonads like *P. fluorescens*, *P. putida*, *P. aeruginosa*, and *P. chlororaphis* produce siderophores. Siderophores produced by *P. aeruginosa* were evaluated for antifungal activity against three phytopathogenic fungi, *Fusarium moniliformae*, *Alternaria solani*, and *Helminthosporium halodes*. Siderophores inhibited the growth of *F. moniliformae* (92%), *A. solani* (89%), and *H. Halodes* (80%) respectively (Sharma et al., 2007; Naik et al., 2016). In addition *P. aeruginosa* produces another siderophore called pyochelin with a lower affinity for iron. They function as biocontrol agents by depriving the pathogens from iron nutrition thus resulting in increased yield of crops. Pyoviridines, pyochelin, salicylic acid, and quinolobactins are the major siderophores produced by fluorescent *Pseudomonas*. Fluorescent yellow green siderophores have been named as pyoverdines or pseudobactins. Hayen and Volmer (2005) studied siderophores including enterobactin, ferrioxamine B, ferrichrome, ferrirhodin, rhodotorulic acid, and coprogen.

10.5.4 Phenazine

Phenazine is a nitrogen containing heterocyclic antimicrobial compound that plays a vital role in the management of soil-borne phytopathogens (Pierson et al., 1995). Phenazines are synthesized by species of *Pseudomonas* and all phenazines exhibited broad spectrum activity against bacteria and fungi (Leisinger and Margraff, 1979; Vinay et al., 2016). The major phenazine derived by *P. aeruginosa* was pyocyanin (1-OH-5-methyl phenazine).

10.5.5 Pyrrolnitrin

Pyrrolnitrin is a broad spectrum antifungal metabolite produced by fluorescent and nonfluorescent *Pseudomonas* strains (Kirner, 1998). A phenyl pyrrol derivative of pyrrolnitrin has been developed as an agricultural fungicide. Schnider et al. (1995) reported that pyrrolnitrin actively persists in soil for at least 30 days. It did not readily diffuse and was released only after lysis of host bacterial cells. This property of slow release facilitated protection against common plant pathogen such as *Rhizoctonia solani* (Vinay et al., 2016; Reshma et al., 2018). Species of *Pseudomonas* (*P. putida* and *P. fluorescens*) having 2,4-DAPG, phenazine, siderophores, and HCN tested the antagonistic activity against *R. solani* (sheath blight of rice), and *F. oxysporum* f. sp. *Ciceris* (chickpea wilt). These secondary metabolites inhibited the growth of mycelium and significantly reduced the germination capacity of the sclerotia produced by pathogens (Rosales et al., 1995; Nandakumar et al., 2002; Mazurier et al., 2009).

10.5.6 Induction of systemic resistance

Induced resistance is a state of enhanced defensive capacity developed by a plant when appropriately stimulated (Van Loon et al., 1998). Inducing the plant's self defense mechanisms by prior application of a biological inducer is thought to be a novel plant protection strategy (Ramamoorthy et al., 2001). Resistance inducing rhizobacteria offer an excellent alternative in providing a natural, effective, safe, persistent, and durable protection (Reshma et al., 2018). One classical biotic inducer is the plant growth promoting bacterium *P. fluorescens* Migula (Iavicoli et al., 2003). Induced resistance is generally systemic and can be triggered by pathogens, certain chemicals, and nonpathogenic rhizosphere bacteria. Inducing the plant's self defense mechanisms by prior application of a biological inducer is thought to be a novel plant protection strategy (Ramamoorthy et al., 2001; Reshma et al., 2018).

10.5.7 Biocontrol

Fluorescent pseudomonads produce different secondary metabolites including a wide range of antibiotic compounds that have been used for biocontrol activity. Antibiotics include a chemically heterogeneous group of small organic molecules of microbial origin, and at low concentrations are deleterious to the growth or metabolic activities of other microorganisms (Thomashow and Weller, 1996). All these secondary metabolites contribute to disease suppression in various pathogen host plant systems. Howell and Stipanovic (1980) first established the importance of antibiotic production in the suppression of seedling pathogens such as *P.ultimum* causing damping off in cotton using *P. fluorescens* (Pf-5).

10.6 *Trichoderma* species

As a group, fungi have a deep impact on human life and ecosystem functionality. Fungi are the principal decomposers in the ecosphere and are essential for recycling nutrients in the environment. *Trichoderma* species are ubiquitous, colonizes on cellulosic materials and also in the rhizosphere of plants (Kubicek et al., 2008; Jaklitsch, 2009). *Trichoderma* species are free living fungi that are present in soil and root ecosystems and other natural habitats especially those containing high organic matter (Harman et al., 2004; Chennappa et al., 2017a,b). *Trichoderma* species have been isolated from decomposed organic matter, root surface of various plants, and decaying barks. The genus *Trichoderma* consists of nine species including *T. aureoviride*, *T. hamatum*, *T. harzianum*, *T. koningii*, *T. longibrachiatum*, *T. piluliferum*, *T. polysporum*, *T. pseudokoningii*, and *T. viride* aggregates defined largely by conidiophore morphology (Fig. 10.1). Among them, the most important species used in the field of biological control are *T. atroviride*, *T. harzianum*, *T. virens*, *T. asperellum*, and *T. reesei* (Benitez et al., 2004; Chennappa et al., 2017a,b).

Among the different groups of fungi, *Trichoderma* has gained immense importance over the last few decades due to its biological control activity and use of secondary metabolites. *Trichoderma* species are ubiquitous and are often the predominant component of the mycoflora in soil ecosystem and play an important role in ecosystem health (Komon-Zelazowska et al., 2007). The ecological adaptability of *Trichoderma* species is evidenced by their widespread distribution, including under different environmental conditions and on various substrates. This physiological flexibility together with the antagonistic action of *Trichoderma* species and the ability of these fungi to promote plant growth have made them attractive biological control agents (Mariola et al., 2007). *Trichoderma* is the only eukaryotic fungi known to possess growth promoting and antagonistic biocompounds against a broad range of soil-borne pathogens such as that of bacteria; that is, *Azotobacter*, *Pseudomonas*, and *Trichoderma* are eco-friendly by nature, which makes them significantly suitable for sustainable organic agriculture practices. The beneficial action of *Trichoderma* species is not limited to fighting with pathogens; they have also been shown to be opportunistic plant symbionts, enhancing systemic resistance of plants (Shoresh et al., 2010).

10.6.1 Secondary metabolites

Species of *Trichoderma* can produce antibiotics (peptaibols), mycotoxins, and more than 100 metabolites with antibiotic activity including polyketides, pyrones, terpenes, amino acids, and polypeptides (Sivasithamparam and Ghisalberti, 1998). Paracelsin is the first characterized secondary metabolite to be isolated and characterized from *Trichoderma* species, which has antibiotic properties; later, different types of peptaibols were identified (Degenkolb et al., 2008; Stoppacher et al., 2008). *Trichoderma asperellum* has the ability to produce 43 different major volatile secondary metabolites such as 1,2-benzene dicarboxylic acid, 2-butoxy-2-oxoethyl butyl ester, 1,2-benzene dicarboxylic acid dibutyl ester, 2H-pyran-2-one, palmitic acid, several phenolic isomers, methyl cyclohexane, 1-methyl-2-nitrobenzene, isoquinoline, dibutyl ester, butyl 8-methylnonyl ester, trimethylsilyl palmitate, trans-9-octa decenoic acid, trimethylsilyl ester, p-aminotoluene, several phenol isomers (p-tert-amylphenol, 4-nonylphenol, 4-dodecylphenol, 4-(1,1,3,3-tetra methylbutyl) Phenol), n-eicosane, 8-propyl quinoline, 5-ethyl-5-propyl undecane, tetrade-cane, heptadecane, cyclooctyl methyl phosphono-fluoridoate, etc. (Nitish and Kumar, 2017).

10.6.2 Antibiotic compounds

Most of the *Trichoderma* species produce antibiotics or diffusible compounds during interactions with pathogens, these compounds inhibit the colonization of pathogenic fungi. The metabolites include harzianic acid, almethicins, tricholin, peptaibols, antibiotics, 6 pentyl−α-pyrone, massoilactone, viridin, glioviridin, gliosporens, and heptelidic acid (Vey et al., 2001; Reino et al., 2008). *T. harzianum* produced pyrone antibiotics, which are effective in controlling the *Gaeumannomyces graminis* var. *tritici*, and *T. virens* showed biocontrol efficiency by producing gliovirin (Howell, 1998). *Trichoderma* species are recognized for numerous mechanisms including competition for space and nutrients, mycoparasitism, antibiosis, inactivation of the pathogen's enzymes, and induced resistance (Sivan and Chet, 1989; Kapulnik and Chet, 2000; Roco and Perez, 2001).

Peptaibols have strong antimicrobial properties against bacteria and fungi. Peptaibols along with cell wall degrading enzymes can inhibit the growth of fungi more effectively. More than 190 peptaibols have been documented that are produced by species of *Trichoderma* (Neuhof et al., 2007). Gliotoxin belongs to nonribosomal peptides and it was the first metabolite isolated from species of *Trichoderma* (Brian, 1944). Gliotoxin is a fungistatic metabolite and it has recorded good antagonistic activity against *Rhizoctonia*. It is effective in cotton disease control (Whilhite and Stanley, 1996). It is also one of the most important antifungal compounds produced by species of *Trichoderma* and it is a very effective biocontrol metabolite. Pyrones have plant growth promoting activities and are especially effective against *Sclerotium rolfsii*, a common plant pathogen (Evidente et al., 2003; Reither et al., 2005).

10.6.3 Mycoparasitism

Biological mechanisms involving *Trichoderma* species operate by way of mycoparasitism, antibiosis, and competition. The mechanism of parasitism includes interaction such as the coiling of hyphae around the pathogen, penetration by haustoria, and release of antibiotics or other chemicals that are harmful to the pathogens and inhibit their growth (Harman et al., 2004). The antagonistic interaction established between two fungal species, which consists of the direct attack of one fungus on another, is known as mycoparasitism and this depends on recognition, attack, penetration, and killing of the host. *Trichoderma* is also one of the best-studied mycoparasitic fungi and it can invade many plant pathogens in a root system. Mycoparasitism is considered as the effective biocontrol strategy and *Trichoderma* species are effective biocontrol agents due to their antagonistic capacities against a broad range of phytopathogenic fungi. Another important factor that makes *Trichoderma* more effective is the production of hydrolytic enzymes and this is the key step in the successful establishment of a mycoparasitic relationship with pathogens. *Trichoderma* species are highly efficient producers of many extracellular enzymes like cellulases, chitinases, glucanases, proteases, etc. They are capable of secreting hydrolytic enzymes and cause mycoparasitism on fungal pathogens of plants. Trichodermin, Dermadin, Sesquierpene, Harzianum A, Harzianolide and Trichodermol, which will antagonize soil-borne pathogens (Nakkeeran et al., 2002; Kucuk and Kyvanc, 2008). They are known to possess cell wall degrading enzymes such as chitinase, β-1, 3-glucanase, β-1, 6-glucanase, cellulose, and protease (Coley-Smith et al., 1991; Bunker and Mathur, 2001; Muhammad and Amusa, 2003; Jayalakshmi et al., 2009).

10.6.4 Cellulolytic enzymes

The cell wall is the first barrier between pathogens and mycoparasites, for example, *Trichoderma*. The degradation of the cell wall is mediated by different set of enzymes such as chitinase, glucanase [β-(1−4), β (1−3) and β (1−6)], N-acetylglucosaminidase, and protease. Chitin is the major structural component of the fungal cell wall and is divided into 1, 4-β acetylglucosaminidases, endochitinases, and exochitinases. The second most important polymer in the cell wall is β-1−3 glucan, which is hydrolyzed by β-1−3 glucanases. Glucanases have been detected and identified in the infected region of the host and are produced by the interaction of pathogens and *Trichoderma*. Overproduction of these enzymes influences biocontrol efficiency, which in turn inhibits the growth and spore germination of many pathogens. Overexpression of the glucanase by *T. harzianum* inhibited the growth of *Botrytis cinerea*, *R. solani*, and *Phytophthora citrophthora* (Ihrmark et al., 2010). Proteases are also important in the degradation of fungal cell wall and *T. atroviride* and *T. virens* have more enzymes than *T. reesei*. *Trichoderma* species produce different types of proteases during mycoparasitism, which include asprtyl and subtilin like groups (Geremia et al., 1993; Seidi et al., 2009). *T. harzianum* inactivated hydrolytic enzymes produced by *B. cinerea* pathogens by secreting proteases and act as a biocontrol agents (Howell, 2003).

10.6.5 Biocontrol

The defensive actions of *Trichoderma* species involve application of lytic enzymes (Kubicek et al., 2001; Viterbo et al., 2002), proteolytic enzymes (Kredics et al., 2005; Suarez et al., 2007; Chen et al., 2009), volatile metabolites (Calistru et al., 1997; Eziashi et al., 2006), and other secondary metabolites (Reino et al., 2008) against pathogens (Benitez et al., 2004). *T. koningii* showed its antifungal activity by producing secondary metabolites such as δ-decanolactone, 6-pentyl-α-pyranone and 6-(4-oxopentyl)-2H-pyran-2-one in the rhizosphere region and inhibited the common soil-borne pathogenic fungi such as *R. solani*, *S. rolfsii*, *Macrophomina phaseolina*, and *F. oxysporum* respectively (Vivek et al., 2014). Similar volatile compounds are also documented from the GCMS analyses; as many as 278 compounds such as cyclohexane, cyclopentane, fatty acids, alcohols, esters, sulfur-containing compounds, simple pyrane, and benzene derivatives from liquid cultures of *T. harzianum*. The compounds isolated from *Trichoderma* species have strong antifungal activity and the compounds include 3-methyl-heptadecanol, methyl cyclohexane, 6-nonylene alcohol, methyl-cyclopentane, 2-methyl heptadecanol, N-methyl pyrollidine, dermadin, ketotriol, koningin-A, palmitic acid, 3-(2′-hydroxy propyl)-4-(hexa-2′-4-dineyl)-2-(5H)-furanone, and 3-(propenone)-4-(hexa-2′-4′-dineyl)-2-(5H)-furanone (Dubey et al., 2011; Siddiquee et al., 2012).

The *Trichoderma* species are able to control ascomycetes, basidiomycetes, oomycetes, and also nematodes (Monte, 2001; Benitez et al., 2004; Kyalo et al., 2007). Species of *Trichoderma viz.*, *T. aureoviride*, *T. virens*, *T. reesei*, *T. hamatum*, *T. atroviride*, *T. hamatum*, *T. harzianum*, *T. koningii*, *T. longibrachiatum*, *T. piluliferum*, *T. polysporum*, *T. pseudokoningii*, and *T. viride* recorded the highest growth inhibition rate against major plant pathogens and a few of them are crop and host specific. Plant pathogens infect all parts of the plant and major infection occurs in the root system and the pathogens include *Phytophthora parasitica* (black eye rot of tomato), *Colletotrichum capsici* (Anthracnose of chili), *F. solani* (wilt of chili), *M. phaseolina* (charcoal rot of sorghum), *A. solani* (early blight), *Rhizoctonia solani* (rice sheath blight/chickpea damping-off), *S. rolfsii* (collar rot of orchids), *Pythium* sp. (damping off), *F. oxysporum* (wilt.), *Aspergillus niger* (collar rot), *Aspergillus flavus* (foot root), *Alternaria porri* (purple blotch), and *S. rolfsii* (bean white mold); many reports are available on biocontrol efficacy (Fig. 10.2) of *Trichoderma* species

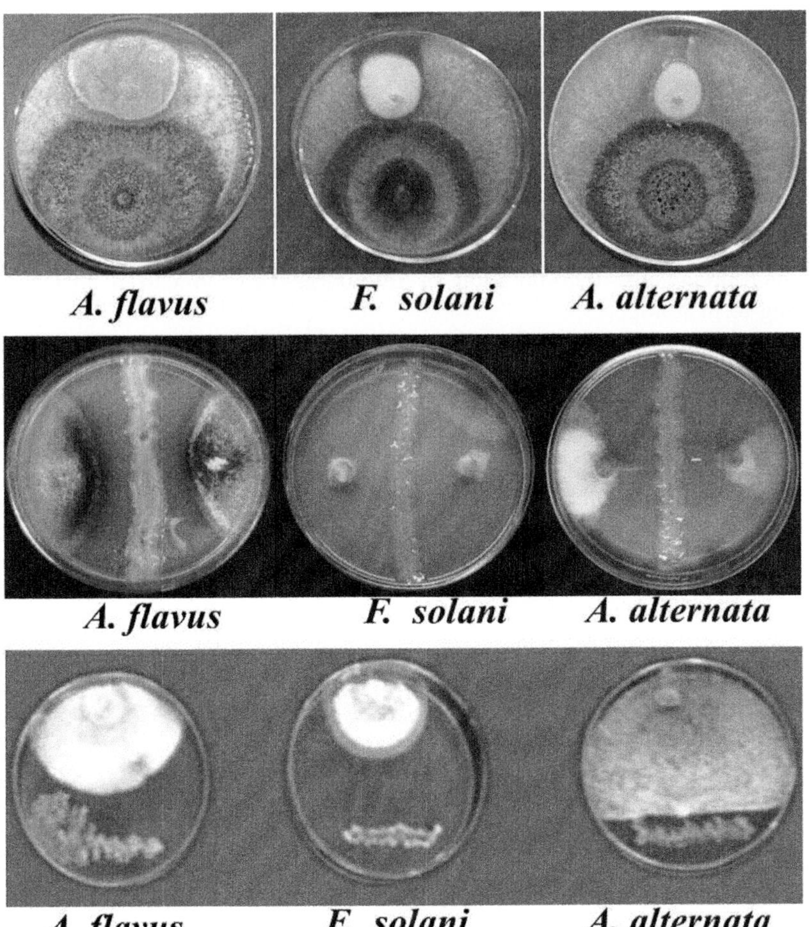

FIGURE 10.2 Biocontrol efficacy of *Trichoderma* (A1), *Azotobacter* (B1), and *Pseudomonas* (C1) isolates against plant pathogens.

(Khan and Sinha, 2007; Kolte and Raut, 2007; Mandhare and Suryawanshi, 2008; Bagwan, 2011; Nainwal and Vishunavat, 2016; Chennappa et al., 2017a,b).

Among all these compounds, 2H-Pyran-2- has been validated in mycotoxin detoxification. Additionally 2H-Pyran-2-, and volatile compounds like diethyl phthalate, 1, 2-benzene dioxylic acid esters, tetra-decanoic acid were reported to be responsible for antagonism against *F. oxysporum* and possesses enhanced biocontrol activity. Studies on the efficacy of volatile metabolites produced by different species of *Trichoderma* tested against *Alternaria alternata* and volatile metabolite (*T. harzianum*- ISO-1) showed a maximum reduction in the mycelial growth (44.60%) of *A. alternata* followed by *T. harzianum* ISO-2 (35.50%) and *T. piluliferum* (36.24%) respectively. *T. viride* has recorded maximum growth inhibition of *Colletotrichum gloeosporioides* (34.37%) followed by *A. alternata* (20%), *F. oxysporum* (15.21%), *Rhizoctonia solani* (12%), and *Curvularia lunata* (13.63%) (Divya, 2015). Similarly, *T. harzianum* has showed highest growth inhibition of *C. gloeosporioides* (35%) followed by *F. oxysporum* (14.80%), *A. alternata* (11.11%), *C. lunata* (9.52%), and *R. solani* (4.54 %) respectively (Sangle et al., 2016). The volatile metabolites assay revealed that the metabolites of *Trichoderma* species had potentially inhibited the mycelia growth of pathogens.

10.7 Common applications formulations and shelf life

Different types of formulations have been developed using *P. fluorescens*, *B. subtillis*, *A. chroococcum*, and *T. viride* either to increase crop health or to manage plant disease depending on the development formulations with suitable carriers that support the survival of bacteria for longer periods. The carriers may be either organic or inorganic in nature, and include peat, talc, lignite, kaolinite, pyrophyllite, zeolite, montmorillonite, alginate, press mud, sawdust, vermiculite, etc. The carriers maintain the survival rate of biocontrol agents by protecting them from desiccation and cell death (Heijnen et al., 1993; Dandurand et al., 1994). The aflotoxin contamination in chili and ground nut was brought down drastically with *P. fluorescens*, which is used for soil enrichment and foliar application (Naik and Sudini, 2011).

The *Trichoderma* species induce various mechanisms to affect growth of plant pathogens, control disease, and improve overall plant health (Singh et al., 2004). The last decade has witnessed a tremendous breakthrough in this aspect, especially in terms of standardization of production techniques of *T. harzianum or T. viride*, *P. florescence* (against fungal diseases) for use against many insect pests and diseases. Biocontrol agents like *Trichoderma* are acclaimed as effective, eco-friendly, and cheap. *Trichoderma* is one of the common fungal biocontrol agents being used worldwide for suitable management of various foliar and soil-borne plant pathogens like *Ceratobasidium*, *Fusarium*, *Rhizoctonia*, *Macrophomina*, *Sclerotium*, *Pythium*, and *Phytophthora* species (Divya, 2015). *Trichoderma* also serve plants as biofertilizers due to their phosphorus solubilizing activity and their ability to decompose organic matter resulting in increased availability of micronutrients to the plants (Singh et al., 2004). Formulations are usually prepared as carrier-based inoculants containing effective microorganisms. Incorporation of microorganisms in carrier material enables easy handling, high effectiveness, and long-term storage.

10.8 Seed biopriming with formulations

Biopriming is a process of seed treatment using liquid formulations or biocontrol agents to protect seeds from diseases and physiological damages. Nowadays, biopriming is gaining wide importance in crop disease management. Biopriming of cowpea seeds with *T. harzianum* reduced root rot incidence at preemergence stage (64%) and postemergence stage (68%) after 40 days of sowing (El-Mohamedy et al., 2006). Similar reports are also recorded on biopriming of faba bean with *T. viride*, *T. harzianum*, *B. subtillis*, and *P. fluorescens*. All the biocontrol agents are effective in protecting the seeds against the incidence of faba bean root rot (El-Mougy and Abdel-Kader, 2008). Maize seeds were bioprimed with *T. harzianum* (1×10^8 spore mL^{-1}) and evaluated against *F. verticillioides* incidence, seed germination, seedling vigor field emergence, yield, and seed weight, and fumonisin production. It was recorded that the *T. harzianum* was more effective in reducing the disease incidence and production of fumonisin by *F. verticillioides* (Chandranayaka et al., 2010). Chickpea seeds on priming with bioagents (*T. viride* and *T. harzianum*) significantly reduced the wilt incidence and increased seed germination and plant growth parameters as compared with chemical fungicides (Kumar et al., 2014). Efficacy of 2,4-DAPG positive fluorescent *Pseudomonas* isolates were tested against chili diseases. The result showed that a consortium (*P. fluorescens* and *T. viride*) recorded the least incidence of damping off (8.50%), wilt (8.60%), followed by the application of *Pseudomonas fluorescens* alone (7.50% and 14%) for damping off (Naik et al., 2014).

10.9 Conclusion

Hence, biodegradation of pesticides and chemical fertilizers plays an important role in agricultural practice to serve safe food to the world and clean environment for the future generations. PGPMs are natural inhabitants of the rhizosphere and have the ability to utilize the abovementioned herbicides and insecticides as a sole carbon source for their energy. Hence, use of PGPMs will help in organic farming to improve soil fertility and quality of the food produce and also to remove environmentally hazardous chemicals from the food chain, thereby ensuring the supply of healthy food for mankind. Therefore this study reveals the application of PGPMs will certainly serve as a safe alternative to chemical pesticides in the management of diseases with their potent plant growth properties.

References

Agarwal, N., Singh, H.P., 2002. Antibiotic resistance and inhibitory effect of *Azotobacter* on soil borne plant pathogens. Indian J. Microbiol. 42, 245–246.

Ahmad, F., Ahmad, I., Khan, M.S., 2005. Indole acetic acid production by the indigenous isolated of *Azotobacter* and fluorescent *Pseudomonas* in the presence and absence of tryptophan. Turkish J. Biol. 29, 29–34.

Ahmad, F., Ahmad, I., Khan, M.S., 2008. Screening of free living rhizospheric bacteria for their multiple growth promoting activities. Microbiol. Res. 173–181.

Aleem, A., Isar, J., Malik, A., 2003. Impact of long term application of industrial wastewater on the emergence of resistance traits of *Azotobacter vinelandii* isolated from rhizosphere soil. Bioresour. Technol. 86, 7–13.

Almon, L., 1958. The vitamin B12 content of *Azotobacter vinelandii*. J. Nutri. 643–648.

Anand, M., Naik, M.K., Ramegouda, G., Rani, D.G.S., 2010. Biocontrol and plant growth promotion activity of indigenous isolates of *Pseudomonas fluorescens*. J. Mycopathol. Res. 48 (1), 45–50.

Antoun, H., Kloepper, J.W., 2001. Plant growth promoting rhizobacteria (PGPR), Encycl. of Genetics. Fitzroy Dearborn, London. 1477–1480.

Bagwan, N.B., 2011. Evaluation of biocontrol potential of *Trichoderma* species against *Sclerotium rolfsii*, *Aspergillus niger* and *Aspergillus flavus*. Int. J. Plant Prot. 4, 107–111.

Bakker, P.A., Lamer, J.G., Bakker, A.W., Schippers, B., 1986. The role of siderophores in potato tuber yield increase by *Pseudomonas putida* in a short rotation of potato. Netherlands J. Plant Pathol. 92, 249–256.

Bangera, M.G., Thomashow, L.S., 1996. Characterization of a genomic locus required for synthesis of the antibiotic 2, 4-diacetylphloroglucinol by the biological control agent *Pseudomonas fluorescens* Q2-87. Mol. Plant Microbiol. Interac. 9 (2), 83–90.

Bano, N., Musarrat, J., 2003. Isolation and characterization of phorate degrading soil bacteria of environmental and agronomic significance. Lett. Microbiol. 36, 349–353.

Barazani, O., Friedman, J., 1999. IAA is the major root growth factor secreted from plant growth mediated bacteria. J. Chem. Ecol. 25, 2397–2407.

Barrera, D.A., Soto, E., 2010. Biotechnological uses of *Azotobacter vinelandii* current state limits and prospects. African J. Biotechnol. 9, 5240–5250.

Becker, J.O., Cook, R.J., 1988. Role of siderophores in suppression of *Pythium* sp. and production of increased growth response of wheat by fluorescent Pseudomonads. Phytopathology 78, 778–782.

Bender, C.L., Rangaswamy, V., Loper, J.E., 1999. Polyketide production by plant associated Pseudomonads. Annu. Rev. Phytopathol. 37, 175–196.

Benitez, T., Rincon, A.M., Limon, M.C., Codon, A.C., 2004. Biocontrol mechanisms of *Trichoderma* strains. Int. Microbiol. 7 (4), 249–260.

Bhosale, H.J., Kadam, T.A., Bobade, A.R., 2013. Identification and production of *Azotobacter vinelandii* and its antifungal activity against *Fusarium oxysporum*. J. Env. Biol. 34, 177–182.

Bowden, K., Broadbent, J.L., Ross, W.J., 1965. Some simple anti-helmintics. Br. J. Pharmacol. 24, 714–724.

Brian, P.W., 1944. Production of gliotoxin by *Trichoderma viride*. Nature. 154, 667–668.

Bunker, R.N., Mathur, K., 2001. Antagonism of local biocontrol agents to *Rhizoctonia solani* inciting dry root rot of chili. J. Mycol. Plant Pathol. 31 (1), 50–53.

Calistru, C., McLean, M., Berjak, P., 1997. In vitro studies on the potential for biological control of *Aspergillus flavus* and *Fusarium moniliforme* by *Trichoderma* species. A study of the production of extracellular metabolites by *Trichoderma* species. Mycopathologia. 137, 115–124.

Carine, L., Gadkar, J.V., Wallon, T., Novinscak, A., Martin, F., 2012. Production of DAPG and HCN by *Pseudomonas* sp. LBUM300 contributes to the biological control of bacterial canker of tomato. Biol. Control. 102 (10), 967–973.

Castillo, J.M., Casas, J., Romero, E., 2011. Isolation of an endosulfan-degrading bacterium from a coffee farm soil: Persistence and inhibitory effect on its biological functions. Sci. Total Environ. 20–27.

Cavaglieri, L.R., Andres, L., Ibanez, M., Etcheverry, M.G., 2005. Rhizobacteria and their potential to control *Fusarium verticillioidies*, effect of maize bacterisation and inoculums density. Antonie Van Leeuwenhoek. 87, 179–187.

Chandranayaka, S., Niranjana, S.R., Shankara, U.A.C., Raj, N.S., Reddy, M.S., Prakash, H.S., et al., 2010. Seed biopriming with novel strain of *Trichoderma harzianum* for the control of toxigenic *Fusarium verticillioides* and fumonisins in maize. Arch. Phytopathol. Plant Prot. 43 (3), 264–282.

Chen, L.L., Liu, L.J., Shi, M., Song, X.Y., Zheng, C.Y., Chen, X.L., et al., 2009. Characterization and gene cloning of a novel serine protease with nematicidal activity from *Trichoderma pseudokoningii* SMF2. FEMS Microbiol. Lett. 299, 135–142.

Chennappa, G., 2015. Impact of Pesticides on *Azotobacter* Species Isolated From Paddy Field Soil Samples. Thesis. University of Mysore.

Chennappa, G., Adkar-Purushothama, C.R., Suraj, U., Tamilvendan, K., Sreenivasa, M.Y., 2013. Pesticide tolerant *Azotobacter* isolates from paddy growing areas of northern Karnataka, India. World J. Microbiol. Biotechnol. 30 (1), 1–7.

Chennappa, G., Adkar-Purushothama, C.R., Naik, M.K., Sreenivasa, M.Y., 2014. Impact of pesticides on PGPR activity of *Azotobacter* sp. isolated from pesticide flooded paddy soils. Greener. J. Biol. Sci. 4 (4), 117–129.

Chennappa, G., Naik, M.K., Adkar-Purushothama, C.R., Amaresh, Y.S., Sreenivasa, M.Y., 2016a. PGPR, abiotic stress tolerant and antifungal activity of *Azotobacter* sp. isolated from paddy soils. Indian J. Expt. Biol. 54, 322–331.

Chennappa, G., Naik, M.K., Sreenivasa, M.Y., 2016b. *Azotobacter*—PGPR activities with special reference to effect of pesticides and biodegradation. In: Singh, D.P. (Ed.). Microbial Inoculants in Sustainable Agricultural Productivity, Vol. 2: Functional Applications, Springer, India. 13(1), pp. 229–244.

Chennappa, G., Naik, M.K., Amaresh, Y.S., Ravikiran, Ashwathanarayan, D.S., Patil, M.G., et al., 2017a. Morphological, molecular characterization of *Trichoderma* species isolated from different rhizosphere soils and its anti-pathogenic properties. Imperial J. Interdisci. Res. 3 (4), 1–12.

Chennappa, G., Naik, M.K., Amaresh, Y.S., Nagaraj, H., Sreenivasa, M.Y., 2017b. *Azotobacter*—a potential bio-fertilizer and bio inoculants for sustainable agriculture. In: Panpatte, D., Jhala, Y.K., Vyas, R., Shelat, H.N. (Eds.), Microorganisms for Green Revolution. Microbes for Sustainable Crop Prodcution, 1. Springer Nature Publishing House, Singapore, pp. 87–106.

Chennappa, G., Sreenivasa, M.Y., Nagaraja, H., 2018a. *Azotobacter salinestris* - A novel pesticide degrading PGPR bacteria. In: Panpatte, D., Jhala, Y.K., Vyas, R., Shelat, H.N. (Eds.), Microorganisms for Green Revolution. Springer Nature Singapore, pp. 23–43. , 2.

Chennappa, G., Nidoni, U., Nagaraja, H., Amaresh, Y.S., Sreenivasa, M.Y., 2018b. *Azotobacter*—a natural resource for bioremediation of toxic pesticides in soil ecosystems. In: Microbial Biotechnology in Agro-environmental Sustainability, Elsevier Publishing House.

Coley-Smith, J.R., Ridout, C.J., Mitchell, C.M., Lynch, J.M., 1991. Control of button rot disease of lettuce (*Rhizoctonia solani*) using preparation of *Trichoderma viride*, *T. harzianum*. Plant Pathol. 40, 359–366.

Dandurand, L.M., Morra, M.J., Chaverra, M.H., Orser, C.S., 1994. Survival of *Pseudomonas* spp. in air dried mineral powders. Soil Biol. Biochem. 26, 1423–1430.

Degenkolb, T., Dieckmann, R., Nielsen, K.F., Grafnhan, T., Theis, C., Zafari, D., et al., 2008. The *Trichoderma brevicompactum* clade: a separate lineage with new species, new peptaiotics and mycotoxins. Mycol. Progress. 7, 177–219.

Divya, N., 2015. Biological and Molecular Characterization of Indigenous Isolates of *Trichoderma* Species. Thesis. University of Agricultural Sciences, Raichur.

Dubey, S.C., Tripathi, A., Dureja, P., Grover, A., 2011. Characterization of secondary metabolites and enzymes produced by *Trichoderma* species and their efficacy against plant pathogenic fungi. Indian. J. Agri. Res. 81 (5), 455–461.

El-Mohamedy, R.S.R., AbdAlla, M.A., Badiaa, R.I., 2006. Soil amendment and seed bio-priming treatments as alternative fungicides for controlling root rot diseases on cowpea plants in Nobaria province. Res. J. Agri. Biol. Sci. 2 (6), 391–398.

El-Mougy, N.S., Abdel-Kader, M.M., 2008. Long-term activity of bio-priming seed treatment for biological control of faba bean root rot pathogens. Australias. Plant Pathol. 37 (5), 464–471.

Evidente, A., Cabras, A., Maddau, L., Serra, S., Andolfi, A., Motta, A., 2003. Virepyronone, a new antifungal 6-substituted 2H pyran-2-one produce by *Trichoderma viride*. J. Agri. Food Chem. 51, 6957–6960.

Eziashi, E.I., Uma, N.U., Adekunle, A.A., Airede, C.E., 2006. Effect of metabolites produced by *Trichoderma* species against *Ceratocystis paradoxa* in culture medium. African J. Biotechnol. 5, 703–706.

Fernando, W.G.D., Nakkeeran, S., Zhang, Y., 2005. Biosynthesis of antibiotics by PGPR and its relation in biocontrol of plant diseases. PGPR: Biocontrol Biofertilization 2, 67–109.

Geremia, R.A., Goldman, G.H., Jacobs, D., Ardiles, W., Vila, S.B., et al., 1993. Molecular characterization of the proteinase encoding gene, *prb1* related to mycoparasitism by *Trichoderma harzianum*. Mol. Microbiol. 8, 603–613.

Ghosh, P.G., Sawant, N.A., Patil, S.N., Aglave, B.A., 2010. Microbial biodegradation of organophosphate pesticides. Int. J. Biotechnol. Biochem. 6, 871–876.

Glick, B.R., 1995. The enhancement of plant growth by free living bacteria. Canadian J. Microbiol. 41, 109–117.

Handelsman, J., Stabb, E.V., 1996. Biocontrol of soil borne plant pathogens. Plant Cell. 8, 1855–1869.

Harman, G.E., Howell, C.R., Viterbo, A., Lorito, I.M., 2004. *Trichoderma* species opportunistic, avirulent plant symbionts. Nat. Rev. 2, 43–56.

Hayen, H., Volmer, D.A., 2005. Rapid identification of siderophores by combined thin-layer chromatography/matrix-assisted laser desorption/ionization mass spectrometry. Rapid Commun. Mass Spectrom 19 (5), 711–720.

Heijnen, C.E., Burgers, S.L.G.E., Van Veer, J.A., 1993. Metabolic activity and population dynamics of rhizobia introduced into un-amended and betonite amended loamy sand. Appl. Env. Microbiol. 59, 743–747.

Helmy, Q., Kardena, E., Nurachman, Z., Wisjnuprapto, 2010. Application of bio-surfactant product by *Azotobacter vinelandii* AV01 for enhanced oil recovery and biodegradation of oil sludge. Int. J. Civil Env. Eng. 10, 7–14.

Howell, C.R., 1998. The role of antibiosis. In: Harman, G.E., Kubicek, C.P. (Eds.), *Trichoderma* and *Gliocladium*, vol. 2. Taylor & Francis, London, pp. 173–184. , Enzymes, biological control, and commercial applications.

Howell, C.R., Stipanovic, R.D., 1980. Suppression of *Pythium ultimum* induced damping off of cotton seedlings by *Pseudomonas fluorescens* and its antibiotic, pyoluteorin. Phytopathology 70, 712–715.

Howell, R.C., 2003. Mechanisms employed by *Trichoderma* species in the biological control of plant diseases: the history and evolution of current concepts. Plant Dis. 87, 4–10.

Iavicoli, A., Boutet, E., Buchala, A., Metraux, J.P., 2003. Induced systemic resistance in *Arabidopsis thaliana* in response to root inoculation with *Pseudomonas fluorescens* CHA0. Mol. Plant Microbe. Interact. 16, 851–858.

Ihrmark, et al., 2010. Comparative molecular evolution of *Trichoderma* chitinases in response to mycoparasitic interactions. Evol. Bioinform. 6, 1–26.

Jaklitsch, W.M., 2009. European species of Hypocrea part I. The green spored species. Stud. Mycol. 63, 1–91.

Javed, A.M., Hafiz, N.A., Shahzad, K., Arshad, M., 2009. Role of plant growth promoting rhizobacteria applied in combination with compost and mineral fertilizers to improve growth and yield of wheat (*Triticum aestivum* L.). Pak. J. Bot. 41 (1), 381–390.

Jayalakshmi, S.K., Raju, S., Usha, R., Benagi, V.I., Sreeramu, K., 2009. *Trichoderma harzianum* L1 as a potential source for lytic enzymes and elicitor of defense responses in chickpea (*Cicer arietinum*) against wilt disease caused by *Fusarium oxysporum* f. sp. ciceri. Aust. J. Crop Sci. 3 (1), 44–52.

Johri, B.N., Sharma, A., Virdi, J.S., 2003. Rhizobacterial diversity in India and its influence on soil and plant health. Adv. Biochem. Engg. Biotechnol. 84, 49–89.

Kannapiran, E., Ramkumar, S.V., 2011. Inoculation effect of nitrogen fixing and phosphate solubilizing bacteria to promote growth of black gram (*Phaseolus mungo* Roxb; Eng). Ann. Biol. Res. 2 (5), 615–621.

Kapulnik, Y., Chet, I., 2000. Induction and accumulation of PR proteins activity during early stages of root colonization by the mycoparasite *T. harzianum* strain T-203. Plant Physiol. Biochem. 38, 863–873.

Kataryan, B.T., Torgashova, G.G., 1976. Spectrum and activity of the herbicidal effect of 2,4-diacetylphloroglucinol. Dokl. Akad. Nauk. Arm. SSR 63, 109–112.

Khan, A., Shaukat, S.S., Islam, S., Khan, A., 2012. Evaluation of fluorescent Pseudomonad isolates for their activity against some plant-parasitic nematodes. American-Eurasian J. Agric. Env. Sci. 12 (11), 1496–1506.

Khan, A.A., Sinha, A.P., 2007. Screening of *Trichoderma* spp. against *Rhizoctonia solani* the causal agent of rice sheath blight. Indian Phytopath. 60 (4), 450–456.

Khan, H.R., Mohiuddin, M., Rahman, M., 2008. Enumeration, isolation and identification of nitrogen-fixing bacterial strains at seedling stage in rhizosphere of rice grown in non-calcareous grey flood plain soil of Bangladesh. J. Faculty. Env. Sci. Technol. 13, 97–101.

Kirner, S., 1998. Functions encoded by pyrrolnitrin biosynthetic genes from *Pseudomonas fluorescens*. J. Bacteriol. 180, 1939–1943.

Kitten, T., Kinscherf, T., McEvoy, G., Willis, D.K., 1998. A newly identified regulator is required for virulence and toxin production in *Pseudomonas syringae*. Mol. Microbiol. 28, 917–929.

Kloepper, J.W., 1992. Plant growth promoting rhizobacteria as biological control agents. In: Marcel Decker (Ed.), Soil Microbial Ecology. Applications in Agriculture, Forestry and Environmental Management. Dekker, New York, pp. 255–274.

Kloepper, J.W., Schroth, M.N., 1980. Plant growth promoting rhizobacteria on radishes. Proceeding International Conference Plant Pathology and Bacteriology 2: 879–882.

Kolte, V.S., Raut, S.P., 2007. Efficacy of *Trichoderma spp.* against *Sclerotium rolfsii* infecting collar rot of orchids (*Dendrobium orchreatum*). Ann. Plant Prot. Scis. 15 (2), 469–539.

Komon-Zelazowska, M., Neuhof, T., Dieckmann, R., Dohren, V.H., Herrera-Estrella, A., 2007. Formation of atroviridin by *Hypocrea atroviridis* is conidiation associated and positively regulated by blue light and the G protein GNA3. Eukaryot. Cell 6, 2332–2342.

Kraepiel, A., Bellenger, J., Wichard, T., Morel, F., 2009. Multiple roles of siderophores in free living nitrogen fixing bacteria. Biometals. 22, 573–581.

Kredics, L., Antal, Z., Szekeres, A., Hatvani, L., Manczinger, L., Vagvolgyi, C., et al., 2005. Extracellular proteases of *Trichoderma* species. A Review. Acta. Microbiol. Immunol. Hung. 52, 169–184.

Kubicek, C.P., Mach, R.L., Peterbuer, C.K., Lorito, M., 2001. *Trichoderma:* from genes to biocontrol. J. Plant. Pathol. 83, 11–24.

Kubicek, C.P., Komon, Z.M., Druzhinina, I.S., 2008. Fungal genes *Hypocrea/Trichoderma*: from barcodes to Biodiversity. J. Zhjiang. Uni. Sci. B. 9 (10), 753–763.

Kucuk, C., Kyvanc, M., 2008. Mycoparasitism in the biological control of *Gibberella zeae* and *Aspergillus ustus* by *Trichoderma harzianum* strains. J. Agri. Technol. 4 (2), 49–55.

Kumar, V., Shahid, M., Singh, A., Srivastava, M., Mishra, A., Srivastava, Y.K., et al., 2014. Effect of biopriming with biocontrol agents *Trichoderma harzianum* (Th. Azad) and *Trichoderma viride* (01pp) on Chickpea Genotype (Radhey). J. Plant Pathol. Microbiol. 5 (5), 252–256.

Kyalo, G., Affokpon, A., Coosemans, J., Coynes, D.L., 2007. Biological control effects of *Pochonia chlamysdosporia* and *Trichoderma* isolates from Benin (West-Africa) on root knot nematodes. Commun. Agri. Appl. Biol. Sci. 72, 219–223.

Leisinger, T., Margraff, R., 1979. Secondary metabolites of fluorescent Pseudomonads. Microbiol. Rev. 43, 422–442.

Lopez, J.G., Toledo, M.V., Reina, S., Salmeron, V., 1981. Root exudates of maize on production of auxins, gibberellins, cytokinins, amino acids and vitamins by *Azotobacter chroococcum* chemically defined media and dialysed soil media. Toxicol. Env. Chem. 33, 69–78.

Lugtenberg, B.J.J., Dekkers, L., Bloemberg, G.V., 2001. Molecular determinants of rhizosphere colonization by Pseudomonads. Annual Rev. Phytopathol. 39, 461–490.

Mali, G.V., Bodhankar, M.G., 2009. Antifungal and phytohormone potential of *Azotobacter chroococcum* isolates from ground nut (*Arachis hypogeal* L.) rhizosphere. Asian J. Exp. Sci. 23, 293–297.

Mandhare, V.K., Suryawanshi, A.V., 2008. Efficacy of some botanicals and *Trichoderma* species against soil borne pathogens infecting chickpea. J. Food. Legumes. 21 (2), 122–124.

Mariola, C.R., Rodriguez-Galan, O., Benitez, T., Sousa, S., Rey, M., Llobell, A., et al., 2007. Microscopic and transcriptome analyses of early colonization of tomato roots by *Trichoderma harzianum*. Int. Microbiol. 10, 19–27.

Martin, X.M., Sumathi, C.S., Kannan, V.R., 2011. Influence of agrochemical and *Azotobacter* spp. application on soil fertility in relation to maize growth under nursery conditions. Eurasian J. Biosci. 5, 19–28.

Maurhofer, M., Keel, C., Schnider, U., Voisard, C., Haas, D., Defago, G., 1992. Influence of enhanced antibiotic production in *Pseudomonas fluorescens* strain CHAO on its disease-suppressive capacity. Phytopathology 82, 190–195.

Maurhofer, M., Hasse, C., Maurwly, D., Metranx, J.P., Defago, G., 1994. Induction of systemic resistance of tobacco to tobacco necrosis virus by the root colonizing *Pseudomonas fluorescens* CHAO: influence of the gac A gene and of pyoverdine production. Phytopathology 84, 136–146.

Mazurier, S., Corberand, T., Lemanceau, P., Raajimakers, J.M., 2009. Phenazine antibiotics produced by fluorescent Pseudomonads contribute to natural soil suppressiveness to *Fusarium* wilt of chick pea. J. I.S.M.E. 3, 977–991.

Mollmann, U., Heinisch, L., Bauernfeind, A., Kohler, T., Ankel-Fuchs, D., 2009. Siderophores as drug delivery agents: application of the Trojan horse strategy. Biometals. 22, 615–624.

Moneke, A.N., Okpala, G.N., Anyanwu, C.U., 2010. Biodegradation of glyphosate herbicide *in vitro* using bacterial isolates from four rice fields. African J. Biotechnol 9, 4067–4074.

Monte, E., 2001. Understanding *Trichoderma*: between biotechnology and microbial ecology. Int. Microbiol. 4, 1–4.

Mrkovacki, N.B., Nikola, A., Cacic, Milic, V.M., 2002. Effects of Pesticides on *Azotobacter chroococcum*. Proceedings. Nat. Sci 23–28.

Muhammad, S., Amusa, N.A., 2003. *In-vitro* inhibition of growth of some seedling blight inducing by manure inhabiting microbes. African J. Biotechnol. 2 (6), 161–164.

Myresiotis, C.K., Vryzas, Z., Mourkidou, E.P., 2012. Biodegradation of soil applied pesticides by selected strains of plant growth promoting rhizobacteria (PGPR) and their effects on bacterial growth. Biodegradation. 23, 297–310.

Nagaraja, H., Chennappa, G., Rakesh, S., Naik, M.K., Amaresh, Y.S., Sreenivasa, M.Y., 2016. Antifungal activity of *Azotobacter nigricans* against trichothecene producing *Fusarium* species associated with cereals. Food. Sci. Biotechnol. 25 (4), 1197–1204.

Naik, M.K., Sudini, H., 2011. Aflatoxin contamination of food commodities and their management. In: Plant Pathology in India: Vision 2030. Indian Phytopathol. Soc. pp. 1–291.

Naik, M.K., Rajalaxmi, K., Amaresh, Y.S., 2013. Search for 2, 4 DAPG positive genes in fluorescent *Pseudomonas* and their exploitation for sustainable disease management. Recent Advances in biofertilizer and bio fungicides (PGPR) for sustainable agriculture. Proc 3rd Asian PGPR Conf. 21–24.

Naik, M.K., Rajalaxmi, K., Rangeshwaran, R., Amaresh, Y.S., Reddy, M.S., 2014. Search for 2, 4 DAPG positive gene in fluorescent *Pseudomonas* and their exploitation for sustainable disease management. Cambridge Scholar 152–163.

Naik, M.K., Chennappa, G., Vinay, J.U., 2016. Potentials antibiotics of fluorescent Pseudomonads in sustainable crop disease management. In: Chowdappa, P., Sharma, P., Singh, D., Misra, A.K. (Eds.), Contemporary Plant Pathology, IPS 6th International Conference on Plant, Pathogens and People, pp. 379–398.

Nainwal, D., Vishunavat, K., 2016. Management of purple blotch and Stemphylium blight of onion in Tarai and Bhabar regions of Uttarakhand. India. J. Appl. Natural Sci. 8 (1), 150–153.

Nakkeeran, S., Krishnamoorthy, A.S., Ramamoorthy, V., Renukadevi, P., 2002. Microbial inoculants in plant disease control. J. Ecobiol 14, 83–94.

Nandakumar, R., Babu, S., Radjacommare, R., Raguchander, T., Samiyappan, R., 2002. *Pseudomonas fluorescens* mediated antifungal activity against *Rhizoctonia solani* causing sheath blight in rice. Phytopathol. Mediterr 41, 109–119.

Neilands, J.B., 1981. Iron absorption and transport in microorganisms. Annu. Rev. Nutr l, 27–46.

Neuhof, T., Dieckmann, R., Druzhinina, I.S., Kubicek, C.P., VonDohren, H., 2007. Intact cell MALDI TOF mass spectrometry analyses of peptaibols formation by the genus *Trichoderma*: can molecular phylogenetic knowledge predict peptaibols structures. Microbiology 153 (10), 3417–3437.

Nitish, R.B., Kumar, J., 2017. Characterization of volatile secondary metabolites from *Trichoderma asperellum*. J. Appl. Nat. Sci. 9 (2), 954–959.

Nowak-Thompson, B., Chaney, N., Wing, J.S., Gould, S.J., Loper, J.E., 1999. Characterization of the pyoluteorin biosynthetic gene cluster of *Pseudomonas fluorescens* Pf-5. J. Bacteriol. 181 (7), 2166–2174.

Page, W., Von Tigerstrom, M., 1988. Aminochelin, a catecholamine siderophore produced by *Azotobacter vinelandii*. J. Gen. Microb. 134, 453–460.

Pal, K.K., McSpadden-Gardener, B., 2006. Biological control of plant pathogens. Plant Health Inst. 10, 1117–1122.

Patten, C.L., Glick, B.R., 2002. Role of *Pseudomonas putida* indole acetic acid in development of the host plant root system. Appl. Env. Microbiol. 68, 3795–3801.

Pierson, L.S., Gaffney, T., Lam, S., Gong, F., 1995. Molecular analysis of genes encoding phenazine biosynthesis in the biological control bacterium *Pseudomonas aureofaciens*. FEMS Microbiol. Lett 134, 299–307.

Raaijmakers, J.M., Vlami, M., de-Souza, J.T., 2002. Antibiotic production by bacterial biocontrol agents. Anton van Leeuw 81, 537–547.

Rajalaxmi, K., Naik, M.K., Amaresh, Y.S., Sunkad, G., Patil, M.G., Hosamani, A.C., 2012. Shelf life of two indigenous isolates of fluorescent Pseudomonads as bioagents in different carrier materials. J. Mycopathol. Res. 50 (1), 85–88.

Ramamoorthy, V., Viswanathan, R., Raguchander, T., Prakasam, V., Samiyappan, R., 2001. Induction of systemic resistance by plant growth promoting rhizobacteria in crop plants against pests and diseases. Crop Protec 20, 1–11.

Reinhardt, E.L., Ramos, P.L., Manfio, G.P., Barbosa, H.R., Pavan, C., Filho, C.A.M., 2008. Molecular characterization of nitrogen fixing bacteria isolated from Brazilian agricultural plants at also Paulo state. Brazilian J. Microb. 39, 414–422.

Reino, J.L., Guerrero, R.F., Hernadez-Galan, R., Collado, I.G., 2008. Secondary metabolites from species of the biocontrol agent *Trichoderma*. Phytochem. Rev. 7, 89–123.

Reither, B., Brunner, K., Schuhmacher, R., Piessi, I., Seidi, V., Krska, R., et al., 2005. The G protein alpha subunit Tga1 of *Trichoderma atraviride* is involved in chitinase formation and differential production of antifungal metabolites. Fungal Genet. Biol. 42, 749–760.

Reshma, P., Naik, M.K., Aiyaz, M., Niranjana, S.R., Chennappa, G., Sheikh, S.S., et al., 2018. Induced systemic resistance by 2, 4-diacetylphloroglucinol positive fluorescent *Pseudomonas* strains against rice sheath blight. Indian J. Exp. Biol. 56, 207–212.

Revillas, J.J., Rodelas, B., Pozo, C., Toledo, M.V., Gonzalez-Lopez, J., 2000. Production of B-group vitamins by two *Azotobacter* strains with phenolic compounds as sole carbon source under diazotrophic and adiazotrophic conditions. J. Appl. Microbiol. 89 (3), 486–493.

Roco, A., Perez, L.M., 2001. *In vitro* biocontrol activity of *Trichoderma harzianum* on *Alternaria alternata* in the presence of growth regulators. Electronic J. Biotechnol. 4 (2), 68–73.

Rosales, A.M., Thomashow, R.J.C., Mew, T.W., 1995. Isolation and Identification of antifungal metabolites produced by rice associated antagonistic *Pseudomonas* spp. Phytopathology 85 (9), 1028–1032.

Sachin, D.N., 2009. Effect of *Azotobacter chroococcum* (PGPR) on the growth of bamboo (*Bambusa bamboo*) and maize (*Zea mays*) plants. Biofrontiers. 1, 24–31.

Sangle, U.R., Kumar, S., Mishra, J.S., 2016. Effect of volatile metabolites released by different *Trichoderma* spp on the growth of pathogen *Sclerotium rolfsii* and effect on *Sclerotia germination*. The Bioscan. 11 (3), 1559–1564.

Schnider, U., Keel, C., Blumer, C., Troxler, J., Defago, G., Haas, D., 1995. Amplification of the house keeping sigma factor in *Pseudomonas fluorescens* CHA0 enhances antibiotic production and improves biocontrol abilities. J. Bacteriol. 177, 5387–5539.

Seidi, V., Song, L., Lindquist, E., Gruber, S., Koptchinskiy, A., et al., 2009. Transcriptomic response of the mycoparasitic fungus *Trichoderma atraviride* to the presence of 546 a fungal prey. BMC Genom. 10, 567.

Sharma, K., Sexena, A., Dak, G., Sharma, R., Agarwal, A., 2007. Isolation and assay of antifungal activity of siderophore producing *Pseudomonas aeruginosa*. J. Mycol. Plant Pathol. 37, 251–253.

Shoresh, M., Harman, G.E., Mastouri, F., 2010. Induced systemic resistance and plant responses to fungal biocontrol agents. Annu. Rev. Phytopathol. 48, 1–23.

Siddiquee, S., Cheong, B.E., Taslima, K., Kausar, H., Hasan, M.M., 2012. Sequence and identification of volatile compounds from liquid cultures of *Trichoderma harzianum* by GCMS using three different capillary columns. J. Chromatographic Sci. 50, 358–367.

Singh, B.K., Walker, A., Morgan, J.A.W., Wright, D.J., 2004. Biodegradation of chloropyrifos by Enterobacter strain B-14 and its use in bioremediation of contaminated soils. Appl. Env. Microbiol. 4855–4863.

Singh, C., Tiwari, S., Gupta, V.K., Singh, J.S., 2018. The effect of rice husk biochar on soil nutrient status, microbial biomass and paddy productivity of nutrient poor agriculture soils. Catena 171, 485–493.

Singh, J.S., 2013. Plant growth promoting rhizobacteria: potential microbes for sustainable agriculture. Resonance 18 (3), 275–281.

Singh, J.S., 2015. Microbes: the chief ecological engineers in reinstating equilibrium in degraded ecosystems. Agric. Ecosyst. Environ. 203, 80–82.

Singh, J.S., Gupta, V.K., 2018. Soil microbial biomass: A key soil driver in management of ecosystem functioning. Sci. Total Environ. 634, 497–500.

Sivan, A., Chet, I., 1989. The possible role of competition between *Trichoderma harzianum* and *Fusarium oxysporum* on rhizosphere colonization. Phytopathology 79, 198–203.

Sivasithamparam, K., Ghisalberti, F.L., 1998. Secondary metabolism *Trichoderma* and *Gliocladium*. In: Kubicek, C.P., Harman, G.E. (Eds.), *Trichoderma* and *Gliocladium*. Taylor and Francis Ltd, London, pp. 139–191.

Spaepen, S., Vanderleyden, J., Remans, R., 2007. Indole 3 acetic acid in microbial and microorganism plant signaling. FEMS Microb. Rev. 31, 425–448.

Stoppacher, N., Zeilinger, S., Omann, M., Lassahn, P.G., Roitinger, A., Krska, R., et al., 2008. Characterisation of the peptaibiome of the biocontrol fungus *Trichoderma atroviride* by liquid chromatography/tandem mass spectrometry. Rapid Commun. Mass Spectrom. 22, 1889–1898.

Suarez, M.B., Vizcaino, J.A., Llobell, A., Mont, E., 2007. Characterization of genes encoding novel peptidass in the biocontrol fungus *Trichoderma harzianum* CECT 2413 using the TrichoEST functional genomics approach. Curr. Genet. 51, 331–342.

Thomashow, L.S., Weller, D.M., 1996, Current concepts in the use of introduced bacteria for biological disease control: mechanisms and antifungal metabolites. In: Stacey, G., Keen, M. (Eds.), Plant-Microbe Interactions. vol. 1, Springer, Boston, MA, pp. 187–235.

Tiwari, P., Singh, J.S., 2017. A plant growth promoting rhizospheric *Pseudomonas aeruginosa* strain inhibits seed germination in *Triticum aestivum* (L) and *Zea mays* (L). Microbiol. Res. 8, 1–7.

Tiwari, S., Singh, C., Singh, J.S., 2018. Land use changes: a key ecological driver regulating methanotrophs abundance in upland soils. Energy Ecol. Environ. 3, 355–371.

Upadhyay, S.K., Singh, D.P., Saikia, R., 2009. Genetic diversity of plant growth promoting rhizobacteria isolated from rhizospheric soil of wheat under saline condition. Cur. Microbiol . Available from: https://doi.org/10.1007/s00284-009-9464-1.

VanLoon, L.C., Bakker, P.A.H.M., Pierterse, C.M.J., 1998. Systemic resistance induced by rhizosphere bacteria. Annu. Rev. Phytopathol. 36, 453–483.

Vasudevan, P., Kavitha, S., Priyadarisini, V.B., Babujee, L., Gnanamanickam, S.S., 2002. Biological control of rice diseases. In: Gnanamanickam, S. S. (Ed.), Biological control of crop diseases. Marcel Dekker, New York, pp. 11–32.

Velusamy, P., Immanuel, J.E., Gnanamanickam, S.S., Thomashow, L., 2006. Biological control of rice bacterial blight by plant associated bacteria diacetyl phloro glucinol. Canadian J. Microbiol. 52 (1), 56–65.

Vey, A., Hoagland, R.E., Butt, T.M., 2001. Toxic metabolites of fungal biocontrol agents. In: Butt, T.M., Jackson, C., Magan, N. (Eds.), Fungi as Biocontrol Agents: Progress, Problems and Potential. CAB Int. Bristol. pp. 311–346.

Vimal, S.R., Patel, V.K., Singh, J.S., 2018. Plant growth promoting *Curtobacterium albidum* strain SRV4: an agriculturally important microbe to alleviate salinity stress in paddy plants. Ecol Indic. (Article in Press).

Vinay, J.U., Naik, M.K., Rangeshwaran, R., Chennappa, G., Shaikh, S.S., Sayyed, R.Z., 2016. Detection of antimicrobial traits in fluorescent pseudomonads and molecular characterization of an antibiotic pyoluteorin. 3 Bitech. 6, 227–238.

Viterbo, A., Ramot, O., Chemin, L., Chet, I., 2002. Significance of lytic enzymes from *Trichoderma* spp. in the biocontrol of fungal plant pathogens. Ant. Van. Leeuw. 81, 549–556.

Vivek, A., Suresh, W., Om, P.S., Jitendra, K., Aditi, K., Jay, S., et al., 2014. Isolation, characterisation of major secondary metabolites of the Himalayan Trichoderma koningii and their antifungal activity. Arch. Phytopathol. Plant Protect. 47 (9), 1063–1071.

Whilhite, S.E., Stanley, D.C., 1996. Timing of gliotoxin biosynthesis in the fungal biological control agent *Gliocladium virens* (*Trichoderma virnes*). Appl. Microbiol. Biotechnol. 45, 513–518.

Yuan, Z., Cang, S., Matsupuji, M., Nakata, K., Nagamatsu, Y., Yoshimoto, A., 1998. High production of pyoluteorin and 2,4-diacetylphloroglucinol by *Pseudomonas fluorescens* S272 grown on ethanol as a sole carbon source. J. Fermentation Bioengg. 86 (6), 559–563.

Zhou, T., Chen, D., Li, C., Sun, Q., Li, L., Liu, F., et al., 2012. Isolation and characterization of *Pseudomonas brassicacearum* J12 as an antagonist against *Ralstonia solanacearum* and identification of its antimicrobial components. Microbiol. Res. 167, 388–394.

Chapter 11

Harnessing the soil microbial wealth for enhancement of plant secondary metabolites in medicinal and aromatic plants

Akanksha Singh, Rupali Gupta, Shruti Chaturvedi and Rakesh Pandey
Department of Microbial Technology and Nematology, CSIR—Central Institute of Medicinal and Aromatic Plants, Lucknow, India

Chapter Outline

11.1 Introduction

Microbial colonizers in the plant rhizosphere play a crucial role in plant overall health by affecting their development and physiology. Though many microbes are beneficial for the plant's well-being, many act as foes by directly harming them and others by acting as true opportunists (Mendes et al., 2013). The assortment and purpose of microbes colonizing the rhizosphere is related to the release of various kinds of root exudates, biogeochemical reactions, and respiratory processes (Narula et al., 2009). The diverse compounds and molecules discharged by the plant roots into the neighborhood of the rhizospheric zone perform multiple functions like attraction of beneficial microbes, inhibition of pathogenic microbes, etc. Like the various agricultural crops, medicinal plants are also affected by the beneficial effects of the soil microbes.

The manipulation of the soil microbiome for optimizing crop productivity is a long-used strategy, however, alteration done with the aim of enhancing the secondary metabolite content in medicinal and aromatic plants (MAPs) is relatively novel (Köberl et al., 2013). Many bacterial genera belonging to *Alcaligenes*, *Arthrobacter*, *Bacillus*, *Pseudomonas*, and *Rhizobium* have been found to demonstrate advantageous effects on the overall well-being of plants (Tilak et al., 2005). The main modes of mechanism adopted by plant growth-promoting rhizobacteria (PGPR) include production of plant hormones, solubilization of minerals like iron and phosphorus, production of iron chelating siderophores, and stimulation of systemic resistance pathway (Bhattacharyya and Jha, 2012). Indirect benefits of PGPR include biocontrol of harmful pathogens whose management aids in better growth of the plants (Podile and Kishore, 2006; Singh et al., 2013a,b) (Fig. 11.1). Medicinal plants, being a potential reservoir of secondary metabolites, maintain an immense assortment of microflora in their own rhizosphere.

Medicinal plants make up a huge percent of the flora that supply raw materials for primary health care, health foods, cosmetics and pharmaceutical industries. Globally, the requirement of industries and people is met exclusively from the

New and Future Developments in Microbial Biotechnology and Bioengineering. DOI: https://doi.org/10.1016/B978-0-12-818258-1.00011-X

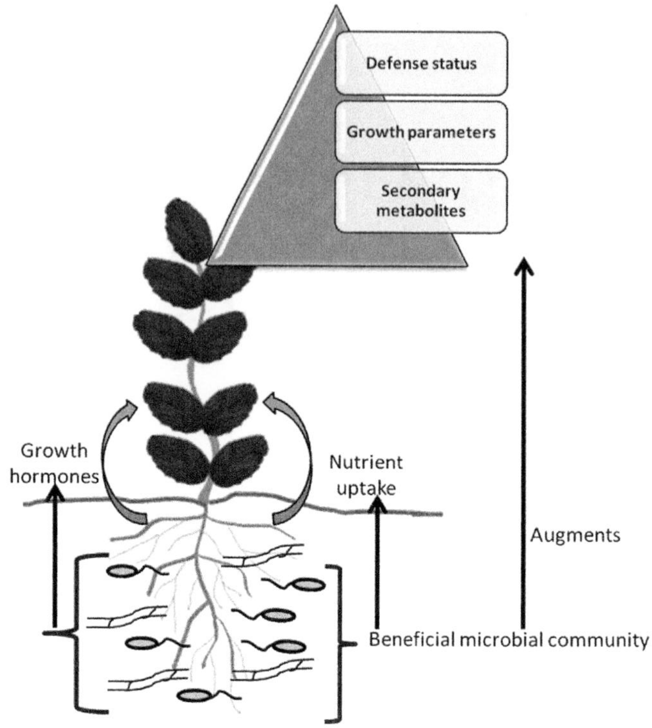

FIGURE 11.1 Diagrammatic representation of the effect of beneficial rhizospheric microbial community on plant health, defense parameters, and secondary metabolite status.

wild natural populations, which often fall short due to lack of raw materials. In this context, utilization of valuable microbes is being looked upon as a potential target for not only enhancing growth but also secondary metabolites of great importance (Gupta et al., 2017a; Singh et al., 2016a). Presently used agricultural methods taken up for managing phytopathogens and improving plant health are largely dependent on the use of hazardous synthetic chemicals. However, this overdependence and indiscriminate use of dangerous fertilizers has severely affected the nontarget organisms, health of human beings and the environment. Thus, environment friendly and attentive farmers have started adopting eco-friendly farming practices for accomplishing the demands of the market.

In the last few decades medicinal plant investigation has principally focused on the bioactive components, however, recently the spotlight has shifted towards the role of microbes in triggering the induction of secondary metabolites, especially the endophytes and rhizospheric microbes. The research shows that considerable quantities of natural products are in fact formed either directly by microbes or indirectly by the interaction of microbial metabolites with the host plants (Gunatilaka, 2006; Singh et al., 2016b). A minute subset of potent microbial cultures have been reported to have phytotherapeutic characteristics (Strobel and Daisy, 2003; Miller et al., 2012). Since leads in the involvement of microbes for production of bioactive phytochemicals are present, the present chapter thus focuses mainly on the production of secondary metabolites by various mechanisms in host–microbe interaction.

11.2 Medicinal and aromatic plants

Their potential applications and effects are excellent reasons for the popularity of herbal treatments for various disorders nowadays (Patel, 2016). For centuries, the most well known and important source in primary health care have been MAPs. All MAPs comprise of chemical compounds of distinct type and concentration. However, their concentration and amount depend on variation in plant species and the plants' age. The optimum utilization of MAPs depends on their presence as well as on their influential treatment ability to control specific disorders.

Fortunately, plant diversity is still obligatory for the well-being of humans, as they provide a considerable number of the cures required in health care (van Seters, 1997). They are also major contributors to biodiversity in specific ecological areas. However, the quality of active components derived from medicinal plants is highly influenced by various abiotic and biotic factors (Solaiman and Anawar, 2015). To choose bioactively rich medicinal plants for cancer treatment and many other ailments is a wise decision and thought to be quite safe in comparison to the synthetic substitutes for human beings and the environment (Toussaint et al., 2007; Nema et al., 2013). Noticeably, there has been increased emphasis on the utilization of materials derived from plants as a source of medicines for an extensive diversity of human ailments.

According to Rasool Hassan (2012), the term "medicinal plant" comprises of various kinds of plants used in herbalism ("herbology" or "herbal medicine"). It has been described in Chinese writings, ancient Unani manuscripts, and Egyptian papyri that since ancient times, plants and herbs have been used for medicinal purposes. Even to date, traditional medicine is used by nearly 80% of the world's inhabitants for their primary health care, as reported by the World Health Organization (WHO). Healing with plants is not an advanced technique, and it probably dates back to the period of *Homo sapiens*' evolution (Máthé, 2015), which proves that the natural environment has been used as a resource of medicinal agents since prehistoric times. Somehow, moderately based on their use in traditional medicine, nowadays natural plant species have been the biggest source for the isolation of a remarkable number of modern-day drugs. These plants contain so-called active ingredients that have a beneficial effect on the metabolic processes of living organisms, including human beings. The trait aromaticity specifies the plants with the fragrant or sweet-smelling aroma; at the same time the word *aroma* is assumed to indicate also the taste of aromatic herbs. Conclusively, on account of easy availability, low cost, and great potential to control disorders, the MAPs are highly useful in human society. Thus, such MAPs urgently need conservation by all means for future generations.

11.3 Secondary metabolites

The rich variety of natural products or secondary metabolites produced by plants has a well-known function in protective actions against various predators and microbial pathogens, based on their toxicity and antagonism to herbivores and microbes. However, some play a role in defense mechanisms against abiotic stresses (e.g., UV-B exposure), while others are necessary for intercommunication between the plants and other organisms (Schäfer and Wink, 2009). However, they are reported to have insignificant effect on growth and developmental processes. Secondary metabolites are mostly dependent on the supply of precursors produced during primary metabolism. Unlike primary ones, secondary metabolites are selectively found to be distributed in the plant kingdom (Ziegler and Taiz, 2006). Also, they have not been found to have direct roles in the processes of respiration, photosynthesis, and other chief metabolic processes of the plants. More recently, numerous secondary metabolites have been reported to have significant ecological functions in plants, for example:

- Protection of plants against biotic stressors.
- Serving as the attractants for pollinators and animals involved in seed dispersal.
- Functioning as intermediaries of plant–plant competition and plant–microbe symbioses.

Earlier, Vining (1992) reported secondary metabolites to be elements having low molecular mass with noncrucial function in normal bacterial function. Hence, he assumed no direct relation to the influence on the development and the growth of microorganisms. Contrary to that, Demain and Fang (2000) showed in their studies that secondary metabolites are necessary for the expansion of rapidly multiplying cells not only as regulators of cellular differentiation processes but cellular inhibitors against competing bacteria as well.

Moreover, a broad range of secondary metabolites have been discovered that have the equivalent precursors divided into a range of compounds resulting in secondary metabolites (Demain and Fang, 2000), while there is a greater divergence among them. When we consider the structural consequence of secondary metabolites, they have discrete characteristics within which they replicate their biological activity. Aside from their involvement in medicine, compelling advantages of these bioactive compounds have been established in agricultural as well as in industrial applications (Table 11.1).

11.4 Plant growth-promoting microbes (PGPMs)

The PGPMs are a group of beneficial microorganisms that are able to colonize the host root to endorse plant growth and health by inducing the production of various plant hormones, siderophores, and phosphate solubilization along with facilitation of efficient nutrients from the surrounding environment. There are several findings on the positive consequences of synergism among various PGPMs and MAPs on enhancement of plant growth and secondary metabolites production (Shrivastava et al., 2015; Egamberdieva and da Silva, 2015). The occurrence of various allelochemicals exudates released by roots such as organic acids, proteins, amino acids and polysaccharides facilitates the microbial

TABLE 11.1 Bioactive secondary metabolites of some medicinal and aromatic plants.

S No	Plant species	Common name	Active agents/drugs
1.	*Catharanthus roseus*	Sadabahar	Alkaloid (vinblastine and vincristine)
2.	*Curcuma longa*	Haldi	Flavonoid (curcumin)
3.	*Withania somnifera*	Ashwaganda	Steroidal lactones, withanolides, notably withaferin A
4.	*Azadirachta indica*	Neem	Limonoids (nimbidinin), Di- and triterpenoids
5.	*Ocimum sanctum*	Tulsi	Apigenin, taxol and ursolic acid, citral
6.	*Artemisia annua*	Maleria buti	Artemisinin
7.	*Bacopa monnieri*	Brahmi	Triterpenoid saponins (bacosides)
8.	*Hyoscyamus niger*	Black henbane	Hyoscyamine, scopolamine, and other tropane alkaloids
9.	*Chlorophytum borivilianum*	Safed musli	Steroidal saponins
10.	*Mentha piperita*	Peppermint	Menthyl acetate, 1,8-cineole, limonene, beta-pinene, and beta-caryophyllene
11.	*Pelargonium graveolens*	Geranium	Citronellol, geraniol, citronellyl formate, gamma-Eudesmol, isomenthone, linalool, geranyl formate, and menthone
12.	*Cassia angustifolia*	Senna	Sennosides A, B, C, and D
13.	*Centella asiatica*	Mandookparni	Glycosides (asiaticoside and madecassoside)
14.	*Cymbopogan martinii*	Palmarosa	β-myrcene, 3-undecyne, neral, geranial, nerol, geranyl acetate, and juniper camphor
15.	*Glycyrrhiza glabra*	Mulethi	Triterpenoid aglycone, glycyrrhetic acid (glycyrrhetinic acid; enoxolone)
16.	*Matricaria chamomilla*	Chamomile	Sesquiterpenes, flavonoids, coumarins, and polyacetylenes
17.	*Phyllanthus amarus*	Bhumi amlaki	Lignans (phyllanthin, hypophyllanthin, nirurin, etc.), flavonoids (quercetin, quercetrin, rutin, etc.), terpenes, tripenes, alkaloids
18.	*Plantago ovata*	Isabgol	Aucubin glycoside (iridoid), sugars, and sterols
19.	*Pogostemon cablin*	Patchouli	Patchouli alcohol, pogostol, seychellene, nor-patchoulinol, patchoulipyridine, methylchavicol, limonene, pinene, and p-methoxy cinnamaldehyde
20.	*Stevia rebaudiana*	Stevia	Stevioside and rebaudioside A
21.	*Vetiveria zizanioides*	Vetiver	Khusimol, α-vetivone and β-vetivone
22.	*Gymnema sylvestre*	Gudmar	Triterpene saponins
23.	*Piper betel*	Betelvine	Piperol-A, piperol-B, and methyl piper betlol

colonization with the host root system (Compant et al., 2010; Pii et al., 2015). The PGPMs are associated with the plant and soil regions including various species of *Bacillus*, *Azotobacter*, *Azospirillum*, *Pseudomonas*, *Bradyrhizobium*, *Flavobacterium*, *Streptomycetes*, *Trichoderma*, etc. that greatly influence the growth and physiology of the host plant (Dojima and Craker, 2016). The MAPs inoculated with PGPMs are benefited with augmented germination rate, biomass production, radical growth, leaf area, physiological parameters like stomatal conductance, photosynthesis rate and transcription, abiotic and biotic tolerance, and other microbial activities (Solaiman and Anawar, 2015; Gupta and Pandey, 2015; Singh et al., 2016a,b; Gupta et al., 2016a). Nowadays, PGPMs are extensively exploited for enhancing the plant growth in diverse areas including agriculture sectors, and also in MAPs cultivation for secondary metabolites production. The commercialization of numerous PGPMs with diverse genera has been done for varied agricultural practices. The rising problems of MAPs cultivation on a commercial scale can be solved by identifying and screening suitable helpful microbes to be used as bioenhancers to promote plant growth without any harmful effect on the environment (Dojima and Craker, 2016). Moreover, PGPMs can modulate the production of bioactive secondary metabolites in MAPs. Nowadays, new research findings are underlined toward the exploitation of PGPMs for the cultivation of MAPs to enhance bioactive secondary metabolites production (Verpoorte et al., 2000; Gupta and Pandey, 2015).

Recently, herbal drugs are attaining much more significance as they are the chief resource of pharmaceuticals for therapeutic use. The bioactive secondary metabolites formed by MAPs are extensively used in various pharmaceutical industries. MAPs are used in traditional medicinal practices and approximately more than 80% of present day medicines are of plant origin (Singh et al., 2016a). Hence, the ever-growing demand for phytotherapeutic compounds in the modern world has compelled MAPs cultivation on a huge magnitude by accepting better agricultural strategies. There are many findings on the beneficial effects of PGPMs when inoculated to MAPs under greenhouse and field conditions. These PGPMs enhanced the plant growth, nutrient uptake, and assembly of various bioactive medicinally useful phytochemical compounds such as flavonoids, phenols, saponins, alkaloids, and tannins in the plant body (Swamy et al., 2016). Recently, microbes in combinational mode with involvement of large numbers of MAPs improved the production of several secondary bioactive metabolites (Rajasekar and Elango, 2011; Singh et al., 2013a,b; Gupta et al., 2016a). Jaleel et al. (2007) studied the effects of *Pseudomonas fluorescens* on yield and ajmalicine production in *Catharanthus roseus* under drought stress. They concluded that among the tested PGPMs, the *P. fluorescens* significantly enhanced the biosynthesis of secondary metabolites such as catharanthine, ajmalicine, serpentine, and vindoline. Likewise, PGPM colonization in the rhizospheric zone improved the major constituents of secondary metabolites in *Origanum majorana* (Banchio et al., 2008). Similarly, plants inoculated with *P. fluorescens* and *Bradyrhizobium* sp. showed an augmentation in the major chemical constituents with respect to control plants. The PGPMs such as *Bacillus*, *Azotobacter*, and *Pseudomonas* when applied to *C. roseus* singly and in combination significantly enhanced the morphological parameters in terms of plant height and weight, nutrient content, along with secondary metabolite content with respect to the control plants (Karthikeyan et al., 2013). Pandey et al. (2011) and Santoro et al. (2011) reported that a large number of microbial strains including *P. fluorescens*, *Bacillus subtilis*, *Azospirillum brasilense*, *Trichoderma harzianum*, *Glomus intraradices*, and *Bacillus megaterium* enhanced the yield and secondary metabolite production in *Mentha piperita* and *Mentha arvensis*, respectively. Santoro et al. (2011) observed twofold increase of monoterpene (pulegone and menthone) production in *P. fluorescens* inoculated plants. Similarly, Singh et al. (2013a,b) observed that *Pseudomonas monteilii* and *Glomus fasciculatum* inoculation alone and in combination with *Coleus forskohlii* enhanced the forskolin content, a key metabolite extensively used in treatment of cancer, respiratory, and heart diseases. In another finding, Ghorbanpour et al. (2013) observed that the inoculation of *Pseudomonas putida* and *P. fluorescens* was able to increase the tropane alkaloids such as hyoscyamine and scopolamine production along with root area in *Hyoscyamus niger* (black henbane) under water-deficit stress conditions. Their results concluded that the inoculation of both these strains was able to produce enhanced level of antioxidant enzymes and proline content. Likewise, *Bacopa monnieri* (Brahmi) treated with *Bacillus pumilus* and *Exiguobacterium oxidotolerans* improved the growth and bacoside A content under saline condition. The plants treated with PGPMs also showed an increased level of proline, catalase, and ascorbate peroxidase activity with reduced lipid peroxidation in comparison to the control plants (Bharti et al., 2013). Likewise, inoculation of antagonistic microbes (*Chitiniphilus* sp. and *Streptomyces* sp.) alone and in combination increased the production of bacoside A in *B. monnieri* under nematode stress (Gupta et al., 2017b,c). Vafadar et al. (2014) studied the effect of *Bacillus polymixa*, *P. putida*, *Azotobacter chroococcum*, and *G. intraradices* on biomass as well as chlorophyll, stevioside, and nitrogen, phosphorus and potassium (NPK) content in *Stevia rebaudiana*. They found that coinoculation with *Glomus* with *Azotobacter* followed by *Glomus* with *Bacillus* and *Azotobacter* with *Pseudomonas* treatments altered the secretion of stevioside content.

Recently, Singh et al. (2016b) compared the effect of inoculation of three different microbial genera viz., *Streptomycetes* sp., *T. harzianum*, and *B. subtilis* cell cultures and their metabolites on withanolide A and withaferin A contents in *Withania somnifera*. They observed the different metabolites of *Streptomyces* sp. and *T. harzianum* significantly enhanced biomass yield and withanolide A. Further, Gupta et al. (2016b) have reported that inoculation of *Streptomyces* sp., *B. megaterium*, and *T. harzianum* to *Artemisia annua* increased the build up of artemisinin content, which is broadly used in malaria without any reputed cases of resistance. According to Gupta et al. (2016c), a host specific response was shown by *T. harzianum* and its metabolites on plant growth and essential oil parameters in *Ocimum basilicum* and *Ocimum sanctum*. They observed a significant improvement in biomass and secondary metabolites in terms of essential oil in the live culture of *T. harzianum* treatment of *O. basilicum*, whereas its metabolites improved biomass and essential oil in *O. sanctum*. Similarly, *Salvia officinalis* when inoculated with various strains of *P. fluorescens* and *P. putida* (singly and in combination) improved the in-plant growth, and secondary metabolites biosynthesis along with defense parameters like total phenolics, and flavonoids content. There was a significant accumulation of *cis*-thujene, camphor, and 1,8-cineol chemical components in the essential oil (Ghorbanpour et al., 2016). Likewise, in a remarkable finding, Kumar et al. (2016) evaluated the effect of *B. subtilis*, *Burkholderia thailandensis*, *Agrobacierium tumefaciens*, *Klebsiella* sp., *Bacillus cereus*, *P. putida*, *P. fluorescens*, and *A. chroococcum* on the growth parameters in *Curcuma longa* and confirmed the positive effect on plant growth parameters as well as the pharmaceutically important curcumin content in the leaves. In another study, Gupta et al. (2016b) reported that the microbes viz. *T. harzianum*, *G. intraradices*, and *B. subtilis* in synergistic mode were able to improve the productivity of crop, secondary metabolite content, and glandular trichome numbers in *Pelargonium graveolens*. This improved yield of secondary metabolites is associated with the enhanced number of glandular trichomes found on the leaves surface. They highlighted a linear correlation between essential oil content and glandular trichomes number in all the microbe treated plants. Currently, more efforts should be made in the direction of enhancing the potential of these PGPMs for the cultivation of MAPs at the commercial level globally. Additionally, the research finding should spotlight the isolation and screening of new and valuable strains for understanding the various mechanisms of plant–microbe interactions that are able to trigger the biosynthetic pathways of bioactive compounds to boost plant secondary metabolites.

11.5 Mechanism of plant–microbe interaction for better plant health and secondary metabolite production

11.5.1 Growth hormones

There are six major groups of plant growth regulators (PGRs) (phytohormones), which are recognized as gibberellins, cytokinins, abscisic acid, ethylene, brassinosteroids, and auxins (Moore, 1979; Salisbury and Ross, 1992). These are well known to have positive effects on the overall growth and development of plants at very low concentrations (less than 1 mM). Several plants like fenugreek (Deore and Bharud, 1990), basil (Arularasu and Sambandamurthi, 1999), and coriander (Badgujar and Warhal, 1988) respectively have been observed with improved herb yield with PGRs. In a greenhouse experiment conducted by Abdelgadir et al. (2009), *Jatropha curcas* plants had increased lateral branching with the exogenous application of cytokinin. In another experiment, with the aim of enhancing the growth and essential oil production using PGRs, culture was done in in vitro in micropropagated plants of peppermint (*M. piperita*), (Santoro et al., 2013).

As per results of an experiment conducted by Nasr and Aali (2016), application of 2,4-D and kinetin altered the yield of callus induction, especially in microsamples of the root. Along with that the best callus induction was obtained in the presence of indole-3-butyric acid (IBA) and kinetin as well. Also, Valizadeh et al. (2007) showed that using the microsample of hypocotyl of caraway in a culturing environment of Murashige and Skoog Medium (MS) or 1 mg L^{-1} of 2,4-D, regeneration of the stem occurs after callus induction and without having to do the culturing. With inclusion, more callus inductions were observed in 0.1 mg L^{-1} concentration of 2,4-D and 2 mg L^{-1} of kinetin (Bagheri et al., 1932). With the aim of finding the best composition of medium for callus induction in *Mentha pulegium* L., Jafari et al. (2016) cultured two explants of *M. pulegium* L. on MS medium amended with NAA, BA, and 2,4-D in varying concentrations in a factorial experiment. However, to their surprise no significant differences ($P < .01$) between explant types for induction of callus and growth rate was observed by the researchers. Whereas, the hormone levels of 1 mg L^{-1} BA and 2 mg L^{-1} NAA (88.00%), 0.5 mg L^{-1} BA and 1 mg L^{-1} NAA (92.00%), 0.5 mg L^{-1} BA and 0.5 mg L^{-1} NAA (86.00%), and BA-free and 1 mg L^{-1} NAA (86.00%) had the most significant effect on percentage of total callus induction. The most effective concentration (0.25 mm d^{-1}) for callus growth rate was observed in BA-free and 1 mg L^{-1} NAA hormone levels. Additionally, the statistical analysis also revealed that for callus induction, among all interactions, the most efficient effect (0.29 mm d^{-1}) was observed with the hypocotyl explant in BA-free and 1 mg L^{-1} NAA. Samantaray et al. (2012) stated that MS medium amended with 2.5 mg L^{-1} 2,4-D was the best composition for callus weight and emergence in *Mentha spicata*. Xu et al. (2009) also investigated and reported the leaf explant with ½ MS medium, 1.5 mg L^{-1} BA, and 1.5 mg L^{-1} NAA as the most proficient mixture for the callus induction in *Mentha haplocalyx*. Similarly, Darvishi et al. (2014) studied the tissue culture of *M. pulegium* with three explant types (stem, leaf, and root) and the MS medium with diverse levels of BAP and 2,4-D for production of callus and reported leaf explants, BAP-free, and 1 mg L^{-1} 2,4-D as the most proficient amalgamation for callus initiation. In addition, the proliferation of the shoot with the elevated BAP was in concurrence with the results obtained by Shukla and Farooqi (1990). Better consequence of utilization of BAP for multiplication of shoot has also been reported for various medicinal plants like *Tanacetum cinerarrfolium* and *Spilanthes acmella*. In continuum, Kollárová et al. (2004) reported induction in callus-like roots in *Karwinskia humboldtiana* with a high concentration of NAA. Similarly, Arockiasamy et al. (2002) and Coelo et al. (2012) found the combination of BA and IAA more beneficial for *Eryngium maritimum* L. and *Thymus lotocephalus* micropropagation, respectively, with respect to BA alone. Additionally, other species of the family Apiaceae, that is, *Crithmum maritimum* (Grigoriadou and Maloupa, 2008), *Thapsia garganica* (Makunga et al., 2003),

Arracacia xanthorrhiza (Sliva et al., 2010), and *Anethum graveolens* (Jana and Shekhawat, 2012) made better response to lower concentration of BA alone as well as combination with NAA, respectively. Bandurski et al. (1995) and Thiem et al. (2013) reiterated that the production of endogenous auxins may be affected by exogenous auxins. Contrary to that, there was no sign of *in vitro* derived shoots rooted on auxin-free MS media in *Centella asiatica* and *A. xanthorrhiza* according to Tiwari et al. (2000) and Sliva et al. (2010). In other study conducted by Liu et al. (2002) the treatment of plants with auxin increased the root dry weight in consequence of repetitive cell division resulting in lateral root formation. IBA treatment (the effective concentration of IBA) in respect to increased numbers of leaves and nodes also enhanced shoot length and fresh weight was shown to be pH dependent (Harbage and Stimart, 1996). The IBA concentrations in the medium, at pH values of 5.6–8.8, were lower and not sufficient to induce root and shoot development.

11.5.2 Alteration in rhizosphere microbiome

The rhizosphere, a nutrient rich zone, is considered as a hotspot for inhabitation of microorganisms belonging to fungi, oomycetes, bacteria, actinomycetes, mycorrhiza, etc. (Hinsinger and Marschner, 2006; Hinsinger et al., 2009; Raaijmakers et al., 2009). Most of the microbial colonizers in the rhizosphere microbiome are components of a food web that exploits huge quantity of nutrients secreted by the host plants. The study of microbes inhabiting the rhizospheric zone of medicinally important plants is essential as they not only have an influence on the overall plant health but also on induction of metabolites of industrial importance (Köberl et al., 2013). On finding the rhizodeposits to be a key factor regulating the diversity of microbes, Cook et al. (1995) hypothesized that plants may adjust the microbiome of the rhizosphere to their own personal profit by carefully choosing stimulation of microbes with characteristics that are advantageous to both growth and defense of the plants. From there, many research groups have worked on the diversity of arbuscular mycorrhiza (AM) fungal community and showed their positive effect on the plant growth and medicinal value molecules (Wubet et al., 2004; Chandra et al., 2010; Zhang et al., 2013). Medicinal plants are known for synthesizing an assortment of biologically active compounds for surviving and flourishing in the natural conditions (Bajguz, 2007; Vardhini and Anjum, 2015). However, interestingly the composition and amount of the bioactive molecules in medicinal plants differ extensively, depending on the type of plant species, soil, and the kind of alliance with the native rhizospheric microbes (Zhao et al., 2011; Morsy, 2014). These secondary metabolites produced by the medicinal plants in turn influence plant–microbial communities associated with the host plants (Qi et al., 2012; Chaparro et al., 2014). Recently, Köberl et al. (2013) found an elevated number of antagonistic bacteria in the root rhizosphere of *Matricaria chamomilla*, *Calendula officinalis*, and *Solanum distichum*. Likewise, it is assumed that the synthesis of phytochemical components of plants is associated either directly or indirectly to the presence of endophytic microbes (Chandra, 2012; Qi et al., 2012). Therefore, comprehending the reaction of root associated microbial communities to shift in the physiochemical status of the rhizosphere may provide useful information about the ecology of microbes associated with plants.

11.5.3 Modulation in secondary metabolite biosynthetic pathway

When the microbes are used as priming agents in plants, a secondary metabolism pathway is kindled in such a manner that when the plant faces any stress in future the defense response generated will be more powerful than the nonprimed counterparts (Singh et al., 2013a,b; 2014). Harnessing the above knowledge and information Awasthi et al. (2011) reported that synergistic interaction between *Glomus mosseae* and *B. subtilis* not only enhanced growth but also yield of artemisinin in *A. annua* L. Likewise, for quite a lot of medicinal plants, fungi and bacteria were identified to produce many active ingredients. A number of biologically active compounds, such as camptothecin, hypericin diosgenin, vinblastin, podophyllotoxin, paclitaxel, etc. have been reported to be produced by various endophytic fungi (Joseph and Priya, 2011; Zhao et al., 2011). In another research finding, the application of synergistic microbial strains namely *Chitiniphilus* sp. MTN22 and *Streptomyces* sp. MTN14 not only mitigated oxidative stress produced by the plant parasitic nematodes but also stimulated total bacoside production by altering the expression of genes concerned with the biosynthesis pathway in *B. monnieri* L. (Gupta et al., 2017a). It has been further reported that not only application of live cultures but also their metabolites affect the secondary metabolite pathway of plants as a study by Singh et al. (2016b) showed enhancement in withanolide A content along with growth of *W. somnifera* (L.) Dunal.

Alkaloids, terpenoids, and phenolic acids are the chief clusters of plant secondary metabolites that find usage in pharmacological and therapeutical purposes. Essential oils derived from MAPs are rich in monoterpenes and phenylpropanoids. A study by Copetta et al. (2006) reported a substantial augmentation in the content of *O. basilicum* essential oil upon inoculation with AM fungi. Similarly, inoculation of *G. intraradices* and *G. mosseae* singly as well as in combination significantly augmented total phenolic content and flowerhead numbers of *Cynara cardunculus* (Ceccarelli et al., 2010). Likewise, inoculation with AM fungi significantly improved plant growth and vinblastine amount in leaves of *Vinca rosea* (Rosa-Mera et al., 2011), and rosmarinic acid in shoots of sweet basil (Toussaint et al., 2007). In another study, plants primed with *Rhizophagus intraradices* and AM fungi mixture showed higher concentration of pseudohypericin and hypericin in the shoots of medicinal value plant *Hypericum perforatum* L. (Zubek et al., 2012). Recently a group of workers showed augmentation in bioactive secondary metabolite apigenin-7-*O*-glucoside upon application of *B. subtilis* Co1-6 and *P. polymyxa* Mc5Re-14 in chamomile plants indicated toward the modulation in the metabolomics of plant along with the microbiome of rhizospheric zone (Schmidt et al., 2014).

11.5.4 Nutrient uptake

PGPRs and their relationship with plants have been utilized commercially for enhancing growth as well as for achieving sustainable agriculture. The rhizospheric microbes are proficient in converting, solubilizing, and mobilizing macro and micronutrients and therefore are considered as key driving forces responsible for recycling nutrients available in the soil (Hayat et al., 2010; Glick, 2012). However, currently importance is being laid toward exploration of biological methods for not only improving crop production but also important useful secondary metabolites through integrated plant nutrient management schemes. Inoculation of plants with phosphate dissolving microorganisms has been proven to augment the yield of many agriculture crops. Dual inoculation of mycorrhiza and phosphate solubilizing bacteria showed tremendous progress in nutrient (N, P, K, Ca, Mg, Fe, Mn, and Zn) uptake in *Marsdenia volubilis* compared with single treatments (Sandhya et al., 2014). Likewise, Khaosaad et al. (2006) showed an increase in essential oil levels in *Origanum* species upon application of arbuscular mycorrhizal fungi. Similarly, a field study conducted by Chaudhary et al. (2008) showed significant augmentation in the herbage production, essential oil concentration, and uptake of micronutrients like P, Zn, and Fe in *A. annua* treated with *Glomus macrocarpum* (GM) and *Glomus fasciculatum* (GF) compared with the control set. In another study, inoculation of *O. basilicum* roots with PGPRs enhanced fresh weight of root, essential oil amount, along with the nitrogen status of the plants with respect to the control plants (Ordookhani et al., 2011). Further, a study conducted by Awasthi et al. (2011) showed enhancement in total biomass, uptake of nutrients, and artemisinin content in *A. annua* under glasshouse conditions when *G. intraradices*, *G. mosseae*, GF, and *Glomus aggregatum* and two free living nitrogen fixing bacteria (NFB) (*B. subtilis* and *Stenotrophomonas* spp.) were inoculated alone or in combinational mode. Similarly, our group's research further showed that plants treated with *B. subtilis* not only reduced root-knot infestation but also showed significant increase in fresh biomass, amount of essential oil, and nutrient uptake as compared with the untreated set in sweet basil under field conditions (Gupta and Pandey, 2015).

11.6 Conclusions

Application of beneficial rhizospheric microbes alters the secondary metabolic pathway thereby positively influencing the quantity of bioactive secondary metabolites in medicinal plants. Since most of the bioactive components of medicinal plants are principally secondary metabolites, the quality of which affects the value of medicinal materials, anything that positively affects their synthesis is accepted with open arms by both the scientific and farming communities. Overall looking into the beneficial and conducive effects of microbes, their application should not only be encouraged for increasing the plant yield but also for the augmented synthesis of bioactive compounds, which in turn would improve the quality of raw materials. Through this chapter we thus hope to draw the desired attention toward the significance and potential of plant–microbe interaction in the field of MAPs to a relatively new field of indirectly manipulating the level of metabolites of pharmaceutical and industrial importance.

Acknowledgments

The authors are thankful to the Director, CSIR-CIMAP, for his encouragement and support. AS and SC are grateful to Department of Science and Technology, Govt. of India for providing fellowship under DST Women Scientist (SR/WOS-B/2017/254-AAS) and (SR/WOS-B/769/2016).

References

Abdelgadir, H.A., Johnson, S.D., Van, S.J., 2009. Promoting branching of a potential biofuel crop *Jatropha curcas* L. by foliar application of plant growth regulators. Plant Growth Regul. 58 (3), 287–295.

Arockiasamy, S., Prakash, S., Ignacimuthu, S., 2002. Direct organogenesis from mature leaf and petiole explants of *Eryngium foetidum* L. Biol. Plant. 45 (1), 129–132.

Arularasu, P., Sambandamurthi, S., 1999. Effect of germination nitrogen and spacing on yield of herbage and oil yield in Tulasi (*Ocimum sanctum* L.). South Indian Hortic. 47 (1/6), 370–372.

Awasthi, A., Bharti, N., Nair, P., Singh, R., Shukla, A.K., Gupta, M.M., et al., 2011. Synergistic effect of *Glomus mosseae* and nitrogen fixing *Bacillus subtilis* strain Daz26 on artemisinin content in *Artemisia annua* L. Appl. Soil Ecol. 49, 125–130.

Badgujar, C.D., Warhal, K.N., 1988. Effect of seed soaking and wrapping on growth and yield of vegetable coriander. J. Maha. Agric. Univ. 13 (3), 344–365.

Bagheri, A.R., Moshiri, F., Khosravinia, S., 1932. Investigation of reaction of micro sample and *in vitro* growth regulators on callus generation, root generation and regeneration of the Iranian Caraway. Sci. Res. J. Eco Tech. Agric. Plants 5, 53–61.

Bajguz, A., 2007. Metabolism of brassinosteroids in plants. Plant Physiol. Biochem. 45 (2), 95–107.

Banchio, E., Bogino, P.C., Zygadlo, J., Giordano, W., 2008. Plant growth promoting rhizobacteria improve growth and essential oil yield in *Origanum majorana* L. Biochem. Syst. Ecol. 36 (10), 766–771.

Bandurski, R.S., Cohen, J.D., Slovin, J.P., Reinecke, D.M., 1995. Auxin biosynthesis and metabolism. In: Davies, P.J. (Ed.), Plant Hormones. Springer, Dordrecht, pp. 39–65.

Bharti, N., Yadav, D., Barnawal, D., Maji, D., Kalra, A., 2013. *Exiguobacterium oxidotolerans*, a halotolerant plant growth promoting rhizobacteria, improves yield and content of secondary metabolites in *Bacopa monnieri* (L.) Pennell under primary and secondary salt stress. World J. Microbiol. Biotechnol. 29 (2), 379–387.

Bhattacharyya, P.N., Jha, D.K., 2012. Plant growth-promoting rhizobacteria (PGPR): emergence in agriculture. World J. Microbiol. Biotechnol. 28 (4), 1327–1350.

Ceccarelli, N., Curadi, M., Martelloni, L., Sbrana, C., Picciarelli, P., Giovannetti, M., 2010. Mycorrhizal colonization impacts on phenolic content and antioxidant properties of artichoke leaves and flower heads two years after field transplant. Plant Soil 335 (1-2), 311–323.

Chandra, S., 2012. Endophytic fungi: novel sources of anticancer lead molecules. Appl. Microbiol. Biotechnol. 95 (1), 47–59.

Chandra, K.K., Kumar, N., Chand, G., 2010. Studies on mycorrhizal inoculation on dry matter yield and root colonization of some medicinal plants grown in stress and forest soils. J. Environ. Biol. 31 (6), 975–979.

Chaparro, J.M., Badri, D.V., Vivanco, J.M., 2014. Rhizosphere microbiome assemblage is affected by plant development. ISME J. 8 (4), 790–803.

Chaudhary, V., Kapoor, R., Bhatnagar, A.K., 2008. Effectiveness of two arbuscular mycorrhizal fungi on concentrations of essential oil and artemisinin in three accessions of *Artemisia annua* L. Appl. Soil Ecol. 40 (1), 174–181.

Coelo, N., Gonçalves, S., González-Benito, M.E., Romano, A., 2012. Establishment of an in vitro propagation protocol for *Thymus lotocephalus*, a rare aromatic species of the Algarve (Portugal). Plant Growth Regul. 66 (1), 69–74.

Compant, S., Clément, C., Sessitsch, A., 2010. Plant growth-promoting bacteria in the rhizo-and endosphere of plants: their role, colonization, mechanisms involved and prospects for utilization. Soil Biol. Biochem. 42 (5), 669–678.

Cook, R.J., Thomashow, L.S., Weller, D.M., Fujimoto, D., Mazzola, M., Bangera, G., et al., 1995. Molecular mechanisms of defense by rhizobacteria against root disease. Proc. Natl. Acad. Sci. 92 (10), 4197–4201.

Copetta, A., Lingua, G., Berta, G., 2006. Effects of three AM fungi on growth, distribution of glandular hairs, and essential oil production in *Ocimum basilicum* L. var. Genovese. Mycorrhiza 16 (7), 485–494.

Darvishi, E., Kazemi, E., Kahrizi, D., Chaghakaboudi, S.R., Khani, Y., 2014. Optimization of callus induction in Pennyroyal (*Mentha pulegium*). J. Appl. Biotechnol. Rep. 1 (3), 97.

De la Rosa-Mera, C.J., Ferrera-Cerrato, R., Alarcón, A., de Jesús Sánchez-Colín, M., Muñoz-Muñiz, O.D., 2011. Arbuscular mycorrhizal fungi and potassium bicarbonate enhance the foliar content of the vinblastine alkaloid in *Catharanthus roseus*. Plant Soil 349 (1-2), 367–376.

Demain, A.L., Fang, A., 2000. The natural functions of secondary metabolites. In: Fiechter, A. (Ed.), History of Modern Biotechnology I. Springer, Berlin, Heidelberg, pp. 1–39.

Deore, B.P., Bharud, R.W., 1990. Growth, yield and storability of fenugreek as influenced by foliar spray of growth substances. J. Maha. Agri. Univ. 15 (2), 208–210.

Dojima, T., Craker, L., 2016. Potential benefits of soil microorganisms on medicinal and aromatic plants. In: Jeliazkov, V.D., Cantrell, C.L. (Eds.), Medicinal and Aromatic Crops: Production, Phytochemistry, and Utilization. ACS Symposium Series American Chemical Society, Washington, DC, pp. 75–90.

Egamberdieva, D., da Silva, J.A.T., 2015. Medicinal Plants and PGPR: A new frontier for phytochemicals. In: Egamberdieva, D., Shrivastava, S., Varma, A. (Eds.), Plant-Growth-Promoting Rhizobacteria (PGPR) and Medicinal Plants. Springer, Cham, pp. 287–303.

Ghorbanpour, M., Ghafarzadegan, R., Khavazi, K., Hatami, M., 2013. Two main tropane alkaloids variations of black henbane (*Hyoscyamus niger*) under PGPRs inoculation and water deficit stress induction at flowering stage. J. Med. Plants 1 (45), 29–42.

Ghorbanpour, M., Hatami, M., Kariman, K., Abbaszadeh Dahaji, P., 2016. Phytochemical variations and enhanced efficiency of antioxidant and antimicrobial ingredients in *Salvia officinalis* as inoculated with different rhizobacteria. Chem. Biodiv. 13 (3), 319–330.

Glick, B.R., 2012. Plant growth-promoting bacteria: mechanisms and applications. Scientifica 15, 1–15.

Grigoriadou, K., Maloupa, E., 2008. Micropropagation and salt tolerance of *in vitro* grown *Crithmum maritimum* L. Plant Cell Tissue Organ Cult. 94 (2), 209.

Gunatilaka, A.L., 2006. Natural products from plant-associated microorganisms: distribution, structural diversity, bioactivity, and implications of their occurrence. J. Nat. Prod. 69 (3), 509–526.

Gupta, R., Pandey, R., 2015. Microbial interference ameliorates essential oil yield and diminishes root-knot infestation in sweet basil under field conditions. Biocontr. Sci. Technol. 25 (10), 1165–1179.

Gupta, R., Singh, A., Srivastava, M., Gupta, M.M., Pandey, R., 2016a. Augmentation of systemic resistance and secondary metabolites by chitinolytic microbes in *Withania somnifera* against *Meloidogyne incognita*. Biocontr. Sci. Technol. 26 (12), 1626–1642.

Gupta, R., Singh, A., Pandey, R., 2016b. Microbe-based technology ameliorates glandular trichomes, secondary metabolites and antioxidants in *Pelargonium graveolens* L'Hér. J. Sci. Food Agric. 96 (12), 4151–4159.

Gupta, R., Singh, A., Kanaujia, R., Kushwaha, S., Pandey, R., 2016c. *Trichoderma harzianum* ThU and its metabolites underscore alteration in essential oils of *Ocimum basilicum* and *Ocimum sanctum*. Proc. Natl. Acad. Sci. India Sect. B: Biol. Sci 1–9.

Gupta, R., Singh, A., Srivastava, M., Singh, V., Gupta, M.M., Pandey, R., 2017a. Microbial modulation of bacoside A biosynthetic pathway and systemic defense mechanism in *Bacopa monnieri* under *Meloidogyne incognita* stress. Sci. Rep. 7, 41867.

Gupta, R., Singh, A., Ajayakumar, P.V., Pandey, R., 2017b. Microbial interference mitigates *Meloidogyne incognita* mediated oxidative stress and augments bacoside content in *Bacopa monnieri* L. Microbiol. Res. 199, 67–78.

Harbage, J.F., Stimart, D.P., 1996. Effect of pH and 1H-indole-3-butyric acid (IBA) on rooting of apple microcuttings. J Am. Soc. Hortic. Sci. 121 (6), 1049–1053.

Hayat, R., Ali, S., Amara, U., Khalid, R., Ahmed, I., 2010. Soil beneficial bacteria and their role in plant growth promotion: a review. Ann. Microbiol. 60 (4), 579–598.

Hinsinger, P., Marschner, P., 2006. Rhizosphere—perspectives and challenges—a tribute to Lorenz Hiltner—Munich, Germany. Plant Soil 283 (1), vii–viii.

Hinsinger, P., Bengough, A.G., Vetterlein, D., Young, I.M., 2009. Rhizosphere: biophysics, biogeochemistry and ecological relevance. Plant Soil 321 (1-2), 117–152.

Jafari, A., Kahrizi, D., Mansouri, M., 2016. Effects of plant growth regulators and explant on callus induction in pennyroyal (*Mentha pulegium* L.). Biharean Biol. 10 (2), 134–136.

Jaleel, C.A., Manivannan, P., Sankar, B., Kishorekumar, A., Gopi, R., Somasundaram, R., et al., 2007. *Pseudomonas fluorescens* enhances biomass yield and ajmalicine production in *Catharanthus roseus* under water deficit stress. Colloids Surf. B: Biointerf. 60 (1), 7–11.

Jana, S., Shekhawat, G.S., 2012. *In vitro* regeneration of *Anethum graveolens*, antioxidative enzymes during organogenesis and RAPD analysis for clonal fidelity. Biol. Plant. 56 (1), 9–14.

Joseph, B., Priya, R.M., 2011. Bioactive compounds from endophytes and their potential in pharmaceutical effect: a review. Am. J Biochem. Mol. Biol. 1 (3), 291–309.

Karthikeyan, B., Sakthivel, U., Narayanan, J.S., 2013. Role of plant growth-promoting rhizobacteria for commercially grown medicinal plants. In: Maheshwari, D.K., Saraf, M., Aeron, A. (Eds.), Bacteria in Agrobiology: Crop Productivity. Springer, Berlin, Heidelberg, pp. 65–76.

Khaosaad, T., Vierheilig, H., Nell, M., Zitterl-Eglseer, K., Novak, J., 2006. Arbuscular mycorrhiza alter the concentration of essential oils in oregano (*Origanum* sp., Lamiaceae). Mycorrhiza 16 (6), 443–446.

Köberl, M., Ramadan, E.M., Adam, M., Cardinale, M., Hallmann, J., Heuer, H., et al., 2013. *Bacillus* and *Streptomyces* were selected as broad-spectrum antagonists against soilborne pathogens from arid areas in Egypt. FEMS Microbiol. Lett. 342 (2), 168–178.

Kollárová, K., Lišková, D., Kákoniová, D., Lux, A., 2004. Effect of auxins on *Karwinskia humboldtiana* root cultures. Plant Cell Tissue Organ Cult. 79 (2), 213–221.

Kumar, A., Singh, M., Singh, P.P., Singh, S.K., Singh, P.K., Pandey, K.D., 2016. Isolation of plant growth promoting rhizobacteria and their impact on growth and curcumin content in *Curcuma longa* L. Biocatal. Agric. Biotechnol. 8, 1–7.

Liu, C., Zhu, J., Liu, Z., Li, L., Pan, R., Jin, L., 2002. Exogenous auxin effects on growth and phenotype of normal and hairy roots of *Pueraria lobata* (Willd.) Ohwi. Plant Growth Regul. 38 (1), 37–43.

Makunga, N.P., Jäger, A.K., Van Staden, J., 2003. Micropropagation of *Thapsia garganica*—a medicinal plant. Plant Cell Rep. 21 (10), 967–973.

Máthé, Á., 2015. Botanical aspects of medicinal and aromatic plants. In: Mathe, A. (Ed.), Medicinal and Aromatic Plants of the World. Springer, Dordrecht, pp. 13–33.

Mendes, R., Garbeva, P., Raaijmakers, J.M., 2013. The rhizosphere microbiome: significance of plant beneficial, plant pathogenic, and human pathogenic microorganisms. FEMS Microbiol. Rev. 37 (5), 634–663.

Miller, K.I., Qing, C., Sze, D.M.Y., Roufogalis, B.D., Neilan, B.A., 2012. Culturable endophytes of medicinal plants and the genetic basis for their bioactivity. Microb. Ecol. 64 (2), 431–449.

Moore, T.C., 1979. Auxins. In: Moore, T.C. (Ed.), Biochemistry and Physiology of Plant Hormones. Springer, New York, pp. 32–89.

Morsy, N., 2014. Phytochemical analysis of biologically active constituents of medicinal plants. Main Group Chem. 13 (1), 7–21.

Narula, N., Kothe, E., Behl, R.K., 2009. Role of root exudates in plant-microbe interactions. J. Appl. Bot. Food Qual. 82, 122–130.

Nasr, N., Aali, H., 2016. Evaluation of the effect of cytokinine and auxin plant growth regulator on callus induction and plant regeneration on medicinal plant *Convolvulus arvensis* L. Biol. Forum Int. J. 8 (2), 29–33.

Nema, R., Khare, S., Jain, P., Pradhan, A., Gupta, A., Singh, D., 2013. Natural products potential and scope for modern cancer research. Am. J. Plant Sci. 4 (06), 1270.

Ordookhani, K., Sharafzadeh, S., Zare, M., 2011. Influence of PGPR on growth, essential oil and nutrients uptake of sweet basil. Adv. Environ. Biol. 5 (4), 672–677.

Pandey, R., Mishra, A.K., Tiwari, S., Singh, H.N., Kalra, A., 2011. Enhanced tolerance of *Mentha arvensis* against *Meloidogyne incognita* (Kofoid and White) Chitwood through mutualistic endophytes and PGPRs. J. Plant Interact. 6 (4), 247–253.

Patel, D.K., 2016. Medicinal and aromatic plants: role in human society. Med. Aromat. Plants 5, e175.

Pii, Y., Mimmo, T., Tomasi, N., Terzano, R., Cesco, S., Crecchio, C., 2015. Microbial interactions in the rhizosphere: beneficial influences of plant growth-promoting rhizobacteria on nutrient acquisition process. Biol. Fert. Soils 51 (4), 403–415.

Podile, A.R., Kishore, G.K., 2006. Plant growth promoting rhizobacteria. In: Gnanamanickam, S.S. (Ed.), Plant-Associated Bacteria. Springer, Amsterdam, pp. 195–230.

Qi, X., Wang, E., Xing, M., Zhao, W., Chen, X., 2012. Rhizosphere and nonrhizosphere bacterial community composition of the wild medicinal plant *Rumex patientia*. World J. Microbiol. Biotechnol. 28 (5), 2257–2265.

Raaijmakers, J.M., Paulitz, T.C., Steinberg, C., Alabouvette, C., Moënne-Loccoz, Y., 2009. The rhizosphere: a playground and battlefield for soilborne pathogens and beneficial microorganisms. Plant Soil 321 (1-2), 341–361.

Rajasekar, S., Elango, R., 2011. Effect of microbial consortium on plant growth and improvement of alkaloid content in *Withania somnifera* (Ashwagandha). Curr. Bot. 2 (8), 27–30.

Rasool Hassan, B., 2012. Medicinal plants (importance and uses). Pharm. Anal. Acta. 3, e139.

Salisbury, F.B., Ross, C.W., 1992. Plant Physiology, 4th Ed Wadsworth Publishing Co, Belmont, CA, pp. 315–405.

Samantaray, A., Sial, P., Kar, M., 2012. Micro-propagation and biochemical analysis of spear mint (*Mentha spicata*). Ind. J. Innov. Dev. 1 (7), 489–493.

Sandhya, A., Vijaya, T., Sridevi, A., 2014. Co-inoculation studies of vesicular arbuscular mycorrhizal fungi (VAM) and phosphate solubilizing bacteria (PSB) on nutrient uptake of *Marsdenia volubilis* (T. Cooke). Ann. Plant Sci. 3 (07), 765–769.

Santoro, M.V., Zygadlo, J., Giordano, W., Banchio, E., 2011. Volatile organic compounds from rhizobacteria increase biosynthesis of essential oils and growth parameters in peppermint (*Mentha piperita*). Plant Physiol. Biochem. 49 (10), 1177–1182.

Santoro, M.V., Nievas, F., Zygadlo, J., Giordano, W., Banchio, E., 2013. Effects of growth regulators on biomass and the production of secondary metabolites in peppermint (*Mentha piperita*) micropropagated *in vitro*. Am. J. Plant Sci. 4 (05), 49.

Schäfer, H., Wink, M., 2009. Medicinally important secondary metabolites in recombinant microorganisms or plants: progress in alkaloid biosynthesis. Biotech. J. Healthc. Nut. Technol. 4 (12), 1684–1703.

Schmidt, R., Köberl, M., Mostafa, A., Ramadan, E.M., Monschein, M., Jensen, K.B., et al., 2014. Effects of bacterial inoculants on the indigenous microbiome and secondary metabolites of chamomile plants. Front. Microbiol. 5, 1–11.

Shrivastava, S., Egamberdieva, D., Varma, A., 2015. Plant growth-promoting rhizobacteria (PGPR) and medicinal plants: the state of the art. In: Egamberdieva, D., Shrivastava, S., Varma, A. (Eds.), Plant-Growth-Promoting Rhizobacteria (PGPR) and Medicinal Plants. Springer, Cham, pp. 1–16.

Shukla, A., Farooqi, A.A., 1990. Utilization of plant growth regulators in aromatic plant production. Curr. Res. Med. Aromat. Plants 12, 152–157.

Singh, A., Jain, A., Sarma, B.K., Upadhyay, R.S., Singh, H.B., 2013a. Rhizosphere microbes facilitate redox homeostasis in *Cicer arietinum* against biotic stress. Ann. Appl. Biol. 163 (1), 33–46.

Singh, R., Soni, S.K., Kalra, A., 2013b. Synergy between *Glomus fasciculatum* and a beneficial *Pseudomonas* in reducing root diseases and improving yield and forskolin content in *Coleus forskohlii* Briq. under organic field conditions. Mycorrhiza 23 (1), 35–44.

Singh, A., Jain, A., Sarma, B.K., Upadhyay, R.S., Singh, H.B., 2014. Rhizosphere competent microbial consortium mediates rapid changes in phenolic profiles in chickpea during *Sclerotium rolfsii* infection. Microbiol. Res. 169 (5), 353–360.

Singh, A., Gupta, R., Saikia, S.K., Pant, A., Pandey, R., 2016a. Diseases of medicinal and aromatic plants, their biological impact and management. Plant Genet. Resour. 14 (4), 370–383.

Singh, A., Gupta, R., Srivastava, M., Gupta, M.M., Pandey, R., 2016b. Microbial secondary metabolites ameliorate growth, *in planta* contents and lignification in *Withania somnifera* (L.) Dunal. Physiol. Mol. Biol. Plants 22 (2), 253–260.

Sliva, Š., Viehmannova, I., Vitamvas, J., 2010. Micropropagation and morphogenesis of arracacha (*Arracacia xanthorrhiza* Bancroft). Agric. Trop. Subtrop. 43 (3), 206–211.

Solaiman, Z.M., Anawar, H.M., 2015. Rhizosphere microbes interactions in medicinal plants. In: Egamberdieva, D., Shrivastava, S., Varma, A. (Eds.), Plant-Growth-Promoting Rhizobacteria (PGPR) and Medicinal Plants. Springer, Cham, pp. 19–41.

Strobel, G., Daisy, B., 2003. Bioprospecting for microbial endophytes and their natural products. Microbiol. Mol. Biol. Rev. 67 (4), 491–502.

Swamy, M.K., Akthar, M.S., Sinniah, U.R., 2016. Response of PGPR and AM fungi toward growth and secondary metabolite production in medicinal and aromatic plants. In: Hakeem, K.R., Akhtar, M.S. (Eds.), Plant, Soil and Microbes. Springer, pp. 145–168.

Thiem, B., Kikowska, M., Krawczyk, A., Więckowska, B., Sliwinska, E., 2013. Phenolic acid and DNA contents of micropropagated *Eryngium planum* L. Plant Cell Tissue Organ Cult. 114 (2), 197–206.

Tilak, K.V.B.R., Ranganayaki, N., Pal, K.K., De, R., Saxena, A.K., Nautiyal, C.S., et al., 2005. Diversity of plant growth and soil health supporting bacteria. Curr. Sci. 136–150.

Tiwari, K.N., Sharma, N.C., Tiwari, V., Singh, B.D., 2000. Micropropagation of *Centella asiatica* (L.), a valuable medicinal herb. Plant Cell Tissue Organ Cult. 63 (3), 179–185.

Toussaint, J.P., Smith, F.A., Smith, S.E., 2007. Arbuscular mycorrhizal fungi can induce the production of phytochemicals in sweet basil irrespective of phosphorus nutrition. Mycorrhiza 17 (4), 291–297.

Vafadar, F., Amooaghaie, R., Otroshy, M., 2014. Effects of plant-growth-promoting rhizobacteria and arbuscular mycorrhizal fungus on plant growth, stevioside, NPK, and chlorophyll content of *Stevia rebaudiana*. J. Plant Interact. 9 (1), 128–136.

Valizadeh, M., Kazemi-Tabar, S.K., Nematzadeh, G.A., 2007. Effect of plant growth regulators on callus induction and regeneration of cumin (*Cuminum cyminum*). Asian J. Agric. Res. 1 (1), 17–22.

van Seters, A.P., 1997. Forest based medicines in traditional and cosmopolitan health care. Med. Plants Forest Conserv. Health Care 5–11.

Vardhini, B.V., Anjum, N.A., 2015. Brassinosteroids make plant life easier under abiotic stresses mainly by modulating major components of antioxidant defense system. Front. Environ. Sci. 2, 67.

Verpoorte, R., van der Heijden, R., Memelink, J., 2000. Engineering the plant cell factory for secondary metabolite production. Transgen. Res. 9 (4-5), 323–343.

Vining, L.C., 1992. Roles of secondary metabolites from microbes. CIBA Foundation Symposium 184–198.

Wubet, T., Weiß, M., Kottke, I., Teketay, D., Oberwinkler, F., 2004. Molecular diversity of arbuscular mycorrhizal fungi in *Prunus africana*, an endangered medicinal tree species in dry Afromontane forests of Ethiopia. New Phytol. 161 (2), 517–528.

Xu, Y., WeiLin, L., Cheng Yuan, L., Yan, L., 2009. Induction and multiplication of leaf callus from two cultivars of M*entha haplocalyx*. J. Plant Resour. Environ 18, 84–88.

Zhang, H.Y., Xue, Q.H., Shen, G.H., Wang, D.S., 2013. Effects of actinomycetes agent on ginseng growth and rhizosphere soil microflora. J. Appl. Ecol. 24 (8), 2287–2293.

Zhao, J., Shan, T., Mou, Y., Zhou, L., 2011. Plant-derived bioactive compounds produced by endophytic fungi. Mini Rev. Med. Chem. 11 (2), 159–168.

Ziegler, E., Taiz, L., 2006. Plant Physiology. Sinauer Associates Inc, Sunderland, MA, p. 764.

Zubek, S., Mielcarek, S., Turnau, K., 2012. Hypericin and pseudohypericin concentrations of a valuable medicinal plant *Hypericum perforatum* L. are enhanced by arbuscular mycorrhizal fungi. Mycorrhiza 22 (2), 149–156.

Chapter 12

Role of plant-associated bacteria in phytoremediation of trace metals in contaminated soils

Pratishtha Gupta[1], Rupa Rani[1], Zeba Usmani[1], Avantika Chandra[2] and Vipin Kumar[1]

[1]*Laboratory of Applied Microbiology, Department of Environmental Science and Engineering, Indian Institute of Technology (ISM), Dhanbad, India,* [2]*Department of Environmental Science and Engineering, Indian Institute of Technology (ISM), Dhanbad, India*

Chapter Outline

12.1 Introduction

Heavy metals are persistent inorganic pollutants that enter into the ecosystem through natural and anthropogenic sources. The natural sources of metal pollution are weathering of rocks, forest fires, release of particles by vegetation, volcanic activities, and erosion. While the processes of smelting, mining, electroplating, fertilizers, and pesticides are anthropogenic sources (Singh et al., 2011; Dixit et al., 2015). The effects of heavy metal pollution are significant on the environment.

Conventional treatment methods include solidification, soil washing, stabilization, vitrification, soil vapor extraction, and electrokinetic remediation. These methods are costly, inadequate, and have negative impacts on physical, chemical, and biological properties of soil (Glick, 2010). Moreover, these approaches do not completely make soil free from metal pollutants, they only convert one form of metal into another (Ullah et al., 2015). In this context, biological means of remediation such as bioremediation and phytoremediation can be a potentially viable option in the treatment of contaminated sites (Girma, 2015). Phytoremediation is broadly defined as the use of potential plants (hyperaccumulators) in removal/extraction of metal pollutants from the contaminated soils. This approach is highly effective, environmentally sustainable, and inexpensive compared with the conventional treatment methods. However, the phytoremediation efficiency of plants can greatly be enhanced by the inoculation of several metal resistant plant growth promoting (PGP) rhizobacterial strains.

Plant growth promoting rhizobacteria (PGPR) can enhance the plant growth by improving plant resistance to several environmental stresses (Gururani et al., 2013). The bacterial strains are able to tolerate heavy metal stress by several resistant mechanisms, including complex formation with thiol-containing molecules, active efflux system, immobilization/mobilization of heavy metals, extra- or intracellular sequestration, and conversion of highly toxic forms of

New and Future Developments in Microbial Biotechnology and Bioengineering. DOI: https://doi.org/10.1016/B978-0-12-818258-1.00012-1

compounds into less toxic ones. These PGPR enhance the plant growth through a variety of mechanisms such as mineral phosphate solubilization, synthesis of siderophores, indole-3-acetic acid (IAA), hydrogen cyanide (HCN), ammonia, antibiotic production, and nitrogen fixation (Khan et al., 2009). These microbes enhance the plant biomass, growth, and metal tolerance ability, resulting in effective microbial-assisted phytoremediation of metal polluted sites. The aim of the present chapter is to describe the effectiveness of PGPR in enhanced phytoremediation of metal polluted sites. This chapter also describes the mode of action for plant growth promotion by PGPR. The mechanism by which microbes' combat metal stress is also comprehensively described in this chapter.

12.2 Plant growth promoting rhizobacteria: the potential root colonizers

Rhizosphere is defined as a small zone of soil immediately surrounding the plant root system (McNear, 2013). Compared with bulk soil, rhizospheric soil is rich in nutrients due to the release of variety of root exudates such as sugars and amino acids. The bacteria inhabiting the rhizospheric zone in plants are termed as rhizobacteria (Ahemad and Kibert, 2013). The PGPR is an association of bacteria that colonize the plant roots and help in plant growth promotion by direct and indirect mechanisms (Singh, 2013).

The PGPR can be divided into two groups: (1) extracellular PGPR and (2) intracellular PGPR (Gupta et al., 2015).

1. The extracellular PGPR are generally confined on the rhizosphere or in the space between the cells of the root cortex. The most common genera of extracellular PGPR are *Azotobacter*, *Bacillus*, *Chromobacter*, *Micrococcus*, and *Pseudomonas* (Ahemad and Kibert, 2013).
2. The intracellular PGPR are located inside the nodular structure of the root cells. The intracellular PGPR belongs to the genera *Allorhizobium*, *Mesorhizobium*, and *Bradyrhizobium* (Ahemad and Kibert, 2013).

12.3 Plant growth promotion: the mechanism of action

PGPR promotes the plant growth due to several PGP properties of rhizobacteria. PGPR promotes the plant growth by two different mechanisms: direct and indirect. The direct plant growth promotion is due to certain metabolites synthesized by bacteria such as phytohormones. The indirect mechanism of plant growth is due to the production of antagonistic substances produced by bacteria or by inducing resistance to pathogens (Beneduzi et al., 2012). A brief description of different indirect and direct methods of plant growth promotion by PGPR is given below.

12.4 Direct mechanism

12.4.1 Nitrogen fixation

Nitrogen is a vital and essential element required by all forms of life. In spite of being abundant (78%) in the atmosphere, it cannot be taken up by the plants. Thus, the atmospheric nitrogen is converted into plant-utilizable forms by a complex enzyme system known as nitrogenase.

The PGPR has an ability to fix the atmospheric form of nitrogen into the forms that can be taken up by the plants. These bacteria convert the atmospheric nitrogen into available forms by two mechanisms: (1) symbiotic and (2) nonsymbiotic.

1. Symbiotic: This is a mutualistic relationship between microbes and plants. In this process, the microbes enter into the roots and form nodules in which nitrogen fixation occurs (Mus et al., 2016). The most common symbiotic nitrogen fixing bacteria include *Rhizobium*, *Bradyrhizobium*, *Mesorhizobium*, and *Frankia*.
2. Nonsymbiotic: This mechanism is carried out by free living bacteria and can enhance nonleguminous plant growth (Bhattacaharya and Jha, 2012). These belong to the genera *Enterobacter*, *Pseudomonas*, *Azotobacter*, *Azospirillum*, and *Burkholderia*.

12.4.2 Phosphate solubilization

Phosphorus is the second most important nutrient for plant growth after nitrogen. Phosphorus plays an important role in all metabolic functions of plants, including respiration, photosynthesis, signal transduction, and macromolecular synthesis. Plants cannot utilize phosphates as they are in insoluble forms. PGPR utilize these insoluble forms of phosphates and convert them into soluble and available forms for plants (Shaikh et al., 2016). The major mechanism by which these PGPR make phosphorus available to plants is by the release of different organic acid, for example gluconic acid, citric acid, succinic acid, and lactic acid (Khan et al., 2014).

12.4.3 Auxins

Auxins are the group of plant hormones that have positive effects on plant growth and development. The main auxin in plant is IAA. The PGPR have the capability to synthesize IAA hence they can increase the internal pool of IAA in plants. The pathway for the IAA synthesis in bacteria is the removal of carboxyl and amino group from tryptophan. IAA plays a significant role in plant–microbe interactions as they act as a signaling molecule affecting gene expression in microorganisms (Spaepen and Vanderleyden, 2011). The rhizobacteria secreted IAA can also lead to the loosening of the plant cell wall, thus facilitating the increased release of root exudates providing additional nutrients to support plant growth (Kumar and Sharma, 2017).

12.4.4 ACC deaminase

The ACC deaminase controls the ethylene production in plant roots and helps in root growth. Hence, the ACC deaminase producing PGPR promotes the plant growth under stress conditions (Arshad et al., 2007).

12.4.5 Hydrogen cyanide

HCN is a secondary metabolite commonly produced by rhizospheric microorganisms (Kamei et al., 2014). HCN production is a common feature of genus *Bacillus* and *Pseudomonas* and it is a potential biocontrol metabolite. The HCN in plants enhances the percentage and rate of seed germination, plant biomass, and nutrient uptake by the plants (Kumar et al., 2015).

12.5 Indirect mechanism

12.5.1 Antibiosis

PGPR is most powerful agents in the production of antibiotics for biocontrol against phytopathogens (Shilev, 2013). A variety of antibiotics are produced by bacteria, such as amphisin, comvcin A, tensin, tropolone produced by *Pseudomonas* sp., oligomysin A, xanthobaccin, and kanosamine produced by *Bacillus* sp., and *Streptomyces* sp.

12.5.2 Siderophores

Siderophores are organic compounds having a low molecular weight are produced by bacteria and fungi. The siderophores bind to Fe^{3+} ions and transport them into cells (Sureshbabu et al., 2016). Fe^{3+} is a biologically important molecule as it is an important constituent of cytochrome, heme and nonheme proteins and cofactor for various enzymes. PGPR, under iron deficiency, synthesize siderophores to sequester iron (Banik et al., 2014; Bharti et al., 2014). Most common forms of siderophores are catecholates (produced by bacteria) and hydroxymates (produced by both bacteria and fungi).

12.5.3 Lytic enzymes

PGPR is an effective biocontrol agent due to production of certain enzymes such as lipases, proteases, phosphatases, dehydrogenases, etc. These enzymes have a significant effect on plant growth as they protect plants from pathogenic fungi and from biotic and abiotic stresses (Kumar and Sharma, 2017) (Table 12.1).

12.6 Microbial mechanism to combat metal stress

Microorganisms have developed various methods such as adsorption, methylation, uptake, oxidation, and reduction to protect themselves from heavy metal toxicity (Rajendran et al., 2003). Adsorption is defined as the nonspecific binding of metal ions to polysaccharides/proteins associated with the bacterial cell wall. In this process, specific membrane proteins like metallothionine form complexes with heavy metals (Ayangbenro and Babalola, 2017). Biomethylation is an

TABLE 12.1 Plant growth promotion activities by rhizospheric bacteria.

PGP bacteria	PGP activities	References
Pseudomonas sp. VRK3	IAA, P-solubilization, siderophores	Hemambika et al. (2013)
Phyllobacterium myrsinacearum	ACC deaminase, siderophores, IAA, P-solubilization	Ma et al. (2013)
Bacillus spp.	IAA, siderophores, P-solubilization, antifungal activity	Karuppiah and Rajaram (2011)
Rahnella sp.	ACC deaminase, P-solubilization, siderophores	He et al. (2013a,b)
Pseudomonas aeruginosa OSG41	P-solubilization, EPS, siderophores, HCN, ammonia	Oves et al. (2013)
Bacillus sp.	P-solubilization, ammonia, siderophores	Sengupta et al. (2015)
Pseudomonas sp. PsA4, *Bacillus* sp. Ba32	IAA, P-solubilization	Rajkumar et al. (2006)
Rhizobacterial strains A3, S32	IAA, siderophores	Rajkumar et al. (2005)
Burkholderia sp.	IAA, siderophores, P-solubilization, ACC deaminase	Jiang et al. (2008)
Pseudomonas sp.	Siderophores, IAA, ammonia, hydrogen cyanide, P-solubilization	Płociniczak et al. (2013)
Mesorhizobium sp. RC3	IAA	Wani et al. (2008)

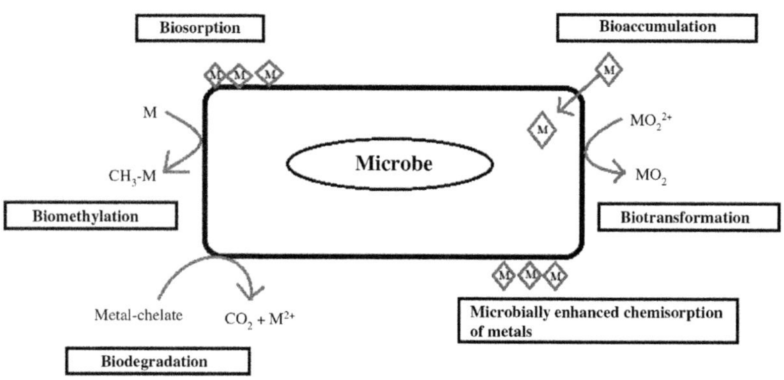

FIGURE 12.1 Mechanism of metal–microbe interaction in the bioremediation process. *Adapted and modified from Tabak, H.H., Lens, P., van Hullebusch, E.D., Dejonghe, W., 2005. Developments in bioremediation of soils and sediments polluted with metals and radionuclides—1. Microbial processes and mechanisms affecting bioremediation of metal contamination and influencing metal toxicity and transport. Rev. Environ. Sci. Biotechnol. 4, 115–156.*

important method that plays a crucial role in heavy metal bioremediation. The stepwise process of methylation results in a completely volatile metal compound (Dixit et al., 2015). For example, several bacterial species, *Alcaligenes faecalis*, *Bacillus* sp., *Pseudomonas* sp., etc. are known that can biomethylate mercury into gaseous methyl mercury.

Reduction of metals can be defined as the process where bacteria utilize metals for anaerobic respiration as terminal electron acceptors. Besides respiration, bacteria also possess a reduction mechanism that imparts resistance to metal stress. For example, reduction of hexavalent chromium to trivalent chromium forms by chromium reducing bacteria. Leaching of different metals such as arsenic, cadmium, copper, cobalt, and zinc, from metal polluted sites is an important feature of metal tolerant bacteria to impart resistance (Gupta and Diwan, 2017). For example, sulfur oxidizing bacteria can indirectly precipitate metals into insoluble sulfides.

12.7 Plant growth promoting rhizobacteria mediated phytoremediation of metal polluted soils

Phytoremediation is the in situ use of green living plants to extract, immobilize, and remove pollutants from surface water, ground water, soil, and sediments (Fig. 12.1). It is an effective, economical, practical, and socially accepted technology to remediate the contaminated sites (Oh et al., 2014). Phytoremediation is classified into several

TABLE 12.2 Case studies of PGP mediated phytoremediation of metal polluted soils.

PGP bacteria	Plant	Heavy metals	Accumulated part	References
Enterobacter cloacae	*B. napus*	As	Root and shoot	Nie et al. (2002)
Microbacterium sp.	*Pisum sativum*	Cr	Root and shoot	Soni et al. (2014)
Bacillus pumilus and *Bacillus* sp.	*Sedum plumbizincicola*	Cd	Root and shoot	Ma et al. (2015)
Bacillus sp.	Cicer	Cr	Root and shoot	Wani and Khan (2010)
Pseudomonas sp.	*Solanum nigrum*	Cd	Root, stem, and leaves	Chen et al. (2014)
Rahnella sp.	*B. napus*	Zn	Root and shoot	He et al. (2013a,b)
Pseudomonas aeruginosa	Chickpea	Cr	Root and shoot	Oves et al. (2013)
Burkholderia sp.	*Lycopersicum esculentum* and *Z. mays*	Pb	Root and shoot	Jiang et al. (2008)
Enterobacter sp.	*Polygonum pubescens*	Cd	Root and shoot	Jiang et al. (2008)
Brucella sp.	Okra	Cr	Root and shoot	Maqbool et al. (2014)

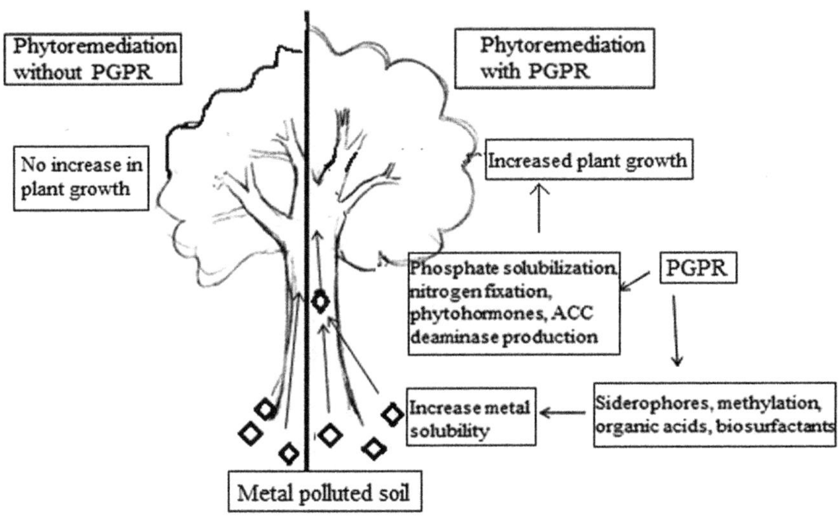

FIGURE 12.2 Mechanism of plant growth promotion by PGPR. *Adapted and modified from Ullah, A., Heng, S., Munis, M.F.H., Fahad, S., Yang, X., 2015. Phytoremediation of heavy metals assisted by plant growth promoting (PGP) bacteria: a review. Environ. Exp. Bot. 117, 28–40.*

forms: phytoextraction, phytostabilization, phytotransformation, phytovolatilization, and rhizoremediation. Out of these several forms, the most common ones in remediation of metal polluted sites are phytoextraction and phytostabilization. Phytoextraction is the process in which plants extract metals from the soil and concentrate them into roots and shoots (Lasat, 2000). Phytostabilization refers to the use of plants to reduce the metal mobility in soil via absorption and precipitation, thus reducing their bioavailability.

The plants that produce a high amount of biomass, grow rapidly, and are able to tolerate and accumulate large concentration of metals are considered as ideal for phytoextraction. The hyperaccumulator plants have extraordinary high metal accumulation ability, but most of these produce low amounts of biomass and have a stunted growth rate under high metal stress conditions. Thus, the inoculation of PGPR in the rhizosphere of these plants could overcome this problem (Glick, 2003) (Table 12.2).

Microbes in the metal polluted soils enhance the mobility and availability of the metals to the plants through the process of chelation, phosphate solubilization, and acidification (Jing et al., 2007). In this context, the PGPR exert synergistic effects on plants by enhancing the plant growth and making nutrients available through a variety of mechanisms such as nitrogen fixation, siderophores, ammonia, and HCN production (Singh, 2015; Tiwari and Singh, 2017; Kumar et al., 2018). The efficiency of phytoremediation process is expected to increase upon inoculation of PGP bacteria (Jing et al., 2007).

The efficiency of the phytoremediation process depends upon the interactions of soil, microbe, metal, and plants. The interaction of plant roots with the microbes determines the extent of phytoremediation (Jing et al., 2007). Microbes play a substantial role in recycling of nutrients, control of pests, detoxification of chemicals, and maintenance of soil structure (Singh et al., 2018; Tiwari et al., 2018; Singh and Gupta, 2018; Vimal et al., 2018). Thus, bacteria can either enhance the phytoremediation capacity of plants or reduce the phytotoxicity of the contaminated soils (Naees et al., 2011). These bacteria can also alter the metal availability and mobility to plants through the release of chelating substances, influencing the redox potential and acidification of the soil environment (Fig. 12.2).

12.8 Conclusions

The ever-increasing concentration of heavy metals in the environment has prompted scientists and researchers to deliver an effective solution for the removal of these pollutants from the biosphere. In consideration of various physical, chemical, and biological means of treatment, phytoremediation of metal polluted sites is an environmental friendly and safe method of treatment. The inoculation of PGP bacteria along the plant rhizosphere is known to enhance the efficiency of the phytoremediation process. These bacteria transform metals and render them into soluble and bioavailable forms, which can be easily taken up by the plants. These bacteria also enhance the plant biomass by making nutrients available to plants under stress conditions. Numerous bacteria with PGP activities have been tested by researchers. The phytoremediation process in association of PGPR can save our planet from the menace of heavy metal stress.

References

Ahemad, M., Kibert, M., 2013. Mechanism and application of plant growth promoting rhizobacteria: current perspective. J. King Saud Univ. Sci. 26, 1–20.

Arshad, M., Saleem, M., Hussain, S., 2007. Perspectives of bacterial ACC deaminase in phytoremediation. Trends Biotechnol. 25, 356–362.

Ayangbenro, A.S., Babalola, O.O., 2017. A new strategy for heavy metal polluted environments: a review of microbial biosorbents. Int. J. Environ. Res. Public Health 14, 94–109.

Banik, S., Das, K., Islam, M., Salimullah, M., 2014. Recent advancements and challenges in microbial bioremediation of heavy metals contamination. JSM Biotechnol. Biomed. Eng. 2, 1035–1043.

Beneduzi, A., Ambrosini, A., Passaglia, L.M.P., 2012. Plant growth-promoting rhizobacteria (PGPR): their potential as antagonists and biocontrol agents. Genet. Mol. Biol. 35, 1044–1051.

Bharti, R.P., Shrivastava, A., Soni, N., Tiwari, A., More, S., Choudhary, J.R., 2014. Phytoremediation of heavy metal toxicity and role of soil in rhizobacteria. Int. J. Sci. Res. Publ. 4, 1–5.

Bhattacharyya, P.N., Jha, D.K., 2012. Plant growth-promoting rhizobacteria (PGPR): emergence in agriculture. World J. Microbiol. Biotechnol. 28, 1327–1350.

Chen, L., Luo, S., Li, X., Wan, Y., Chen, J., Liu, C., 2014. Interaction of Cd hyperaccumulator *Solanum nigrum* L. and functional endophyte *Pseudomonas* sp. Lk9 on soil heavy metals uptake. Soil Biol. Biochem. 68, 300–308.

Dixit, R., et al., 2015. Bioremediation of heavy metals from soil and aquatic environment: an overview of principles and criteria of fundamental processes. Sustainability 7, 2189–2212.

Girma, G., 2015. Microbial bioremediation of some heavy metals in soils: an updated review. Indian J. Sci. Res 6, 147–161.

Glick, B.R., 2003. Phytoremediation: synergistic use of plants and bacteria to clean up the environment. Biotechnol. Adv. 21, 383–393.

Glick, B.R., 2010. Using soil bacteria to facilitate phytoremediation. Biotechnol. Adv 28, 367–374.

Gupta, P., Diwan, B., 2017. Bacterial exoploysaccharides mediated heavy metal removal: a review on biosynthesis, mechanism and remediation strategies. Biotechnol. Rep. 13, 58–71.

Gupta, G., Parihar, S.S., Ahirwar, N.K., Snehi, S.K., Singh, V., 2015. Plant growth promoting rhizobacteria (PGPR): current and future prospects for development of sustainable agriculture. J. Microbiol. Biochem. Technol. 7, 096–102.

Gururani, M., Upadhyaya, C., Baskar, V., Venkatesh, J., Nookaraju, A., Park, S., 2013. Plant growth-promoting rhizobacteria enhance abiotic stress tolerance in *Solanum tuberosum* through inducing changes in the expression of ros-scavenging enzymes and improved photosynthetic performance. J. Plant Growth Regul. 32, 245–258.

He, H., et al., 2013a. Characterization of endophytic *Rahnella* sp. JN6 from *Polygonum pubescens* and its potential in promoting growth and Cd, Pb, Zn uptake by *Brassica napus*. Chemosphere 90, 1960–1965.

He, J., et al., 2013b. Cadmium tolerance in six poplar species. Environ. Sci. Pollut. Res. 20, 163–174.

Hemambika, B., Balasubramanian, V., Kannan, V.R., James, R.A., 2013. Screening of chromium-resistant bacteria for plant growth-promoting activities. Soil Sediment Contam. Int. J. 22, 717–736.

Jiang, C., Sheng, X., Qian, M., Wang, Q., 2008. Isolation and characterization of a heavy metal-resistant *Burkholderia* sp. from heavy metal-contaminated paddy field soil and its potential in promoting plant growth and heavy metal accumulation in metal-polluted soil. Chemosphere 72, 157–164.

Jing, Y., He, Z., Yang, X., 2007. Role of soil rhizobacteria in phytoremediation of heavy metal contaminated soils. J. Zhejiang Univ. Sci. B 8, 192–207.

Kamei, A., Dolai, A.K., Kamei, A., 2014. Role of hydrogen cyanide secondary metabolite of plant growth promoting rhizobacteria as biopesticides of weeds. Global J. Sc. Front. Res. D. Agric. Vet. 14I, 109–112.

Karuppiah, P., Rajaram, S., 2011. Exploring the potential of chromium reducing *Bacillus* sp. and their plant growth promoting activities. J. Microbiol. Res. 1, 17–23.

Khan, A.A., Jilani, G., Akhtar, M.S., Naqvi, S.M.S., Rasheed, M., 2009. Phosphate solubilizing bacteria: occurrence, mechanism and their role in crop production. J. Agric. Biol. Sci. 1, 48–58.

Khan, M.S., Zaidi, A., Ahmad, E., 2014. Mechanism of phosphate solubilization and physiological functions of phosphate-solubilizing microorganisms. In: Khan, M., Zaidi, A., Musarrat, J. (Eds.), Phosphate Solubilizing Microorganisms. Springer, Cham.

Kumar, A., et al., 2015. Does a plant growth promoting rhizobacteria enhance agricultural sustainability? J. Pure Appl. Microbiol. 9, 715–724.

Kumar, V., Sharma, N., 2017. Plant growth promoting rhizobacteria as growth promoters for wheat: a review. Agric. Res. Tech.: Open Access J. 12, 555857. Available from: https://doi.org/10.19080/ARTOAJ.2017.12.555857.

Kumar, A., Kaushal, S., Saraf, S.A., Singh, J.S., 2018. Microbial biofuels: a solution to carbon emissions and energy crisis. Front. Biosci. (Landmark Ed) 1 (23), 1789–1802.

Lasat, M.M., 2000. Phytoextraction of metals from contaminated soil: a review of plant/soil/metal interaction and assessment of pertinent agronomic issues. J. Hazard. Subst. Res. 2, 1–25.

Ma, Y., Rajkumar, M., Luo, C.Y., Freitas, H., 2013. Phytoextraction of heavy metal polluted soils using *Sedum plumbizincicola* inoculated with metal mobilizing *Phyllobacterium myrsinacearum* RC6b. Chemosphere 93, 1386–1392.

Ma, Y., Oliveira, R.S., Nai, F., Rajkumar, M., Luo, Y., Rocha, I., et al., 2015. The hyperaccumulator *Sedum plumbizincicola* harbors metal-resistant endophytic bacteria that improve its phytoextraction capacity in multi-metal contaminated soil. J. Environ. Manag. 156, 62–69.

Maqbool, Z., Asghar, H.N., Shahzad, T., Hussain, S., Riaz, W., Ali, S., et al., 2014. Isolating, screening and applying chromium reducing bacteria to promote growth and yield of okra (*Hibiscus esculentus* L.) in chromium contaminated soils. Ecotoxicol. Environ. Saf. Available from: https://doi.org/10.1016/j.ecoenv.2014.07.007.

McNear Jr., D.H., 2013. The rhizosphere—roots, soil and everything in between. Nat. Educ. Knowl. 4, 1.

Mus, F., Crook, M.B., Garcia, K., Garcia Costas, A., Geddes, B.A., Kouri, E.D., et al., 2016. Symbiotic nitrogen fixation and the challenges to its extension to nonlegumes. Appl. Environ. Microbiol. 82, 3698–3710. Available from: https://doi.org/10.1128/AEM.01055-16.

Naees, M., Ali, Q., Shahbaz, M., Ali, F., 2011. Role of rhizobacteria in phytoremediation of heavy metals: an overview. Int. Res. J. Plant Sci. 2, 220–232.

Nie, L., Shah, S., Burd, G.I., Dixon, D.G., Glick, B.R., 2002. Phytoremediation of arsenate contaminated soil by transgenic canola and the plant growth- promoting bacterium *Enterobacter cloacae* CAL2. Plant Physiol. Biochem. 40, 355–361.

Oh, K., Cao, T., Li, T., Cheng, H., 2014. Study on application of phytoremediation technology in management and remediation of contaminated soils. J. Clean Energy Technol. 2, 216–220.

Oves, M., Khan, M.S., Zaidi, A., 2013. Chromium reducing and plant growth promoting novel strain *Pseudomonas aeruginosa* OSG41 enhances chickpea growth in chromium amended soils. Eur. J. Soil Biol. 56, 72–83.

Płociniczak, T., Kukla, M., Watroba, R., Piotrowska-Seget, Z., 2013. The effect of soil bioaugmentation with strains of *Pseudomonas* on Cd, Zn and Cu uptake by *Sinapis alba* L. Chemosphere 91, 1332–1337.

Rajendran, P., Muthukrishnan, J., Gunasekaran, P., 2003. Microbes in heavy metal remediation. Indian J. Exp. Biol. 43, 935–944.

Rajkumar, M., Nagendran, R., Lee, K.J., Lee, W.H., 2005. Characterization of a novel Cr^{6+} reducing *Pseudomonas* sp. with plant growth-promoting potential. Curr. Microbiol. 50, 266–271.

Rajkumar, M., Nagendran, R., Lee, K.J., Kim, S.Z., 2006. Influence of plant growth promoting bacteria and Cr^{6+} on the growth of Indian mustard. Chemosphere 62, 741–748.

Sengupta, K., Maiti, T.K., Saha, P., 2015. Degradation of 4-nitrophenol in presence of heavy metals by a halotolerant *Bacillus* sp. strain BUPNP2, having plant growth promoting traits. Symbiosis 65, 157–163.

Shaikh, S.S., Sayyed, R.Z., Reddy, M.S., 2016. Plant growth-promoting rhizobacteria: an eco-friendly approach for sustainable agroecosystem. In: Hakeem, K., Akhtar, M., Abdullah, S. (Eds.), Plant, Soil and Microbes. Springer, Cham.

Shilev, S., 2013. Soil rhizobacteria regulating the uptake of nutrients and undesirable elements by plants. In: Arora, N.K. (Ed.), Plant Microbe Symbiosis: Fundamentals and Advances. Springer, India, pp. 147–150.

Singh, J.S., 2013. Plant growth promoting rhizobacteria a potential microbe for sustainable agriculture. Resonance 18, 275. Available from: https://doi.org/10.1007/s12045-013-0038-y.

Singh, J.S., 2015. Microbes: the chief ecological engineers in reinstating equilibrium in degraded ecosystems. Agric. Ecosyst. Environ. 203, 80–82.

Singh, J.S., Gupta, V.K., 2018. Soil microbial biomass: a key soil driver in management of ecosystem functioning. Sci. Total Environ. 634, 497–500.

Singh, R., Gautam, N., Mishra, A., Gupta, R., 2011. Heavy metals and living systems: an overview. Indian J. Pharmacol. 43, 246–253.

Singh, C., Tiwari, S., Gupta, V.K., Singh, J.S., 2018. The effect of rice husk biochar on soil nutrient status, microbial biomass and paddy productivity of nutrient poor agriculture soils. Catena 171, 485–493.

Soni, S.K., Singh, R., Awasthi, A., Kalra, A., 2014. In vitro Cr (VI) reduction by cell-free extracts of chromate-reducing bacteria isolated from tannery effluent irrigated soil. Environ. Sci. Pollut. Res. Int. 21, 1971–1979.

Spaepen, S., Vanderleyden, J., 2011. Auxin and plant–microbe interactions. Cold Spring Harb. Perspect. Biol. Available from: https://doi.org/10.1101/cshperspect.a001438.

Sureshbabu, K., Amaresan, N., Kumar, K., 2016. Amazing multiple function properties of plant growth promoting rhizobacteria in the rhizosphere soil. Int. J. Curr. Microbiol. Appl. Sci. 5, 661–683.

Tabak, H.H., Lens, P., van Hullebusch, E.D., Dejonghe, W., 2005. Developments in bioremediation of soils and sediments polluted with metals and radionuclides—1. Microbial processes and mechanisms affecting bioremediation of metal contamination and influencing metal toxicity and transport. Rev. Environ. Sci. Biotechnol. 4, 115–156.

Tiwari, P., Singh, J.S., 2017. A plant growth promoting rhizospheric *Pseudomonas aeruginosa* strain inhibits seed germination in *Triticum aestivum* (L) and *Zea mays* (L). Microbiol. Res. 8, 1–7.

Tiwari, S., Singh, C., Singh, J.S., 2018. Land use changes: a key ecological driver regulating methanotrophs abundance in upland soils. Energy Ecol. Environ. 3, 355–371.

Ullah, A., Heng, S., Munis, M.F.H., Fahad, S., Yang, X., 2015. Phytoremediation of heavy metals assisted by plant growth promoting (PGP) bacteria: a review. Environ. Exp. Bot. 117, 28–40.

Vimal, S.R., Patel, V.K., Singh, J.S., 2018. Plant growth promoting *Curtobacterium albidum* strain SRV4: an agriculturally important microbe to alleviate salinity stress in paddy plants. Ecol. Indic. (Article in press).

Wani, P.A., Khan, M.S., 2010. Bacillus species enhance growth parameters of chickpea (*Cicer arietinum* L.) in chromium stressed soils. Food Chem. Toxicol. 48, 3262–3267. Available from: https://doi.org/10.1016/j.fct.2010.08.035.

Wani, P.A., Khan, M.S., Zaidi, A., 2008. Chromium reducing and plant-growth promoting *Mesorhizobium* improves chickpea growth in chromium amended soil. Biotechnol. Lett. 30, 159–163.

Further reading

Jing, Y.X., et al., 2013. Characterization of bacteria in the rhizospheric soils of *Polygonum pubescens* and their potential in promoting growth and Cd, Pb, Zn uptake by *Brassica napus*. Int. J. Phytoremed. 16, 321–333.

Chapter 13

Use of microbial biofertilizers technology in agro-environmental sustainability

Himani Meena and Siddhardha Busi
Department of Microbiology, School of Life Sciences, Pondicherry University, Kalapet, India

Chapter Outline

13.1 Introduction

Agriculture is one of the largest industries and provides a huge economic balance to the world's economic system. One-third of the global gross domestic product (GDP) is contributed by agriculture products, which boosts the world's economic condition. India has the second largest population after China, with 1.3 billion people and a growth rate of 1.26%, and it is highly dependent on agriculture for food and survival. Development in agriculture can be an useful tool for reduction of poverty and augmentation of prosperity in developing countries to feed billions of people. The increased demand for food leads to production of genetically modified crops, climate change, biofuel production, and elevated transportation rates, creating high risk of food security in the population (Singh et al., 2018; Tiwari et al., 2018; Singh and Gupta, 2018; Vimal et al., 2018). The agriculture field plays a crucial role in development of economic growth, livelihoods, food security, poverty reduction, and rural development. Climate change is one of the major problems faced by the farmers during crop cultivation, which results in insignificant crop production and less nutritional food product. Several technologies are available to achieve the goal of maximum crop production without economical loss. Use of fertilizer on agricultural fields is one of the main strategies to enhance plant growth and soil fertility, and to increase crop yield (Singh, 2013, 2015; Tiwari and Singh, 2017; Kumar et al., 2018). Fertilizer is divided into two different categories based on its synthesis procedure, that is, synthetic or inorganic fertilizer and manure or organic fertilizer. These are further classified into micronutrient and macronutrient suppliers, which provide essential nutrients based on soil requirements.

New and Future Developments in Microbial Biotechnology and Bioengineering. DOI: https://doi.org/10.1016/B978-0-12-818258-1.00013-3

13.2 Chemical fertilizer

Developing countries such as India are suffering from a food crisis due to increased population and rapid urbanization. Food security is one of the major problems arising as a result of decreasing soil fertility and reduced crop production. Use of chemical or inorganic fertilizer was advised to recover from food crises and for economic reasons. The Chinese government formulated a policy about the appropriate use of chemical fertilizer through its agriculture extension program. Pana and Zhang (2018) proposed three basic knowledges that contribute to inappropriate fertilizer management: effectiveness knowledge, procedural knowledge, and declarative knowledge. Insufficient knowledge about the appropriate use of fertilizer in a particular sized field with high impact comes under effectiveness knowledge, whereas procedural knowledge addresses the problem low crop yield. Declarative knowledge describes the adequate knowledge of fertilizer use on agricultural land. Emmanuel et al. (2016) performed a survey on an agriculture extension service that works as a helping unit for farmers for optimized use of chemical fertilizer. Studies on the new concept of Ghana, South Africa and applied some parameters such as age and sex of the farmers, farm size and the exact amount of the chemical fertilizer with rice productivity. They found that utilization of chemical fertilizer was maximized among male farmers compared with female farmers. Irrigation of the paddy field was also considered as an important factor for increased rice production. A training program was required to train farmers for proper utilization of fertilizer to enhance plant growth and control the food crisis for population volume. Based on different variables, selection bias and probability instrumental variables calculated and immediate boost in rice production and adoption of chemical fertilizer in agriculture fields was observed. Fertilizer is highly recommended for fulfilling the requirement of essential elements on agricultural land. Straight fertilizer such ammonium nitrate (NH_4NO_3), calcium ammonium nitrate ($Ca\ (NO_3)_2 \cdot NH_4NO_3 \cdot 10H_2O$) superphosphate ($P_2O_5$), and triple superphosphate fulfill single element (nitrogen, phosphorus, and potassium) requirements in land soil. Multinutrient fertilizer provides various nutrient elements at same time and can be divided into three groups: binary fertilizer that provides phosphorus and nitrogen together (water soluble, monoammonium phosphate, and diammonium phosphate), whereas the NPK provides nitrogen, phosphorus, and potassium with single fertilizer application. Other elemental requirements, that is, cobalt and boron, are achieved by applying micronutrient fertilizer, ethylene-diamine-tetra-acetic acid (EDTA). Chemical fertilizer can be subdivided into three major classes based on the requirements of the agriculture land: nitrogen fertilizer, phosphate fertilizer, and potassium fertilizer. Some examples of nitrogen fertilizer, that is, sodium nitrate, potassium nitrate, calcium nitrate, and urea (decomposed by microbial enzyme urease), are highly water soluble and can percolate into soil and then convert into nitrate from gaseous and ammonium forms. Ammonium fertilizers (ammonium chloride, ammonium phosphate) are resistant to the leaching process and provide nitrogen to slow growing plants by transforming NH_4 into NO_3 (Balasubramanian et al., 2017). Based on the solution solubility, phosphate fertilizers are classified as water soluble phosphatic group (single, double, and triple super phosphate), insoluble (rock phosphate), and citric acid soluble compounds (H_3PO_4). Potassic fertilizer, that is, potassium sulfate and muriate of potash, adds potassium and nitrogen to the land. These chemical fertilizers, due to their high nutrient concentration, increase the plant growth and soil fertility and increase crop yield in agriculture fields. Chemical fertilizers are easily available with suitable transport facility but excessive and prolonged use of the chemical fertilizer is a growing problem with some hazardous effects on the environment, that is, decreased soil fertility, soil structure, and productivity. The nonbeneficial effects of chemical fertilizer can be minimized by adopting manure or organic fertilizer.

13.3 Biofertilizers

Chemical fertilizer possesses a high amount of phosphorus and nitrogen, which have tremendous negative effect on ground and surface water and lead to water resources related problems. Draining of elemental composition from the soil level to the ground water contaminates the natural habitat through formation of particulate matter and enhancing unwanted microbial growth. Water resources are highly influenced by presence of phosphorus and nitrogen, which boosts algal growth in water bodies, increases the cost of water cleaning processes, and magnifies the risk of health issues. Manure (animal droppings, which are highly nitrogenous) or organic matter (decomposed plant material) can be utilized as fertilizer and can reduce the terrible side effects of chemical fertilizer. Animal droppings or organic fertilizer have numerous benefits; they are rich in nutrients, provide moisture to the soil, and hold water in the soil to fight against drought conditions for slow growing plants. Percolation of nutrients from agricultural land to the terrestrial land and ground water is the biggest challenge for ecologists. Nutrient ratio, nutrient concentration, availability of nutrient, uniformity, and proper timing are some of the important factors that promote appropriate use of manure. Using animal manure may escalate nutrient availability, generate huge amount of soil organic matter, and activate the soil microbial community. Hui et al. (2017) conducted a study in maize fields in Gongzhuling, Jilin Province, China (42°57′N,

148°57′E) over a 30-year period. Chemical fertilizers containing individual nitrogen, phosphorus, and potassium (NPK fertilizer) and manure were used in the agriculture and changes in the soil profile were noted. The experimental results showed an immediate increase in maize crop yield and soil organic carbon (SOC) for an initial 12 years in the chemical fertilizer treated field compared with the manure treated field. But a significant decrease in crop productivity and SOC amount was observed over a long period of chemical fertilizer use. Agriculture fields treated with manure displayed consequential increase in maize production and maintained a specific level of SOC in the soil. At threshold level of SOC (41.96 Mg C ha^{-1}), an induced maize production was observed in the manure treated field, which clearly indicates the need for replacing the chemical fertilizer with manure partially. Use of manure in agriculture for fertilization can increase the carbon storage of the soil, which is highly recommended for sequestering the carbon into the soil to access site specific conditions depending on the regional site. Chaudhary et al. (2017) examined the prolonged use of NPK fertilizer and organic manure [farmyard manure (FM)], green manure (GM), and straw incorporation (SI) on rice-wheat field (*Typic Ustochrepts* soil) in Punjab, India. Single fertilizers and combinations of different fertilizers were used and SOC, labile fractions of SOC (LFSOC), C stabilization, C sequestration, and C management index (CMI) were calculated over a 15-year period. Soil density (from 7.3% to 16.6%) was minimized over the duration whereas drastic increase in soil porosity index (from 11.7% to 26.2%) was observed in the experiment. Labile fraction of SOC, that is, labile carbon (LC), water soluble carbon (WSC), hot water soluble carbon (HWSC), and particulate organic C (POC) were determined to perform qualitative analysis of the soil. Remarkable changes were recorded in LC (51%−85%), WSC (67%−131%), POC (64%−159%), and HWSC (38%−131%) in comparison to SOC (29%−59%), which was a good sign of protracted practice of the nutrient management system.

13.4 Benefits of biofertilizer over chemical fertilizer

Excessive use of chemical fertilizer in agriculture causes severe environmental pollution such as water pollution, soil pollution, and air pollution. The presence of heavy metals and radionuclides in chemical fertilizer has increased the threat of accumulating inorganic matter. Particulate matter formation and greenhouse gas emission produced due to chemical fertilizer have contributed to the environmental issues such as eutrophication of water sources, loss of crop yield, and crop quality deterioration. Several studies have been conducted that explain the effects of using chemical fertilizers in detail.

13.5 Influences on agroecosystem

13.5.1 Crop productivity and soil fertility

Inappropriate use of chemical fertilizer utilization in agriculture has deleterious effect on crop productivity by influencing the soil properties and other climatic parameters, that is, water quality, primary, secondary, and micronutrient availability, seasonal factors, and pest incidence. Due to less nutrient content, manure is unable to maintain the specific level of nutrient in the soil required for optimum production of crops. Significant amount of chemical fertilizer should be applied onto the field to enhance the crop production and quality. Legumes play an important role in maintaining the soil structure and soil fertility and help in fixing atmospheric nitrogen. Rurangwa et al. (2018) explained a method to enhance the crop yield with rotational crop production on agro-ecosystem zone by supplying chemical fertilizer with manure. Legume cultivation using chemical fertilizer with organic manure promotes mycorrhizal growth with enhanced phosphorus uptake and facilitate movement of insoluble P and minimize pest or insect infestation to prevent the declining efficiency of field crops. Elouear et al. (2016) applied sheep manure and potassium fertilizer to the contaminated soil (lead, zinc, cadmium) while growing alfalfa (*Medicago sativa* L.). The reports stated that the sheep manure acts as good phytostabilizing agent whereas the potassium fertilizer was found to be an advanced phytoextracting agent. Bovine manure escalates the SOC accumulation, increases soil fertility, and improves yield response, nutrient uptake, soil health status, and fertilizer efficiency (Zavattaro et al., 2017).

13.5.2 Soil antibiotic resistance

Soil is known as treasure trove of antibiotic molecules that play an important role in the pharmaceutical field. Extreme chemical fertilizer utilization and tillage causes increased soil erosion, vitiates the agroecosystem, increases water demand, and slows down crop productivity. Availability of bacterial communities and antibiotic resistance genes in their genome isolated from soil have magnified the production of antibiotics in the medical research area. Xie et al. (2018) examined the long-term effect of chemical fertilizers and manure alone or in combination on the soil microbiome. During 25 years of experiment, the researchers presented reports based on high throughput screening and found out that the use of manure instead of chemical fertilizer can enhance the diversity in bacterial community by horizontal gene transfer. Clusters of antibiotic resistance gene and movable transposons were found in the newly occupied group of microbes by using high throughput quantitative PCR and Illumina HiSeq sequencing. Microbial diversity shifts under an unbalanced diet of nitrogen:phosphorus:potassium ratio and unstable pH conditions. Manure application enhances the SOC, which provides suitable conditions for the substrate specific microbial community and reduces the leaching of trace nutrients from soil to the water sources. Diversification of soil microbiota, enhanced microbial growth, microbial activity (catabolic utilization profiles), and production of antibiotics are responsible for overall increase in plant growth and induced crop production (Wei et al., 2017). However, researchers have suggested that supplementing chemical fertilizer with biofertilizer on the agro-ecosystem may increase the soil biological function and the dominant microbiota can suppress disease caused by foreign pathogenic strains (Tao et al., 2017). Chen et al. (2017) reported the abundance of bacterial communities and their effects on soil chemical properties of early rice/late rice *Astragalus sinicus* L. on red clay soil near Nanchang, Jiangxi province, China after application of manure plus chemical fertilizers. Microbial communities play a prominent role in decreased acidification, incorporation of nutrient, regulation of nutrient chain, and detoxifying the xenobiotic and accelerating the plant growth stimulation properties.

13.5.3 Influence on water resources

Micronutrients leaching into water resources through the agriculture ecosystem have numerous negative impacts on aquaculture and aquatic animals. Accretion in nitrogen concentration in aquaculture has increased the risk of eutrophication, a phase in which the oxygen availability is much less for consumption of marine organisms and has a detrimental effect on aquatic life. The mechanism behind the less O_2 availability is governed by elevated concentration of nitrogen, which promotes the growth of algae; at crowded levels the bloom explodes and marine bacteria utilize the huge amount of dissolved oxygen for degradation of the algae.

13.5.4 Influence on air

Anthropogenic nitrous oxide (N_2O) emissions are one of the major concerns in ecosystem management and as per the annual report of the Agriculture Management System, enhanced agriculture practices to satisfy the global food demand, are contributing the foremost part of N_2O emission. Fundamental research showed that the increased N_2O emission is caused by the presence of residual N, which remains unutilized during the plant growth. Lazcano et al. (2016) demonstrated N_2O emission through experiments for summer and winter season forage crops using 15N-labeled urea ammonium nitrate (UAN) provided to the soil. They reported that the N_2O emissions were found to be 0.76%−1.30% and proved that these are highly influenced by climatic parameters, that is, high soil moisture, inorganic N present in the soil, and temperature using multiple regression analysis. Chemical fertilization also contributes to release of greenhouse gas (methane) in the environment and manure rectified the problem by providing large amounts of sequestered carbon and can thus be used as potential candidate for eco-farming systems (Liu et al., 2015).

13.6 Microbial biofertilizer

Developing countries depend on agricultural land to overcome their food crises and the current scenario of commercial fertilizer has boomed very fast as it provides an instant relief to the scarcity of food and protection from harmful microbes. Tremendous use of chemical fertilizer on agro-ecosystems has caused hazardous effects on the environment, which has led to the discovery of an alternative for fertilizer. Utilization of plant microbial consortia as fertilizer minimizes contamination of commercially available fertilizers and supports plant growth stimulation and availability of nutrient direct into the soil, provides resistance against abiotic stress, and also acts as a biocontrol agent. Existing microbial consortia are classified based on their particular role played during the fertilization process, that is, phosphate-solubilizing microbe, nitrogen fixing microbe, siderophore producing microbe, and exopolysaccharide secretors (Panda et al., 2016). Native and nonnative microbes can be used as microbial biofertilizers and their activity relies on the persistence of the microbial cells in a new place. Adequate knowledge should be available about the inoculum, that is, biological function, and well recognized features such as environmental adaptions and mechanisms, which are important parameters before application of microbial consortia on plant soil and and root zones to ensure the coevolution takes place without any lethal consequences. Horizontal gene transfer plays an important role in successive applications of microbial biofertilizer by transmitting the resistant genes to the intrinsic microorganisms. Comprehensive studies of crop microbial consortia are required to distinguish among detrimental and constructive microbes to understand the mechanism behind the natural selection tool. The National Centre of Organic Farming, India has submitted an annual report on excessive production of chemical fertilizer due to increased demand for synthetic fertilizer in agricultural land. Microbial biofertilizer can be individual or an admixture of microbial strains and can be comprised of diverse microbes, for example, *Acetobacter*, *Azatobacter*, *Bacillus*, *Rhizobacteria*, and plant growth promoting bacteria (vesicular arbuscular mycorrhiza). Studies on plant associated microbiome demonstrate untapped microbial consortia that can be used as useful tools due to their increased genetic pool and immense production of secondary metabolite production with antimicrobial activity. Microbial biofertilizers are divided into different groups based on their mechanisms to combat the problem.

13.6.1 Rhizobacteria group

The *Rhizobacteria* are a complex cluster of bacteria that regulate fixation of atmospheric nitrogen N_2 and possess a symbiotic relationship with leguminous plants, located in root nodules. Each leguminous plant adopts its own *rhizobacterial* species based on their involvement in plant growth, plant nutrition management, and soil fertility. *Rhizobacteria* can be a promising candidate to employ as biofertilizer without any negative effect on environmental conditions. Plant growth promoting bacteria can provide benefits to the plants in many ways, which include P solubilization, mineralization of organic phosphate, atmospheric N_2 fixation, exudation of beneficial plant hormones (gibberellin 3, auxin, and cytokinin), and modulation of production of various antimicrobial peptides (antibiotic, enzymes, and siderophore), and they have a negative impact on surrounding harmful microbes as a biocontrol agent. *Bacillus* sp. is well known plant growth promoting rhizobacteria, which stimulates plant growth by inhibiting the ROS generation and hydrogen peroxide and enhances secondary metabolite production. Ge et al. (2016) discovered a new bacterial species, *Bacillus methylotrophicus* strain NKG1 from Changbai Mountain, China, and applied it on the *Solanum lycopersicum* plant as a biofertilizer and biocontrol agent. *B. methylotrophicus* application on seeds escalated the seedling fresh weight, root length, and seedling length by 27.4%, 57.7%, and 12.5%, respectively, which was much more than the control sample. In the field experiment, *B. methylotrophicus* showed positive effect on tomato plant height, crown width, stem diameter, and maximum fruit diameter, and antagonistic effect on the causative agent of gray tomato mold, *Botrytis cinerea*. Brown alga produce a higher amount of alginate oligosaccharide, which contributes to enhanced crop production. A *Bacillus litoralis* strain, *Bacillus* M3, isolated from decayed kelp fermented seaweed *Sargassum horneri* for 12 hour. Alginate lyase produced by *Bacillus* M3 degraded alginate showed substantial effect on plant growth, *Lycopersicon esculentum* L. *B. litoralis* can be used as a potential candidate as biofertilizer and biocontrol agent (Wang et al., 2016). *Bacillus* strain synthesizes 1-aminocyclopropane-1-carboxylate (ACC) deaminase enzyme that cleaves ACC to produce 2-oxobutanoate and NH_3 as by-product. ACC acts as precursor in ethylene production, which reduces plant growth and increases susceptibility toward pathogenic strain at excessive level. Xu et al. (2014) examined the effect of ACC deaminase producing *Bacillus subtilis* (HYT-12-1) on seeds of *Lycopersicum esculentum* Mill. and bacterial traits involved in plant growth promotion. *B. subtilis* (HYT-12-1) showed highest enzymatic activity at 112.02 nmol a-ketobutyrate mg^{-1} protein h^{-1}. Plant growth promoting activities were recorded as siderophore production (24%), nitrogen fixation (85%), P solubilization (37%), and ACC deaminase activity (6%). Production of gibberellin from seedborne *Bacillus amyloliquefaciens* RWL-1 was demonstrated by Shahzad et al. (2016) as well as its effect on growth of *Oryza sativa*. *B. amyloliquefaciens* is well known Gram-positive microorganism involved in the production of various antibiotics and phytohormones responsible for the plant growth promotion. Upregulated gibberellin production by *B. amyloliquefaciens* was recorded as GA1 (17.54 $\pm$ 2.40 ng), GA4 (310 $\pm$ 5.41 ng), GA7 (192.60 $\pm$ 3.32 ng), and GA9 (19.04 $\pm$ 2.49 ng). An increased synthesis of endogenous salicylic acid (1615.06 $\pm$ 10.81 μg) was noticed and decreased production of endogenous abscisic acid (ABA) (23.31 $\pm$ 2.76 ng) and jasmonic acid (25.51 $\pm$ 4.20 ng) was measured. Qin et al. (2015) performed complete sequence analysis of *B. amyloliquefaciens* L-H15, based on its PGPR activity. They reported the presence of 3,864,316 bp circular chromosome with (46.6%) G + C content, 3931 coding sequences (CDSs), 52 tRNA genes, and 8 rRNA operons, which were involved in phytohormone production (IAA), biofilm formation, and Iturin A production and may contribute to its biocontrol properties. The sequencing of *B. amyloliquefaciens* L-H15 genome was achieved using Illumina Hiseq 2000 platform (2 $\times$ 100 bp). Another genome study was conducted by Choi et al. (2014) on *B. amyloliquefaciens* GB03 and the presence of nonribosomal peptide synthetase and polyketide synthase gene clusters highly recommends *B. amyloliquefaciens* GB03 as a biocontrol agent. Genome sequencing unveiled genes involved in antibiotic production, that is, surfactin, fengycin, bacillibactin, bacilysin, macrolactin, and difficidin in both GB03 and FZB42. Gene clusters for root colonization, swarming, biofilm formation, biofertilization, and phytohormone production were found in both strains.

13.6.2 Azotobacter group

The *Azotobacter* are free living bacteria that can fix atmospheric nitrogen and can be used as biofertilizer in underprivileged areas with low soil fertility and organic fertilizer. The *Azotobacter* do not colonize on plant roots unlike *rhizobacteria* and prefer to grow on upland crop systems. Supplementing *Azotobacter* is beneficial to the soil due to high productivity of plant growth hormone and antibiotic production to fight against pathogenic strains (Noar and Bruno-Bárcena, 2018). *Azatobacter vinelandii* generates siderophores called azotobactin in iron deficiency. The *A. vinelandii* possess three forms of nitrogenase enzyme, which contains specific metal ion groups for iron (Fe), molybdenum (Mo), and vanadium (V). It is composed of two subunits: dinitrogenase reductase and dinitrogenase. Genes encoding dinitrogenase contain two subunits, $\alpha_2\beta_2$ unit in Mo dinitrogenase and δ unit in of V and Fe nitrogenases. Wang et al. (2018) treated *Azotobacter chroococcum* with carbon dots and evaluated its impact on nitrogen fixation activity. Carbon dots (4 μg/mL) applied on *A. chroococcum* altered nitrogenase structural configuration and enhanced the biocatalytic activity of nitrogen fixation. Carbon dots increased the nitrogen fixation process by 158% compared with the control.

13.6.3 Azospirrilum group

The *Azospirillum* are associated with nitrogen fixation and are recommended for wetland conditions as biofertilizer. This species exists within the plant roots and can produce different plant growth promoting substrates under basic pH conditions and moistened soil texture. The *Azospirillum* enhances plant growth by governing many factors such as phytohormone production, providing a shield against abiotic stress, P solubilization, and nitrogen fixation. Cassan and Zorita (2016) explained three ways to overcome the nitrogen deficiency, by increasing availability of nitrogen in the soil, enhancing competition among soil microbes for nitrogen content utilization, and boosting nitrogen balance in the plant. Fukami et al. (2018a,b) reviewed the role of *Azospirillum* against abiotic and biotic stress during crop cultivation. They explained how the phytohormones control different signaling pathways involved in induced systemic resistance (ISR) and systemic acquired resistance. NPR1 protein triggers stimulation of a defense gene that is an end-product of two different pathways, initiated by specific components such salicylic acid and phytohormones for jasmonic acid/ethylene pathways deprived of salicylic acid. In an experiment conducted by Fukami et al. (2018a,b), *Azospirillum brasilense* was applied on *Zea mays* L. and the effect on secondary metabolite production was recorded. Upregulation of catalase (CAT), superoxide dismutase (SOD), and malondialdehyde (MDA), and ascorbate peroxidase (APX) was observed in the maize, which helps to induce production of salicylic acid (leaves and root) and ABA (leaves). This report concludes that application of *A. brasilense* boosts plant defense system to fight against pathogenic strains and survives under unfavorable condition. Iron-regulated transporter 1 (IRT1) present in wild-type *Arabidopsis* Col-0 plants, conquers heavy metal ions from the soil, and reduces plant fertility due to metal toxicity. *B. subtilis* and *A. brasilense* were used in combination on *Arabidopsis* Col-0 plants to cease metal uptake and increase plant production. The ABA supplemented by *B. subtilis* and *A. brasilense* suppressed *itr1* gene expression and diminished the Cd toxicity in the plant (Xu et al., 2018). Grouzdev et al. (2018) sequenced methylotrophic *Azospirillum* sp. B2 strain and denoted the presence of 7059 genes (6821 were protein-coding genes, 66 coded for tRNAs, and 4 were noncoding RNAs) by NCBI Prokaryotic Genome Annotation Pipeline (PGAP).

13.6.4 Cyanobacterial group (*Azolla-Anabena*)

Cyanobacteria are a phylum of anaerobic, photosynthetic blue green algae that produce energy through photosynthesis and generate oxygen. They require basic soil conditions for proper growth and survival. They can be utilized as a biofertilizer in the paddy field during sapling season for nitrogen fixation. Nitrogen fixing water floating fern, *Azolla* sp., can be used as biofertilizer on wetland cropping systems and also utilized for the production of livestock feed. *Azolla* is highly rich in nitrogen and able to manufacture micronutrients, that is, protein, amino acids, and lipids. *Azolla* has the fastest growing rate and can minimize the weed growth in paddy field to reduce plant diseases (Bhuvaneshwari and Singh, 2015). *Azolla* and *Anabena* possess an endosymbiotic relationship and have beneficial effect on crop systems. Zhu et al. (2018) examined capability of *Anabena flos-aquae* to remove nitrogen and phosphate via production of dehydrogenase enzyme (DH) and extracellular polymeric substances (EPS). Nitrogen and phosphorus removal was quantified and found to be 94.9% and 96.8%, from ammonium form nitrogen group whereas 97.7% of phosphorus from orthophosphate form phosphorous group.

13.6.5 Phosphate-solubilizing microbes

Availability of insoluble phosphorus in soil is a major concern for farmers during the cropping season. Microbial species provide soluble phosphorus for plant growth and development by solubilizing the insoluble phosphorus in the soil. Microbial biofertilizer (*Pseudomonas* and *Bacillus*) applies three basic mechanisms to solubilize the soil phosphate: secretion of specific complex to solubilize excessive phosphorus, mineralization through released cellular enzyme, and substrate degradation with by-product P (Sharma et al., 2013). Mukhtar et al. (2017) isolated six phosphate-solubilizing microbial species from three different sites: 77-NS2 (*Bacillus endophyticus*), 77-CS-S1 (*Bacillus sphaericus*), and 77-NS5 (*Enterobacter aerogenes*) from sugarcane; PSB5 (*Bacillus safensis*) and PSB12 (*Bacillus megaterium*) from wheat; and halophilic phosphate-solubilizing strain AT2RP3 (*Virgibacillus* sp.) from *Atriplex amnicola* and applied these on *Triticum aestivum* L. Among the six species, PSB12 (*B. megaterium*) solubilized a maximum of 148.1, 305.6, 217.2 μg/mL and phosphate, grown on medium containing glucose, sucrose, and maltose respectively. *Pseudomonas* sp. is a Gram-negative microorganism that can be used as potential phosphate solubilizer. Hariprasad et al. (2014) explained the potential of *Pseudomonas aeruginosa* strain 2 apa to solubilize the raw phosphate and production of numerous microbial secondary metabolites under harsh condition such as auxin, salicylic acid, siderophore production, phenolics, and jasmonic acid.

13.6.6 Mycorrhizal fungi

Mycorrhizal fungi are among those fungal groups that grow in collaboration with plant roots and survive based on their symbiotic relationship. Mycorrhizal fungi promote root uptake of micronutrients and water from the soil and enhance plant growth and development. They exist in the exterior part of the root as root extension and help in mobilizing the micronutrients from vertical to lateral within root zones (Abd-Alla et al., 2013). Arbuscular mycorrhizal fungi (AMF) are natural rhizospheric inhabitants and are well recognized due to the presence of transporters (phosphate transporters), which mediate the transfer of mineral nutrients from soil to the plant and promote plant root and stem development. Hoseinzade et al. (2016) examined the efficiency of AMF with endophytic *Herbaspirillum seropedicae* for consumption of micronutrient and make it available for plant consumption. Microbial biofertilizers are listed below based on their mechanism involved in fertilization process (Table 13.1).

TABLE 13.1 Microbial biofertilizer and their beneficial roles.

S. no.	Microbial biofertilizer	Microorganisms	Beneficial traits	References
1.	Rhizobacteria group	*Bacillus methylotrophicus* *B. litoralis* *B. subtilis* *B. s amyloliquefaciens*	Biofertilizer, biocontrol, alginate lyase production and plant growth promotion phytohormones production	Ge et al. (2016), Wang et al. (2016), Xu et al. (2014), Shahzad et al. (2016), Qin et al. (2015), Choi et al. (2014)
2.	*Azotobacter* group	*Azatobacter vinelandii* *A. chroococcum*	Nitrogen fixation and soil fertility increase	Noar and Bruno-Bárcena (2018), Wang et al. (2018)
3.	*Azospirrilum* group	*Azospirrilum brasilense* Methylotrophic *Azospirillum* sp.	Plant growth promotion phytohormones production, salicylic acid and abscisic acid production	Cassan and Zorita (2016), Xu et al. (2018), Fukami et al. (2018a,b), Grouzdev et al. (2018)
4.	Cyanobacteria group (*Azolla- Anabaena*)	*Azolla-Anabena*	Nitrogen fixation and phosphate solubilization	Bhuvaneshwari and Singh (2015), Zhu et al. (2018)
5.	Phosphate-solubilizing microbes	*Bacillus endophyticus* *Bacillus sphaericus* *Enterobacter aerogenes* *Bacillus safensis* *Bacillus megaterium* *Virgibacillus* sp. *Pseudomonas aeruginosa*	Inorganic phosphate solubilization and organic phosphate mineralization	Sharma et al. (2013), Mukhtar et al. (2017), Hariprasad et al. (2014)
6.	Mycorrhizal fungi	Arbuscular mycorrhizal fungi	Plant growth promotion and phytohormones production	Hoseinzade et al. (2016)

13.7 Beneficial traits of microbial biofertilizer

13.7.1 Phosphate solubilization

Solubilization of mineral phosphates through microbial acid phosphatase enzyme and production of low molecular organic acid fulfills the basic need for phosphates in agricultural land to maximize crop yield and can be achieved by inoculation of microbial strains capable of providing the needs for plant growth and stimulation. Insufficient supply of phosphorus in soil can arrest cell division, metabolic pathways regulation, and chlorophyll production; restrict the mechanism of photosynthesis; hamper energy transfer (ATP, GTP, and NADPH); and reduce manufacturing of polysaccharides, lipids, and sugars. Plants can utilize two forms of soluble phosphorus: the monobasic ($H_2PO_4^-$) and the diabasic (HPO_4^{2-}). Phosphate precipitation with naturally occurring cations iron (Fe^{+3}) and aluminum (Al^{+3}) lead to unavailability of mineral phosphate in the soil in acidic environment (Panda et al., 2016). Batool and Iqbal (2018) used phosphate-solubilizing bacteria as a substitute to the commercial fertilizer during the crop season of *T. aestivum* (Var. Galaxy 2013). They isolated phosphate-solubilizing *rhizobacteria* from soil samples from various sites of Punjab, Pakistan. *Rhizobacteria* were selected based on their phosphate-solubilizing activity and plant growth promoting features (phytohormones, siderophore, ammonium, and hydrogen peroxide). Selected *Rhizobacteria* showed P solubilization index on a scale from 4 to 7 and 30–246 μg/mL solubilization in liquid culture. Panda et al. (2016) isolated a phosphate-solubilizing bacteria consortium from the Eastern Himalayan region, Sikkim, India and the plants selected for isolation were maize, rice, ginger, and cardamom. From three different zones, 42 phosphate-solubilizing bacteria were isolated and tested for phosphate solubilization activity on solid medium and liquid Pikovskaya's medium. The isolates belong to the bacterial genera *Micrococcus*, *Bacillus*, *Pseudomonas*, *Delftia*, *Staphylococcus*, *Microbacterium*, among which *B. amyloliquefaciens* exhibited antifungal activity against soil- and root-borne pathogenic strains *Rhizoctonia*, *Phythium*, and *Fusarium* that was higher than the leaf pathogens. Phosphate solubilization was higher when tricalcium phosphate used as substrate and solubilized P concentration varying from 30.2 to 203.7 mg L^{-1}. *Aspergillus*, *Penicillium*, *Klebsiella*, *Burkholderia*, and *Staphylococcus* are among phosphate solubilizers and were isolated from oil palm *Elaeis guineensis* Jacq. They showed higher P solubilization activity on SRS medium supplemented with tricalcium phosphate as substrate and released a significant amount of organic components (gluconic acid, citric acid, succinic acid, and acetic acid). These organic molecules contribute to decreasing the soil pH for better P solubilization (Acevedo et al., 2014). Microbial biofertilizer can be an alternative for the important crops that grows on low P content soil as presented in Fig. 13.1.

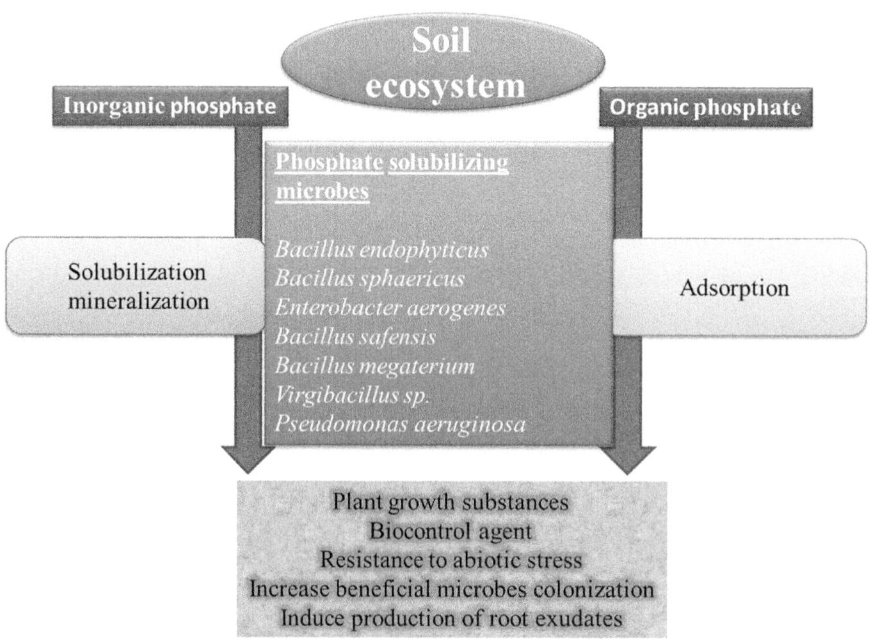

FIGURE 13.1 Mechanism of phosphate solubilization regulated by phosphate-solubilizing microorganisms.

13.7.2 Nitrogen fixation

Diazotrophs occupy a dominant gene encoding for production of ammonia by fixing atmospheric nitrogen by the nitrogenase enzyme. Diazotrophs are specifically found in legumes, and some typical examples of diazotrophs are *Rhizobium*, *Bradyrhizobium*, *Azorhizobium*, *Sinorhizobium*, and *Mesorhizobium*. Moreover, biological N_2 fixation (BNF) was signified as an economically beneficial and environmentally sound alternative to chemical fertilizers. Pham et al. (2017) isolated an endospheric fungal strain, *Pseudomonas stutzeri* A15, from rice rhizosphere and analyzed it for its plant growth promoting activity with nitrogen fixing ability. Plant growth promotion activity was independent of nitrogen fixing ability and secretion of specific enzyme ACC deaminase contributes to the N_2 fixing activity. *Phanerochaete chrysosporium* and *Bacillus thuringiensis* supply nitrogen to the soil and are subsequently responsible for nicotine reduction in tobacco plant (Shang et al., 2017). Cyanobacteria are classified as photosynthetic blue green algae, are able to produce oxygen, and are present in freshwater lakes, streams, oceans, damp soil, and wet rocks. Cyanobacteria possess important traits that play a specific role in agriculture as plant growth promoting rhizobacteria and biofertilizer. *A. flos-aquae* and *Anabena azotica* are well known photosynthetic cyanobacteria and have the ability to fix atmospheric nitrogen and convert it into small peptides and ammonia to increase the soil fertility (Shen et al., 2017). *A. chroococcum* is a heterotrophic microorganism that acts as a treasure of lytic enzymes (phosphatase, nitrogenase) and various phytohormones (gibberellin, auxin, and cytokinin). *Azotobacter* plays an important role in fixing the soil nitrogen and provides ammonia for plant growth and development (Abd El-Fattah et al., 2013) as explained in Fig. 13.2.

13.7.3 Plant growth promotion

Rhizomicrobes are complex microbes residing in the rhizosphere. Plants secrete root exudates that attract microorganisms and mediate their host–microbe interaction. Root exudates enhance microorganisms' biological activities, which has a direct and indirect influence on plant growth and development. Rhizospheric microorganisms regulate the transportation of nutrients to the plant, solubilizing the precipitated P and providing tolerance toward abiotic/biotic stresses, and minimizing the environmental impact of chemical intensive farming practices.

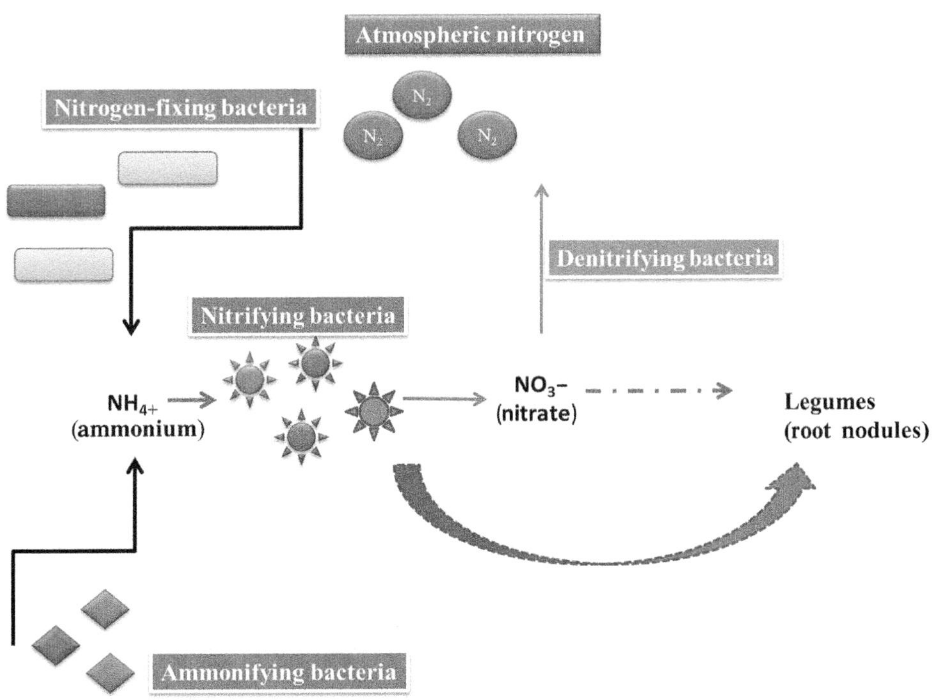

FIGURE 13.2 Nitrogen fixation processes for fixing atmospheric nitrogen.

13.7.4 Siderophore production

Fe^{3+} is the richest form of iron present in the environment but the competition between various life forms for iron is high due to higher solubility of the small molecule. Plant growth promoting rhizobacteria produce an ample amount of low molecular iron chelators that can aggregate the soluble Fe^{3+} in stable form and use it for plant growth and development (Kumar et al., 2014). Siderophores mediate uptake of heavy metals Al, Cd, Cu, Ga, In, Pb, and Zn from the soil and can be used in bioremediation.

13.7.5 Phytohormones secretion

Plants produce various kinds of phytohormones (gibberellin, auxin, cytokinin, and ethylene) that affect plant growth and provide tolerance against negative environmental condition. Plant growth promoting rhizobacteria produce higher amount of auxin hormone, which has a positive impact on plant roots and enhances root development.

13.7.6 Biocontrol

Plant growth promoting rhizobacteria have a negative impact on pathogenic strains and act as antagonists toward foreign microorganisms (Mohandas et al., 2013; Chowdhury et al., 2015). Competing for nutrient availability or residing in the niche, bioactive compound production with antimicrobial activity and induced systemic resistance are some of the mechanisms used for suppressing growth of the disease causing pathogenic strains.

Mohandas et al. (2013) isolated eight actinomycetes from the *Glomus mosseae* spores, identified as *Streptomyces fradiae*, *S. avermitilis*, *S. cinnamonensis*, *Leifsonia poae*, and *Streptomyces canus* based on RNA polymerase-subunit gene sequencing. Among all the microorganisms, 10.1 μg/mL indole acetic acid (IAA) and 12.0 μg/mL of gibberellic acid 3 (GA3) concentration of phytohormones were produced by *S. canus* whereas *S. avermitilis*, *S. canus*, and *L. poae* displayed a high level activity of phosphate solubilization, siderophore production, and chitin degrading activity and can be used as bioinoculant for growth promotion, nutrient mobilization, and as a biocontrol agent in guava seedling production. All the actinomycetes exhibited antifungal activity against *Fusarium oxysporum* and *Alternaria solani*. Chowdhury et al. (2015) reviewed biocontrol mechanism controlled by root associated *B. amyloliquefaciens* FZB42. The FZB42 is a commercially available biofertilizer that produces volatile compounds and cyclic

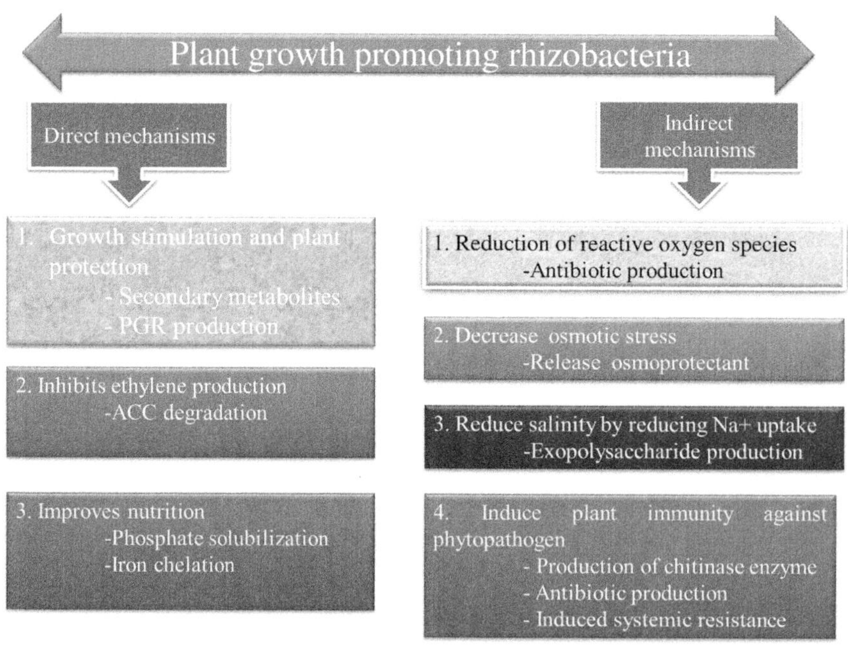

FIGURE 13.3 Graphical representation of plant growth promoting rhizobacteria and its beneficial traits.

lipopeptides eliciting ISR. About 10% of the whole genome is responsible for bioactive compounds production and parallel gene profile. Kumar et al. (2014) used a combination of microbial strains *B. megaterium*, *Arthrobacter chlorophenolicus*, and *Enterobacter* sp. isolated from soil at certain time intervals of the cropping season. They estimated the nutrient concentration and growth of *Triticum aestuvum* L. under in vitro and in vivo conditions. Plant height, grain yield, straw yield, and test weight were taken as amendatory parameters for concluding the overall improvement after application of the triple consortium. During pot assay, intensified plant growth with 17.5%, 79.8% of grain yield, 78.6% of straw yield, and 26.7% in test weight was recorded. Results from the field experiment showed 29.4% increase in plant height with 27.5% grain yield, 29.5% straw yield, and 17.6% test weight. Indigenous microbes displayed a wide range of biological activities such as hydrogen cyanide (HCN) production, auxin production, and nitrogen fixation (15.0 mg N g^{-1} carbon source, *A. chlorophenolicus* BHU5 and 26.4 μg/mL) and siderophore production (Fig. 13.3).

13.8 Conclusions

Excessive use of synthetic fertilizer has increased the amount of heavy metals in the ground and surface water with drastic side effects. It minimizes the crop production, crop variety, and crop quality. Agriculture land is amazing due to the presence of a magnified number of microbes that are expedient for enhancing the overall performance of crop system in the field. Microbial consortia are divided into specific categories based on their particular nature. The *Anabena*, *Azatobacter*, *Pseudomonas*, *Bacillus*, *Klebsiella*, and mycorrhizal fungi are well recognized microorganisms that help in improving the field operations with augmented nutritional values, phytohormone production, nitrogen fixation, phosphate and potassium solubilization, plant growth and promotion, stimulation, and siderophore production. Single and combined microbial consortia can replace the abusive and irregular application of chemical fertilizer and support the agriculture land for greater crop yield with higher quality.

References

Abd El-Fattah, D.A., Eweda, W.E., Zayed, M.S., Hassanein, M.K., 2013. Effect of carrier materials, sterilization method, and storage temperature on survival and biological activities of *Azotobacter chroococcum* inoculant. AOAS 58 (2), 111–118.

Acevedo, E., Castaneda, T.G., Prada, F., Navia, M., Romero, H.M., 2014. Phosphate-solubilizing microorganisms associated with the rhizosphere of oil palm (*Elaeis guineensis Jacq.*) in Colombia. Appl. Soil Ecol. 80, 26–33.

Balasubramanian, S., Nelson, A., Koloutsou-Vakakis, S., Lin, J., Rood, M.J., Myles, L., et al., 2017. Evaluation of denitrification decomposition model for estimating ammonia fluxes from chemical fertilizer application. Agric. For. Meteorol. 237–238, 123–134.

Batool, S., Iqbal, A., 2018. Phosphate solubilizing rhizobacteria as alternative of chemical fertilizer for growth and yield of *Triticum aestivum* (Var. Galaxy 2013). Saudi Biol. Sci . Available from: https://doi.org/10.1016/j.sjbs.2018.05.024 (In press).

Bhuvaneshwari, K., Singh, P.K., 2015. Response of nitrogen-fixing water fern Azolla biofertilization to rice crop. 3 Biotech 5, 523–529.

Cassan, F., Zorita, M.D., 2016. *Azospirillum* sp. in current agriculture: from the laboratory to the field. Soil. Biol. Biochem. 103, 117–130.

Chaudhary, S., Dheri, G.S., Brar, B.S., 2017. Long-term effects of NPK fertilizers and organic manures on carbon stabilization and management index under rice-wheat cropping system. Soil Till. Res. 166, 59–66.

Chen, D., Yuan, L., Liu, Y., Ji, J., Hou, H., 2017. Long-term application of manures plus chemical fertilizers sustained high rice yield and improved soil chemical and bacterial properties. Eur. J. Agron. 90, 34–42.

Choi, S.K., Jeong, H., Kloepper, J.W., Ryu, C.M., 2014. Genome sequence of *Bacillus amyloliquefaciens* GB03, an active ingredient of the first commercial biological control product. Genom. Announc. 2 (5), e01092-14.

Chowdhury, S.P., Hartmann, A., Gao, X.W., Borriss, R., 2015. Biocontrol mechanism by root-associated *Bacillus amyloliquefaciens* FZB42—a review. Front. Microbiol. 6, 780.

Elouear, Z., Bouhamed, F., Boujelben, N., Bouzid, J., 2016. Application of sheep manure and potassium fertilizer to contaminated soil and its effect on zinc, cadmium and lead accumulation by alfalfa plants. Sustain. Environ. Res. 26, 131–135.

Emmanuel, D., Sekyere, E.O., Owusu, V., Jordaan, H., 2016. Impact of agricultural extension service on adoption of chemical fertilizer: implications for rice productivity and development in Ghana. NJAS—Wageningen J. Life Sci. 79, 41–49.

Fukami, J., Cerezini, P., Hungria, M., 2018a. *Azospirillum*: benefits that go far beyond biological nitrogen fixation. AMB Expr. 8, 73.

Fukami, J., Ollero, F.J., Osa, C.D., Fernandez, R.V., Nogueira, M.A., Megías, M., et al., 2018b. Antioxidant activity and induction of mechanisms of resistance to stresses related to the inoculation with *Azospirillum brasilense*. Arch. Microbiol. Available from: https://doi.org/10.1007/s00203-018-1535-x.

Ge, B., Liu, B., Nwet, T.T., Zhao, W., Shi, L., Zhang, K., 2016. *Bacillus methylotrophicus* strain NKG-1, isolated from Changbai mountain, China, has potential applications as a biofertilizer or biocontrol agent. PLoS One 11 (11), e0166079.

Grouzdev, D.S., Tikhonova, E.N., Krutkina, M.S., Kravchenko, I.K., 2018. Genome sequence of methylotrophic *Azospirillum* sp. strain B2, isolated from a raised *Sphagnum* bog. Genom. Announc. 6, e00492-18.

Hariprasad, P., Chandrashekar, S., Singh, S.B., Niranjana, S.R., 2014. Mechanisms of plant growth promotion and disease suppression by *Pseudomonas aeruginosa* strain 2apa. J. Basic Microbiol. 54, 792–801.

Hoseinzade, H., Ardakani, M.R., Shahdi, A., Rahmani, H.A., Noormohammadi, G., Miransari, M., 2016. Rice (*Oryza sativa* L.) nutrient management using *mycorrhizal fungi* and endophytic *Herbaspirillum seropedicae*. J. Integr. Agric. 15 (6), 1385–1394.

Hui, L., Wen-ting, F., Xin-hua, X., Ping, Z., Hong-jun, G., Nan, S., et al., 2017. Chemical fertilizers could be completely replaced by manure to maintain high maize yield and soil organic carbon (SOC) when SOC reaches a threshold in the Northeast China Plain. J. Integr. Agric. 16 (4), 937–946.

Kumar, A., Maurya, B.R., Raghuwanshi, R., 2014. Isolation and characterization of PGPR and their effect on growth, yield and nutrient content in wheat (*Triticum aestivum* L.). ISBAB 3, 121–128.

Kumar, A., Kaushal, S., Saraf, S.A., Singh, J.S., 2018. Microbial biofuels: a solution to carbon emissions and energy crisis. Front. Biosci. (Landmark Ed) 1 (23), 1789–1802.

Lazcano, C., Tsang, A., Doane, T.A., Pettygrove, G.S., Horwath, W.R., Burge, M., 2016. Soil nitrous oxide emissions in forage systems fertilized with liquid dairy manure and inorganic fertilizers. Agric. Ecosyst. Environ. 225, 160–172.

Liu, H., Li, J., Li, X., Zheng, Y., Feng, S., Jiang, G., 2015. Mitigating greenhouse gas emissions through replacement of chemical fertilizer with organic manure in a temperate farmland. Sci. Bull. 60 (6), 598–606.

Mohandas, S., Poovarasan, S., Panneerselvama, P., Saritha, B., Upreti, K.K., Kamal, R., et al., 2013. Guava (*Psidium guajava* L.) rhizosphere *Glomus mosseae* spores harbor actinomycetes with growth promoting and antifungal attributes. Sci. Hort. 150, 371–376.

Mukhtar, S., Shahid, I., Mehnaz, S., Malik, K.A., 2017. Assessment of two carrier materials for phosphate solubilizing biofertilizers and their effect on growth of wheat (*Triticum aestivum* L.). Am. J. Microbiol. Res. 205, 107–117.

Noar, J.D., Bruno-Bárcena, J.M., 2018. *Azotobacter vinelandii*: the source of 100 years of discoveries and many more to come. Nat. Microbiol. 164, 421–436.

Pana, D., Zhang, N., 2018. The role of agricultural training on fertilizer use knowledge: a randomized controlled experiment. Ecol. Econ. 148, 77–91.

Panda, B., Rahman, H., Panda, J., 2016. Phosphate solubilizing bacteria from the acidic soils of Eastern Himalayan region and their antagonistic effect on fungal pathogens. Rhizosphere 2, 62–71.

Pham, V.T.K., Rediers, H., Ghequire, M.G.K., Nguyen, H.H., Mot, R.D., Vanderleyden, R., et al., 2017. The plant growth-promoting effect of the nitrogen-fixing endophyte *Pseudomonas stutzeri* A15. Arch. Microbiol. 199, 513–517.

Qin, Y., Han, Y., Shang, Q.M., Li, P., 2015. Complete genome sequence of *Bacillus amyloliquefaciens* L-H15, a plant growth promoting rhizobacteria isolated from cucumber seedling substrate. J. Biotechnol. 200, 59–60.

Rurangwa, E., Vanlauwe, B., Giller, K.E., 2018. Benefits of inoculation, P fertilizer and manure on yields of common bean and soybean also increase yield of subsequent maize. Agric. Ecosys. Environ. 261, 219–229.

Shahzad, R., Waqas, M., Khan, A.L., Asaf, S., Khan, M.A., Kang, S.M., et al., 2016. Seed-borne endophytic *Bacillus amyloliquefaciens* RWL-1 produces gibberellins and regulates endogenous phytohormones of *Oryza sativa*. Plant Physiol. Biochem. 106, 236–243.

Shang, C., Chen, A., Chen, G., Li, H., Guan, S., He, J., 2017. Microbial biofertilizer decreases nicotine content by improving soil nitrogen supply. Appl. Biochem. Biotechnol. 181, 1–14.

Sharma, S.B., Sayyed, R.Z., Trivedi, M.H., Gobi, T.A., 2013. Phosphate solubilizing microbes: sustainable approach for managing phosphorus deficiency in agricultural soils. Springer Plus 2, 587.

Shen, J.Y., Liao, J.Z., Guo, L.L., Su, R.F., 2017. Effects of monosulfuron-ester on metabolic processes of nitrogen-fixing cyanobacteria *Anabaena flos-aquae* and *Anabaena azotica*. Braz. J. Microbiol. 48 (3), 544–550.

Singh, J.S., 2013. Plant growth promoting rhizobacteria: potential microbes for sustainable agriculture. Resonance 18 (3), 275–281.

Singh, J.S., 2015. Microbes: the chief ecological engineers in reinstating equilibrium in degraded ecosystems. Agric. Ecosyst. Environ. 203, 80–82.

Singh, J.S., Gupta, V.K., 2018. Soil microbial biomass: a key soil driver in management of ecosystem functioning. Sci. Total Environ. 634, 497–500.

Singh, C., Tiwari, S., Gupta, V.K., Singh, J.S., 2018. The effect of rice husk biochar on soil nutrient status, microbial biomass and paddy productivity of nutrient poor agriculture soils. Catena 171, 485–493.

Tiwari, P., Singh, J.S., 2017. A plant growth promoting rhizospheric *Pseudomonas aeruginosa* strain inhibits seed germination in *Triticum aestivum* (L) and *Zea mays* (L). Microbiol. Res. 8, 1–7.

Tiwari, S., Singh, C., Singh, J.S., 2018. Land use changes: a key ecological driver regulating methanotrophs abundance in upland soils. Energy Ecol. Environ. 3, 355–371.

Vimal, S.R., Patel, V.K., Singh, J.S., 2018. Plant growth promoting *Curtobacterium albidum* strain SRV4: an agriculturally important microbe to alleviate salinity stress in paddy plants. Ecol. Indic. Available from: https://doi.org/10.1016/j.ecolind.2018.05.014 (in press).

Wang, M., Chen, L., Liu, Z., Zhang, Z., Qin, S., Yan, P., 2016. Isolation of a novel alginate lyase-producing *Bacillus litoralis* strain and its potential to ferment *Sargassum horneri* for biofertilizer. MicrobiologyOpen. 5 (6), 1038–1049.

Wang, H., Li, H., Zhang, M., Song, Y., Huang, J., Huang, H., et al., 2018. Carbon dots enhance the nitrogen fixation activity of *Azotobacter Chroococcum*. ACS Appl. Mater. Interfaces 10, 16308–16314.

Wei, M., Hu, G., Wang, H., Bai, E., Lou, Y., Zhang, A., et al., 2017. 35 years of manure and chemical fertilizer application alters soil microbial community composition in a fluvo-aquic soil in Northern China. Eur. J. Soil Biol. 82, 27–34.

Xie, W.Y., Yuan, S.T., Xu, M.G., Yang, X.P., Shen, Q.R., Zhang, W.W., et al., 2018. Long-term effects of manure and chemical fertilizers on soil antibiotic resistome. Soil Biol. Biochem. 122, 111–119.

Xu, M., Sheng, J., Chen, L., Men, Y., Gan, L., Guo, S., et al., 2014. Bacterial community compositions of tomato (*Lycopersicum esculentum* Mill.) seeds and plant growth promoting activity of ACC deaminase producing *Bacillus subtilis* (HYT-12-1) on tomato seedlings. World J. Microbiol. Biotechnol. 30, 835–845.

Xu, Q., Pan, W., Zhang, R., Lu, Q., Xue, W., Wu, C., et al., 2018. Inoculation with *Bacillus subtilis* and *Azospirillum brasilense* produces abscisic acid that reduces Irt1-mediated cadmium uptake of roots. J. Agric. Food Chem. 66, 5229–5236.

Zavattaro, L., Bechini, L., Grignani, C., Evert, F.K., Mallast, J., Spiegel, H., et al., 2017. Agronomic effects of bovine manure: a review of long-term European field experiments. Eur. J. Agron. 90, 127–138.

Zhu, Y., Tu, X., Chai, X.S., Wei, Q., Guo, L., 2018. Biological activities and nitrogen and phosphorus removal during the *Anabaena flos-aquae* biofilm growth using different nutrient form. Bioresour. Technol. 251, 7–12.

Chapter 14

Microbe-mediated management of degraded and marginal lands

Ashmita Tandon[1,4], Sonal Srivastava[1,4], Touseef Fatima[2,4], Udit Yadav[1,4], Anshu[3,4], Sanjeev Kumar[1,4], Subhi Srivastava[3,4], Ratna Katiyar[3], Suchi Srivastava[1,4] and Poonam C. Singh[1,4]

[1]*Academy of Scientific and Innovative Research (AcSIR), Ghaziabad, India,* [2]*Integral University, Lucknow, India,* [3]*Lucknow University, Lucknow, India,* [4]*CSIR-National Botanical Research Institute, Lucknow, India*

Chapter Outline

14.1 Introduction

Soil is a non replenishable and indispensable asset that is critical to sustain vital functions, on which rests the nexus of food and nutritional security, which further impacts climate and economy (UNCCD, 2015; Rojas et al., 2016). Land degradation (LD) has been variously defined as "a long-term loss of ecosystem functions and services, caused by disturbances from which the system cannot recover unaided" (UNEP, 2007); and "[a]ny reduction or loss in the biological or economic productive capacity of the land resource caused by human activities, exacerbated by natural processes and often magnified by and closely intertwined with climate change and biodiversity loss" (UNCCD, 2014). The major causes of LD are intensive cropping, water erosion, wind erosion, salinity/alkalinity, water logging, mass movement (in cold areas), frost heaving (in cold areas), frost shattering, anthropogenic activities (mining, brick kiln, urban waste, industrial effluents, etc.) (Rojas et al, 2016). Urbanization has decreased cultivated agricultural land by 4.2 M ha reducing scope for land expansion (MoA, 2016). Depletion of soil fertility is a problem of global dimension that has been triggered by indiscriminate usage and chronic mismanagement. It is a matter of concern due to the increasing global population demanding 70% increased food grain for a population of 9.1 billion across world and 1.6 billion in India by 2050 (FAO, 2011a,b; UNCCD, 2014). The projected 113 million tonnes (Mt) of additional food grains for India, 377 Mt more from current production of 264 Mt, demands 50% more land on which to grow the crops (ICAR, 2010). However the current rate of land degradation transmutes an annual loss of 20 million tones (Mt) in food grain production, which if continued, would reduce global food production by 12% (IFPRI, 2012). Temporary gains through

New and Future Developments in Microbial Biotechnology and Bioengineering. DOI: https://doi.org/10.1016/B978-0-12-818258-1.00014-5

forced production than the maximum land capacity using more irrigation water and heavy nutrient inputs will further strain the land leading to an inevitable and permanent loss of fertility. It suggests encroachment and exploitation of forest areas, which will have ecological implications. Therefore, the United Nations has ranked land degradation among the world's greatest environmental challenges and threats to human society (http://faostat.fao.org/site/698/default.aspx). The "Zero Net Land Degradation" goal set at the Rio20 conference has been widely accepted and taken into action (UNCCD, 2014). The United Nations declared 2015 as the "International Year of Soil" to generate awareness for soil health to support food security and ecosystem functions. The United Nations Convention to Combat Desertification designated "Inclusive cooperation for achieving land degradation neutrality (LDN)" to combat desertification and drought as its global theme on The World Day (June 17, 2018). The UN has incorporated LDN into its sustainable development goals (SDGs). SDG 15 specifically focuses on the restoration of degraded and desertified lands and assists nations in adopting voluntary LDN targets. However, planning remedial measures and developing techniques to offset land degradation requires scaling and mapping the extent of degraded lands and their physicochemical and biological analysis.

14.2 Current status of land degradation

The progressive agricultural soil degradation is a "silent emergency" that has built gradually over the years, affecting some 1.9 M ha of global land (UNCCD, 2015; FAO–ITPS, 2015). Estimates indicate that up to 25% of all land worldwide is highly degraded, and 36% is slightly or moderately degraded (FAO, 2011a). The land degradation hotspots comprise 29% of land area covering all major agroclimatic zones (IFPRI, 2012). The major portion (78%) of land degradation affects humid regions while the rest (22%) occurs in arid, semiarid, dry subhumid, and hyperarid regions of the world (Caspari et al., 2015). Land Degradation Assessment in Drylands has estimated that in total 33% of land and 20% of cropland is degraded (FAO, 2011a). In the past two decades more than 50% of all fertile, food-producing soil has been degraded to some extent (UNCCD, 2015).

India has 328.73 M ha of geographical area with 181.95 M ha agricultural land that comprises 139.9 M ha cultivated area, 3.2 M ha land with miscellaneous tree cover, and 12.6 M ha culturable wasteland. The remaining 26.3 M ha agricultural land is fallow or barren and susceptible to severe degradation (MoA, 2016). The Department of Land Resources, Ministry of Rural Development initiated the National Wasteland Updation Project in collaboration with the National Remote Sensing Centre. As a part of this initiative the entire country was mapped for degraded/wastelands using Resources at-1- LISS III satellite and a wasteland atlas showing state and district-wise spatial distribution of wastelands across the country was released (DoLR, 2011; SAC, 2016). Real-time land degradation monitoring and analysis carried out through various government agencies in India show a cumulative increase of 1.87 M ha area undergoing process of desertification/land degradation in India constituting 0.57% of the total geographical area of the country in a period of 10 years (2003–2005 to 2011–2013) (SAC, 2016, ICAR, 2010). The arid, semiarid, and subhumid regions in India have 60.11%, 28.68%, and 30.86% of the degraded land area (MoEF, 2014; SAC, 2016). The occurrence of these degraded lands also shows potential to increase the arable land following their reclamation.

14.3 Characteristics of degraded lands

The soil characteristics are defined by their pH, salt concentration, texture, clay content, etc. Terms used to describe the soil characteristics in the chapter are listed below.

- *Soil pH:* Soil pH is a measure of the acidity or basicity of a soil. It is measured in slurry of soil mixed with water (or a salt solution, such as 0.01 M $CaCl_2$). Acid soils have a pH below 7 and alkaline soils have a pH above 7. Ultraacidic soils have pH $<$ 3.5 and very strongly alkaline soils have pH $>$ 9.
- *Electrical conductivity (EC):* The electrical conductivity of the saturated soil paste extract (ECe) is a reciprocal of resistance that increases with the soluble salts and total salt concentrations in soil expressed in deci Siemens per meter (dS m^{-1}) at 25°C.
- *Exchangeable sodium percentage (ESP):* The amount of sodium as a proportion of all cations in soil is termed as the exchangeable sodium percentage.
- *Sodium absorption ratio (SAR):* Sodium absorption ratio is the ratio of the concentration of sodium (Na^+) cations to the combined concentration of calcium (Ca^{2+}) and magnesium (Mg^{2+}) cations in soil. It gives relative concentrations of Na^+, Ca^{2+}, and Mg^{2+} ions in saturated soil extract. A soil with SAR value below 2 is desirable for

agricultural purposes. The SAR values above 12 deteriorate the physical conditions of soil to an extent that plants have difficulty in absorbing water.

- *Slaking:* Clay particles swell upon wetting, which builds pressure that compresses the air entrapped within the pore spaces, which breaks soil aggregates. This breaking of soil aggregates into smaller particles is called slaking.

An account of different types of land degradation and their causes is detailed below and summarized in Table 14.1. Table 14.2 summarizes the challenges in management of the various forms of degraded land.

14.3.1 Salt affected soil (sodic/alkali soils)

Salinity is the major form of land degradation and affects many areas, especially the arid and semiarid regions. Soils with ECe greater than 4 dS m^{-1} are termed saline. An ESP value of 15 is a limit above which the soil are characterized as sodic/alkali (Chhabra, 2004). Most soils show a sharp deterioration in physical properties around or above this ESP limit. A soil with SAR value below 2 is desirable for agricultural purposes. The SAR values above 12 deteriorate the physical conditions of soil to an extent that plants have difficulty in absorbing water. The commonly accepted criteria for salinity thresholds are (1) saline soils: ECe > 4 dS m^{-1} pH < 8.5; (2) sodic soils: ESP > 15SAR > 13, ECe < 4 dS m^{-1}, and pH 8.5 or greater; and (3) saline–sodic soils: ESP > 15, SAR > 13, and ECe $\geq$ 4 dS m^{-1} (Chhabra, 2004; Yan et al., 2015). Saline–sodic soils are the most degraded form of soil due to the combination of both salinity and sodicity. Pedogenic sources or weathering of natural salt bearing soil minerals contribute to soil salinity. Saline soil also transforms into saline–sodic or sodic when irrigated with water containing high amounts of carbonates and bicarbonates of Na and Ca, and leaching of soluble NaCl and $CaCl_2$ salts causing accumulation of Na^+ ions in soil profile (Chaganti et al., 2015). The wet and dry monsoonal cycles and prevailing flood irrigation practices increase upward

TABLE 14.1 Causes and properties of various forms of land degradation.

Land degradation	Causes	Properties
Salinity and sodicity	Pedogenic sources (weathering of natural salt bearing soil minerals) High temperature, low rainfall and high evaporation in arid and semiarid regions Surface salts blown by winds and deposited on soil surface at other places Ground water containing high amount of salts	Faulty irrigation practices Injudicious use of fertilizers Capillary rise from shallow water table Water-logging/flooding Ingression of sea water (in coastal areas) Excessive seepage
Acidity	Mineralization/decomposition of organic matter Soil respiration Acid rain Formation of nitric (HNO_3) and sulfuric (H_2SO_4)acids in rainwater due to dissolved ammonia (NH_3)and sulfur dioxide (SO_2) gases respectively Heavy rainfall resulting in on-site leaching of base cations Irrigation with rain water (acidic pH)	Bacterial oxidation of elemental sulfur and SO_2 to H_2SO_4 Proton extrusion/excretion from plant roots accompanied with excess uptake of base cations Atmospheric N_2fixation by leguminous plants Application of acidic fertilizers. Elevated N deposition in soil due to application of N fertilizers Deforestation and removal of crop residues
Water logging	Poor drainage Hard pans of $CaCO_3$ Surface crusting Low water table/poor soil depth	Clayey texture with low OC Low aggregation Excess rainfall
Nutrient depletion	Intensive crop cultivation Erosion/run-off Leaching Atmospheric volatilization Fertilizer inefficiency	Crop residue removal Removal of organic matter Denitrification Nitrate leaching Unbalanced fertilizer doses

TABLE 14.2 Challenges in management of various forms of degraded land.

Degraded/marginal lands	Problems to be managed
Saline and sodic soils	High pH; salt accumulation; drought; Na and salt toxicity; low atmospheric CO_2-C sequestration; N mineralization and N loss; ammonia volatilization; moisture deficit; low availability of Zn, Mn, Fe, and Al; insoluble P complexes with Ca and Mg; poor physical structure; reduced microbial activity; low OC and IC; water-logging
Acid soils	Low pH; Al, Fe, Mn, and Cu toxicity; deficiency of Ca, Mg, S, B, and Mo; nitrate leaching and N loss; poor soil structure
Water logging/flooding	Poor aeration (anoxia), accumulation of toxic elements and plant pathogens, increased salt build-up, sulfide toxicity, emission of GHGs (CH_4 and N_2O), N loss by denitrification and nitrate leaching, change in soil redox status, low CO_2-C sequestration, loss of OC
Nutrient depletion	Deficiency of multiple nutrients, accumulation of heavy metals and fertilizer residues, lowering of soil pH, secondary pollution and eutrophication in water bodies
Erosion/desertification	Surface loss of soil; loss of N, P, and C; increased salt build up, moisture deficit (arid, semiarid, and hyperarid regions); off-site pollution of landscape and water bodies; loss of OM; increase in CO_2 emissions (surface loss of carbonates)

moisture flux owing to high evaporation rates in absence of crop cover favoring heavy build up of salts. Continuous agriculture also causes naturally occurring salts in the subsoil to be drawn up to the soil surface. The excess Na^+ on the cation exchange sites repels soil particles and increases the interparticle distance causing dispersion and breakdown of soil aggregates. Severe slaking occurs within minutes of wetting in sodic soil. When swelling exceeds the attractive forces of the clay particles they are dispersed. Clay dispersion breaks soil aggregates and translocation of individual soil particles causes blockage of pores in soil. Rain and/or irrigation weakens soil aggregates and blocks more than 50% of soil pores reducing water and air movement in saline–sodic soil. Surface seals are formed after aggregate dispersion and lower the soil hydraulic conductivity, reduce water infiltration/rate, and increase surface runoff. Surface seals develop into thick crusts of high strength upon drying, which makes soil susceptible to extreme erosion and water-logging. Exchange of Na^+ ions with other cations in soil destabilizes the soil structure and deteriorates its hydraulic properties, which increases surface crusting. The high clay content with low aggregate stability reduces water infiltration, aeration, water permeability, and soil hydraulic conductivity. Reduced infiltration and hydraulic conductivity increases surface run-off, puddling, and water-logging, and reduces plant-available water, which impedes seed bed preparation, root penetration, seedling growth, and plant establishment. The high density and low hydraulic conductivity of soil surface crusts reduce the flow of water, air, and heat through soil layers and increase susceptibility of soil to surface erosion. It reduces the ability of plant cells to take up water from soil, which has negative implications on crop production (Rengasamy, 2010). Hard setting occurs when the slaked soil dries. It is similar to surface crust, but its formation takes place in the lower depths where it creates an impermeable compact subsoil layer. It can be either limited to the upper few mm of soil or extend through the entire soil profile. Hard compact crust formed on soil surface by slaking, clay dispersion, and clay swelling destroys the soil structure and reduces air and water movement. Like Na^+, presence of K^+ ions also causes clay swelling and dispersion. Excess salts lower the amount of water available to the plants, which reduces the non-limiting soil water range and impedes ability of plants to uptake water from soil (Rojas et al., 2016). Low porosity, permeability, and high strength of soil restrict water storage and transport and also growth of plants in soil.

Salt affected soils are deficient in major plant nutrients including nitrogen (N), phosphorus (P), and potassium (K) (NPK) while high pH also reduces the availability of micronutrients such as Fe, Al, Zn, Mn, and Cu. The organic carbon (OC) content of these soils is lower than normal agricultural soil (Setia et al., 2013). Dispersion of aggregates caused by elevated sodium-salts and low soil water content fosters organic C loss that decreases primary productivity of soil. Global salt-affected soils have lost 3.47 t of OC per hectare and with an increase in the extent of saline soils the loss of 6.8 Pg in soil OC is expected by the year 2100 (Setia et al., 2013). Repeated wetting and drying of sodic soils differently influence C accessibility. When wet, dispersion of aggregates exposes organic matter and increases its availability to microbes, which accelerates C loss from soil as compared with dry conditions (Wong et al., 2010). Also salts in saline and sodic soil pose ion toxicity and have osmotic effects on plants and microbes, which over a period of time reduces soil organic C and N content (Yan et al., 2015). The changes in physical and chemical properties of soil and deficiency of nutrients and OM reduce soil microbial populations, microbial biomass, and enzymes. The reduced enzyme activities further affect nutrient contents and biogeochemical processes in soil. Thus, the physicochemical conditions of salt-affected soils are deteriorated to an extent that causes severe reduction in plant growth and yield.

Therefore, to reclaim such soils we need to manage high pH; salt accumulation; drought; Na and salt toxicity; low atmospheric CO_2-C sequestration; N mineralization and N loss; ammonia volatilization; moisture deficit; low availability of Zn, Mn, Fe, and Al; insoluble P complexes with Ca and Mg; poor physical structure; reduced microbial activity; low OC and IC; and water-logging.

14.3.2 Acid soil

Acid soils occur in many areas of the world. Soils of the humid tropics and forest areas in temperate zones are predominantly acidic. Based on the pH values the acid soils are strongly acidic (pH < 4.5); moderately acidic (pH 4.5–5.5), and slightly acidic (pH 5.5–6.5) (ICAR, 2010). India has 6.24 M ha strongly acidic and 24.41 M ha moderately acidic soils, which occur over vast stretches in the northeastern states namely Nagaland, Manipur, Tripura, Mizoram, Assam, Sikkim, and Meghalaya and also over some portions in Chhattisgarh, Jharkhand, Kerala, and Goa (ICAR, 2010; Maji et al., 2012). Acid soils occur with the build-up of hydrogen (H^+) and aluminum (Al^{3+}) cations in the soil, which replace the base cations such as potassium (K^+), calcium (Ca^{2+}), magnesium (Mg^{2+}), and sodium (Na^+) on their leaching (FAO–ITPS, 2015; ICAR, 2010; Guo et al., 2010). Rain water with pH 5–5.6 is slightly acidic due to dissolved CO_2 and when soil is exposed to rain, it attains the same equilibrium pH as the rain water. The effect of rain water on soil pH is more pronounced when it contains SO_2 and nitrogen oxides (NO_x), NH_3, and other oxidizing agents. These atmospheric gases form nitric acid (HNO_3) and sulfuric acid (H_2SO_4) with rain water increasing the precipitation of acid rain and acidification of the soil and water bodies. The CO_2 released during microbial and plant root respiration gets dissolved in water to form carbonic acid (H_2CO_3), which dissociates to form H^+ ions, acidifying the soil. Excessive rainfall followed by low surface evaporation due to high relative humidity and low temperature acidify the soil. Faulty agricultural practices lead to acid soil horizons in semiarid regions. Input of nitrogen fertilizers and excess deposition of nitrogen from atmosphere cause leaching of Ca and Mg base cations reducing the soil buffering capacity. Nitrogen fertilizer amount exceeding 5 g m^{-2} $year^{-1}$ significantly decreases the soil pH. The hydrogen ions released during nitrification of ammonium-based fertilizers in soil acidify the soil when NO_3^- is leached from the root zone. Uptake of base cations from nitrogenous fertilizers by plants also generates protons increasing the soil acidity (Guo et al., 2010; Chen et al., 2016). Fertilizer inputs in cereal cropping systems release 20–221 kmol H^+ ha^{-1} and uptake of base cations by plants generates 15–20 kmol H^+ ha^{-1} annually in soil (Guo et al., 2010). Extended episodes of drought in soil cause oxidation of sulfur and sulfur compounds that decrease soil pH.

The implications of soil acidity include reduced availability of many minerals and nutrients and on microbial flora. In acidic soil phosphorus is fixed as Fe and Al oxides that reduce its availability in soil (Behera and Shukla, 2015). Acidic conditions in soil mobilize toxic cations, particularly Al, Fe, and Mn ions, which pose toxic effects on plants (Bojórquez Quintal et al., 2017). Lowering of soil pH adversely affects the growth of microorganisms and causes shift in microbial population from bacteria to fungi. Acidity damages plant root hairs and affects moisture and nutrient uptake by plants (http://www.krishisewa.com/articles/soil-fertility/381-soil-acidity.html). The acidic conditions in the soil decrease microbial respiration and microbial biomass (Chen et al., 2016). The acidity in soil causes clay mineral dissolution and reduces soil cation exchange capacity (CEC), which produces nonreversible changes in the soil structure. Soil acidification extends deep into subsoil if preventive measures are not taken, which then becomes difficult to rectify. Therefore, to reclaim acidic soils we need to manage low pH; poor soil structure; deficiency of Ca, Mg, S, B and Mo Al, Fe, Mn, and Cu toxicity; nitrate leaching; and N loss.

14.3.3 Eroded and desertified soil

Soil erosion is one of the most severe forms of land degradation caused by removal of surface soil by erosive agents, commonly water or wind. Water and wind account for 87% of erosion, while excessive grazing pressure, loss of vegetation cover, and the lack of adequate soil conservation practices also increase the susceptibility of soils to erosion (FAO–ITPS, 2015). The upper fertile soil layer is naturally lost at the rate of one millimeter per year. Modern farming practices strip soil of carbon and other nutrients making it less robust and subsequently increasing the rate of loss by 10–40 times higher than the rate at which it could be naturally replenished. Soil particles mobilized during tillage are lost at the rate of 5 Pg annually. Water and winds annually erode 20–30 Pg $year^{-1}$ and 2 Pg $year^{-1}$ soil across the globe. Winds in dry arid and semiarid areas mobilize soil particles that are carried as dust in the atmosphere. The unsustainable land management practices annually drive 75 billion tonnes of top soil from croplands, which causes an estimated loss of $400 billion to agricultural production. Erosion by water and wind affects 93.68 and 9.48 M ha land in India (ICAR, 2010). Water removes about 16.4 ton soil ha^{-1} resulting in an annual loss of 5.3 billion tons of soil. Nearly 29% of eroded soil is lost to the sea, while 61% is transported to other places, and the remaining 10% is deposited in reservoirs (Bhattacharyya et al., 2015). Soil in dry lands is more susceptible to loss by wind erosion and the loss is expected to increase with reduction in precipitation and other changes in climate (Plaza-Bonilla et al., 2015).

Organic carbon having low density and loosely bound in the upper soil surface is preferentially removed by surface runoff and wind, severely depleting the soil OC pool. The soil OM, which is usually protected and stabilized as the integral part of soil aggregates, becomes exposed to microbial processes and subsequently lost due to mineralization. About 20% of OC displaced along with eroded sediments is mineralized to CO_2 under aerobic conditions and CH_4 under anaerobic environments (FAO—ITPS, 2015). The OC loss from soil can be as high as 0.5%—3.0% in areas affected by erosion and surface runoff. Therefore, to reclaim such soils we need to manage surface loss of soil; loss of N, P, and C; increased salt build up; moisture deficit (arid semiarid and hyper arid regions); off-site pollution of landscape and water bodies; loss of OM; and increase in CO_2 emissions (surface loss of carbonates).

14.3.4 Flooded/water-logged soil

Water-logging/flooding is a type of land degradation in which the soil profile is saturated with waters either temporarily or permanently (FAO—ITPS, 2015). Water-logging is often associated with soils affected by salt, high levels of ground water table, high water inflow and low water outflow, low infiltration rate, low hydraulic conductivity, and poor drainage (Bhattacharyya et al., 2015). India has in total 14.30 M ha water-logged area, which occurs in Andhra Pradesh, Assam, Madhya Pradesh, Maharashtra, and many parts of Rajasthan, Uttar Pradesh, Bihar, West Bengal, and Tamil Nadu. The prevailing climatic and geological conditions and anthropogenic activities like faulty irrigation and poor drainage lead to water-logging in these areas (ICAR, 2010; SAC, 2016). Flooding excludes air from soil pore spaces that are displaced by water, blocking the pores and restricting normal air circulation in soil. Adequate aeration is critical for rhizospheric processes, nutrient and water uptake, and heat movement through soil. Soil organisms require oxygen for respiration and nutrient cycling, and biogeochemical transformations. Water-logged soils have insufficient oxygen in the root zone, which prevents plant root respiration and restricts plant growth. The oxygen concentration of $>1\%$ with diffusion rate of 0.2 $\mu g\ cm^{-2}\ min^{-1}$ are required to satisfy the water and nutrient needs of the plants (Morales-Olmedo et al., 2015). Below this oxygen deficiency (hypoxia) or its complete absence (anoxia) develops, which retards growth and causes considerable damage to the roots of cultivated plants. Continuous water stagnation destroys the soil structure and makes the soil surface compact, which impedes seedling emergence, root growth, and plant establishment. The anaerobic conditions in water-logged soils lead to denitrification (reduction of NO_3 or NO_2 to N_2O and N_2), which accounts for more than 40% of the N losses on application of ammonium fertilizers (Zhu et al., 2011). Some N is also lost through leaching and as NH_3 volatilization. Water inundation decreases soil redox potential, and more iron (III) oxides and manganese (IV) oxides are reduced to iron (II) oxides and manganese (II) oxides, which increases their toxicity in soil (Zengin, 2013; Otoidobiga et al., 2015). Water-logged soils release volatile organic sulfur compounds, such as hydrogen sulfide (H_2S), carbon sulfide (CS_2), carbonyl sulfide (COS), and methyl mercaptan (CH_3SH) and dimethyl disulfide (DMDS) during microbiological processes, which are lost to the atmosphere (Irfan et al., 2010; Jing et al., 2017). The sulfate reducing bacteria produce toxic sulfide ions that prevent plant sulfur uptake and cause sulfide toxicity in soil (Otoidobiga et al., 2015). The Zn and Cu are immobilized and unavailable for plant uptake particularly in the calcareous soils with high pH and organic matter (Das, 2014; Ramesh et al., 2014). The organic acids, CH_4, CO_2, and ethylene gases accumulate in the plant rhizosphere during water-logging, which restricts plant root growth (Irfan et al., 2010). The excessive moisture content lowers the soil temperature, which retards bacteriological activities in soil. Salts present in lower layers are brought up to the surface by capillary action of water and deposited at the surface, increasing the salt build up. Water-logging and flooding thus creates anoxic conditions that are nonconducive conditions for plant growth and crop production, thereby decreasing the economic value of the land. Therefore, to reclaim water-logged soils we need to manage poor aeration (anoxia), accumulation of toxic elements and plant pathogens, increased salt build-up, sulfide toxicity, emission of GHGs (CH_4 and N_2O), N loss by denitrification and nitrate leaching, change in soil redox status, low CO_2-C sequestration, and loss of OC.

14.3.5 Nutrient depleted soil

Depletion of nutrients from soil is another form of land degradation common in many areas of the world. Plants receive primary, secondary, and micronutrients vital for their growth from soil. The nutrient depletion occurs when nutrients removed from soil exceed the amount that is being replenished. In India the cropping systems have created a deficit of primary and secondary nutrients in soil. Majority of the Indian soils show 8–10 Mt deficit of N, P, K along with the deficiencies in nutrients like sulfur (S) and micronutrients like iron (Fe), zinc (Zn), manganese (Mn), copper (Cu), boron (B), nickel (Ni), and molybdenum (Mo). The primary nutrients like nitrogen (N) are relatively mobile and thus readily lost from the soil through volatilization, leaching, and denitrification while other elements like potassium (K) and phosphorus (P) being nonmobile remain in soil nutrient reserves for longer periods. Statewise deficiencies in soil nutrients have been well documented by Shukla et al. (2014). Across the country N, P, K, S, Zn, B, Mo, Fe, Mn, and Cu deficiency has reached 89%, 80%, 50%, 41%, 49%, 33%, 13%, 12%, 5%, and 3%, respectively (Jones et al., 2013). The synthetic fertilizer supplements fail to effectively replenish the lost nutrients requiring repeated applications for subsequent crops (Jones et al., 2013; Shukla et al., 2014). The physicochemical constraints and unbalanced use of fertilizers reduce nutrient use efficiency in soil. Excessive use of chemical fertilizers further aggravate multinutrient deficiency in soil and also add to contaminants such as silver, nickel, selenium, thallium and vanadium, mercury, lead, cadmium, and uranium, which pose hazards for human health (MoA, 2016). Thus, the widespread nutrient deficiency generated in soil constrains plant growth, deteriorates soil health, and makes soil too weak and fragile to sustain long-term crop production. Therefore, to minimize nutrient deficiency in soil we need to manage deficiency of multiple nutrients, accumulation of heavy metals and fertilizer residues, lowering of soil pH, and secondary pollution and eutrophication in water bodies.

14.4 Land degradation and soil carbon and nutrient reserves

The three major reservoirs that regulate the carbon cycle on Earth are the oceans, the atmosphere, and terrestrial systems. Soils comprise 80% of total terrestrial carbon C and have 3.3 times more C than the atmosphere and 4.5 times more C. Of the total terrestrial carbon OC is of importance for crop productivity, and improving soil organic carbon (SOC) holds immense potential to enhance soil and atmosphere quality (FAO, 2017). The estimates of SOC are 684 Pg C in 0–30 cm soil, which is of agricultural importance. The total soil C reserves in India are 13.69 Pg in 0–30 cm and 63.90 Pg in 0–1.5 m comprising 9.55 Pg OC in 0–30 cm and 29.92 Pg OC in 0–1.5 m (Pal et al., 2015). The soils in arid and semiarid regions mostly have IC that constitutes 78% of the total C stock (Pal et al., 2015). Several practices like faulty drainage, plowing, biomass burning/removal, deforestation, erosion, salinization, nutrient depletion, soil structural decline and compaction, and acidification influence soil C reserves and aggravate OC depletion from soil. The global SOC loss estimates vary from 44 to 537 Pg (FAO–ITPS, 2015). The OC being loosely held in the superficial soil layers is susceptible to loss, which is more rapid in the arid and semiarid and dry subhumid tropical regions (Pal et al., 2015). The intensive agriculture adversely affects microbial biomass in topsoil, which depletes soil C reserves (Pabst et al., 2013). Soil is also a huge reservoir for nitrogen (N), and contains 63–67 Pg N in 0–30 cm, 133–140 Pg N, and 136 (92–140) Pg N in 0–1 m of upper soil surface, globally (Oertel et al., 2016). Cultivated regions contain high amount of N, which exceeds 3.5 Mg (Megagram) (1 Mg = 10^6 g) N per hectare probably due to addition of N fertilizers and increase in soil microbial activity. The soil N pool is present in four major forms: (1) soil OM (plant material and humus), (2) microbial biomass, (3) ammonium ions (NH_4^+) held by clay particles and organic matter, and (4) mineral N, which includes free NH_4^+ in solution, nitrate (NO_3^-), and nitrite (NO_2^-) ions (Cameron et al., 2013). Most N (97%–99%) present in the upper 0–30 cm of soil surface is bound with organic matter, mineralization of which releases 8–120 kg N ha^{-1} in plant available forms, NO_3^- and/or NH_4^+ ions (Chen et al., 2014). The mineral-N pool is prone to loss through (1) ammonia volatilization, (2) leaching, and (3) denitrification. Faulty agricultural practices like application of excess doses of N fertilizers also facilitate loss of nitrogen from soil (Cameron et al., 2013). The N loss from soil is usually high in the arid and semiarid regions and increases with aridity, creating N deficiency in soil (Fisk et al., 2015). Soil total phosphorus (P) is 400–4000 kg ha^{-1} in the upper 0–30 cm, which occurs as organic P (OP) and inorganic P (IP) (Yadav and Verma, 2012). Organic P is about 30%–80% (between 4% and 90% in most soils) comprising organophosphorous compounds such as inositol phosphates (10%–50% of OP), phospholipids (1%–5% of OP), and nucleic acids (0.2%–2.5% of OP), which are available to plants after mineralization (Arcand and Schneider, 2006; Yadav and Verma, 2012). The microbial biomass contains about 3%–24% of the total OP (Arcand and Schneider, 2006). Phytin (Ca–Mg salt of phytic acid) is the most abundant soil OP that is not directly available to the plants. Inorganic P occurs as Ca–phosphate in calcareous or normal soils and Fe^{3+} and Al^{3+} phosphates in acidic soil (Yadav and Verma, 2012). The total P content in arid regions (alkaline and/or calcareous soils) of the country is 560–900 kg P ha^{-1}, of which the available form is only 15–25 kg ha^{-1}, and the rest is fixed on clay minerals and with $CaCO_3$ as Ca–phosphates (Yadav and Verma, 2012). The chemical P fertilizers added are precipitated as dibasic calcium phosphate dihydrate ($CaHPO_4 \cdot 2H_2O$) or Octa-calcium phosphate ($Ca_4H(PO_4)_3 \cdot 3H_2O$), which are converted into insoluble apatites (hydroxyapatite, $Ca_{10}[OH]_2[PO_4]_6$) and fluorapatite ($Ca_{10}F_2[PO_4]_6$) (Imas, 2000). Plants uptake P as orthophosphate anions $H_2PO_4^{-1}$, HPO_4^{-2} and PO_4^{-3} (in the order of preference $H_2PO_4^{-1} > HPO_4^{-2} > PO_4^{-3}$) from soil, which is dependent upon soil pH (Arcand and Schneider, 2006; Yadav and Verma, 2012). In most soils only 0.1%–1% of the total P reserves are available for plants and the major fraction is immobilized. Like N and P, potassium (K) is an essential nutrient required by plants. Soils contain K ranging between 10,000 and 50,000 kg ha^{-1} in the upper 0–0.2 m of the soil profile (Kafkafi et al., 2001). In agricultural soils K content is between 10 and 20 g K kg^{-1} soil (Kafkafi et al., 2001). The K content in soil across India varies from less than 0.5%–3% with an average of 1.52% (Meena et al., 2014). The soil K exists in four major forms: (1) water soluble K, (2) exchangeable K, (3) nonexchangeable/fixed K, and (4) structural/mineral K (Meena et al., 2014). The water soluble K constitutes 0.1%–0.2% of total soil K in the upper soil layers and is immediately available to plants and microbes. The exchangeable K and fixed K constitute 1%–2% and 1%–10% of the total soil K and replenish the K lost from soil. The remaining (90%–98%) K is present in crystalline structures of mineral particles (e.g., muscovite, orthoclase, biotite, feldspar, illite, mica, quartz, smectite, kaolinite), which are usually insoluble thus unavailable for the plants (Meena et al., 2014; Rashid et al., 2016). Bacteria and fungi in the rhizospheric region have ability to mobilize and decompose mineral particles releasing K from dense crystalline layers in water soluble, freely available form for plant and microbial uptake (Rashid et al., 2016). Table 14.3 summarizes the microbe-mediated restoration of nutrients (NPK) in stressed soils.

TABLE 14.3 Microbe-mediated restoration of soil fertility in degraded lands.

Bacteria and/or fungal strains	Major role	Region/crop
Shewanella putrefaciens and *Bacillus* sp.	Increased iron availability	Calcareous soils
Bacillus megaterium and AM fungi	Provides N, P, K and drought tolerance	Arid and Semiarid regions
Pseudomonas aeruginosa BS8	Provides N, P, and Fe with biocontrol (antifungal) activity	Variable conditions
Bradyrhizobium diazoefficiens and *Bradyrhizobium japonicum* strains	Enhanced N fixation Increased yield	Soybean
Pseudomonas aeruginosa mixed culture *Azotobacter chroococcum* + *Azospirillum brasilense*	Increase NPK Promote plant growth	Rice
Bradyrhizobium diazoefficiens and *B. japonicum*	Increase Nitrogen Promote yield	Soybean
Bacillus cereus, B. megaterium, Burkholderia caryophylli, Pseudomonas cichorii and *Pseudomonas syringae*	Increase P	Subtropical flooded and temperate nonflooded soils
Micrococcus sp.	Increase P, Fe and organic content, reduce pH Promote plant growth	Effective with cowpea crop
Rhizophagus irregularis and *Pseudomonas alcaligenes*	Increase soil P and its plant uptake	Soils with high organic P (phytate)
Rhizophagus irregularis and *Bradyrhizobium vigna*	Increase P by hydrolysis from organic P Increase soil organic matter, enzyme activity and pH	Subtropical red soi(Soils with Fe and Al phosphates)
A. brasilense and *Pantoeadispersa*	Increase organic matter, microbial biomass C, nutrients Increase plant growth and nutrient uptake	Soils of semiarid regions
B. megaterium, Enterobacter sp. and *Bacillus thuringiensis*	Improve NPK Favor plant establishment and nutrient uptake	Degraded soil of semiarid region
Rhizophagus intraradices, B. megaterium and *B. thuringiensis*	Increase nutrients Alleviate drought stress	Arid soil under drought conditions

14.5 Microbes as indicators of soil health and its productivity

Soil quality is "the capacity of a reference soil to function, within natural or managed ecosystem boundaries, to sustain plant and animal productivity, maintain or enhance water and air quality, and support human health and habitation." It encompasses interactions of soil with surrounding environment, including its implications on human and animal health (Katyal, et al., 2016). Soil is not just a growing medium, but rather a living system whose function is mediated by diverse living organisms and is of prime importance in soil functioning (Katyal, et al., 2016). A healthy soil has good structure with high water and nutrient-holding capacity to maintain a diverse community of organisms that form beneficial associations with plant roots, control plant diseases, and recycle nutrients. The assessment of soil health and quality is necessary to evaluate its degradation status, which is usually measured indirectly using several physical, chemical, and biological indicators of which biological indicators related to functional processes are most sensitive to environmental changes (Cardoso et al., 2013).

The microbial biomass, though it occupies a tiny fraction (<0.5%) of the total soil volume and constitutes less than 10% of the total organic matter in soil, is the major driver of soil biological functions, which influence soil quality. The metabolic characteristics and diversity of microorganisms responsible for N, P, S, and C cycles in soil are sensitive to physicochemical, soil use, and management changes in soil and show rapid shift during disturbance, which can be monitored. Thus, a healthy soil has a reserve of quiescent microorganisms that quickly respond to foreign substances to provide higher resilience or biological buffering against disturbances and maintain soil homeostasis to sustain various ecological processes. High microbial diversity and functional redundancy indicate good soil quality and enable soil to recover from stress. The parameters describing the microbial functions and their diversity provide an early indication of improvement or warning of soil degradation. These microbial markers have been discussed with respect to their role as markers and their effect on soil health.

14.6 Microbial intervention in restoration of degraded soil

The goal of LDN seems achievable by sustainable land management (SLM) practices and their proper application. The response hierarchy of LDN is Avoid > Reduce > Reverse land degradation (UNCCD-SPI, 2015). The applied methods are envisaged to help soil health amelioration, enhance its buffering capacity against degradation forces, and impart other socioeconomic benefits. The various SLM approaches include organic farming, crop rotation, and mixed cropping, which are useful to restore soil properties in degraded lands by (1) boosting soil functions, (2) managing soil erosion, (3) maintaining SOC and nutrients, (4) improving soil structure and long-term fertility, and (5) enhancing soil biodiversity (Singh, 2015; Rojas et al., 2016). The role of microbes in all these processes of SLM is imperative. Therefore, the introduction of these practices improves diversity of microbes in soil. Their cosmopolitan survival, massive efficient genetic pool, catabolic versatility, and stress tolerance potential make them important ecological engineers to restore degraded soil (Singh, 2015).

14.6.1 Nutrient management

Several problems like micro- and macronutrient deficiency created due to fertilizer imbalances and nutrient mining can be managed using microbes and help reduce chemical inputs in soil (Jones et al., 2004; Miransari, 2013). The AM fungi and many organic acid producing rhizospheric microbes have been extensively used in management of nutrients like P, K, and Si in soil (Smith and Smith, 2012; Liu et al., 2016).

14.6.2 Soil structure

The arrangement of soil particles into aggregates and the distribution of pore spaces present within or between the aggregates describe the soil structure (Bronick and Lal, 2005). The soil structure affects water mobility, diffusion of air, biological activity, seed germination, and root growth. Healthy soil comprises of well formed stable aggregates protected from disintegrating forces, which depends upon the architectural traits of microbes and their interactions with plant roots (Rillig et al., 2015). The microbes and their biochemical activities increase formation and stabilization of soil aggregates, which improves soil structure. Bacteria and fungi produce several substances that work as gluing agents for silt and clay particles causing their aggregation in soil. Bacteria dominate in microaggregates while fungi in macroaggregates help in formation and stabilization of soil aggregate. Inside soil aggregates, water content and proximity to substrates are increased and microbes are protected from attack by predators (Ruamps et al., 2011). The various bacterial and fungal constituents with various roles in soil aggregation and stabilization include the following.

14.6.3 Cell wall components and their residues

The microbial cell walls and their residues form electrostatic bonds with minerals, which are stable against decomposition and persist after the death of the organism in the soil (Miltner et al., 2012). These serve in structural organization of soil aggregates and are excellent microbial markers indicating their presence in soil. The role of various cell wall components of bacteria and fungi in soil is well evident.

14.6.3.1 Extracellular polymeric substances

The EPS and mucilage are gel-like, water-rich macromolecules forming a complex mixture of biopolymers polysaccharides, proteins, glycoproteins, nucleic acids, lipids, and humic substances that aids in formation and stabilization of microaggregates. Its hygroscopic nature improves cellular hydration, which increases microbial tolerance to water stress in dry soil (Sandhya et al., 2009). EPS retain and protect enzymes and their substrates, which improves efficiency of soil enzymatic activity (Flemming and Wingender, 2010). The EPS matrix embeds the microbial biofilms and helps them to closely adhere to surfaces (soil and root), bridging the gap between substrates, which eases the availability and uptake of water and nutrients. (Flemming and Wingender, 2010; Redmile-Gordon et al., 2014). Bacteria and polysaccharides associated with clay particles serve as nucleus to initiate formation of "composite building units" for microaggregates in soil (Schutter and Dick, 2002). The large number of free functional groups like carboxyl, amide, and phosphoryl present in EPS attach microorganisms to minerals, causing aggregation of soil particles (Omoike and Chorover, 2006). The pores and networks in biofilm allow the flow of nutrients and oxygen from soil (Li and Yuan, 2002). The cross-linked structure of EPS enables it to shrink and swell yet remain saturated across a wide range of soil metric potentials (Or et al., 2007). The hydrophilic nature of mucilage in soil rhizosphere allows it to retain water even when bulk soil water is scarce, which improves water uptake in plant roots (Carminati, 2012). Besides structural improvement in soil EPS producing plant growth promoting microbes such as *Bacillus* and *Enterobacter* species reduce Na^+ uptake by plants (Upadhyay et al., 2011) and *Pseudomonas* increase drought tolerance in plants (Sandhya et al., 2009), which makes them effective for application in degraded salt affected soils of arid and semiarid regions. Biofilm forming microbial communities raise microbial diversity through breaking dormancy of the soil microbial seed bank (Seneviratne and Kulasooriya, 2013). It has also been suggested that cyanobacteria induce soil aggregation and water permeability, and are quite useful in improving quality of poor structured soils of arid or subarid regions. Cyanobacterial cells form crusts on the surface of soils in arid and semiarid regions that are a major source of organic C and well adapted to drylands because of extracellular polysaccharide (EPS) secretions that stabilize soil structure and provide resistance to desiccation. Thus polysaccharides/EPS from bacteria, fungi, algae, or root exudates interact strongly with the soil components to form and stabilize soil aggregates and contribute to formation of soil structure.

14.6.3.2 Fungal hyphae

Fungal hyphae enmesh microaggregates to form macroaggregates, and soil aggregation increases with increased hyphal density. They improve aggregate stability by reorientation of clay particles, binding particles with extracellular polysaccharides, and enmeshing particles. Hyphae with extracellular polysaccharides and enzymes form "sticky string bags" that enmesh, cross-link, adhere, and glue soil particles and aggregates that form and stabilize soil structure (Six et al., 2004; Caesar-Tonthat, 2002; Daynes et al., 2012). The fungi involved in soil structure improvement have been extensively reported. The AM fungi are the most important biotic components to influence soil aggregation (Rillig et al., 2002; Rillig et al., 2015). The macroporous soil structure formed by hyphae of AM, ectomycorrhizal, and saprophytic fungi allows easy penetration of water and air and prevents soil erosion (Rillig et al., 2002; Zheng et al., 2014). It facilitates transport of nutrients and water for fungal uptake and protects the fungal hyphae and spores from drought and microbial attack (Driver et al., 2005; Rashid et al., 2016). The AM hyphal networks stabilize soil aggregates more due to their long residence time and low palatability than saprobic fungi (Rillig et al., 2002).

14.6.3.3 Glomalin

Like EPS glomalins are glue-like hydrophobic glycoproteins produced by AM fungi, which closely adhere to hyphal walls forming a protective sheath (Miller and Jastrow, 2000; Rashid et al., 2016). The sticky glomalin serves as an adhesive agent that attaches fungal hyphae to sand, silt, and clay particles and organic matter forming clumps of soil aggregates (Borie et al., 2008; Rashid et al., 2016). This structure is stable to resist wind and water erosion, but porous to allow movement of air, water, and roots through it. It also harbors more beneficial microbes, holds more water, and prevents soil crusting. The recalcitrant and hydrophobic nature of glomalin increases its soil concentration, which protects other aggregating agents stabilizing soil aggregates and improving soil structure (Rillig et al., 2002; Rashid et al., 2016).

14.6.3.4 Hydrophobic substances

Hydrophobins are a class of small hydrophobic proteins produced by fungi that help in attachment of mycelia to surfaces, alter surface properties, lower water tension and are useful in exploring large distances and transporting nutrients and water for fungi (Linder et al., 2005, Diehl, 2013). These hydrophobic substances often bind to soil particles and promote formation of water stable aggregates (Ambriz et al., 2010). *Agaricus bisporus* produces surface active hydrophobic proteins that promote formation of water stable aggregates in soil (Spohn and Rillig, 2012).

14.6.4 Microbes affecting soil physicochemical properties

Sulfur-oxidizing microbes convert elemental sulfur into sulfuric acid (sulfate), which facilitates dissolution of calcite mineral to release calcium in soil, which in turn replaces sodium ions. It lowers soil pH, which increases availability of P and K in the soil (Stamford et al., 2015). Application of sulfur dusts inoculated with sulfur oxidizing bacteria such as *Thiobacillus* (also called *Acidothiobacillus*) are useful for reclamation of saline–sodic soils (Stamford et al., 2015; Zhao et al., 2017). Mycorrhizal fungi are effective in managing toxicities of Mn, Fe, B, and Al, a prominent feature in acidic soils, by binding to structures such as spores, hyphae, and vesicles or its immobilization by the glomalin-related proteins, which confer Al tolerance to plants (Seguel et al., 2013; Aguilera et al, 2011). Microbes produce siderophores that increase uptake of Fe and form complexes with other heavy metals (Pb, In, Cu, Ga, Cd and Al) (Schalk et al., 2011; Rashid et al., 2016). Purple nonsulfur bacteria (PNPB) *Rhodopseudomonas palustris* improves rice growth and yield from flooded soils. PNPB produce 5-aminolevulinic acid that reduces salt stress for plants (Nunkaew et al., 2015).

14.6.5 Microbe-mediated amelioration of nutrient deficiency

Physicochemical conditions of soils affect availability of nutrient such as P, K, Zn, Fe, Mn, and Cu, which are deficient in most soils either because of their low levels or the low solubility. A major fraction of nutrients in soil exists in bound mineral forms. Microbes interact with mineral surfaces in soil and rocks to dissolve these minerals and release the constituent elements in available form for their growth (Uroz et al., 2009). Minerals are solubilized by the production of organic acids such as succinic, citric, gluconic, ketogluconic, and oxalic acids or by production of protons (H^+), which decrease the local pH and cause acidification of soil microenvironment. *Bradyrhizobium japonicum*, *Cupriavidus necator*, *Ralstonia solanacearum*, *Dechloromonas agitate*, and *Nocardioides* sp. have an ability to oxidize phyllosilicate minerals to release K (Wu et al., 2005; Meena et al., 2014). Mineral forms of K and P are also solubilized by fungi through release of organic acid anions citrate, malate, and oxalate (Han and Lee, 2005; Wu et al., 2005). AM fungi *Glomus mosseae* and *G. intraradices* increase K uptake in corn plants. *Aspergillus terreus* and *Aspergillus niger* solubilize K from insoluble feldspar and potassium aluminum silicates and P from rock phosphates through organic acids production (Li et al., 2016). Zinc solubilizing microbes such as *Pseudomonas aeruginosa* (Bapiri et al., 2012), *Ralstonia picketti* (Gontia-Mishra et al., 2017), *Burkholderia cepacia* (Vaid et al., 2014), *Klebsiella pneumoniae* (Gontia-Mishra et al., 2016), *Gluconacetobacter diazotrophicus* (Sarathambal et al., 2010), and *Bacillus aryabhattai* (Ramesh et al., 2014) are known to solubilize zinc from insoluble zinc compounds and increase their uptake in plants. Bluish vivianite crystals have been observed on the roots of rice plants grown in flooded soil, which contain 2.0–4.5 g P kg^{-1} vivianite on an oven-dry basis (Nanzyo et al., 2010). Vivianite is a ferrous iron phosphate mineral, crystals of which are formed by metal reducing bacteria in organic and iron-rich sediments of waterlogged soil (Rothe et al., 2016). Microbes not only manage microdeficiency but also help to ameliorate high nutrient concentrations in soil and their adverse affects on plants (Miransari, 2013). Major microbial mechanisms to increase soil nutrient availability and uptake in plants are summarized in Table 14.4.

14.6.6 Microbe-mediated soil sequestration of carbon and nutrient (NPK)

Plant roots and rhizospheric microorganisms retain 50%–70% of soil C (Clemmensen et al., 2013). Microbial interactions in soil maintain and regulate many nutrient cycles to conserve long term soil C, N, and nutrient reserves and replenish lost nutrients (Kaiser et al., 2015). Microbes grow on plant residues and incorporate plant-derived substrates and C in their biomass. After death microbial biomass residues release C, which remains in soil (Miltner et al., 2009;

TABLE 14.4 Major microbial mechanisms to increase soil nutrient availability and uptake in plants.

Mechanisms	Effects
Bacteria	
1. Lowering of pH by release of organic acids and protons (H^+) 2. Chelation of the cations 3. Acidolysis, that is, acidification of their own cells, rhizosphere and surroundings of K minerals with discharge of H^+, inorganic and organic acids	Directly dissolve K and P from inorganic sources Mobilize or solubilize insoluble form and structural unavailable forms of K and P compounds in to soil solution Chelate Si and Al ions associated with K minerals
4. Production of siderophores	Chelate and form high affinity complexes with Fe^{3+} Membrane mediated uptake of complexes into cells reduce Fe^{3+} to Fe^{2+} Solubilize Fe from minerals or organic compounds
5. Release of several phenazines like Phenazine-1-carboxamide and the glycopeptidic antibiotic bleomycin (*Pseudomonas*) 6. Flavins and quinone-containing humic substances (*Shewanella oneidensis*)	Shuttles electrons Mineral oxide reduction Dissolve mineral oxides crystals Stimulate mineral reduction (Fe-dissimilation) Soluble electron shuttles Interact with outer membrane cytochromes Enhance extracellular electron transfer Metal reduction (dissimilation)
7. Biological N-fixation 8. Nitrification	Fix atmospheric N_2 in the form of NH_3 Aerobic oxidation of NH_4^+ to NO_2^- by ammonium-oxidizing bacteria (*Nitrosomonas*, *Nitrosospira*, and *Nitrosococcus* species) and NO_2^- to NO_3^- by nitrite-oxidizing bacteria (*Nitrobacter*, *Nitrospina*, *Nitrococcus*, and *Nitrospira* species)
9. Extracellular depolymerases (proteases, chitinases and peptidoglycan hydrolases) 10. Urease, amino acid oxidase	Hydrolysis of complex N containing molecules such as large proteins and polypeptide, Chitin, Peptidoglycan into small organic molecules Extracellular mineralization (Ammonification) of small organic molecules
11. Production of H_2S	React with ferric phosphate to yield ferrous sulfate with concomitant release of phosphate
12. Exopolysaccharides (EPSs)	Solubilize tricalcium phosphate
13. Siderophores	Mineral dissolution Enhance P availability
Fungi	
1. Soluble K and P uptake and transport in mycelia 2. Release of H^+ (Proton extrusion), CO_2 and organic acids	Mobilize P and K in soil Solubilize P and K and/or enhance their solubilization from inorganic sources
3. Bacteria present in mycelia	Translocate C and P from roots to the associated bacteria Promote N fixation by bacteria
VAM fungi	
1. Symbiotic association of bacteria and fungi with legumes 2. Association of AM fungi with Rhizobia	Provide high amount of P for Nitrogen fixation. Provide plant derived C to increase the potential ability of nonsymbiotic N_2 fixers Fix atmospheric N_2 through N-fixing bacteria Mycelia protect the intracellular bacteria of the genus Burkholderia
3. Extraradical hyphae of AM fungi 4. Extensive hyphae in soil and around roots of AM fungi	Exploration of soil for nutrients Allows the fungus to penetrate into small soil pores Extend beyond P, S, and Zn depletion zone in soil Enables host plant to access distant inorganic nutrients and increase their mobility to plants

(*Continued*)

TABLE 14.4 (Continued)

Mechanisms	Effects
Bacteria	
5. Production of hydrolytic enzymes 6. Pectinase, xyloglucanase, and cellulase	Decompose soil organic matter Better exploit N resources Improve uptake of organic N molecules like amino acids.
7. Enhanced plant growth, altered root exudates and signaling pathways 8. Provide organic C for soil microbes	Affects the activities of heterotrophic soil microbes. Compete with heterotrophic soil microbes for mineral N Increase mineralization of organic N, increase N uptake by the plant
9. Production of organic acids (anions) 10. Release of protons (H^+) during NH_4^+ assimilation 11. Carbonic acid formation during respiration (CO_2 + soil water) 12. Release of inorganic acids such as H_2SO_4, HCl, HNO_3	Directly dissolve mineral P (insoluble P complexes with Ca, Al, Fe) Chelation of cations Organic acids (anions) exchanged for P on adsorption sites (ligand exchange) Form soluble complexes with metal ions associated with
13. Production of phosphatases, phytases, phosphonatases and C–P lyases	Mineralize (hydrolyze) organic P compounds
14. Drying–rewetting (flush-events)	Microbial cell lysis {About 30%–45% of microbial P (0.8–1 mg kg^{-1}) released from microbial cells during initial flush event in soil}
15. Extensive branched, extraradical hyphae 16. Small hyphal diameter	Extend beyond P depletion zone in soil Exploration of soil for nutrients Allows the fungus to penetrate into small soil pores
17. Higher substrate affinity into hyphae (low Km) 18. Chemical modification of adsorbed P	Improved P uptake Orthophosphate is scavenged and delivered to plants Higher (P) influx rates per unit surface area Increased (P) uptake
19. Alter rhizospheric microenvironment 20. Increased proton efflux or pCO_2	Greater access to soil P Lower pH Solubilization of P in neutral and calcareous soil
21. Produce carboxylates	Ligand exchange Dissolution of P-bearing Fe/Al oxides P mobilization Induces alteration in surface charges on clays
22. Storage as polyphosphates	The internal Pi concentration relatively low Allows an efficient transfer of P from the extra radical to the intraradical mycelium
23. Chelating compounds and metabolites 24. Metallophores, Ethylene diamine tetraacetate	Form complexes with Zn Reduce their reaction in soil Chelation of Zn Zn chelates move towards roots Release chelating ligand at root surface
25. Changes in root architecture	Improves root growth, root surface area, and absorptive area
26. Sulfatases and Sulfate transport system, Intracellular mono-oxygenase multienzymes, Laccases and peroxidases	Sulfur transformations Mineralize soil S pools to plant available inorganic sulfates Mineralization and mobilization of Organo-S (sulfate esters and sulfonates) Facilitate uptake of sulfate from microbes
27. Infection needle with type III secretion system (T3SS)	Putatively transfer nutrients via this infection needle

Miltner et al., 2012). In a study with ^{13}C-labeled bacterial cells 50% of the biomass C was returned to soil majority (40%) of which becomes part of recalcitrant soil OM after 224 days of incubation (Miltner et al., 2012). In soil with organic matter inputs total bacterial necromass is usually more than living microbial biomass that stores C, which constitutes soil OC (Miltner et al., 2012). Thus, more than 80% of soil OC is of microbial origin (Liang and Balser, 2011). Soil fungal communities accumulate more OC in their biomass and express a broader suite of enzymes that decompose complex OM and create a balance between microbial biomass to regulate soil C reserves (Malik et al., 2016; Waring et al., 2013).

The extraradical mycelia of AM fungi form extensive networks in soil and constitute 20%–30% of the total microbial biomass. It receives up to 20% of C in the form of carbohydrates from the host plant (Leake et al., 2004) and about 30% of the C retained in the root system is incorporated and processed into the rhizospheric soil and made available to other soil organisms (Zhang et al., 2017). AM fungi release large amounts of chitin and glomalin-related proteins, which are relatively stable after turnover, which enhances soil OC reserves. Glomalin and glomalin related proteins are major components of soil organic matter that contribute to the soil C and N pool (Zhang et al., 2017). The increase in soil OC by AM fungi improves soil structure even in compact clayey soils (Xu et al., 2015).

The soil mineral N pool is immobilized in the microbial biomass, producing a labile pool of organic N that exists in equilibrium with a stable N pool associated with the humus. This equilibrium is shifted toward immobilization (assimilation) by increasing the input of C relative to N and toward mineralization (decomposition) by increasing the input of N relative to carbon (Fisk et al., 2015). The proteins are hydrolyzed to release amino acids, which release ammonia after ammonification and are converted into nitrite (NO_2^-) and nitrate (NO_3^-) after oxygen incorporation by two specialized groups of chemoautotrophic bacteria *Nitrosomonas* and *Nitrobacter*. Microbes extract N from recalcitrant organic matter (having high C/N ratio) in the form of amino acids to increase soil N availability (Redmile-Gordon et al., 2014). AM symbiosis increases N uptake and its assimilation in plants preventing denitrification, which reduces N_2O emissions (Lazcano et al., 2014). AM fungi acquire N from both inorganic and organic N sources in soil and its symbiotic association increases N uptake NO_3^- and/or NH_4^+ by plants (Whiteside et al., 2012; Hodge and Storer, 2015). Cyanobacteria (blue-green algae) are autotrophic prokaryotes that can survive stress conditions in degraded lands and promote C, N, and P levels in soil. Cyanobacteria synthesize carbohydrates during photosynthesis and diazotrophic cyanobacteria also fix atmospheric N in soil (Mager, 2010; Singh et al., 2016). In this way they regulate both C and N status of soils. Cyanobacteria release organic acids, and chelator compounds that solubilize inorganic phosphates such as $(Ca)_3(PO_4)_2$, $FePO_4$, $AlPO_4$, and hydroxyapatite $[Ca_5(PO_4)_3OH]$ present in soils (Singh et al., 2016). It is not only the obligate symbiotic associations between fungus and plant roots that facilitate nutrient acquisition in plants but there are certain entomophytes such as *Metarhizium* and *Beauveria*, which have reported ability to transfer N to plants. *Metarhizium robertsii* is a soil-borne fungus with tripartite interaction (fungus/insect/plant) that trades insect N to plants in exchange for plant photosynthates (Behie et al., 2017).

14.7 Microbial markers and their significance in soil

Microbial biomass is an important building block of soil organic matter. Microbial cells and their exudates contain several molecules such as amino sugars, ergosterol, phospholipid fatty acids (PLFAs), and polysaccharides, which are markers of their presence and eventually become an indispensible part of soil organic matter after turnover. Microbial cell wall materials and outer membrane components are important sources for long-term soil carbon and nitrogen reserves (Liang et al., 2011) and serve as microbial markers (Kögel-Knabner, 2002). The structural molecules such as amino sugars, peptidoglycan chitin, melanin pigments, and phospholipid fatty acids in membrane and cell walls of bacteria and fungi are major constituents of soil organic matter that trace soil microbial dynamics. Some important microbial markers in soil are described in the following sections.

14.7.1 Amino sugars

Glucosamine, muramic acid, galactosamine, and mannosamine are prominent amino sugars in soil (Amelung et al., 2008). Fungal glucosamine and bacterial muramic acid accumulate as microbial residues in soil. Glucosamine is mainly obtained from chitin, a polymer of *N*-acetyl-D-glucosamine and a structural building block of fungal cell walls (Chantigny et al., 1997). *N*-acetyl glucosamine and muramic acid polymerize in bacterial cell walls forming peptidoglycan, which consists of muramic acid and glucosamine in a molar ratio of 1:2 (Engelking et al., 2007). Cell walls of Gram-positive bacteria consist of 90% peptidoglycan, whereas in Gram-negative bacteria it is only 5%–20%. Most soil glucosamine is of fungal origin (Amelung et al., 2008). Galactosamine derived from microbes mainly of bacterial origin accounts for 30%–50% of amino sugar content in soil (Engelking et al., 2007). These microbial amino sugars are relatively stable and persist in soil after cell death. Therefore, the proportion of total amino sugars to total soil C reflects the relative contribution of the microbial community to soil C sequestration.

14.7.2 Ergosterol

The amount of ergosterol is a marker for fungal biomass that determines its activity in soil. Ergosterol is synthesized and present in the cell membrane of fungi belonging to basidiomycota, ascomycota, except in some zygomycota, and all glomeromycota (Klamer and Bååth, 2004). It is important for the fungal growth and controls membrane permeability, viscosity, and activity of membrane-bound enzymes. Since ergosterol has rapid turnover it specifies the growth and presence of living fungal biomass in soil. The average amount of ergosterol present in soil, aquatic, and plant inhabiting fungi is 4.5 $\mu g\ mg^{-1}$ dry mass of mycelia (Wallander et al., 2013). However, the free ergosterol present in fungal cell membrane (it is esterified in cytosol) is converted into relatively stable esterified form, which indicates dry or inactive fungal biomass in OM (Wallander et al., 2010).

14.7.3 Chitin

The chitin is an unbranched polymer of *N*-acetyl-D-glucosamine and one of the most abundant polymers in soil. It is a dominant polysaccharide that constitutes 10%–15% of the original fungal biomass (Duo-Chuan, 2006). Chitinases are hydrolytic enzymes produced by microbes to degrade chitin in soil organic matter for use as N and C, which is incorporated into their biomass.

14.7.4 Phospholipid fatty acids

PLFAs are the major membrane components of bacteria, actinomycetes, and fungi that are produced during growth (Zelles, 1999). They provide information on viable soil microbial biomass as they are rapidly degraded after cell death (Zelles, 1999; Frostegård et al., 2011). Different fatty acids serve as markers for specific microbial groups and variations in their composition predict the relative amount of bacteria, actinomycetes, and fungi in soil (Zelles, 1999). For example hydroxylated and cyclopropyl fatty acids are specific for Gram-negative bacteria, branched fatty acids on positions other than iso and anteiso for Gram-positive bacteria, ester-linked 10-methyl branched saturated fatty acids for actinomycetes (Zelles, 1999), oleic acid (18:1 ω 9) for Zygomycetes (Klamer and Bååth, 2004), and 16:1 ω 5 for AM fungi while 18:2 ω 6, 9 is specific for Ascomycetes, Basidiomycetes, and Zygomycetes (Klamer and Bååth, 2004). The ratio of fungal-specific FAs to bacterial-specific FAs estimates the ratio of fungi and bacteria and their effects on soil (Frostegard et al., 2011). In stress conditions such as high temperature, starvation, osmotic stress, low pH, and heavy metal toxicity microbial adaptations are increased in trans unsaturated fatty acids and cyclopropyl fatty acids to maintain the living cell membrane by minimizing membrane lipid loss or changes in membrane fluidity (Kaur et al., 2005). These adaptations cause changes in PLFA patterns, which also occur due to shifts in microbial species composition (Kaur et al., 2005; Frostegård et al., 2011). Thus PLFAs are useful biomarkers that generate fingerprints of soil microbial community and assess the physiological status of microbes during stress conditions in soil.

14.8 Prospects of microbe based technologies

Productive land decreases with every new episode of land degradation, which has far reaching consequences. The strategies used to reclaim soil mostly rely on the use of fertilizers, pesticides, and chemical ameliorants, which have several limitations. Firstly, they are costly and unaffordable for poor and marginal farmers. Secondly, their doses have to be repeated over a long duration, which increases their dependency in soil. Thirdly, they usually target one or two aspects of soil health and treat a single problem while neglecting others for which some other ameliorant is usually required. Fourthly, they are dependent on the region, climate, and soil type so they have to be tailored for each situation. Lastly, they are required in large amounts, which need inputs to develop the raw materials required in their production. These limitations have encouraged new sustainable eco-friendly strategies, application of microorganisms being one of them.

Direct application of microbes enhances the microbial diversity in the agro-ecosystems by breaking dormancy of the inactive soil microbial pool to establish a functional equilibrium. The microbes and their products are safe and produce no toxic substances that would accumulate in soil and enter the food chain. Microbes self-replicate to maintain their numbers, which prevents repetitive applications. Microbe-mediated mobilization of surplus inorganic mineral reserves reduces chemical fertilizer dependency and the hazards of injudicious and unbalanced use of NPK fertilizers in the soil (Meena et al., 2014; Rashid et al., 2016). It has ability to degrade and compost complex organic matter, which releases nutrients, stimulating plant growth in soil. Microbe primed organic matter is decomposed rather quickly, which supplements nutrients and carbon to soil. Composted crop residues like organic manures replenish nutrients in soil and are promising alternatives for sustainable agro-waste and land management (Indoria et al., 2018; Siddiquee et al., 2017). Microbes provide effective biocontrol against pathogenic soil organisms and pests that rarely acquire resistance and also do not harm the native beneficial flora (Wu et al., 2005). Microbial services in soil are confined not only to increasing nutrient uptake and improving plant health but their symbiotic associations with plants also impart tolerance to various abiotic stress. Fungal endophyte *Curvularia protuberate* is responsible for thermotolerance of monocot *Dichanthelium lanuginosum* (panic grass), growing in the geothermal soils of Yellowstone National Park (Rodriguez et al., 2008). Thus interactive association of fungal endophytes to confer stress tolerance to plants is a novel strategy for growing crop plants in degraded soils (http://www.sciencemag.org/news/2006/08/fungi-helps-grass-beat-heat). The combinations of organic matter (compost), living plant roots, and AM fungi are effective in improving soil physicochemical properties (porosity, bulk density, soil water retention characteristics and water-stable aggregation concentration, and content of soil OC, nitrogen, CEC, and pH) (Daynes et al., 2013). Various studies using organic amendments with microbial combinations in restoration of soil health have been summarized in Table 14.5. Biofilm forming biofertilizers effectively restore the microbial diversity and increase nutrient use efficiency in intensively cropped degraded lands (Seneviratne and Kulasooriya, 2013). Application of cyanobacteria having natural soil binding agents is used to prevent surface soil erosion in deserts (Park et al., 2017). Combinations of ecologically diverse microbial strains (microbial consortium) with related functions are preferred as bioinoculants to improve the efficiency. Wu et al. (2005) coinoculated N-fixer (*Azotobacter chroococcum*), P solubilizer (*Bacillus megaterium*) K solubilizer (*Bacillus mucilaginous*), and AM fungus (*Glomus mosseae* or *Glomus intraradices*) to improve soil properties and nutrient content. Likewise, biofertilization with microbial coinoculants along with organic fertilizers, animal manure, compost, plant and crop waste, plant roots, and other associated factors like growing of cover crops provides organic matter to meet the nutrient demands of microbes that also synergize C sequestration in soil.

TABLE 14.5 Microbial interventions with organic amendments used in restoration of soil health.

Study	Major observation	Reference
Bacillus subtilis and two strains of *B. mucilaginosus* + vermicompost	PGP, increased soil N and C content in tomato by spinach rotation system; dose dependent synergistic effect of vermicompost and PGPR	Song et al. (2015)
AM fungal consortia + *Bacillus megaterium* + compost (Sugar beet residue treated with *Aspergillus niger*)	Increased plant growth, water content, and C, K, N, P, Ca, and Mg in soil, which was more pronounced with addition of composted residue	Armada et al. (2014)
Enterobacter sp. + sugar beet residue *Bacillus thuringiensis* + sugar beet residue *Bacillus megaterium* + sugar beet residue	Improved NPK in plant biomass; crop residue increased available P in rhizosphere soil by 29%; *Enterobacter* sp. with sugar beet residue improved total N by 46% as well as microbiological and biochemical properties	Mengual et al. (2014a)
Azospirillum brasilense + *Pantoeadispersa* + organic olive residue	Stimulated growth of *Pinushalepensis* Mill. and increased nutrient uptake; enhanced enzyme activities, total carbohydrates, and microbial biomass C and nutrients in soil	Mengual et al. (2014b)
Pyrene mineralizing bacteria + farmyard manure	Enhanced bioremediation of polyaromatic hydrocarbons in soil	Adam et al. (2017)
Native AM fungi + *Yarowia lipolytica* + agro-wastes (dry olive cake + sugar beet)	Increased soil nutrient availability; microbial interactions with agro-wastes better modify soil physicochemical, biological and fertility properties, which enhances plant performance	Medina et al. (2004)

14.9 Conclusion

Land degradation is caused by intensive agriculture stripping soil of vital nutrients, carbon reserves, and beneficial organisms, which reduces its ability to sustain crop cultivation. Soils show signs of chronic nutrient and carbon deficiency. Excessive nutrient mining followed by unbalanced replenishment with fertilizers and contamination with hazardous chemical residues and heavy metals released from agro-chemicals used during crop cultivation have deteriorated soil health and quality. Degraded soil is nonconducive to support microflora and fauna and thus loses its viability to provide necessary ecosystem services. It takes over 500−1000 years to regenerate an inch of soil from rocks, which suggests that allowing soil to degrade is much more expensive than to restore soil or at least halt land degradation (Stewart, 2015). It is urgent to evaluate land management strategies and remodify them so as to sustain good soil health for a long time. Microbes are the most dynamic and versatile components of the soil ecosystem. Their immense metabolic diversity, unique qualities, and adaptability to survive extreme environmental conditions make them likely candidates for improvement and restoration of health of degraded and marginal lands. The management strategies with emphasis on improving upon microbe-driven sustainable carbon and nutrient sequestration potential of soil are needed to strengthen the nexus between soil and agricultural productivity.

References

Adam, I.K., Duarte, M., Pathmanathan, J., Miltner, A., Brüls, T., Kästner, M., 2017. Microbial communities in pyrene amended soil−compost mixture and fertilized soil. AMB Express 7 (1), 7.

Aguilera, P., Borie, F., Seguel, A., Cornejo, P., 2011. Fluorescence detection of aluminum in arbuscular mycorrhizal fungal structures and glomalin using confocal laser scanning microscopy. Soil Biol. Biochem. 43 (12), 2427−2431.

Ambriz, E., Báez-Pérez, A., Sánchez-Yáñez, J., Moutoglis, P., Villegas, J., 2010. Fraxinus−Glomus−Pisolithus symbiosis: plant growth and soil aggregation effects. Pedobiologia 53 (6), 369−373.

Amelung, W., Brodowski, S., Sandhage-Hofmann, A., Bol, R., 2008. Combining biomarker with stable isotope analyses for assessing the transformation and turnover of soil organic matter. Adv. Agron. 100, 155−250.

Arcand, M.M., Schneider, K.D., 2006. Plant-and microbial-based mechanisms to improve the agronomic effectiveness of phosphate rock: a review. An. Acad. Bras. Ciên. 78 (4), 791−807.

Armada, E., Portela, G., Roldán, A., Azcón, R., 2014. Combined use of beneficial soil microorganism and agrowaste residue to cope with plant water limitation under semiarid conditions. Geoderma 232, 640−648.

Bapiri, A., Asgharzadeh, A., Mujallali, H., Khavazi, K., Pazira, E., 2012. Evaluation of zinc solubilization potential by different strains of fluorescent Pseudomonads. J. Appl. Sci. Environ. Manage. 16 (3).

Behera, S.K., Shukla, A.K., 2015. Spatial distribution of surface soil acidity, electrical conductivity, soil organic carbon content and exchangeable potassium, calcium and magnesium in some cropped acid soils of India. Land Degrad. Dev. 26 (1), 71−79.

Behie, S.W., Moreira, C., Sementchoukova, I., Barelli, L., Zelisko, P., Bidochka, M., 2017. Carbon translocation from a plant to an insect-pathogenic endophytic fungus. Nat. Commun. 8, 14245.

Bhattacharyya, R., Ghosh, B.N., Mishra, P.K., Mandal, B., Rao, C.S., Sarkar, D., et al., 2015. Soil degradation in India: challenges and potential solutions. Sustainability 7 (4), 3528−3570.

Bojórquez Quintal, E., Escalante-Magaña, C., Echevarría-Machado, I., Martínez-Estévez, M., 2017. Aluminum, a friend or foe of higher plants in acid soils. Front. Plant Sci. 8, 1767.

Borie, F., Rubio, R., Morales, A., 2008. Arbuscular mycorrhizal fungi and soil aggregation. J. Soil Sci. Plant Nutr. 8 (2), 9−18.

Bronick, C.J., Lal, R., 2005. Soil structure and management: a review. Geoderma 124 (1), 3−22.

Caesar-Tonthat, T.C., 2002. Soil binding properties of mucilage produced by a basidiomycete fungus in a model system. Mycol. Res. 106 (8), 930−937.

Cameron, K.C., Di, H.J., Moir, J.L., 2013. Nitrogen losses from the soil/plant system: a review. Ann. Appl. Biol. 162 (2), 145−173.

Cardoso, E.J., Vasconcellos, R.L., Bini, D., Miyauchi, M.Y., Santos, C.A., Alves, P.R., et al., 2013. Soil health: looking for suitable indicators. What should be considered to assess the effects of use and management on soil health? Sci. Agric. 70 (4), 274−289.

Carminati, A., 2012. Model of root water uptake coupled with rhizosphere dynamics. Vadose Zone J. 11 (3).

Caspari, T.M., van Lynden, G.W.J., Bai, Z.G., 2015. Land Degradation Neutrality: An Evaluation of Methods. Report commissioned by German Federal Environment Agency (UBA) (62). Umweltbundesamt.

Chaganti, V.N., Crohn, D.M., Šimůnek, J., 2015. Leaching and reclamation of a biochar and compost amended saline−sodic soil with moderate SAR reclaimed water. Agric. Water Manage 158, 255−265.

Chantigny, M.H.D.A., Angers, D., Prevost, L., Vezina, F., 1997. Chali- the physical protection, bioavailability and composition of particufour. Soil aggregation and fungal and bacterial biomass under late organic matter. Biol. Fertil. Soils 32, 360−367.

Chen, B., Liu, E., Tian, Q., Yan, C., Zhang, Y., 2014. Soil nitrogen dynamics and crop residues. A review. Agron. Sustain. Dev. 34 (2), 429−442.

Chen, D., Li, J., Lan, Z., Hu, S., Bai, Y., 2016. Soil acidification exerts a greater control on soil respiration than soil nitrogen availability in grasslands subjected to long-term nitrogen enrichment. Funct. Ecol. 30 (4), 658−669.

Chhabra, R., 2004. Classification of salt-affected soils. Arid Land Res. Manage. 19 (1), 61–79.

Clemmensen, K.E., Bahr, A., Ovaskainen, O., Dahlberg, A., Ekblad, A., Wallander, H., et al., 2013. Roots and associated fungi drive long-term carbon sequestration in boreal forest. Science 339 (6127), 1615–1618.

Das, S.K., 2014. Role of micronutrient in rice cultivation and management strategy in organic agriculture—a reappraisal. Agric. Sci. 5 (09), 765.

Daynes, C.N., Zhang, N., Saleeba, J.A., McGee, P.A., 2012. Soil aggregates formed in vitro by saprotrophic Trichocomaceae have transient water-stability. Soil Biol. Biochem. 48, 151–161.

Daynes, C.N., Field, D.J., Saleeba, J.A., Cole, M.A., McGee, P.A., 2013. Development and stabilisation of soil structure via interactions between organic matter, arbuscular mycorrhizal fungi and plant roots. Soil Biol. Biochem. 57, 683–694.

Diehl, D., 2013. Soil water repellency: dynamics of heterogeneous surfaces. Colloids Surf. A: Physicochem. Eng. Aspects 432, 8–18.

DoLR, 2011. Wastelands Atlas of India 2011 *(Change Analysis based on temporal satellite data of 2005-06 and 2008-09)*. Government of India, Ministry of Rural Development Department of Land Resources-New Delhi.

Driver, J.D., Holben, W.E., Rillig, M.C., 2005. Characterization of glomalin as ahyphal wall component of arbuscular mycorrhizal fungi. Soil Biol. Biochem. 37 (1), 101–106.

Duo-Chuan, L., 2006. Review of fungal chitinases. Mycopathologia 161, 345e360.

Engelking, B., Flessa, H., Joergensen, R.G., 2007. Shifts in amino sugar and ergosterol contents after addition of sucrose and cellulose to soil. Soil Biol. Biochem. 39, 2111–2118.

FAO, 2011a. Land degradation assessment in drylands (LADA). Global and Argentina, China, Cuba, Senegal, South Africa, Tunisia. Project findings and recommendations, Rome.

FAO, 2011b. The state of the world's land and water resources for food and agriculture (SOLAW) - managing systems at risk. Rome and Earthscan. Food and Agriculture Organization of the United Nations, London.

FAO–ITPS, 2015. Status of the World's Soil Resources (SWSR) – Main Report. Food and Agriculture Organization of the United Nations and Intergovernmental Technical Panel on Soils, Rome, Italy.

FAO, 2017. Soil Organic Carbon: The Hidden Potential. Food and Agriculture Organization of the United Nations, Rome.

Fisk, L.M., Barton, L., Jones, D.L., Glanville, H.C., Murphy, D.V., 2015. Root exudate carbon mitigates nitrogen loss in a semiarid soil. Soil Biol. Biochem. 88, 380–389.

Flemming, H.C., Wingender, J., 2010. The biofilm matrix. Nat. Rev. Microbiol. 8 (9), 623–633.

Frostegård, Å., Tunlid, A., Bååth, E., 2011. Use and misuse of PLFA measurements in soils. Soil Biol. Biochem. 43 (8), 1621–1625.

Gontia-Mishra, I., Sapre, S., Sharma, A., Tiwari, S., 2016. Amelioration of drought tolerance in wheat by the interaction of plant growth-promoting rhizobacteria. Plant Biol. 18 (6), 992–1000.

Gontia-Mishra, I., Sapre, S., Tiwari, S., 2017. Zinc solubilizing bacteria from the rhizosphere of rice as prospective modulator of zinc biofortification in rice. Rhizosphere 3, 185–190.

Guo, J.H., Liu, X.J., Zhang, Y., Shen, J.L., Han, W.X., Zhang, W.F., et al., 2010. Significant acidification in major Chinese croplands. Science 327 (5968), 1008–1010.

Han, H.S., Lee, K.D., 2005. Phosphate and potassium solubilizing bacteria effect on mineral uptake, soil availability and growth of eggplant. Res. J. Agric. Biol. Sci. 1 (2), 176–180.

Hodge, A., Storer, K., 2015. Arbuscular mycorrhiza and nitrogen: implications for individual plants through to ecosystems. Plant Soil. 386 (1-2), 1–9.

ICAR, 2010. Degraded and Wastelands of India: Status and Spatial Distribution. Indian Council of Agricultural Research (ICAR), New Delhi.

IFPRI, 2012. Global Food Policy Report. International Food Policy Research Institute, Washington, D.C.

Imas, P., 2000. Integrated nutrient management for sustaining crop yields in Calcareous soils. In: GAUPRII-IPI National Symposium, International Potash Institute, Gujarat, India. September 19–22.

Indoria, A.K., Sharma, K.L., Sammi Reddy, K., Ch Srinivasarao, K., Srinivas, S.S., Balloli, M., Osman, G., Pratibha, Raju, N.S., 2018. Alternative sources of soil organic amendments for sustaining soil health and crop productivity in India–impacts, potential availability, constraints and future strategies. Curr. Sci. 115 (11), 2052.

Irfan, M., Hayat, S., Hayat, Q., Afroz, S., Ahmad, A., 2010. Physiological and biochemical changes in plants under waterlogging. Protoplasma 241 (1-4), 3–17.

Jing, W.W., Li, N., Li, X.F., Li, D.Q., Wang, L.L., 2017. Exchange fluxes of VOSCs between rice paddy fields and the atmosphere in the oasis of arid area in Xinjiang, China. J. Atmos. Chem. 75 (1), 17–32.

Jones, D.L., Hodge, A., Kuzyakov, Y., 2004. Plant and mycorrhizal regulation of rhizodeposition. New Phytol. 163 (3), 459–480.

Jones, D.L., Cross, P., Withers, P.J., DeLuca, T.H., Robinson, D.A., Quilliam, R.S., et al., 2013. Nutrient stripping: the global disparity between food security and soil nutrient stocks. J. Appl. Ecol. 50 (4), 851–862.

Kafkafi, U., Xu, G., Imas, P., Magen, H., Tarchitzky, J., 2001. Potassium and Chloride in Crops and Soils: The Role of Potassium Chloride Fertilizer in Crop Nutrition. International Potash Institute.

Kaiser, C., Franklin, O., Richter, A., Dieckmann, U., 2015. Social dynamics within decomposer communities lead to nitrogen retention and organic matter build-up in soils. Nat. Commun. 6, 8960.

Katyal, J.C., Chaudhari, S.K., Dwivedi, B.S., Biswas, D.R., Rattan, R.K., Majumdar, K., 2016. Soil health: concept, status and monitoring. Bull. Indian Soc. Soil Sci. 30, 1–98.

Kaur, A., Chaudhary, A., Kaur, A., Choudhary, R., Kaushik, R., 2005. Phospholipid fatty acid–a bioindicator of environment monitoring and assessment in soil ecosystem. Curr. Sci. 10, 1103–1112.

Klamer, M., Bååth, E., 2004. Estimation of conversion factors for fungal biomass determination in compost using ergosterol and PLFA 18: 2ω6, 9. Soil Biol. Biochem. 36 (1), 57–65.

Kögel-Knabner, I., 2002. The macromolecular organic composition of plant and microbial residues as inputs to soil organic matter. Soil Biol. Biochem. 34 (2), 139–162.

Lazcano, C., Barrios-Masias, F.H., Jackson, L.E., 2014. Arbuscular mycorrhizal effects on plant water relations and soil greenhouse gas emissions under changing moisture regimes. Soil Biol. Biochem. 74, 184–192.

Leake, J., Johnson, D., Donnelly, D., Muckle, G., Boddy, L., Read, D., 2004. Networks of power and influence: the role of mycorrhizal mycelium in controlling plant communities and agroecosystem functioning. Can. J. Bot. 82 (8), 1016–1045.

Li, Z., Bai, T., Dai, L., Wang, F., Tao, J., Meng, S., et al., 2016. A study of organic acid production in contrasts between two phosphate solubilizing fungi: *Penicillium oxalicum* and *Aspergillus niger*. Sci. Rep. 6, 25313.

Li, X., Yuan, Y., 2002. Settling velocities and permeabilities of microbial aggregates. Water Res. 36, 3110–3120.

Liang, C., Balser, T.C., 2011. Microbial production of recalcitrant organic matter in global soils: implications for productivity and climate policy. Nat. Rev. Microbiol. 9 (1), 75.

Liang, C., Cheng, G., Wixon, D.L., Balser, T.C., 2011. An absorbing markov chain approach to understanding the microbial role in soil carbon stabilization. Biogeochemistry 106 (3), 303–309.

Linder, M.B., Szilvay, G.R., Nakari-Setälä, T., Penttilä, M.E., 2005. Hydrophobins: the protein-amphiphiles of filamentous fungi. FEMS Microbiol. Rev. 29, 877–896.

Liu, W., Zhang, Y., Jiang, S., Deng, Y., Christie, P., Murray, P.J., et al., 2016. Arbuscular mycorrhizal fungi in soil and roots respond differently to phosphorus inputs in an intensively managed calcareous agricultural soil. Sci. Rep. 6, 24902.

Mager, D.M., 2010. Carbohydrates in cyanobacterial soil crusts as a source of carbon in the southwest Kalahari, Botswana. Soil Biol. Biochem. 42 (2), 313–318.

Maji, A.K., Obi Redyy, G.P., Sarkar, D., 2012. Acid Soils of India – Their Extent and Spatial Distribution. NBSS & LUP, Nagpur, pp. 1–138, NBSS Publication No. 145.

Malik, A.A., Chowdhury, S., Schlager, V., Oliver, A., Puissant, J., Vazquez, P.G., et al., 2016. Soil fungal: bacterial ratios are linked to altered carbon cycling. Front. Microbiol. 7, 1247.

Medina, A., Vassileva, M., Caravaca, F., Roldán, A., Azcón, R., 2004. Improvement of soil characteristics and growth of *Dorycnium pentaphyllum* by amendment with agrowastes and inoculation with AM fungi and/or the yeast *Yarowia lipolytica*. Chemosphere 56 (5), 449–456.

Meena, V.S., Maurya, B.R., Verma, J.P., 2014. Does a rhizospheric microorganism enhance K^+ availability in agricultural soils? Microbiol Res. 169, 337–347.

Mengual, C., Schoebitz, M., Azcón, R., Roldán, A., 2014a. Microbial inoculants and organic amendment improves plant establishment and soil rehabilitation under semiarid conditions. J. Environ. Manage. 134, 1–7.

Mengual, C., Roldán, A., Caravaca, F., Schoebitz, M., 2014b. Advantages of inoculation with immobilized rhizobacteria versus amendment with olive-mill waste in the afforestation of a semiarid area with Pinushalepensis Mill. Ecol. Eng. 73, 1–8.

Miller, R.M., Jastrow, J.D., 2000. Mycorrhizal fungi influence soil structure. Arbuscular Mycorrhizas: Physiology and Function. Springer, Netherlands, pp. 3–18.

Miltner, A., Kindler, R., Knicker, H., Richnow, H.H., Kästner, M., 2009. Fate of microbial biomass-derived amino acids in soil and their contribution to soil organic matter. Org. Geochem. 40 (9), 978–985.

Miltner, A., Bombach, P., Schmidt-Brücken, B., Kästner, M., 2012. SOM genesis: microbial biomass as a significant source. Biogeochemistry 111 (1-3), 41–55.

Miransari, M., 2013. Soil microbes and the availability of soil nutrients. Actaphysiol. Plant. 35 (11), 3075–3084.

MoA, 2016. State of Indian Agriculture (2015-16). Government of India, Ministry of Agriculture & Farmers Welfare, Department of Agriculture, Cooperation & Farmers Welfare Directorate of Economics and Statistics-New Delhi.

MoEF, 2014. Elucidation of the Fifth National Report Submitted to UNCCD Secretariat (2012) Ministry of Environment and Forests, Government of India Paryavaran Bhawan, CGO Complex, Lodi Road, New Delhi- 110 003. Published by: Indian Council of Forestry Research & Education (ICFRE), P.O. New Forest, Dehradun.

Morales-Olmedo, M., Ortiz, M., Sellés, G., 2015. Effects of transient soil waterlogging and its importance for rootstock selection. Chil. J. Agric. Res. 75, 45–56.

Nanzyo, M., Yaginuma, H., Sasaki, K., Ito, K., Aikawa, Y., Kanno, H., et al., 2010. Identification of vivianite formed on the roots of paddy rice grown in pots. Soil Sci. Plant Nutr. 56 (3), 376–381.

Nunkaew, T., Kantachote, D., Nitoda, T., Kanzaki, H., 2015. Selection of salt tolerant purple nonsulfur bacteria producing 5-aminolevulinic acid (ALA) and reducing methane emissions from microbial rice straw degradation. App. Soil Ecol. 86, 113–120.

Oertel, C., Matschullat, J., Zurba, K., Zimmermann, F., Erasmi, S., 2016. Greenhouse gas emissions from soils-a review. Chemie der Erde-Geochem. 76 (3), 327–352.

Omoike, A., Chorover, J., 2006. Adsorption to goethite of extracellular polymeric substances from *Bacillus subtilis*. Geochim. Cosmoch. Acta. 70 (4), 827–838.

Or, D., Phutane, S., Dechesne, A., 2007. Extracellular polymeric substances affecting pore-scale hydrologic conditions for bacterial activity in unsaturated soils. Vadose Zone J. 6, 298–305.

Otoidobiga, C.H., Keita, A., Yacouba, H., Traore, A.S., Dianou, D., 2015. Dynamics and activity of iron-reducing bacterial populations in a West African rice paddy soil under subsurface drainage: case study of kamboinse in burkinafaso. Agric. Sci. 6 (08), 860.

Pabst, H., Kühnel, A., Kuzyakov, Y., 2013. Effect of land-use and elevation on microbial biomass and water extractable carbon in soils of Mt. Kilimanjaro ecosystems. Appl. Soil Ecol. 67, 10–19.

Pal, D.K., Wani, S.P., Sahrawat, K.L., 2015. Carbon sequestration in Indian soils: present status and the potential. Proc. Natl. Acad. Sci., India Sect. B: Biol. Sci. 85 (2), 337–358.

Park, C.H., Li, X.R., Zhao, Y., Jia, R.L., Hur, J.S., 2017. Rapid development of cyanobacterial crust in the field for combating desertification. PLoS One 12 (6), e0179903.

Plaza-Bonilla, D., Arrúe, J.L., Cantero-Martínez, C., Fanlo, R., Iglesias, A., Álvaro-Fuentes, J., 2015. Carbon management in dryland agricultural systems. A review. Agron. Sustain. Dev. 35 (4), 1319–1334.

Ramesh, A., Sharma, S.K., Sharma, M.P., Yadav, N., Joshi, O.P., 2014. Inoculation of zinc solubilizing *Bacillus aryabhattai* strains for improved growth, mobilization and biofortification of zinc in soybean and wheat cultivated in Vertisols of central India. Appl. Soil Ecol. 73, 87–96.

Rashid, M.I., Mujawar, L.H., Shahzad, T., Almeelbi, T., Ismail, I.M., Oves, M., 2016. Bacteria and fungi can contribute to nutrients bioavailability and aggregate formation in degraded soils. Microbiol. Res. 183, 26–41.

Redmile-Gordon, M.A., Brookes, P.C., Evershed, R.P., Goulding, K.W., Hirsch, P.R., 2014. Measuring the soil-microbial interface: extraction of extracellular polymeric substances (EPS) from soil biofilms. Soil Biol. Biochem. 72, 163–171.

Rengasamy, P., 2010. Soil processes affecting crop production in salt-affected soils. Funct. Plant Biol. 37 (7), 613–620.

Rillig, M.C., Wright, S.F., Eviner, V.T., 2002. The role of arbuscular mycorrhizal fungi and glomalin in soil aggregation: comparing effects of five plant species. Plant Soil. 238, 325–333.

Rillig, M.C., Aguilar-Trigueros, C.A., Bergmann, J., Verbruggen, E., Veresoglou, S.D., Lehmann, A., 2015. Plant root and mycorrhizal fungal traits for understanding soil aggregation. New Phytol. 205 (4), 1385–1388. 1.

Rodriguez, R.J., Henson, J., Van Volkenburgh, E., Hoy, M., Wright, L., Beckwith, F., et al., 2008. Stress tolerance in plants via habitat-adapted symbiosis. ISME J. 2 (4), 404–416.

Rojas, R.V., Achouri, M., Maroulis, J., Caon, L., 2016. Healthy soils: a prerequisite for sustainable food security. Environ. Earth Sci. 75 (3), 180.

Rothe, M., Kleeberg, A., Hupfer, M., 2016. The occurrence, identification and environmental relevance of vivianite in waterlogged soils and aquatic sediments. Earth-Sci. Rev. 158, 51–64.

Ruamps, L.S., Nunan, N., Chenu, C., 2011. Microbial biogeography at the soil pore scale. Soil Biol. Biochem. 43 (2), 280–286.

SAC, 2016. Desertification and Land Degradation Atlas of India (Based on IRS AWiFS data of 2011-13 and 2003-05). Space Applications Centre, ISRO, Ahmedabad, 219p.

Sandhya, V., Skz, A., Grover, M., Reddy, G., Venkateswarlu, B., 2009. Alleviation of drought stress effects in sunflower seedlings by the exopolysaccharides producing *Pseudomonas putida* strain GAP-P45. Biol. Fertil. Soils 46, 17–26.

Sarathambal, C., Thangaraju, M., Paulraj, C., Gomathy, M., 2010. Assessing the zinc solubilization ability of Gluconacetobacterdiazotrophicus in maize rhizosphere using labelled 65 Zn compounds. Indian J. Microbiol. 50, 103–139.

Schalk, I.J., Hannauer, M., Braud, A., 2011. New roles for bacterial siderophores in metal transport and tolerance. Environ. Microbiol. 13, 2844–2854.

Schutter, M.E., Dick, R.P., 2002. Microbial community profiles and activities among aggregates of winter fallow and cover-cropped soil. Soil Sc. Soc. Am. J. 66 (1), 142–153.

Seguel, A., Cumming, J.R., Klugh-Stewart, K., Cornejo, P., Borie, F., 2013. The role of arbuscular mycorrhizas in decreasing aluminiumphytotoxicity in acidic soils: a review. Mycorrhiza. 23, 167–183.

Seneviratne, G., Kulasooriya, S.A., 2013. Reinstating soil microbial diversity in agroecosystems: the need of the hour for sustainability and health. Agric. Ecosyst. Environ. 164, 181–182.

Setia, R., Gottschalk, P., Smith, P., Marschner, P., Baldock, J., Setia, D., et al., 2013. Soil salinity decreases global soil organic carbon stocks. Sci. Total Environ. 465, 267–272.

Shukla, A.K., Tiwari, P.K., Prakash, C., 2014. Micronutrients deficiencies vis-a-vis food and nutritional security of India. Ind. J. Fertil. 10, 94–112.

Siddiquee, S., Shafawati, S.N., Naher, L., 2017. Effective composting of empty fruit bunches using potential Trichoderma strains. Biotech. Rep. 13, 1–7.

Singh, J.S., 2015. Microbes: the chief ecological engineers in reinstating equilibrium in degraded ecosystems. Agric. Ecosyst. Environ. 203, 80–82.

Singh, J.S., Kumar, A., Rai, A.N., Singh, D.P., 2016. Cyanobacteria: a precious bioresource in agriculture, ecosystem, and environmental sustainability. Front. Microbiol. 7, 529.

Six, J., Bossuyt, H., Degryze, S., Denef, K., 2004. A history of research on the linkbetween (micro) aggregates, soil biota, and soil organic matter dynamics. Soil Till. Res. 79, 7–31.

Smith, S.E., Smith, F.A., 2012. Fresh perspectives on the roles of arbuscular mycorrhizal fungi in plant nutrition and growth. Mycologia 104, 1–3.

Song, X., Liu, M., Wu, D., Griffiths, B.S., Jiao, J., Li, H., et al., 2015. Interaction matters: synergy between vermicompost and PGPR agents improves soil quality, crop quality and crop yield in the field. Appl. Soil Ecol. 89, 25–34.

Spohn, M., Rillig, M.C., 2012. Temperature-and moisture-dependent soil water repellency induced by the basidiomycete *Agaricus bisporus*. Pedobiologia 55, 59–61.

Stamford, N.P., Figueiredo, M.V., da Silva Junior, S., Freitas, A.D., Santos, C.E., Junior, M.A., 2015. Effect of gypsum and sulfur with Acidithiobacillus on soil salinity alleviation and on cowpea biomass and nutrient status as affected by PK rock biofertilizer. Sci. Hortic 192, 287–292.

Stewart N. The value of land: Prosperous lands and positive rewards through sustainable land management. United Nations University-Institute for Water, Environment and Health; 2015.

UNCCD, 2014. Land Degradation Neutrality: Resilience at local, national and regional levels, s.l. United Nations Convention to Combat Desertification.

UNCCD, 2015. Desertification, Land Degradation & Drought (DLDD): Some Global Facts and Figures. United Nations Conventions to Combat Desertification. Available from http://www.unccd.int/Lists/SiteDocumentLibrary/WDCD/DLDD%20Facts.pdf

UNCCD-SPI, 2015. PivotC Carbon. Science Policy Brief. 01. UNCCD-SPI.

UNEP, 2007. GEO-4: Global Environmental Outlook. Environment for Development. United Nations Environ. Programme, Nairobi.

Upadhyay, S.K., Singh, J.S., Singh, D.P., 2011. Exopolysaccharide-producing plant growth-promoting rhizobacteria under salinity condition. Pedosphere 21, 214–222.

Uroz, S., Calvaruso, C., Turpault, M.P., Frey-Klett, P., 2009. Mineral weathering by bacteria: ecology, actors and mechanisms. Trends Microbiol. 17 (8), 378–387.

Vaid, S.K., Kumar, B., Sharma, A., Shukla, A.K., Srivastava, P.C., 2014. Effect of Zn solubilizing bacteria on growth promotion and Zn nutrition of rice. J. Soil Sci. Plant Nutr. 14, 889–910.

Wallander, H., Johansson, U., Sterkenburg, E., Brandström, Durling, M., Lindahl, B.D., 2010. Production of ectomycorrhizal mycelium peaks during canopy closure in Norway spruce forests. New Phytol. 187, 1124–1134.

Wallander, H., Ekblad, A., Godbold, D.L., Johnson, D., Bahr, A., Baldrian, P., et al., 2013. Evaluation of methods to estimate production, biomass and turnover of ectomycorrhizal mycelium in forests soils—a review. Soil Biol. Biochem. 57, 1034–1047.

Waring, B.G., Averill, C., Hawkes, C.V., 2013. Differences in fungal and bacterial physiology alter soil carbon and nitrogen cycling: insights from meta-analysis and theoretical models. Ecol. Lett. 16, 887–894.

Whiteside, M.D., Garcia, M.O., Treseder, K.K., 2012. Amino acid uptake in arbuscular mycorrhizal plants. PLoS One 7, e47643.

Wong, V.N., Greene, R.S., Dalal, R.C., Murphy, B.W., 2010. Soil carbon dynamics in saline and sodic soils: a review. Soil UseManage. 26, 2–11.

Wu, S.C., Cao, Z.H., Li, Z.G., Cheung, K.C., Wong, M.H., 2005. Effects of biofertilizer containing N-fixer, P and K solubilizers and AM fungi on maize growth: a greenhouse trial. Geoderma 125, 155–166.

Xu, P., Liang, L.Z., Dong, X.Y., Shen, R.F., 2015. Effect of arbuscular mycorrhizal fungi on aggregate stability of a clay soil inoculating with two different host plants. Acta Agric. Scand. Sect. B-Soil Plant Sci. 65, 23–29.

Yadav, B., Verma, A., 2012. Phosphate solubilization and mobilization in soil through microorganisms under arid ecosystems. In: Ali, M. (Ed.), The Functioning of Ecosystems. In Tech, Croacia Capítulo, pp. 99–108. , Chapter 6.

Yan, N., Marschner, P., Cao, W., Zuo, C., Qin, W., 2015. Influence of salinity and water content on soil microorganisms, Int. Soil Water Conserv. Res, 3. pp. 316–323.

Zelles, L., 1999. Fatty acid patterns of phospholipids and lipopolysaccharides in the characterisation of microbial communities in soil: a review. Biol. Fertil. Soils. 29, 111–129.

Zengin, F., 2013. Biochemical and physiological effect of excess manganese (Mn) in bean (*Phaseolus vulgaris* L. cv. Strike). Proc. Natl. Acad. Sci 83, 651–657.

Zhang, J., Tang, X., Zhong, S., Yin, G., Gao, Y., He, X., 2017. Recalcitrant carbon components in glomalin-related soil protein facilitate soil organic carbon preservation in tropical forests. Sci. Rep. 7, 2391.

Zhao, C., Gupta, V.V., Degryse, F., McLaughlin, M.J., 2017. Abundance and diversity of sulfur-oxidising bacteria and their role in oxidising elemental sulfur in cropping soils. Biol. Fert. Soils. 53, 159–169.

Zheng, W., Morris, E.K., Rillig, M.C., 2014. Ectomycorrhizal fungi in association with Pinussylvestris seedlings promote soil aggregation and soil water repellency. Soil Biol. Biochem. 78, 326–331.

Zhu, T.B., Zhang, J.B., Cai, Z.C., 2011. The contribution of nitrogen transformation processes to total N_2O emissions from soils used for intensive vegetable cultivation. Plant Soil 343, 313–327.

Further reading

Kumar, A., Kaushal, S., Saraf, S.A., Singh, J.S., 2018. Microbial biofuels: a solution to carbon emissions and energy crisis. Front. Biosci. (Landmark ED) 1 (23), 1789–1802.

Meena, V.S., Maurya, B.R., Verma, J.P., Aeron, A., Kumar, A., Kim, K., et al., 2015. Potassium solubilizing rhizobacteria (KSR): isolation, identification, and K-release dynamics from waste mica. Ecol. Eng. 81, 340–347.

Singh, J.S., 2013. Plant growth promoting rhizobacteria: potential microbes for sustainable agriculture. Resonance 18 (3), 275–281.

Singh, J.S., Gupta, V.K., 2018. Soil microbial biomass: a key soil driver in management of ecosystem functioning. Sci. Total Environ. 634, 497–500.

Singh, C., Tiwari, S., Gupta, V.K., Singh, J.S., 2018. The effect of rice husk biochar on soil nutrient status, microbial biomass and paddy productivity of nutrient poor agriculture soils. Catena 171, 485–493.

Tiwari, P., Singh, J.S., 2017. A plant growth promoting rhizospheric *Pseudomonas aeruginosa* strain inhibits seed germination in *Triticum aestivum* (L) and *Zea mays* (L). Microbiol. Res. 8, 1–7.

Tiwari, S., Singh, C., Singh, J.S., 2018. Land use changes: a key ecological driver regulating methanotrophs abundance in upland soils. Energy Ecol. Environ. 3, 355–371.

Vimal, S.R., Patel, V.K., Singh, J.S., 2018. Plant growth promoting *Curtobacterium albidum* strain SRV4: an agriculturally important microbe to alleviate salinity stress in paddy plants. Ecol Indic article in press.

Chapter 15

Trichoderma—a new strategy in combating agriculture problems

Y.S. Amaresh[1], G. Chennappa[2], S. Avinash[1], M.K. Naik[1] and M.Y. Sreenivasa[3]

[1]*Department of Plant Pathology, College of Agriculture, UAS, Raichur, India,* [2]*Department of Processing and Food Engineering, CAE, UAS, Raichur, India,* [3]*Department of Studies in Microbiology, University of Mysore, Mysore, India*

Chapter Outline

15.1 Introduction

Chemical control of plant diseases can be impressive but this is comparatively a short term measure and additionally, the accumulation of harmful chemical residues sometimes causes serious ecological imbalance (Singh, 2014, 2015). In recent years, the increasing use of potentially hazardous chemicals in agriculture has resulted in growing concern for both the environment and public health (Singh and Gupta, 2018; Vimal et al., 2018). Moreover use of such chemicals entails a substantial cost to the nation and developing countries. Fungi are the principal decomposers in the ecosphere and are essential for recycling nutrients in the environment. Some groups of fungi are infectious agents and can cause a wide variety of diseases in animals, plants, and humans (Idnurm and Heitman, 2005). Plant diseases result in severe losses of agricultural and horticultural crops every year and thus need to be controlled to maintain the quality and abundance of food and feed. Biological control involves the use of beneficial microorganisms (Singh and Strong, 2016; Tiwari and Singh, 2017), it offers an eco-friendly approach to the management of plant disease, and can be incorporated into cultural and physical controls thereby resulting in limited chemical usage for an effective integrated pest management system (Monte, 2001; Singh, 2016). Various microbes are being used as biocontrol agents for the control of plant pathogens. *Trichoderma*, a fungus commonly available in soil and root ecosystems, has gained immense importance over the last few decades due to its biological control activity against several plant pathogens. The *Trichoderma* species being ubiquitous and often predominant component of the mycoflora in native and agricultural soils play an important role in ecosystem health (Komon-Zelazowska et al., 2007). The ecological adaptability of *Trichoderma* species is evidenced by their widespread distribution, including under different environmental conditions and on various substrates. This physiological flexibility together with the antagonistic action of *Trichoderma* species against phytopathogenic fungi and the ability of these fungi to promote plant growth have made them attractive biological control agents (Chennappa et al., 2017; Mariola et al., 2007). The *Trichoderma* species that are widely marketed as biopesticides, bioinoculants and soil amendments are well studied, due to their ability to protect plants, enhance vegetative growth, and control pathogen populations (Vinale et al., 2008). The *Trichoderma* species are recognized for the numerous mechanisms that are involved in attacking other fungi. These mechanisms include competition for space and nutrients (Sivan and Chet, 1989), mycoparasitism (Haran et al., 1996), antibiosis (Ghisalberti and Sivasithamparam, 1991), inactivation of the pathogen's enzymes (Roco and Perez, 2001), and induced resistance (Kapulnik and Chet, 2000).

New and Future Developments in Microbial Biotechnology and Bioengineering. DOI: https://doi.org/10.1016/B978-0-12-818258-1.00015-7

The secondary metabolites are small molecules that are not directly essential for growth but have important roles in signaling, development, and interaction with other organisms (Hoffmeister and Keller, 2007; Osbourn, 2010; Singh et al., 2016). The production of secondary metabolites is strain dependent and varies in relation to the equilibrium between elicited biosynthesis and biotransformation rates (Vinale et al., 2009). Based upon chemical and analytical reports, *Trichoderma* species are prolific producers of secondary metabolites (peptaibols, nonribosomal peptides, volatile and nonvolatile terpenes, pyrones, siderophores, and nitrogen containing compounds) as 373 different molecules have been identified (Reino et al., 2008; Mukherjee et al., 2012a,b; Crutcher et al., 2013). Hence, these fungi are promising agents that can be applied for remediation of polluted soil and water by using these biocontrol agents (Harman et al., 2004). Species of *Trichoderma* can produce antibiotics, mycotoxins, and more than100 metabolites have been documented with antibiotic activity including polyketides, pyrones, terpenes, amino acids, and polypeptides (Sivasithamparam and Ghisalberti, 1998).

15.2 Properties of *Trichoderma*

The *Trichoderma* species (Fig. 15.1) are ubiquitous, colonize on cellulosic materials and in the rhizosphere of plants (Kubicek et al., 2008; Jaklitsch, 2009), and are free living fungi that are present in soil and root ecosystems and other natural habitats, especially those containing high organic matter (Harman et al., 2004). *Trichoderma* species have been isolated from decomposed organic matter, root surface of various plants, and decaying barks.

The genus *Trichoderma* consists of nine species: *Trichoderma aureoviride*, *T. hamatum*, *T. harzianum*, *T. koningii*, *T. longibrachiatum*, *T. piluliferum*, *T. polysporum*, *T. pseudokoningii*, and *T. viride* aggregates defined largely by conidiophore morphology (Fig. 15.2). Among them, the most important species that are used in the field of biological control are *T. atroviride*, *T. harzianum*, *T. virens*, *T. asperellum*, and *T. reesei* (Rifai, 1969; Benitez et al., 2004).

15.3 Secondary metabolites

Most of the *Trichoderma* species produce antibiotics or diffusible compounds during interactions with pathogens, and these compounds inhibit the colonization of plant pathogens in the rhizosphere. The metabolites include harzianic acid, almethicins, tricholin, peptabols, antibiotics, 6 pentyl-α-pyrone, massoilactone, viridian, glioviridin, gliosporens, and heptelidic acid (Vey et al., 2001; Reino et al., 2008). *T. harzianum* produces pyrone antibiotics, which are effective in controlling the *Gaeumanno myces graminis* var. *tritic*, and *T. virens* shows biocontrol efficiency by producing gliovirin

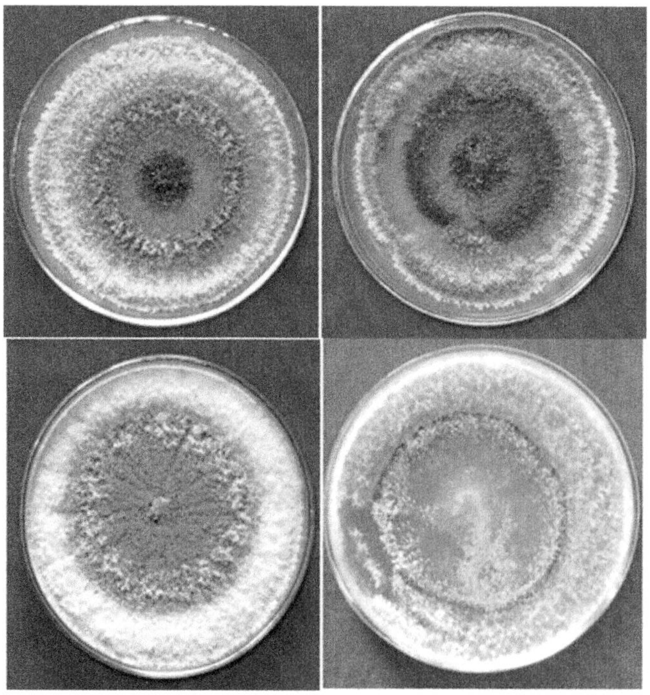

FIGURE 15.1 Morphological and cultural variation of *Trichoderma* spp.

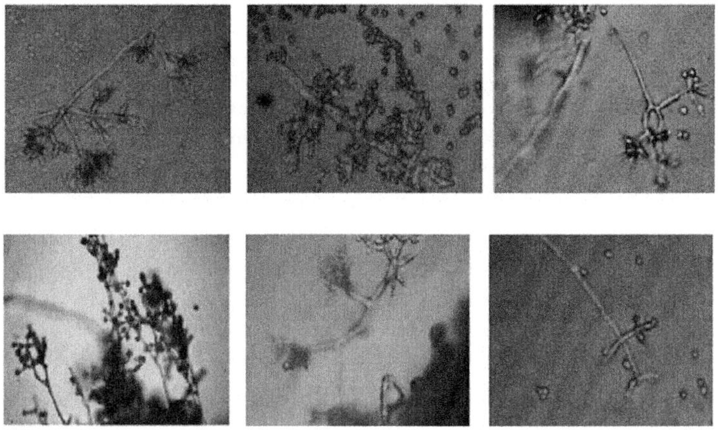

FIGURE 15.2 Sporulation (conidia and conidiophore) patterns of *Trichoderma* spp.

(Howell, 1998). *Trichoderma asperellum* has the ability to produce different major volatile secondary metabolites such as 1,2-benzenedicarboxylic acid, 2-butoxy-2-oxoethyl butyl ester, 1,2-benzenedicarboxylic acid dibutyl ester, 2H-pyran-2-one, palmitic acid, several phenolic isomers, methyl cyclohexane, 1-methyl-2-nitrobenzene, isoquinoline, dibutyl ester, butyl 8-methylnonyl ester, trimethyl silylpalmitate, trans-9-octadecenoic acid, trimethylsilyl ester, p-aminotoluene, several phenol isomers (p-tert-amylphenol, 4-nonylphenol, 4-dodecylphenol, 4-(1,1,3,3-tetramethylbutyl) phenol), n-eicosane, 8-propylquinoline, 5-ethyl-5-propylundecane, tetradecane, heptadecane, cyclooctyl methyl phosphono-fluoridoate, etc. (Nitish and Kumar, 2017). Similar volatile compounds are also documented—as many as 278 compounds such as cyclohexane, cyclo-pentane, fatty acids, alcohols, esters, sulfur-containing compounds, and simple pyrane and benzene derivatives from liquid cultures of *T. harzianum*. The compounds isolated from *Trichoderma* species have strong antifungal activity and the compounds include 3-methyl-heptadecanol, methyl cyclohexane, 6-nonylene alcohol, methyl-cyclopentane, 2-methyl heptadecanol, *N*-methyl pyrollidine, dermadin, ketotriol, koningin-A, palmitic acid, 3-(2′-hydroxypropyl)-4-(hexa-2′-4-dineyl)-2-(5H)-furanone, and 3-(propenone)-4-(hexa- 2′-4′-dineyl)-2-(5H)-furanone (Siddiquee et al., 2012; Dubey et al., 2011).

Peptaibols have strong antimicrobial properties against bacteria and fungi. Peptaibols along with cell wall degrading enzymes can inhibit the growth of fungi more effectively. More than 190 peptaibols have been documented that are produced by species of *Trichoderma* (Neuhof et al., 2007). Paracelsin was the first characterized secondary metabolite isolated and characterized from *Trichoderma* species, which has antibiotic properties, and later different types of peptaibols were identified (Degenkolb et al., 2008; Stoppacher et al., 2008). Gliotoxin belongs to the nonribosomal peptides and it was the first metabolite isolated from species of *Trichoderma* (Brian, 1944). Gliotixin is a fungistatic metabolite and it has good antagonistic activity against *Rhizoctonia*. It is effective in cotton disease control (Whilhite and Stanley, 1996). It is also one of the most important antifungal compounds produced by species of *Trichoderma* and it is the most effective biocontrol metabolite (Reither et al., 2005). Pyrones have plant growth promoting activities and are effective against *S. rolfsii*, a common plant pathogen (Evidente et al., 2003).

15.4 Defensive actions of *Trichoderma* enzymes

The defensive actions of *Trichoderma* species involve application of lytic enzymes (Kubicek et al., 2001; Viterbo et al., 2002), proteolytic enzymes (Kredics et al., 2005; Suarez et al., 2007; Chen et al., 2009), volatile metabolites (Calistru et al., 1997; Eziashi et al., 2006), and other secondary metabolites (Reino et al., 2008) against pathogens (Benitez et al., 2004). Another important factor that makes *Trichoderma* more effective is the production of hydrolytic enzymes and this is the key step in the successful establishment of a mycoparasitic relationship with pathogens. *Trichoderma* species are highly efficient producers of many extracellular enzymes (chitinases, glucanases, and proteases). They are capable of secreting hydrolytic enzymes and cause mycoparasitism on fungal pathogens of plants. The cell wall is the first barrier between pathogen and mycoparasite, that is, *Trichoderma*. The degradation of cell wall is mediated by different set of enzymes including chitinase, glucanase [β-(1−4), β(1−3), and β(1−6)], *N*-acetyl glucosaminidase, and protease.

Chitin is the major structural component of the fungal cell wall and chitinases have the ability to degrade chitin. Chitinases are group of enzymes including 1,4-β acetylglucosaminidases, endochitinases, and exochitinases that are involved in the breakdown of chitin to simpler components. Similarly the second most important polymer in the cell wall is β-1−3 glucan, which is hydrolyzed by β-1−3 glucanases. Glucanases have been detected and identified in the infected region of the host and are produced by interaction of pathogens and *Trichoderma*. Overproduction of these enzymes influences the biocontrol efficiency, which in turn inhibits the growth and spore germination of many pathogens. Overexpression of the glucanase by *T. harzianum* inhibited the growth of *Botrytis cinerea*, *R. solani*, and *Phytophthora citrophthora* (Ihrmark et al., 2010). Further, proteases are also important in the degradation of fungal cell wall and *T. atroviride* and *T. virens* have more enzymes than *T. reesei*. *Trichoderma* species produce different types of proteases during mycoparasitism, which include asprtyl and subtilin like groups (Geremia et al., 1993; Seidi et al., 2009). *T. harzianum* inactivated hydrolytic enzymes produced by *B. cinerea* pathogen by secreting proteases and acts as a biocontrol agent (Howell, 2003).

The antagonistic interaction established between two fungal species, where the direct attack of one fungus on another is known as mycoparasitism, depends on recognition, attack, penetration, and killing of the host. Mechanisms of parasitism include interaction with pathogens like coiling of hyphae around the cell, penetration by haustoria, and *Trichoderma* species release antibiotics or other chemicals that are harmful to the pathogens and inhibit their growth (Harman et al., 2004). *Trichoderma* is also one of the best-studied mycoparasitic fungi and it can invade many plant pathogens in a root system. Mycoparasitism is considered as an effective biocontrol strategy and *Trichoderma* species are effective biocontrol agents due to their antagonistic capacities against a broad range of phytopathogenic fungi.

15.5 Compatibility with pesticides

There are reports where biocontrol agents that can tolerate fungicides up to a certain level were mixed with fungicides resulting in eradication of diseases (De Cal et al., 1994). The combination of fungicide tolerant biological control agents with reduced levels of fungicide under integrated management strategies would promote a degree of disease suppression similar to that achieved with full dosage of fungicides (Monte, 2001). The *Trichoderma* species have played an important role in integrated disease management practices. Seed treatment and soil application of *Trichoderma* along with compatible fungicides, insecticides, and weedicides are a common practice among farmers for the economic and effective management of seed- and soil-borne plant disease. Therefore, it's crucial to screen out popularly used fungicides, insecticides, and weedicides for their compatibility with efficient *Trichoderma* isolates, so they can be commercially exploited under a biointensive Integrated Pest Management (IPM) program. Samuels (1996) reported that *Trichoderma* strain alone or in combination with fungicides have been used for management of several seed- and soil-borne plant pathogens.

Species of *Trichoderma* showed strong tolerance to fungicides at different concentration and the tolerance may vary with species to species. These species, including *T. viride*, *T. virens*, and *T. harzianum*, recorded the maximum tolerance to carboxin, captan, tebuconazole, thiram, mancozeb, copper oxychloride, vitavax powder (Carboxin 37.5% + Thiram 37.5% WP), and carbendazim. The evaluated fungicides have shown compatibility with *Trichoderma* and *Trichoderma* can be used for the management of many plant diseases along with fungicides. Some of the fungicides such as Bordeaux mixture and azoxystrobin are moderately compatible with *Trichoderma* isolates (Ansari et al., 2011; Arunasri et al., 2011; Bagwan, 2010; Madhavi et al., 2011; Patil and Gupta, 2011; Roberti et al., 2002; Sachin-Upmanyu and Shyam, 2002). *Trichoderma* was most sensitive and not compatible with to benomyl, tebuconazole, propiconazole, chlorothalonil and hexaconazole, metalaxyl and thiophanate methyl (Shukla, 2011; Rangnathswamy et al., 2012; Bheemaraya., Patil et al., 2012; Nandeesha et al., 2013; Ortellado, 2013). The intolerance or sensitivity towards the fungicides depends on the physiological properties and environmental factors. Insecticides such as thiamethoxam, acetamiprid, chlorpyriphos, indoxacarb, quinalphos, bavistin, phorate, and carbofuran caused maximum inhibition of mycelia growth of *Trichoderma* species including *T. harzianum*, *T. viride*, and *T. hamatum*. Hyphal growth decreased with increase in concentration of insecticides. However, copper oxychloride, imidacloprid and chlorantraniliprole are compatible with *Trichoderma* species and these could be safely applied with *Trichoderma* for the management of insect pests and nematodes (Jayaraj and Ramabadran, 1996; Jebakumar et al., 2000; Bhai and Thomas, 2010; Madhavi et al., 2011; Shukla, 2011; Singh et al., 2012; Bheemaraya., Patil et al., 2012; Saravanan et al., 2013; Thiruchchelvan et al., 2013).

15.6 Application of *Trichoderma*

The *Trichoderma* species induce various mechanisms to affect growth of plant pathogens, control disease, and improve overall plant health (Singh et al., 2004). The last decade has witnessed a tremendous breakthrough in this respect, especially on standardization of production techniques of *Trichoderma* species to use them against many insect pests and diseases. Biocontrol agents like *Trichoderma* are acclaimed as effective, eco-friendly, and cheap. *Trichoderma* is one of the common fungal biocontrol agents being used worldwide for suitable management of various foliar and soil-borne plant pathogens (*Ceratobasidium*, *Fusarium*, *Rhizoctonia*, *Macrophomina*, *Sclerotium*, *Pythium*, and *Phytophthora* species). *Trichoderma* also serve as bioinoculants due to their plant growth promotion properties and their ability to decompose organic matter resulting in increased availability of micronutrients to the plants (Singh et al., 2004).

15.7 *Trichoderma* as biocontrol agents

The *Trichoderma* species are among the most promising biocontrol agents as compared with the other genera of fungi. *Trichoderma* species are able to control ascomycetes, basidiomycetes, and oomycetes (Monte 2001; Benitez et al., 2004) and also nematodes (Kyalo et al., 2007). Species of *Trichoderma* including *T. aureoviride*, *T. hamatum*, *T. harzianum*, *T. koningii*, *T. longibrachiatum*, *T. piluliferum*, *T. polysporum*, *T. pseudokoningii*, and *T. viride* recorded the highest growth inhibition rate against major plant pathogens, and a few of them are crop and host specific. Plant pathogens infect all parts of the plant and major infection occurs in the root system and the pathogens such as *Rhizoctonia solani*, *Sclerotium rolfsii*, *Pythium* sp., *R. solani*, *Fusarium oxysporum*, *Aspergillus niger*, *Aspergillus flavus*, *Alternaria porri*, and *S. rolfsii*, and many reports are available on biocontrol efficacy of *Trichoderma* species (Khan and Sinha, 2007; Kolte and Raut, 2007; Mandhare and Suryawanshi, 2008; Bagwan, 2011; Nainwal and Vishunavat, 2016). The *Trichoderma koningii* showed its antifungal activity by producing secondary metabolites including δ-decanolactone, 6-pentyl-α-pyranone, and 6-(4-oxopentyl)-2H-pyran-2-one in the rhizosphere region and inhibited the common soil-borne pathogenic fungi such as *R. solani*, *Sclerotium rolfsii*, *Macrophomina phaseolina*, and *F. oxysporum*, respectively (Vivek et al., 2014). Additionally 2H-pyran-2-, a volatile compound, is responsible for antagonism against *F. oxysporum* and possesses enhanced biocontrol activity.

Many reports are available on the effect of secondary metabolites produced by *Trichoderma* species against common plant pathogens causing different types of diseases in agriculture and horticulture crops. Among many plant pathogens *S. rolfsii* is the most studied pathogen against the *Trichoderma* species. Different *Trichoderma* isolates have been screened for the production of volatile compounds and their effect on the growth of *S. rolfsii*. It is revealed that the growth inhibition of *S. rolfsii* ranged between 69% by *T. viride* and 18% by *T. piluliferum*. Volatile metabolites produced by *T. viride*, *T. harzianum*, and *T. longibrachiatum* resulted in reduction of mycelial growth of *S. rolfsii* to the extent of up to 60.8%. Antagonists' efficacy of *Trichoderma* species evaluated against *S. rolfsii* and *T. harzianum* showed the highest percent reduction of mycelial growth (36%) followed by *T. viride* (33%). Antagonism efficacy of native isolate of *Trichoderma* species was evaluated against *S. rolfsii* and maximum inhibition of the radial growth of *S. rolfsii* was recorded by *Trichoderma* isolate with 70% over the control. Effect of volatile compounds produced by *Trichoderma* species tested on *R. bataticola*, *T. viride* shown maximum inhibition of growth (37.09%) followed by *T. hamatum* (32.81%). In case of *S. rolfsii* maximum growth inhibition was recorded by *T. viride* (54.21%) and least inhibition was recorded by *T. virens* (17.10%). Species of *Trichoderma* showed growth inhibitory effect on *S. rolfsii* which is due to the release different types of volatile metabolites and *T. harzianum* found more growth inhibitory effect (89%) as compared to other species (Rao and Kulkarni, 2003; Shaigan et al., 2008; Rekha et al., 2012; Padmaja et al., 2013; Nair, 2015; Joao et al., 2017; Sangle et al., 2016). In examining *T. viride*, *T. harzianum*, and *T. virens* against *F. oxysporum*, it is found that *T. viride* and *T. harzianum* have shown the highest ability to inhibit the growth of pathogens with the range of 60%. Volatile substances derived from *T. viride* and *T. harzianum* inhibited mycelial growth of the *F. oxysporum* to an extent of 72% and 56%, respectively. Antagonistic action of volatile metabolites of *T. harzianum* and *T. viride* was evaluated against *F. solani*. The growth of pathogen was inhibited by *T. harzianum* (upto 38%) and *T. viride* (35%). *T. pseudokoningii* was tested against *F. oxysporum* and caused maximum inhibition of the mycelial growth of *F. oxysporum* (54%). *T. longibrachiatum* and *T. viride* were ranked next in respect to growth inhibition of *F. oxysporum* (43%). Effect of volatile metabolites produced by *Trichoderma* species was tested against fungal plant pathogens. *T. viride* inhibited the mycelial growth of *F. oxysporum* (41% inhibition) followed by *T. viride* and *T. harzianum* (35% and 30% inhibition) over control (Sumana and Devaki, 2012; Dubey and Suresh, 2006; Bokhari and Perveen, 2012; Rini and Sulochana, 2007; Faheem et al., 2010). *T. pseudokoningii* was tested against *R. solani* and caused maximum inhibition of the mycelial growth of *R. solani* (54%). *T. longibrachiatum* and *T. viride* were ranked next in respect to growth inhibition of *R. solani* (41%). The secondary metabolites produced by the culture filtrate of *T. harzianum* caused maximum growth inhibition of *R. bataticola* (91%) followed by *T. viride* (90%) and *T. virens* (88%). Effect of volatile compounds produced by *Trichoderma* species was tested on *R. bataticola*, and *T. viride* showed maximum inhibition of growth (37.09%) followed by *T. hamatum* (32.81%). As regards the antagonistic capability of *T. harzianum* isolates against some pathogenic soil-borne fungi, maximum growth inhibition was noticed against *Rhizoctonia* sp. by *T. harzianum* (53.6%).

In terms of antagonistic action of *Trichoderma* strains against *Sclerotinia sclerotiorum* and *S. cepivorum*, the maximum inhibition effect of volatile compounds was occasioned by *T. longibrachiatum* with 31.5% and 59.2% on *S. sclerotiorum* and *S. cepivorum*, respectively. Efficacy of volatile metabolites producing *T. harzianum* and *Trichoderma* sp. against *Alternaria alternata* was tested and it was found that the *T. harzianum* inhibited mycelial growth by 59.47% over control. Volatile metabolites produced from *T. harzianum* ISO-1 showed a maximum reduction (44.60%) in the mycelial growth of *A. alternata* followed by *T. harzianum* ISO-2 (35.50%) and *T. piluliferum* (36.24%). The effect of volatile metabolites recorded that *Trichoderma* inhibited the growth of pathogens, that is, *R. solani* (100%), *F. graminearum* (88%), and *P. ultiumum* (86.6%). The effect of *Trichoderma* species was tested on major pathogens of red gram and maize and growth inhibition was shown by *T. harzianum* on *R. bataticola* (69.7%), *F. udum* (69.07%), the *T. hamatum* on *Exerohilum turcicum* (70.33%), and *T. virens* was more efficient on *Colletotrichum truncatum* (64.38%). Volatile compounds released by *Trichoderma* species were tested on the growth of seed-borne pathogens. *T. viride* recorded the maximum growth inhibition for *Colletotrichum gloeosporioides* (34.37%) followed by *A. alternata* (20.0%), *F. oxysporum* (15.21%), *R. solani* (12.0%), and the minimum for *Curvularia lunata* (13.63%) whereas for *T. harzianum* the highest growth inhibition was observed for *C. gloeosporioides* (35.0%) followed by *F. oxysporum* (14.80%), *A. alternata* (11.11%), *C. lunata* (9.52%), and lowest for *R. solani* (4.54%), respectively (Castillo et al., 2011; AL-Saeedi and AL-Ani 2014; Rini and Sulochana, 2007; Dubey et al. 2011; Nair, 2015; Arzanlou et al., 2014; Thakur and Harsh 2014; Raut et al., 2014; Prabhavathi et al., 2015; Tapwal et al., 2015). Similarly, *Trichoderma* species isolates were tested for their bioefficacy with fungal pathogens including *M. phaseolina*, *R. bataticola*, *R. solani*, and *S. rolfsii*. The highest percent inhibition of *M. phaseolina* (81.11%) was recorded by Tri-12 isolate and dual culture assay against *R. solani* revealed that among all the *Trichoderma* isolates tested, Tri-12 was significantly superior with recorded 82.59% inhibition of growth of *R. solani*. Isolate Tri-2 had shown the highest percent inhibition of mycelial growth of *S. rolfsii* (47.78%), respectively (Nair, 2015).

15.8 *Trichoderma* formulations

The *Trichoderma* have been formulated with different carrier materials such as talc, lignite, vermiculite wheat bran, pesta granules, alginate prills, and press mud, and agriculture waste based application of *Trichoderma* increases the crop health and manages plant diseases by inhibiting the plant pathogens. *Trichoderma* can be applied in various ways such as soil application, furrow application, aerial spraying, seed treatment root dipping, and even a combination of all, and the application depends on the problems and diseases. Talc based formulations of *Trichoderma* were developed at Tamil Nadu Agricultural University, Coimbatore for seed treatment of pulse crops and rice (Kumar et al., 2014). Initially *Trichoderma* culture was grown in liquid medium and after 1 week of incubation period, liquid culture was mixed with talc powder in the ratio of 1:2 and dried to 8% moisture under shade. The talc formulation of *Trichoderma* has a shelf life of up to 6 months. It has become popular in India for management of several soil-borne diseases of various crops through seed treatment (4–5 g/kg seed).

15.9 Conclusions

Three isolates, *T. virens*, *T. viride*, and *T. hamatum*, performed better against major soil-borne pathogens. The majority of the *Trichoderma* isolates are compatible with fungicides and insecticides. Among all the isolates, *T. hamatum* has shown highest compatibility with fungicides followed by *T. virens* and *T. viride*. The species of *Trichoderma* are strong biocontrol weapons against a wide range of plant pathogens for sustainable agriculture production and management.

References

AL-Saeedi, S., AL-Ani, B.M., 2014. Study of antagonistic capability of *Trichoderma harzianum* isolates against some pathogenic soil borne fungi. Agric. Biol. J. N. Am 5 (1), 15–23.

Ansari, M.M., Mirza, S., Gupta, G.K., Srivastava, S.K., 2011. Characterization of local isolates of *Trichoderma* in vitro evaluation against *Sclerotium rolfsii* (causal organism of collar rot of soybean) and compatibility with seed dressing fungicides. Soybean Res. 9, 123–135.

Arunasri, P., Chalam, T.V., Reddy, N.P., Reddy, S.T., 2011. Investigations on fungicidal sensitivity of *Trichoderma* spp. and *Sclerotium rolfsii* in crossandra. Int. J. Appl. Biol. Pharm. Technol. 2 (2), 104–110.

Arzanlou, M., Khodaei, N.A., Babai, A.A., Azar, A.M., 2014. Inhibitory effect of *Trichoderma* isolates on growth of *Alternaria alternata*, the causal agent of leaf spot disease on sunflower, under laboratory conditions. Arch. Phytopathol. Plant. Prot 47 (13), 1592–1599.

Bagwan, N.B., 2010. Evaluation of *Trichoderma* compatibility with fungicides, pesticides, organic cakes and botanicals for integrated management of soil borne diseases of soybean [*Glycine max* (L.) Merril]. Int. J. Plant Prot. 3 (2), 206–209.

Bagwan, N.B., 2011. Evaluation of biocontrol potential of *Trichoderma* species against *Sclerotium rolfsii*, *Aspergillus niger* and *Aspergillus flavus*. Int. J. Plant Prot. 4, 107–111.

Benitez, T., Rincon, A.M., Limon, M.C., Antonia, C., 2004. Biocontrol mechanisms of *Trichoderma* strains. Int. Microbiol. 7, 249–260.

Bhai, R.S., Thomas, J., 2010. Compatibility of *Trichoderma harzianum* with fungicides insecticides and fertilizers. Indian Phytopathol. 63 (2), 145–148.

Bheemaraya Patil, M.B., Ramesh, Vendan, K.T., Amaresh, Y.S., Rao, K., 2012. Compatibility of *Trichoderma* spp. with commonly used fungicides, insecticides and plant extracts. Indian J. Plant Prot. 40 (2), 118–122.

Bokhari, N.A., Perveen, K., 2012. Antagonistic action of *Trichoderma harzianum* and *Trichoderma viride* against *Fusarium solani* causing root rot of tomato. Afr. J. Microbiol. Res. 6 (44), 7193–7197.

Brian, P.W., 1944. Production of gliotoxin by *Trichoderma viride*. Nature 154, 667–668.

Calistru, C., McLean, M., Berjak, P., 1997. In vitro studies on the potential for biological control of *Aspergillus flavus* and *Fusarium moniliforme* by *Trichoderma* species. A study of the production of extracellular metabolites by Trichoderma species. Mycopathologia 137, 115–124.

Castillo, F.D.H., Padilla, A.M.B., Morales, G.G., Siller, M.C., Herrera, R.R., Cristobal, N., et al., 2011. In vitro antagonist action of *Trichoderma* strains against *Sclerotinia sclerotiorum* and *Sclerotium cepivorum*. Am. J. Agric. Biol. Sci. 6 (3), 410–417.

Chen, L.L., Liu, L.J., Shi, M., Song, X.Y., Zheng, C.Y., Chen, X.L., et al., 2009. Characterization and gene cloning of a novel serine protease with nematicidal activity from *Trichoderma pseudokoningii* SMF2. FEMS Microbiol. Lett. 299, 135–142.

Chennappa, G., Naik, M.K., Amaresh, Y.S., Ravikiran, Ashwathanarayan, D.S., Patil, M.G., et al., 2017. Morphological, molecular characterization of *Trichoderma* species isolated from different rhizosphere soils and its anti-pathogenic properties. Imp. J. Interdiscip. Res. 3 (4), 1–12.

Crutcher, F.K., Parich, A., Schuhmacher, R., 2013. A putative terpene cyclase, vir4, is responsible for the biosynthesis of volatile terpene compounds in the biocontrol fungus *Trichoderma virens*. Fungal Genet. Biol 56, 67–77.

De Cal, A.S., Pascual, A., Melgarejo, P., 1994. In vitro studies on the effects of fungicides on beneficial fungi of peach twig mycoflora. Mycopathologia 126 (1), 15–20.

Degenkolb, T., Dieckmann, R., Nielsen, K.F., Grafnhan, T., Theis, C., Zafari, D., et al., 2008. The *Trichoderma brevicompactum* clade: a separate lineage with new species, new peptaibiotics and mycotoxins. Mycol. Prog. 7, 177–219.

Dubey, S.C., Suresh, M., 2006. Randomly amplified polymorphic DNA markers for *Trichoderma* species and antagonism against *Fusarium oxysporum* f. sp. *ciceris* causing chickpea wilt. J. Phytopathol. 154, 663–669.

Dubey, S.C., Aradhika, T., Dureja, P., Anita, G., 2011. Characterization of secondary metabolites and enzymes produced by *Trichoderma* species and their efficacy against plant pathogenic fungi. Indian J. Agric. Sci. 81, 455–461.

Evidente, A., Cabras, A., Maddau, L., Serra, S., Andolfi, A., Motta, A., 2003. Virepyronone, a new antifungal 6- substituted 2H pyran-2-one produce by *Trichoderma viride*. J. Agric. Food Chem. 51, 6957–6960.

Eziashi, E.I., Uma, N.U., Adekunle, A.A., Airede, C.E., 2006. Effect of metabolites produced by *Trichoderma* species against *Ceratocystis paradoxa* in culture medium. Afr. J. Biotechnol. 5, 703–706.

Faheem, A.V.K., Razdan, F.A., Mohiddin, F.A., Bhat, K.A., Sheikh, P.A., 2010. Effect of volatile metabolites of *Trichoderma* species against seven fungal plant pathogens *invitro*. J. Phytol. 2 (10), 34–37.

Geremia, R.A., Goldman, G.H., Jacobs, D., Ardiles, W., Vila, S.B., et al., 1993. Molecular characterization of the proteinase encoding gene, *prb1* related to mycoparasitism by *Trichoderma harzianum*. Mol. Microbiol. 8, 603–613.

Ghisalberti, E.L., Sivasithamparam, K., 1991. Antifungal antibiotics produced by *Trichoderma* spp. Soil Biol. Biochem. 23, 1011–1020.

Haran, S., Schickler, H., Oppenheim, A., Chet, I., 1996. Differential expression of *Trichoderma harzianum* chitinases during mycoparasitism. Phytopathology 86, 980–985.

Harman, G.E., Petzoldt, R., Comis, A., Chen, J., 2004. Interactions between *Trichoderma harzianum* strain T22 and maize inbred line M017 and effects of these interactions on diseases by *Pythium ultimum* and *Collectotrichum graminicola*. Phytopathology 94, 147–153.

Hoffmeister, D., Keller, N.P., 2007. Natural products of filamentous fungi: enzymes, genes and their regulation. Nat. Prod. Rep. 24, 393–416.

Howell, C.R., 1998. The role of antibiosis. In: Harman, G.E., Kubicek, C.P. (Eds.), *Trichoderma* and *Gliocladium*. Vol. 2. Enzymes, Biological Control, and Commercial Applications. Taylor & Francis, London, pp. 173–184.

Howell, C.R., 2003. Mechanisms employed by *Trichoderma* species in the biological control of plant diseases: the history and evolution of current concepts. Plant Dis. 87, 4–10.

Idnurm, A., Heitman, J., 2005. Light controls growth and development via a conserved pathway in the fungal kingdom. PLoS Biol. 3, 95.

Ihrmark, K., et al., 2010. Comparative molecular evolution of *Trichoderma* chitinases in response to mycoparasitic interactions. Evol. Bioinform. 6, 1–26.

Jaklitsch, W.M., 2009. European species of Hypocrea part I. The green spored species. Stud. Mycol. 63, 1–91.

Jayaraj, J., Ramabadran, P., 1996. Effect of certain soil insecticides on in vitro growth, sporulation and cellulose production by *T. harzianum*. J. Biol. Control 10, 111–115.

Jebakumar, R.S., Anandaraj, M., Sarma, Y.R., 2000. Compatibility of phorate and chlorpyriphos with *Trichoderma harzianum* (Rifai.) applied for integrated disease management in black pepper. J. Spices Aromat. Crops 9 (2), 111–115.

Joao, M., Silva, D., Raiza, D.R., Teixeira, O., Jessica, R., Rocha, D., et al., 2017. In vitro and in vivo antagonism of *Screrotium rolfsii* Sacc. by strains of *Trichoderma* spp. Int. J. Agric. Environ. Biotechnol 2 (1), 60–67.

Kapulnik, Y., Chet, I., 2000. Induction and accumulation of PR proteins activity during early stages of root colonization by the mycoparasite *T. harzianum* strain T-203. Plant Physiol. Biochem. 38, 863–873.

Khan, A.A., Sinha, A.P., 2007. Screening of *Trichoderma* spp. against *Rhizoctonia solani* the causal agent of rice sheath blight. Indian Phytopathol. 60 (4), 450–456.

Kolte, V.S., Raut, S.P., 2007. Efficacy of *Trichoderma* spp. against *Sclerotium rolfsii* infecting collar rot of orchids (*Dendrobium orchreatum*). Ann. Plant Prot. Sci. 15 (2), 469–539.

Komon-Zelazowska, M., Neuhof, T., Dieckmann, R., Dohren, V.H., Herrera-Estrella, A., 2007. Formation of atroviridin by *Hypocrea atroviridis* is conidiation associated and positively regulated by blue light and the G protein GNA3. Eukaryotic Cell 6, 2332–2342.

Kredics, L., Antal, Z., Szekeres, A., Hatvani, L., Manczinger, L., Vagvolgyi, C., et al., 2005. Extracellular proteases of *Trichoderma* species. A review. Acta Microbiol. Immunol. Hung. 52, 169–184.

Kubicek, C.P., Mach, R.L., Peterbuer, C.K., Lorito, M., 2001. *Trichoderma:* from genes to biocontrol. J. Plant. Pathol. 83, 11–24.

Kubicek, C.P., Komon, Z.M., Druzhinina, I.S., 2008. Fungal genes *Hypocrea/Trichoderma*: from barcodes to biodiversity. J. Zhejiang Univ. Sci. B 9 (10), 753–763.

Kumar, S., Thakur, M., Rani, A., 2014. *Trichoderma*: mass production, formulation, quality control, delivery and its scope in commercialization in India for the management of plant diseases. Afr. J. Agric. Res 9 (53), 3838–3852.

Kyalo, G., Affokpon, A., Coosemans, J., Coynes, D.L., 2007. Biological control effects of *Pochonia chlamysdosporia* and *Trichoderma* isolates from Benin (West-Africa) on root knot nematodes. Commun. Agric. Appl. Biol. Sci. 72, 219–223.

Madhavi, G.B., Bhattiprolu, S.L., Reddy, V.B., 2011. Compatibility of biocontrol agent *Trichoderma viride* with various pesticides. J. Hortic. Sci. 6 (1), 71–73.

Mandhare, V.K., Suryawanshi, A.V., 2008. Efficacy of some botanicals and *Trichoderma* species against soil borne pathogens infecting chickpea. J. Food Legumes 21 (2), 122–124.

Mariola, C.R., Rodriguez-Galan, O., Benitez, T., Sousa, S., Rey, M., Llobell, A., et al., 2007. Microscopic and transcriptome analyses of early colonization of tomato roots by *Trichoderma harzianum*. Int. Microbiol. 10, 19–27.

Monte, E., 2001. Understanding *Trichoderma*: between biotechnology and microbial ecology. Int. Microbiol. 4, 1–4.

Mukherjee, P.K., Buensanteai, N., Moran Diez, M.E., 2012a. Functional analysis of non ribosomal peptide synthetases (NRPSs) in *Trichoderma virens* reveals a polyketide synthase (PKS)/NRPS hybrid enzyme involved in the induced systemic resistance response in maize. Microbiology 158, 155–165.

Mukherjee, P.K., Horwitz, B.A., Kenerley, C.M., 2012b. Secondary metabolism in *Trichoderma* a genomic perspective. Microbiology 158, 35–45.

Nainwal, D., Vishunavat, K., 2016. Management of purple blotch and Stemphylium blight of onion in Tarai and Bhabar regions of Uttarakhand, India. J. Appl. Nat. Sci. 8 (1), 150–153.

Nair, D., 2015. Biological and molecular characterization of indigenous isolates of *Trichoderma* species. Thesis. University of Agricultural Sciences, Raichur.

Nandeesha, B.S., Kumar, M.R., Reddy, N.P., 2013. Evaluation of different fungicides and their compatibility with potential *Trichoderma* spp. for the management of *Aspergillus niger*, incitant of collar rot of groundnut. Asian J. Biol. Sci. 2 (1), 59–63.

Neuhof, T., Dieckmann, R., Druzhinina, I.S., Kubicek, C.P., VonDohren, H., 2007. Intact cell MALDI TOF mass spectrometry analyses of peptaibols formation by the genus *Trichoderma*: can molecular phylogenetic knowledge predict peptaibols structures. Microbiology 153 (10), 3417–3437.

Nitish, R.B., Kumar, J., 2017. Characterization of volatile secondary metabolites from *Trichoderma asperellum*. J. Appl. Nat. Sci. 9 (2), 954–959.

Ortellado, M.F., 2013. In vitro compatibility of native *Trichoderma* spp. with fungicides for seed treatment. Investig. Agrar 15 (1), 15–22.

Osbourn, A., 2010. Secondary metabolic gene clusters: evolutionary tool kits for chemical innovation. Trends Genet. 26, 449–457.

Padmaja, M., Narendra, K., Swathi, J., Sowjanya, K.M., Babu, P.J., Satya, A.K., 2013. *Invitro* antagonism of native isolates of *Tricoderma* spp. against *Sclerotium rolfsii*. Int. J. Res. Pharm. Biomed. Sci. 4 (3), 886–891.

Patil, R.A., Gupta, V.R., 2011. Different substrate of *Trichoderma viride* affecting its efficacy and compatibility with fungicide. J. Plant Dis. Sci. 16 (2), 141–144.

Prabhavathi, N.M., Amaresh, Y.S., Yadahalli, P., 2015. In vitro evaluation of selected bioagent isolates isolated from conservation agriculture ecosystem of red gram and maize. Trends Biosci. 8 (21), 5990–5993.

Rangnathswamy, M., Patibanda, A.K., Chandrasekhar, G.S., Sandeep, D., Mallesh, S.B., Kumar, H.B.H., 2012. Compatibility of *Trichoderma* isolates with selected fungicides in vitro. Int. J. Plant Prot. 5 (1), 12–15.

Rao, N.S., Kulkarni, S., 2003. Effect of *Trichoderma* spp. on the Growth of *Sclerotium rolfsii* Sacc. J. Biol. Control 181–184.

Raut, I., Badea Doni, M., Calin, M., Oancea, F., Vasilescu, G., Sesan, T.E., et al., 2014. Effect of volatile and non volatile metabolites from *Trichoderma* spp. against important phytopathogens. Rev. Chim. (Bucharest) 65 (11), 1285–1288.

Reino, J.L., Guerrero, R.F., Hernandez, G.R., 2008. Secondary metabolites from species of the biocontrol agent *Trichoderma*. Phytochem. Rev. 7, 89–123.

Reither, B., Brunner, K., Schuhmacher, R., Piessi, I., Seidi, V., Krska, R., et al., 2005. The G protein alpha subunit Tgal of *Trichoderma atraviride* is involved in chitinase formation and differential production of antifungal metabolites. Fungal Genet. Biol. 42, 749–760.

Rekha, D., Patil, M.B., Shetty, S.P., Swamy, K.M., Rajini, B.G., 2012. In vitro screening of native *Trichoderma* isolates against *Sclerotium rolfsii* causing collar rot of ground nut. Int. J. Sci. Nat. 3 (1), 117–120.

Rifai, M.A., 1969. Revision of genus *Trichoderma*. Mycol. Pap 116, 1–56.

Rini, C.R., Sulochana, K.K., 2007. Usefulness of *Trichoderma* and *Pseudomonas* against *Rhizoctonia solani* and *Fusarium oxysporum* infecting tomato. J. Trop. Agric. 45 (1), 21–28.

Roberti, R., Badiali, F., Cesari, A., 2002. Compatibility of different strains of *Trichoderma* and *Gliocladium* with fungicides and herbicides. J. Plant Pathol. 84, 192.

Roco, A., Perez, L.M., 2001. In vitro biocontrol activity of *Trichoderma harzianum* on *Alternaria alternata* in the presence of growth regulators. Electron. J. Biotechnol. 4 (2), 68–73.

Sachin-Upmanyu, S.K., Shyam, K.R., 2002. Innovative approaches for the management of root rot and web blight (*Rhizoctonia solani*) of French bean. J. Mycol. Plant Pathol. 32 (3), 317–331.

Samuels, G.J., 1996. *Trichoderma*: a review of biology and systematics of the genus. Mycol. Res. 100 (8), 923–935.

Sangle, U.R., Kumar, S., Mishra, J.S., 2016. Effect of volatile metabolites released by different *Trichoderma* spp. on the growth of pathogen *Sclerotium rolfsii* and effect on sclerotia germination. Bioscan. 11 (3), 1559–1564.

Saravanan, L., Kalidas, P., Phanikumar, T., Deepthi, P., Babu, K.R., 2013. In vitro compatibility of *Trichoderma viride* with agrochemicals. Ann. Plant Prot. Sci. 22 (1), 190–239.

Seidi, V., Song, L., Lindquist, E., Gruber, S., Koptchinskiy, A., et al., 2009. Transcriptomic response of the mycoparasitic fungus *Trichoderma atraviride* to the presence of 546 a fungal prey. BMC Genom 10, 567.

Shaigan, S., Seraji, A., Moghaddam, S.A.M., 2008. Identification and investigation on antagonistic effect of *Trichoderma* spp. on tea seedlings white foot and root rot (*Sclerotium rolffsi* Sacc.) in vitro condition. J. Biol. Sci. 11, 3246–3350.

Shukla, H.S., 2011. Compatibility of *Trichoderma viride* with botanicals and synthetic pesticide, insecticide and herbicide. Asian J. Chem. Environ. Res. 4 (2), 62–63.

Siddiquee, S., Cheong, B.E., Taslima, K., Kausar, H., Hasan, M.M., 2012. Sequence and identification of volatile compounds from liquid cultures of *Trichoderma harzianum* by GCMS using three different capillary columns. J. Chromatogr. Sci. 50, 358–367.

Singh, J.S., 2014. Cyanobacteria: a vital bio-agent in eco-restoration of degraded lands and sustainable agriculture. Clim. Change Environ. Sustain. 2, 133–137.

Singh, J.S., 2015. Microbes: the chief ecological engineers in reinstating equilibrium in degraded ecosystems. Agric. Ecosyst. Environ 203, 80–82.

Singh, J.S., 2016. Microbes play major roles in ecosystem services. Clim. Change Environ. Sustain. 3, 163–167.

Singh, J.S., Strong, P.J., 2016. Biologically derived fertilizer: a multifaceted bio-tool in methane mitigation. Ecotoxicol. Environ. Saf. 124, 267–276.

Singh, J.S., Gupta, V.K., 2018. Soil microbial biomass: a key soil driver in management of ecosystem functioning. Sci. Total Environ. 634, 497–500.

Singh, B.K., Walker, A., Morgan, J.A.W., Wright, D.J., 2004. Biodegradation of chloropyrifos by Enterobacter strain B-14 and its use in bioremediation of contaminated soils. Appl. Environ. Microbiol. 4855–4863.

Singh, P., Srivastava, S., Shrivastava, S.K., Singh, H.B., 2012. Compatibility of different insecticides with *Trichoderma harzianum* under in vitro condition. Plant Pathol. J. 11, 73–76.

Singh, J.S., Abhilash, P.C., Gupta, V.K., 2016. Agriculturally important microbes in sustainable food production. Trends Biotechnol. 34, 773–775.

Sivan, A., Chet, I., 1989. The possible role of competition between *Trichoderma harzianum* and *Fusarium oxysporum* on rhizosphere colonization. Phytopathology 79, 198–203.

Sivasithamparam, K., Ghisalberti, F.L., 1998. Secondary metabolism Trichoderma and Gliocladium. In: Kubicek, C.P., Harman, G.E. (Eds.), Trichoderma and Gliocladium. Taylor & Francis Ltd, London, pp. 139–191.

Stoppacher, N., Zeilinger, S., Omann, M., Lassahn, P.G., Roitinger, A., Krska, R., et al., 2008. Characterisation of the peptaibiome of the biocontrol fungus *Trichoderma atroviride* by liquid chromatography/tandem mass spectrometry. Rapid Commun. Mass Spectrom. 22, 1889–1898.

Suarez, M.B., Vizcaino, J.A., Llobell, A., Mont, E., 2007. Characterization of genes encoding novel peptidass in the biocontrol fungus *Trichoderma harzianum* CECT 2413 using the TrichoEST functional genomics approach. Curr. Genet. 51, 331–342.

Sumana, K., Devaki, N.S., 2012. In vitro evaluation of some bioagents against tobacco wilt pathogen. J. Biopest. 5 (1), 18–22.

Tapwal, A., Tyagi, A., Thakur, G., Chandra, S., 2015. In vitro evaluation of *Trichoderma* species against seed borne pathogens. Int. J. Chem. Biochem. Sci. 1, 2349–2724.

Thakur, S., Harsh, N.S.K., 2014. Efficacy of volatile metabolites of phylloplane fungi of *Rauwolfia serpentina* against *Alternaria alternata*. Curr. Res. Environ. Appl. Mycol. 4 (2), 152–156.

Thiruchchelvan, N., Mikunthan, G., Thirukkumaran, G., Pakeerathan, K., 2013. Effect of insecticides on bio-agent *Trichoderma harzianum* rifai under in vitro condition. Am. Eurasian J. Agric. Environ. Sci 13 (10), 1357–1360.

Tiwari, P., Singh, J.S., 2017. A plant growth promoting rhizospheric *Pseudomonas aeruginosa* strain inhibits seed germination in *Triticum aestivum* (L) and *Zea mays* (L). Microbiol. Res 8, 1–7.

Vey, A., Hoagland, R.E., Butt, T.M., 2001. Toxic metabolites of fungal biocontrol agents. In: Butt, T.M., Jackson, C., Magan, N. (Eds.), Fungi as Biocontrol Agents: Progress, Problems and Potential. CAB Int. Bristol. pp. 311–346.

Vimal, S.R., Patel, V.K., Singh, J.S., 2018. Plant growth promoting *Curtobacterium albidum* strain SRV4: an agriculturally important microbe to alleviate salinity stress in paddy plants. Ecol. Indic. (Article in press).

Vinale, F., Sivasithamparam, K., Ghisalberti, E.L., Marra, R., Barbetti, M.J., Li, H., et al., 2008. A novel role for *Trichoderma* secondary metabolites in the interactions with plants. Physiol. Mol. Plant Pathol. 72, 80–86.

Vinale, F., Ghisalberti, E.L., Sivasithamparam, K., Marra, R., Ritieni, A., Ferracane, R., et al., 2009. Factors affecting the production of *Trichoderma harzianum* secondary metabolites during the interaction with different plant pathogens. Lett. Appl. Microbiol. 48, 705–711.

Viterbo, A., Ramot, O., Chemin, L., Chet, I., 2002. Significance of lytic enzymes from *Trichoderma* spp. in the biocontrol of fungal plant pathogens. Ant. Van. Leeuw. 81, 549–556.

Vivek, A., Suresh, W., Om, P.S., Jitendra, K., Aditi, K., Jay, S., et al., 2014. Isolation, characterisation of major secondary metabolites of the Himalayan *Trichoderma koningii* and their antifungal activity. Arch. Phytopathol. Plant Prot. 47 (9), 1063–1071.

Whilhite, S.E., Stanley, D.C., 1996. Timing of gliotoxin biosynthesis in the fungal biological control agent *Gliocladium virens* (*Trichoderma virnes*). Appl. Microbiol. Biotechnol 45, 513–518.

Further reading

Tiwari, S., Singh, C., Singh, J.S., 2018. Land use changes: a key ecological driver regulating methanotrophs abundance in upland soils. Energy Ecol. Environ. 3, 355–371.

Chapter 16

Phycoremediation of industrial effluents contaminated soils

Praveen Solanki[1], M.L. Dotaniya[2], Neha Khanna[3], S. Udayakumar[4], C.K. Dotaniya[5], Shiv Singh Meena[6], Maitreyie Narayan[1] and R.K. Srivastava[1]

[1]*Department of Environmental Science, GBPUA&T, Pantnagar, India,* [2]*ICAR-Indian Institute of Soil Science, Bhopal, India,* [3]*Department of Agricultural Chemistry and Soil Science, Dr. B.R. Ambedkar University, Agra, India,* [4]*Department of Agronomy, Horticulture, and Plant Science, South Dakota State University, Brookings, SD, United States,* [5]*College of Agriculture, SKRAU, Bikaner, India,* [6]*Department of Soil Science, GBPUA&T, Pantnagar, India*

Chapter Outline

16.1 Introduction

Phycoremediation is algae-based remediation of contaminated soil, in which different kinds of algae are used with respect to the target pollutant to be removed (Romera et al., 2007; Olguin and Sanchez-Galvan, 2012). In addition to the reclamation of contaminated soil, algal biomass is also produced by the process called *carbon sequestration*; this biomass further could be used for the production of "green fuel" (Rawat et al., 2011; Bala et al., 2016; Singh, 2013; Singh, 2015; Tiwari and Singh, 2017; Singh et al., 2018; Tiwari et al., 2018; Singh and Gupta, 2018; Vimal et al., 2018). Therefore deposition of algae into the soil leads to increase in organic matter and carbon content, which enhances the soil's biological as well as physical properties (Prajapati et al., 2013; Dotaniya et al., 2014; Solanki et al., 2017a). However, currently the major challenges for application of phycoremediation technology are quantity and quality of targeted pollutants and adaptation of this technique with respect to the different climatic conditions (Pittman et al., 2011; Srivastava et al., 2015), hence indigenous species of algae should be preferred. Therefore the growth of algae and efficiency for the phycoremediation process are regulated by characteristics of industrial effluents to be removed and climatic condition of the targeted site (Arora et al., 2006; Dubey et al., 2013; Khatoon et al., 2017). Microalgae namely *Spirulina platensis*, *Desmodesmus* sp., *Haematococcus pluvaris*, *Nodularia* sp., *Chlorella vulgaris*, *Cyanothece* sp., *Nannochloropsis oculata*, *Botryococcus sudeticus*, *Dunaliella* sp., *Scenedesmus*, sp., *Phormidium* sp., *Chlamydomonas* sp., *Oscillatoria* sp., *Arthrospira* sp., *Nostoc* sp., etc. are now been producing extensively and applied for phycoremediation of environmental pollutants and production of biomass for various applications (Dubey et al., 2013; Pacheco et al., 2015). Microalgae for phycoremediation can be cultured and further harvesting can be done using different techniques such as filtration, flocculation, magnetic separation, and centrifugation (Coward et al., 2014; Kumar et al., 2018).

New and Future Developments in Microbial Biotechnology and Bioengineering. DOI: https://doi.org/10.1016/B978-0-12-818258-1.00016-9

TABLE 16.1 Classification of degraded land in India (Sehgal and Abrol, 1994).

Types of degradation	Area (million hectares)
Water erosion	148.9
Wind erosion	13.5
Saline and alkali soils	10.1
Water-logging	11.6
Decline if soil fertility	3.7

In this modern era of rapid industrialization and development, the greatest challenge is the safe disposal of effluents in their existing form, for example, solid or liquid (De Bashan and Bashan, 2010). Though the quantitative generation of industrial effluents is linearly proportional to the economic growth and development of the particular country, the safe management and disposal of these effluents have been more closely examined recently (Anirudhan and Sreekumari, 2011). Each industry has a common drainage point also known as *the point source of pollution* for discharging of industrial effluents that may consist of several organic and inorganic materials, pollutants, and heavy metals, depending on the type of industry (Zinicovscaia and Cepoi, 2016). In addition to this, human activities with rising living standards have resulted in the tremendous production of undesirable wastes. The wide range of raw materials is used in industrial manufacturing units and after processing the unused materials or effluents are discharged into sewage system or deposited into landfills, which leads to contamination of water as well as soil resources (Abdel-Halim et al., 2013).

India is the second most populous country and supports nearly 16% of the world's human population as well as approximately 20% of the world's livestock on a tiny land of 328.73 million hectares (2.5% of the world's total geographical area). Beside the tremendous pressure on this small land resource to yield enough agricultural production and to support the huge population, most of the agricultural land, approximately 187.8 mha (57% of 328.73 mha), has been severely degraded (Sehgal and Abrol, 1994; Khanna and Solanki, 2014). In India, to feed the huge population it has become essential to use agrochemicals and chemical fertilizers for modern agriculture, however, extensive application in an unsustainable manner resulted in many environmental issues including contamination of agricultural soil through heavy metal and other toxic substances (Prakash and Kumar, 2013; Qari and Hassan, 2014; Solanki et al., 2017b). The classification of degraded land in India is depicted in Table 16.1. This chapter mainly focuses on the phycoremediation of industrial effluents contaminated soils and the by-products of algae for further production of green fuels.

16.2 Environmental risk of soil pollution with heavy metals

Developmental and industrial activities in many developing countries including India have been pursued extensively and are growing significantly without much concern paid to environmental issues, which has imposed pressure on finite environmental resources. Elements having high atomic weight (usually between 63.5 and 200.6) and density (usually >5 g/mL) are known as *heavy metals* (Lapedes, 1974), which are highly known for their long *persistence* into the environment; hence they are highly resistant to degradation. Beside the long persistence of heavy metal, they are highly fat soluble and sequentially accumulate into fat tissues by the process called *biomagnification* (Srivastava et al., 2015). Heavy metals are classified by Valls and Lorenzo (2002) into harmful (when the concentrations are high, e.g., Fe, Cu, Zn, Mn, Ni, Co, Cr, etc.) and highly toxic (even when the concentrations is very low, e.g., Hg, Cd, Pb, etc.). Some metals are essential for soil microbial activities as well as for plant development (e.g., Fe, Mn, Cu, Ni, Mo, Zn, etc.), which are well known as "trace elements" since they are required in trace quantity (Tarley et al., 2004; Solanki and Debnath, 2014; Solanki, 2014). In contrast to trace elements, others such as Al, Sn, Au, Cd, Pb, Sr, Ti, Hg, etc. have no essentiality for microbial and plant biological activities and are stated as *toxic metals* (Prasad and Freitas, 2000; Chamarthy et al., 2001; Barrera et al., 2006). Recently more deposition of industrial effluents into landfills has led to an environmental risk of soil pollution, as they consist of several heavy metals such as chromium (Cr) and lead (Pb) in tannery effluents (Vajpayee et al., 2001; Dotaniya et al., 2016). Though no reliable investigations are available to measure the degree of soil degradation, it is accepted that the problem with regard to disposal of industrial effluents is extensive and significant.

16.3 Characterization of industrial effluents

More than 10,000 types of dyes are produced synthetically and extensively used for textile coloring, dyestuff manufacturing, paper printing, food coloring additives, cosmetics, etc. (Zollinger, 1987; Robinson et al., 2001; Lim et al., 2010). In general, coloring efficiency of dyes is up to 90% and the rest is discharged as industrial effluent due to the inefficiency of the coloring process. Jin et al. (2007) reported that about 2.8×10^5 tons of various textile dye effluents were released into environmental segments per annum, which represents a critical source of soil and water pollution (Ratna and Padhi, 2012). Since India is one of the most emerging countries for industrial and infrastructural development, it generates a significant quantity of industrial effluents, which are disposed unscientifically on the land and in water bodies. Furthermore, quantitative generation of industrial effluents is correlated with the industrial and economic growth of the country. In India, a considerable amount of effluents are generated from various industries, for example, petrochemicals, pesticides, pharmaceuticals, paint and dye, fertilizers, petroleum, asbestos, inorganic chemicals, caustic soda, tannery, general engineering industries, and so on (Vajpayee et al., 1995; Dotaniya et al., 2016). Discharged effluents from these industries consist of many pollutants namely pesticides, various heavy metals, cyanides, complex aromatic compounds, dye chemicals, and several other chemicals elements or their compounds, which are flammable, reactive, toxic, and corrosive, or have explosive properties (Farinella et al., 2007; Lesage et al., 2007). Jais et al. (2017) analyzed wet market wastewater, which was acidic in nature with pH of 4.4–6.3 while the BOD and COD were 295 and 763 mg/L, respectively.

16.4 Safe disposal of industrial effluents

Safe disposal of industrial effluents is one of the most emerging issues globally since most of the water and soil resources have been already lost their quality to protect biotic and abiotic component of the environment. In addition to this, discharge of industrial effluents into nearby water bodies and land causes several diseases in human beings and aquatic animals (Faragallah et al., 2009; Migahed et al., 2017; Dotaniya et al., 2018). Therefore mostly developed as well as underdeveloping countries including India have adopted different environmental regulations such as the Environmental (Protection) Act–1986 (EPA, 1986) in India for proper management and safe disposal of industrial effluents to protect all four segments (*atmosphere*, *lithosphere*, *hydrosphere*, and *biosphere*) of the environment. Many regulations have been accepted throughout the world such as ECOLEX (a multination's environmental law), which has been adopted by many international organizations namely IUCN (International Union for Conservation of Nature and Natural Resources), FAO (Food and Agriculture Organization), UNEP (United Nation Environment Programme), etc. and stated that wastewater and other effluents are subjected to treatment prior to their final disposal (Solanki et al., 2017c). National and international standards (ECOLEX) for industrial wastewater effluents discharge into the environment are depicted in Table 16.2 (EPA, 1986; ECOLEX, 2002; Jais et al., 2017).

Various medical waste and infected materials should be safely disposed of through properly installed incinerator and the remaining ash should be dumped into the sanitary landfill. Similarly toxic and hazardous wastes should be dumped into the scientifically constructed landfill. Discharge of various wastewaters from respective industries should be remediated or treated up to the desired level of pollution load before being drained into the nearest water body (Solanki et al., 2017d, 2018b).

Some autotrophic organisms, for example, cyanobacteria or algae (BGA), being ubiquitous, have been reported for decolorization of many dyes (Baptista and Vasconcelos, 2006; Mulbry et al., 2008). Decolorization of dyes through different algae species namely *Oscillatoria* and *Chlorella* occurs mainly by carbon sequestration into biomass, CO_2, and H_2O transformation of molecules to noncolored and adsorption of chromophores into its biomass (Mohan et al., 2002; Acuner and Dilek, 2004; Daeshwar et al., 2007). Table 16.3 represents the phycoremediation of different industrial effluents using various algal species.

The phycoremediation process for various industrial effluents takes 3–26 days based on pollution load and required pollutant reduction efficiency (De Bashan and Bashan, 2010; Liu et al., 2013); however, *Scenedesmus* sp. took 40 days for phycoremediation of swine industry wastewater in Korea (Kim et al., 2007) and the highest percentage (%) reduction for total nitrogen, total phosphorus, and total carbon was accordingly 87.0, 83.2, and 12.9. Similarly, Mata et al. (2012) investigated phycoremediation of brewery industry wastewater using *Scenedesmus obliquus* for remediation of chemical oxygen demand (COD) and total nitrogen (TN) and concluded that nine days retention time for phycoremediation, was optimum for maximum removal of COD and TN (57.50% and 20.80%, respectively). While Kothari et al. (2012) investigated phycoremediation of dairy industry effluents using *C. pyrenoidosa* and concluded that the maximum removal efficiency of 49.90% for nitrate (NO_3^-), 79.06% for nitrite (NO_2^-), 83.23% for phosphate (PO_4^-) and 32.00%

TABLE 16.2 Effluents discharge standards (mg/L, except pH or otherwise stated).

Parameter	Maximum permissible limits	
	Indian standards	ECOLEX
pH	6.0–9.0	5.0–9.0
Temperature (°C)	<5°C from receiving water body	–
Electrical conductivity (EC, μS cm^{-1})	500.0	–
Biochemical oxygen demand (BOD_5)	100.0	40.0
Chemical oxygen demand (COD)	250.0	120.0
Suspended solids (SS)	100.0	35.0
Total dissolved solids (TSS)	500.0	–
Oil and grease	10.0	10.0
Sulfide (SO_4)	5.0	0.002
Cadmium (Cd)	0.05	0.010
Chromium (Cr)	2.00	NA
Copper (Co)	3.00	0.500
Zinc (Zi)	15.0	2.00
Lead (Pb)	0.10	0.05
Manganese (Mn)	2.00	0.20
Nickel (Ni)	3.00	0.10
Aluminum (Al)	NA	5.00
Iron (Fe)	3.00	2.00

for iron (Fe) ions were observed. An investigation was done by Hultberg et al. (2013) for phycoremediation of tomato cultivated greenhouse wastewater in Sweden using *C. vulgaris* and reported that *C. vulgaris* shows more phosphorus remediation efficiency (99.70%) than that of *C. pyrenoidosa*, which was reported by Kothari et al. (2012). However, for nitrogen reduction, *C. pyrenoidosa* was observed more efficient (79.06%) as compared with *C. vulgaris* (20.70%).

Application of *C. vulgaris* was investigated for phycoremediation of leather processing effluents, which consist of heavy metals, residual pigments, casein, and other chemicals (Hanumantha Rao et al., 2011; Solanki et al., 2018c). The results show the significant reduction in BOD, COD, free ammonia, nitrite, calcium, and magnesium by 22%, 38%, 80%, 89%, 63%, and 50%, respectively. Similarly the application of different algae species namely *Chlorella*, *Scenedesmus*, *Gloeocystis*, *Chlamydomonas*, and *Cyanobacteria* was studied for phycoremediation of carpet industrial effluents (Chinnasamy et al., 2010). While *Nostoc* sp. was employed for phycoremediation of dairy effluents (Mulbry et al., 2008; Kotteswari et al., 2012; Brar et al., 2017). Pathak et al. (2014) studied the application of *C. pyrenoidosa* for phycoremediation of textile effluents and reported 62% and 87% reduction of nitrate and phosphate, respectively.

16.5 Industrial effluents and their effect on soil quality

Many industrial effluents are directly deposited into landfills since it is cheaper than other effluent management techniques namely incineration, pyrolysis, gasification, etc., however, leachates from landfills along with storm water are contaminating the agricultural land in many ways. Mining activities for the abandoned digging of elemental ore are highly responsible for heavy metal contamination of nearby lands, which is scientifically known as *mine spoil* (Clemente et al., 2012; Yang et al., 2012; Fellet et al., 2014). Furthermore, it causes several health risks and ecological complications to the nearby existing environmental resources (Kabata-Pendias and Pendias, 2001; Ali et al., 2013). When these types of industrial effluents are discharged on agricultural soil, heavy metals and various organic compounds showing

TABLE 16.3 Phycoremediation methodology for various industrial effluents.

Industrial effluents	Types of microalgae	Method	Reactor scale	Parameters	Removal efficiency	Countries	References
Drainage discharge from greenhouse	*Chlorella vulgaris*	Photobioreactor (batch culture)	Sample 200 mL, continuously aerated, 16 h/ 8 h day/ night, 20°C, light intensity of 100 μmol/m^2/s conducted for 9 days	TP TN	 99.70 20.70	Sweden	Hultberg et al. (2013)
Dairy wastewater	*Chlorella pyrenoidosa*	Batch culture	Sample 1000 mL with various concentration (0%, 25%, 50%, 75%, 100%) of dairy wastewater + 2 mL of homogeneous algal suspension conducted for 10 days	NO_3^- NO_2^- P Fe	 49.09 79.06 83.23 32.00	India	Kothari et al. (2012)
Dairy wastewater	*Pithophora* sp.	Photobioreactor (batch culture)	Sample 100 mL, 27°C ± 2°C, intensity of 2000 lx Conducted for 20 days	NH_3 NO_3^- P	 99.01 84.56 97.98	India	Silambarasan et al. (2012)
Wet market wastewater	*Scenedesmus* sp.	Batch culture	Sample 625 mL, added with 1 × 10^5 cell/mL of microalgae, natural condition conducted for 8 days	TOC TN PO_4^{3-} Zn Fe Cu	 71.73 73.01 87.60 79.65 59.33 100.0	Malaysia	Jais et al. (2015)
Swine wastewater	*Scenedesmus obliquus*	Photobioreactor (batch culture)	Sample 250 mL + 2 mL homogeneous algal suspension, agitation at 150 rpm, light intensity 40 μmol/m^2/s conducted for 21 days	TN TP TC	 59.00 24.00 27.00	Korea	Abou-Shanab et al. (2013)

phytotoxicity even at comparatively low concentration can negatively affect soil fertility as well as soil productivity (Yewalkar et al., 2007).

Since the 19th century, applications of polythene and plastic have greatly increased and various products have been extensively made. The term *white pollution* includes solid waste such as plastic bags and polythene products, which are disposed into the landfilling and soil environment and adversely affect the soil properties and its biological ecosystem (Steinbuchel, 2001). The raw materials for manufacturing of plastic products include polymers of polypropylene, polyvinyl chloride, polystyrene, etc. that are known for high resistance to degradation by any methods, and hence leading to several urban environmental issues. The environmental issues with regard to white pollution have opened many challenges for researchers in finding safe disposal methodologies or invention for such plastics, which are susceptible to biological degradation (Shivlata and Satyanarayana, 2015; Narayan et al., 2018). Narayana and Babu (1993) assessed existing data on soil loss, degradation and summarized that, in India, the soil was being eroded at the rate of 16.35 t/ha annually.

16.6 Mechanism of phycoremediation

The main mechanism used by algae for phycoremediation of contaminated soils is very much related to its surface properties and its cell-wall constituents, which contain polymers, namely pectin, hemicelluloses, cellulose, lignin, and arabinogalactan proteins (Domozych et al., 2012). Besides these, functional groups, for example, hydroxyl, carboxyl, and the amino groups are also present in algae biomass, which affect its phycoremediation efficiency. The negative charges of these functional groups, found the potential to bind heavy metals (Chen and Hao, 1998; Wang and Chen, 2009; Abdel Monem et al., 2010; Al-Gheethi et al., 2014). Various microalgae remove heavy metals using two basic processes, namely bioaccumulation and biosorption. In biosorption metal ions are bound on functional groups attached to the cell surface through ion exchange, chelation, complexation, and microprecipitation (Akar and Tunali, 2006; Zinicovscaia and Cepoi, 2016), while in bioaccumulation, the heavy metals are transported and translocated through the cell membrane into different tissues by the active–passive transport system (Shaikh and Bhosle, 2011; Olguin and Sanchez-Galvan, 2012). Many living cells adopt a detoxification process to reduce the toxicity of different heavy metals, which is followed by precipitation of metals as phosphate, carbonate, or sulfide (Hamdy, 2000; Radway et al., 2001). Jaishankar et al. (2014) reported the causes of detoxification mechanisms as putative entrapment in extracellular polymeric substance, volatilization, and participation.

The phycoremediation potential of algae to accumulate and absorb heavy metals into its biomass has been investigated (Gupta and Rastogi, 2008; Jeyakumar and Chandrasekaran, 2014). Mechanisms such as chemisorption, ion exchange, surface precipitation, and covalent bonding were observed in phycoremediation (Munoz and Guieyssea, 2006). Scenedesmus *incrassatulus* was used for phycoremediation of Cu (copper), Cr (chromium), and Cd (cadmium) from artificial wastewater with removal efficiency of 31.7%, 52.7%, and 24.1%, respectively (Pena-Castro et al., 2004), while Jacome Pilco et al. (2009) reported 91% Cr reduction from synthetic industrial effluents using *S. incrassatulus*. However, Chen et al. (2012) reported a complete remediation of Cd_2^+ from aqueous solution through phycoremediation using *S. obliquus*. Gurbuz et al. (2009) reported 46% and 50% phycoremediation efficiency for Zn (Zinc) and Fe, respectively, from industrial effluents. Probably all living organisms require a little fraction of heavy metals, namely iron, copper, manganese, zinc, cobalt, nickel, etc. for their morphological growth and development (Park et al., 2006), however, these metals also become toxic to organisms beyond their maximum permissible limits. In Fig. 16.1 bioaccumulation and biosorption pathways for various heavy metals removal are represented.

16.7 Major products from algae

16.7.1 Algal biomass

Harvested biomass of algae is a source of various industrial products including green fuel (Rawat et al., 2011). The pigments such as astaxanthin and carotenoids obtained from different kind of algae are potentially used as natural colorants and antioxidants, while polyunsaturated fatty acids are efficiently applicable as health supplements (Pangestuti and Kim, 2011). The production of algal biomass, different types of bioreactors, for example, open ponds (Boussiba et al., 1988; Tredici and Materassi, 1992), flat-plate photobioreactor (Milner, 1953), tubular bioreactors (Lee and Low, 1991; Ugwu et al., 2002), vertical-column photobioreactors (Kaewpintong et al., 2007), internally illuminated photobioreactors (Ogbonna et al., 1999), high rate ponds (HRPs), and continuously stirred tank reactors (CSTRs) have been successfully operated. Since they are autotrophic in nature, algae synthesize their own food and produce enough biomass by

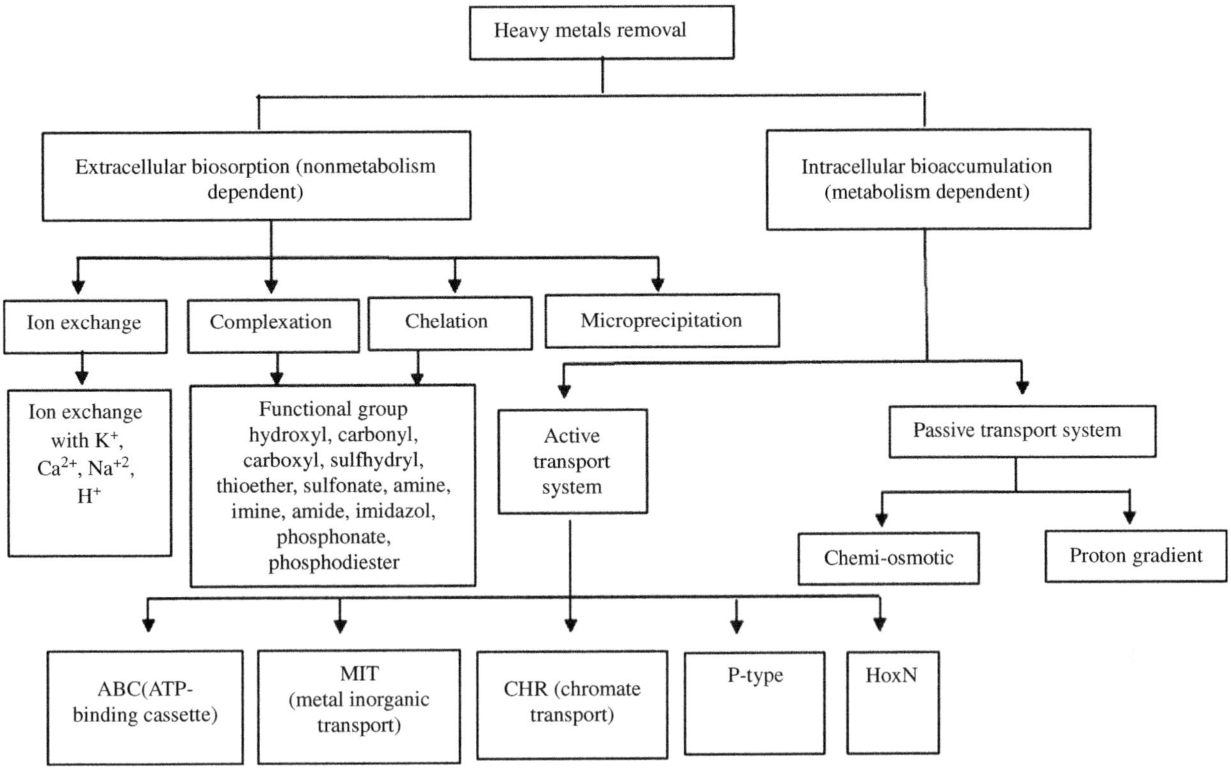

FIGURE 16.1 Different mechanisms for heavy metals removal by various microorganisms based on previous investigations (Fath and Kolter, 1993; Fagan and Saier, 1994; Saier et al., 1994; Saier, 1994; Nies et al., 1998; Regine et al., 2000; Abdel Monem et al., 2010; Al-Gheethi et al., 2015; Jais et al., 2017).

utilizing sunlight and carbon dioxide (CO_2) through the process called *photosynthesis* (Yaakob et al., 2011; Singh and Gupta, 2016). Higher absorption of CO_2 from the atmosphere and deposition into algal biomass leads to higher *carbon sequestration* (Abdelaziz et al., 2014; Singh et al., 2016). The biomass yield and carbon sequestration efficiency of few algae species are depicted in Table 16.4. Beside these, they remove organic and inorganic pollutants as well as many micronutrients such as Cu, Zn, Fe, and Mg as supplements for their multiplication and growth (Mcelwee et al., 2006; Rahman et al., 2012; Zulkifli et al., 2012). Furthermore, different heavy metals present in wastewater have shown toxicity for various algae. Therefore phycoremediation ability of microalgae is restricted to the toxicity level of different heavy metals. The *Euglena* sp., *Scenedesmus* sp., and *Chlorella* sp. were found to be a suitable and common species to use for phycoremediation since they had high tolerance and removal efficiency with respect to pollution load. Fig. 16.2 represents the most common types of raw media, biomass harvesting methods, quantification methods, and various applications of algal biomass.

16.7.2 Green fuel

Harvested algal biomass obtained through the mass cultivation of different algae contributes to mitigation of global warming through carbon dioxide sequestration from the atmosphere (Ross et al., 2008; Grierson et al., 2009). Besides this, the biomass contains high quality of oil and produced 72 times more than that of *Jatropha* per hectare. This oil can be processed and various types of green fuel namely biofuel, nutritional, ethanol (Parjo and Razak, 2015), etc. could be produced as depicted in Fig. 16.3. Furthermore, the quality of contained oil in algae can be significantly improved through manipulating the culture media, which is not possible in the higher plants such as *Pongamia* and *Jatropha*, commonly used plants for biodiesel production (Park et al., 2011; Mata et al., 2010). Such positive and different green products from algae make it a potential candidate for carbon sequestration and climate change mitigation in an eco-friendly manner (Chang et al., 2011).

TABLE 16.4 Relative comparison of biomass yield and CO_2 sequestration of various microalgae with respect to different CO_2, NO_x, SO_x concentration and temperature.

Microalgal species	CO_2 (%)	Temperature (°C)	NO_x/SO_x (mg/L)	Biomass yield (mg/L/d)	CO_2 sequestration rate (mg/L/d)	References
Nannochloris sp.	15	25	0/50	350	658	Negoro et al. (1991)
Nannochloropsis sp.	15	25	0/50	300	564	Negoro et al. (1991)
Chlorella sp.	50	35	60/20	950	1790	Maeda et al. (1995)
Chlorella sp.	20	40	ND	700	1316	Sakai et al. (1995)
Chlorella sp.	50	25	ND	386	725	Sung et al. (1999)
Chlorella sp.	15	25	0/60	1000	1880	Lee et al. (2002)
Chlorella sp.	50	25	ND	500	940	Yue and Chen (2005)
Chlorogloeopsis sp.	5	50	ND	40	20.45	Ono and Cuello (2007)
Chlorococcum sp.	50	22	ND	44	82.0	Ota et al. (2009)

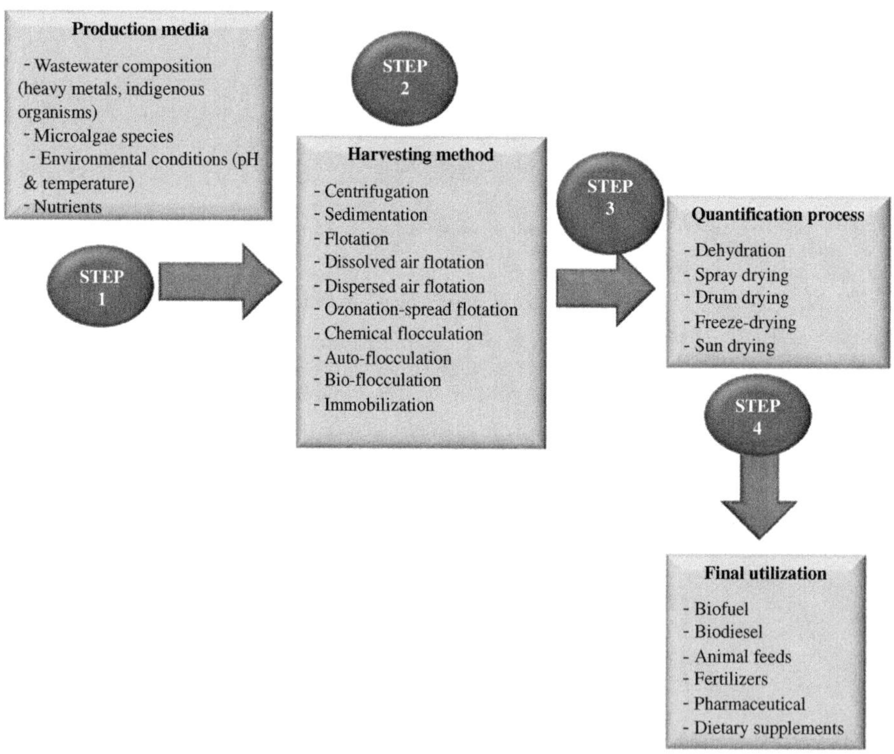

FIGURE 16.2 Various methods of algal biomass harvesting from wastewater and its utilization for different fuels. *Adapted from Jais, N.M., Mohamed, R.M.S.R., Al-Gheethi, A.A., Amir Hashim, M.K., 2017. The dual roles of phycoremediation of wet market wastewater for nutrients and heavy metals removal and microalgae biomass production. Clean Technol. Environ. Policy 19, 37–52.*

16.7.3 Algal biochar

In biological origin, biochar is known as *charcoal*. Algal biochar has been reported by Lehmann and Joseph (2009) as a potential candidate for carbon sequestration as well as a *soil ameliorant* with the capacity to improve the water holding capacity (WHC), organic matter, total organic carbon (TOC), and soil fertility status of many soils (Lehmann et al.,

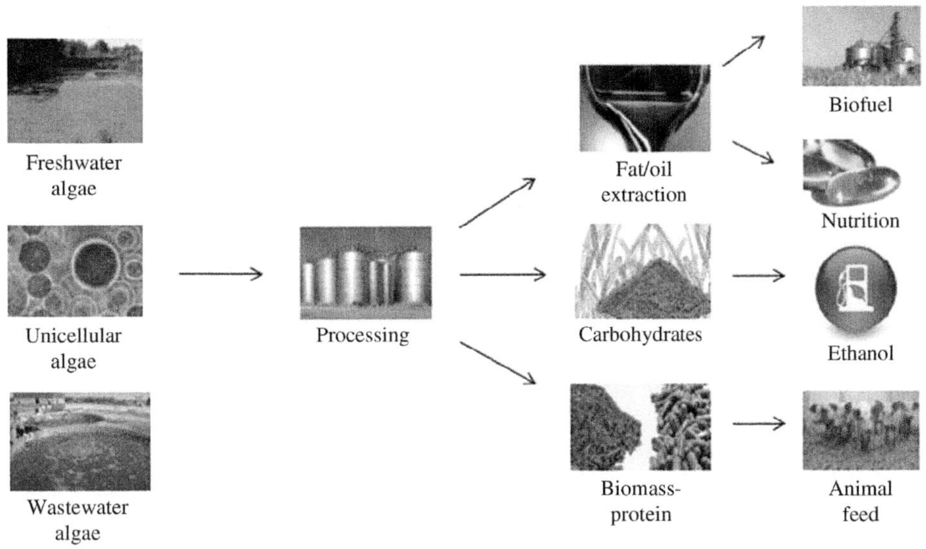

FIGURE 16.3 Potential by-products of harvested microalgae biomass. *Courtesy: Mielke, J.A., Ma, Y., Saqui-Salces, M., Urriola, P.E., Chen-Shurson, G.C., 2016. Potential use of microalgae products in swine diets. http://www.extension.umn.edu/agriculture/swine/potential-use-ofmicroalgae. Retrieved on April 8, 2016 and Jais, N.M., Mohamed, R.M.S.R., Al-Gheethi, A.A., Amir Hashim, M.K., 2017. The dual roles of phycoremediation of wet market wastewater for nutrients and heavy metals removal and microalgae biomass production. Clean Technol. Environ. Policy 19, 37–52.*

2011; Paz-Ferreiro et al., 2014). Thies and Rilliz (2009) reported on enhanced microbial activity in soil treated with algal biochar, since it provides suitable nutritive substrates for healthy microbial growth and activities. Beesley et al. (2011) suggested application of biochar for remediation of contaminated soil, while Choppala et al. (2012) reported that heavy metals mobility can be reduced and their redox state can be altered (from Cr^{+6} to Cr^{+3}) using biochar.

16.8 Future line of work

Since there is more focus on eco-developmental technologies to achieve the United Nations Sustainable Development Goals, 2030 (UNSDG-2030), in this regard algae is one of the most promising candidates as it shows a significant role in phycoremediation of contaminated soil and industrial effluents, and can produce a considerable amount of biomass, which is processed for the production of green fuels, feed for various animal species, and also used as nutritional supplements. Besides this, the atmospheric carbon sequestration potential of autotrophic algae is an additional achievement to mitigate the considerable increase in the *carbon emission* across the world (Mata et al., 2010; Rawat et al., 2011). Furthermore, genetically engineered and more advanced algal strains with changed in the physiological and morphological properties through the incorporation of *catabolic genes* into algal species would be other promising platforms in the future for scientific research and development (R&D) activities to enhance the above-said efficiencies of various algal species.

16.9 Conclusions

Algae have potential applications in phycoremediation of contaminated soil and wastewater, as well as harvested algal biomass, and show a significant potential in producing a different kind of green fuels and other by-products for their multipurpose uses. In future to achieve sustainable development goals, applications of different algal species could play a very crucial role, since most of its mechanisms are more eco-friendly, sustainable, and cost-effective in nature. However, since algae is a part of the biotic component, its overall efficiency is supposed to be operated by the various climatic phenomena and anthropogenic pollutants. The atmospheric carbon sequestration through various microalgae is considered as eco-friendly harvesting of CO_2 from the atmosphere into algal biomass and latter deposition of this carbon reaches from algae to the agricultural soil leading to significant positive changes in soil physical and microbial properties, thus improving *soil fertility* status.

Acknowledgments

The authors are thankful to Department of Environmental Science, G.B. Pant University of Agriculture and Technology (GBPUA&T), Pantnagar, Uttarakhand, India and University Grants Commission (UGC), New Delhi, India for providing the academic platform while writing this chapter.

References

Abdel Monem, M.O., Al-Zubeiry, A.H., Al-Gheethi, A.A., 2010. Biosorption of nickel by *Pseudomonas cepacia* 120S and *Bacillus subtilis* 117S. Water Sci. Technol. 61, 2994–3007.

Abdelaziz, A.E., Leite, G.B., Belhaj, M.A., Hallenbeck, P.C., 2014. Screening microalgae native to Quebec for wastewater treatment and biodiesel production. Bioresour. Technol. 157, 140–148.

Abdel-Halim, S.H., Shehata, A.M.A., El-Shahat, M.F., 2003. Removal of lead ions from industrial waste water by different types of natural materials. Water Res. 37, 1678–1683.

Abou-Shanab, R.A.I., Ji, M., Kim, H., Paeng, K., Jeon, B., 2013. Microalgal species growing on piggery wastewater as a valuable candidate for nutrient removal and biodiesel production. J. Environ. Manag. 115, 257–264.

Acuner, E., Dilek, F.B., 2004. Treatment of tectilon yellow 2G by *Chlorella vulgaris*. Proc. Biochem. 39 (5), 623–631.

Akar, T., Tunali, S., 2006. Biosorption characteristics of *Aspergillus flavus* biomass for removal of Pb (II) and Cu (II) ions from an aqueous solution. Bioresour. Technol. 97, 1780–1787.

Al-Gheethi, A.A., Norli, I., Lalung, J., Megat Azlan, A., Nur Farehah, Z.A., Kadir, M.O., 2014. Biosorption of heavy metals and cephalexin from secondary effluents by tolerant bacteria. Clean Technol. Environ. Policy 16, 137–148.

Al-Gheethi, A.A., Mohamed, R.M., Efaq, A.N., Amir, H.K., 2015. Reduction of microbial risk associated with grey water utilized for irrigation. J. Water Health 14 (3), 379–398.

Ali, H., Khan, E., Sajad, M.A., 2013. Phytoremediation of heavy metals—concepts and applications. Chemosphere 91, 869–881.

Anirudhan, T.S., Sreekumari, S.S., 2011. Adsorptive removal of heavy metal ions from industrial effluents using activated carbon derived from waste coconut buttons. J. Environ. Sci. 23, 1989–1998.

Arora, A., Saxena, S., Sharma, D.K., 2006. Tolerance and phytoaccumulation of chromium by three *Azolla* species. World J. Microbiol. Biotechnol. 22, 97–100.

Bala, J.D., Lalung, J., Al-Gheethi, A.A., Norli, I., 2016. A review on biofuel and bioresources for environmental applications. Renewable Energy and Sustainable Technologies for Building and Environmental Applications. Springer, Switzerland, pp. 205–225.

Baptista, M.S., Vasconcelos, M.T., 2006. Cyanobacteria metal interactions: requirements, toxicity, and ecological implications. Crit. Rev. Microbiol. 32, 127–137.

Barrera, H., Urena-Nunez, F., Bilyeu, B., Barrera-Diaz, C., 2006. Removal of chromium and toxic ions presents in mine drainage by ectodermis of *Opuntia*. J. Hazard. Mater. 136, 846–853.

Beesley, L., Moreno-Jimenez, E., Gomez-Eyles, J.L., Harris, E., Robinson, B., Sizmur, T., 2011. A review of biochars' potential role in the remediation, revegetation and restoration of contaminated soils. Environ. Pollut. 159, 3269–3282.

Boussiba, S., Sandbank, E., Shelef, G., Cohen, Z., Vonshak, A., Ben-Amotz, A., et al., 1988. Outdoor cultivation of the marine microalgae *Isochrysis galbana* in open reactors. Aquaculture 72, 247–253.

Brar, A., Kumar, M., Vivekanand, V., Pareek, N., 2017. Photoautotrophic microorganisms and bioremediation of industrial effluents: current status and future prospects. 3 Biotech 7, 18.

Chamarthy, S., Seo, C.W., Marshall, W.E., 2001. Adsorption of selected toxic metals by modified peanut shells. J. Chem. Technol. Biotechnol. 76, 593–597.

Chang, R.L., Ghamsari, L., Manichaikul, A., Hom, E.F., Balaji, S., Fu, W., et al., 2011. Metabolic network reconstruction of *Chlamydomonas* offers insight into light-driven algal metabolism. Mol. Syst. Biol. 7, 518–531.

Chen, J.M., Hao, O.J., 1998. Microbial chromium (VI) reduction. Crit. Rev. Environ. Sci. Technol. 28, 219–251.

Chen, C.Y., Chang, H.W., Kao, P.C., Pan, J.L., Chang, J.S., 2012. Biosorption of cadmium by CO_2-fixing microalga *Scenedesmus obliquus* CNW-N. Bioresour. Technol. 105, 74–80.

Chinnasamy, S., Bhatnagar, A., Hunt, R.W., Das, K.C., 2010. Microalgae cultivation in a wastewater dominated by carpet mill effluents for biofuel applications. Bioresour. Technol. 101, 3097–3105.

Choppala, G.K., Bolan, N.S., Megharaj, M., Chen, Z., Naidu, R., 2012. The influence of biochar and black carbon on reduction and bioavailability of chromate in soils. J. Environ. Qual. 41, 1175–1184.

Clemente, R., Walker, D.J., Pardo, T., Martinez-Fernandez, D., Bernal, M.P., 2012. The use of a halophytic plant species and organic amendments for the remediation of a trace elements contaminated soil under semi-arid conditions. J. Hazard. Mater. 15, 223–224.

Coward, T., Lee, J.G.M., Caldwell, G.S., 2014. Harvesting microalgae by CTAB-aided foam flotation increases lipid recovery and improves fatty acid methyl ester characteristics. Biomass Bioenergy 67, 354–362.

Daeshwar, N., Ayazloo, M., Khataee, A.R., Pourhassan, M., 2007. Biological decolorization of dye solution containing malachite green by microalgae *Cosmarium* sp. Bioresour. Technol. 98 (6), 1176–1182.

De Bashan, L.E., Bashan, Y., 2010. Immobilized microalgae for removing pollutants: review of practical aspects. Bioresour. Technol. 101, 1611–1627.

Domozych, D.S., Marina, C.M., Fangel, J.U., Mikkelsen, M.D., Ulvskov, P., Willats, W., 2012. The cell walls of green algae: a journey through evolution and diversity. Front. Plant Sci. 3, 82.

Dotaniya, M.L., Saha, J.K., Meena, V.D., Rajendiran, S., Coumar, M.V., Kundu, S., et al., 2014. Impact of tannery effluent irrigation on heavy metal build up in soil and ground water in Kanpur. Agrotechnology 2, 77.

Dotaniya, M.L., Meena, V.D., Rajendiran, S., Coumar, M.V., Saha, J.K., Kundu, S., et al., 2016. Geoaccumulation indices of heavy metals in soil and groundwater of Kanpur India under long term irrigation of tannery effluent. Bull. Environ. Contam. Toxicol. 98, 706–711.

Dotaniya, M.L., Rajendiran, S., Dotaniya, C.K., Solanki, P., Meena, V.D., Saha, J.K., et al., 2018. Microbial assisted phytoremediation for heavy metal contaminated soils. In: Kumar, V., Kumar, M., Prasad, R. (Eds.), Phytobiont and Ecosystem Restitution. Springer, Singapore, pp. 295–317.

Dubey, S.K., Dubey, J., Mehra, S., Tiwari, P., Bishwas, A., 2013. Potential use of cyanobacterial species in bioremediation of industrial effluents. Afr. J. Biotechnol. 10, 1125–1132.

ECOLEX, 2002. Standard for effluent discharge regulations. General Notice No. 44. http://faolex.fao.org/docs/texts/mat52519.doc. Retrieved on January 28, 2018.

EPA, 1986. Environmental (Protection) Act India.

Fagan, M.J., Saier, M.H.J., 1994. P-type ATPases of eukaryotes and bacteria: sequence comparisons and construction of phylogenetic trees. J. Mol. Evol. 38 (1), 57–99.

Faragallah, H., Askar, A., Okbah, M., Moustafa, H., 2009. Physicochemical characteristics of the open Mediterranean sea water for about 60 km from Damietta harbor. Egypt J. Ecol. Nat. Environ. 1, 106–119.

Farinella, N.V., Matos, G.D., Arruda, M.A.Z., 2007. Grape bagasse as a potential biosorbent of metals in effluent treatments. Bioresour. Technol. 98, 1940–1946.

Fath, M.J., Kolter, R., 1993. ABC-transporters: the bacterial exporters. Microbiol. Rev. 57 (4), 995–1017.

Fellet, G., Marmiroli, M., Marchiol, L., 2014. Elements uptake by metal accumulator species grown on mine tailings amended with three types of biochar. Sci. Total Environ. 468–469, 598–608.

Grierson, S., Strezov, V., Herbertson, J., Ellem, G., Mc Gregor, R., 2009. Thermal characterization of microalgae under slow pyrolysis conditions. J. Anal. Appl. Pyrol. 85, 118–123.

Gupta, V.K., Rastogi, A., 2008. Biosorption of lead from aqueous solutions by green algae Spirogyra species: kinetics and equilibrium studies. J. Hazard. Mater. 152, 407–414.

Gurbuz, F., Ciftci, H., Akcil, A., 2009. Biodegradation of cyanide containing effluents by *Scenedesmus obliquus*. J. Hazard. Mater. 162 (1), 74–79.

Hamdy, A.A., 2000. Biosorption of heavy metals by marine algae. Curr. Microbiol. 41, 232–238.

Hanumantha Rao, P., Ranjith Kumar, R., Raghavan, B., Subramanian, V., Sivasubramanian, V., 2011. Application of phycoremediation technology in the treatment of wastewater from a leather processing chemical manufacturing facility. Water SA 37 (1), 7–14.

Hultberg, M., Carlsson, A.S., Gustafsson, S., 2013. Treatment of drainage solution from hydroponic greenhouse production with microalgae. Bioresour. Technol. 136, 401–406.

Jacome Pilco, C.R., Cristiani-Urbania, E., Flores-Cotera, L.B., Velasco-Garcia, R., Ponce-Noyola, T., Canizares-Villanueva, R.O., 2009. Continuous Cr (VI) removal by *Scenedesmus incrassulatus* in an airlift photobioreactor. Bioresour. Technol. 100, 2388–2391.

Jais, N.M., Mohamed, R.M.S.R., Apandi, W.A., Matias-Peralta, H., 2015. Removal of nutrients and selected heavy metals in wet market wastewater by using microalgae *Scenedesmus* sp. Appl. Mech. Mater. 773–774, 1210–1214.

Jais, N.M., Mohamed, R.M.S.R., Al-Gheethi, A.A., Amir Hashim, M.K., 2017. The dual roles of phycoremediation of wet market wastewater for nutrients and heavy metals removal and microalgae biomass production. Clean Technol. Environ. Policy 19, 37–52.

Jaishankar, M., Tseten, T., Anbalagan, N., Mathew, B.B., Beeregowda, K.N., 2014. Toxicity, mechanism and health effects of some heavy metals. Interdiscip. Toxicol. 7 (2), 60–72.

Jeyakumar, R.S., Chandrasekaran, V., 2014. Adsorption of lead (II) ions by activated carbons prepared from marine green algae: equilibrium and kinetics studies. Int. J. Ind. Chem. 5, 1–9.

Jin, X., Liu, G., Xu, Z., Tao, W., 2007. Decolourisation of a dye industry effluent by *Aspergillus fumigatus* XC6. Appl. Microbiol. Biotechnol. 74, 239–243.

Kabata-Pendias, A., Pendias, H., 2001. Trace Elements in Soil and Plants, 3rd edition CRC Press, p. 403.

Kaewpintong, K., Shotipruk, A., Powtongsook, S., Pavasant, P., 2007. Photoautotrophic high-density cultivation of vegetative cells of *Haematococcus pluvialis* in airlift bioreactor. Bioresour. Technol. 98, 288–295.

Khanna, N., Solanki, P., 2014. Role of agriculture in the global economy. Agrotechnology 2 (4), 221.

Khatoon, H., Solanki, P., Narayan, M., Tewari, L., Rai, J.P.N., 2017. Role of microbes in organic carbon decomposition and maintenance of soil ecosystem. Int. J. Chem. Stud. 5 (6), 1648–1656.

Kim, M.K., Chang, M.U., Acreman, J., 2007. Enhanced production of *Scenedesmus* spp. (green microalgae) using a new medium containing fermented swine wastewater. Bioresour. Technol. 98, 2220–2228.

Kothari, R., Pathak, V.V., Kumar, V., Singh, D.P., 2012. Experimental study for growth potential of unicellular alga *Chlorella pyrenoidosa* on dairy wastewater: an integrated approach for treatment and biofuel production. Bioresour. Technol. 116, 466–470.

Kotteswari, M., Murugesan, S., Ranjith Kumar, R., 2012. Phycoremediation of dairy effluent by using the microalgae *Nostoc* sp. Int. J. Environ. Res. Dev. 2, 35–43.

Kumar, A., Kaushal, S., Saraf, S.A., Singh, J.S., 2018. Microbial biofuels: a solution to carbon emissions and energy crisis. Front. Biosci. (Landmark Ed.) 1 (23), 1789–1802.

Lapedes, D.N., 1974. Dictionary of Scientific and Technical Terms. McGraw Hill, New York, p. 674.

Lee, Y.K., Low, C.S., 1991. Effects of photobioreactor inclination on the biomass productivity of an outdoor algal culture. Biotechnol. Bioeng. 38, 995–1000.

Lee, J.S., Kim, D.K., Lee, J.P., Park, S.C., Koh, J.H., Cho, H.S., 2002. Effects of SO_2 and NO on growth of *Chlorella* sp. KR-1. Bioresour. Technol. 82, 1–4.

Lehmann, J., Joseph, S., 2009. Biochar for Environmental Management: Science and Technology. Earthscan, London and Sterling, VA, USA.

Lehmann, J., Rillig, M.C., Thies, J., Masiello, C.A., Hockaday, W.C., Crowley, D., 2011. Biochar effects on soil biota a review. Soil Biol. Biochem. 43, 1812–1836.

Lesage, E., Mundia, C., Rousseau, D.P.L., Van de Moortel, A.M.K., Du Laing, G., Meers, E., et al., 2007. Sorption of Co, Cu, Ni and Zn from industrial effluents by the submerged aquatic macrophyte *Myriophyllum spicatum* L. Ecol. Eng. 30, 320–325.

Lim, S.L., Chu, W.L., Phang, S.M., 2010. Use of *Chlorella vulgaris* for bioremediation of textile wastewater. Bioresour. Technol. 101, 7314–7322.

Liu, G., Xu, X., Zhu, L., Xing, S., Chen, J., 2013. Biological nutrient removal in a continuous anaerobic–aerobic–anoxic process treating synthetic domestic wastewater. Chem. Eng. J. 225, 223–229.

Maeda, K., Owada, M., Kirmura, N., Omata, K., Karube, I., 1995. CO_2 fixation from the flue gas on coal fired thermal power plant by microalgae. Energy Convers. Manag. 36, 717–720.

Mata, T.M., Martins, A.A., Caetano, N.S., 2010. Microalgae for biodiesel production and other applications: a review. Renew. Sust. Energy Rev. 14, 217–232.

Mata, T.M., Melo, A.C., Simoes, M., Caetano, N.S., 2012. Parametric study of a brewery effluent treatment by microalgae *Scenedesmus obliquus*. Bioresour. Technol. 107, 151–158.

Mcelwee, K., Baker, J., Clair, D., 2006. Pond Fertilization: Ecological Approach and Practical Application. Oregon State University, Aquaculture Collaborative Research Support Program.

Mielke, J.A., Ma, Y., Saqui-Salces, M., Urriola, P.E., Chen-Shurson, G.C., 2016. Potential use of microalgae products in swine diets. http://www.extension.umn.edu/agriculture/swine/potential-use-ofmicroalgae. Retrieved on April 8, 2016.

Migahed, F., Abdelrazak, A., Fawzy, G., 2017. Batch and continuous removal of heavy metals from industrial effluents using microbial consortia. Int. J. Environ. Sci. Technol. 14, 1169–1180.

Milner, H.W., 1953. Rocking tray. In: Burlew, J.S. (Ed.), Algal Cultures From Laboratory to Pilot Plant. Carnegie Institution, Washington, DC, p. 108.

Mohan, S.V., Rao, C.N., Prasad, K.K., Karthikeyan, J., 2002. Treatment of simulated reactive yellow 22 azo dye effluents using *Spirogyra* species. Waste Manag. 22 (6), 575–582.

Mulbry, W., Kondrad, S., Pizarro, C., Kebede-Westhead, E., 2008. Treatment of dairy manure effluent using freshwater algae: algal productivity and recovery of manure nutrients using pilot-scale algal turf scrubbers. Bioresour. Technol. 99, 8137–8142.

Munoz, R., Guieyssea, B., 2006. Algal-bacterial processes for the treatment of hazardous contaminants: a review. Water Res. 40, 2799–2815.

Narayan, M., Solanki, P., Srivastava, R.K., 2018. Treatment of sewage (domestic wastewater or municipal wastewater) and electricity production by integrating constructed wetland with microbial fuel cell. Sewage. Intech Open. Available from: https://cdn.intechopen.com/pdfs/60245.pdf.

Narayana, V.V.D., Babu, R., 1993. Estimation of soil erosion in India. In: Narayan, V.V.D. (Ed.), Soil and Water Conservation Research in India. ICAR Publication, New Delhi.

Negoro, M., Shioji, N., Miyamoto, K., Yoshiharu, M., 1991. Growth of microalgae in high CO_2 gas and effects of SOx and NOx. Appl. Biochem. Biotechnol. 28–29, 877–886.

Nies, D.H., Koch, S., Wachi, S., Peitzsch, N., Saier, M.H.J., 1998. CHR, a novel family of prokaryotic proton motive force-driven transporters probably containing chromate/sulfate transporters. J. Bacteriol. 180, 5799–5802.

Ogbonna, J.C., Soejima, T., Tanaka, H., 1999. An integrated solar and artificial light system for internal illumination of photobioreactors. J. Biotechnol. 70, 289–297.

Olguin, E.J., Sanchez-Galvan, G., 2012. Heavy metal removal in phytofiltration and phycoremediation: the need to differentiate between bioadsorption and bioaccumulation. New Biotechnol. 30, 3–8.

Ono, E., Cuello, J.L., 2007. Carbon dioxide mitigation using thermophilic cyanobacteria. Biosyst. Bioeng. 96, 129–134.

Ota, M., Kato, Y., Watanabe, H., Watanabe, M., Sato, Y., Smith, R.L., 2009. Fatty acid production from a highly CO_2 tolerant alga *Chlorococcum littorale* in the presence of inorganic carbon and nitrate. Bioresour. Technol. 100, 5237–5248.

Pacheco, M.M., Hoeltz, M., Moraes, M.S., Schneider, R.C., 2015. Microalgae: cultivation techniques and wastewater phycoremediation. J. Environ. Sci. Health Tox. Hazard. Subst. Environ. Eng. 50, 585–601.

Pangestuti, R., Kim, S.K., 2011. Biological activities and health benefit effects of natural pigments derived from marine algae. J. Funct. Foods 3, 255–266.

Parjo, K., Razak, R.A., 2015. Phycoremediation of wastewaters and potential hydrocarbon from microalgae: a review. Adv. Environ. Biol. 9, 1–8.

Park, D., Yun, Y.S., Jo, J.H., Park, J.M., 2006. Biosorption process for treatment of electroplating wastewater containing Cr (VI): laboratory-scale feasibility test. Ind. Eng. Chem. Res. 45, 5059–5065.

Park, J., Craggs, R., Shilton, A., 2011. Wastewater treatment high rate algal ponds for biofuel production. Bioresour. Technol. 102, 35–42.

Pathak, V.V., Singh, D.P., Kothari, R., Chopra, A.K., 2014. Phycoremediation of textile wastewater by unicellular microalga *Chlorella pyrenoidosa*. Cell. Mol. Biol. 60, 35–40.

Paz-Ferreiro, P., Lu, H., Fu, S., Mendez, A., Gasco, G., 2014. Use of phytoremediation and biochar to remediate heavy metal polluted soils: a review. Solid Earth 5, 65–75.

Pena-Castro, J.M., Martinez-Jeronimo, F., Esparza-Garcia, F., Canizares-Villanueva, R.O., 2004. Heavy metals removal by the microalga *Scenedesmus incrassatulus* in continuous cultures. Bioresour. Technol. 94 (2), 219–222.

Pittman, J.K., Dean, A.P., Osundeko, O., 2011. The potential of sustainable algal biofuel production using wastewater resources. Bioresour. Technol. 102, 17–25.

Prajapati, S.K., Kaushik, P., Malik, A., Vijay, V.K., 2013. Phycoremediation and biogas potential of native algal isolates from soil and wastewater. Bioresour. Technol. 135, 232–238.

Prakash, B.S., Kumar, S.V., 2013. Batch removal of heavy metals by biosorption onto marine algae-equilibrium and kinetic studies. Int. J. Chem. Technol. Res. 5, 1254–1262.

Prasad, M.N.V., Freitas, H., 2000. Removal of toxic metals from solution by leaf, stem and root phytomass of *Quercus ilex* L. (holly oak). Environ. Pollut. 110, 277–283.

Qari, H.A., Hassan, I.A., 2014. Removal of pollutants from waste water using *Dunaliella* algae. Biomed. Pharmacol. J. 7, 147–151.

Radway, J.C., Wilde, E.W., Whitaker, M.J., Weissman, J.C., 2001. Screening of algal strains for metal removal capabilities. J. Appl. Phycol. 13, 451–455.

Rahman, A., Ellis, J.T., Miller, C.D., 2012. Bioremediation of domestic wastewater and production of bioproducts from microalgae using waste stabilization ponds. J. Bioremed. Biodegrad. 3 (6), 6199.

Ratna, D., Padhi, B.S., 2012. Pollution due to synthetic dyes toxicity and carcinogenicity studies and remediation. Int. J. Environ. Sci. 3, 941–955.

Rawat, I., Ranjith Kumar, R., Mutanda, T., Bux, F., 2011. Dual role of microalgae: phycoremediation of domestic wastewater and biomass production for sustainable biofuels production. Appl. Energy 88, 411–3424.

Regine, H.S., Vieira, F., Volesky, B., 2000. Biosorption: a solution to pollution. Int. Microbiol. 3, 17–24.

Robinson, T., McMullan, G., Marchant, R., Nigam, P., 2001. Remediation of dyes in textile effluent: a critical review on current treatment technologies with a proposed alternative. Bioresour. Technol. 77, 247–255.

Romera, E., Gonzalez, F., Ballester, A., Blazquez, M.L., Munoz, J.A., 2007. Comparative study of biosorption of heavy metals using different types of algae. Bioresour. Technol. 98, 3344–3353.

Ross, A., Jones, J.M., Kubacki, M.L., Bridgeman, T.G., 2008. Classification of macroalgae as fuel and its thermochemical behavior. Bioresour. Technol. 99, 6494–6504.

Saier, M.H.J., 1994. Computer-aided analyses of transport protein sequences: gleaning evidence concerning function, structure, biogenesis, and evolution. Microbiol. Rev. 58, 71–93.

Saier, M.H., Tam, R., Reizer, A., Reizer, J., 1994. Two novel families of bacterial membrane proteins concerned with nodulation, cell division and transport. Mol. Microbiol. 11 (8), 41–847.

Sakai, N., Sakamoto, Y., Kishimoto, N., Chihara, M., Korube, I., 1995. *Chlorella* strains from hot springs tolerant to high-temperature and high CO_2. Energy Convers. Manag. 36, 693–696.

Sehgal, J., Abrol, I.P., 1994. Soil Degradation in India: Status and Impact. Oxford and IBH, New Delhi, 80 pp.

Shaikh, P.R., Bhosle, A.B., 2011. Bioaccumulation of chromium by aquatic macrophytes *Hydrilla* sp. and *Chara* sp. Adv. Appl. Sci. Res. 2, 214–220.

Shivlata, L., Satyanarayana, T., 2015. Thermophilic and alkaliphilic actinobacteria: biology and potential applications. Front. Microbiol. 6, 1014.

Silambarasan, T., Vikramathithan, M., Dhandapani, R., Mukesh, D.J., Kalaichelvan, P.T., 2012. Biological treatment of dairy effluent by microalgae. World J. Sci. Technol. 2 (7), 132–134.

Singh, J.S., 2013. Plant growth promoting rhizobacteria: potential microbes for sustainable agriculture. Resonance 18 (3), 275–281.

Singh, J.S., 2015. Microbes: the chief ecological engineers in reinstating equilibrium in degraded ecosystems. Agric. Ecosyst. Environ. 203, 80–82.

Singh, J.S., Gupta, V.K., 2016. Degraded land restoration in reinstating CH_4 sink. Front. Microbiol. 7 (923), 1–5.

Singh, J.S., Kumar, A., Rai, A.N., Singh, D.P., 2016. Cyanobacteria: a precious bioresource in agriculture, ecosystem and environmental sustainability. Front. Microbiol. 7 (529), 1–19.

Singh, J.S., Gupta, V.K., 2018. Soil microbial biomass: a key soil driver in management of ecosystem functioning. Sci. Total Environ. 634, 497–500.

Singh, C., Tiwari, S., Gupta, V.K., Singh, J.S., 2018. The effect of rice husk biochar on soil nutrient status, microbial biomass and paddy productivity of nutrient poor agriculture soils. Catena 171, 485–493.

Solanki, P., 2014. Effect of Sewage Sludge on Marigold and Golden rod. Thesis Acharya N. G. Ranga Agricultural University, Rajendranagar, Hyderabad. pp. 172.

Solanki, P., Debnath, P., 2014. Role of biosolids in sustainable development. Agrotechnology 2 (4), 220.

Solanki, P., Meena, S.S., Narayan, M., Khatoon, H., Tewari, L., 2017a. Denitrification process as an indicator of soil health. Int. J. Curr. Microbiol. Appl. Sci. 6, 2645–2657.

Solanki, P., Sharma, S.H.K., Akula, B., 2017b. Sewage sludge and its impact on soil property. Environ. Ecol. 35 (4C), 3186–3195.

Solanki, P., Narayan, M., Meena, S.S., Srivastava, R.K., 2017c. Floating raft wastewater treatment system: a review. Int. J. Pure Appl. Microbiol. 11, 1113–1116.

Solanki, P., Narayan, M., Srivastava, R.K., 2017d. Effectiveness of domestic wastewater treatment using floating rafts a promising phyto-remedial approach: a review. J. Appl. Nat. Sci. 9 (4), 1931–1942.

Solanki, P., Narayan, M., Rabha, A.K., Srivastava, R.K., 2018b. Assessment of cadmium scavenging potential of *Canna indica* L. Bull. Environ. Contam. Toxicol. 101 (4), 446–450.

Solanki, P., Narayan, M., Meena, S.S., Srivastava, R.K., Dotaniya, M.L., Dotaniya, C.K., 2018c. Phytobionts of wastewater and restitution. In: Kumar, V., Kumar, M., Prasad, R. (Eds.), Phytobiont and Ecosystem Restitution. Springer, Singapore.

Srivastava, S., Agrawal, S.B., Mondal, M.K., 2015. A review on progress of heavy metal removal using adsorbents of microbial and plant origin. Environ. Sci. Pollut. Res. 22, 15386–15415.

Steinbuchel, A., 2001. Perspectives for biotechnological production and utilization of biopolymers: metabolic engineering of polyhydroxyalkanoate biosynthesis pathways as a successful example. Macromol. Biosci. 1, 1–24.

Sung, K.D., Lee, J.S., Shin, C.S., Park, S.C., 1999. Isolation of a new highly CO_2 tolerant freshwater microalgae *Chlorella* sp KR-1. Renew. Energy 16, 1019–1022.

Tarley, C.R.T., Ferreira, S.L.C., Arruda, M.A.Z., 2004. Use of modified rice husks as a natural solid adsorbent of trace metals: characterization and development of an on-line preconcentration system for cadmium and lead determination by FAAS. Microchem. J. 77, 163–175.

Thies, J.E., Rilliz, M.C., 2009. Characteristics of biochar: biological properties. In: Lehmann, S., Joseph, J. (Eds.), Biochar for Environmental Management. Earthscan, Virginia, pp. 85–106.

Tiwari, P., Singh, J.S., 2017. A plant growth promoting rhizospheric *Pseudomonas aeruginosa* strain inhibits seed germination in *Triticum aestivum* (L) and *Zea mays* (L). Microbiol. Res. 8, 1–7.

Tiwari, S., Singh, C., Singh, J.S., 2018. Land use changes: a key ecological driver regulating methanotrophs abundance in upland soils. Energy Ecol. Environ. 3, 355–371.

Tredici, M.R., Materassi, R., 1992. From open ponds to vertical alveolar ponds: the Italian experience in the development of reactors for the mass cultivation of photoautotrophic microorganisms. J. Appl. Phycol. 4, 221–231.

Ugwu, C.U., Ogbonna, J.C., Tanaka, H., 2002. Improvement of mass transfer characteristics and productivities of inclined tubular photobioreactors by installation of internal static mixers. Appl. Microbiol. Biotechnol. 58, 600–607.

Vajpayee, P., Rai, U.N., Sinha, S., Tripathi, R.D., Chandra, P., 1995. Bioremediation of tannery effluent by aquatic macrophytes. Bull. Environ. Contam. Toxicol. 55, 546–553.

Vajpayee, P., Rai, U.N., Ali, M.B., Tripathi, R.D., Yadav, V., Sinha, S., et al., 2001. Chromium-induced physiologic changes in *Vallisneria spiralis* L. and its role in phytoremediation of tannery effluent. Bull. Environ. Contam. Toxicol. 67, 246–256.

Valls, M., Lorenzo, V., 2002. Exploiting the genetic and biochemical capacities of bacteria for the remediation of heavy metal pollution. FEMS Microbiol. Rev. 26, 327–328.

Vimal, S.R., Patel, V.K., Singh, J.S., 2018. Plant growth promoting *Curtobacterium albidum* strain SRV4: an agriculturally important microbe to alleviate salinity stress in paddy plants. Ecol. Indic. (in press).

Wang, J., Chen, C., 2009. Biosorbents for heavy metals removal and their future. Biotechnol. Adv. 27, 195–226.

Yaakob, Z., Fakir, Z.Y., Ali, E., Abdullah, S.R.S., Takriff, M.S., 2011. An overview of microalgae as a wastewater treatment. In: Jordan International Energy Conference, Amman.

Yang, D., Zeng, D.H., Li, L.J., Mao, R., 2012. Chemical and microbial properties in contaminated soils around a magnesite mine in Northeast China. Land Degrad. Dev. 23, 256–262.

Yewalkar, S.N., Dhumal, K.N., Sainis, J.K., 2007. Chromium (VI)-reducing *Chlorella* sp. isolated from disposal sites of paper-pulp and electroplating industry. J. Appl. Phycol. 19, 459–465.

Yue, L.H., Chen, W.G., 2005. Isolation and determination of cultural characterization of a new highly CO_2 tolerant freshwater microalgae. Energy Conserv. Manag. 46, 1868–1876.

Zinicovscaia, I., Cepoi, L., 2016. Cyanobacteria for Bioremediation of Wastewaters. Springer, Berlin.

Zollinger, H., 1987. Colour Chemistry—Synthesis, Properties and Applications of Organic Dyes and Pigments. VCH, New York, pp. 92–102.

Zulkifli, A.R., Roshadah, H., Tunku Khalkausar, T.F., 2012. Control of water pollution from nonindustrial premises. In: A Conference Bayview Hotel, Langkawi, Kedah, November 5, 2012.

Further reading

Dotaniya, M.L., Rajendiran, S., Meena, V.D., Saha, J.K., Coumar, M.V., Kundu, S., et al., 2017. Influence of chromium contamination on carbon mineralization and enzymatic activities in Vertisol. Agric. Res. 6, 91–96.

Solanki, P., Rabha, A.K., Narayan, M., Srivastava, R.K., 2018a. Relative comparison for phytoremediation potential of Canna and Pistia for wastewater recycling. Environ. Ecol. 36A, 316–320.

Chapter 17

Microbial services in agro-environmental management

Antra Chatterjee[1], Madan Kumar Mandal[2] and Neha Chaurasia[2]

[1]*Center of Advanced Study in Botany, Institute of Science, Banaras Hindu University, Varanasi, India,* [2]*Department of Biotechnology and Bioinformatics, North Eastern Hill University, Shillong, India*

Chapter Outline

17.1 Introduction

Agricultural land systems cover about 40%–50% of the Earth's land surface (Singh et al., 2018) on which inhabitants depend for secure crop production. An analysis of global population growth rate says that the global population is expected to rise from 7.2 to 9.6 billion by 2050 (UN-DESA, 2015). Thus it will increase demand for food and fodder necessities as well as feedstock. In 2015, 17 sustainable development goals were adopted by the UN to eradicate hunger and extreme poverty by 2030, while at the same time protecting the environment and global climate (Smith et al., 2007; UN, 2013). More sustainable crop production could be achieved by improving plant fertilization strategies and management.

Fertilizer plays a vital role in maintaining soil fertility, enhancing crop yield, and improving harvest quality (Kumar et al., 2018; Tiwari et al., 2018). Fertilizer includes inorganic and organic (biofertilizer) fertilizer that can supply essential nutrients to plants for their growth and development. Inorganic fertilizers contain one or more chemical elements that provide essential nutrients to crop plant for maintaining growth and development (Martínez-Alcántara et al., 2016). It includes single nutrient fertilizer like nitrogen fertilizer (N), phosphate fertilizer (P), potash fertilizer (K), and multinutrient fertilizer such as NK, NP, NK, and NPK. While biofertilizers include crop waste, animal manure, and compost as well as microbes. Application of biofertilizer in soil increases soil organic matter, which provides a steady release of nutrients to the crop plants and improves soil chemical and physiological properties (Barakat et al., 2012). It also increases nutrient retention and water holding capacity of soil, promoting aggregation and improvement in soil structure, moisture, and temperature (Lal, 2004). Water holding capacity of soil is one of the most imperative soil factors for plant growth and development, controlling carbon allocation, nutrient cycle, and the rate of photosynthesis. Studies

New and Future Developments in Microbial Biotechnology and Bioengineering. DOI: https://doi.org/10.1016/B978-0-12-818258-1.00017-0

have shown that soil water holding capacity influences crop yield (Yang et al., 2014; Williams et al., 2016). These factors are also suggested as a means to buffer yields against future variable environmental conditions (Singh, 2015a,b,c,d; Vimal et al., 2016). Ameliorated soil physical conditions increase root growth, which facilitates higher nutrient uptake. Moreover, the amendment of biofertilizer has attracted more concern not only because of its impact on soil chemical and physical properties but also due to its role in mitigation of methane emission and carbon sequestration (Favoino and Hogg, 2008; Baldi et al., 2010).

Biochar is the charcoal produced by the thermal decomposition of biomass of plant, chopstick, agricultural waste, etc. under a limited oxygen supply (Lehmann, 2007b). Amendment of biochar in soil improves soil fertility by enhancing both nutrient retention and water holding capacity of soil (Lehmann, 2007a,b). Muhammad et al. (2017) reported that the addition of biochar to the paddy field significantly enhanced the soil pH from 4.2 to 6.2, total nitrogen by 135% organic carbon by 90% and accumulation of Mn, Mo, Na, and Zn compared with control soil, which leads to increase in rice crop yield. Biochar has also a great capability of carbon sequestering in soil and significantly reduces CH_4 emissions in paddy field (Liu et al., 2011; Sohi, 2012). Biochar in paddy fields significantly promotes the growth of paddy plants, which leads to the development of aerobic environment in the vicinity of the rhizosphere. As methanogenic bacteria are generally anaerobic organisms, the aerobic environment inhibits methanogenic activity, thus reducing CH_4 production in paddy fields (Watanabe et al., 1997; Dong et al., 2013).

Various studies have suggested that microbes play a vital role in maintaining soil quality and promoting uptake of nutrients by plant (Kuritz, 1999; Han et al., 2006; Wani et al., 2007; Singh, 2009; Akoijam et al., 2015; Cuellar-Bermudez et al., 2017). The rhizospheric microbes are known to fix atmospheric nitrogen and make it available for the plant and accumulate various essential metals for plant growth such as Fe, Zn, Mn, and Cu (Whiting et al., 2001; Comte et al., 2008; Braud et al., 2009a; Chu et al., 2010; Hider and Kong, 2010). Rhizospheric microbes improve the rate of metal uptake in plants by various mechanisms such as production of extracellular polymeric substances (EPS) and siderophores along with rhizosphere acidification, metal oxidation, and reduction. Among various mechanisms, EPS play a vital role in biosorption of heavy metals. Exopolymers that are secreted by rhizospheric bacteria are able to bind strongly with heavy metals and form organometallic complexes. Joshi and Juwarkar (2009) reported that the EPS produced by *Azotobacter* sp. was capable of binding with CrO_4^{2+} and Cd^{2+}. Comte et al. (2008) demonstrated that the rhizospheric pH also effect on the heavy metal adsorption process by EPS. At pH 7–8, affinity of EPS toward heavy metal ions are Pb > Cu > Cd at pH 6 and Cu > Pb > Cd. It was demonstrated that siderophores secreted by *Pseudomonas aeruginosa* were capable of chelating Ag^{+}, Al^{3+}, Cd^{2+}, Co^{2+}, Cr^{2+}, Cu^{2+}, Eu^{3+}, Ga^{3+}, Hg^{2+}, Mn^{2+}, Ni^{2+}, Sn^{2+}, Tb^{3+}, and Zn^{2+} (Braud et al., 2009a). Hence, it enhances their bioavailability for plant in the vicinity of the rhizosphere.

Among the microalgae, cyanobacteria are distinctive as they have the ability to carry out photosynthesis and nitrogen fixation, and grow in extreme environments including wastewater and polluted habitats (Kuritz, 1999; Singh, 2009; Akoijam et al., 2015; Cuellar-Bermudez et al., 2017). Nitrogen fixing cyanobacteria are ideal candidates to produce eco-friendly biofertilizer as they can fix atmospheric nitrogen and make it available for the plant. Cyanobacteria also have ability to degrade pesticides, which has attracted further research to evaluate their pesticide degrading ability. Various articles have reported the ability of cyanobacteria to biodegrade the pesticides includes methyl-parathion, monosulfuron-ester, lindane, glyphosate, etc. (Ibrahim et al., 2014; Shen et al., 2017; Tiwari et al., 2018). El-Bestawy et al. (2007) demonstrated the potential of lindane degradation by *Oscillatoria* sp. R12. *Anabaena sphaerica* also showed a higher growth at $5\,\mu g\,mL^{-1}$ malathion (organophosphorus pesticide) concentration in medium without any adverse effect on physiological and biochemical indices compared with standard medium (Chakraborty et al., 2017). Anees et al. (2014) also showed the ability of *Nostoc muscorum* to degrade butachlor. Hence, microbes that have plant growth promoting and pesticide degrading ability would be a sustainable solution for the problems related to high crop production and pesticide degradation. This chapter illustrates the importance of microbial biofertilizer utilization to increase crop production and degradation of pesticides.

17.2 Role of phosphate solubilizing bacteria in sustainable agricultural practices

Among the essential macronutrients necessary for plant growth and development phosphorus (P) finds its place next to nitrogen (N). Phosphorus is present in micromolar quantities as compared with other elements, which are present in millimolar quantities (Ozanne, 1980). Inorganic phosphorus (Pi) usually accounts for 35%–70% and organic phosphorus (Po) accounts for just 15%–80% of the total phosphorus in soils. Inorganic P, which accounts for 35%–70% of the total phosphorus content of soil, mainly includes Ca–P, Fe–P, and Al–P whereas that of organic P in soil is synthesized by microorganisms and plants in the form of inositol phosphates (soil phytate), phosphonates, and active forms as orthophosphate diesters (phospholipids and nucleic acid), labile orthophosphate monoesters, and organic polyphosphates (Condron et al., 2005). Plants can uptake phosphate as orthophosphate (Ranjan et al., 2013). Low availability of phosphorus results in browning of leaves, smaller leaf development, weaker stems, and hence lower yield (Mahantesh and Patil, 2011). Applications of fertilizers with high phosphorus content are almost becoming mandatory for plant growth and development due to the fact that phosphorus is unavailable to the plants due to its fixation with the other elements in soil. Although, the imprudent application of phosphatic fertilizer resulted in enhancement of crop yield but with an additional adverse impact on the ecosystem including disposal of gypsum, emission of highly poisonous HF, and accumulation of heavy metal such as Cd (Sharma et al., 2013). In agricultural soils concentration of phosphates is enough to sustain crop yields worldwide for about 100 years (Goldstein, 1996). In the present scenario advanced biotechnological incorporation in understanding the mechanisms responsible for solubilization together with improvement of crop production should go in hand to hand with managing a sustainable environment system. Sustainable agriculture employing phosphate solubilizing microorganisms (PSMs) is an alternative source of phosphate to the plants hence a better way of management of nutritional security together with the environment health. Organisms such as bacteria, fungi, algae, actinomycetes, and mycorrhizae, having properties to solubilize soil phosphate via production of organic acids and enzymes, come under PSMs. Application of PSMs includes availability of soluble phosphate, enhancement in biological nitrogen fixation, increasing the availability of trace elements such as iron and zinc, and by production of plant growth promoting regulators (Kucey et al., 1989; Ponmurugan and Gopi, 2006). The PSM registered their presence in rhizosphere, although their numbers vary from soil to soil. Division of the total population of PSM depicts 1%–50% of bacteria and 0.5%–0.1% of fungi (Kucey, 1983). Important genera in the bacterial kingdom include *Bacillus* and *Pseudomonas* whereas that of fungi include *Aspergillus* and *Penicillium* (Illmer and Schinner, 1992). Although endophytic actinobacteria *Streptomyces* species also occur in phosphate mobilizing sites in abundance. They possess additional antimicrobial activities hence also act as plant growth promoting microbial agents. Other than bioavailability of phosphates to plants and increasing their yield, PSMs may also be useful in the bioremediation of soil polluted with heavy metals (Ahemad, 2015; Monica and Harshada, 2015) or for bioleaching of rare elements for mined ores (Singh, 2015a). Stress tolerant phosphate solubilizing bacteria are found to be present in maximum in rhizoplane region followed by the rhizosphere and root-free soil in alkaline soils (Johri et al., 1999). Because of their marked presence under such stressed conditions, PSM strains can be used as an excellent model for understanding the physiological, biochemical, and molecular mechanisms of phosphate solubilization under stressful environment (Khan et al., 2013). In the rhizospheric zone, the population of highly metabolically active PSMs is more pronounced as compared with nonrhizospheric zones (Vazquez et al., 2000). Another functionality of PSM that can be utilized is bioremediation of soil polluted with heavy metals (Ahemad, 2015; Monica and Harshada, 2015) as well as for bioleaching of rare elements for mined ores (Shin et al., 2015). Formation of a clear halo zone can be seen around the colonies formed by phosphate solubilizing bacterial and fungal strains on an agar plate media having inorganic phosphate as a single source of phosphorus (Vimal et al., 2017; Singh and Seneviratne, 2017a,b; Tiwari and Singh, 2017; Vimal et al., 2018). For the detection of phosphate solubilization bacteria are inoculated on agar plate containing Pikovskaya media (Pikovskaya, 1948). Phosphate solubilizing bacteria generally lose their phosphate solubilizing activity upon repeated subculturing; although phosphate solubilizing fungi do not lose such activity (Kucey, 1983). The PSMs include both aerobic and anaerobic strains of bacteria; however, greater number of aerobic strains are present in submerged soils. The bacterial species have potential to solubilize tricalcium phosphate, dicalcium phosphate, hydroxyapatite, and rock phosphate (Rodríguez and Fraga, 1999). Phosphate solubilizing bacterial genera include *Pseudomonas* sp., *Bacillus* sp., *Burkholderia* sp., *Rhizobium* sp., *Achromobacter* sp., *Agrobacterium* sp., *Aerobacter* sp., *Micrococcus* sp., *Flavobacterium* sp., *Erwinia* sp., *Rhodococcus* sp., *Arthrobacter* sp., *Serratia* sp., *Enterobacter* sp., *Chryseobacterium* sp., *Gordonia* sp., *Phyllobacterium* sp., *Delftia* sp., *Azotobacter* sp., *Xanthomonas* sp., *Klebsiella* sp., *Pantoea* sp., *Vibrio proteolyticus*, and *Xanthobacter agilis* (Kumar et al., 2001). Nitrogen fixing bacteria *Rhizobium leguminosarum* bv. *trifolii* (Abril et al., 2007) and *Rhizobium* sp. (Sridevi et al., 2007) showed inorganic and organic phosphate mobilizing activity. The PSM among fungus showed greater solubilizing property as compared with bacteria because of their ability to produce acids in mass (Venkateswarlu et al., 1984). Yeasts that can solubilize inorganic phosphate include *Torula thermophila*, *Saccharomyces cerevisiae*, *Rhodotorula minuta*, *Yarrowia lipolytica*, *Schizosaccharomyces pombe*, and *Pichia fermentans* (Varsha-Narsian et al., 1994; Vassilev et al., 2001). Arbuscular mycorrhizal fungi have also been reported to increase the uptake of phosphate (Duponnois et al., 2005). The interactive application of inoculation of different PSMs together is an area of budding research that needs proper investigation to be done under different agroclimatic conditions in agricultural fields (Widada et al., 2007).

17.3 Role of nitrogen fixing cyanobacteria in pesticide degradation

The ample use of pesticides in crop fields and household systems has polluted different environments all over the world. The high population growth across the world increases the demand for large-scale food production. The high crop production needs suitable application of fertilizers and pesticides. Pesticides include various types of chemical products with particular functions or targets (Table 17.1). The use of pesticides has as its main positive aspects enhanced crop production and reduction of vector-borne diseases (Damalas, 2009; Agrawal et al., 2010). But, only 0.1% of total pesticides are used in the field to target pests and the residual 99.9% is distributed throughout the environment, which impinges on the nontarget organisms such as green algae, cyanobacteria, fishes, reptiles, birds, and mammals (Kumar et al., 2011; Bhardwaj and Sharma, 2013; Ravoet et al., 2015). Thus degradation of these residual pesticides from the environment is needed to shield nontarget organisms from exposure to pesticides.

Although a variety of physical and chemical processes have been established to remove excess pesticides from the environment, the use of biological organisms has attracted significant consideration. In this perspective, autotrophic macrophytes, green algae, and cyanobacteria are strong candidates for the bioremediation of pesticide-contaminated environments (El-Bestawy et al., 2007). Among them, cyanobacteria are very unique as they have the ability to carry out photosynthesis and nitrogen fixation, and grow in extreme environments including wastewater and polluted habitats (Kuritz, 1999; Singh, 2009; Akoijam et al., 2015; Cuellar-Bermudez et al., 2017). However, the role of cyanobacteria in biodegradation has been stimulated after the isolation of filamentous cyanobacteria like *Microcoleus chthonoplastes* and *Phormidium corium* from mats formed around the massive oil spill caused by the 1991 war in the Persian Gulf (Kuritz 1999). Various articles have reported the ability of cyanobacteria to biodegrade the naturally occurring aromatic compounds including naphthalene, phenanthrene, phenol, and catechol (Cerniglia et al., 1980; Narro et al., 1992; Shashirekha et al., 1997). El-Bestawy et al. (2007) investigated the potential of lindane degradation by *Oscillatoria* sp. R12. *Oscillatoria* sp. R12 showed 12.62-fold and 8.71-fold higher growths compared with control at 5 and 10 ppm lindane concentration in medium, respectively. The *A. sphaerica* also showed higher growth at 5 $\mu g\ mL^{-1}$ malathion (organophosphorus pesticide) concentration in medium without any adverse effect on physiological and biochemical indices compared with standard medium (Chakraborty et al., 2017). Hence, cyanobacteria, which have both nitrogen fixing and pesticide degrading ability, would be utilized for sustainable high crop production and pesticide degradation.

TABLE 17.1 Classification of pesticides based on chemical composition and their characteristics.

Pesticides	Main component	Characteristics	Example
Organochlorine	Chlorine is the central atom in the molecule	They are lipophilic; they accumulate in fatty tissue of animals; toxic to a variety of animals; long-term persistent	DDT, aldrin, lindane, chlordane, mirex, endosulfan, isobenzan, toxaphene, chloro propylate
Organophosphates	Phosphorus is the central atom in the molecule	Some of them affect the central nervous system. They are absorbed by plants and transferred in the food chain accordingly. They are less persistent than chlorinated hydrocarbons	Malathion, methyl-parathion, diazinon, dimefox, mipafox, ronnel, enitrothion, bidrin
Carbamates	Chemical structure based on a plant alkaloid *Physostigma venenosum*	Carbamates are acid derivatives; kill a limited spectrum of insects; highly toxic to vertebrates; relatively low persistence	Methyl-carbaryl, thio-vernolate, dithio-methane
Pyrethroids	Compounds similar to the synthetic pyrethrins (alkaloids obtained from petals of *Chrysanthemum cinerariaefolium*)	Affect the nervous system; are less persistent than other pesticides; are the safest in terms of their use, some are used as household insecticides	Tetramethrin, ptrethrin cyclethrin Alphamethrin, decamethrin, cypermethrin
Trazines	Trazines are group of nitrogen-containing heterocycles	They block electron transfer by preventing plastoquinone (QB) to bind with D1 protein; hence inhibit PS II system	Atrazine, simazine, ametryn, atratone, chlorazine, cynazine

17.4 Strategies of cyanobacteria in degrading pesticides

Organophosphates and the organochlorine class of pesticides are most commonly used in crop fields to control insects and weeds. The organophosphate group of pesticides possesses a very strong C–P bond, which is highly stable and resistant to chemical hydrolysis, thermal decomposition, and photolysis. Mainly pesticides are used in crop fields to protect crop plants from pest attacks but they also affect cyanobacteria, which are beneficial for crop plants. It also has been reported that nitrogen fixing cyanobacteria were more tolerant to 2,4-dinitrophenol compared with non–nitrogen fixing strains under field conditions (Singh et al., 2016). Some nitrogen fixing cyanobacteria have developed strategies to degrade these pesticides as follows (Fig. 17.1).

Nitrogen fixing cyanobacteria have enzymes like C–P lyase (EC 4.99.1) and alkaline phosphatase (EC 3.1.3.1) that are used to hydrolyze the C–P bond of the organophosphates group of pesticides. The free phosphate groups derived from hydrolysis of the C–P bond of organophosphates group of pesticides are used as a nutrient source for their growth and development (Balakumar and Ravi, 2001; Subramanian et al., 1994). A 140% and 348% increase in C–P lyase was noticed in *Nostoc* sp. and *Westiellopsis* sp. respectively upon exposure to glyphosate pesticide (Balakumar and Ravi, 2001). While increase in alkaline phosphatase was recorded in *Aulosira fertilissima* ARM 68 under presence of various pesticides such as monocrotophos, malathion, dichlorvos, phosphomidon, and quinolphos (Subramanian et al., 1994). The λ-hexachlorocyclohexane (lindane) is an organochlorine pesticide used for rice crop protection. Nitrogen fixing cyanobacteria such as *Nostoc ellipsosporum* and *Oscillatoria* sp. are found to be tolerant to lindane, which degrades it into 1,2,3- and 1,2,4-trichlorobenes via pentachlorocyclohexane intermediate (Kuritz, 1999; El-Bestawy et al., 2007). Nitrophenolic pesticides are a class of pesticides that possess nitro functional group, such as dinoseb, nitrofurans, nitroimidazoles, and 2,4-dinitrophenol. Nitrogen fixing cyanobacteria detoxify these pesticides by reducing the nitro functional group using nitroreductase enzyme. Table 17.2 illustrates the pesticide-tolerant cyanobacterial strains and their mechanisms of pesticide degradation.

Very few studies have been performed to evaluate molecular aspects pesticide degradation by cyanobacteria. The gene drgA codes for nitroreductase, which is used by nitrogen fixing cyanobacteria to reduce nitro functional group of nitrophenolic pesticides (Leganes et al., 1992). It has been suggested that the *nir operon* is responsible for lindane degradation by nitrogen fixing cyanobacteria (Flores et al., 1983; Omata, 1995). Kuritz et al. (1997) demonstrated that *Anabaena* sp. mutants in different genes of *nir operon* (such as *nirA*, *nirB*, *nirC*, *nirD*) were unable to degrade lindane. Kuritz and Wolk (1995) cloned *linA* gene to *Anabaena* sp. and *N. ellipsosporum* and studied the lindane degrading ability. They found that due to the presence of *linA* gene, *Anabaena* sp. and *N. ellipsosporum* did not depend on *nir operon* for degradation of lindane.

17.5 Application of biochar in methane mitigation and paddy crop production

Although many factors contribute to global warming, increasing greenhouse gas (GHG) emissions is the main driver. Methane (CH_4) is the second most imperative GHG after CO_2 (Han et al., 2016). The major anthropogenic sources of CH_4 emission are rice cropping systems. Estimates of the annual emission of CH_4 from paddy fields range from 31 to 112 Tg year^{-1}, which is approximately 9%–19% of total global CH_4 emissions (Tokida et al., 2011). Still, farmers have been pouring fertilizers and rice straws into paddy field, both of which lead to enhanced emissions of CH_4 and worsen global warming (Ye et al., 2015). Hence, control of CH_4 emission from paddy field is essential now and in the future.

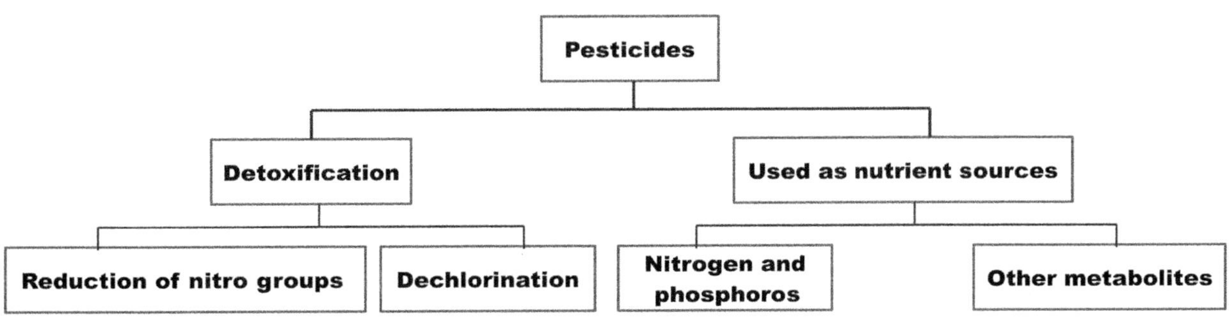

FIGURE 17.1 Strategies followed by nitrogen fixing cyanobacteria for degradation of pesticides.

TABLE 17.2 List of pesticide-tolerant nitrogen fixing cyanobacterial strains and their mechanisms of pesticide degradation.

Pesticide	Tolerant strains	Mechanisms	References
Methyl-parathion	*Fischerella* sp.	Used as phosphorus source	Tiwari et al. (2018)
Monosulfuron-ester	*Anabaena azotica*	Tolerant up to 101.3 nmol L^{-1}	Shen et al. (2017)
Malathion	*Nostoc muscorum*	Used as phosphorus source	Ibrahim et al. (2014)
	A. oryzae		
Butachlor	*N. muscorum*	Used as metabolites	Anees et al. (2014)
Glyphosate	*Anabaena* sp.	Used as phosphorus source	Forlani et al. (2008), Balakumar and Ravi (2001)
	Nostoc punctiforme		
	Westiellopsis sp. L. ACW 101		
Lindane	*Anabaena* sp. PCC 7120	Dechlorination of lindane	El-Bestawy et al. (2007), Kuritz (1999)
	N. ellipsosporum		
	Oscillatoria sp. R12		
2,4-Dinitrophenol	*Anabaena* UAM 202	Detoxification by the reduction of nitro groups	Leganés and Fernández-Valiente (1992)
	Nostoc UAM 205		
DDT	*Anabaena* sp.	Metabolized to DDD and DDE	Lal et al. (1987)

Biochar is the charcoal deriving from the thermal decomposition of biomass of plant, agricultural waste, chopsticks, etc. under a limited oxygen supply (Lehmann, 2007b). It improves soil fertility by enhancing nutrient retention and water holding capacity of soil (Lehmann, 2007a,b). It has also a great capability of carbon sequestering in soil and significantly reduces CH_4 emissions in paddy fields (Liu et al., 2011; Sohi, 2012).

17.6 Biochar production

Biochar is the solid carbon reach product derived from biomass of organic material by pyrolysis. Biomass of organic material is heated at 400°C–500°C under oxygen limited conditions in a reactor. At these temperatures, biomass endures exothermic processes and releases gaseous components, heat, and biochar (coproduct) as shown in Fig. 17.2.

17.7 Application of biochar in paddy fields to improve paddy crop production

Rice is one of the major food crops used by half of the world's population as a staple food. Many studies have been recorded that biochar amendment in paddy fields improves soil fertility and paddy crop production (Khan et al., 2013; Muhammad et al., 2017; Zheng et al., 2012). The amendment of biochar to paddy fields increases nutrient concentrations, cation exchange ability, pH, and water holding ability (Khan et al., 2013; Zheng et al., 2012). Muhammad et al. (2017) noticed that the amendment of biochar to the paddy field significantly enhanced the soil pH from 4.2 to 6.2, total nitrogen by 135% organic carbon by 90% and accumulation of Mn, Mo, Na, Zn compared with control soil, which leads to increased rice crop yield. Khan et al. (2013) also reported that the addition of sewage sludge biochar to the soil improves soil pH, nitrogen, and carbon content that in turn increased shoot biomass by 71.3%–92.2% and rice yield by 148.8%–175.1% compared with control. It also reported that the amendment of biochar at the rate of 10 and 40 t ha^{-1} increases the production of rice by 12% and 14%, respectively compared with control (Zhang et al., 2010). Thus amendment of biochar to paddy field improves soil fertility and paddy crop production as well as mitigating CH_4 emission. However, further studies are required to evaluate long-term effects of biochar addition on soil fertility and crop production. The biochar application in paddy fields mitigates the CH_4 emission mainly by the following three ways:

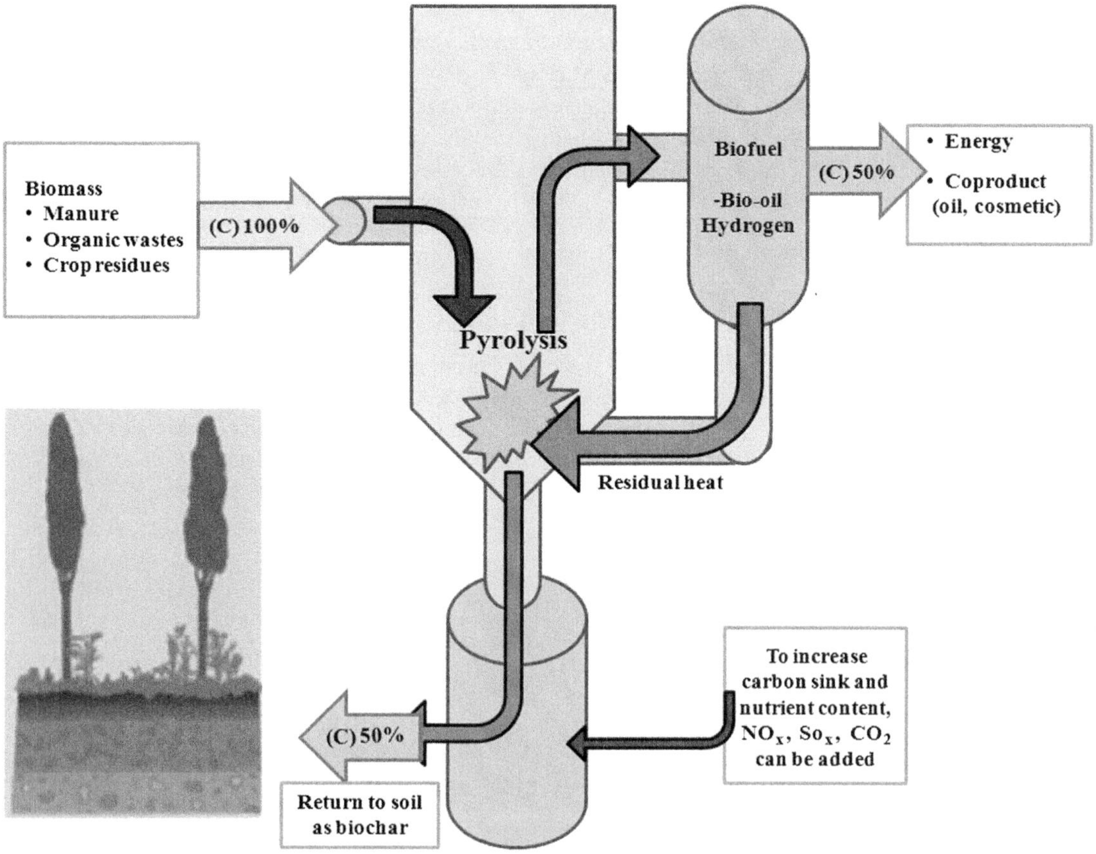

FIGURE 17.2 Pyrolysis of organic material to produce bioenergy and biochar. *Modified from Lehmann, J., 2007b. Bioenergy in the black. Front. Ecol. Environ. 5(7), 381–387.*

1. Methanogenic bacteria uses soil dissolves organic carbon as a substrate to produce CH_4.

$$CO_2 + 4H_2 \xrightarrow{\text{Methanogenic bacteria}} CH_4 + 2H_2O$$

   Biochar has great potential to sequester soil carbon, which leads to less availability of substrate to methanogenic bacteria for CH_4 production, hence reduces CH_4 emission in paddy fields (Liu et al., 2011).
2. Biochar amendment in paddy fields significantly promotes the growth of paddy plants. The stimulated plants could bring more oxygen availability to the aerenchyma tissues of rhizosphere. As methanogenic bacteria are generally anaerobic organisms, a large availability of oxygen inhibits methanogenic activity, thus reduces CH_4 production in paddy fields (Dong et al., 2013; Watanabe et al., 1997).
3. The large surface area and high porosity of biochar may enhance the absorption of CH_4 in paddy field that increases the substrate for methanotrophic bacteria. A methanotrophic bacterium uses CH_4 as their only carbon and energy sources hence reduces the emission of CH_4 in paddy fields (Dong et al., 2013; Liu et al., 2011).

$$CH_4 + O_2 \xrightarrow{\text{Methanotropic bacteria}} HCOOH$$

17.8 Role of rhizospheric microbes in essential heavy metal uptake

Essential heavy metals are metals with high density and atomic number greater than 20 (Raskin et al., 1994). Some metals such as Fe, Zn, Co, Mo, and Cu are required in minute amounts by organisms for their growth and development, known as trace elements. However, heavy metals such as Pb, As, Cd, and Hg do not have any role in growth and development of living organisms, but they cause harmful effects and are called toxic metals (Raskin et al., 1994). Even excessive amounts of trace elements also cause hazardous effects to the plants and other organisms (Vimal et al., 2018).

Excessive application of pesticides, fertilizers, and improper waste management practices led to heavy metal accumulation in both soil and water. Therefore they easily accumulate into the plant and animal bodies and lead to bioaccumulation. Humans, animals, and plants are suffering from various diseases and disorders caused by bioaccumulation of heavy metals (Vimal et al., 2018). Since soil heavy metal remediation is a global concern, a variety of physical, chemical, and biological methods are used for remediation of heavy metals. In bioremediation, both microbes and plants are used for removal of pollutants. The branch of bioremediation that uses plants for the remediation of pollutants from contaminated soils is called phytoremediation. The rhizospheric microbes that live in the vicinity of root zone of the plant play a vital role in the phytoremediation process.

17.9 Rhizospheric microbes and heavy metal uptake

Rhizosphere zone is the immediate surrounding area of the root. The microbes found in the vicinity of root zone are rhizospheric microbes such as *Bacillus cereus*, *P. aeruginosa*, *Azotobacter* sp., *Rhizobium meliloti*, *Aspergillus niger*, and *Azospirillum* sp. It is well evaluated that plant–microbe interactions determine the effectiveness of metal uptake. *Thlaspi caerulescens* plants inoculated with rhizospheric microbes showed highest accumulation of Zn (Whiting et al., 2001). Various mechanisms involved in improving the rate of heavy metal uptake in plants are as follows:

1. EPS production,
2. rhizosphere acidification,
3. siderophore production,
4. plant growth promotion, and
5. metal oxidation and reduction.

17.10 Role of microbial extracellular polymeric substances on heavy metal uptake

Among various mechanisms, EPS plays a vital role in biosorption of heavy metals. Exopolymers that are secreted by rhizospheric bacteria are able to bind strongly with heavy metals and form organometallic complexes. The EPS are a mixture of polysaccharides, mucopolysaccharides, proteins, humic substances, uric acid, lipids, and glycoproteins that depend on the microbes and culture conditions (Sheng et al., 2010; Ahemad and Kibret, 2013). For example, EPS produced by Gram-negative bacteria are neutral or polyanionic while by some Gram-positive bacteria are cationic (Sutherland, 2001; Vu et al., 2009). These substances have functional groups such as carboxyl, hydroxyl, amine, and phosphoric groups, which facilitate EPS to bind with heavy metals (Liu and Fang, 2002; Seneviratne et al., 2014). EPS produced by *Azotobacter* sp. was capable of binding with CrO_4^{2+} and Cd^{2+} (Joshi and Juwarkar, 2009).

17.11 Rhizosphere acidification and heavy metal mobilization

Soil microbes enhance the solubility of heavy metal ions by secretion of different organic acids and by decomposition of organic matter. Both rhizospheric bacteria and plant roots are capable of producing organic acids such as succinate, malate, gluconate, citrate, 2-oxoglutarate, and oxalate. These organic acids create a low pH condition in the rhizosphere and increase the heavy metal mobilization and uptake (Rozycki and Strzelczyk, 1986; Seneviratne et al., 2017). Metal ions can bind with organic acids by complexation reaction in soil solution and increase their mobility for plant uptake. It was reported that *Gluconacetobacter diazotrophicus* produce 5-ketogluconic acid, which leads to the solubilization of Zn compounds (Wani et al., 2007). It was reported that acetic and malic acids stimulate Cd uptake by maize roots (Han et al., 2006). *Pseudomonas fluorescens* G10 and *Microbacterium* sp. G16 have also been noticed to increase Pb uptake in *Brassica napus* by excretion of organic acid (Sheng et al., 2008).

17.12 Siderophore production by rhizospheric microbes

Siderophores are iron chelating compounds (>10 kDa) that are generally produced under Fe limited conditions by bacteria, fungi, and plants to facilitate iron uptake (Chu et al., 2010; Hider and Kong, 2010). Iron from minerals or organic compounds are solubilized by siderophores under low iron conditions. Siderophores can also form complexes with other heavy metals (Glick and Bashan, 1997; Rajkumar et al., 2010). It was reported that siderophores excreted by *P. aeruginosa* were able to chelate Ag^{+}, Al^{3+}, Cd^{2+}, Co^{2+}, Cr^{2+}, Cu^{2+}, Eu^{3+}, Ga^{3+}, Hg^{2+}, Mn^{2+}, Ni^{2+}, Sn^{2+}, Tb^{3+}, and Zn^{2+} metals (Braud et al., 2009), hence increasing their bioavailability for the plant in the rhizosphere.

The functional groups of siderophores that coordinate Fe are classified into three main groups: hydroxyl-carboxylate, catecholate, and hydroxamate (Raymond and Dertz, 2004). Siderophores production by the microbes in the vicinity of the rhizosphere promotes the growth of plants. These siderophores solubilize unavailable forms of heavy metal–bearing minerals and form metal–siderophore complexes. Plants can uptake metals from the metal–siderophore complex by chelate degradation and release of metals or by a ligand exchange reaction (Braud et al., 2009b). Siderophores production has also been studied in some mycorrhizal fungi. It has been demonstrated that *Scleroderma verrucosum*, *Suillus luteus*, and *Rhizopogon luteolus* (ectomycorrhizal fungi) produce catecholates and hydroxamates siderophores under iron limited conditions (Goodell et al., 1997; Machuca et al., 2007).

17.13 Plant growth promotion via rhizospheric microbes

It has been known that most rhizospheric microbes synthesize and release plant growth regulators such as IAA. IAA promotes seed and tuber germination, enhances the root development, and stimulates lateral and adventitious root formation and also helps to develop resistance to stress conditions in plants. The bacterial IAA enhances root length and surface area that increase the soil nutrient uptake by plant (Glick, 2012). It has been reported that *Pseudomonas syringae*, *Bradyrhizobium* sp., *Rhizobium* sp., *Azospirillum* sp., *Klebsiella* sp., *Agrobacterium tumefaciens*, and cyanobacteria (*Synechocystis* sp.) have capability to synthesize and release IAA (Spaepen and Vanderleyden, 2011; Patten and Glick, 1996; Ahemad and Kibret, 2013). Plant growth promoting rhizospheric bacteria also protect plants from various abiotic stresses such as drought, salinity, flooding, and heavy metals by reducing plant ethylene hormone level in plants through 1-aminocychlopropane-1-carboxylate deaminase enzyme activity. The high level of ethylene hormone leads to senescence and abnormal root growth in plants. Production of ethylene hormone in roots gets accelerated under stresses like drought, salinity, flooding, and heavy metals, which causes abnormal plant growth (Raghuwansi and Prasad, 2018).

17.14 Metal oxidation and reduction

The bioavailability of essential metals in the vicinity of the rhizosphere also depends on oxidation or reduction reactions carried out by rhizospheric microbes. Metal oxidation by rhizospheric bacteria is also an important method for phytoextraction of heavy metals (Seneviratne et al., 2017). It was noticed that sulfur-oxidizing rhizosphere bacteria such as *Thiobacillus ferrooxidans* and *Thiobacillus thiooxidans* are capable to increase Cu mobilization and its uptake in plant tissue. It has been reported that the sulfur-oxidizing bacteria are able to decrease the pH of rhizospheric soil by reducing sulfur to sulfates, which increases the Cu availability for plant uptake (Shi et al., 2011).

17.15 Microalgae playing major role in lindane degradation

Lindane is an isomer of hexachlorocyclohexane that was first synthesized in 1825 but its insecticidal property was discovered by Teunis van der Linden (Hardie, 1964). It is synthesized by chlorination of benzene under ultraviolet light. Actually, lindane is the mixture of α, β, γ, and δ isomers of hexachlorocyclohexane (Kutz et al.,1991). Lindane isomers differ based on the number of axial (ax) positioned chlorine atoms (Mertens, 2006).

Lindane is a widely used pesticide due to its low cost and effectiveness in controlling diseases, enhancing agricultural production, and protecting livestock (Salam and Das, 2012). Due to its high affinity for lipids and chemical stability, it accumulates in living organisms and causes toxicity (Benimeli et al., 2007). Hence, degradation of lindane from environment becomes an essential need to protect nontarget organisms. Microalgae are very unique as they have the capability to carry out photosynthesis and grow in extreme environments including polluted habitats (Kuritz, 1999; Singh, 2009; Akoijam et al., 2015; Cuellar-Bermudez et al., 2017). Microalgae have the potential to degrade pesticides and have attracted further research to evaluate lindane degrading ability. Gonzalez et al. (2012) noticed that *Scenedesmus intermedius* develops high lindane-removal ability upon previous exposure.

17.16 Mechanism of lindane degradation by microalgae

It has been reported that different microorganisms are capable of using halogenated compounds as nutrients for their growth and development. The main reaction during microalgal degradation of lindane is the removal of the halogen atom (dehalogenation) from lindane (Fig. 17.3). The halogen atoms are usually responsible for the toxic character of lindane. During the dehalogenation step, the halogen atoms are replaced by hydrogen or a hydroxyl group.

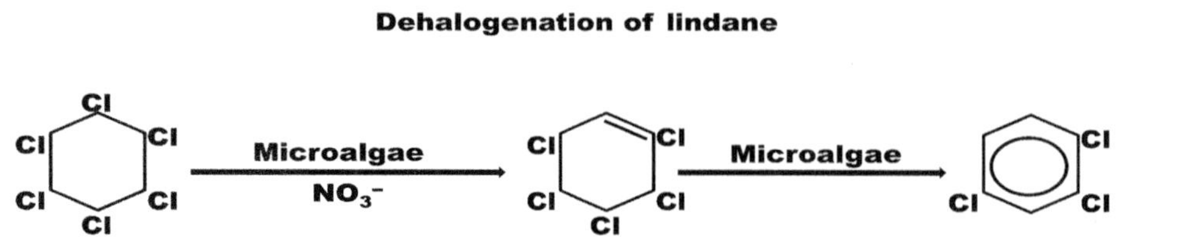

Lindane **γ-Pentachlorocyclohexene** **Trichlorobenzene**

FIGURE 17.3 Mechanism followed by microalgae for dehalogenation of lindane.

Kuritz and Wolk (1995) reported that *Anabaena* sp. strain PCC 7120 and *N. ellipsosporum* converted lindane first to γ-pentachlorocyclohexene and then to a mixture of 1,2,3- and 1,2,4-trichlorobenzene. This dehalogenation process was cometabolic and depended on the availability of nitrate in the medium.

Lindane is one of the most efficient pesticides and is generally used for rice crop protection (Abdullah et al., 1997). Singh (1973) first reported that *Cylindrospermum* sp., *A. fertilissirna*, and *Plectonema boyanurn* were tolerant up to 80 $\mu g\ mL^{-1}$ concentrations of lindane in growth medium. Lindane was also degraded by many microalgal strains including *Anabaena* sp. strain P30, *Fischerella muscicola* UT1829, *Fisherella* sp. CALU926, *Calothrix* sp. ATCC29112, *Nostoc pameloides* UT1627, *N. muscorum* UT387, *Nostoc* sp. strains GSV39, 40, and 236, *Plectonema boryanum*, *Phormidium uncinatum*, *Plectonema* sp., and *Synechococcus* sp. PCC7942 (Kuritz et al., 1997). It was reported that lindane dehalogenation by *Anabaena* sp. PCC 7120 and *N. ellipsosporum* occurred only in the presence of nitrate. This process was inhibited by both ammonium and darkness (Kuritz, 1999). Kuritz (1999) also demonstrated that genetically engineered *Anabaena* sp. PCC7120 and *N. ellipsosporum* with *linA* operon from *Pseudomonas paucimobilis* were capable of dechlorination of lindane without nitrate requirement. Gonzalez et al. (2012) noticed that *S. intermedius* developed high lindane-removal ability by rare mutation prior to lindane exposure. Pérez-Legaspi et al. (2016) also studied lindane-removal ability of *Nannochloris oculata* upon culturing under 0.1 and 0.5 $mg\ L^{-1}$ lindane supplemented medium and observed 73% and 68.2% decreased lindane content in medium.

17.17 Conclusion

With the advent of the green revolution, intensive agricultural practices that have come into existence include use of high-yielding, disease-resistant crop varieties, and constant input of agrochemicals such as chemical fertilizers and pesticides. Application of such chemicals adversely affects the dynamic equilibrium of soil and affects agro-biodiversity by destroying nontarget, useful soil flora and fauna. One such approach is sustainable agricultural practice, including application of biofertilizer to enhance the soil nutrient content, which further results in better productivity. Filamentous, heterocystous, nitrogen fixing, photosynthetic cyanobacteria are part of the tropical paddy field ecosystem assumed as an excellent source of global nitrogen economy of rice fields and embraced as a better alternative to agrochemicals with significant economic and environmental benefits. It is a well-known fact that plants are host and partner to abundant microbiota including bacteria and fungi and their association both outside and inside (within tissues) guides us to understand and nurture novel ways to enhance adaptability and hence productivity against different abiotic stresses in a sustainable manner. The plant-breeding and genetic engineering technologies are presently exploiting the plant–microbe interactions in the rhizosphere to enhance food production under different abiotic stresses. Combination of different species of algae and fungi, fungi and bacteria is a recent move toward yielding the maximum benefit of microbial biofertilizer technology and taking a novel step toward agro-environmental sustainability.

References

Abdullah, A.R., Bajet, C.M., Matin, M.A., Nhan, D.D., Sulaiman, A.H., 1997. Ecotoxicology of pesticides in the tropical paddy field ecosystem. Environ. Toxicol. Chem. 16 (1), 59–70.

Agrawal, A., Pandey, R.S., Sharma, B., 2010. Water pollution with special reference to pesticide contamination in India. J. Water Resour. Prot. 02 (05), 432–448.

Ahemad, M., Kibret, M., 2013. Recent trends in microbial biosorption of heavy metals. Biochem. Mol. Biol 1 (1), 19–26.

Akoijam, C., Langpoklakpam, J.S., Chettri, B., Singh, A.K., 2015. Cyanobacterial diversity in hydrocarbon-polluted sediments and their possible role in bioremediation. Int. Biodeterior. Biodegrad. 103, 97–104.

Anees, S., Suhail, S., Pathak, N., Zeeshan, M., 2014. Potential use of rice field cyanobacterium *Nostoc muscorum* in the evaluation of butachlor induced toxicity and their degradation. Bioinformation 10 (6), 365–370.

Balakumar, T., Ravi, V., 2001. Catalytic degradation of the herbicide glyphosate by the paddy field isolates of cyanobacteria. In: Chen, F., Jiang, Y. (Eds.), Algae and Their Biotechnological Potential. Marcel Dekker, Inc., New York, pp. 195–196.

Baldi, E., Toselli, M., Eissenstat, D.M., Marangoni, B., 2010. Organic fertilization leads to increased peach root production and lifespan. Tree Physiol. 30 (11), 1373–1382.

Barakat, M.R., Yehia, T.A., Sayed, B.M., 2012. Response of Newhall Naval orange to bioorganic fertilization under newly reclaimed area conditions I: vegetative growth and nutritional status. ISHS Acta Hort. 4 (1), 18–25.

Benimeli, C.S., Castro, G.R., Chaile, A.P., Amoroso, M.J., 2007. Lindane uptake and degradation by aquatic streptomyces sp. strain M7. Int. Biodeterior. Biodegrad. 59, 148–155.

Bhardwaj, T., Sharma, J.P., 2013. Impact of pesticides application in agricultural industry: an Indian scenario. Int. J. Agric. Food Sci. Technol. 4 (8), 2249–3050.

Braud, A., Hannauer, M., Mislin, G.L.A., Schalk, I.J., 2009a. The *Pseudomonas aeruginosa* pyochelin-iron uptake pathway and its metal specificity. J. Bacteriol. 191, 3517–3525.

Braud, A., Jezequel, K., Bazot, S., Lebeau, T., 2009b. Enhanced phytoextraction of an agricultural Cr- and Pb-contaminated soil by bioaugmentation with siderophore-producing bacteria. Chemosphere 74 (2), 280–286.

Cerniglia, C.E., Baalen, C.V., Gibson, D.T., 1980. Metabolism of naphthalene by the cyanobacterium. J. Gen. Microbiol. 116, 485–494.

Chakraborty, S., Tiwari, B., Shila, S., Kumar, A., 2017. Biocatalysis and agricultural biotechnology differential physiological, oxidative and antioxidative responses of cyanobacterium *Anabaena sphaerica* to attenuate malathion pesticide toxicity. Biocatal. Agric. Biotechnol. 11, 56–63.

Chu, B.C., Garcia-Herrero, A., Johanson, T.H., Krewulak, K.D., Lau, C.K., Peacock, R.S., et al., 2010. Siderophore uptake in bacteria and the battle for iron with the host; a bird's eye view. Biometals 23 (4), 601–611.

Comte, S., Guibaud, G., Baudu, M., 2008. Biosorption properties of extracellular polymeric substances (EPS) towards Cd, Cu and Pb for different pH values. J. Hazard Mater. 151 (1), 185–193.

Condron, L.M., Turner, B.L., Cade-Menun, B.J., 2005. Chemistry and dynamics of soil organic phosphorus. Soil Sci. Soc. Am. J. 46, 87.

Cuellar-Bermudez, S.P., Aleman-Nava, G.S., Chandra, R., Garcia-Perez, J.S., Contreras-Angulo, J.R., Markou, G., et al., 2017. Nutrients utilization and contaminants removal. A review of two approaches of algae and cyanobacteria in wastewater. Algal Res. 24, 438–449.

Damalas, C., 2009. Understanding benefits and risks of pesticide use. Sci. Res. Essays. 4 (10), 945–949.

Dong, D., Yang, M., Wang, C., Wang, H., Li, Y., Luo, J., et al., 2013. Responses of methane emissions and rice yield to applications of biochar and straw in a paddy field. J. Soils Sedim. 13 (8), 1450–1460.

Duponnois, R., Colombet, A., Hien, V., Thioulouse, J., 2005. The mycorrhizal fungus *Glomus* intraradices and rock phosphate amendment influence plant growth and microbial activity in the rhizosphere of *Acacia holosericea*. Soil Biol. Biochem. 37, 1460–1468.

El-Bestawy, E.A., El-Salam, A.Z.A., Mansy, A.E.R.H., 2007. Potential use of environmental cyanobacterial species in bioremediation of lindane-contaminated effluents. Int. Biodeterior. Biodegrad. 59, 180–192.

Favoino, E., Hogg, D., 2008. The potential role of compost in reducing greenhouse gases. Waste Manag. Res. 26 (1), 61–69.

Flores, E., Guerrero, M.G., Losada, M., 1983. Photosynthetic nature of nitrate uptake and reduction in the cyanobacterium *Anacystis nidulans*. BBA Bioenergetics. 722 (3), 408–416.

Forlani, G., Pavan, M., Gramek, M., Kafarski, P., Lipok, J., 2008. Biochemical bases for a widespread tolerance of cyanobacteria to the phosphonate herbicide glyphosate. Plant Cell Physiol. 49 (3), 443–456.

Glick, B.R., 2012. Plant growth-promoting bacteria: mechanisms and applications. Scientifica 2012, 15.

Glick, B.R., Bashan, Y., 1997. Genetic manipulation of plant growth-promoting bacteria to enhance biocontrol of phytopathogens. Biotechnol. Adv. 15 (2), 353–378.

Goldstein, A.H., 1996. Involvement of the quinoprotein glucose dehydrogenase in the solubilization of exogenous phosphates by Gram-negative bacteria. In: Torriani-Gorini, A., Yagil, E., Silver, S. (Eds.), Phosphate in Microorganisms: Cellular and Molecular Biology. ASM, Washington, DC, pp. 197–203.

Gonzalez, R., Garcia-Balboa, C., Rouco, M., Lopez-Rodas, V., Costas, E., 2012. Adaptation of microalgae to lindane: a new approach for bioremediation. Aquat. Toxicol. 109, 25–32.

Goodell, B., Jellison, J., Liu, J., Daniel, G., Paszczynski, A., Fekete, F., 1997. Low molecular weight chelators and phenolic compounds isolated from wood decay fungi and their role in the fungal biodegradation of wood. J. Biotechnol. 53, 133–162.

Han, F., Shan, X., Zhang, S., Wen, B., Owens, G., 2006. Enhanced cadmium accumulation in maize roots—the impact of organic acids. Plant Soil 289 (1), 355–368.

Han, X., Sun, X., Wang, C., Wu, M., Dong, D., Zhong, T., et al., 2016. Mitigating methane emission from paddy soil with rice-straw biochar amendment under projected climate change. Sci. Rep. 6, 1–10.

Hardie, D.W.F., 1964. Benzene hexachloride. In: Kirk, R.E., Othmer, D.F. (Eds.), Encyclopedia of Chemical Technology. John Wiley & Sons, New York, pp. 267–281.

Hider, R.C., Kong, X., 2010. Chemistry and biology of siderophores. Nat. Prod. Rep. 27 (5), 637–657.

Ibrahim, W.M., Karam, M.A., El-Shahat, R.M., Adway, A.A., 2014. Biodegradation and utilization of organophosphorus pesticide malathion by cyanobacteria. BioMed Res. Int. 2014, 6.

Illmer, P.A., Schinner, F., 1992. Solubilization of inorganic phosphates by microorganisms isolated from forest soil. Soil Biol. Biochem. 24, 389–395.

Jayaraj, R., Megha, P., Sreedev, P., 2016. Review article. Organochlorine pesticides, their toxic effects on living organisms and their fate in the environment. Interdiscip. Toxicol. 9 (3–4), 90–100.

Khan, S., Chao, C., Waqas, M., Arp, H.P.H., Zhu, Y.G., 2013. Sewage sludge biochar influence upon rice (*Oryza sativa* L) yield, metal bioaccumulation and greenhouse gas emissions from acidic paddy soil. Environ. Sci. Technol. 47 (15), 8624–8632.

Kucey, R.M.N., 1983. Phosphate solubilizing bacteria and fungi in various cultivated and virgin Alberta soils. Can. J. Soil. Sci. 63, 671–678.

Kucey, R.M.N., Janzen, H.H., Legget, M.E., 1989. Microbial mediated increases in plant available phosphorus. Adv. Agron. 42, 199–228.

Kumar, V., Behl, R.K., Narula, N., 2001. Establishment of phosphate-solubilizing strains of *Azotobacter chroococcum* in the rhizosphere and their effect on wheat cultivars under greenhouse conditions. Microbiol. Res. 156, 87–93.

Kumar, A., Nagpure, N.S., Trivedi, S.P., Kumar, R., Kushwaha, B., 2011. Profenofos induced DNA damage in freshwater fish, *Channa punctatus* (Bloch) using alkaline single cell gel electrophoresis. Mutat. Res. 726 (2), 209–214.

Kumar, A., Kaushal, S., Saraf, S.A., Singh, J.S., 2018. Microbial biofuels: a solution to carbon emissions and energy crisis. Front. Biosci. (Landmark Ed) 1 (23), 1789–1802.

Kuritz, T., 1999. Cyanobacteria as agents for the control of pollution by pesticides and chlorinated organic compounds. J. Appl. Microbiol 85 (Suppl. 1), 186S–192S.

Kuritz, T., Wolk, C.P., 1995. Use of filamentous cyanobacteria for biodegradation of organic pollutants. Appl. Environ. Microbiol 161 (1), 234.

Kuritz, T., Bocanera, L.V., Rivera, N.S., 1997. Dechlorination of lindane by the cyanobacterium *Anabaena* sp. strain PCC7120 depends on the function of the *nir operon*. J. Bacteriol. 179 (10), 3368–3370.

Kutz, F.W., Wood, P.H., Bottimore, D.P., 1991. Organochlorine pesticides and polychlorinated biphenyls in human adipose tissue. Rev. Environ. Contam. Toxicol. 120, 1–82.

Lal, R., 2004. Soil carbon sequestration impacts on global climate change and food security. Science. 304, 1623–1627.

Lal, S., Lal, R., Saxena, D.M., 1987. Bioconcentration and metabolism of DDT, fenitrothion and chlorpyrifos by the blue-green algae *Anabaena* sp. and *Aulosira fertilissima*. Environ. Pollut. 46 (3), 187–196.

Laura, M., Snchez-Salinas, E., Dantn Gonzlez, E., Luisa, M., 2013. Pesticide biodegradation: mechanisms, genetics and strategies to enhance the process. In: Biodegradation—Life of Science, pp. 251–289.

Leganés, F., Fernández-Valiente, E., 1992. Effects of phenoxy acetic herbicides on growth, photosynthesis, and nitrogenase activity in cyanobacteria from rice fields. Arch. Environ. Contam. Toxicol. 22 (1), 130–134.

Lehmann, J., 2007a. A handful of carbon. Nature 447 (7141), 143–144.

Lehmann, J., 2007b. Bioenergy in the black. Front. Ecol. Environ. 5 (7), 381–387.

Liu, H., Fang, H.H.P., 2002. Characterization of electrostatic binding sites of extracellular polymers by linear programming analysis of titration data. Biotechnol. Bioeng. 80 (7), 806–811.

Liu, Y., Yang, M., Wu, Y., Wang, H., Chen, Y., Wu, W., 2011. Reducing CH_4 and CO_2 emissions from waterlogged paddy soil with biochar. J. Soils Sediments 11 (6), 930–939.

Machuca, A., Pereira, G., Aguiar, A., Milagres, A.M.F., 2007. Metal-chelating compounds produced by ectomycorrhizal fungi collected from pine plantations. Lett. Appl. Microbiol. 44 (1), 7–12.

Mahantesh, P., Patil, C.S., 2011. Isolation and biochemical characterization of phosphate solubilizing microbes. Int. J. Microbiol. Res. 3, 67–70.

Martínez-Alcántara, B., Martínez-Cuenca, M.R., Bermejo, A., Legaz, F., Quiñones, A., 2016. Liquid organic fertilizers for sustainable agriculture: nutrient uptake of organic versus mineral fertilizers in citrus trees. PLoS One 11 (10), 1–20.

Mertens, B., 2006. Microbial Monitoring and Degradation of Lindane in Soil (PhD thesis). Ghent University, Belgium. ISBN 90-5989-126-0.

Muhammad, N., Aziz, R., Brookes, P.C., Xu, J., 2017. Impact of wheat straw biochar on yield of rice and some properties of Psammaquent and Plinthudult. J. Soil Sci. Plant Nutr. 17 (3), 808–823.

Narro, M.L., Cerniglia, C.E., Baalen, C.V.A.N., Gibson, D.T., 1992. Metabolism of phenanthrene. Metab. Clin. Exp. 58 (4), 1351–1359.

Omata, T., 1995. Structure, function and regulation of the nitrate transport system of the cyanobacterium *Synechococcus* sp. PCC7942. Plant Cell. Physiol. 36 (2), 207–213.

Ozanne, P.G., 1980. Phosphate nutrition of plants—a general treatise. In: Khasawneh, F.E., Sample, E.C., Kamprath, E.J. (Eds.), The Role of Phosphorus in Agriculture. America Soil Science Society of America, Madison, WI, pp. 559–589.

Patten, C.L., Glick, B.R., 1996. Bacterial biosynthesis of indole-3-acetic acid. Can. J. Microbiol. 42 (3), 207–220.

Pérez-Legaspi, I.A., Ortega-Clemente, L.A., Moha-León, J.D., Ríos-Leal, E., Gutiérrez, S.C.R., RubioFranchini, I., 2016. Effect of the pesticide lindane on the biomass of the microalgae *Nannochloris oculata*. J. Environ. Sci. Health B. 51 (2), 103–106.

Ponmurugan, P., Gopi, C., 2006. In vitro production of growth regulators and phosphatase activity by phosphate solubilizing bacteria. Afr. J. Biotechnol. 5, 348–350.

Raghuwanshi, R., Prasad, J.K., 2018. Perspectives of rhizobacteria with ACC deaminase activity in plant growth under abiotic stress. In: Giri, B., Prasad, R., Varma, A., (Eds.), Root Biology, pp. 303–321.

Rajkumar, M., Ae, N., Prasad, M.N.V., Freitas, H., 2010. Potential of siderophore-producing bacteria for improving heavy metal phytoextraction. Trends Biotechnol 28 (3), 142–149.

Ranjan, A., Mahalakshmi, M.R., Sridevi, M., 2013. Isolation and characterization of phosphate- solubilizing bacterial species from different crop fields of Salem, Tamil Nadu, India. Int. J. Nutr. Pharmacol. Neurol. Dis. 3, 29.

Raskin, I., Kumar, P.N., Dushenkov, S., Salt, D.E., 1994. Bioconcentration of heavy metals by plants. Curr. Opin. Biotechnol. 5 (3), 285–290.

Ravoet, J., Reybroeck, W., De Graaf, D.C., 2015. Pesticides for apicultural and/or agricultural application found in Belgian honey bee wax combs. Bull. Environ. Contam. Toxicol. 94 (5), 543–548.

Raymond, K.N., Dertz, E.A., 2004. Biochemical and physical properties of siderophores. In: Crosa, J., Mey, A., Payne, S. (Eds.), Iron Transport in Bacteria. ASM Press, Washington, DC, pp. 3–17.

Rodríguez, H., Fraga, R., 1999. Phosphate solubilizing bacteria and their role in plant growth promotion. Biotechnol. Adv. 17, 319–339.

Rózycki, H., Strzelczyk, E., 1986. Organic acids production by *Streptomyces* spp. isolated from soil, rhizosphere and mycorrhizosphere of pine (*Pinus sylvestris* L.). Plant Soil 96 (3), 337–345.

Salam, J.A., Das, N., 2012. Remediation of lindane from environment—an overview. Int. J. Adv. Biotechnol. Res. 2 (1), 9–15.

Seneviratne, M., Seneviratne, G., Madawala, H.M.S.P., Iqbal, M.C.M., Rajakaruna, N., Bandara, T., et al., 2014. A preliminary study of the role of bacterial—fungal co-inoculation on heavy metal phytotoxicity in serpentine soil. Aust. J. Bot. 63 (4), 261–268.

Seneviratne, M., Seneviratne, G., Madawala, H., Vithanage, M., 2017. Role of rhizospheric microbes in heavy metal uptake by plants. In: Singh, J.S., Seneviratne, G. (Eds.), Agro-Environmental Sustainability: Volume 2: Managing Environmental Pollution. Cham. Springer, pp. 147–163.

Sharma, S.B., Sayyed, R.Z., Trivedi, M.H., Gobi, T.A., 2013. Phosphate solubilizing microbes: sustainable approach for managing phosphorus deficiency in agricultural soils. Springerplus 2, 587–601.

Shashirekha, S., Uma, L., Subramanian, G., 1997. Phenol degradation by the marine cyanobacterium *Phormidium valderianum* BDU 30501. J. Ind. Microbiol. Biotechnol. 19 (2), 130–133.

Shen, J.Y., Liao, J.Z., Guo, L.L., Su, R.F., 2017. Effects of monosulfuron-ester on metabolic processes of nitrogen-fixing cyanobacteria *Anabaena* flos-aquae and *Anabaena azotica*. Braz. J. Microbiol. 48 (3), 544–550.

Sheng, G., Yu, H., Li, X., 2010. Extracellular polymeric substances (EPS) of microbial aggregates in biological wastewater treatment systems: a review. Biotechnol. Adv. 28 (6), 882–894.

Sheng, X.-F., Xia, J.-J., Jiang, C.-Y., He, L.-Y., Qian, M., 2008. Characterization of heavy metal-resistant endophytic bacteria from rape (*Brassica napus*) roots and their potential in promoting the growth and lead accumulation of rape. Environ. Pollut. 156 (3), 1164–1170.

Shi, J.Y., Lin, H.R., Yuan, X.F., Chen, X.C., Shen, Y.X., 2011. Enhancement of copper availability and microbial community changes in rice rhizospheres affected by sulfur. Molecules 16 (2), 1409–1417.

Singh, B.K., 2009. Organophosphorus-degrading bacteria: ecology and industrial applications. Nat. Rev. Microbiol. 7 (2), 156–164.

Singh, D., Khattar, J., Kaur, G., Singh, Y., 2016. Toxicological impact of herbicides on cyanobacteria. Annu. Res. Rev. Biol. 9 (4), 1–39.

Singh, J.S., 2015a. Biodiversity: current perspective. Clim. Change Environ. Sustain. 3 (1), 71–72.

Singh, J.S., 2015b. Microbes: the chief ecological engineers in reinstating equilibrium in degraded ecosystems. Agric. Ecosyst. Envron. 203, 80–82.

Singh, J.S., 2015c. Plant-microbe interactions: a viable tool for agricultural sustainability. Appl. Soil Ecol. 92, 45–46.

Singh, J.S., 2015d. Microbes play major roles in the ecosystem services. Clim. Change Environ. Sustain. 3, 163–167.

Singh, J.S., Seneviratne, G., 2017a. Agro-Environmental Sustainability: Volume 2: Managing Environmental Pollution, 2. Springer, Dordrecht, The Netherlands, p. 229.

Singh, J.S., Seneviratne, G., 2017b. Agro-Environmental Sustainability: Volume 1: Managing Crop Health. Springer, Dordrecht, The Netherlands, p. 316.

Singh, C., Tiwari, S., Gupta, V.K., Singh, J.S., 2018. The effect of rice husk biochar on soil nutrient status, microbial biomass and paddy productivity of nutrient poor agriculture soils. Catena 171, 485–493.

Singh, P.K., 1973. Effect of pesticides on blue-green algae. Arch. Mikrobiol. 89 (4), 317–320.

Smith, P., Martino, D., Cai, Z., Gwary, D., Janzen, H., Kumar, P., et al., 2007. Policy and technological constraints to implementation of greenhouse gas mitigation options in agriculture. Agric. Ecosyst. Environ. 118 (1–4), 6–28.

Sohi, P.S., 2012. Carbon storage with benefits can one cell influence cancer. Science. 338, 1034–1036.

Spaepen, S., Vanderleyden, J., 2011. Auxin and plant-microbe interactions. Cold Spring Harb. Perspect. Biol 3 (4).

Sridevi, M., Mallaiah, K.V., Yadav, N.C.S., 2007. Phosphate solubilization by *Rhizobium* isolates from Crotalaria species. J. Plant Sci. 2, 635–639.

Subramanian, G., Sekar, S., Sampoornam, S., 1994. Biodegradation and utilization of organophosphorus pesticides by cyanobacteria. Int. Biodeterior. Biodegrad. 33 (2), 129–143.

Sutherland, I., 2001. Biofilm exopolysaccharides: a strong and sticky framework. Microbiology 147 (1), 3–9.

Tiwari, B., Verma, E., Chakraborty, S., Srivastava, A.K., Mishra, A.K., 2018. Tolerance strategies in cyanobacterium *Fischerella* sp. under pesticide stress and possible role of a carbohydrate-binding protein in the metabolism of methyl parathion (MP). Int. Biodeterior. Biodegrad. 127, 217–226.

Tiwari, P., Singh, J.S., 2017. A plant growth promoting rhizospheric *Pseudomonas aeruginosa* strain inhibits seed germination in *Triticum aestivum* (L) and *Zea mays* (L). Microbiol. Res. 8 (7233), 73–79.

Tiwari, S., Singh, C., Singh, J.S., 2018. Land use changes: a key ecological driver regulating methanotrophs abundance in upland soils. Energy Ecol. Environ 3, 355–371.

Tokida, T., Adachi, M., Cheng, W., Nakajima, Y., Fumoto, T., Matsushima, M., et al., 2011. Methane and soil CO_2 production from current-season photosynthates in a rice paddy exposed to elevated CO_2 concentration and soil temperature. Global Change Biol. 17 (11), 3327–3337.

UN, 2013. World Population Projected to Reach 9.6 Billion by 2050. United Nations (UN), Department of Economy and Social Affairs. <https://www.un.org/en/development/desa/news/population/un-report-world-population-projected-to-reach-9-6-billion-by-2050.html>.

UN-DESA, 2015. World Population Prospects: The 2015 Revision. <https://www.un.org/en/development/desa/news/population/2015-report.html>.

Varsha-Narsian, J., Thakkar, J., Patel, H.H., 1994. Inorganic phosphate solubilization by some yeast. Ind. J. Microbiol. 35, 113–118.

Vassilev, N., Vassileva, M., Azcon, R., Medina, A., 2001. Preparation of gel-entrapped mycorrhizal inoculum in the presence or absence of *Yarrowia lipolytica*. Biotechnol. Lett. 23, 907–909.

Venkateswarlu, B., Rao, A.V., Raina, P., Ahmad, N., 1984. Evaluation of phosphorus solubilization by microorganisms isolated from arid soil. J. Ind. Soc. Soil. Sci. 32, 273–277.

Vimal, S.R., Patel, V.K., Singh, J.S., 2018. Plant growth promoting *Curtobacterium albidum* strain SRV4: an agriculturally important microbe to alleviate salinity stress in paddy plants. Ecol. Indic (in press).

Vimal, S.R., Singh, J.S., Arora, N.K., Singh, D.P., 2016. PGPR: an effective bioagent in stress agricultural management. In: Sarma, B.K., Jain, A. (Eds.), Microbial Empowerment in Agriculture: A Key to Sustainability and Crop Productivity. Biotech Books, New Delhi, pp. 81–108.

Vimal, S.R., Singh, J.S., Arora, N.K., Singh, S., 2017. Soil-plant-microbe interactions in stressed agriculture management: a review. Pedosphere 27 (2), 177–192.

Vu, B., Chen, M., Crawford, R.J., Ivanova, E.P., 2009. Bacterial extracellular polysaccharides involved in biofilm formation. Molecules. 14 (7), 2535–2554.

Wani, P.A., Khan, M.S., Zaidi, A., 2007. Chromium reduction, plant growth-promoting potentials, and metal solubilization by *Bacillus* sp. isolated from alluvial soil. Curr. Microbiol. 54 (3), 237–243.

Watanabe, I., Hashimoto, T., Shimoyama, A., 1997. Methane-oxidizing activities and methanotrophic populations associated with wetland rice plants. Biol. Fertil. Soils. 24 (3), 261–265.

Whiting, S.N., de Souza, M.P., Terry, N., 2001. Rhizosphere bacteria mobilize Zn for hyperaccumulation by *Thlaspi caerulescens*. Environ. Sci. Technol. 35 (15), 3144–3150.

Widada, J., Damarjaya, D.I., Kabirun, S., 2007. The interactive effects of arbuscular mycorrhizal fungi and rhizobacteria on the growth and nutrients uptake of sorghum in acid soil. In: Rodriguez- Barrueco, C., Velazquez, E. (Eds.), First International Meeting on Microbial Phosphate Solubilization. Springer, Dordrecht, The Netherlands, pp. 173–177.

Williams, A., Hunter, M.C., Kammerer, M., Kane, D.A., Jordan, N.R., Mortensen, D.A., et al., 2016. Soil Water holding capacity mitigates downside risk and volatility in US rainfed maize: time to invest in soil organic matter? PLoS One. 11 (8).

Yang, F., Zhang, G.L., Yang, J.L., Li, D.C., Zhao, Y.G., Liu, F., et al., 2014. Organic matter controls of soil water retention in an alpine grassland and its significance for hydrological processes. J. Hydrol. 519, 3010–3027.

Ye, R., Doane, T.A., Morris, J., Horwath, W.R., 2015. The effect of rice straw on the priming of soil organic matter and methane production in peat soils. Soil Biol. Biochem. 81, 98–107.

Zhang, A., Cui, L., Pan, G., Li, L., Hussain, Q., Zhang, X., et al., 2010. Effect of biochar amendment on yield and methane and nitrous oxide emissions from a rice paddy from Tai Lake plain, China. Agric. Ecosyst. Environ. 139 (4), 469–475.

Zheng, R.L., Cai, C., Liang, J.H., Huang, Q., Chen, Z., Huang, Y.Z., et al., 2012. The effects of biochars from rice residue on the formation of iron plaque and the accumulation of Cd, Zn, Pb, As in rice (*Oryza sativa* L.) seedlings. Chemosphere. 89 (7), 856–862.

Further reading

Consuelo Gomez, K.B., 1999. Leaching heavy metals from contaminated soil by using *Thiobacillus ferrooxidans* or *Thiobacillus thiooxidans*. Geomicrobiol. J. 16 (3), 233–244.

Singh, J.S., 2013. Plant growth promoting rhizobacteria: potential microbes for sustainable agriculture. Resonance 18 (3), 275–281.

Singh, J.S., Gupta, V.K., 2018. Soil microbial biomass: a key soil driver in management of ecosystem functioning. Sci. Total Environ. 634, 497–500.

Chapter 18

Rhizosphere microbiome and plant probiotics

Jintu Dutta[1] and Utpal Bora[1,2]

[1]*Centre for the Environment, Indian Institute of Technology Guwahati, Guwahati, India,* [2]*Department of Biosciences and Bioengineering, Indian Institute of Technology Guwahati, Guwahati, India*

Chapter Outline

18.1 Introduction

Microbes are one of the most crucial components to all life forms on Earth. The omnipresence of microbes makes them important to study their habitat to understand their long existence. One gram of soil contains 4000 different bacterial "genomic units" as evident from DNA−DNA reassociation (Torsvik et al., 1990). There is an intensive interaction between plant roots and microbes. Diversity of the microbes is found to be very numerous in the plant root region, also known as the rhizosphere zone. The rhizosphere is the narrow dynamic region where plant−microbe interaction is intense and complex (Hinsinger and Marschner, 2006; Perret et al., 2007; Jones and Hinsinger, 2008; Hinsinger et al., 2009; Raaijmakers et al., 2009). But when we look into the rhizosphere region, there are different categories of microbes present. Some of them are good, some of them are bad, and even some of them are ugly. These good, bad and ugly microbes refers here to the good microorganisms, plant pathogens and opportunistic human pathogens respectively. Rhizosphere organisms that have been well studied for their beneficial effects on plant growth and health are the nitrogen-fixing bacteria, mycorrhizal fungi, plant growth promoting rhizobacteria (PGPR), biocontrol microorganisms, mycoparasitic fungi, and protozoa. Rhizosphere organisms that are deleterious to plant growth and health include the pathogenic fungi, oomycetes, bacteria, and nematodes. A third group of microorganisms that can be found in the rhizosphere are the human pathogens (Mendes et al., 2013). Plants release some rhizodeposits (e.g., exudates, border cells, mucilage) that regulate the microbial diversity and in return microbes also help the plants to survive in various stress conditions. Plant roots provide a niche for the proliferation of soil bacteria that flourish on root exudates and lysates. Root exudates have been grouped into two classes, that is, low molecular weight compounds, such as amino acids, organic acids, sugars, phenolic compounds, volatile compounds (VOCs) and other secondary metabolites, and high molecular weight compounds, such as polysaccharides and proteins (Narasimhan et al., 2003; Bais et al., 2006; Badri and Vivanco, 2009; Huang et al., 2013). This exudation is plant specific and is considered to regulate the evolution or specific physiological adaptation of rhizobacteria to particular soil habitat conditions (Crowley and Rengel, 1999). The nature of root exudates is crucial for the ecosystem distribution and niche-specificity of certain plants (Dakora and Phillips, 2002). The compounds are released by the roots via at least two potential mechanisms, for example, root exudates are transported across the cellular membrane and secreted into the surrounding rhizosphere and plant products are also released from root border cells and root border-like cells, which separate from roots as they grow (Hawes et al., 2000; Vicre et al., 2005). There may be almost every compound present in the root products and rhizodepositions except some specific compounds involved in photosynthesis.

New and Future Developments in Microbial Biotechnology and Bioengineering. DOI: https://doi.org/10.1016/B978-0-12-818258-1.00018-2

Rhizosphere microbes (Fig. 18.1) are responsible for many functions and are central to crop fertility, recycling nutrients, detoxifying pollutants, regulating carbon storage, and controlling the production and absorption of greenhouse gases such as methane and nitrous oxides (Singh, 2013, 2015; Tiwari and Singh, 2017). These microorganisms are now considered as key engineers that may be manipulated to reestablish biodiversity and functions in degraded ecosystems (Young et al., 2005; Heneghan et al., 2008; Calderón et al., 2016). These microbes have a very important role in the growth and in ecological fitness of their host plant (Buée et al., 2009). Moreover, the microbes play a vital role to maintain in both natural and managed agriculture soil ecosystems because they are involved in different key processes such as formation of soil structure, organic matter decomposition, removal of toxin, degrading xenobiotic compounds like pesticides, and the cycling of carbon, bioremediation of toxic compounds, phosphorus, sulfur, etc. Furthermore, microbes residing in the rhizosphere soil help in solubilizing and mobilizing the nutrients to plants and also suppress the soil-borne plant diseases (Dey et al., 2012; Ahemed and Kibret, 2014). The genetic diversity of PGPR has been also reported in different crops such as wheat (Beneduzi et al., 2008a,b; Upadhyay et al., 2009), common bean (Remans et al., 2008), rice (Beneduzi et al., 2008a,b), maize (Cordero et al., 2012), and tea (Dutta and Thakur, 2017). A diverse range of PGPR genera have been previously documented such as *Arthrobacter*, *Azospirillum*, *Azotobacter*, *Bacillus*, *Pseudomonas*, *Klebsiella*, *Burkholderia*, *Erwinia*, *Flavobacterium*, *Micrococcous*, *Enterobacter*, *Xanthomonas*, *Chromobacterium*, *Serratia*, and *Caulobacter* (Bal et al., 2013).

In recent times, advances in technology have facilitated the process of understanding the other 99% of microbes that are not culturable to date in our laboratory conditions. The "meta-omics" approaches like metagenomics, metatranscriptomics, metaproteomics, and metabolomics are making us reach a point that was previously undiscovered. A new era of microbial ecology was initiated when sequencing of ribosomal RNAs and the genes encoding them was introduced to

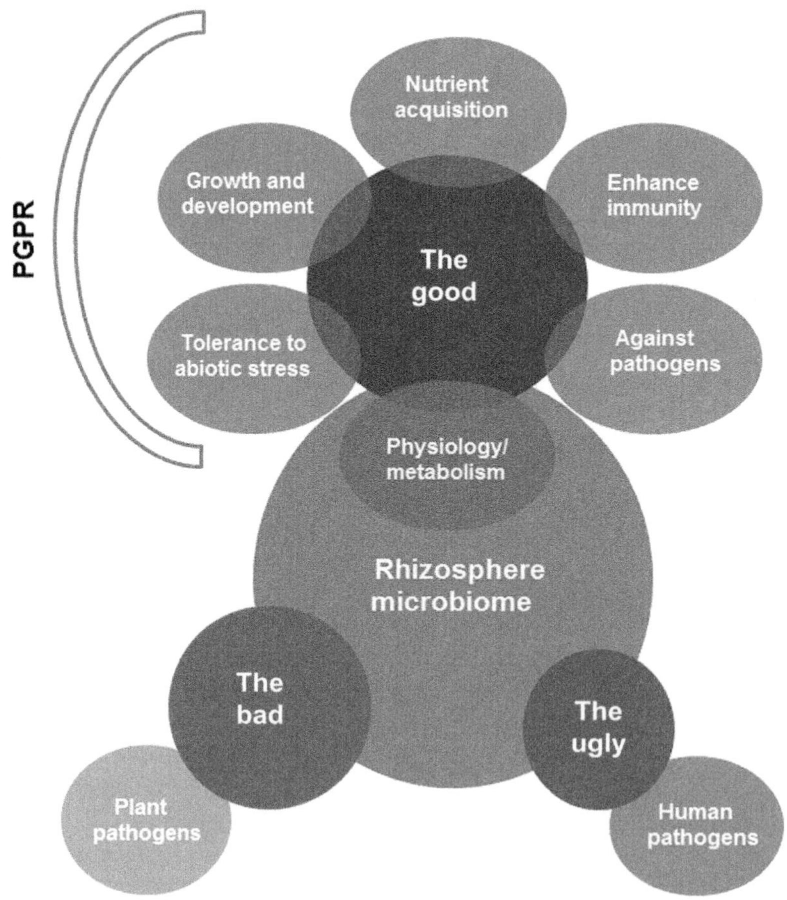

FIGURE 18.1 The schematic overview of rhizosphere microbiome and the functions of the different microbial group present in the rhizosphere soil. Here in the figure, "the good" depicts the plant growth promoting rhizobacteria (PGPR); "the bad" depicts the plant pathogenic; and "the ugly" depicts the opportunistic human pathogens. *Modified from Mendes, R., Garbeva, P., Raaijmakers, J.M., 2013. The rhizosphere microbiome: significance of plant beneficial, plant pathogenic, and human pathogenic microorganisms. FEMS Microb. Rev. 37, 634–663.*

describe uncultured bacteria in the environment. Apart from the microbial structure and diversity, the other major goal of metagenomic studies is to recognize the functional potential of the complex microbial communities.

18.2 Rhizosphere microbiome

Since Hiltner (1904) coined the term *rhizosphere* and defined the rhizosphere as the narrow soil zone influenced by plant roots, thereafter this hotspot for microbial colonization and interaction with plant has received attention from scientists across the globe. From the research on the rhizosphere, it was observed that the microbial activity in that region is essential for plant functioning as it assists the plant in nutrient uptake and offers protection against different pathogens (Berendsen et al., 2012). The bacterial population in the rhizosphere is higher than in bulk soil and up to 15% of the root surface may be covered by a variety of microbial colonies because such bacteria possess metabolic versatility to acclimatize and utilize root exudates efficiently (Van Loon, 2007). Mendes et al. (2011) documented more than 33,000 bacterial and archaeal species in the rhizosphere. The bacteria present in the rhizosphere utilize the nutrients released from the host plant and they also secrete metabolites into the rhizosphere. The understanding of rhizosphere interaction has developed from studies on agricultural or horticultural crop plants and model species such as *Arabidopsis thaliana* and *Medicago truncatula*. In addition, substantial effort is also being made to understand the microbial ecology of the rhizosphere of noncultivated plant species in natural ecosystems (Bever et al., 2012) and how microorganisms influence resource allocation, biodiversity, and aboveground interactions with herbivores and their natural enemies (Pineda et al., 2010). In addition to a comprehensive phylogenetic analysis of the rhizosphere microbiome, there is a strong need to go beyond cataloging microbial communities and to determine which microorganisms are active during the different developmental stages of plant/root growth and which functions and pathways are displayed in time and space (Singh and Gupta, 2018; Vimal et al., 2018). The major limitation in the microbiological studies in the soil environment is that the largest proportion of the microbes present in the soil is unculturable under the present laboratory conditions (Amann et al., 1995; Kent and Triplett, 2002; Doornbos et al., 2012). However, recent advances in technologies like metagenomics where we can extract the total DNA of the soil to analyze the microbial population provide us with deeper insight and a more complete picture of the rhizosphere microbiome (Leveau, 2007; Sorensen et al., 2009; Hirsch and Mauchline, 2012; Bakker et al., 2013). Thus the microbial players in the rhizosphere are on their way to being exposed and, perhaps the present era of "omics" including transcriptomics and proteomics studies of the microbiome has been initiated to reveal microbial activities in complex environments (Urich et al., 2008; Gosalbes et al., 2012; Jansson et al., 2012).

Who is there in the rhizosphere and what they are doing is the important question to understand the rhizosphere microbiome (Fig. 18.2) The meta-analysis of 19 clone libraries obtained from the rhizosphere of 14 plant species revealed more than 1200 distinguishable bacterial taxa from 35 different taxonomic orders, with the Proteobacteria as the most dominant phylum (Hawkes et al., 2007). Uroz et al. (2010) demonstrated the species richness and spatial diversity of oak rhizosphere soil. They obtained 5619 operational taxonomic units (OTU) in the rhizosphere of oak and showed that the bacterial community was dominated by the Actinobacteria and Proteobacteria. Despite the harsh abiotic conditions in Antarctic soils, up to 732 OTUs were detected in the rhizosphere of two vascular plants (Teixeira et al., 2010). For most of the rhizosphere samples from the Antarctic, Firmicutes were the most abundant phylum, whereas in many other rhizosphere studies, the Proteobacteria are commonly more abundant. Weinert et al. (2011) showed that the number of OTUs found in the rhizosphere of three potato cultivars grown at two distant field sites ranged from 1444 to 2015. The dominant phylum detected was the Proteobacteria (46%), followed by Firmicutes (18%), Actinobacteria (11%), Bacteroidetes (7%), and Acidobacteria (3%). The metagenomic study of rhizosphere revealed that the rhizosphere soil microbial communities and abundances of the classes *Bacteroidetes*, *Betaproteobacteria*, *Chlorobi*, *Dehalococcoidetes*, and *Methanobacteria*. These bacterial classes include strains that potentially promote plant growth and phytic acid utilization, and some gene clusters relating to phytic acid utilization, such as alkaline phosphatase and citrate synthase, with the phytic acid utilization status of the plant (Unno and Shinano, 2013). The high throughput sequencing to explore the rhizosphere microbial diversity in the alfalfa and barley planted oil contaminated soil samples analysis of 16s rRNA sequences showed the *Proteobacteria* to be the most enriched (45.9%) followed by *Bacteriodetes* (21.4%) and *Actinobacteria* (10.4%) phyla (Kumar et al., 2018a,b). Another study by Bai et al (2014) showed the microbial richness and diversity in rhizosphere soil is higher than in influent water, supporting the proposition that rhizosphere soil supports significant microbial growth. Bacteria were the dominant domain (83.3%), with Gammaproteobacteria being the most abundant phylum in the rhizosphere soil. For microbial function annotation, a number of biodegradation pathways of xenobiotic compounds were observed in rhizosphere soil (Kumar et al., 2018a,b; Singh et al., 2018; Tiwari et al., 2018). Nitrification and denitrification processes were the primary transformation route

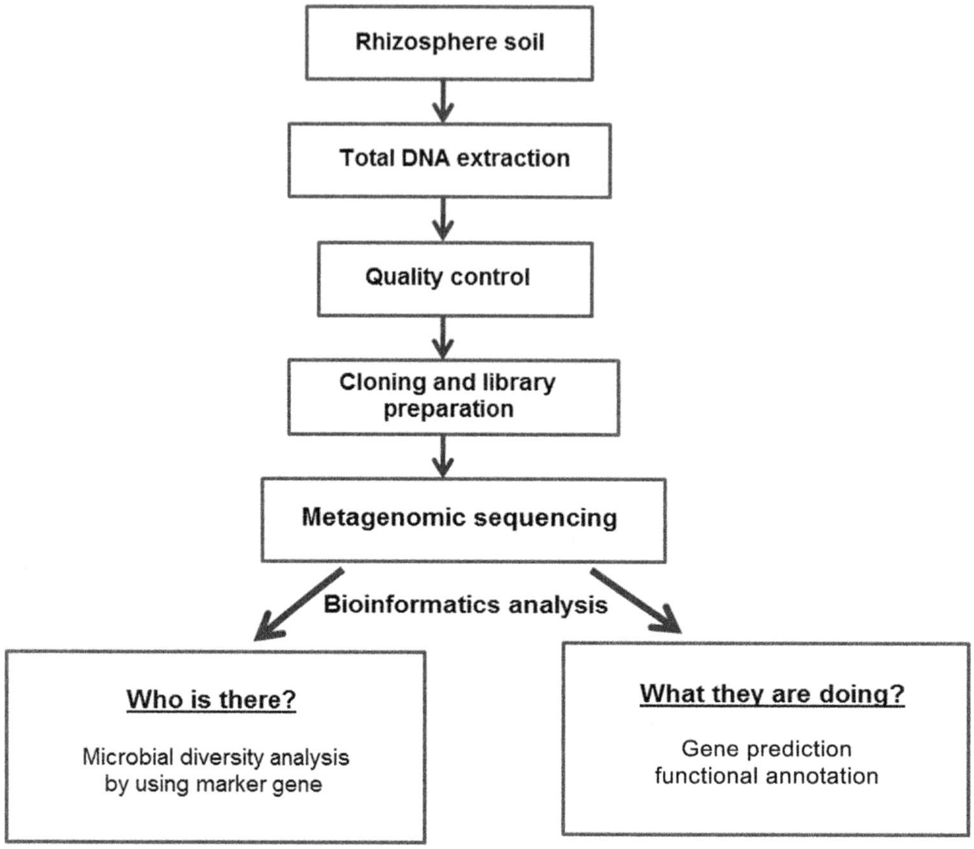

FIGURE 18.2 Overview of the methodological workflow of metagenomic sequencing.

in the N cycle. This study found that soluble $Fe^{2+}$and Mn^{4+} from influent water were converted to Fe^{3+} and Mn^{4+} in rhizosphere soil by chemical oxidation (Fe^{2+}) and biological oxidation (Mn^{2+}, employing multicopper oxidase), respectively.

18.3 Plant probiotics

Probiotics are defined as live microorganisms that when administered in adequate amounts, confer a health benefit on the host (Hill et al., 2014; Neef and Sanz, 2013). The beneficial microbes having multifarious traits to promote plant growth are plant probiotics. The PGPR are the group of beneficial bacteria that are considered as plant probiotics and promote plant growth directly or indirectly. They mainly reside in the rhizosphere zone and improve growth, nutrition, and health of plants in different agro-ecosystems (Philippot et al., 2013). In the last decade, several studies have been carried out to understand the functioning of rhizosphere as an ecological niche, which eventually gained attention in sustainable agricultural practice. PGPR are the potential tools to contribute in the development of sustainable agricultural systems (Schippers et al., 1995) (Table 18.1).

Commonly, PGPR contribute to the plant health in three different ways (Glick, 1995) that is, producing particular compounds for the plants (Dobbelaere et al., 2003; Zahir et al., 2004), facilitating the uptake of certain nutrients from the soil (Lucas García et al., 2004 a,b; Çakmakçi et al., 2006), and reducing or preventing the microorganisms that can cause plant disease (Jetiyanon and Kloepper, 2002; Guo et al., 2004; Saravanakumar et al., 2008). A putative rhizobacteria qualifies as PGPR when it is able to produce a positive effect on the plant upon inoculation, hence demonstrating good competitive skills over the existing rhizosphere communities (Fig. 18.3). In the rhizosphere, generally about 2%–5% of bacteria are PGPR and the majority of plausible PGPR bacterial genera including *Arthrobacter*, *Azospirillum*, *Azotobacter*, *Acinetobacter*, *Bacillus*, *Brevibacillus*, *Aneurinibacillus*, *Sporosarcina*, *Enterobacter*, *Burkholderia*, *Pseudomonas*, *Klebsiella*, *Bradyrhizobium*, *Rhizobium*, *Erwinia*, *Flavobacterium*, *Micrococcous*, *Xanthomonas*, *Chromobacterium*, *Serratia*, *Caulobacter*, *Frankia*, and *Thiobacillus* have been documented to promote

TABLE 18.1 Plant probiotics and their host plant.

Plant probiotics (genus level)	Host plant	References
Bacillus	Rice, tea, potato, common bean	Beneduzi et al. (2008a,b), Dutta et al. (2015), Ahmed and Hasnain (2010) and Ortiz-Castro et al. (2008)
Burkholderia	Rice, tea	Estrada et al. (2013) and Dutta et al. (2015)
Pseudomonas	Tea, tomato, rice	Dutta et al. (2015), Hu et al. (2016), and Habibi et al. (2014)
Phyllobacterium	Strawberry	Flores-Félix et al. (2015)
Enterobacter	Tea, wheat, sugarcane	Dutta et al. (2015), Rana et al. (2011), and Taulé et al. (2012)
Paenibacillus	Rice, barley, wheat	Beneduzi et al. (2008a,b), Bakaeva et al. (2017)
Streptomyces	Cocoa, wheat, chickpea	Boudjeko et al. (2017), Jog et al. (2014), and Gopalakrishnan et al. (2015)
Azospirillum	Sugarcane, bean, soybean, rice	Tejera et al. (2005), Yadegari et al. (2010), Hungria et al. (2013), and Sahoo et al. (2014a,b)
Azotobacter	Rice, cotton, wheat	Sahoo et al. (2014a,b) and Kumar et al. (2007)
Micromonospora sp.	Pine	Valdes et al. (2005)
Rhodococcus sp.	Mustard	Belimov et al. (2005)
Trichoderma sp.	Tomato, canola	Vinale et al. (2008)

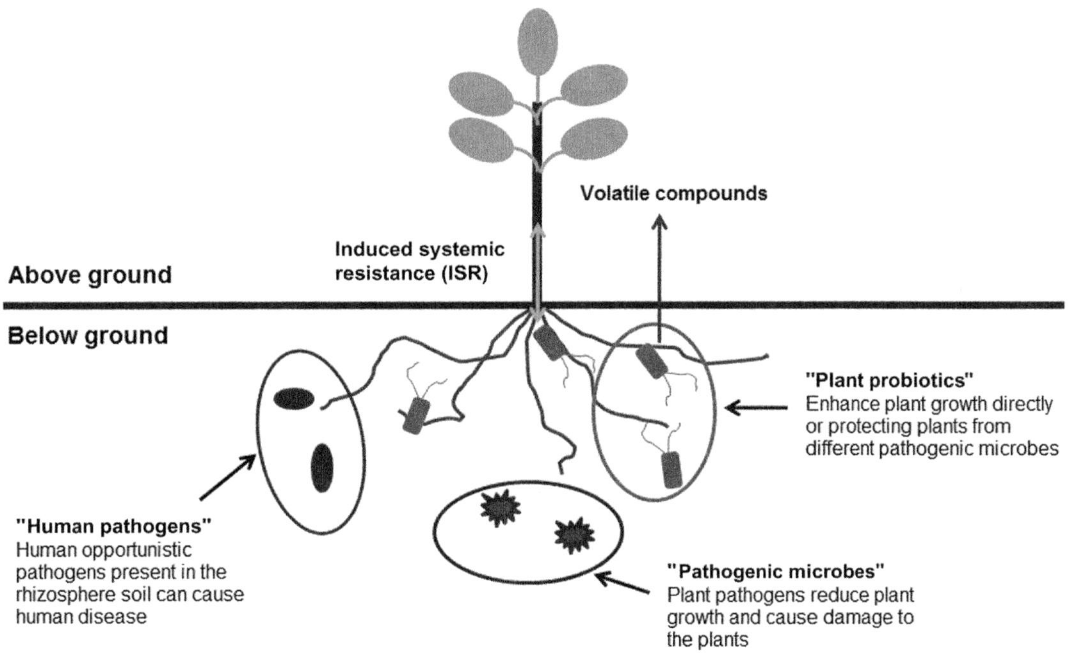

FIGURE 18.3 The belowground–aboveground interaction of the microbes with the host plant and the functional attributes of the microbes in the rhizosphere soil.

plant growth (Glick, 1995; Vessey, 2003; Bal et al., 2013; Dutta et al., 2015, 2017). The probiotic pseudomonads isolated from rhizospheric soil form protective biofilms on the root surface and produce the metabolites or enzymes to boost plant health (Kim and Anderson, 2018). The probiotic *Psedumonas* species was also reported to enhance rhizosphere microbiome function and the suppression of the bacterial plant pathogen *Ralstonia solanacearum* in the tomato rhizosphere microbiome (Hu et al., 2016). Flores-Félix et al. (2015) reported that the genus *Phyllobacterium* produced biofilms and well colonized with the roots of strawberry plants. They also observed the increase in yield and quality of strawberry plants with significantly higher content in vitamin C, one of the most interesting bioactive compounds in strawberries. Due to its beneficial attributes, plant probiotics could be used as an alternative to different agro-chemicals inputs such as fertilizers and pesticides in agriculture. These probiotic products could manage the sustainable agriculture and quality agricultural product at low cost (Tikhonovich and Provorov, 2011).

18.4 Conclusions

Current available methodologies to culture microbes in the laboratory are struggling to describe the microbial diversity present in the rhizosphere soil or any other environmental samples. It is known as the great plate count anomaly and most of the microbes present in the environment are unculturable at present laboratory conditions. Meanwhile, all the techniques continue to advance and in recent times the metagenomic technique has been developed, where we can study the total DNA present in a particular environmental sample. This technology was accepted instantly and widely to understand how to scale up our methods to understand the uncultured microbes in the environment. The metagenomic study of the different rhizosphere soil samples showed the microbial richness and diversity in the rhizosphere. Previously unknown microbes and their function in the rhizosphere soil provide a clue to understand the plant–microbe interaction in the rhizosphere zone more closely. Moreover, the construction of metagenomic libraries captures the phylogenetic and genetic diversity in environmental samples. Future advances in understanding the differences among communities or environments will be derived from comparative metagenomics in which libraries prepared from different sites or at different times can be compared. Moreover, the genomics of environmental DNA can lead to the discovery of important physiological processes in uncultured microorganisms. However, it is important to sequence the target DNA very carefully due to the large size of the organisms present in the environmental sample. Some strong correlations between the presence of an organism with microbial processes or soil structure and function would be helpful in making wise choices prior to making large investments in such sequencing projects. Our fundamental task is to find the linkage of specific organisms to the ecosystem function over time and space and this research work must go forward if we want to understand the role of these uncultured organisms in diverse ecosystems.

References

Ahemad, M., Kibret, M., 2014. Mechanisms and applications of plant growth promoting rhizobacteria: current perspective. J. King Saud Univ. Sci. 26, 1–20.

Ahmed, A., Hasnain, S., 2010. Auxin-producing Bacillus sp.: Auxin quantification and effect on the growth of *Solanum tuberosum*. Pure Appl. Chem 82, 313–319.

Amann, R.I., Ludwig, W., Schleifer, K.H., 1995. Phylogenetic identification and in situ detection of individual microbial cells without cultivation. Microbiol. Rev. 59, 143–169.

Badri, D.V., Vivanco, J.M., 2009. Regulation and function of root exudates. Plant Cell Environ. 32, 666–681.

Bai, Y., Liang, J., Liu, R., Hu, C., Qu, J., 2014. Metagenomic analysis reveals microbial diversity and function in the rhizosphere soil of a constructed wetland. Environ. Technol. 35, 2521–2527.

Bais, H.P., Weir, T.L., Perry, L.G., Gilroy, S., Vivanco, J.M., 2006. The role of root exudates in rhizosphere interactions with plants and other organisms. Ann. Rev. Plant Biol. 57, 233–266.

Bakaeva, M.D., Chetverikov, S.P., Korshunova, T.Y., Loginov, O.N., 2017. The new bacterial strain *Paenibacillus* sp. IB-1: a producer of exopolysaccharide and biologically active substances with phytohormonal and antifungal activities. Appl. Biochem. Microbiol. 53, 201–208.

Bakker, P.A.H.M., Berendsen, R.L., Doornbos, R.F., Wintermans, P.C.A., Pieterse, C.M.J., 2013. The rhizosphere revisited: root microbiomics. Front. Plant Sci. 4, 165.

Bal, H.B., Nayak, L., Das, S., Adhya, T.K., 2013. Isolation of ACC deaminase producing PGPR from rice rhizosphere and evaluating their plant growth promoting activity under salt stress. Plant Soil 366, 93–105.

Belimov, A.A., Hontzeas, N., Safronova, V.I., Demchinskaya, S.V., Piluzza, G., Bullitta, S., et al., 2005. Cadmium-tolerant plant growth-promoting bacteria associated with the roots of Indian mustard (Brassica juncea L. Czern.) 37, 241–250.

Beneduzi, A., Peres, D., Beschoren da Costa, P., Zanettini, M.H.B., Passaglia, L.M.P., 2008a. Genetic and phenotypic diversity of plant-growth promoting bacilli isolated from wheat fields in southern Brazil. Res. Microbiol. 159, 244–250.

Beneduzi, A., Peres, D., Vargas, L.K., Bodanese-Zanettini, M.H., Passaglia, L.M.P., 2008b. Evaluation of genetic diversity and plant growth promoting activities of nitrogen-fixing bacilli isolated from rice fields in South Brazil. Appl. Soil Ecol. 39, 311–320.

Berendsen, R.L., Pieterse, C.M.J., Bakker, P.A.H.M., 2012. The second genome and plant health. Trends Plant Sci. 17, 478–486.

Bever, J.D., Platt, T.G., Morton, E.R., 2012. Microbial population and community dynamics on plant roots and their feedbacks on plant communities. Annu. Rev. Microbiol. 66, 265–283.

Boudjeko, T., Tchinda, R.A.M., Zitouni, M., Nana, J.A.V.T., Lerat, S., Beaulieu, C., 2017. *Streptomyces cameroonensis* sp. nov., a Geldanamycin producer that promotes *Theobroma cacao* growth. Microb. Environ. 32, 24–31.

Buée, M., De Boer, W., Martin, F., van Overbeek, L., Jurkevitch, E., 2009. The rhizosphere zoo: an overview of plant-associated communities of microorganisms, including phages, bacteria, archaea, and fungi, and of some of their structuring factors. Plant Soil 321, 189–212.

Çakmakçi, R., Dönmez, F., Aydın, A., Sahin, F., 2006. Growth promotion of plants by plant growth-promoting rhizobacteria under greenhouse and two different field soil conditions. Soil Biol. Biochem. 38, 1482–1487.

Calderón, K., Spor, A., Breuil, M.C., Bru, D., Bizouard, F., Violle, C., et al., 2016. Effectiveness of ecological rescue for altered soil microbial communities and functions. ISME 16, 1751–7362.

Cordero, P., Cavigliasso, A., Príncipe, A., Godino, A., Jofré, E., Mori, G., et al., 2012. Genetic diversity and antifungal activity of native *Pseudomonas* isolated from maize plants grown in a central region of Argentina. Syst. Appl. Microbiol. 35, 342–351.

Crowley, D.E., Rengel, Z., 1999. Biology and chemistry of nutrient availability in the rhizosphere. In: Rengel, Z. (Ed.), Mineral Nutrition of Crops: Fundamental Mechanisms and Implications. Food Products Press, New York, pp. 1–40.

Dakora, F.D., Phillips, D.A., 2002. Root exudates as mediators of mineral acquisition in low-nutrient environments. Plant Soil 245, 35–47.

Dey, R., Pal, K.K., Tilak, K.V.B.R., 2012. Influence of soil and plant types on diversity of rhizobacteria. Proc. Natl. Acad. Sci. USA 82, 341–352.

Dobbelaere, S., Vanderleyden, J., Okon, Y., 2003. Plant growth promoting effects of diazotrophs in the rhizosphere. Crit. Rev. Plant Sci. 22, 107–149.

Doornbos, R.F., Van Loon, L.C., Bakker, P.A.H.M., 2012. Impact of root exudates and plant defense signaling on bacterial communities in the rhizosphere. Agron. Sustain. Dev. 32, 227–243.

Dutta, J., Thakur, D., 2017. Evaluation of multifarious plant growth promoting traits, antagonistic potential and phylogenetic affiliation of rhizobacteria associated with commercial tea plants grown in Darjeeling, India. PloS One 12, e0182302.

Dutta, J., Handique, P.J., Thakur, D., 2015. Assessment of culturable tea rhizobacteria isolated from tea estates of Assam, India for growth promotion in commercial tea cultivars. Front. Microbiol. 6, 1252.

Estrada, G.A., Baldani, V.L.D., de Oliveira, D.M., et al., 2013. Selection of phosphate-solubilizing diazotrophic *Herbaspirillum* and *Burkholderia* strains and their effect on rice crop yield and nutrient uptake. Plant Soil 369, 115–129.

Flores-Félix, J.D., Silva, L.R., Rivera, L.P., Marcos-García, M., García-Fraile, P., Martínez-Molina, E., et al., 2015. Plants probiotics as a tool to produce highly functional fruits: the case of *Phyllobacterium* and vitamin C in strawberries. PLoS One 10, e0122281.

Glick, B.R., 1995. The enhancement of plant growth by free-living bacteria. Can. J. Microbiol. 41, 109–117.

Gopalakrishnan, S., Srinivas, V., Alekhya, G., Prakash, B., Kudapa, H., Rathore, A., et al., 2015. The extent of grain yield and plant growth enhancement by plant growth-promoting broad-spectrum *Streptomyces* sp. in chickpea. SpringerPlus 4. Available from: https://doi.org/10.1186/s40064-015-0811-3.

Gosalbes, M.J., Abellan, J.J., Durban, A., Perez-Cobas, A.E., Latorre, A., Moya, A., 2012. Metagenomics of human microbiome: beyond 16s rDNA. Clin. Microbiol. Infect. Dis. 18, 47–49.

Guo, J.H., Qi, H.Y., Guo, Y.H., Ge, H.L., Gong, L.Y., Zhang, L.X., et al., 2004. Biocontrol of tomato wilt by plant growth-promoting rhizobacteria. Biol. Control 29, 66–72.

Habibi, S., Djedidi, S., Prongjunthuek, K., Mortuza, M.F., Ohkama-Ohtsu, N., Sekimoto, H., et al., 2014. Physiological and genetic characterization of rice nitrogen fixer PGPR isolated from rhizosphere soils of different crops. Plant Soil 379, 51–66.

Hawes, M.C., Gunawardena, U., Miyasaka, S., Zhao, X., 2000. The role of root border cells in plant defense. Trends Plant Sci. 5, 128–133.

Hawkes, C.V., DeAngelis, K.M., Firestone, M.K., 2007. Root interactions with soil microbial communities and processes. The Rhizosphere. Elsevier, New York.

Heneghan, L., Miller, S.P., Baer, S., Callaham Jr., M.A., Montgomery, J., Pavao-Zuckerman, M., et al., 2008. Integrating soil ecological knowledge into restoration management. Restor. Ecol 16, 608–617.

Hill, C., Guarner, F., Reid, G., Gibson, G.R., Merenstein, D.J., Pot, B., et al., 2014. Expert consensus document: the international scientific association for probiotics and prebiotics consensus statement on the scope and appropriate use of the term probiotic. Nat. Rev. Gastroenterol. Hepatol. 11, 506–514.

Hiltner, L., 1904. Uber neue erfahrungen und problem auf dem gebiete der bodenbakteriologie. Arbeit. Deut. Landwirt. Gesell. 98, 59–78.

Hinsinger, P., Marschner, P., 2006. Rhizosphere 2004 conference-perspectives and challenges—a tribute to Lorenz Hiltner. Plant Soil 283, 7–8.

Hinsinger, P., Bengough, A.G., Vetterlein, D., Young, I.M., 2009. Rhizosphere: biophysics, biogeochemistry and ecological relevance. Plant Soil 321, 117–152.

Hirsch, P.R., Mauchline, T.H., 2012. Who's who in the plant root microbiome? Nat. Biotechnol 30, 961–962.

Hu, J., Wei, Z., Friman, V.P., Gu, S.H., Wang, X.F., Eisenhauer, N., et al., 2016. Probiotic diversity enhances rhizosphere microbiome function and plant disease suppression. mBio 7, e01790-16.

Huang, L.F., Song, L.X., Xia, X.J., Mao, W.H., Shi, K., Zhou, Y.H., et al., 2013. Plant-soil feedbacks and soil sickness: from mechanisms to application in agriculture. J. Chem. Ecol. 39, 232–242.

Hungria, M., Nogueira, M.A., Araujo, R.S., 2013. Co-inoculation of soybeans and common beans with rhizobia and *Azospirilla*: strategies to improve sustainability. Biol. Fert. Soil. 49, 791–801.

Jansson, J.K., Neufeld, J.D., Moran, M.A., Gilbert, J.A., 2012. Omics for understanding microbial functional dynamics. Environ. Microbiol 14, 1–3.

Jetiyanon, K., Kloepper, J.W., 2002. Mixtures of plant growth-promoting rhizobacteria for induction of systemic resistance against multiple plant diseases. Biol. Control 24, 285–291.

Jog, R., Pandya, M., Nareshkumar, G., Rajkumar, S., 2014. Mechanism of phosphate solubilization and antifungal activity of *Streptomyces* spp. isolated from wheat roots and rhizosphere and their application in improving plant growth. Microbiology 160, 778–788.

Jones, D.L., Hinsinger, P., 2008. The rhizosphere: complex by design. Plant Soil 312, 1–6.

Kent, A.D., Triplett, E.W., 2002. Microbial communities and their interactions in soil and rhizosphere ecosystems. Annu. Rev. Microbiol. 56, 211–236.

Kim, Y.C., Anderson, A.J., 2018. Rhizosphere pseudomonads as probiotics improving plant health. Mol. Plant Pathol. 19, 2349–2359.

Kumar, R., Bhatia, R., Kukreja, K., Behl, R.K., Dudeja, S.S., Narula, N., 2007. Establishment of *Azotobacter* on plant roots: chemotactic response, development and analysis of root exudates of cotton (*Gossypium hirsutum* L.) and wheat (*Triticum aestivum* L.). J. Basic Microbiol. 47, 436–439.

Kumar, A., Kaushal, S., Saraf, S.A., Singh, J.S., 2018a. Microbial bio-fuels: a solution to carbon emissions and energy crisis. Front. Biosci. (Landmark Ed) 1 (23), 1789–1802.

Kumar, V., AlMomin, S., Al-Aqeel, H., Al-Salameen, F., Nair, S., Shajan, A., 2018b. Metagenomic analysis of rhizosphere microflora of oil-contaminated soil planted with barley and alfalfa. PLoS One 13 (8), e0202127.

Leveau, J.H.J., 2007. The magic and menace of metagenomics: prospects for the study of plant growth promoting rhizobacteria. Eur. J. Plant Pathol. 119, 279–300.

Lucas García, J.A., Domenech, J., Santamaría, C., Camacho, M., Daza, A., Gutierrez Mañero, F.J., 2004a. Growth of forest plants (pine and holm-oak) inoculated with rhizobacteria: relationship with microbial community structure and biological activity of its rhizosphere. Environ. Exp. Bot. 52, 239–251.

Lucas García, J.A., Probanza, A., Ramos, B., Barriuso, J., Gutierrez Mañero, F.J., 2004b. Effects of inoculation with plant growth promoting rhizobacteria (PGPRs) and *Sinorhizobium fredii* on biological nitrogen fixation, nodulation and growth of *Glycine max* cv. Osumi. Plant Soil 267, 143–153.

Mendes, R., Kruijt, M., de Bruijn, I., Dekkers, E., van der, V.M., Schneider, J.H., et al., 2011. Deciphering the rhizosphere microbiome for disease-suppressive bacteria. Science 332, 1097.

Mendes, R., Garbeva, P., Raaijmakers, J.M., 2013. The rhizosphere microbiome: significance of plant beneficial, plant pathogenic, and human pathogenic microorganisms. FEMS Microb. Rev. 37, 634–663.

Narasimhan, K., Basheer, C., Bajic, V.B., Swarup, S., 2003. Enhancement of plant-microbe interactions using a rhizosphere metabolomics-driven approach and its application in the removal of polychlorinated biphenyls. Plant Physiol. 132, 146–153.

Neef, A., Sanz, Y., 2013. Future for probiotic science in functional food and dietary supplement development. Curr. Opin. Clin. Nutr. Metab. Care 16, 679–687.

Ortiz-Castro, R., Valencia-Cantero, E., López-Bucio, J., 2008. Plant growth promotion by *Bacillus megaterium* involves cytokinin signaling. Plant Signal. Behav. 3, 263–265.

Perret, J.S., Al-Belushi, M.E., Deadman, M., 2007. Nondestructive visualization and quantification of roots using computed tomography. Soil Biol. Biochem 39, 391–399.

Philippot, L., Raaijmakers, J.M., Lemanceau, P., Putten, W.H.V., 2013. Going back to the roots: the microbial ecology of the rhizosphere. Nat. Rev. Microbiol. 11, 789–799.

Pineda, A., Zheng, S.J., van Loon, J.J.A., Pieterse, C.M.J., Dicke, M., 2010. Helping plants to deal with insects: the role of beneficial soil-borne microbes. Trends Plant Sci. 15, 507–514.

Raaijmakers, J.M., Paulitz, T.C., Steinberg, C., Alabouvette, C., Moënne-Loccoz, Y., 2009. The rhizosphere: a playground and battlefield for soilborne pathogens and beneficial microorganisms. Plant Soil 321, 341–361.

Rana, A., Saharan, B., Joshi, M., Prasanna, R., Kumar, K., Nain, L., 2011. Identification of multi-trait PGPR isolates and evaluating their potential as inoculants for wheat. Ann. Microbiol. 61, 893–900.

Remans, R., Beebe, S., Blair, M., Manrique, G., Tovar, E., Rao, I., et al., 2008. Physiological and genetic analysis of root responsiveness to auxin-producing plant growth promoting bacteria in common bean (*Phaseolus vulgaris* L.). Plant Soil 302, 149–161.

Sahoo, R.K., Ansari, M.W., Dangar, T.K., Mohanty, S., Tuteja, N., 2014a. Phenotypic and molecular characterisation of efficient nitrogen-fixing *Azotobacter* strains from rice fields for crop improvement. Protoplasma 251, 511–523.

Sahoo, R.K., Ansari, M.W., Pradhan, M., Dangar, T.K., Mohanty, S., Tuteja, N., 2014b. Phenotypic and molecular characterization of native *Azospirillum* strains from rice fields to improve crop productivity. Protoplasma 251, 943–953.

Saravanakumar, D., Lavanya, N., Muthumeena, B., Raguchander, T., Suresh, S., Samiyappan, R., 2008. *Pseudomonas fluorescens* enhances resistance and natural enemy population in rice plants against leaf folder pest. J. Appl. Entomol. 132, 469–479.

Schippers, B., Scheffer, R.J., Lugtenberg, B.J.J., Weisbeck, P.J., 1995. Biocoating of seeds with plant growth promoting rhizobacteria to improve plant establishment. Outlook Agric. 24, 179–185.

Singh, J.S., 2013. Plant growth promoting rhizobacteria: potential microbes for sustainable agriculture. Resonance 18 (3), 275–281.

Singh, J.S., 2015. Microbes: the chief ecological engineers in reinstating equilibrium in degraded ecosystems. Agric. Ecosyst. Environ. 203, 80–82.

Singh, J.S., Gupta, V.K., 2018. Soil microbial biomass: a key soil driver in management of ecosystem functioning. Sci. Total Environ. 634, 497–500.

Singh, C., Tiwari, S., Gupta, V.K., Singh, J.S., 2018. The effect of rice husk biochar on soil nutrient status, microbial biomass and paddy productivity of nutrient poor agriculture soils. Catena 171, 485–493.

Sorensen, J., Hauberg Nicolaisen, M., Ron, E., Simonet, P., 2009. Molecular tools in rhizosphere microbiology—from single-cell to whole-community analysis. Plant Soil 321, 483–512.

Taulé, C., Mareque, C., Barlocco, C., Hackembruch, F., Reis, V.M., Sicardi, M., et al., 2012. The contribution of nitrogen fixation to sugarcane (*Saccharum officinarum* L.), and the identification and characterization of part of the associated diazotrophic bacterial community. Plant Soil 356, 35–49.

Teixeira, L.C.R.S., Peixoto, R.S., Cury, J.C., Sul, W.J., Pellizari, V.H., Tiedje, J., et al., 2010. Bacterial diversity in rhizosphere soil from Antarctic vascular plants of Admiralty Bay, maritime Antarctica. ISME J. 4, 989–1001.

Tejera, N., Lluch, C., Martínez-Toledo, M.V., Gonzàlez-López, J., 2005. Isolation and characterization of *Azotobacter* and *Azospirillum* strains from the sugarcane rhizosphere. Plant Soil 270, 223–232.

Tikhonovich, I.A., Provorov, N.A., 2011. Microbiology is the basis of sustainable agriculture: an opinion. Ann. Appl. Biol. 159, 155–168.

Tiwari, P., Singh, J.S., 2017. A plant growth promoting rhizospheric *Pseudomonas aeruginosa* strain inhibits seed germination in *Triticum aestivum* (L) and *Zea mays* (L). Microbiol. Res. 8, 1–7.

Tiwari, S., Singh, C., Singh, J.S., 2018. Land use changes: a key ecological driver regulating methanotrophs abundance in upland soils. Energy Ecol. Environ. 3, 355–371.

Torsvik, V., Goksøyr, J., Daae, F.L., 1990. High diversity in DNA of soil bacteria. Appl. Environ. Microbiol. 56, 782–787.

Unno, Y., Shinano, T., 2013. Metagenomic analysis of the rhizosphere soil microbiome with respect to phytic acid utilization. Microb. Environ. 28, 120–127.

Upadhyay, S.K., Singh, D.P., Saikia, R., 2009. Genetic diversity of plant growth promoting rhizobacteria isolated from rhizospheric soil of wheat under saline condition. Curr. Microbiol. 59, 489–496.

Urich, T., Lanzen, A., Qi, J., Huson, D.H., Schleper, C., Schuster, S.C., 2008. Simultaneous assessment of soil microbial community structure and function through analysis of the meta-transcriptome. PLoS One 3, e2527.

Uroz, S., Buée, M., Murat, C., Frey-Klett, P., Martin, F., 2010. Pyrosequencing reveals a contrasted bacterial diversity between oak rhizosphere and surrounding soil. Environ. Microbiol. Rep. 2, 281–288.

Valdes, M., Perez, N.O., Estrada-de los Santos, P., CaballeroMellado, J., Pena-Cabriales, J.J., Normand, P., et al., 2005. Non-Frankia actinomycetes isolated from surface-sterilized roots of *Casuarina equisetifolia* fix nitrogen. Appl Environ Microbiol 71, 460–466.

Van Loon, L.C., 2007. Plant responses to plant growth-promoting rhizobacteria. Eur. J. Plant Pathol. 119, 243–254.

Vessey, J.K., 2003. Plant growth promoting rhizobacteria as biofertilizers. Plant Soil 255, 571–586.

Vicre, M., Santaella, C., Blanchet, S., Gateau, A., Driouich, A., 2005. Root border-like cells of *Arabidopsis*. Microscopical characterization and role in the interaction with rhizobacteria. Plant Physiol. 138, 998–1008.

Vimal, S.R., Patel, V.K., Singh, J.S., 2018. Plant growth promoting *Curtobacterium albidum* strain SRV4: an agriculturally important microbe to alleviate salinity stress in paddy plants. Ecol. Indic (in press).

Vinale, F., Sivasithamparam, K., Ghisalberti, E.L., et al., 2008. A novel role for *Trichoderma* secondary metabolites in the interactions with plants. Physiol. Mol. Plant Pathol. 72, 80–86.

Weinert, N., Piceno, Y., Ding, G.C., Meincke, R., Heuer, H., Berg, G., et al., 2011. PhyloChip hybridization uncovered an enormous bacterial diversity in the rhizosphere of different potato cultivars: many common and few cultivar-dependent taxa. FEMS Microbiol. Ecol. 75, 497–506.

Yadegari, M., Rahmani, H.A., Noormohammadi, G., Ayneband, A., 2010. Plant growth promoting rhizobacteria increase growth, yield and nitrogen fixation in *Phaseolus vulgaris*. J. Plant Nutr. 33, 1733–1743.

Young, T.P., Petersen, D.A., Clary, J.J., 2005. The ecology of restoration: historical links, emerging issues and unexplored realms. Ecol. Lett. 8, 662–673.

Zahir, A.Z., Arshad, M., Frankenberger, W.T., 2004. Plant growth promoting rhizobacteria: applications and perspectives in agriculture. Adv. Agron. 81, 97–168.

Chapter 19

Microbial resources mediated bioremediation of persistent organic pollutants

Siddharth Boudh[1], Jay Shankar Singh[1] and Preeti Chaturvedi[2]

[1]*Department of Environmental Microbiology, Babasaheb Bhimrao Ambedkar University, Lucknow, India,* [2]*Aquatic Toxicology Laboratory, Environmental Toxicology Division, CSIR—Indian Institute of Toxicology Research, Lucknow, India*

Chapter Outline

19.1 Introduction

Persistent organic pollutants (POPs) are bioaccumulative, toxic chemicals persistent in nature that do not break down easily in the environment. The POPs are synthetic chemicals that can cause harmful effects on human health and animals. POPs are chemically stable and do not readily degrade in the environment. POPs are lipophilic (have an affinity for fats) and easily soluble in fat and biomagnified as they move up through the food chain. They can travel long distances on wind and water and can be found in regions far away from their points of origin or use. These compounds are semivolatile in nature and because of their physicochemical property they can occur either in vapor phase or can be absorbed in atmospheric particles. These properties help them in their wide ranging transportation in the atmosphere. POPs have become widespread environmental contaminants and a global problem, and due to rapid industrial development and human activities POPs have become lethal to mankind. Some major organic chemical compounds that cause contamination in environment are organochlorine pesticides (OCPs), polycyclic aromatic hydrocarbons (PAHs), polychlorinated biphenyls (PCBs), dioxins, and dibenzofurans (Ersekova et al., 2014).

Various remediation technologies are being developed to treat these pollutants. These compounds stay intact within the environment massively for a long time period because of their nonbiodegradable nature, that is, they resist photolytic, chemical, and biological degradation. In addition, these compounds are of great concern because of long-range transport, persistent nature, and bioaccumulation tendency; once they enter into the food chain they get accumulated in the fatty tissue of the human body and pose a risk to cause adverse effects to the environment, wildlife, and human

New and Future Developments in Microbial Biotechnology and Bioengineering. DOI: https://doi.org/10.1016/B978-0-12-818258-1.00019-4

health (Kallenborn et al., 2012). Information on the concentration of POPs is increasing rapidly. Taken together the data and availability of the information on POPs have allowed an understanding of the environmental fate.

19.2 Characteristic and structural properties of persistent organic pollutants

Pure aldrin is white whereas technical grades are tan to dark brown. It is used for termite control. It is toxic to human but much less toxic to plants. Dieldrin use is restricted to control wood borers, termites, and textile pests. Dieldrin is highly toxic for fish but moderately toxic for warm blooded organisms. Endrin is white crystalline solid, used as insecticide and rodenticide. It is highly persistent in soil with a half-life of up to 12 years in some cases. It is highly toxic for fish, aquatic invertebrates, and phytoplankton. Pure chlordane is water insoluble has a melting point of 104°C. Chlordane is highly persistent in soil with an approximate half-life of 4 years. The chemical properties of chlordane are low water solubility, high stability, and semivolatility so it shows long-range transport, and has been detected in arctic air, water, and organisms. Pure heptachlor is a white crystalline solid that is insoluble in water. It is a persistent dermal insecticide with some fumigant action. It is nonphytotoxic at insecticidal concentrations and shows moderate toxicity for mammals. Dichlorodiphenyltrichloroethane (DDT) was first synthesized by Austrian chemist Othmar Zeidler, but its insecticidal properties were identified by Paul Muler in 1939. It is available in odorless, colorless crystalline, or white powder form and shows bioaccumulation in adipose tissues and biomagnification. It is mainly used on cotton crops; it is also used to control mosquito vectors of malaria in numerous countries. DDT is highly persistent in soil and its half-life extends up to 15 years, and to 7 days in air. DDT is highly toxic to fish and affects their behavior. DDT also affects temperature selection in fish. DDT is acutely toxic to birds by adversely affecting their reproduction. In humans DDT exhibits estrogen-resembling and probable carcinogenic properties. The International Agency for Research on Cancer (IARC) has classified DDT as a possible human carcinogen.

Toxaphene is a waxy solid that is yellow to amber color with a chlorine/turpentine-like odor. It is most heavily used OCP. Toxaphene is highly toxic for fish and shows carcinogenic effect on rats and mice but it is nontoxic to plants. IARC has classified toxaphene and Mirex as a possible human carcinogen. Hexachlorobenzene (HCB) is a chlorinated monocyclic aromatic compound in which the benzene ring is fully substituted by chlorine. HCB is a white crystalline solid. It has been used as fungicide, in the formulation of military pyrotechnic products and tracer bullets, for impregnating paper, and as wood preservatives. It shows toxicity in aquatic animals and is a possible human carcinogen. PCBs and hexachlorocyclohexane are possible human carcinogens while polychlorinated dibenzo-p-dioxins (PCDDs) and polychlorinated dibenzofurans (PCDFs) are toxic for humans in some cases.

There are several conventions held in different places on the globe to eliminate or restrict the production and use of POPs, such as the Stockholm Convention on Persistent Organic Pollutants, Convention on Long-Range Transboundary Air Pollution (CLRTAP), Rotterdam Convention on the prior informed consent procedure for certain hazardous chemicals and pesticides in international trade, and the Basel Convention on the control of transboundary movements of hazardous wastes and their disposal. Molecular structure of different types of POPs in the environment is given in Table 19.1.

The Stockholm Convention is an international environmental treaty to eliminate or restrict the production and use of POPs so that the environment and human health stay protected. Initially, there were 12 POPs listed under Stockholm convention, which were grouped under Annex A, Annex B, Annex C (Table 19.2).

The new list of 16 POPs, according to the Stockholm Convention on Persistent Organic Pollutants (POPs) June (2017), is shown in Table 19.3.

19.3 Principle of microbial degradation of persistent organic pollutants

Microbial degradation by the use of any kind of microorganism including bacteria, fungi, yeast, algae, and actinomycetes requires three vital components: an effective microbe, foodstuff, and nutrient. These three components are known as the "bioremediation triangle." Microorganisms easily find their food from the contaminated environment as they consume pollutants or contaminants as a source of carbon, which provides the nutrients to grow and survive in such harsh environmental conditions. Microorganisms obtain their energy by breaking chemical bonds and transporting electrons away from the pollutant. This process of transporting electrons is termed as an oxidation—reduction reaction, which changes the toxic chemical state of the pollutant. The pollutant that loses an electron is oxidized and the chemical that gains the electron is reduced. The energy released in this oxidation and reduction process is utilized by microorganisms along with carbon to increase the number of microbial cells for the degradation of pollutant more rapidly. Generally, microbes use molecular oxygen for accepting electrons but nitrate, sulfate, carbon dioxide, and iron also serve as

TABLE 19.1 Structure of different types of POPs.

Name	Molecular formula
Aldrin	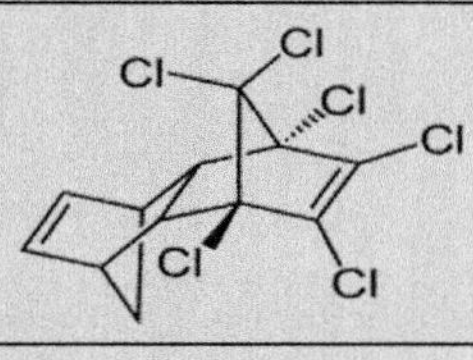
Chlordane	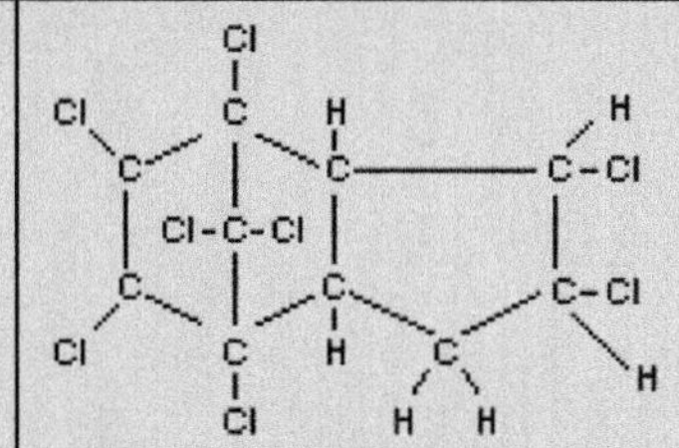
Chlordecone	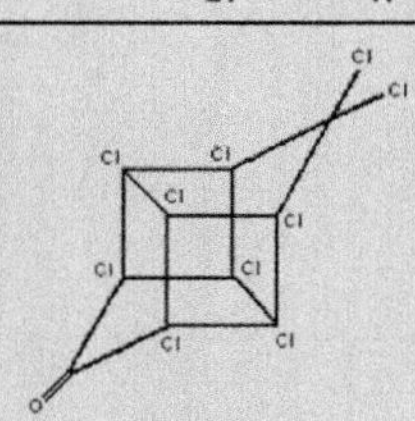
Dichlorodiphenyltrichloroethane (DDT)	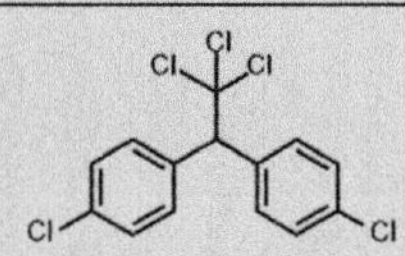
Dieldrin	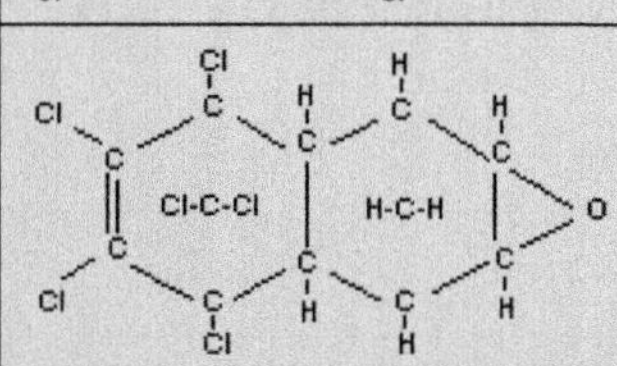
Endrin	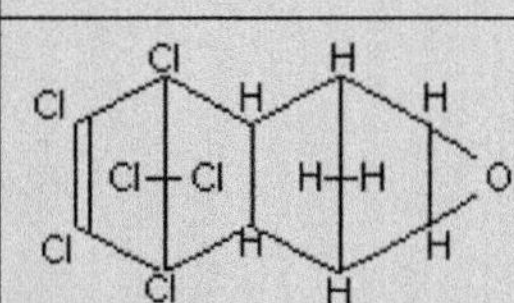
Endosulfane	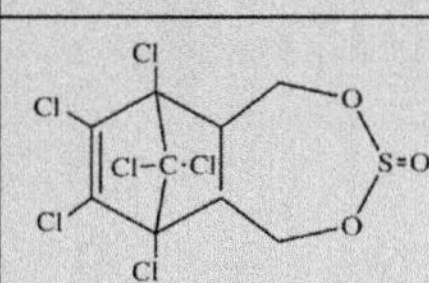
Heptachlor	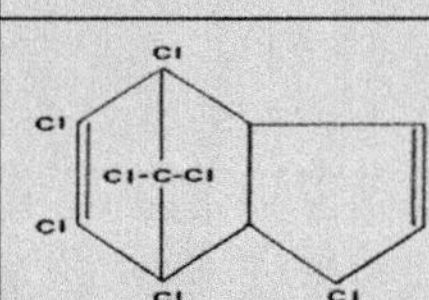
Hexachlorobenzene	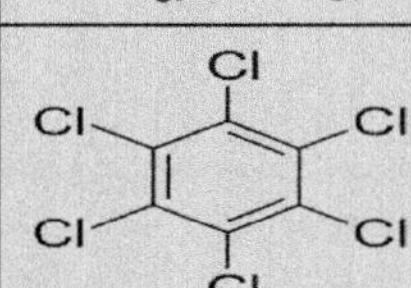

(Continued)

TABLE 19.1 (Continued)

Name	Molecular formula
Hexachlorocyclohexane	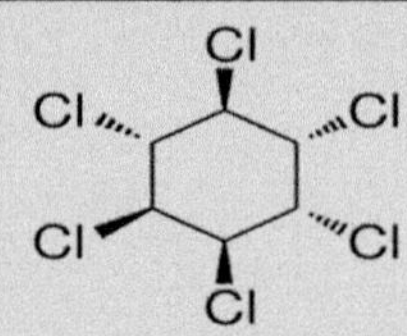
Hexabromobiphenyl	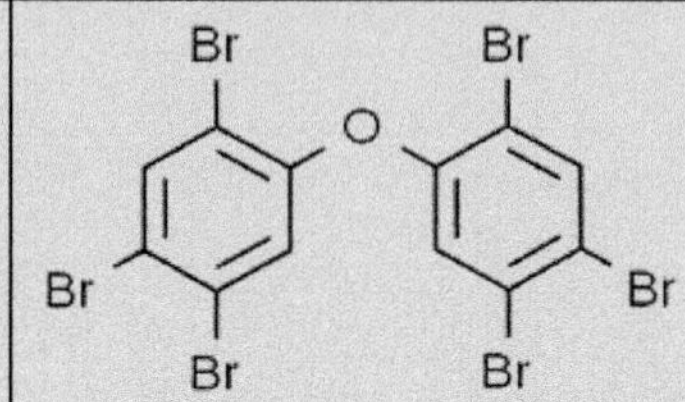
Mirex	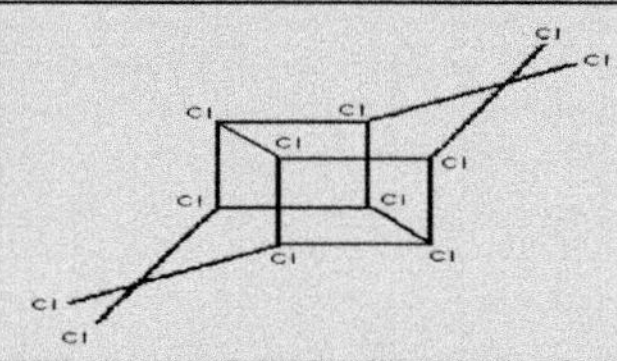
Polychlorinated dibenzofurans (PCDFs)	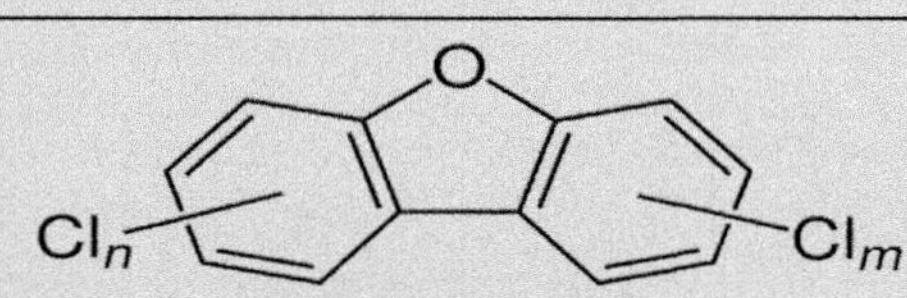
Polychlorinated dibenzo-*p*-dioxins (PCDDs)	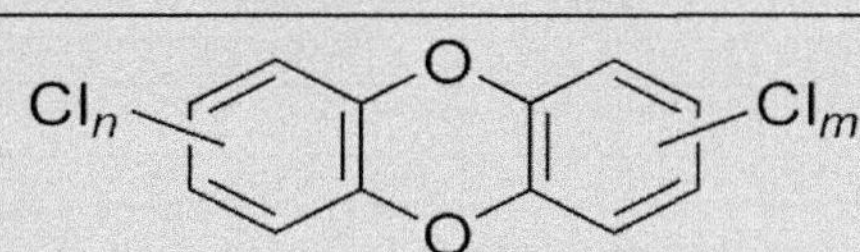
Pentachloroenzene	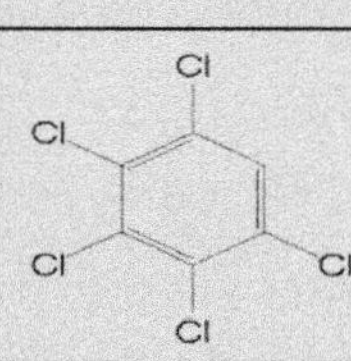
Perfluorooctanoates (PFOs)	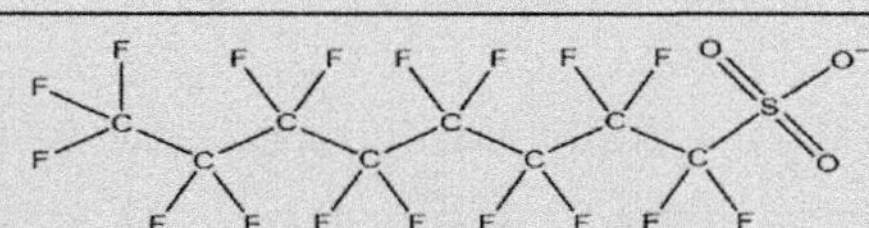
Texaphene	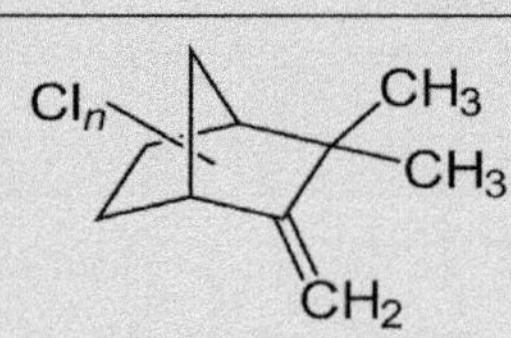

TABLE 19.2 The POPs listed under Stockholm Convention.

Annex A: Pesticides	Aldrin, chlordane, dieldrin, endrin, heptachlor, hexachlorobenzene, mirex, toxaphene, and polychlorinated biphenyl
Annex B: Industrial chemical	dichlorodiphenyltrichloroethane
Annex C: By-product	Polychlorinated dibenzo-*p*-dioxins, dibenzofurans, and hexachlorobenzene

TABLE 19.3 Details of 16 new POPs.

S. no.	Chemical	Annex	Specific exemptions/acceptable purposes
1.	Alpha hexachlorocyclohexane	A	Production: None Use: None
2.	Beta hexachlorocyclohexane	A	Production: None Use: None
3.	Chlordecone	A	Production: None Use: None
4.	Hexabromobiphenyl	A	Production: None Use: None
5.	Hexabromocyclododecane	A	Production: As allowed by the parties listed in the Register of specific exemptions. Use: Expanded polystyrene and extruded polystyrene in buildings in accordance with the provisions of the Part VII of Annex A
6.	Hexabromodiphenyl ether and heptabromodiphenyl ether (commercial octabromodiphenyl ether)	A	Production: None Use: Articles in accordance with the provisions of Part IV of Annex A
7.	Hexachlorobutadiene	A and C	Production: None Use: None
8.	Lindane	A	Production: None Use: Human health pharmaceutical for control of head lice and scabies as second-line treatment
9.	Pentachlorobenzene	A and C	Production: None Use: None
10.	Pentachlorophenol and its salts and esters	A	Production: As allowed for the parties listed in the Register in accordance with the provisions of Part VIII of Annex A Use: Pentachlorophenol for utility poles and cross-arms in accordance with the provisions of Part VIII of Annex A
11.	Perfluorooctane sulfonic acid (PFOS), its salts and perfluorooctane sulfonyl fluoride (PFOSF)	A	Production: For the use below Use: Acceptable purposes and specific exemptions in accordance with Part III of Annex B
12.	Polychlorinated naphthalenes	A and C	Production: For the use below Use: Production of polyfluorinated naphthalenes, including octafluoronaphthalene
13.	Technical endosulfan and its related isomers	A	Production: As allowed for the parties listed in the Register Use: Crop–pest complexes as listed in accordance with the provisions of Part VI of Annex A
14.	Tetrabromodiphenyl ether and pentabromodiphenyl ether (commercial pentabromodiphenyl ether)	A	Production: None Use: Articles in accordance with the provisions of Part V of Annex A
15.	Decabromodiphenyl ether (Commercial mixture, c-DecaBDE)	A	Production: As allowed for the parties listed in the Register Use: Vehicles, aircraft, textile, additives in plastic housings, etc., polyurethane foam for building insulation, in accordance with Part IX of Annex A
16.	Short-chain chlorinated paraffins (SCCPs)	A	Production: As allowed for the parties listed in the Register Use: Additives in transmission belts, rubber conveyor belts, leather, lubricant, additives, tubes for outdoor decoration, bulbs, paints, adhesives, metal, processing, plasticizers

effective electron acceptors. On the basis of the removal or degradation of a pollutant from the contaminated environment, there are basically two important systems, which are as follows.

In situ remediation practices are applied for the treatment of pollutants at the contaminated site where they are found. This technique was originally developed as a cheap, more reliable, and standard alternative method to clean up water bodies and soil sediments contaminated with organic and inorganic pollutants such as hydrocarbons, pesticides, chemical solvents, and heavy metals (Boudh and Singh, 2019). It involves an aerating system to stimulate microorganisms for the degradation of an organic contaminant in the unsaturated zone, that is called as "bioventing" or in saturated soils and groundwater. It is a highly superior method for the management of contaminated environment and enhances chemostatic abilities and catabolic activities of microorganisms. The chemotaxis increases the bioavailability of pollutants under the influence of chemical gradient to the microbial cell and thus plays a crucial role in the transformation of the chemical pollutant (Jain et al., 2005).

It is a biological process that involves the removal of contaminated soil and water from the contaminated site, which is then treated at another place to enhance microbial degradation of organic pollutant with help of indigenous microbial population and supplying nutrients. It requires the bioreactor and nutrients to speed up the breakdown of organic pollutants into their by-products by the process of biostimulation.

The continuous entrance and subsequent activities of POPs form industrial and agricultural sources into the environment, unceasingly deteriorating the quality of environmental flora and fauna. Due to high exposure and induced toxicity, POPs have become the major contaminants in the list of priority pollutants. Over the past few years, various conventional treatment technologies including incineration, land filling, solvent extraction, chemical reduction, solidification alkali metal reduction, and stabilization have been utilized for the removal and degradation of POPs from industrial wastes and soil sources (Perelo, 2010). But, due to the expensive cost and maintenance of sophisticated instruments, and requirement of a high range of chemicals and manpower for the eradication of POPs, such traditional methods are not very effective or reliable for complete degradation or removal of such pollutants regarding their complexity and chemical toxicity in terms of generating an enormous amount of sludge and secondary pollutants, as well as their lethal effects on the working and local area pollution (Chandra and Kumar, 2015). Hence, to overcome these problems of traditional technologies, microbial degradation technology is emerging as a sustainable and cost-effective tool in the field of bioremediation, which explored the new ways and potential of microorganisms (including bacteria, fungi, algae, and actinomycetes) for the significant remediation of POPs from the environment (Fig. 19.1).

Microbial technologies have a number of advantages over chemical technologies due to their nontoxicity and safety from an environmental point of view. Microorganisms are able to utilize minerals, hydrocarbons, lindane, pesticides, and metals by converting or changing their physical and chemical nature into simpler and less or nontoxic products and to use for their growth, nutrition, and metabolic activities (Boudh and Singh, 2019; Gadd, 2010). Microbial communities have the ability to biotransform highly toxic and hazardous states of organic pollutants into a nontoxic product by developing different kinds of resistant mechanism and enzymatic reactions (Ortiz-Hernandez et al., 2013). Different types of microbial enzymes like laccase, lignin peroxidases (LiP), manganese peroxidases (MnP), hydrolases, glutathione (GSH) *S*-transferases (GST), and oxidoreductases produced from bacterial, algal, and fungal sources have been well reported for the detoxification and degradation/removal of POP by catalyzing various enzymatic reaction of

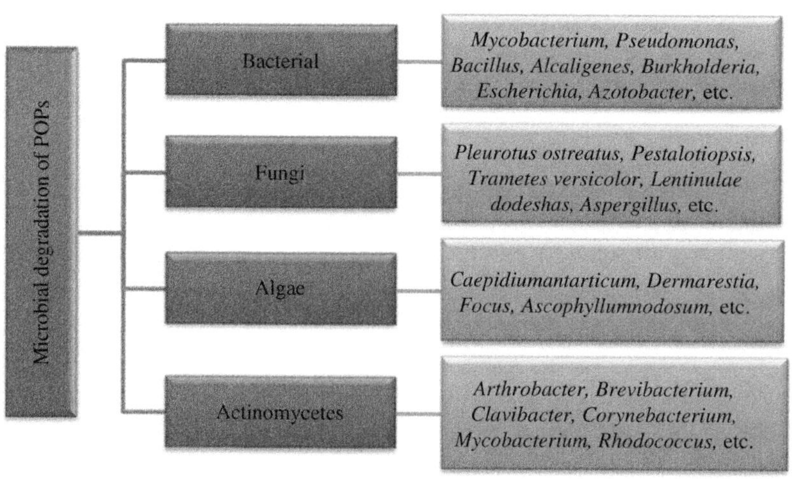

FIGURE 19.1 Microbes involved in degradation of POPs.

oxidation of aromatic compounds, hydrolysis and ring cleavage, reduction of nitro and amino groups, dechlorination, and addition of functional groups to convert their chemical structure (Ramakrishnan et al., 2011). Thus bioremediation is the most advantageous and worthwhile process by which microorganisms are stimulated to quickly reduce or remove hazardous organic pollutants and contaminants to environmentally safe levels in soil sediments and other matrices.

19.4 Bacterial mediated degradation of different types of persistent organic pollutants

Bacteria are one of the most diverse and largest representatives of a taxonomic unit that is ubiquitously distributed into the environment and plays an important role in the field of bioremediation for the degradation and detoxification of POPs and xenobiotics compounds from the contaminated site in an in situ and ex situ manner. Microbial remediation of POPs by the use of bacterial species is a viable biodegradation technology to transform and mineralize harmful offensive organic pollutants into inoffensive metabolites. It is a feasible remedial practice that exploits the metabolic flexibility and adaptability of a microorganism to survive under such a contaminated environment and degrade hazardous pollutants.

The biodegradation process through microbes can take place either aerobically or anaerobically and degrade pollutants into their final products of biomass, carbon dioxide, and methane. Aerobic degradation process involves a variety of aerobic bacteria such as *Pseudomonas*, *Bacillus*, *Micrococcus*, and *Escherichia*, which are able to grow in the presence of oxygen and utilize molecular oxygen as a terminal electron acceptor and cosubstrate for the conversion or biotransformation of the pollutant into of carbon dioxide. The anaerobic process is completed in the absence of oxygen with help of anaerobic bacterial strains of *Desulfovibrio*, *Methanococcus*, *Methanospirillum*, *Desulfotomaculum*, etc., which react with the functional group of organic pollutants and withdraw the electron from it and cause breakdown or degradation of such pollutant leading to the formation of methane and inorganic salts. Anaerobic microbial degradation is more useful for transformation or mineralization of tetrachloro-ethylene, chlorinate phenols, and nitro-substituted aromatics in comparison to the aerobic process. Anaerobic degradation also involves methanotrophic and methanogenic bacteria due to the formation of methane gas as by-product of anaerobic digestion. Methanotrophs are able to utilize this methane for their metabolic activity and this simultaneously results in the formation of carbon dioxide as a final end product (Jha et al., 2011). During methanogenesis, these microbes oxidize methane gas by the use of oxygen and convert it into carbon dioxide. Methanotrophs also produce methane monooxygenase enzyme, which can react with methane and degrade an extensive range of chlorinated hydrocarbons. In addition to methanotrophs, aerobic and anaerobic degradation processes also utilize nitrifying and denitrifying, sulfate reducers, and ferric ion reducers bacteria act as an electron acceptor that favors the complete mineralization of pollutant depending on their electron acceptor availability (Diaz, 2004). Many bacteria have been recognized and reported for their degradative potential to utilize many organic pollutants including PAHs, PCBs, DDT, etc.

19.4.1 Degradation of pesticides

Pesticides are extensively used in agricultural practices and continuously accumulate into soil systems, which persists for years due to their problematic degradation (Boudh and Singh, 2019). Generally, aldrin, dieldrin, DDT, benzene hexachloride (BHC), endrin, carbamates, organochlorines, etc. are used, which are highly toxic and harmful for nontargeted organisms. Bacterial reduction of these pesticides represents a safe and natural way for environmental management. A variety of bacterial strains have been reported from different habitats such as soil, wastewater, activated sludge, and animal feces that can easily metabolize DDT and convert it into dichlorodiphenyldichloroethane (DDD) under both aerobic and anaerobic conditions (Lai and Saxena, 1982). Most pesticides are degraded via enzymatic pathway due to the chemical complexity of pesticides. Enzymatic degradation of pesticides involved two-stage metabolism processes. There are mainly three types of enzymes responsible for the proper degradation and detoxification of pesticides. In first stage metabolism are hydrolases, esterases, and mixed function oxidases (MFO), whereas, second stage metabolism requires GST. Hydrolases commonly work in the absence of redox cofactors. Esterases catalyze the hydrolysis of carboxyl esters, amides and phosphate esters, etc. MFO catalyzes both oxidation and reduction reactions by incorporating one oxygen molecule into the substrate while the other one is reduced to water. MFO comprises two membrane protein enzymes: cytochrome P450 and NADPH-dependent cytochrome P450 reductase.

The *Alcaligenes eutrophus* A5 was found capable for the aerobic degradation of DDT (Nadeau et al., 1994). This bacterium possess dioxygenase enzyme that can cause oxidation of DDT at *ortho* and *meta* position resulting in an unstable intermediate dihydrodiol compound. This compound further degraded into two hydroxylated compounds by dehydrogenase and finally through *meta*-cleavage results in a terminal end product of 4-CBA (chlorobenzoate).

The OCPs are highly toxic and carcinogenic and their extensive use in the agricultural practices create a dangerous threat to the environment. The *Alcaligenes faecalis* JBW4 was identified and reported (Kong et al., 2013) for the effective degradation of potential for endosulfan residue of organochlorine.

19.4.2 Degradation of polycyclic aromatic hydrocarbons

PAHs are ubiquitous, stable recalcitrant pollutants in the environment with high hydrophobicity and low solubility and therefore are not easily eliminated from the atmosphere (Brassington et al., 2007). A large number of bacteria that can easily metabolize PAHs completely into carbon dioxide and other metabolic intermediates have been identified (*Pseudomonas*, *Mycobacterium*, *Alcaligenes*) (Harayama, 1997). Several low molecular weight PAHs (naphthalene and anthracene) are easily degraded by the soil microorganisms. *Streptomyces* sp. ERI-CPDA-1 isolated from oil contaminated soil was able to degrade 99.14% naphthalene, 17.5% phenanthrene, and 98.25% diesel oil in just 7 days (Balachandran et al., 2012). *Brevibacterium* sp. DPO1361 and *Pseudomonas* sp. F274 is reported for the effective reduction of fluorine via monooxygenation resulting in the formation of phthalate followed by converting into protocatechuate. Enzymatic degradation of phenanthrene was observed by *Pseudomonas mendocina* with help of two active enzymes: one is PAH dioxygenase and the other is catechol-2-3 oxygenase (Tian et al., 2002). Dioxygenases enzymes are mainly responsible for the degradation of PAH by oxidizing its benzene ring and from *cis*-dihydrodiols. These dihydrodiols are further catalyzed by dehydrogenase to form dihydroxylated intermediates that can finally be metabolized into carbon dioxide and water through catechols. Sometime complex bi- and tricyclic PAHs are degraded by the oxidation of multicomponent enzyme system of naphthalene dioxygenase that contained NADH oxidoreductase, ferredoxin, and an oxygenase component. Biotransformation of naphthalene was easily carried out by bacteria as it is the most simple and soluble form of PAH catalyzed by naphthalene dioxygenase (Goyal and Zylstra, 1997), and attacks the aromatic ring to form *cis*-(1R, 2S) dihydroxy-1,2-dihydronaphthalene. This intermediate is simultaneously dehydrogenated to 1,2-dihydronaphthalene via *cis*-dehydrogenase and subsequently metabolized to salicylate hydroxyl-2H-chromene-2-carboxylic acid, *cis-o*-hydroxybenzyl pyruvate, and 2-hydroxybenzaldehyde. The salicylate is subsequently converted to catechol through decarboxylation followed by ring fission in *meta*- and *ortho*-pathways.

19.4.3 Degradation of polychlorinated biphenyls

PCBs are hazardous pollutants and extremely toxic due to their long-term persistence. They include chlorinated organic pollutants such as pentachlorophenol (PCP), tributyltin, tetrachlorophenol, and atrazine. These PCB compounds can be degraded by several aerobic and anaerobic bacterial strains such as *Pseudomonas*, *Bacillus*, *Rhodococcus*, *Burkholderia* (Ahmed and Focht, 1973). These bacteria transform PCBs into their metabolic products by utilizing biphenyl as a sole carbon source for their growth and nutrition. PCBs are mostly degraded by microbial biphenyl dioxygenase via *meta*-cleavage pathway, which results in two major intermediate compounds: trichloroacetic acid (TCA) and CBA. Various types of biphenyl (Bph) dioxygenase (BphA, BphB, BphC, and BphD) are reported for the catabolic transformation of PCBs via biphenyl upper (Bph) pathway and lower pathway. Bph dioxygenase is a three-component system enzyme having an electron transport chain (that catalyzes the transportation of electron from NADH to terminal dioxygenase) ferredoxin and ferredoxin reductase (Furukawa et al., 2004). *Pseudomonas pseudo alcaligenes* KF707 was investigated for the efficient degradation potential of PCBs with reduced accumulation of CBA (Ohmori et al., 2011). It follows an upper pathway enzymatic system. The degradation of PCB starts with the formation of 2,3-dihydro-2,3-diol by action of 2,3 dioxygenase (BphA) by introducing molecular oxygen into biphenyl molecule followed by the action of dihydrodiol dehydrogenase (BphB) leading to the dehydrogenation of 2,3-dihydro-2,3-diol into 2,3-dihydroxybiphenyl. This 2,3-dihydroxybiphenyl is cleaved by 2,3-dihydroxybiphenyl 1,2 dioxygenase (BphC) at meta position and results in the formation of 2-hydroxy-6-oxo-6-phenylhexa-2,4-dienoic acid. This intermediate is catalyzed by 2-hydroxy-6-oxo-6-phenylhexa-2,4-dieneoate hydrolase (BphD) and finally results in an end product of chlorinated benzoic acid (Ohmori et al., 2011).

19.5 Fungi mediated degradation of persistent organic pollutants

Fungi are the most diverse and primitive saprophytic, eukaryotic organisms having the ability to grow on different organic and inorganic substrates present in the environment. They play an important role in the decomposition of efficient decomposers of lignin and cellulose. Lignin and cellulose are long chain organic pollutants that are difficult to degrade. The use of fungi for biodegradation of such long chain organic compounds has intensified in recent years.

Other recalcitrant organic pollutants including phenols, PCBs, petroleum, PAHs, oils, wood, textile dyes, pesticides, and leather can also be converted into harmless products by use of fungi.

The process of utilization of fungal species for the degradation and biotransformation or remediation of toxic pollutants is known as mycoremediation or mycodegradation. The increasing use of such remediation technology provides a new opportunity for the biodegradative pathway to decontaminate the environment. Mycodegradation is recognized as very useful, cost-effective, and eco-friendly way for the removal of POPs. White rot fungi are most probably used for degradation and reduction of such pollutants by using a wide array of intracellular as well as extracellular enzymes includes laccases, lipases, LiP, MnP, versatile peroxidase, and sometimes cytochrome P450 (Bhattacharya et al., 2012). These enzymes have high oxidative potential and ability to oxidize bulky substrates into a valuable product through their flexible machinery conjoining with secondary metabolites. The white rot fungi have the significant competency to reduce complex polysaccharide, lignin, PAHs, and PCBs by involving enzymatic biotransformation from soil sediments and wastewaters into a new simpler form. *Phanerochaete chrysosporium* is a member of the white rot fungi and is one of the best strains that has been extensively reported for the enzymatic reduction of fluorine and benzopyrene into quinone through secreting LiP and MnP. Several other fungal strains like *Pleurotus ostreatus*, *Pestalotiopsis*, *Trametes versicolor*, *Lentinula edodes*, *Aspergillus*, and certain molds can greatly extend the process of remediation of insecticide endosulfan, imazalil, thiophanate, ortho-phenyl phenol, atrazine, chlorpyrifos, starch, pectins, fats, and keratin even more profitably via oxidative and reductive processes of ligninolytic enzymes (Chandra and Chowdhary, 2015).

Ligninolytic enzymes play a crucial role in the delignification of lignocellulosic waste materials, degradation of toxic pollutants, and removal of xenobiotic substances, wastewater treatment and decolorization of dyes, and conversion of high molecular weight coal fraction to low molecular weight coal fractions (Yadav and Yadav, 2015).

19.5.1 Lignin peroxidase

It is an extracellular heme peroxide of white rot basidiomycetes belongs to the family of oxidoreductases and its various isoforms are known to exist from other fungal sp. also. Aromatic rings are oxidized by LiP enzymes, which are moderately activated via electron donating substitutes. It is mainly responsible for the oxidative cleavage of $C\alpha{-}C\beta$ bonds in nonphenolic aromatic substrates to form highly reactive aryl cation radicals (Ruiz-Dueñas et al., 2008).

19.5.2 Manganese peroxidase

MnP is a glycoprotein dependent H_2O_2 oxidize Mn (II) to Mn (III) and catalyzes the degradation of phenolic substrates with the formation of phenoxy radicals (Wong, 2009). Mn (III) acts as a chelating agent with organic acids having high redox potential and oxidizes phenolic compounds.

19.5.3 Laccase

Laccases are multinuclear, multicopper containing enzymes that catalyze the oxidation of various organic substrates including phenol with the simultaneous reduction of molecular oxygen to water (Alcade, 2007). They are not only produced from white rot fungi, but also from the members belong to the family Basidiomycetes, Ascomycetes, and Deuteromycetes. Laccases have been used for various applications in organic synthesis for example in the oxidation of functional groups, the coupling of phenol and steroids and medical agents, that is, anesthetics, antiinflammatory, antibiotics, etc. (Chandra and Chowdhary, 2015). Rather than these enzymes, there are several other extracellular oxidoreductases and auxiliary oxidases like glucose oxidase, glyoxal oxidase, aryl alcohol oxidase, pyranose 2-oxidase, and cellobiose/quinone oxidoreductase that may also contribute into the biodegradation of organic pollutants (Wong, 2009).

19.6 Algae mediated degradation of persistent organic pollutants

The utilization of micro- and macroalgae for the treatment of POPs, including recalcitrants, xenobiotics and pesticides, phenolics, and heavy metal compounds from wastewater and soil is known is phycoremediation. Phycoremediation is gaining greater importance and attention from the environmental perspective due to its excellence in photosynthesis, energy metabolism, and survival capability under harsh stress condition of extremely toxic pollutants. In marine and aquatic ecosystem algae, cyanobacteria, and protozoan species are effectively involved in the degradation process of pesticides, polyaromatic compounds, and hydrocarbons. Various algal species like *Rhodococcus*, *Caepidium antarcticum*, *Dermarestia* sp., *Fucus* sp., and *Ascophyllum nodosum* are investigated and reported for efficient degradation of hydrocarbons by producing certain hydrocarbon-degrading enzymes (Chekroun et al., 2014). Generally, hydrocarbons, lindane, pesticides, organ chlorines, organophosphates, carbonates, nitro phenols, cyclic aromatic, and xenobiotic compounds severely affect the environment by their widespread use in agricultural and industrial management practices (Boudh and Singh, 2019). These compounds dangerously produce hazardous health effects in all living beings due to their toxic persistent nature. Therefore phycoremediation is considered as the most desirable technology for the removal and detoxification of harmful products into a harmless product through bioaccumulation strategies or by doing CO_2 fixation.

Degradation of 2,4,6-trinitrotoluene (TNT) and 1,3,5-trinitrobenzene was investigated by (Cruz-Uribe et al., 2007) by the use of three different algal species *Acrosiphonia coalita* (green alga), *Porphyra yezoensis* (red alga), and *Portieria hornemannii* and observed that all the three species are capable to reduce explosive toxic TNT compound to 2-amino,4,6 2,4-Dinitrotoluene (DNT) and 4-amino, 2,6, dinitrotoluene under axenic conditions. Algae and cyanobacteria utilize organic pollutants as their nutritive and growth source. An investigation of gene expression of algae *Chlamydomonas reinhardtii* under the higher concentration of TNT was reported by Patel et al. (2004). According to their investigation, ~158 responsive genes were differentially expressed in the presence of TNT stress, resulting in the upregulation of the genes responsible for energy metabolism and photosynthetic ability.

The most common feature of algae and cyanobacteria is their photosynthetic quality because they contain chlorophyll pigment machinery. This photosynthetic ability plays an important role in the degradation of organic pollutants. The mechanism of degradation of any organic or xenobiotic pollutant can be processed in three phase route by algae and other photosynthetic organisms (Torres et al., 2008). Phase I process involves the oxidation, reduction, and hydrolysis reaction by addition of reactive functional groups, which transforms lipophilic recalcitrants or xenobiotics into water-soluble compound and NADPH with the help of Cyt P450 and Cytb5. In Phase II large and polar groups like GSH and glucuronic acid are added to xenobiotics and finally in Phase III followed by Phase I and II, xenobiotic compounds are compartmentalized into cell walls or vacuoles (Torres et al., 2008). Like, bacterial and fungal organisms, algal species also produce some valuable enzymes, which are very useful for eradication of POPs (Chekroun et al., 2014).

19.7 Actinomycetes mediated degradation of persistent organic pollutants

Actinomycetes belong to the phylum of Gram-positive bacteria having great economic importance in respect to agriculture and forestland due to their valuable involvement in the soil ecosystem. These microorganisms have a prodigious potential for the degradation and detoxification of a variety of organic pollutants like pesticides, hydrocarbons, and aromatic compounds. Biodegradation of contaminated pesticides like organophosphates, acetanilide, organochlorines, carbamates, sulfonylurease, and s-triazines is very possible with the use of actinomycetes (De Schrijver and De Mot, 1999). Actinomycetes also play an important role for the degradation of complex polymers and help in the bioremediation of terrestrial and aquatic soil ecosystem through the recycling of contaminated toxic organic pollutants (Goodfellow and Williams, 1983).

Actinomycetes strains produce a variety of extracellular peroxidase enzymes, cellulose, and hemicellulose degrading enzymes, which are found to most suitable for the solubilization of lignin and its related complex polysaccharide compounds into the simple form of sugars (mono- and disaccharides) by using hydrolases (Monson et al., 2001). Currently, petroleum hydrocarbons are becoming a major environmental soil contaminant, which raises serious concerns regarding their extensive use in industries and households in the form of energy and chemical substances. Some useful research investigations explored the importance of Streptomyces flora as a good bioagent for the biotransformation and biodegradation of chlorinated hydrocarbons (Barabas et al., 2001). Actinomycetes microorganisms related to the genus *Arthrobacter*, *Brevibacterium*, *Clavibacter*, *Corynebacterium*, *Mycobacterium*, *Rhodococcus*, *Streptomyces*, etc. have been reported for their extreme potential for the remediation of pesticides through cometabolism. These microbes could utilize substrates of pesticides as a source of energy for their growth and metabolizing into a nutrient source of carbon, phosphorus, nitrogen, and sulfur (De Schrijver and De Mot, 1999).

Polychlorinated phenols, DDT, aldrin, and s-triazines have been found to detoxified and metabolized into their intermediates under aerobic as well as an anaerobic environment with help of actinomycetes. Most PCPs are degraded by removing chlorine substituents and cleaving ring whereas mono- and dichlorophenols are degraded only after ring cleavage.

Cytochrome P450s are imperative enzymes that are mainly responsible for biodegradation mechanism of actinomycetes. These enzymes are related to monooxygenases and having broad substrate specificity in actinomycetes catalyzes the oxidation of substrate for the remediation purpose of pollutants. *Mycobacterium chlorophenolicum* PCP-1 is reported for degradation of PCP by sequential dechlorination catalyzed by microbial enzyme action (Apajalahti and

Salkinoja-Salonen, 1987). It can easily cultivate at higher concentration PCP because it could consume PCP as its sole carbon source for its growth and energy. Dechlorination starts with *para-* and *ortho*-hydroxylation of PCP catalyzed by membrane-bound P450 and cytoplasmic hydroquinone dehalogenase enzymes yielding small intermediates and finally degraded into maleic acid (Uotila et al., 1991). Biodegradation of DDT to DDD and Pentachloronitrobenzene (PCNB) to pentachloro aniline was also reported by *Nocardia* and *Streptomyces* strains. However, actinomycetes are a considerable source of treatment for POPs, but, a keen advancement is needed for their better exploitation in the field of bioremediation for the complete mineralization and detoxification of harmful organic chemicals.

19.8 Conclusion and future challenges

Contamination by POPs is widespread because POPs are chemically stable, bioaccumulate, and circulate globally by means of atmosphere, oceans, and other pathways. This article is about the bioremediation of POPs by microbial resources and future research for the next upcoming years. By upgrading the basic source inventories, national and international supplemented broad area coverage of POPs concentration and its bioremediation reliability improvements may be beneficial databases for POPs and its bioavailability from groundwater, soil, and sediments.

Expert research communities who have studied more traditional POPs can shift their studies to newer classes of POPs, possibly more polar persistent compound metabolites, and their biodegradation, toxicity mechanism, trace gas/particles, and other contaminants, which cause a serious threat to the entire globe. Hence more detailed studies could be done that can help in understanding the important aspects about controlling this global problem of POPs and development in quantitative assessment can be undertaken in the exposure and air exchange of the POPs. Further improvement is required and it is very challenging to the researchers to develop new techniques in identification and quantification of new POPs. Government bodies need to make stricter rules and regulations on releasing and exposing the harmful contaminants into the environment. Anthropogenic activities have caused huge pollution to the environment. The POPs and organic pollutants such as PAHs, pesticides, industrial chemicals, and by-products like PCDFs are toxic to the environment and represent a threat to humans and wildlife because they bioaccumulate in adipose tissue as they are soluble in fat and insoluble in water and persistent in nature, making POPs extremely dangerous. Bioremediation is a cost-effective, safe, and remarkable alternative to traditional physicochemical practices for the remediation of these POPs at a contaminated site, as it can selectively degrade the pollutants without harming the site or its indigenous flora and fauna. However, despite being an effective solution to the contaminated environmental problems bioremediation technologies have limited applications due to the substrate and environmental variability and limited biodegradative potential of microorganism. Fundamental to these concerns is the management of POPs (POPs) including various other harmful chemicals that are lethal to humans and wildlife. Therefore the health problems, environmental concerns, and various techniques for biodegradation of POPs need to be addressed before their extensive implementation.

Acknowledgments

The authors thank the Head, Department of Environmental Microbiology, Babasaheb Bhimrao Ambedkar University, Lucknow, for providing facilities and encouragements. Siddharth Boudh is thankful to UGC, New Delhi, for financial support from Rajiv Gandhi National Fellowship (Award Letter No. F.1-17.1/2013-14/RGNF-2013-14-SC-UTT-37387).

References

Ahmed, M., Focht, D.D., 1973. Degradation of polychlorinated biphenyls by two species of *Achromobacter*. Can. J. Microbiol. 19, 47–52.

Alcade, M., 2007. Laccase: biological functions structure and industrial applications. In: Polaina, J., Maccabe, A.P. (Eds.), Industrial Enzymes: Structure, Function and Applications, 26. Springer, The Netherlands, pp. 461–476.

Apajalahti, J.H.A., Salkinoja-Salonen, M.S., 1987. Dechlorination and *para*-hydroxylation of polychlorinated phenols by *Rhodococcus chlorophenolicus*. J. Bacteriol. 169, 675–681.

Balachandran, C., Duraipandian, V., Balakrishnan, K., Ignacimuthu, S., 2012. Petroleum and polycyclic aromatic hydrocarbons (PAHs) degradation and naphthalene metabolism in *Streptomyces* sp. (ERI-CPDA-1) isolated from oil contaminated soil. Bioresour. Technol. 112, 83–90.

Barabas, G., Vargha, I.M., Szabo, A., Penyige, S., Damjanovich, J., Szollosi, J., et al., 2001. N-alkane uptake and utilization by *Streptomyces* strains. Ant. Van. Leeu. 79, 269–276.

Bhattacharya, S., Angayarkanni, J., Das, A., Palaniswamy, M., 2012. Mycoremediation of benzo [A] pyrene by *Pleurotus ostreatus* isolated from Wayanad district in Kerala, India. Int. J. Pharm. Bio. Sci. 2, 84–93.

Boudh, S., Singh, J.S., 2019. Pesticide contamination: environmental problems and remediation strategies. In: Bharagava, R., Chowdhary, P. (Eds.), Emerging and Eco-Friendly Approaches for Waste Management. Springer, Singapore, pp. 245–269.

Brassington, K.J., Hough, R.L., Paton, G.L., Semple, K.T., Risdon, G.C., Crossley, J., et al., 2007. Weathered hydrocarbon wastes: a risk management primer. Crit. Rev. Environ. Sci. Technol. 37, 199–232.

Chandra, R., Chowdhary, P., 2015. Properties of bacterial laccases and their application in bioremediation of industrial wastes. Environ. Sci. Process. Impacts 17, 326–342.

Chandra, R., Kumar, V., 2015. Biotransformation and biodegradation of organophosphates and organohalides. In: Chandra, R. (Ed.), Environmental Waste Management. CRC Press, Boca Raton, FL, pp. 475–527.

Chekroun, K.B., Sánchez, E., Baghour, M., 2014. The role of algae in bioremediation of organic pollutants. Int. J. Environ. Res. Pub. Environ. Health 1, 19–32.

Cruz-Uribe, O., Cheney, D.P., Rorrer, G.L., 2007. Comparison of TNT removal from seawater by three marine macroalgae. Chemosphere 67, 1469–1476.

De Schrijver, A., De Mot, R., 1999. Degradation of pesticide by actinomycetes. Crit. Rev. Microbiol. 25 (2), 85–119.

Diaz, E., 2004. Bacterial degradation of aromatic pollutants: a paradigm of metabolic versatility. Int. Microbiol. 7, 173–180.

Ersekova, A., Hilscherova, K., Klanova, J., Giesy, J.P., Novak, J., 2014. Effect-based assessment of passive air samples from four countries in Eastern Europe. Environ. Monit. Assess. 186, 3905–3916.

Furukawa, K., Suenaga, H., Goto, M., 2004. Biphenyl dioxygenases: functional versatilities and directed evolution. J. Bacteriol. 186, 5189–5196.

Gadd, G.M., 2010. Metals, minerals and microbes. Geomicrobiol. Bioremed. Microbiol. 156, 609–643.

Goodfellow, M., Williams, S.T., 1983. Ecology of actinomycetes. Annu. Rev. Microbiol. 37, 189–216.

Goyal, A.K., Zylstra, G.J., 1997. Genetics of naphthalene and phenanthrene degradation by *Comamonas testosteroni*. J. Ind. Microbiol. Biotechnol. 19, 401–407.

Harayama, S., 1997. Polycyclic aromatic hydrocarbons bioremediation design. Curr. Opin. Biotechnol 8, 268–273.

Jain, R.K., Kapur, M., Labana, S., Lal, B., Sarma, P.M., Bhattacharya, D., et al., 2005. Microbial diversity: application of micro-organisms for the biodegradation of xenobiotics. Curr. Sci. 89, 101–112.

Jha, A.K., Singh, K., Sharma, C., Singh, S.K., Gupta, P.K., 2011. Assessment of methane and nitrous oxide emission from livestock in India. J. Earth Sci. Clim. Change 1, 107.

Kallenborn, R., Halsall, C., Dellong, M., Carlsson, P., 2012. The influence of climate change on the global distribution and fate processes of anthropogenic persistent organic pollutants. J. Environ. Monit 14, 2854–2869.

Kong, L., Zhu, S., Zhu, L., Xie, H., Wei, K., Yan, T., et al., 2013. Colonization of *Alcaligenes faecalis* strain JBW4 in natural soils and its detoxification of endosulfan. Appl. Microbiol. Biotechnol. 98 (3), 1407–1416.

Lai, R., Saxena, D.M., 1982. Accumulation, metabolism, and effects of organochlorine insecticides on microorganisms. Microbiol. Rev. 46, 95–127.

Monson, M.G., Ball, A.S., Reeder, B.J., Silkstone, G., Nicholls, P., Wilson, M.T., 2001. Extracellular heme peroxidases in actinomycetes: a case of mistaken identity. Appl. Environ. Microbiol. 67, 4512–4519.

Nadeau, L.J., Menn, F.M., Breen, A., Sayler, G.S., 1994. Aerobic degradation of 1,1,1-tricholor-2,2-bis(4-chlorophenyl)ethane (DDT) by *Alcaligenes eutrophus* A5. Appl. Environ. Microbiol. 60, 51–55.

Ohmori, T., Morita, H., Tanaka, M., Miyauchi, K., Kasai, D., Furukawa, K., et al., 2011. Development of a strain for efficient degradation of polychlorinated biphenyls by patchwork assembly of degradation pathways. J. Biosci. Bioeng. 111, 437–442.

Ortiz-Hernandez, L.M., Sanchez-Salinas, E., Dantan Gonzalez, E., Godinez, M.L.C., 2013. Pesticide biodegradation: mechanism, genetics and strategies to enhance the process. Intech. Open Sci. 38, 251-287.

Patel, N., Cardoza, V., Christensen, E., Rekapalli, B., Ayalew, M., Stewart, C.N., 2004. Differential gene expression of *Chlamydomonas reinhardtii* in response to 2,4,6-trinitrotoluene (TNT) using microarray analysis. Plant Sci. 167, 1109–1122.

Perelo, L.W., 2010. Review: in situ and bioremediation of organic pollutants in aquatic sediments. J. Hazard. Mater. 177, 81–89.

Ramakrishnan, B., Megharaj, M., Ventateshwarlu, K., Sethunathan, N., Naidu, R., 2011. Mixtures of environmental pollutants: effects of microorganisms and their activities in soils. Rev. Environ. Poll. Contam. Toxicol 211, 63–120.

Ruiz-Dueñas, F.J., Morales, M., Mate, M.J., Romero, A., Martínez, M.J., Smith, A.T., 2008. Site-directed mutagenesis of the catalytic tryptophan environment in *Pleurotus eryngii* versatile peroxidase. Biochemistry 47, 1685–1695.

Tian, L., Ma, P., Zhong, J.J., 2002. Kinetics and key enzyme activities of phenanthrene degradation by *Pseudomonas mendocina*. Process Biochem. 37, 1431–1437.

Torres, M.A., Barros, M.P., Campos, S.C.G., Pinto, E., Rajamani, S., Sayre, R.T., et al., 2008. Biochemical biomarkers in algae and marine pollution: a review. Ecotoxicol. Environ. Saf 71, 1–15.

Uotila, J.S., Salkinoja-Salonen, M.S., Apajalahti, J.H.A., 1991. Dechlorination of pentachlorophenol by membrane bound enzymes of *Rhodococcus chlorophenolicus* PCP-I. Biodegradation 25–31.

Wong, D.W.S., 2009. Structure and action mechanism of ligninolytic enzymes. Appl. Biochem. Biotechnol. 157, 174–209.

Yadav, M., Yadav, H., 2015. Applications of ligninolytic enzymes to pollutants, wastewater, dyes, soil, paper, and polymers. Environ. Chem. Lett. 13, 309–318.

Further reading

Boudh, S., Tiwari, S., Singh, J.S., 2017. Microbial-mediated lindane bioremediation. In: Singh, J.S., Seneviratne, G. (Eds.), Agro-Environmental Sustainability. Springer, The Netherlands, pp. 213–233.

Chapter 20

Microbial approaches in management and restoration of marginal lands

Umesh Pankaj, Geetu Singh and Rajesh Kumar Verma

Divion of Agronomy and Soil Science, CSIR-Central Institute of Medicinal and Aromatic Plants, Lucknow, India

Chapter Outline

20.1 Introduction

Available facts show that total lands have shrunk noticeably due to the severe struggle between different parts of economy for land use, unfortunate anthropogenic modifications, and salinization, leaving little scope for agricultural extension (Singh, 2013, 2015; Tiwari and Singh, 2017; Kumar et al., 2018). On the other hand the population of the world is continuously increasing at a shocking rate of around 1.09% per annum. To confirm food security for such a large population, countries needs to produce huge quantities of food grain, feedstuff, and fuel. Marginal lands have received great attention, which could help to address these food challenges. Literature suggests that large areas of marginal land are available in Asian countries, which may hold potential for the production of food crops. The available marginal lands in Asian countries are at about 4 million km^2 or 6.5% of total land area. The term "marginal land" has been used quite loosely without a concrete definition. Marginal lands are defined as having poor physical characteristic, limited rainfall, high salt content, and poorly and imperfectly drained soils. Plants are vital for the survival of life on this planet. Plants are affected by different environmental stresses during their life cycle and the agricultural crop production is globally restricted by various abiotic and biotic stresses. Biotic factors include pathogenic microorganisms like plant viruses, bacteria, fungi, and some insects that damage the plants (Singh et al., 2018; Tiwari et al., 2018). There are numerous abiotic stresses engendered by different environmental conditions such as UV radiations, extreme temperatures, freezing, drought, salinity, acidity, flooding, and different heavy metals (Singh and Gupta, 2018; Vimal et al., 2018). Salinity is one of the major threats that affect plant growth and productivity worldwide (Mahajan and Tuteja, 2005; Mittler, 2006). So, this demands intensive effort on agro-technology-driven productivity enhancement in degraded lands like drought affected land and salt-affected soil (sodic and alkaline soil). Sodium chloride is the most soluble salt and most of the soil's salinity is developed due to natural geological, hydrological, and pedological processes. Globally, 75 countries have been recognized to have large areas of salt-affected lands. Martinez-Beltran and Manzur (2005) estimated that nearly 831 million hectares of land are salt-affected worldwide. Across many countries, about 95 million

New and Future Developments in Microbial Biotechnology and Bioengineering. DOI: https://doi.org/10.1016/B978-0-12-818258-1.00020-0

hectares of soil are under primary salinization (salt accumulation through natural processes in soils and water) whereas 77 million hectares suffer from secondary salinization (as a result of human activities and ever-rising groundwater) (Metternicht and Zink, 2003; Munns, 2005). It is estimated that every day between 2000 and 4000 ha (Shabala and Cuin, 2008; Qadir et al., 2014) of irrigated land in arid and semiarid areas across the globe are degraded by salinity and become unsuitable for crop production. Salt accumulation in soil is a major threat to agricultural production and ecosystem sustainability because it reduces plant growth and increases the risk of soil erosion (Giri et al., 2003; Al-Karaki, 2006). The high salt concentration negatively affects soil microbial activity as well as soil chemical and physical properties, thus causing a decline in soil productivity. When exposed to biotic and abiotic stress plant metabolism is disrupted (Swarbrick et al., 2006; Massad et al., 2012), which leads to the reduction in proficiency and ultimately productivity (Shao et al., 2008). After recognition, the plant's constitutive basal defense mechanisms (Andreasson and Ellis, 2010) lead to an activation of complex signaling cascades of defense varying from one stress to another (Abou Qamar et al., 2009; Chinnusamy et al., 2004).

20.2 Problems associated with marginal lands

Drought is a major problem in agriculture worldwide. For example, Europe experienced an extreme drought event in 2003, exacerbated by high summer temperatures, which led to a dramatic reduction in primary productivity (Olesen et al., 2011). With the predicted increase in reduced rainfall and heat events due to global warming, plant productivity in temperate regions is threatened; water scarcity may lead to reduced plant development, leaf wilting, unbalanced fruit composition, and seed maturation (Ciais et al., 2005).

The second major threat in environment is salt-affectedness, which refers to soil that has higher concentration of various types of salts and has become saline or sodic (Szabolcs, 1989) in nature. A sodic soil has too much sodium associated with the negatively charged clay particles. Soil salinity occurs through natural or anthropogenic processes that lead to the accumulation of salts in the soil water to a high level that hinders plant growth. Natural salinity is a result of salt accumulation over a long period of time, and is the result of weathering of rocks with the release of soluble salts including chloride of sodium, calcium, and magnesium, and to a lesser level of sulfates and carbonates. In addition, deposition of oceanic salt carried by wind and rain also leads to salinity. Anthropogenic causes of salinity are (1) land clearing and the replacement of perennial vegetation with annual crops, and (2) irrigation of land with salt-rich water and insufficient drainage system (Munns and Tester, 2008).

Soil salinization is a serious global environmental problem associated to agriculture. Excessive salts have negative impact on living organisms in a specific environment. Salts in soil can adversely affect the performance or individual physiology of the plants in a significant way. High salt concentrations like sodium, calcium, magnesium, chloride, carbonate, bicarbonate, and sulfate potentially affect the agricultural crops' growth and yield. Abundance of soluble ions in soil solution can decrease the activity of essential nutrients in the soil and can lead to reduction in accessibility and uptake of some nutrients by the plants. It is known that salt stress causes reduction in phosphorus accumulation in plants, which developed P-deficiency symptoms. The direct effects of salts on plant growth are nutrient imbalance, which causes a decrease in the nutrient uptake and carriage to the shoot leading to ion deficiencies (Munns, 2002).

Salt-affected soils are generally classified on the basis of electrical conductivity of the saturated extract (ECe), sodium adsorption ratio (SAR) and exchangeable sodium percentage (ESP), and pH (Rengasamy, 2010). On the basis of the above parameters salt-affected soils are classified as (1) saline soils, containing high levels of soluble salts (ECe > 4.0 dS m^{-1}); (2) sodic soils, having high levels of exchangeable sodium (SAR > 13.0 and ESP > 15.0); and (3) saline–sodic soils, in which both soluble salts and exchangeable sodium are high (Ghasemi et al., 1995).

20.3 Harmful effects of drought and salinity on plant

Drought and salinity obstructs the photosynthetic pathway and increases photorespiration, changing the cells' homeostasis with increased production of reactive oxygen species (ROS) such as super oxide radical, hydrogen peroxide, and hydroxyl radical (Das and Roychoudhury, 2014). Under optimal growth conditions, low levels of ROS are generated in cell organelles such as chloroplasts, peroxisomes, and mitochondria (Apel and Hirt, 2004). The increased production of ROS during stress induces the activation of stress-responsive and plant defense pathway genes (Pitzschke et al., 2006). Wide research has been done regarding stress tolerance of plants, mainly on water relations, photosynthesis, and accumulation of various inorganic ions and organic metabolites.

20.4 Management of marginal land

To fulfill the needs of the rapidly multiplying population and to keep the land resources in good physical shape and increase the productivity for the future generations, it is imperative to closely look into the reasons for land degradation and development of salinity and adopt innovative measures to reverse these processes. Putting botanical means into practice depending upon the nature of the issue could be highly helpful for mitigating the abiotic stress.

20.5 Phyto-restoration of marginal land

During drought, unbalanced uptake of macro- and micronutrients that may precipitate due to lack of water availability (Vassilev et al., 2012; Qi and Zhao, 2013) can dramatically exacerbate the already compromised health status of plants. Agricultural fields are generally used for cultivation of food crops and thus, to meet the current demand of food security, it would be sensible to restore marginal land that is generally not suited for most of the agricultural crops. Drought and salinity tolerant plants are widely used for the restoration of marginal lands in a sustainable manner. Plants respond to dry conditions in several ways, including modification of root architecture (shallow versus deep rooting) and leaf shape. Such responses can differ between perennial trees and annual plants, such as cereals. Although drought is an increasing problem in agriculture, the contribution of the root-associated bacterial microbiome to plant adaptation to water stress is poorly studied. Nowadays, medicinal and aromatic plants (MAPs) have drawn the attention of researchers for the cultivation on marginal land due to not only their aroma, flavor/taste, and clinical use but also they are proven to be good cash crops and a boon for improving the income of farmers. MAPs are cultivated for a wide variety of products derived from plant crude materials and processed for their use as pharmaceuticals, herbal remedies, cosmetics, sweets, dietary supplements, polishes, insecticides, and the aroma industry (Lange, 2004). Earlier, studies indicated that grasses have a unique capability to cope with drought as well salt stress conditions especially aromatic grasses (family Poaceae) such as palmarosa (*Cymbopogon martinii*), lemongrass (*Cymbopogon flexuosus*), and vetiver (*Chrysopogon zizanioides*) (Patra and Singh, 1998; Tomar and Minhas, 2004).

20.6 Microbial approach for management of marginal lands

Drought imposes adaptive constraints on both plants and their associated microorganisms. The microbes have the ability to show resistance to abiotic stresses typically associated with drought, such as resistance to salinization, osmotic stress, and growth at high and low temperatures. Interestingly, some microbes are osmotolerant and showed the ability to grow under low moisture conditions and are of great benefit to plants. PGP bacteria can directly enhance micronutrient uptake and affect phytohormone homeostasis, or indirectly stimulate the plant immune system against phytopathogens (Balloi et al., 2010) and improve soil texture and structure (Mapelli et al., 2012). Various types of beneficial properties of plant growth-promoting rhizobacteria (PGPR) are shown in Fig. 20.1. For instance, some PGP bacteria are able with the 1-aminocyclopropane-1-carboxylate (ACC) deaminase enzyme (Glick et al., 2007) to cleave the plant ethylene precursor ACC, thereby lowering the level of ethylene in developing or stressed plants (Glick, 2004). The majority of

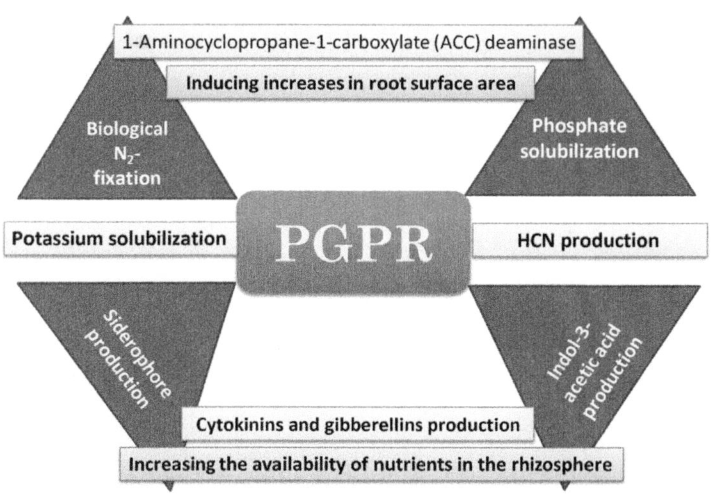

FIGURE 20.1 Beneficial and effective role of plant growth-promoting rhizobacteria under salt-affected soil for a sustainable agriculture.

PGPRs had the potential to contribute to plant nutrition by siderophore release and solubilization of inorganic phosphate compounds. Furthermore, the PGP bacteria are able to secrete mucilaginous material, possibly positively affecting bacteria root adhesion and colonization. Such an exopolymeric matrix contributes to soil stabilization through (1) an increase in the amount of root adhering soil, (2) an improvement in water-holding capacity and reduced water loss during desiccation due to its hydrophilic properties, (3) a stimulation of root exudation, and (4) protection of roots from the mechanical effects of soil hardness (Rolli et al., 2015).

Salinity is a major soil constraint that affects the physiological traits and yield of crops worldwide. Salt-affected soil reduces crop productivity and leads to abandonment of many agricultural areas. Deleterious effect of salinity includes nutrient imbalance, osmotic stress, and membrane destabilization contributing toward lowering of growth and yield of crops (Zhu, 2007). For sustainable crop production in salt-affected areas it has been vital to mitigate these deleterious effects of salts over the cultivated crops. Saline—alkaline soil exhibits a specific type of extreme habitat for the adaptation of a halophilic microbial community that grows luxuriantly at high pH and high salt concentrations (Qin et al., 2016). A well-established fact is that rhizobacteria colonizing roots improve the growth and productivity of the plants and also can be useful to alleviate this worldwide problem. For successful performance of plant growth-promoting rhizobacteria (PGPR) in salt-affected soil, both abilities of salt-tolerant and plant growth promotion (phosphate solubilization, production of siderophore and auxin) are required for their better survival and performance in natural field conditions (Bharti et al., 2013). An indirect mechanism of plant growth promotion is protecting the plant against soil-borne diseases by producing hydrogen cyanide, particularly caused by pathogenic fungal spp. (Lugtenberg and Kamilova, 2009).

Rhizobacteria inhabiting within the rhizosphere region may also enter into the root cortex (endoroot), found on the surface of root and endobacteria (found within the cortical cells). An efficient PGPR must possess properties like (1) capable to colonize on the surface the root, (2) capable to grow and proliferate within rhizosphere, and (3) compatible with other microbiota in the rhizosphere (Kloepper et al., 1991). PGPR promotes plant growth directly by providing essential nutrients to plants (nitrogen, phosphorus, and essential minerals) and modulating plant hormone levels. Earlier studies have reported various PGPR involved in drought and salinity amelioration activities are exemplified in Table 20.1.

TABLE 20.1 Some examples of PGPR and AMF inoculation for plant growth promotion under different environmental stress condition.

Crops	Type of stress	Bacteria/AMF	Response	Reference
Retama (*Retama sphaerocarpa*)	Drought	*Bacillus thuringiensis*; *Glomus intraradices*	Enhanced root development, reduced water required to produce shoot biomass	Marulanda et al. (2006)
Medicago spp. (*M. nolana, M. rigidula, M. rotata*)	Drought	*Sinorhizobium meliloti* (wild type and genetically modified derivative	Plants inoculated with mycorrhiza and *Sinorhizobium* strains are less affected by water stress; mycorrhizal plants modulated by genetically modified *Sinorhizobium* proved better	Vazquez et al. (2001)
Trifoliate orange (*Poncirus trifoliate*)	Drought	*Funneliformis mosseae*	Greater root-hair growth and auxin synthesis	Liu et al. (2018)
Tomato (*Lycopersicon esculentum*)	P-deficient environment	*Pseudomonas putida, Azotobacter chroococcum; Azosprillum lipoferum; Glomus* sp.	Inoculation enhanced lycopene and antioxidant activity, shoot and fruit potassium content	Ordookhani et al. (2010)
Red clover (*Trifolium pratense* L.)	Heavy metal	*Brevibacillus* sp.	Single as well as dual inoculation caused positive effect under lead contamination	Vivas et al. (2003)
Lettuce (*Lactuca sativa*)	Salinity	*Pseudomonas mendocina; Glomus mosseae*	Enhanced plant biomass; however, aggregate stability decreased under salinity even with inoculation	Kohler et al. (2009)
Maize (*Zea mays* L.)	Coastal saline soil	*Massilia* sp. *RK4* & *Rhizophagus intraradices*	Proline content was significantly reduced, which is a stress indicator produced by plants	Krishnamoorthy et al. (2016)

PGPR, Plant growth-promoting rhizobacteria; *AMF*, arbuscular mycorrhizal fungi.

20.7 Role of arbuscular mycorrhizal fungi in management of marginal land

In 1881 the Polish botanist, Franciszek Kamienski discovered a mutualistic association of fungus and Indian pine plant roots (*Monotropa hypopitys* L.). In 1885 Frank named the symbiotic process between the fungi and roots using the Greek word *mykorrhizen*, meaning "mycorhiza (fungus-root)," formerly also called endomycorrhiza, or endotrophic mycorrhizas. Amongst the mycorrhizal associations, the arbuscular mycorrhizal fungi (AMF) association is the most common one (Sjöberg, 2005; Tahat and Sijam, 2012). AMF reproduce asexually and there is no known sexual state yet. About 80% of plants on Earth are colonized with AMF.

Various studies have shown that AMF are integral components of a natural ecosystem and have potential to survive in extreme environmental conditions like high salinity and drought prone environments (Giri et al., 2003; Evelin et al., 2009). AMF ameliorates the adverse effects of salinity and promotes plant growth by different mechanisms (Fig. 20.2). AMF colonized plants alleviate the detrimental effects of soil salinity (Abdel Latef et al., 2016). AMF symbiosis promotes plant growth under salt stress by enhancing the water and nutrient uptake (phosphorus and nitrogen) by plants (Al-Khaliel, 2010). AMF also helps in the seed germination and physiological traits of the plants (switch grass, *Panicum virgatum* L. and chili, *Capsicum annuum* L.) under abiotic stress (Rueda-Puente et al., 2010). AMF improves the salt and drought tolerant capacity of food crops like maize, tomato, lettuce, etc. (Jahromi et al., 2008; Ruiz-Lozano and Azcón, 1995). AMF also protects plants from stress through the following processes: adjustment of cellular osmotic, detoxification of reactive oxygen species, maintenance of membrane integrity, and enzymes and proteins stabilization (Abdel Latef and Chaoxing, 2011). Furthermore, some of these solutes are called osmoprotectants because they protect cellular components from dehydration damage. These solutes include proline, soluble sugars, polyols, trehalose, and quaternary ammonium compounds such as proline, betaine, alaninebetaine, glycinebetaine, pipecolatebetaine, and hydroxyprolinebetaine (Hamdia et al., 2004).

The mycorrhizosphere is a unique arena of plant–microbe and microbe–microbe interactions. It is found that AMF and PGPRs may interact synergistically or antagonistically in the mycorrhizosphere (Barea et al., 2002; Awasthi et al., 2011). These microbial interactions in the mycorrhizosphere may need to be studied and synergistic interactions between them could be exploited and employed for environmentally friendly low input agriculture (Krishnamoorthy et al., 2016). A better understanding of interactions among soil microorganisms with plants is a crucial step for the sustainable management of soil fertility and crop production (Johansson et al., 2004). The plant growth-promoting activities of microorganisms may also vary with soil type, strains, host plant, and cultivation practices. Besides, some extent of specificity also exists among plants and their associated microorganisms such that plants actively participate in selecting and promoting the particular microbial consortia to ameliorate the particular environmental stress (Mendes et al., 2011).

20.8 Interaction among plant, arbuscular mycorrhizal fungi and plant growth-promoting rhizobacteria under abiotic stressed environment

Plants can withstand different environmental stress by interacting with soil microorganisms including AMF and PGPR. Some previous studies conducted under different stress environments for crop improvement by coinoculation of different PGPR and AMF have been listed in Table 20.1. AMF ameliorates stress to plants by improving the rate of nutrient

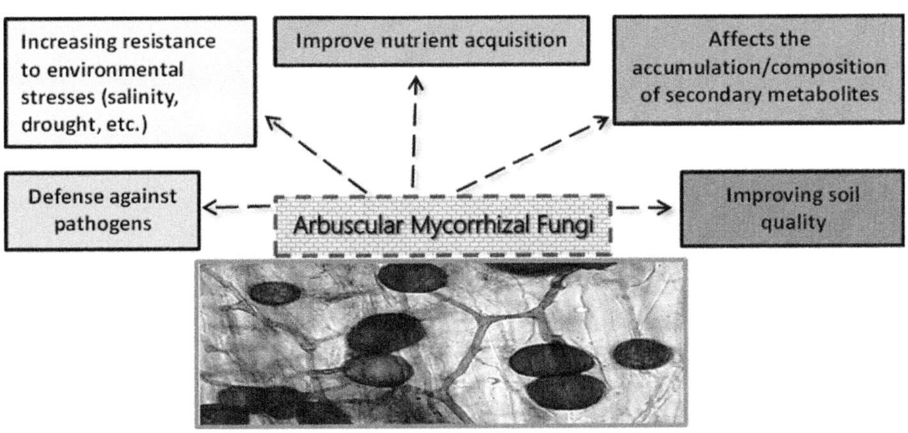

FIGURE 20.2 Schematic presentation of arbuscular mycorrhizal fungi role in plants and soil.

uptake, water absorption, enhanced photosynthesis, and accumulation of compatible organic solutes (Abdel Latef and Chaoxing, 2011). The inoculation of *Rhizophagus intraradices* improved the tolerance levels of maize plant by improving the nutrients uptake and reducing stress levels evident by lower proline accumulation within mycorrhizae treated plants (Estrada et al., 2013). Iron is playing a measurable role in plant growth and development through participation in metabolic process such as DNA synthesis, respiration, and photosynthesis. Some graminaceous plants release iron specific chelating compounds called phytosiderophores. Additionally, other plants can exude organic acids that form complexes with iron ions. Under stressed soil, the presence has been shown of certain groups of PGPR that produce siderophores and promote iron scavenging, and the iron–siderophores complexes are easily accessible to plants (Kloepper et al., 1980). Moreover, siderophores are effective in desorption of metal such as Cd and Pb from a kaolinite clay at a moderate pH (Hepinstall et al., 2005).

The microbial consortium is defined as two or more groups of microorganisms that live symbiotically. It can be endosymbiotic or ectosymbiotic. The model of consortium was first developed by Johannes Reinke in 1872. Nowadays most of the research findings are focusing on consortium development to tackle the salinity or drought problems (Krishnamoorthy et al., 2016). An overview of the interactions between roots and associated microbes in environmental stress conditions is shown in Fig. 20.3 (Qin et al., 2016). In the mechanistic process shown in the image, the red arrows indicate the upregulation of plant genes or chemical signals or osmolytes under stress soil, while the blue arrows indicate the downregulation. Roots provide three distinctive compartments (endosphere, rhizoplane, and rhizosphere) for colonization by microbes. Both AMF fungi and bacteria can produce ACC-deaminase and a series of phytohormones to influence the plant endogenous hormone levels and ultimately reprogram the hormone signaling in plants. The ACC-deaminase producing microbes often reduce the excessive ethylene production in plants caused by different stress.

To maintain ion homeostasis within plants subjected to stress conditions, bacterial exopolysaccharides bind the toxic Na^+ in case of sodic soil and restrict Na^+ influx into the roots. Volatile organic compounds produced by PGPR can trigger induction of high-affinity K^+ transporter (HKT1) in shoots and reduction of HKT1 in roots, limiting Na^+ entry into roots and facilitating shoot-to-root Na^+ recirculation (Fig. 16.3). AMF increase the K^+/Na^+ ratio by selectively

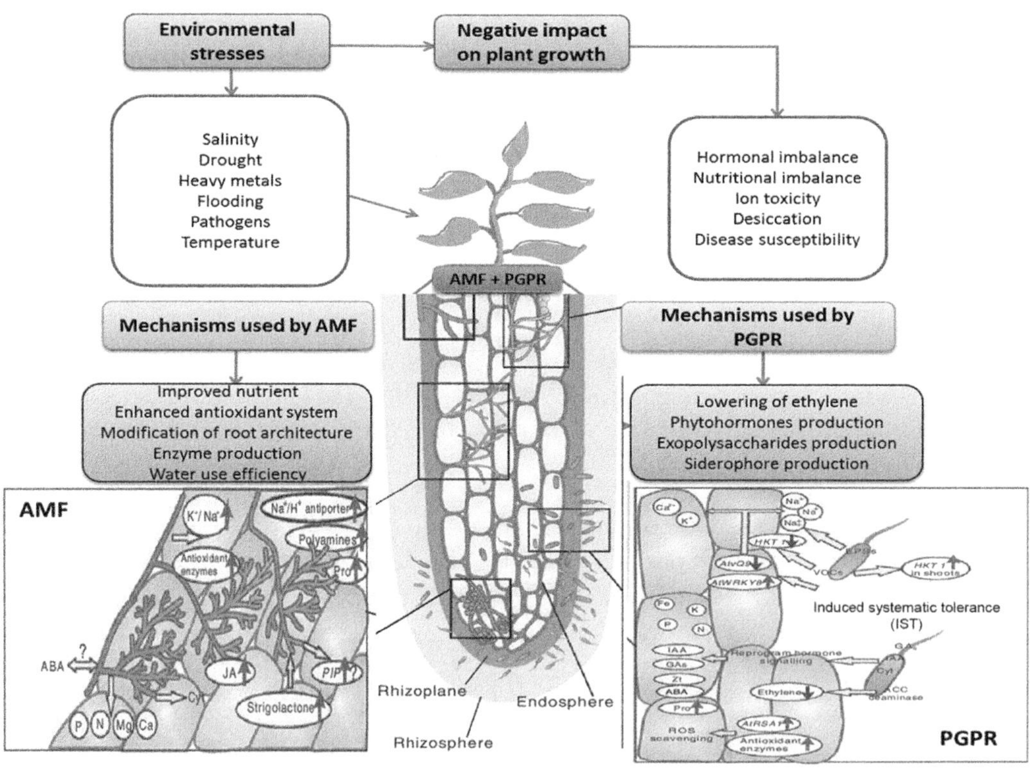

FIGURE 20.3 Combined effects of PGPR and AMF for the alleviation of environmental stresses. Image showing interaction and mechanism of PGPR and AMF. *ABA*, Abscisic acid; *GAs*, gibberellic acids; *HKT1*, high-affinity K^+ transporter; *IAA*, indole-3-acetic acid; *MS*, methionine synthase; *Pro*, proline; *PIPs*, plasma membrane intrinsic proteins; *PGPR*, Plant growth-promoting rhizobacteria; *AMF*, arbuscular mycorrhizal fungi. *Courtesy: Qin, Y., Druzhinina, I.S., Pan, X., Yuan, Z., 2016. Microbially mediated plant salt tolerance and microbiome-based solutions for saline agriculture. Biotechnol. Adv. 34, 1245–1259.*

enhancing K^+ and Ca^{2+} uptake and avoiding translocation of toxic N^+ under salinity stress (Elhindi et al., 2016). PGPR and AMF generally positively regulate the expression of genes encoding the plasma membrane integral proteins to aquaporin activity for efficient water uptake in drought and salt-stressed plants. Almost all microbes are capable of increasing the antioxidative systems in plants for reactive oxygen species scavenging. Some important osmolytes (proline and polyamines) are also promoted in microbe-infected plants. Endophytes are proposed to produce some effectors (plant hormones, signaling molecules, and microRNAs) for activating the signaling pathways. Endophytes often significantly promote metabolic efficiency and alter the ratio of upregulation and downregulation (UR/DR) genes and they also produce some bioactive compounds (melanin, mannitol, and trehalose) in vitro and in plant to alleviate plant abiotic stress.

20.9 Use of microbes with organic fertilizer

Organic matter content of soil promotes plant growth by improving soil water-holding capacity. Moreover, organic matter improves soil aeration, increases absorption and release of nutrients, and prevents soil erosion and leaching (Sekhon and Meelu, 1994; Reijntjes et al., 1992). Amendments of organic fertilizers with suitable bioinoculants (Fig. 20.4) improved the physicochemical and biological parameters of postharvest soil of different crops in field condition (Singh et al., 2013). Hussain et al. (2001) showed that amendments of salt-affected soil with farm yard manure (FYM) improved the physicochemical properties of soil including hydraulic conductivity, soil bulk density, and soil porosity. The ocimum (*Ocimum basilicum*) plant growth under salt stress soil was improved by application of bioinoculants like *Dietzia natronolimnaea* and *Glomus intraradices* along with organic manure, for example, vermicompost (Bharti et al., 2016). Similar report was published by Krishnamoorthy et al. (2016) in which the growth of maize was significantly improved by coinoculation of *Massilia* spp. and *R. intraradices.* The AMF play a vital role in the carbon cycling, consequently increasing carbon flow into soil and improving poor nutrient soil health (Cardoso and Kuyper, 2006). Interaction between host plant and AMF indirectly affects soil carbon storage (Zhu and Miller, 2003); in particular carbon flow from host to AMF can be lost into the mycorrhizosphere up to 70% as organic carbon (Jakobsen and Rosendahl, 1990). In addition, AMF produces glomalin (soil iron-containing glycoproteinaceous), which is helpful in soil aggregation and storage of soil carbon (Driver et al., 2005). Glomalin concentrations are also correlated with soil aggregate water stability in the soil system.

Soil enzymes are considered as highly sensitive and make changes in soils, and have been anticipated as indicators to determine the degree of soil stresses (Schmidt et al., 2011). Enzymes play important roles in soil nutrient cycling, and their activity represents soil microbial activity or an indication of soil quality. The interactions between soil properties and enzyme activities under different amendment such as organic and inorganic fertilizers were investigated by several researchers. The soil enzymes like dehydrogenase, catalase, protease, urease, and acid and alkaline phosphatase are important parameters for accessing soil quality (Mariela et al., 2016). Alkaline phosphatase is the main enzyme involved in the cycling of P because it can transform organic P into inorganic P, which is available to the plants (Dick and Burns, 2011). Alkaline phosphatase is very sensitive to soil environmental conditions and an important indicator for the mineralization of the organic P and biological activity of soil (Zhang et al., 2014). Soil enzyme activity is

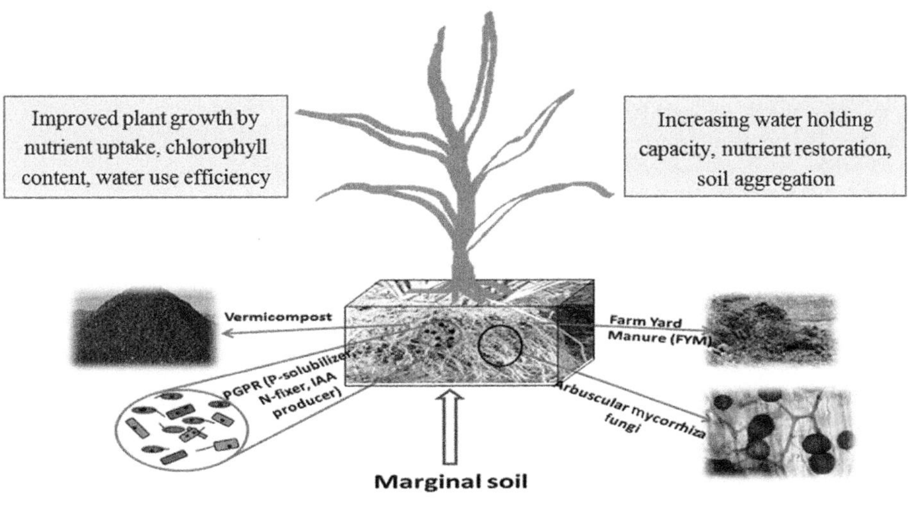

FIGURE 20.4 Application of organic fertilizer with coinoculation of microbes (PGPR and AMF) for the management of marginal lands. *PGPR*, Plant growth-promoting rhizobacteria; *AMF*, arbuscular mycorrhizal fungi.

improved by application of various PGPR with organic fertilizers (FYM/vermicompost) under marginal lands (Lin-lin et al., 2017; Singh, 2016). AMF and PGPRs are also considered as a source of soil enzymes for biochemical reactions since they can increase the activity of soil enzymes under various environmental stress as well poor nutrient soils (Alguacil et al., 2009).

20.10 Conclusion

Microbes are being considered as environmentally friendly potential tools for the resolution to agro-environmental problems since valuable microorganisms have the capability to enhance plant growth, nutrient availability, and uptake, and support plant health under harsh conditions. PGPRs and AMF are able to alleviate environment stress in plants, and increase germination, growth, and yield of plants. Under high abiotic stress plant growth is improved through biological applications and this can significantly influence the plant enzymes activities and nutrients acquisition. Thus, PGPR and AMF with inoculation of organic manure (FYM/Vermicompost) can solve the problems of low productivity under marginal or avoided lands like drought-stressed and saline soil. It is clear that the amelioration of abiotic stress soils using biological means, including growing of salt/drought tolerant genotype, addition of organic matters, and development of integrated nutrients systems, will lead to the desired results and provide a greater profit.

References

Abdel Latef, A.A., Chaoxing, H., 2011. Effect of arbuscular mycorrhizal fungi on growth, mineral nutrition, antioxidant enzymes activity and fruit yield of tomato grown under salinity stress. Sci. Hortic. 127, 228–233.

Abdel Latef, A.A.H., Hashem, A., Rasool, S., Allah, E.F.A., Alqarawi, A.A., Egamberdieva, D., et al., 2016. Arbuscular mycorrhizal symbiosis and abiotic stress in plants: a review. J. Plant Biol. 59, 407–426.

Abou Qamar, S., Luo, H., Laluk, K., Mickelbart, V.M., Mengiste, T., 2009. Crosstalk between biotic and abiotic stress responses in tomato is mediated by AIM1 transcription factor. Plant J. 58, 1–13.

Alguacil, M.M., Caravaca, F., Azcón, R., Roldan, A., 2009. Changes in biological activity of a degraded Mediterranean soil after using microbially-treated dry olive cake as a biosolid amendment and arbuscular mycorrhizal fungi. Eur. J. Soil Biol. 44, 347–354.

Al-Karaki, G.N., 2006. Nursery inoculation of tomato with arbuscular mycorrhizal fungi and subsequent performance under irrigation with saline water. Sci. Hortic. 109, 1–7.

Al-Khaliel, A.S., 2010. Effect of salinity stress on mycorrhizal association and growth response of peanut infected by *Glomus mosseae*. Plant Soil Environ. 56 (7), 318–324.

Andreasson, E., Ellis, B., 2010. Convergence and specificity in the Arabidopsis MAPK nexus. Trends Plant Sci. 15, 106–113.

Apel, K., Hirt, H., 2004. Reactive oxygen species: metabolism, oxidative stress, and signal transduction. Annu. Rev. Plant Biol. 55, 373–399.

Awasthi, A., Bharti, N., Nair, P., Singh, R., Shukla, A.K., Gupta, M.M., et al., 2011. Synergistic effect of *Glomus mosseae* and nitrogen fixing *Bacillus subtilis* strain Daz26 on artemisinin content in *Artemisia annua* L. Appl. Soil Ecol. 49, 125–130.

Balloi, A., Rolli, E., Marasco, R., Mapelli, F., Tamagnini, I., Cappitelli, F., et al., 2010. The role of microorganisms in bioremediation and phytoremediation of polluted and stressed soils. Agrochimica 54, 353–369.

Barea, J.M., Pozo, M.J., Azcón, R., Azcón-Aguilar, C., 2002. Microbial cooperation in the rhizosphere. J. Exp. Bot. 56 (417), 1761–1778.

Bharti, N., Yadav, D., Barnawal, D., Maji, D., Kalra, A., 2013. *Exiguobacterium oxidotolerans*, a halotolerant plant growth promoting rhizobacteria, improves yield and content of secondary metabolites in *Bacopa monnieri* (L.) Pennell under primary and secondary salt stress. World J. Microbiol. Biotechnol. 29 (2), 379–387.

Bharti, N., Pandey, S.S., Barnawal, D., Patel, V.K., Kalra, A., 2016. Plant growth promoting rhizobacteria *Dietzia natronolimnaea* modulates the expression of stress responsive genes providing protection of wheat from salinity stress. Sci. Rep. 6, 347–368.

Cardoso, I.M., Kuyper, T.W., 2006. Mycorrhizas and tropical soil fertility. Agric. Ecosyst. Environ. 116, 72–84.

Chinnusamy, V., Schumaker, K., Zhu, J.K., 2004. Molecular genetics perspectives on cross-talk and specificity in abiotic stress signalling in plants. J. Exp. Bot. 55, 225–236.

Ciais, P., Reichstein, M., Viovy, N., Granier, A., Ogee, J., Allard, V., et al., 2005. Europe-wide reduction in primary productivity caused by the heat and drought in 2003. Nature 437, 529–533.

Das, K., Roychoudhury, A., 2014. Reactive oxygen species (ROS) and response of antioxidants as ROS-scavengers during environmental stress in plants. Front. Environ. Sci. Available from: https://doi.org/10.3389/fenvs.2014.00053.

Dick, R.P., Burns, R.G., 2011. A Brief History of Soil Enzymology Research. Soil Science Society of America, Madison, pp. 1–34.

Driver, J.D., Holben, W.E., Rillig, M.C., 2005. Characterization of glomalin as a hyphal wall component of arbuscular mycorrhizal fungi. Soil Biol. Biochem. 37, 101–106.

Elhindi, K.M., El-din, A.S., Elgorban, A.M., 2016. The impact of arbuscular mycorrhizal fungi in mitigating salt-induced adverse effects in sweet basil (*Ocimum basilicum* L.). Saudi J. Biol. Sci. 24 (1), 170–179.

Estrada, B., Aroca, R., Maathuis, F.J., Barea, J.M., Ruiz-Lozano, J.M., 2013. Arbuscular mycorrhizal fungi native from a Mediterranean saline area enhancemaize tolerance to salinity through improved ion homeostasis. Plant Cell Environ. 36, 1771–1782.

Evelin, H., Kapoor, R., Giri, B., 2009. Arbuscular mycorrhizal fungi in alleviation of salt stress: a review. Ann. Bot. 104, 1263–1280.

Ghasemi, F., Jakeman, A.J., Nix, H.A., 1995. Salinisation of Land and Water Resources: Human Causes, Extent, Management, and Case Studies. NSW University Press, Sydney.

Giri, B., Kapoor, R., Mukerji, K.G., 2003. Influence of arbuscular mycorrhizal fungi and salinity on growth, biomass and mineral nutrition of *Acacia auriculiformis*. Biol. Fertil. Soil 38, 170–175.

Glick, B.R., 2004. Bacterial ACC deaminase and the alleviation of plant stress. Adv. Appl. Microbiol. 56, 291–312.

Glick, B.R., Cheng, Z., Czarny, J., Duan, J., 2007. Promotion of plant growth by ACC deaminase-producing soil bacteria. Eur. J. Plant Pathol. 119, 329–339.

Hamdia, M.A.E.S., Shaddad, M.A.K., Doaa, M.M., 2004. Mechanisms of salt tolerance and interactive effects of Azospirillum brasilense inoculation on maize cultivars grown under salt stress conditions. Plant Growth Regulation 44 (2), 165–174.

Hepinstall, S.E., Turner, B.F., Maurice, P.A., 2005. Effects of siderophores on Pb and Cd adsorption to kaolinite. Clays Clay Mineral 53, 557–563.

Hussain, N., Hassan, G., Arshadullah, M., Mujeeb, F., 2001. Evaluation of amendments for the improvement of physical properties of sodic soil. Int. J. Agric. Biol. 3, 319–322.

Jahromi, F., Aroca, R., Porcel, R., Ruiz-Lozano, J.M., 2008. Influence of salinity on the in vitro development of *Glomus intraradices* and on the in vivo physiological and molecular responses of mycorrhizal lettuce plants. Microb. Ecol. 55, 45–53.

Jakobsen, I., Rosendahl, L., 1990. Carbon flow into soil and external hyphae from roots of mycorrhizal *cucumber* plants. New Phytol. 115, 77–83.

Johansson, J.F., Paul, L.R., Finlay, R.D., 2004. Microbial interactions in the mycorrhizosphere and their significance for sustainable agriculture. FEMS Microbiol. Ecol. 48, 1–13.

Kloepper, J.W., Leong, J., Teintze, M., Schroth, M.N., 1980. Enhancing plant growth by siderophores produced by plant growth-promoting rhizobacteria. Nature 286, 885–886.

Kloepper, J.W., Rodrguez-kbana, R., Mcinroy, J.A., Collins, D.J., 1991. Analysis of populations and physiological characterization of microorganisms in rhizospheres of plants with antagonistic properties to phytopathogenic nematodes. Plant Soil. 136, 95–102.

Kohler, J., Hernandez, J.A., Caravaca, F., Roldan, A., 2009. Introduction of antioxidant enzyme is involved in the greater effectiveness of a PGPR versus AM fungi with respect to increasing the tolerance of lettuce to severe salt stress. Environ. Exp. Bot. 65, 245–252.

Krishnamoorthy, R., Kim, K., Subramanian, P., Senthilkumar, M., Anandham, R., Sa, T., 2016. Arbuscular mycorrhizal fungi and associated bacteria isolated from salt-affected soil enhances the tolerance of maize to salinity in coastal reclamation soil. Agric. Ecosyst. Environ. 231, 233–239.

Kumar, A., Kaushal, S., Saraf, S.A., Singh, J.S., 2018. Microbial bio-fuels: a solution to carbon emissions and energy crisis. Front. Biosci. (Landmark ED) 1 (23), 1789–1802.

Lange, D., 2004. Medicinal and aromatic plants: trade, production, and management of botanical resources. Acta Hortic. 629, 177–197.

Lin-lin, J., Guang-ming, H., Lan, Y., Liu, S., Gao, J., Yang, X., et al., 2017. Corn cob biochar increases soil culturable bacterial abundance without enhancing their capacities in utilizing carbon sources in Biolog Eco-plates. J. Integr. Agric. 16 (03), 713–724.

Liu, C.Y., Zhang, F., Zhang, D.J., Srivastava, A.K., Wu, Q.S., Zou, Y.N., 2018. Mycorrhiza stimulates root-hair growth and IAA synthesis and transport in trifoliate orange under drought stress. Sci. Rep. 8, 1978. Available from: https://doi.org/10.1038/s41598-018-20456-4.

Lugtenberg, B., Kamilova, F., 2009. Plant-growth-promoting rhizobacteria. Annu. Rev. Microbiol. 63, 541–556.

Mahajan, S., Tuteja, N., 2005. Cold, salinity and drought stresses: an overview. Arch. Biochem. Biophys. 444 (2), 139–158.

Mapelli, F., Marasco, R., Balloi, A., Rolli, E., Cappitelli, F., Daffonchio, D., et al., 2012. Mineral-microbe inter-actions: biotechnological potential of bioweathering. J. Biotechnol. 157, 473–481.

Mariela, F.P., Iván Pável, M.E., Luis Manuel, S.R., María Jesús, F.G., Reyes, L.O., 2016. Dehydrogenase and mycorrhizal colonization: tools for monitoring agrosystem soil quality. Appl. Soil Ecol. 100, 144–153.

Martinez-Beltran, J., Manzur, C.L., 2005. Overview of salinity problems in the world and FAO strategies to address the problem. Paper Presented at the International Salinity Forum, Riverside.

Marulanda, A., Barea, J.M., Azcon, R., 2006. An indigenous drought tolerant strain of *Glomus intraradices* associated with a native bacterium improves water transport and root development in *Retama sphaerocarpa*. Microb. Ecol. 52, 670–680.

Massad, T.J., Dyer, L.A., Vega, C.G., 2012. Cost of defense and a test of the carbon-nutrient valance and growth-differentation balance hypotheses for two co-occurring classes of plant defense. PLoS One 7, e7554.

Mendes, R., Kruijt, M., de Bruijn, I., et al., 2011. Deciphering the rhizosphere microbiome for disease-suppressive bacteria. Science 332, 1097–1100.

Metternicht, G., Zink, J., 2003. Remote sensing of soil salinity: potentials and constraints. Remote Sens. Environ. 85, 1–20.

Mittler, R., 2006. Abiotic stress, the field environment and stress combination. Trends Plant Sci. 11, 15–19.

Munns, R., 2002. Comparative physiology of salt and water stress. Plant Cell Environ. 25 (2), 239–250.

Munns, R., 2005. Salinity stress and its impact. In: Blum A, ed. Plant Stress. http://www.plantstress.com/Articles/index.asp.

Munns, R., Tester, M., 2008. Mechanisms of salinity tolerance. Annu. Rev. Plant Biol. 59, 651–681.

Olesen, J.E., Trnka, M., Kersebaum, K.C., Skjelvag, A.O., Seguin, B., Peltonen-Sainio, P., et al., 2011. Impacts and adaptation of European crop production systems to climate change. Eur. J. Agron. 34, 96–112.

Ordookhani, K., Khavazi, K., Moezzi, A., Rejali, F., 2010. Influence of PGPR and AMF on antioxidant activity, lycopene and potassium contents in tomato. Afr. J. Agric. Res. 5, 1108–1116.

Patra, P., Singh, D.V., 1998. Medicinal and aromatic crops. In: Tyagi, N.K., Minhas, P.S. (Eds.), Agricultural Salinity Management in India. Central Soil Salinity Research Institute, Karnal, pp. 499–506.

Pitzschke, A., Fornazi, C., Hirt, H., 2006. Reactive oxygen species signalling in plants. Antioxid. Redox Signal. 8, 1757–1764.

Qadir, M., Quillérou, E., Nangia, V., Murtaza, G., Singh, M., Thomas, R.J., et al., 2014. Economics of salt-induced land degradation and restoration. Nat. Resour. Forum. 38 (4), 282–295.

Qi, W.Z., Zhao, L., 2013. Study of the siderophore-producing *Trichoderma asperellum* Q1 on cucumber growth promotion under salt stress. J. Basic Microbiol. 53, 355–364.

Qin, Y., Druzhinina, I.S., Pan, X., Yuan, Z., 2016. Microbially mediated plant salt tolerance and microbiome-based solutions for saline agriculture. Biotechnol. Adv. 34, 1245–1259.

Reijntjes, C., Haverkort, B., Waters-Bayer, A., 1992. Farming for the Future: An Introduction to Low-External-Input and Sustainable Agriculture. Macmillan Press L, London.

Rengasamy, P., 2010. Soil processes affecting crop production in salt affected soils. Funct. Plant Biol. 37, 613–620.

Rolli, E., Marasco, R., Vigani, G., Ettoumi, B., Mapelli, F., Deangelis, M.L., et al., 2015. Improved plant resistance to drought is promoted by the root-associated microbiome as a water stress-dependent trait. Environ. Microbiol. 17 (2), 316–331.

Rueda-Puente, E.O., et al., 2010. Effect of plant growth promoting bacteria and mycorrhizal on *Capsicum annuum* L. var. aviculare ([Dierbach] D'Arcy and Eshbaugh) germination under stressing abiotic conditions. Plant Physiol. Biochem. 48, 724–730.

Ruiz-Lozano, J.M., Azcón, R., 1995. Hyphal contribution to water uptake in mycorrhizal plants as affected by the fungal species and water status. Physiol. Plant. 95, 472–478.

Schmidt, M., Torn, W.I., Abiven, M.S., Dittmar, S., Guggenberger, T., Janssens, G., et al., 2011. Persistence of soil organic matter as an ecosystem property. Nature 478 (7367), 49–56.

Sekhon, G.S., Meelu, O.P., 1994. Organic matter management in relation to crop production in stressed rainfed systems. In: Virmani, S.M., Katyal, J.C., Eswaran, H., Abrol, I.P. (Eds.), Stressed Ecosystems and Sustainable Agriculture. Oxford University Press and IBH Publishing, New Delhi.

Shabala, S., Cuin, T.A., 2008. Potassium transport and plant salt tolerance. Physiol. Plant. 133, 651–669.

Shao, H.B., Chu, L.Y., Shao, M.A., Abdul Jaleel, C., Hong-Mei, M., 2008. Higher plant antioxidants and redox signalling under environmental stresses. Comp. Rend. Biol. 331, 433–441.

Singh, J.S., 2013. Plant growth promoting rhizobacteria: potential microbes for sustainable agriculture. Resonance 18 (3), 275–281.

Singh, J.S., 2015. Microbes: the chief ecological engineers in reinstating equilibrium in degraded ecosystems. Agric. Ecosyst. Environ. 203, 80–82.

Singh, K., 2016. Microbial and enzyme activities of saline and sodic soils. Land Degrad. Dev. 27 (3), 706–718.

Singh, J.S., Gupta, V.K., 2018. Soil microbial biomass: a key soil driver in management of ecosystem functioning. Sci. Total Environ. 634, 497–500.

Singh, R., Soni, S.K., Patel, R.P., Kalra, A., 2013. Technology for improving essential oil yield of *Ocimum basilicum* L. (sweet basil) by application of bioinoculant colonized seeds under organic field conditions. Ind. Crop Prod. 45, 335–342.

Singh, C., Tiwari, S., Gupta, V.K., Singh, J.S., 2018. The effect of rice husk biochar on soil nutrient status, microbial biomass and paddy productivity of nutrient poor agriculture soils. Catena 171, 485–493.

Sjöberg, J., 2005. Arbuscular Mycorrhizal Fungi-Occurrence in Sweden and Interaction With a Pathogenic Fungus in Barley (Ph.D. Thesis). Swedish Univ. Agri. Sci., Uppsala, pp. 1–5.

Swarbrick, P.J., Schulze-Lefert, P., Scholes, J.D., 2006. Metabolic consequences of susceptibility and resistance in barley leaves challenged with powdery mildew. Plant Cell Environ. 29, 1061–1076.

Szabolcs, I., 1989. Salt-Affected Soils. CRC Press Inc, Boca Raton, FL.

Tahat, M.M., Sijam, K., 2012. Arbuscular mycorrhizal fungi and plant root exudates bio-communications in the rhizosphere. Afr. J. Microbiol. Res. 6, 7295–7301.

Tiwari, P., Singh, J.S., 2017. A plant growth promoting rhizospheric *Pseudomonas aeruginosa* strain inhibits seed germination in *Triticum aestivum* (L) and *Zea mays* (L). Microbiol. Res. 8, 1–7.

Tiwari, S., Singh, C., Singh, J.S., 2018. Land use changes: a key ecological driver regulating methanotrophs abundance in upland soils. Energy Ecol. Environ. 3, 355–371.

Tomar, O.S., Minhas, P.S., 2004. Relative performance of some aromatic grasses under saline irrigation. Ind. J. Agron. 49 (3), 207–208.

Vassilev, N., Eichler-Lobermann, B., Vassileva, M., 2012. Stress-tolerant P-solubilizing microorganisms. Appl. Microbiol. Biotechnol. 95, 851–859.

Vazquez, M.M., Azcon, R., Barea, J.M., 2001. Compatibility of a wild type and its genetically modified *Sinorhizobium* strain with two mycorrhizal fungi on *Medicago* species as affected by drought stress. Plant Sci. 161, 347–358.

Vimal, S.R., Patel, V.K., Singh, J.S., 2018. Plant growth promoting *Curtobacterium albidum* strain SRV4: an agriculturally important microbe to alleviate salinity stress in paddy plants. Ecol Indic. in press.

Vivas, A., Azcon, R., Biro, B., Barea, J.M., Ruiz-Lozano, J.M., 2003. Influence of bacterial strains isolated from lead-polluted soil and their interactions with arbuscular mycorrhizae on the growth of *Trifolium pratense* L. under lead toxicity. Can. J. Microbiol. 49, 577–588.

Zhang, T.B., Kang, Y.H., Liu, S.H., Liu, S.P., 2014. Alkaline phosphatase activity and its relationship to soil properties in a saline-sodic soil reclaimed by cropping wolfberry (*Lycium barbarum* L.) with drip irrigation. Paddy Water Environ. 12, 309–317.

Zhu, J.K., 2007. Plant salt stress. Encyclopedia of Life Sciences. John Wiley and Sons, Ltd, pp. 1–3.

Zhu, Y.G., Miller, R.M., 2003. Carbon cycling by arbuscular mycorrhizal fungi in soil-plant systems. Trends Plant Sci. 8, 407–409.

Further reading

Hamdia, M.A.E.S., Shaddad, M.A.K., Doaa, M.M., 2004. Mechanisms of salt tolerance and interactive effects of *Azospirillum brasilense* inoculation on maize cultivars grown under salt stress conditions. Plant Growth Regul. 44 (2), 165–174.

Liu, A., Hamel, C., Hamilton, R.I., Smith, D.L., 2000. Mycorrhizae formation and nutrient uptake of new corn (*Zea mays* L.) hybrids with extreme canopy and leaf architecture as influenced by soil N and P levels. Plant Soil. 221, 157–166.

Rengasamy, P., Greene, R.S.B., Ford, G.W., Mehanni, A.H., 1984. Identification of dispersive behaviour and management of red-brown earths. Aust. J. Soil Res. 22, 413–431.

Chapter 21

Host–endophytes cross-talk: an essential prerequisite for plant ecosystem functioning

Shatrupa Ray[1], Prachi Singh[1], Jyoti Singh[1,2] and Harikesh Bahadur Singh[1]

[1]*Department of Mycology and Plant Pathology, Institute of Agricultural Sciences, Banaras Hindu University, Varanasi, India,* [2]*Department of Botany, Institute of Sciences, Banaras Hindu University, Varanasi, India*

Chapter Outline

21.1 Introduction

An incessant rise in global population combined with a steep reduction in arable land has led to a significant loss in agricultural production. In addition, use of poor land management practices as well as an increased use of chemical fertilizers and pesticides have resulted in a drastic decline of the agro-economy. Use of plant growth promoting bacteria (PGPBs) as biofertilizers and biopesticides offers a promising alternative to ensure sustainability in agriculture. However, a full-fledged exploitation of PGPBs in agriculture requires a thorough understanding of the host–bacterial interaction in an applied setting, for example, under stress conditions. In line to the aforementioned perspective, plants have evolved a variety of mechanisms over millions of years to cope with various forms of environmental stresses. Hosting of endophytes within the plants' interior is one such mechanism to successfully cope with environmental stresses as well as augment growth and productivity.

Endophytes are particular microbes carrying out the entire or a part of their life cycle within the host interior. Though a complete understanding of endophytic benefit for adapting to such a lifestyle has not yet been comprehended, endophyte mediated host benefits, particularly under adverse conditions, have been much documented. Conrath et al. (2015) reported that endophytic microbes systemically prime host plants for an augmented broad spectrum defense response against phytopathogens. Similarly, reports validate endophyte mediated host defense enhancement against a variety of abiotic stresses (Ali et al., 2014). An interesting report by Beltran-Garcia et al. (2014) demonstrated digestion of endophytic bacteria by *Agave tequilana* for fulfillment of nitrogen requirement. In yet another report, Coutinho et al. (2015) reported an efficient interaction of endophytic microbes with host plants as compared with the rhizospheric microbes under diverse environmental conditions. Thus, the scope of exploitation of the intimate association of endophytic microbiota with host in terms of agricultural productivity under diverse conditions cannot be ignored.

New and Future Developments in Microbial Biotechnology and Bioengineering. DOI: https://doi.org/10.1016/B978-0-12-818258-1.00027-3

Though the endophytic diversity particularly within the root tissues has been much elucidated, the gradual change in phytomicrobiota with space and time has still remained a matter of research. In the current chapter, the authors aim to thoroughly focus on the (1) latest updates to reveal the phytomicrobiota with precision, (2) types of bacteria thriving in endophytic lifestyle, (3) modes of transmission of endophytic microbiota within hosts, (4) host mediated phytohormone production and their role in endophytic colonization, as well as (5) host–endophyte interaction for enhancing host productivity and defense. The chapter will be a minor attempt towards a better comprehension of host–endophyte interaction, which may further lead to development of new strategies for sustainable agriculture.

21.2 Techniques to study the phytomicrobiome

Assessment of the plant microbiome according to the classical technique involves a series of culture dependent as well as independent techniques. While the culture dependent mechanisms involve growth of microbes under different nutrient, temperature, and pH conditions, the technique often misses out on the microbial diversity, particularly the obligate ones (Turner et al., 2013). On the contrary, the culture independent techniques, such as 16S rRNA polymerase chain reaction (PCR) amplification allow species as well as strain specific taxonomic identification, particularly of bacterial cultures. Similarly, fungal identification primarily involves 18S rRNA region PCR amplification as well as amplification of the hypervariable internally transcribed spacer regions. However, in spite of the high demand for the above methods, PCR amplification based identification involves designing of specific primers. Thus this technique is more or less biased towards the identification of target organisms (Pinto and Raskin, 2012). In this context, use of high throughput sequencing (e.g., 454-pyrosequencing) enables the identification of diverse sequences within a sample thereby confirming presence of rare microbial species (Lundberg et al., 2012).

A thorough comprehension of inter- and intraspecies cross-talk of microbial species with host systems requires a complete overview of the microbial diversity associated (Barea et al., 2005). Metagenomics enables the qualitative as well as quantitative revelation of genes associated with a metabolic process thereby disclosing the functional potential of a microbiome. For instance, through metagenomic analysis of different plant species, identification of various novel microbial rhodopsins, particularly in the phyllospheric region, is made possible (Atamna-Ismaeel et al., 2012). Similarly, stable isotope probing (SIP) allows identification of particular substrates during a physiological process conducted by microbes. The SIP combined with DNA profiling enables a deeper comprehension of the path of carbon flow during host–microbe cross-talk at the rhizospheric interface (Chen and Murrell, 2010). For instance, the fact that plant derived carbon was actually metabolized by microbes was identified by incubating host plants with $^{13}CO_2$ (Radajewski et al., 2000). Further, verification of transport of the labeled $^{13}CO_2$ from host to the rhizospheric microbes was done by purification of the heavy ^{13}C from rhizobacterial chromosomal DNA from the light nonlabeled DNA fraction by CsCl density gradient centrifugation technique. This technique further enables the identification of labeled rhizomicrobiota whose population is stimulated due to exudation by host (Liu et al., 2017b). In line to the above context, gene expression analysis of SIP-mRNA will enable identification of endophytes utilizing particular substrates (Dumont et al., 2011). However, certain drawbacks of the aforementioned technique involve an increased biasness towards Gram-positive bacteria owing to a higher GC content (Rasche et al., 2009) as well as trouble in distinguishing active from nonactive cells (Schmid et al., 2006). Proteogenomic analyses enable revelation of species with particular metabolic requirements thereby implicating those compounds as a primary nutrition source in the region of concern (Lu et al, 2006; el Zahar Haichar et al., 2008). For instance, Knief et al. (2012) reported the abundance of *Methylobacterium* spp. and other methylotrophs in the phyllospheric region of hosts depending on the expression of porins, TonB dependent receptors, ATP binding cassette (ABC) transporters, etc. Moreover, the report also confirmed that the phyllospheric *Methylobacterium* spp. derived their nutrition from methanol obtained from plant pectin.

21.3 Host–endophytes interaction

The surrounding environment plays a vital role in determining the dominant phytomicrobiota. For instance, a low soil pH is reported to harbor a larger population of acidobacteria as compared with the soil of higher pH (Müller et al., 2016). Rapid emergence of next-generation sequencing strategies, such as fluidic force microscopy, revealed that the basic microbial genera inhabiting the host interior belong to proteobacteria, mostly from the alphaproteobacteria class, actinobacteria, and bacteroidetes, with *Methylobacterium*, *Pseudomonas*, and *Sphingomonas* being among the most abundant genera (Stiefel et al., 2013).

The exquisite account of plant–bacterial interactions includes rhizobial communication with leguminous plants. Nitrogen fixing bacterial species, such as *Azorhizobium*, *Bradyrhizobium*, *Ensifer*, *Mesorhizobium*, *Rhizobium*, and

Sinorhizobium particularly included in "rhizobia" induce a range of differentiating structures in roots of leguminous plants, such as *Fabaceae* and *Parasponia*, thereby leading to nodule formation (Wang et al., 2012; Vaishnav et al. 2017). A sequence of plant bacterial signal exchange leads to gradual changes in root structures, such as root curling and the formation of infection thread (Bonafante and Anca, 2009). Following complete infection, the transformed bacterial cells (bacteroids) become surrounded within plant membrane thereby forming the symbiosome, a structure similar to that of the mycorrhizal arbuscules and meant for atmospheric nitrogen fixation (Kolton et al., 2014).

Contrary to the rhizobial strains involved in nitrogen fixation within leguminous plants, endophytes remain in the intra- and/or intercellular spaces of plant cells while retaining their original structure. In this context, obligate endophytes, such as *Balansia*, *Epichloë*, and *Neotyphodium*, from the family *Clavicipitaceae* (*Ascomycota*) complete their entire life cycle within the plant interior (Card et al., 2016). Simultaneously, facultative endophytes, such as *Pseudomonas Azospirillum* and fungi of the genera *Hypocrea* and *Trichoderma*, particularly flourish in the external environment with intermittent entry into the plant tissues (Hardoim et al., 2008). In this context, it is quite precise that the facultative endophytes consume host nutrients thereby probably reducing host ecological stability. Interestingly, a matter of significant debate still exists as to whether the facultative endophytes selectively recruit their hosts for dissemination or vice versa (Santoyo et al., 2016; Vaishnav et al. 2018).

21.4 Endophytic transmission into hosts

Endophytic colonization within the host system occurs either horizontally through the environment or vertically from the parent to the next generation. In some cases, endophytic transmission may also be by mixed modes (Bright and Bulgheresi, 2010). Vertical transmission is more common in fungal endophytes, such as the genera of *Epichloe* and *Neotyphodium*, an ecological adaptation to ensure secure transmission of the symbionts from one parent to another (Moran, 2006). On the contrary, horizontal mode of recruitment of bacterial endophytes is primarily selected (Foster and Wenseleers, 2006) by host plants owing to their sessile nature thereby providing a successful means of adapting to the changing environment. The endogenous microbiota recruited by horizontal mode depends majorly on the host's environment and not on the genotype (Peiffer et al., 2013; Schlaeppi and Bulgarelli, 2015). Primarily, endophytes recruited by horizontal mode are a subpopulation of the rhizospheric microbiota. However, once recruited, the microbial metabolism gets significantly altered so as to adapt within the host interior (Sessitsch et al., 2012). In this context, it may be suggested that not all rhizospheric bacteria have the traits to become endophytic. Only the microbes having characteristic traits distinct from rhizomicrobiota become endophytic (Compant et al., 2010). Fig. 21.1 elaborates the various routes of transmission of endophytic bacteria within the host interior followed by establishment thereby leading to various host benefits.

21.4.1 Entry from soil

The earliest entry zone begins at the spermosphere, the area of soil surrounding the seeds. Seed exudates in form of proteins, sugars, and fatty acids the immediate soil microbiota, which colonizes the seeds. An ecological adaptation by the host plants, the spermospheric microbiota differ characteristically from the rhizospheric microbiota as these microbes colonize the seeds within the first few hours of sowing (Nelson, 2004).

Another prominent mode of entry through soil is through the rhizosphere. This region is richly endowed with root exudates comprising of chiefly organic acids, sugars, fatty acids, amino acids, growth factors, hormones, vitamins, as well as antimicrobial compounds. Reports attest the prominent role of root exudates in shaping the rhizosphere microbiome (Bertin et al., 2003). According to Lebeis et al. (2015), defense hormone signaling of the host enables selective recruitment of specific microbes from the available microbiota. However, it is also believed that endophyte may have camouflaged their microbe associated molecular patterns (MAMPs), such as bacterial flagellin, lipopolysaccharide, or translation elongation factor (Tu) so as to escape the defense mechanism of the host (Bulgarelli et al., 2013). Similarly, other evidence also justifies the amelioration of root exudates by the rhizospheric microbiota (Ray et al., 2018a). The route followed by endophytes during entry from the rhizospheric region involves initial colonization on the rhizoplane region followed by gradual colonization on root internal tissues finally leading to their entry in the xylem vessels of shoot and leaf (Ray et al., 2016). Release of cell wall degrading enzymes, particularly endoglucanase and endopolygalaturonase, enables their entry and colonization through the root surface as well as within, though reports also justify their gradual entry through the root cracks formed during emergence of root hairs (Compant et al., 2005).

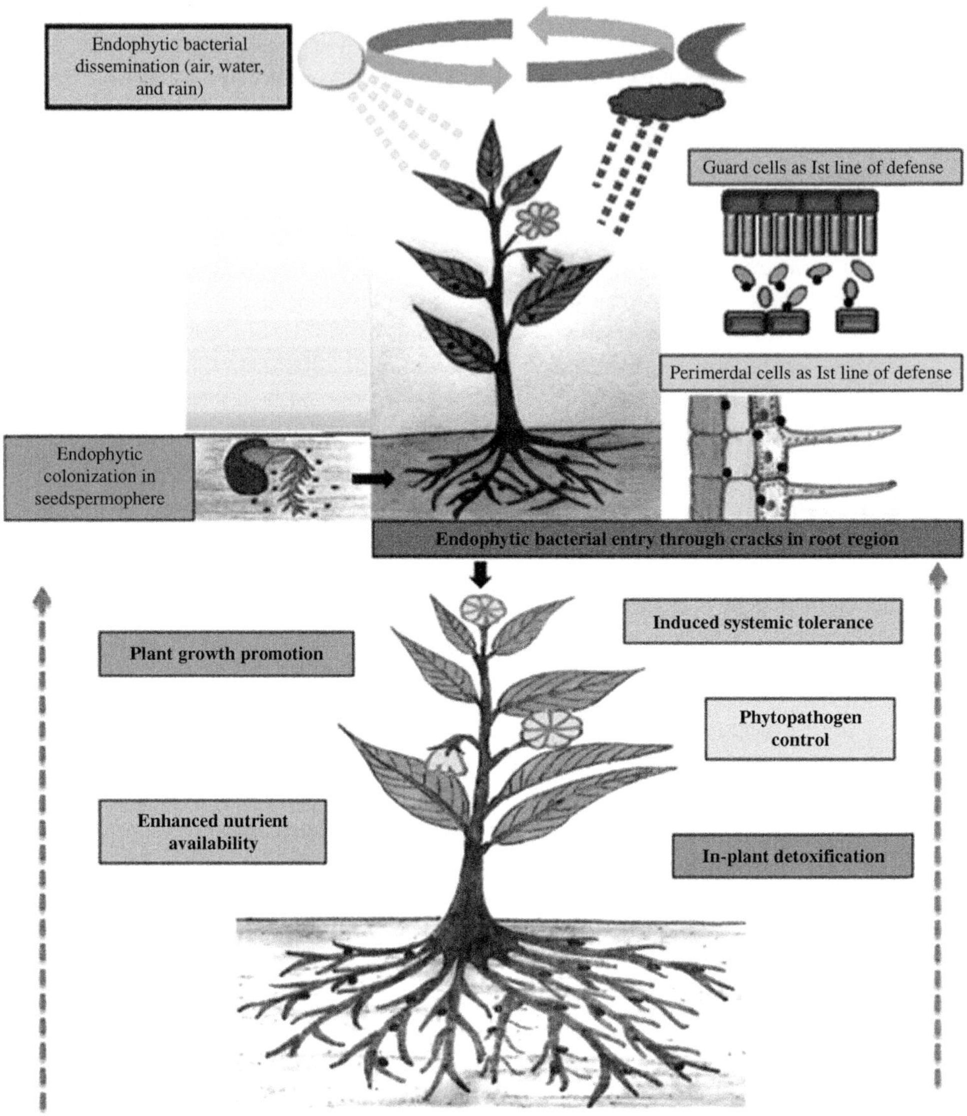

FIGURE 21.1 Schematic representation of endophytic bacterial entry into host tissues followed by successful colonization, which leads to several host benefits.

21.4.2 Entry from air

Plausible sources of bacterial environments include air, rain, water, as well as insects. In this context, entry of bacterial endophytes apart from the root surface includes the stomata, though this route is majorly utilized by phytopathogens (Melotto et al., 2008). The guard cells form the first line of defense thereby distinguishing pathogens from endophytes in a MAMP mediated screening (Arnaud and Hwang, 2015). According to Baldotto et al. (2011), *Herbaspirillum seropedicae* was found to colonize the intercellular spaces of the leaf mesophyll through the stomata. Similarly, White et al. (2014) reported the presence of *Bacillus amyloliquefaciens* in the guard cells of the stomatal region of *Amaranthus caudatus*. In yet another report, Fukami et al. (2016) demonstrated the presence of bacteria on the interior of leaves, possibly the stomatal region, upon being inoculated by leaf spray method.

21.5 Role of phytohormones in endophytic colonization

Phytohormones play a crucial role in modulating the phytomicrobiota, particularly for advanced plant nutrition and resistance against environmental stress conditions. For instance, Iniguez et al. (2005) reported ethylene mediated suppression of endophytic population in *Medicago truncatula* while the ethylene insensitive *M. truncatula* expressed hypercolonization of the endophytic population. Apart from ethylene, jasmonic acid is also reported to play a pivotal role in modulation of endophytomicrobiota (Miché et al., 2006). In this context, it may be comprehended that phytohormonal level is basically adjusted by the host plant so as to maintain an optimized concentration of plant beneficial microbiota.

Apart from concentration optimization of the endophytic microbiota, phytohormones also play a major role in managing the diversity of endosphere microbiome. For instance, Liu et al. (2017a) reported suppression of endophytic diversity in wheat roots by jasmonic acid mediated signaling pathway while the shoot endomicrobiota and the rhizomicrobiota remained unaffected. In this context, Lebeis et al. (2015) particularly stated that root cells of different isogenic *Arabidopsis* mutants differentially their endophytic community in response to salicylic acid mediated signaling. Another significant report by Kniskern et al. (2007) stated the specific role of salicylic acid in lowering the diversity of endophytic bacteria in leaves of *Arabidopsis* while the endophytic diversity remained quite unaffected by jasmonic acid.

In addition to salicylic acid and jasmonic acid, ethylene signaling also plays a major role in maintaining the density and diversity of the endophytic microbiome (Liu et al., 2017b). Long et al. (2010) reported that the diversity of culturable root endophytes was lowered in response to ethylene mediated signaling in isogenic transformed *Nicotiana attenuate* plants as compared with their wild type counterparts. Thus, it may be concluded that plant defense signaling pathway, particularly the phytohormones, significantly affect the population as well as diversity of the endophytic microbiome.

21.6 Endophytes mediated host benefit

21.6.1 Phytohormone production

Endophytes are well documented for production of various phytohormones, such as auxins (Zúñiga et al., 2013), gibberellins (GA) (Shahzad et al., 2016), cytokinins (Bhore et al., 2010), etc. (Tables 21.1 and 21.2). Indole acetic acid (IAA) being the most abundant member of the auxin family is biosynthesized primarily from tryptophan following multiple pathways. However in plant associated bacteria, predominantly two major routes are followed, that is, indole-3-acetamide and indole-3-pyruvate pathways though indole-3-acetaldehyde pathway is also reported (Ray et al., 2017, 2018b). According to a report by Khan et al. (2014), IAA and GA released by *Sphingomonas* sp. LK11 enhanced growth of tomato seedlings. Similarly, *S. mutabilis* strain IA1 collected from the Saharan desert region produced IAA and GA, which augmented growth of wheat seedlings as well as enhanced host resistance against *Fusarium culmorum* (Toumatia et al., 2016).

Another phytohormone of notable importance is 1-aminocyclopropane-1-carboxylate (ACC) deaminase. Under stress conditions, ACC gets accumulated in root tissues followed by systemic transfer to aerial tissues where the molecule is converted to ethylene (ET) by ACC-oxidase present in leaves (Bulgarelli et al., 2013). Endophytic bacteria enable optimization of ethylene levels produced within the host by production of ACC deaminase from *acdS* gene. The enzyme utilizes ACC, the immediate precursor of ET, as a carbon and nitrogen source thereby producing α-ketobutyrate and ammonia and blocking the ET biosynthesis pathway (Ali et al., 2014). According to Argueso et al. (2007), auxin production by endomicrobiota stimulates ACC synthase thereby increasing ACC concentration, which further triggers ACC deaminase activity. In this context, Qin et al. (2014) reported ACC deaminase production by the endomicrobiota of *Limonium sinense*, which enhanced overall plant growth under salinity stress conditions. Thus, it may be predicted that augmenting phytohormone production through use of endophytic microbiota is a feasible approach towards sustainable agriculture under stress conditions (Vaishnav et al., 2018).

21.6.2 Siderophore production

Endophytic microbes have an augmented ability of producing organic acid and siderophores (low molecular weight iron chelating compounds) particularly under iron limiting conditions (Ahmed and Holmström, 2014). In some reports, it has been documented that endophytes have a higher affinity for Fe^{3+} ions than phytosiderophores. In some cases, microbes have even been reported to degrade phytosiderophores. However, plant hosts not associated with microbes have demonstrated considerable iron deficiency (Carvalhais et al., 2013). Sessitsch et al. (2012) reported significant upregulation of genes related to siderophore production, reception, and iron storage in the root endophytic microbiota of rice plants. Apart from providing iron to hosts, siderophores are also reported to play a significant role in biocontrol as they quench the bioavailable iron thereby depriving phytopathogens of their access (Aznar et al., 2015).

TABLE 21.1 Phytohormones released by endophytic bacteria.

S. no.	Endophytic bacteria	Phytohormones	Host plant	References
1.	*Pseudomonas putida, P. fluorescens, P. stutzeri, P. synxantha, P. thivervalensis*	IAA, zeatin	*Solanum nigrum, Nicotiana attenuate, Prosopis strombulifera, Echinacea, Pinus sylvestris*	Lata et al. (2006), Long et al. (2008), Sgroy et al. (2009), Pirttilä et al. (2004)
2.	*Agrobacterium tumefaciens, A. rhizogenes*	IAA, zeatin	*S. nigrum, N. attenuata*	Long et al. (2008), Pirttilä et al. (2004)
3.	*Bacillus pumilus, B. subtilis, B. cereusB. licheniformis, B. axargiensis,* and *B. malacitensis*	IAA, gibberellic acid, zeatin, abscisic acid (ABA)	*P. strombulifera,* sugarbeet, *Echinacea,* soyabean	Lata et al. (2006), Sgroy et al. (2009), Shi et al. (2009), Senthilkumar et al. (2009)
4.	*Pantoea* spp.	IAA	*S. nigrum*	Long et al. (2008)
5.	*Aeromonas* spp.	IAA	*S. nigrum*	Long et al. (2008)
6.	*Acinetobacter johnsonii*	IAA, gibberellic acid, zeatin, ABA	Sugarbeet	Shi et al. (2009)
7.	*Chryseobacterium indologene*	IAA, gibberellic acid, zeatin, ABA	Sugarbeet	Shi et al. (2009)
8.	*Sphingomonas* spp.	IAA, gibberellic acid	*Tephrosia apollinea*	Khan et al. (2014)
9.	*Enterobacter* spp.	IAA	*Piper nigrum*	Jasim et al. (2013)
10.	*Klebsiella* spp.	IAA	*P. nigrum, Saccharum* sp.	Jasim et al. (2013), Govindarajan et al. (2007)
11.	*Burkholderia vietnamiensis*	IAA	*Saccharum* sp.	Govindarajan et al. (2007)
12.	*Methylobacterium extorquens*	Zeatin	*P. sylvestris*	Pirttilä et al. (2004)
13.	*Brevibacterium halotolerans*	Gibberellic acid	*P. strombulifera*	Sgroy et al. (2009)
14.	*Achromobacter xylosoxidans*	Gibberellic acid	*P. strombulifera*	Sgroy et al. (2009)
15.	*Lysinibacillus fusiformis*	Gibberellic acid	*P. strombulifera*	Sgroy et al. (2009)

21.6.3 Nitrogen metabolism

Nitrogen metabolism is thoroughly reported in the genomes of rhizospheric and endophytic microbiota. According to Sessitsch et al. (2012), the metagenome of rice roots revealed microbial genes related to nitrification and ammonia oxidation, which further clarifies the role of endophytic microbiota in the nitrogen cycling process. Straub et al. (2013) reported the presence of nitrogen fixation and denitrification genes in some endophytic bacteria. Similarly, foliar endophytes have also been reported to active roles in nitrogen cycling process (Moyes et al., 2016). In yet another report, Puri et al. (2016) endophytic nitrogen fixing *Paenibacillus polymyxa* P2b-2R from rhizospheric as well as endospheric tissues of maize. Though significant research on endophyte mediated nitrogen metabolism has been done to date, the fraction of total nitrogen contributed by endophytes to the host remains to be investigated (Liu et al., 2017b).

21.6.4 Quorum quenching of phytopathogens

Quorum quenching is the process of disruption of a signaling system responsible for upregulation of genes related to virulence, biofilm formation and maturation, as well as competence (Miller and Bassler, 2001). While endophytic bacteria mediated quenching of violcein pigment released by biosensor *Chromobacterium violaceum* is well documented (Kusari et al., 2014), *Bacillus* and *Pseudomonas* sp. having *carAB* gene, responsible for quenching the virulence factor of phytopathogenic *Xanthomonas* sp. and *Xyllela fastidiosa* are also reported by Newman et al. (2008). Similarly, endophytic *B. amyloliquefaciens* was reported to express *aiiA* gene coding for quorum quenching lactonase AiiA to inhibit phytopathogenic *B. dothidea* (Yin et al., 2011). Endophytic microbes, being within the host, are constantly under the

TABLE 21.2 Phytohormones released by endophytic fungi.

S. no.	Endophytic fungi	Phytohormones	Host plant	References
1.	*Aspergillus fumigatus, A. niger*	Indole acetic acid (IAA), gibberellic acid, ABA	Soybean (*Glycine max*), *Cannabis sativa*	Khan et al. (2011a), Lubna Asaf et al. (2018)
2.	*Penicillium chrysogenum, P. crustosum, P. funiculosum, P. commune, P. citrinum*	IAA, gibberellic acid	*Teucrium polium*; cucumber (*Cucumis sativus*), soybean (*G. max*); *Sesamum indicum*; *Ixeris repenes*	Choi et al. (2005), Hassan (2017), Khan et al. (2008), Waqas et al. (2012), Khan et al. (2011b)
3.	*Fusarium* spp.	IAA, gibberellic acid	*Euphorbia pekinensis*	Dai et al. (2008)
4.	*Exophiala* spp.	Gibberellic acid	*C. sativus*	Khan et al. (2011c)
5.	*Paecilomyces formosus*	Gibberellic acid	*C. sativus*	Khan et al. (2012a)
6.	*Rhodotorula minuta*	Zeatin	Scots pine (*Pinus sylvestris*)	Pirttilä et al. (2004)
7.	*Galactomyces geotrichum*	IAA, gibberellic acid	*Trapa japonica*	Waqas et al. (2014)
8.	*Pestalotiopsis aff. neglecta*	IAA	*Platycladus orientalis*	Hoffman et al. (2013)
9.	*Phoma* spp., *P. glomerata, P. herbarum*	IAA, gibberellic acid	Cucumber (*Cucumis sativum*), soybean (*G. max*)	Waqas et al. (2012), Hamayun et al. (2009a)
10.	*Chaetomium globosum*	IAA, gibberellic acid	Pepper (*Capsicum annum*)	Khan et al. (2012b)
11.	*Cadophora malorum*	Gibberellic acid	*Calystegia soldanella*	You et al. (2012)
12.	*Arthrinium phaeospermum*	Gibberellic acid	*Carex kobomugi*	
13.	*Cladosporium sphaerospermum*	Gibberellic acid	Soybean (*G. max*)	Hamayun et al. (2009b)
14.	*Chrysoporium pseudomerdarium*	Gibberellic acid	Soybean (*G. max*)	Hamayun et al. (2009c)
15.	*Scolecobasidium tshawytschae*	Gibberellic acid	Soybean (*G. max*)	Hamayun et al. (2009d)
16.	*Gliomastix murorum*	Gibberellic acid	*Elymus mollis*	Khan et al. (2009)

influence of various chemomicrobial interactions. A thorough understanding of these interactions will enable an optimal release of quorum quenching substances active against phytopathogens (Mookherjee et al., 2018).

21.6.5 Biocontrol activity against phytopathogens

Biocontrol activity by endophytic bacteria, particularly by the application of antimicrobials, has remained a promising mechanism to date. For instance, exogenous application of *B. amyloliquefaciens* released iturins on cotton plants induced an MAMP triggered defense pathway thereby causing ROS burst and disruption of phytopathogenic fungal cell wall integrity (Han et al., 2015). Similarly, endophytic *Pseudomonas poae* was reported to release lipopeptide poaeamide with antifungal activity against *Rhizoctonia solani* (Zachow et al., 2015). Apart from lipopeptides, bacterial endophytes are also reported to release volatile antimicrobials, such as 2,3-butanediol from *Enterobacter aerogenes*, which express resistance against northern corn leaf blight fungus *Setosphaeria turcica* (D'Alessandro et al., 2014). In spite of the promising scope of antimicrobial volatiles of endophytic bacteria, field application offering large scale contribution to agriculture is still in its infancy.

21.7 Conclusions

Sculpting of root endophytic microbiome by host plants is an essential phenomenon during endophytic recruitment. In this context, metaomic technologies and high throughput DNA sequencing play a pivotal role in assessment of endophytic composition in a culture independent manner. However, despite the plethora of information available on host–endophyte relationship, certain facets still remain untouched (Liu et al., 2017). For instance, rare presence of Gemmatimonadetes, Acidobacteria, and Archaea in the plant's endosphere demands critical analysis. Further, the extent of endophytic colonization required for enhancing growth promotion and defense in host is a realm entailing a thorough investigation.

Enhancing agricultural productivity by exploiting the phytomicrobiota is an exciting arena. However, manipulation of the endophytic microbiota for augmenting host growth and productivity has still remained quite challenging. Further, most of the research to date has been focused on the basis of presumption of a similar mechanism in rhizobacteria as that of the endophytic bacteria. Thus research needs to be directed towards elucidation of plant growth promotional and defense related traits restricted to endophytes. In line to the above concept, metaomic approaches should be considered to precisely comprehend host–endophyte relationships.

Research is required to understand how beneficial endophytic bacteria are recruited, optimized, and maintained within the host during various growth stages. While several genes demonstrating the endophytic lifestyle have been characterized within the genome, transposon mutagenesis may provide a scope for analysis of endophytic genes hitherto unknown. In this context, engineering of bacterial endophytes may augment their potential for enhanced plant growth promotion and defense. Though much work remains to be done, it may be predicted that bacterial endophytes may behave as partial or total substitutes of chemical fertilizers in the near future.

References

Ahmed, E., Holmström, S.J., 2014. Siderophores in environmental research: roles and applications. Microb. Biotechnol 7 (3), 196–208.

Ali, S., Charles, T.C., Glick, B.R., 2014. Amelioration of high salinity stress damage by plant growth-promoting bacterial endophytes that contain ACC deaminase. Plant Physiol. Biochem 80, 160–167.

Argueso, C.T., Hansen, M., Kieber, J.J., 2007. Regulation of ethylene biosynthesis. J. Plant Growth Regul. 26 (2), 92–105.

Arnaud, D., Hwang, I., 2015. A sophisticated network of signaling pathways regulates stomatal defenses to bacterial pathogens. Mol. Plant. 8 (4), 566–581.

Atamna-Ismaeel, N., Finkel, O.M., Glaser, F., Sharon, I., Schneider, R., et al., 2012. Microbial rhodopsins on leaf surfaces of terrestrial plants. Environ. Microbiol 14 (1), 140–146.

Aznar, A., Chen, N.W., Thomine, S., Dellagi, A., 2015. Immunity to plant pathogens and iron homeostasis. Plant Sci. 240, 90–97.

Baldotto, L.E.B., Olivares, F.L., Bressan-Smith, R., 2011. Structural interaction between GFP-labeled diazotrophic endophytic bacterium Herbaspirillum seropedicae RAM10 and pineapple plantlets' Vitória'. Braz. J. Microbiol. 42 (1), 114–125.

Barea, J.M., Pozo, M.J., Azcon, R., Azcon-Aguilar, C., 2005. Microbial co-operation in the rhizosphere. J. Exp. Bot. 56 (417), 1761–1778.

Beltran-Garcia, M.J., White Jr, J.F., Prado, F.M., Prieto, K.R., Yamaguchi, L.F., Torres, M.S., et al., 2014. Nitrogen acquisition in Agave tequilana from degradation of endophytic bacteria. Sci. Rep. 4, 6938.

Bertin, C., Yang, X.H., Weston, L.A., 2003. The role of root exudates and allelochemicals in the rhizosphere. Plant Soil. 256 (1), 67–83.

Bhore, S.J., Ravichantar, N., Loh, C.Y., 2010. Screening of endophytic bacteria isolated from leaves of Sambung Nyawa [*Gynura procumbens* (Lour.) Merr.] for cytokinin-like compounds. Bioinformation 5 (5), 191–197.

Bonafante, P., Anca, I.A., 2009. Plants, mycorrhizal fungi, and bacteria: a network of interactions. Annu. Rev. Microbiol. 63, 363–383.

Bright, M., Bulgheresi, S., 2010. A complex journey: transmission of microbial symbionts. Nat. Rev. Microbiol. 8 (3), 218–230.

Bulgarelli, D., Schlaeppi, K., Spaepen, S., van Themaat, E.V.L., Schulze-Lefert, P., 2013. Structure and functions of the bacterial microbiota of plants. Annu. Rev. Plant Biol. 64, 807–838.

Card, S., Johnson, L., Teasdale, S., Caradus, J., 2016. Deciphering endophyte behaviour: the link between endophyte biology and efficacious biological control agents. FEMS Microbiol. Ecol. 92 (8), 1–19.

Carvalhais, L.C., Muzzi, F., Tan, C.H., Hsien-Choo, J., Schenk, P.M., 2013. Plant growth in *Arabidopsis* is assisted by compost-soil derived microbial communities. Front. Plant Sci. 4, 235.

Chen, Y., Murrell, J.C., 2010. When metagenomics meets stable-isotope probing: progress and perspectives. Trends Microbiol. 18 (4), 157–163.

Choi, W.Y., Rim, S.O., Lee, J.H., Lee, J.M., Lee, I.J., et al., 2005. Isolation of gibberellins producing fungi from the root of several *Sesamum indicum* plants. J. Microbiol. Biotechnol. 15 (1), 22–28.

Compant, S., Reiter, B., Sessitsch, A., Nowak, J., Clément, C., et al., 2005. Endophytic colonization of *Vitis vinifera* L. by plant growth-promoting bacterium *Burkholderia* sp. strain PsJN. Appl. Environ. Microbiol. 71 (4), 1685–1693.

Compant, S., Clément, C., Sessitsch, A., 2010. Plant growth-promoting bacteria in the rhizo-and endosphere of plants: their role, colonization, mechanisms involved and prospects for utilization. Soil Biol. Biochem. 42 (5), 669–678.

Conrath, U., Beckers, G.J., Langenbach, C.J., Jaskiewicz, M.R., 2015. Priming for enhanced defense. Ann. Rev. Phytopathol 53, 97–119.

Coutinho, B.G., Licastro, D., Mendonca-Previato, L., Cámara, M., Venturi, V., 2015. Plant influenced gene expression in the rice endophyte *Burkholderia kururiensis* M130. Mol. Plant Microb. Interact. 28 (1), 10–21.

D'Alessandro, M., Erb, M., Ton, J., Brandenburg, A., Karlen, D., et al., 2014. Volatiles produced by soil-borne endophytic bacteria increase plant pathogen resistance and affect tritrophic interactions. Plant Cell Environ 37 (4), 813–826.

Dai, C.C., Yu, B.Y., Li, X., 2008. Screening of endophytic fungi that promote the growth of *Euphorbia pekinensis*. Afr. J. Biotechnol. 7 (19), 3505–3510.

Dumont, M.G., Pommerenke, B., Casper, P., Conrad, R., 2011. DNA-, rRNA- and mRNA- based stable isotope probing of aerobic methanotrophs in lake sediment. Environ. Microbiol 13 (5), 1153–1167.

el Zahar Haichar, F., Marol, C., Berge, O., Rangel-Castro, J.I., Prosser, J.I., et al., 2008. Plant host habitat and root exudates shape soil bacterial community structure. ISME J. 2 (12), 1221.

Foster, K.R., Wenseleers, T., 2006. A general model for the evolution of mutualisms. J. Evol. Biol. 19 (4), 1283–1293.

Fukami, J., Nogueira, M.A., Araujo, R.S., Hungria, M., 2016. Accessing inoculation methods of maize and wheat with *Azospirillum brasilense*. AMB Expr. 6 (1), 3.

Govindarajan, M., Kwon, S.W., Weon, H.Y., 2007. Isolation, molecular characterization and growth-promoting activities of endophytic sugarcane diazotroph Klebsiella sp. GR9. World J. Microbiol. Biotechnol. 23 (7), 997.

Hamayun, M., Khan, S.A., Khan, A.L., Rehman, G., Sohn, E., et al., 2009a. *Phoma herbarum* as a new gibberellin-producing and plant growth-promoting fungus. J. Microbiol. Biotechnol. 19 (2), 1244–1249.

Hamayun, M., Khan, S.A., Ahmad, N., Tang, D.S., Kang, S.M., et al., 2009b. *Cladosporium sphaerospermum* as a new plant growth-promoting endophyte from the roots of *Glycine max* (L.) Merr. World J. Microbiol. Biotechnol. 25 (4), 627–632.

Hamayun, M., Khan, S.A., Iqbal, I., Na, C.I., Khan, A.L., et al., 2009c. *Chrysosporium pseudomerdarium* produces gibberellins and promotes plant growth. J. Microbiol. 47 (4), 425–430.

Hamayun, M., Khan, S.A., Kim, H.Y., Chaudhary, M.F., Hwang, Y.H., et al., 2009d. Gibberellin production and plant growth enhancement by newly isolated strain of *Scolecobasidium tshawytschae*. J. Microbiol. Biotechnol. 19 (6), 560–565.

Han, Q., Wu, F., Wang, X., Qi, H., Shi, L., Ren, A., et al., 2015. The bacterial lipopeptide iturins induce *Verticillium dahliae* cell death by affecting fungal signalling pathways and mediate plant defence responses involved in pathogen associated molecular pattern-triggered immunity. Environ. Microbiol. 17 (4), 1166–1188.

Hardoim, P.R., van Overbeek, L.S., van Elsas, J.D., 2008. Properties of bacterial endophytes and their proposed role in plant growth. Trends Microbiol. 16 (10), 463–471.

Hassan, S.E.D., 2017. Plant growth-promoting activities for bacterial and fungal endophytes isolated from medicinal plant of *Teucrium polium* L. J. Adv. Res. 8 (6), 687–695.

Hoffman, M.T., Gunatilaka, M.K., Wijeratne, K., Gunatilaka, L., Arnold, A.E., 2013. Endohyphal bacterium enhances production of indole-3-acetic acid by a foliar fungal endophyte. PLoS One. 8 (9), e73132.

Iniguez, A.L., Dong, Y., Carter, H.D., Ahmer, B.M., Stone, J.M., et al., 2005. Regulation of enteric endophytic bacterial colonization by plant defenses. Mol. Plant Microb. Int. 18 (2), 169–178.

Jasim, B., John Jimtha, C., Jyothis, M., Radhakrishnan, E.K., 2013. Plant growth promoting potential of endophytic bacteria isolated from *Piper nigrum*. Plant Growth Regul 71 (1), 1–11.

Khan, S.A., Hamayun, M., Yoon, H., Kim, H.Y., Suh, S.J., et al., 2008. Plant growth promotion and *Penicillium citrinum*. BMC Microbiol. 8 (1), 231.

Khan, S.A., Hamayun, M., Kim, H.Y., Yoon, H.J., Lee, I.J., et al., 2009. Gibberellin production and plant growth promotion by a newly isolated strain of *Gliomastix murorum*. World J. Microb. Biotech. 25 (5), 829–833.

Khan, A.L., Hamayun, M., Kim, Y.H., Kang, S.M., Lee, J.H., et al., 2011a. Gibberellins producing endophytic *Aspergillus fumigatus* sp. LH02 influenced endogenous phytohormonal levels, isoflavonoids production and plant growth in salinity stress. Process Biochem. 46 (2), 440–447.

Khan, A.L., Hamayun, M., Kim, Y.H., Kang, S.M., Lee, I.J., 2011b. Ameliorative symbiosis of endophyte (*Penicillium funiculosum* LHL06) under salt stress elevated plant growth of *Glycine max* L. Plant Physiol. Biochem. 49 (8), 852–861.

Khan, A.L., Hamayun, M., Ahmad, N., Waqas, M., Kang, S.M., et al., 2011c. *Exophiala* sp. LHL08 reprograms *Cucumis sativus* to higher growth under abiotic stresses. Physiol. Plant. 143 (4), 329–343.

Khan, A.L., Hamayun, M., Kang, S.M., Kim, Y.H., Jung, H.Y., et al., 2012a. Endophytic fungal association via gibberellins and indole acetic acid can improve plant growth under abiotic stress: an example of *Paecilomyces formosus* LHL10. BMC Microbiol. 12 (1), 3.

Khan, A.L., Shinwari, Z.K., Kim, Y., Waqas, M., Hamayun, M., et al., 2012b. Role of endophyte *Chaetomium globosum* lk4 in growth of *Capsicum annuum* by production of gibberellins and indole acetic acid. Pak. J. Bot. 44 (5), 1601–1607.

Khan, A.L., Waqas, M., Kang, S.M., Al-Harrasi, A., Hussain, J., et al., 2014. Bacterial endophyte *Sphingomonas* sp. LK11 produces gibberellins and IAA and promotes tomato plant growth. J. Microbiol. 52 (8), 689–695.

Knief, C., Delmotte, N., Chaffron, S., Stark, M., Innerebner, G., et al., 2012. Metaproteogenomic analysis of microbial communities in the phyllosphere and rhizosphere of rice. ISME J. 6 (7), 1378.

Kniskern, J.M., Traw, M.B., Bergelson, J., 2007. Salicylic acid and jasmonic acid signaling defense pathways reduce natural bacterial diversity on *Arabidopsis thaliana*. Mol. Plant Microb. Interact. 20 (12), 1512–1522.

Kolton, M., Frenkel, O., Elad, Y., Cytryn, E., 2014. Potential role of flavobacterial gliding-motility and type IX secretion system complex in root colonization and plant defense. Mol. Plant Microb. Interact. 27 (9), 1005–1013.

Kusari, P., Kusari, S., Lamshöft, M., Sezgin, S., Spiteller, M., et al., 2014. Quorum quenching is an antivirulence strategy employed by endophytic bacteria. Appl. Microbiol. Biotechnol. 98 (16), 7173–7183.

Lata, H., Li, X.C., Silva, B., Moraes, R.M., Halda-Alija, L., 2006. Identification of IAA-producing endophytic bacteria from micropropagated *Echinacea* plants using 16S rRNA sequencing. Plant Cell Tissue Organ Cult. 85 (3), 353–359.

Lebeis, S.L., Paredes, S.H., Lundberg, D.S., Breakfield, N., Gehring, J., et al., 2015. Salicylic acid modulates colonization of the root microbiome by specific bacterial taxa. Science. 349 (6250), 860–864.

Liu, H., Carvalhais, L.C., Schenk, P.M., Dennis, P.G., 2017a. Effects of jasmonic acid signalling on the wheat microbiome differ between body sites. Sci. Rep. 7, 41766.

Liu, H., Carvalhais, L.C., Crawford, M., Singh, E., Dennis, P.G., et al., 2017b. Inner plant values: diversity, colonization and benefits from endophytic bacteria. Front Microbiol. 8, 2552.

Long, H.H., Schmidt, D.D., Baldwin, I.T., 2008. Native bacterial endophytes promote host growth in a species-specific manner; phytohormone manipulations do not result in common growth responses. PLoS One. 3 (7), e2702.

Long, H.H., Sonntag, D.G., Schmidt, D.D., Baldwin, I.T., 2010. The structure of the culturable root bacterial endophyte community of Nicotiana attenuata is organized by soil composition and host plant ethylene production and perception. New Phytol 185 (2), 554–567.

Lu, Y., Rosencrantz, D., Liesack, W., Conrad, R., 2006. Structure and activity of bacterial community inhabiting rice roots and the rhizosphere. Environ. Microbiol. 8 (8), 1351–1360.

Lubna Asaf, S., Hamayun, M., Gul, H., Lee, I.J., Hussain, A., 2018. *Aspergillus niger* CSR3 regulates plant endogenous hormones and secondary metabolites by producing gibberellins and indole acetic acid. J. Plant Interact. 13 (1), 100–111.

Lundberg, D.S., Lebeis, S.L., Paredes, S.H., Yourstone, S., Gehring, J.S., et al., 2012. Defining the core *Arabidopsis thaliana* root microbiome. Nature. 488 (7409), 86–90.

Melotto, M., Underwood, W., He, S.Y., 2008. Role of stomata in plant innate immunity and foliar bacterial diseases. Annu. Rev. Phytopathol. 46, 101–122.

Miché, L., Battistoni, F., Gemmer, S., Belghazi, M., Reinhold-Hurek, B., 2006. Upregulation of jasmonate-inducible defense proteins and differential colonization of roots of *Oryza sativa* cultivars with the endophyte *Azoarcus* sp. Mol. Plant Microb. Interact. 19 (5), 502–511.

Miller, M.B., Bassler, B.L., 2001. Quorum sensing in bacteria. Annu. Rev. Microbiol. 55 (1), 165–199.

Mookherjee, A., Singh, S., Maiti, M.K., 2018. Quorum sensing inhibitors: can endophytes be prospective sources? Arch. Microbiol. 200 (2), 355–369.

Moran, N.A., 2006. Symbiosis. Curr. Biol. 16, 866–871.

Moyes, A.B., Kueppers, L.M., Pett-Ridge, J., Carper, D.L., Vandehey, N., et al., 2016. Evidence for foliar endophytic nitrogen fixation in a widely distributed subalpine conifer. New Phytol. 210 (2), 657–668.

Müller, D.B., Vogel, C., Bai, Y., Vorholt, J.A., 2016. The plant microbiota: systems-level insights and perspectives. Annu. Rev. Genet. 50, 211–234.

Nelson, E.B., 2004. Microbial dynamics and interactions in the spermosphere. Annu. Rev. Phytopathol. 42, 271–309.

Newman, K.L., Chatterjee, S., Ho, K.A., Lindow, S.E., 2008. Virulence of plant pathogenic bacteria attenuated by degradation of fatty acid cell-to-cell signaling factors. Mol. Plant Microb. Interact. 21 (3), 326–334.

Peiffer, J.A., Spor, A., Koren, O., Jin, Z., Tringe, S.G., et al., 2013. Diversity and heritability of the maize rhizosphere microbiome under field conditions. Proc. Natl. Acad. Sci. USA. 110 (16), 6548–6553.

Pinto, A.J., Raskin, L., 2012. PCR biases distort bacterial and archaeal community structure in pyrosequencing datasets. PloS One. 7 (8), e43093.

Pirttilä, A.M., Joensuu, P., Pospiech, H., Jalonen, J., Hohtola, A., 2004. Bud endophytes of Scots pine produce adenine derivatives and other compounds that affect morphology and mitigate browning of callus cultures. Physiol. Plant. 121 (2), 305–312.

Puri, A., Padda, K.P., Chanway, C.P., 2016. Seedling growth promotion and nitrogen fixation by a bacterial endophyte *Paenibacillus polymyxa* P2b-2R and its GFP derivative in corn in a long-term trial. Symbiosis. 69 (2), 123–129.

Qin, S., Zhang, Y.J., Yuan, B., Xu, P.Y., Xing, J., et al., 2014. Isolation of ACC deaminase-producing habitat-adapted symbiotic bacteria associated with halophyte *Limonium sinense* (Girard) Kuntze and evaluating their plant growth-promoting activity under salt stress. Plant Soil. 374 (1-2), 753–766.

Radajewski, S., Ineson, P., Parekh, N.R., Murrell, J.C., 2000. Stable-isotope probing as a tool in microbial ecology. Nature. 403 (6770), 646.

Ray, S., Singh, S., Sarma, B.K., Singh, H.B., 2016. Endophytic *Alcaligenes* isolated from horticultural and medicinal crops promotes growth in okra (*Abelmoschus esculentus*). J. Plant Growth Regul. 35 (2), 401–412.

Ray, S., Singh, V., Bisen, K., Keswani, C., Singh, S., Singh, H.B., 2017. 11 Endophytomicrobiont: a multifaceted beneficial interaction. In: Singh, H. B., Sarma, B.K., Keswani, C. (Eds.), Advances in PGPR Research. CABI, pp. 218–232.

Ray, S., Mishra, S., Bisen, K., Singh, S., Sarma, B.K., Singh, H.B., 2018a. Modulation in phenolic root exudate profile of *Abelmoschus esculentus* expressing activation of defense pathway. Microbiol. Res. 207, 100–107.

Ray, S., Singh, J., Rajput, R., Singh, H., Singh, S., 2018b. Endophytic bacteria: an essential requirement of phyto nutrition. Nutr. Food Sci. Int. J. 5 (2), 55–57.

Rasche, F., Lueders, T., Schloter, M., Schaefer, S., Buegger, F., et al., 2009. DNA based stable isotope probing enables the identification of active bacterial endophytes in potatoes. New Phytol. 181 (4), 802–807.

Santoyo, G., Moreno-Hagelsieb, G., del Carmen Orozco-Mosqueda, M., Glick, B.R., 2016. Plant growth-promoting bacterial endophytes. Microbiol. Res. 183, 92–99.

Schlaeppi, K., Bulgarelli, D., 2015. The plant microbiome at work. Mol. Plant Microb. Interact. 28 (3), 212–217.

Schmid, M.D., Elesi, M.S., Othballer, M.R., Chloter, N.S., Ee, E.L., et al., 2006. Localization and visualization of microbial community structure and activity in soil microhabitats. In: Schulz, B., Boyle, C., Sieber, T. (Eds.), Intestinal Microorganisms of Termites and Other Invertebrates. Springer, Berlin, Germany, pp. 439–461.

Senthilkumar, M., Swarnalakshmi, K., Govindasamy, V., Lee, Y.K., Annapurna, K., 2009. Biocontrol potential of soybean bacterial endophytes against charcoal rot fungus, *Rhizoctonia bataticola*. Curr. Microbiol. 58 (4), 288.

Sessitsch, A., Hardoim, P., Döring, J., Weilharter, A., Krause, A., et al., 2012. Functional characteristics of an endophyte community colonizing rice roots as revealed by metagenomic analysis. Mol. Plant Microb. Interact. 25 (1), 28–36.

Sgroy, V., Cassán, F., Masciarelli, O., Del Papa, M.F., Lagares, A., et al., 2009. Isolation and characterization of endophytic plant growth-promoting (PGPB) or stress homeostasis-regulating (PSHB) bacteria associated to the halophyte Prosopis strombulifera. Appl. Microbiol. Biotechnol 85 (2), 371.

Shahzad, R., Waqas, M., Khan, A.L., Asaf, S., Khan, M.A., et al., 2016. Seed-borne endophytic *Bacillus amyloliquefaciens* RWL-1 produces gibberellins and regulates endogenous phytohormones of *Oryza sativa*. Plant Physiol. Biochem. 106, 236–243.

Shi, Y., Lou, K., Li, C., 2009. Promotion of plant growth by phytohormone-producing endophytic microbes of sugar beet. Biol. Fertil. Soils 45 (6), 645.

Stiefel, P., Zambelli, T., Vorholt, J.A., 2013. Isolation of optically targeted single bacteria by application of fluidic force microscopy to aerobic anoxygenic phototrophs from the phyllosphere. Appl. Environ. Microbiol 79 (16), 4895–4905.

Straub, D., Rothballer, M., Hartmann, A., Ludewig, U., 2013. The genome of the endophytic bacterium H. frisingense GSF30(T) identifies diverse strategies in the Herbaspirillum genus to interact with plants. Front. Microbiol 4, 168.

Toumatia, O., Compant, S., Yekkour, A., Goudjal, Y., Sabaou, N., et al., 2016. Biocontrol and plant growth promoting properties of *Streptomyces mutabilis* strain IA1 isolated from a Saharan soil on wheat seedlings and visualization of its niches of colonization. S. Afr. J. Bot 105, 234–239.

Turner, T.R., James, E.K., Poole, P.S., 2013. The plant microbiome. Genome Biol 14 (6), 209.

Vaishnav, A., Hansen, A.P., Agrawal, P.K., Varma, A., Choudhary, D.K., 2017. Biotechnological perspectives of Legume–Rhizobium symbiosis. In: Hansen, A.P., Choudhary, D.K., Agrawal, P.K., Varma, A. (Eds.), Rhizobium Biology and Biotechnology. Springer, Cham, pp. 247–256.

Vaishnav, A., Shukla, A.K., Sharma, A., Kumar, R., Choudhary, D.K., 2018. Endophytic bacteria in plant salt stress tolerance: current and future prospects. J. Plant Growth Regul. 1–19.

Wang, D., Yang, S., Tang, F., Zhu, H., 2012. Symbiosis specificity in the legume: rhizobial mutualism. Cell Microbiol 14 (3), 334–342.

Waqas, M., Khan, A.L., Kamran, M., Hamayun, M., Kang, S.M., et al., 2012. Endophytic fungi produce gibberellins and indoleacetic acid and promotes host-plant growth during stress. Molecules 17 (9), 10754–10773.

Waqas, M., Khan, A.L., Kang, S.M., Kim, Y.H., Lee, I.J., 2014. Phytohormone-producing fungal endophytes and hardwood-derived biochar interact to ameliorate heavy metal stress in soybeans. Biol. Fert. Soils 50 (7), 1155–1167.

White, J.F., Torres, M.S., Sullivan, R.F., Jabbour, R.E., Chen, Q., et al., 2014. Occurrence of *Bacillus amyloliquefaciens* as a systemic endophyte of vanilla orchids. Microsc. Res. Tech 77 (11), 874–885.

Yin, X.T., Xu, L.N., Xu, L., Fan, S.S., Liu, Z.Y., et al., 2011. Evaluation of the efficacy of endophytic *Bacillus amyloliquefaciens* against *Botryosphaeria dothidea* and other phytopathogenic microorganisms. Afr. J. Microbiol. Res 5, 340–345.

You, Y.H., Yoon, H., Kang, S.M., Woo, J.R., Choo, Y.S., et al., 2012. *Cadophora malorum* Cs-8-1as a new fungal strain producing gibberellins isolated from *Calystegia soldanella*. J. Basic Microbiol 53 (7), 630–634.

Zachow, C., Jahanshah, G., de Bruijn, I., Song, C., Ianni, F., et al., 2015. The novel lipopeptide poaeamide of the endophyte *Pseudomonas poae* RE *1-1-14 is involved in pathogen suppression and root colonization. Mol. Plant Microb. Interact 28 (7), 800–810.

Zúñiga, A., Poupin, M.J., Donoso, R., Ledger, T., Guiliani, N., et al., 2013. Quorum sensing and indole-3-acetic acid degradation play a role in colonization and plant growth promotion of *Arabidopsis thaliana* by *Burkholderia phytofirmans* PsJN. Mol. Plant Microb. Interact. 26 (5), 546–553.

Chapter 22

Tapping soil biodiversity for enhancing resource use efficiency

Ranjan Paul[1], Sonalika Sahoo[1], Avijit Ghosh[2] and Gobinath R[3]

[1]*Division of Soil Resource Studies, ICAR-National Bureau of Soil Survey and Land Use Planning, Nagpur, India,* [2]*ICAR-Indian Grassland and Fodder Research Institute, Jhansi, India,* [3]*ICAR-Indian Institute of Rice Research, Rajendranagar, Hyderabad, India*

Chapter Outline

22.1 Introduction

Soil is one of the most diverse habitats on Earth and contains the most diverse assemblages of living organisms. Soil biodiversity is the variety of all living organisms found within the heterogeneous and dynamic soil system. The Convention on Biological Diversity (CBD) describes it as the variation in soil life, from genes to communities, and the ecological complexes of which they are part. Soil biodiversity comprises not only the aboveground but the belowground diversity of animal and plant lives ranging from the micro- to macrolevel. Biodiversity is measured in the simplest way in terms of species richness (number of distinct species present) and evenness (proportional abundance of species

New and Future Developments in Microbial Biotechnology and Bioengineering. DOI: https://doi.org/10.1016/B978-0-12-818258-1.00028-5

present in a system). Biological activity in soils is largely concentrated in the topsoil. The biological components occupy a tiny fraction (<0.5%) of the total soil volume and make up less than 10% of the total soil organic matter. This living component consists of plant roots and soil organisms. Soil microorganisms are responsible for a large part of biological activity (60%–80%), which is associated with processes regulating nutrient cycles and decomposition of organic residues.

22.2 Soil biodiversity: provider of essential ecosystem goods and services

Soil and soil biodiversity are offering multiple services for the benefit of the nature's ecosystem; which is driven by multiple soil workers such as microorganisms and insects. The relationship between these services and soil biodiversity are positive and highly complex in nature. The European Commission (2010) classified the ecosystem services into seven categories and which are followed:

22.2.1 Soil organic matter recycling and fertility

Soil organisms contribute to modifying soil structure and creating new habitats. Soil organic matter is an important building block for soil structure, contributing to soil aeration, and enabling soils to absorb water and retain nutrients. All soil microbes are involved in the formation and decomposition of soil organic matter, and thus contribute to structuring the soil. For example, some species of fungi produce a protein that plays an important role in soil aggregation due to its sticky nature. The decomposition of soil organic matter by soil organisms releases nutrients in forms usable by plants and other organisms. The residual soil organic matter forms humus, which serves as the main driver of soil quality and fertility. As a result, soil organisms indirectly support the quality and abundance of plant primary production. It should be underlined that soil organic matter as humus can only be produced by the diversity of life that exists in soils and it cannot be manmade. When the soil organic matter recycling and fertility service is impaired, all life on Earth is threatened, as all life is either directly or indirectly reliant on plants and their products, including the supply of food, energy, nutrients (e.g., nitrogen produced by the rhizobium bacteria in synergy with the legumes), construction materials, and genetic resources. This service is crucial in all sorts of ecosystems, including agriculture and forestry. Plant biomass production also contributes to the water cycle and local climate regulation, through evapotranspiration.

22.2.2 Regulation of carbon flux and climate control *via* the carbon storage

Soil is estimated to contain about 2500 billion tonnes of carbon to 1 m depth. The soil organic carbon pool is the second largest carbon pool on the planet and is formed directly by soil biota or by the organic matter (e.g., litter, aboveground residues) that accumulates due to the activity of soil biota. Every year, soil organisms process 25,000 kg of organic matter (the weight of 25 cars) in soil in a surface area equivalent to a soccer field. Soil organisms increase the soil organic carbon pool through the decomposition of dead biomass, while their respiration releases carbon dioxide (CO_2) to the atmosphere. Carbon can also be released to the atmosphere as methane, a much more powerful greenhouse gas than CO_2, when soils are flooded or clogged with water. In addition, part of the carbon may leak from soils to other parts of the landscape or to other pools, such as the aquatic pool. Peat lands and grasslands are among the best carbon storage systems in Europe, while land-use change, through the conversion of grasslands to agricultural lands, is responsible for the largest carbon losses from soils.

22.2.3 Water cycle

Regulation, infiltration, storage, purification, transfer to aquifers and surface effluents, erosion prevention, and regulation of flows in effluents (flooding or drying out of rivers). Soil ecosystem engineers affect the infiltration and distribution of water in the soil, by creating soil aggregates and pore spaces. Soil biodiversity may also indirectly affect water infiltration, by influencing the composition and structure of the vegetation, which can shield off the soil surface, influence the structure and composition of litter layers, and influence soil structure by rooting patterns. It has been observed that the elimination of earthworm populations due to soil contamination can reduce the water infiltration rate significantly, in some cases even by up to 93%. The diversity of microorganisms in the soil contributes to water purification, nutrient removal, and to the biodegradation of contaminants and pathogenic microbes. Plants also play a key role in cycling of water between soil and atmosphere through their effects on (evapo-) transpiration.

22.2.4 Decontamination and bioremediation

The decontamination and bioremediation of contaminant sites are reclaimed through physical, chemical and biological neutralization of contaminants present in the soil. Microbes play a key role in bioremediation, by accumulating pollutants in their bodies, degrading pollutants into smaller, nontoxic molecules, or modifying those pollutants into useful metabolic molecules (e.g., taking several months in the case of hydrocarbons, but much more for other molecules). Humans often use these remediation capacities of soil organisms to directly engineer bioremediation, whether in situ or ex situ, or by promoting microbial activity. Phytoremediation, which is indirectly mediated by soil organisms, is also useful to remove persistent pollutants and heavy metals.

22.2.5 Pest control

Biological control of pests and pathogens of plants, animals, and humans is discussed here. Efficient pest control is essential to the production of healthy crops, and the impairment of this service can have important economic costs, as well as food safety costs. Ensuring efficient natural pest control avoids having to use engineered control methods, such as pesticides, which have both huge economic and ecological costs. The use of pesticides, for instance, can be at the origin of a loss of more than 8 billion dollars per year due to environmental and societal damages. In natural ecosystems, the loss of pathogenic and root-feeding soil organisms will cause a loss of plant diversity and will enhance the risk of exotic plant invasions. Changes in vegetation also influence aboveground biodiversity. Loss of this ecosystem service, therefore, will cause loss of biodiversity in entire natural ecosystems.

22.2.6 Human health

This includes both direct (e.g., provisioning of pharmaceutical molecules) and indirect services (e.g., avoided impacts linked to the nonprovisioning of the above mentioned services). Soil organisms, with their astonishing diversity, are an important source of chemical and genetic resources for the development of new pharmaceuticals. For instance, many antibiotics used today originate from soil organisms, for example penicillin, isolated from the soil fungus *Penicillium notatum* by Alexander Fleming in 1928, and streptomycin, derived in 1944 from a bacteria living in tropical soil. Given that antibiotic resistance develops fast, the demand for new molecules is unending. Soil biodiversity can also have indirect impacts on human health. Land-use change, global warming, or other disturbances to soil systems can release soil-borne infectious diseases and increase human exposure to those diseases. Finally, disturbed soil ecosystems may lead to more polluted soils or less fertile crops, all of which, if they reach large proportions, can indirectly affect human health, for example through intoxication of contaminated food or massive migrations. Loss of soil biodiversity, therefore, could reduce our capacity to develop novel antibiotic compounds, it could enhance the risk of infectious diseases, and it could increase the risk for humans to ingest toxic or contaminated food.

22.2.7 Sustainable agriculture

Given the escalating population growth, land degradation, and increasing demands for food, achieving sustainable agriculture and viable agricultural systems is critical to the issue of food security and poverty alleviation in most, if not all, developing countries. It is fundamental to the sustained productivity and viability of agricultural systems worldwide. Sustainable agriculture (including forestry) involves the successful management of agricultural resources to satisfy human needs while maintaining or enhancing environmental quality and conserving natural resources for future generations. The sustained use of the earth's land and water resources and thereby plant, animal and human health is dependent upon maintaining the health of the living biota that provide critical processes and ecosystem services. However, current technologies and development support for increased agricultural production have largely ignored this vital management component. Improvement in agricultural sustainability requires, alongside effective water and crop management, the optimal use and management of soil fertility and soil physical properties. Both rely on soil biodiversity and soil biological processes. This calls for the widespread adoption of management practices that enhance soil biological activity and thereby build up long-term soil productivity and health. FAO considers the issue of soil biodiversity and soil ecosystem management of great importance to the achievement of sustainable, resource-efficient, and productive agriculture. Soil biodiversity has been identified as an area requiring particular attention under the program of work on agricultural biodiversity of the Conference of the Parties to the CBD (Table 22.1).

TABLE 22.1 Soil biodiversity for enhancing resource use efficiency.

Essential services	Link with enhanced resource use efficiency
Supplementation and recycling of plant nutrients	Better utilization of inherent source of N, P, K, S, and micronutrients present within soil organic matter and minerals
Scavenging CO_2 from air	Sequestering of C within huge biomass in terrestrial and aquatic ecosystem forming humus, long-term reservoir of soil fertility
Modification of soil physical structure and maintain water regimes by organic matter recycling and humification	Increase water use efficiency (more crop per drop) by infiltration, storage, purification, transfer to aquifers
Decontamination and bioremediation	Protection of environmentally vulnerable areas for better resource use efficiency
Biological control of pest and pathogens	Reduction in use of pesticides and chemicals

22.3 Soil biodiversity and agricultural sustainability

Soil and its living organisms are an integral part of agricultural and forestry ecosystems, playing a critical role in maintaining soil health, ecosystem functions, and productivity. Of primary importance is the contribution of soil organisms to a wide range of essential services and to the sustainable function of all ecosystems, by acting as the primary driving agents of nutrient cycling; regulating the dynamics of soil organic matter, soil carbon sequestration, and greenhouse gas emissions; modifying soil physical structure and water regimes; enhancing the amount and efficiency of nutrient acquisition by the vegetation; and enhancing plant health. These services are not only essential to the functioning of natural ecosystems but constitute an important resource for agricultural production and food security as well as the sustainable management of agricultural systems.

It is well known that land management practices alter soil conditions and the soil community of micro-, meso-, and macroorganisms. However, the relationship between specific practices and soil functions is less clear. In general, the structure of soil communities is largely determined by ecosystem characteristics and land use systems. For example, arid systems have few earthworms, but have termites, ants, and other invertebrates that serve similar functions. On the other hand, the level of activity of different species depends on specific management practices as these affect the microenvironment conditions, including temperature, moisture, aeration, pH, pore size, and type of food sources.

Management strategies including tillage, crop rotations, and use of plant residues and manure change soil habitats and the food web and alter soil quality, or the capacity of the soil to perform its functions. For example, soil compaction, poor vegetation cover, and/or lack of plant litter covering the soil surface tend to reduce the number of soil arthropods. Farming practices that minimize soil disturbance (plowing) and return plant residues to the soil, such as no-tillage farming and crop rotation, allow slowly rebuilding and restoring soil organic matter. Reducing tillage tends to also result in increased growth of fungi, including mycorrhizal fungi.

Farming communities are concerned with land management issues such as water availability to plants, access to sources of fuel and fodder, control of soil erosion, and land degradation, especially avoiding soil nutrient depletion and pollution of air, soil, and water resources. Nonetheless, farmers are essentially driven not by environmental concerns, but by economics, by issues of costs and returns and efficiency in terms of labor and energy and use of external inputs. A central paradigm for the farmer for the maintenance and management of soil fertility, without undue reliance on costly and often risky external inputs, is to utilize his or her management practices to influence soil biological populations and processes in such a way as to improve and sustain land productivity.

Options whereby farmers can actually manage soil biodiversity to enhance agricultural production can be classified into direct and indirect interventions:

- Direct methods of intervening in the production system aim to alter the abundance or activity of specific groups of organisms through inoculation and/or direct manipulation of soil biota. Inoculation with soil beneficial organisms, such as nitrogen-fixing bacteria, Mycorrhiza, and earthworms, have been shown to enhance plant nutrient uptake, increase heavy metal tolerance, improve soil structure and porosity, and reduce pest damage.
- Indirect interventions are means of managing soil biotic processes by manipulating the factors that control biotic activity (habitat structure, microclimate, nutrients, and energy resources) rather than the organisms themselves.

Examples of indirect interventions include most agricultural practices such as the application of organic material to soil (for example through composting), tillage, irrigation, green manuring and liming, as well as cropping system design and management. These must not be conducted independently, but in a holistic fashion, because of the recurrent interactions between different management strategies, hierarchical levels of management, and different soil organisms.

Direct and indirect benefits of improving soil biological management in agricultural systems can be assessed in terms of economic, environmental, and food security benefits:

- *Economic benefits:* Soil biological management reduces input costs by enhancing resource use efficiency (especially decomposition and nutrient cycling, nitrogen fixation, and water storage and movement). Less fertilizer may be needed if nutrient cycling becomes more efficient and less fertilizer is leached from the rooting zone. Fewer pesticides are needed where a diverse set of pest-control organisms is active. As soil structure improves, the availability of water and nutrients to plants also improves. It is estimated that the value of "ecosystem services" (e.g., organic waste disposal, soil formation, bioremediation, N_2 fixation and biocontrol) provided each year by soil biota in agricultural systems worldwide may exceed US $1542 billion (Pimentel et al, 1997).
- *Environmental protection:* Soil organisms filter and detoxify chemicals and absorb the excess nutrients that would otherwise become pollutants when they reach groundwater or surface water. The conservation and management of soil biota help to prevent pollution and land degradation, especially through minimizing the use of agro-chemicals and maintaining/enhancing soil structure and cation exchange capacity. Excessive reduction in soil biodiversity, especially the loss of keystone species or species with unique functions, for example, as a result of excess chemicals, compaction, or disturbance, may have catastrophic ecological effects leading to loss of agricultural productive capacity. The mix of soil organisms in the soil also partially determines soil resilience, the desirable ability of a given soil to recover its functions after a disturbance such as fire, compaction, and tillage.
- *Food security:* Soil biological management can improve crop yield and quality, especially through controlling pests and diseases and enhancing plant growth. Soil biodiversity determines the resource use efficiency, as well as the sustainability and resilience of low-input agro-ecological systems, which ensure the food security of much of the world's population, especially the poor. In addition, some soil organisms are consumed as an important source of protein by different cultures and others are used for medicinal purposes. At least 32 Amerindian groups in the Amazon basin use terrestrial invertebrates as food, and especially, as sources of animal protein, a strategy that takes advantage of the abundance of these highly renewable elements of the rainforest ecosystem.
- Ecological approaches and opportunities to promote soil biota for sustainable agriculture.

As noted above, soil biota may be beneficial, neutral, or detrimental to plant growth. Thus soil biota and their ecological interactions must be effectively managed for maximum productivity. Land managers need unbiased information that will enable them to develop biologically based management strategies to control or manipulate soil stabilization, nutrient cycling, crop diseases, pest infestations, and detoxification of natural and manmade contaminants. These strategies will require improved understanding of the effects on soil biota of habitats, food sources, host interactions, and the soil physical and chemical environment. Understanding the ecology regulating both beneficial and detrimental organisms is essential to harnessing and controlling their activity in agro-ecosystems with a view to promoting viable, productive, and sustainable systems.

Capturing the benefits of soil biological activity for agricultural production requires adhering to the following ecological principles:

- *Supply organic matter:* Each type of soil organism occupies a specific niche in the web of life and favors a different substrate and nutrient source. A rich and varied source of organic matter will tend to support a wide variety of soil organisms.
- *Increase plant varieties:* Crops should be mixed and their spatial–temporal distribution varied to create a greater diversity of niches and resources that stimulate soil biodiversity. Through crop rotation and intercropping it is possible to encourage a wider variety of organisms, improve nutrient cycling, and natural processes of pest and disease control.
- *Protect the habitat of soil organisms:* The activity of soil biodiversity can be stimulated by improving soil living conditions such as aeration, temperature, moisture, nutrient quantity, and quality. Reduced soil tillage and minimized compaction are of particular note.

Adaptation and further development of integrated soil biodiversity management into sustainable land management practices require solutions that pay adequate consideration to the synergies between the soil ecosystem and its productive capacity and agroecosystem health. The ecosystem approach is a strategy for the integrated management of land, water, and living resources that promotes conservation and sustainable use in an equitable way. It addresses the essential processes, functions, and interactions among organisms and their environment. It recognizes that humans, with their cultural diversity, are an integral component of ecosystems. There are several practical examples of holistic agricultural management systems that promote and enhance agroecosystem health, including biodiversity, biological cycles, and soil biological activity.

22.3.1 Organic agriculture

This includes the management of locally available resources to optimize competition for food and space between different plant and animal species. The manipulation of the temporal and spatial distribution of biodiversity is the main productive "input" of organic farmers. By refraining from using mineral fertilizers and synthetic pesticides, pharmaceuticals, and genetically modified seeds and breeds, biodiversity is relied upon to maintain soil fertility and to prevent pests and diseases. Scientific research has demonstrated that organic agriculture significantly increases the density and species richness of indigenous invertebrates, specialized endangered soil species, beneficial arthropods, earthworms, symbionts, and microbes. Suitable conditions for soil fauna and flora, as well as soil forming and conditioning and nutrient cycling, are encouraged by organic practices such as manipulation of crop rotations and strip-cropping, green manuring and organic fertilization (animal manure, compost, crop residues), minimum tillage, and of course, avoidance of synthetic pesticide and herbicide use.

22.3.2 Conservation agriculture

This aims to maintain and improve crop yields and resilience against drought and other hazards, while at the same time protecting and stimulating the biological functioning of the soil. Essential principles of conservation agriculture are no-tillage (and direct seeding) or reduced tillage, the maintenance of a cover of live or dead vegetal material on the soil surface and the use of crop rotations. Crop sequences are planned over several seasons to minimize the buildup of pests or diseases and to optimize plant nutrient use by synergy between different crop types. Management practices that affect the placement and incorporation of residues influence the capacity of soil organisms to recycle nutrients. Tillage, for example, affects soil porosity and the placement of residues by collapsing the pores and tunnels constructed by soil animals, affecting the water holding, gas, and nutrient exchange capacities of the soil. The placement of residues influences soil surface temperature, rate of evaporation and water content, and nutrient loading and rate of decay. Conservation tillage, and particularly no tillage, reduces soil disturbance, increases organic matter content, improves soil structure, buffers soil temperatures, and allows soils to trap and retain more rainwater. These soils are more biologically active and biologically diverse, have higher nutrient loading capacities, and release nutrients more continuously.

22.4 Utilization of soil biodiversity for enhancing resource use efficiency

Soil biodiversity is doing a major role in the utilization of natural resource, enhance the use of resource effectively and environmental remediation in both direct and indirect measures.

Direct Measures:

- Atmospheric N fixation
- Phosphorus solubilization
- Phosphorus mobilization
- Potassium (K) solubilization and mobilization
- Siderophore production for increased Fe uptake
- Sulphur oxidization
- Organic waste decomposition

Indirect Measures:

- Aggregate stabilization (by glomalin and mycellia of fungi) and improved soil physical properties
- Enhancing infiltration rate by earthworm, ants, and termites
- Decontaminate polluted soil/water and make them productive for agricultural uses

22.4.1 Atmospheric N fixation

Biological nitrogen fixation (BNF), the conversion of atmospheric dinitrogen to reduced form by bacteria and archaea, is an important ecosystem service that microbes provide to eukaryotes. BNF contributes 65% of the total input of N on Earth, or 96% of the N input derived from natural processes; hence, it is considered to be the second major contributor to life after photosynthesis (Ormeño-Orrillo et al., 2013). It is estimated that the annual global input of BNF is approximately 55 Tg N, whereas synthetic N-fertilizers contribute 79 Tg N (Crews and Peoples, 2005). A diverse group of organisms have the ability to fix nitrogen, but only a very small proportion of species are able to do so; about 87 species in 2 genera of archaea, 38 genera of bacteria, and 20 genera of cyanobacteria have been identified as diazotrophs or organisms that can fix nitrogen (Sprent, 2009; Zhang et al., 1991). There are plenty of bacterial species that are able to establish nitrogen-fixing symbiosis. They are classified into four groups: symbiotic nodule formers, rhizospheric associative, endophytic, and soil free-living. The nodule-forming bacteria (rhizobia) that associate with legumes are probably the most efficient way to exploit N_2 fixation in agriculture. The associative (lives on root surface) and endophytic (lives in inner plant parts) bacteria may establish intimate and mutual relationships with plants, usually resulting in plant growth promotion.

22.4.1.1 Symbiotic nodule formers

22.4.1.1.1 Rhizobia

Rhizobia induce the formation of nodules on the roots and sometimes on the stems of legumes. Rhizobia have been used as an efficient nitrogen fixer for many years. Rhizobia enter the root hairs, multiply there, and form nodules, which are the seat of nitrogen fixation. The legume–Rhizobium symbiosis is the most important symbiotic association in terms of BNF, producing roughly 200 million tons of fixed nitrogen annually (Ferguson et al., 2010). Rhizobium inoculants in different locations and soil types were reported to significantly increase the grain yields of Bengal gram (Patil and Medhane, 1974); pea, alfalfa, and sugar beet rhizosphere (Ramachandran et al., 2011); lentil (Rashid et al., 2012); soybean (Grossman et al., 2011); ground nut (Sharma et al., 2011); and berseem (Hussain et al., 2002). Along with legumes, Rhizobium isolates obtained from wild rice have also been found to supply nitrogen to the rice plant (Peng et al., 2008).

Rhizobia are bacteria allocated to the α- and β-proteobacteria orders. Here is the list and description of the main genera, but the number of rhizobial genera and species will vary as the studies of rhizobial phylogeny and diversity progress.

22.4.1.1.2 Rhizobium

Rhizobium was the first genus described (Rhizobium leguminosarum Frank 1889), and means "root living" bacteria. Some examples of nodulating Rhizobium are *R. etli, R. gallicum, R. leguminosarum, R. tropici, R. freirei, R. paranaense*. Rhizobium is the largest group of rhizobia in terms of the number of species. A ubiquitous genus, isolated worldwide, it elicits nodules in a broad spectrum of grain legume species (Sprent, 2009).

22.4.1.1.3 Mesorhizobium

Mesorhizobium is the second largest genus in number of rhizobial species. It is the usual symbiont of *Lotus* spp., but nodulates many other legumes (Laranjo et al., 2014). Mesorhizobium has been recognized as an excellent biological model to understand the process of rhizobial symbiosis (Laranjo et al., 2014).

22.4.1.1.4 Sinorhizobium (or *Ensifer*)

Chen et al. (1988) isolated from soybean nodules some unusual fast-growing rhizobia called Sinorhizobium. In addition to *Bradyrhizobium, Ensifer* microsymbionts of soybean have been isolated in China, initially from primitive genotypes (Keyser et al., 1982; Ruiz-Sainz et al., 1984). It is also reported as indigenous to Brazilian soils of the Cerrados in the Midwest regions (Souza, 2016), indicating that the genus may nodulate a number of other legume species.

22.4.1.1.5 Bradyrhizobium

Bradyrhizobium can nodulate in a broad range of legume species (Parker, 2002; Menna et al., 2009; Bautista et al., 2010). It is usually found in nodules of *Acacia* spp., *Lupin* spp., cowpea, and soybean (Sprent, 2009). The genus represents the main bacteria nodulating Acacia spp. in Australia and other countries of Asia (Perrineau et al., 2014) and cowpea in African and Asian soils (Radl et al., 2014). Bradyrhizobium (*B. japonicum, B. elkanii, B. diazoefficiens*) represents the main rhizobia nodulating soybean. It is noteworthy that, in Brazil, recently described Bradyrhizobium species, including *B. manausense* (Silva et al., 2014), *B. tropiciagri, B. embrapense* (Delamuta et al., 2015), *B. viridifuturi* (Helene et al., 2015), and *B. stylosanthis* (Delamuta et al., 2016), reveal the great diversity of this genus in that country. A similar situation occurs in China, with several new species described in the past few years.

22.4.1.1.6 Allorhizobium

The genus *Allorhizobium* and the species *Allorhizobium undicola* were first proposed for a group of strains capable of efficient N_2-fixing symbiosis with *Neptunia natans*, an indigenous stem-nodulated tropical legume found in waterlogged areas of Senegal (de Lajudie et al., 1998).

22.4.1.1.7 Azorhizobium

The genus *Azorhizobium* is composed of three species: *A. caulinodans* (Dreyfus et al., 1988), *A. doebereinerae* (Moreira et al., 2006), and *A. oxalatiphilum* (Lang et al., 2013). In comparison with other rhizobial genera (e.g., *Rhizobium, Bradyrhizobium*), *Azorhizobium* is often regarded as a more specific bacterium for symbiosis, particularly with *Sesbania virgata* (Moreira et al., 2006). *Azorhizobium* is one of the few exceptions of rhizobia able to fix N_2 as a free-living soil bacterium (Dixon and Kahn, 2004).

22.4.1.1.8 Microvirga

Microvigra is identified by Ardley et al. (2012) from nodules of *Listia angolensis* and *Lupinus texensis*. They identified three novel species belonging to the genus *Microvirga*, naming them as *Microvirga lupini*, *Microvirga lotononidis*, and *Microvirga zambiensis*. Then Radl et al. (2014) identified a novel species of *Microvirga* from cowpea nodules in the semiarid region of Brazil.

22.4.1.1.9 Methylobacterium

Methylobacterium is generally found in *Crotalaria* nodules (Sy et al., 2001; Jourand et al., 2004). It can grow on media containing methanol as a carbon source, which is probably unique among rhizobia. Among 51 species of the genus *Methylobacterium* only a few form nodules. *Methylobacterium* has also been isolated from nodules of field beans (*Vicia faba*), cowpea, black gram (*Vigna mungo*), soybean, and *Sesbania* sp.

22.4.1.1.10 Phyllobacterium

Phyllobacterium has been isolated from legume nodules, root surfaces, and the rhizosphere of several legume and non-legume plant species (Mantelin et al., 2006). *Phyllobacterium* strains has also been isolated from nodules of common bean *Lotus corniculatus* and *Sophora flavescens* (Jiao et al., 2015).

22.4.1.1.11 Actinorhizal symbiosis

The ability to fix nitrogen with legumes has spread to hundreds of species of rhizobia, while the ability to fix nitrogen with actinorhizal plants has been restricted to the Frankia. Actinorhizal symbioses are between Gram-positive soil actinobacteria of the genus *Frankia* and 24 genera of dicotyledonous plants, from eight families belonging to three different orders (Franche et al., 2009), collectively called actinorhizal plants. The well-known genera include *Alnus* (alder), *Eleagnus* (autumn olive), *Hippophae* (sea buckthorn), and *Casuarina* (beef wood) (Franche et al., 2009). Based on their host plants, ten species, that is, *Frankia alni*, *F. elaeagni*, *F. brunchorstii*, *F. discariae*, *F. casuarinae*, *F. ceanothi*, *F. coriariae*, *F. dryadis*, *F. purshiae*, and *F. cercocarpi*, were assigned (Becking, 1970). One remarkable characteristic of *Frankia* is the presence of two unique developmental structures that are critical to its survival: vesicles and spores. Vesicles are the site for actinorhizal nitrogen fixation. Actinorhizal symbiosis is a major contributor to nitrogen inputs in forests, wetlands, fields, and disturbed sites of temperate and tropical regions. The contributions of actinorhizal symbioses are comparable to those of Rhizobium–legume interactions. Typical annual contributions by Alnus associations are 12–200 kg of N ha^{-1}, and those by Hippophae associations are 27–179 kg of N ha^{-1} (Baker and Mullin, 1992).

22.4.1.2 *Associative and endophytic bacteria*

Rhizospheric associative bacteria proliferate on the root surface, nourished on root exudates. Sometimes bacteria of this group proliferate in the intercellular spaces of the root cortex and may survive in the soil (Baldani et al., 1997). The most studied genera of associative diazotrophic bacteria are *Acetobacter*, *Azoarcus*, *Azospirillum*, *Burkholderia*, and *Herbaspirillum* (Baldani and Baldani, 2005) but, without question, the genus *Azospirillum* (with emphasis on *A. brasilense* and *A. lipoferum*) is the most important agronomically (Baldani and Baldani, 2005; Hungria et al., 2016), and is marketed as inoculants in many countries. Both associative and endophytic bacteria may establish intimate and reciprocal relationships with plants, usually resulting in plant growth promotion. The benefits of these diazotrophic bacteria as plant growth promoting bacteria extrapolates to the role in BNF, including other features such as plant hormone production and phosphate solubilization (Giller, 2001; Hungria et al., 2016).

Inputs from the associative and endophytic diazotrophic bacteria towards the nutrient supply to agriculture are comparatively lower than those by rhizobial symbiosis. However, associative and endophytic diazotrophic bacteria may also promote plant growth by the synthesis of phytohormones (Bashan and De-Bashan, 2010). Among the several strains of associative and endophytic diazotrophic bacteria that have been investigated as potential inoculants, some bacteria may behave as an endophyte in one plant species, and as associative in another. Therefore, we have listed associative and endophytic diazotrophic bacteria together. Here is the list of some of the most prominent strains.

22.4.1.3 Free living N - fixers

The soil free-living group includes the genera *Azotobacter*, *Bacillus*, *Beijierinckia*, *Burkholderia*, *Clostridium*, *Desulfovibrio, Derxia*, *Enterobacter*, *Klebsiella*, *Paenibacillus*, *Serratia* (e.g., Baldani and Baldani, 2005; Silva et al., 2011) and many others, most of which decompose soil organic matter. The free-living group also includes cyanobacteria (e.g., *Nostoc* and *Anabaena*) and phototrophic sulfur bacteria. In addition to those in the free-living group, cyanobacteria may live in symbiosis with fungi (forming lichens) or with plants (e.g., *Nostoc* with briophytes, a few gymnosperms and angiosperms, and *Anabaena* with the aquatic fern *Azolla*). Considering cropping systems of agronomic importance, the *Anabaena–Azolla* association is broadly used mainly by means of incorporation and mineralization of the resulting biomass into rice paddy fields in China and Vietnam, contributing as much as 80 kg ha^{-1} to the soil (Giller, 2001).

Azotobacter plays an important role in the nitrogen cycle in nature as it possesses a variety of metabolic functions (Sahoo et al., 2014). Besides playing role in nitrogen fixation, *Azotobacter* has the capacity to produce vitamins such as thiamine and riboflavin and plant hormones such as indole acetic acid (IAA), gibberellins (GA) and cytokinins (CK) (Abd-El-Fattah et al., 2013). *A. chroococcum* improves the plant growth by enhancing seed germination and advancing the root architecture (Gholami et al., 2009) by inhibiting pathogenic microorganisms around the root systems of crop plants. This genus includes diverse species, namely, *A. chroococcum, A. vinelandii, A. beijerinckii, A. nigricans, A. armeniacus*, and *A. paspali.* It is used as a biofertilizer for different crops, such as wheat, oat, barley mustard, sesame, rice, linseed, sunflower, castor, maize, sorghum, cotton, jute, sugar beets, tobacco, tea, coffee, rubber, and coconuts (Wani et al., 2013).

22.4.1.4 Cyanobacteria

The free-living group also includes cyanobacteria (e.g., *Nostoc* and *Anabaena*). In addition to those in the free-living group, cyanobacteria may live in symbiosis with fungi (forming lichens) or with plants (e.g., *Nostoc* with briophytes, a few gymnosperms and angiosperms, and *Anabaena* with the aquatic fern *Azolla*). Some cyanobacterial members are endowed with the specialized cells known as heterocysts, thick-walled modified cells that are considered the site of nitrogen fixation by nitrogenase enzyme. Several cyanobacterial species such as *Anabaena variabilis, Nostoc muscorum, Aulosira fertissima*, and *Tolypothrix tenuis* were found to be effective biofertilizers. Considering cropping systems of agronomic importance, the *Anabaena–Azolla* association is broadly used mainly by means of incorporation and mineralization of the resulting biomass into rice paddy fields in China and Vietnam, contributing as much as 80 kg ha^{-1} to the soil (Giller, 2001).

22.4.2 Phosphorus solubilization

Organisms with phosphate solubilizing abilities under diverse soil and agro-climatic conditions have become economically important and found to be an alternative to utilize the fixed or occluded phosphorus in the soil P stress environment. Most important phosphorus solubilizers/mobilizers found in the soil environment are;

1. Phosphorus Solubilizing bacteria (PSB) and fungi (PSF) (solubilize insoluble mineral P)
2. Arbuscularmycorrhizae (AM) for rendering unavailable P sources accessible to the plant

22.4.2.1 PSB and PSF

Important bacterial genera include *Bacillus* and *Pseudomonas* (Illmer and Schinner, 1992) and fungal genera include *Aspergillus* and *Penicillium* (Motsara et al., 1995). Efficient PSM cultures are mass produced for supply to the farmers as microphos. In India, IARI microphos culture (Gaur, 1990) has been developed that contained two efficient PSB (*Pseudomonas striata* and *Bacillus polymyxa*) and three PSF (*Aspergillus awamori*, *A. niger*, and *Penicillium digitatum*). Earlier studies on phosphate solubilization/plant growth promotion by pseudomonads include *P. fluorescens* (Di Simine et al., 1998); *P. corrugata* (Pandey and Palni, 1998); *P. aeruginosa* (Musarrat et al., 2000); *P. stutzeri* (Vazquez et al., 2000); *P. putida* (Pandey et al., 2008); *P. rhizosphaerae* (Peix et al., 2003); *P. lutea* (Peix et al., 2004); *P. trivialis*, *P. poae* (Gulati et al., 2008), and *P. lurida* (Selvakumar et al., 2008)

22.4.2.2 Mycorrhizae

Mycorrhizae are symbiotic association of fungi and plant roots of vascular plants, which supplies food materials to the fungi, while hyphae network of the fungi enhances the ability of the roots to absorb nutrients like P and Zn nutrition by mobilization, solubilization, and mineralization of inherent sources in soil, and improve water use and drought tolerance. Their mechanisms include:

1. Better absorption of available nutrients (changing root morphology)
2. Increase availability (producing H_2CO_3 and phosphatase)
3. Increase mobility (intracellular mobility)

There are two types: ectomycorrhizal fungi (Amenita, Boletus, etc.) and endomycorrhizal fungi (Arbusculermycorrhiza/AM fungi/VAM), Glomus, Endogene, VAM, Gigaspora, etc.

In edible fruit plant *Diospyroblancoi* some AMF are reported such as: *Acaulosporalongula, A. scrobiculata, A. tuberculata* (Ningsih et al., 2013).

22.4.2.3 *Phosphate solubilization and growth promotion by* Pseudomonas fragi, *a psychrotolerant bacterium*

Selvakumar et al. (2009) isolated Pseudomonas *fragi* from the high altitude Himalayan rhizosphere soil. The isolated species having an inherent ability to grow and solubilize the occluded P or fixed P at different temperatures ranging from 4°C to 300°C and the higher production of indole acetic acid (IAA) and hydrogen cyanide sustain the PGPR activities at the wide varied temperatures. Fig. 22.1 shows an initial increase in the soluble P levels, followed by a subsequent drop. This decline in the soluble P levels observed at all incubation temperatures can be attributed to the various phenomena, such as the depletion of nutrients, production of toxic metabolites in the growth medium, or to the autolysis of cells. The decline in the pH of the broth can be attributed to the production of organic acids by the bacterium as a result of its glucose metabolism.

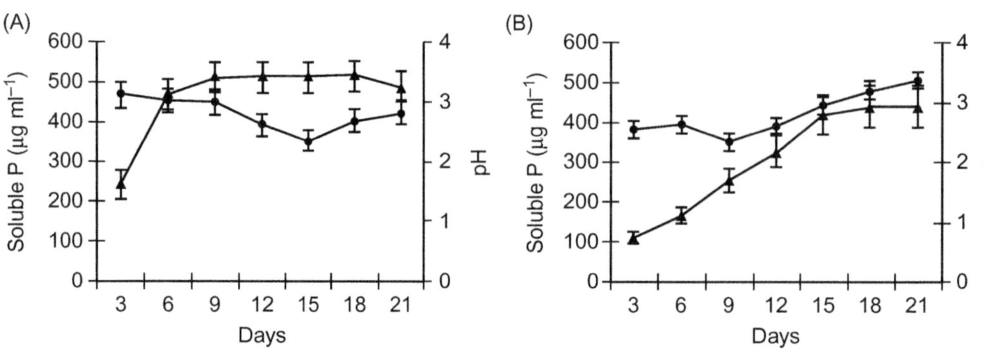

FIGURE 22.1 Effect of incubation temperatures on the tricalcium phosphate solubilization by *Pseudomonas fragi* CS11RH1 at (A) 30°C and (B) 15°C. *Adapted from Selvakumar et al., 2009.*

22.4.2.4 Phosphate solubilization by Penicillium *spp. isolated from soil samples of Indian Himalayan region*

Among the *Penicillium* isolates from Indian Himalayan region, *P. citrinum* had the highest P solubilization efficiency (Tables 22.2 and Table 22.3) (Pandey et al., 2008).

22.4.3 Potash solubilizers

Potassium solublisers are having the ability to secrete the organic acids like acetic acid, citric acid and maleic acid to solublise the K from non exangeable form of K minerals such as muscovite,orthoclase and microcline. Agriculturally important potassium solubilisers are *Bacillus mucilagenous* (Friedrich et al., 1991) and *Frateuria aurentia* (Chandra et al., 2005).

22.4.4 Siderophore producer

Pseudomonas putida is an iron source for dicot and monocot plants by two bacterial siderophores (pseudobactin and ferrioxamine B) (Bar Ness et al., 1991). Spingobacterium, Pseudomonas poae and Enterobacter endosymbiont are reported from tobacco rhizosphere as Fe source in low Fe soils (Tian et al., 2009).

22.4.5 Effective microorganism based

These include a multiculture of major microbes and a consortium of photosynthetic bacteria, lactic acid bacteria, yeasts, actinomycetes, and fermenting fungi, which are useful in recycling of organic wastes.

TABLE 22.2 Effect of inoculation of *Pseudomonas fragii* on the nutrient uptake parameters of wheat seedlings.

Treatments	Root length (cm)	Shoot length (cm)	Dry root biomass (g)	Dry shoot biomass(g)	N uptake (mg plant^{-1})	P uptake (mg plant^{-1})	K uptake (mg plant^{-1})
Control	25.3	24.1	0.12	0.09	0.013	0.17	2.78
P. fragi	29.4	29	0.22	0.19	0.030	0.44	6.27
LSD ($P = .05$)	3.3	3.8	0.04	0.02	0.005	0.01	0.02

Adapted from Pandey et al., 2008.

TABLE 22.3 Growth characteristics and phosphate solubilization on Pikovskaya medium after 1 week of incubation, for *Penicillium* sp.

Penicillium sp.	Temperature range (°C)	pH range	Salt tolerance (%)	Zone of solubilization (mm)
P. aurantiogriseum	4–35 opt 21	3.0–12.0 opt. 4.0–5.0	20	6.3
P. citrinum	9–50 opt. 28	3.0–12.0 opt. 6.0–9.0	20	8.6
P. janthinellum	9–50 opt. 28	3.0–12.0 opt. 5.0	15	5.1
P. oxalicum	4–35 opt. 21	3.0–12.0 opt. 4.0–6.0	15	7.1
P. pinetonum	9–42 opt. 28	3.0–12.0 opt. 5.0–6.0	15	7.3
P. pinophilum	4–35 opt. 28	3.0–12.0 opt. 6.0	15	8.2
P. purpurogenum	9–50 opt. 28	3.0–12.0 opt. 4.0	05	7.6
P. raistrickii	4–35 opt. 21	3.0–12.0 opt. 4.0–5.0	15	5.8

Adapted from Pandey et al., 2008.

22.4.6 Sulfur oxidizing bacteria

Thiobacillus thioxidans was isolated from composts of soil, sulfur, and rock phosphate. (Waksman and Joffe, 1921). It oxidizes elemental sulfur to sulfuric acid, deriving the necessary carbon from the CO_2 of the atmosphere. *Sulfolobus* sp. was isolated from acidic thermal habitat in the United States (Brock et al., 1972). It is a thermophilic facultative autotroph (55°C −80°C).

Bacterial strains CVO and FWKO B were isolated from produced brine in an oil field in Saskatchewan, Canada (Gevertz et al., 2000) that can oxidize elemental S and sulfide to H_2SO_4. Both strains are members of the epsilon subdivision of the division Proteobacteria, with CVO most closely related to *Thiomicrospira denitrifcans*. FWKO B is most closely related to members of the genus Arcobacter. *Sulfurimonas autotrophicagen* is a novel S oxidizing proteobacterium isolated from hydrothermal habitat in the Mid Okinawa region (Inagaki et al., 2003).

22.4.7 Diversity of growth promoting bacteria associated with rice under deepwater and salinity

In northeastern India, traditionally grown paddy varieties in low lying fields were investigated. Endophytic bacterial communities *Pantoea, Citrobacter, Klebsiella, Pseudomonas,* and *Microbacterium* were found to release pectinase, cellulase. *Pantoea, Citrobacter, Klebsiella* were able to fix atmospheric N (Verma et al., 2001)

Swaminathania salitolerans (a novel bacterium) can fix N and solubilize phosphate. It was isolated from rhizosphere, roots, and stems of salt tolerant, mangrove associated wild rice (*Porteresia coarctata* Tateoka) (Loganathan and Nair, 2004).

In the saline soils along the coastline of Tamil Nadu two important PGPRs, *Pseudomonas alcaligens* and *Pseudomonas pseudoalcaligens* (Rangarajan et al., 2002), were isolated.

Other bacterial community in the same soil include *Ochrobactrum anthropi, Serratia marcescens,* and *Pseudomonas auruginosa* (Tripathi et al., 2002)

22.4.8 Organic waste decomposition and use as nutrient source

The vermicomposts have more available nutrients per weight than the organic waste from which they are produced. Various earthworm species involved in this process are *Eisenia fetida, Lumbricus rubellus* (temperate), and *Eudrilus eugeniae, Perionyx excavatus* (tropical).

22.4.9 Isolation and identification of cellulose degrading bacteria from mangrove soil and their cellulase production ability

Cellulose degrading bacteria (CDB) were isolated from mangrove soil of the Mahanadi river delta, Odisha, India to evaluate their cellulase production ability (Behera et al., 2014). In total 15 CDB were isolated based on their halo zone formation on Congo red agar medium. From the morphological and biochemical characterization, the isolates were identified as *Micrococcus* spp. (CDB-4, 6, 15), *Bacillus* spp. (CDB-1, 2, 9, 12, 14), *Pseudomonas* spp. (CDB-10, 11, 13), *Xanthomonas* spp. (CDB-3, 7), and *Brucella* spp. (CDB-5, 8) (Table 22.4).

The cellulase production ability of the bacterial isolates in the production medium were in the following order: CDB-10 (13.591 U mL^{-1} min^{-1}) < CDB-1 (17.321 U mL^{-1} min^{-1}) < CDB-9 (18.53 U mL^{-1} min^{-1}) < CDB-3 (22.24 U mL^{-1} min^{-1}) < CDB-6 (24.71 U mL^{-1} min^{-1}) < CDB-15 (27.298 U mL^{-1} min^{-1}) < CDB-13 (29.653 U mL^{-1} min^{-1}) < CDB-11 (32.124 U mL^{-1} min^{-1}) < CDB-7 (34.971 U mL^{-1} min^{-1}) < CDB-8 (44.48 U mL^{-1} min^{-1}) < CDB-4 (58.701 U mL^{-1} min^{-1}) < CDB-2 (59.307 U mL^{-1} min^{-1}) < CDB-14 (63.013 U mL^{-1} min^{-1}) < CDB-5 (96.374 U mL^{-1} min^{-1}) < CDB-12 (98.253 U mL^{-1} min^{-1}).

22.4.10 Isolation and identification of sulfur oxidizing bacteria and their sulfur oxidizing ability

Twelve isolates from mangrove soil of Mahanadi river delta (Behera et al., 2014) efficiently reduced the pH of the medium to 4.2 from the initial pH 8.0.

From the morphological and biochemical characterization, the isolates were identified as *Micrococcus* sp. (SOB1, 8), *Bacillus* sp. (SOB2, 6, 9, 10, 11), *Pseudomonas* sp. (SOB3, 4, 5, 12), and *Klebsiella* spp (SOB-7) (Fig. 22.2).

TABLE 22.4 Evaluation of cellulase activity of some cellulose degrading bacteria in carboxy methyl cellulose agar plate through halo zone formation.

Isolates	Colony size (cm)	Zone size (cm)	Ratio of clear zone diameter to colony diameter
CDB-1	0.9	1.8	2
CDB-2	1.1	1.8	1.636
CDB-3	0.8	1.4	1.75
CDB-4	1	2	2
CDB-5 (*Brucella* spp.)	0.8	2	2.5
CDB-6	1	2.1	2.1
CDB-7	1.4	1.8	1.28
CDB-8	1.1	2	1.818
CDB-9	1	1.3	1.3
CDB-10	1	1.7	1.7
CDB-11	0.7	0.9	1.285
CDB-12	1	2.1	2.1
CDB-13	1	2.1	2.1
CDB-14	0.9	1.3	1.444
CDB-15	1.2	1.5	1.25

CDB, Cellulose degrading bacteria.
Adapted from Behera et al., 2014.

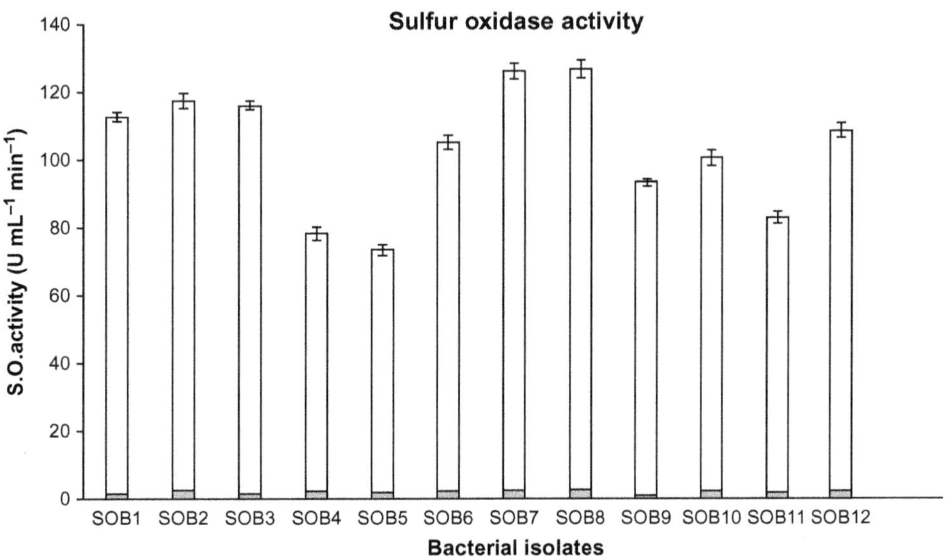

FIGURE 22.2 SOB-7 showed maximum sulfate ion followed by the SOB-8. Maximum sulfate ion production from 14 to 150 mg mL^{-1} by a *Thiobacillus* spp. Isolated from sludge amended soil. Source: *Behera et al., 2014.*

22.4.11 Modification of soil physical structures and maintaining water regimes

Glomalin, a type of protein produced by AM fungi, has the ability to modify soil physical structures and water regimes with the help of its glue-like property that fixes soil particles and favors the formation of stable aggregates. Commonly found genera are *Acaulospora, Gigaspora,* etc. (Wright and Upadhyaya, 1998). Earthworms promote soil fragmentation and aeration, and bring about soil turning and dispersion in farmlands. Termites increase water infiltration rate in crusted soil, so water use efficiency is also increased (Mando et al., 1996).

22.4.12 Increasing aggregate stability by glomalin produced by AM fungi

Result from the study conducted by Wright and Upadhyaya (1998) showed that aggregate stability is highly correlated with the glomalin content in soil (Fig. 22.3).

22.4.13 Biodegradation of xenobiotics

Commonly found xenobiotics in soil system include nitro aromatic compounds, poly aromatic hydrocarbons (PAHs), and chlorinated hydrocarbons. Bacterial strains like *Arthrobacter protophormiae* and *A. ralstonia* (Chauhan and Jain, 2000) have been proven to have bioremediation capabilities. Likewise a microbial consortium of *Klebsiella pneumonae* and *Pseudomonas fluorescens* (Srivastava and Thakur, 2007) has also been used for the same.

22.4.14 Biological treatment of tannery wastewater using novel microbial consortium BM-S-1

Tannery wastewater was treated by using a novel microbial consortium BM-S-1 (Kim et al., 2014) which was enriched from natural soils and composed of following microbes: *Lactobacillus vini* (47.7%), *Lactobacillus ghanensis* (28.3%), *Acetobacter lovaniensis* (7.6%), *Lactobacillus nagelii* (5.4%), *Gluconacetobacter liquefaciens* (1.1%), *Sporolactobacillus* (0.9%), and *Clostridia sp.* (0.2%). Chemical oxygen demand (COD), total nitrogen (TN), total P (TP), chromium (Cr), and suspended solids (MLSS) were measured in the influent as well as in the effluent and resulting in 98.3%, 98.6%, 93.6%, and 88.5% removal efficiencies for COD, TN, TP, and Cr respectively (Table 22.5).

22.4.15 Effect of termites on infiltration into crusted soil

To show this effect two treatments were made (Mando et al., 1996). Treatment I included application of a mulch mixture of wood and straw without insecticide (termite plots) and treatment II included application of the same mulch and the insecticide Dursban (nontermite plots).

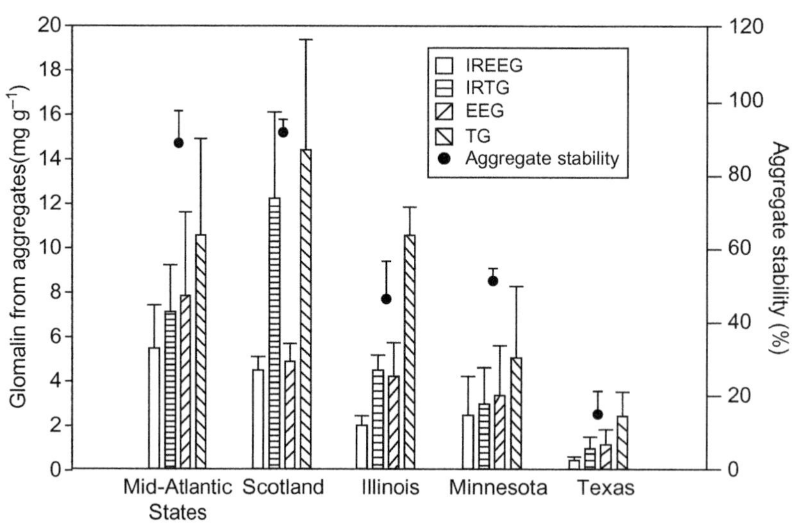

FIGURE 22.3 Means and standard deviations of fractions of glomalin and aggregate stability in soils from four areas of the United States and one area of Scotland. *IREEG*, Immunoreactive easily extractable glomalin; *EEG*, easily extractable glomalin; *IRTG*, immunoreactive total glomalin; *TG*, total glomalin. **Source:** *Wright and Upadhyaya, 1998.*

TABLE 22.5 Concentrations for COD, TN, TP, and MLSS in the full scale, tannery wastewater undergoing bioaugmentation with BM-S-1.

Treatment stage	COD (mg L^{-1})	TN (mg L^{-1})	TP (mg L^{-1})	MLSS (mg L^{-1})
Influent	6910.1	487.2	5.54	3210
B	796.4	92.6	1.39	11,200
PA	634.4	87.7	2.20	9500
PS	226.4	22.9	1.57	
SA	163.9	31.2	0.47	10,830
Effluent	117.1	10.5	0.36	
SD	1730	81.5	3.05	27,740
% Removal	98.3	98.6	93.6	

COD, Chemical oxygen demand; *TN*, total nitrogen; *TP*, total P; *Cr*, chromium; *MLSS*, suspended solids.
Adapted from Kim et al., 2014.

22.4.16 Soil biodiversity and ecosystem multifunctionality related or not-experimental evidence

Wagg et al. (2014) investigated whether reductions of biodiversity in soil communities belowground have consequences for the overall performance of an ecosystem; this question remains unresolved. It is important to investigate this in view of recent observations that soil biodiversity is declining and that soil communities are changing upon land use intensification. They established soil communities differing in composition and diversity and tested their impact on eight ecosystem functions in model grassland communities. They showed that soil biodiversity loss and simplification of soil community composition impairs multiple ecosystem functions, including plant diversity, decomposition, nutrient retention, and nutrient cycling. The average response of all measured ecosystem functions (ecosystem multifunctionality) exhibited a strong positive linear relationship to indicators of soil biodiversity, suggesting that soil community composition is a key factor in regulating ecosystem functioning. Their results indicate that changes in soil communities and the loss of soil biodiversity threaten ecosystem multifunctionality and sustainability.

22.5 Economic and environmental benefits of soil biodiversity

22.5.1 Economic

Soil biodiversity takes care of the management of soil health, structure, and composition. This in turn provides the needed base for successful plant life and therefore the food source for most human and animal life. One can conclude that it is therefore economically "priceless." However, various studies have been made to estimate the economic value of the different services soil biodiversity provides. Recycling of organic wastes is considered to be one of the most important uses of soil biodiversity. Mankind produces more than 38 billion metric tons of organic waste on a global scale annually. Were it not for the decomposing/recycling activity of soil organisms, much of the globe's land surface would be literally covered with organic debris. The economic value of this service represents approximately 50% of the total benefits of soil biotic activity worldwide (Table 22.6).

22.5.2 Environment

22.5.2.1 Structuring the soil and contributing to climate regulation

Soil organisms work the sand, clay, or silt, forming new structures and habitats that aerate the soil and allow water to permeate through it. Some species of fungi, for example, produce a sticky protein that binds soil particles together, thus stabilizing soil aggregates, while larger creatures like termites drive tunnels through the soil. The work done by soil organisms also enables soil to store and release carbon, helping to regulate the flux of greenhouse gases and thus the global climate system. This has a direct impact on human health, crop productivity, water resources, and food security.

TABLE 22.6 Worldwide total estimated economic benefits of biodiversity.

Activity	World economic benefits ($\times$ 10^9)
Waste recycling (38 billion ton wastes)	760
Soil formation	25
Nitrogen fixation (140–170 mt)	90
Bioremediation	121
Biotechnology	6
Biocontrol	160
Pollination	200
Wild food	180

Adapted from Pimentel et al., 1997.

Soil stores carbon mainly in the form of organic matter, and is the second largest carbon pool on Earth, after the oceans. The more organic matter there is in soil, the better a carbon sink it can be. A well-managed soil can thus be an important buffer against climate change. Different types of soil have different carbon storage capacities. Peatland soils, for example, covers only a fraction of Europe's land area, but stores 20% of all soil carbon in Europe. Grasslands and forests accumulate carbon in their soil, while croplands often tend to release it. In Europe, the largest emissions of CO_2 from soil are due to land-use change from grasslands to arable land, and to intensive tillage without the addition of organic matter. Soil organisms play a major role in processing the organic matter in soil; some even incorporate it into the soil themselves. For example, dung beetles are able to bury dead bodies of small animals in the soil, thus making their organic matter available as a food source for their own larvae as well as for other soil organisms. Earthworms can mix litter from surface layers through the soil underneath.

22.5.2.2 Storing and purifying water

A similarly vital role of the below-ground factory of life is to purify and store water. As water infiltrates the ground, contaminants including bacteria and viruses are absorbed by soil particles, making the water both clean and safe. However, this purification capacity depends on the soil being rich in microorganisms, which perform the work. The more biodiversity in soil, the better this function can be performed. Meanwhile, channels, nests, and galleries created by earthworms, ants, and termites all promote water absorption, while vegetation, with its leaf litter and root systems, helps to capture water and to structure the below-ground soil. Cutting back vegetation, for example by deforestation, does the opposite, allowing soil to be washed away. Without a vibrant soil community, the soil becomes poor in structure and water runoff increases, leading to erosion and flooding. If the soil's ability to absorb, cleanse, and store water is compromised, groundwater will be impaired, and more water treatment facilities will be required. Maintaining the soil's ability to process and cleanse water will save money and safeguard health and wellbeing.

Few people are aware that soil organisms have the remarkable ability to clean up certain types of pollution, or at the very least dilute their impact. In a procedure called bioremediation, microbes in the soil are able to decompose some organic pollutants and convert them to nontoxic molecules. Bioremediation is a natural process that has been frequently harnessed by humans. It is the cheapest method of soil decontamination and has proved effective in numerous cases. One famous example was the clean-up of the Exxon Valdez oil tanker spill in Alaska in 1989. As part of the efforts to clean crude oil from 2000 km of coastline, a mix of nutrients and fertilizers that encouraged bacterial growth was applied to the contaminated sand and sediment. The bacterial activity led to a fivefold increase in the rate of oil degradation, and an efficient site clean-up, although the oil spill caused the death of many marine and shoreline animals. Toxins that can be removed from the soil by natural bioremediation include chemicals used in wood treatments, solvents used in dry cleaning, agricultural pesticides, and even polychlorinated biphenyls now banned substances that were formerly used in plastics and electrical components. A soil rich in biodiversity is essential to obtain the best results from bioremediation. While microorganisms work on the chemical pollutants, other organisms that control the structure and porosity of soil help it to absorb, disperse, and degrade contaminants. Natural soil decontamination does have its limits, of course. The process can take years, if not decades, since some persistent pollutants cannot be broken down and sometimes the contamination load is simply too great for the soil. In addition, heavy metals such as cadmium, lead, and mercury cannot be degraded and have been found to accumulate in the food chain or to contaminate ground water. Soil may offer its services as a natural detoxifier, but we cannot expect it to work miracles.

22.5.2.3 *Protecting soil biodiversity*

Protection of individual species has historically been the goal of conservation bodies and it has been relatively easy to define which species are under threat. For example the CITES Convention (Convention on International Trade in Endangered Species Flora) clearly defines which species should be protected. Recently, the focus has shifted away from individual species to conservation of habitats through the use of "keystone" species, which act as indicator organisms within a particular environment or ecosystem. However, in cases where the habitat contains species that are unknown, unidentified, or difficult to study, a different approach needs to be adopted. In soil systems it is unlikely that we may ever know the true number and identity of all the species and it is virtually impossible to concentrate on a particular species because the same species may be found elsewhere (in an area not under threat) and they may interact more closely than in aboveground habitats. In these cases, the strategy has been to focus on protecting the function of the ecosystem so-called ecosystem services. This allows a multitiered approach to looking at soil that will embrace the different interaction levels occurring in soil. This complexity in soil, lack of knowledge, and technical difficulties present a unique challenge to conserving soil habitats.

In 2006 the European Commission adopted the Thematic Strategy for Soil Protection of European Soils. The proposed Directive "lays down a framework for the protection and sustainable use of soil based on the principles of integration of soil issues into other policies, preservation of soil functions within the context of sustainable use, prevention of threats to soil and mitigation of their effects, as well as restoration of degraded soils to a level of functionality consistent at least with the current and approved future use of the land."

The key elements of the directive as proposed by the commission are:

- A requirement for central and local Government to consider the impacts that new policies will have on soils whilst they are being developed (Article 3);
- A duty on all land users to prevent or minimize harm to soils (Article 4);
- A requirement to limit or mitigate the effects of soil sealing (the covering of the soil surface with an impermeable material such as concrete) (Article 5);
- A requirement to reduce the risks relating to soil erosion, organic matter decline, compaction, salinization, and landslides, by identifying risk areas, and deciding on a program of measures to address these risks (Articles 6–8);
- A requirement to prevent soil contamination, compile an inventory of contaminated sites, and remediate those sites listed on the inventory (Articles 9–14); and
- A requirement to raise awareness of soils issues, report to the Commission, and exchange information (Articles 15–17).

Action is required at the global level because soil is a nonrenewable natural resource of common interest to Europe because of the crucial functions it performs for society and the ecosystems. European environmental legislation is incomplete without soil policy, hampering the objective to reach a high level of environmental protection in Europe. Differences among Member States in dealing with soil problems may distort competition within the single market. Most of the costs of soil degradation are not borne by the land users, who are responsible for the degradation, but by the taxpayers. Soil degradation has transboundary consequences, as soil contamination may affect the quality of food and feed products. The health of the European population can be impaired as a result of soil degradation.

22.5.2.4 *Problems in conserving soil biodiversity*

Many of the problems in conserving biodiversity are associated with the lack of recognition of the importance it plays in agricultural production. Although many farmers and the farming community have a profound knowledge of their agriculture, training and education are often needed to highlight the roles of the soil biota at various levels of the ecosystem/landscape. Soil quality assessments such as chemical and physical properties provide some knowledge of resources but should be supplemented with information on resources (human and organic such as composts) and biological indicators of soil quality and function.

To overcome any limitations to agricultural production, Swift (2004) proposed a series of potential "entry points" at which management practices could be improved. These include both direct interventions such as inoculation for disease and pest control and soil fertility improvement (such as rhizobia, actinomycetes, mycorrhizae, diazotrophs) and indirect interventions through, for example, cropping system design, organic matter management, and genetic control of soil function (manipulating resistance to disease, organic matter, and root exudates). A potential set of improvements could be tested together using an "adaptive" experimentation approach whose results feedback over a number of cropping cycles. This would involve other members of the farming community such as extension agents and local community facilitators and be evaluated according to local agricultural, climatic, soil, socioeconomic, and cultural conditions as long as the farmers, etc. can identify problems that may lead to the failure of the adopted system. Any system undertaken must be flexible to meet the needs and priorities of those concerned.

The final decision of whether to adopt the practice is by no means certain as the farmer may choose to revert back to the traditional management strategy. The selection of best practice is a long-term process and requires a level of commitment, for example monitoring, and the appropriate incentives so that the improvements in agricultural production and human wellbeing can be shown and sustained.

22.6 Conclusion

The available evidence suggests that soil biodiversity confers ecosystem functioning. Under resource limitation both water and nutrient use efficiencies are increased in the presence of soil fauna and associated with increased plant and mycorrhizal diversities and aboveground production. With higher functional dissimilarity of the soil fauna, the net diversity effect on ecosystem processes is higher. This will be an important finding in explaining the contribution of soil biodiversity to the efficiency of resource use, but need to be substantiated. Although the functional aspects of agroecosystems, which are important to farmers are much more numerous than those treated in this chapter, it is clear that soil biodiversity will only be meaningful, when integrated with aboveground biodiversity to sustain ecosystem functioning. We feel that the knowledge gained and to be developed will be useful only, if inspired by, and combined with farmers' knowledge, perceived problems, and opportunities for application. The value of soil biodiversity is also to be recognized by society at large. We suggest that identifying the value of soil biodiversity in terms of economic benefits is a meaningful step in a research programme aimed at sustaining soil biodiversity and its use and as part of a wider strategy of conserving and using agrobiodiversity. The transformation of agricultural production from one of the greatest threats to global biodiversity and ecosystem services to a major contributor to ecosystem integrity is unquestionably a key challenge of the 21st century. Many elements of soil biodiversity could also help to achieve the critical goals of agricultural sustainability, resilience of food systems, and adaptation to climate change. To realize these potentials, the agricultural and conservation research and policy communities will need to re-evaluate and coordinate their priorities and strategies.

References

Abd-El-Fattah, A.A., Sadik, N.A.H., Shaker, O.G., Aboulftouh, M.L., 2013. Differential microRNAs expression in serum of patients with lung cancer, pulmonary tuberculosis, and pneumonia. Cell Biochem. Biophys. 67 (3), 875–884.

Ardley, J.K., Parker, M.A., De Meyer, S.E., Trengove, R.D., O'Hara, G.W., Reeve, W.G., et al., 2012. *Microvirga lupini* sp. nov., *Microvirga lotononidis* sp. nov., and *Microvirga zambiensis* sp. nov. are Alphaproteobacterial root nodule bacteria that specifically nodulate and fix nitrogen with geographically and taxonomically separate legume hosts. Int. J. Syst. Evol. Microbiol. 62 (11), 2579–2588.

Baker, D.D., Mullin, B.C., 1992. Actinorhizal symbioses. Biol. Nitrogen Fixation 259–291.

Baldani, J.I., Baldani, V.L., 2005. History on the biological nitrogen fixation research in graminaceous plants: special emphasis on the Brazilian experience. An. Acad. Bras. Cienc. 77 (3), 549–579.

Baldani, J., Caruso, L., Baldani, V.L., Goi, S.R., Döbereiner, J., 1997. Recent advances in BNF with nonlegume plants. Soil Biol. Biochem. 29 (5–6), 911–922.

Bar-Ness, E., Chen, Y., Hadar, Y., Marschner, H., Römheld, V., 1991. Siderophores of *Pseudomonas putida* as an iron source for dicot and monocot plants. Plant Soil 130, 231–241.

Bashan, Y., De-Bashan, L.E., 2010. How the plant growth-promoting bacterium Azospirillum promotes plant growth—a critical assessment, Advances in Agronomy, vol. 108. Academic Press, pp. 77–136.

Bautista, V.V., Monsalud, R.G., Yokota, A., 2010. *Devosia yakushimensis* sp. nov., isolated from root nodules of *Pueraria lobata* (Willd.) Ohwi. Int. J. Syst. Evol. Microbiol. 60 (3), 627–632.

Becking, J.H., 1970. Frankiaceae fam. nov.(Actinomycetales) with one new combination and six new species of the genus Frankia Brunchorst 1886, 174. Int. J. Syst. Evol. Microbiol. 20 (2), 201–220.

Behera, B.C., Patra, M., Dutta, S.K., Thatoi, H.N., 2014. Isolation and characterisation of sulphur oxidising bacteria from mangrove soil of Mahanadi River Delta and their sulphur oxidising ability. J. Appl. Environ. Microbiol. 2, 1–5.

Brock, T.D., Brock, K.M., Belly, R.T., Weiss, R.L., 1972. *Sulfolobus:* a new genus of sulfur-oxidizing bacteria living at low pH and high temperature. Arch. Mikrobiol 84, 54–68. Available from: https://doi.org/10.1007/BF00408082.

Chandra, K., Greep, S., Ravindranath, P., Sivathsa, R.S.H., 2005. Liquid biofertilizers. Regional Center for Organic Farming Hebbal, Bangalore, India.

Chen, W.X., Yan, G.H., Li, J.L., 1988. Numerical taxonomic study of fast-growing soybean rhizobia and a proposal that *Rhizobium fredii* be assigned to *Sinorhizobium* gen. nov. Int. J. Syst. Evol. Microbiol. 38 (4), 392–397.

Crews, T.E., Peoples, M.B., 2005. Can the synchrony of nitrogen supply and crop demand be improved in legume and fertilizer-based agroecosystems? A review. Nutr. Cycl. Agroecosyst. 72 (2), 101–120.

de Lajudie, P., Willems, A., Nick, G., Moreira, F., Molouba, F., Hoste, B., et al., 1998. Characterization of tropical tree rhizobia and description of *Mesorhizobium plurifarium* sp. nov. Int. J. Syst. Evol. Microbiol. 48 (2), 369–382.

Delamuta, J.R.M., Ribeiro, R.A., Ormeño-Orrillo, E., Parma, M.M., Melo, I.S., Martínez-Romero, E., et al., 2015. *Bradyrhizobium tropiciagri* sp. nov. and *Bradyrhizobium embrapense* sp. nov., nitrogen-fixing symbionts of tropical forage legumes. Int. J. Syst. Evol. Microbiol. 65 (12), 4424–4433.

Delamuta, J.R.M., Ribeiro, R.A., Araújo, J.L.S., Rouws, L.F.M., Zilli, J.É., Parma, M.M., et al., 2016. *Bradyrhizobium stylosanthis* sp. nov., comprising nitrogen-fixing symbionts isolated from nodules of the tropical forage legume Stylosanthes spp. Int. J. Syst. Evol. Microbiol. 66 (8), 3078–3087.

Di Simine, C.D., Sayer, J.A., Gadd, G.M., 1998. Solubilization of zinc phosphate by a strain of *Pseudomonas fluorescens* isolated from a forest soil. Biol. Fertil. Soils 28, 87–94.

Dixon, R., Kahn, D., 2004. Genetic regulation of biological nitrogen fixation. Nat. Rev. Microbiol. 2 (8), 621.

Dreyfus, B., Garcia, J.L., Gillis, M., 1988. Characterization of *Azorhizobium caulinodans* gen. nov., sp. nov., a stem-nodulating nitrogen-fixing bacterium isolated from *Sesbania rostrata*. Int. J. Syst. Evol. Microbiol. 38 (1), 89–98.

Ferguson, B.J., Indrasumunar, A., Hayashi, S., Lin, M.H., Lin, Y.H., Reid, D.E., et al., 2010. Molecular analysis of legume nodule development and autoregulation. J. Integr. Plant Biol. 52 (1), 61–76.

Franche, C., Lindström, K., Elmerich, C., 2009. Nitrogen-fixing bacteria associated with leguminous and nonleguminous plants. Plant Soil 321 (1–2), 35–59.

Friedrich, S., Platonova, N.P., Karavaiko, G.I., Stichel, E., Glombitza, F., 1991. Chemical and microbiological solubilization ofsilicates. Acta Biotechnol 11, 187–196.

Gaur, A.C., 1990. Phosphate solubilizing microorganisms as biofertilizers. Omega Scientific Publisher, New Delhi, p. 176.

Gevertz, D., Telang, A.J., Voordouw, G., Jenneman, G.E., 2000. Isolation and characterization of strains CVO and FWKO B, two novel nitrate-reducing, sulfide-oxidizing bacteria isolated from oil field brine. Appl. Environ. Microbiol. 66, 2491–2501. Available from: https://doi.org/10.1128/AEM.66.6.2491-2501.2000.

Gholami, A., Shahsavani, S., Nezarat, S., 2009. The effect of plant growth promoting rhizobacteria (PGPR) on germination, seedling growth and yield of maize. Int. J. Biol. Life Sci. 5 (1), 35–40.

Giller, K.E., 2001. Nitrogen Fixation in Tropical Cropping Systems. Cabi.

Grossman, J.M., Schipanski, M.E., Sooksanguan, T., Seehaver, S., Drinkwater, L.E., 2011. Diversity of rhizobia in soybean [Glycine max (Vinton)] nodules varies under organic and conventional management. Appl. Soil Ecol. 50, 14–20.

Gulati, A., Rahi, P., Vyas, P., 2008. Characterization of phosphate solubilizing fluorescent *Pseudomonads* from the rhizosphere of seabuckthorn growing in the cold deserts of Himalayas. Curr. Microbiol. 56, 73–79.

Helene, L.C.F., Delamuta, J.R.M., Ribeiro, R.A., Ormeño-Orrillo, E., Rogel, M.A., Martínez-Romero, E., et al., 2015. *Bradyrhizobium viridifuturi* sp. nov., encompassing nitrogen-fixing symbionts of legumes used for green manure and environmental services. Int. J. Syst. Evol. Microbiol. 65 (12), 4441–4448.

Hungria, M., Nogueira, M.A., Araujo, R.S., 2016. Inoculation of Brachiaria spp. with the plant growth-promoting bacterium *Azospirillum brasilense*: an environment-friendly component in the reclamation of degraded pastures in the tropics. Agric. Ecosyst. Environ. 221, 125–131.

Hussain, N., Mujeeb, F., Tahir, M., Khan, G.D., Hassan, N.M., Bari, A., 2002. Effectiveness of Rhizobium under salinity stress. Asian J. Plant Sci. 1 (1), 12–14.

Illmer, P.A., Schinner, F., 1992. Solubilization of inorganic phosphates by microorganisms isolated from forest soil. Soil Biol Biochem 24, 389–395. Available from: https://doi.org/10.1016/0038-0717(92)90199-8.

Inagaki, F., Takai, K., Kobayashi, H., Nealson, K.H., Horikoshi, K., 2003. *Sulfurimonas autotrophica*gen. nov., sp. nov., a novel sulfur-oxidizing ε-proteobacterium isolated from hydrothermal sediments in the Mid-Okinawa Trough. Int. J. Syst. Evol. Microbiol. 53, 1801–1805.

Jiao, Y.S., Yan, H., Ji, Z.J., Liu, Y.H., Sui, X.H., Zhang, X.X., et al., 2015. *Phyllobacterium sophorae* sp. nov., a symbiotic bacterium isolated from root nodules of *Sophora flavescens*. Int. J. Syst. Evol. Microbiol. 65 (2), 399–406.

Jourand, P., Giraud, E., Bena, G., Sy, A., Willems, A., Gillis, M., et al., 2004. *Methylobacterium nodulans* sp. nov., for a group of aerobic, facultatively methylotrophic, legume root-nodule-forming and nitrogen-fixing bacteria. Int. J. Syst. Evol. Microbiol. 54 (6), 2269–2273.

Keyser, H.H., Bohlool, B.B., Hu, T.S., Weber, D.F., 1982. Fast-growing rhizobia isolated from root nodules of soybean. Science 215 (4540), 1631–1632.

Kim, I.S., Ekpeghere, K.I., Ha, S.Y., Kim, B.S., Song, B., Kim, J.T., et al., 2014. Full-scale biological treatment of tannery wastewater using the novel microbial consortium BM-S-1. J. Environ. Sci. Health A 49 (3), 355–364.

Laranjo, M., Alexandre, A., Oliveira, S., 2014. Legume growth-promoting rhizobia: an overview on the *Mesorhizobium* genus. Microbiol. Res. 169 (1), 2–17.

Loganathan, P., Nair, S., 2004. *Swaminathania salitolerans* gen. nov.sp. nov, a salt-tolerant nitrogen-fixing and phosphate solubilizing bacterium from wild rice (*Porteresiacoarctata* Tateoka). Int. J. Syst. Evol. Microbiol. 54, 1185–1190.

Mando, A., Stroosnijder, L., Brussaard, L., 1996. Effects of termites on infiltration into crusted soil. Geoderma 74, 107–113.

Mantelin, S., Fischer-Le Saux, M., Zakhia, F., Béna, G., Bonneau, S., Jeder, H., et al., 2006. Emended description of the genus Phyllobacterium and description of four novel species associated with plant roots: *Phyllobacterium bourgognense* sp. nov., *Phyllobacterium ifriqiyense* sp. nov., *Phyllobacterium leguminum* sp. nov. and *Phyllobacterium brassicacearum* sp. nov. Int. J. Syst. Evol. Microbiol. 56 (4), 827–839.

Menna, P., Barcellos, F.G., Hungria, M., 2009. Phylogeny and taxonomy of a diverse collection of Bradyrhizobium strains based on multilocus sequence analysis of the 16S rRNA gene, ITS region and glnII, recA, atpD and dnaK genes. Int. J. Syst. Evol. Microbiol. 59 (12), 2934–2950.

Motsara, M.R., Bhattacharyya, P., Srivastava, B., 1995. Biofertilizer technology, marketing, and usage- A source book-cum-glossary. Fertilizer Development and Consultation Organisation, New Delhi.

Musarrat, J., Bano, N., Rao, R.A.K., 2000. Isolation and characterization of 2, 4-dichlorophenoxyacetic acid-catabolizing bacteria and their biodegradation efficiency in soil. World J. Microbiol. Biotechnol. 16, 495–497.

Ningsih, D.R., Kramadibrata dan, K., Gunawan, A.W., 2013. Arbuscular mycorrhizal fungi in bisbul trees (Diospyros blanco) in Bogor. Biotropia 20 (2), 112–121.

Ormeño-Orrillo, E., Hungria, M., Martinez-Romero, E., 2013. Dinitrogen-fixing prokaryotes. The Prokaryotes: Prokaryotic Physiology and Biochemistry, pp. 427–451.

Pandey, A., Palni, L.M.S., 1998. Isolation of *Pseudomonas corrugate* from Sikkim Himalayas. J. Microbiol. Biotechnol. 14, 411–413.

Pandey, A., Das, N., Kumar, B., Rinu, K., Trivedi, P., 2008. Phosphate solubilization by *Penicillium* spp. isolated from soilsamples of Indian Himalayan region. World J. Microbiol. Biotechnol. 24, 97–102.

Parker, M.A., 2002. Bradyrhizobia from wild *Phaseolus*, *Desmodium*, and *Macroptilium* species in northern Mexico. Appl. Environ. Microbiol. 68 (4), 2044–2048.

Patil, P.L., Medhane, N.S., 1974. Seed inoculation studies in gram (*Cicer arietinum* L.) with different strains of Rhizobium sp. Plant Soil 40 (1), 221–223.

Peix, A., Rivas, R., Mateos, P.F., Martinez-Molina, E., Ro-drigueBarrueco, C., Velazquez, E., 2003. *Pseudomonas rhizosphaerae* sp. nov., a novel species that actively solubilizes phosphatein vitro. Int. J. Syst. Evol. Microbiol. 53, 2067–2072.

Peix, A., Rivas, R., Santa-Regina, R., Mateos, P., Martinez-Molina, E., Rodríguez-Barrueco, C., et al., 2004. *Pseudomonas lutea* sp. nov., a novel phosphate-solubilizing bacterium isolated from the rhizosphere of grasses. Int. J. Syst. Evol. Microbiol. 54, 847–850.

Peng, S., Khush, G.S., Virk, P., Tang, Q., Zou, Y., 2008. Progress in ideotype breeding to increase rice yield potential. Field Crops Res. 108 (1), 32–38.

Perrineau, M.M., Zelzion, E., Gross, J., Price, D.C., Boyd, J., Bhattacharya, D., 2014. Evolution of salt tolerance in a laboratory reared population of *Chlamydomonas reinhardtii*. Environ. Microbiol. 16 (6), 1755–1766.

Pimentel, D., Wilson, C., McCullum, C., Huang, R., Dwen, P., Flack, J., et al., 1997. Economic and environmental benefits of biodiversity. Bio Science 47 (11), 747–757.

Radl, V., Simões-Araújo, J.L., Leite, J., Passos, S.R., Martins, L.M.V., Xavier, G.R., et al., 2014. *Microvirga vignae* sp. nov., a root nodule symbiotic bacterium isolated from cowpea grown in semi-arid Brazil. Int. J. Syst. Evol. Microbiol. 64 (3), 725–730.

Rangarajan, S., Saleena, L.M., Nair, S., 2002. Diversity of Pseudomonas spp. isolated from rice rhizosphere populations grown along a salinity gradient. Microb. Ecol. 43, 280–289.

Ramachandran, V.K., East, A.K., Karunakaran, R., Downie, J.A., Poole, P.S., 2011. Adaptation of *Rhizobium leguminosarum* to pea, alfalfa and sugar beet rhizospheres investigated by comparative transcriptomics. Genome Biol. 12 (10), R106.

Rashid, M.H.O., Schäfer, H., Gonzalez, J., Wink, M., 2012. Genetic diversity of rhizobia nodulating lentil (*Lens culinaris*) in Bangladesh. Syst. Appl. Microbiol. 35 (2), 98–109.

Ruiz-Sainz, J.E., Beringer, J.E., Gutierrez-Navarro, A.M., 1984. Effect of the fungicide captafol on the survival and symbiotic properties of *Rhizobium trifolii*. J. Appl. Bacteriol 57 (2), 361–367.

Sahoo, R.K., Ansari, M.W., Dangar, T.K., Mohanty, S., Tuteja, N., 2014. Phenotypic and molecular characterisation of efficient nitrogen-fixing Azotobacter strains from rice fields for crop improvement. Protoplasma 251 (3), 511–523.

Selvakumar, G., Kundu, S., Joshi, P., Nazim, S., Gupta, A.D., Mishra, P.K., et al., 2008. Characterization of a cold tolerant plant growth-promoting bacterium *Pantoeadispersa 1A* isolated from a sub-alpine soil in the North Western Indian Himalayas. World J. Microbiol. Biotechnol. 24, 955–960.

Selvakumar, G., Joshi, P., Nazim, S., Mishra, P.K., Bisht, J.K., Gupta, H.S., 2009. Phosphate solubilization and growth promotion by *Pseudomonas fragi* CS11RH1 (MTCC 8984), a psychrotolerant bacterium isolated from a high altitude Himalayan rhizosphere. Biologia 64, 239–245.

Sharma, P., Sardana, V., Kandola, S.S., 2011. Response of groundnut (*Arachishypogaea* L.) to Rhizobium inoculation. Libyan Agric. Res. Centre J. Int. 2, 101–104.

Souza, C., 2016. Constitutional Engineering in Brazil: The Politics of Federalism and Decentralization. Springer.

Sprent, J.I., 2009. Legume Nodulation: A Global Perspective. John Wiley and Sons.

Srivastava, S., Thakur, I.S., 2007. Evaluation of biosorption potency of Acinetobacter sp. for removal of hexavalent chromium from tannery effluent. Biodegradation 18 (5), 637–646.

Swift, M.J., Izac, A.M., van Noordwijk, M., 2004. Biodiversity and ecosystem services in agricultural landscapes—are we asking the right questions? Agriculture, Ecosystems & Environment 104 (1), 113–134.

Tian, F., Ding, Y., Zhu, H., Yao, L., Du, B., 2009. Genetic diversity of siderophore-producing bacteria of tobacco rhizosphere. Braz. J. Microbiol 40 (2), 276–284.

Tripathi, A.K., Verma, S.C., Ron, E.Z., 2002. Molecular characterization of a salt-tolerant bacterial community in the rice rhizosphere. Res. Microbiol 153, 579–584.

Vazquez, P., Holguin, G., Puente, M.E., Lopez-Cortez, A., Bashan, Y., 2000. Phosphate-solubilizing microorganisms associated with the rhizosphere of mangroves in a semiarid costal lagoon. Biol. Fertil. Soils 30, 460–468.

Verma, A., Kukreja, K., Pathak, D.V., Suneja, S., Narula, N., 2001. In vitro production of plant growth regulators (PGRs) by *Azorobacter chroococcum*. Indian J. Microbiol 41, 305–307.

Wagg, C., Bender, S.F., Widmer, F., van der Heijden, M.G., 2014. Soil biodiversity and soil community composition determine ecosystem multifunctionality. Proc. Natl. Acad. 111 (14), 5266–5270.

Wani, S.A., Chand, S., Ali, T., 2013. Potential use of *Azotobacter chroococcum* in crop production: an overview. Curr. Agric. Res. J. 1 (1), 35–38.

Waksman, S.A., Joffee, J., 1921. The oxidation of sulfur by microorganisms. Proc. SoC. Exp. Biol. Med. 18, 1–3.

Wright, S.F., Upadhyaya, A., 1998. A survey of soils for aggregate stability and glomalin, a glycoprotein produced by hyphae of arbuscularmycorrhizal fungi. Plant Soil 198, 97–107.

Chapter 23

Application of plant growth promoting rhizobacteria in remediation of pesticides contaminated stressed soil

Rupa Rani[1], Vipin Kumar[1], Pratishtha Gupta[1] and Avantika Chandra[2]

[1]*Laboratory of Applied Microbiology, Department of Environmental Science and Engineering, Indian Institute of Technology (ISM), Dhanbad, India,*
[2]*Department of Environmental Science and Engineering, Indian Institute of Technology (ISM), Dhanbad, India*

Chapter Outline

23.1 Introduction

Pesticides are a diverse group of chemicals used worldwide as herbicides, insecticides, fungicides, rodenticides, molluscides, and nematicides. In agriculture, these pesticides are applied to prevent or control pest diseases, weeds, and other plant pathogens in an effort to maintain high agricultural production or reduce yield losses (Damalas and Eleftherohorinos, 2011). About 45% of annual food production is lost due to pests, thus the need to increase the world food production in the rapidly growing population is well recognized (Schreinemachers and Tipraqsa, 2012; Abhilash and Singh, 2009). To meet the food demand, it has resulted in a dramatic increase in the consumption of both fertilizers and pesticides. Unlike fertilizers, the intensive use of pesticides does not directly contribute to better crop yields but simply helps to control the potential losses caused by animal pests (such as insects, mites, nematodes, and rodents) (Verma et al., 2014), so the usage of pesticides has become essential in agriculture. Pesticides can be classified in several ways, but the most common are based on their chemical composition (Table 23.1).

23.2 Scenario of pesticide consumption

Change in regional climate and the increase in the demand for agro-products has resulted in an increase in consumption and application rate of pesticides (Shetty et al., 2008). Some countries exhibit maximum consumption of total pesticide according to their agricultural lands. The rate of total pesticide consumption in several countries was dependent on agricultural lands. For instance Turkey, Colombia, India, Bolivia, Ecuador, Germany, and Chile showed maximum pesticide consumption as well as agricultural area (Verma et al., 2014) (Fig. 23.1). According to the Directorate of Plant Protection, Quarantine and Storage, Government of India, the maximum usage of pesticides (39,948 MT) has been found in Uttar Pradesh as compared with the other top six pesticide using states of India, that is, Punjab (29,235 MT),

New and Future Developments in Microbial Biotechnology and Bioengineering. DOI: https://doi.org/10.1016/B978-0-12-818258-1.00029-7

TABLE 23.1 Classification of pesticides according to their chemical composition and their general characteristics.

Groups	Main composition	Characteristics	Examples
Organochlorine	Carbon atoms, chlorine, hydrogen, and occasionally oxygen	Lypophilic; they accumulate in fatty tissue of animals, are transferred through the food chain, toxic to a variety of animals, long-term persistent	Endosulfan, DDT, aldrin, lindane, chlordane, mirex
Organophosphate	Possess central phosphorus atom in the molecule; the organophosphate pesticides can be aliphatic, cyclic, and heterocyclic	The pesticide is soluble in organic solvents and water, less persistent than chlorinated hydrocarbons, affect the central nervous system and transferred to leaves and stems for leaf-eating insects and pests	Malathion, methyl, parathion, diazinon
Biological	Viruses, microorganisms or their metabolic products	Only *Bacillus thuringiensis* (Bt) and its subspecies are used with some frequency; these are applied against forest pests and crops, particularly against butterflies; also affect other caterpillars	Dispel, foray, thuricide
Pyrethroids	Compounds similar to the synthetic pyrethrins (alkaloids obtained from petals of *Chrysanthemum cinerariifolium*)	Affect the nervous system; are less persistent than other pesticides; are the safest in terms of their use; some are used as household insecticides	Pyrethrins
Carbamates	Chemical structure based on a plant alkaloid *Physostigma venenosum*	Carbamate acid derivatives; kill a limited spectrum of insects, but are highly toxic to vertebrates; relatively low persistence	Sevin, carbaryl

Source: Courtesy Badii, M.H., Landeros, J., 2015. Pesticides That Affect Human Health and Sustainability. CULCyT (Badii and Landeros, 2015); Ortiz-Hernández, M.L., Sánchez-Salinas, E., Dantán-González, E., Castrejón-Godínez, M.L., 2013. Pesticide biodegradation: mechanisms, genetics and strategies to enhance the process. In: Biodegradation—Life of Science. Intech (Ortiz-Hernandez et al., 2013).

Haryana (21,908 MT), Maharashtra (16,480 MT), Rajasthan (15,239 MT), Gujarat (13,430 MT), and Tamil Nadu (12,851 MT) for the duration of 2005–10. In India, the yearly consumption of pesticide is increasing due to rapid agricultural production (Verma et al., 2014).

23.3 Organochlorine pesticides

The organochlorine pesticides have been extensively used for controlling pests and parasites in agricultural practices and in malaria control programs and are recognized to be highly persistent in the environment. Their strong lipophilicity, resistance to environmental degradation, stability, biomagnification ability through the food chain, long half-life in the environment, and potential for long-range transport are the main reasons for their persistence in the environment (Fisher, 1999). They mainly include aldrin, dieldrin, DDT, endrin, endosulfan, heptachlor, heptachlor epoxide, and hexachlorocyclohexane. Characteristics, applications, and effects of organochlorine pesticides are listed in Table 23.2. Their persistence and ubiquitous distribution in the environment have resulted in adverse effects to human health and environment. Although most organochlorine pesticides were banned in the 1970s and 1980s in Europe and North America, large amounts of these pesticides were applied in developing countries of the tropical and subtropical region as late as in the 1990s (Kannan et al., 1997) because of their low cost, broad spectrum, and versatility against various pests (FAO, 2005).

23.4 Microbial degradation of organochlorine pesticides

Bioremediation is a process that involves removal of pesticides from the contaminated soil using microbial strains. It may be carried out in several ways such as biotransformation, bioaccumulation, biomineralization, biodegradation, and cometabolism. There are several sources of microorganisms having the ability to degrade pesticides. As pesticides are mainly used for agricultural purposes, soil is the medium that frequently acquires these pesticides, besides wastewater, activated sludge, sewage sludge, pesticide industry effluent, sediments, and natural water. Around the world, in many

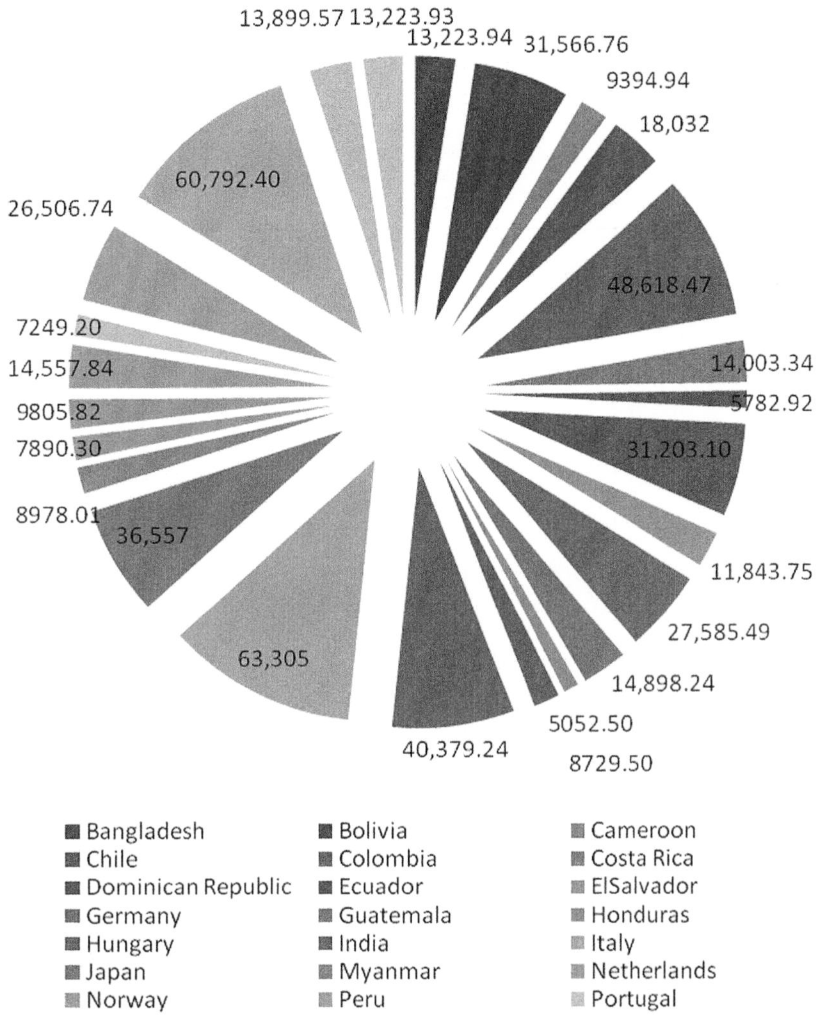

FIGURE 23.1 Worldwide scenario of pesticide consuming countries in the year 2010 (tonnes year^{-1}). *Courtesy Schreinemachers, P., Tipraqsa, P., 2012. Agricultural pesticides and land use intensification in high, middle and low income countries. Food Policy 37 (6), 616–626; FAO, 2010. Food and Agricultural Organization of the United Nations. http://faostat.fao.org/site/424/default.asp#ancor (FAO, 2010); World Bank, 2010. World Development Indicators; Rising Global Interest in Farmland. Can It Yield Sustainable and Equitable Benefits? Washington. http://data.worldbank.org/data-catalog/world-developmentindicators (World Bank, 2010); Verma, J.P., Jaiswal, D.K., Sagar, R., 2014. Pesticide relevance and their microbial degradation: a-state-of-art. Rev. Environ. Sci. Biotechnol. 13 (4), 429–466.*

laboratories, there are several microorganisms characterized by their identification, growth, and degradation of pesticides. The isolation and characterization of microorganisms having the potential for degradation of pesticides gives the possibility for the removal of pesticides from the contaminated sites (Ortiz-Hernández et al., 2011). Several studies have reported that different microbial strains have different potential for degradation of pesticides (Table 23.3).

23.5 Endosulfan

Endosulfan (6,7,8,9,10,10-hexachloro-1,5,5a,6,9-ahexahydro-6,9-methano-2,3,4-benzodioxyanthiepin-3-oxide), an organochlorine pesticide, is a mixture of two stereoisomers: 70% alpha-endosulfan and 30% beta-endosulfan. It is used extensively to control pests in a wide range of crops like sugarcane, fruits, vegetables, cotton, tobacco, and for the control of mites, tsetse flies, cabbage worms, and home garden pests, as well as a wood preservative (Mitton et al., 2016). It is used as an insecticide under different trade names like Thiodan, Endox, Thiomul, Beosit, Endocell, Malix, Thionex, Insecto, and Tiovel (Weber et al., 2010). The persistence of endosulfan is due to the presence of relatively reactive cyclic sulfite diester group, which makes it moderately persistent as compared with other organochlorine compounds (Giri and Rai, 2012); also, due to its hydrophobic nature, it tends to get adsorbed to soil particles

TABLE 23.2 Characteristics, applications, and effects of organochlorine pesticides.

Organochlorine pesticides	Scientific name	Characteristics	Applications	Effects on human	References
Aldrin	1,2,3,4,10,10-Hexachloro-1,4,4a,5,8,8ahexahydro-1,4-endo-exo-5,8-dimethanonaphthalene	Colorless solid; before the ban it was heavily used as pesticides to treat seeds and soil, easily enters the insect's body and quickly converts to dieldrin	Controls weeds, insect pests, and fungal diseases; has been routine in agriculture for the past century	Nervous system damage, including convulsions, headaches, irritability, and nausea	Bandala et al. (2002), Wrobel et al. (2015)
Dieldrin	1,2,3,4,10,10-Hexachloro-6,7-epoxy-1,4,4a,5,6,7,8,8aoctahydro-endo-1,4-exo-5,8-dimethanonaphthalene	White crystals with faint chemical smell; low solubility in water, but is highly soluble in organic solvents	Control of soil insects and several insect vectors of disease	Headaches, dizziness, uncontrollable muscle movements; irritability, and vomiting; convulsion and kidney damage	Wrobel et al. (2015), Stern (2014)
DDT	1,1,1-Trichloro-2,2-bis(4-chlorophenyl)ethane	Colorless, crystalline, tasteless and almost odorless, highly hydrophobic, insoluble in water, but has good solubility in most organic solvents	Used mainly to control mosquito-borne malaria, agriculture	Breast cancers, male infertility, miscarriages and nervous system and liver damage	Aislabie et al. (1997), Kamanavalli and Ninnekar (2004)
Endrin	1,2,3,4,10,10-Hexachloro-6,7 epoxy-1,4,4a,5,7,8,8a-octahydro-1,4-endo, endo-5, 8-dimethanonaphthalene	Solid, white, almost odorless substance	Used against a wide range of agricultural pests and as rodenticide	Central nervous system injury, convulsions, headaches, dizziness, nausea, vomiting	Wu et al. (2014)
Endosulfan	6,7,8,9,10,10-Hexachloro-1,5,5a,6,9,9a-hexahydro-6,9-methano-2,4,3-benzodioxathiepine-3-oxide	Brown crystals, slight sulfur dioxide odor	Used mainly to control insect pests on a cotton, coffee, fruits, oil seeds	Causes headaches, nausea and vomiting, seizures, and in extreme cases, unconsciousness and death	Shivaramaiah and Kennedy (2006), Bhattacharjee et al. (2014)
Heptachlor	1,4,5,6,7,8-Heptachloro-3a,4,7,7a-tetrahydro-4-7-methanoindene	It is lipophilic and poorly soluble in water	Used against soil insects, termites, controlling malaria mosquitoes	Affects nervous system, neurological effects including irritability, salivation, and dizziness	Wu et al. (2014)
Heptachlor Epoxide	1,4,5,6,7,8,8-Heptachloro-3a,4,7,7a-tetrahydro-4,7-methano-1*H*-indene	Looks like a white powder, dissolves more easily in water, and can stay in the soil and water for many years	Used in agriculture and public health programs	Damage the liver and kidney tissue, increased red blood cells	Purnomo et al. (2013)
Hexachlorocyclohexane	(1,2,3,4,5,6-hexachlorocyclohexane)	White solid that may evaporate into the air as a colorless vapor with a slightly musty odor	Used to protect against grasshoppers, cohort insects, rice insects	Blood disorders, dizziness, headache, seizures, and changes in the levels of sex hormones	Lal et al. (2008)

TABLE 23.3 Degradation of organochlorine pesticides by microorganisms.

Pesticides	Incubation conditions	PGPR	Degradation		References
Endosulfan	21 days at 30°C	*Staphylococcus equorum*	81.45%		Bhattacharjee et al. (2014)
		Enterobacter sp.	82.87%		
	pH 8.0	*Bacillus subtilis*	83.69%		
	3 days	*B. subtilis*	71.0%		Kumar et al. (2014)
	7 days		94.2%		
	pH 6.5				
	Temperature 30°C				
	14 days	*Pseudomonas* sp.	98.02 ± 0.18%		Harikumar et al. (2013)
	119.48 ± 0.53 μg L^{-1}	*Bacillus* sp.	50 ± 0.08%		
	36 h	*Pseudomonas* sp.			Siva and Rajam (2013)
	3 days at room temperature	*Bacillus* sp.	50%		Shivaramaiah and Kennedy (2006)
	5 days	Mixed bacterial culture	α-80%, β-75%		Odukkathil and Vasudevan (2013)
			99% endosulfan		
		Arthrobacter sp.	94% endosulfate		
	15 days	*Pseudomonas fluorescens*	α-92.80%		Giri and Rai (2012)
			β-79.35%		
	20 days	Pseudomonas *aeruginosa*	α-50.25%		Kalyani et al. (2009)
			β-69.77%		
	30 days	*Ochrobactrum* sp.	α-57%	β-04%	Kumar et al. (2008)
		Arthrobacter sp.	88%	60%	
		Burkholderia sp.	91%	67%	
	pH 8.5	*P. aeruginosa*	α-86%	β-60%	Jayashree and vasudevan (2007)
	pH 7		47%	31%	
	14 days at 28°C ± 2°C	*Staphylococcus* sp.	12.2%		Kumar and Philip (2006a)
	28 days at 28°C ± 2°C	*Bacillus circulans*	11.9%		
		Mixed culture	23.96%		
Hexachlorocyclohexane	15 days at 28°C	*P. aeruginosa*	α- and γ- > 98%		Kumar et al. (2006)
			β-17%		
			δ-76%		
	8 days	*B. circulans*	α-80.7%		Gupta et al. (2000)
	(28 days for β-isomer)	Bacillus *brevis*	β-39.7%		
			γ-94.6%		
			δ-80.6%		
			α-94.9%		
			β-39.4%		
			γ-98.4%		
			δ-76.8%		

(Continued)

TABLE 23.3 (Continued)

Pesticides	Incubation conditions	PGPR	Degradation	References
DDT	30 days	*B. circulans*	77%	Kaushik (1991)
	135 days	*B. brevis*	91%	
	8 days	*B. circulans*	α-81%–95%	Kaushik (1989)
		B. brevis	γ-95%–100%	
	5 days	*Bacillus* sp.	89.3%	Pant et al. (2013)
	6 days	*Pseudoxanthomonas*	60%	Wang et al. (2010)

PGPR, Plant growth promoting rhizobacteria.

(Harikumar et al., 2014). The persistence of endosulfan depends on physicochemical properties like moderately high octanol/water partition coefficient (3.55–0.362) and low water solubility (0.322 mg L^{-1}, 25°C). As per the Bureau of Indian Standards, the permissible limit of endosulfan in drinking water is 0.4 μg L^{-1} (BIS, 2012) and as per the US Environmental Protection Agency, the permissible limit of endosulfan in rivers, lakes, and streams is 74 μg L^{-1} (USEPA, 2002). Endosulfan is highly lipophilic and exposure to it can lead to bioaccumulation and biomagnification, resulting in adverse effects on target and nontarget organisms. Contamination and recalcitrance of endosulfan in soil and aquatic environments results in accumulation in crops, tea, phytoplankton, zooplankton, fishes, and vegetables (Chowdhury et al., 2007; Chen et al., 2002; Arrebola et al., 2001; Ramaneswari and Rao, 2000). Human exposure to endosulfan can affect the central nervous system, cause liver and kidney damage, nervous disorders, and have teratogenic and mutagenic reproductive effects (ATSDR Agency for Toxic Substances and Disease Registry, 2000; Lu et al., 2000). More than two decades of aerial spraying of endosulfan on the cashew plantations in the government owned Plantation Corporation of Kasargod district, Kerala, South India, has been linked to massive health effects including birth defects, physical deformities, cerebral palsy, vision loss, infertility cancers, and deaths in the communities (NIOH, 2002; Abhilash and Singh, 2009).

In 2001 endosulfan was listed as a persistent toxic pollutant by the Agency for Toxic Substances and Disease Registry and in 2011 it was listed as a persistent organic pollutant (POP) by the Stockholm Convention because of its bioaccumulation potential, toxicity, and persistence. However, POP convention relaxed the ban on endosulfan use for a few crop–pest compounds with a 5-year phase-out period (UNEP/POPS/POPRC, 2008). Endosulfan is considered as a health and environmental concern due to its ubiquitous and persistent nature, which has led to an interest in the degradation of endosulfan. Endosulfan degradation through biological treatment is considered to be a more feasible option as compared with existing methods such as incineration and landfill (Sutherland et al., 2000; Siddique et al., 2003) because the latter involve high treatment costs and possibilities of causing secondary pollution problems.

23.6 Mechanism of microbial degradation of endosulfan

Microorganisms play a substantial role in the degradation of endosulfan, which involves the enzymatic breakdown of endosulfan into nontoxic compounds on the contaminated sites. Many authors have reported the isolation of bacterial cocultures (Awasthi et al., 1997) and mixed bacterial cultures (Sutherland et al., 2000; Kumar and Philip, 2006b; Kumar et al., 2008) having potential for degradation of endosulfan. Bajaj et al. (2010) reported the formation of metabolites in the course of the degradation of endosulfan into ES diol, ES lactone, and ES ether based on the retention time by GC–mass spectral analysis. Kumar et al. (2014) reported the microbial strain *Bacillus subtilis* MTCC 8561 to degrade endosulfan as well as its equally toxic metabolite endosulfan sulfate to endodiol and endosulfan lactone (nontoxic metabolites) up to 94.2% within 7 days. From the metabolites identified after degradation, it has been concluded that the bacterium *B. subtilis* MTCC 8561 metabolizes endosulfan and endosulfan sulfate by directly hydrolyzing it to endosulfan diol followed by the formation of endosulfan lactone through oxidation. The maximum amount of endosulfan diol, noticed after bacterial degradation, indicated that the bacterium hydrolyzed the compound, which releases a sulfite group from endosulfan and sulfate group from endosulfan sulfate, and used them as sulfur source in growth and metabolism. Both the endosulfan diol and endosulfan lactone formed are nontoxic in nature. The proposed degradation

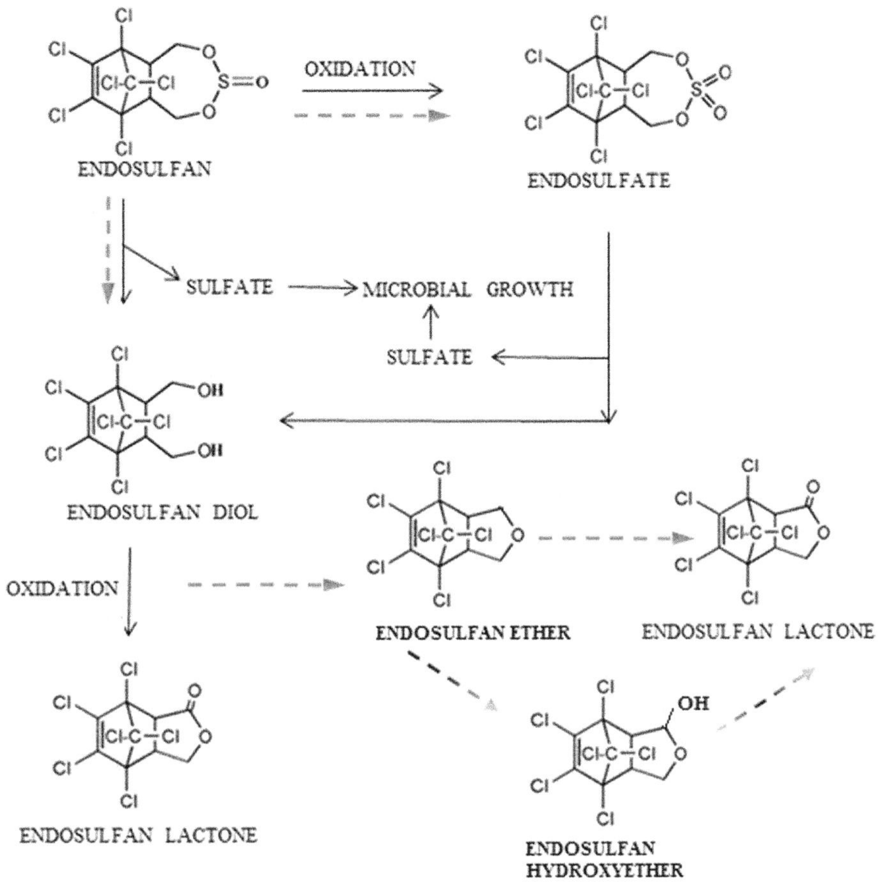

FIGURE 23.2 Proposed pathway for metabolism of endosulfan. *Courtesy Dotted lines indicate pathway of metabolism by* Pseudomonas *sp. strain IITR01 (Bajaj, A., Pathak, A., Mudiam, M.R., Mayilraj, S., Manickam, N., 2010. Isolation and characterization of a* Pseudomonas *sp. strain IITR01 capable of degrading α-endosulfan and endosulfan sulfate. J. Appl. Microbiol. 109 (6), 2135–2143); solid lines indicate pathway of metabolism by* Bacillus subtilis *MTCC 8561 (Kumar, A., Bhoot, N., Soni, I., John, P.J., 2014. Isolation and characterization of a* Bacillus subtilis *strain that degrades endosulfan and endosulfan sulfate. 3 Biotech. 4 (5), 467–475).*

pathway of endosulfan by *Pseudomonas* sp. strain IITR01 (Bajaj et al., 2010) and *B. subtilis* MTCC 8561 (Kumar et al., 2014) is represented in Fig. 23.2.

23.7 Plant growth promoting rhizobacteria

The plant growth promoting rhizobacteria (PGPR) are a group of bacteria that proficiently colonize in plant roots (rhizosphere) and enhance plant growth and development (Kloepper, 1994). The rhizosphere is the narrow zone of soil directly surrounding the root system where the maximum microbial activities occur due to the presence of high nutrient levels (Walker et al., 2003). PGPR facilitates resource acquisition (nitrogen, phosphorus, and essential minerals) or modulate plant hormone levels of their host plants through a number of proposed mechanisms (Glick, 2012). The mechanisms of PGPR mediated enhancement of plant growth include biological nitrogen fixation, the mobilization of recalcitrant soil nutrient, antagonism against phytopathogenic bacteria by producing siderophores, and producing numerous plant growth regulators and vitamins (da Luz, 2001; Kravchenko et al, 2002). PGPR that mediate plant growth promotion through the production of various substances are listed in Table 23.4. The PGPR can exhibit beneficial traits for plant growth as follows:

23.7.1 Phosphate solubilization

Microorganisms coupled with phosphate solubilizing activity, often called phosphate solubilizing microorganisms, may provide the available forms of P to the plants and hence are a viable substitute to chemical phosphate fertilizers

TABLE 23.4 Examples of plant growth promoting rhizobacteria.

Scientific name	Microorganisms	Effects to plants	References
Pisum sativum	*Enterobacter cloacae*	Production of IAA is 112 μg mL^{-1}	Khalifa et al. (2016)
		Increase in root dry weight is 39 mg $plant^{-1}$	
		Increase in lateral roots is 20.57 cm	
Peucedanum japonicum	*Bacillus* sp.	Increase in shoot growth by 10.9%	Hong and Lee (2014)
		Increase in root dry weight by 30.8%	
Vigna unguiculata	*Azotobacter* sp.	Increase in yield by 30%	Sivasankari and Anandharaj. (2014)
	Bacillus sp.	Phosphate solubilization index 38.55	
		Produced IAA 24.2 μg mL^{-1}	
Malus domestica	*B. circulans*	Increase in seed germination by 22.32%	Mehta et al. (2015)
		Increase in shoot length by 15.91%	
		Increase in root length by 25.10%	
		Increase in shoot dry weight by 52.92%	
		Increase in root dry weight by 31.4%	
		Increases nitrogen by 18.75%	
		Increases potassium by 57.69%	
		Increases phosphorus by 22.22%	
Zea mays	*Arthrobacter* sp.	Production of IAA is 135.22 μg mL^{-1}	Kavamura et al. (2013)
Zea mays	*Enterobacter* sp.	Production of IAA is 20 μg mL^{-1}	Ogbo and Okonkwo (2012)
Alyssum serpyllifolium and *Brassica juncea*	*Pseudomonas* sp. A3R3	Increased the plant biomass	Ma et al. (2011)
Solanum lycopersicum	*Bacillus* sp.	Produced IAA 20.3 μg mL^{-1}	Banerjee et al. (2010)
		Produced IAA 3 μg mL^{-1}	
	Arthrobacter sp.	Solubilization of phosphate (165.37 μg mL^{-1})	
Arabidopsis thaliana	*B. cereus*	Increase in shoot weight by 68.8%	Niu et al. (2011)
		Increase in root weight by 57.2%	
		Increase in biomass by 67.4%	
		Increase in seed production by 34.8%	
Cicer arietinum	*Bacillus* sp. PSB10	Improved growth, nodulation, chlorophyll, leghaemoglobin, seed yield and grain protein reduced the uptake of chromium in roots, shoots, and grains	Wani and Khan (2010)
Zea mays	*P. aeruginosa*	Promoted plant growth facilitated soil metal mobilization enhanced Cr and Pb uptake	Braud et al. (2009)
	Pseudomonas fluorescens		
	Ralstonia metallidurans		
Cicer arietinum	*Pseudomonas* sp.	Enhanced fresh and dry weight of plants	Tank and Saraf (2009)
Brassica juncea and *Cucurbita mixta*	*P. aeruginosa*	Stimulated plant growth	Sinha and Mukherjee (2008)
		Reduced Cd uptake	

(Continued)

TABLE 23.4 (Continued)

Scientific name	Microorganisms	Effects to plants	References
Vigna mungo	*P. aeruginosa* MKRh3	Less cadmium accumulation extensive rooting and enhanced plant growth	Ganesan (2008)
Solanum lycopersicum, *Abelmoschus esculentus*, *Spinacia oleracea*	*Bacillus subtilis*, *P. aeruginosa*	Increased dry biomass 31% for tomato	Adesemoye et al. (2008)
		Increased dry biomass 36% for okra	
		Increased dry biomass 83% for African spinach	
Lactuca sativa L.	Pseudomonas *putida* CC-R2-4	Increase in shoot length and root length	Rekha et al. (2007)
	B. subtilis CC-g104		
Lycopersicon esculentum, *Hordeum vulgare*	*Methylobacterium oryzae*, *Burkholderia* sp., *Bacillus* sp.	Increased root weight up to 16.7% and shoot weight up to 347%	Canbolat et al. (2006)
Brassica juncea	*B. subtilis* SJ101	Facilitated Ni accumulation	Zaidi et al. (2006)
Vigna radiate	*P. putida* KNP9	Reduced Pb and Cd uptake	Tripathi et al. (2005)
Glycine max	*Pseudomonas* sp.	Phosphate solubilization index 35.22	Bandala et al. (2002)
		Produced IAA 19.6 $\mu g\ mL^{-1}$	

(Khan et al., 2007). Phosphorus is available in soils in both organic and inorganic forms but the amount of available forms to plants is very low. This low availability of phosphorus to plants is due to the fact that the majority of P is present in insoluble forms in soil, while the plants absorb it only in two soluble forms, the monobasic ($H_2PO_4^-$) and the diabasic (HPO_4^{2-}) ions. Bacterial genera like *Azotobacter*, *Bacillus*, *Beijerinckia*, *Burkholderia*, *Enterobacter*, *Erwinia*, *Flavobacterium*, *Microbacterium*, *Pseudomonas*, *Rhizobium*, and *Serratia* are reported as the most significant phosphate solubilizing bacteria (Bhattacharyya and Jha, 2012).

23.7.2 Indole acetic acid

Indole acetic acid (IAA) is the most common natural auxin found in plants and its positive effect on root growth was also observed (Sharma et al., 2003). Rhizobacteria can synthesize IAA and enhance cell proliferation and stimulate the host's uptake of minerals and nutrients from the soil (Zahran, 2001). IAA has a positive effect on plant cell division, extension, and differentiation; stimulates seed and tuber germination; increases the rate of xylem and root development; controls processes of vegetative growth; initiates lateral and adventitious root formation; mediates responses to light, gravity, and fluorescence; affects photosynthesis, pigment formation, biosynthesis of various metabolites, and resistance to stressful conditions (Miransari and Smith, 2014).

23.7.3 Siderophore production

In the aerobic environment, iron presents principally as Fe^{3+} and is likely to form insoluble hydroxides and oxyhydroxides, consequently making it less accessible to plants and microorganisms (Rajkumar et al., 2010). Commonly, bacteria acquire iron by the secretion of low molecular mass iron chelators denoted as siderophores, which have high association constants for complexing iron. Plants assimilate iron from bacterial siderophores by several mechanisms, for instance, chelation and release of iron, the direct uptake of siderophore–Fe complexes, or by a ligand exchange reaction (Schmidt, 1999). Many studies have reported the plant growth promotion by siderophore-mediated Fe-uptake as a result of siderophore producing rhizobacterial inoculations (Rajkumar et al., 2010).

23.7.4 Aminocyclopropane-1-carboxylate deaminase

Ethylene is an essential plant growth hormone and is produced endogenously by more or less all plants. Besides being a plant growth regulator, ethylene has also been known as a stress hormone (Saleem et al., 2007). Under stress conditions, for example, salinity, drought, water logging, heavy metals, and pathogenicity, the endogenous level of ethylene is significantly increased, which adversely affects the plant growth. PGPR that possess the enzyme 1-aminocyclopropane-1-carboxylate (ACC) deaminase facilitate plant growth and development by reducing ethylene levels, inducing salt tolerance, and decreasing drought stress in plants (Nadeem et al., 2007; Zahir et al., 2008).

23.8 Conclusion

Investigations into the microbial degradation of organochlorine pesticides are useful for the development of methods for the remediation of contaminated soils. Biodegradation is primarily a strategy of microorganisms that involves the removal of organochlorine pesticides from the contaminated site for their own survival. Biodegradation of organochlorine pesticides with specific microorganisms is economical and environmentally and socially acceptable. PGPR contribute significantly to the removal of toxic pesticides. The vast majority of studies that investigate the mode of action of PGPR indicated pesticides degradation by *Azospirillum*, *Azotobacter*, *Bacillus*, *Enterobacter*, *Gordonia*, *Klebsiella*, *Paenibacillus*, *Pseudomonas*, and *Serratia*. The use of PGPR as a tool for pesticide degradation on the contaminated sites needs to be adopted at larger scale. Therefore it is important that efforts are made for the disposal of pesticides and for the remediation of contaminated sites. The use of microbial strains having plant growth activities is ecologically superior over the chemical methods. Hence, microbial degradation of organochlorine pesticides in soil by using PGPR can be used as an alternative and eco-friendly approach for the bioremediation of pesticide contaminated sites.

References

Abhilash, P.C., Singh, N., 2009. Pesticide use and application: an Indian scenario. J. Hazard. Mater. 165 (1–3), 1–12.

Adesemoye, A.O., Obini, M., Ugoji, E.O., 2008. Comparison of plant growth-promotion with *Pseudomonas aeruginosa* and *Bacillus subtilis* in three vegetables. Braz. J. Microbiol. 39 (3), 423–426.

Aislabie, J.M., Richards, N.K., Boul, H.L., 1997. Microbial degradation of DDT and its residues—a review. N. Z. J. Agric. Res. 40 (2), 269–282.

Arrebola, F.J., Egea-González, F.J., Moreno, M., Fernández-Gutiérrez, A., Hernández-Torres, M.E., Martínez-Vidal, J.L., 2001. Evaluation of endosulfan residues in vegetables grown in greenhouses. Pest Manage. Sci. 57 (7), 645–652.

ATSDR (Agency for Toxic Substances and Disease Registry), 2000. Toxicological Profile for Endosulfan (Update). US Department of Health and Human Services, Atlanta, GA, pp. 11–81.

Awasthi, N., Manickam, N., Kumar, A., 1997. Biodegradation of endosulfan by a bacterial coculture. Bull. Environ. Contam. Toxicol. 59 (6), 928–934.

Badii, M.H., Landeros, J., 2015. Pesticides That Affect Human Health and Sustainability. CULCyT.

Bajaj, A., Pathak, A., Mudiam, M.R., Mayilraj, S., Manickam, N., 2010. Isolation and characterization of a *Pseudomonas* sp. strain IITR01 capable of degrading α-endosulfan and endosulfan sulfate. J. Appl. Microbiol 109 (6), 2135–2143.

Bandala, E.R., Gelover, S., Leal, M.T., Arancibia-Bulnes, C., Jimenez, A., Estrada, C.A., 2002. Solar photocatalytic degradation of Aldrin. Catal. Today 76 (2–4), 189–199.

Banerjee, S., Palit, R., Sengupta, C., Standing, D., 2010. Stress induced phosphate solubilization by *Arthrobacter* sp. and *Bacillus* sp. isolated from tomato rhizosphere. Aust. J. Crop. Sci. 4 (6), 378.

Bhattacharjee, K., Banerjee, S., Bawitlung, L., Krishnappa, D., Joshi, S.R., 2014. A study on parameters optimization for degradation of endosulfan by bacterial consortia isolated from contaminated soil. Proc. Natl. Acad. Sci. India B 84 (3), 657–667.

Bhattacharyya, P.N., Jha, D.K., 2012. Plant growth-promoting rhizobacteria (PGPR): emergence in agriculture. World. J. Microbiol. Biotechnol. 28, 1327–1350.

BIS, 2012. Indian Standard Drinking Waters Specification, second revision Bureau of Indian Standards, Manak Bhavan, New Delhi.

Braud, A., Jézéquel, K., Bazot, S., Lebeau, T., 2009. Enhanced phytoextraction of an agricultural Cr-and Pb-contaminated soil by bioaugmentation with siderophore-producing bacteria. Chemosphere 74 (2), 280–286.

Canbolat, M.Y., Bilen, S., Çakmakçı, R., Şahin, F., Aydın, A., 2006. Effect of plant growth-promoting bacteria and soil compaction on barley seedling growth, nutrient uptake, soil properties and rhizosphere microflora. Biol. Fertil. Soils 42 (4), 350–357.

Chen, Y., Zhang, X., Liu, H., Wang, Y., Xia, X., 2002. Study on *Pseudomonas* sp. WBC-3 capable of complete degradation of methylparathion. Wei. Sheng. Wu. Xue. Bao. 42 (4), 490–497.

Chowdhury, A.G., Das, C., Kole, R.K., Banerjee, H., Bhattacharyya, A., 2007. Residual fate and persistence of endosulfan (50 WDG) in Bengal gram (*Cicer arietinum*). Environ. Monit. Assess. 132 (1–3), 467–473.

Damalas, C.A., Eleftherohorinos, I.G., 2011. Pesticide exposure, safety issues, and risk assessment indicators. Int. J. Environ. Res. Public Health 8 (5), 1402–1419.

FAO, 2005. Proceedings of the Asia Regional Workshop, Regional Office for Asia and the Pacific, Bangkok.

FAO, 2010. Food and Agricultural Organization of the United Nations. http://faostat.fao.org/site/424/default.asp#ancor.

Fisher, B.E., 1999. Most unwanted. Environ. Health Perspect. 107 (1), A18.

Ganesan, V., 2008. Rhizoremediation of cadmium soil using a cadmium-resistant plant growth-promoting rhizopseudomonad. Curr. Microbiol. 56 (4), 403–407.

Giri, K., Rai, J.P.N., 2012. Biodegradation of endosulfan isomers in broth culture and soil microcosm by *Pseudomonas fluorescens* isolated from soil. Int. J. Environ. Stud. 69 (5), 729–742.

Giri, K., Rai, J.P.N., 2012. Plant growth-promoting bacteria: mechanisms and applications. Scientifica 1–15.

Gupta, A., Kaushik, C.P., Kaushik, A., 2000. Degradation of hexachlorocyclohexane (HCH; α, β, γ and δ) by *Bacillus circulans* and *Bacillus brevis* isolated from soil contaminated with HCH. Soil Biol. Biochem. 32 (11–12), 1803–1805.

Harikumar, P.S., Jesitha, K., Sreechithra, M., 2013. Remediation of endosulfan by biotic and abiotic methods. J. Environ. Prot. 4 (05), 418.

Harikumar, P.S., Jesitha, K., Megha, T., Kokkal, K., 2014. Persistence of endosulfan in selected areas of Kasaragod district, Kerala. Curr. Sci. 1421–1429.

Hong, S.H., Lee, E.Y., 2014. Vegetation restoration and prevention of coastal sand dunes erosion using ion exchange resins and the plant growth-promoting rhizobacteria *Bacillus* sp. SH1RP8 isolated from indigenous plants. Int. Biodeter. Biodegr. 95, 262–269.

Jayashree, R., Vasudevan, N., 2007. Effect of tween 80 added to the soil on the degradation of endosulfan by *Pseudomonas aeruginosa*. Int. J. Environ. Sci. Technol. 4 (2), 203–210.

Kalyani, S., Sharma, S., Singh, J., Dureja, S., Lata, P., 2009. Enrichment and isolation of endosulfan-degrading microorganism from tropical acid soil. J. Environ. Sci. Health B 44 (7), 663–672.

Kamanavalli, C.M., Ninnekar, H.Z., 2004. Biodegradation of DDT by a *Pseudomonas* species. Curr. Microbiol. 48 (1), 10–13.

Kannan, K., Tanabe, S., Giesy, J.P., Tatsukawa, R., 1997. Organochlorine pesticides and polychlorinated biphenyls in foodstuffs from Asian and oceanic countries. Rev. Environ. Contam. Toxicol 152, 1–55.

Kaushik, C.P., 1989. Loss of HCH from surface soil layers under subtropical conditions. Environ. Pollut. 59 (3), 253–264.

Kaushik, C.P., 1991. Persistence and metabolism of HCH and DDT in soil under subtropical conditions. Soil Biol. Biochem. 23 (2), 131–134.

Kavamura, V.N., Santos, S.N., da Silva, J.L., Parma, M.M., Ávila, L.A., Visconti, A., et al., 2013. Screening of Brazilian cacti rhizobacteria for plant growth-promotion under drought. Microbiol. Res 168 (4), 183–191.

Khalifa, A.Y., Alsyeeh, A.M., Almalki, M.A., Saleh, F.A., 2016. Characterization of the plant growth promoting bacterium, *Enterobacter cloacae* MSR1, isolated from roots of non-nodulating *Medicago sativa*. Saudi J. Biol. 23 (1), 79–86.

Khan, M.S., Zaidi, A., Wani, P.A., 2007. Role of phosphate-solubilizing microorganisms in sustainable agriculture—a review. Agron. Sustain. Dev. 27 (1), 29–43.

Kloepper, J.W., 1994. Plant growth-promoting rhizobacteria (other systems). In: Okon, Y. (Ed.), *Azospirillum*/Plant Associations. CRC Press, Boca Raton, FL, pp. 111–118.

Kravchenko, L.V., Makarova, N.M., Azarova, T.S., Provorov, N.A., Tikhonovich, I.A., 2002. Isolation and phenotypic characterization of plant growth–promoting rhizobacteria with high antiphytopathogenic activity and root-colonizing ability. Microbiology 71 (4), 444–448.

Kumar, M., Philip, L., 2006a. Endosulfan mineralization by bacterial isolates and possible degradation pathway identification. Bioremediat. J. 10 (4), 179–190.

Kumar, M., Philip, L., 2006b. Bioremediation of endosulfan contaminated soil and water—optimization of operating conditions in laboratory scale reactors. J. Hazard. Mater. 136 (2), 354–364.

Kumar, M., Gupta, S.K., Garg, S.K., Kumar, A., 2006. Biodegradation of hexachlorocyclohexane-isomers in contaminated soils. Soil. Biol. Biochem. 38 (8), 2318–2327.

Kumar, M., Lakshmi, C.V., Khanna, S., 2008. Biodegradation and bioremediation of endosulfan contaminated soil. Bioresour. Technol. 99 (8), 3116–3122.

Kumar, A., Bhoot, N., Soni, I., John, P.J., 2014. Isolation and characterization of a *Bacillus subtilis* strain that degrades endosulfan and endosulfan sulfate. 3 Biotech 4 (5), 467–475.

Lal, R., Dadhwal, M., Kumari, K., Sharma, P., Singh, A., Kumari, H., et al., 2008. *Pseudomonas* sp. to *Sphingobium indicum*: a journey of microbial degradation and bioremediation of hexachlorocyclohexane. Indian. J. Microbiol. 48 (1), 3–18.

Lu, Y., Morimoto, K., Takeshita, T., Takeuchi, T., Saito, T., 2000. Genotoxic effects of alpha-endosulfan and beta-endosulfan on human HepG2 cells. Environ. Health Perspect. 108 (6), 559.

Luz, W.C.D., 2001. Evaluation of plant growth-promoting and bioprotecting rhizobacteria on wheat crop. Fitopatol. Bras. 26, 597–600.

Ma, Y., Rajkumar, M., Luo, Y., Freitas, H., 2011. Inoculation of endophytic bacteria on host and nonhost plants—effects on plant growth and Ni uptake. J. Hazard. Mater. 195, 230–237.

Mehta, P., Walia, A., Kulshrestha, S., Chauhan, A., Shirkot, C.K., 2015. Efficiency of plant growth-promoting P-solubilizing *Bacillus circulans* CB7 for enhancement of tomato growth under net house conditions. J. Basic Microbiol. 55 (1), 33–44.

Miransari, M., Smith, D.L., 2014. Plant hormones and seed germination. Environ. Exp. Bot. 99, 110–121.

Mitton, F.M., Gonzalez, M., Monserrat, J.M., Miglioranza, K.S., 2016. Potential use of edible crops in the phytoremediation of endosulfan residues in soil. Chemosphere 148, 300–306.

Nadeem, S.M., Zahir, Z.A., Naveed, M., Arshad, M., 2007. Preliminary investigations on inducing salt tolerance in maize through inoculation with rhizobacteria containing ACC deaminase activity. Can. J. Microbiol. 53, 1141–1149.

NIOH, 2002. Final Report of the Investigation of Unusual Illnesses Allegedly Produced by Endosulfan Exposure in Padre Village of Kasargod District (N. Kerala). National Institute of Occupational Health, Indian Council of Medical Research, Ahmedabad.

Niu, D.D., Liu, H.X., Jiang, C.H., Wang, Y.P., Wang, Q.Y., Jin, H.L., et al., 2011. The plant growth–promoting rhizobacterium *Bacillus cereus* AR156 induces systemic resistance in *Arabidopsis thaliana* by simultaneously activating salicylate-and jasmonate/ethylene-dependent signaling pathways. Mol. Plant Microbe Interact. 24 (5), 533–542.

Odukkathil, G., Vasudevan, N., 2013. Enhanced biodegradation of endosulfan and its major metabolite endosulfate by a biosurfactant producing bacterium. J. Environ. Sci. Health B 48 (6), 462–469.

Ogbo, F., Okonkwo, J., 2012. Some characteristics of a plant growth promoting *Enterobacter* sp. isolated from the roots of maize. Adv. Microbiol. 2 (3), 368–374.

Ortiz-Hernández, M.L., Sánchez-Salinas, E., Olvera-Velona, A., Folch-Mallol, J.L., 2011. Pesticides in the environment: impacts and its biodegradation as a strategy for residues treatment. Pesticides-Formulations, Effects, Fate. InTech.

Ortiz-Hernández, M.L., Sánchez-Salinas, E., Dantán-González, E., Castrejón-Godínez, M.L., 2013. Pesticide biodegradation: mechanisms, genetics and strategies to enhance the process. Biodegradation-Life of Science. Intech.

Pant, G., Mistry, S.K., Sibi, G., 2013. Isolation, identification and characterization of p, p′-DDT degrading bacteria from soil. J. Environ. Sci. Technol. 6 (3), 130–137.

Purnomo, A.S., Mori, T., Putra, S.R., Kondo, R., 2013. Biotransformation of heptachlor and heptachlor epoxide by white-rot fungus *Pleurotus ostreatus*. Int. Biodeter. Biodegr. 82, 40–44.

Rajkumar, M., Ae, N., Prasad, M.N.V., Freitas, H., 2010. Potential of siderophore-producing bacteria for improving heavy metal phytoextraction. Trends Biotechnol. 28 (3), 142–149.

Ramaneswari, K., Rao, L.M., 2000. Bioconcentration of endosulfan and monocrotophos by *Labeo rohita* and *Channa punctata*. Bull. Environ. Contam. Toxicol. 65 (5), 618–622.

Rekha, P.D., Lai, W.A., Arun, A.B., Young, C.C., 2007. Effect of free and encapsulated *Pseudomonas putida* CC-FR2-4 and *Bacillus subtilis* CC-pg104 on plant growth under gnotobiotic conditions. Bioresour. Technol. 98 (2), 447–451.

Saleem, M., Arshad, M., Hussain, S., Bhatti, A.S., 2007. Perspective of plant growth promoting rhizobacteria (PGPR) containing ACC deaminase in stress agriculture. J. Indian Microbiol. Biotechnol. 34, 635–648.

Schmidt, W., 1999. Mechanisms and regulation of reduction-based iron uptake in plants. New Phytol. 141 (1), 1–26.

Schreinemachers, P., Tipraqsa, P., 2012. Agricultural pesticides and land use intensification in high, middle and low income countries. Food Policy 37 (6), 616–626.

Sharma, A., Johri, B.N., Sharma, A.K., Glick, B.R., 2003. Plant growth-promoting bacterium *Pseudomonas* sp. strain GRP3 influences iron acquisition in mung bean (*Vigna radiata* L. Wilczek). Soil Biol. Biochem. 35 (7), 887–894.

Shetty, P.K., Murugan, M., Sreeja, K.G., 2008. Crop protection stewardship in India: wanted or unwanted. Curr. Sci. 457–464.

Shivaramaiah, H.M., Kennedy, I.R., 2006. Biodegradation of endosulfan by a soil bacterium. J. Environ. Sci. Heal B 41 (6), 895–905.

Siddique, T., Okeke, B.C., Arshad, M., Frankenberger, W.T., 2003. Enrichment and isolation of endosulfan-degrading microorganisms. J. Environ. Qual. 32 (1), 47–54.

Sinha, S., Mukherjee, S.K., 2008. Cadmium–induced siderophore production by a high Cd-resistant bacterial strain relieved Cd toxicity in plants through root colonization. Curr. Microbiol. 56 (1), 55–60.

Siva, G.V., Rajam, A., 2013. Degradation of endosulfan using *Pseudomonas* sp. ED1 isolated from pesticide contaminated soil. J. Acad. Ind. Res. 2 (3), 170.

Sivasankari, B., Anandharaj, M., 2014. Isolation and molecular characterization of potential plant growth promoting *Bacillus cereus* GGBSTD1 and *Pseudomonas* spp. GGBSTD3 from vermisources. Adv. Agric. Available from: https://doi.org/10.1155/2014/248591.

Stern, A.H., 2014. Hazard identification of the potential for dieldrin carcinogenicity to humans. Environ. Res. 131, 188–214.

Sutherland, T.D., Horne, I., Lacey, M.J., Harcourt, R.L., Russell, R.J., Oakeshott, J.G., 2000. Enrichment of an endosulfan-degrading mixed bacterial culture. Appl. Environ. Microbiol. 66 (7), 2822–2828.

Tank, N., Saraf, M., 2009. Enhancement of plant growth and decontamination of nickel-spiked soil using PGPR. J. Basic Microbiol. 49 (2), 195–204.

Tripathi, M., Munot, H.P., Shouche, Y., Meyer, J.M., Goel, R., 2005. Isolation and functional characterization of siderophore-producing lead-and cadmium-resistant *Pseudomonas putida* KNP9. Curr. Microbiol. 50 (5), 233–237.

UNEP/POPS/POPRC, 2008. 4/14 Stockholm convention on persistent organic pollutants, consideration of chemicals newly proposed for inclusion in annexes A, B or C of the convention: endosulfan. In: 4th Meeting, Geneva.

USEPA, 2002. Reregistration Eligibility Decision for Endosulfan. United States Environmental Protection Agency, Washington, DC, EPA 738-R-02-013.

Verma, J.P., Jaiswal, D.K., Sagar, R., 2014. Pesticide relevance and their microbial degradation: a-state-of-art. Rev. Environ. Sci. Biotechnol. 13 (4), 429–466.

Walker, T.S., Bais, H.P., Grotewold, E., Vivanco, J.M., 2003. Root exudation and rhizosphere biology. Plant Physiol. 132, 44–51.

Wang, G., Zhang, J., Wang, L., Liang, B., Chen, K., Li, S., et al., 2010. Co-metabolism of DDT by the newly isolated bacterium, *Pseudoxanthomonas* sp. wax. Braz. J. Microbiol. 41 (2), 431–438.

Wani, P.A., Khan, M.S., 2010. *Bacillus species* enhance growth parameters of chickpea (*Cicer arietinum* L.) in chromium stressed soils. Food Chem. Toxicol. 48 (11), 3262–3267.

Weber, J., Halsall, C.J., Muir, D., Teixeira, C., Small, J., Solomon, K., et al., 2010. Endosulfan, a global pesticide: a review of its fate in the environment and occurrence in the Arctic. Sci. Total Environ. 408 (15), 2966–2984.

World Bank, 2010. World Development Indicators; Rising Global Interest in Farmland. Can It Yield Sustainable and Equitable Benefits? Washington. <http://data.worldbank.org/data-catalog/world-developmentindicators>.

Wrobel, M.H., Grzeszczyk, M., Mlynarczuk, J., Kotwica, J., 2015. The adverse effects of aldrin and dieldrin on both myometrial contractions and the secretory functions of bovine ovaries and uterus in vitro. Toxicol. Appl. Pharmacol. 285 (1), 23–31.

Wu, C., Luo, Y., Gui, T., Huang, Y., 2014. Concentrations and potential health hazards of organochlorine pesticides in shallow groundwater of Taihu Lake region, China. Sci. Total Environ. 470, 1047–1055.

Zahir, Z.A., Munir, A., Asghar, H.N., Shaharoona, B., Arshad, M., 2008. Effectiveness of rhizobacteria containing ACC-deaminase for growth promotion of pea (*Pisum sativum*) under drought conditions. J. Microbiol. Biotechnol. 18, 958–963.

Zahran, H.H., 2001. Rhizobia from wild legumes: diversity, taxonomy, ecology, nitrogen fixation and biotechnology. J. Biotechnol. 91 (2–3), 143–153.

Zaidi, S., Usmani, S., Singh, B.R., Musarrat, J., 2006. Significance of *Bacillus subtilis* strain SJ-101 as a bioinoculant for concurrent plant growth promotion and nickel accumulation in *Brassica juncea*. Chemosphere 64 (6), 991–997.

Further reading

John Bullied, W., Buss, T.J., Kevin Vessey, J., 2002. *Bacillus cereus* UW85 inoculation effects on growth, nodulation, and N accumulation in grain legumes: field studies. Can. J. Plant Sci. 82 (2), 291–298.

Index

Note: Page numbers followed by "*f*" and "*t*" refer to figures and tables, respectively.

B

C

D

E

Q

R

CPI Antony Rowe
Eastbourne, UK
September 25, 2019

Table of Contents

INTRODUCTION

Robert Gleave

It is a pleasure to introduce this admirable and ground-breaking translation of this important work by the great Imāmī Shīʿī theologian and jurist, Ḥasan b. Yūsuf Ibn al-Muṭahhar al-Ḥillī, commonly referred to as al-ʿAllāma al-Ḥillī (d.726/1325). The *Tabṣirat al-mutaʿallimīn fī aḥkām al-dīn* ("Enlightening Students in the Laws of Religion") is a work of *fiqh* (literally "understanding", but usually translated as "jurisprudence"). A work of *fiqh* is an attempt by a jurist to outline his or her doctrines on a set list of Islamic legal issues. The format of a work of *fiqh* is quite standard and predictable. *Fiqh* works invariably begin with a discussion of ritual purity, move on to ritual duties such as prayer, fasting, alms-tax, pilgrimage; after this, the chapters are in a less standard order but they include contracts and commercial transactions, marriage and divorce, international relations (though the term is anachronistic, the chapter refers to the relations between the Islamic state and others) and judicial procedure. A quick glance at the contents of the *Tabṣirat al-mutaʿallimīn* shows that it follows this format closely, which was common across the schools of law – Sunnī, Shīʿī and Ibādī.

In this introduction to the work and its author, I divide my comments into three sections: (1) The Reputation and Legacy of al-ʿAllāma al-Ḥillī; (2) The Life and Works of al-ʿAllāma al-Ḥillī; and (3) His Methodology in the *Tabṣirat al-mutaʿallimīn*. One may ask the question: "Why discuss al-ʿAllāma's life and legacy before his life and works?" The reason being that his great intellectual, political and societal impact on Imāmī Shīʿī history has led to a disputed legacy. All the facts and details of al-ʿAllāma's life are, to an extent, refracted through the subsequent tradition of scholarship around him and in response to his achievements. To understand his importance, one needs first to recognize how he has been interpreted and presented in the extant literary sources.

The Reputation and Legacy of al-ʿAllāma al-Ḥillī

The author of *Tabṣirat al-mutaʿallimīn* ("the book between your hands" as the Arabic expression goes) is famous for many things within the Imāmī

Shīʿī tradition. He is certainly one of the most prolific Imāmī Shīʿī writers, authoring over a hundred books in most genres of the religious sciences.[1] His theological works became teaching texts and fundamental to the training of Imāmī Shīʿī religious experts; his works in legal theory (*uṣūl al-fiqh*) and law (of which the *Tabṣirat al-mutaʿallimīn* is but one example) became standard reference books and the subject of subsequent commentary. He is lauded as the jurist who most effectively combined the rationalist theological premises inherent in the Imāmī Shīʿī adoption of Muʿtazilism with the textual requirements of Islamic law (namely, reference to the Quran and Sunna). These, amongst other contributions, form the bedrock of his reputation in the Imāmī Shīʿī tradition.

To expand on this one can cite examples. In this work, in the section on divorce of the *Tabṣirat al-mutaʿallimīn*, al-ʿAllāma al-Ḥillī writes:

> The divorce is complete upon his stating, "You are divorced" (*ṭāliq*), devoid of any conditions or descriptions, and it is conditional upon being heard by two just male witnesses.[2]

In holding this doctrine, al-ʿAllāma was not unusual. The view that the validity of any divorce depends on the husband's divorce pronouncement being witnessed by two just men is so widely held within the Imāmī Shīʿī school that from the earliest times, the jurists were claiming a consensus of the community (*ijmāʿ al-firqa*) on this point. Al-ʿAllāma's adherence to this doctrine, along with his fellow Imāmī Shīʿī *fuqahāʾ* was, according to one report, to play a role in one of the better known episodes in al-ʿAllāma's life – namely, his involvement in the conversion of the Mongol Sultan Uljeytu (r. 1304–1316) from Sunnism to Shīʿism. The earliest account of the intertwining of al-ʿAllāma, this doctrine and the conversion of Uljeytu is to be found recounted by Muḥammad Taqī al-Majlisī (d.1070/1659, henceforth Majlisī I) in his *Rawḍat al-muttaqīn* – and it is worth recounting in full as it exemplifies how Imāmī Shīʿism has promoted al-ʿAllāma's role in Islamic history:

1 Muḥsin al-Amīn, *Aʿyān al-Shīʿa* (Beirut: Dār al-Taʿāruf, 1403/1983), 5:402 says, "It is said that he wrote over one hundred works (*muʾallafāt*), even though what has come down to us in terms of names does not exceed ninety five, as you will see..." He then proceeds to list all the book titles by al-ʿAllāma known to him. Sabine Schmidtke, *The theology of al-Allama al-Hilli (d. 726/1325)* (Berlin : K. Schwarz, 1991), 74ff lists 126 known titles.

2 See below, p. 354 and al-ʿAllāma al-Ḥillī, *Tabṣirat al-muta ʾallimīn* (Tehran: Intishārāt-e faqīh, 1368Sh/[1989]), 190.

The reason why Sulṭān Muḥammad Uljeytu turned to the [Shīʿī] Faith is because [Sulṭān Muḥammad] had become angry with his wife and he said to her, "You are divorced" three times. Then he softened [his position] and gathered together the scholars. They said, "There is no escaping the need for a *muḥallil* [that is, a man who will marry the wife, and then divorce her, so she can remarry her former husband].

He said, "You have different opinions on every question [of law] – on this you don't have any difference of opinion?" They said not.

One of his ministers said, "There is a scholar in Ḥilla who will say that this divorce is invalid." So he sent a letter to al-ʿAllāma, and bid him come.

When he had sent for him, the Sunni scholars all said, "He has an invalid school [of law], and the Rawāfiḍ are irrational. It is not fitting for the King to call on a man of such inferior rationality. The King said, "Wait until he arrives".

When ʿAllāma arrived, the King sent for all the scholars of the four schools and gathered them together. When al-ʿAllāma entered, he took his shoes in his hands, and entered the gathering. He said, "Peace be with you" (*al-salāmu ʿalaykum*), and then sat down with the King.

They all pointed out to the King, "Didn't we say to you that they had weak intellects?"

The King replied, "Ask him about what he did."

So they asked him, "Why didn't you bow down before the king, and why did you abandon good etiquette?"

Al-ʿAllāma said, "The Messenger of God was a king, and he was always greeted with *al-salāmu ʿalaykum*. God says, "...when you enter houses, give greetings of peace upon each other – a greeting from Allah, blessed and good..." (Q24:61). There is no difference between us and you – for all of us are not permitted to bow down to anyone but God."

So they asked him, "Why did you sit down next to the King?"

And he said, "There was nowhere else to sit."

He said everything in Arabic; a translator was conveying it to the King.

They asked him, "Why did you take your sandals with you? This is not fitting for a rational being, never mind a man."

He replied, "I was afraid that the Hanafis would steal them, just like Abū Ḥanīfa stole the sandals of the Prophet."

At this, the Ḥanafīs screamed out, "God forbid – never! When was Abū Ḥanīfa present during the time of the Prophet!?! He was born 100 years after the Messenger of God's death."

So al-ʿAllāma said, "Oops, I forgot – perhaps the thief was al-Shāfiʿī."

And then the Shāfiʿīs exclaimed, saying, "Al-Shāfiʿī was born on the day when Abū Ḥanīfa died, and was 4 years in the belly of his mother, and he didn't come out, due to respect for Abū Ḥanīfa – so when he died, he came out. His appearance was in the second century after the death of the Messenger of God."

So al-ʿAllāma said, "Oh! Perhaps it was Mālik."

And the Mālikīs cried in the same way as the Ḥanafīs. And then he said, "Perhaps it was Aḥmad b. Ḥanbal..." and they responded in the same way as the Shāfiʿīs.

Then al-ʿAllāma turned to the King and said, "King – you know that none of the leaders of the four schools was alive at the time of the Prophet, and not even during the time of the companions. This is one of their innovations: They prefer to follow the *mujtahid*s of these Four Schools. Even if one of them is greater than the other by a single degree, they are not allowed to do their *ijtihād* and come to conclusions which oppose the opinions of the others."

The King said, "Not a single one of them was around at the time of the Prophet and the companions?" And everyone said, "No."

Then al-ʿAllāma said, "We are the Shīʿa, followers of the Prince of Believers, soul of the Messenger of God, his brother, the son of his uncle, and his inheritor." [He continued:] "... and in any case, the divorce which the King initiated is invalid, because he did not fulfil the conditions, amongst which is the witnessing of two just individuals. Did the King make the declaration in the presence of two such people?"

The King said, "No."

Al-ʿAllāma began to debate with the Sunni scholars, until he was completely victorious over all of them. The King himself then became Shīʿī, and sent out a letter to the regions and provinces saying that all should mention the Twelve Imams in the sermons, and their names should be written on mosques and prayer halls. The one written in Isfahan is still there today in the Old Jāmiʿ mosque, written in [Sulṭan Muḥammad's] time, in three places. Also, the one on the minaret of the Dār al-Siyyāda still exists, which Sulṭan Muḥammad completed after what had happened to his brother Ghāzān.[3]

The source quoted here, *Rawḍat al-muttaqīn*, is a commentary on the famous hadith collection of Muḥammad b. ʿAlī Ibn Bābawayh (d. 381/991), *Man lā yaḥḍuruhu al-faqīh*. The hadith which prompts Majlisī to recount this story is:

3 Muḥammad Taqī al-Majlisī, *Rawḍat al-muttaqīn* (Qom: Bunyād-e Farhang-e Islāmī, 1389/[1969]), 1:30–32.

Al-Qāsim b. Muḥammad al-Jawharī relates from ʿAlī b. Abī Ḥamza, who said:

> Abū ʿAbdallāh [Imam Jaʿfar al-Ṣādiq] said, "The only [valid] divorce is according to the Sunna. ʿAbdallāh b. ʿUmar divorced [his wife] three times in one session, and his wife was menstruating at the time. The Messenger of God refused to recognise his divorce and said, 'Anything which opposes the Book of God should be referred to the Book of God.'"[4]

From this hadith, Majlisī I embarks on an extended discussion of how declaring the divorce formula (such as "You are divorced" or "I divorce you" or some other explicit expression of divorce) three times in one session is against proper practice (*sunna*), and declaring it whilst your wife is menstruating also makes the divorce invalid. There is some examination of the chain of transmission (*isnād*) of this report, and the citation of other reports from the Prophet and Imams which support the discouragement of triple divorce pronouncements in single sessions. In one of the reports cited by Majlisī (found not in *Man lā yaḥḍuruhu al-faqīh*, but in al-Kulaynī's [d. 329/941] *al-Kāfī*), Imam Muḥammad al-Bāqir lists invalid divorces, amongst which is: "If he divorces her during her waiting period but without two just witnesses, then his divorce is not actually a divorce."[5] It is this report from Imam al-Bāqir which triggers Majlisī I to tell the story translated above. The whole story appears as one of Majlisī I's asides – it is not his main concern. The story is, it might be argued, a bit of light relief in the midst of the heavy, technical language of Majlisī I's hadith commentary.

Now, the *Rawḍat al-muttaqīn* is, of course, a late source, composed some three hundred years after al-ʿAllāma's death; and I make no claims for the story's historicity here. It is interesting that the story does not feature in the various versions of Uljetu's conversion listed in an important article by Judith Pfeiffer. Pfeiffer does reference the presence and influence of al-ʿAllāma (and his son) in the court of Uljeytu.[6] The story itself is most likely an elaboration, reflecting the tendency amongst some Safavid jurists to overplay the influence of pre-Safavid jurists in the political realm.

What the story reveals though is the high standing and pride which much

4 Ibn Bābawayh, *Man lā yaḥḍuruhu al-faqīh* (Qom: Mu'assasat al-Nashr al-Islāmī, 1404/[1983]), 3:496.

5 "*wa-in ṭalaqahā li-l-ʿidda bi-ghayr shāhiday ʿadal fa-laysa ṭalāqahu bi-ṭalāqin...*" Muḥammad b. Yaʿqūb al-Kulaynī, *al-Kāfī* (Tehran: Dār al-Kutub al-Islāmiyya, 1367Sh/[1988]), 6:61.

6 Judith Pfeiffer, "Conversion Versions: Sultan Öljeytü's Conversion to Shi'ism (709/1309) in Muslim Narrative Sources." *Mongolian Studies* 22 (1999): 35-67

of subsequent Imāmī Shīʿī tradition displays towards al-ʿAllāma al-Ḥillī, his life and his achievements. Amongst the so-called Uṣūlī school of jurisprudence, al-ʿAllāma is seen as the initiator of a more integrated and nuanced form of jurisprudence, and a scholar whose erudition prompted much reaction, even from non-Shīʿī authors. The best-known example of this is Ibn Taymiyya's *Minhāj al-sunna al-nabawiyya* – an extensive critique of al-ʿAllāma's summa of Imāmī Shīʿī doctrine *Minhāj al-karāma fī maʿrifat al-imāma.*[7]

Respect for al-ʿAllāma in the Shīʿī tradition was not universal. In the 17th century, there emerged a virulent, Shīʿī rejection of al-ʿAllāma's "Uṣūlī" model of jurisprudence. The so-called Akhbārī school, from the work of Muḥammad Amīn al-Astarābādī (d. 1036/1626) onwards, would target al-ʿAllāma, accusing him of incorporating Sunni analytical methods and doctrine into the Imāmī Shīʿī milieu. Amongst his so-called "deviations" was to introduce into Imāmī Shīʿism techniques of hadith evaluation derived from Sunnī thinking. Specifically, al-ʿAllāma divided reports into four categories of varying reliability, which mirrored those employed by Sunni hadith scholars. The Akhbārīs viewed this as an unjustified innovation and departure from the way of the Imams. Al-ʿAllāma also laid out a full theory of *ijtihād* – that is, he gave to the jurist the power to deduce and infer from the original sources, and to produce his own opinion concerning a legal issue. Indeed, underpinning the whole Akhbārī critique of al-ʿAllāma was his acceptance of an epistemology in which legal doctrines which were known with less than absolute certainty (*ʿilm, qaṭʿ*) retained a form of probative force (*ḥujjiyya*). Reports of uncertain pedigree, inferences and implicit meanings of texts, the jurists' own view as to the appropriateness of a ruling – all of these were not certain knowledge, but mere opinion (*ẓann*). The acceptance in law of these "less than certain" pieces of evidence was, for the Akhbārīs, a compromise on the certainty they felt could be attained from the direct legal statements of the Imams. Hence, for Akhbārīs, al-ʿAllāma was their *bête noire*, a traitor to the cause of Imāmī Shīʿism. The result was that he received abuse and vitriol from the Akhbārīs, and the juristic method he introduced was branded as heretical and dangerous.[8]

7 This work has been translated and published in this series dedicated to the works of al-ʿAllāma; see al-ʿAllāma al-Ḥillī, *The Way of Nobility: Knowledge of the Imam (Minhāj al-karāma fī maʿrifat al-imāma). A parallel English-Arabic text translated, introduced and annotated by Saiyad Nizamuddin Ahmad* (Birmingham: AMI Press, 2023). See also Tariq Al-Jamil, "Ibn Taymiyya And Ibn Al-Mutahhar Al-Hilli: Shi'i Polemics And The Struggle For Religious Authority In Medieval Islam" in Shahab Ahmed and Yossef Rapoport, *Ibn Taymiyya and his times* (Oxford: OUP, 2010), 229–246.

8 I deal will the Akhbārī critique of al-ʿAllāma in my *Scripturalist Islam: The History and Doc-*

Regardless of whether one assesses his achievements negatively (like the Akhbārīs) or positively (as did the Uṣūlīs), Western scholarship has generally identified in the works al-ʿAllāma a major advance in Imāmī theology and law.[9] His incorporation of Muʿtazilī theology was more thorough-going than previous Imāmī writers, and his incorporation of *ijtihād* into *uṣūl al-fiqh* was a game-changer which set the scene for all major subsequent advances in Imāmī legal theory. Whether he is vilified or lauded, or simply recognized as the genius synthesizer and erudite polymath that he was, his central position in Imāmī intellectual history has been iterated and reiterated both within the tradition and outside of it.

The Life and Works of al-ʿAllāma al-Ḥillī

There are a number of highly competent accounts of al-ʿAllāma's life and works available in the secondary literature. The modern introductions to nearly all of his works contain the standard set of comments: al-ʿAllāma's life and travels, his teachers, his pupils, and his works. Sabine Schmidtke has given a detailed English language account of al-ʿAllāma's life, following a similar pattern, but supplemented by some critical historical analysis and evaluation.[10] Bearing in mind the legacy and reputation of al-ʿAllāma mentioned above, there is always a tendency for biographers to emphasize some elements of his life over others. Any aim of producing a pure, unbiased account of the life and works of al-ʿAllāma (or anyone else) is doomed to failure. Even the most basic account (as I am setting out below) is not neutral, but geared to a particular purpose. In this case, my purpose is to give the reader my account of how *Tabṣirat al-mutaʿallimīn* fits into al-ʿAllāma's life and works.

Al-ʿAllāma's entry in his own work of scholarly biography *Khulāṣat al-aqwāl fī maʿrifat aḥwāl al-rijāl* is very short on detail – as the demands of the genre dictate.[11] It consists of his names, place of birth (and date) followed by a list of works. He does though mention in this entry that he was born in Hilla

trines of the Akhbārī Shīʿī School (Leiden: Brill, 2007), particularly that of al-Astarābādī.

9 Hossein Modarressi, "Rationalism and Traditionalism in Shīʿī Jurisprudence: A Preliminary Survey." *Studia Islamica*, no. 59 (1984), 144; Norman Calder, "Doubt and Prerogative: The Emergence of an Imāmī Shīʿī Theory of Ijtihād." *Studia Islamica*, no. 70 (1989), 65ff.

10 Schmidtke, *The theology of al-Allama al-Hilli*, 9–41.

11 al-ʿAllāma al-Ḥillī, *Khulāṣat al-aqwāl fī maʿrifat aḥwāl al-rijāl* (Qom: Muʾassasat Nashr al-Fiqāha, 1417/[1996]), 109.

on 19th Ramadan 648 (as in the printed edition, but 29th Ramadan in other manuscripts, and al-Ḥillī himself records it as 27th Ramadan elsewhere). All of these dates correspond to December 1250. His family appears to have been learned, and his father in particular wrote works in the religious sciences. Indeed, it would be unusual for someone of al-ʿAllāma's stature not to have come from a religious family: by this time in Hilla, the scholarly leadership was coalescing into families (the familial relations of the scholars of the period are clear from the records of scholarly licenses, or *ijāzas*). The importance of the family, and his father in particular, is reflected in the story al-ʿAllāma tells of his father's role as the representative of Hilla in the negotiations with the Mongol invaders. His sisters and brothers, and their linkages with the scholarly community in Hilla and elsewhere, are recorded in the Shīʿī biographical literature and confirm the interconnected nature of learned families, a feature which continues to this day in Imāmī Shīʿī seminary teaching contexts.

His teachers also were drawn from his relatives. Aside from his father (who appears to have taught him theology, Quranic exegesis and grammar) the most famous of al-ʿAllāma's teachers was the principal scholar of the day, Jaʿfar b. Muḥammad al-Ḥillī (d. 676/1277), known as al-Muḥaqqiq al-Awwal (hereon Muḥaqqiq I). In much of al-ʿAllāma's work, one sees a continuation of the themes first introduced by Muḥaqqiq I. For example, on the issue of *ijtihād*, Muḥaqqiq I appears to introduce the novel position that it might be permitted in certain areas of the law; this is continued by al-ʿAllāma and expanded to cover all areas of legal investigation when carried out by a qualified person.[12] Muḥaqqiq I taught him law (*fiqh*) and legal theory (*uṣūl al-fiqh*), and the influence of Muḥaqqiq I's work on the *Tabṣirat al-mutaʿallimīn* is discussed in the next section. There were a host of other teachers mentioned by al-Ḥillī in his own *ijāzas*. After this initial study under these teachers, al-ʿAllāma mentions the famous Naṣīr al-Dīn al-Ṭūsī (d. 672/1274), whose observatory in Maragha had become a centre for Islamic learning and science by the time al-Ḥillī encountered Naṣīr al-Dīn. His studies with these scholars was in the areas of philosophy, theology and logic. He may have visited Maragha to study with Naṣīr al-Dīn, and if so, he most likely left after al-Ṭūsī's death in 672/1274. From then on, al-ʿAllāma was based in Hilla (with occasional visits to Baghdad), writing and teaching.

It was during this time that Uljaytu came to the throne, and pushed the

12 See Calder, "Doubt and Prerogative," 65–67.

Islamic nature of Mongol rule to the fore. The encounter with al-ʿAllāma – a version of which is given above – and Uljaytu's conversion to Shīʿī Islam is dated to 709/1310, at a time when al-ʿAllāma was around sixty years of age. There are numerous stories of al-ʿAllāma's debates with the other (Sunni) legal scholars at Uljaytu's court. Apparently he was lauded at court, considered part of the favoured group of scholars, and had good relationships with the courtiers and Uljaytu himself. Al-ʿAllāma had a special position in Uleytu's "travelling seminary" (*madrasa sayyāra*) which moved from place to place with the Īlkhān, providing education for Uljaytu's retinue. Schmidtke considers it likely al-ʿAllāma left the court in 716/1316 before Uljaytu's death later that year. By most accounts, he returned to Hilla where he taught for the remaining years of his life. He died in 726/1325. His most immediate legacy was his students, of which the most renowned was his son, Fakhr al-Muḥaqqiqīn (d. 771/1369) who had travelled with al-ʿAllāma to Uljeytu's court and made quite an impression there, it seems. Al-ʿAllāma's nephews ʿAmīd al-Din and Ḍiyāʾ al-Dīn al-Aʿrājī were also well-respected scholars. Of these, only Fakhr al-Muḥaqqiqīn became a major contributor to the history of Imāmī jurisprudence according to later Shīʿī tradition.

In terms of literary output, al-ʿAllāma was a prolific author. The bio-bibliographical sources record 126 works attributed to him though we do not have them all, even in manuscript form; he himself lists 66 of his works up to 1320/1294 (i.e. 5 years before his death) in his short autobiographical notice in his *Khulāṣat al-aqwāl*. Most of these, at the time of him writing the entry are still in the process of composition:

> *wa-hādhihi al-kutub fīhā kathīr lam yatumm*
> ("and these books there is much unfinished")[13]

The phrasing here is revealing – clearly the idea that there is in the books much which is unfinished reveals that he worked on several books simultaneously, rarely bringing any one of them to completion. Most genres of religious literature have a set format – a prime example being works of *fiqh*, mentioned above, but it is not an inaccurate characterization of other genres as well. Given this generic predictability, there were clearly chapters in these works which al-ʿAllāma feels should be there, but are not yet composed, or not yet finished. One should be alert to this possibility when

13 al-ʿAllāma al-Ḥillī, *Khulāṣat al-aqwāl*, 109.

considering the *Tabṣirat al-mutaʿallimīn*, since it is quite possible that what we have is merely the state of the work at the death of al-ʿAllāma. He may have originally planned for additional chapters on other topics. Whilst some works in the list receive a short comment from al-ʿAllāma, for the *Tabṣirat al-Mutaʿallimīn* he gives the title, but makes no comment. We therefore do not know what he himself thought of his work.

As Schmidkte points out, there are issues with all these sources, as works are given two titles, or misattributed to him. Even given these caveats, though, we can be quite certain of a large proportion of the works listed there, and that these works cover various Islamic sciences including logic, theology and philosophy; in the transmitted sciences he worked on exegesis, hadith reporters and scholarly biography; his work included contributions in grammar and mysticism. The works which concern us here are those on law – seven recorded on legal theory (*uṣūl*) many of which have survived and are edited and published, and nineteen recorded on jurisprudence, as well as shorter treatises on specific areas of law. The *Tabṣirat al-mutaʿallimīn* is one of these nineteen.[14]

Al-ʿAllāma's Methodology in the Tabṣirat al-Mutaʿallimīn

Within the genre of *fiqh* there are different sub-genres which occurred and reoccurred in Muslim legal literature. The *Tabṣirat al-mutaʿallimīn* is an excellent example of a *mukhtaṣar* (literally – "an abbreviation"). *Mukhtaṣars* are brief, succinct statements of the law, with little or no argumentation, and which summarize the position of the law school (and the jurist) at that point in time. They have played an important role in the history of Islamic law as they are seen by some commentators as one element of the fully formed *madhhab*.[15] That is, they summarize where the legal thinking of the group is at a point it time, producing a condensed expression of the favoured positions. The brevity of such works meant they invited commentary and critique – indeed, one could argue they were written in such an elliptical abbreviated style that one of their main functions was to be the subject of commentary.[16]

14 Schmidtke, *The theology of al-Allama al-Hilli*, 74ff.

15 See C. Melchert, *The Formation of the Sunni Schools of Law: 9th-10th centuries CE* (Leiden: Brill, 1997), passim.

16 See N. Calder, *Islamic Jurisprudence in the Classical Era* (Cambridge: CUP, 2010), 22–73 for an analysis of Ḥanafī *muhtaṣars*.

The other feature of their role in the development of a legal school was that they were memorized by novices in order to familiarize themselves fully in the doctrine of the school. These features can be ascribed to the *Tabṣirat al-mutaʿallimīn* – and hence it is a *mukhtaṣar* for the Imāmī school.

It is not the first such text – indeed one of the most successful of the Imāmī *mukhtaṣar*s – *al-Mukhtaṣar al-nāfiʿ* ("The Useful Breviary") – was composed by al-ʿAllāma's teacher, Muḥaqqiq I. The success of a work like this is measured in the commentaries it attracted: *al-Mukhtaṣar al-nāfiʿ* attracted 94 commentaries recorded in the bio-bibliographical tradition, including one by al-ʿAllāma; by contrast, the *Tabṣirat al-mutaʿallimīn*, though no slouch, attracted only 30 commentaries (according to the authoritative Āghā Buzurg al-Ṭihrānī), and it is fair to say that these only began to emerge in the 19th century CE, indicating that the work itself was not the focus of study for some centuries, but, for some reason, became popular as a subject of commentary in the post-Bihbahānī era.[17] I suspect that the reason for its sudden popularity is linked to the rise in the production of *fiqh* and *uṣūl* literature generally in the post-Akhbārī environment. The collapse of the Akhbārī school, for whom commentary was restricted to the Quran and hadith primarily, meant scholars were seeking to revive the commentarial tradition. The standard texts were, of course, the subject of renewed commentary (*al-Lumʿa al-Dimashqiyya* of al-Shahīd al-Awwal and *al-Mukhtaṣar al-nāfiʿ*). But in order to show an element of originality, scholars might choose lesser known texts, and it is into this category that the *Tabṣirat al-mutaʿallimīn* falls. Since then, the text itself has been the subject of extensive commentary, perhaps the most prominent being that of Ayatallah Ṣādiq Ḥusaynī al-Shirāzī, part of the controversial Shiraziyya movement.

Al-ʿAllāma's methodology in the *Tabṣirat al-mutaʿallimīn* is typical of that of a jurist writing a *mukhtaṣar* work. Abbreviation takes precedence over comprehensibility, and sentences regularly need to be explained before they can be understood. Incidentally, this makes translation into English a particular challenge for such works, and why the translation in this volume is such a momentous effort. What the style demonstrates is that the text was written to be taught rather than read, commented upon rather than used as an immediate source of guidance. Having said that, the *Tabṣirat al-mu-*

17 For the *shurūh* titles of these commentaries, see Āghā Buzurg al-Ṭihrānī, *al-Dharīʿa ilā taṣanīf al-Shīʿa* (Beirut: Dār al-Aḍwā', 1983), 14:57ff for *al-Mukhtaṣar al-nāfiʿ* and 13:133ff for *Tabṣirat al-mutaʿallimīn*; for the *ḥāshiya* references, see al-Ṭihrānī, *al-Dharīʿa*, 6:193 for *al-Mukhtaṣar al-nāfiʿ* and 6:31ff for *Tabṣirat al-mutaʿallimīn*.

ta'allimīn is more comprehensible than most *mukhtaṣars*, as this translation ably demonstrates.

One can give an example of how al-ʿAllāma carries out his task by explicating the example given above: namely that a valid divorce requires two just male witnesses. The rule itself is particular to the Imāmī Shīʿa, as other schools do not make such a requirement. The source of the rule is presented in statements of the Imams, such as:

> Aḥmad b. Muḥammad b. Abī Naṣr said to Abū l-Ḥasan [Imam ʿAlī al-Riḍā]: "How does one do a *sunna* (acceptable, legitimate) divorce?" And he said, "Divorce her when she is pure from her period, before you have intercourse with her, in the presence of two just witnesses, just as God lays out in his book... And I said to him, "If he divorces her when she is pure, not having had sexual intercourse with her, but with one man and two women?" And he said, "The testimony of women is not allowed in divorce, and it might be possible for them to witness in relation to other women concerning menstrual blood and whether it is present." I said, "If two anti-Shīʿī men are the witnesses to the divorce, is it valid?" And he said, "He who was born with a good innate disposition can testify to a divorce, once you know him to be good."[18]

The last phrase is subject to a debate as to whether the Imam means a non-Shīʿī can be a just witness, or whether the Imam is, in effect, ruling out non-Shīʿī believers.

More significantly, it is a specifically Shīʿī interpretation of the Quranic verse:

> O Prophet, when you [Muslims] divorce women, divorce them for [the commencement of] their waiting period and keep count of the waiting period, and fear Allah, your Lord. Do not turn them out of their [husbands'] houses, nor should they [themselves] leave [during that period] unless they are committing a clear immorality. And those are the limits [set by] Allah. And whoever transgresses the limits of Allah has certainly wronged himself. You know not; perhaps Allah will bring about after that a [different] matter. And when they have fulfilled their term, either retain them according to acceptable terms or part with them according to acceptable terms. And bring to witness two just men from among you and establish the testimony for God.[19]

18 Kulaynī, *al-Kāfī*, 6:68.

19 Quran 65:1–2.

Some exegetes understand the phrase "bring to witness two just men from among you" as referring to the phrase "retain them" – that is, witnesses are only needed if the husband takes back the wife. The Shīʿa though understand the phrase to refer to the "part with them" element of the verse – that is, the witnesses are needed to enact the divorce. Al-Ṭūsī and Shīʿī jurists after him argue that such phrases always refer to the nearest element, and the nearest element in the verse is the "part with them" phrase.[20]

From all of this, al-ʿAllāma produces his abbreviated statement of the law:

> The divorce is complete upon his stating, "You are divorced" (*ṭāliq*), devoid of any conditions or descriptions, and it is conditional upon being heard by two just male witnesses.[21]

Here we can see his presentation is even more abbreviated that of his teacher Muḥaqqiq I in his *al-Mukhtaṣar al-nāfiʿ*. The exposition there is a little more expansive, with a discussion on the meaning of "just" (*ʿadl*) when applied to the witnesses, and an explicit mention of women's testimony being ineligible. Al-ʿAllāma avoids the latter explicit mention of women by using the phrase "the hearing of two just *men*". Furthermore, the word "hearing" used by al-ʿAllāma here reveals his position on the question of whether the husband has to call them to be official witnesses to this event. This was a point of discussion, but al-ʿAllāma states they need only hear the words of divorce – that is sufficient for the condition to be met.

Techniques such as this demonstrate how in the *Tabṣirat al-mutaʿallimīn*, al-ʿAllāma's aim is two-fold. Of course, one aim is brevity – since that is valued in itself as a demonstration of his skill as a wordsmith. But more than this al-ʿAllāma wishes to establish a dominant school doctrine through subtle use of words (such as "hearing" rather than "witnessing" on the above example). The result is a highly abbreviated text which is, perhaps, too successful in the task he sets himself. That the text failed to recognize significant difference of opinion within the Shīʿī school could be one of the reasons it was not such a popular book for commentary. The tradition generally thrived on exploring juristic difference, whilst al-ʿAllāma in this text wishes to flatten out the diversity in order to consolidate the school.

20 Muḥammad b. al-Ḥasan al-Ṭūsī, *al-Tibyān fī tafsīr al-Qurʾān* (al-Najaf: Maktabat al-Amīn, 1964–69), 27:27ff.

21 See the translation below, p. 354.

Conclusions

It is, then, a privilege to introduce the translation of the *Tabṣirat al-mutaʿallimīn* laid out below and expertly executed by Nazmina Dhanji. The Arabic in works like this is terse and sometimes challenging, and translation needs to reflect this so that the reader gains a sense of the style. More importantly, the translation here enables the reader to understand the text itself. A translation is, in some degree, a commentary on the translated text; in a text of *fiqh*, the translator is required to tie up the loose ends and ensure that the reader, with some concentration, can fully grasp the intended meaning. A *mukhtaṣar* work is a most useful source book for personal religious life. But a *mukhtaṣar* is also a demonstration of scholarly prowess on the author's part, and one sees this in al-ʿAllāma al-Ḥillī's attempt at extreme compression. When rhetorically asking why al-ʿAllāma took on the burden of writing all the works of *fiqh* that he did, Āyatallāh Jaʿfar al-Ṣubḥānī lists them one by one, explaining their distinct qualities. The first work he names is *Tabṣirat al-mutaʿallimīn*. As you read this splendid translation, you may understand why Ṣubḥānī describes it thus:

> The first is *Tabṣirat al-mutaʿallimīn*: a comprehensive, condensed, complete work of *fiqh*, without any commentary or argumentation. In this book, al-ʿAllāma lays out his legal opinions and his edicts on all the [topics of law]... [A]nd because of the book's concision, and the fluency of its expressions, the jurists have awarded it a position of great importance, from the time of its writing until the present day. They have adorned it with commentaries and explications, thereby making it, from the earliest time, a book to be studied seriously.[22]

22 *wa-qad kān fī sālif al-zamān kitāban dirāsiyyan.* Jaʿfar al-Subḥānī, "Introduction" in al-ʿAllāma al-Ḥillī, *Taḥrīr al-aḥkām*, ed. Ibrāhīm al-Bahādurī (Qom: Iʿtimād, 1420/[1999]), 1:9.

Enlightening Students in the Laws of Religion

تبصرة المتعلّمين في أحكام الدين

In the Name of Allah,
the Infinitely Compassionate,
the Ever-Merciful.

All praise be to God, Whose authority is eternal, Whose prestige is sublime, and Whose proof is lucid; Who has blessed His servants by sending them His prophets, and maintained His grace upon them through the bestowal of responsibilities that lead to His good reward. May God send his blessings upon the chief of His prophets in the world, Muḥammad, the chosen one, and his pure progeny.

We have authored this book titled *Tabṣirat al-mutaʿallimīn fī aḥkām al-dīn* (*Enlightening Students in the Laws of Religion*) to guide novices and benefit students, and we seek God's help and succour in the process, for He is the Kindest of all givers and the most Generous of all benefactors. We will proceed in order of importance.[1]

1 This translation, and the Arabic transcription opposite, are based on the following edition: al-ʿAllāma al-Ḥillī. *Tabṣirat al-Mutaʿallimīn fī aḥkām al-dīn*, ed. Muḥammad Hādī al-Gharawī. (Tehran : Intishārāt-i Faqīh, 1368Sh/[1989]).

بِسْمِ اللهِ الرَّحْمنِ الرَّحِيْمِ

الحمد لله القديم سلطانه، العظيم شأنه، الواضح برهانه، المنعم على عباده بإرسال أنبيائه، المتطول عليهم بالتكليف المؤدى إلى حسن جزائه، وصلى الله على سيد رسله في العالمين، محمد المصطفى وعترته الطاهرين.

أما بعد: فهذا الكتاب الموسوم ب (تبصرة المتعلمين في أحكام الدين)، وضعناه لإرشاد المبتدئين وإفادة الطالبين، مستمدين من الله المعونة والتوفيق، إنه أكرم المعطين، وأجود المسؤولين.
ونبدأ بالأهم فالأهم:

The Book of Ritual Purity*

* Lexically the term *ṭahāra* refers to physical cleanliness, but in Islamic legal terminology it is a term applied to the appropriate state required for performing prayers (*ṣalāt*), such as ablution (*wuḍū'*), its substitute dry ablution (*tayammum*) and major ablution (*ghusl*).

كتاب الطهارة

Chapter 1: Water

Water (as a cleansing agent) is of two types: unmixed (*muṭlaq*) and mixed (*muḍāf*). Unmixed water is that which is appropriate to be called water and nothing else, whereas mixed water is the opposite. Therefore, unmixed water is clean in and of itself, and is also cleansing.

Regarding its encountering an impurity, it is divided into:

(a) running water – such as the water in rivers, for example, which does not become impure due to an impurity that falls into it, as long as its colour, taste and smell remain unchanged. If they do change, then only that water that has changed becomes impure, and not that which flows before or after it. The same ruling applies to rainwater during a downpour, and water in the public bath that is connected to a source.[1]

(b) stagnant water – such as in pits and containers. If the amount is as much as a *kurr* – which is 1200 Iraqi *raṭl*,[2] or the amount that a container of three and a half hand spans in height, width and length can hold[3] – then the water does not become impure when an impurity falls in, as long as one of its [three] qualities remain unchanged. If they do change, then it becomes impure, and can be purified by adding a *kurr* amount of water to it all at once until the changes cease to exist. If the amount is less than a *kurr*, then it becomes impure upon contact with an impurity, even if its qualities do not change. It can be purified by adding a *kurr* amount of water to it all at once.

(c) well-water – which, if its qualities change due to an impurity falling in, becomes pure again when it is extracted and the change ceases. If not, then it is considered to remain in its original state of purity.

1 The water source would need to be flowing water or a large container exceeding the amount of *kurr*.

2 This is equivalent to an imperial British pound or 449.28 grams.

3 376 litres.

الباب الأوّل: في المياه

الماء ضربان: مُطلَق ومُضاف، فالمطلق ما يستحق إطلاق اسم الماء عليه ولا يمكن سلبه عنه، والمضاف بخلافه. فالمطلق طاهِرٌ مُطَهِّرٌ.

وباعتبار وقوع النجاسة فيه ينقسم أقساماً:

الأول: الجاري، كمياه الأنهار، ولا ينجس لما [يقع] فيه من النجاسة ما لم يتغيّر لونه أو طعمه أو ريحه بها، فإن تغيّر نجس المتغيّر خاصةً دون ما قبله وبعده. وحكم ماء الغيث حال نزوله وماء الحمام إذا كانت له مادةٌ حكمه.

الثاني: الواقف، كمياه الحياض والأواني، إن كان مقداره كُرًا - وحدّه ألف ومائتا رطل بالعراقي، أو كان كل واحد من طوله وعرضه وعمقه ثلاثة أشبار ونصفا بشبر مستوى الخلقة - لم ينجس بوقوع النجاسة فيه ما لم تغيّر أحد أوصافه. فإن غيّرته نجس، ويطهر بإلقاء كرّ دفعةً عليه حتى يزول تغيّره. وإن كان أقلّ من كرّ نجس بوقوع النجاسة فيه وإن لم تغيّر أوصافه، ويطهر بإلقاء الكرّ دفعةً عليه.

الثالث: ماء البئر، إن تغيّر بوقوع النجاسة فيه نجس، وطهر بزوال التغيّر بالنزح وإلّا فهو على أصل الطهارة.

A group of our scholars have ruled that well-water becomes impure when an impurity falls into it even if the water does not change in quality, and that it is obligatory to extract all of it if an inebriating drink, beer, semen, menstrual, postpartum or other female blood drips therein, or if an animal dies inside it. If this is not possible, then four men working in pairs alternately must keep draining it for a full day. If a camel, cow or the like dies in it, then a *kurr* amount of water must be extracted out. If a human being dies in it, then seventy buckets must be extracted out, and fifty for soft faeces and a considerable quantity of blood,[4] besides the three bloods (seen by women). Forty buckets are to be extracted if a dog, bird of prey, pig, fox or rabbit dies, or if a man urinates therein. Ten buckets are to be drawn out for dry faeces and little blood, and seven if a bird or mouse dies and decomposes therein; or if a child has urinated in it, or a person in the state of *janāba* (ritual impurity after sexual intercourse or discharge of semen) has washed himself or performed major ablution (*ghusl*) in it, or if a dog fell in but came out alive. Seven buckets are to be drawn out for chicken droppings, three for a mouse or snake, and a single bucket for a sparrow and similar species, or if a suckling baby, not yet on solid food, were to pass urine therein. In my opinion, these are all recommendations.

(d) water leftover from that previously consumed by animals. They are all pure except for the leftovers of a dog, pig or disbeliever.[5]

As for mixed water, it is that which has been extracted from other entities[6] or mixed and diluted with other substances such that it is no longer just water without added considerations, such as rose water or broth. It becomes impure as soon as an impurity comes into contact with it, whether it is a small or large amount. This mixed water cannot be used for cleansing an impurity, ritual purification or ablution purposes, even if it is clean.

4 A considerable quantity of blood, as opposed to a small, negligible amount, is the amount of blood which exceeds the area of an average coin.

5 The impurity of the body of a disbeliever being akin to that of a dog or a pig is unique to Imāmī *fiqh*, and other schools of thought do not consider this to be the case. Other manuscripts of this work also add the term *nāṣibī* (one who harbours enmity towards the Ahl al-Bayt) to the list after the word *kāfir*.

6 Such as fruits, vegetables and the like.

وجماعة من أصحابنا حكموا بنجاستها بوقوع النجاسة فيها وإن لم يتغيّر ماؤها ، وأوجبوا نزح الجميع بوقوع المسكّر أو الفقاع أو المني أو دم الحيض أو الاستحاضة أو النفاس فيها أو موت بَعِير فيها . فإن تعذر تراوح أربعة رجال عليها مثنى يوماً ، ونزح كرّ لموت الجمل والبقرة وشبههما ، ونزح سبعين [دلواً] لموت الانسان ، وخمسين للعذرة الذائبة والدم الكثير غير الدماء الثلاثة ، وأربعين لموت الكلب والسنور والخنزير والثعلب والأرنب وبول الرجل ، ونزح عشرة للعذرة اليابسة وللدم القليل ، وسبع لموت الطير والفأرة إذا تفسخت أو انتفخت ، وبول الصبي واغتسال الجنب وخروج الكلب منها حياً ، وخمس لذرق الدجاج ، وثلاثة للفأرة والحية ، ودلو للعصفور وشبهه وبول الرضيع . وعندي أن ذلك - أي كلّها - مُستحبّ .

الرابع : أسعار الحيوان ، كلّها طهارة إلّا الكلب والخنزير والكافر .

وأما المضاف فهو المعتصر من الاجسام ، أو الممتزج بها مزجاً يسلبه الإطلاق كماء الورد والمرق ، وهو ينجس بكلّ ما يقع فيه من النجاسة سواء كان قليلاً أو كثيراً . ولا يجوز رفع الحدث به ولا الخبث وإن كان طاهراً .

Other Rulings

1) Water that has previously been used for ablution remains pure and can be used as a purifying agent.

2) Water that has been used to remove a material impurity[7] becomes impure, whether its qualities have changed due to the impurity or not, except for water used after using the lavatory (*istinjā'*).

3) The dirty wash-water in a public bathroom is impure, unless impurities are known to have been removed.

4) Impure water cannot be used to purify anything, neither in cleansing physical impurities nor ritual impurities, nor must it be used for drinking except in case of necessity.

7 Such as semen, menstrual blood, etc.

مسائل:

الأولى: الماء المستعمل في رفع الحدث طاهِرٌ مُطَهِّرٌ.

الثانية: المستعمل في إزالة النجاسة نجس، سواء تغيّر بالنجاسة أو لم يتغيّر، عدا ماء الاستنجاء.

الثالثة: غسالة الحمام نجسة ما لم يعلم خلوها من النجاسة.

الرابعة: الماء النجس لا يجوز استعماله في الطهارة، ولا إزالة النجاسة، ولا الشرب إلّا مع الضرورة.

Chapter 2: Minor (Wet) Ablution

This chapter is divided into several sections.

Section One: Causes Necessitating Ritual Ablution

Ritual ablution is obligatory after passing urine and faeces, passing wind from the usual place, sleep that overcomes the senses of hearing and sight, any other loss of consciousness, and irregular light bleeding seen by women (*istiḥāḍa*). Nothing else makes it obligatory.

Section Two: Etiquette of Using the Lavatory

It is obligatory for the one wishing to relieve himself to be discreet and cover his private parts. He must not face *qibla* (direction of the Kaʿba in Mecca) nor have his back towards it, both indoors and outdoors.

It is recommended for him to enter the lavatory with his left foot first, to exit with his right, to cover his head therein, to invoke God's name (*Bismillāh...*), to perform *istibrāʾ*,[8] to supplicate whilst entering and exiting, to perform *istinjāʾ*,[9] to leave promptly and to cleanse himself using both stones and water.

It is reprehensible to relieve oneself in the streets, near places where people come to collect water, in places that would upset or embarrass people, under fruit-bearing trees or places that are frequented by the public, and facing the sun or the moon. It is likewise reprehensible to urinate on hard ground, in animal dens, in water or against the direction of the wind. It is also reprehensible to eat, drink, brush teeth, talk – except in the remembrance of God or due to necessity – and to cleanse oneself with the right hand, or even with the left if one is wearing a ring with the name of God, His prophets or one of the Imams engraved on it.

It is obligatory to cleanse oneself after having relieved oneself, by washing the private parts after passing urine, and washing the anus after a bowel movement. In the absence of water, one may suffice with three clean stones, or three pieces of cloth.

8 *Istibrāʾ*: the process of pressing or jerking the male urinary organ up to three times to ensure that no urine is left in the tract.

9 *Istinjāʾ*: the process of wiping and cleaning oneself after opening the bowels with an agent other than water, historically stones or broken pieces of earthenware.

الباب الثاني: في الوضوء

وفيه فصول

الفصل الأول: في موجبه

إنما يجب بخروج البول والغائط والريح من المعتاد والنوم الغالب على السمع وما في معناه والاستحاضة القليلة الدم، ولا يجب بغير ذلك.

الفصل الثاني: في آداب الخلوة

ويجب ستر العورة على طالب الحدث، ويحرم عليه استقبال القبلة واستدبارها في الصحاري والبنيان.

ويستحبّ له تقديم اليسرى عند الدخول إلى الخلاء واليمنى عند الخروج، وتغطية الرأس، والتسمية، والاستبراء، والدعاء عند الدخول والخروج، والاستنجاء، والفراغ، والجمع بين الأحجار والماء.

ويكره الجلوس في الشوارع، والمشارع، ومواضع اللعن، وتحت الأشجار المثمرة، وفيئ النزال، واستقبال الشمس والقمر، والبول في الأرض الصلبة، ومواطن الهوام، وفي الماء، واستقبال الريح به، والأكل، والشرب، والسواك، والكلام إلاّ بذكر الله تعالى أو للضرورة، والاستنجاء باليمين، وباليسار وفيها خاتم عليه اسم الله تعال أو أنبيائه عليهم السلام أو أحد الأئمة عليهم السلام.

ويجب عليه الاستنجاء، وهو غسل مخرج البول معه خاصةً، وغسل مخرج الغائط مع التعدي، وبدونه يجزي ثلاثة أحجار طاهرة، أو ثلاث خرق.

Section Three: Method of Performing Ablution With Water

There are seven components that are obligatory:

(1) The intention: made whilst washing the face or whilst washing the hands at the beginning, which is recommended; and to maintain the intention until one completes it.

(2) Washing the face: from the hairline until the tip of the beard or chin lengthwise, and the span between the thumb and the middle finger breadthwise.

(3) Washing the arms: from the elbows to the fingertips. It is not allowed to wash them in reverse.

(4) Wiping the scalp at the front of the head or the hair: with the wetness leftover [on the hand], without using fresh water, with the minimum span sufficing for it to be called wiping.

(5) Wiping the skin of the feet: from the tips of the toes to the ankles, and it is allowed in reverse.

(6) Order: it must be performed in the order stated.

(7) Succession: the actions must be performed one after the other without delay.

It is recommended to wash the hands before submerging them in the vessel of water [to perform ablution], once after waking from sleep or urinating, twice after having relieved the bowels, and thrice after a state of sexual impurity. It is also recommended to take water from the vessel with the right hand, to invoke God's name, to rinse the mouth and the nose three times, and to perform the washings twice. When washing the arms, it is recommended to start by pouring water on the back of the arm, and for women on the inner arm, and to do the reverse for the second washing; and to supplicate during each action.

Drying oneself afterwards with a towel is reprehensible, as is using assistance.

Appointing someone to perform ablution on one's behalf or on him is not allowed.

الفصل الثالث: في كيفيته

ويجب فيه سبعة أشياء:

١. النية مقارنة لغسل الوجه أو لغسل اليدين المستحب، واستدامتها حكما حتى يفرغ.

٢. وغسل الوجه من قصاص شعر الرأس إلى محادر شعر الذقن طولاً، وما اشتملت عليه الابهام والوسطى.

٣. وغسل اليدين من المرفقين إلى أطراف الأصابع، ولو عكس لم يجز.

٤. ومسح بشرة مقدم الرأس أو شعره بالبلل من غير استئناف ماء جديد، بأقل ما يقع عليه اسم المسح.

٥. ومسح بشرة الرجلين الأصابع إلى الكعبين، ويجوز منكوسا.

٦. والترتيب على ما قُلناه.

٧. والموالاة وهي متابعة الأفعال بعضها لبعض من غير تأخير.

ويستحب فيه غسل اليدين قبل إدخالهما الإناء، مرة من حدث النوم والبول ومرتين من الغائط وثلاثاً من الجنابة، ووضع الإناء على اليمين، والاغتراف بها، والتسمية، والمضمضة، والاستنشاق ثلاثاً، وتثنية الغسلات، ووضع الماء في غسل اليدين على ظهر الذراعين والمرأة على باطنهما، وبالعكس في الثانية، والدعاء عند كل فعل.

ويكره التمندل والاستعانة.

ويحرم التولية.

Other Rulings

1) A person in the state of ritual impurity (*ḥadath*) [i.e. not in a state of minor ablution] is not allowed to touch the script of the Quran.

2) If one is certain of having been in an impure state, but doubts whether he has purified himself through ablution, then he must purify himself, whereas the opposite situation does not obligate purification.

3) He who doubts in his having performed any of the parts of ablution whilst he is still performing it can go back and perform it, then continue. If he doubts after having completed it, he can ignore his doubt.

مسائل:

الأولى: لا يجوز للمحدث مس كتابة القرآن.

الثانية: لو تيقن الحدث وشك في الطهارة تطهر، وبالعكس لا يجب الطهارة.

الثالثة: لو شك في شيء من أفعال الوضوء وهو على حاله أتى به وبما بعده، ولو انصرف لم يلتفت.

Chapter 3: Major Ablution

Major ablution becomes obligatory after sexual impurity (*janāba*), menstruation, irregular female bleeding, postpartum bleeding, touching a dead body after it has gone cold but before it has been washed with a major ablution, and upon death. Recommended major ablutions will follow later.

Section One: Janāba

The state of *janāba* is caused by the discharge of semen however it occurs (e.g. through sexual activity, a wet dream, masturbation etc.), and by penetrative sexual intercourse – such that the head of the penis is no longer visible – be it vaginal or anal, and regardless of whether there has been an ejaculation of semen or not. Major ablution becomes obligatory upon entering the state of *janāba*.

The obligations of major ablution are: to make the intention whilst washing the hands or the head [if one is performing the ablution in sequential parts (*tartībiyyan*)] and to keep that intention until the ablution is finished [i.e. the person cannot change his intention midway], to wash the whole body, manually washing the parts that cannot be reached by pouring water alone. One must begin with the head, then the right-hand side of the body, then the left.[10]

If the body is submerged in water all at once (*irtimās*), then there is no need to perform it in order.

It is recommended to perform *istibrā'* (urethral cleansing) either by passing urine or by manually clearing the urethra of any liquid; to rinse the mouth and the nose; to wash using at least one *ṣāʿ* of water or more,[11] and to manually wash even those parts that the water can reach.

Before having performed the major ablution, one is not allowed to recite any of the four chapters containing obligatory prostrations,[12] nor to touch the script of the Quran, nor an inscription of the name 'Allāh' or the names of His messengers and Imams, nor to enter a mosque unless it is to pass through it – with the exception of the Sacred Mosque or the Prophet's Mosque[13] – nor to place anything therein.

10 This is the first of two methods of performing the major ablution and is known as a sequentially ordered ablution (*al-ghusl al-tartībī*). The other, described immediately after it, is ablution by full submersion into water, and is known as *al-ghusl al-irtimāsī*.

11 *Ṣāʿ*: approximately 3 kilograms.

12 Sūra al-Sajda, Sūra Fuṣṣilat, Sūra al-Najm and Sūra al-ʿAlaq.

13 One cannot even pass through the two mosques in a state of *janāba*.

الباب الثالث: في الغسل

ويجب بالجنابة، والحيض، والاستحاضة، والنفاس، ومس الأموات بعد بردهم وقبل تطهيرهم بالغسل – وللموت. ويستحب لما يأتي. فهاهنا فصول:

الفصل الأول: في الجنابة

وهي تحصل بإنزال الماء مطلقاً، وبالجماع في الفرج حتى تغيب الحشفة – سواء القبل والدبر وإن لم ينزل، ويجب فيه الغسل.

والواجب فيه النية عند غسل اليدين أو الرأس مستدامة الحكم، واستيعاب الجسد بالغسل، وتخليل ما لا يصل إليه الماء إلا به، والبدأة بالرأس ثم بالجانب الأيمن ثم الأيسر.

ويسقط الترتيب مع الارتماس.

ويستحب فيه الاستبراء بالبول، أو الاجتهاد، والمضمضة، والاستنشاق، والغسل بصاع فما زاد، وتخليل ما يصل إليه الماء.

ويحرم عليه قبل الغسل قراءة العزائم، ومس كتابة القرآن، أو شيء عليه اسمه تعالى، أو أسماء أنبيائه أو أحد الأئمة عليهم السلام، ودخول المساجد إلا اجتيازاً – إلا المسجد الحرام ومسجد الرسول ﷺ، ووضع شيء فيها.

It is reprehensible to recite more than seven verses of the Quran, to touch a physical copy of the Quran (*muṣḥaf*), to eat or drink except after having rinsed the mouth and nose, to sleep except after having performed the minor ablution (*wuḍūʾ*) and to dye one's hair whilst in that state.

If one does anything to breach the state of purity whilst performing the major ablution,[14] then he must start afresh.

Section Two: Menstruation

This is usually dark, thick blood that gushes out with a burning sensation and warmth.

The blood that a woman sees after the age of fifty (if she is not a Qurayshi nor a Nabaṭī[15]) or sixty (if she is), and before the age of nine, is not considered menstrual blood.

Menstrual blood must flow for a minimum of three consecutive days, and last no more than ten days. The days in between are termed a cycle.

If menstrual blood is seen for more than ten days and the woman has a regular cycle, then she resorts to the days of her cycle. If a woman sees blood for the first time, or has irregular cycles, she acts per the signs characteristic of the blood.[16] If, on the other hand, she is unable to distinguish between the blood of *ḥayḍ* and the blood of *istiḥāḍa*, then she resorts to the cycles of her immediate female relatives. If she has none, or they differ from each other's, then to the cycles of her peers, and if she has none or they differ from each other's, then she considers her cycle to be seven days each month, or three days in one month and ten in the next, alternately. The woman with irregular cycles similarly considers seven days to be her menstrual cycle, or three and ten days alternately every two months.

A menstruating woman is forbidden from entering mosques, unless it is to pass through them, except for the two sacred mosques; as well as from reciting the chapters with obligatory prostrations, and from touching the script of the Quran.

Her husband is forbidden from having intercourse with her, and if he does then he is punished and made to pay an atonement as a recommendation.

14 Such as passing wind, urine or faeces.

15 These are women whose lineage either traces back to the Qurayshi tribe from her father's side, or to the Nabaṭī tribe who used to reside in a place called Nabaṭ between Kufa and Basra.

16 That is, if her menstrual blood and her irregular bleeding have distinct characteristics she will act in accordance with those.

ويكره قراءة ما زاد على سبع آيات، ومس المصحف، والأكل، والشرب إلا بعد المضمضة والاستنشاق، والنوم إلا بعد الوضوء، والخضاب.

ولو أحدث في أثناء الغسل أعاد.

الفصل الثاني: في الحيض

وهو في الأغلب دم أسود غليظ يخرج بحرقة وحرارة.

وما تراه بعد خمسين سنة إن لم تكن قرشية ولا نبطية، أو بعد ستين سنة إن كانت أحدهما، أو قبل تسع سنين مطلقاً، فليس بحيض.

وأقله ثلاثة أيام متواليات وأكثره عشرة أيام، وما بينهما بحسب العادة.

ولو تجاوز الدم العشرة فإن كانت المرأة ذات عادة مستقرة رجعت إليها، وإن كانت مبتدئة أو مضطربة ولها تميز عملت عليه، ولو فقدته رجعت المبتدئة إلى عادة أهلها. فإن فقدن فإلى أقرانها. فإن فقدن أو كن مختلفات تحيضت في كل شهر سبعة أيام، أو ثلاثة من الأول وعشرة من الثاني. والمضطربة تتحيض بالسبعة أو الثلاثة والعشرة في الشهرين.

ويحرم عليها دخول المساجد إلا اجتيازا، عدا المسجدين، وقراءة العزائم ومسّ كتابة القرآن.

ويحرم على زوجها وطؤها، ولو وطأ عزّر وكفّر مستحباً.

She is not obligated to perform the daily prayer, nor to fast, nor to purify herself ritually from impurities that necessitate ritual ablution (*wuḍūʾ*). She cannot circumambulate the Kaʿba, nor retreat in a mosque (*iʿtikāf*), and it is not valid to divorce her during this time. She is not obligated to make up for her missed prayers but must make up for missed fasts.

Reciting any part of the Quran, apart from the chapters with obligatory prostrations is disapproved,[17] as is touching the physical copy of the Quran (*muṣḥaf*) or carrying it, dyeing the hair, having penetrative sexual intercourse before having performed the major ablution, and sexual foreplay involving the parts between the navel and the knees.

It is recommended for her to perform the minor ritual ablution at every prescribed daily prayer and to sit in her usual place of prayer, remembering God for the duration of the prayer.

Section Three: Irregular Bleeding in Women

This type of blood is usually yellowish, cool and light, and is seen by a woman after the days of menstrual or postpartum bleeding, or after the menopause.

If there is a little blood, which is when it stains the cotton but does not seep through, then she must change her pad and renew her *wuḍūʾ* for every prayer. If there is a lot of blood, i.e. it seeps into the cotton but does not flow through, she is obligated in addition to changing her pad to perform major ablution (*ghusl*) before the morning prayer. And if there is even more blood, whereby it drips through the pad, then she is obligated to perform two more major ablutions: one for the midday (*ẓuhr*) and afternoon (*ʿaṣr*) prayers, and one for the evening (*maghrib*) and night (*ʿishāʾ*) prayers, performing both prayers together. The two major ablutions are performed just like that for menstruation.

When she does what we have just mentioned, her status is as one who is ritually pure.

17 Since in the case of the *ʿazāʾim* she is prohibited from reciting them.

ولا ينعقد لها صلاة ولا صوم، ولا طهارة رافعة للحدث، ولا طواف، ولا اعتكاف، ولا يصح طلاقها، ولا يجب عليها قضاء الصلاة، ويجب قضاء الصوم.

ويكره لها قراءة ما عدا العزائم، ومسّ المصحف وحمله، والخضاب، والوطيء قبل الغسل، والاستمتاع منها بما بين السرة والركبة.

ويستحب لها الوضوء لكل صلاة فريضة، والجلوس في مصلاها ذاكرة بقدر صلاتها.

الفصل الثالث: في الاستحاضة

وهو في الأغلب دم أصفر بارد رقيق تراه بعد أيام الحيض، أو أيام النفاس، أو بعد اليأس.

فإن كان الدم قليلا – وهو أن يظهر على القطنة ولا يغمسها – وجب عليها تغيير القطنة وتجديد الوضوء لكل صلاة. وإن كان كثيراً – وهو أن يغمس القطنة ولا يسيل – وجب عليها مع ذلك تغيير الخرقة والغسل لصلاة الغداة. وإن كان أكثر منه – وهو أن يسيل – وجب عليها مع ذلك غسلان: غسل للظهر والعصر تجمع بينهما، وغسل للمغرب والعشاء تجمع بينهما. وغسلها كغسل الحائض.

وإذا فعلت ما قلناه صارت بحكم الطاهر.

Section Four: Postpartum Bleeding

This is the bleeding that a woman experiences straight after or accompanying childbirth.

There is no minimum number of days that it lasts, but it does not exceed ten days. The same rulings apply to her as the menstruating woman.

Section Five: Performing Major Ablution on The Deceased

There are five parts to this discussion:

1. At the Point of Death:

The dying person must be made to face the *qibla* by lying him on his back, with his face and the soles of his feet facing it.

It is recommended to prompt him to declare the two testimonies of faith, to profess his belief in the Imams and in the awaited saviour; to recite the Quran, to close his eyes and his mouth, to open out his hands, to inform the believers, and to make swift preparations for his funeral unless there is any discrepancy.

It is disapproved for a person in the state of menstruation or sexual impurity to be near him at that time, or for any metal to be placed on his abdomen.

2. The Major Ablution:

The body must be washed three times: first with cedar berry water, secondly with camphor water, and thirdly with pure water, just like the major ablution.

If it is feared that his flesh will disintegrate as a result of the washing, then dry ablution (*tayammum*) can be performed instead.

It is recommended for the person washing to stand on the right side of the body, to press the belly gently during the first two washings [so as to empty the deceased's bowels of any stored faeces], to remember God, to seek forgiveness, to ensure that the water drains into a hole, to wash the body under a shelter, to make it face the *qibla*, to wash the head and body with the foam of the cedar berry water and the private parts with *ushnān* (soda plant powder used as a soap substitute), and to perform a minor ablution on the deceased body [before washing it with the three aforementioned washings]. It is disapproved to sit the body upright, to clip its fingernails or shave its hair.

الفصل الرابع: في النفاس

وهو الدم الذي تراه عقيب الولادة أو معها .

ولا حد لأقلّه ، وأكثره عشرة أيام. وحكمها حكم الحائض في جميع الأحكام.

الفصل الخامس: في غسل الأموات

ومباحثه خمسة:

المبحث الأول: الاحتضار

يجب فيه استقبال الميت بالقبلة بأن يلقى على ظهره ويجعل وجهه وباطن رجليه إليها .

ويستحب تلقينه الشهادتين ، والاقرار بالأئمة عليهم السلام ، وكلمات الفرج ، وقراءة القرآن ، وتغميض عينيه وإطباق فيه ، ومد يديه ، وإعلام المؤمنين ، وتعجيل أمره إلا مع الاشتباه.

ويكره أن يحضره جنب أو حائض ، أو يجعل على بطنه حديد .

المبحث الثاني: الغسل

ويجب تغسيله ثلاث مرات : الأولى بماء السدر ، والثانية بماء الكافور ، والثالثة القراح كغسل الجنابة.

ولو خيف تناثر لحمه يمّم.

ويستحب وقوف الغاسل على يمينه ، وغمز بطنه في الغسلتين الأولتين ، والذكر ، والاستغفار ، وإرسال الماء إلى حفيرة ، وتغسيله تحت سقف ، واستقبال القبلة به ، وغسل رأسه وجسده برغوة السدر ، وفرجه بالأشنان ، وأن يوضّأ . ويكره إقعاده ، وقصّ أظفاره ، وترجيل شعره .

3. Shrouding:

It is obligatory to shroud the body in three pieces of cloth: a loincloth [that covers the region between the belly button to the knees], a tunic [that covers the region from the shoulders to half way down the lower leg] and a sheet [that covers the entire body from head to toe], and to apply camphor to the parts of the body that touch the ground in prostration.

For a deceased male, it is recommended to add a Yemeni covering without any gold or silver embroidery, another piece to cover his thighs and a turban to wrap around his head, with the tail end left untucked. For a female, it is recommended to add another sheet to cover her breasts, a *namaṭ* (a woollen shawl with contrasting stripes to cover the whole body over the shroud), and a scarf instead of a turban.

It is recommended also that the shroud should be made of cotton and perfumed with *dharīra* (a fragrant powder), and two branches of a date palm tree [placed on either side]. Then the deceased person's name should be written on the cover sheet, the tunic, the loincloth and the two palm branches, and that he has declared the two testimonies of faith, along with the names of the Imams. The camphor must weigh the equivalent of 13.3 dirhams' value.[18]

It is abominable to shroud the body with black garments, to insert camphor into the ears and eyes, and to fragrance shrouds by burning incense.

4. Prayers for the Deceased:

It is obligatory to pray over every deceased Muslim, or one who enjoys the same rights as a Muslim over the age of six years for a deceased child, be they male or female, free or slave.

The prayer is recommended for a child under the age of six.

Those whose right to inherit from the deceased is greatest have the greater right to perform the prayer, and the husband is more entitled [to pray over his wife] than any others [from her family such as her father]. A Hāshimī (descendant of Banū Hāshim) is the worthiest of leading the prayer if the next of kin requests him to do so, and it is recommended for him to do so, provided certain conditions are met. And the Imam is the one who is most worthy of leading the prayer. As for its necessity, it is a communal obligation.

18 One dirham equals 2.420 grams of silver.

المبحث الثالث: التكفين

ويجب تكفينه في ثلاثة أثواب: مئزر، وقميص، وأزار، ومساس مساجده بالكافور.

ويستحب أن يزاد الرجل حبرة غير مطرزة بالذهب، وخرقة لفخذيه، وعمامة يعمم بها محنكاً، ويزاد المرأة لفافة أخرى لثدييها، ونمطاً، وتعوض عن العمامة بقناع.

والتكفين بالقطن، وتطييبه بالذريرة، وجريدتان من النخل، وأن يكتب على اللفافة والقميص والإزار والجريدتين اسمه وأنه يشهد الشهادتين و (أسماء) الأئمة عليهم السلام، وأن يكون الكافور ثلاثة عشر درهماً وثلثاً.

ويكره التكفين في السواد، وجعل الكافور في سمعه وبصره، وتجمير الأكفان.

المبحث الرابع: الصلاة عليه

وهي تجب على كل ميت مسلم أو بحكمه - ممن بلغ ست سنين من أولادهم - ذكرا كان أو أنثى، حرا أو عبدا.

وتستحب على من نقص سنه عن ذلك.

وأولاهم بالصلاة عليه أولاهم بالميراث، والزوج أولى من غيره، والهاشمي أحق إذا قدمه الولي - ويستحب له تقديمه مع الشرائط - والإمام أولى من غيره. ووجوبها على الكفاية.

Method: To make the intention, then to recite "*Allāhu akbar*" (God is greater) 5 times, with supplications between each one, the best of these being: to recite the first *Allāhu akbar,* followed by the two testimonies of faith; then the second followed by invoking blessings on the Prophet and his household; then the third followed by supplicating for all the believers; then the fourth followed by supplicating for the deceased person if he's a believer, damning him if he was a hypocrite, supplicating for him if he was weak [in faith]; in the case of a child, asking God to make him a forerunner for his parents into Paradise, and asking God to raise him with those whom he adhered to in the case of someone he does not know. Then he ends by reciting "*Allāhu akbar*" a fifth time. There is no recitation of chapters from the Quran in the prayer, nor concluding with blessings of peace (*taslīm*).

Whilst ritual purification is recommended for the prayer, it is not obligatory.

Other Rulings

1) Prayers can only be performed after washing and shrouding the body.

2) It is disapproved to perform the prayer for the deceased twice.[19]

3) If the prayer was not performed for a deceased person, then it must be performed at his graveside within a day and a night.

4) It is recommended for the prayer leader to stand in line with the deceased male's waist, and the deceased female's chest. If there are two bodies of each gender, he must stand closest to the man.

5) The head of the deceased must be placed towards the right side of the prayer leader.

5. Burial:

The minimum obligation is to hide the body in the ground to make it inaccessible to animals and predators, and to suppress its smell from reaching people. The body must be buried on its right side facing the *qibla*.

19 That is, it is not recommended for a single individual to pray over the deceased person's body twice.

وكيفيتها : أن يكبّر بعد النية خمساً بينها أدعية، أفضلها أن يكبّر ويتشهّد الشهادتين، ثم يصلّي على النبي وآله عليهم السلام بعد الثانية، ثم يدعو للمؤمنين بعد الثالثة، ثم يدعو للميّت إن كان مؤمناً وعليه إن كان منافقاً وبدعاء المستضعفين إن كان منهم في الرابعة، ولو كان طفلاً سأل اللّه تعالى أن يجعله لأبويه فرطاً ، وإن لم يعرفه سأل اللّه تعالى أن يحشره مع من يتولّاه، ثم يكبّر الخامسة وينصرف - بعد رفع الجنازة - ولا قراءة فيها ولا تسليم.

ويستحب فيها الطهارة وليست شرطاً .

مسائل:

الأولى: لا يصلّى عليه إلا بعد تغسيله وتكفينه.

الثانية: يكره الصلاة على الجنازة مرّتين.

الثالثة: لو لم يصلّ على الميّت صلّى على قبره يوماً وليلة.

الرابعة: يستحب أن يقف الإمام عند وسط الرجل وصدر المرأة، ولو اتفقا جعل الرجل مما يليه.

الخامسة: يجب أن يجعل رأس الميّت عن يمين المصلّي.

المبحث الخامس: الدفن

والواجب ستره في الأرض عن الهوام والسباع، وطم رائحته عن الناس على جانبه الأيمن موجهاً إلى القبلة.

It is recommended to walk behind the bier while it is being carried, or to have two pallbearers carrying it on either side, or four, one at each corner. The body should be placed at the foot of the grave first if it is a male,[20] and in front of it closest to the direction of the *qibla* if it is a female. A male body should be lowered to the grave head-first, while the body of a female should be laid to rest on the ground on her back [without either the head or the feet being lowered first]. It is recommended that the grave be dug as deep as the full height of a person or up to his collarbone, and a proper grave is preferable to an existing hole in the ground, which should be at least the height of a seated man. It is recommended to be in constant remembrance (of God) whilst taking and lowering the body into the grave, for the next of kin to descend into the grave barefoot, with his head uncovered and buttons undone, to place the body in. The knots of the shroud should be undone, and the cheek placed on the soil. It is recommended that some soil from Karbala be buried with the body. It is recommended to prompt the deceased (*talqīn*) with the recitation of the two testimonies of faith and the names of the Imams as it is being lowered in; to cover the body with bricks; and for the next of kin to exit the grave from the foot end of the grave. Those attending the burial should push soil into the grave with the back of their hands. It should be filled with its own earth and raised above the ground by a span of four fingers. Then water should be poured onto the grave, all around it. Then those attending should place their hands on the grave and ask God to have mercy on the deceased. After everyone has left, the next of kin should again recite the prompting of the testimony of faith (*talqīn*).

It is discouraged for family members to descend into the grave except when the deceased is a female, and for them to cover the grave with soil. It is abominable to put an enclosure around the grave unnecessarily or to plaster it or decorate it. It is also abominable to bury two bodies in the same grave, or to transfer a body from the grave to another location, except for holy places of visitation.

If a person dies at sea, the body must be weighed down and sunk into the sea.

Non-Muslims must not be buried in Muslim burial grounds, except for a *dhimmī*[21] woman who dies carrying the child of a Muslim man inside her, but she must be buried with her back to the *qibla*.

20 That is, if the body is that of a deceased male, then it should be placed on the ground near the end of the grave at which the body's feet will be once it has been lowered into the ground.

21 *Dhimmī*: protected non-Muslim living in Muslim lands

ويستحب إتباع الجنازة أو مع أحد جانبيها ، وتربيعها ، ووضعها عند القبر إن كان رجلاً ، وقدامه مما يلى القبلة إن كان امرأة ، وأخذ الرجل من قبل رأسه والمرأة عرضاً ، وحفر القبر قدر قامة أو إلى الترقوة ، واللحد أفضل من الشق بقدر ما يجلس فيه الجالس ، والذكر عند تناوله وعند وضعه في اللحد ، والتحفي ، وحل الأزرار ، وكشف الرأس ، وحل عقد الأكفان ، ووضع خده على التراب ، ووضع شيء من التربة معه ، وتلقينه الشهادتين والإقرار بالأئمة (عليهم السلام) ، وشرج اللبن ، والخروج من قبل رجليه ، وإهالة الحاضرين التراب بظهور الأكف ، وطم القبر ، وتربيعه ، وصب الماء عليه دوراً ، ووضع اليد عليه ، والترحّم ، وتلقين الولي بعد الانصراف .

ويكره نزول ذي الرحم ، وإهالته التراب ، وفرش القبر بالساج من غير حاجة وتجصيصه ، وتجديده ، ودفن ميتين في قبر واحد ، ونقله إلى غير المشاهد .

والميت في البحر يثقل ويرمى فيه .

ولا يدفن في مقبرة المسلمين غيرهم ، إلا الذميّة الحامل من المسلم فيستدبر بها القبلة .

Other Rulings

1) A martyr does not need to be washed or shrouded. Rather he is buried in his own clothes after performing the funeral prayers for him.

2) The same rules apply to the dismembered limbs of a deceased person as they would to the whole body, as long as there is bone in them they are to be washed, shrouded and buried, as would a miscarried foetus of four months and above. If there is no bone, then the piece of flesh is to be wrapped in a cloth and buried, as is the miscarried foetus of less than four months.

3) The shroud should be paid for from the deceased's own assets before any dues are paid from them. The husband is responsible for paying for the shroud of his wife, even if she was a woman of her own means.

4) The same rulings apply to an illegitimate person as one born in wedlock, apart from camphor, which must not be used on him.

5) Whoever touches a dead body after it has gone cold and before it has been washed with the major ablution, whether he touches the body itself or a dismembered limb, is obligated to perform his own major ablution (*ghusl mass al-mayyit*). If one touches a dismembered part devoid of bone or a corpse that is not human, he need only wash his hands.

مسائل:

الأولى: الشهيد لا يغسل ولا يكفن بل يصلّى عليه وهو في ثيابه.

الثانية: صدر الميّت كالميت في أحكامه، وغيره إن كان فيه عظم غسل وكفن ودفن، وكذا السقط لأربعة أشهر، وإلاّ دفن بعد لفه في خرقة، وكذا السقط لدون أربعة.

الثالثة: يؤخذ الكفن من أصل التركة قبل الديون، وكفن المرأة على زوجها وإن كانت موسرة.

الرابعة: الحرام كالحلال إلاّ في الكافور فلا يقربه.

الخامسة: من مسّ ميّتاً من الناس بعد برده بالموت وقبل تطهيره بالغسل أو مسّ قطعة منه فيها عظم قطعت من حي أو ميت وجب عليه الغسل، ولو خلت القطعة من عظم أو كان الميّت من غير الناس غسل يده خاصّةً.

Section Six: Recommended Major Ablutions

They are: The Friday *ghusl*, whose timing is from sunrise until noon; the *ghusl* on the eve of the first night of the month of Ramadan, the middle of the same month, the seventeenth, nineteenth, twenty-first, and twenty-third of Ramadan; the *ghusl* on the eve of Eid al-Fiṭr; during the day on the two Eids;[22] the eve of the middle of Rajab; the eve of the middle of Shaʿbān; the Day of Mabʿath;[23] the Day of Ghadīr;[24] the Day of Mubāhila;[25] the *ghusl* before entering the state of *iḥrām* [before performing the pilgrimage] and before visiting the Prophet and the Imams; the *ghusl* before making up the missed Prayer of Signs (*ṣalāt al-āyāt*) for one who intentionally did not pray it during a total solar eclipse; the *ghusl* for repentance; the *ghusl*s before the prayer for fulfilment of needs and the prayer when soliciting a good outcome (*istikhāra*); the *ghusl* before entering the Sanctuary of Mecca; before entering the Sacred Mosque in Mecca and the Kaʿba; before entering Medina and the Prophet's mosque; and the *ghusl* given to a new-born.

22 Eid al-Fiṭr and Eid al-Aḍḥā.

23 This is twenty-seventh day of Rajab, and commemorates the start of the Prophet's mission.

24 18th Dhū l-Ḥijja.

25 24th Dhū l-Ḥijja.

الفصل السادس: في الأغسال المسنونة

وهي: غسل يوم الجمعة ووقته من طلوع الفجر إلى الزوال، وأول ليلة من رمضان، وليلة النصف منه، وسبع عشرة، وتسع عشرة، وإحدى وعشرين، وثلاث وعشرين، وليلة الفطر، ويومي العيدين، وليلة نصف رجب، وليلة نصف شعبان، ويوم مبعث، والغدير، والمباهلة، وغسل الإحرام، وزيارة النبي والأئمة عليهم السلام، وقضاء الكسوف مع الترك عمداً واحتراق القرص كله، وغسل التوبة، وصلاة الحاجة، والاستخارة، ودخول الحرم، والمسجد الحرام والكعبة، والمدينة، ومسجد النبي ﷺ وغسل المولود.

Chapter 4: Dry Ablution

Dry ablution as a substitute for ritual ablution becomes obligatory when water is unavailable, or cannot be used due to an illness, extreme cold, fear or thirst, or if there is no vessel or instrument available to access it, or if the high cost of water would cause immediate financial loss, or even if there is no loss to be incurred as such but it is unreasonably high.

It is obligatory to search for water first, as far as the area covered by an arrow's throw on all four sides on a rocky plain, or two arrows' throw in a desert land.

If a person needs to remove an impurity from himself and there is no surplus water, he must use the water to remove it, and perform dry ablution [instead of his ritual ablution].

Dry ablution can only be done using clean soil. One is allowed to perform it with chalky, limey or stony soil.

It is abominable to use manure and sand, but they may be used if there is nothing other than mud available with which to perform dry ablution.

Method: One must strike the earth with both palms whilst making the intention, then shake off the hands [of excess soil] and wipe them over the face, from the hairline until the tip of the nose. Then he must wipe the back of the right hand with his left palm, from the wrist to the fingertips, and then the back of the left hand with the right palm.

If the dry ablution substitutes a major ablution (*ghusl*), then he must strike the earth twice: once for the face, and once more for the two hands, observing the correct order.

The same things that nullify ritual ablution (*wuḍūʾ*) nullify dry ablution, in addition to the availability of water and the possibility of using it. If water becomes available before he has started his prayer, he must use it for purification, but if it becomes available during his prayer, he can complete it and does not need to repeat whatever he had prayed previously with dry ablution.

Dry ablution is not valid when performed before the time for prayer has set in, and is only allowed to be performed as a last resort,[26] and there are two opinions regarding its permissibility when there is ample time left.

26 That is, when there is little time left for the prayer to be performed.

الباب الرابع: في التيمّم

ويجب عند فقد الماء، أو تعذر استعماله لمرض أو برد أو خوف عطش أو عدم آلة يتوصل بها إليه أو ثمن يضر في الحال، ولو لم يضره وجب وإن كثر.

ويجب الطلب غلوة سهم في الحزنة وسهمين في السهلة من جوانبه الأربع.

ولو كان عليه نجاسة ولا يفضل الماء عن إزالتها تيمّم وأزالها به.

ولا يصح إلا بالتراب الخالص، ويجوز بأرض النورة والجص والحجر.

ويكره بالسبخة والرمل، ولو لم يجد إلا الوحل تيمّم به.

وكيفيته: أن يضرب بيديه على الأرض ناوياً، وينفضّهما، ويمسح بهما وجهه – من قصاص الشعر إلى طرف الأنف –، ثم يمسح ظهر كفه الأيمن ببطن الأيسر، ثم ظهر الأيسر ببطن الأيمن – من الزند إلى طرف الأصابع.

ولو كان بدلاً من الغسل ضرب ضربتين: ضربة للوجه وأخرى لليدين ويجب الترتيب.

وينقضه كل نواقض الطهارة، ويزيد [عليها] وجود الماء مع التمكن من استعماله، ولو وجده قبل شروع الصلاة تطهر، ولو وجده في الأثناء أتم صلاته، ولا يعيد ما صلى بتيمّمه.

ولا يجوز قبل دخول الوقت، ويجوز مع الضيق، وفي حال السعة قولان.

Chapter 5: Impurities

There are ten impurities: the urine and faeces of creatures whose flesh is impermissible to eat and whose blood gushes out with force; the semen, remains and blood of all creatures whose blood gushes out, dogs, pigs, disbelievers, intoxicants[27] and beer.

It is obligatory to remove these impurities from one's body and clothing for the purpose of prayer, except for blood which is less than the area of an average coin, not including the three types of blood,[28] or the blood of something intrinsically impure [such as a dog or a pig].

The blood on boils and wounds, even if bleeding, and that which can only be removed with great difficulty is excused, as well as the impurity on items of clothing that one can pray without, such as belts, socks or headgear.

In the case of a woman looking after a baby, who has only one garment,[29] it would suffice for her to wash it once each day.

When one knows the exact location of the impurity on his clothing, he must remove it. If he does not know its location, he must wash the whole garment.

If the impure garment gets muddled up with another one,[30] he must pray wearing each of them in turn.

If one is unable to wash the garment, then he can pray without clothes on if he has nothing else to wear. If he dreads the cold, then he may pray with them on, and does not have to repeat the prayer later.

If one intentionally prays with an impurity on him, he must repeat the prayer within its timeframe [if there is still time to perform the prayer] or afterwards [if the time of prayer has lapsed]. If he forgot about the impurity and remembered during his prayer, he must repeat it within its time frame. If, however, he remembered only after having completed his prayer, he does not have to repeat it.

The sun's rays purify any urine or other impurity on the ground, that dries because of it, including buildings, mats and flooring. The earth purifies the soles of shoes.[31]

27 That is, wines and spirits

28 Since these have their own rulings.

29 That is, a garment that becomes impure during the day due to the child.

30 Such that it is impossible to determine which one of them is impure.

31 Provided the earth is clean and the shoes are dry.

الباب الخامس: في النجاسات

وهي عشرة: البول والغائط مما لا يؤكل لحمه من ذي النفس السائلة، والمني من ذي النفس السائلة مطلقاً، وكذا الميتة والدم منه، والكلب والخنزير، والكافر، والمسكر، والفقاع.

ويجب إزالتها عن الثوب والبدن للصلاة - عدا ما نقص عن الدرهم البغلي من الدم، غير الدماء الثلاثة ودم نجس العين.

وعفي عن دم القروح والجروح مع السيلان ومشقة الإزالة، وعن نجاسة ما لا يتم الصلاة فيه كالتكة والجورب والقلنسوة.

ويكفي المربية للصبي إذا لم يكن لها إلا ثوب واحد غسله في اليوم مرة واحدة.

ويجب إزالة النجاسة مع علم موضعها، ولو جهل غسل جميع الثوب.

ولو اشتبه الثوب بغيره صلى في كل واحد منهما مرة.

ولو لم يتمكن من غسل الثوب صلى عرياناً إذا لم يجد غيره، ولو خاف البرد صلى فيه، ولا إعادة.

ولو صلى في النجس مع العلم أعاد في الوقت وخارجه، ولو نسي حالة الصلاة أعاد في الوقت، ولو لم يتقدم العلم حتى فرغ فلا إعادة.

وتطهر الشمس ما تجففه من البول وغيره على الأرض، والأبنية، والحصر والبواري.

والأرض باطن الخف.

If a container [e.g. cooking utensils] becomes impure, it must be washed: thrice, of a dog's saliva after rubbing it with soil first; seven times of contact with a pig; thrice of wine or a mouse, though seven times is better; and from all other impurities once, or preferably thrice.

The use of gold and silver utensils for eating and other things is forbidden, and the use of gold or silver-plated utensils is disapproved.

The utensils used by polytheists are considered pure so long as their having touched them with wetness has not been ascertained with certainty.

ولو نجس الإناء وجب غسله، فيغسل من ولوغ الكلب ثلاثاً أولاهن بالتراب، ومن الخنزير سبعاً، ومن الخمر والفأرة ثلاثاً والسبع أفضل، ومن غير ذلك مرة والثلاث أفضل.

ويحرم استعمال أواني الذهب والفضة في الأكل وغيره، ويكره المفضض.

وأواني المشركين طاهرة ما لم يعلم مباشرتهم لها برطوبة.

The Book of Prayer

كتاب الصلاة

Chapter 1: Preliminaries

This chapter contains several sections.

Section One: Number of Prayers

There are five obligatory daily prayers: the noon prayer (*ẓuhr*), which consists for four units whilst in one's home town or two when travelling; the afternoon prayer (*ʿaṣr*), which is the same; the evening prayer (*maghrib*) which has three units in both cases; the night prayer (*al-ʿishāʾ*), which is the same as the noon prayer; and the dawn prayer (*fajr*) which has two units in both cases.

The supererogatory prayers consist of thirty-four units in total when at home: eight units before the noon prayer, eight units before the afternoon prayer, four units after the evening prayer, and two units prayed seated after the night prayer, which are counted as one unit. Then eight units of prayer for the vigil prayer in the middle of the night (*ṣalāt al-layl*), followed by two even units (*ṣalāt al-shafʿ*) and one odd unit (*ṣalāt al-witr*), and two units before the prayer at dawn.

When travelling, there is no need to perform the daytime supererogatory prayers, nor the seated prayer after the night prayer.

The other obligatory prayers are: the Friday prayer, the two Eid prayers, the prayer during an eclipse or an earthquake, the prayer of signs (*ṣalāt al-āyāt*), the prayer after circumambulation of the Kaʿba, the prayer over the deceased, prayers made as an oath to God and their like.

All other prayers are recommendations.

وفيه أبواب:

الباب الأول: في المقدمات

وفيه فصول

الفصل الأول: في أعدادها

الصلاة الواجبة في كل يوم وليلة خمس: الظهر أربع ركعات في الحضر وفي السفر ركعتان، والعصر كذلك، والمغرب ثلاث فيهما، والعشاء كالظهر، والصبح ركعتان فيهما.

والنوافل اليومية أربع وثلاثون في الحضر: ثمان ركعات قبل الظهر، وثمان بعدها للعصر، وأربع ركعات بعد المغرب، وركعتان من جلوس بعد العشاء تعدان ركعة، وثمان ركعات صلاة الليل، وركعتا الشفع، وركعة الوتر، وركعتا الفجر.

وتسقط في السفر نوافل النهار والوتيرة خاصة.

ومن الصلوات الواجبة: الجمعة، والعيدان، والكسوف، والزلزلة، والآيات، والطواف، والجنائز، والمنذور، وشبهه.

وما عدا ذلك مسنون.

Section Two: Timings of the Prayer

When the sun begins its descent past its zenith, the time for the noon prayer (*ẓuhr*) sets in and lasts for the length of time it would take to pray four units. When this elapses, the remaining time is a combined time for the noon and afternoon (*ʿaṣr*) prayer until just enough time remains before sunset to pray the four units for the afternoon prayer, which is exclusively for that prayer. When the sun has set to the point where all redness has disappeared from the eastern horizon, the time for the evening prayer (*maghrib*) sets in and lasts for the length of time it would take to perform it. Then it is a combined time for both the evening and night prayer (*ʿishāʾ*) until just enough time remains before the middle of the night to pray four units, which is exclusively for the *ʿishāʾ* prayer. When the second [or true] dawn appears, the time for the dawn (*fajr*) prayer sets in and lasts until sunrise.

The timings of the supererogatory prayers (*nawāfil*) are as follows: the one preceding the noon prayer should be prayed from the time the sun begins its descent until shadows are as long as their objects. If one has not prayed any of the supererogatory units by then, he must attend to his obligatory prayer. If, however, he has performed even one supererogatory unit, he may complete it quickly followed immediately by the obligatory prayer. The time of the supererogatory prayers for *ʿaṣr* sets in as soon as one has completed the *ẓuhr* prayer, and lasts until the time when shadows are twice the length of their objects. If he has started a supererogatory unit by then, he may finish it quickly then pray the obligatory thereafter. But if not, then he goes on to his obligatory prayer. The time for the evening supererogatory prayer is straight after *maghrib* until any redness disappears from the western horizon. If he has not prayed before it disappears, he moves on to pray the *ʿishāʾ* prayer. The time for the *ʿishāʾ* supererogatory prayer is straight after *ʿishāʾ* and lasts for the same amount of time. The time for the night vigil prayer sets in after the middle of the night, though to pray it as close as possible to dawn is preferable. If dawn sets in, and he has only prayed four units of it, he may complete the rest quickly, then pray the *fajr* prayer straight after. And if not, then he can make up for them afterwards. The time for the two supererogatory units before the *fajr* prayer sets in immediately after the night vigil prayer, though it is better to pray them once dawn has broken, and it is better to pray them even if dawn has broken, up until the redness appears on the eastern horizon.

الفصل الثاني: في أوقاتها

إذا زالت الشمس دخل وقت الظهر حتى يمضي مقدار أربع ركعات، ثم يشترك الوقت بينها وبين العصر إلى أن يبقى لغروب الشمس مقدار أربع ركعات فيختص بالعصر. وإذا غربت الشمس – وحده غيبوبة الحمرة المشرقية – دخل وقت المغرب إلى أن يمضي مقدار أدائها، ثم يشترك الوقت بينها وبين العشاء إلى أن يبقى لانتصاف الليل مقدار أربع فيختص بالعشاء. وإذا طلع الفجر الثاني دخل وقت الصبح إلى أن تطلع الشمس.

وأما النوافل: فوقت نافلة الظهر إذا زالت الشمس إلى أن يصير ظل كل شيء مثله، فإذا صارت كذلك ولم يصل شيئًا من [النافلة] اشتغل بالفريضة، ولو تلبس بركعة من النافلة زاحم بها الفريضة. ووقت نافلة العصر بعد الظهر إلى أن يصير ظل كل شيء مثليه، ولو خرج وقد تلبس بركعة زاحم بها وإلا فلا، ووقت نافلة المغرب بعدها إلى أن تذهب الحمرة المغربية، ولو ذهبت ولم يكملها اشتغل بالعشاء. ووقت الوتيرة بعد العشاء وتمتد بامتداد وقتها. ووقت نافلة الليل بعد انتصافه، وكلما قرب من الفجر كان أفضل، ولو طلع وقد تلبس بأربع زاحم بها الصبح وإلا قضاها، ووقت ركعتي الفجر عند الفراغ من صلاة الليل، وتأخيرها إلى طلوعه أفضل، وإذا طلع الفجر زاحم بها ولو إلى طلوع الحمرة المشرقية.

Other Rulings

1) The obligatory prayers and their reparative prayers (*qaḍāʾ*) can be prayed at any time, as long as they do not encroach on the prayers stipulated for that particular time frame. The same rule applies to supererogatory prayers, as long as they do not encroach on the obligatory prayers.

2) It is disapproved to begin praying supererogatory prayers at sunrise, sunset, between high noon and the start of the sun's descent (except for Friday), and straight after the morning prayer or the afternoon prayer, unless they are for specific purposes.[1]

3) It is better to perform the prayers as soon as their times set in, except in certain situations,[2] and one is not allowed to delay the prayers beyond their stipulated times or to pray them earlier.

Section Three: The Direction of Prayer

This is the Kaʿba, if possible, or in its direction from afar.

One who is praying in the Kaʿba itself may face whichever of its walls he desires, and if he prays on its roof, then he must stand in such a way that part of it is visible in front of him.

Every community faces the corner corresponding to their locality: the Iraqis face the Iraqi corner, the Yemenis face the Yemeni corner, the North Africans face the North African corner, and the People of Shām (the Levant) face the Shāmī corner.

For those facing the Iraqi corner, the way to find *qibla* is to stand with the dawn parallel to the left shoulder, and for dusk to be parallel with the right shoulder. At noon, once the sun has reached its highest point and begins its descent, it is to face the direction whereby it shines directly on the right eyebrow; and at night, to have the Pole star behind the right shoulder.

If there are no outward signs, then one must pray in all four directions

1 Such as the prayer for visitation of a sacred place, the prayer for the fulfilment of a need, the prayer for seeking a good outcome, the prayer for soliciting rainfall, the prayer to express gratitude, the prayer for the respect of a mosque, the prayer to welcome the beginning of a new month, etc.

2 If, for example, he is waiting for an impediment to pass, or in the case of a fasting person whose company at the table is anticipated, or the fasting person who needs to eat first, or the pilgrim descending from ʿArafāt to al-Mashʿar.

مسائل:

الأولى: تصلى الفرائض في كل وقت أداءً وقضاءً ما لم تتضيق الحاضرة، والنوافل ما لم تدخل الفريضة.

الثانية: يكره ابتداء النوافل عند طلوع الشمس، وغروبها، وقيامها نصف النهار إلى أن تزول - إلا يوم الجمعة -، وبعد الصبح والعصر عدا ذي السبب.

الثالثة: تقديم كل صلاة في أول وقتها أفضل - إلا في مواضع - ولا يجوز تأخير الصلاة عن وقتها، ولا تقديمها عليه.

الفصل الثالث: في القبلة

وهي الكعبة مع القدرة، وجهتها مع البعد.

والمصلي في الكعبة يستقبل أي جدرانها شاء، وعلى سطحها يبرز بين يديه بعضها.

وكل قوم يتوجهون إلى ركنهم: بالعراقي لأهل العراق، واليماني لأهل اليمن، والمغربي لأهل المغرب، والشامي لأهل الشام.

وعلامة العراق جعل الفجر محاذياً لمنكبه الأيسر والشفق لمنكبه الأيمن، وعين الشمس - عند الزوال - على طرف الحاجب الأيمن مما يلي الأنف والجدي خلف المنكب الأيمن.

if one is at liberty to do so, but if constrained, he may pray in any direction he wishes.

If he intentionally does not face the *qibla*, then he must repeat his prayer.

If he assumed the wrong direction by mistake or forgot, having performed his prayer between the east and the west, then he need not repeat the prayer. But if he prayed exactly facing either direction, then he must repeat it [by ensuring he prays in a direction between the east and west], if there is still time. If he prayed with his back to the *qibla*, then he must repeat it in any case.

One is not allowed to perform the prayer on horseback out of choice, except for the supererogatory prayers.

Section Four: Clothing for Prayer

It is obligatory to clothe one's coverable parts (*ʿawra*), either with cotton or linen, or with plants that grow from the earth, or the fur of four-legged woodland animals [such as a hare], or the wool, fur and feathers of animals whose flesh is fit for consumption.

Praying whilst clad in the hide of a dead animal is not allowed, even if it has been tanned; nor in the hide, wool, fur and feathers of animals that are not fit for consumption, even if they have been slaughtered and tanned; nor in pure silk for men, where he has a choice of other garments. It is, however, allowed during a battle, for women, on riding saddles and as bedspreads. One is also not allowed to pray in usurped clothing, nor to have the soles of the feet covered by anything that is not an extension of trousers and the like.

Wearing black clothing is disapproved, except for the turban and footwear, as is wearing one's shirt tucked in to the trousers; having any visible metal on one's person; covering the face; wearing outer garments tightly fastened (except during a battle), and covering oneself with a single cloth drawn under the arms and held to one side.

The clothing worn for prayer must be pure – apart from the dispensations mentioned previously – and rightfully belonging to oneself or considered as such.

A man's coverable parts (*ʿawra*) are his genitals, front and back, whereas for a woman it is her entire body. She is permitted to keep her face, two hands and two feet uncovered. The slave-woman and the minor girl can keep the head uncovered.

It is recommended for a man to cover his entire body, especially with a long cloak (*ridāʾ*), and for a woman to cover herself with three garments: a

ومع فقد الإمارات يصلي إلى أربع جهات مع الاختيار، ومع الضرورة إلى أي جهة شاء.

ولو ترك الاستقبال عمداً أعاد.

ولو كان ظاناً أو ناسياً وكان بين المشرق والمغرب فلا إعادة، ولو كان إليهما أعاد في الوقت.

ولو كان مستدبراً أعاد مطلقاً.

ولا يصلي على الراحلة اختياراً إلا نافلة.

الفصل الرابع: في اللباس

يجب ستر العورة أما بالقطن، أو الكتان، أو ما أنبتته الأرض من أنواع الحشيش، أو بالخز الخالص، أو بالصوف والشعر والوبر مما يؤكل لحمه أو جلده مع التذكية.

ولا يجوز الصلاة في جلد الميتة وإن دبغ، ولا جلد ما لا يؤكل لحمه وإن ذكي ودبغ، ولا صوفه وشعره ووبره، ولا الحرير المحض للرجال مع الاختيار ويجوز في الحرب وللنساء، وللركوب، والافتراش له - ولا في المغصوب، ولا ما يستر ظهر القدم إذا لم يكن له ساق.

ويكره في الثياب السود - إلا العمامة والخف - وأن يأتزر فوق القميص وأن يستصحب الحديد ظاهراً، واللثام، والقباء المشدود - في غير الحرب - واشتمال الصماء.

ويشترط في الثوب الطهارة إلا ما عفي عنه مما تقدم - والملك أو حكمه.

وعورة الرجل قبله ودبره، وجسد المرأة عورة، وسوغ لها كشف الوجه واليدين والقدمين، وللأمة والصبية كشف الرأس.

ويستحب للرجل ستر جميع جسده، والرداء وللمرأة ثلاثة أثواب: قميص ودرع وخمار.

tunic, a long loose over-garment, and a head covering.

If a man has nothing to cover himself with, he may pray standing up if he is certain of being safe from onlookers, but if not, then he must pray seated and gesture the actions.

Section Five: The Place for Prayer

One can perform the prayer in any place that he owns or has permission to use, but it becomes invalid if he learns that the place was usurped.

The place of prostration is required to be pure.

It is recommended to offer the obligatory daily prayer in the mosque and the supererogatory ones at home.

Praying in the bathroom is disapproved, as is praying in the valleys of Ḍajnān, al-Shaqra and al-Bayḍā', on wet clay, between graves, on sand, in swamps, in camel-pens, on ant-hills, on the downward slope of a valley, in the middle of the road, inside the Ka'ba – especially the obligatory prayer – in the houses of Magians or fire-worshippers. One should also not pray where a woman is either praying in front of him or beside him, nor in an open doorway, nor facing a person, nor a blazing fire and nor a wall where urine drains past.

One is only allowed to place the forehead on the earth during prostration, or on that which grows out of the earth, being neither edible nor wearable. It must also be owned or permissible for use, and free of impurity. One is not allowed to knowingly prostrate on a usurped or impure place.

It is not necessary for the ground beneath the rest of the parts of the body during prostration to be pure.

Prostration on anything other than earth is not allowed, such as skins, or anything that may be extracted from the ground such as minerals.

If there is no earth available to prostrate on, then one may do so on snow, tar or anything else. If the ground is too hot, then one may prostrate on clothing, or barring that, one one's hand.

ولو لم يجد ساتراً صلى قائماً بالإيماء إن أمن إطلاع غيره، وإلا قاعداً مؤمياً.

الفصل الخامس: في المكان

كل مكان مملوك أو مأذون فيه يجوز فيه الصلاة، وتبطل في المغصوب مع علم الغصب.

ويشترط طهارة موضع الجبهة.

ويستحب الفريضة في المسجد، والنافلة في المنزل.

وتكره الصلاة في الحمام، ووادي ضجنان، والشقرة، والبيداء، وذات الصلاصل، وبين المقابر، وأرض الرمل، والسبخة، ومعاطن الإبل، وقرى النمل، وجوف الوادي، وجواد الطريق، والفريضة جوف الكعبة، وبيوت المجوس والنيران، وأن يكون بين يديه أو إلى أحد جانبيه امرأة تصلي، وإلى باب مفتوح، أو إنسان مواجه، أو نار مضرمة، أو حائط ينزمن بالوعة.

ولا يجوز السجود إلا على الأرض، أو ما أنبتته الأرض - مما لا يؤكل ولا يلبس - إذا كان مملوكاً أو في حكمه خالياً من نجاسة، ولا يجوز على المغصوب مع العلم ولا على نجاسة.

ولا يشترط طهارة مساقط بقية أعضاء السجود.

ولا يجوز السجود على ما ليس بأرض كالجلود، أو ما خرج عنها بالاستحالة كالمعادن.

ويجوز مع عدم الأرض السجود على الثلج والقير وغيرهما، ومع الحر على الثوب، فإن فقد فعلى اليد.

Section Six: The Public (adhān) and Individual (iqāma) Calls to Prayer

These two calls are both recommended for the five daily prayers, whether prayed on time or made up in lieu afterwards, whether praying by oneself or congregationally, man or woman, as long as the latter recites them without her voice being heard.

Their recommendation is emphasised for the prayers that are recited aloud, especially the dawn and dusk prayers.

The *adhān* is as follows:

Allāhu akbar (God is greater) – repeated four times
Ash-hadu an lā ilāha illa Allāh (I bear witness that there is no god but God) – repeated twice
Ash-hadu anna Muḥammadan rasūl Allāh (I bear witness that Muḥammad is the messenger of God) – repeated twice
Ḥayya ʿalā al-ṣalāt (Hasten to the prayer) – repeated twice
Ḥayya ʿalā al-falāḥ (Hasten to success) – repeated twice
Ḥayya ʿalā khayr al-ʿamal (Hasten to the best of deeds) – repeated twice
Allāhu akbar (God is greater) – repeated twice
Lā ilāha illa Allāh (There is no god but God) – repeated twice

The *iqāma* is similar, except that two of the repetitions of *Allāhu akbar* are dropped from the beginning; *Lā ilāha illa Allāh* is only recited once at the end; and the phrase *Qad qāmat al-ṣalāt* (Indeed the prayer has been established) is recited twice after *Ḥayya ʿalā khayr al-ʿamal.* In total, there are thirty-five phrases in both together.

Adhān is only recited once the time for prayer has set in, except for the dawn prayer,[3] where it is recommended to recite it again once the time has set in. They must both be recited in the correct order.

The public caller to the prayer (*mu'adhdhin*) must be a veracious individual of good repute, observant of the times of prayer. He must purify himself, stand at a height facing *qibla*, raise his voice and recite the *adhān* in a slow, measured tone and the *iqāma* at a slightly faster pace, leaving a short pause between them to sit or prostrate or to pace a little.

3 When it is recited earlier to announce the prayer.

الفصل السادس: في الأذان والإقامة

وهما مستحبان في الصلوات الخمس أداءاً وقضاءاً، للمنفرد والجامع، رجلاً كان أو امرأة، بشرط أن تسر.

ويتأكدان في الجهرية، خصوصاً في الغداة والمغرب.

وصورة الأذان:

الله أكبر، الله أكبر، الله أكبر، الله أكبر،

أشهد أن لا إله إلا الله، أشهد أن لا إله إلا الله،

أشهد أن محمدا رسول الله، أشهد أن محمدا رسول الله،

حي على الصلاة، حي على الصلاة،

حي على الفلاح، حي على الفلاح،

حي على خير العمل، حي على خير العمل،

الله أكبر، الله أكبر،

لا إله إلا الله، لا إله إلا الله.

والإقامة مثله إلا التكبير فإنه يسقط منه مرتان في أوله، والتهليل يسقط مرة واحدة في آخره، ويزيد «قد قامت الصلاة» مرتين بعد «حي على خير العمل». فجميع فصولهما خمسة وثلاثون فصلاً.

ولا يؤذن قبل دخول الوقت إلا في الصبح، ويستحب إعادته بعد دخوله.

ويشترط فيهما الترتيب.

ويستحب كون المؤذن عدلاً، صيتاً، بصيراً بالأوقات، متطهراً، قائماً على مرتفع، مستقبلاً للقبلة، رافعاً صوته، مرتلاً للأذان، محدراً للإقامة، فاصلاً بينهما بجلسة أو سجدة أو خطوة.

It is disapproved for him to be walking or riding whilst reciting if he is able to stand still, to pronounce the vowels at the end of the phrases without stopping, to talk during the *adhān* and *iqama*, and to insert stresses and echoes in the wrong places.

It is prohibited to say the phrase: *al-ṣalātu khayrun min al-nawm* (Prayer is better than sleep).

ويكره أن يكون ماشياً أو راكباً مع القدرة والإعراب أواخر الفصول، والكلام في خلالهما، والترجيع لغير الأشعار.

ويحرم قول «الصلاة خير من النوم».

Chapter 2: The Actions in the Prayer

There are obligatory as well as recommended actions, and are dealt with in separate sections below:

Section One: The Eight Obligatory Actions

1) Intention: this is made with the opening *takbīr* (*takbīrat al-iḥrām*). One's intention must be to draw near [to God]; it must specify the name of the prayer and whether it is an obligatory or recommended prayer, whether it is being prayed on time or in lieu of a missed prayer; and it must be maintained until the end.

2) The Opening *Allāhu akbar*: This, like the intention, is a cornerstone of the prayer, and must be uttered in Arabic as: *Allāhu akbar*, and not in translation, if possible. One who does not know how to pronounce it must learn to do so properly, though one who is mute should gesture its meaning using sign language and with proper intent in his heart. It must also be recited whilst in a standing position, wherever physically possible.

It is recommended to raise both palms up to the earlobes whilst reciting it.

3) Standing: This is also a cornerstone of the prayer, wherever physically possible. If one is unable to stand, he may lean on something, and if he cannot do that either, then he may perform the prayer seated. If he cannot sit, then he can pray in a reclining position whilst gesturing with his limbs, and if he cannot do that either, then he may pray lying down.

4) Recitation of the Quran: It is obligatory to recite the Opening Chapter (Sūrat al-Ḥamd) and another chapter, in both units of a two-unit prayer, and in the first two units of all other prayers. It is not permissible to recite them in translation. One must learn to recite them well if he does not know them, and if unable to recite well, then to recite them as best as he can. If he does not know any of it, then he may recite *Allāhu akbar* and *Lā ilāha illa Allāh* instead. The mute person may move his tongue and recite it in his heart. In the third and fourth units, one can choose between the recitation of the Quran and four pronouncements glorifying God as follows: *Subḥānallāhi wal-Ḥamdu lillāhi wa Lā ilāha illā Allāhu wa Allāhu akbar* (Glory be to God, and all praise be to God, and there is no god but God, and God is greater).

الباب الثاني: في أفعال الصلاة

وهي واجبة ومندوبة، فها هنا فصول:

الفصل الأول: الواجبات ثمانية

الأول: النية، مقارنة لتكبيرة الإحرام. ويجب نية القربة، والتعيين، والوجوب أو الندب، والأداء أو القضاء، واستدامة حكمها إلى الفراغ.

الثاني: تكبيرة الإحرام، وهي ركن - وكذا النية - وصورتها: «الله أكبر» ولا يكفي الترجمة مع القدرة. ويجب التعلم، والأخرس يشير بها مع عقد قلبه. وشرطها القيام مع القدرة. ويستحب رفع اليدين بها إلى شحمتي الأذنين.

الثالث: القيام وهو ركن مع القدرة، ولو عجز اعتمد، فإن تعذر صلى قاعداً، ولو عجز صلى مضطجعاً بالإيماء، ولو عجز صلى مستلقياً.

الرابع: القراءة، ويجب الحمد والسورة في الثنائية، والأوليين من غيرها، ولا يجزي الترجمة، ويجب التعلم لو لم يحسن مع المكنة، ومع العجز يصلي بما يحسن، وإن لم يحسن شيئاً كبر الله وهلّله، والأخرس يحرك لسانه ويعقد بها قلبه، ويتخير في الثالثة والرابعة بينها وبين التسبيح أربعاً، وصورته «سبحان الله والحمد لله ولا إله إلا الله والله أكبر».

It is obligatory to recite audibly in the dawn prayer, the first two units of the evening prayer, and the first two units of the night prayer, and to recite quietly in the rest of them.

The chapters of the Quran that contain obligatory prostrations are not allowed to be recited in the obligatory daily prayers. One must also not recite chapters that would cause the time of prayer to elapse, nor two chapters after Sūrat al-Ḥamd.

It is recommended to recite the *basmala* at the beginning of the chapters audibly, even in the units where one recites quietly. It is also recommended to recite Sūrat al-Jumuʿa and Sūrat al-Munāfiqūn in the Friday prayer and the two noon prayers that follow it.

It is prohibited to recite *āmīn*,[4] and doing so invalidates the prayer.

5) Bowing: It is obligatory to bow once in every unit, since it is a cornerstone, except for the prayer that is performed during an eclipse and the prayer of signs. One must bend at the waist to the extent whereby the palms reach the knees. If unable to, he bends as far as possible, and if unable to do so, he can gesture with his limbs. He must hold the position until he has completed the glorification, of which once is obligatory, as follows: *Subḥāna rabbi al-ʿaẓīm wa-biḥamdih* (Glory be to my Lord, the Great, and praise is due unto Him). Then he must stand up straight and still.

It is recommended to say *Allāhu akbar* before bowing and to raise one's palms whilst doing so; to open out the fingers whilst placing them on the knees and push them back gently; to keep the back straight and stretch out the neck; to supplicate and increase the glorification; and to recite: *Samiʿa Allāhu li-man ḥamidah* (God hears the one who praises Him) upon standing again.

Bowing whilst keeping the two hands under one's clothes is disapproved.

6) Prostration: It is obligatory to prostrate twice in every unit as they are a cornerstone of the prayer. The prostration must be performed with seven parts touching the ground: the forehead, both hands, both knees, and both big toes. The place where one prostrates the forehead cannot be higher than the level of one's feet by more than a brick. If one is unable to prostrate, then he can gesture, or raise a surface up to a level that he can prostrate on and do so. One must hold the position until he has completed the glorification, which is obligatory to recite once as follows: *Subḥāna rabbī al-aʿlā wa-biḥamdih* (Glory be to my Lord, the Highest, and praise is due unto Him). He must sit up straight and still between the two prostrations, and he must place his forehead only on that which is valid to prostrate on.

4 After the recitation of Sūra al-Ḥamd, the first chapter of the Quran.

ويجب الجهر في الصبح، وأولتي المغرب، وأولتي العشاء، والإخفات في البواقي.

ولا يجوز قراءة العزائم في الفرائض، ولا ما يفوت الوقت بقراءته، ولا قراءة سورتين بعد الحمد.

ويستحب الجهر بالبسملة في الإخفات، وقراءة الجمعة والمنافقين في الجمعة وظهريها.

ويحرم قول «آمين»، ويبطل.

الخامس: الركوع، ويجب في كل ركعة مرة - إلا في الكسوف والآيات - وهو ركن، ويجب أن ينحني قدراً تصل كفاه إلى ركبتيه، ولو عجز أتى بالممكن، وإلا أومأ، وأن يطمئن بقدر التسبيح، وأن يسبح مرة واحدة، صورتها «سبحان ربي العظيم وبحمده»، وأن ينتصب قائماً مطمئناً.

ويستحب التكبير له، ورفع اليدين به، ووضع يديه على ركبتيه مفرجات الأصابع، وردهما إلى خلفه، وتسوية ظهره، ومد عنقه، والدعاء، وزيادة التسبيح وأن يقول بعد رفع رأسه: «سمع الله لمن حمده». ويكره أن يركع ويداه تحت ثيابه.

السادس: السجود، ويجب في كل ركعة سجدتان، وهما ركن، ويجب في كل سجدة السجود على سبعة أعضاء: الجبهة واليدين والركبتين وإبهامي الرجلين، وعدم علو موضع السجود على القيام بأزيد من لبنة، ولو تعذر السجود أومأ، أو رفع شيئاً وسجد عليه، وأن يطمئن بقدر التسبيح، وأن يسبح مرة واحدة، صورتها: «سبحان ربي الأعلى وبحمده» وأن يجلس بينهما مطمئنا وأن يضع جبهته على ما يصح السجود عليه.

It is recommended to say *Allāhu akbar* before it as well as after lifting the head up; to place the hands down on the ground first; to touch the nose to the ground; to supplicate and increase the glorification; to hold oneself still after raising the head from the second prostration; to supplicate between the two prostrations; and to stand up by using the hands as supports on the grounds whilst lifting the knees up first.

Prostrating in a slouching manner is disapproved.

7) The Testimony of faith whilst seated: This is obligatory to perform once in a two-unit prayer, and twice in a three and four-unit prayer. It is obligatory to sit down, where possible, and declare the two testimonies of faith, followed by invoking blessings on the Prophet and his household (peace be upon them). The minimum one must say is: *Ash-hadu an lā ilāha illa Allāh wa ash-hadu anna Muḥammadan rasūl Allāh, Allāhumma ṣalli ʿalā Muḥammadin wa āli Muḥammad* (I testify that there is no god but God, and I testify that Muḥammad is the messenger of God, O God bless Muḥammad and the household of Muḥammad).

It is recommended to sit slightly leaning onto the left thigh, with the right foot resting on the sole of the left foot, and to supplicate after the obligatory recitation.

8) Greeting of Peace: There is a difference of opinion regarding its obligation. One recites it as follows: *As-salāmu ʿalaynā wa ʿalā ʿibādillāhi al-ṣāliḥīn* (Peace be upon us all and upon God's righteous servants), or *As-salāmu ʿalaykum wa raḥmatullāhi wa barakātuh* (May the peace, mercy and blessings of God be upon you all).

It is recommended for the one praying by himself to address the greeting of peace facing the *qibla* first, then to gesture with the corner of his eyes towards his right. The imam of a congregation should turn his face slightly (in addition to gesturing to the right with his eyes), whereas the follower should gesture by turning his face to the right and to the left, if there is anyone sitting to his left.

Section Two: Recommended Actions in the Prayer

There are five such actions recommended:

1) To begin the prayer with seven *takbīrāt* including the *takbīrat al-iḥrām*, with three instances of supplications between them.

2) *Qunūt* (beseeching God with hands raised): This is performed in the second unit, after the recitation and before bowing. If one forgets, he can perform it in lieu after standing up from bowing.

ويستحب التكبير له وعند رفع الرأس منه، والسبق بيديه إلى الأرض، والإرغام بالأنف، والدعاء والتسبيح الزائد، والطمأنينة عقيب رفعه من الثانية، والدعاء، بينهما، والقيام معتمداً على يديه سابقًا برفع ركبتيه.

ويكره الإقعاء.

السابع: التشهد، ويجب في كل ثنائية مرة، وفي الثلاثية والرباعية مرتين، ويجب فيه الجلوس بقدره، والشهادتان، والصلاة على النبي وآله عليهم السلام، وأقله: «أشهد أن لا إله إلا الله وأشهد أن محمدا رسول الله، اللّهم صل على محمد وآل محمد».

ويستحب أن يجلس متوركاً، وأن يدعو بعد الواجب.

الثامن: التسليم، وفي وجوبه خلاف، وصورته: «السلام علينا وعلى عباد الله الصالحين»، أو: «السلام عليكم ورحمة الله وبركاته». ويستحب أن يسلم المنفرد إلى القبلة ويومئ بمؤخر عينيه إلى يمينه، والإمام (يومئ إلى يمينه) بصفحة وجهه، والمأموم (يومئ بصفحة وجهه) إلى يمينه ويساره - إن كان على يساره أحد.

الفصل الثاني: في مندوبات الصلاة

وهي خمسة:

الأول: التوجه بسبع تكبيرات بينها ثلاثة أدعية، واحدة منها تكبيرة الإحرام.

الثاني: القنوت، وهو في كل ثانية قبل الركوع وبعد القراءة ويقضيه لو نسيه بعد الركوع.

3) To keep the gaze fixed down on the place of prostration when standing, on the palms during *qunūt*, on the space between his feet when bowing, on the tip of his nose when prostrating, and on his lap whilst seated.

4) To place the hands on the thighs, parallel with the knees, when standing; to place them in front of the face when beseeching in *qunūt*; on the knees when bowing; parallel to the ears when prostrating; and on the thighs whilst seated.

5) Post-Prayer Rituals: The minimum recommendation is to recite the Glorification of Fāṭima al-Zahrā',[5] and there is no maximum. It is recommended to recite that which has been mentioned in narrations.

Section Three: Actions That Invalidate the Prayer

The prayer is invalidated by all the things that invalidate one's state of purity, even if it is unintentional, as well as by intentionally looking behind one's back, uttering two letters or more that do not constitute a supplication or recitation from the Quran, laughing loudly, making significant movements that are not part of the prayer, crying over worldly matters, and folding the arms during the prayer.

Looking left and right during the prayer is disapproved, as is yawning, stretching, cracking one's knuckles, fidgeting, slouching, clearing phlegm, spitting, blowing on the place of prostration, moaning, or suppressing the urge to urinate or defecate.

Breaking one's prayer unnecessarily is prohibited. There is a difference of opinion regarding the validity of the prayer performed by a man with braided hair.

Blessing someone who has sneezed,[6] replying to a *salām* greeting and supplicating for legitimate desires within the prayer are allowed.

5 This consists of reciting *Allāhu akbar* 34 times, *al-ḥamdu lillāh* 33 times, and *subḥān Allāh* 33 times.

6 By saying the customary phrase: *yarḥamuk Allāh* (May God have mercy on you)

الثالث: نظره في حال قيامه إلى موضع سجوده، وفي حال قنوته إلى باطن كفيه، وفي ركوعه إلى بين رجليه، وفي سجوده إلى طرف أنفه، وفي جلوسه إلى حجره.

الرابع: وضع اليدين قائماً على فخذيه بحذاء ركبتيه، وقانتاً تلقاء وجهه، وراكعاً على ركبتيه، وساجداً بحذاء أذنيه، وجالساً على فخذيه.

الخامس: التعقيب، وأقله تسبيح الزهراء عليها السلام، ولا حصر لأكثره، ويستحب أن يأتي فيه بالمنقول.

الفصل الثالث: في قواطع الصلاة

ويبطلها كل نواقض الطهارة - وإن كان سهواً -، وتعمد الالتفات إلى ما ورائه، والكلام بحرفين فصاعدا - مما ليس بدعاء ولا قرآن - والقهقهة، والفعل الكثير الخارج عنها، والبكاء لأمور الدنيا، والتكفير.

ويكره الالتفات يميناً وشمالاً، والتثاؤب، والتمطي، والفرقعة، والعبث، والإقعاء، والتنخم، والبصاق، ونفخ موضع السجود، والتأوه، ومدافعة الأخبثين.

ويحرم قطع الصلاة لغير ضرورة، وفي عقص الشعر للرجل قولان. ويجوز تسميت العاطس، ورد السلام، والدعاء بالمباح.

Chapter 3: Other (Obligatory) Prayers

This chapter contains several sections.

Section One: Friday Prayers

They consist of two units instead of the noon prayer, to be performed from when the sun begins its descent until the time when shadows are the same lengths as their objects.

Its Prerequisites: The just authority, or whoever he appoints (to lead the prayer). The number of people must be at least five, including the imam. The two sermons must both include praise of God and invocation of blessings on the Prophet and his household, admonition and recitation of a short chapter from the Quran. The congregation must be such that there is not another Friday congregational prayer taking place within a three-mile radius.

The Friday prayer, when all its conditions have been met, is obligatory upon every responsible free male of sound mind and body, neither blind nor disabled, nor too aged to attend, nor a traveller.

If the Friday prayer is more than two *farsakh* away,[7] then one is not obliged to attend. If the time for it elapses, then the *ẓuhr* prayer becomes obligatory.

The sermons must be delivered before noon, such that the time for prayer sets in as they end. The imam must stand to deliver them wherever possible. It is recommended for him to be in a state of purity, eloquent, diligent with his prayers, wearing a long cloak (*ridāʾ*) and to lean on something whilst speaking; and it is recommended to listen to him attentively.

7 One *farsakh* equals approximately 5.5 kilometres.

الباب الثالث: في بقية الصلوات

وفيه فصول

الفصل الأول: في الجمعة

وهي ركعتان عوض الظهر، ووقتها من زوال الشمس إلى أن يصير ظل كل شيء مثله.

وشروطها: السلطان العادل، أو من نصبه، والعدد - وهو خمسة نفر أحدهم الإمام - والخطبتان -، وهما حمد الله تعالى والصلاة على النبي وآله والوعظ وقراءة سورة خفيفة من القرآن - والجماعة، وأن لا يكون هناك جمعة أخرى بينهما أقل من ثلاثة أميال.

وتجب مع الشروط على كل مكلف حر ذكر سليم من المرض والعمى والعرج، وأن لا يكون هماً، ولا مسافراً.

ولو كان بينه وبين الجمعة أزيد من فرسخين لم يجب الحضور. ولو فاتت وجبت الظهر.

ويجب إيقاع الخطبتين بعد الزوال قبلها، وقيام الخطيب مع القدرة. ويستحب فيهما الطهارة، وأن يكون الخطيب بليغاً مواظباً على الصلاة، مرتدياً، معتمداً على شيء، والإصغاء إليه.

Other Rulings

1) To recite *adhān* a second time is an innovation.

2) It is prohibited to transact after the call to prayer, and any transaction must be annulled.

3) If it is possible to gather as a congregation during the occultation of the twelfth Imam, the Friday prayer is recommended.

4) It is recommended to precede the prayer with twenty units of supererogatory prayers, to shave the head, clip the fingernails, trim the moustache, to walk with dignity and composure, to clean the body and perfume it, to supplicate, and to recite the Quran audibly.

Section Two: The Prayer on the Two Eids

It is obligatory for it to be prayed in congregation, with the same prerequisites as the Friday prayer. If the prerequisites are not met, it is recommended to pray it normally in congregation as well as individually. The time for its performance is between sunrise and noon, and it does not need to be made up in lieu if the time elapses.

مسائل:

الأولى: الأذان الثاني بدعة.

الثانية: يحرم البيع بعد النداء، وينعقد.

الثالثة: لو أمكن الاجتماع حال الغيبة استحبت الجمعة.

الرابعة: يستحب التنفل بعشرين ركعة، وحلق الرأس، وقص الأظفار، وأخذ الشارب، والمشي بسكينة ووقار، وتنظيف البدن، والتطيب، والدعاء والجهر بالقراءة.

الفصل الثاني: في صلاة العيدين

وهي واجبة جماعة بشروط الجمعة، ومع فقدها تستحب جماعة وفرادى، ووقتها بعد طلوع الشمس إلى الزوال، ولا تقضى لو فاتت.

It consists of two units: Sūrat al-Ḥamd and Sūrat al-Aʿlā are recited in the first unit, followed by five *takbīr* (*Allāhu akbar*) with *qunūts* between them. Then one recites *Allāhu akbar* a sixth time and bows down, then prostrates twice and stands for the second unit. He recites Sūrat al-Ḥamd and Sūrat al-Shams, then recites four *takbīrs* with *qunūts* between them, followed by a fifth *takbīr* before bowing again.

It is recommended to perform the prayer out in the open, and to walk to the prayer barefoot with dignity and composure, to eat before the prayer on Eid al-Fiṭr, and after the prayer on Eid al-Aḍḥā from the slaughtered meat. It is also recommended to recite the special *takbīr*[8] after each of the four prayers between maghrib on the eve of Eid until after the Eid prayer on Eid al-Fiṭr, and after fifteen prayers during Eid al-Aḍḥā, the first of which is after the *ẓuhr* prayer on the Day of Eid for whoever is in Mina, and after ten prayers for whoever is not.

Other Rulings

1) Performing supererogatory prayers before and after it is disapproved, except in the Prophet's mosque (peace be upon him), before leaving for the prayer.

2) It is said that it is obligatory to perform an extra *takbīr* and *qunūt* in the prayer.

3) The two sermons follow the prayer.

4) Travelling after sunrise and before having prayed is prohibited, and doing so before sunrise is disapproved.

8 The special takbīr for Eid al-Fiṭr is as follows: *Allāhu akbar, Allāhu akbar, Allāhu akbar, Lā ilāha illa Allāh, wa-Llāhu akbar, Allāhu akbar, wa-lillāhi al-ḥamd. Allāhu akbar ʿalā mā hadānā* (Allah is greater, Allah is greater, Allah is greater, there is no god but Allah and Allah is greater. Allah is greater and to Allah belongs all praise. Allah is greater over all that He has guided us). For Eid al-Aḍḥā, it is: *Allāhu akbar, Allāhu akbar, Allāhu akbar, Lā ilāha illa Llāh wa-Llāhu akbar, Allāhu akbar ʿalā mā hadānā, Allāhu akbar ʿalā mā razaqanā min bahīmat al-anʿām* (Allah is greater, Allah is greater, Allah is greater, there is no god but Allah and Allah is greater. Allah is greater for all that He has guided us, Allah is greater for all the livestock that he has given us provision from).

وهي ركعتان، يقرأ في الأولى الحمد والأعلى، ثم يكبر خمساً يقنت بينها، ثم يكبر السادسة للركوع، ويسجد السجدتين، ثم يقوم فيقرأ الحمد والشمس، ثم يكبر أربعاً ويقنت بينهما، ثم يكبر الخامسة للركوع. ويستحب الإصحار بها، والخروج حافياً بسكينة ووقار، وأن يطعم قبل خروجه في الفطر وبعده في الأضحى مما يضحى به، والتكبير عقيب أربع صلوات: أولها المغرب، وآخرها العيد في الفطر، وفي الأضحى عقيب خمسة عشرة: أولها ظهر العيد لمن كان بمنى، وفي غيرها عقيب عشر.

مسائل:

الأولى: يكره التنفل قبلها وبعدها إلا في مسجد النبي ﷺ قبل خروجه.

الثانية: قيل التكبير الزائد واجب، وكذا القنوت.

الثالثة: الخطبتان بعدها.

الرابعة: يحرم السفر بعد طلوع الشمس قبلها، ويكره قبله.

Section Three: The Prayer for Eclipses and Other Natural Events

It becomes obligatory during a solar eclipse, a lunar eclipse, an earthquake, terrifyingly strong winds and other such fear-inducing natural phenomena. It is a two-unit prayer, where each unit consists of five bowings and two prostrations.

Method: One must make the intention and say *Allāhu akbar*, then recite Sūrat al-Ḥamd and a second *sūra*, or part of a *sūra*. Then he bows and stands erect again. If he had recited the second *sūra* in full, he recites Sūrat al-Ḥamd a second time and a *sūra*, or [if not, then] the next part of the previous *sūra* after it. He continues in the same manner until the fifth bowing. If he had not completed the second *sūra* in the first cycle, then it suffices for him to recite parts of it to completion, instead of Sūrat al-Ḥamd. Once he has completed five bowings, he says *Allāhu akbar*, prostrates twice, then stands up for the second unit and does the same thing that he did in the previous unit, then completes it with *tashahhud* and *taslīm*.

It is recommended to recite long *sūras* in it; to spend as long in the bowing position as when standing; to pray it in congregation, even if one should have to pray it again, provided there is enough time remaining; to say *Allāhu akbar* when standing back up from bowing, except after the fifth and tenth bowing, when he should say: *samiʿa Allāhu li-man ḥamidah* (God hears the one who praises Him); and to raise the hands in *qunūt* five times.

The time for the prayer of natural events sets in from the beginning of the event until the sun or moon starts coming out of the eclipse; for other natural events, for their entire duration; and for earthquakes, until the end of one's life.

If he misses the prayer, intentionally or forgetfully, he must pray in lieu later. If he was unaware of it during a total eclipse, he must pray in lieu when he finds out, but not for a partial eclipse.

If the time for the eclipse prayer coincides with the time for the daily prayer, he may choose which one to pray first as long as there is no time constraint. If the time is short, then he must prioritize the daily obligatory prayer, and there is no obligation to pray the other in lieu there was not enough time to pray both on time.

الفصل الثالث: في صلاة الكسوف

وتجب - عند كسوف الشمس، وخسوف القمر، والزلزلة، والرياح المخوفة، وغيرها من أخاويف السماء - ركعتان، تشتمل كل ركعة على خمس ركوعات وسجدتين.

وكيفيتها: أن ينوي ويكبر، ويقرأ الحمد وسورة أو بعضها، ثم يركع، ثم ينتصب، فإن كان أتم السورة قرأ الحمد ثانياً وسورة أو بعضها، وهكذا إلى أن يركع خمساً، وإن لم يكن أتمها اكتفى بتمامها عن الفاتحة، فإذا ركع خمساً كبر وسجد سجدتين، ثم قام وصنع ثانياً كما صنع أولاً، وتشهد وسلم.

ويستحب أن يقرأ فيها السور الطوال، ومساواة الركوع للقيام، والجماعة، والإعادة مع بقاء الوقت، والتكبير عند الانتصاب من الركوع - إلا في الخامس والعاشر فإنه يقول: «سمع الله لمن حمده» والقنوت خمس مرات.

ووقت الكسوف والخسوف من حين ابتدائه إلى ابتداء الانجلاء، وفي غيرهما مدته، وفي الزلزلة مدة العمر.

ولو فاتته عمداً أو نسياناً قضاها، ولو كان جاهلاً فإن كان قد احترق القرص كله قضى وإلا فلا.

ولو اتفقت وقت حاضرة تخير ما لم تتضيق إحداهما، ولو تضيقتا قدم الحاضرة، ولا قضاء مع عدم التفريط.

Chapter 4: Recommended Prayers

Among these are:

1. The Prayer for Rain (*ṣalāt al-istisqā'*), which is encouraged to be prayed during a drought. Its performance is the same as the Eid Prayer, except that one beseeches and implores God in the *qunūt* for an abundance of rain and mercy through it. Furthermore, it is recommended to beseech God in the prayer through transmitted supplications and to fast for three days. On the third day, which should be either a Monday or Friday, children should be separated from their mothers [so as to intensify the crying and sense of need], the imam should remove his cloak [and place it over his right shoulder]; and upon completion of the prayer, he should recite *Allāhu akbar* one hundred times facing *qibla*, *Subḥānallāh* one hundred times to his right, *Lā ilāha illa Allāh* one hundred times to his left, and *al-Ḥamdu lillāh* one hundred times towards the people, and for them to follow him in this. It is also recommended to repeat the prayer if the response is delayed.
2. The Prayer Recommended during the Nights of Ramadan: This consists of a thousand units in total, whereby twenty units are performed on each night. An extra hundred units are prayed on the odd nights,[9] and an extra ten on the last ten nights.
3. Various Other Recommended Prayers: On the eve of Eid al-Fiṭr,[10] on the Day of al-Ghadīr,[11] on the eve of the 15th Shaʿbān,[12] on the eve and Day of the Mabʿath (the start of the Prophet's mission);[13] the prayer of ʿAlī,[14] of Fāṭima,[15] and of Jaʿfar.[16]

9 This refers to the odd nights that may possibly encompass the Night of Power (*laylat al-qadr*), namely the 19th, 21st and 23rd nights of Ramadan.

10 This is a two-unit prayer in which one recites Sūrat al-Ḥamd once and Sūrat al-Tawḥīd 1,000 times in the first unit, and Sūrat al-Ḥamd and Sūrat al-Tawḥīd once each in the second.

11 This is a two-unit prayer to be offered half an hour before noon.

12 This is four-unit prayer.

13 This is a twelve-unit prayer in which one recites Sūrat al-Ḥamd and Sūrat Yāsīn in each unit.

14 This is a four-unit prayer with two *tashahhuds* and two *salāms* (two cycles of two units), and in which once recites Sūrat al-Ḥamd once and Sūrat al-Tawḥīd fifty times in each unit.

15 This is a two-unit prayer in which one recites Sūrat al-Ḥamd once and Sūrat al-Qadr one hundred times in the first unit, followed by Sūrat al-Ḥamd once and Sūrat al-Tawḥīd one hundred times in the second unit.

16 This is a four-unit prayer in two cycles of two units. In the first unit, one recites Sūrat al-Ḥamd and Sūrat al-Zilzāl once each, followed by *Subḥān Allāh wa-l-ḥamdu li-Llāhi wa-lā ilāha illa Llāhu wa-Llāhu akbar* fifteen times. This phrase is then repeated in each action of the prayer: the bowing, standing, the two prostrations, and the sitting after them. In the second unit, one recites Sūrat al-Ḥamd and Sūrat al-ʿĀdiyāt once each, in the third, and Sūrat al-Naṣr, and

الباب الرابع: في الصلوات المندوبة

فمنها صلاة الاستسقاء، وهي مؤكدة عند قلة المياه.

وكيفيتها مثل صلاة العيد، إلا أنه يقنت لسؤال توفير المياه والاستعطاف به - ويستحب بالمأثور - وأن يصوم الناس ثلاثاً، والخروج يوم الإثنين أو الجمعة والتفريق بين الأطفال وأمهاتهم، وتحويل الرداء، وتكبير الإمام بعدها مائة مستقبل القبلة، والتسبيح كذلك يميناً، والتهليل يساراً، والتحميد تلقاء الناس، ومتابعتهم له، والمعاودة مع تأخير الاجابة.

ومنها نافلة رمضان، وهي ألف ركعة، في كل ليلة عشرين، وفي ليالي الإفراد زيادة مائة، وفي العشر الأواخر زيادة عشر.

ومنها صلاة ليلة الفطر، ويوم الغدير، وليلة نصف شعبان، وليلة المبعث ويومه، وصلاة عليّ وفاطمة وجعفر - عليهم السلام.

Chapter 5: Lapses in the Prayer

Whoever intentionally leaves out one of the obligatory actions in the prayer, his prayer is void, even if he was ignorant of his obligation to perform that action. The only exception to this is the tone of the voice, for which he is excused if he was ignorant of which ones are to be recited audibly. The same ruling applies to intentionally including something that is not part of the prayer. If one forgets to perform an action, if it is a cornerstone of the prayer, he may perform it as soon as he remembers if he is still in its place[17], but if not, then he must repeat the prayer.

If he performs an additional bowing, intentionally or mistakenly, he must repeat the prayer. If he mistakenly misses out a whole unit or two and only remembers after the end of the prayer, having turned away from *qibla* or spoken, then he must repeat the prayer. If he prayed on usurped land or in usurped or impure clothing, or knowingly prostrated on an impure space, he must repeat the prayer. He must repeat the prayer in any case if he prayed without ritual purification, before the time had set in, or with his back to the *qibla*.

If the lapse concerns an action that is not a cornerstone of the prayer, then there are several scenarios:

1) Where there is no ruling either way as in the case of someone who forgets to recite anything at all until the bowing, or recites audibly instead of quietly and vice versa, or forgets to recite the glorification whilst bowing, or during the pauses before and after each action, or the glorification in prostration, or fails to touch one of the seven parts to the ground, or to sit still for a while after raising his head from prostration, or during *tashahhud*.

in the fourth, Sūrat al-Ḥamd and Sūrat al-Tawḥīd. Every unit also contains the above-mentioned glorification as per the first unit, so that it is recited seventy-five times in each unit, making a total of three-hundred times in the whole prayer.

17 That is, he has not moved on to the next cornerstone.

الباب الخامس: في السهو

من ترك شيئاً من واجبات الصلاة عمداً بطلت صلاته وإن كان جاهلاً، عدا الجهر والإخفات فقد عذر لو جهلهما، وكذلك لو فعل ما يجب تركه عمداً. أما الناسي فإن ترك ركناً أتى به إن كان في محله وإلا أعاد.

ولو زاد ركوعا عمداً أو سهواً أعاد، ولو نقص من الصلاة ركعة أو ركعتين سهواً ولم يذكر حتى تكلم أو استدبر القبلة أعاد. ولو صلّى على مكان مغصوب أو في ثوب مغصوب أو نجس، أو سجد عليه مع العلم أعاد، ولو صلى بغير طهارة أعاد مطلقًا، أو قبل الوقت، أو مستدبر القبلة أعاد.

وإن كان غير ركن فله أقسام:

الأول: ما لا حكم له، وهو من نسي القراءة حتى ركع، أو الجهر، أو الإخفات، أو تسبيح الركوع أو طمأنينته حتى ينتصب، أو رفع الرأس منه، أو طمأنينته، أو تسبيح السجود، أو طمأنينته، أو إحدى الأعضاء السبعة أو رفع الرأس منه، أو طمأنينته في الرفع منهما، أو طمأنينة الجلوس في التشهد.

2) When one is obligated to salvage the prayer as in the case of one who remembers during the second *sūra* that he forgot to recite Sūrat al-Ḥamd. He can go back and recite Sūrat al-Ḥamd, then repeat the second *sūra*. He who remembers having forgotten to bow before he goes into prostration can go ahead and bow. He who remembers after having stood up that he forgot to prostrate can sit back down, prostrate and continue – performing two extra prostrations in lieu of lapses (*sajdatay al-sahw*) afterwards – and similarly for one who forgot to perform the *tashahhud*. If he remembers after the *taslīm* that he had forgotten the *tashahhud* or the invocation of blessings on the Prophet (peace be upon him), he may recite them in lieu there and then.

3) Doubts: If one has a doubt about which unit he is currently praying in a two or three-unit prayer, or in the first two units of a four-unit prayer, he must start the prayer again; and similarly if he does not know how many units he has prayed. If he has a doubt regarding the performance of an action that he has already moved on from, he need not entertain the doubt; but if he has not moved on, he must perform it. If he then remembers that he had indeed performed it previously, he must start the prayer again if it was a cornerstone action, otherwise not. If he has a doubt about which unit he is currently on beyond the first two units in a four-unit prayer, and he is unsure either way, he can assume it is the later unit, then pray an extra unit thereafter as a precaution.

If he doubts between the second and third unit, as to whether he has prayed two units or three, or between the third and fourth unit, he must take it as the higher number, then perform an extra precautionary unit straight after the *taslīm*, prayed either as one unit standing or as two units seated.

If his doubt is whether he is on the second or the fourth unit, he must take is as the fourth, then pray two extra precautionary units standing.

If he doubts as to whether he is on the second, third or fourth unit, he must take it as the fourth, then pray two extra precautionary units standing and two whilst seated, after the prayer.

الثاني : ما يوجب التلافي ، فمن ذكر أنه لم يقرأ الحمد وهو في السورة قرأ الحمد وأعاد السورة ، ومن ذكر ترك الركوع قبل السجود ركع ، ومن ذكر بعد القيام ترك سجدة قعد وسجد - ويسجد سجدتي السهو - ، وكذا لو ذكر ترك التشهد ، ولو ذكر التسليم ترك التشهد أو الصلاة على النبي ﷺ قضاه .

الثالث : الشك ، إن كان في عدد الثنائية أو الثلاثية أو الأولين من الرباعية أعاد . وكذا إذا لم يعلم كم صلى ، وإن كان في فعل قد انتقل عنه لم يلتفت وإلا أتى به ، فإن ذكر أنه قد فعله استأنف إن كان ركناً وإلا فلا ، فلو شك فيما زاد على الأولين في الرباعية ولا ظنّ بنى على الزائد واحتاط .

فمن شك بين الإثنين والثلاث أو بين الثلاث والأربع بنى على الأكثر ، فإذا سلم صلى ركعة من قيام أو ركعتين من جلوس .

ومن شك بين الإثنين والأربع بنى على الأربع وصلى ركعتين من قيام .

ومن شك بين الإثنين والثلاث والأربع بنى على الأربع وصلى ركعتين من قيام وركعتين من جلوس .

Other Rulings:

1) Doubts by a person who is prone to doubt and constant indecisiveness are to be ignored. Similarly, the lapse by an imam in congregation can be ignored as long as the follower salvages it and vice-versa. There is no need to entertain a doubt that crops up within the remedial measure (i.e. the two remedial prostrations in lieu of the lapse or extra precautionary prayers).

2) If he doubts in the units of a supererogatory prayer, he should assume the lower unit, but he can assume the higher.

3) If he mistakenly talks during the prayer or stands up when he should be sitting, or vice versa, or recites the *taslīm* before the end of the prayer, he must perform the two remedial prostrations at the end. They are also obligatory if he has a doubt as to whether he is on the fourth or fifth unit, in which case he assumes it is the fourth, then performs the two prostrations.

4) The two remedial prostrations are to be performed straight after the prayer. In both, he should recite: *Bismillāhi wa-billāhi, Allāhumma ṣalli ʿalā Muḥammadin wa-āli Muḥammad* (In the Name of God and by God; O God, bless Muḥammad and his household) or *Assalamu ʿalayka ayyuhan nabiyyu wa-raḥmatullāhi wa-barakātuh* (Peace, mercy and blessings be upon you, O Prophet). Then he recites the basic *tashahhud* and *taslīm*.

5) If a responsible person misses his prayer, mistakenly or intentionally, or if he missed it due to oversleeping or being in a state of intoxication, and if he is a Muslim, he must pray in lieu of it. He need not pray in lieu if he was unconscious for the whole duration of the prayer or for prayers that he missed while he was a disbeliever. The apostate, however, must repay missed prayers [if he becomes Muslim again]. If someone cannot find any water or earth to purify himself with, then he is absolved of that prayer and need not pray in lieu of it.

6) If the time for the daily obligatory prayer sets in, and he is also obligated to perform a prayer in lieu, he may choose which one he prays first. If the time for the obligatory prayer becomes restricted, then it is prioritized [and the prayer in lieu should be delayed].

7) Prayers made up in lieu must be performed in the correct order.

8) If a person missed a daily prayer, not knowing which one he missed, he must pray a three-unit prayer, a four-unit prayer and a two-unit prayer in lieu of it.[18]

18 His intention for all three would be in lieu of a missed prayer. If he is travelling and does not know which one he missed, he need only perform a two-unit and a three-unit prayer in lieu.

مسائل:

الأولى: لا سهو على من كثر سهوه وتواتر، ولا على الإمام والمأموم إذا حفظ عليه الآخر، ولا سهو في سهو.

الثانية: من سهى في النافلة بنى على الأقل، وإن بنى على الأكثر جاز.

الثالثة: من تكلم ساهياً، أو قام في حال القعود، أو قعد في حال القيام، أو سلم قبل الإكمال، وجب عليه سجدتا السهو، وكذا يجبان على من شك بين الأربع والخمس فإنه يبني على الأربع ويسجدهما.

الرابعة: سجدتا السهو بعد الصلاة، ويقول فيهما: «بسم الله وبالله، اللّهمّ صلّ على محمد وآل محمد»، أو «السلام عليك أيها النبي ورحمة الله وبركاته»، ثم يتشهد خفيفاً ويسلم.

الخامسة: المكلف إذا أخل بالصلاة عمداً أو سهواً أو فاتته بنوم أو بسكر وكان مسلماً قضى، وإن كان مغمى عليه جميع الوقت أو كان كافراً فلا قضاء، والمرتدّ يقضي، ولو لم يجد ما يتطهر به من الماء والتراب سقطت أداءً وقضاءً.

السادسة: إذا دخل وقت الفريضة وعليه فائتة تخير بينهما، وإن تضيقت الحاضرة تعينت.

السابعة: الفوائت تترتّب كالحواضر.

الثامنة: من فاتته فريضة ولم يعلم ما هي صلى ثلاثاً وأربعاً وإثنين.

9) When he is in his hometown, he may shorten the prayer in lieu of what he missed while travelling; and when travelling, the traveller performs the full prayer in lieu of whichever prayer he missed while at home.

10) It is recommended to offer prayers in lieu of missed daily supererogatory prayers (*nawāfil*). If he missed them due to illness, he may offer 750 grams of grain in charity per two units missed, or if he cannot afford to, then per day's prayers missed.

التاسعة: الحاضر يقضي ما فاته في السفر قصراً، والمسافر يقضي ما فاته في الحضر تماماً.

العاشرة: يستحب قضاء النوافل المرتبة، ولو فاتت بمرض أستحب أن يتصدق عن كل ركعتين بمدّ، فإن لم يتمكن فعن كل يوم.

Chapter 6: The Congregational Prayer

It is obligatory to pray the Friday and Eid prayers in congregation if the prerequisites are met, and recommended in all of the remaining obligatory prayers, as well as the two Eid prayers when their prerequisite conditions cannot be met, and the prayer for rain. A congregation is formed of two or more people, and is invalidated if there is a barrier between the imam and the follower that prevents him from being seen – except between women – or if the imam is standing on an unusually higher level above the congregation. The opposite is allowed, however. The followers must not stand at an irregular distance behind the imam, nor outside of the rows. If the follower joins the prayer with the imam in the state of bowing, then he may count it as his first unit, otherwise not. The follower does not recite the *sūras* when praying behind a suitable imam,[19] and he must not precede him in his actions.

He must make an intention to follow the imam, and it is permitted for him to intend to pray a different prayer to the one intended by the imam [e.g. the imam makes an intention to pray the afternoon prayer while the follower joins the prayer behind the imam but intends to pray the noon prayer].

If there is only one follower, he is recommended to stand to the right of the imam, but if there are more people then they stand behind him, unless they are all naked, in which case he sits between them in one row. The same rulings apply to women, but if they are praying with the men, then they must stand behind them.

The imam must be of age of jurisprudential responsibility, of legitimate birth, and veracious. A man praying sitting down should not lead one who is standing, nor should an illiterate man lead a literate one, nor someone with a speech impediment lead one who pronounces correctly, and nor should a woman or hermaphrodite lead a man.

A descendant of the Banū Hāshim and the resident imam of the mosque take priority in leading. Priority should be given to the one who is better at reciting the Quran, then the most knowledgeable among them, then one who emigrated [for the sake of God] first, then the oldest, then the most presentable.

19 That is, someone who follows the same school of jurisprudence and sect as him.

الباب السادس: في صلاة الجمعة

وهي واجبة في الجمعة والعيدين بالشرائط، ومستحبة في الفرائض الباقية، والعيدين مع اختلاف الشرائط، والاستسقاء.

وتنعقد بإثنين فصاعداً، ولا تصح مع حائل بين الإمام والمأموم يمنع المشاهدة - إلا في المرأة -، ولا مع علو الإمام في المكان بما يعتد به، ويجوز العكس، ولا يتباعد المأموم بالخارج عن العادة من دون صفوف.

ولو أدرك الإمام راكعاً أدرك الركعة وإلا فلا، ولا يقرأ المأموم مع المرضى ولا يتقدمه في الأفعال.

ولا بد من نية الائتمام، ويجوز اختلافهما في الفرض.

وإذا كان المأموم واحداً استحب أن يقف عن يمينه، وإن كانوا جماعة فخلفه، إلا العاري فإنه يجلس وسطهم.

وكذا المرأة، ولو صلين مع الرجال تأخرن عنهم.

ويعتبر في الإمام التكليف، والعدالة، وطهارة المولد.

ولا يؤم القاعد القائم، ولا الأمّي القارئ، ولا المؤف اللسان صحيحه، ولا المرأة رجلاً ولا خنثى.

والهاشمي وصاحب المسجد أولى.

ويقدم الأقرأ، فالأفقه، فالأقدم هجرة، فالأسن، فالأصبح.

It is disapproved for residents of a place to be led by a traveller, or for those purified by *wuḍū'* to be led by one who purified himself with *tayammum* (dry ablution), or for the able-bodied to be led by a disabled person or a leper. A man who has been disciplined with a fixed penalty for a crime after his repentance is also discouraged from leading, as is the uncircumcised man, someone who is disliked by the followers, and nomadic travellers leading those who have migrated and settled in a place.

Other Rulings:

1) If the imam breaks his state of purity in the prayer, he must step aside and appoint someone else to deputize for him and be prayer leader. If he dies or faints, those following him must appoint someone to complete the prayer.

2) If someone entering the congregational prayer fears that he will miss that unit of the prayer [because of his distance from the congregation and the time it would take to join them] he [may recite the first *takbīrat al-iḥrām* and begin the prayer from his position and] bow down into the bowing position, then walk [towards the congregation] and join with them.

3) If the imam begins the prayer[20] while one is in the middle of a supererogatory prayer, he must break it. If he is in the middle of an obligatory prayer, then he must complete it as a supererogatory one,[21] then follow him. If the imam were to be an actual Imam,[22] then he would have to break the prayer, and follow him [without the need for him to change his intention to a supererogatory prayer and completing two units].

4) If one misses some of the prayer, he joins in with the imam and considers the point at which he joined to be the beginning of his prayer. When the imam concludes the prayer with *taslīm*, he stands to complete his own prayer.

20 Some manuscripts begin this sentence with the verb *aḥrama* instead of *dakhala*. Both however indicate the same meaning that the imam begins his prayer, since *aḥrama* means that the imam has performed the first *takbīrat al-iḥrām*.

21 That is, he changes his intention from an obligatory prayer to a supererogatory one, and performs only two units of prayer before joining the congregrational prayer. See Sayyid Ṣādiq al-Shīrāzī, *Sharḥ Tabṣirat al-muta'allimīn*, vol. 1, p. 97.

22 That is, one of the twelve Imams from the progeny of the Prophet.

ويكره أن يأتمّ الحاضر بالمسافر ، والمتطهّر بالمتيمّم ، والسليم بالأجذم والأبرص والمحدود بعد توبته والأغلف . ويكره إمامة من يكرهه المأمومون ، والأعرابي بالمهاجرين .

مسائل:

الأولى : لو أحدث الإمام استناب ، ولو مات أو أغمي عليه قدموا إماماً .

الثانية : لو خاف الداخل فوات الركعة ركع ومشى ولحق بهم .

الثالثة : إذا دخل الإمام وهو في نافلة قطعها ، ولو كان في فريضة أتمها نافلة ، ولو كان إمام الأصل قطعها وتابعه .

الرابعة : لو فاته بعض الصلاة دخل مع الإمام وجعل ما يدركه أول صلاته ، فإذا سلم الإمام قام وأتم الصلاة .

الخامسة : يستحب عمارة المساجد مكشوفة ، والميضاة على أبوابها ، والمنارة مع حائطها ، والإسراج فيها ، وإعادة المستهدم .

5) It is recommended to build mosques with open roofs, to keep he ablution areas near its entrances and the minaret in line with its walls, to place lights therein and to repair whatever is dilapidated.

The equipment of a mosque [that has been damaged] is allowed to be used in another.

It is prohibited to decorate them, to adorn them with images, to take ownership of them or use them as roads, to bring any kind of impurity therein, to remove even their pebbles from them, and if they are taken, then they must be replaced.

It is disapproved to build mosques up to be very high; to have balconies and prayer niches built into the walls; to use them as walkways; to trade or transact in them; to use them for public announcements or meting out punishments; to recite poetry in them; to use them as workshops; to sleep or spit in them; to use them to detain mentally unstable people; and to carry out executive orders therein.

It is recommended to enter a mosque with the right foot, and to exit with the left, to supplicate whilst entering and leaving, and to sweep them.

ويجوز استعمال آلته في غيره منها .

ويحرم زخرفتها ، ونقشها بالصور ، وأخذها أو بعضها في ملك أو طريق ، وإدخال النجاسة إليها ، وإخراج الحصى منها وتعاد لو أخرج.

ويكره تعليتها ، والشرف والمحاريب في حائطها ، وجعلها طريقاً ، والبيع فيها والشراء ، والتعريف ، وإقامة الحدود ، وإنشاد الشعر ، وعمل الصنائع ، والنوم ، والبصاق ، وتمكين المجانين ، وإنفاذ الأحكام.

ويستحب تقديم الرجل اليمنى دخولاً ، واليسرى خروجاً ، والدعاء فيهما ، وكنسها .

Chapter 7: The Prayer During a Fearful Situation

This is a shortened prayer regardless of whether [the battle] occurs in one's hometown or not, or whether it is being prayed in congregation or individually. It has three conditions: a) that the group of Muslims be large enough to be divided into two groups whereby one resists enemy attack while the other prays, b) that the enemy camp is sufficiently large enough to incite fear, and c) that the enemy position is opposite to the direction of *qibla* [i.e. behind the congregation when they are praying].

Method: The imam should lead the first group in the first unit,[23] then stands for the second, whilst waiting for them to finish the rest of their prayer (individually) with the *taslīm*,[24] whereupon the second group joins him while he prays the second unit. He pauses during the *tashahhud* so that they can catch up with him, and he completes the prayer alongside them with *taslīm*. In a three-unit prayer, he would pray the first unit with the first group, and the next two with the second group, or vice versa.

It is obligatory to stay armed if necessary, as long as the weapons are not an impediment to any obligatory action in the prayer. In an intensely threatening situation, one may pray standing, walking or riding, to the best of one's ability. He can prostrate on the saddlebow of his saddle or gesture. He should face *qibla* wherever possible. If he is unable to pray with gestures, then he may pray by reciting the four pronouncements glorifying God: *Subḥānallāhi wal-ḥamdu lillāhi wa lā ilāha illa Allāhu wallāhu akbar* in place of every unit.

One who is sinking into quicksand or drowning can pray by gesturing.

The prayer can only be shortened thus during travel and fear.

23 That is, while the second group is countering the enemy.

24 That is, when they would leave to counter the enemy, whilst the second group pray.

الباب السابع: في صلاة الخوف

وهي مقصورة سفراً وحضراً جماعة وفرادى، وشروطها ثلاثة: أن يكون في المسلمين كثرة يمكنهم الافتراق قسمين يقاوم كل قسم العدو، وأن يكون في العدو كثرة يحصل معها الخوف، وأن يكون العدو في خلاف جهة القبلة.

وكيفيتها: أن يصلي الإمام بالأولى ركعة ويقف في الثانية حتى يتموا ويسلموا فيجيئ الباقون فيصلي بهم الثانية ويقف في التشهد حتى يلحقوه فيسلم بهم، وإن كانت ثلاثية صلى بالأولى ركعة وبالثانية ركعتين أو بالعكس.

ويجب أخذ السلاح ما لم يمنع شيئاً من الواجبات فيؤخذ مع الضرورة. وصلاة شدة الخوف بحسب الإمكان واقفاً أو ماشياً أو راكباً، ويسجد على قربوس سرجه وإلا أومأ، ويستقبل القبلة ما أمكن، ولو لم يتمكن من الايماء صلى بالتسبيح عوض كل ركعة: «سبحان الله والحمد لله ولا إله إلا الله والله أكبر».

والموتحل والغريق يصليان إيماءاً، ولا يقصران إلا مع السفر أو الخوف.

Chapter 8: Prayers Performed by a Traveller

During a journey, every four-unit prayer is shortened to two, provided five conditions are met:

1) The distance intended covers the distance of eight *farsakh*, or four if one intends to return the same day.

2) That his journey is not to a place where he has a house or place of residence and has lived therein for six months or more, or where he intends to stay for more than ten days. If that is the case, but the distance there fulfils the conditions, then he may shorten his prayers during the journey.

3) That the journey is undertaken for legitimate purposes. If it is to commit a sin, then he is not allowed to shorten his prayer.

4) That his travelling does not exceed his time spent at home, such as sailors, transporters, shepherds, Bedouins, and travelling tradesmen. The yardstick is that he must not be staying in his town of residence for more than ten days at a time. If he stays in his own town or in another town for more than ten days, then he must shorten his prayers when he leaves.

5) That the enclosures of his town should have disappeared from his line of sight, and he should no longer be able to hear the call to prayer in his hometown, before he can shorten his prayers.

Once these conditions are met, it is obligatory for him to shorten his prayers, except in the sanctuary of God [Mecca], the Prophet's mosque in Medina (peace be upon him), the mosque of Kufa and Karbala, whereby he has a choice. If he intentionally prays the full prayer anywhere else, he must repeat it. If he was ignorant of the ruling, he need not repeat it, and if he forgot, then he must repeat it if he remembers within the time, but if not, then not.

If he sets off on his journey after the time for prayer has set in, he must shorten the prayer if he is performing the prayer in its time. And if he returns home from the journey after the time for prayer has set in, then he must pray in full.

If a traveller intends to stay somewhere for more than ten days, he must pray in full. If he did not intend to do so initially, but ends up having to extend his stay, then he may shorten his prayer up to thirty days extended in total, then he must pray in full.

الباب الثامن: في صلاة المسافر

يسقط في السفر من كل رباعية ركعتان بشروط خمسة:

أحدها: قصد المسافة، وهي: ثمانية فراسخ، أو أربعة مع قصد العود في يومه.

الثاني: أن لا ينقطع سفره ببلد له فيه ملك قد استوطنه ستة أشهر فصاعداً أو عزم على إقامة عشرة أيام، ولو قصد المسافة وله على رأسها منزل قصر في طريقة خاصة.

الثالث: إباحة السفر، فلو كان عاصياً بسفره لم يقصر.

الرابع: أن لا يكون سفره أكثر من حضره كالملاح والمكاري والراعي والبدوي والذي يدور في تجارته. والضابط: من لا يقيم في بلده عشرة أيام، ولو أقام أحد هؤلاء في بلده أو بلد غير بلده عشرة قصر إذا خرج.

الخامس: أن يتوارى عنه جدران بلده أو يخفى أذان مصره، فلا يترخص قبل ذلك.

ومع حصول الشرائط يجب التقصير، إلا في حرم الله وحرم رسوله ﷺ ومسجد الكوفة والحائر - على ساكنه السلام - فإنه يتخير، ولو أتم في غيرها عمداً أعاد، والجاهل لا يعيد، والناسي يعيد في الوقت لا خارجه.

ولو سافر بعد دخول الوقت قصر مع بقاء الوقت، ولو دخل من السفر بعد دخول الوقت أتم.

ولو نوى المسافر إقامة عشرة أيام أتم، ولو لم ينو قصر إلى ثلاثين يوما ثم يتم.

The Book of Alms Tax

كتاب الزكاة

This book consists of two sections: income tax and the alms of *fiṭra*.

Chapter 1: Prerequisites and Timing of the Obligation

The alms tax on income is obligatory on every responsible free person who has reached the age of responsibility and is the owner of the taxable amount that he has access to.

It is recommended for the guardians who invest on behalf of a child to pay the *zakāt* from it. If one has no access to his assets, and it is impossible for him to pay out of them, then it is not obligatory for him. When he does recover them after years, it is recommended for him to pay one year's *zakāt* from them.

There is no *zakāt* payable on a loan.

The borrower must pay the *zakāt* on the loan if he has left it as it is (un-invested) for a year.

The *zakāt* becomes obligatory on the first day of the twelfth Islamic month, provided the necessary conditions have been fulfilled up until the completion of that year. One is not allowed to postpone payment if he can afford to pay, and he remains liable until he does. Nor is he allowed to pay in advance of the obligation. If he does so, it is treated as a loan [to the recipient] to be returned to him. However, if the obligation sets in and the recipient [to whom the payments was made] continues to be someone to whom *zakāt* is rightfully paid, then it can be counted as *zakāt*.

It is not allowed to send the *zakāt* outside of one's hometown if there are deserving needy people within it, and he remains liable [were he to send money elsewhere]. If there are no needy people, then it can be sent out and he is no longer liable. It is incumbent to make the intention when remitting the payment.

There are two conditions to the liability: being a Muslim and having the capacity to pay. The disbeliever who embraces Islam is absolved from making up for the time when he was a disbeliever and did not pay. He who cannot pay it despite it being obligatory upon him is not liable to make up for it later, if the taxable amount is destroyed or squandered.

وهي قسمان: زكاة المال، وزكاة الفطرة. وهنا أبواب:

الباب الأول: في شرائط الوجوب ووقته

إنما تجب زكاة المال على البالغ العاقل الحر المالك للنصاب المتمكن من التصرف فيه.

ويستحب لمن اتجر في مال الطفل من أوليائه إخراجها عنه.

والمال الغائب إذا لم يتمكن صاحبه منه لا تجب فيه. ولو مضت عليه أحوال كذلك استحب إخراج زكاة حول عنه بعد وجوده.

ولا زكاة في الدين.

وزكاة القرض على المقترض إن تركه بحاله حولاً.

ومع هلال الثاني عشر تجب مع بقاء الشرائط في كمال الحول، ولا يجوز التأخير مع المكنة فيضمن، ولا تقديمها قبل وقت الوجوب. فإن دفع كان قرضاً له استعادته واحتسابه منها مع بقائه على الاستحقاق وتحقق الوجوب.

ولا يجوز نقلها عن بلدها مع وجود المستحق فيه، ويضمن، ولو عدم: نقل ولا ضمان. ولا بد من النية عند الإخراج.

وأما الضمان فشرطه اثنان: الإسلام، وإمكان الأداء. فالكافر يسقط عنه بعد إسلامه، ومن لم يتمكن من إخراجها مع الوجوب إذا تلفت لم يضمنها.

Chapter 2: Taxable Things From Which *Zakāt* Must Be Paid

There are nine commodities and no more, contained within the three groupings below:

1. *Livestock*

Zakāt is obligatory to be paid out of three categories of livestock: camels, cows and sheep, upon the fulfilment of four conditions: the taxable amount [for each commodity has been reached]; that they are domesticated, bred or farmed; that they have been owned for a whole year, and that they are not kept solely to work.

There are twelve taxable limits for camels: For five camels, one sheep is given as *zakāt*. For ten camels, two sheep are given. For fifteen camels, three sheep are given. For twenty camels, four sheep are given, then five sheep for twenty-five camels. For twenty-six camels or more, a one-year old she-camel is given as *zakāt*. For thirty-six camels or more, a two-year old she-camel is given. For forty-six camels or more, a three-year old she-camel is given, until the limit of sixty-one camels, when a she-camel over five years old is given. For seventy-six camels or more, two she-camels over two years old are given, and for ninety-one camels or more, two three-year old she-camels are given. The next limit is one-hundred-and-one camels, for which a three-year old she-camel is given for every additional fifty camels, and a two-year old she-camel for every additional forty camels.

Cattle only have two taxable limits: For thirty cows and above, a calf over one year old must be given. For forty cows and above, a two-year old calf is given.

As far as sheep are concerned, there are five taxable limits: When forty sheep are owned, one must be given in *zakāt*. Then for one-hundred and twenty-one sheep, two sheep are given. Then, for two-hundred and one sheep, three are given; and four for three hundred and one sheep, after which he gives one sheep for every additional one-hundred sheep that he owns, starting at four hundred.

Zakāt need not be paid on the additional numbers between two taxable bands, which are termed *shanaq* for extra camels, *waqṣ* for extra cows, and *ʿafw* for extra sheep.

الباب الثاني: فيما تجب فيه الزكاة

وهي تسعة أصناف لا غير، وينضمّها ثلاثة فصول:

الفصل الأول: زكاة النعم

تجب الزكاة في النعم الثلاثة: الإبل والبقر والغنم، بشروط أربعة: النصاب والسوم والحول وأن لا تكون عوامل.

فنصاب الإبل اثنا عشر: خمس وفيها شاة، ثم عشر وفيها شاتان، ثم خمس عشرة وفيها ثلاث شياه، ثم عشرون وفيها أربع شياه، ثم خمس وعشرون وفيها خمس شياه، ثم ست وعشرون وفيها بنت مخاض، ثم ست وثلاثون وفيها بنت لبون، ثم ست وأربعون وفيها حقة، ثم إحدى وستون وفيها جذعة، ثم ست وسبعون وفيها بنتا لبون، ثم إحدى وتسعون وفيها حقتان، ثم مائة وواحدة ففي كل خمسين حقة وفي كل أربعين بنت لبون بالغاً ما بلغ.

وأما البقر فلها نصابان: أحدهما ثلاثون وفيه تبيع أو تبيعة، والثاني أربعون وفيه مسنة.

وأما الغنم ففيها خمسة نصب: أربعون وفيها شاة، ثم مائة وإحدى وعشرون ففيها شاتان، ثم مائتان وواحدة ففيها ثلاث شياه، ثم ثلاثمائة وواحدة ففيها أربع شياه، ثم أربعمائة ففي كل مائة شاة، بالغاً ما بلغت.

وما لا يتعلق به الزكاة - وهو ما بين النصابين - في الإبل شنقاً، وفي البقرة وقصاً، وفي الغنم عفواً.

The condition that animals be grazing (i.e. domesticated, farmed or bred) applies to all of them, throughout the year. If animals stay inside of their own accord, eating fodder, or are kept inside and fed fodder by their owners, the year begins afresh from when they resume grazing.

Ownership of the animals for a year is a condition that applies to all, and the year consists of twelve months, whereby *zakāt* becomes payable as the twelfth month begins. If the taxable number of animals falls below the number required for taxation before the year is completed, then the obligation is lifted, even if the owner of livestock deliberately intends to avoid taxation by keeping their number below the limit [e.g. by giving some of his livestock away]. If it is after a year, however, then it is not lifted.

Other Rulings:

1) The ewe that is to be given in *zakāt* should be at least seven months old if it is the tax for sheep, and at least six months old if it is for goats. It suffices to give either a male or female sheep [as *zakāt*].

One year-old camels are termed *bint al-makhāḍ* and *tabīʿ*. Two-year old camels are called *bint al-labūn* and *al-masanna*. A camel that has completed three years and entered into its fourth is termed *ḥiqqa*, and one that has entered into its fifth year, a *jadhaʿa*.

2) An animal cannot be given in *zakāt* if it is sick, too old, has recently given birth,[1] has physical defects, is not considered a good feeder, or is a stallion used to impregnate females. The [owner's] male camels can be taken from for *zakāt*, if they are all sick.

3) If the owner is obligated to give a one-year old camel[2] but he only has a two-year old camel, then he must give that one, and two ewes or twenty dirhams are to be returned to him instead. If it is the opposite situation,[3] then he must give the one-year old camel along with two ewes or twenty dirhams. The same formula applies in the case of three and four-year old camels. A two-year old male camel is valued the same as one-year old she-camel.

4) It is not obligatory to give away the animal itself, rather one may give its monetary value instead.

1 That is, up to fifteen days.

2 This is the *zakāt* which becomes obligatory for someone who owns 26 camels.

3 That is, when the obligation upon him is to give a two-year old camel as *zakāt* for 36 camels but he only has a one-year old camel

وأما السوم، فهو شرط في الجميع طول الحول، فلو اعتلف في أثناء الحول من نفسها، أو أعلفها مالكها، استأنف الحول بعد العود إلى السوم.

وأما الحول: فهو شرط في الجميع، وهو اثنا عشر شهرا، وبدخول الثاني عشر تجب الزكاة. ولو ثلم النصاب قبل الحول سقط الوجوب ولو قصد الفرار، ولو كان بعده لم يسقط.

مسائل:

الأولى: الشاة المأخوذة في الزكاة أقلها الجذع من الضأن، والثني من المعز، ويجزئ الذكر والانثى. وبنت المخاض والتبيع: هو الذي كمل حولاً. وبنت اللبون والمسنة: ما كمل الحولين. والحقة: ما كملت ثلاثاً ودخلت في الرابعة. والجذعة: ما دخلت في الخامسة.

الثانية: لا تؤخذ المريضة، ولا الهرمة، ولا الوالدة، ولا ذات العوار، ولا تعد الأكولة، ولا فحل الضراب. ولو كانت إبله مراضاً أخذ منها.

الثالثة: من وجب عليه بنت مخاض وعنده بنت لبون، دفعها واستعاد شاتين أو عشرين درهماً، ولو كان بالعكس دفع بنت مخاض ومعها شاتين أو عشرين درهماً، وكذا الحقة والجذعة، وابن اللبون يساوي بنت المخاض.

الرابعة: لا يجب إخراج العين، بل يجوز دفع القيمة.

2. *The Zakāt Payable on Gold and Silver*

Zakāt on these two becomes obligatory upon fulfilment of [three] conditions, namely: ownership for a year, which we have dealt with; taxable amounts; and that they be coins and minted.

The taxable amount for gold is twenty dinars' worth,[4] from which half a dinar must be paid, then two *qayrāṭ* (one tenth of a dinar) for every four dinars owned on top. There is no *zakāt* payable on less than twenty dinars' worth of gold, nor on amounts in between which are less than four dinars.

The taxable amount for silver is two hundred dirhams' worth, from which five dirhams is payable, then a dinar for every forty dirhams' worth thereafter. There is nothing payable for amounts less than two hundred dirhams, nor on amounts less than forty dirhams in between bands, nor on ingots and jewellery, even if they are kept in this state to evade tax if it is before a year has elapsed, but if it is after, then it is payable on them.

3. *The Zakāt Payable on Crops*

Zakāt is payable on four categories of crops: wheat, barley, dates and raisins. It is not obligatory for any other crops. The obligation rests on the fulfilment of two conditions:

Firstly, the taxable amounts, which is five *wasaq* for each crop.[5] One *wasaq* is equivalent to sixty *ṣāʿ*, one of which in turn equals four *mudd*. One *mudd* is equivalent to two and a quarter Iraqi *raṭl*[6]. One-tenth of this amount then becomes payable as *zakāt*, provided the crops are watered by irrigation channels, rainfall or groundwater. If they are watered manually by water drawn from wells by oxen, or with buckets and watering cans, then half of a tenth of the total is payable (i.e. one-twentieth). The extra weight (over five *wasaq*) is treated similarly and the fraction of *zakāt* paid out of it, even if it is a little, after all costs and taxes have been deducted from the total. If the crops are watered with both running water as well as manually with buckets and the like, one must consider which method is used more often, and calculate the *zakāt* accordingly. If they are watered equally with both means, then they are to be divided as such (i.e. one-tenth is payable on half of the crops and one-twentieth on the other half).

4 Twenty dinars' worth is the same as twenty *mithqāl* in weight. One *mithqāl* is equal to 3.456 grains.

5 One *wasaq* is 185 kilograms.

6 One *raṭl* is 449.28 grams, which is equivalent to a pound in imperial weight.

الفصل الثاني: في زكاة الذهب والفضة

تجب الزكاة فيهما بشروط: الحول وقد مضى، والنصاب، وكونهما مضروبين بسكة المعاملة.

ونصاب الذهب: عشرون ديناراً ففيه نصف دينار، ثم أربعة دنانير ففيها قيراطان، وهكذا دائماً. ولا يجب فيما نقص عن عشرين ولا عن أربعة شيء.

ونصاب الفضة: مائتا درهم ففيها خمسة دراهم، ثم أربعون ففيها درهم ولا شيء فيما نقص عن المائتين، ولا عن الأربعين، ولا السبائك، ولا الحلي وإن قصد الفرار قبل الحول وبعده تجب.

الفصل الثالث: في زكاة الغلات

تجب الزكاة في أربعة أجناس منها، وهي: الحنطة، والشعير، والتمر، والزبيب. ولا تجب فيما عداها.

وإنما تجب فيها بشرطين:

الأول: النصاب، وهو في كل واحد منها خمسة أوسق، كل وسق ستون صاعاً، كل صاع أربعة أمداد، كل مد رطلان وربع بالعراقي، فيجب العشر إن سقى سيحاً أو بعلاً أو عذيا وإن سقى بالقرب والدوالي والنواضح فنصف العشر، ثم كل ما زاد وإن قلّ، بعد إخراج المؤون من بذر وغيره. ولو سقى بهما اعتبر بالأغلب، ولو تساويا قسط.

Secondly, that they must be growing on his property at his disposal. If they have been transferred to him through trade or as a gift, then *zakāt* is not obligatory on them if they came to him after having been harvested, but if it was before, then it is.

The *zakāt* on crops is payable as soon as they are matured in strength, and on fruits as soon as they are fit for consumption. The *zakāt* must be given out as soon as the crops have been winnowed and the fruit plucked.

If there are different types of crops, the taxable amounts must be calculated from each one separately, and not combined as one.

4. *Commodities on Which the Payment of Zakāt Is Recommended*

It is recommended to pay *zakāt* on the capital one uses for trade, provided that one year has elapsed, that the initial capital stays the same or increases over the course of the year, and that the value of his capital meets the taxable amount. He may pay in either of the two currencies.

Zakāt is recommended to be paid on horses provided that he has owned them for a year, they are domesticated and grazing, and they are female. He should pay two dinars for an *ʿatīq*[7] and one dinar for a *birdhūn*.[8]

It is recommended to pay *zakāt* on whatever grows out of the ground besides the four obligatory crops, provided that the same conditions that apply to obligatory crops are fulfilled and they are harvested in the same way.

7 A superior breed of horse.
8 An ordinary horse.

الثاني: أن ينمو في ملكه، فلو انتقلت إليه بالبيع أو الهبة أو غيرهما لم تجب الزكاة إن كان نقلها بعد بدو الصلاح، وإن كان قبله وجبت.

ويتعلق الزكاة بالغلات إذا اشتدّت، وفي الثمار إذا بدا صلاحها. ووقت الإخراج عند التصفية وجذ الثمرة. وإن اجتمعت أجناس مختلفة ينقص كل جنس عن النصاب، لم يضم بعضه إلى بعض.

الفصل الرابع: فيما يستحب فيه الزكاة

يستحب الزكاة في مال التجارة بشرط: الحول، وأن يطلب برأس المال أو بزيادة في الحول كله، وبلوغ قيمته النصاب، ويقوم بالنقدين.

ويستحب في الخيل بشرط: الحول، والسوم، والأنوثة. فيخرج عن العتيق ديناران، وعن البرذون دينار واحد. ويستحب فيما تخرج الأرض عدى الأجناس الأربعة من الحبوب، بشرط حصول شرائط الوجوب في الغلات، ويخرج كما يخرج منها.

Chapter 3: Those Entitled to Receive *Zakāt*

There are eight groups of people entitled to receive *zakāt*:

1) and 2) are the poor and the destitute, and these are people who do not possess enough to support themselves and their dependants for a year, and are unable to make ends meet even after working. A man who owns a house to live in, or has a slave to serve him or an animal to ride can be given it [i.e. a poor person is not required to sell his possessions to make ends meet before he can be considered a recipient of *zakāt*].

3) *zakāt* collectors,

4) those who are to be encouraged to take part in *jihād* [to defend Muslims territories], even if they are non-Muslims.

5) to manumit slaves, which would be the *mukātab* slaves [contracted to earn their freedom] and those undergoing extreme difficulty.

6) those who are in financial debt, not due to having committed acts of disobedience to God.

7) for the cause of God, which includes any noble action for proximity to God, such as *jihād*, pilgrimage, building mosques and reservoirs.

8) to wayfarers, who have left their hometowns and have either become stranded or do not have sufficient funds during their travel, even if they were wealthy in their own town, including the guest, as long as their travel is for a legitimate purpose.

The first two groups of people must be believers, and the children of believers may also be entitled. If a non-Twelver Shiʿa gives *zakāt* to another non-Twelver, then he must be give his *zakāt* again [to another Twelver Shiʿa Muslim] after they have converted to the Twelver *madhhab*.

They must also not be one's own dependants who are mandatorily his own responsibility, such as his parents, grandparents, children, grandchildren, spouse or slave.

They must also not be Hashimīs, if the one giving is a non-Hashimī himself, since they are provided for from the *khums* tax. Hashimīs can, however, receive other recommended alms, and one can give *zakāt* to their protégés. One is allowed to give all his *zakāt* to a single person, though it is recommended to divide it equally amongst people in all the different categories.

The minimum amount of *zakāt* to be given to the poor is what is obligatory within the first tax band, and there is no maximum amount.

الباب الثالث: في المستحق للزكاة

وهم ثمانية أصناف:

الأول والثاني: الفقراء والمساكين، وهم الذين لا يملكون قوت سنتهم لهم ولعيالهم، ويكون عاجزاً عن تحصيل الكفاية بالصنعة. ويعطى صاحب دار السكنى وعبد الخدمة وفرس الركوب.

الثالث: العاملون، وهم السعاة للصدقات.

الرابع: المؤلفة قلوبهم، وهم الذين يستمالون للجهاد وإن كانوا كفاراً.

الخامس: في الرقاب، وهم المكاتبون والعبيد الذين في الشدة.

السادس: الغارمون، وهم المدينون في غير معصية الله.

السابع: في سبيل الله، وهو كل مصلحة أو قربة، كالجهاد، والحج، وبناء المساجد والقناطر.

الثامن: أبن السبيل، وهو المنقطع به في الغربة، وإن كان غنياً في بلده، والضيف إذا كان سفرهما مباحاً.

ويعتبر في الأولين الايمان، ويعطى أولاد المؤمنين، ولو أعطي المخالف مثله أعاد مع الاستبصار.

وأن لا يكونوا واجبي النفقة عليه، من الأبوين وإن علوا، والأولاد وإن نزلوا، والزوجة، والمملوك.

وأن لا يكونوا هاشميين إذا كان المعطي من غيرهم وتمكنوا من الخمس.

وتحل للهاشمي المندوبة، ويجوز إعطاء مواليهم. ويجوز تخصيص واحد بها أجمع.

والمستحب تقسيطها على الاصناف.

وأقل ما يعطى الفقير ما يجب في النصاب الأول، ولا حد للكثرة.

Chapter 4: *Zakāt Al-Fiṭra*

It is obligatory every year upon any free person of age of responsibility who is financially independent inasmuch as he can provide for himself [and his dependants, if he has any] for the period of one year. It must be paid at the very onset of the month of Shawwāl before the Eid al-Fitr prayer.

One can pay it earlier in Ramadan, but it must not be delayed beyond the Eid, unless there is a legitimate excuse [such as forgetfulness].

If payment on time is missed, it must be paid later. If he had set it aside but it ended up being used up without negligence or excess, then there is no obligation to pay it. It is not permissible to send the money out of one's hometown if there are deserving people within it.

The amount that is payable: is nine *raṭl*[9] *of* either wheat, barley, dates, raisins, rice, or cooked yogurt whey; or four *raṭl* of milk. The best thing to give is dates, then raisins, then whatever is most frequently consumed in a year [by the one paying the *zakāt*]. It is permissible to give the value of the foodstuff instead.

It is obligatory to give the *zakāt* on one's own behalf, and on behalf of each of one's dependants, be they Muslim or non-Muslim, free or slave, young or old, even if he is maintaining them voluntarily [and not mandatorily].

He must make the intention and remit the *zakāt* to one who is entitled to receive it. The best thing to do is to give it to the Imam (peace be upon him), and during his occultation, to the Shīʿī jurists [to distribute to those who are entitled].

A poor person must not be given less than one *ṣāʿ*,[10] and there is no upper limit.

It is recommended to prioritise one's own needy relatives, then one's poor neighbours before giving to others. It is recommended for the poor to pay the *zakāt* too.

9 Approximately three kilograms.
10 Approximately three kilograms.

الباب الرابع: في زكاة الفطرة

وهي واجبة على المكلف الحر الغني، وهو مالك سنته، في كل سنة، عند هلال شوال، وتتضيق عند صلاة العيد.

ويجوز تقديمها في رمضان، ولا تؤخر عن العيد إلا لعذر.

ولو فاتت قضيت، ولو عزلها ثم تلفت من غير تفريط فلا ضمان. ولا يجوز نقلها عن بلده مع وجود المستحق.

وقدرها: تسعة أرطال [بالعراقي]، من الحنطة والشعير والتمر والزبيب والأرز والأقط، ومن اللبن أربعة أرطال بالمدني.

وأفضلها: التمر، ثم الزبيب، ثم ما يغلب على قوت السنة. ويجوز إخراج القيمة.

ويجب أن يخرجها عن نفسه وعن من يعوله من مسلم وكافر، حر وعبد، صغير وكبير، وإن كان متبرعاً بالعيلولة.

ويجب فيها النية، وإيصالها إلى مستحق زكاة المال.

والأفضل صرفهما إلى الإمام عليه السلام، ومع غيبته إلى المأمون من فقهاء الامامية.

ولا يعطى الفقير أقل من صاع، ولا حد لأكثره.

ويستحب اختصاص القرابة بها ثم الجيران. ويستحب للفقير إخراجها.

Chapter 5: The One-Fifth Tax

It is obligatory to be paid out of the spoils of war, minerals, treasures that have been discovered by unearthing them from the ground, valuables obtained by diving, profits made from business, industry and agriculture, the land of a *dhimmī* if he had bought it from a Muslim, and any commodity that is owned that has become mixed with the same commodity gained unlawfully whereby it is not possible to distinguish between the two.

The taxable amount for minerals and treasures is twenty dinars' worth, and for those obtained by diving, one dinar's worth. The taxable amount for profits from business, industry and agriculture is considered after the deduction of one year's necessary expenditure for himself and his family, after which it is obligatory on the remainder.

The tax becomes payable upon acquisition of these things.

The *khums* tax is divided into six parts: one part for God, one part for His Messenger, one part for his close relatives – these three all fall under the Imam's share. Then one share for the needy among those who are Hāshimī, one share for their orphans and one share for wayfarers from among them.

It is not to be sent out of town if there are eligible people within it, and one can prioritise between the latter three groups with respect to their shares. Those who are entitled to receive it must be believers, and the orphans must be poor.

Spoils of war are considered as being: any derelict place whose inhabitants have vacated it; any land that has been acquired through means other than war; any land whose inhabitants have willingly embraced Islam without resistance; mountain summits; valleys; unoccupied arid land; crown land and assets [previously belonging to kings which have not been misappropriated]; possessions and property of those with no heirs; and spoils of war obtained without the Imam's consent. These are all regarded as belonging to the Imam.

In his absence, we have been permitted to deal with properties, businesses and marriage of female captives.

الباب الخامس: في الخمس

وهو واجب في غنائم دار الحرب، والمعادن، والغوص، وأرباح التجارات والصناعات والزراعات، وأرض الذمي إذا اشتراها من مسلم، والحرام الممتزج من الحلال ولم يتميز.

ويعتبر في المعادن والكنوز عشرون ديناراً، وفي الغوص دينار، وفي أرباح التجارات والصناعات والزراعات الزيادة عن مؤنة السنة له ولعياله بقدر الاقتصاد فيجب في الزائد.

ووقت الوجوب: وقت حصول هذه الأشياء.

ويقسم الخمس ستّة أقسام: سهم لله، وسهم لرسوله، وسهم لذي القربى فهذه الثلاثة للإمام. وسهم للمساكين من الهاشميين، وسهم لأيتامهم، وسهم لأبناء سبيلهم.

ولا يحمل عن البلد مع وجود المستحق فيه، ويجوز اختصاص بعض الطوائف الثلاثة بنصيبهم.

ويعتبر فيهم الإيمان، وفي اليتيم الفقر.

والأنفال: كل أرض خربة باد أهلها، وكل أرض لم يوجف عليها بخيل ولا ركاب، وكل أرض أسلمها أهلها من غير قتال، ورؤوس الجبال، وبطون الاودية، والموات التي لا أرباب لها، والآجام، وصوافي الملوك وقطائعهم غير المغصوبة، وميراث من لا وارث له، والغنائم المأخوذة بغير إذن الإمام. فهذه كلها للإمام.

وأبيح لنا المساكن، والمتاجر، والمناكح.

The Book of Fasting

كتاب الصوم

Chapter 1: Introduction to Fasting

Fasting is to intentionally abstain from various things which break the fast. If one is fasting specifically in the month of Ramadan, then the intention of proximity to God suffices. If not, then he needs to specify the reason. The intention must be made the night before, or up until noon is allowed. After noon, the time for making an intention has elapsed. It is obligatory to abstain from those things in both Ramadan fasts and other specified fasts, as well as fasts in lieu of them.

The intention of fasting the whole month of Ramadan can be made at the beginning of the month, and one is even allowed to make it prior to the onset of the month. He should fast a recommended fast on the thirtieth day of Shaʿbān, known as the day of doubt, and if it turns out that it is the first day of Ramadan, it will count as such. If he wakes up intending not to fast but has not done anything to make the fast invalid, and it turns out to be the first of Ramadan, he has until noon to make his intention to fast. If it is after noon, then he must continue to abstain and then fast in lieu of it later.

The duration of the fast is in the daytime, from the second dawn until sunset.

فيه أبواب:

الباب الأول: الصوم

الصوم هو الامساك عن المفطرات مع النية، فإن تعين الصوم كرمضان كفت فيه نية القربة، وإلا افتقر إلى التعيين. ووقتها الليل، ويجوز تجديدها إلى الزوال، فإذا زالت الشمس فات وقتها ووجب الإمساك في رمضان والمعين، ثم قضاه.

ويجزي في رمضان نية عن الشهر في أوّله، ويجوز تقديم النية عليه.

ويوم الشك يصام - ندباً - عن شعبان، فإن اتفق أنه من رمضان أجزأ. ولو أصبح بنية الإفطار ولم يفطر ثم تبين أنه من رمضان جدد النية إلى الزوال، ولو كان بعد الزوال أمسك واجباً وقضى.

ومحل الصوم النهار، من طلوع الفجر الثاني إلى الغروب.

Chapter 2: Things That One Must Abstain From

These fall into two categories: obligatory and recommended.

The obligatory things to abstain from are: eating, drinking, sexual intercourse (both frontal and rear-entry), wilful ejaculation of semen, intentionally allowing thick smoke to enter the throat, intentionally remaining in the state of sexual impurity until dawn, and going back to sleep in that state after two wakings (without performing *ghusl*) until dawn. These seven things lead to both having to make up fasts in lieu of those missed (*qaḍā'*) as well as paying the atonement penalty (*kaffāra*).

The things that require reparative fasts to be made up in lieu are: inadvertently continuing to eat after dawn under the assumption that it is still night time, whist having the ability to be mindful of it, even if someone else had told him that there was still time left; eating before nightfall under the mere assumption of darkness of the sky – though if he had considerable belief that night had fallen, then found out that it had not, then he need not make up the fast – or by following others in assuming nightfall when in reality it was not yet time; going back to sleep in the state of sexual impurity after one waking without performing *ghusl* until dawn; inducing vomiting; unintentionally swallowing water whilst washing the mouth to cool oneself down except for water that is swallowed naturally when performing ablution for prayers; and injecting liquid enemas.

It is obligatory to refrain from lying about Almighty God, His Messenger and the Imams, peace be upon them. There are two opinions regarding submerging one's head in water.[1] As for abstaining from everything else that is generally prohibited aside from what we have already listed, it would apply even more so during the fast.

The recommended things to abstain from are: taking medicines through the nose, applying eyeliner or kohl containing musk or aloe; blood-letting; taking a bath if it leads to weakness; smelling narcissus or other fragrant flowers; using suppositories; wearing wet clothes to keep cool; kissing, petting, or being intimate with one's spouse with desire; and for women to sit in water.

The fast is not broken by keeping the stone of a ring in the mouth, chewing mastic, tasting food and spitting it out, feeding a bird by mouth, and for men to sit in water.

1 According to one group, if a person immerses their head in water intentionally then his fast becomes nullified and as a result must be made up for at a later date. The other opinion holds that it is impermissible for someone fasting to immerse their head in water, but doing so does not invalidate their fast. See Sayyid Ṣādiq al-Shīrāzī, *Sharḥ Tabṣirat al-muta'allimīn*, 1: 185.

الباب الثاني: فيما يمسك عنه

وهو ضربان: واجب، وندب.

فالواجب: الأكل، والشرب، والجماع في القبل والدبر، والاستمناء، وإيصال الغبار الغليظ إلى الحلق متعدياً، والبقاء على الجنابة متعمداً حتى يطلع الفجر، ومعاودة النوم بعد انتباهتين حتى يطلع الفجر. وهذه السبعة توجب القضاء والكفارة.

ويجب القضاء: بالإفطار بعد الفجر مع ظن بقاء الليل وترك المراعاة مع القدرة عليها، ولو أخبره غيره ببقاء الليل، وقبل الغروب للظلمة الموهمة - ولو غلب على الظن دخول الليل فلا قضاء - وتقليد الغير في دخول الليل ولم يدخل، ومعاودة النوم بعد انتباهة واحدة قبل الغسل حتى يطلع الفجر، وتعمد القيء، ودخول الماء إلى الحلق للتبرّد - دون ماء المضمضة للصلاة - والحقنة بالمائعات.

ويجب الإمساك عن الكذب على الله تعالى وعلى رسوله وعلى الائمة عليهم السلام.

وفي الارتماس في الماء قولان، وكذا الإمساك عن كل محرم سوى ما ذكرناه، ويتأكد في الصوم.

والمندوب: [ترك] السعوط، والكحل بما فيه صبر أو مسك، وإخراج الدم، ودخول الحمام للضعفان، وشم النرجس والرياحين، والحقنة بالجامد وبل الثوب على الجسد، والقبلة والملاعبة والمباشرة بشهوة، وجلوس المرأة في الماء.

ولا يفسد الصوم بمص الخاتم، ومضغ العلك، وذوق الطعام إذا لفظه، وزق الطائر، واستنقاع الرجل في الماء.

Other Rulings:

1) Payment of the atonement is only obligatory for breached Ramadan fasts, fasts due to specific solemn vows, overdue reparative (*qaḍā'*) fasts, and breaching the *i'tikāf* mosque retreat [during the month of Ramadan].

Breaches in non-specific fasts for general oaths, reparative fasts that are made up on time, and recommended fasts are not liable for the atonement fine.

2) The atonement for specified fasts is either to free a slave, or fast for two consecutive months, or to feed sixty needy people.

The atonement for an overdue reparative fast is to feed ten needy people, and if unable to do so, then to fast three days instead.

If a person intentionally does not fast for two days, then the atonement is doubled. A person who refuses to fast [without just cause] must be reprimanded and killed if he insists on the correctness of his action.

3) A man who forces his wife to have sex with him is responsible for paying the atonement, whereas the wife who voluntarily consents [to intercourse] must pay her own atonement.

مسائل:

الاولى: الكفارة لا تجب إلا في رمضان والنذر المعين، وقضاء رمضان بعد الزوال، والاعتكاف على وجه.

وما لا يتعين صومه كالنذر المطلق وقضاء رمضان قبل الزوال، والنافلة، لا يجب بإفساده شيء.

الثانية: كفارة المتعين: عتق رقبة، أو صيام شهرين متتابعين، أو إطعام ستين مسكينا.
وكفارة قضاء رمضان بعد الزوال: إطعام عشرة مساكين، فإن عجز صام ثلاثة أيام.
ولو تكرر الإفطار في يومين تكررت الكفارة. ويعزّر المفطر، ولو كان مستحلاً قتل.

الثالثة: المكره لزوجته يتحمل عنها الكفارة، والمطاوعة تكفر عن نفسها.

Chapter 3: Types of Fasts

There are four types of fasts: obligatory, recommended, disapproved and prohibited fasts.

Obligatory Fasts: Ramadan fasts, atonement fasts, fasts as penalties for unintentional bloodshed during the Hajj, solemn vows and the like, fasting during the *iʿtikāf* retreat according to one point of view, and reparative fasts in lieu of missed Ramadan fasts. Other fasts in lieu will be dealt with in due course.

The onset of the month of Ramadan is determined by sighting the crescent moon, or upon completion of thirty days of the month of Shaʿbān, or upon evidence of the crescent having been sighted.

There are seven prerequisites to fasting, which are: to have reached the age of responsibility, to have full mental capacity, to be of good health, to not be travelling or under similar duress, and to not be in a state of menstrual or post-natal bleeding.

The prerequisites for reparative fasts are: to have reached the age of responsibility, to have full mental capacity and to be a Muslim.

The apostate must make up for fasts that he missed during his period of apostasy.

The person fasting reparative fasts in lieu of Ramadan may choose to do so anytime until noon, by which he must specify his intention.

الباب الثالث: في أقسامه

وهي أربعة: واجب، ومندوب، ومكروه، ومحظور.

والواجب شهر رمضان، والكفارات، ودم المتعة والنذر، وشبهه، والاعتكاف على وجه، وقضاء الواجب بغير رمضان يأتي في أماكنه.

وأما شهر رمضان: فعلامته رؤية الهلال، أو مضي ثلاثين من شعبان، أو قيام البينة بالرؤية.

وشرائط وجوبه سبعة: البلوغ، وكمال العقل، والسلامة من المرض، والإقامة، أو حكمها، والخلو من الحيض، والنفاس.

وشرائط القضاء: البلوغ، وكمال العقل، والإسلام.

والمرتدّ يقضي ما فاته من زمان ردّته.

ويتخير قاضي رمضان في إتمامه إلى الزوال، فيتعين.

Recommended Fasts: One may fast any day of the year except for forbidden days. There are sixteen occasions that have special emphasis: the first Thursday of every month, the first Wednesday in the second-third of the month, the last Thursday from the last third of the month, the Day of Ghadīr,[2] the Day of Mubāhila,[3] the Day of Mab'ath,[4] the birthday of the Prophet,[5] the day celebrating the creation of the earth,[6] the Day of 'Āshūrā' by way of mourning,[7] the Day of 'Arafa as long as it does not weaken him from supplicating,[8] the first of Dhū l-Ḥijja, the first of Rajab, the whole of Rajab, the whole of Sha'bān, the luminous days [of full-moons],[9] every Thursday and Friday.

It is recommended for those who are temporarily excused from fasting to refrain from eating the rest of the day in situations such as when the traveller arrives home after noon, or beforehand, but is not fasting; the sick person if he starts to recover; women in the state of menstruation or postpartum bleeding who are clean (but have yet to purify themselves ritually); the disbeliever who embraces Islam during a day of Ramadan; the child who becomes religiously responsible; someone with diminished mental capacity or who was unconscious and then regains consciousness.

The voluntary fast of a guest without the permission of his host is not valid, nor the wife's voluntary fast without her husband's permission, nor the child's without his father's permission, and nor the slave's without his owner's permission.

Disapproved Fasts: supererogatory fasts whilst travelling or when invited for a meal; on the Day of 'Arafa if it makes one too weak to supplicate; when one is doubtful about it being the first of Ramadan.

2 The Day of Ghadīr: 18th Dhū l-Ḥijja.

3 The Day of Mubāhila: 25th Dhū l-Ḥijja.

4 The Day of Mab'ath: 27th Rajab.

5 The birthday of the Prophet: either the 12th or 17th Rabī' al-Awwal.

6 The day celebrating the creation of the earth (*daḥw al-arḍ*): 24th Dhū l-Ḥijja.

7 This is the tenth day of Muḥarram. The idea is to abstain from food and drink in grief and mourning for Ḥusayn ibn 'Alī and his family, and one must break the fast in the afternoon before sunset, for he is not actually fasting but rather abstaining out of grief and mourning.

8 The Day of 'Arafa: 9th Dhū l-Ḥijja.

9 The 11th, 12th and 13th of every month.

والمندوب: جميع أيام السنة إلا المنهي عنه. والمؤكّد ستة عشر قسماً: أوّل خميس من كل شهر، وأوّل أربعاء من العشر الثاني، وآخر الخميس من الثالث، ويوم الغدير، والمباهلة، ويوم المبعث، ومولد النبي ﷺ، ويوم دحو الأرض، وعاشوراء على وجه الحزن، وعرفة لمن لا يضعفه عن الدعاء، وأوّل ذي الحجة، وأوّل رجب، ورجب كله، وشعبان كله، وأيام البيض، وكل خميس، وجمعة.

ويستحب الإمساك - وإن لم يكن صوماً - للمسافر القادم بعد الزوال أو قبله وقد أفطر، والمريض إذا برئ كذلك، والحائض والنفساء إذا طهرتا، والكافر إذا أسلم، والصبي إذا بلغ، والمجنون إذا أفاق، وكذا المغمى عليه.

ولا يصح صوم الضيف تطوّعاً بدون إذن المضيف، ولا المرأة بدون إذن الزوج، ولا الولد بدون إذن الوالد، ولا المملوك بدون إذن المولى.

والمكروه: النافلة سفراً، والمدعو إلى طعام، وعرفة مع ضعفه عن الدعاء أو شك الهلال.

Prohibited Fasts: Fasting on the two Eids; the days of *tashrīq* during the Hajj for those who are in Mina; the day of doubt as to whether the month of Ramadan has set in [with the intention of Ramadan]; fasting as an oath to commit a sin; a fast of silence; fasting consecutive days without a break; fasting obligatory fasts during a journey, except for: fasting vows specific to that end, fasts as atonement to be done during the Hajj for having shed blood or intentionally departed from ʿArafāt before sunset, or the fasting of a frequent traveller who travels more than he stays at home, which is someone who does not remain more than ten days at a stretch in his town of residence.

Other Rulings:

1) The obligatory fasts, with respect to timing, are divided into three categories:

Those with fixed timing: Ramadan fasts, fasts in lieu of them, solemn vows and fasting during *iʿtikāf.*

Those with flexible timing: the fast as atonement for injury while shaving the head in Hajj, the atonement for breached Ramadan fasts, and the atonement for hunting.

Consecutive timing: fast for the atonement of a breached oath, atonement for unintentional killing, atonement for *ẓihār* [a pre-Islamic form of divorce], atonement for bloodshed in Hajj, and the atonement of reparative fasts of Ramadan.

2) All these fasts must be completed consecutively apart from the general oaths and the like, reparative fasts, atonement for hunting, and the seven fasts in lieu of a sacrificial animal in Hajj.

3) When obligated to fast consecutively, if one does not do so for a good reason, he may resume where he left off, but if not, he must start again from the beginning, except for someone who is obligated to fast two months consecutively, in which case he must fast the first month consecutively and at least one day from the second. He who is obligated to fast a month consecutively must fast at least fifteen days (so as not to have to start again). One who must fast three days in lieu of the sacrificial animal for *ḥajj al-tamattuʿ*, may fast the two consecutive days of *tarwiya* and *ʿarafa*,[10] and the third after the days of *tashrīq*.[11]

10 The 8th and 9th Dhū l-Ḥijja.

11 The 11th, 12th and 13th days of Dhū l-Ḥijja are known as the days of *tashrīq*.

والمحرم: صوم العيدين، وأيام التشريق لمن كان بمنى، ويوم الشك على أنه من رمضان، وصوم نذر المعصية، وصوم الصمت، والوصال، والواجب في السفر إلا النذر المقيد به، وبدل دم المتعة، والبدنة لمن أفاض من عرفات قبل الغروب عامداً، أو يكون سفره أكثر من حضره، وهو كل من ليس له في بلده مقام عشرة أيام.

مسائل:

الاولى: الصوم الواجب ينقسم إلى:

مضيق، وهو رمضان، وقضاؤه، والنذر، والاعتكاف.

ومخير، وهو صوم كفارة أذى حلق الرأس، وكفارة رمضان، وجزاء الصيد.

ومرتب، وهو صوم كفارة اليمين، وقتل الخطأ، والظهار، ودم الهدي، وكفارة قضاء رمضان.

الثانية: كل الصوم يجب فيه التتابع إلا النذر المطلق وشبهه، والقضاء، وجزاء الصيد، والسبعة في بدل الهدي.

الثالثة: كل ما يشترط فيه التتابع إذا أفطر لعذر، بنى، وإن كان لغيره استأنف، إلا من وجب عليه شهران فصام شهراً ومن الثاني ولو يوماً، ومن وجب عليه شهر فصام خمسة عشر يوماً، والثلاثة في بدل هدي التمتع إذا صام يومي التروية وعرفة صام الثالث بعد أيام التشريق.

Chapter 4: Those Who Are Exempt From Fasting

When a woman begins menstruating or bleeds postpartum, her fast is invalidated regardless of the time of day. She must make up fasts in lieu of those missed. If she purifies herself after dawn (*fajr*), she is recommended to refrain from eating that day, and make up the missed fast later.

If a boy becomes religiously responsible or if a mentally incapacitated person regains consciousness before dawn, they must both mandatorily fast that day. But if it is after, then no.

If a sick person recovers or a traveller returns home before noon, and they have not done anything to invalidate the fast, they must mandatorily fast that day, and it counts as a fast. But if it is after noon, then not. If an illness continues until the following Ramadan, then the obligation to fast in lieu ceases, and he must give one dry measure (*mudd*) of food in charity for every fast missed instead. If he recovers between the two Ramadans for long enough to be able to make them up, then he must do so, but there is no penalty to atone. If, however, he takes them lightly, he must make them up in lieu as well as pay a penalty of one *mudd* per missed fast. In the case of more than two Ramadans' worth of fasts, the ruling is restricted to two Ramadans' worth.

The sick and the traveller must both mandatorily not fast, and even if they do, they are not considered valid. The same conditions that necessitate the prayer to be shortened whilst travelling equally apply to the fast.

The old man and woman who cannot fast due to old age can give one *mudd* of food in charity for each fast instead, as can someone suffering from polydipsia, who must make them up in lieu as and when he recovers.

The pregnant woman in the last stages of her pregnancy and the nursing mother whose milk is at risk of drying up may not fast, and instead give in charity and make them up in lieu later.

If an ill person dies due to his illness, then it is recommended for his next of kin to make up reparative fasts on his behalf. If, however, someone dies whilst having had the opportunity to make up his fasts, or having missed them due to travel and the like, then it is obligatory upon the next of kin – the oldest male child – to make them up. If there are two next of kin, they divide them up between themselves. A deceased woman must also have her fasts made up in lieu, but if the oldest child is female, then the obligation to make them up is lifted, and she must instead pay one *mudd* of food in charity for every fast left. If the deceased had two months' worth of fasts to be made up, his next of kin must fast one month in lieu and give in charity for the other month from the deceased's estate.

الباب الرابع: في المعذورين

إذا حاضت المرأة أو نفست، أي وقت كان من النهار، بطل صومها وتقضيه ولو طهرت بعد الفجر أمسكت استحباباً وقضته.

ولو بلغ الصبي أو أفاق المجنون قبل الفجر صاما ذلك اليوم واجباً، وإلا فلا.

والمريض إذا برئ أو قدم المسافر قبل الزوال ولم يفطرا أمسكا واجباً وأجزأهما، وإلا فلا.

ولو استمر المرض إلى رمضان آخر سقط القضاء وتصدّق عن الماضي لكل يوم بمدّ، ولو برئ بينهما وكان عازماً على الصوم قضاه ولا كفارة، وإن تهاون قضى وكفر عن كل يوم بمدّ، وحكم ما زاد على رمضانين حكم رمضانين.

ويجب الافطار على المريض والمسافر، فلو صاما لم يجزهما، وشرائط قصر الصلاة شرائط قصر الصوم.

والشيخ والشيخة مع عجزهما، يتصدقان عن كل يوم بمدّ، وكذا ذو العطاش، ويقضي مع البرء.

والحامل المقرب والمرضعة القليلة اللبن تفطران وتقضيان مع الصدقة.

ولو مات المريض في مرضه استحب لوليّه القضاء عنه، ولو مات بعد استقرار الصوم والفوات بسفر وغيره قضى الولي - وهو أكبر أولاده الذكور - واجباً، ولو كان وليان تحاصا. ويقضى عن المرأة، ولو كان الأكبر أنثى فلا قضاء، وتصدق من التركة عن كل يوم بمد، ولو كان عليه شهران قضى الولي شهراً، وتصدق من مال الميت عن الآخر.

Chapter 5: Ritual Retreat

This involves staying in the sacred mosque in Mecca, the Prophet's mosque in Medina, the central mosques of Kufa or Basra especially for worship.

Its conditions are: to make the intention, fast and spend at least three days in the mosque. It is made obligatory by making an oath or a vow to that effect; otherwise it is recommended and voluntary. If two days [in retreat] have been completed, then staying for a third becomes obligatory.

One must not leave the mosque except out of necessity[12] or for an act of obedience such as attending a brother's funeral, visiting a sick person, participating in funeral prayers or to give testimony.

If he does leave, he must not walk under a shade, sit down, or pray anywhere else, unless it is in Mecca. It is recommended for him to stipulate with God the condition that his retreat in the mosque can be annulled and cancelled, so that if he wishes he can terminate his stay in the mosque as long as two complete days have not yet passed. He is forbidden from enjoying the company of women,[13] transacting and trading, smelling fragrances and disputing.

It is invalidated by the same things that invalidate the fast. If he engages in sexual intercourse during the retreat, he must pay the atonement penalty for breaking a fast in the month of Ramadan, even if it was during the night. If it was in the daytime in the month of Ramadan, then he must pay double the penalty.[14]

If he breaks his fast during the *iʿtikāf* with something other than sexual intercourse (e.g. he eats or drinks), if it is an obligatory retreat due to a specific solemn vow, then he must pay the atonement, but if not, then not, unless it is the third day.

If a woman starts menstruating or a person falls ill during the retreat, they should leave, but they must make up those days in lieu another time.

12 That is, in order to relieve oneself and visit the lavatory.

13 This ruling is ambiguous, in that it could either be referring to sexual intercourse specifically or to any form of enjoyment such as kissing and touching.

14 The atonement penalty (*kaffāra*) for intentionally breaking a fast in the month of Ramadan are either to free a slave, to fast for two consecutive months, or to feed sixty poor people. See the Book of Fasting, Chapter Two, Other Rulings.

الباب الخامس: في الاعتكاف

وهو اللبث للعبادة في مسجد مكة، أو مسجد النبي ﷺ، أو جامع الكوفة أو البصرة خاصة.

وشرطه: النية، والصوم، وإيقاعه ثلاثة أيام فما زاد.

وهو واجب وندب: فالواجب ما أوجب بالنذر وشبهه، والندب ما تبرع به، فإذا مضى يومان وجب الثالث.

ولا يخرج عن المسجد إلا لضرورة أو طاعة كتشييع أخ أو عيادة مريض وصلاة جنازة وإقامة شهادة.

ومع الخروج لا يمشي تحت الظلال ولا يجلس ولا يصلي إلا بمكة. ويستحب الاشتراط.

ويحرم عليه الاستمتاع بالنساء، والبيع والشراء، وشم الطيب، والجدال.

ويفسده ما يفسد الصوم.

ولو جامع فيه كفر مثل كفارة رمضان وإن كان ليلاً، وفي نهار رمضان تتضاعف الكفارة.

ولو أفطر بغيره مما يوجب الكفارة، فإن وجب بالنذر المعين كفر، وإلا فلا، إلا في الثالث.

ولو حاضت المرأة أو مرض المعتكف خرجا وقضيا مع وجوبه.

The Book of Pilgrimage

كتاب الحج

This book consists of several chapters:

Chapter 1: Criteria for the Pilgrimage

They are: the obligatory pilgrimage as a pillar of Islam; the pilgrimage that becomes obligatory as the result of a solemn vow or the like; that which one performs when commissioned to do so on behalf of another person; and that which is performed in lieu of a thwarted pilgrimage. The pilgrimage as a pillar of Islam is essentially obligatory for every male, female and transgender person, provided they fulfil six criteria: the age of religious responsibility, mental capacity, freedom [from slavery], financial means, transportation and accessibility [to Mecca, and freedom from danger en route].

If a young boy or girl undertakes the pilgrimage, it does not count [since there is as yet no obligation upon him or her to perform the pilgrimage] unless they attain the age of religious maturity at one of the two obligatory stations; and the same thing applies to a slave [unless he is freed]. However, the state of *iḥrām* (religious consecration) of a young boy is perfectly correct, even if he has not yet reached the age of moral discernment (*mumayyaz*), as is that of someone with diminished mental capacity or the slave, with his owner's permission.

If a poor person is employed to carry out some work or fulfil some duties that happen to take him on the pilgrimage this would not count for him. However, if he attains the ability to perform the Hajj therefore, the pilgrimage then becomes obligatory upon him. If someone has the financial means but is ill, then there is no obligation on him to deputize someone.

Once the criteria have been met, it is obligatory to make haste in undertaking it without delay, and if one neglected it until his death despite having been able to, it is obligatory to pay someone from his assets - before they are divided - to undertake it on his behalf from the closest location to Mecca. This applies even if he has left behind only enough to cover the expenses of hiring someone.

وفيه أبواب :

الباب الأول: في أقسامه

وهي : حجة الإسلام، وما يجب بالنذر وشبهه ، وبالاستيجار ، والإفساد . فحجة الإسلام واجبة بأصل الشرع مرة واحدة على الذكور والإناث والخناثى ، بشروط ستة : البلوغ، والعقل ، والحرية ، والزاد ، والراحلة ، وإمكان المسير .

فلو حج الصبي لم يجزه إلا إذا أدرك أحد الموقفين بالغاً ، وكذا العبد . ويصح الاحرام بالصبي غير المميز وبالمجنون ، ومن العبد بإذن المولى .

ولو تسكع الفقير لم يجزئه بعد الاستطاعة . ولو كان المتمكن مريضاً لم يجب الاستنابة .

ويجب مع الشرائط على الفور ، ولو أهمل مع الاستقرار حتى مات ، قضي من صلب ماله من أقرب الاماكن ولو لم يخلف غير الأجرة .

It is not permissible for someone who has not yet performed the obligatory pilgrimage to either perform it voluntarily as a recommended act or as a deputy on behalf of someone else. A woman requires neither a *maḥram* (male relative) nor her husband's permission for the obligatory pilgrimage, but she does for a supererogatory one.

A deputy hired to undertake it must be: Muslim and of sound mental capacity. They must not be under obligation to perform their own pilgrimage. If the deputy does not have their own obligation to perform the pilgrimage, then they are permitted to undertake it on behalf of someone else, even if they themselves are a first-timer or a woman. If someone voluntarily undertakes it on behalf of a deceased person (for whom the pilgrimage was obligatory but was not performed by them), then the deceased would have their religious duty fulfilled and their responsibility absolved.

ولا يجوز لمن وجب عليه الحج أن يحج تطوّعاً ولا نائباً . ولا يشترط في المرأة المحرم ولا إذن الزوج، ويشترط في الندب.

أما النائب، فشرطه: الإسلام، والعقل، وأن لا يكون عليه حج واجب، ولو لم يكن جاز ولو كان صرورة أو امرأة، ولو تبرّع عن الميت برئت ذمته.

Chapter 2: The Different Types of Pilgrimage

There are three different types: *ḥajj al-tamattuʿ*, *ḥajj al-qirān* and *ḥajj al-ifrād*.

Ḥajj al-tamattuʿ: consists of entering the state of *iḥrām*[1] (for *ʿumra*) from the *mīqāt,*[2] circumambulating the Kaʿba (*ṭawāf*) seven times, reciting a two-unit prayer at the Station of Abraham (*maqām Ibrāhīm*), briskly walking between Ṣafā and Marwa seven times (*saʿy*), and clipping the nails and hair; then re-entering the state of *iḥrām* a second time in Mecca for the Hajj, staying in ʿArafāt on the 9th Dhū l-Ḥijja until sunset, departing for al-Mashʿar [Muzdalifa] and staying there after the dawn prayer, throwing stones at *Jamrat al-ʿUqba,*[3] then making the sacrifice, then shaving the head on the Day of Sacrifice in Mina, then performing the Hajj circumambulation, the two-unit prayer and the *saʿy* thereafter, performing the circumambulation of women (*ṭawāf al-nisāʾ*) and the two-unit prayer after it, spending the nights of the 11th and 12th Dhūl-Ḥijja in Mina, stoning the three columns during those two days and on the 13th if one chooses to stay.

This type is obligatory for whoever resides at a distance of 24 km or more from Mecca in every direction.

Ḥajj al-ifrād: the Hajj pilgrimage is undertaken first, and then the *ʿumra* is performed straight after leaving the state of *iḥrām* for Hajj.

Ḥajj al-qirān: is the same, except that the sacrificial animal is brought along by the pilgrim from when he enters the state of *iḥrām*.

The conditions for *ḥajj al-tamattuʿ* are: making the intention, undertaking it in the months of Hajj, which are Shawwāl, Dhū l-Qaʿda and Dhū l-Ḥijja, performing the Hajj and *ʿumra* components within the same year, and entering the state of *iḥrām* from Mecca itself.

The conditions for the other two types of Hajj are: making the intention, undertaking them in the months of Hajj, entering the state of *iḥrām* from the *mīqāt* or from one's own home if it is closer than the *mīqāt*.

1 *iḥrām*: state of obligatory religious consecration that the pilgrim enters upon engaging in the process of the pilgrimage.

2 *mīqāt* (pl. *mawāqīt*): special places designated for the pilgrim to enter the state of *iḥrām*.

3 *Jamrat al-ʿUqba*: the last of the three columns symbolising Satan appearing to Hajar in three different instances. The first column is called *Jamrat al-Ūlā*, and the middle one, *Jamrat al-Wusṭā*.

الباب الثاني: في أنواعه

وهي ثلاثة: تمتُّع بالعمرة إلى الحج، وقِران، وإفراد.

أما التمتّع: فصورته الإحرام من الميقات، والطواف بالبيت سبعاً، وصلاة ركعتين في مقام إبراهيم عليه السلام، والسعي بين الصفا والمروة سبعاً، والتقصير. والإحرام ثانياً من مكة بالحج، والوقوف بعرفات تاسع ذي الحجة إلى الغروب، والإفاضة إلى المشعَر والوقوف به بعد الفجر، ورمي جمرة العقبة، ثم الذبح، ثم الحلق يوم النحر بمنى، وطواف الحج، وركعتاه، وسعيه، وطواف النساء، وركعتاه، والمبيت بمنى ليلة الحادي عشر والثاني عشر، ورمي الجمار الثلاث في اليومين، ثم إن أقام الثالث عشر رمى.

وهذا فرض من نأى عن مكة اثنا عشر فما زاد من كل جانب.

والمفرد: يقدم الحج ثم يعتمر عمرة مفردة بعد الاحلال.

والقارن: كذلك، لكنه يسوق الهدي عند إحرامه.

وشرط التمتّع: النية، ووقوعه في أشهر الحج، وهي شوال وذو القعدة وذو الحجة، وإتيان الحج والعمرة في عام واحد، وإنشاء إحرام الحج من مكة.

وشرط الباقيين: النية، ووقوعه في أشهر الحج، وعقد الإحرام من الميقات أو من منزله إن كان دون الميقات.

In both, it is permissible to perform the circumambulation before going to ʿArafāt, but it is recommended to recite the *talbiya*[4] during each round of the circumambulation to renew his state. The *ḥajj al-tamattuʿ* pilgrim is obliged to offer the sacrificial animal [on the day of Eid], whereas the other two types of pilgrims are not obligated to do so.

4 *talbiya*: the obligatory declaration of obedience and submission to God by pronouncing "*labbayk, allāhumma labbayk, labbayk, lā sharīka laka labbayk...*", which means "Here I am at Your service, O God, yes, at Your service. Responding to Your call, there is no partner to You..."

ويجوز لهما الطواف قبل المضي إلى عرفات، لكنهما يجددان التلبية عند كل طواف استحباباً، ويجب على المتمتع الهدي، ولا يجب على الباقين.

Chapter 3: The State of *Iḥrām*

One can only enter the state of *iḥrām* from the *mīqāt*, of which there are six designated places:

For the people of Iraq: al-ʿAqīq, the best point being al-Maslakh, then Ghamra, and the last point, Dhāt ʿIrq. One must not pass this point without having entered the state of *iḥrām*.[5]

For the people of Medina: Masjid al-Shajara, or in case of emergency, al-Juḥfa, which is the designated *mīqāt* for the people of Greater Syria.

For the people of Yemen: Yalamlam.

For the people of Ṭā'if: Qarn al-Manāzil.

For *ḥajj al-tamattuʿ*, Mecca is the *mīqāt*.

Whoever resides close to the *mīqāt* may consider his own home as his *mīqāt*.

Children may enter it from Fakh.[6]

If someone is travelling in from elsewhere to perform the Hajj, he must enter the state of *iḥrām* from the designated place for his townsfolk.

One is not allowed to enter *iḥrām* prior to these *mīqāt*, and if he intentionally delays it beyond the *mīqāt*, he must return and enter *iḥrām* from there. If he is unable to return, then his Hajj is void. However, if he forgot or was ignorant of the ruling, he must return if possible, but if not then he can enter it from wherever he is. If he forgot the *iḥrām* until after the Hajj rituals had been completed, his Hajj is valid, according to one narration.

The obligatory parts of entering the state of *iḥrām* are: making the intention and maintaining it throughout, reciting the four-part *talbiya* declaration for the *tamattuʿ* and *ifrād* pilgrims, and the same in addition to marking and garlanding the sacrificial animal for the *qirān* pilgrim. The *talbiya* is as follows: "*Labbayk allāhumma labbayk, labbayk. Innal-ḥamda wan-niʿmata wal-mulk laka, lā sharīka laka labbayk*" – "Here I am at Your service, O God, yes, at Your service. Surely all praise, bounty and sovereignty belong to You; there is no partner to You; at Your service." He must also wear two garments appropriate to pray in.

5 *muḥrim*: one who has entered the state of *iḥrām*.

6 Fakh: the name of a well close to Mecca. This is an allowance made for children due to their possible frailty and inability to withstand the elements.

الباب الثالث: في الإحرام

وإنما يصح من الميقات، وهي ستة:

لأهل العراق: العقيق، وأفضله المسلخ، وأوسطه غمرة، وآخره ذات عرق. فلا يجوز عبورها إلا محرما.

ولأهل المدينة: مسجد الشجرة، وعند الضرورة الجحفة، وهي ميقات أهل الشام اختياراً.

ولليمن. يلملم.

وللطائف: قرب المنازل. ولحج التمتع مكة.

ومن كان منزله أقرب من الميقات فمنزله ميقاته. وفخ للصبيان.

ومن حج على طريق أحرم من ميقات أهله.

ولا يجوز الإحرام قبل هذه المواقيت، ولو تجاوزها متعمداً رجع وأحرم منها، وإن لم يتمكن بطل حجه، وإن كان ناسياً أو جاهلاً رجع مع المكنة، وأحرم من موضعه إن لم يتمكّن. ولو نسي الإحرام حتى كملت مناسكه صح حجه على رواية.

والواجب في الإحرام: النية، واستدامتها حكماً، والتلبيات الأربع للمتمتّع والمفرِد، وهي والإشعار والتقليد للقارِن، وصورتها «لَبَّيكَ اللَّهُمَّ لَبَّيْكَ لَبَّيْك، إنّ الحَمدَ والنِّعْمَةَ والمُلْكَ لك، لا شَرِيكَ لَكَ لَبَّيْك»، ولبس الثوبين مما يصح فيه الصلاة.

The recommended parts are: for the *tamattuʿ* pilgrim to let his hair grow from the beginning of Dhū l-Qaʿda; to clean the body; to trim the nails and the moustache; to remove the hair from the pubic region and the armpits; to perform a *ghusl* before wearing the *iḥrām*; and to do so either straight after the noon prayer, or any obligatory prayer, or after reciting six units of prayer, or just two units; to recite the *talbiya* loudly as soon as his mount reaches within a mile of al-Baydāʾ en route from Medina; to supplicate and to express what type of Hajj he intends to perform, and whether it is for himself or as someone's deputy; for the *tamattuʿ* pilgrim to continuously recite the *talbiya* until he sees the buildings of Mecca, for the *ifrād* and *qirān* pilgrim to do so until noon on the Day of ʿArafa, and for the *ʿumra* pilgrim to do so until he enters the sanctuary (*al-ḥaram*). The *iḥrām* clothing must be unstitched cotton. Women's *iḥrām* is the same as men's except in the prohibition of wearing stitched clothing; and menstruation does not prevent them from entering the state of *iḥrām*.

والمندوب: توفير شعر الرأس للمتمتّع من أوّل ذي القعدة، وتنظيف الجسد، وقصّ الأظفار والشارب، وأخذ العانة والإبطين بالنورة، والغسل أمامه، والإحرام عقيب الظهر، أو فريضة، أو ست ركعات، أو ركعتين، ورفع الصوت بالتلبية إذا علت راحلته البيداء على طريق المدينة، والدعاء والتلفظ بالنوع والاشتراط وتكرار التلبية إلى أن يشاهد بيوت مكة للمتمتع، وإلى عند الزوال يوم عرفة للمفرد والقارن، وإذا دخل الحرم للمعتمر، والإحرام في قطن محض.

وإحرام المرأة كإحرام الرجل إلا في تحريم المخيط، ولا يمنعها الحيض منه.

Chapter 4: Counterindications in the State of *Iḥrām*

It is obligatory to refrain from fourteen actions: hunting on land, capturing an animal for consumption, trapping it, pointing it out for someone else to hunt, consuming the meat of a hunted animal, slaughtering an animal; having sexual intercourse, kissing, being intimate with, and looking lustfully at women; marrying, arranging a marriage or witnessing it; masturbating, using or smelling fragrance; for men to wear stitched clothing, covering the tops of the feet, telling lies, swearing argumentatively by saying "No, by God!" and "Yes, by God!"; killing insects on one's body, unnecessarily removing hair, using dye, covering the head for men, intentionally walking under a shelter, clipping the fingernails, cutting trees or pulling up herbage that does not belong to oneself except plucking fruits, Meccan lemongrass and dates from palms.

Using black kohl is disapproved, as is looking in the mirror, wearing a ring for adornment purposes, blood-letting through cupping, rubbing the body, carrying weapons unnecessarily – or not at all according to another opinion; for a woman to cover her face, wearing *iḥrām* garments that are dirty or that stand out, using henna for cosmetic purposes, entering public baths, responding to someone calling his name by saying "*labbayk*", and using aromatic herbs and fragrances.

Scratching the body and brushing one's teeth are allowed as long as they do not cause bleeding.

الباب الرابع: في تروك الإحرام

والواجب منها أربعة عشر تركاً: صيد البر، وإمساكه، وأكله، والإشارة إليه، والإغلاق عليه، وذبحه، والنساء: وطئاً وتقبيلاً ولمساً ونظراً بشهوة، وعقداً له ولغيره وشهادة عليه، والاستمناء، والطيب، والمخيط للرجال، وما يستر ظهر القدم، والفسوق وهو الكذب، والجدال وهو قول لا والله وبلى والله وقتل هوام الجسد، وإزالة الشعر مع غير الضرورة، واستعمال الدهن، وتغطية الرأس للرجال، والتظليل سائراً، وقصّ الأظفار، وقطع الشجر والحشيش النابت في غير ملكه إلا الفواكه والإذخر والنخل.

ويكره الاكتحال بالسواد، والنظر في المرآة، ولبس الخاتم للزينة، والحجامة، ودلك الجسد، ولبس السلاح اختياراً، على أحد القولين في ذلك كله، والنقاب للمرأة، والإحرام في الثياب الوسخة والمعلمة، والحناء للزينة، ودخول الحمام وتلبية المنادي، واستعمال الرياحين.

ويجوز حك الجسد والسواك ما لم يدم.

Chapter 5: Atonement Penalties (*Kaffāra*) for Breaching the State of *Iḥrām*

This chapter contains two sections:

Section One: Atonement Penalties for Hunting

This includes any land animal that is customarily allowed to be hunted but prohibited in *iḥrām*. Fishing of animals who lay their eggs and hatch in water is allowed, as are pheasants.

If the *muḥrim* hunts an ostrich, he must provide a sacrificial animal (*badana*).[7] If this is not possible, then he must spend the value of the sacrificial animal on staple foods to feed up to sixty people two *mudds* each. Whatever is left over after having fed sixty people, he may keep for himself, and he is not obligated to make up any shortfall. If he cannot afford to feed people, he must fast a day for every two *mudd* [i.e. every person unfed]. If he cannot fast that many days, then he must fast for eighteen days.

If he hunts and kills a wild ox or donkey, then he must provide an ox for sacrifice. If this is not possible, then he must spend the value of the sacrificial animal on staple foods to feed up to thirty people two *mudds* each. If it feeds fewer people, he does not have to make up the rest, but he is allowed to keep the surplus. If he is not able to feed people, he must fast a day per person. If he cannot fast that many days, then he must fast for nine days.

If he hunts and kills a deer, fox or rabbit, then he must provide a sheep for sacrifice. If he cannot do so, then he must spend the value of the sacrificial animal on staple foods to feed up to ten people two *mudds* each. He may keep any surplus and does not have to make up for the shortfall. If he is unable to feed people, then he must fast a day per person. If he cannot fast that many days, then he must fast for three days.

If he breaks an ostrich egg and the chick inside was moving, he must provide a young virgin camel for every egg. If the chick inside was not moving, then he must facilitate the breeding of as many camels as eggs destroyed, and the calves are to be presented as offerings to the House of God. If he is unable to do this, then he must sacrifice a sheep for every egg. And if unable to do that, then he must fast for three days.

7 *badana*: this is usually a camel or a cow.

الباب الخامس: في كفّارات الإحرام

وفيه فصلان

الفصل الأول: في كفّارات الصيد

وهو الحيوان المحلّل الممتنع في البرّ، ويجوز صيد البحر وهو ما يبيض ويفرخ فيه، والدجاج الحبشي.

ففي النعامة (بدنة)، ومع العجز يفض ثمن البدنة على البر ويطعم لكل مسكين مدان، وما زاد عن ستين له، ولا يجب عليه ما نقص عنه. ولو عجز صام عن كل مدين يوماً، فإن عجز صام ثمانية عشر يوماً.

وفي بقرة الوحش وحماره (بقرة)، فإن لم يجد فض ثمنها على البر وأطعم ثلاثين مسكيناً لكل واحد مدان، ولا يجب عليه التتميم، والفاضل له، وإن عجز صام عن كل مدين يوماً، فإن عجز صام تسعة أيام.

وفي الظبي والثعلب والأرنب (شاة)، فإن عجز فض ثمنها على البر وأطعم عشرة لكل مسكين مدان، والفاضل له، ولا يجب عليه التتميم، فإن عجز صام عن كل مدين يوماً، فإن عجز صام ثلاثة أيام.

وفي كسر بيض النعام إذا تحرك الفرخ، لكل بيضة (بكرة) من الإبل، وإن لم يتحرك أرسل فحولة الإبل في إناث بعددها فالناتج هدي لبيت الله، فإن عجز فعن كل بيضة شاة، فإن عجز صام ثلاثة أيام.

In the case of broken eggs of sand grouse or partridge, if the chick inside was moving, then he must offer a young lamb for sacrifice, but if it was not moving, then he must facilitate for the breeding of as many lambs as broken eggs and present the lambs as offerings to the House of God. If he is unable to do that, then the same ruling as for ostrich eggs applies.

If he kills a dove, he must sacrifice a sheep; and for its chick, a young lamb, and for each egg, a dirham [in charity].

The non-*muḥrim* who hunts it within the sanctuary of the *ḥaram* must pay a dirham for every dove hunted, half a dirham for a chick and a quarter for an egg. A *muḥrim* who hunts within the sanctuary must pay both the atonement penalty and the fine.

The atonement for killing a lizard, hedgehog or gerbil is a young goat; for a sand grouse, francolin or similar bird, a young weaned camel; and for birds such as sparrows, larks, wagtails and the like, he must give away a *mudd* of foodstuff. For every locust and louse that he throws off from his body, he must give away a handful of foodstuff, and if he kills many locusts, then he must offer a sheep for sacrifice, but he is not obligated if he did it unavoidably so.

If a pilgrim eats that which he killed, then he must pay double the atonement penalty, and if he eats what someone else slaughtered, then he pays just the once. If a group of pilgrims kill an animal together, then they must each pay the atonement fine. Whoever has game with him ceases to be its owner upon entering the state of *iḥrām*, and he must set it free. If he keeps it, he is held liable for it.

Other Rulings:

1) The *muḥrim* who kills an animal outside the sanctuary of the *ḥaram* must atone with the sacrifice, whereas the *non-muḥrim* who kills it within the *ḥaram* must pay its value. The *muḥrim* who kills within the *ḥaram* must therefore pay both types of atonement for all offences up until the sacrifice of a camel, after which they do not accrue further.

2) The pilgrim who has killed an animal remains liable for it, regardless of whether it was done intentionally, accidentally or inadvertently. If he repeats the unintentional offence, he must pay the atonement again, just as he must do if he repeats it intentionally.

3) If he is compelled by circumstances to eat the meat of the hunted game or dead animal, then he may eat only the game meat, then offer a sacrifice if he can afford to. If he cannot, or faces extenuating circumstances, then he may eat the meat of the dead animal.

وفي بيض القطا والقبج إذا تحرك الفرخ، لكل بيضة (من صغار الغنم)، وإن لم يتحرك أرسل فحولة الغنم في إناث بعددها والناتج هدي للبيت، ولو عجز كان كبيض النعام.

وفي الحمامة (شاة)، وفي فرخها (حمل)، وفي بيضها (درهم).

وعلى المحل في الحرم عن الحمامة درهم، وعن الفرخ نصف، وعن البيضة ربع، ويجتمعان على المحرم في الحرم.

وفي الضب والقنفذ واليربوع، وفي القطاة والدراج وشبهه (جمل فاطم)، وفي العصفور والقنبرة والصعوة (مد)، وفي الجرادة والقملة يلقيها عن جسده (كف من طعام)، وفي الجراد الكثيرة (شاة) ولو لم يتمكن من التحرز لم يكن عليه شيء.

ولو أكل ما قتله كان عليه فداءان، ولو أكل ما ذبحه غيره ففداء واحد، ولو اشترك جماعة في قتله فعلى كل واحد فداء، وكل من معه صيد يزول ملكه عنه بالإحرام. ويجب عليه إرساله. فإن أمسكه ضمنه.

مسائل:

الاولى: المحرم في الحل يجب عليه الفداء. والمحل في الحرم القيمة. ويجتمعان على المحرم في الحرم ما لم يبلغ بدنة فلا يتضاعف.

الثانية: القاتل يضمن الصيد بالقتل عمداً وسهواً وجهلاً. ولو تكرّر خطأً تكرّرت الكفّارة. وكذا العمد.

الثالثة: لو اضطرّ إلى أكل الصيد والميتة أكل الصيد وفداه مع المكنة. وإلا أكل الميتة.

4) The atonement fee due for killing a domesticated animal goes to its owner, and if was not owned by anyone, then it is to be given away in charity. The fine for killing doves of the *ḥaram* sanctuary is to be spent on feed for the doves within it.

5) The atonement sacrifice that a *muḥrim* becomes liable for during the Hajj must be slaughtered at Mina. If he is an *'umra* pilgrim, then he must do so in Mecca, in the designated place known as al-Ḥazūra.

6) The boundary of the *ḥaram* sanctuary is calculated as the area within one day's journey from the Sacred Mosque, and whoever hunts within it is held liable.

Section Two: Remaining Contraindications

There are several matters to be dealt with.

1) The *muḥrim* who intentionally has sexual intercourse with his wife by penetration before the two ritual stays, knowing of its prohibition, his Hajj is invalidated, though he must complete it, then undertake another reparative one the following year, and offer a sacrificial camel as atonement. This applies, regardless of whether the Hajj was obligatory or supererogatory. The same applies to her too if she did it voluntarily, and they must separate immediately, which means that they must not be alone together without other people present, from when they committed the sin until after the Hajj rituals have been completed, as well as when they undertake the pilgrimage in future.

If he coerced her, then her pilgrimage is valid, and he bears her share of the atonement too. If the act was committed after the two ritual stays, then their pilgrimage is valid, but they must each sacrifice a camel.

If he did it before the circumambulation upon first entering Mecca (*ṭawāf al-ziyāra*), then he must sacrifice a camel, but if he cannot afford one, then an ox or a sheep. If he did it before the circumambulation of women (*ṭawāf al-nisā'*), then he must sacrifice a camel. If he had already completed five rounds of that circumambulation, then there is no atonement due. If he had sexual intercourse in the state of *iḥrām* during his *'umra* pilgrimage before the *sa'y* ritual, then his *'umra* is invalidated but he must complete the pilgrimage regardless, sacrifice a camel, as well as undertake a reparative *'umra* in lieu.

If a man looks lustfully at someone other than his wife and ejaculates as a result, he must sacrifice a camel. If he cannot afford it, then an ox, and if not, then a sheep.

الرابعة: فداء الصيد المملوك لصاحبه، وغير المملوك يتصدق به، وحمام الحرم يشتري بقيمته علف لحمامه.

الخامسة: ما يلزمه في إحرام الحج ينحره أو يذبحه بمنى، وإن كان معتمراً فبمكة في الموضع المعروف بالحزورة.

السادسة: حد الحرم بريد في بريد، من أصاب فيه صيداً ضمنه.

الفصل الثاني: في بقية المحظورات

وفيه مسائل:

الأولى: من جامع امرأته قبل أحد الموقفين قبلاً أو دبراً عامداً عالماً بالتحريم بطل حجه، وعليه إتمامه والقضاء من قابل، وبدنة. سواء كان الحج فرضاً أو نفلاً، وعليها مثل ذلك إن طاوعته، وعليهما الافتراق، وهو أن لا ينفردا بالاجتماع إن حجا في القابل من موضع المعصية إلى أن يفرغا من المناسك.

ولو أكرهها صح حجها ويحمل عنها الكفارة، ولو كان بعد الموقفين صح الحج ووجبت البدنة على كل واحد منهما.

ولو جامع قبل طواف الزيارة لزمه بدنة، فإن عجز فبقرة أو شاة، ولو جامع قبل طواف النساء لزمه بدنة.

ولو كان قد طاف منه خمساً فلا كفارة. ولو جامع في إحرام العمرة قبل السعي بطلت. وعليه بدنة وقضاؤها وإتمامها.

ولو نظر إلى غير أهله فأمنى كان عليه بدنة، فإن عجز فبقرة، وإن عجز فشاة.

If he looks at his wife without lust but ends up ejaculating, then there is no atonement obligatory, but if it was done lustfully and he ejaculated, then he must sacrifice a young camel, as would be the case if he ejaculated whilst having intimate sexual contact with her.

If a *muḥrim* contracts a marriage with a fellow *muḥrim* and they consummate the marriage, they are both liable to pay double the atonement penalty.

2) The *muḥrim* who perfumes himself must sacrifice a sheep, regardless of whether it was through anointing himself with it, daubing it, using it as incense or consuming it. There is no problem with smelling the cloth of the Kaʿba.

3) For clipping the nails in *iḥrām*, he must offer a *mudd* of foodstuff for every nail. If he cut the nails of both hands and feet in one sitting, then he must sacrifice a sheep, but if he did it in more than one sitting, then two sheep. The person who issues a *fatwa* (ruling) allowing the clipping of nails is held liable if his follower clips his nails causing his finger to bleed, and he too must then sacrifice a sheep.

4) Wearing stitched clothing also obligates sacrificing a sheep, even if it was done out of necessity.

5) A man who shaves his head must sacrifice a sheep or feed a *mudd* of food to ten needy people, or fast for three days, even if he had been compelled to do so.

6) A man who removes hairs from his armpits must sacrifice a sheep, or feed three needy people. If hair from his head or his beard fall out because of his touching it, then he must offer a handful of foodstuffs in charity for it, but if it was while performing his ablution, then he does not have to pay anything.

7) A man who takes shelter while walking must sacrifice a sheep, as he must do if he covers his head, even if done out of necessity.

8) He who swears an oath argumentatively three times over something that is true must sacrifice a sheep – the same as someone who swears once over something false. If the latter repeats the offense over something false, then he must sacrifice an ox, and a camel if he does it thrice.

9) If a *muḥrim* oils his body with a scented oil or pulls out a tooth, he must sacrifice a sheep.

10) If he cuts down a big tree, he must offer an ox; a sheep for a small tree and the equivalent value for any parts of trees or bushes that he cuts or uproots.

11) The atonement penalty applies anew for every repeated act of sexual intercourse, wearing stitched clothing or using perfume.

12) The ignorant person or the one who forgot need not pay the atonement penalty, except in the case of hunting.

ولو نظر إلى أهله بغير شهوة فأمنى فلا شيء عليه، وإن كان بشهوة فأمنى فجزور، وكذا لو أمنى عند الملاعبة.

ولو عقد المحرم لمحرم فدخل كان عليهما كفارتان.

الثانية: من تطيب لزمه شاة، سواء الصبغ والإطلاء والبخور والأكل، ولا بأس بخلوق الكعبة.

الثالثة: في تقليم كل ظفر مدّ من طعام، وفي يديه ورجليه شاة مع اتّحاد المجلس، ولو تعدّد فشاتان. وعلى المفتي إذا قلم المستفتي فأدمى إصبعه شاة.

الرابعة: في لبس المخيط شاة وإن كان لضرورة.

الخامسة: في حلق الشعر شاة، أو إطعام عشرة مساكين لكل مسكين مد، أو صيام ثلاثة أيام وإن كان مضطراً.

السادسة: في نتف الإبطين شاة، وفي أحدهما إطعام ثلاثة مساكين، ولو سقط من رأسه أو لحيته شيء بمسه تصدّق بكف من طعام، وإن كان في الوضوء فلا شيء.

السابعة: في التظليل سائراً شاة، وكذا في تغطية الرأس وإن كان لضرورة.

الثامنة: في الجدال صادقاً ثلاثاً شاة، وكذا في الكاذب مرة، ولو ثنى فبقرة، ولو ثلث فبدنة.

التاسعة: في الدهن الطيب وقلع الضرس شاة.

العاشرة: في الشجرة الكبيرة بقرة، وفي الصغيرة شاة، وفي أبعاضها قيمته.

الحادية عشرة: تتكرّر الكفارة بتكرّر الوطيء، واللبس، مع اختلاف المجلس، والطيب كذلك.

الثانية عشرة: لا كفارة على الجاهل والناسي إلا في الصيد.

Chapter 6: Circumambulating the Kaʿba

It is obligatory to perform once for the *ʿumra tamattuʿ* pilgrimage, and twice for the Hajj of *tamattuʿ*. For each of the other two *ʿumra* and Hajj [*ifrād* and *qirān*] it must be performed twice.

The pilgrim must be in a state of ritual purity for it and must have removed any impurity from his body and clothing. It is obligatory for men to have been circumcised beforehand.

It is obligatory to make the intention, then to circle around the Kaʿba seven times starting from the Black Stone and ending thereat, to keep the Kaʿba to one's left, to ensure that he touches the Black Stone at some point during it, to remain between the Kaʿba and the Station of Abraham (*maqām Ibrāhīm*), and to pray the two-unit prayer for circumambulation behind the Station of Abraham thereafter.

It is recommended to supplicate upon entering Mecca and the Sacred Mosque, to chew the Meccan lemongrass, to enter Mecca from its highest point barefoot and with solemness and dignity, to perform a *ghusl* at Biʾr Maymūn or Fakh, to greet and kiss the Black Stone in every round, or at least to gesture towards it whilst supplicating, to supplicate during the circumambulation, to place the hands on the ground at the foot of the Kaʿba and supplicate whilst touching the cheek or the belly to the ground, to greet the Rukn al-Yamānī corner and the other corners when passing them. It is recommended to perform three-hundred and sixty *ṭawāf*, but if one is not able to, then three-hundred and sixty rounds.

Ṭawāf is a cornerstone of the Hajj and one who omits it intentionally invalidates his Hajj. If he forgets to perform it, then he must do so as soon as he remembers, and if he physically cannot perform it, then he must deputise someone to do it on his behalf.

If someone has a doubt about the number of rounds that he completed after having completed the *ṭawāf*, he need not entertain his doubt. If his doubt occurs during the *ṭawāf*, then he must start all over again if the number of rounds is below seven. If it is above seven, then he must stop wherever he is.

If he remembers during an obligatory *ṭawāf* that he is not in a state of purity, then he must start again. If one joins another *ṭawāf* onto an obligatory circumambulation, without a break in between, it is invalidated. Joining recommended *ṭawāf*s together is disapproved.

الباب السادس: في الطواف

وهو واجب مرة في العمرة المتمتّع بها ، ومرتين في حجه ، وفي كل واحد من عمرة الباقيين مرتين، وكذا في حجهما .

ويشترط فيه الطهارة ، وإزالة النجاسة عن الثوب والبدن ، والختان في الرجل .

ويجب فيه النية ، والطواف سبعة أشواط ، والابتداء بالحجر والختم به ، وجعل البيت على يساره ، وإدخال الحجر فيه ، ويكون بين المقام والبيت ، وصلاة ركعتيه في مقام إبراهيم عليه السلام .

ويستحب فيه الدعاء عند الدخول إلى مكة والمسجد ، ومضغ الإذخر ودخول مكة من أعلاها حافياً بسكينة ووقار ، والغسل من بئر ميمون أو فخ ، واستلام الحجر في كل شوط ، وتقبيله أو الايماء إليه ، والدعاء عند الاستلام وفي الطواف ، والتزام المستجار ووضع الخد عليه والبطن ، والدعاء ، واستلام الركن اليماني وباقي الأركان ، والطواف ثلاثمائة وستين طوافاً ، فإن لم يتمكّن فثلاثمائة وستين شوطاً .

والطواف ركن من تركه عمداً بطل حجه ، وناسياً يأتي به ، ومع التعذّر يستنيب .

ولو شك في عدده بعد الانصراف لم يلتفت ، وفي الأثناء يعيد إن كان فيما دون السبعة ، وإلا قطع .

ولو ذكر في طواف الفريضة عدم الطهارة أعاد . ولو قرن في طواف الفريضة بطل ، ويكره في النافلة .

If he inadvertently does an eighth round, then he must complete fourteen rounds in total [by adding on six more] and pray the obligatory two-unit prayer before the *sa'y*, and an extra two-unit recommended prayer after it. If he had inadvertently completed fewer rounds [remembering later], if he had completed more than half, then he can resume and complete the rest. If he remembered after returning home, then he must deputise someone to complete the rest on his behalf. If he had completed less than half to begin with, then he must start over. The same applies to someone who had to break his *ṭawāf* out of necessity or to pray a supererogatory prayer.

It is not permissible to perform the *ṭawāf* and *sa'y* of the Hajj before having completed the compulsory ritual stays, except for the woman who fears that she will start her menstrual period. If she does indeed start menstruating before it, then she must wait to perform it until the last minute before going to stay at 'Arafāt. If she is not cleansed by then, her Hajj of *tamattu'* is invalidated and it becomes a Hajj of *ifrād*, and she must perform an *'umra* in lieu afterwards. If she starts menstruating during the *ṭawāf* and has already completed half of it, she can leave the rest and continue with the rest of the Hajj rituals; then complete the remaining rounds of the *ṭawāf* after she is cleansed of her menses. If she had completed less than half of the *ṭawāf* when she started menstruating, then she is as one who has not done the *ṭawāf* at all.

The woman who bleeds irregularly (*istiḥāḍā*) can perform the *ṭawāf* as one who is ritually pure, provided she cleanses herself as obligatorily required of her.

ولو زاد سهواً أكمل أسبوعين، وصلى ركعتي الواجب قبل السعي والمندوب بعده. ولو نقص من طوافه وقد تجاوز النصف أتم، ولو رجع إلى أهله استناب، ولو كان أقل استأنف، وكذا من قطع الطواف لحاجة أو صلاة نافلة.

ولا يجوز تقديم طواف حج التمتع وسعيه على الوقوف إلا لخائفة الحيض ولو حاضت قبله انتظرت الوقوف، فإن لم تطهر بطل متعتها وصارت حجتها مفردة، وتقضي العمرة بعد ذلك. ولو حاضت خلاله فإن تجاوزت النصف تركت بقية الطواف وفعلت بقية المناسك، ثم قضت الفائت بعد طهرها، وإلا فحكمها حكم من لم تطف.

والمستحاضة إذا فعلت ما يجب عليها كانت كالطاهرة.

Chapter 7: Brisk Walk Between Ṣafā and Marwa

This is obligatory to do once for every instance of being in *iḥrām*. It is obligatory to make the intention, then to begin at Ṣafā and end at Marwa. The *sa'y* consists of seven laps, whereby walking from Ṣafā and back counts as two laps.

It is recommended to be in a state of purity, to greet the Black Stone, to drink water from the well of Zamzam, to pour water over oneself from the bucket opposite the Black Stone, to leave from the gate of Ṣafā and to climb to the top of the hillock and pause there, to face the Black Stone corner and recite *Allāhu Akbar* and *Lā ilāha illa Allāh* seven times, to supplicate and then walk out from either side of it. It is recommended to jog from the point parallel to the minaret to Zuqāq al-'Attārīn for that is part of Wādī Muḥassar, and to walk the rest of the *sa'y*.

Sa'y is a cornerstone of the pilgrimage, and it is invalidated if it is intentionally omitted, but not inadvertently so, in which case one would return and perform it. If he is physically unable to do it, then he must deputize someone to do it on his behalf. If he intentionally walks more than seven laps, the *sa'y* is void, but not if it was done by mistake. If he realized that he had done fewer than seven laps, then he must return and start again. If he has to stop in the middle out of necessity or for the obligatory prayer, then he can continue where he left off. If after having assumed completion of the rituals, coming out of the state of *iḥrām*, sleeping with his wife and cutting his nails, he remembers that he had forgotten to walk a lap, then he must go back and do it, and sacrifice a cow as atonement.

When he has completed the *sa'y* for the *'umra* pilgrimage, he must perform *taqṣīr*, which is to clip his nails and cut a small lock of his hair. He must not shave his head, and if he does so then he must sacrifice an animal as atonement, as he must if he forgets to do *taqṣīr* before entering the state of *iḥrām* again for the Hajj pilgrimage. After *taqṣīr*, everything that had been contraindicated in the state of *iḥrām* is now permissible, except for hunting as long as he remains in the vicinity of the *ḥaram* sanctuary. It is recommended for him to continue wearing unstitched clothing to look like the other *muḥrims*.

الباب السابع: في السعي

وهو واجب في كل إحرام مرة، وتجب فيه النية، والبداءة بالصفا والختم بالمروة، والسعي سبعة أشواط من الصفا إليه شوطان.

ويستحب فيه الطهارة، واستلام الحجر، والشرب من زمزم، والاغتسال من الدلو المقابل للحجر، والخروج من باب الصفا والصعود عليه، واستقبال ركن الحجر بالتكبير والتهليل سبعاً، والدعاء والمشي طرفيه، والهرولة من المنارة إلى زقاق العطارين فإنه من وادي محسر، والسعي ماشياً.

وهو ركن يبطل الحج بتركه عمداً لا سهواً، ويعود لأجله. فإن تعذّر استناب، ولو زاد على السبع عمداً بطل، لا سهواً. ويعيده لو لم يحصل عدد أشواطه، ولو قطعه لقضاء حاجة أو صلاة فريضة تممه. ولو ظن الاتمام فأحل وواقع أهله وقلّم الأظفار ثم ذكر نسيان شوط أتمّ ويكفر ببقرة.

وإذا فرغ من سعي العمرة قصّر، وأدناه أن يقصّ أظفاره أو شيئاً من شعره، ولا يحلق رأسه، فإن فعل كان عليه دم، وكذا لو نسيه حتى أحرم بالحج، ومع التقصير يحل من كل شيء أحرم منه إلا الصيد ما دام في الحرم، ويستحب له أن يتشبّه بالمحرمين في ترك لبس المخيط.

Chapter 8: The Rituals of the Hajj

This chapter contains several sections.

Section One: Entering the State of Iḥrām for Hajj

When the pilgrim has completed his *'umra*, he must enter the state of *iḥrām* again from Mecca itself, and it is recommended to do so on the Day of *tarwiya* at midday under the rainwater spout of the Ka'ba.

The method is the same as previously mentioned, except that he now makes the intention for the *iḥrām* of Hajj, and stops reciting the *talbiya* at noon on the Day of 'Arafa.

If he forgets to put on the *iḥrām* until he reaches the plains of 'Arafāt, then he does so there if he is not able to return to Mecca. And if he does not remember until after he has completed his rituals, then he need not atone in any way.

Section Two: Staying in 'Arafāt

This is a cornerstone of the Hajj and invalidates it if it is intentionally not adhered to. If one forgets to do it within the designated time frame and does not reach al-Mash'ar as a result, then his Hajj is void.

It is obligatory to make the intention, and to stay at 'Arafāt until sunset on the Day of 'Arafa. If he is physically unable to stay there during the day, then he must do so at night, even if it just before dawn. If he is not able to do that either or he forgets to do it until daybreak, then he must stay at Muzdalifa, and that would suffice him. If one departs from 'Arafāt before sunset, he must sacrifice a camel in atonement, and if he cannot, then he must fast for eighteen days if he knowingly departed. If it was done in ignorance or forgetfulness, then he need not atone.

Namira, Thawiyya, Dhū l Majāz, 'Urana and Arāk are all places bordering 'Arafāt and staying in those places does not suffice.

الباب الثامن: في أفعال الحج

وفيه فصول

الفصل الأول: في إحرام الحج

إذا فرغ من العمرة وجب عليه الإحرام بالحج من مكة، ويستحب أن يكون يوم التروية عند الزوال من تحت الميزاب.

وكيفيته كما تقدّم، إلا أنه ينوي إحرام الحج، ويقطع التلبية يوم عرفة عند الزوال.

ولو نسيه حتى يحصل بعرفات أحرم بها إذا لم يتمكن من الرجوع، ولو لم يتذكر حتى يقضي مناسكه لم يكن عليه شيء.

الفصل الثاني: في الوقوف بعرفات

وهو ركن في الحج، يبطل الإخلال به عمداً. ولو تركه ناسياً حتى فات وقته ولم يحصل بالمشعر بطل حجه.

ويجب فيه النية، والوقوف بعرفات إلى غروب الشمس من يوم عرفة. ولو لم يتمكن من الوقوف نهاراً وقف ليلاً ولو قبل الفجر، ولو لم يتمكن أو نسي حتى طلع الفجر، وقف بالمشعر وأجزأه، ولو أفاض منها قبل الغروب وجب عليه بدنة، ولو عجز صام ثمانية عشر يوماً إن كان عالماً، وإن كان جاهلاً أو ناسياً فلا شيء عليه.

ونمرة، وثوية، وذو المجاز، وعرنة، والأراك: حدود لا يجزي الوقوف بها.

It is recommended to leave for Mina on the Day of *tarwiya* [the day before *'arafa*] after midday, though the Imam must pray his midday prayers in Mina. Then he is recommended to remain there until dawn on the Day of 'Arafa, and he must not go beyond Wādī Muḥassar until the sun has risen. It is recommended to supplicate upon arriving in Mina, when departing from there and en route; to stop at the foot of the hill towards the left of it and to supplicate standing; and to pray his noon (*ẓuhr*) and afternoon (*'aṣr*) prayers together with a single *adhān* and two *iqāma*.

It is discouraged to stay at the top of the hill, to sit down or to be mounted on an animal.

Section Three: Staying at Al-Mash'ar

When the sun sets on the Day of 'Arafa, the pilgrim must set out for al-Mash'ar.

It is recommended to walk purposefully and slowly, to supplicate upon reaching al-Kathīb al-Aḥmar, to delay the evening prayers (*maghrib* and *'ishā'*) until he can pray them both together when he reaches, even if a quarter of the night has already passed, and to combine them together with a single *adhān* and two *iqāma*.

It is obligatory to make the intention and to remain there between dawn and sunrise, and if he misses that window of time for some emergency, then he may stay until midday. If he knowingly and intentionally leaves before dawn, then he must sacrifice a sheep in atonement, though his Hajj remains valid if he stayed at 'Arafāt. Women and those who fear danger are allowed to depart before dawn.

Al-Mash'ar covers the area between al-Ma'zamayn up to al-Ḥiyāḍ and Wādī Muḥassar.

This stay is a cornerstone of the Hajj and whoever intentionally omits it, during the night or in the daytime, his Hajj is void. If he inadvertently misses it but managed to stay at 'Arafāt, then his Hajj is still valid.

ويستحب أن يخرج إلى منى يوم التروية بعد الزوال والإمام يصلي بها ، ثم يثبت بها إلى فجر عرفة ، ولا يجوز وادي محسر حتى تطلع الشمس ، ويدعو عند نزولها والخروج منها وفي الطريق ، وأن يقف مع السفح في ميسرة الجبل داعياً قائماً ، وأن يجمع بين الظهرين بأذان وإقامتين. ويكره الوقوف في أعلى الجبل ، وقاعداً ، وراكباً .

الفصل الثالث: في الوقوف بالمشعر

وإذا غربت الشمس من يوم عرفة أفاض إلى المشعر .

ويستحب أن يقتصد في المسير ، ويدعو عند الكثيب الأحمر ، ويؤخر العشائين حتى يصلّيهما فيه ولو صار ربع الليل ، ويجمع بينهما بأذان وإقامتين.

وتجب فيه النية ، والكون فيه بين طلوع الفجر إلى طلوع الشمس ، ولو فاته لضرورة فإلى الزوال ، ولو أفاض قبل الفجر عالماً عامداً كفر بشاة وصح حجه إن وقف بعرفة. ويجوز للمرأة والخائف الإفاضة قبله.

وحد المشعر : بين المأزمين إلى الحياض إلى وادي محسر .

وهذا الوقوف ركن ، من تركه ليلاً ونهاراً عمداً بطل حجه ، ولو كان ناسياً وأدرك عرفات صح حجه.

Other Rulings

1) The recommended customary time frame to stay in ʿArafāt is from noon on the Day of ʿArafa until sunset, or until dawn in extenuating circumstances.

The recommended customary time frame for staying at al-Mashʿar is from the break of dawn on the Day of Sacrifice until sunrise, or until noon in extenuating circumstances.

If he manages to stay at one of the two places within the recommended time frame, but misses the other due to some emergency, then his Hajj is still valid. If he managed to stay at both places but only in the extended window, then his Hajj is void according to one opinion, but the scholars are unanimous that it would be void if he only managed one stay in the extended window.

2) Whoever misses these Hajj rituals need not complete the rest of them, and he comes out of the state of *iḥrām* after an *ʿumra mufrida*. He must then obligatorily undertake a reparative Hajj in lieu the following year.

3) It is recommended to stay at these places after praying and supplicating, for a first-timer to walk to al-Mashʿar on foot, to climb to the top of the hill known as Qazaḥ and to remember God much there.

4) It is recommended to pick up the pebbles for the stoning rituals from al-Mashʿar, but one can pick them up from anywhere within the vicinity of the *ḥaram* apart from the mosques.

Section Four: The Stay in Mina

There are three things that are obligatory to do on the Day of Sacrifice in Mina:

First: to stone the *Jamrat al-ʿUqba* column with seven pebbles freshly collected from the vicinity of the *ḥaram*, with the intention to strike the column with his action, which is known as *ramy* [aiming and throwing].

It is recommended that the pebbles be rough and mottled, the size of fingertips, [picked up as pebbles and] not broken from stones, and not smooth and hard. It is recommended to supplicate before throwing each pebble, to be in a state of purity, to throw from a distance of ten to fifteen arm lengths (*dhirāʿ*) away, to flick the pebbles with the thumb and nail of the forefinger, to be facing this particular column with one's back towards the *qibla* but to face the *qibla* whilst stoning the rest. It is permissible to throw pebbles on behalf of someone who is incapacitated.

مسائل:

الاولى : وقت الوقوف الاختياري بعرفة من زوال الشمس يوم عرفة إلى غروبها ، والاضطراري إلى الفجر .

ووقت الوقوف الاختياري بالمشعر من طلوع الفجر يوم النحر إلى طلوع الشمس ، والاضطراري إلى الزوال .

فإن أدرك أحد الموقفين اختياراً وفاته الآخر لضرورة صح حجه ، وإن أدرك الاضطراريين معاً فاته الحج على قول ، أما لو أدرك أحدهما فإنه يبطل حجه إجماعاً .

الثانية من فاته الحج سقطت عنه أفعاله ، ويحل بعمرة مفردة ، ويقضي الحج في القابل مع الوجوب

الثالثة : يستحب الوقوف بعد الصلاة والدعاء ، ووطأ المشعر بالرجل للصرورة ، والصعود على قزح ، وذكر الله عليه .

الرابعة : يستحب التقاط حصى الرمي منه ، ويجوز من أي جهات الحرم كان ، عدى المساجد .

الفصل الرابع: في نزول منى

ويجب يوم النحر بمنى ثلاثة :

أحدها : رمي جمرة العقبة بسبع حصيات ملتقطة من الحرم أبكاراً ، مع النية ، وإصابة الجمرة بفعله بما يسمّى رمياً .

ويستحب أن تكون رخوة برشا قدر الانملة ، ملتقطة ، لا مكسّرة ولا صلبة ، والدعاء عند كل حصاة ، والطهارة ، والتباعد بمقدار عشرة أذرع إلى خمسة عشر ذراعاً ، والرمي خذفاً وأن يستقبل هذه الجمرة ويستدبر القبلة ، وفي غيرها يستقبلها . ويجوز الرمي عن العليل .

Second: to slaughter the sacrificial offering. It is obligatory to sacrifice the animal after the stoning ritual, in that order, especially for the *tamattuʿ* pilgrim, for both obligatory and voluntary pilgrimages. The master whose slave is doing Hajj may either make him fast or provide an animal for him to sacrifice on his behalf. If he is manumitted before the two stays, then he himself must provide an animal for sacrifice if he can afford to do so, but if not, then he can fast instead.

It is obligatory to make the intention, to slaughter it in Mina on the Day of Sacrifice, to not share the same animal with anyone for an obligatory sacrifice. It is obligatory for the animal to be of the right age: camels should be five years old and over, cows and sheep should be over a year old; lambs under one year can suffice. They must be healthy[8] and not so emaciated that there is no fat on their kidneys.

It is recommended that the animal be fat, and that it accompanied the pilgrim at ʿArafāt. Female animals are recommended for camels and cows, and male for sheep and goats. It is recommended to supplicate whilst slaughtering, to keep a third of the meat for oneself to consume, to gift a third of it [to friends], and to feed a third to the needy and poor.

If someone has no access to a sacrificial animal but can afford its value, he must leave its payment with someone who can buy it and slaughter it for him at any time until the end of Dhū l-Ḥijja. If he is not able to afford it either, then he must fast three consecutive days during the Hajj and seven when he returns home. He may fast the three days in advance from the first of Dhū l-Ḥijja but not prior to that. If he departs home without having fasted the three days, then he must appoint someone to sacrifice an animal in Mecca the following year.

As for the *qirān* pilgrim, he must slaughter or sacrifice the animal in Mina if he is performing the Hajj first, and in Mecca if he is taking it with him to perform his *ʿumra*. He is allowed to ride his animal and drink of its milk as long as it does not cause harm to it or to its offspring. If the *qirān* pilgrim's sacrificial animal dies en route, he is not obliged to substitute it with another, unless it was by way of liability. The only obligatory sacrifices that must be donated in charity [and not kept for consumption] are the ones from a solemn vow (*nadhr*). The butcher must not be gifted the meat from an obligatory sacrifice.

8 That is, they must not be obviously lame, blind, too old, or with broken horns or ears, and not diseased.

الثاني: الذبح، ويجب بعد الرمي مرتباً، وهو الهدي، على المتمتع خاصة، في الفرض والنفل. وللمولى إلزام المملوك بالصوم، أو أن يهدي عنه، فإن عتق قبل أحد الموقفين لزمه الهدي مع القدرة وإلا صام.

وتجب فيه النية، وذبحه بمنى يوم النحر، وعدم المشاركة في الواجب، وأن يكون من النعم ثنياً قد دخل في السادسة إن كان من البدن، وفي الثانية إن كان من البقر والغنم، ويجزي من الضأن الجزع لسنة، تامّاً غير مهزول بحيث لا يكون على كليتيهما شحم.

ويستحب أن تكون سمينة قد عرف بها، إناثاً من الإبل والبقر، وذكراناً من الضأن والمعز، والدعاء عند الذبح، وأن يأكل ثلثه ويهدي ثلثه ويطعم القانع والمعتر ثلثه.

ولو فقد الهدي ووجد ثمنه خلفه عند من يشتريه ويذبحه طول ذي الحجة، ولو فقده صام ثلاثة أيام متتابعة في الحج وسبعة إذا رجع، ويجوز تقديم الثلاثة من أول ذي الحجة ولا يجوز تقديمها عليه، فإن خرج ولم يصمها تعين الهدي في القابل بمنى.

وأما هدي القِران: فيجب ذبحه أو نحره بمنى إن قرن بالحج، وبمكة إن قرن بالعمرة. ويجوز ركوب الهدي وشرب لبنه ما لم يضر به وبولده، وإذا هلك هدي القِران لم يلزمه بدله إلا أن يكون مضموناً، ولا يتعين للصدقة إلا بالنذر، ولا يعطى الجزار من الهدي الواجب.

The timing recommended to slaughter the sacrifices is on the Day of Sacrifice (*yawm al-naḥr*) and the next three days if they are being slaughtered in Mina, or the next two days if it is elsewhere. The obligatory sacrifice of Hajj *tamattuʿ* suffices, and if one does not have a sacrificial animal, he can give its value in charity. Sacrificing an animal that one has raised from young is disapproved, as is giving the animal hides to the butcher [by way of payment].

Third: to shave the head. It is obligatory to shave the head or to do *taqṣīr* (cutting a lock of hair and clipping the nails) on the Day of Sacrifice after having slaughtered the animal in Mina, though shaving the head is preferable and especially emphasised for first-timers and those with thick, matted hair. Women must only do *taqṣīr*.

If the pilgrim departs before having shaved his head or done *taqṣīr*, he must return and do one of them. If he is unable to return, he must obligatorily do it wherever he is, and it is recommended for him to send his hair to Mina to be buried there. He who does not have any hair on his head can pass a razor across his scalp.

The pilgrim is not allowed to visit the Kaʿba before doing *taqṣīr*, and if he intentionally circumambulates without having done so, he must atone by sacrificing a sheep. The one who does so forgetfully need not pay anything, but he must redo his *ṭawāf* after *taqṣīr*.

Once he has shaved his head or done *taqṣīr*, the pilgrim is no longer in the state of *iḥrām* and all the previous contraindications are now permissible, except for perfume and intimacy with women. Once he has completed the *ṭawāf al-ziyāra*,[9] perfume is permissible, and upon completing *ṭawāf al-nisāʾ*, women become permissible.

Section Five: The Remaining Rituals

Upon completion of the rituals at Mina, he must proceed to Mecca – on the same day or the next day for the *tamattuʿ* pilgrim, or at any time until the end of Dhū l-Ḥijja for the *qirān* and *ifrād* pilgrim – to perform the *ṭawāf* of Hajj and the two-unit prayer after it, the *saʿy* for Hajj, the *ṭawāf al-nisāʾ*: also seven rounds, and the two-unit prayer after it. The method of all these rituals is outlined in the section on *ʿumra* rituals. *Ṭawāf al-nisāʾ* is obligatory upon every pilgrim.

9 Also known as *ṭawāf al-ḥajj*.

وأما الأضحية : فمستحبة يوم النحر ، وثلاثة بعده بمنى ، ويومان في غيرها ، ويجزي هدي التمتع عنها ، فلو فقدها تصدّق بثمنها . ويكره التضحية بما يربيه ، وإعطاء الجزار الجلود .

الثالث : الحلق ، ويجب يوم النحر بعد الذبح الحلق أو التقصير بمنى ، والحلق أفضل ، ويتأكد للصرورة والملبد . ويتعين في المرأة التقصير .

ولو رحل قبل الحلق أو التقصير رجع وفعل أحدهما ، فإن تعذّر حلق أو قصر أين كان - وجوباً - وبعث شعره إلى منى ليدفن بها استحباباً . ومن ليس على رأسه شعر يمر الموسى عليه .

ولا يزور البيت قبل التقصير ، فإن طاف قبله عمداً كفّر بشاة ، ولا شيء على الناسي ، ويعيد طوافه .

فإذا حلق أو قصر أحل مما عدا الطيب والنساء ، فإذا طاف طواف الزيارة حل الطيب ، ويحل النساء بطوافهن .

الفصل الخامس : في بقية المناسك

فإذا تحلل بمنى مضى - ليومه أو غده إن كان متمتعاً ، ويجوز للقارن والمفرد طول ذي الحجة - إلى مكة لطواف الحج ، ويصلي ركعتيه ثم يسعى للحج ، ثم يطوف للنساء ، كل ذلك سبعاً ، ثم يصلي ركعتيه ، وصفة ذلك كما قلنا في أفعال العمرة ، وطواف النساء واجب على كل حاج .

When he has completed these rituals, he must return to Mina and mandatorily spend the nights of the 11th and 12th Dhū l-Ḥijja there. He must stone all three columns on each of the two days spent there, with seven pebbles each. He must start from the first column, aiming at it from its left-hand side, whilst reciting *Allāhu Akbar* and supplicating. Then he stones the second in the same manner, and then the third. If he stones them in reverse order or lapses in some way, he must go back and recommence to observe the correct sequence.

Stoning can be done at any time between sunrise and sunset. It is not permissible to stone at night, except for those who have good reason to do so, such as those who fear danger in the daytime, shepherds, and slaves. If someone spends the third day in Mina, he must stone on that day too, and if not, then he must bury the remainder of his pebbles there. If he spends the two nights elsewhere other than in Mina, he must obligatorily sacrifice a sheep per night in atonement, unless he spent those nights in Mecca engaged in worship the whole night. One can leave after half the night has passed.

The only people allowed to leave Mina on the 12th Dhū l-Ḥijja before sunset as a first cohort are those who abstained from hunting and intimacy with women. Anyone else may not leave until the next day, and if they do, then they must sacrifice a sheep in atonement. The first cohort can leave after midday, and the second on the following day before midday.

If someone forgot to perform the ritual of stoning on any of the days, he must make up for it and do it on the following day. If he forgot to stone a column but does not know which one it was, he must stone all three again. If he forgot to stone and remembered after entering Mecca, he must return to Mina and stone. If, however, he is unable to, then he can continue with the rest of his rituals, but it is recommended for him to come back and stone the following year, or to deputise someone to do it on his behalf.

It is recommended to stay in Mina all three days of *tashrīq*.

When all these rituals have been performed, the Hajj is complete. The pilgrim is now recommended to go to Mecca to perform the farewell *ṭawāf*, and to enter the Kaʿba, especially the first-timer. He is recommended to offer prayers at each of its corner, between the two pillars, and on the stone of the red marble. It is also recommended to visit Masjid al-Ḥasba, to pray in it, and to take a short rest there, lying on one's back. Similarly, visiting Masjid al-Khīf is also encouraged, and to exit it from Bāb al-Ḥanāṭin, to prostrate at that gate and to supplicate. It is commendable to buy a dirham's worth of dates and to give them out in charity.

فإذا فرغ من هذه المناسك رجع إلى منى وبات بها ليلتي الحادي عشر والثاني عشر من ذي الحجة واجباً ، ويرمى في اليومين الجمار الثلاث ، كل جمرة في كل يوم ، بسبع حصيات ، يبدأ بالجمرة الأولى ويرميها عن يسارها مكبراً داعياً ، ثم الثانية كذلك ، ثم الثالثة ، ولو نكس أعاد على ما يحصل معه الترتيب .

ووقت الرمي : ما بين طلوع الشمس إلى غروبها .

ولا يجوز الرمي ليلاً إلا للمعذور كالخائف والرعاة والعبيد ، فإن أقام اليوم الثالث رماها أيضاً وإلا دفن حصاه بمنى ، ولو بات الليلتين بغير منى وجب عليه عن كل ليلة شاة ، إلا أن يبيت بمكة مشتغلاً بالعبادة . ويجوز أن يخرج بعد نصف الليل .

ويجوز النفر الأول لمن اتقى [الصيد والنساء] إذا لم تغرب الشمس في الثاني عشر بمنى ، ولا يجوز لغيره ، فإن نفر كان عليه شاة ، والنافر في الأول يخرج بعد الزوال ، وفي الثاني يجوز قبله .

ولو نسى رمي يوم قضاه من الغد مقدماً ، ولو نسى جمرة وجهل عينها رمى الثلاث ، ولو نسى الرمي حتى دخل مكة رجع ورمى ، فإن تعذر مضى ورمى في القابل أو استناب مستحباً .

ويستحب الإقامة بمنى أيام التشريق .

فإذا فرغ من هذه المناسك تمّ حجه ، واستحب له العود إلى مكة لطواف الوداع ودخول الكعبة خصوصا للصرورة ، والصلاة في زواياها ، وبين الأسطوانتين ، وعلى الرخامة الحمراء ، ودخول مسجد الحصبة والصلاة فيه ، والاستلقاء على قفاه ، وكذلك مسجد الخيف ، ويخرج من المسجد من باب الحناطين ، ويسجد عند باب المسجد ويدعو ، ويشتري بدرهم تمراً يتصدق به .

A prolonged stay in Mecca after the Hajj is disapproved but staying in Medina is encouraged.

A menstruating woman can only bid farewell at the gate of the Sacred Mosque.

The pilgrim is then highly encouraged to proceed to Medina to visit the Prophet (peace be upon him), to visit Fāṭima (peace be upon her) from the *rawḍa*,[10] to visit the Imams buried in Baqīʿ cemetery, the martyrs at Uḥud, especially Ḥamza, and to spend three days in spiritual retreat inside the Prophet's mosque.

10 *rawḍa*: The small area between the pulpit and the Prophet's tomb, which he had termed a garden (*rawḍa*) from among the gardens of Paradise.

ويكره أن يجاور بمكة، ويستحب بالمدينة.

والحائض تودع من باب المسجد.

ثم يأتي المدينة لزيارة النبي ﷺ استحباباً مؤكداً، وزيارة فاطمة عليها السلام من الروضة، وزيارة الائمة عليهم السلام بالبقيع، وزيارة الشهداء خصوصاً حمزة بأحد، والاعتكاف ثلاثة أيام بها.

Chapter 9: The *Umra* Rituals

The *ʿumra* pilgrimage, just like the Hajj, becomes obligatory upon the fulfilment of certain conditions and factors.

Its rituals consist of making the intention, entering the state of *iḥrām*, *ṭawāf* and the two-unit prayer thereafter, *saʿy*, *ṭawāf al-nisāʾ* and its two-unit prayer, and *taqṣīr* or shaving the head.

The *tamattuʿ* pilgrim who undertakes the *ʿumra* first does not need to perform a separate *ṭawāf al-nisāʾ* for the *ʿumra*.

The *ʿumra mufrida* pilgrimage [unlinked to the Hajj] can be undertaken at any time during the year and is especially recommended in the month of Rajab. The *qirān* and *ifrād* pilgrims perform it after their Hajj, whereas the *tamattuʿ* pilgrim completes it beforehand as part of his Hajj.

الباب التاسع: في العمرة

وهي فريضة مثل الحج بشرائطه وأسبابه.

وأفعالها: النية، والإحرام، والطواف، وركعتاه، والسعي، وطواف النساء، وركعتاه، والتقصير أو الحلق.

وليس في المتمتع بها طواف النساء.

ويجوز المفردة في جميع أيام السنة، وأفضلها رجب، والقارن والمفرد يأتي بها بعد الحج، والمتمتع بها يجزي عنها.

ولو اعتمر في أشهر الحج جاز أن ينقلها إلى التمتع، ويجوز في كل شهر، وأقله في كل عشرة أيام، ولا حد لها عند السيد المرتضى (قده).

Chapter 10: Those Who Are Taken Ill During the Hajj (*Maḥṣūr*) and Those Who Are Prevented from It (*Maṣdūd*)

Someone who is prevented from performing the Hajj (*maṣdūd*) is one who is blocked by enemies. If such a person is already in *iḥrām* then he must slaughter his sacrificial animal wherever he is, and all contraindications become permissible once more. The rulings in this category apply to someone who is specifically prevented from entering Mecca or the two obligatory stays (ʿArafāt and al-Mashʿar). The obligatory Hajj remains binding upon him, but if it was an optional pilgrimage, he need not redo it. One can only come out of the state of *iḥrām* by arranging for an animal to be sacrificed on his behalf [in the designated place] at an agreed time, upon which he would make the intention to be released from *iḥrām*. If he cannot do this, then sacrificing the animal that he has brought along with him [wherever he is detained] suffices. The same rulings apply to a detained *ʿumra* pilgrim.

The *maḥṣūr* is one who is taken ill during the Hajj and prevented from completing it. He must arrange for an animal to be sacrificed in the designated place on his behalf if he has not brought one along. Otherwise, he can send his animal with others, and when it is due to have reached the designated place – Mina for a Hajj pilgrim and Mecca for an *ʿumra* pilgrim – he does his *taqṣīr* and comes out of the state of *iḥrām*, whereupon everything becomes permissible for him again, except for women, until he manages to undertake the Hajj himself the following year if this was an obligatory Hajj. Alternatively, he can deputize someone to perform the *ṭawāf al-nisāʾ* on his behalf if it is a supererogatory pilgrimage. If the illness or impediment ceases, he must join in the Hajj rites. If he manages to complete one of the two obligatory stays on time, then his Hajj is valid, but if not, then it is void.

الباب العاشر: في المحصور والمصدود

المصدود: الممنوع بالعدو، فإن تلبّس بالإحرام نحر هديه وأحل من كل شيء أحرم منه. وإنما يتحقّق الصد بالمنع عن مكة أو عن الموقفين ولا يسقط الواجب، ويسقط المندوب. ولا يصح التحلّل إلا بالهدي ونية التحلل، ويجزي هدي السياق عنه، والمعتمر المصدود كالحاج.

والمحصّر: هو الممنوع بالمرض، فيبعث هديه إن لم يكن قد ساق. وإلا اقتصر على هدي السياق، فإذا بلغ محله – وهو منى إن كان حاجاً، ومكة إن كان معتمراً – قصر، وأحل إلا من النساء حتى يحج في القابل إن كان واجباً، أو يطاف طواف النساء عنه إن كان ندباً، ولو زال الحصر التحق، فإن أدرك أحد الموقفين صح حجه، وإلا فلا.

The Book of Struggle and Combat

كتاب الجهاد

This book contains various sections.

Section One: Those Who Are Obliged to Engage in It

Jihād is an Islamic obligation as a collective responsibility,[1] provided that nine conditions are fulfilled: that the obliged be of age of religious responsibility, of sound mental capacity, free, male, neither aged, nor disabled, nor blind, and nor too ill to engage in combat. Fighting the enemy through *jihad* is incumbent when summoned by the Imam, or one who has been appointed by him.

Undertaking *jihād* under an oppressive government is not allowed, unless the Muslim community is in greater danger at the hands of a feared enemy, in which case they must defend themselves and not intend it as a *jihād* in aiding an oppressive government. The one who is physically unable to engage in *jihād* must appoint someone else to fight instead of him if he can, and it is allowed to appoint someone else in other circumstances besides incapacity.

It is recommended to defend Islamic territories (*murābaṭa*)[2] from three to forty days. If it takes longer than that, then it is considered *jihād*. *Murābaṭa* becomes obligatory in instances where someone makes a solemn vow (*nadhr*) to do so.

1 A collective responsibility (*wājib kifā'ī*) is one whereby if sufficient, efficient people undertake to fulfil this responsibility, the rest of the community is absolved from the obligation.

2 *Murābaṭa* is to spy or keep watch over the enemy by taking up a position at the edge enemy city walls so as to inform the Muslims of the enemy's movements.

وفيه فصول

الفصل الأول: فيمن يجب عليه

وهو فرض على الكفاية بشروط تسعة: البلوغ، والعقل، والحرية، والذكورة، وأن لا يكون هماً، ولا مقعداً، ولا أعمى، ولا مريضاً يعجز عنه، ودعاء الإمام أو من نصبه إليه.

ولا يجوز مع الجائر إلا أن يدهم المسلمين عدو يخشى عليه منه فيدفعه ولا يقصد معونة الجائر، والعاجز يجب أن يستنيب مع القدرة ويجوز لغير العاجز.

ويستحب المرابطة ثلاثة أيام إلى أربعين فإن زادت كانت جهاداً، ويجب بالنذر [وشبهه].

Section Two: Those Who Must Be Fought

There are three categories of people who must be fought:

1) Jews, Christians and Magians. They must be fought until they either embrace Islam or abide by the conditions of *dhimma*, which are: acceptance of the *jizya* poll tax, that they do not harass the Muslims, that they do not openly engage in prohibited acts such as drinking alcohol in public, that they do not build new churches nor ring church bells, and that they abide by the laws of the Muslim lands.

If they adhere to these, then they must be left alone. The *jizya* is not a fixed amount and lies at the discretion of the Imam. It must not be taken from minors, people of diminished mental capacity, incompetent people and women. The tax can either be charged per capita or per area of land owned. If they embrace Islam, it no longer applies. If someone who is due to pay the *jizya* dies before the end of the tax year, it must be taken from his assets; and it may even be taken from the value of prohibited things that he owned. The beneficiaries of it are the people who defend the Muslim territories (*mujāhidūn*). Whilst they live in Muslim lands, *dhimmīs* must not build new churches or synagogues on them, though they may renovate existing ones, provided that they do not tower over the buildings of the Muslims. If they purchase a building from a Muslim which overlooks other buildings owned by the neighbouring Muslims, then they are not required to demolish the part that exceeds the other heights. They are not allowed to enter mosques.

الفصل الثاني: فيمن يجب جهادهم

وهم ثلاثة أصناف:

الأول: اليهود والنصارى والمجوس، وهؤلاء يُقتلون إلى أن يسلموا أو يلتزموا شرائط الذمة، وهي: قبول الجزية، وأن لا يؤذوا المسلمين، وأن لا يتظاهروا بالمحرمات كشرب الخمر، وأن لا يحدثوا كنيسة ولا يضربوا ناقوساً، وأن يجري عليهم أحكام المسلمين.

فإن التزموا بهذه كف عنهم، ولا حد للجزية بل بحسب ما يراه الإمام ولا تؤخذ من الصبيان، والمجانين والبله والنساء، ويجوز وضعها على رؤوسهم وأراضيهم، ولو أسلموا سقطت، ولو مات الذي بعد الحول أخذت من تركته ويجوز أخذها من ثمن المحرمات. ومستحقها المجاهدون. وليس لهم استيناف بيعة ولا كنيسة في دار الإسلام، ويجوز تجديدهما، ولا يجوز أن يعلو الذمي على بناء المسلمين، ويقر ما ابتاعه من مسلم على حاله، ولا يجوز أن يدخلوا المساجد.

2) Apart from these, all other types of disbelievers must be fought without exception, unless they embrace Islam. The ones who are nearest and present the gravest danger must be fought first. They can only be fought after having been invited to accept Islam first by either the Imam or one whom he has appointed. If they refuse, then it is permissible to fight them, though in the event of interest, negotiating a temporary truce is allowed with the Imam's permission. Individuals [less than ten people] from among the Muslims – even slaves – may extend protection to individuals from the polytheists. If a non-Muslim enters Muslim lands [at a time of war] believing it is safe to do so, he is returned to his place of safety and fought in that land.

One is not allowed to flee from a battle if the enemies outnumber the Muslims by a ratio of two to one, unless he is turning aside for the sake of combat or withdrawing to side with another wing of the army. One is allowed to fight with all methods and means of warfare except throwing poison into their lands. If the enemies use children, women or even Muslims as human shields, and victory is only possible by killing them, then they may be killed. Women must not be fought – even if they are helping the menfolk – unless one is compelled to do so. If someone embraces Islam but still lives in enemy territory, his life must be protected, as must his minor children from being captured. All his liquid assets must be protected and moved to Islamic territories, and any land that he owns is considered spoils of war belonging to the Muslims. If a slave embraces Islam before his master and leaves, he is treated as a free man.

3) Outlaws, or anyone who revolts against a legitimate Imam, must be fought when summoned to do so by the Imam or one whom he has appointed, by way of collective responsibility, unless they retreat. Such rebels fall under two categories: those who are part of a group, in which case their wounded can be detained, their escapees pursued and their captives killed; and those who do not belong to a group, in which case the wounded one must not be detained, the escapee must not be pursued, nor the captive killed. In addition, it is not permissible to capture the children, women or property of either group.

الثاني: من عدا هؤلاء من الكفار يجب جهاده ولا يقبل منه إلا الإسلام، ويبدأ بقتال الأقرب والأشد خطراً. وإنما يحاربون بعد الدعاء من الإمام أو من نصبه إلى الإسلام فإن امتنعوا أحل قتالهم، ويجوز المهادنة مع المصلحة بإذن الإمام، ويمضي ذمام آحاد المسلمين - وإن كان عبداً - لآحاد المشركين، ويُرد من دخل بشبهة الأمان إلى مأمنه ثم يقاتل، ولا يجوز الفرار إذا كان العدو على الضعف من المسلمين إلا لمتحرف لقتال أو متحيز إلى فئة، ويجوز المحاربة بسائر أنواع الحرب إلا إلقاء السم في بلادهم. ولو تترسوا بالصغار والنساء أو المسلمين ولم يمكن الفتح إلا بقتلهم جاز، ولا يقتل النساء - وإن عاون - إلا مع الضرورة. ومن أسلم في دار الحرب حقن دمه وولده الصغار من السبي وماله من الأخذ مما ينقل ويحول، وأما الأرضون فمن الغنائم. ولو أسلم العبد قبل مولاه وخرج ملك نفسه.

الثالث: البغاة، وهم كل من خرج على إمام عادل، ويجب قتاله مع دعاء الإمام أو من نصبه، على الكفاية، إلا أن يرجعوا، وهم قسمان: من له فئة، فيُجهز على جريحهم ويُتبع مدبرهم ويُقتل أسيرهم. ومن لا فئة له، فلا يجهز على جريحهم ولا يتبع مدبرهم ولا يقتل أسيرهم. ولا يحل سبى ذراري الفريقين ولا نساؤهم ولا أموالهم.

Section Three: The Division of the Spoils of War

Everything that is collected as war booty and taken out from polytheistic lands is at the Imam's disposal, such as treasures, artefacts, prizes and whatever he chooses for himself. Then he must divide the rest into fifths. Four-fifths of moveable assets belong to the fighters who were present in the battle, even if they may not have fought at the time. Cavaliers receive twice as much as foot soldiers, and those who provided more horses, three times as much. Someone who was born in the land of *jihād* after its occupation but before the division of booty also receives a share, as do those who join the Muslims to help them. No one must be given preferential treatment in favour of their class, level of suffering or anything else. Spoils of war collected on ships are divided in the same manner, but only the cavalry receive a share. Cavaliers are considered as such upon their entering the land and not the battle. Bedouin nomads do not receive a share, even if they fought.

Women and children taken as prisoners of war may be owned as captives. Adult males above the age of puberty, if captured in the midst of the battle, must be killed as long as they have not submitted to Islam, and the Imam may choose to either decapitate them or to cut off their arms and legs on alternate sides and leave them to die. If, however, they were captured after the end of the battle, they must not be killed, and the Imam can choose to either release them from captivity without recompense, ransom them, or take them as slaves.

As for captured lands: all cultivated land belongs to the Muslim community as a whole, and not just the combatants, and it is for the Imam to oversee them. They must not be sold, donated, gifted or owned by anyone in particular, rather the Imam must distribute their yields for public benefit. Lands that are barren at the time of capture belong to the Imam, and no one can use them without his permission. These are the rulings pertaining specifically to conquered lands. As for lands belonging to people who have contracted a peace treaty with the Muslims, they continue to belong to their owners. If a landowner sells it, the *jizya* due on the land is transferred to the new owner. If one embraces Islam, then the *jizya* that is paid due to land-ownership is relinquished [just as *jizya* upon the individuals themselves is relinquished]. If the land was specified in the treaty as belonging to the Muslims, then it is treated as land from spoils of war.

الفصل الثالث: في قسمة الغنائم

جميع ما يغنم من بلاد الشرك يخرج منه ما شرطه الإمام كالجعائل والرضخ والأجر وما يصطفيه، ثم يخمس الباقي، وأربعة الأخماس الباقية إن كان مما ينقل ويحول فللمقاتلة ومن حضر القتال وإن لم يقاتل خاصة، للراجل سهم وللفارس سهمان، ولذي الأفراس ثلاثة، ومن وُلِد بعد الحيازة قبل القسمة أسهم له، وكذا من يلحقهم للمعونة، ولا يفضل أحد على غيره لشرفه أو شدة بلائه، ويقسم ما يغنم في المراكب هذه القسمة، ولا يسهم لغير الخيل، والاعتبار بكونه فارساً عند الحيازة لا بدخول المعركة، ولا نصيب للأعراب وإن جاهدوا.

والأسارى من الإناث والأطفال يملكون بالسبي. والذكور البالغون: إن أُخِذوا قبل أن تضع الحرب أوزارها وجب قتلهم ما لم يسلموا، ويتخير الإمام بين ضرب أعناقهم وقطع أيديهم وأرجلهم من خلاف ويتركهم حتى ينزفوا ويموتوا، وإن أُخِذوا بعد انقضاء الحرب لم يجز قتلهم، ويتخير الإمام بين المن والفداء والاسترقاق.

وأما الأرضون: فما كان حياً فللمسلمين كافة لا يختص بها الغانمون، والنظر فيها إلى الإمام، ولا يصح بيعها ولا وقفها ولا هبتها ولا ملكها على الخصوص، بل يصرف الإمام حاصلها في المصالح. والموات وقت الفتح للإمام، لا يتصرف فيها إلا بإذنه. هذا حكم الأرض المغنومة وأما أرض الصلح: فلأربابها، ولو باعها المالك انتقل ما عليها من الجزية إلى رقبته، ولو أسلم سقط ما على أرضه أيضاً ولو شرطت الأرض للمسلمين كانت كالمغنومة.

The land belonging to people who willingly embraced Islam is left to their owners, and they only have to pay the *zakāt* (alms) tax due off the land upon fulfilling the criteria. Any land whose maintenance has been abandoned by its owner can be let out by the Imam on the basis of a share of the produce and income paid by the developer to its proprietor. Whoever develops uncultivated land with the Imam's permission has the right of tenure, and if the land has a proprietor, then he must pay him its rent. If it does not, then to the Imam, and during his occultation he may keep it himself and remit it to him upon his reappearance.

The criteria for transfer of ownership of land for the sake of cultivation is that it must not already belong to a Muslim, nor have been secured previously by a developer, nor be a sacred place dedicated to worship, nor a fief [previously granted to another individual by an infallible], and nor a place where the Muslims have set up a military hospital or infirmary. The cultivation of the land must be undertaken in accordance with regular customs; and surrounding a region of land with stones [for temporary settlement] does not necessitate transfer of ownership, just a greater deservingness [to own the land].

وأما أرض من أسلم عليها أهلها طوعاً فلأربابها ، وليس عليهم سوى الزكاة مع الشرائط .

وكل أرض ترك أهلها عمارتها فالإمام يقبلها ويدفع طسقها من المتقبل إلى أربابها ، وكل من أحيى أرضاً مواتاً بإذن الإمام فهو أحق بها ، ولو كان لها مالك كان عليه طسقها له ، وإلا فللإمام ، ومع غيبته فهو أحق ، ومع ظهوره له رفع يده .

وشرط التملك بالإحياء : أن لا يكون في يد مسلم ، ولا حريماً لعامر ، ولا مشعراً لعبادة ، ولا مقطّعاً ولا محجراً .

والإحياء بالعادة ، والتحجير لا يفيد التمليك بل الأولوية .

Section Four: Encouraging Good and Forbidding Wrongdoing

These are both rational obligations upon the community as a whole, with four pre-conditions: that the one undertaking this duty knows what is good and what is bad; that he considers his admonishing having a positive impact; that the person being admonished does not already show signs of relinquishing their actions; and that there is no risk of harm ensuing from these actions.

There are two types of good deeds: those that are obligatory, in which case it is obligatory to encourage them; and those that are commendable, whereby it is in turn commendable to enjoin them. Acts of wrongdoing, however, are all abominable, and thus forbidding them is obligatory.

First, one must disapprove of the wrongdoing in his heart, then discourage it verbally, and lastly physically. If physical injury is necessary, then he is not allowed to inflict it without prior permission from the Imam.

Fixed legal punishments can only be implemented under his authority.

A man is allowed to mete out a fixed legal punishment on his own slave, child or wife, if it is safe to do so.

During the occultation of the Imam, the jurists may implement them, provided it is safe to do so. It is obligatory for the people to assist them in this task, for they are authorized to issue edicts and adjudicate court cases among the people with the necessary conditions. A jurist is not allowed to implement the laws of a non-Imami legal school, and if he is compelled to do so, he may act in dissimulating his faith (*taqiyya*) as long as it does not constitute killing anybody.

Taking up governmental posts is allowed when appointed by the legitimate authority, and if insisted upon, it becomes obligatory. On the other hand, it is prohibited to accept the same positions under a tyrannical government, if he is certain that he will be prevented from enjoining good-doing and forbidding wrong-doing as a result. If he is compelled even without this certainty, then he is allowed to take it up, but he must strive to uphold and implement rulings with the truth.

الفصل الرابع: في الأمر بالمعروف والنهي عن المنكر

وهما يجبان عقلاً على الكفاية بشروط أربعة: أن يعلم المعروف والمنكر، وأن يجوز تأثير الإنكار وأن لا يظهر أمارة الإقلاع، وانتفاء المفسدة.

والمعروف قسمان: واجب وندب، فالأمر بالواجب واجب، وبالمندوب مندوب. وأما المنكر فكله قبيح فالنهي عنه واجب.

وينكر أولاً بالقلب، ثم باللسان، ثم باليد. ولو افتقر إلى الجراح لم يفعله إلا بإذن الإمام.

والحدود لا يقيمها إلا بأمره.

ويجوز للرجل إقامة الحد على عبده وولده وزوجته إذا أمن الضرر.

وللفقهاء إقامتها حال الغيبة مع الأمن، ويجب على الناس مساعدتهم. ولهم الفتوى والحكم بين الناس مع الشرائط المبيحة للفتيا، ولا يجوز الحكم بمذهب أهل الخلاف، فإن اضطرّ عمل بالتقية ما لم يكن قتلاً.

ويجوز الولاية من قبل العادل، ولو ألزمه وجبت، ويحرم من الجائر ما لم يعلم تمكنه من الأمر بالمعروف والنهي عن المنكر، ولو أكره بدونه جاز، ويجتهد في إنفاذ الحكم بالحق.

The Book of Trade, Commerce and Transactions

كتاب المتاجر

This book contains various sections.

Section One: Trading

Trading may be obligatory for a person who has no other means of income beside it so long as the trading in question is permissible in itself. It may even be considered commendable if by means of it one seeks comfort for one's dependants; or disapproved if one uses it to monopolize[1] the market. Generally, it is permissible when trading is neither a necessity nor poses a danger. It is only prohibited when dealing with prohibited commodities or services, which are detailed below:

1) Earning a livelihood by selling essentially impure things is forbidden: the likes of alcohol, all intoxicants, beer, dead animals, blood, dogs apart from hunting dogs, sheepdogs, guard-dogs and farm dogs, and impure fat [derived from pigs or animals not slaughtered ritually] used for outdoor lighting.

2) Earning a livelihood by selling prohibited instruments or tools is also forbidden, such as lutes and mandolins, idols and crosses, gambling aids such as chess, backgammon, and gambling chips.

3) Trading in anything that supports prohibited acts is also forbidden, such as selling arms and ammunition to enemies of the religion, selling and letting out property or transport for prohibited practices, selling grapes for the production of wine or wood for idols to be made thereof. Both of these are discouraged to be sold to the makers of such things.

4) It is also forbidden to trade in things that have absolutely no benefit, like selling animals that were originally humans such as monkeys, bears, catfish, turtles and eels, though there is no problem in selling beasts of prey.

5) It is equally forbidden to earn a livelihood through prohibited practices themselves, such as sculpting living creatures; singing in any context other than rightfully at a wedding; wailing and lamenting a person's death with false expressions; mocking people of faith; safeguarding, trading in and transcribing corruptive literature for any other purpose than to refute them; learning and teaching magic, witchcraft, sorcery and fortune-telling; gambling; fraud; cross-dressing for men[2]; adorning mosques and copies of the Quran[3]; supporting oppressors in their oppressive ways; and any earning from any form of illicit sexual relations, including prostitution.

1 A *muḥtakir* is one who monopolizes by deliberately hoarding a commodity in the anticipation of its price rising and his gains increasing.

2 By wearing garments usually worn by women, or wearing gold or silk.

3 That is, with images and sculptures of living creatures.

وفيه فصول

الفصل الأول: التجارة

قد تجب إذا لم يكن للإنسان معيشة سواها وكانت مباحة، وقد تستحب إذا أراد التوسعة على عياله، وقد تكره كالمحتكر، وقد تباح بأن لا يحتاج إليها ولا ضرر في فعلها، وقد تحرم إذا كانت في محرم. وهي أصناف:

الأول: يحرم التكسب ببيع الأعيان النجسة، كالخمر وكل مسكر، والفقاع، والميتة، والدم، والكلب إلا كلب الصيد والماشية والحائط والزرع، والدهن النجس للاستصباح به تحت السماء.

الثاني: يحرم التكسب بالآلات المحرمة، كالعود، والمِزمر، والأصنام والصلبان، وآلات القمار كالشطرنج والنرد والأربعة عشر.

الثالث: يحرم التكسب بما يقصد به المساعدة على الحرام كبيع السلاح لأعداء الدين والمساكن للمحرمات، والحمولة لها، وبيع العنب ليعمل خمراً، والخشب ليعمل صنماً، ويكره بيعهما على من يعمل ذلك من غير شرط.

الرابع: ما لا ينتفع به يحرم التكسب به، كالمسوخ البرية كالقردة والدب، والبحرية كالجري والسلاحف والطافي، ولا بأس بالسباع.

الخامس: يحرم التكسب بما يحرم عمله، كعمل الصور المجسمة، والغناء في غير العرس بالحق، وهجاء المؤمنين، وحفظ كتب الضلال ونسخها لغير النقض، وتعلم السحر والقيافة والكهانة والشعبدة، والقمار، والغش، وتزيين الرجل بالمحرم، وزخرفة المساجد والمصاحف، ومعونة الظالمين في ظلمهم، وأجر الزانية.

6) Charging a fee and earning a livelihood by charging for actions that are public collective obligations is forbidden, such as washing, shrouding and burying the deceased; charging a fee for passing a legal judgment, and accepting bribes for doing so. It is permissible to accept a salary from the public funds for the profession of a judge, public administrator or *muezzin*.

Earning one's livelihood from the following professions is disapproved: money-exchanging, selling shrouds, handling foodstuffs, slave-trading, slaughtering animals, gold-smithery, blood-cupping by stipulating a price, weaving clothing; as is making a living from breeding animals, transcribing or teaching the Quran, and midwifery by stipulating a price.

The government's collection of revenue by way of shares or taxes is permissible, even if it is not rightfully entitled to it.

Accepting gifts from an oppressive ruler is, however, forbidden if one knows that those specific things have been usurped, but if not, then it is permissible.

If one has been told to distribute a certain amount of money to a group of people in a specific manner, he is not allowed to counter the instructions. If left unspecified, then he may take some for himself equivalent to the share of his counterparts, if he belongs to that group, according to one opinion.

السادس: ما يجب فعله يحرم التكسب به كأجرة تغسيل الموتى وتكفينهم ودفنهم، والأجرة على الحكم، والرشا فيه. ويجوز أخذ الرزق من بيت المال، وكذا الأذان.

وأما المكروه: فالصرف، وبيع الأكفان، والطعام، والرقيق، والذباحة، والصياغة، والحجامة مع الشرط، والحياكة، وأجرة الضراب، وأجرة تعليم القرآن ونسخه، وكسب القابلة مع الشرط.

وما يأخذه السلطان باسم المقاسمة أو الزكاة حلال وإن لم يكن مستحقا له.

وجوائز الظالم حرام إن علمت بعينها وإلا حلت.

ومن أمر بصرف مال إلى قبيل وعين له لم يجز التعدي، وإلا جاز أن يتناول منه مثل غيره إذا كان منهم، على قول.

Section Two: The Code of Conduct in Trading and Commerce

It is recommended to become well-versed in this code so that one knows the difference between valid and corrupt trades and remains safe from usury. It is also recommended to treat any two buyers equally, to take goods back if requested by a buyer wishing to return them, to utter the two testimonies of faith and saying *Allāhu Akbar* when contracting a trade, and to always accept less in weight when buying but to give more when selling.

It is abominable to flatter the vendor and to criticize the goods or the buyer of them; to conceal any defects of goods one is selling; to swear oaths in order to make a sale; to trade in the dark; to make a profit from a believer by charging him extra, or to someone whom he has promised to give a good deal; to trade between dawn and sunrise; to be the first to enter the marketplace before anybody else; to conduct business with low, despicable and corrupt people; to continue haggling after the deal has been concluded; to offer a lower price for goods called out at a higher price to another buyer; to use weights and measures in spite of not being competent with them; to gazump and interfere in someone else's transaction; for a city dweller to appoint a rural person to trade on his behalf; to go out of town to receive tradesmen or buyers [before they have had a chance to ascertain the price of goods in the town], and this applies up to a boundary of four *farsakh* from the town.

One has the option to withdraw from the transaction or return the commodity when there is fraud or deception, whereby the price of something is being increased by the vendor, simply for the sake of pushing prices up for others, after secret collusion with a buyer who does not intend to purchase the goods. Monopoly is the hoarding of wheat, barley, dates, raisins, oil, and salt in order to sell them for a higher price when there is a lack of them available to buy. Such a trader can be forced to sell the goods, and he is not allowed to fix their price.

الفصل الثاني: في آداب التجارة

يستحب التفقّه فيها ليعرف صحيح البيع وفاسده ويسلم من الربا ، وأن يسوي بين المبتاعين ، ويقيل المستقيل ، ويشهد الشهادتين عند العقد ويكبر الله تعالى ، ويأخذ الناقص ويعطي الراجح .

ويكره مدح البائع وذم المشتري وكتمان العيب والحلف على البيع ، والبيع في المظلم ، والربح على المؤمن ، وعلى الموعود بالإحسان ، والسوم بين طلوع الفجر وطلوع الشمس ، وأن يدخل السوق قبل غيره ، ومعاملة الأدنين ، وذوي العاهات والأكراد ، والاستحطاط بعد الصفقة ، والزيادة وقت النداء والتعرض للكيل والوزن مع عدم المعرفة ، والدخول على سوم أخيه ، وأن يتوكّل حاضر لباد ، وتلقى الركبان ، وحده أربعة فراسخ فما دون .

ويثبت الخيار مع الغبن الفاحش والنجش ، وهو زيادة لزيادة من واطاه البائع ، والاحتكار وهو حبس الحنطة والشعير والتمر والزبيب والسمن والملح للزيادة في الثمن مع عدم غيره ، ويجبر على البيع ، ولا يسعر عليه .

Section Three: Contractual Agreements of Sale

This is simply a confirmation of sale, with a statement such as "I have sold... to you", and the acceptance of it with "I have bought ... from you". The validity of the contract depends on its issuance from the actual owner, and his being of responsible legal age; or if it is a minor, by one who holds guardianship over him, such as the father, grandfather, governor, agent, representative or lawyer. A contract made by anyone else apart from these would require permission from them.

If a vendor sells his own goods combined with those belonging to someone else, the contract will only apply to his goods, and the owner of the rest is responsible for his part. The buyer has the right of withdrawal if that owner rescinds the transaction.

Goods that are traded using measures, weights or number must be sold as such, and the itemised value must be made known. A joint purchase of a proportion of bulk goods is allowed provided their ratio is known. Estimating the weight by comparing similar-sized containers is allowed.

Everything that is for sale must either be available for viewing or have their specifications sufficiently detailed as to remove any ambiguity thereof. If the buyer finds it as described, he may buy it, otherwise, he has the right of withdrawal.

If full knowledge of the goods is not possible without testing them, the vendor is still allowed to sell them by their description, and if they turn out to be contrary to the description, the buyer may withdraw. If his testing them leads to the goods being damaged, then he should buy the goods. If they turned out to be faulty from the onset, then he can ask for the difference between the perfect and the faulty one. If the faulty or broken goods have no value, then he can ask for a refund.

One is not allowed to sell fish that is still in the net, nor milk that is still in the udder, nor animals that are still in their mothers' wombs. They are only allowed if other things are purchased alongside them. Similarly, one is not allowed to sell a horse that is still being cross bred. One is allowed to sell the musk that is still in the civet's gland, even if it has not yet ruptured, as well as the wool on sheep's backs.

The price and specification of the goods must be made known, either by viewing or description. It is not permissible to sell goods for dinars and dirhams interchangeably if there is a risk that they will be mistaken for one another.

الفصل الثالث: في عقد البيع

وهو الإيجاب، كقوله «بعتك» والقبول وهو «اشتريت».

وإنما يصح إذا صدر عن مكلف مالك، أو بحكمه كالأب والجد والحاكم وأمينه والوصي والوكيل، ويقف عقد غيرهم على الإجازة.

ولو جمع بين ملكه وغيره مضى في ملكه وتخير المالك في الآخر، وللمشتري مع فسخ المالك الخيار.

ويشترط في المكيل والموزون والمعدود معرفة المقدار بأحدها، ويجوز ابتياع بعض الجملة مشاعاً إذا علمت نسبته، ويجوز الإندار للظروف بما يقاربها.

ويشترط في كل مبيع أن يكون مشاهداً أو موصوفاً بما يرفع الجهالة، فإن وجد على الوصف وإلا كان له الخيار.

ولو افتقرت معرفته إلى الاختبار جاز بيعه بالوصف أيضاً، ويتخير مع خلافه ولو أدى اختباره إلى الإفساد جاز شراؤه، فإن خرج معيباً أخذ أرشه، وإن لم يكن له قيمة بعد الكسر أخذ الثمن.

ولا يجوز بيع السمك في الأجمة، ولا اللبن في الضرع، ولا ما في بطون الأنعام، ويجوز لو ضم معها غيرها. ولا ما يلقح الفحل، ويجوز بيع المسك في فأره وإن لم يفتق، وبيع الصوف على ظهور الغنم.

ولا بد أن يكون الثمن معلوماً قدراً ووصفاً بالمشاهدة أو الصفة، يبيع بدينار غير درهم نسيئة ولا نقداً مع جهل نسبته إليه.

The goods for sale must be at the vendor's disposal to sell, so it is not permissible to sell a fugitive slave. If, however, he combined it with another item, the transaction is valid. Similarly, he cannot sell a bird in the sky.

The keeper of goods [i.e. their current vendor] is liable for any illegally obtained goods [that the original owner is reclaiming]. If he had assumed them to be legal and sold them as such, or painted them and consequently increased their value, he must return the value to their rightful owner with the increase. And if the value has reduced, then he must make up for the loss in original value.

If the two parties of such a transaction cannot agree on the value, then the vendor's word holds more weight if the goods remain or are in his possession. And the buyer's word holds more weight if the goods have already been sold or are in his possession.

ويشترط أن يكون مقدوراً على تسليمه، فلا يصح بيع الآبق منفرداً ولو ضم إليه غيره صح، ولا الطير في الهواء.

وكل بيع فاسد فإنه مضمون على قابضه. ولو علمه صنعة، أو صبغه فزادت قيمته رجع بالزيادة، ولو نقص ضمن النقصان كالأصل.

وإذا اختلف المتبايعان في قدر الثمن فالقول قول البائع إن كان باقياً، وقيل إن كان في يده، وقول المشتري إن كان تالفاً، وقيل إن كان في يده.

Section Four: Right of Withdrawal From a Transaction

They consist of seven types:

1) The right of withdrawal at the point of sale (*khiyār al-majlis*): Both the vendor of an item as well as the buyer have the right to withdraw from a transaction before its conclusion as long as they have not physically parted and left the place where the transaction is taking place, and as long as they have not waived this right either before or after the contract. This does not apply to anything other than the transaction between them.

2) The right to return purchased animals (*khiyār al-ḥaywān*): Anyone who buys an animal has an automatic grace period of three days from the point of sale whereby he can rescind the transaction if he wishes, as long as they do not waive this right in the contract and as long as the buyer does not use the animals during this period. If there is something wrong with the animal during this period, or even after it, then the seller is responsible as long as the buyer has not done anything to cause the injury or illness. Faults that are discovered after the grace period, even without any mishandling by the buyer, do not entitle him to return it.

3) Extended right to return goods (*khiyār al-sharṭ*): This can be stipulated for any goods for which an extended right of withdrawal is required and is not limited to a particular time frame. Rather it remains up to them both to stipulate its terms, as long as the term is clearly defined. The terms can be stipulated by either party, both, and even a third party, defining a term within which the vendor must refund the value of the goods if they are returned. If the goods have been delivered but the full price has not been paid, the buyer remains responsible for purchasing them. The buyer is responsible for any damage to the goods during this period and gets to keep any accretion.

4) The right of withdrawal upon having been overcharged (*khiyār al-ghaban*): This right is exercised when a vendor sells goods without a standard price or the buyer is made to pay more than the standard value, without knowing the customary price that people would pay for it without feeling cheated. In this case the buyer who has been overcharged may withdraw from the sale.

الفصل الرابع في الخيار

وأقسامه سبعة:

الأول: خيار المجلس، فمن باع شيئاً ثبت له وللمشتري الخيار ما لم يتفرّقا، أو يشترطا سقوطه قبل العقد أو بعده، ولا يثبت في غير البيع.

الثاني: خيار الحيوان، وكل من اشترى حيواناً ثبت له الخيار خاصة ثلاثة أيام من حين العقد، إن شاء الفسخ فيها فسخ، ما لم يشترطا سقوطه أو يتصرف المشتري فيه، فإن تلف في هذه المدة قبل القبض أو بعده فمن البائع ما لم يحدث المشتري فيه حدثاً، والعيب الحادث من غير تفريطه لا يمنع الرد بالسابق.

الثالث: خيار الشرط، وهو يثبت في كل مبيع اشترط الخيار فيه، ولا يتقدر بمدة معينة، بل لهما أن يشترطا مهما شاءا بشرط أن تكون المدة مضبوطة، ويجوز اشتراطه لأحدهما أو لهما أو لثالث واشتراط مدة يرد فيها البائع الثمن ويرتجع المبيع، فإن خرجت ولم يأت بالثمن كاملاً لزم البيع، والتلف من المشتري في المدة والنماء له.

الرابع: خيار الغبن. وهو أن يبيع بدون ثمن المثل أو يشتري بأكثر منه، ولا يعرف القيمة مما لا يتغابن الناس فيه فيختار المغبون الفسخ.

5) If a vendor sells something without taking the money, and has yet to deliver the goods, without stipulation of the delay, he has a three-day grace period within which to conclude the transaction. If the buyer comes to collect the goods within the three days, he is entitled to receive them. But if he comes after the three days, the vendor has the right to cancel the transaction. If the goods are damaged or lost whilst in the vendor's possession during the grace period, they remain his responsibility. Goods that have a short expiry date have a one-day grace period.

6) The right of withdrawal on sight (*khiyār al-ru'ya*): One who buys something as per its described specifications, without having seen it, has the right to withdraw from the transaction if he finds that it does not meet the specifications. If the vendor had not seen the goods either, selling them by their description, and they turn out to be of a better quality [than what the buyer paid for], then he has the right to withdraw from the transaction.

7) The right of withdrawal due to defects (*khiyār al-'ayb*): this will be dealt with below in its own section.

The right of withdrawal from a transaction is inherited. If goods are damaged or lost before they are collected, the vendor is responsible for them, and if any parts are missing, the buyer has the option of either returning them or being refunded the difference between the defective item and a perfect one.

الخامس: من باع شيئاً ولم يقبض الثمن ولا سلم السلعة ولم يشترط التأخير، لزم البيع ثلاثة أيام، فإن جاء المشتري فهو أحق بالسلعة، وإن مضت كان للبائع الفسخ، ولو تلفت السلعة كانت من مال البائع على كل حال، وما لا بقاء له يثبت الخيار فيه يوماً.

السادس: خيار الرؤية، فمن اشترى موصوفاً غير مشاهد كان للمشتري خيار الفسخ إذا وجده دون الوصف، ولو لم يشاهده البائع وباعه بالوصف فظهر أجود كان الخيار للبائع.

السابع: خيار العيب، وسيأتي.

والخيار موروث. والمبيع إذا تلف قبل القبض كان من مال البائع، وإن تغيب تخير المشتري بين الرد والامساك بالأرش.

Section Five: Defective Goods

This refers to anything that is more or less than the standard quality expected. Once the two parties conclude the transaction or stipulate for the goods to be in perfect condition, it is necessary for them to be such. If the buyer has waived any such stipulation for them to be free of defects, buying them as they are, then he is not held liable. Otherwise, if a defect or flaw appears, the buyer has the option of either returning it or to keeping it along with a compensation of the difference between a perfect item and the defective one, as long as he has not used it. If he has used it, or the defect occurred whilst in his possession, he can only be refunded the difference in compensation; and if he purchased it knowing that it was defective, then he has no right to compensation.

If a vendor sells two items together as a single deal, and one of them turns out to be defective, the buyer can either ask for compensation of the difference between the defective item and a perfect one, or return both items. He may not return only the faulty item. If two people share in the purchase of an item as a single deal, neither one has the right to return his share as being defective, unless the other agrees. Using a defective item invalidates one's right to return it, except in the case of having had sexual intercourse with a pregnant slave-girl, in which case he may return her with five percent of her value in addition. A misrepresented un-milked ewe (*muṣarrāt*)[4] can similarly be returned along with the value of the milk that has been taken from it, if he cannot return it in exactly the same condition.

If the vendor had previously claimed that his goods were in perfect condition, but has no evidence of this, [and they turn out to be defective,] then the buyer's word supersedes accompanied by an oath of honesty. If the buyer claims that the goods were defective before he bought them, then the vendor's word supersedes, with an oath of honesty to that effect.

4 *muṣarrāt*: this refers to a milk-producing ewe that has not been milked or suckled by lamb for two days or more, to give the false impression that it is full and productive.

الفصل الخامس: في العيوب

وهو كل ما زاد أو نقص عن المجرى الطبيعي. فإن أطلق المتبايعان البيع أو اشترطا الصحة اقتضى الصحة، وإن تبرء المشتري من العيوب فلا ضمان. وبدونه إذا ظهر عيب تخير المشتري بين الرد والإمساك بالأرش ما لم يتصرف، فإن كان قد تصرف أو حدث فيه عيب عنده ثبت الأرش خاصة، ولو علم بالعيب ثم اشتراه فلا أرش أيضاً.

ولو باع شيئين صفقةً وظهر العيب في أحدهما كان للمشتري الأرش أوردّ الجميع لا المعيب وحده، ولو اشترى اثنان صفقة لم يكن لأحدهما رد حصته بالعيب إلا إذا وافقه الآخر، والتصرف يبطل رد المعيب إلا في الوطي في الحامل فيردها مع نصف عُشر القيمة، والحلب في الشاة المصراة فيردها مع قيمة اللبن إن تعذر المثل.

ولو ادعى البائع التبري من العيوب ولا بينة فالقول قول المشتري مع يمينه، ولو ادعى المشتري تقدم العيب على العقد فالقول قول المشتري مع يمينه.

Section Six: Transactions in Cash, Credit and Finance

Making a transaction dictates that the value of the goods must be paid. It is valid for both parties to stipulate for it to be delayed for a specified term, but not if the term is left unspecified or vague. Neither is it valid if the vendor sells it for one price if payment is made immediately, and a higher price for delayed payment.

If someone sells something for future payment, then buys it back before the agreed term for either more or less than the original price, immediately or later on, the transaction is valid only if there were no conditions stipulated. If he buys it back after the agreed term, then it is valid in any case even if the price is different to the original price. Some have said that it is not valid if the price is different, but the more likely opinion is the opposite.

The buyer is under no obligation to pay the full price before the agreed term [even upon early delivery], nor on the vendor to accept payment before it. Once the agreed date has arrived and he makes the payment, the seller must accept it, and if he refuses, then liability for perished goods rests with the one who has them in his possession.

If someone buys goods for future payment to then sell on for a profit, he must give his buyer the same period of time to pay him. If he conceals this fact, then the buyer has the right to reject the transaction or to keep it at the current value for which the vendor had bought it. If he sells them for a profit, he must not add on the profit as a fixed percentage of the price, but rather based on the value of the goods. If he bought several items in bulk at a certain price, he is not allowed to then sell them individually on the basis of percentage profit unless he makes this clear to the buyer.

الفصل السادس: في النقد والنسيئة والمرابحة

إطلاق العقد يقتضي حلول الثمن، فإن شرطا تأجيله مدة معينة صح، ويبطل في المجهولة، وكذا لو باعه بثمن حالاً وبأزيد مؤجّلاً.

وإذا باع نسيئة ثم اشتراه قبل الأجل بزيادة أو نقصان من جنس الثمن وغيره حالاً ومؤجلاً صح مع عدم الشرط، ولو اشتراه بعد حلوله جاز بغير الجنس مطلقاً، وبه: قيل لا يجوز مع التفاوت، والأقرب خلافه.

ولا يجب دفع الثمن قبل الأجل ولا قبضه قبله، ولو حل ودفع وجب القبض، فإن امتنع كان هلاكه من صاحب الحق.

ولو اشترى نسيئة وجب أن يخبر بالأجل إذا باعه مرابحة، فإن أخفى تخير المشتري بين الرد والإمساك بالثمن حالاً. وإذا باع مرابحة نسب الربح إلى السلعة لا إلى الثمن. ولو اشترى أمتعة صفقة بثمن لم يجز بيع أفرادها مرابحة بالتقويم إلا بعد الإعلام.

Section Seven: What Is Automatically Included in the Sale

Whoever sells a piece of land, any palms or trees are included in it if stipulated as such. If not, then they are not included, unless he makes a statement to the effect of, "I am selling it to you along with all that is enclosed within its gates", in which case, everything from the top to the bottom of that land is usually included, apart from its inhabitants.

If someone sells a palm tree that has already been pollinated, then the fruits [of that year] belong to the seller, but if it had not yet been pollinated, then the fruits are the buyer's. The fruits of trees are not automatically included in the sale without stipulation. If a vendor has excluded a palm tree from the sale of a plantation [for his own use], he has right of way to it, to pick its fruit, and to let its branches grow over the land.

الفصل السابع: فيما يدخل في المبيع

من باع أرضاً دخل فيها النخل والشجر مع الشرط، وإلا فلا، ويدخل لو قال: «بعتُكها وما أغلق عليه بابها» ويدخل في الدار الأعلى والأسفل إلا أن يستقل بالسكنى عادة.

ولو باع نخلاً مؤبراً فالثمرة للبائع، ولو لم يؤبر فالثمرة للمشتري. ولا يدخل الحمل في الابتياع من غير شرط، فلو استثنى نخلة كان له المدخل إليها والمخرج منها ومدى جرائدها في الأرض.

Section Eight: Delivery of Goods

This entails the release of immovable and non-transportable goods; the weighing out and measuring goods that are customarily weighed and measured; the counting out of goods that are normally counted; and the transfer of animals.

It is obligatory on the vendor to submit the goods, and on the buyer to pay the value, and they must be compelled to do so if they refuse. It is obligatory to deliver them as soon as possible.

Selling something in advance that is not yet in one's possession is allowed, except for foodstuffs, which should only be sold if they are available for assessment. The vendor's word in stating any lack or defect in the goods must be taken, provided that he weighs or measures in the buyer's presence, swears an oath of honesty and there is no evidence to the contrary. The buyer's word supersedes if he does not weigh or measure in his presence.

It is valid to make the purchase conditional upon certain plausible and permissible services at the time of contracting the transaction, but one is not allowed to place an impossible condition that is outside the vendor's remit, such as making the crops flourish, for example. It is valid to stipulate that a slave be emancipated as a condition.

If someone stipulates a condition that is not feasible, or that the slave should not be emancipated, or that the slave-girl must be a virgin, the condition is automatically void, and there are strong grounds for the nullification of the whole transaction.

If a specific amount was stipulated as a condition, and the product falls short, the buyer has the right to either return the goods or keep them at a recalculated price, regardless of whether the missing amounts are of the same type or not. If he chooses to keep only the balance of the goods, the vendor has the right to re-value them, but if he insists on the whole amount, then the vendor has no choice [but to provide him with the missing amount or to refund him]. If there is a surplus amount to what was originally stipulated, from the same type of product, the vendor takes back the extra, and the buyer then has the right to accept the transaction or not. If, however, the surplus is from a different product, then the transaction is void at face value, but it is possible to apply both the rulings for an advanced sale and the sale of differing goods together in one transaction.

الفصل الثامن: في التسليم

وهو التخلية فيما لا ينقل ويحول، والكيل والوزن فيما يكال أو يوزن والقبض باليد في الأمتعة، والنقل في الحيوان.

وهو واجب على البائع في البيع وعلى المشتري في الثمن، ويجبران معاً لو امتنعا . ويجب التسليم مفرغاً .

ويجوز بيع ما لم يقبض قبله إلا أن يكون طعاماً فلا يبيعه إلا تولية .

والقول قول البائع في عدم النقصان مع حضور المشتري الكيل والوزن مع يمينه وعدم البينة، وقول المشتري مع عدم حضوره.

ويصح في حال العقد اشتراط ما يسوغ ويدخل تحت القدرة، ولا يجوز اشتراط ما ليس بمقدور كصيرورة الزرع سنبلًا، ويصح اشتراط العتق.

ولو اشترط ما لا يسوغ أو عدم العتق أو عدم وطأ الأمة بطل الشرط، وفي إبطال البيع وجه قوي.

ولو شرط مقداراً فنقص تخير المشتري بين الرد والإمساك بالقسط من الثمن، سواء كانت أجزاءه متساوية أو مختلفة، فإن أخذ بالقسط تخيّر البائع، ولو أخذه بالجميع فلا خيار، ولو زاد متساوي الأجزاء أخذ البائع الزائد فيتخير المشتري حينئذٍ، ولو زاد المختلف فالوجه البطلان، ويجوز أن يجمع بين سلف وبيع مختلفين صفقة.

Section Nine: Usury

This is well-known to be expressly forbidden in Islam, and entails the selling of a commodity that is traded by weights and measures for extra from another commodity of the same type, either in kind, such as selling a certain amount of product for two amounts of the same product, for example; or in principle, such as selling a certain amount of the product for the same amount for future payment.

Usury applies only to trades within the same commodity, and to commodities that are traded by weights and measures.

Bartering two of the same commodity of equal commercial value with immediate exchange of goods is allowed, but not on the basis of the exchange being delayed. Similarly, different commodities prone to usury may be bartered with a surplus (in ratios or measures), preferably with immediate exchange of goods, though a delayed transaction is allowed but discouraged. Commodities that are not prone to usury are also allowed to be bartered thus, except in the case of the two currencies (dinars and dirhams).

Wheat and barley are treated as one category of commodity here, and the same goes for commodities deriving from the same origin such as sesame and sesame oil, ghee and butter, as well as the good quality item and the one of poorer quality from the same commodity. Meats differ based on which animal they come from, as do different types of fat.

If a commodity is sold by amount in one place, and by weight in another, then the ruling pertaining to that locality prevails. Selling unripe dates for ripe ones is not allowed, even if they are identical in weight; and selling meat for an animal is disapproved.

It is acceptable to sell a dirham and a *mudd*'s worth of dates for two dirhams or two *mudds*.

Whoever has consumed usury inadvertently is not at fault but must return whatever he received as usury to its rightful owner if he is able to locate him, or to his next of kin. If not, then he must give it away in charity on his behalf.

There is no usury between father and son, nor between master and slave, nor between husband and wife, and nor between the Muslims and those at war with them. It is, however, not allowed between Muslims and protected non-Muslims (*dhimmīs*).

الفصل التاسع: في الربا

وهو معلوم التحريم بالضرورة من الشرع، وهو بيع أحد المثلين بآخر مع زيادة عينية كبيع قفيز بقفيزين، أو حكمية كبيع قفيز بقفيز نسيئة.

وشرطه أمران: الاتّحاد في الجنس، والكيل أو الوزن.

ويجوز بيع المثلين متساوياً نقداً، ولا يجوز نسيئة، وكل ربوي يجوز بيعه بمخالفه نقداً متفاضلاً ونسيئةً على كراهية، وكذا غير الربوي، إلا أن يكون أحد العوضين من الأثمان.

والشعير والحنطة جنس واحد هنا، وكذا كل شيء مع أصله كالسمسم والشيرج، وكل فرعين من أصل واحد كالسمن والزبد، والجيد والرديء. واللحوم تختلف باختلاف الحيوان، وكذا الأدهان.

ولو كان الشيء جزافاً في بلدة وموزوناً في أخرى فلكل بلد حكم نفسه، ولا يباع الرطب بالتمر وإن تساويا، ويكره اللحم بالحيوان.

ولو باع درهماً ومدّ تمر بدرهمين أو مدين صح.

ومن ارتكب الربا بجهالة فلا أثم عليه، ويعيد ما أخذ منه على مالكه إن وجده أو ورثته، ولو جهل تصدّق به عنه.

ولا ربا بين الوالد وولده، ولا بين السيد وعبده، ولا بين الرجل وزوجته، ولا بين المسلم والحربي، ويثبت بينه وبين الذمي.

As for exchanging currencies, the condition is that the exchange takes place in the same sitting. If the two currencies are the same, then the amounts must be the same, and if not, then not. If one receives at least some of the amount, only that part of the transaction is valid. If the two parties leave the place whilst maintaining each other's company, then exchange the currencies, the transaction is valid.

Gold may be sold for silver and vice versa. Coins mixed with impure metals (*ad-darāhim al-maghshūsha*) may be spent if they are known to be in circulation, but if not, then transacting them is not allowed, until their state is made clear.

Items that are moulded from a mixture of silver and gold must not be sold as either purely silver or purely gold if the exact proportions of each can be separated. If not, then it is sold for the lesser mineral therein. If they are present in equal amounts, it may be sold for either one. The dust [of the minerals] leftover after moulding must be given away in charity.

It is allowed to borrow currency and to stipulate that it be repaid in another place. It is also allowed to buy a dirham for another dirham on condition that a ring is made to go with it or something similar [to avoid usury], though this cannot be applied to anything else.

وأما الصرف : فشرطه التقابض في المجلس ، فإن تساوى الجنس وجب تساوي المقدار ، وإلا فلا . ولو قبض البعض صح فيه خاصة ، ولو فارقا المجلس مصطحبين ثم تقابضا صح .

ومعدن الذهب يباع بالفضة وبالعكس ، والدراهم المغشوشة إذا كانت معلومة الصرف جاز إنفاقها ، وإلا فلا ، إلا أن يبين حالها .

والمصاغ من الجوهرين إن أمكن تخليصه لم يبع بأحدهما قبله ، وإلا بيع بالناقص ، ومع التساوي يباع بهما ، وتراب الصاغة يتصدق به .

ويجوز أن يقرضه ويشترط الإقباض بأرض أخرى ، وأن يشتري درهماً بدرهم ويشترط صياغة خاتم على أشكال . ولا ينسحب على غيره .

Section Ten: The Transaction of Fruits

Selling fruits before their appearance [on the trees] is not allowed. Once they have appeared, it is allowed, even if their quality cannot be ascertained yet, provided that one is able to know clearly what is being sold, or they are being sold together with something else, or they are being sold two years in advance. If none of these conditions are fulfilled, then there are two opinions about the validity of the transaction.

If some of the fruits in an orchard are suitable for sale, the sale of the whole lot is allowed. The same applies to the sale of several orchards if one is known to yield fruit. Fruits may be sold in their husks, as are other crops in their stalks, chaff and fodder. The buyer must pay the seller extra for cutting it earlier [if used as fodder], or for the extra use of land if he wishes it to remain uncut, after which the buyer must cut it for him.

Vegetables can be sold after they have grown to size, when produced in stages, yield after yield, as well as leaves of trees that may be plucked or stripped, in several stages.

One is allowed to exclude a specified public portion of the yield from the sale, or even a palm tree, specific fruit tree or specific weights' worth. If the exclusion is not adhered to, then it must be recovered from his balance of produce.

Muḥāqala, which is the sale of grain futures and bi-products from the same piece of land, is forbidden, as is *muzābana*, which is the sale of dates still on the palm for dry dates from the same palm, unless they are bare palm (*'ariyya*) being appraised. If two people jointly own a fruit tree, one of them can accept his partner's share of fruit for a specific weight of another commodity in exchange.

Whoever passes fruit trees unintentionally is allowed to eat fruit from them, but not to take it away with him and nor to waste it.

الفصل العاشر: في بيع الثمار

لا يجوز بيع الثمرة قبل ظهورها ، ويجوز بعده وإن لم يبد صلاحها ، بشرط القطع أو مع الضميمة أو عامين، ولو فقد الجميع فقولان.

ولو أدرك بعض البستان جاز بيع الجميع ، وكذا يجوز بيع البستانين إذا أدرك أحدهما ، وبيع الثمرة في كمامها ، والزرع قائماً وحصيداً وقصيلاً ، وعلى المشتري قطعه، فإن تركه طالبه البائع بأجرة الأرض مدة التبقية ، وللبائع قطعه.

ويجوز بيع الخضر بعد انعقادها لقطة ولقطات ، وما يجز أو يخرط جزة وجزات وخرطة وخرطات.

ويجوز استثناء حصة مشاعة أو نخلاً أو شجراً معيناً أو أرطالاً معلومة ، فإن خاست سقط من الثنيا بحسابه.

والمحاقلة حرام، وكذا المزابنة إلا العرية ويجوز أن يتقبل أحد الشريكين بحصة صاحبه بوزن معلوم.

ومن مر بثمرة نخل لا قصداً جاز أن يأكل من غير استصحاب ولا أضرار.

Section Eleven: Trading in Animals

Any owned animal can be sold whereupon it then becomes the property of the buyer, except for the runaway animal or slave that has escaped from its owner, and the slave mother (*umm al-walad*) whilst her child is with her, in which case he can either delay purchasing her or buy them both.

It is also not permissible to purchase one's own father as a slave, nor one's son, or any other family members related to him by blood or milk, and the same goes for a woman. If one does own a relative, then he must manumit him. Neither is it permissible for a Muslim to be owned as a slave by a disbeliever or given away as a *waqf*. If a husband owns his wife as a slave or vice versa, the ownership prevails, and the marriage is nullified.

The joint purchase of parts of animals is allowed. If one of the partners in a joint transaction stipulates the head or hide of the animal, for example, he may have them per his share of the transaction, and not in accordance with the parts stipulated. If someone tasks another with the purchase of an animal as a joint transaction between them, this is valid and he is liable for paying half of the cost, but if he has stipulated the amount of capital, then he is not bound to pay half.

One who sells a slave-girl must ensure that she is clear of her menses before he sells her if she has been menstruating, and if not, then he must wait forty-five days. If he does not ascertain that, then it is obligatory on the buyer to do so [to ensure that she is not pregnant]. This is not necessary for a woman who has entered menopause, an underage girl, one who has already been cleared for pregnancy, and the slave-girl owned by a woman. The new master is not allowed to have frontal sexual intercourse with a pregnant slave-girl until four months and ten days have passed of her pregnancy. If he does, then he must detach himself away, for if he does not, then he is highly discouraged from selling her child in the future.

It is recommended to change the slave's name, to feed him something sweet, to give four dirhams in charity for him, and to not disclose to him how much he was purchased for. Separating a mother from her child before he reaches seven years of age is disapproved.

الفصل الحادي عشر: في بيع الحيوان

كل حيوان مملوك يصح بيعه ويستقر ملك المشتري عليه، إلا الآبق منفرداً ، وأم الولد مع وجود ولدها وإيفاء ثمنها أو القدرة عليه، إلا أن يكون العبد أباً للمشتري وإن علا، أو ابناً وإن نزل، أو واحدة من المحرمات عليه نسباً ورضاعاً ، وكذا المرأة في العمودين، فيعتق عليه لو ملكه، أو يكون المشتري كافراً والعبد مسلماً ، أو يكون موقوفاً . ولو ملك أحد الزوجين صاحبه استقر الملك وبطل النكاح.

ويجوز ابتياع أبعاض الحيوان المشاعة، ولو شرط أحد الشريكين الرأس والجلد بما له، كان له بنسبة ما له لا ما شرط، ولو أمره بشراء حيوان أو غيره بشركته صح ولزمه نصف الثمن، ولو شرط رأس المال لم يلزمه.

وعلى البائع استبراء الأمة قبل بيعها بحيضة إن كانت تحيض، وإلا فخمسة وأربعين يوماً ، ولو لم يستبرئ وجب على المشتري، ويسقط في اليائسة والصغيرة والمستبراة وأمة المرأة، ولا يطأ الحامل قبلاً إلا بعد مضي أربعة أشهر وعشرة أيام، فإن فعل عزل، ولو لم يعزل كره له بيع ولدها .

ويستحب تغيير اسمه، وإطعامه شيئاً من الحلاوة، والصدقة عنه بأربعة دراهم، ولا يريه ثمنه في الميزان. ويكره التفرقة بين الأم والولد قبل سبع سنين.

If after having impregnated a newly-purchased slave-girl, it turns out that she had been sold unlawfully [and belonged to someone other than the vendor], her original master may take her away. If she had been a virgin, then the buyer must pay her full market value, and if not, then half of it, as well as the value of the child once she gives birth to it alive. Then he may recover all that from the one who sold her to him, if he did not know that she was already owned by another at the point of sale.

It is permissible to buy slaves that have been captured in war by the oppressors, as it is to buy the daughter, sister or other relatives of a disbeliever.

If someone buys a slave-girl who has stolen something from a land that has contracted a peace treaty with the Muslims (*arḍ al-ṣulḥ*), he can return her to the vendor, who must refund the purchase price. If the vendor died without leaving an heir, then he must send her to the governor.

ولو ظهر استحقاق الأمة بعد حملها انتزعها المالك، وعلى المشتري قيمتها إن كانت بكراً وإلا فنصفه وقيمة الولد يوم سقوطه حياً، ويرجع بذلك كله على البائع إن لم يكن علم بالغصب وقت البيع.

ويجوز شراء ما يسبيه الظالمون، وكذا بنت [الكافر] وأخته وغيرهما من أقاربه.

ومن اشترى جارية سرقت من أرض الصلح ردها على البائع واسترجع الثمن، وإن مات ولا عقب له دفعها إلى الحاكم. ولو دفع إلى مملوك غيره [الـ] مأذون مالاً ليعتق نسمة ويحج عنه فاشترى أباه، ثم ادعى كل من الثلاثة شراءه من ماله، فالقول قول سيد المملوك مع عدم البينة.

ولو وطأ الشريك جارية الشركة، حد بنصيب غيره، فإن حملت قومت عليه وانعقد الولد حراً، وعليه قيمة حصص الشركاء منه عند سقوطه، ولو اشترى كل من المأذونين صاحبه ولا سبق، بطل العقدان.

Section Twelve: Advanced Transactions

These are allowed on condition that all details about the item such as its type and specifications are mentioned to remove any ambiguity about it, and the price is paid in full before parting ways. If the vendor only takes part of the payment, the balance is void. Goods that are usually sold by weight and measure must be evaluated as such. The time of delivery must be fixed from the onset, and any possibility of delay declared. If the vendor is unable to deliver the goods on time, the buyer has the right to either cancel the transaction or wait.

If the goods delivered do not match the specifications, or are delivered before time, then the buyer is not obliged to accept them, but he must do so if they are delivered on time and as specified, or even better than specified.

It is permissible to stipulate feasible conditions, but not that they be from a certain crop in a certain field, or the fabric spun by a particular woman, or the fruit of one specific tree.

The vendor bears the cost of hiring someone to weigh or measure the goods on his behalf, whereas the buyer is responsible for the cost of an evaluator or someone to weigh or measure the goods on his behalf upon receipt of them. If an agent volunteers to undertake these, then there is no hire fee payable [by the buyer].

The broker is not liable for the quality of the goods or any wastage that occurs whilst they are in his possession if there has been no negligence on his part, and he has the last word in a dispute if he swears an oath denying any negligence and the vendor has no evidence against him. The same applies if the dispute concerns the original value of the goods, even if negligence has been proven.

الفصل الثاني عشر: في السلف

وشروطه: ذكر الجنس والوصف الرافع للجهالة، وقبض الثمن قبل التفرق، ولو قبض البعض بطل الباقي، وتقدير المبيع ذي الكيل والوزن بمقداره، وتعيين أجل مضبوط، وإمكان وجوده بعد الحلول، فإن تعذر تخير المشتري بين الفسخ والصبر.

ولو دفع دون الصفة أو أكثر أو قبل الأجل لم يجب القبول، بخلاف ما لو دفعه في وقته بصفته أو أزيد منها.

ويجوز اشتراط ما هو سائغ، ولا يجوز أن يشترط من زرع أرض بعينها، أو غزل امرأة بعينها أو ثمرة نخلة بعينها.

وأجرة الكيال ووزان المتاع وبائع الأمتعة على البائع، وأجرة الناقد ووزان الثمن ومشتري الأمتعة على المشتري. ولو تبرّع الواسطة فلا أجرة.

ولا ضمان على الدلال في الجودة ولا التلف في يده إذا لم يفرط، والقول قوله في التفريط مع اليمين وعدم البينة، وفي القيمة لو ثبت التفريط.

Section Thirteen: The Right of Pre-Emption

When a joint owner of something decides to sell his share of the property, the other partner has the right of pre-emption or refusal, with certain conditions: that the property be divisible or suitable to own shares in; that the share is transferable through sale; that if the sale goes ahead, it will be a joint transaction at the point of sale. This right also applies to sharing the right of way to streams and water rights; but does not apply when there are more than two partners. The other partner may only exercise his right to buy it first if he has the means to do so, and to that the value be payable immediately upon purchase.

If the owner of a piece of freehold land sells his share, the administrator of *waqf* property is allowed to assume the right of pre-emption. The *dhimmī* has no right of pre-emption over a Muslim, though conversely a Muslim does over him. The partner with the right of pre-emption can only purchase what is being contractually sold for a fixed price, even though it may exclude him from some share of it, and if it is not fungible, then he must accept its value instead. If he claims to not have the cash at hand, he is given a three-day grace period, and if the cash is in another town, he is given enough time to go and procure it in addition to the three days, provided it does not cause any loss to the seller. The absentee partner has the same right of pre-emption [as a present partner], which he may exercise in person, as do the incompetent person, the minor and the mentally incapacitated. They may exercise their right themselves when their condition ceases, or through a guardian. The absentee partner [who finds his share sold] may purchase it back from the buyer, and the latter is responsible for recovering his money.

If the property is being sold on credit at a later date, the partner can only exercise his right in the present, otherwise he must appoint a guarantor if he is not able to procure the funds on time.

In a dispute about the price of the property between the partner [wishing to exercise his right of pre-emption] and the buyer [who has already bought the share], the buyer has the last word when accompanied by an oath of honesty, if the partner has no evidence.

The right of pre-emption is inherited in the way that other properties and assets are inherited.

If one's right of pre-emption is forfeited before a sale, it is not invalidated, as opposed to when the partner knew about the sale, witnessed it and congratulated the seller but did not exercise his right.

الفصل الثالث عشر: في الشفعة

إذا باع أحد الشريكين حصته في ملكه كان للآخر الشفعة، بشروط: أن يكون الملك مما يصح قسمته، وأن ينتقل الحصة بالبيع، وأن يكون البيع مشاعاً مع الشفيع حال البيع، أو يكون شريكاً في الطريق والنهر والساقية، وأن لا يزيد الشركاء على إثنين، وأن يكون الشريك قادراً عليه، وأن يطالب على الفور مع المكنة.

ولو باع صاحب الشقص الطلق نصيبه جاز لصاحب الوقف الأخذ بالشفعة، ولا يثبت لذمي على مسلم ويثبت للمسلم عليه. ويأخذ الشفيع بما وقع عليه العقد وإن أبرأه من بعضه، ولو لم يكن مثلياً أخذ بقيمة الثمن، ولو ذكر غيبة الثمن أجل ثلاثة أيام، وينظر لو كان في بلد آخر بما يمكن وصوله إليه مع ثلاثة أيام ما لم يستضر المشتري. ويثبت للغائب ويطالب مع حضوره، وللسفيه والصبي والمجنون، ويطالبون مع زوال الأوصاف، أو الولي. والشفيع يأخذ من المشتري ودركه عليه.

ولو كان الثمن مؤجلاً أخذ الشفيع في الحال، وألزم بكفيل إذا لم يكن ملياً على إيفاء الثمن عند الأجل.

والقول قول المشتري مع يمينه في كمية الثمن إذا لم يكن للشفيع بينة. والشفعة تورث كالأموال.

ولو أسقط الشفعة قبل البيع لم تبطل، بخلاف ما لو بارك أو شهد على أشكال.

The Book of Hiring, Tenancy and Letting

كتاب الإجارة

(والوديعة وتوابعهما)

This book contains various sections.

Section One: Hiring

Hiring has six conditions that must be fulfilled: a) there must be a contract, consisting of the proposal and the acceptance indicating that the utility of a thing is transferred [from the possession of the owner to the one hiring], for a specified term in exchange for an agreed and stipulated compensation; b) that the contract be conducted by someone having legitimate right of utility [i.e. the one hiring must either be the owner of the item being leased or have the authority to hire it on behalf of its owner]; c) that full specifications of the item are known in weight or measure, and it suffices for items like these [i.e. items usually weighed or measured before sale] and others beside them [i.e. other kinds of commodities that are usually sold without first being weighted or measured] that they are traded by observation [i.e. they do not need to be physically weighed or measured but can be hired on the basis that both the hirer and the leaser agree on the amount of a commodity visible in front of them]; d) that the utility of the item is stipulated either by length of time or by usage; e) that the utility be owned or considered as such; and f) that the exact term be fixed beforehand, such that there is no possibility of the period increasing or decreasing.

The contract of lease is binding and cannot be nullified except by mutual agreement, and not through sale of the property or death. The leaser is a custodian [of the item] and held liable upon any infringement.

For the contract to become binding, the hire charge must be paid in advance in full, unless it is stipulated to be paid in specified installments or after the term, which is valid.

The leaser is permitted to hire out the commodity or utility that he has hired to a third party for a fee that is higher or lower than what he paid as long as there is no stipulation in the original lease contract that prevents him from doing so. If the owner prevents the leaser from using the property or if it perishes before he can use it, the agreement is void. If an oppressor prevented him from using the property after having obtained it, the rental remains valid and the tenant can recover compensation from the oppressor.

If the housing is damaged through no fault of the tenant, he may either nullify the contract and ask for a refund of rent that has been paid in advance, or he can compel the landlord to repair it.

وفيه فصول

الفصل الأول: في الإجارة

وشروطها ستة: العقد، وهو الإيجاب والقبول الدالان بالوضع على تمليك المنفعة مدة من الزمان بعوض معلوم. وأن يكون ممن هو جائز التصرف. والعلم بالأجرة كيلاً أو وزناً، ويكفي فيهما وفي غيرهما المشاهدة. وأن تكون المنفعة معلومة بالزمان أو العمل، ومملوكة أو في حكمها. وضبط المدة بما لا يزيد وينقص.

وهي لازمة لا تبطل إلا بالتراضي، لا بالبيع ولا بالموت. والمستأجر أمين يضمن مع التعدي.

وإطلاق العقد يقتضي تعجيل الأجرة، ولو شرط دفعها نجوماً معينة أو بعد المدة صح.

وللمستأجر أن يؤجر بأكثر أو أقل إن لم يشترط عليه المباشرة، ولو منعه المؤجر من العين أو هلكت قبل القبض بطلت، ولو منعه ظالم بعد القبض صحت ورجع المستأجر على الظالم.

ولو انهدم المسكن من غير تفريط فسخ المستأجر ورجع بنسبة المتخلف من الأجرة، أو ألزم المالك بالعمارة.

If, in a dispute, two parties disagree on whether or not there has been a hire contract agreement between them, then the party denying the contract has the last word if the claimant cannot produce evidence to support his claim. The tenant has the last word when it comes to the amount of rent charged, claims of neglect and the value of the original property [with evidence], and the owner has the last word in the restitution of the property and the duration of the lease.

In any situation, whereby the amount of rent payable is nullified, the market rent for a similar property applies. Renting a non-distinguishable portion of a commodity is also permissible.

All skilled craftspeople are liable for damage caused to items handed in their care, even if they are highly skilled, such as a launderer who tears the garment he is given to wash, for example.

والقول قول منكر الإجارة مع عدم بينة المدعي، وقول المستأجر في قدر الأجرة والتفريط وقيمة العين، وقول المالك في رد العين وقدر المستأجر.

وكل موضع يبطل فيه الإجارة يثبت فيه أجرة المثل. ويصح أجرة المشاع.

ويضمن الصانع ما يجنيه وإن كان حاذقاً، كالقصار يخرق الثوب.

Section Two: Sharecropping and Letting Orchards

These are both binding contractual agreements that can only be voided by mutual annulment.

Sharecropping: Five conditions must be adhered to: a) that there be a contract; b) that the yield from the crops must be shared; c) that the term be stipulated; d) that the exact division of shares of the joint ownership is specified; and e) that the land is arable and beneficial.

The tenant can farm the land himself, or have someone else farm it on his behalf, or work on it together with others, as long as it has not been stipulated in the contract that it must be farmed by the tenant himself directly. He may grow whatever he wishes as long as specific crops have not been stipulated in the contract. The landowner is responsible for paying the land tax unless stated otherwise in the contract. Both sides are allowed to estimate the yield of their crops, and if they come to an agreement, they must accept it assuming the crops harvest. If the sharecropping did not go to plan, or the tenant farmer did not grow his produce, the customary ground rent is still payable.

Sharecropping to produce wheat and barley is disapproved, as is stipulating gold or silver in addition to the share of crops.

If the land is flooded before the tenant farmer can harvest it, the contract is nullified. If only part of it is flooded, he may annul the contract and pay rent only for what he has farmed, and no more. The same applies to land that one rents.

Letting Orchards: Six conditions must be adhered to: a) that a contract is made with the owner; b) that the term is stipulated; c) that the orchard is capable of producing fruits; d) that either the exact section of the orchard to be let is specified; e) or that the share of the yield from the entire orchard that the contract applies to is specified; and f) that the orchard has firm roots whose fruits would be edible upon its continuance.

These conditions are valid both before the fruit has appeared on the trees and afterwards, upon greater yield through farming. For it to become binding, the contract necessitates that the tenant farmer undertakes all the farming necessary for maximum yield of the crops, whilst the owner is responsible for securing the orchard, irrigating it and paying the land tax.

Upon nullification of the contract, the tenant farmer must be paid an equivalent salary, and the produce belongs to the owner.

It is disapproved for the owner to ask for payment in gold or silver on top of the shares he receives of the produce, and it is obligatory to adhere to the contract if the fruit is edible.

الفصل الثاني: في المزارعة والمساقاة

وهما عقدان لازمان لا يبطلان إلا بالتفاسخ.

أما المزارعة فشروطها خمسة: العقد، وأن يكون النماء مشاعا، والأجل المعلوم، وتعيين الحصة بالجزء المشاع، وكون الأرض مما ينتفع بها.

وله أن يزرع بنفسه أو بغيره أو بالشركة ما لم يشترط المباشرة. ويزرع ما شاء مع عدم التخصيص في العقد. والخراج على المالك ما لم يشترط عليه. والخرص جائز من الطرفين، فإن اتفقا كان مشروطاً بالسلامة، وإذا بطلت المزارعة أو لم يزرع العامل يثبت أجرة المثل.

ويكره إجارة الأرض بالحنطة والشعير، وأن يشترط مع الحصة ذهباً أو فضة.

ولو غرقت الأرض قبل القبض بطلت، ولو غرق بعضها تخير العامل في الفسخ والإمضاء، وكذا لو استأجرها.

وأما المساقاة فشروطها ستة: العقد من أهله، والمدة المعلومة، وإمكان حصول الثمرة فيها، وتعيين الحصة، وشياعها، وأن يكون على أصل ثابت له ثمرة ينتفع بها مع بقائه.

وتصح قبل ظهور الثمرة وبعدها مع الاستزادة بالعمل، وإطلاق العقد يقتضي قيام العامل بكل ما يستزاد به الثمرة، وعلى المالك بناء الجدران وعمل الناضح والخراج.

ومع بطلانها يثبت للعامل أجرة المثل، والنماء لربه.

ولو شرط على العامل مع الحصة ذهباً أو فضة كره، ووجب الوفاء مع سلامة الثمرة.

Section Three: Payment of Reward

The contract must consist of a proposal and acceptance, to the effect of: "Whoever finds and brings back my slave or does such and such a task shall have *x* amount of reward." The acceptance need not be uttered verbally.

Any type of permissible and intentional work carried out can be entitled to payment of a reward, even if the nature of it is undefined. If the exact compensation is known, then it must be paid for that work, and if not, then whatever would be a typical wage for that work, except in the case of someone hired to find a lost camel or runaway slave. If they are found within the city, then he must be paid one dinar for each, and four if found outside the city.

If someone volunteers himself for a job, then he need not be paid, whether others doing the same job are being paid or not. If a person offers a reward for work on behalf of another person, it is binding upon him to pay it when the work has been fulfilled.[1]

The reward is merited upon delivery [of the goods or service], and once work has begun on a task, the proposer cannot breach the contract and not pay the reward for work completed. If a person pledges two different rewards, then the most recent pledge is the one that he is held accountable to upon completion of the work.

If a reward is offered for a task that would necessitate some work from several individuals, then they all have a share of the reward. And if each of them produced the complete work independently, then each one must receive the full reward. If a reward was offered to fetch something from a particular distance away, and the person fetched it from a part of that distance, then he receives the relative proportion of the reward.

The owner [proposing the reward] has the last word in a dispute involving the amount of the reward [when the owner claims it was unspecified], determining exactly what the reward was being offered for, as well as the amount. In case of the latter, the lesser of the typical reward and the claimed reward must be paid, with no further pursuit.

1 For example, if Zayd says "I will pay a reward to anyone who fixes Khalid's car", then Zayd is required to pay the reward when Khalid's car gets fixed.

الفصل الثالث: في الجُعالة

ولا بد فيها من الإيجاب والقبول، كقوله «من رد عبدي أو فعل كذا فله كذا» ولا يفتقر إلى القبول لفظاً.

ويجوز على كل عمل محلل مقصود وإن كان مجهولاً، فإن كان العوض معلوماً لزم بالفعل، وإلا فأجرة المثل، إلا في البعير والآبق يوجدان في المصر فعن كل واحد دينار وفي غير المصر أربعة.

ولو تبرّع فلا أجرة سواء جعل لغيره أو لا، ولو تبرع الأجنبي بالجعل لزمه مع العمل.

ويستحق الجعل بالتسليم، ومع التلبس بالعمل ليس للجاعل الفسخ بدون أجرة ما عمل. ويعمل بالمتأخر من الجعالتين.

ولو جعل لفعل يصدر عن كل واحد بعضه فللجميع الجعل، ولو صدر من كل واحد فلكل واحد جعل، ولو جعل للرد من مسافة فرد من بعضها فله النسبة.

والقول قول المالك في عدم الجعل، وفي تعيين المجعول فيه، وفي القدر. فيثبت فيه الأقل من أجرة المثل والمدعي، وعدم السعي.

Section Four: The Prize Money in Racing and Archery

Both of these must consist of an offer and acceptance, and are valid for competitions involving arrows, lances and swords, as well as camels, elephants, horses, mules and donkeys.

The reward may either be an amount of money that is owed to the winner by the other participants or a prize in the form of some material commodity, and can be financed either by a third-party, or by one of the two contestants, or from the public treasury. It is paid to the winner or to the legitimator of the competition (*muḥallil*).[2] The presence of a *muḥallil* is not a pre-requisite.

In races, the distance, the prize-money and the animal must all be determined before the race, each having a fair chance of winning as the other.

In archery competitions, the type of throwing-contest needs to be determined in advance, as do the number of throws, the style of throwing, the distance, the target, the prize-money, and for the instruments to all be of the same type. There is no need to identify individual arrows or bows.

The following scenarios are permissible: for two contenders to race alongside the *muḥallil*, and the winner receives the stakes (of the two contenders). If it is a draw, they each keep their own stake; if it is a draw between a contender and the *muḥallil*, the winning contender keeps his own stake as well as half of his opponent's, and the rest goes to the *muḥallil*. If the contract is corrupted, then there is no prize.

If the prize has been deservedly claimed, the financier must either pay the exact prize or its value.

Winning in a horse-race is determined by the neck and nape of the horse to be ahead. There is no requirement to stipulate a certain number of steps that it must be ahead of the other contender to win, nor a certain number of targets hit out of a total [in archery].

2 *muḥallil*: (a.), literally, "someone who makes a thing legal, legaliser, legitimator", the figure who, in classical Islamic law acts as something like a dummy or a "man of straw", in order to authenticate or make permissible some legal process otherwise of doubtful legality or in fact prohibited. It thus forms part of the mechanisms and procedures subsumed under *ḥiyal*, legal devices, often used for evading the spirit of the law whilst technically satisfying its letter. Thus the *muḥallil* is found in gambling, racing for stakes, e.g. with horses or pigeons, and archery cont. (Encyclopedia of Islam).

الفصل الرابع: في السبق والرماية

ولا بُدّ فيهما من إيجاب وقبول، وإنما يصحان في السهام والحراب والسيوف، والإبل والفيلة والخيل والبغال والحمير خاصة.

ويجوز أن يكون العوض ديناً وعيناً، وأن يبذله أجنبي أو أحدهما أو من بيت المال، وجعله للسابق منهما أو للمُحلِّل، وليس المحلّل شرطاً.

ولا بُدّ في المسابقة من تقدير المسافة والعوض وتعيين الدابة، وتساويهما في احتمال السبق.

ويفتقر الرمي إلى تقدير الرشق وعدد الإصابة وصفتها وقدر المسافة والغرض والعوض وتماثل جنس الآلة، ولا يشترط تعيين السهم ولا القوس.

ولو قالا «من سبق منا ومن المحلّل فله العوضان» فمن سبق من الثلاثة فهما له، فإن سبقا فلكل ما له، وإن سبق أحدهما والمحلّل فللسابق ما له ونصف الآخر والباقي للمحلّل. ولو فسد العقد فلا أجرة.

ولو كان العوض مستحقاً فعلى الباذل مثله أو قيمته.

ويحصل السبق بالتقدم بالعنق والكتد ولا يشترط ذكر المحاطة والمبادرة.

Section Five: Partnership

Partnership is valid only for assets and not in offering services – as each must receive the wage for his own work – nor in the joint purchase and sale of goods on credit upon their own liability, and nor as a collective undefined partnership (*mufāwaḍa*).

It becomes a partnership when two or more partners enter it as a single venture, or through the merging of possessions such that when they are mixed each party's own share is no longer distinguishable from the other party's share. Each of them owns a share of the profit and loss proportionate to their share of the capital.

If they agree to an equal share of the profits and losses despite unequal shares of capital investment, then it is acceptable and vice versa. It is not valid, however, for one of the partners to act without the other's permission, and he must limit his remit to what is permitted by the other.

If there is no risk of damage or loss by terminating the partnership and dividing it equally, the partner being prevented from continuing the partnership may demand his share of the assets back. If they both own equal shares, then drawing lots suffices, though it is more prudent for a mediator to fairly divide the capital between them. This is not, however, an essential requirement. The partner himself is a trustee.

A contract for future partnership is not valid. The partnership is automatically terminated with the death or mental incapacity of a partner.

Entering into partnership with disbelievers is disapproved. Neither partner has the right to demand his capital in cash, but rather his share of the goods by mutual agreement. Dividing the shares in an endowment fund is not valid, except upon dissolution of the company.

الفصل الخامس: في الشركة

إنما يصح في الأموال دون الأعمال - فلكل أجرة عمله - والوجوه والمفاوضة.

ويتحقق باستحقاق الشخصين - فما زاد - عيناً واحدة، أو بمزج المتساويين بحيث يرتفع الامتياز بينهما. ولكل منهما في الربح والخسران بقدر ماله.

ولو اشترطا التساوي مع اختلاف المالين أو بالعكس جاز ولا يصح تصرف أحدهما بدون إذن الآخر، ويقتصر على المأذون.

ومع انتفاء الضرر بالقسمة يجبر الممتنع عنها مع المطالبة. ويكفي القرعة في تحقّق القسمة مع تعديل السهام، والأحوط حضور قاسم وليس شرطاً. والشريك أمين.

ولا تصح مؤجلة وتبطل بالموت والجنون.

ويكره مشاركة الكفار، وليس لأحد الشركاء المطالبة بإقامة رأس المال. وإنما تصح القسمة بالتراضي.

ولا تصح قسمة الوقف، ويجوز قسمته مع الطلق.

Section Six: The Brokered Transaction

This is when a person pays someone else to transact for a commission of the profit, and is only allowed for cash assets [i.e. the investor must provide finances or funds while the broker provides the technical know-how to manage the investment]. They both have a share in the profit. The broker must be given his stipulated share of the profits, but if there is a problem with the transaction, he is paid a customary fee and the profit goes to the owner of the capital. The brokered transaction is not binding, but may be terminated by either party as and when they please.

The broker must act as instructed [i.e. he must use the finances to transact in those business activities agreed upon with the investor], but if there is no specified commodities or business activities stipulated by the investor, then the employee can use the funds in any way he pleases so long as he acts in the interest of both parties, but he bears responsibility for any loss if he acts contrary to the stipulations made by the investor. The contract is terminated upon death, and its validity depends on both parties knowing the exact amount of funds or finances. The broker has a right to keep his share of any surplus profit made and is not responsible for any loss if there was no negligence on his part. He has the last word in a dispute defending his actions, the amount of initial capital, and any wastage or loss. The owner has the last word if the broker refuses to return the capital.

If the broker uses the capital to buy his own father out of slavery, he must use his own share of the profit towards his manumission, and the father can work towards the rest.

The broker is allowed to spend as much as is necessary from the capital on his travel expenses. He is not allowed to have sexual relations with the slave-girl of the investor without his permission.

For the transaction to be launched, the broker needs to invest the capital itself and the value must be representative. If the owner pulls out of the brokered transaction, the broker must still receive his wage for the time he spent on it.

الفصل السادس: في المضاربة

وهي أن يدفع الإنسان مالاً إلى غيره ليعمل فيه بحصة من ربحه.

وإنما تصح بالأثمان الموجودة، والشركة في الربح، وللعامل ما شرط له، ولو وقعت فاسدة فله أجرة المثل والربح لصاحب المال. وليست لازمة.

ويقتصر على المأذون، ولو أطلق تصرف كيف شاء مع اعتبار المصلحة، ويضمن لو خالف. وتبطل بالموت. ويشترط العلم بمقدار المال.

ويملك العامل حصته من النماء بالظهور، ولا خسران عليه بدون التفريط. والقول قوله في عدمه وفي قدر رأس المال والتلف والخسران، وقول المالك في عدم الرد.

ولو اشترى العامل أباه عتق نصيبه من الربح فيه وسعى الأب في الباقي. وينفق العامل من الأصل في السفر قدر كفايته.

ولا يطأ جارية القراض من دون إذن، والإطلاق يقتضي الشراء بعين المال وثمن المثل، ولو فسخ المالك المضاربة فللعامل أجرته إلى ذلك الوقت.

Section Seven: Deposits in Trust

This is a non-mandatory, valid agreement between two parties [where one deposits something as a trust in the custody of the other]. The custodian must look after it as per convention. If the owner stipulates a specific manner or place where it must be kept, then he must adhere to it. If he does not, he is responsible for any damage or loss, unless there was threat of danger.

If a person entrusts another with an animal then it is his responsibility to provide the food and drink. The remainder of these provisions must be returned [to the owner when the animal is given back]. The entrusted party is only liable for damage or loss if he was negligent, not otherwise. The dissolution of the contract only occurs once he returns the entrusted item or is absolved of responsibility by the owner. If he is entrusted with something out of fear that it might otherwise be usurped by an oppressor, then he is permitted to willfully deny its possession if confronted by the oppressor. If necessary, he is permitted to swear an oath to the oppressor in this regard. If, however, he confesses to having the wealth of another in his possession, he is liable [to pay it back].

Reason dictates that the entrusted item be mandatorily returned to the depositor, or to his heirs after his death, unless he had usurped it, in which case it must be returned to its rightful owner. If one does not know who the rightful owner is, it may be treated as lost property to be given away in charity after some time, if he wishes. The only time he must return the usurped item to an oppressor is if it is mixed with other legitimate belongings of his, in which case he must return the whole lot to him.

In a dispute, the custodian has the last word accompanied by an oath of honesty if the owner accuses him of damage, negligence, not having returned the trust, or its value, without evidence. The owner has the last word in case of damages if he claims that it was a loan and not a trust.

الفصل السابع: في الوديعة

وهي عقد جائز من الطرفين، ويجب حفظها بمجرى العادة، ولو عين المالك حرزاً تعين، فلو خلف ضمن إلا مع الخوف.

ويجب على الودعي علف الدابة وسقيها، ويرجع به [على المالك]، ويضمن المستودع مع التفريط لا بدونه، ولا يزول إلا بالرد إلى المالك أو الإبراء. ويحلف للظالم ويوري، ولو أقر له لم يضمن.

ويجب ردها عقلاً على المودع أو إلى ورثته بعد موته، إلا أن يكون غاصباً فيردها على مالكها، ومع الجهل لقطة يتصدق بها إن شاء، إلا أن يمتزج بمال الظالم فيردها عليه.

والقول قول الودعي في التلف وعدم التفريط والرد والقيمة مع يمينه، وقول المالك أنه دين لا وديعة مع التلف.

Section Eight: Lending and Borrowing

Any owned item that is fit for use and remains intact after use can be borrowed, on condition that the lender has right of disposal.

The borrower must use it in an ordinary customary manner and is not held liable unless he is stipulated as such in the contract, or he has misused it, or it is a high-value item [such as gold or silver]. If the item decreases through regular conventional use, the borrower is not held liable, unless he borrowed a usurped item from someone who had taken it wrongfully. If he was unaware and later came to know, he must return the item to the lender. The borrower must also limit himself to use what he has borrowed according to what the lender permits [and for no other purpose].

In a dispute, the borrower has the last word with an oath of honesty if he is accused of negligence or a reduction in the value without evidence; and the lender has the last word if the borrower is not returning the item. Pawned or mortgaged lending is valid, but the lender retains the right to reclaim and stop the mortgage after the stipulated time.

الفصل الثامن: في العارية

كل عين مملوك يصح الانتفاع بها مع بقائها صح إعارتها، بشرط كون المعير جائز التصرف.

وينتفع المستعير على العادة، ولا يضمن مع التلف بدون التضمين أو التعدي، أو كون العين أثماناً، ولو نقصت بالاستعمال المأذون، فيه لم يضمن، ولو استعار من الغاصب ضمن، فإن كان جاهلاً رجع على المعير بما يؤخذ منه. ويقتصر المستعير على المأذون.

والقول قول المستعير مع يمينه في عدم التفريط والقيمة معه، وقول المالك في الرد. ويصح الإعارة للرهن، وله المطالبة بالافتكاك بعد المدة.

Section Nine: Lost Property

Someone who finds lost children must be of age of legal responsibility and a Muslim. He must have the permission of his master if he himself is a slave. If the child is found within Muslim lands, he is considered free, and if not, then he is considered a slave (*riqq*).

The lost child's first next of kin is the Imam, in the absence of an actual next of kin, and he is responsible for paying the blood-money owed by him if he causes injury to another person. When the child reaches maturity and confesses to being a slave, then his word is accepted and the [legitimate] sovereign must take care of his maintenance. If the [legitimate Imam] is unavailable [e.g. he is in occultation], then some believers can support him together; and if they cannot, then the finder can provide for him himself, with a mind to ask the child to pay him back when he grows up, but he must have this intention to begin with, and not impose it later. If he has a father, grandfather or previous guardian, they must be compelled to take him.

If he is a slave, then he must be returned to his owner. If he runs away, or dies before he can be returned, the finder is not held liable. Taking in a lost child is a communal obligation, and whatever possessions are found on him at the time of his finding belong to him, the child.

It is discouraged to take in lost animals unless they face an imminent threat. Therefore, a camel found near pasture and water must not be taken, unless it has been purposely abandoned in a barren place. Taking it in results in ownership of it. A sheep found on pasture-less land may be taken in without liability [e.g. if it is eaten]. He is responsible for its upkeep in the absence of a [legitimate] sovereign and must return it if its owner comes to claim it. If he benefits from it (i.e. its milk), then that is considered his compensation. If a whole year passes without the owner claiming the lost animal, then he may opt to either carry on looking after it and is then not liable for any loss or damage, or he may opt to keep it as his own, in which case he is fully responsible for it.

الفصل التاسع: في اللقطة

يشترط في ملتقط الصبي: التكليف، والإسلام، وإذن المولى في المملوك فإن كان في دار الإسلام فهو حر، وإلا فرق.

ووارث الأول الإمام مع عدم الوارث وهو عاقلته.

ولو بلغ رشيداً فأقر بالرقية قبل، وينفق عليه السلطان، فإن تعذر فبعض المؤمنين، فإن تعذر أنفق الملتقط ويرجع مع نيته لا بدونها، ولو كان له أب أو جد أو ملتقط قبله أجبر على أخذه.

ولو كان مملوكاً رده على مولاه، فإن أبق أو تلف من غير تفريط فلا ضمان.

وأخذ اللقيط واجب على الكفاية، وهو مالك لما يده عليه.

ويكره أخذ الضوال إلا مع التلف، فلا يؤخذ البعير في كلاء وماء، ويؤخذ في غيره إذا ترك من جهد، ويملكه الأخذ، وتؤخذ الشاة في الفلاة مضمونة وينفق مع تعذر السلطان ويرجع بها، ولو انتفع تقاص، وإذا حال الحول على الضالة ونوى الاحتفاظ فلا ضمان، ولو نوى التملك ضمن.

Picking up lost property is disapproved. If one does pick something up that is worth less than a dirham, he may keep it. If it is worth a dirham or more, then he must announce that it has been found for a whole year. If he had found it in the sanctuary of Mecca, then he must give it in charity after the year has passed and he would not be liable if he chooses to keep it in his possession in the hope of finding its owner. If the item had been found elsewhere, he is allowed to keep it as his own or to give it away in charity. He is liable [to the owner if he demands his property to be returned to him]. If, however, he chooses to continue looking after it [until the owner is found] then he is not held liable for any loss or damage. If the lost item is perishable, he may use it after valuing it, and is liable for paying its value to the owner if he comes to reclaim it, or he may remit it to the governor, and is then no longer liable for it.

Picking up lost property that is worth little but may be of benefit is also disapproved. That which is found in wastelands or derelict places may be kept by the finder. If something is found in an owned place, the owner must be notified. If he acknowledges it as his own, then it is his, but if not, then the finder may keep it. The same applies to something that is found in the belly of an animal that one has recently bought.

The guardian is responsible for making known any property that is found by a child or mentally deranged person. It suffices for a slave to announce on behalf of his master. Whoever finds something can either announce it himself or appoint a representative. It is not necessary for the notification process to be done continuously. It does not suffice to simply describe the lost property or person, but rather evidence must be furnished [by the one wishing to claim it as their property]. The finder of lost property is to be trusted [in his claims].

ويكره أخذ اللقطة، فإن أخذها وكانت دون الدرهم ملكها ، وإن كانت درهماً فما زاد عرفها حولاً ، فإن كانت في الحرم تصدّق بها بعده ولا ضمان إذا استبقاها أمانة، وإن كانت في غيره فإن نوى التملك جاز ويضمن، وكذا إن تصدّق بها ، ولو نوى الحفظ فلا ضمان، ولو كانت مما لا يبقى انتفع بها بعد التقويم وضمن القيمة، أو يدفعها إلى الحاكم فلا ضمان.

ويكره أخذ ما يقل قيمته ويكثر نفعه وما يوجد في فلاة أو خربة فلو أجده ولو كان في مملوكة عرف المالك، فإن عرفه فهو له وإلا فللواجد ، وكذا ما يوجد في جوف الدابة.

ويتولى الولي التعريف لو التقط الطفل أو المجنون، ويكفي تعريف العبد في تملك المولى، وله أن يعرف بنفسه وإن يستنيب. ولا يشترط فيه التوالي، ولا يكفي الوصف بل لا بدَّ من البينة، والملتقط أمين.

Section Ten: Usurped Property

This is prohibited by dictate of reason, and is someone's unlawful taking and controlling another's property, even if it is unoccupied; and he remains liable for any loss or damage until he releases it.

If the usurper forcefully lives in the house with the owner, then he is liable for half of it. If he usurped a pregnant animal, he must also return the offspring. If he prevents the owner from receiving the animal that has been dispatched to him or from sitting on his rug, he is not held liable [unless damage can be directly attributed to him]. If the property is usurped from a previous usurper, the owner may choose which one of the two to demand compensation from.

The usurper is not liable for a free man that he has wrongfully seized unless he is a minor, nor for paying the wage of a labourer if he prevents him from working. If, however, he employs him, then he must pay him a wage. The usurper is liable for any loss or damage resulting from removal of a restraint from a usurped slave, mentally incapacitated person or horse. If he leaves a door open and a thief steals the usurped goods, the thief is held liable for them. One can be held liable for wine and pork usurped from a non-Muslim *dhimmī*, or for their value if they are hidden from public sight, but not if they belong to a Muslim.

الفصل العاشر: في الغصب

وهو حرام عقلاً. ويتحقق بالاستيلاء على مال الغير ظلماً وإن كان عقاراً، ويضمن بالاستقلال.

ولو سكن الدار قهراً مع المالك ضمن النصف، ولو غصب حاملاً ضمن الحمل، ولو منع المالك من إمساك الدابة المرسلة أو من القعود على بساطه لم يضمن، ولو غصب من الغاصب تخير المالك في الاستيفاء ممن شاء.

ولا يضمن الحر إلا أن يكون صغيراً، ولا أجرة الصانع لو منعه عنها، ولو استعمله فعليه أجرة عمله، ولو أزال القيد عن العبد المجنون أو الفرس ضمن، ولو فتح باباً فسرق غيره المتاع ضمن السارق، ويضمن الخمر والخنزير للذمي، وبقيمتهما عندهم - مع الاستتار -، لا للمسلم.

The usurped property must be returned, and if it is damaged then he is liable for the repairs. If he is unable to do so, he must replace it with a similar one, and if he cannot do that, then he must pay its current value. If the value has changed from the time that he usurped it until the time it became damaged, he is liable for paying the higher value. If its market value has increased, he is not liable for the difference if he returns the same item. If the market value of the item has increased due to a change in its properties, then he is liable for the difference.[3] If he improved the item in a way that has no bearing on its value, then he in not liable for the difference. If the market value increases due to his having removed a part of it,[4] then he is liable for paying the difference in market value. If the original form of the item has changed irreversibly, the usurper must return it as well as any compensation resulting from a loss in value. He is not due a refund from any increase in value of the item due the change in its original form.

If someone usurps a slave and inflicts harm on him equivalent to the market value of the slave, he must return him as well as the compensation due, according to one opinion.

If the misappropriated goods have been mixed with similar goods, or those of better quality, and the resulting mixture cannot be separated, then he must return an amount similar to what was usurped from that mixture.[5] If they have been mixed with lower quality goods or cannot be differentiated, they must be replaced with the same.

Any profit that is made from the misappropriated goods belongs to the owner.[6] If someone mistakenly buys something, unaware that they were stolen goods, he has the right to return them to the usurper for a refund, as well as any penalty or loss of profit that he incurs as a result, even in the past, though this may be problematic. If he bought them knowingly, then he has no right to return them.

3 For example, if at the beginning of the month someone usurps some unripe fruit which ripens during the course of the month, and then returns the fruit back to its rightful owner but only after it has passed its best quality, then he is liable to pay the loss of profit the owner suffers as a result of not being able to sell his fruit at its highest market value.

4 For instance, in the case of a male slave whose genitals are cut off by the usurper, thus increasing his value due to sexual impotence and him no longer being a threat to the women of the household.

5 For example, if someone usurps one litre of apple juice and mixes it with another litre of apple juice of better quality, then the usurper must return one litre from the resulting mixture.

6 For example, the offspring of livestock born whilst in the possession of the usurper belong to the original owner.

ويجب رد المغصوب، فإن تعيب ضمن الأرش، فإن تعذر ضمن مثله، فإن تعذر فقيمته يوم المطالبة، ولو لم يكن مثلياً ضمنه بأعلى القيم من حين الغصب إلى حين التلف على إشكال، ولو زاد للسوق لم يضمنه مع الرد، ولو زاد للصفة ضمنها، ولو تجددت صفة لا قيمة لها لم يضمنها، ولو زادت القيمة لنقص بعضه كالحب فعليه الأرش، ولو زادت العين بأثرها رجع الغاصب بها وعليه أرش النقصان وليس له الرجوع بأرش نقصان عينه.

ولو غصب عبداً وجنى [عليه] بكمال قيمته رده مع الأرش على قول، ولو امتزج المغصوب بمساويه أو بأجود رده، ولو كان بأدون ضمن المثل.

وفوائد المغصوب للمالك. ولو اشتراه جاهلاً بالغصب رجع بالثمن على الغاصب وبما غرم عوضاً عما لا نفع في مقابلته، أو كان، على إشكال. ولو كان عالماً فلا رجوع بشيء.

If someone cultivates usurped land, the crops belong to him, but he is liable for the ground rent of the land. In case of a dispute with the owner of the usurped property about the amount payable, the usurper's word accompanied by an oath of honesty holds weight if the owner has no evidence for his claim.

ولو زرع [في] المغصوب كان الزرع له وعليه الأجرة، والقول قول الغاصب في القيمة مع اليمين وتعذر البينة.

Section Eleven: Developing Uncultivated Land

One is not allowed to handle the property of anyone else, such as a pathway, waterway or pasture, without his permission, even if it be to improve it.

Newly-build pathways on public lands must have a right of way, in disputes, of seven *dhirāʿ*.[7] The distance between watering wells must be kept to forty *dhirāʿ*, and between irrigation wells, sixty *dhirāʿ*. The distance between two springs should be one thousand *dhirāʿ* on soft land, and five hundred on hard ground.

Water from sources flowing down from higher altitudes may be retained by cultivators up to ankle level for palm plantations, and up to dorsal level for other crops. Then cultivators lower down may do the same. One is allowed to protect the pasture land which is within his own property, and the Imam has this right absolutely.

The owner of water rights to a river is not allowed to change its course without the permission of the owner of any watermills along the river. It is discouraged to sell water rights from underground canals or rivers.

It is allowed to have skylights or portholes to the sides of buildings on thoroughfares, as long as they do not affect passers-by. Windows overlooking others require permission as do doors opening out onto them.

This applies equally to both: already-existing windows overlooking an empty plot as well as newly made windows overlooking existing buildings, and the first door to open out onto the street is allowed to remain. Later openings such as a door opposite an existing door may be ordered to be shut. Existing doorways have a greater right to remain open than later ones.

If someone erects a roof porthole facing a thoroughfare, the people living opposite cannot stop him from doing so, even if it spans the width of the path. If it falls, and the people opposite erect one, the original person cannot stop them from doing so.

One is recommended to erect a parapet between the next-door neighbour's wall and one's own if need be. If it is to be used by a muezzin, then it is advisable to consult them before erecting it, otherwise a dispute thereafter may be settled by compensation.

7 One *dhirāʾ* is equivalent to the distance of one arm span (i.e. from the tip of the middle finger to the end of the elbow).

الفصل الحادي عشر: في إحياء الموات

لا يجوز التصرف في ملك الغير بغير إذنه ولو فيما فيه صلاحه، كالطريق والنهر والمراح.

وحد الطريق المبتكر في المباحة مع المشاحة سبعة أذرع، وحريم بئر المعطن أربعون ذراعاً، والناضح ستون، والعين في الرخوة ألف، وفي الصلبة خمسمائة.

ويحبس النهر للأعلى إلى الكعب في النخل، وللزرع إلى الشراك، ثم كذلك لمن هو دونه، وللمالك أن يحمي المرعى في ملكه، وللإمام مطلقاً.

وليس لصاحب النهر تحويله إلا بإذن صاحب الرحى المنصوبة عليه. ويكره بيع الماء في القنوات والأنهار.

ويجوز إخراج الرواشن والأجنحة في الطرق النافذة ما لم تضر المارة، ومع الإذن في المرفوعة، وكذا فتح الأبواب.

ويشترك المتقدم والمتأخر في المرفوعة إلى الباب الأول وصدر الدرب، ويختص المتأخر بما بين البابين، ولكل منهما تقديم بابه لا تأخيرها.

ولو أخرج الرواشن في النافذ فليس لمقابله منعه وإن استوعب عرض الدرب، ولو سقط فبادر مقابله لم يكن للأول منعه.

ويستحب للجار وضع خشب جاره على حائطه مع الحاجة، ولو أذن جاز الرجوع قبل الوضع، وأما بعده فبالأرش.

If neighbours dispute about sole ownership of a wall between them, the one who is prepared to testify in court has the last word if the other refuses to do so. If they both agree to testify or both refuse, then it belongs to them jointly. If it is attached to a building belonging to either one of them or he has an element belonging to him within the wall [i.e. a recess or door] then it is his, if accompanied by an oath of honesty.

The joint-owner of a wall, mill, well or river does not have the right of disposal over it without the permission of his co-owner, and in case of damage, he cannot coerce his co-owner to rebuild it.

For neighbours who live in the same building, in case of a dispute, the occupier of the ground floor has the last word regarding the walls of the house, and the one on the top floor regarding the roof, the walls of upstairs rooms and the staircase. The basement is both their responsibility, as is the loft or access to it between them both. The rest is up to the occupier of the ground floor.

The neighbour has right over the branches of a tree that bend into his property, and if they are a nuisance, he may cut them back from his property.

In a dispute, the rider of an animal has greater right over it than the one who holds its reins. In a dispute [about being overlooked], the occupier of the ground floor of a property, in keeping the door of his room open, has the last word over his neighbour, after swearing an oath of honesty and in the absence of evidence.

ولو تداعيا جداراً مطلقاً فهو للحالف مع نكول الآخر، ولو حلفا أو نكلا فلهما، ولو اتّصل بناء أحدهما أو كان له عليه طرح فهو له مع اليمين.

ولا يتصرف الشريك في الحائط والدولاب والبئر والنهر بغير إذن شريكه، ولا يجبر الشريك على العمارة.

والقول قول صاحب السفل في جدران البيت، وقول صاحب العلو في السقف وجدران الغرفة والدرجة، وأما الخزانة تحتها فلهما، وطريق العلو في الصحن بينهما، والباقي للأسفل.

وللجار عطف أغصان الشجرة، فإن تعذّر قطعها عن ملكه.

وراكب الدابة أولى من قابض لجامها. وصاحب الأسفل أولى بالغرفة المفتوح بابها إلى غيره مع التنازع واليمين وعدم البينة.

The Book of Debts

كتاب الديون

This book contains various sections.

Section One

Taking a loan whilst one can avoid doing so is disapproved, and if he does take a loan he must make an intention to repay it. The reward for giving someone a loan is equivalent to the reward of giving charity.

It is prohibited to stipulate an increase in amount or in characteristics as a condition of the repayment, though one may accept repayment with increase without this being stipulated as condition of the loan. If a particular place has been designated for the repayment, then that must be adhered to.

Anything that has a set value and characteristics is suitable for lending. The borrower is responsible for returning the same item, one identical or its value at the time of repayment. It is not obligatory to return exactly the same item unless the lender stipulates so, nor must he delay the term of repayment. It is acceptable to ask for repayment of a debt earlier than the contracted term in exchange for waiving some of the amount.

If the lender disappears and there is no news of him, it is obligatory for the borrower to maintain his intention to repay him, and to stipulate it in his will before death. If after some time has passed, he remains unaware of his whereabouts and the likelihood of someone like him being alive still is impossible, then he must remit the loan to his heirs. If he cannot find them, then he must donate it as charity on his behalf, however the better opinion is that it belongs to the Imam.

It is not valid for two partners to divide a debt [from their shared capital].[1] It is acceptable to sell what one lent to another person, for cash or in exchange for something of a lesser value if it is not of the same type [i.e. a crop], and as long as it is not usury. One is not allowed to exchange one debt for another.

A Muslim is allowed to accept the repayment of his loan from a *dhimmī*, with money made from the sale of prohibited items. If, however, the *dhimmī* embraced Islam after he sold the items, he is entitled to demand payment [if he has not yet been paid from the sale of those items].

1 If, for example, two partners agree to give half of their shared capital to Zayd and another half to Khalid, then it is not possible for them to divide the debt such that they agree that one party will be owed by Zayd and the other party will be owed by Khalid. Rather if they agree to loan their wealth, they must both be considered creditors of the debt.

وفيه فصول

الفصل الأول: في الدين

يكره الدين مع القدرة، ولو استدان وجب نية القضاء، وثواب القرض ضعف ثواب الصدقة.

ويحرم له اشتراط زيادة القدر أو الصفة، ويجوز قبولها من غير شرط، ولو شرط موضع التسليم لزم.

وكل ما ينضبط وصفه وقدره صح قرضه، وذو المثل يثبت في الذمة مثله وغيره قيمته وقت التسليم. ولا يجب إعادة العين بدون اختيار المقترض، ولا يتأجل الحال، ويصح تعجيل المؤجل بإسقاط بعضه.

ولو غاب المدين وانقطع خبره وجب على المستدين نية القضاء والوصية به عند الوفاة، فإن جهل خبره ومضت مدة لا يعيش مثله إليها غالباً سلم إلى ورثته، ومع فقدهم يتصدّق به عنه، والأولى أنه للإمام.

ولو اقتسم الشريكان الدين لم يصح، ويصح بيع الدين بالحاضر وإن كان أقل منه إذا كان من غير جنسه أولم يكن ربوياً، ولا يصح بدين مثله.

وللمسلم قبض دينه من الذمي من ثمن ما باعه من المحرمات ولو أسلم الذمي بعد البيع استحق المطالبة.

A slave may not take a loan without the permission of his master, and if he does so, he is responsible for its repayment if he is manumitted. Otherwise, he is absolved of it [if he has not been released and dies]. If the master does permit him, then he is responsible for its repayment and not the slave, even after his manumission. The creditor of a slave is in reality the creditor of the master. If he gave him permission to engage in trade, and the slave as a result takes out a business loan of his accord, the master is responsible for its repayment. If, however, it was a loan for another purpose, he remains responsible for it after his manumission.

وليس للعبد الاستدانة بدون إذن المولى، فإن فعل تبع به إن عتق وإلا سقط، ولو أذن له لزمه دون المملوك وإن عتق. وغريم المملوك كغرماء المولى. ولو أذن له في التجارة فاستدان لها لزم المولى، وإن كان لغيرها تبع به بعد العتق.

Section Two: Pawnship

It is incumbent here that there be an offer and an acceptance from its owner, and there is a difference of opinion regarding the stipulation of terms for reclaiming it. The pre-requisite is that the pawn be an owned and available item, fit for seizure and sale, in exchange for a fixed right that is owed, whether the pawn be kept as it is or used for profit.

Pawning something that one does not own is dependent upon the permission of the owner. If the debtor joins something of his with something belonging to someone else as a pawn, he is only bound with respect to what belongs to him [and of the whole if he has the permission of the owner of the part that does not belong to him]. The agreement is binding upon the debtor [i.e. he cannot demand the creditor to return the pawn unless he first repays the original debt].

Pawning a pregnant female slave does not include her foetus as a pawn, even if she was impregnated anew after having been pawned. Any income generated from the pawned item belongs to the mortgager. A pawn held in security for one of two loans cannot be used for the other. It is valid, however, to take out a second loan and designate the pawn for the first loan as a security for both loans. The legal guardian [of a minor, for example] can pawn his property if there is benefit to his ward, in doing so.

Both the mortgager and the mortgagee are prohibited from selling or using the property without permission of its owner, and if the creditor has been given right of disposal over it as an agent, then this remains in effect as long as the mortgager is alive, and if appointed as the executor of a will, then he is liable for the property. The pawnship is inheritable.

The mortgagee is protected and cannot be held liable without infringement, and he is thus liable to produce the like of the item if it is exactly replaceable or its value on the day of reclamation. He has the last word with regard to its value – not the value of the loan – when accompanied by an oath of honesty, and in the absence of any evidence to the contrary and any negligence on his part. He has greater right over it than other creditors, and if there is any surplus after the loan has been repaid, then he has a share in the surplus. If the loan amount is of a higher value than the pawn whereby he has a debt unsecured by a pawn, then all creditors have an equal right to it.

الفصل الثاني: في الرهن

ولا بدَّ فيه من الإيجاب والقبول من أهله، وفي اشتراط الإقباض إشكال. ويشترط فيه أن يكون عيناً مملوكاً يمكن قبضه ويصح بيعه، على حق ثابت في الذمة عيناً كان أو منفعة.

ويقف رهن غير المملوك على الإجازة، ولو ضمنها لزم في ملكه، ويلزم من جهة الراهن.

ورهن الحامل ليس رهناً للحمل وإن تجدّد، وفوائد الرهن للراهن. ورهن أحد الدينين ليس رهناً على الآخر. ولو استدان آخر وجعل الرهن على الأول رهناً عليهما صح. وللولي الرهن مع مصلحة المولى عليه.

وكل من الراهن والمرتهن ممنوع من التصرف بغير إذن صاحبه، ولو شرط وكالة المرتهن لم ينعزل ما دام حياً. ولو وصى إليه لزم، والرهانة موروثة.

والمرتهن أمين لا يضمن بدون التعدي، فيضمن به مثله إن كان مثلياً وإلا قيمته يوم القبض. والقول قوله مع يمينه - في قيمته - وعدم بينة التفريط، لا قدر الدين. وهو أحق به من باقي الغرماء، ولو فضل من الدين شيء شارك في الفاضل، ولو فضل من الرهن وله دين بغير رهن تساوي الغرماء فيه.

If the mortgagee disposes of the property without permission, he remains liable for it and must pay any fees. If the mortgager had authorized the sale before the term of the contract, and he sold it, he has no right of disposal over the value except after the term.

If the mortgagee fears that the next of kin will disclaim the debt and has no evidence for his case, he is allowed to redeem it from the value of the pawn without his permission. In a dispute over the item, the owner who claims it was a deposit left in trust has the last word over another who claims it was a pawn.

ولو تصرّف المسترهن بدون إذن ضمن وعليه الأجرة، ولو أذن الراهن في البيع قبل الأجل فباع لم يتصرّف في الثمن إلا بعده، ولو خاف جحود الوارث ولا بينة جاز أن يستوفي من الرهن من تحت يده، والقول قول المالك مع ادعاء الوداعة وادعاء الآخر الرهن.

Section Three: Limitation of Legal Competence

There are six possible causes for this:

1) Juvenility. The minor does not have the right of disposal over his own property except upon reaching the age of responsibility and maturity. The first is established either through the growth of hair [facial and pubic] or nocturnal emission or the attainment of fifteen years of age for boys, and nine for girls; and the second through testing him in the correct handling of his possessions whereby he can be safe from being swindled, and he is competent in his general comportment.

If he lacks one of these two pre-requisites, the limitation on him remains even if he has come of age. In men, maturity is identified by his contemporaries who would attest to it, and in women either by their contemporaries' attestation (in which case four women must testify) or by men's (in which case two men must testify).

2) Mental incapacity. The one who is mentally incapacitated cannot handle his own property except at times when he is mentally sound.

3) Incompetence. Such a person must have legal limitations, however only with respect to his property.

4) Slavery. The slave's actions have no legal effect without his master's permission, and even if he gives him ownership over something, he does not actually own it according to the most correct opinion.

5) Terminal illness. He only has the right of disposal over one third of his estate, especially those of his assets that he wishes to donate. The same applies if he was to die of his illness.

6) Bankruptcy. The bankrupt person becomes legally limited upon fulfilling four conditions: a) that his debts have been confirmed by the state, b) that they have become due for repayment, c) that his assets are valued much lower than his total debt, and d) that his creditors have demanded that he be legally curtailed.

If the state declares him bankrupt, he loses the right of disposal over his property as long as the limitation on him remains, so if he borrows money after that or buys something on credit, the lender and the vendor do not have the same rights over his assets as his original creditors.[2] If he squanders someone else's wealth, that person also has a share in them, as would a previous lender if he admits to a previous debt.

2 If they were unaware of his bankruptcy.

الفصل الثالث: في الحجر

وأسبابه ستة:

الأول: الصغر. فالصغير ممنوع من التصرّف إلا مع البلوغ والرشد، ويعلم الأول بالإنبات أو الاحتلام أو بلوغ خمس عشرة سنة في الذكور وتسع في الأنثى، والثاني بإصلاح ما له عند اختباره بحيث يسلم من المغابنات وتقع أفعاله على الوجه الملائم.

ولا يزول الحجر مع فقد أحدهما وإن طعن في السن، ويثبت في الرجال بشهادة أمثالهم، وفي النساء بشهادتهن أو بشهادة الرجال.

الثاني: الجنون. ولا يصح تصرّف المجنون إلا في أوقات إفاقته.

الثالث: السفه. ويحجر عليه في ماله خاصة.

الرابع: الملك. فلا ينفذ تصرّف المملوك بدون إذن مولاه، ولو ملكه شيئاً لم يملكه على الأصح.

الخامس: المريض. تمضي وصيته في الثلث خاصة، ومنجزاته المتبرّع بها كذلك إذا مات في مرضه.

السادس. الفلس. ويحجر عليه بشروط أربعة: ثبوت ديونه عند الحاكم، وحلولها، وقصور أمواله عنها، ومطالبة أربابها الحجر.

وإذا حجر عليه الحاكم بطل تصرّفه في ماله ما دام الحجر، فلو اقترض بعده أو اشترى في الذمة لم يشارك المقرض والبائع الغرماء، ولو أتلف مال غيره شارك صاحبه، وكذا لو أقر بدين سابق.

If the bankrupt person admits to a current debt for an item in his possession, it is said that it must be paid to the claimant, and he has the option to either sell the item or to annul the transaction. If he finds the item itself, he has the right to take it back, even if it has become mixed with items of equal or lesser value. If there is nothing else but that item, he must divide its original value, minus any increase, between all the creditors.

A creditor cannot single out items that are owed to him from a deceased person's estate if it falls short of covering his debts. Grains, eggs, crops and hatcheries are excluded from specification in any case.

The executor of the estate must take out the portion owed, and the vendor must divide the value between all the creditors.

Other Rulings:

1) If he is bankrupt and still owes the cost of a slave-mother [whose child he has fathered], she must be sold [but not her child] or returned to the vendor.

2) Demanding or forcing a bankrupt person to pay back his debts for the sake of making profit is not permissible, nor to make him sell his house or his servant.

3) A postponed debt is not covered by the declaration of bankruptcy, and it remains payable even if the debtor passes away. If the creditor passes away, then it is no longer due.

4) The bankrupt person can support himself and his family from his own wealth until the day when his assets are divided between his creditors. If he dies in the meantime, his shroud [for burial] is to be bought from it first.

5) Assets are to be divided equally to cover existing debts. If an existing debt becomes known later, the original division is nullified, and it must be included in the total. Once his assets have been divided, he is released, and the bankruptcy is lifted upon repayment of the debts.

6) Legal guardianship over the property of the minor and the mentally incapacitated is held by the father and the grandfather, and in their absence, their successor. If there is no such person, then the state is the guardian. Guardianship over the property of an incompetent or a bankrupt person lies solely with the state.

ولو أقر بعين - قيل - يدفع إلى المقر له وله إجازة بيع الخيار وفسخه ومن وجد عين ماله كان له أخذها - ولو خلطها بالمساوي والأدون، وإن لم يكن سواها ، دون نمائها والضرب مع الغرماء .

ولا اختصاص في [مال] الميت مع قصور التركة، ويخرج الحب والبيض بالزرع والاستفراخ عن الاختصاص.

وللشفيع أخذ الشقص، ويضرب البائع مع الغرماء .

مسائل:

الأولى: لو أفلس بثمن أم الولد بيعت أو أخذها البائع.

الثانية: لا تحل مطالبة المعسر أو إلزامه بالتكسب ولا بيع دار سكناه ولا عبد خدمته.

الثالثة: لا يحل بالحجر الدين المؤجل، ولو مات من عليه حل، ولا يحل بموت صاحبه.

الرابعة: ينفق عليه من ماله إلى يوم القسمة وعلى عياله، ولو مات قدم الكفن.

الخامسة: يقسم المال على الديون الحالة بالتقسيط ، ولو ظهر دين حال نقضت وشاركهم، ومع القسمة يطلق ويزول الحجر بالأداء .

السادسة: والولاية في مال الطفل والمجنون للأب والجد له، فإن فقد فالوصي ، فإن فقد فالحاكم، وفي مال السفيه والمفلس للحاكم خاصة.

Section Four: Guarantee

The guarantee is valid when it is initiated by someone qualified to do so, and must necessarily be agreed to by both the guarantor and the lender, upon which the borrower is released from liability for it, even if he refuses, and responsibility for what is owed is transferred to the guarantor. The guarantee is valid upon acceptance by the lender, regardless of the financial means of the guarantor, and even if the lender does so knowing of the guarantor's straitened circumstances at the time of the contract. If he did not know, then he has the right to rescind.

A future guarantee for a current debt is valid, and vice versa. The guarantor may only demand the borrower to pay him back what he settled if he stood as guarantor upon his request, but not otherwise.

It is not a pre-requisite to know the exact amount that is guaranteed, but he is liable for whatever can be proven to have been borrowed through records.

If a slave stands as guarantor without the permission of his master, he remains liable for it after his manumission.

The claimed right must be accompanied by proof, whether it is already incumbent or due upon fulfilment of certain conditions.

If someone guarantees the contracted payment or delivery of goods, he remains liable for it even if the contract is nullified, but not if the transaction is purposely rescinded.

Transfer of Debt

The transferral of a debt must be agreed to by all three parties, though the acceptance of it is not mandatory. If accepted, however, it becomes binding and the original debtor is absolved of his debt, and responsibility for paying it is transferred to the third person. The contract remains in place whether he is wealthy or whether it is known at the time of the contract that his financial circumstances are straitened, but if this was not known, then he can rescind it.

If the third person demands repayment for the debt that he had paid off, and the original debtor maintains that the former is still responsible for the debt, the third person has the last word with an oath of honesty.

If a buyer transfers the debt owed by him for the cost of goods, and then returns the goods, the transferred debt will also be nullified, though there is debate around this. The buyer must go back to the vendor for a refund.

الفصل الرابع: في الضمان

وإنما يصح إذا صدر عن أهله ولا بدَّ من رضا الضامن ولمضمون له، ويبرأ المضمون عنه وإن أنكره، وينتقل المال على الضامن، فإن كان ملياً أو علم المضمون له بإعساره وقت الضمان صح وإلا كان له الفسخ.

ويصح مؤجّلاً وإن كان الدين حالاً، وبالعكس. ويرجع الضامن على المضمون عنه بما أداه إن ضمن بسؤاله [وإلا فلا].

ولا يشترط العلم بقدر المضمون، ويلزمه ما تقوم به البينة خاصة.

ولو ضمن المملوك بغير إذن مولاه تبع به بعد العتق.

ولا بدَّ في الحق من الثبوت، سواء كان لازماً أو آيلاً إليه.

ولو ضمن عهدة الثمن لزمه مع بطلان العقد لا تجدّد فسخه.

وأما الحوالة:

فيشترط فيها رضا الثلاثة، ولا يجب قبولها، ومعه يلزم ويبرأ المحيل، وينتقل المال إلى ذمة المحال عليه إن كان ملياً أو علم بإعساره، وإلا فله الفسخ.

ولو طالب المحال عليه بما أداه فادّعى المحيل ثبوته في ذمته فالقول قول المحال عليه مع يمينه.

ولو أحال المشتري بالثمن ثم فسخ بطلت الحوالة على إشكال، ويرجع المشتري على البائع مع قبضه.

If the vendor transfers responsibility of goods that he owes, over to a third party, but the goods are returned, the transferral remains in place. If the entire transaction is cancelled, then it is cancelled for both the original vendor and the third-party seller.

Surety

The condition of surety is the agreement of both the bondsman and the claimant specifically. There is a difference of opinion regarding stipulation of a term, and identifying the defendant for whom surety is being guaranteed. The bondsman is responsible for presenting the defendant or paying the bail due for him.

If someone forcefully sets free a defendant who is in the custody of his claimant, he must return him or pay the bail due for him. If he is a murderer, then he must return him or pay the blood-money compensation [due to the victim's family].

The bondsman is absolved of his responsibility upon the death of the defendant, upon his delivering him, upon the defendant's surrendering himself, or if the claimant releases him from liability.

If the bondsman and the claimant stipulate a place where the defendant is to be brought, it must be adhered to. If not, then the default location is the town where the surety was contracted.

ولو أحال البائع أجنبياً ثم فسخ لم تبطل الحوالة، ولو بطل البيع بطلت فيهما .

وأما الكفالة:

فيشترط فيها رضا الكفيل والمكفول له خاصة، وفي اشتراط الأجل قولان، وتعيين المكفول، وعلى الكافل دفع المكفول أو ما عليه.

ومن أطلق غريماً عن يد صاحبه قهراً ألزم بإعادته أو ما عليه، ولو كان قاتلاً دفعه أو الدية.

ولو مات المكفول أو دفعه الكفيل أو سلم نفسه أو أبرأه المكفول له برئ الكفيل.

ولو عينا موضعاً للتسليم لزم، وإلا انصرف إلى بلد الكفالة.

Section Five: The Compromise

This is allowed in both situations where there is admission [of a previous right owed] as well as those where this is denied, provided that it does not lead to permitting what is prohibited and vice versa, whether the two parties are aware of the particulars or not, and whether it concerns debts or commodities. Such an agreement can only be dissolved with the consent of both parties, or when a breach in either of the two constituents merits it.

It is permissible for two partners to compromise in the dissolution of their partnership allowing one to take both profit and loss, and the other his capital.

If two people are in possession of a dirham each, and one of them claims that they both belong to him whereas the other claims that they possess one each, one dirham is given to the first claimant and the other is shared equally between them both. Similarly, in the case where someone has entrusted two dirhams with a person, and another person has entrusted a third dirham with the same trustee, who then loses one of the two dirhams through no negligence on his part, [the owner of the two dirhams receives a dirham and a half, and the latter the remaining half dirham].

If two identical items of clothing [of differing values] held in trust get mixed up, they must be sold and the amount divided between the two owners in proportion to their values.

Seeking a compromise does not automatically entail confirmation [of having a right to something], as opposed to [contracts] where one may say: "Sell it to me", "Make me its legal owner", "Gift it to me", "Defer it for me" or "You have fulfilled it".

الفصل الخامس: في الصلح

وهو جائز مع الإقرار والإنكار، إلا ما حلل حراماً أو بالعكس، مع علم المصطلحين بالمقدار أو جهلهما، ديناً (أو) عيناً ولا يبطل إلا برضاهما أو استحقاق أحد العوضين.

ولو اصطلح الشريكان على أنّ لأحدهما الربح والخسران وللآخر رأس المال صح.

ولو أدعى أحدهما درهمين في يدهما والآخر أحدهما أعطى الآخر نصف درهم. وكذا لو أودع أحدهما درهمين والآخر ثالثاً وتلف أحدهما بغير تفريط [فلصاحب الإثنين درهم ونصف وللآخر ما بقي].

ولو اشتبه الثوبان بيعا وقسم الثمن على نسبة رأس مالهما.

وليس طلب الصلح إقراراً، بخلاف [ما إذا قال] بعني أو ملكني أو هبني أو أجلني أو قضيت.

Section Six: Admission (Of Previous Liability)

This entails informing of a previous right owed, and does not have to be specified verbally but can be indicated by other means known to convey admission. If one were to say, for example: "Yes" or "Aye" in response to the question, "Do you owe such and such?" then that counts as an admission. Similarly, a positive answer to a negative question such as, "Do you not owe this amount?" Merely saying, "I admit" without stating what it is he is admitting to does not suffice. His admission is not valid if it is conditional. If he were to say, "If so-and-so testifies to it then he is truthful" [then it would count as an admission] and he is bound by it even if that person does not testify to it.

The one who admits liability must be of the age of responsibility and free. A slave's admission is followed up after he has been manumitted.

The creditor must be the legal owner, but if the admission is in favour of a slave, then it belongs to his master. If a person admitting liability details the amount he has in his possession, then his explanation is accepted even if the amount detailed be very small; but if he refuses to specify what he owes, he must be detained. If he says, "a thousand and one dirhams", his specification of a thousand is accepted.[3] If he says, "a thousand, and three dirhams" or "one hundred, and twenty dirhams", it is assumed that the whole compound number is included in the dirhams. If he says, "a few dirhams" then it is assumed to be twenty. If he says, "a sum of dirhams" then it is one hundred dirhams. If he says, "such and such number of dirhams" then it is assumed to be eleven dirhams; and "some and a few dirhams" is twenty-one dirhams. This is only in those cases where the one making admission is aware of these technicalities [i.e. he knows, for example, that saying *kadhā dirham* means 20 dirhams], otherwise he must specify the nature of it. If he says, "a hundred as a future debt", or "the cost of wine", or "an item I never received", or "something I bought with the right to return it", then it is the creditor's word that supersedes accompanied with an oath of honesty.

3 It will be assumed that by "a thousand and one dirham" he means one thousand and one dirhams.

الفصل السادس: في الإقرار

وهو إخبار عن حق سابق. ولا يختص لفظاً، ويصح بالإشارة المعلومة. ولو قال: «نعم» أو «أجل» [في] جواب «عليك كذا» فهو إقرار، وكذا «بلى» عقيب «أليس عليك» بخلاف نعم. ولو قال «أنا مُقر» فليس بإقرار إلا أن يقول به، ولو علقه بشرط بطل، ولو قال «إن شهد فلان فهو صادق» لزمه وإن لم يشهد.

ويشترط في المُقِرّ: التكليف، والحرية، ويتبع العبد بإقراره بعد العتق.

وفي المُقَرّ له: أهلية التملك، ولو أقر للعبد فهو لمولاه، وأن فسر المقر به بما يملك قبل وإن قل، ولو لم يفسر حُبس عليه. ولو قال ألف درهم قُبِل تفسيره في الألف، ولو قال ألف وثلاثة دراهم أو مائة وعشرون درهماً فالجميع دراهم، ولو قال كذا درهماً فعشرون، ولو قال كذا درهم فمائة، و[لو قال] كذا كذا درهماً أحد عشر، وكذا [و] كذا درهماً أحد وعشرون. هذا مع معرفته وإلا فله التفسير، ولو قال مائة مؤجلة، أو من ثمن خمر، أو مبيع لم أقبضه، أو ابتعت بخيار، فالقول قول الغريم مع اليمين.

An exceptive statement is considered valid if the exception is of the same nature as the whole,[4] but not if it is different.[5] If he says, "Ten minus three minus three", he must pay four. If he says "I owe so-and-so a dirham and a dirham, except a dirham", then the exception is invalid [so he owes two dirhams not one]. If he says, "Ten minus five bar three", he must pay eight. If he says, "Ten take away one", it is not acceptable. If he says, "This belongs to *x* rather to *y*", then it is considered as belonging to the first person, and the second is owed its value.

Debts owed as currencies, weights and measures must be referred back to the custom of the locality, and if the commodity in question has many types, then it is the admitter's responsibility to clarify which is meant. If he admits to something measured by vessels, the vessels are not included in the debt. If he admits to a dry measure (*qafīz)* of wheat, then changes it to the same measure of barley, he must pay both. If he admits to one measure of wheat, and then changes it to two measures, he must pay two measures.

If he admits to a debt of one thousand owed to someone at the start of the next month, for example, he remains liable for it at that time, as opposed to one that is conditional upon the arrival of Zayd, for example. If the exact amount due is ambiguous because he admits to an amount using the plural "some dirhams" [but he doesn't specify how many], the lesser amount is considered [i.e. three dirhams]. If he cannot identify the creditor between two contenders, they must each swear an oath of honesty in the absence of any evidence. If he does not know who the creditor is, then is able to identify him, but the creditor denies it, the judge can either seize the item owed from him or keep it on file as a debt still owed if he swears an oath of honesty.

If the creditor denies ownership of a slave that is supposedly owed to him, in Shaykh Ṭūsī's opinion, he must be manumitted, though there is an alternative opinion. If he maintains this claimed ownership, contrary to witnesses, then he must swear an oath of honesty.

4 For example, if someone were to admit "I owe 100 dirhams minus 10 dirhams", then the amount owed is assumed to be 90 dirhams.

5 For example, if someone were to admit "I owe 100 dirhams minus 10 dates", then the price of 10 dates would first have to be determined and then subtracted from 100. So if, for example, 10 dates costs 5 dirhams, then the amount he admits to owing is 95 dirhams.

ويحكم بما بعد الاستثناء المتصل والمنفصل، ويسقط بقدر قيمة المنفصل.

ولو قال عشرة إلا ثلاثة إلا ثلاثة لزمه أربعة، والوجه بطلان الاستثناء في درهم ودرهم إلا درهماً. ولو قال عشرة إلا خمسة إلا ثلاثة لزمه ثمانية، ولو قال عشرة ينقص واحداً لم يقبل.

ولو قال هذا لفلان بل لفلان كان للأول وغرم للثاني القيمة.

ويرجع في النقد والوزن والكيل إلى عادة البلد، ومع التعذر إلى تفسيره.

ولو أقر بالمظروف لم يدخل الظرف. ولو قال قفيز حنطة بل قفيز شعير لزمه القفيزان، ولو قال قفيز حنطة بل قفيزان لزمه اثنان.

ولو قال إذا جاء رأس الشهر فله عليَّ ألف أو بالعكس لزمه، بخلاف إن قدم زيد. ولو أبهم الجمع حمل على أقله. ولو أبهم المُقَرّ له كانا خصمين ولهما اليمين على عدم العلم، ولو أبهم المُقَرّ له ثم عيّن فأنكر المُقَرّ له انتزعه الحاكم أو أقرّه في يده بعد يمينه.

ولو أنكر المقر له بالعبد (قال الشيخ): يعتق، وفيه نظر. ولو ادّعى المواطاة على الأشهاد كان له الإحلاف.

Other Rulings

1) A person's admission to being a child's parent has three conditions: (a) the possibility of fathering that child [given their respective ages], (b) that the child's father is unknown, and (c) that there are no other individuals claiming to be that child's father. It is not necessary to seek corroboration from a minor child, and his denial of it after the age of maturity can be disregarded. Corroboration must be sought from an older child, or if the admission pertains to relations other than a child.

If the admission of a relative, other than a child, entails that the two are each other's sole heirs, they can inherit each other, but the inheritance does not extend to others besides them. If there are existing heirs present, then the newly admitted relative is not included among them.

2) If an heir admits to a relative more entitled than himself, his share is transferred to him. If they both have equal entitlement, then the newly admitted relative receives an amount equivalent to his share of the original amount.

If a person admits to two relatives, who deny each other's right, their denial is disregarded. If he admits to a relative who is more entitled than himself, then to a third who is more entitled than the creditor, if the creditor corroborates his admission, his share is transferred to the third person, and if not, then to the second person, and he remains indebted to the third.

If a child admits to the existence of a sibling, and they both admit to the existence of a third, but the third person denies the second one, the third person receives half of the inheritance, and the second receives a sixth. If their kinship can be confirmed, then his denial is disregarded.

3) Kinship can only be corroborated by the testimony of two just male witnesses, not a man and two women nor a man swearing an oath of honesty.

If two brothers bear witness that the deceased has a son, and they are both just, he has more entitlement to the inheritance than them, and the kinship is confirmed. If they are corrupt, his share of the inheritance is confirmed but not the kinship.

مسائل:

الأولى: يشترط في الإقرار بالولد إمكان البنوة والجهالة وعدم المنازع. ولا يشترط تصديق الصغير، ولا يلتفت إلى إنكاره بعد البلوغ، ويشترط في الكبير وفي غير الولد.

ومع تصديق غير الولد ولا وارث يتوارثان، ولا يتعدّ التوارث إلى غيرهما، ولو كان له ورثة مشهودون لم يقبل في النسب.

الثانية: لو أقر الوارث بأولى منه دفع ما في يده إليه، ولو كان مساوياً دفع بنسبة نصيبه من الأصل.

ولو أقر بإثنين فتناكرا لم يلتفت إلى تناكرهما، ولو أقر بأَولى منه ثم بأَولى من المقر (له) فإن صدقه دفع إلى الثالث وإلا إلى الثاني وغرم للثالث.

ولو أقر الولد بآخر ثم أقرا بثالث وأنكر الثالث الثاني كان للثالث النصف وللثاني السدس، ولو كانا معلومي النسب لم يلتفت إلى إنكاره.

الثالثة: يثبت النسب بشهادة عدلين لا برجل وامرأتين ولا برجل ويمين. ولو شهد الإخوان بابن [للميت] وكانا عدلين كان أَولى منهما وثبت النسب، ولو كانا فاسقين ثبت الميراث دون النسب.

Section Seven: Agency

It is incumbent for it to consist of a proposal and an agreement – either immediately or after some time, and must not be attached to any conditions.[6] It is not binding for either party [i.e. both parties have the right to cancel the arrangement at any time they wish].

When an agent is dismissed by the one who deputized him, his agency is terminated upon his knowledge of the dismissal. It is also terminated by death, mental incapacity, loss of consciousness, expiry of the thing over which he was to act as agent, and any conduct of the principal which affects the transaction. Agency is valid for anything which the law (*shar'*) does not stipulate one's direct and personal implementation.[7]

The agent must not exceed beyond his permitted remit, unless it pertains to preferring a particular market, but if the remit of his agency is general then it is permissible to do so provided it is to the advantage of the principal, except in admissions of liability.

When there are no qualifications made by the deputiser the agent sells at the current time for a representative value, in the local currency, purchases good quality goods, and delivering the goods if selling, or paying the price if buying, on time, and that he returns them if faulty.

Government agency does not necessarily entail collection of funds.

Both types of agency require one to have the capacity to act and the freedom to do so. It is acceptable to appoint a slave as an agent, or for him to deputize an agent with the permission of his master.

An agent may not deputize another agent to act on his behalf without permission of his principal. It is the state's responsibility to deputize agents on behalf of legally incompetent and simple-minded people. It is recommended for people in outstanding positions to deputize others to act on their behalf.

A *dhimmī* may not be deputized to stand against a Muslim. The agent cannot be held liable unless there has been an infringement, but his agency is not terminated upon doing so. His word is to be taken when accompanied by an oath of honesty in a dispute, in the absence of any evidence proving non-fault, or regarding his dismissal, his knowledge of his dismissal, any loss incurred, or regarding his conduct. There are two opinions regarding the role of the agent in returning goods.[8]

6 For example, one cannot say, "I appoint you as a deputy over my wealth if it rains tomorrow."

7 Such as the daily prayer, which each individual must perform themselves.

8 In reference to when the agent and his principal disagree about whether the latter's goods have been returned back to him by the agent.

الفصل السابع: في الوكالة

ولا بدَّ فيها من الإيجاب والقبول - وإن كان فعلاً أو متأخراً - والتنجيز. وهي جائزة من الطرفين. ولو عزله المُوَكِّل بطل تصرّفه مع علمه بالعزل. وتبطل بالموت والجنون والإغماء وتلف متعلّقها وفعل المُوَكّل، وتصح فيما لم يتعلق غرض الشرع بإيقاعه مباشرة.

ولا يتعدى الوكيل المأذون إلا في تخصيص السوق، ولو عمّم التصرف صح مع المصلحة إلا في الإقرار.

والإطلاق يقتضي البيع حالاً بثمن المثل بنقد البلد، وابتياع الصحيح وتسليم المبيع في البيع وتسليم الثمن في الشراء، والرد بالعيب.

ولا يقتضي وكالة الحكومة القبض.

ويشترط أهلية التصرف فيهما، والحرية. ولو توكّل العبد أو وكّل بإذن مولاه صح.

ولا يوكّل الوكيل بغير إذن، وللحاكم التوكيل عن السفهاء والبله ويستحب لذوي المروّات.

ولا يتوكّل الذمي على المسلم. ولا يضمن الوكيل إلا بتعدٍّ ولا تبطل وكالته به. والقول قوله مع اليمين وعدم البينة في عدمه، وفي العزل والعلم به والتلف والتصرّف، وفي الرد قولان.

In a dispute regarding the agency itself, its denier has the final word. If the agent claims in a dispute that he had been given permission to sell an item at a specified price, then the principal's word holds more weight. If the same item can be found, then it is to be recovered. If it is lost or can no longer be recovered, then he must pay the representative value or its actual value if it is unique.

If an agent marries someone to a woman but the principal denies having deputized him as an agent, then the principal must swear an oath of honesty and the agent is responsible for paying the bridal gift (*mahr)*, or half, according to some. If the principal was found to be lying, he must divorce her.

If someone deputizes two agents to act on his behalf, neither of them can act individually unless he has permitted them to do so, and this can only be proven by two witnesses.

If the agent delays delivery of goods, after having been requested and despite being able to do so, he is held liable.

والقول قول مُنكِرِ الوكالة، وقول الموكِّل لو ادّعى الوكيل الإذن في البيع بثمن معين، فإن وجدت العين استعيدت، وإن فقدت أو تعذرت فالمثل أو القيمة إن لم يكن مثلياً.

ولو زوّجه فأنكر الموكّل الوكالة حلف، وعلى الوكيل المهر وقيل نصفه، ويجب على الموكّل طلاقها مع كذبه.

ولو وكّل إثنين لم يكن لأحدهما الانفراد بالتصرف إلا أن يأذن لهما، ولا تثبت إلا بشاهدين.

ولو أخر الوكيل التسليم مع القدرة والمطالبة ضمن.

The Book of Charitable Gifts and That Which Follows It

كتاب الهبات وتوابعه

This book contains several sections.

Section One: Gifts

It is only valid to gift capital assets, even if they are jointly owned, through an offer, acceptance and receipt by a legally responsible and free subject. If one were to gift somebody that which he owes him, then it is a remission of his debt.

The consent of the person gifting is required for receipt of the gift, unless he is gifting what is already in his keep.

The father and grandfather have guardianship over gifts received by the minor and the mentally incapacitated. The one who gifts something to somebody else is not allowed to ask for it to be returned after the item has been transferred to the person to whom it was gifted if that person is a blood relative, or after the item has been consumed and is no longer available, or if it was exchanged for something else, though there is a difference of opinion about the type of usage [that may warrant returning it]. It has been said that [gifts by] spouses are treated as those of blood relatives [and thus cannot be requested back if gifted]. In other than those instances, one is permitted to retract the gift. If there is a fault with it, there is no compensation due. If the gift undergoes growth whereby the growth is inseparable from the original, it too must be returned. If not, then the beneficiary can keep it.

Other Rulings

1) One is not allowed to take back a charitable donation after receipt, even if it was to someone unrelated. If, however, the latter took it without permission of the owner, then it is not remitted to him.

2) The charitable donation must be given with the intention of proximity to God.

3) Charity can be given to the *dhimmī*, even if he is not related.

4) Charity given out of the public's sight is better, unless there is risk of false accusation.

وفيه فصول

الفصل الأول: في الهبة

الهبة إنما تصح في الأعيان المملوكة وإن كانت مشاعة، بإيجاب وقبول وقبض من المكلف الحر، ولو وهبه ما في ذمته كان إبراءاً.

ويشترط في القبض إذن الواهب إلا أن يهبه ما في يده.

وللأب والجد ولاية القبول والقبض عن الصغير والمجنون وليس له الرجوع بعد الإقباض إن كانت لذي الرحم بعد التلف أو التعويض، وفي التصرف خلاف، وقيل الزوجان كالرحم. وله الرجوع في غير ذلك. فإن عاب فلا أرش، وإن زادت زيادة متصلة تبعت وإلا فللموهوب (له).

مسائل:

الأولى: لا يجوز الرجوع في الصدقة بعد الإقباض وإن كانت على الأجنبي، ولو قبضها من غير إذن المالك لم تنتقل إليه.

الثانية: لا بدَّ في الصدقة من نية القربة.

الثالثة: يجوز الصدقة على الذمي وإن كان أجنبياً.

الرابعة: صدقة السر أفضل إلا مع التهمة.

Section Two: Endowments

The most explicit way of declaring an endowment is to say "I have endowed such and such". Others expression may be used but require contextual indicators to show that the donor is making an endowment. The conditions of an endowment are: the acceptance, the intention of proximity to God, and the receipt [by the beneficiary]. The guardian receives the endowment on behalf of a child, and the trustee overseeing the affairs of a certain group receives it on their behalf. Other conditions for the validity of endowments include immediate implementation [i.e. the endowment is not conditional upon anything], its perpetuity [i.e. one cannot give something as endowment for a certain period of time, e.g. one year], and that the endowment be given to other than oneself.

If the donor stipulates that it be returned to him at some point, it is considered a retention. Similarly, if he stipulates a time frame or endows it to people who would eventually pass away, the endowment returns to the donor's heirs thereafter.

The endowment must also be a capital asset that is fit for use as long as it stands, even if it is jointly owned [i.e. a person is permitted to give away his shares in a joint property as an endowment].

The donor must also be legally permitted to transact or utilise the property he is giving as an endowment [i.e. otherwise the endowment is invalid]. The beneficiary must already be living, defined and have legal capacity to own property [hence a slave cannot be an endowment because he does not have the right of ownership in Islamic law]. It is also a requirement that the endowed property be used for legitimate purposes [e.g. one cannot endow a building if that building will be used for the production of alcoholic beverages].

The donor is allowed to appoint himself as a trustee over his own endowment, and if he does not appoint a trustee, then it is left to the beneficiaries to oversee it. Non-existent people [i.e. unborn grandchildren] can benefit from the endowment after those who are living.

The endowment can be spent on charity towards the poor and other such projects with the intention of proximity to God.

الفصل الثاني: في الوقف

وصريح ألفاظه «وقفت»، والباقي بقرينة.

وشروطه: القبول، والتقرب، والإقباض. ويتولى الولي القبض عن الطفل، والناظر في المصالح القبض عنها، والتنجيز والدوام، وإخراجه عن نفسه.

ولو شرط عوده كان حبساً، ولو جعله إلى أمد أو لمن ينقرض غالباً رجع إلى ورثة الواقف.

وأن يكون عيناً مملوكة ينتفع بها مع بقائها وإن كانت مشاعة.

وجواز تصرف الواقف، ووجود الموقوف عليه وتعيينه، وأهليته للتملك، وإباحة منفعة الوقف على الموقوف عليه.

وله جعل النظر لنفسه، فإن أطلق كان لأربابه، ويصح الوقف على المعدوم تبعاً للموجود.

ويصرف الوقف على البرّ إلى الفقراء ووجوه القرب.

If a Muslim were to make an endowment for synagogues and churches, it would be invalid, as opposed to if a non-believer did so. Similarly, it is invalid to endow to one who is at war [with Islam][1] even if he is a relative, but not to a *dhimmī*, even if he is a non-relative.

A Muslim's endowment donated for the benefit of the poor must be spent on poor Muslims, and a non-Muslim's on the poor of his creed. That which is donated for the benefit of the Muslims must be spent on those who pray towards the *qibla*; that which is donated for the benefit of the believers or the Imāmīs, must be spent on the Twelver Shīʿīs, and similarly for the descendants or adherents of a specified person or faction. If one makes an endowment to the father of a household, then it will be passed down through the lineage of his sons, and there is a difference of opinion regarding the daughters. If it is stipulated to be for both, then it is divided equally among them, as long as he does not stipulate a preferred ratio.

[A general endowment made for the benefit of:] "the people" entails those that speak the same language, "the clan" means those who are the most closely related to him, "the neighbours" entails those living within a forty *dhirāʿ* radius from his house, "in the way of God" means anything by means of which God's proximity is sought, "the protégés" (*al-mawālī*) includes both the ex-slaves who have manumitted themselves as well as those whom he has freed.

A general endowment for "the poor" does not include every single pauper, but rather the poor people of his town and those who have come to him for help. If the endower himself becomes poor later, he is allowed to benefit along with the rest of them.

Other Rulings

1) If the original benefit intended by the endowment ceases to exist, it is spent on a good cause.

2) It is valid to include those who will be born in the future along with existing beneficiaries. It is not acceptable to endow something only to change the terms of it later. If one were to stipulate the remittance of an endowment to a whole group but exclude somebody, the endowment is void.

1 Such a person is referred to in Imāmī *fiqh* as *al-kāfir al-ḥarbī*.

ولو وقف المسلم على البِيع والكنائس بطل بخلاف الكافر ، ويبطل على الحربي وإن كان رحماً ، لا الذمي وإن كان أجنبياً .

وينصرف وقف المسلم على الفقراء إلى المسلمين ، والكافر إلى فقراء ملته ، وعلى المسلم إلى المصلي إلى القبلة والمؤمنين أو الإمامية إلى الاثنا عشرية ، وكذا كل منسوب إلى من انتسب إليه ، ولو نسب إلى أب كان لمن انتسب إليه بالأبناء ، وفي البنات قولان ، ولو شرك استوى الذكور والإناث ما لم يفضل .

والقوم أهل اللغة ، والعشيرة الأقرب في النسب ، والجيران لمن يلي داره إلى أربعين ذراعاً ، وسبيل الله كل ما يتقرب به إليه ، والموالي الأعلون والأدنون .

ولا يتبع كل فقير في الوقف على الفقراء ، بل يعطي أهل البلد منهم ومن حضره ، ومن صار منهم جاز له أن يأخذ معهم .

مسائل:

الأولى : إذا بطلت المصلحة الموقوف عليها صرف إلى البر .

الثانية : لو شرط إدخال من يوجد مع الموجود صح ، ولو أطلق وأقبض لم يصح ، ولو شرط نقله بالكلية أو إخراج من يريد بطل الوقف .

3) The maintenance costs for an [endowed] slave are borne by the beneficiary, but if he is disabled, he is freed and responsible for his own upkeep. If the endowed [slave] commits a crime, the endowment would only be nullified upon his death by capital punishment. If he is a victim of a crime, the compensation due belongs to the beneficiary [who owns him].

4) If one makes an endowment to his grandchildren, both the children of his sons and his daughters, male and female receive equal shares. If, however, he says "my descendants", then it entails his son's children specifically, based on one opinion.

5) Any legitimate clause that the donor stipulates is enforceable.

6) A property given for someone to reside in, both temporarily (*suknā*) as well as life-long (*'umrā*), needs a contractual proposal, acceptance and receipt. The property is non-transferable, and if a specific term is stipulated, then it must be adhered to, even if the landlord dies. If the donor said, "You can live here until the end of your life", the contract ceases only upon the resident's death. If he said, "until the end of my life", then it ceases upon his own death, and in this case, if the resident dies before he does, the property passes on to his heirs during his lifetime. If he did not specify a term, however, the landlord may evict him whenever he chooses.

If the property is sold, the temporary tenancy agreement is not invalidated. The resident may live there himself, along with those who would customarily live with him such as his children, his wife and his servant. He is not allowed to let anyone else reside there without permission, nor to sublet it.

Anything that is suitable to endow can be refurbished or improved, such as property, slaves or furniture. One who donates the use of his horse or his slave in the service of places of worship, or in the way of God, must do so permanently until the object remains alive.

الثالثة: نفقة المملوك على الموقوف عليه، ولو أقعد انعتق وكانت نفقته على نفسه. ولو جنى الموقوف لم يبطل الوقف إلا بقتله قصاصاً، ولو جني عليه كانت القيمة للموقوف عليه.

الرابعة: لو وقف على أولاد أولاده، اشترك أولاد البنين والبنات الذكور والإناث، ولو قال «من انتسب إلي» فهو لأولاد البنين خاصة، على قول.

الخامسة: كل ما يشترطه الواقف من الأشياء السائغة لازم

السادسة: يفتقر «السكنى» و»العمرى» إلى إيجاب وقبول وقبض، وليست ناقلة، فإن عين مدة لزمت ولو مات المالك، وكذا لو قال له «عمرك» فإن مات الساكن بطلت، ولو قال «مدة حياتي» بطلت بموته، ولو مات الساكن قبله انتقل الحق إلى ورثته مدة حياته، ولو لم يعين كان للمالك إخراجه متى شاء.

ولو باع المسكن لم تبطل السكنى، وللساكن أن يسكن بنفسه ومن جرت عادته به كالولد والزوجة والخادم، وليس له إسكان غيره من دون إذن، ولا إجارته. وكل ما يصح وقفه يصح إعماره كالملك والعبد والأثاث، ولو حبس فرسه أو غلامه في خدمة بيوت العبادة وسبيل الله لزم ما دامت العين باقية.

Section Three: Wills

Making a will is obligatory and must consist of an offer and acceptance. In the absence of verbal communication, a written directive or non-verbal communication of one's intent suffices. It is not obligatory to act upon every directive found in the deceased's handwriting, rather only those that are legitimate,[2] so if a Muslim were to will for a church to be built after his death, it would not be valid, and the executors of the will would have to retract it.

The person making the will (testator) must be of sensible conduct. The beneficiary must be alive at the time. The executor of the will must be a Muslim and a legally responsible subject, and the estate being distributed must be possessed [by the testator].

If the testator fatally harms himself, then makes a will, it is not valid, but if the will preceded his death, then it is valid. A will made for the benefit of a child in-utero is valid, as long as it is born alive. Likewise, a will made for a *dhimmī*, but not for one who is at war with Islam. One can make a will for the benefit of his slave, the slave-mother of his child, his slave who would be automatically freed upon his death (*mudabbar*), and his slave having the contractual option to buy his own freedom (*mukātab*) – leaving him enough to manumit himself. It is not valid to make a will for other people's slaves. If the amount that the testator leaves his slave is enough to manumit him, then he is freed and receives nothing. If it is more than the amount necessary, then he is entitled to the surplus. And if it is less than the amount needed, then he is employed in work that would make up the deficit. The same applies to the slave-mother of his child, and she does not automatically inherit from the stipulated shares [of inheritance].

If the testator leaves money in his will to manumit his slave and to pay off a debt, payment of the debt takes precedence. If manumission of the slave has already been implemented, it remains in order if his price is twice that of the debt. He must then be employed in the service of the debtor as payment for half the value of the debt, and the heirs of the third are responsible for the rest.

If he makes a bequest for both male and female relatives, they receive equal shares unless he expresses a preferential division. The same applies to paternal and maternal uncles.

2 On condition that one knows with full confidence that it is indeed his handwriting, and has not been changed.

الفصل الثالث: في الوصايا

وهي واجبة، ولا بدَّ فيها من إيجاب وقبول، ويكفي الإشارة والكتابة مع الإرادة والتعذر لفظاً، ولا يجب العمل بما يوجد بخطه.

وإنما تصح في السائغ. فلو أوصى المسلم ببناء كنيسة لم تصح، وله الرجوع فيها.

ويشترط صحة تصرف الموصي، ووجود الموصى له، والتكليف والإسلام في الوصي والملك في الموصى به.

ولو جرح نفسه بالمهلك ثم أوصى لم تصح، ولو تقدمت الوصية صحت.

وتصح الوصية للحمل بشرط وقوعه حياً، وللذمي دون الحربي، ولمملوكه وأم ولده ومدبّره ومكاتبه، لا مملوك الغير، وللمكاتب فيما تحرّر منه، فإن كان ما أوصى به لمملوكه بقدر قيمته عتق ولا شيء له، وإن زاد أُعطي الفاضل، وإن نقص استسعى فيه. وأم الولد كذلك، لا من النصيب.

ولو أوصى بالعتق وعليه دين قدم الدين، ولو نجز العتق صح إذا كانت قيمته ضعف الدين، وسعى للديان في نصف قيمته وللورثة في الثلث.

ولو أوصى لذكور وإناث تساووا إلا مع التفضيل، وكذا الأعمام والأخوال.

If he makes a general bequest to his relatives, then they are the ones who are immediately related to him. His clan, neighbours, and money donated in the way of God, for good causes and to the poor are all treated like endowments (*waqf*).

If the beneficiary stipulated in a will dies before the testator, and the latter did not change his will, his share goes to his heirs, and if he does not have any heirs, then to the heirs of the testator. A will made for the benefit of an unborn child in utero is valid.

It is recommended that the executor of a will be a relative, even if he is an heir. If someone appoints a morally upright person, who then behaves immorally, the appointment is void.

It is valid to appoint a woman as the executor of a will, as well as a child, and a slave with the permission of his master. In the case of a child, he must form part of a collective, which would then execute the will [for immediate matters]. Once the child attains maturity, they may execute it jointly. A legitimate decision made previously [before the child attained maturity] cannot be retracted after he has attained maturity.

It is valid for a disbeliever to appoint another disbeliever to execute a will.

If someone appoints two people as executors and either stipulates that they jointly execute it or leaves it unstipulated, neither of them has the right to execute it independently of the other, and the judge can enforce that they execute it jointly. If they refuse to cooperate without just cause, the judge can substitute them with two other people instead. If one of them is incapable, he may jointly execute it with someone else. If the testator stipulated that they execute it independently, then they may do so, each acting at his own discretion. They are allowed to divide the estate amongst themselves to execute.

If the testator comes to know that the executor he has appointed has refused, his refusal is in order. If he does not come to know of it, then the executor must go ahead and execute the will. If he behaves perfidiously, the judge may replace him. The executor is not liable for any loss, unless there has been negligence on his part. He has the right to get his debt owed to him by the testator repaid, or to raise a loan subject to creditworthiness, or to conduct his own business, taking a customary fee if need be. He may appoint another executor on his behalf with prior authorization, and not without, and he must not exceed that which has been authorized. The authority in charge of Muslim affairs oversees the appointment of an executor for one who has none.

ولو أوصى لقرابته فهم المعروفون بنسبه، والعشيرة والجيران والسبيل والبرّ والفقراء كالوقف.

ولو مات الموصى له قبله ولم يرجع كانت لورثته، فإن لم يكن وارث فلورثة الموصي.

وتصح الوصية بالحمل.

ويستحب للقريب وإن كان وارثاً. وإذا أوصى إلى عدل ففسق بطلت.

ويصح أن يوصي إلى المرأة والصبي بشرط انضمامه إلى الكامل، وإلى المملوك بإذن مولاه، فيمضي الكامل الوصية إلى أن يبلغ ثم يشتركان، ولا ينقض بعد بلوغه ما تقدم مما هو سائغ.

ولو أوصى الكافر إلى مثله صح.

ولو أوصى إلى اثنين وشرط الاجتماع أو أطلق فليس لأحدهما الانفراد، ويجبرهما الحاكم على الاجتماع. ولو تشاحا فإن تعذر استبدل، ولو عجز أحدهما ضم إليه، ولو شرط الانفراد جاز وتصرف كل واحد منهما، ويجوز الاقتسام.

وإذا بلغ الموصى رد الموصى إليه صح الرد، وإلا فلا، ولو خان استبدل به الحاكم. ولا يضمن الوصي إلا مع التفريط، وله أن يستوفي دينه أو يقترض مع الملاءة، أو يقوم على نفسه، ويأخذ أجرة المثل مع الحاجة، وأن يوصي مع الإذن لا بدونه ولا يتعدّى المأذون، ويتولى الحاكم من لا وصي له.

The bequest applies to up to a third of the total estate. If [dispensation of] it exceeds a third, the surplus is subject to authorization from the heirs. If some of the heirs allow it, it is paid out only from their share. If they authorized it before [the testator's] death, it remains valid [and they cannot retract it].

The beneficiary comes to own his share of inheritance after the death [of the testator] and his acceptance of it. The obligatory dues take precedence in being paid out of the one-third, then the remainder, in order of priority for non-obligatory dues. If they are of equal importance, they are treated equally.

If a man makes a bequest for "a portion" (*juz'*), it is to be taken as one-seventh; if he says "a share" (*sahm*), it signifies one-eighth, and if he says "a part" (*shay'*), it is to be taken as one-sixth.

It is valid to make a bequest to leave a beneficiary as much as a share of one of his heirs from the one-third. If it does not exceed the one-third, or the heirs authorize him, then he is treated as one of them. If he makes a bequest to leave as much as the share of his son to someone, but has nothing other than that, the named beneficiary receives half of it, with permission from the heirs, but a third without. If the man has two sons, then the beneficiary receives a third. If his children are both male and female, he is given the equivalent of the lesser share, unless the testator had specified the greater amount.

If the executor forgets one of the dispensations [of the one-third], it reverts to the inheritance. If there are two conflicting dispensations, he implements the latest, and if they do not conflict, he implements them both. If the one-third is insufficient to implement them all, he executes them in order of priority.

A person's claim to wealth in a bequest must be corroborated by two just [male] witnesses, or one male witness and two female witnesses, or one male witness and an oath of honesty, or four female witnesses. The testimony of a single female witness affords him one-quarter of the claimed amount, and that of two females a half. A person's claim to guardianship can only be corroborated by two male witnesses.

وتمضي الوصية بالثلث فما دون، ولو زادت وقف الزائد على إجازة الورثة، ولو أجاز بعض مضى في قدر حصته، ولو أجازوا قبل الموت صح.

ويملك الموصى به بعد الموت والقبول. ويقدم الواجب من الأصل والباقي من الثلث، ويبدأ بالأول فالأول في غير الواجب، ولو جمع تساووا. ولو أوصى بجزء ماله فالسُبع، والسهم الثُمن، والشيء السُدس.

ولو أوصى بمثل نصيب أحد الورثة صحت من الثُلث، فإن لم يزد أو أجازوا كان الموصى له كأحدهم، فلو أوصى بمثل نصيب ابنه وليس له سواه أعطى النصف مع الإجازة والثُلث بدونها، ولو كان [له] ابنان فالثلث، ولو اختلفوا أعطى الأقل إلا أن يعين الأكثر.

ولو نسي الوصي وجهاً رجع ميراثاً. ويعمل بالأخير من المتضادّين، فإن لم يتضادّا عمل بهما. ولو قصر الثلث بدئ بالأول فالأول.

وتثبت الوصية بالمال بشاهدين، وشاهد وامرأتين، وشاهد ويمين، وأربع نساء، وتقبل الواحدة في الربع، والاثنتان في النصف. ولا تثبت الولاية إلا برجلين.

If a man makes a bequest for his slave to be manumitted, but has not left anything towards that, one-third of the slave's price is manumitted. If he makes a bequest towards part of his slave's manumission, but has double the amount set aside for it, it can be used for full manumission. If he makes a bequest for all his slaves to be manumitted but has not left a specific amount towards that, a third of them may be freed [from the one-third] by drawing lots. If he had ranked them, then they are manumitted in that order. As for the slave whose freedom he owes as expiation [for a crime], naming the slave suffices, and if he says that it is a believing slave, then it is obligatory to free him. If a believing slave cannot be found, then he must free one whose religious affiliation is unknown. If it comes to be known afterwards that his religious affiliation conflicted [with Islam], the manumission is still valid.

The dispensations made by a terminally ill person are considered as being from the one-third, even if they have already been discharged.

The same applies to admissions of debts if they are claimed from him while he is ill. If not, then they are calculated from his estate. This ruling applies to any terminal illness, even if it is not imminent.

Compensations due as penalties for crimes and blood-money are taken from the estate. If the testator had defined the cost for the manumission of a believing slave, but the money was not found with him, the executor must wait for it to be found. If found, the slave must be manumitted with the minimum amount, and be given the surplus.

A bequest made in favour of anyone over whom the testator holds guardianship is valid, and even in the absence of such, it is valid for the fulfilment of rights due upon him.

A bequest made to exclude one of his children from his share of the inheritance is invalid.

ولو أعتق عبده ولا شيء له عتق ثُلثه، ولو أعتق بعضه وله ضعفه عتق كله، ولو أعتق مماليكه ولا شيء سواهم عتق ثُلثهم بالقرعة، ولو رتّبهم بدئ بالأول فالأول. ويجزي في الرقبة مسمّاها، ولو قال مؤمنة وجب، فإن لم يوجد عتق من لا يعرف بنصب، ولو بانت بالخلاف بعد العتق صح.

وتصرّفات المريض من الثُلث وإن كانت منجزة.

أما الإقرار فإن كان متهماً فكذلك، وإلا فمن الأصل.

وهذا حكم يتعلق بمطلق المرض الذي يحصل به الموت وإن لم يكن مخوفاً.

ويحتسب من التركة أرش الجناية والدية.

ولو عيّن ثمن الرقبة ولم توجد به توقع الوجود، فإن وجد بأقل أعتق وأعطي الفاضل.

وتصح الوصية على كل من للموصى عليه ولاية، ولو انتفت صحت في إخراج الحقوق عنه.

ولو أوصى بإخراج بعض ولده من الميراث لم يصح.

The Book of Marriage

كتاب النكاح

This book contains several sections.

Section 1: Introduction

Marriage is of three types: permanent, fixed-term, and slave concubinage (*milk al-yamīn*).

The first type requires a contract, which is a proposal from the eligible person and its acceptance, in the past tense, from the other eligible party. If one [an agent] were to say, "I marry [have married] your daughter 'so-and-so' to 'so-and-so'", and the other says, "Yes", it suffices as a proposal. In case a verbal pronouncement [in Arabic] is not possible, translation and signalling can suffice. It is valid for a woman to offer herself in marriage. [Permission from] a legal guardian is not a pre-condition if one has attained the age of legal responsibility and maturity, and nor are witnesses.

Any claim of marriage without stipulated evidence (*bayyina*) and assent can be ignored. If a man claims that a certain woman is his wife but her sister claims to be his wife, the ruling is in favour of his evidence to the contrary, unless her claim pre-dates his, or with evidence of his having consummated with her.

The father's choice of one of his many daughters – without naming which one – to marry to a suitor is valid, provided the latter has seen them all. If not, then the contract is invalid.

It is recommended to choose for marriage a virgin woman, chaste and of noble lineage. [It is also recommended] to offer two units of prayer, to have the marriage witnessed and announced, to recite the sermon before the contract, to consummate the marriage at night, to offer a two-unit prayer before consummation of the marriage, to supplicate before consummating the marriage, to encourage her to do the same and to request God for male offspring.

وفيه فصول

الفصل الأول: في النكاح

النكاح ثلاثة: الدائم، والمنقطع، وملك اليمين.

ويفتقر الأول إلى العقد، وهو الإيجاب من أهله والقبول بلفظ الماضي من أهله، ولو قيل «زوجت بنتك (فلانة) من فلان؟» فقال «نعم» كفى في الإيجاب، ويجزئ مع العجز الترجمة والإشارة. ولو زوّجت المرأة نفسها صح. ولا يشترط الولي مع البلوغ والرشد ولا الشهود.

ولا يلتفت إلى دعوى الزوجية بغير بينة أو تصديق. ولو ادعت أخت الزوجة زوجيته، حكم لبينته إلا مع تقديم تاريخها أو دخوله بها.

والقول قول الأب في تعيين المعقود عليها بغير تسمية مع رؤية الزوج للجميع، وإلا بطل العقد.

ويستحب أن يتخير البكر العفيفة الكريمة الأصل، وصلاة ركعتين، والإشهاد والإعلان، والخطبة أمام العقد، وإيقاعه ليلاً، وصلاة ركعتين عند الدخول، والدعاء، وأمرها بمثله، وسؤال الله الولد.

Performing the marriage when the moon is in the phase of the scorpion is disapproved, as is marrying off a barren person; consummating the marriage on the night of a lunar eclipse, on the day of a solar eclipse, at high-noon, at sunset and before the disappearance of the red glow from the evening sky, on a moonless night, between dawn and sunrise, and on the first night of every lunar month – except for the month of Ramadan; the middle night of every lunar month, during an earthquake, hurricane or cyclone, facing or with the back towards the *qibla*, on a ship, fully-naked, straight after a nocturnal emission and before having done a *ghusl* (major ablution) or *wuḍū'* (ritual ablution). Looking at the woman's genitals is also disapproved, as is talking during intercourse unless it is in remembrance of God; penetration into the anus; unnecessary abstention from a free woman without her permission; and for a traveller to sleep with his wife at night when returning home.

Consummating a marriage with girl younger than nine years old is prohibited.

Looking, without lust, at a woman that one wishes to marry or purchase, and at *dhimmī* women is permissible.

ويكره إيقاع العقد والقمر في العقرب، وتزويج العقيم، والجماع ليلة الخسوف ويوم الكسوف وعند الزوال وعند الغروب وقبل ذهاب الشفق وفي المحاق وبعد الفجر حتى تطلع الشمس وفي أول ليلة كل شهر - إلا رمضان - وليلة النصف وعند الزلزلة والريح الصفراء والسوداء ومستقبل القبلة ومستدبرها وفي السفينة وعارياً وعقيب الاحتلام قبل الغسل أو الوضوء، والنظر إلى فرج المرأة، والكلام بغير الذكر، والوطي في الدبر، والعزل عن الحرة بغير إذنها، وأن يطرق المسافر أهله ليلاً.

ويحرم الدخول بامرأة قبل تسع سنين. ويجوز النظر إلى من يريد التزويج بها أو شراءها، وإلى أهل الذمة بغير تلذّذ.

Section Two: Guardianship (For Marriage)

Guardianship to marry is held by the father, or any paternal grandfather, the legal guardian (*waṣī*) and the [just] ruler charged with the affairs of the Muslims. The father acts as the guardian for minors and those with diminished mental capacity. When either of these two states ceases, neither [i.e. the minor or the mentally impaired] has the option [to refuse the marriage], whereas the mature and legally responsible have no legal guardian to oversee them [in their marital affairs]. The legal guardian and the [just] ruler have guardianship over someone who is mentally incapacitated, even if they are of legal age, whether male or female, so long as they act in their best interests. Marrying others besides them [i.e. those of minor age and the mentally incapacitated] depends on their consent, and in the case of a virgin girl her silence is sufficient of her consent.

The master holds unconditional guardianship over his slave, be it male or female. The mother cannot be a guardian for marriage.

It is recommended for the legally responsible woman to seek her father's consent, and in his absence, to appoint her brother to act on her behalf. The agent appointed to act on her behalf cannot conduct her marriage to himself without her consent.

If two minors are betrothed by their fathers, [if either one dies before consummation] the other inherits from him/her. If they had been betrothed by other than their fathers, the validity rests upon their consent upon coming of age, so if one of them dies before then, the marriage is invalidated. If one of them comes of age and consents to the marriage, then dies, the other must swear an oath of honesty upon coming of age, denouncing that his/her consent was solely for the purpose of inheriting. After that he/she may inherit, otherwise, they do not inherit from each other.

الفصل الثاني: في الأولياء

إنما الولاية للأب وإن علا، والوصي، والحاكم. فالأب على الصغيرين والمجنونين، ولا خيار [لهما] بعد زوال الوصفين والبالغ الرشيد لا ولاية عليه - ذكراً كان أو أنثى - والحاكم والوصي على المجنون البالغ ذكراً وأنثى مع المصلحة. ويقف عقد غيرهم على الإجازة، ويكفي فيها سكوت البكر.

وللمولى الولاية على مملوكه ذكراً وأنثى مطلقاً. ولا ولاية للأم.

ويستحب للبالغة أن تستأذن أباها، وأن توكّل أخاها مع فقده.

وليس للوكيل أن يزوّجها من نفسه بغير إذنها.

ولو زوّج الصغيرين الأبوان توارثا، ولو كان غيرهما وقف على الإجازة فإن مات أحدهما قبل البلوغ بطل، وإن بلغ أحدهما وأجاز ثم مات حلف الثاني بعد بلوغه على انتفاء الطمع وورث، وإلا فلا.

Section Three: Women Whom One Is Forbidden to Marry

These are divided into two main categories: those who are biologically related, and those forbidden due to external factors.

Those that are biologically related include one's mother, grandmother and further up the line, daughter, granddaughter and further down the line, sister and her daughters, and their daughters, paternal aunt or great-aunt, maternal aunt or great-aunt, brother's daughters, and their daughters.

There are several external considerations that cause women to be forbidden to marry:

1. *Those Who Are Forbidden Due to Marital Relationships:*

A man who has sexual relations with a woman through a marriage contract or by concubinage, her mother, grandmother and upwards in that line, are forbidden to him forever, as are her daughters, granddaughters and down that line, whether they existed prior to consummation of their marriage or were born afterwards.

The concubine with whom marriage has been consummated is forbidden to her husband's father, grandfather and upwards, as well as his sons, grandsons and downwards.

If a man contracts a marriage with a woman, her mother is forbidden to him forever, even if that marriage was not consummated. Her daughter is also forbidden to him, as long as she is in the marriage contract. If he divorces her before consummation, he may marry her daughter thereafter, but if he consummated the marriage, then she too is forbidden to him forever.

It is forbidden to be married to the sister of one's wife at the same time [as being married to her], but not so if he married them at different times. The same goes for his wife's niece – both her brother's and sister's daughter – unless he has permission from the girl's paternal or maternal aunt [i.e. his wife]. If he contracts the marriage without her consent, it is invalid.

The one who fornicates with his paternal or maternal aunt, their daughters are forbidden to him forever. If a man owns two sisters as slave-girls, and copulates with one of them, the other is forbidden for him [to copulate with] at the same time. If he does, he is considered as having sinned, but the first one does not become forbidden to him.

الفصل الثالث: في المحرمات

وهي قسمان: نسب وسبب. فالنسب: الأم وإن علت، والبنت وإن سفلت، والأخت وبناتها وإن نزلن، والعمة والخالة وإن علتا، وبنات الأخ وإن نزلن.

وأما السبب فأمور:

الأول: ما يحرم بالمصاهرة

فمن وطأ امرأة بالعقد أو الملك حرمت عليه أمّها وأن علت، وبنتها وإن نزلت، تحريما مؤبدا، سواء سبقن على الوطء أو تأخرن عنه.

وتحرم الموطوءة بالملك أو العقد على أب الواطئ وإن علا، وعلى أولاده وإن نزلوا.

ومن عقد على امرأة ولم يدخل بها حرمت عليه أمّها أبداً، وبنتها ما دامت الأم في عقده، فإن طلّقها قبل الدخول جاز له العقد على بنتها، ولو دخل حرمت أبداً.

وتحرم أخت الزوجة جمعاً لا عيناً، وكذا بنت أختها وبنت أخيها إلا مع إذن العمة والخالة، ولو عقد من دون إذنهما بطل.

ومن زنا بعمته أو خالته حرمت عليه بناتهما أبداً. ولو ملك الأختين فوطأ إحداهما حرمت الأخرى جمعاً، فلو وطأها أثم ولم تحرم الأولى.

A free man is forbidden from permanently marrying more than four free women, or more than two slave-girls. He may be married to two free women and two slave-girls at the same time, or three free women and one slave-girl. A male slave may not be permanently married to more than four slave-girls at one time, or two free women at any one time. He may be married to one free woman and two slave-girls at the same time.

A married man is not allowed to permanently marry a slave-girl without the prior consent of his free wife, and if he does contract the marriage without it, it is invalid. If he marries a free woman after already being married to a slave-girl, without her knowledge of this, she has the choice [to remain in the contract or not]. If he marries them both together in one contract, it is the free woman's marriage that holds validity.

It is forbidden to marry a woman who already has a husband, and a woman who is in her prescribed waiting period (*ʿidda*) [post-divorce or widowhood]. If a man unwittingly marries a woman who is in her waiting period, the contract is invalid. If he had consummated that marriage, she is forbidden to him forever, any resulting offspring are his, and she keeps the *mahr* (bridal gift). She must complete the first waiting period, then begin the second after that one. If he married her knowing that she is in her waiting period, she is forbidden for him to marry forever [even if the marriage had not been consummated].

Other Rulings:

1) A man who commits sodomy with another male, his mother, his sister and his daughter are permanently forbidden to him to marry. If any of them were already married to him prior to that, then they are not forbidden.

2) If a man has sexual intercourse with a girl who has not yet reached nine years of age, and causes injury to her in the process,[1] she is permanently forbidden for him to marry, and he remains responsible for her upkeep forever.

3) If a man fornicates with a woman, marrying her is not prohibited for him. If he fornicates with a woman married to another man, or a woman who is in her waiting period of revocable divorce, then he is forever prohibited from marrying her.

1 By rupturing her hymen.

ويحرم على الحر في الدائم ما زاد على أربع حرائر، وفي الإماء ما زاد على أمتين، وله أن يجمع بين حرتين وأمتين أو ثلاث حرائر وأمة، وعلى العبد ما زاد على أربع إماء، وفي الحرائر ما زاد على حرتين، وله أن ينكح حرة وأمتين.

ولا يجوز نكاح الأمة على الحرة إلا بإذنها، ولو عقد بدونه كان باطلاً. ولو أدخل الحرة على الأمة ولم تعلم فلها الخيار، ولو جمعهما في عقد صح على الحرة.

ويحرم العقد على ذات البعل، والمعتدّة ما دامت كذلك، ولو تزوّجها في عدّتها جاهلاً بطل العقد، فإن دخل حرمت أبداً والولد له والمهر للمرأة، وتتم عدّة الأول وتستأنف للثاني، ولو عقد عالماً حرمت أبداً بالعقد.

مسائل:

الأولى: من لاط بغلام فأوقبه حرم عليه أم الغلام وأخته وبنته أبداً، ولو سبق عقدهنّ لم يحرمن.

الثانية: لو دخل بصبية لم تبلغ تسعاً فأفضاها حرمت أبداً ولم تخرج من حباله.

الثالثة: لو زنا بامرأة لم يحرم نكاحها ولو زنا بذات بعل أو في عدّة رجعية حرمت أبداً.

4) If a man contracts a marriage whilst in the state of *iḥrām*, knowing that doing so is prohibited, she is permanently forbidden for him to marry. And if he did it in ignorance, the contract is void, but she is not forbidden to him.

5) There are no restrictions on the number of women one may contract fixed-term marriages with or take as slave concubines.

6) If a free woman has been divorced three times, she becomes forbidden to he who divorced her thrice until she has married another husband [and been divorced by him], even if it be a slave. If a slave-girl is divorced twice, she is forbidden for the same man to marry again until she has married another man, even if it be a free man.

7) The woman who has been divorced nine times with just as many waiting periods, having been married to two other men in between, is permanently forbidden to the man who originally divorced her.

8) If a man divorces one of his four wives with a revocable divorce, he is not allowed to permanently marry another woman in her place until after she has completed her waiting period, but he is allowed if the divorce is absolute. If a man with three wives contracts to marry two more women at the same time, both marriages are void. If he marries one after the other, the second one is void. The same ruling applies to marrying two sisters together.

2. *Those Related Through Wet-Nursing:*

Those related through wet-nursing are forbidden to marry just as those related by blood, when the wet-nurse is producing milk because of a valid marriage contract [i.e. the wet-nurse is not producing milk because of *zinā*]. [The minimum term of suckling that leads to this prohibition is] a day and a night [i.e. twenty-four hours]; or the milk that nourished a child is enough to grow flesh and strengthen bones; or fifteen full consecutive sucklings from the breast without being suckled by another woman in between. Additionally, the breastfeeding must have taken place within the first two years of the child's life. There are two opinions about whether the wet-nurse's own children [as a result of which she is producing milk] should be within the first two year's. The milk that is fed must be the result of impregnation by one man.[2]

2 She may nurse the same child with the milk produced from two consecutive pregnancies, but not by different men.

الرابعة: لو عقد المحرم عالماً بالتحريم حرمت أبداً، ولو كان جاهلاً بطل العقد ولم تحرم.

الخامسة: لا تنحصر المتعة وملك اليمين في عدد.

السادسة: لو طُلّقت الحرة ثلاثاً حرمت حتى تنكح زوجاً غيره وإن كانت تحت عبد، ولو طُلّقت الأمة طلقتين حرمت حتى تنكح زوجاً غيره وإن كانت تحت حر.

السابعة: المطلّقة تسعاً للعدّة ينكحها بينها رجلان تحرم على المطلِّق أبداً.

الثامنة: لو طلّق إحدى الأربع رجعياً لم يجز أن ينكح بدلها حتى تخرج من العدّة، ويجوز في البائن. ولو عقد ذو الثلاث على اثنين دفعة بطلا، ولو ترتّب بطل الثاني، وكذا الحكم في الأختين.

الثاني: الرضاع

ويحرم منه ما يحرم بالنسب، إذا كان عن نكاح، يوماً وليلة، أو ما أنبت اللحم وشدّ العظم، أو كان خمس عشرة رضعة كاملة من الثدي لا يفصل بينها رضاع أخرى، وأن يكون في الحولين بالنسبة إلى المرتضع، وفي ولد المرضعة قولان، وأن يكون اللبن لفحل واحد.

If two wives of the same man nurse two children who are not their own [i.e. one child each], they will be prohibited from marrying one another. But if a woman nurses two children with the milk from pregnancies by two different men, they will not be prohibited to each other.

Fulfilment of the above conditions results in the wet-nurse becoming a mother to the suckled child, the man who impregnated her[3] [to become] his father, their brothers and sisters his paternal and maternal uncles and aunts, and their children his siblings. All biological and milk-children of the 'milk-father' are prohibited for that suckled child to marry. As for the milk-mother, her biological children are prohibited to marry, but not her milk-children [from another man's milk].

The biological father of the suckled child cannot marry the children of his child's milk-father: neither his biological children nor his milk-children. His other children who have not been nursed from that milk may marry the children of the milk-mother and father.

If the older of a man's two wives were to suckle the younger one, they both become prohibited to him, if her milk had resulted from him impregnating her. If not, then only the [older] wife who suckled becomes prohibited.[4] If it transpires that a man's milk-mother also nursed his wife, she becomes prohibited to him. The milk-mother of a man's [biological] child is not prohibited for him to marry, though she may be biologically prohibited.

It is recommended to choose an intelligent, chaste, clean, Muslim lady to suckle one's child.

3. *Mutual Repudiation of Spouses Resulting From Liʿān*[5]

By its enactment both spouses are forever prohibited from each other, as does a man's falsely accusing his mute or deaf wife of adultery.

3 Literally known as "owner of the milk" (*dhū l-laban*).

4 As she technically becomes the younger one's "milk-mother".

5 *liʿān*: Mutual repudiation. A Quranic institution governing cases in which a husband accuses his wife of adultery without supplying witnesses. Quran 24:6–9 explicitly instructs the husband to swear four times that his accusation is true, followed by a fifth oath in which he invokes the wrath of God upon himself if he is lying. The wife may then neutralize this claim by responding with four oaths of her own, the fifth of which calls upon her the wrath of God if her husband is telling the truth. If she refuses to take the oath, she is presumed guilty and subject to the punishment for adultery. If she takes the oath, she is declared innocent and permanently divorced from her husband. Her husband, in turn, forfeits any paternity claims over children born subsequent to their sworn oaths.

فلو أرضعت امرأتان [صبيين] بلبن فحل واحد نشر الحرمة بينهما، ولو أرضعت امرأة صبيين بلبن فحلين لم ينشر الحرمة.

ومع الشرائط تصير المرضعة أمّاً وذو اللبن أباً وأخوتهما أخوالاً وأعماماً وأولادهما إخوة.

ويحرم أولاد صاحب اللبن ولادةً ورضاعاً على المرتضع وأولاد المرضعة ولادةً لا رضاعاً.

ولا ينكح أبو المرتضع في أولاد صاحب اللبن ولادةً ورضاعاً، ولا في أولاد زوجة المرضعة ولادةً لا رضاعاً، ولأولاده الذين لم يرتضعوا من هذا اللبن النكاح في أولاد المرضعة والفحل.

ولو أرضعت كبيرة الزوجتين صغيرتهما حرمتا إن كان دخل بالمرضعة وإلا فالمرضعة. ولو أرضعت الأم من الرضاعة الزوجة حرمت ولا تحرم أم الولد من الرضاع على أبيه وإن حرمت من النسب.

ويستحب اختيار المسلمة الوضيئة العفيفة العاقلة للرضاع.

الثالث: اللعان

ويثبت به التحريم المؤبّد، وكذا قذف الزوج امرأته الصماء الخرساء.

4. *Disbelief:*

A Muslim man is not permitted to marry non-Muslims other than women of the Book by consensus, though there are two opinions; and a Muslim woman is only allowed to marry a Muslim man.

If one of the spouses [from a Muslim couple] apostatizes before consummation of the marriage, it is annulled with immediate effect. If it is after consummation, then it is annulled after the woman completes her waiting period, unless the husband is a born-Muslim apostate (*murtadd fiṭrī*), in which case it is annulled immediately.[6]

The waiting period for a woman whose born-Muslim husband apostatizes is the same as a widow's period; for all other types, it is the waiting period of divorce.

If the husband from a couple from the People of the Book converts to Islam, the marriage remains in order; but if the wife converts to Islam and he does not, the marriage is annulled immediately if it is before consummation, and subject to the waiting period if not. If he too converts during the waiting period, he may remain married to her.

If one of two disbelieving spouses both without *dhimmī* status [i.e. they are *kāfir ḥarbī*] converts to Islam before consummation of the marriage, it is annulled with immediate effect, and if it is after, then they are no longer married after completion of the waiting period.

If a non-*dhimmī* unbeliever having more than four unbelieving wives converts to Islam, and they do too, he must choose four from among them to keep as wives, and the rest of them are annulled. If a *dhimmī* having four wives converts to Islam, he may remain married to them. If he has more than four wives, he must choose four and the rest of his marriages are annulled.

6 Since he would be killed for his apostasy.

الرابع: الكفر

ولا يجوز للمسلم أن ينكح غير الكتابية إجماعاً، وفيها قولان، ولا للمسلمة أن تنكح غير المسلم.

ولو ارتدّ أحد الزوجين قبل الدخول انفسخ في الحال، ويقف بعده على انقضاء العدّة، إلا أن يرتدّ الزوج عن فطرة فينفسخ في الحال.

وعدّة المرتدّ عن فطرة عدّة الوفاة، وعن غيرها عدّة الطلاق.

ولو أسلم زوج الكتابية ثبت عقده، ولو أسلمت دونه قبل الدخول انفسخ العقد، وبعده يقف على العدّة. فإن أسلم فيها كان أملك بها.

ولو كان الزوجان حربيين وأسلم أحدهما قبل الدخول انفسخ النكاح في الحال، ولو كان بعده وقف على انقضاء العدّة.

[ولو أسلم الزوج الحربي على أكثر من أربع حربيات وأسلمن فاختار أربعاً انفسخ نكاح البواقي] ولو أسلم الذمي وعنده أربع ثبت عقده عليهنّ، ولو كن أزيد تخير أربعاً وبطل نكاح البواقي.

Other Rulings:

1) A believing woman [i.e. a follower of the twelve Imams] is not allowed to marry a man from the opposing sect [i.e. a Sunnī], though the opposite is allowed. Marrying an immoral person is disapproved.

2) *shighār* marriages are invalid, and this is where one's marriage to one woman is designated as the bridal gift (*mahr*) for another.

3) It is permissible to get a free woman married to a slave, a Hāshimī woman to a non-Hāshimī man, and an Arab woman to a non-Arab man, and vice versa. It is obligatory to respond favourably to the marriage proposal of a believing man having the means [to provide for his wife].

مسائل:

الأولى: لا يجوز للمؤمنة أن تتزوج بالمخالف، ويجوز العكس ويكره تزويج الفاسق.

الثانية: نكاح الشغار باطل، وهو جعل نكاح امرأة مهراً لأخرى.

الثالثة: يجوز تزويج الحرة بالعبد والهاشمية بغيره والعربية بالعجمي وبالعكس، ويجب إجابة المؤمن القادر [على النفقة].

Section Four: Fixed-Term Marriage

The prerequisites of this are a proposal and acceptance from those eligible, stipulation of the bridal gift (*mahr*) and mandatory stipulation of the term fixed for the marriage. The contract is invalidated if the bridal gift is not stipulated, and it is most likely invalidated if the term is not stipulated.

It is prohibited [to contract fixed-term marriages] with disbelievers, except for women of the Book; with a slave-girl whilst already married to a free woman without her consent; with one's wife's niece (brother's or sister's daughter) without the wife's consent.

Marrying a licentious woman, or a virgin without her father's permission, is disapproved.

There is no defined amount for the bridal gift. If the man decides to gift back the remaining time of the marriage to his temporary wife before consummation, he must pay her half of the bridal gift. If she does not make herself available for any part of the marriage, then an amount in proportion to the time she is unavailable is deducted from her bridal gift. If the contract turned out to be invalid before consummation, the bridal gift is not payable, but if it was after and she was ignorant of the cause for invalidation, then she may receive half of it. Any child born is the man's, even if he is absent. If he denies him, this does not imply there is *liʿān* (mutual repudiation).

Neither divorce, *liʿān*, nor *ẓihār*[7] apply to fixed-term marriages, nor are the spouses entitled to inherit from one another, even if they stipulate so in the contract. The waiting period after the term is concluded is two menstrual cycles or forty-five days, and after demise of the husband, four months and ten days.

7 A pre-Islamic form of divorce where a man would repudiate his wife by likening her to his mother's back, which deprived the wife of her marital rights, yet prevented her from marrying again.

الفصل الرابع: في المتعة

ويشترط فيها الإيجاب والقبول من أهله، وذكر المهر، ولا بدَّ فيه من ذكر الأجل المعيّن. ولو لم يذكر المهر بطل، ولو لم يذكر الأجل فالأقرب البطلان.

ويحرم غير الكتابية من الكفار، والأمة على الحرة من دون إذنها، وبنت الأخ والأخت من دون إذن العمة والخالة.

ويكره الزانية والبكر من دون إذن الأب.

ولا حدّ للمهر ولو وهبها المدّة قبل الدخول ثبت نصفه، ولو أخلت ببعض المدة لها أسقط بنسبته. ولو ظهر بطلان العقد فلا مهر قبل الدخول، وبعده لها المهر مع جهلها. ويلحق به الولد وإن عزل، ولو نفاه فلا لعان.

ولا يقع بها طلاق ولا لعان ولا ظهار، ولا ميراث لها وإن شرط وتعتدّ بعد الأجل بحيضتين، أو بخمسة وأربعين يوماً، وفي الموت بأربعة أشهر وعشرة أيام.

Section Five: Marrying Slaves

A male and female slave may not contract themselves in marriage without the permission of their master, and if they do, its validity rests upon their master's consent. If the master consents to his male slave's marriage, he is responsible for providing his slave's bridal gift and the maintenance of his wife, and the female slave's master receives the bridal gift on her behalf, which is payable upon consummation. If neither of them had permission from their masters, any ensuing offspring belongs to the masters jointly. If one of them permits the marriage, then the child belongs to the other. If one of the spouses is free, then the child is also free, as long as the master did not stipulate for it to be bonded.

If a free man knowingly marries a slave without the permission of her master, then he is considered to have fornicated, and the child born of them is also a slave. If he did so, ignorant of the ruling, then he is not punishable for the crime of fornication, but he must pay the bridal gift and the cost of manumitting the child when it is born. The same ruling applies if she had misled him to believe that she was free.

The father is responsible for freeing his children born into slavery, and their master is obligated to release them to him. If he cannot afford to do so, he may work towards the cost [of their freedom].

The bridal gift is only payable upon consummation of the marriage.

If a free woman knowingly marries a male slave [without permission from his master], she is not entitled to the bridal gift, and any ensuing offspring is born a slave. If she did so in ignorance, the child will be free and there will be nothing due to pay. The slave will remain responsible for the bridal gift upon consummation, after his manumission.

If a free man or male slave fornicates with a female slave, the child belongs to her master. Even if the man [i.e. the fornicator] buys a part of the female slave's freedom, the contract is invalid, and according to one opinion she still cannot become lawful for him. If she manumits herself, she has the right to annul the marriage.

الفصل الخامس: في نكاح الإماء

ولا يجوز للعبد والأمة أن يعقدا لأنفسهما بغير إذن المولى، فإن فعل أحدهما وقف على الإجازة، ولو أذن المولى للعبد ثبت مهر عبده عليه ونفقة زوجته، وثبت لمولى الأمة مهر أمته. ويستقر بالدخول.

ولو لم يأذنا فالولد لهما، ولو أذن أحدهما فالولد للآخر. ولو كان أحد الزوجين حراً فالولد مثله ما لم يشترط المولى الرقية.

ولو تزوج الحر من دون إذن المولى عالماً فهو زان والولد رق، ولو كان جاهلاً سقط الحد دون المهر وعليه قيمة الولد يوم سقوطه حياً، ولو ادّعت الحرية فكذلك.

وعلى الأب فكّ أولاده، ويلزم المولى دفعهم إليه، ولو عجز سعى في القيمة، ومع عدم الدخول لا مهر.

ولو تزوّجت الحرة بعبد عالمةً فلا مهر والولد رق ومع الجهل حر ولا قيمة، وعلى العبد المهر يتبع به بعد العتق مع الدخول.

ولو زنى الحر أو العبد بمملوكة فالولد لمولاها، ولو اشترى جزءاً من زوجته بطل العقد ولم تحل بالتحليل على قول، ولو أعتقت الأمة كان لها فسخ النكاح.

It is permissible [for a master to marry his slave girl] and to make her bridal gift her manumission, if the manumission precedes, or the marriage contract according to another opinion. The slave-mother who begets her master's child is still a slave albeit that she cannot be sold while the child exists, unless the master has no other slaves with which to pay off a debt equivalent in value to her price. She is automatically freed with the demise of her master from her child's share of inheritance. If it does not cover her price, then she may work for her price.

When a slave-girl is sold, her buyer has the right to annul her [previous] marriage instantly, and the master of her [slave] husband would have the right to do the same; and vice versa [if he is sold].

If her buyer annulled her marriage before her husband had consummated the marriage, then the bridal gift is not payable. If it was allowed to be consummated before the sale, then he is entitled to the bridal gift, and if it was consummated after the sale was concluded, then the seller keeps the bridal gift.

The male slave holds the right to divorce. If the couple both belong to one master, then he has the right to annul the marriage.

The master who gets his slave-girl married is prohibited from sleeping with her, touching her or looking at her lustfully thereafter, as long as she is the wife of another man. Neither one of two partners who jointly own a female slave has the right to sleep with her.

It is obligatory for the buyer of a slave-girl to wait for her to complete one menstrual cycle [before having intercourse with her],[8] and if he manumits her, then it is permissible for him to sleep with her by marrying her without waiting. If anyone else wishes to marry her, however, [upon his manumission of her], then they must wait until she has completed the waiting period of a free woman.

If a master permits someone else to use his slave-girl, she becomes permissible for him, even if she belongs to one of his male slaves, but she can only be used for those things permitted by the master. Any resulting offspring would be free [if the person who was permitted to use her is a free man].

8 To ensure she is menstruating, not pregnant, of good health, etc.

ويجوز جعل العتق مهراً لمملوكته إذا قدم العتق أو النكاح على خلاف. وأمّ الولد رق لا يجوز بيعها مع وجوده إلا في ثمن رقبتها إذا لم يكن غيرها ، وتنعتق بموت المولى من نصيب الولد ، ولو عجز سعت.

وإذا بيعت الأمة كان للمشتري على الفور فسخ النكاح ولصاحب العبد أيضاً ، وكذا العبد.

ومع فسخ مشتري الأمة قبل الدخول لا مهر ، ولو أجاز قبله فله المهر ، وبعده فللبائع.

وطلاق العبد بيده، ولو كانا لواحد كان للمولى فسخه.

ويحرم لمن زوّج أمته وطيها ولمسها والنظر بشهوة ما دامت في حباله. وليس لأحد الشريكين وطئ المشتركة بالملك.

ويجب على مشتري الجارية استبراؤها ولو أعتقها حل له وطيها بالعقد من غير استبراء ، ولا بدَّ لغيره من عدّة الحرة.

ولو حلل أمته على غيره حلت له ولو كان لمملوكه، ولا يحل غير المأذون، وينعقد الولد حراً.

Section Six: Defects (That Have a Bearing on the Validity of the Marriage)

There are four such defects pertaining to the man: mental incapacity, having no testicles, impotence and castration of the male organ; and seven pertaining to the woman: mental incapacity, a missing limb, leprosy, problems with her private parts, not being a virgin, blindness, and disability.

The marriage cannot be annulled if defects appear for the first time after the marriage has been contracted, except for impotence. There is an opinion that allows annulment upon the onset of mental incapacity. In the case of pre-existing defects, a decision to keep the marriage or annul must be made immediately [i.e. before consummation] upon realisation. Annulling a marriage is not a divorce. Cases of impotence must be ruled by the [just] ruler in charge of Muslim affairs.

The bridal gift is not payable if the marriage is annulled before consummation, and if it is after, then a nominal sum is payable which the groom may recover from the one who swindled him.

If the woman annuls the marriage then she is not entitled to the bridal gift if it is annulled before consummation, unless it is a case of impotence, whereby she is entitled to half of it. If it is after consummation, then she receives a nominal amount. In a dispute [in the absence of evidence], the denier of the defect has the last word.

If the case is taken to court, the judge may give the husband who has been accused of impotence but denies it up to a year, and if he manages to have intercourse with his wife or another woman in that time, then the marriage cannot be annulled. If not, then it is annulled, and she is entitled to half of the bridal gift.

If a man marries a woman, believing her to be a free woman, and it transpires that she is a slave, the marriage is annulled and she is not entitled to the bridal gift unless it is after consummation, which the man may recover from the person who swindled him. Similarly, if he had stipulated for her to be the daughter of a *mahr* [i.e. permanent free wife], and she turned out to be daughter of a concubine.

If a woman marries a man, believing him to be a free man and he turns out to be a slave, she has the right to annul the marriage, and she is entitled to the bridal gift if she discovers after consummation, but not before.

الفصل السادس: في العيوب

وهي أربعة في الرجل: الجنون، والخصاء، والعنن، والجب.

وسبعة في المرأة: الجنون، والجذام، والبرص، والقرن، والإفضاء، والعمى، والإقعاد.

ولا فسخ بالمتجدّد بعد العقد في غير العنة، وفي الجنون المتجدّد قول بالفسخ. والخيار على الفور، ليس بطلاق. ولا بدَّ من الحاكم في العنة خاصة.

ولا مهر في الفسخ قبل الدخول من الرجل، وبعده المسمى، ويرجع الزوج على المدلس.

ومن المرأة لا مهر لها قبل الدخول إلا في العنة فيثبت نصفه، وبعده المسمى. والقول قول المنكر للعيب.

ويؤجل الحاكم العنين مع المرافعة سنة، فإن وطأها أو غيرها فلا فسخ، وإلا فسخت ولها نصف المهر.

ولو تزوّجها حرة فبانت أمة فسخ ولا مهر إلا مع الدخول، فيرجع على المدلس. وكذا لو شرطت بنت مهيرة فخرجت بنت أمة.

ولو تزوجته حراً فبان عبداً فلها الفسخ، والمهر بعد الدخول لا قبله.

Section Seven: The Dowry

This is the price in lieu of the vulva and becomes the property of the woman through the contract. Half of it is written off if they divorce before consummation. If the marriage is consummated through penetration, either frontal or rear-entry, her ownership of the dowry continues.

It is valid for it to be something physical (e.g. a ring or a house), a debt [to be paid by the husband in the future], or it can be a benefit. There is no fixed amount to the dowry [i.e. the Shariah does not specify a fixed amount], be it large or small, though it must be described or seen [by the bride].

If the dowry was not fixed at the time of marriage, the contract is still in order, but he must give her a customary dowry upon consummation, which should not exceed the traditional amount.[9] If it does exceed that amount, it must be reduced to the traditional amount. In such a case [where the dowry was unspecified], upon divorce [before consummation], she is entitled to receive provision commensurate with his financial capacity: if he is prosperous, she is entitled to a high-quality garment and ten dinars; if he is of average means, then five dinars; and if he is poor, then a ring and one dirham.

If he marries her, and the dowry is left to either one of them to decide, the marriage is in order and the wishes of the deciding party must be adhered to, as long as the amount that the woman fixes does not exceed the traditional amount, if she is the one who is deciding. If the deciding party were to die before making a decision on the amount for the dowry, she is entitled to the equivalent provision. If he marries her with the dowry stipulated to be a servant or a house or a property without specification, then she is entitled to the average of those things. If the bridal gift is left to be 'as tradition dictates' then it is five hundred dirhams.

If two non-Muslim *dhimmīs* marry upon the dowry of wine, the marriage is in order. If one of them converts to Islam before receipt of the dowry, she must receive its value instead. If a Muslim got married with wine as the dowry, the marriage is said to be valid, but the traditional dowry must be given instead upon consummation. Another opinion states that the contract is void. If a man gifts her a *mudabbar* slave [who would be automatically freed upon his death], the terms of the slave's manumission by *tadbīr* is nullified.

If conditions are placed on the contract that involve forbidden things or acts, only the conditions are void, [but the contract remains valid]. If it is stipulated as a condition that he must not move her out of her hometown, then this must be adhered to.

9 *Mahr al-sunna*: traditionally recommended amount of five hundred gold dirhams.

الفصل السابع: في المهر

وهو عوض البضع، وتملّكه المرأة بالعقد، ويسقط نصفه بالطلاق قبل الدخول، ولو دخل قُبلاً أو دبراً استقرّ.

ويصح أن يكون عيناً أو ديناً أو منفعةً، ولا يتقدر قلة وكثرة، ولا بدَّ فيه من الوصف أو المشاهدة.

ولو لم يتعيّن صح العقد، وكان لها مع الدخول مهر المثل ما لم يتجاوز السنّة، فإن تجاوز ردّ إليها، ومع الطلاق لها المتعة، للموسر بالثوب المرتفع أو عشرة دنانير، والمتوسط بخمسة، والفقير بخاتم أو درهم.

ولو تزوّجها بحكم أحدهما صح، ويلزم ما يحكم به صاحب الحكم ما لم تتجاوز المرأة مهر السنّة إن كانت الحاكمة، ولو مات الحاكم قبله فلها المتعة.

ولو تزوّجها على خادم مطلقاً أو دار أو بيت كان لها وسط ذلك، ولو قال «على السنّة» فخمسمائة درهم.

ولو تزوّج الذميان على خمر صح، فإن أسلم أحدهما قبل القبض فلها القيمة، ولو تزوّج المسلم عليه قيل يصح ويثبت على الدخول مهر المثل، وقيل يبطل العقد، ولو أمهر المدبّر بطل التدبير.

In a dispute about the amount of the dowry, the husband has the last word. If he denies the amount after having consummated the marriage, then she is entitled to the traditional amount. If the dispute is about consummation and she claims that the marriage has been consummated, he has the last word accompanied by an oath of honesty, though this is debatable.

A father who betroths his minor son must guarantee the dowry in case of his poverty.

The woman has the right to refuse sex if the marriage has not yet been consummated until she receives the dowry in hand.

ولو شرط في العقد المحرم بطل الشرط خاصة، ولو اشترط أن لا يخرجها من بلدها لزم.

والقول قول الزوج في قدر المهر، ولو أنكره بعد الدخول فالوجه مهر المثل، ولو ادّعت المواقعة فالقول قوله مع يمينه على إشكال.

ولو زوّج الأب الصغير ضمن المهر مع فقره، وللمرأة الامتناع قبل الدخول حتى تقبض المهر.

Section Eight: Division of Time, and Disobedience

A permanent wife has the right to have [her husband spend] one night of every four with her. Two wives are entitled to two nights out of four between them, and three to three nights out of four between them. If he has four wives, then he must spend one night out of four with each, but if one of them gives up her turn, he can spend it wherever he wishes. If she gifts her turn to a specific co-wife, then he must spend it with her.

The obligation upon him is simply to sleep next to her, and not necessarily to have sexual intercourse. The free woman is entitled to two nights as opposed to one each for the slave-girl and the non-Muslim *kitābī* wife. The new virgin bride may have her husband spend up to seven nights with her consecutively upon consummation of the marriage, and a bride who had been married previously up to three days.

It is recommended for the husband to provide equally for all his wives.

It is obligatory for the wife to make herself available to him and to remove anything repulsive from herself. He is allowed to strike his disobedient wife (*nāshiza*) after having turned his back to her when sleeping in the marital bed and thereafter sleeping in a separate bed. If he behaves cruelly or neglects his duties towards her, she may either demand [her rights] from him, or she may relinquish part of her own right or all of it to win back his favours. He is allowed to accept this.

If both of them despise each other, the judge may appoint two arbitrators from her family or unrelated, and if they see a way for them to reconcile, then they facilitate a reconciliation. If, however, they consider a separation best, they consult them both regarding divorce and due payment. No decision is to be made if the two arbitrators have differing opinions.

الفصل الثامن: في القسم والنشوز

للزوجة دائماً ليلة من أربع، وللزوجتين ليلتان، وللثلاث ثلاث، ولو كنّ أربع فلكل واحدة ليلة.

ولو وهبته إحداهن وضع ليلتها حيث شاء، ولو وهبت الضرّة بات عندها.

والواجب المضاجعة ليلاً لا المواقعة، وللحرة ليلتان، وللأمة والكتابية ليلة، وتختص البكر عند الدخول بسبع، والثيب بثلاث.

ويستحب التسوية في الإنفاق. ويجب على الزوجة التمكين وإزالة المنفر. وله ضرب الناشزة بعد وعظها وهجرها، ولو نشز طالبته، ولها ترك بعض حقها أو كله استمالة له، ويحل قبوله.

ولو كره كل منهما صاحبه أنفذ الحاكم حكَمين من أهلها أو أجنبيين، فإن رأيا الصلح أصلحا، وإن رأيا الفرقة راجعاهما في الطلاق والبذل، ولا حكم مع اختلافهما.

Section Nine: Rulings Pertaining to Children

The child is considered as belonging to the husband in a permanent marriage following its consummation and the passage of at least six months from the time of intercourse until delivery of the child after gestation of between six and ten months. If the husband has been absent or has abstained from his wife for more than ten months, the child she gives birth to may not be considered his.

In a dispute [about paternity] where he denies having penetrated her, he has the final word. If he admits to having slept with her but denies paternity of the child, he must necessarily go through a process of *liʿān* [mutual swearing] in court. He is not allowed to accept paternity of a child born out of wedlock as his own.

If a woman remarries after divorcing her first husband, and gives birth to a child within six months, it belongs to the first husband. If six months or more have passed, then it belongs to the new husband. If it has been less than six months since her new marriage but more than ten months since her divorce from the first husband, then it belongs to neither of them. The same applies to a slave-girl who is sold after her master has slept with her.

If a master admits to having fathered the child borne by his slave-girl or with a fixed-term wife, then the child is his, and he cannot deny paternity after that. If a slave-girl is used for sexual intercourse by her master as well as another man, the child born is considered the master's. If the child bears features that preclude it from being the offspring of the master, he is neither allowed to accept paternity over the child nor to deny it. Rather, he is recommended to leave him something by way of inheritance in his will.

If a group of people have intercourse with a slave-girl, and they all shift the responsibility onto each other, the child's father is determined by drawing lots, and the rest of them are responsible for paying equal shares towards her freedom and the freedom of the child on the day that he is born alive.

If a man slept with a woman mistakenly [assuming her to be his wife or concubine], from which a child then ensues, the child is his. If it transpires that she has a husband, when she had previously assumed herself to be widowed or abandoned, she must be returned to him immediately upon separation from the second man, and fulfilling a waiting period.

الفصل التاسع: في أحكام الأولاد

يلحق الولد في الدائم مع الدخول، ومضي ستة أشهر من حين الوطيء ووضعه لمدة الحمل، وهي ستة أشهر إلى عشرة، فلو غاب أو اعتزل أكثر من عشرة أشهر ثم ولدت لم يلحق به.

والقول قوله في عدم الدخول، ولو اعترف به وأنكر الولد لم ينتف إلا باللعان، ولا يجوز له إلحاق ولد الزنا به.

ولو تزوّجت بآخر بعد طلاق الأول وأتت بولد لأقل من ستة أشهر فهو للأول، وإن كان لستة أشهر فصاعداً فهو للأخير، ولو كان لأقل من ستة أشهر من وطيء الثاني أو أكثر من عشرة أشهر من طلاق الأول فليس لهما، وكذا الأمة لو بيعت بعد الوطيء.

ولو اعترف بولد أمته أو المتعة ألحق به، ولا يقبل نفيه بعد ذلك.

ولو وطأها المولى وأجنبي فالولد للمولى، ومع إمارة الانتفاء لا يجوز إلحاقه ولا نفيه، بل يستحب أن يوصي له بشيء.

ولو وطأها المشتركون فتداعوه ألحق بمن تخرجه القرعة، ويغرم للباقين حصصهم من قيمة الأمة وقيمته يوم سقوطه حياً.

ولو وطأ بالشبهة لحق به الولد، فإن كان لها زوج وظنّت خلوها ردت عليه بعد العدّة من الثاني، ويجب عند الولادة استبدار النساء أو الزوج بالمرأة.

During childbirth, the woman must be attended to by other women or by her husband. It is recommended to bathe the new-born baby; to recite the *adhān* in his right ear and the *iqāma* in his left; to anoint him with a bit of soil from the grave of Imam Ḥusayn (peace be upon him) and with water from the river Euphrates; to name him by one of the prophets or Imams' names (peace be upon them) and to give him an agnomen (*kunya*). One should not give the *kunya* Abū l-Qāsim to a child named Muḥammad. It is also recommended to shave the baby's head on the seventh day, to sacrifice a sheep (*'aqīqa*) thereafter, to donate the equivalent amount of gold or silver as the baby's hair to charity, to pierce his ear, to circumcise a boy on the seventh day. Circumcision becomes obligatory upon attaining puberty. Snipping the private parts of baby girls is recommended.

It is recommended to sacrifice a ram for a baby boy, and an ewe for a baby girl, with the same qualities as the sacrificial animal for pilgrimage. The parents must not eat from its meat, and nor must any of its bones be broken.

The best person to breastfeed the child is its mother; if she is a free woman then she is entitled to take a fee from the child's father. If the father dies, then the fee is to be taken from the property of the child. The mother cannot be forced to breastfeed her child, but a slave-girl can.

The term for breastfeeding is two years, or a minimum of twenty-one months. The mother has greater priority to breastfeed her child if she agrees to the same fee as another woman charges or does it voluntarily. She has priority of custody of a male child whilst she is breastfeeding him if she is a free Muslim woman, and a female child until the age of seven. She loses her right to custody if she remarries. If the father dies, is a slave or a disbeliever, then the mother has the right of custody.

ويستحب غسل المولود، والأذان في أذنه اليمنى، والإقامة في اليسرى، وتحنيكه بتربة الحسين عليه السلام وبماء الفرات، وتسميته باسم أحد الأنبياء أو الأئمة عليهم السلام والكنية، ولا يكنّى محمداً بأبي القاسم، وحلق رأسه يوم السابع، والعقيقة بعده، والتصدّق بوزنه ذهباً أو فضةً، وثقب أذنه، وختانه فيه، ويجب بعد البلوغ. وخفض الجواري مستحب.

ويستحب له أن يعق عن الذكر بذكر وعن الأنثى بأنثى، بصفات الأضحية، ولا يأكل الأبوان منها، ولا يكسر شيء من عظامها. وأفضل المراضع الأمّ. وللحرة الأجرة على الأب، ومع موته من مال الرضيع، ولا تجبر على إرضاعه، وتجبر الأمة.

وحدّ الرضاع حولان، وأقله أحد وعشرون شهراً، والأمّ أحق بإرضاعه إذا رضيت بما يطلب غيرها من أجرة أو تبرع وأحق بحضانة الذكر مدة الرضاع إذا كانت حرة مسلمة، وبالأنثى إلى سبع سنين، وتسقط الحضانة لو تزوّجت. ولو مات الأب أو كان مملوكاً أو كافراً فالأمّ أولى.

Section Ten: Maintenance

The wife obligatorily has the right to be maintained, by way of food, clothing and shelter, provided she is in a permanent marriage and makes herself available to her husband whenever she can, and even if she is a *dhimmī* or a slave. If a woman is divorced irrevocably or her husband dies, she is not entitled to any maintenance unless she is pregnant, she is owed back whatever is not paid to her of her maintenance by her husband.

As for other relatives, it is obligatory to maintain one's parents and grandparents up the line, and children and grandchildren down the line, on condition that they are poor and incapable of earning. The father is responsible for providing for his child. If he lacks the means or cannot work, then the paternal grandfather is responsible. If they are both unable, then it is the mother's responsibility, and if she is unable to, then her parents.

As for the slave, it is his master's responsibility to provide for him, and to allow him to sufficiently earn his provision. If he cannot, then the master must top it up. It is obligatory to provide for one's animals and if he refuses to do so he is forced [by the just ruler] to sell them, slaughter them if they are permissible to slaughter, or give them away.

الفصل العاشر: في النفقات

أما الزوجة فيجب لها النفقة: من الاطعام والكسوة والسكنى، مع العقد الدائم والتمكين التام مع القدرة وإن كانت ذمّية أو أمة. فإن طُلّقت بائناً أو مات الزوج فلا نفقة مع عدم الحمل وتقضي مع الفوات.

وأما الأقارب، فيجب للأبوين وإن علوا والأولاد وإن نزلوا خاصة، بشرط الفقر، والعجز عن التكسّب. وعلى الأب نفقة الولد، فإن فقد أو عجز فعلى أب الأب وهكذا، فإن فقدوا فعلى الأمّ، فإن فقدت فآباؤها.

وأما المملوك: فتجب نفقته على مولاه، وله أن يجعلها في كسبه مع الكفاية، وإلا أتمه المولى.

ويجب للبهائم، فإن امتنع أجبر على البيع، أو الذبح إن كانت مذكاة أو الإنفاق.

The Book of Separation

كتاب الفراق

This book contains several sections.

Section One: Divorce

The man divorcing [his wife] must be of mature age, sane, acting of his own volition, and intent upon divorcing his wife. A legal guardian may issue a divorce on behalf of his client if he is a mentally incapacitated person, but not if his client is a minor or in an inebriated state.

The woman being divorced must be a permanent wife and clear of her menses or postpartum bleeding if her husband is present and has had sexual intercourse with her. If he is absent, then she must have experienced one full menstrual cycle for the divorce to be valid, even if she is menstruating at the time. If he divorces her while she is clear of bleeding, he must not have had intercourse with her in that cycle – unless she is a minor [and not yet menstruating], menopausal, pregnant, or one whose status is unsettled [whereby she has constant irregular bleeding] – in which case she must wait three months.

The divorce is complete upon his stating, "You are divorced," devoid of any conditions or descriptions, and it is conditional upon being heard by two just male witnesses.

وفيه فصول

الفصل الأول: في الطلاق

ويشترط في المطلِّق: البلوغ، والعقل، والاختيار، والقصد. وللولي أن يطلّق عن المجنون لا الصغير والسكران.

وفي المطلَّقة: دوام الزوجية، وخلوها عن الحيض والنفاس إن كان حاضراً ودخل بها، ولو كان غائباً يقدر انتقالها من طهر إلى آخر صح طلاقه ولو كانت حائضاً، وأن يطلّقها في طهر لم يقربها فيه بجماع إلا في الصغيرة واليائسة والحامل، والمسترابة تصبر ثلاثة أشهر.

ولا يقع إلا بقوله «طالق» مجرداً عن الشرط والصفة، ويشترط سماع رجلين عدلين.

Section Two: Types of Divorce

They are the innovated forms of divorce and the divorce based on prophetic guidance. The former is the divorcing of a menstruating woman who is not pregnant or a woman who is bleeding postnatally, if her husband is present,[1] or divorcing a woman who bleeds irregularly without waiting three months to do so and pronouncing the divorce three times consecutively. All of these are invalid. The latter – based on prophetic guidance - is further categorised into irrevocable and revocable.

The first [irrevocable divorce] is that of the menopausal woman, the young girl who is not yet menstruating, the woman with whom the marriage has not been consummated, the woman who seeks divorce from her husband against compensation (*khulʿ*), the woman who is divorced by mutual consent (*mubārāt*) as long as they both continue on to dissolution of the marriage, and the woman who is being divorced a third time, having been divorced revocably twice before that.

The second [revocable divorce] includes the rest of women, whereby the man has the right to take her back.

Ṭalāq al-ʿidda is where the man retains the right to take back his wife during her waiting period and sleeps with her, then divorces her again after she is clear of her menses. The maximum number of times the same man may do this is nine times, between which two other men must contract a permanent marriage with her [and divorce her]. Apart from that, she is prohibited to him [upon expiry of the waiting period and] after the third time, until she marries someone else.

The conditions of the *muḥallil*[2] are: that he be an adult, and that he consummates the marriage with her through frontal intercourse after a valid permanent marriage contract. Just as this marriage would nullify the [prohibition resulting from the] three divorces, so does it serve to nullify less than that.

1 That is, not absentee, imprisoned or travelling for some time, and can hence ascertain her being clear of bleeding.

2 *muḥallil*: the man who marries the thrice-divorced woman, and upon whose death, or divorce from him, she once again becomes permissible (*ḥalāl*) for her original husband. To expressly assume the role of a *muḥallil* in marrying women specifically for the purpose of facilitating reconciliation with their ex-husbands is prohibited.

الفصل الثاني: في أقسامه

وهو بدعة، وسنة.

فالأول: طلاق الحائض الحائل أو النفساء مع حضور الزوج والمسترابة قبل ثلاثة أشهر، وطلاق الثلاث مرسلاً والكل باطل.

والثاني بائن، ورجعي.

فالأول: طلاق اليائسة، والصغيرة، وغير المدخول بها والمختلعة والمبارءة مع استمرارهما على البذل، والمطلّقة ثلاثاً بينها رجعتان.

والثاني: ما عداه مما للرجل المراجعة فيه.

وطلاق العدّة من أحد هذه: ما يراجع في العدّة ويواقع ثم يطلّق بعد الطهر فهذه تحرم بعد تسع ينكحها بينها رجلان مؤبداً. وما عداه تحرم في كل ثالثة حتى تنكح غيره.

ويشترط في المحلّل: البلوغ، والوطي قبلاً بالعقد الصحيح الدائم. وكما يهدم الثلاث يهدم ما دونها.

Both a verbal statement as well as a gesture revoking the divorce are valid, and they need not be witnessed. The woman's word in stating completion of her waiting period upon starting her [third] menstrual period is accepted.

It is disapproved for a terminally ill man to divorce his wife, and if he does, then she would still inherit from him, even if it is an irrevocable divorce, up until a year thereafter, as long as he does not die after that. If he dies, even a moment after the year has passed, or she remarries in that time, or he is cured from his illness, then she does not inherit. He inherits from her only in a revocable divorce during the waiting period. His contracting a marriage [whilst terminally ill] is valid if it is consummated, but not otherwise.

ويصح الرجعة نطقاً وفعلاً، ولا يجب فيها الإشهاد. ويقبل قول المرأة في انقضاء العدّة بالحيض.

ويكره طلاق المريض، ويقع، لكن ترثه المرأة - وإن كان بائناً إلى سنة - ما لم يمت بعدها ولو بلحظة أو تتزوّج هي، أو يبرأ من مرضه. وهو يرثها في الرجعى في العدّة، ونكاحه صحيح مع الدخول، وإلا فلا.

Section Three: Waiting Periods

Minor girls under the age of menstruation, menopausal women and those women with whom the marriage has not been consummated do not have to adhere to a waiting period.

The waiting period of women who have regular periods is three menstrual cycles for free women, and two for slave women. A girl who is of the age of menstruation but has not yet started her periods must observe a waiting period of three months if she is free, and one month and a half if not.

The waiting period of a pregnant woman continues until she delivers, even if it is a miscarriage.

The waiting period of a free woman whose husband has died is four months and ten days, regardless of whether she is a minor, menopausal or otherwise, and whether the marriage was consummated or not. If she is pregnant, then it is the longer of the two terms [i.e. until delivery or four months and ten days]. Both types of women must observe a state of mourning.[3] For a slave-woman, the waiting period is two months and five days, or the longer of the two periods if she is pregnant.

The slave-mother who has a child with her master (*umm al-walad*) must observe the waiting period of a free woman upon his death, but others must wait as a slave-woman.

If a slave-woman's husband dies and she is subsequently manumitted, she must observe the waiting period as a free woman. If he manumits his slave-woman after having slept with her, she must observe a waiting period of three menstrual cycles.

If a husband dies after having given his wife a revocable divorce, both the free woman and the slave observe the waiting period of bereavement. If it is an irrevocable divorce, then they observe the waiting period of divorce.

The husband is not allowed to expel his revocably-divorced wife from the marital home until she finishes her waiting period, unless she commits an indecent act; and nor is she allowed to leave the house in the middle of the night unnecessarily, and if she does she must return before dawn. He is obligated to maintain her during this waiting period.

The waiting period of a divorcee commences from when he issues it, and of a widow from when she receives news of his death.

3 Known as *al-ḥadād*, this involves refraining from adorning oneself with colourful clothing, make-up, applying kohl or henna, and anything that women usually beautify themselves with, and this may vary according to era, geographical location, and types of women.

الفصل الثالث: في العدد

لا عدّة في الطلاق على الصغيرة، واليائسة، وغير المدخول بها .

والمستقيمة الحيض عدّتها ثلاثة أقراء إن كانت حرة وإلا فقرءان. وإن كانت في سنّ من تحيض ولا حيض لها فعدّتها ثلاثة أشهر إن كانت حرة، وإلا فشهر ونصف.

والحامل عدّتها وضع الحمل وإن كان سقطاً . وعدّة الحرة المتوفّى عنها زوجها أربعة أشهر وعشرة أيام، صغيرة أو يائسة أو غيرهما ، دخل أو لا ، ولو كانت حاملاً فأبعد الأجلين، وعليهما الحداد ولو كانت أمة فشهران وخمسة أيام، والحامل بأبعد الأجلين.

وأم الولد تعتدّ من وفاة الزوج كالحرة، وغيرها كالأمة. ولو مات زوج الأمة ثم أعتقت اعتدّت كالحرة، ولو أعتق أمته بعد وطيها اعتدّت بثلاثة أقراء .

ولو مات بعد الطلاق رجعياً اعتدّت الحرة والأمة للوفاة، ولو كان بائناً أتمّت عدّة الطلاق.

ولا يجوز للزوج أن يخرج الرجعية من بيت الطلاق حتى يخرج عدّتها إلا أن تأتي بفاحشة، ولا لها أن تخرج إلا مع الضرورة بعد نصف الليل وترجع قبل الفجر، وعليه نفقة عدّتها .

وتعتدّ المطلّقة من وقت إيقاعه، والمتوفّى عنها من حين البلوغ.

Section Four: *Khul'* and *Mubārāt*

Based on one opinion, the divorce initiated by the wife (*khulʿ*) is not effective until it is followed through with the actual divorce (*ṭalāq*). It must be accompanied by payment of compensation, which can be anything ownable, on condition that it is defined and within the wife's remit. The man has the right to take more than what she offers him.

The husband granting the *khulʿ* must be: of age of legal responsibility, acting of his own accord with the intention of separation. The wife – assuming the marriage has been consummated – must be clear of her menses in a cycle whereby they have not had sexual intercourse despite him having been in town; and she must not be pregnant even though she be capable of conceiving a child; specify her reason as being her aversion of him. There must be two just male witnesses present, and the *khulʿ* must be devoid of any conditions originally stipulated in the nuptial contract. If there is no aversion on her part, then it is void.

He must not already be in possession of the compensation, and she has the right to return to the marriage if the *khulʿ* was in lieu of compensation as long as she is still within her waiting period. If she does return [within that time], he must give back the compensation, but if not, then no. They do not inherit from each other while she is in her waiting period. If it turns out that the compensation has not been paid, it is said that the *khulʿ* is void. If a slave-wife has spent her bridal gift with permission, then the *khulʿ* is still in order [without payment], but if it was without consent, then she remains responsible for paying it.

If a Muslim man is paid wine as compensation for *khulʿ*, if he follows through by divorcing her, the divorce remains revocable. If a man grants his wife a *khulʿ* divorce in compensation for a thousand but does not specify the thing of which he is demanding a thousand, then the divorce is void. If he granted the *khulʿ* for vinegar as compensation, but it turned out to be wine, the *khulʿ* remains in order, but he remains entitled to the same amount of vinegar.

الفصل الرابع: في الخلع والمباراة

ولا يقع الخلع بمجرده ما لم يتبع بالطلاق على قول. ولا بدَّ فيه من الفدية، وهي ما يصح تملّكه بشرط التعيين واختيار المرأة. وله أن يأخذ أزيد مما أعطاها.

ويشترط في الخالع: التكليف، والاختيار، والقصد. وفي المرأة مع الدخول الطهر الذي لم يقربها فيه بجماع مع حضوره، وانتفاء الحمل، وإمكان الحيض، واختصاصها بالكراهية، وحضور شاهدين عدلين، وتجريده عن شرط لا يقتضيه العقد. ويبطل لو انتفت الكراهية منها.

ولا يملك الفدية، ولها الرجوع في الفدية ما دامت في العدّة، وإذا رجعت كان له الرجوع في البضع، وإلا فلا. ولا توارث بينهما في العدّة.

ولو بانت الفدية مستحقة - قيل - يبطل الخلع. ولو بذلت الأمة مع الإذن صح، وبدونه تتبع به.

ولو كانت فدية المسلم خمراً فإن أتبع بالطلاق كان رجعياً. ولو خالعها على ألف ولم يعيّن بطل، ولو خالع على خل فبان خمراً صح، وله بقدره خل.

If a man divorces his wife (*ṭalāq*) in exchange for compensation, the divorce is irrevocable, even if the wording of *khulʿ* was not used, provided she initiated it, saying, "Divorce me in exchange for *x* amount", and he responded immediately. If he divorced her later, then there is no compensation due and the divorce is revocable.

The conditions of *mubārāt* (divorce due to mutual aversion) are the same as those for *khulʿ*, except that here there is mutual aversion, and its wording would take the form of: "I have absolved you of *x* amount so you are divorced". It is irrevocable, as long as she does not return to work on the marriage during her waiting period. He is not allowed to give her anything in addition to what [i.e. the bridal gift] he has already given her.

ولو طلّق بفدية كان بائناً وإن تجرّد عن لفظ الخلع، ولو قالت «طلقني بكذا» كان الجواب على الفور، فإن تأخر فلا فدية، وكان رجعياً.

وشروط المباراة كالخلع، إلا أن الكراهية منهما، وصورتها «بارأتك بكذا فأنت طالق» وهي بائن ما لم ترجع في البذل في العدّة، ولا يحل له الزائد على ما أعطاها.

Section Five: *Ẓihār*[4]

This type of divorce is prohibited, and the wording would constitute him saying to his wife, "You are to me as the back of my mother" or any of his other immediate female relatives forbidden to marry.

Its validity requires the following conditions: that it be heard by two just male witnesses, that the husband be of sound mind, issuing the divorce of his own volition and intending it, that it be issued whilst the wife is in a state of cleanliness whereby he has not slept with her that cycle despite having been in town, and that she be of those who normally menstruate. There is a difference of opinion about [its validity with regard to] the temporary wife, the slave wife, the wife with whom he has not consummated the marriage, and about the inclusion of a condition. It cannot take place upon having married a second wife, and nor as a solemn oath.

When he wishes to have intercourse with her again, he must pay a penalty (*kaffāra*), meaning that intercourse with her is forbidden until he pays the penalty, so if he divorces her in this manner, then takes her back during the waiting period, she is not permissible for him until he pays the penalty. If she leaves, or the divorce is irrevocable and they remarried in the waiting period, or one of them passed away or apostatised, then there is no penalty payable. If he intentionally has intercourse with her before paying the penalty, he must may a double penalty, with a penalty added on each time that he has intercourse with her. If he cannot afford to pay it, then seeking forgiveness suffices.

If the wife escalates the matter and takes him to court, the judge may give him three months from the date of the hearing to either pay the penalty or divorce her.

If a man issues his slave wife with a *ẓihār* divorce, then buys her and has intercourse with her consequent to his ownership over her, then there is no penalty due.

4 *ẓihār*: a pre-Islamic form of divorce or separation whereby the husband would liken his wife to his mother or other immediate female relative forbidden to marry.

الفصل الخامس: في الظهار

وهو حرام، وصورته أن يقول لزوجته «أنت عليّ كظهر أمي» أو إحدى المحرمات.

وشرطه: سماع شاهدي عدل، وكمال المظاهر، والاختيار، والقصد، وإيقاعه في طهر لم يجامعها فيه إذا كان حاضراً ومثلها تحيض. وفي المتمتع بها، والأمة، وغير المدخول بها، ومع الشرط قولان ولا يقع في إضرار ولا يمين.

ومع إرادة الوطيء يجب الكفارة، بمعنى تحرم الوطيء حتى يكفّر.

فإن طلّق وراجع في العدّة لم تحل حتى يكفّر، ولو خرجت أو كان بائناً فاستأنف في العدّة أو مات أحدهما أو ارتدّ فلا كفّارة، ولو وطأ قبل التكفير عامداً [لزمته] كفّارتان، ويتكرّر بكل وطيء كفّارة، ولو عجز أجزأه الاستغفار.

وإذا رافعته أنظره الحاكم ثلاثة أشهر من حين المرافعة فيضيق عليه بعدها حتى يكفّر أو يطلّق.

ولو ظاهر زوجته الأمة ثم اشتراها ووطأها بالملك فلا كفّارة.

Section Six: *Īlāʾ*[5]

Īlāʾ can only have effect when sworn in the name of God, and for a reason other than having taken a second wife, by a man of sound mind doing it of his own volition and intention, even he if he is a slave, castrated or a eunuch.

The woman must be a permanent wife with whom the marriage has been consummated. He may either swear the *īlāʾ* indefinitely or for a specific duration more than four months.

If she takes him to court for this, the judge may give him four months' respite [to either go back to her or to divorce her properly]. If he decides to go back, he must pay the penalty[6] and return to her. If he does not do so in that time, then the judge gives him a choice of either divorcing her or accepting God's law by paying the penalty, and imprisons him without access to food and drink until he chooses one of the two options. The divorce issued as a result is revocable.

If he swore an *īlāʾ* for a specific duration and held out from having intercourse with her until expiry, he need not pay a penalty. He must, however, pay the penalty if he had intercourse with her before the stipulated time. In a dispute about whether he has had intercourse with her or not, he has the last word when accompanied by an oath of honesty.

Violation of the oath constitutes frontal sexual intercourse by an able man, and in an impotent man, the display of a resolve to have sexual intercourse were he able to do so. The penalty is not charged repeatedly upon repetition of the oath.

5 *īlāʾ*: a form of divorce or separation whereby a man takes a solemn vow not to have sexual relations with his wife for more than four months.

6 For having violated a solemn oath.

الفصل السادس: في الايلاء

ولا ينعقد بغير اسم الله تعالى، ولا لغير إضرار من كامل مختار قاصد، وإن كان عبداً أو خصياً أو مجبوباً.

ولا بدَّ أن تكون المرأة منكوحة بالدائم، مدخولاً بها، يولي مطلقاً أو أزيد من أربعة أشهر.

وإذا رافعته أنظره الحاكم أربعة أشهر، فإن رجع وكفّر وإلا ألزمه الطلاق أو الفيئة والتكفير، ويضيق عليه في المطعم والمشرب حتى يفعل أحدهما. ويقع الطلاق رجعياً.

ولو آلى مدة فدافع حتى خرجت فلا كفّارة، وعليه الكفّارة لو وطأ قبلـ[ـها] ولو ادّعى الإصابة فالقول قوله مع يمينه.

وفئة القادر الوطيء قبلاً، وفئة العاجز إظهار العزم على الوطيء مع القدرة. ولا يتكرّر الكفّارة بتكرّر اليمين.

Section Seven: *Li'ān*[7]

Its grounds: Accusation of adultery against the wife by the husband claiming to have witnessed it, but with no other evidence; and the denial of paternity to a child outwardly purported to be his.

Both the husband and the wife who are cursing each other must be: of legal age of responsibility and in a permanent marriage. The wife must not be deaf or mute. There is a difference of opinion regarding consummation of the marriage.

It takes the form of the husband saying four times: "I bear witness by God that I am telling the truth in what I have stated about this woman." Then the judge admonishes him, and if he retracts his statement, he sentences him to the prescribed fixed punishment of someone who has made a false testimony (*hadd al-qadhf*). If not, then he says, "May God's curse be upon the liars." Then the wife must repeat four times, "I bear witness by God that he is a liar." The judge then admonishes her, and if she admits to what she is being charged with, he sentences her to stoning; and if not, he says, "May God's wrath overtake her if he is telling the truth." She then becomes permanently forbidden to him.

The testimony must be uttered verbally whilst they are both standing, beginning with the man, and the woman must be identified. It must be uttered in Arabic if possible, but other languages are allowed if not. In the case of the man, the testimonies precede and are then followed by the invocation of God's curse on him, and for the woman, the testimonies precede followed by the invocation of God's wrath on her.

It is recommended for the judge to sit with his back to the *qibla* with the man standing to his right and the woman to his left [facing the judge]; for whoever wishes to hear the *li'ān* to be present; and for the judge to admonish them before the invocation of curse and wrath.

If the husband does admit to having lied after the *li'ān*, he is sentenced to the punishment for false accusation [of a chaste woman], and she remains forbidden to him. The child would inherit from him upon his acknowledgement that it is his after the *li'ān*, but the father and his relatives would not inherit from him in turn.

7 *li'ān*: divorce on the grounds of adultery denied by the wife.

الفصل السابع: في اللعان

وسببه: قذف الزوجة بالزنا مع ادّعاء المشاهدة وعدم البينة وإنكار ولد يلحق به ظاهراً.

ويشترط في الملاعن والملاعنة: التكليف، وسلامة المرأة من الصمم والخرس، ودوام النكاح. وفي اشتراط الدخول قولان.

وصورته: أن يقول الرجل «أشهد بالله أني لمن الصادقين فيما قلته عن هذه المرأة» أربع مرات. ثم يعظه الحاكم فإن رجع حدّه، وإلا قال «إنّ لعنة الله عليه إن كان من الكاذبين» ثم تقول المرأة أربع مرّات «أشهد بالله أنه لمن الكاذبين»، ثم يعظها الحاكم، فإن اعترفت رجمها وإلا قالت «إن غضب الله عليها إن كان من الصادقين» فتحرم أبداً.

ويجب: التلفظ بالشهادة، وقيامها عند التلفظ، وبداءة الرجل، وتعيين المرأة، والنطق بالعربية مع القدرة، ويجوز غيرها مع العذر، والبدأة بالشهادات ثم باللعن في الرجل، وفي المرأة تبدأ بالشهادات ثم بالغضب.

ويستحب: جلوس الحاكم مستدبر القبلة، ووقوف الرجل عن يمينه والمرأة عن يساره، وحضور من يسمع اللعان، والوعظ قبل اللعن والغضب.

ولو أكذب نفسه بعد اللعان حدّ للقذف، ولم يزل التحريم، ويرثه الولد مع اعترافه بعد اللعان، ولا يرثه الأب ولا من يتقرب به.

If the wife admits to the charge after the pronouncement of the *liʿān* four times, she must be punished according to one opinion. If the divorced woman claims to be pregnant with his child but he denies having had intercourse with her, and she provides evidence of them having been together in private, the most appropriate solution is for the *liʿān* to be suspended until consummation can be ascertained.

ولو اعترفت المرأة بعد اللعان أربعاً ، قيل تحدّ . ولو ادّعت المرأة المطلقة الحمل منه فأنكر الدخول فأقامت بينة بإرخاء الستر ، فالأقرب سقوط اللعان ما لم يثبت الدخول .

The Book of Manumission

كتاب العتق

This book contains several sections.

Section One: Enslavement

Enslavement is restricted to captives of war, or *dhimmīs* who violate the conditions [of their *dhimmī* status]. A person who admits to being a slave is is treated as such if his admission is not under duress, but a person's claim to freedom is not accepted if he is being sold in the marketplace unless he has evidence.

A man or woman cannot enslave either of their parents or grandparents, nor their own children or grandchildren; and neither can a man enslave his immediate female relatives who are forbidden to him to marry. If he does own any of them, he must manumit them. The same rules apply to those who are related through milk as through blood.

وفيه فصول:

الفصل الأول: في الرقّ

يختصّ الرق بأهل الحرب، أو بأهل الذمة إن أخلوا بالشرائط.

ويحكم على المقر بالرقية مختاراً، ولا يقبل قول مدعي الحرية إذا كان يباع في الأسواق إلا ببينة.

ولا يملك الرجل ولا المرأة أحد الأبوين وإن علوا، والأولاد وإن نزلوا، ولا [يملك] الرجل بالمحارم بالنسب من النساء، ولو ملك أحد هؤلاء عتق. وحكم الرضاع حكم النسب.

Section Two: Manumission

One must utter the statement by which he frees his slave explicitly by saying "You are free". As for the utterance "You are manumitted" (*lafẓ al-ʿitq*) there is a difference of opinion. Nevertheless, the manumission of a slave will not occur unless one of these utterances has been made, nor will it take place through indication or writing when the master is capable of uttering the declaration of his slave's freedom. It cannot be conditional nor part of a vow, but if one was to stipulate an act of service or the like alongside the manumission, he is allowed to do so.

The manumitter must be of age of legal responsibility, acting of his own volition with the intention of seeking proximity to God, and the slave must be a Muslim. Freeing a slave who is a non-Twelver Imami [i.e. Sunnis, Ismailis, Waqifis etc.] is disapproved, but if one made a solemn vow to free him [a Sunni, for example] or a disbeliever, it remains valid. It is recommended to free a slave who has spent seven years under one's mastership.

If one makes a solemn vow to free every "old" slave of his, he must free every slave that he has owned for more than six months. If he vows to free the first slave that he ever owned, but he came to own them as a group, the slave must be chosen by drawing lots.

The slave cannot own anything, even if his master gives him possession over certain things, so when he manumits him and he has any property in his hand, it still belongs to the master, even if he knows about it and does not take it away from him.

If a man commits to manumitting "a third" of his slaves, he must select them randomly. If he commits to manumitting "a part" of his slave, he must manumit all of him. If he owned them jointly with a partner, he is responsible for manumitting his partner's share of slaves. If he cannot afford to do so, he may employ the slave to work for a share of his freedom.

If someone manumits a pregnant slave-woman, her unborn child is not automatically freed unless he contractually specifies its manumission.

A slave's blindness, leprosy, disability and his master's torture of him are all grounds for manumitting him, as is his embracing of Islam and his relinquishment of another creed before his previous masters.

If a wealthy man dies, leaving behind a sole heir who is a slave, he must be bought from his master, manumitted and given the remainder of the inheritance.

الفصل الثاني: في العتق

والصريح «أنت حر» ، وفي لفظ العتق إشكال، ولا يقع بغيرهما ، ولا بالإشارة والكتابة مع القدرة، ولا يقع مشروطاً ولا في يمين ولو شرط مع العتق شيئاً من خدمة وغيرها جاز.

وشرطه: تكليف المعتق والاختيار، والقصد ، والقربة، وإسلام العبد. ويكره [اعتاق] المخالف. ولو نذر عتقه أو عتق الكافر صح.

ويستحب أن يعتق من مضى [له] في ملكه سبع سنين.

ولو نذر عتق كل عبد له قديم عتق من ملكه ستة أشهر فصاعداً ، ولو نذر عتق أول مملوك يملكه فملك جماعة استخرج بالقرعة - على خلاف.

والعبد لا يملك شيئاً وإن ملكه مولاه على الأقوى، فلو أعتقه وبيده مال فالمال للمولى وإن علم به ولم يستثنه.

ولو أعتق ثلث عبيده استخرج بالقرعة. ولو أعتق بعض عبده عتق كله، ولو كان له شريك قوم عليه حصة شريكه واعتقت، ولو كان معسراً سعى العبد في النصيب.

ولو أعتق الحبلى فالوجه: عدم عتق الحمل إلا أن يعتقه بالنصوصية.

وعمى المملوك، وجذامه، وتنكيل المولى به، والإقعاد: أسباب في العتق، وكذا إسلام العبد وخروجه قبل مواليه.

ولو مات ذو المال وله وارث مملوك لا غير اشتري من مولاه وأعتق وأعطي الباقي.

Section Three: Manumission Upon the Master's Death

This is when a man of sound mind, intentionally says to his slave, "You are my slave during my lifetime and freed upon my death." Thereupon, the slave is manumitted from the one-third of the estate like the bequest. He may retract his declaration whenever he wishes. Manumission is to be implemented after payment of [the master's] debts.

If a pregnant slave-woman is manumitted upon her master's death, her freedom is specific to her and not her child. If, however, the pregnancy resulted from another slave after the master's declaration, then the child is manumitted as well.

If a person retracts his declaration of future manumission for his slave who is a mother, it is said that his retraction of her manumission would not automatically apply to her children's manumission being retracted. The more accurate opinion is that his retraction of the mother's manumission [after his death] is specific to her, and therefore does not apply to her children. If he were to retract his future manumission of them both, it would be valid.

The male child of a male slave [who is to be manumitted upon the death of his master] born from a female slave is also manumitted through his father's manumission, and his future manumission is not invalidated upon the death of his father preceding the death of his master. Slaves are to be manumitted from the one-third of the bequest, and if it falls short, they may be employed for their freedom.

A slave's attempted escape invalidates the declaration of future manumission.

الفصل الثالث: التدبير

وهو أن يقول «أنت رق في حياتي حر بعد وفاتي»، من الكامل القاصد، فينعتق من الثلث بعد الوفاة كالوصية، وله الرجوع متى شاء، وهو متأخر عن الدين.

ولو دبّر الحبلى اختصت بالتدبير دون الحمل، أما لو تجدد الحمل من مملوك بعد التدبير فإنه يكون مدبراً.

ولو رجع في تدبير الأم قيل: لا يصح رجوعه في تدبير الأولاد، والأقرب أن رجوعه في تدبير الأم خاصة ليس رجوعاً في تدبير الأولاد، ولو رجع في تدبيرهما صح الرجوع.

وولد المدبّر من مملوكه مدبّر، ولا يبطل تدبير الولد بموت أبيه قبل مولاه، وينعتقون من الثلث، فإن عجز استسعوا.

وإباق المدبّر إبطال للتدبير.

Section Four: Contractual Manumission Between Master and Slave

This is of two types: unrestricted and conditional.

The unrestricted contract is for the master to tell his slave, man or woman, "I contract your manumission upon *x* amount, payable at such and such time." It can either be paid in one instalment or several. The latter would say, "I accept." It is said that this type of agreement is contingent upon the master saying, "When you have paid, you are free" so the slave is freed in proportion to the amount that he pays. The master cannot cancel the contract, even if the slave is unable to make the payments. The Imam is obligated to bail him out from the share of the treasury designated for emancipation of slaves if he is unable to make his payments.[1]

If he fathers children with a slave-woman, his children are free in proportion to his own freedom. If he dies without having paid off any part of his freedom, his master inherits everything that he leaves behind. If he had managed to buy a proportion of his freedom, his master receives an amount from his inheritance proportionate to his bondage, and the rest goes to his heirs, who then remit the remainder of the amount stipulated in the contract.

If there is no money to pay the remainder, his children may work to pay the remainder of their father's manumission fee. Upon fulfilment, they too are manumitted, and they inherit in proportion to their freedom.

If a *mukātab* slave makes a bequest or is left something in a bequest, the amount is determined in proportion to his freedom, and similarly, if he is liable for a prescribed punishment, [he is sentenced in proportion to his freedom], as would be the case if the master had intercourse with his partly-freed slave-woman [in an unrestricted contract], he would be punishable in proportion to her freedom.

As for the conditional contract, this is when it includes a statement to the effect of: "If you are unable to make the payments, you will return to being a slave." In this situation, the installments do not contribute towards partial freedom, and he is only manumitted when the full amount has been paid. If he is unable to make the payments – at most a full instalment late – he goes back to being a slave, though it is recommended for the master to be patient with him [especially when the payment is delayed toward the end of the contractual period.]

1 Such monies are collected from the *zakāt* fund and specifically allocated for the emancipation of slaves. See Chapter 3: Those Entitled to Receive *Zakāt*.

الفصل الرابع: في الكتابة

وهي قسمان: مُطلَقة، ومشروطة.

فالمطلقة: أن يقول لعبده أو أمته «كاتبتك على كذا على أن تؤديه في نجم كذا» أما في نجم واحد أو نجوم متعدّدة، فيقول «قبلت». وقيل: يفتقر إلى قول: «فإذا أديت فأنت حر» فهذا يتحرر منه بقدر ما يؤدّي، وليس لمولاه فسخ الكتابة وإن عجز، ويفكه الإمام من سهم الرقاب وجوباً مع العجز.

فإن أولد من مملوكة تحرّر من أولاده بقدر ما فيه من الحرية، وإن مات ولم يتحرّر منه شيء كان ميراثه للمولى، وإن تحرّر منه شيء كان لمولاه من ماله بقدر الرقية، ولورثته بقدر الباقي، ويؤدون منه ما بقي من مال الكتابة.

ولو لم يكن مال سعى الأولاد فيما بقي على أبيهم، ومع الأداء ينعتق الأولاد ويرث بقدر نصيب الحرية.

ولو أوصى أو أوصي له بشيء صح بقدر نصيب الحرية، وكذا لو وجب عليه حد، ولو وطأ المولى المطلَقة حد بنصيب الحرية.

وأما المشروطة: فأن يقول بعد ذلك «فإن عجز فأنت رد في الرق». وهذا لا يتحرّر منه شيء إلا بأداء جميع ما عليه، فإن عجز وحده أن يؤخر نجماً عن وقته ردّ في الرق، ويستحب للمولى الصبر عليه.

The manumission price must mandatorily be a fixed debt to be paid by means of legitimate assets. The master is discouraged from fixing a price higher than the market value [of the slave].

The conditional contract of manumission is void upon the death of the *mukātab* slave, and his wealth and children all belong to his master. The *mukātab* slave does not have the right of disposal over his property that he has not earned himself, except with his master's consent, and the master's right of disposal over his property may cease through arrangements other than the full remittance [of his manumission price].

If the master compels his *mukātab* slave-woman to have sexual intercourse with him, she is entitled to a bridal gift. She does not have the right to get married without the consent of her master. If her children are not already free, then the terms of her contract apply equally to them and they are manumitted as and when she is, whether the contract is unrestricted or conditional.

The children of slaves who have been partially manumitted through an unrestricted contract are also manumitted proportionally. Their earnings are their own to keep if they are manumitted but belong to the master if they are still in bondage.

If a *mukātab* mother is at risk of missing her payments and her master threatens to cancel the contract, she may resort to using her children's money, and God knows best the correct course of action.

ولا بدَّ في العوض من كونه ديناً مؤجلاً معلوماً مما يصح تملّكه. ويكره أن يتجاوز به القيمة.

وإذا مات المشروط بطلت الكتابة، وكان ماله وأولاده لمولاه. وليس للمكاتب أن يتصرّف في ماله بغير الاكتساب إلا بإذن المولى، وينقطع تصرف المولى عن ماله بغير الاستيفاء.

ولو وطأ مكاتبته مكرهاً فلها المهر، وليس لها أن تتزوّج بدون إذن المولى، وأولادها بعد الكتابة إذا لم يكونوا أحراراً حكمهم حكمها ينعتقون بعتقها مشروطة كانت أو مطلقة.

ولو انعتق من المطَلقة بعضها انعتق من الولد بقدره، وكسبهم إن عتقوا فلهم وإن رقوا فللمولى.

ولو أشرفت الأم على العجز وهمّ المولى بالفسخ استعانت به. والله أعلم بالصواب.

The Book of Oaths

كتاب الأيمان

This book contains several sections.

Section One: Oaths

An oath cannot be taken without invoking the Names of God, most High, neither can one take an oath by disassociating themselves from God or any one of the Prophets or Imams (peace be upon them).[1]

The conditions necessary for the one swearing the oath are: legal responsibility, intention and volition. The oath sworn by a disbeliever is valid.

Oaths are only binding in relation to acts that are obligatory or recommended, or in respect of a neutral act whose performance is considered more preferable or given equal importance, or to abandon a forbidden, discouraged or neutral act, or in respect of a neutral act whose non-performance is considered more preferable.

If the action connected to the oath and its non-fulfilment hold equal importance, both religiously as well as in one's worldly life, it is obligatory to act as per the dictates of the oath.

An oath cannot be sworn to dictate anyone else's actions, nor actions in the past and nor for impossible actions.

If a previously possible action turns out to be impossible to fulfil, the oath ceases to be binding. One is allowed to swear an oath contrary to the truth if it guarantees a favourable outcome if he knows of a way to dissimulate the truth.

If the person swearing an oath makes it conditional upon the will of God (by saying *inshā'Allāh*), the oath is rendered void. The father, husband and master have the right to nullify an oath made by their child, wife or slave respectively for any non-obligatory action.

Payment of a penalty becomes obligatory if one neglects to do what is mandatorily binding upon him through the oath, or by doing what he swore he would not do and which became mandatory for him due to the oath, unless the oath was sworn knowingly upon a lie [in which case there is no penalty for desisting from it].

One is not allowed to swear an oath [attesting to something] without knowing [the facts].

The oath becomes binding by saying words to the effect of: "By God, I will do *x* thing" or "By God's eternity..." or "I swear by God..." or "I swear by the Lord of the Book..." – but not "By God's right".

1 For example, if one were to take an oath by saying, "If I do not perform such and such action, then I will disassociate myself from God and His Messengers." This would not count as a legitimate oath. In fact, many scholars including al-ʿAllāma al-Ḥillī consider this type of oath impermissible in the Sharia.

وفيه فصول:

الفصل الأول: في اليمين

لا ينعقد اليمين بغير أسماء الله تعالى، ولا بالبراءة منه أو من أحد الأنبياء أو الأئمة عليهم السلام.

ويشترط في الحالف: التكليف، والقصد، والاختيار. ويصح من الكافر.

وإنما ينعقد على فعل الواجب أو المندوب أو المباح مع الأولوية أو التساوي أو ترك الحرام أو ترك المكروه أو ترك المباح مع الأولوية.

ولو تساوى متعلق اليمين وعدمه في الدين والدنيا وجب العمل بمقتضى اليمين.

ولا يتعلّق بفعل الغير، ولا بالماضي ولا بالمستحيل.

ولو تجدّد العجز عن الممكن انحلّت اليمين. ويجوز أن يحلف على خلاف الواقع مع تضمّن المصلحة والتورية إن عرفها.

ولو استثنى بالمشيئة انحلّت اليمين. وللوالد والزوج والمولى حل يمين الولد والزوجة والعبد في غير الواجب.

وإنما تجب الكفّارة بترك ما يجب فعله أو فعل ما يجب تركه باليمين، لا بالغموس.

ولا يجوز أن يحلف إلا مع العلم.

وينعقد لو قال: والله لأفعلنّ، أو بالله، أو تالله، أو أيم الله، أو لعمر الله، أو أقسم بالله، أو أحلف بربّ المصحف. دون: وحق الله.

Section Two: Solemn Vows and Promises

The conditions necessary of a person making a solemn vow are: legal responsibility, volition, intention and Islam. Additionally, for non-obligatory acts, the wife and the slave require the consent of the husband and master respectively.

It can either be to do a good deed, such as saying for example, "If I am granted a child, then I will take upon myself to do *x* for God", or by way of gratitude, such as, "If this sick person is cured, then I will take upon myself to do *x* for God", or by way of restraint, such as, "If I do this forbidden act, then I will take upon myself to do *x* for God" or "If I do not perform this obligatory act, then I will take upon myself to do *x* for God". If he takes something upon himself as a vow but does not specify that it is for God, then it is not binding.

The dictate of the vow should be an act of obedience for God that is within the person's capacity to perform. If he simply vows to perform "an act of obedience" without specifying what, he must either donate something to charity, perform a two-unit prayer or fast for a day.

If he vows to fast for "an undefined period of time", then he must fast for six months, and if he says, "for some time", then it is five months. If he vows to give "a lot of money" in charity, then he should give eighty dirhams. If he is unable to give the full amount at once, then he may donate it little by little until he has fulfilled it. There is no time constraint if the vow was unconditional, but if he had defined it by a certain time or place, then he must adhere to it.

If he makes a vow to fast on a specific day, and it coincides with a journey that he must make, he may not fast on that day and instead fast in lieu later. The same applies if a woman is menstruating or bleeding postnatally on that day. If it coincides with an Eid, he must not fast and instead fast in lieu later. The same applies if he is incapable of fasting that day.

A Promise is where one would say, "I have made a promise to God" or "I promise God that when *x* happens, I will do *y*". This is binding and the same rulings as the oath apply to it.

Vows and promises can only be sworn through verbal utterance.

If one designates his animal, his slave or slave-girl as an offering to God's Sacred House [or to any other sacred place of visitation], it must be sold and its value spent for the benefit of the intended Sacred House or place of visitation, and as provision for the pilgrims and visitors thereof.

الفصل الثاني: في النذر والعهود

ويشترط في الناذر: التكليف، والاختيار، والقصد، والإسلام، وإذن الزوج والمولى في الزوجة والعبد في غير الواجب.

وهو أما برّ كقوله «إن رزقت ولداً فللّهِ عليّ كذا»، أو شكر كقوله «إن برئ المريض فللّهِ عليَّ كذا» أو زجر كقوله «إن فعلت محرما فللّهِ عليَّ كذا»، أو «إن لم أفعل الطاعة فللّهِ عليَّ كذا»، أو تبرّع كقوله «لله عليَّ كذا». ولو قال «عليَّ» ولم يقل «لله» لم يجب.

ومتعلّق النذر يجب أن يكون طاعة لله مقدوراً للناذر، ولو نذر فعل طاعة ولم يعيّن تصدّق بشيء أو صلّى ركعتين أو صام يوماً.

ولو نذر صوم حين كان عليه ستة أشهر، ولو قال زماناً فخمسة.

ولو نذر الصدقة بمال كثير فثمانون درهماً ولو عجز ناذر الصدقة بماله قومه وتصدّق شيئاً فشيئاً حتى يوفّى، ومع الإطلاق لا يتقيد بوقت، ولو قيده بوقت أو مكان لزم.

ولو نذر صوم يوم بعينه فاتّفق له السفر أفطر وقضاه، وكذا لو حاضت المرأة أو نفست، ولو كان عيداً أفطر ولا قضاء، وكذا لو عجز عن صومه.

والعهد: أن يقول «عاهدت الله»، أو «عليَّ عهد الله أنه متى كان كذا فعليَّ كذا». وهو لازم وحكمه حكم اليمين.

ولا ينعقد النذر والعهد إلا باللفظ.

ولو جعل دابته أو عبده أو جاريته هدياً لبيت الله تعالى [أو أحد المشاهد] بيع وصرف ثمنه في مصالح البيت أو المشهد الذي جعل له، وفي معونة الحاج والزائرين.

Section Three: Penalties

These are either (fixed) gradational, choice-based, a combination of these two, or all penalties charged together.

The gradational penalty is charged for *ẓihār* divorce and accidental murder. In both cases, it is to free a slave in expiation. If one is unable to do that, then he must fast for two months consecutively; and if he cannot do that either, then he must feed sixty poor people. The penalty for breaking a fast in lieu of a Ramadan fast after midday is to feed ten poor people. If unable to do so, he can fast for three days consecutively instead.

The choice-based penalty is charged to someone who breaks a fast during the day in the month of Ramadan, or breaks a defined solemn vow, or, according to one opinion, breaks any solemn promise or vow. He has the choice to either free a slave, or fast for two months consecutively, or feed sixty poor people.

There are two instances where both the above penalty types apply: the breaking of an oath, in which case he must free a slave or feed ten poor people or clothe them. If he cannot afford to do any of them, then he may fast for three days consecutively. The same penalty applies to divorce by *īlāʾ*.

The crime for which all penalties are combined is the intentional and wrongful murder of a believer, in which case one must free a slave, fast for two months consecutively and feed sixty poor people.

It has been said that whoever swears an oath disavowing God must pay the penalty of *ẓihār*, and if he cannot do so, then the penalty for breaking an oath. If a woman cuts off her hair in her grief over a deceased person, she must pay the same penalty as a Ramadan fast missed. And if she pulls it out or scratches her face, or if a man tears his clothes over the bereavement of his wife or son, then the same penalty as breaking an oath is charged.

If a man marries a woman in her waiting period, he must separate from her immediately and pay a penalty fine of five *ṣāʿ*[2] of flour. He who misses the *ʿishāʾ* prayer (the last prayer of the day) by oversleeping past its time limit must fast from when he wakes up [in expiation]. One who is unable to fast on the day that he had vowed to do so must donate two *mudd* of food in charity to a poor person.

2 One *ṣāʿ* is equivalent to 3,700 grams.

الفصل الثالث: في الكفارات

وهي: مُرتّبة، ومُخيّرة، وما يجتمع فيه الأمران، وكفّارة الجمع.

فالمُرتّبة: كفّارة الظهار، وقتل الخطأ. ويجب فيهما عتق رقبة، فإن عجز صام شهرين متتابعين، فإن عجز أطعم ستين مسكيناً. وكفّارة من أفطر يوماً من قضاء شهر رمضان بعد الزوال: إطعام عشرة مساكين، فإن عجز صام ثلاثة أيام متتابعات.

والمُخيّرة: كفّارة من أفطر يوماً من شهر رمضان، أو من نذر معين، أو خالف عهداً أو نذراً على قول. وهي: عتق رقبة، أو صيام شهرين متتابعين، أو إطعام ستين مسكيناً.

وما يجتمع فيه الأمران: كفّارة اليمين: عتق رقبة، أو إطعام عشرة مساكين، أو كسوتهم، فإن عجز صام ثلاثة أيام متتابعات. وكذا الايلاء.

وكفّارة الجمع: في قتل المؤمن عمداً ظلماً: عتق رقبة، وصيام شهرين متتابعين، وإطعام ستين مسكيناً.

وقيل: من حلف بالبراءة فعليه كفّارة ظهار، فإن عجز فكفّارة اليمين. وفي جز المرأة شعرها في المصاب كفّارة رمضان، وفي نتفه أو خدش وجهها أو شق الرجل ثوبه في موت ولده أو زوجته: كفّارة يمين.

ولو تزوّج امرأة في عدّتها فارقها وكفّر بخمسة أصوع من دقيق.

ولو نام عن العشاء الآخرة حتى خرج الوقت أصبح صائماً.

ولو عجز عن صوم يوم نذره تصدّق بمدين على مسكين.

Other Rulings:

1) He who has sufficient funds to free a slave or can raise the funds by selling something must do so. The slave must be a believing Muslim, and can be a runaway slave, a slave-mother, or a *mudabbar* slave.

2) One who cannot find a slave to free or finds one but does not have the funds to free him, can then have the option of the next penalty in the gradational sequence. The clothes on one's body, one's servant and one's home cannot be sold to pay a penalty.

3) The penalty charged to a slave in the case of *ẓihār* and accidental murder – pertaining to the fasting option – is half that of the free man.

4) If he is unable to fast in the gradational penalty system, then it is obligatory for him to feed two *mudd* of food to every poor person. If he is unable to find the stipulated number of people, then he may feed the same people numerous times, and it must constitute their main meal. It is recommended to provide bread. The best food to give alongside it is meat, or vinegar as a moderate option, and salt as the lowest. One cannot feed only children unless they are included with the adults, in which case two children are counted as one adult.

5) Clothing the poor means to provide two garments if one can afford it, but if not, then one.

6) The person liable to pay the penalty must make his intention to do so with the proximity of God, and define the penalty. He must be a legally responsible Muslim.

مسائل:

الأولى: من وجد الثمن وأمكنه الشراء فقد وجد الرقبة، ويشترط فيها الإيمان، ويجزئ الآبق، وأم الولد، والمدبّر.

الثانية: من لم يجد الرقبة، أو وجدها ولم يجد الثمن انتقل إلى الصوم في المرتّبة، ولا يباع ثياب بدنه ولا خادمه ولا مسكنه.

الثالثة: كفّارة العبد في الظهار وقتل الخطأ - في الصوم - نصف كفّارة الحر.

الرابعة: إذا عجز عن الصيام في المرتّبة وجب الإطعام لكل مسكين مدّ من طعام، ولو تعذّر العدد جاز التكرار، ويطعم غالب قوته. ويستحب الإدام، وأعلاه اللحم، وأوسطه الخل، وأدناه الملح. ولا يجوز إطعام الصغار إلا منضمّين إلى الرجال، وإن انفردوا احتسب الاثنان بواحد.

الخامسة: الكسوة لكل فقير ثوبان مع القدرة، وإلا فواحد.

السادسة: لا بدَّ من نية القربة والتعيين، والتكليف في المكفّر، وإسلامه.

The Book of Hunting and Related Matters

كتاب الصيد وتوابعه

This book contains several sections.

Section One: Hunted Animals Fit for Consumption

Animals may be hunted by two means: hunting dogs and arrows.

If a hunting dog kills some game, whereof it has been prevented from partaking, it is permissible to consume it provided six conditions are met: a) that the dog has been trained to hunt whereby it goes when its owner dispatches him and stays back when its owner restrains him, b) that the dog is not accustomed to consuming that which it is sent to hunt, and may only do this on rare occasions, c) that the person dispatching the dog is a Muslim or one treated as such, d) that he dispatches the dog with the intention [of hunting game for consumption], e) that he pronounces God's name when he dispatches it, and f) that he does not lose sight of the game whilst it is alive.

If he forgets to pronounce God's name, believing it to be obligatory, it is still permissible to consume. If someone other than the dispatcher pronounced it, it is not permissible. It is similarly impermissible if the dog of a disbeliever assisted his dog even if God's name was pronounced, or if it was dispatched by someone who did not pronounce God's name or did not make the intention.

Hunting by arrows includes other weapons such as swords, spears, and crossbows as long as they tear through the animal. Anything killed by such a weapon is permissible for consumption if the hunter has pronounced the name of God and is a Muslim or treated as such. If he shoots a metal-pointed arrow and accidentally kills an animal that is unable to escape,[1] it is allowed to be consumed. If an arrow or a dog kill a chick, it is not permissible for consumption.

1 For example, when an animal enters a state of shock or is rooted to the ground.

وفيه فصول:

الفصل الأول: فيما يؤكل صيده

وهو أمران: الكلب، والسهم.

أما الكلب: فإذا قتل صيداً وهو الممتنع حل أكله بشروط ستة: أن يكون الكلب معلماً يسترسل إذا أرسله وينزجر إذا زجره، وأن لا يعتاد أكل ما يصيده وإلا اعتبار بالنادر، وأن يكون المرسِل مسلماً أو في حكمه، قاصداً لإرسال الكلب، وأن يسمّي عند إرساله، وأن لا يغيب عن العين حياً.

فلو نسي التسمية - وكان يعتقد وجوبها - حلّ الأكل. ولو سمّى غير المرسِل لم يحلّ. وكذا لا يحلّ لو شاركه كلب الكافر إن سمّى، أو من لم يسمّ، أو لم يقصد.

وأما السهم: فيدخل فيه السيف والرمح والسهم والمعراض إذا خرق، فيؤكل ما يقتله أحدها إذا سّمى المرسِل وكان مسلماً أو بحكمه، ولو قتل ما فيه حديدة معترضاً حلّ، ولو قتل السهم أو الكلب فرخاً لم يحلّ.

If he shoots it with an arrow but it falls off a cliff or in water and then dies, it is not allowed to be consumed. If a sword splits an animal into two, both parts are permissible if they are either both still moving or both not. If only one part is still moving in a manner whereby it may remain alive, it must be slaughtered before it is permissible to consume [and the other immobile part is considered carrion], but if both parts are immobile, then they are both permissible together. If a net cuts off a part of the animal, then that part is considered carrion.

If he aims at some game but hits another animal instead, it is permissible to consume. If he aimed at it without the intention to hunt it [for consumption] then it is not permissible.

Other means used for hunting such as panthers, nets or other such things do not make the animal permissible for consumption until it has been ascertained as fit for consumption, i.e. still alive, then slaughtered for permissible consumption.

ولو رماه بسهم فتردّى من جبل أو وقع في الماء فمات لم يحلّ، ولو قدّه السيف بنصفين حلاّ إن تحرّكا أو لم يتحرّكا ، ولو تحرّك أحدهما حركة ما حياته مستقرّة حل بعد التذكية خاصة، وإلا حلاّ معاً . ولو قطعت الحبالة بعضه فهو ميتة.

ولو رمى صيداً فأصاب غيره حلّ، ولو رماه لا للصيد فأصاب لم يحلّ.

وباقي آلات الصيد كالفهود والحبالة وغيرهما لا يحلّ ما لم يدرك ذكاته - وهو المستقرّ حياته - ويذكيه.

Section Two: Slaughtering for Consumption

The person slaughtering must be a Muslim or treated as such, even if he is someone who denies the leadership of the Imams of the *ahl al-bayt*, but if a *dhimmī* or *nāṣibī*[2] slaughters an animal, it is not permissible to consume.

Slaughtering must be carried out with a sharp metal blade if available, but if not then any sharp tool that can sever the arteries may be used if necessary.

It is obligatory to slit the oesophagus, the two jugular veins and the throat. For animals slaughtered by *naḥr*,[3] it suffices to thrust the blade into the throat at the juncture of the neck and chest.

When performing the slaughter it is necessary to make the animal face toward the *qibla* and pronounce God's name over it. If either of these are omitted wittingly, then the slaughtered animal is impermissible for consumption; but if either of them are omitted due to forgetfulness then the animal is permissible for consumption.

Camels must mandatorily be slaughtered by *naḥr*, and other animals by regular slaughter (*dhibḥ*), and they must all still be moving as living creatures move after the cut, the least of which would be a movement of the tail or the corner of the eye, or the gushing out of blood. If these are absent, then it is not permissible to consume.

It is recommended for sheep to have all but one of their legs tied, for cows to have their tails hanging free, for camels' hooves to be tied to their armpits, and for birds to be hanging forwards.

Meat that is sold in Muslim markets can be assumed to be permissible and fit for consumption as long as one does not know any different. If it is not possible to slaughter an animal regularly or by *naḥr*, such as rebellious animals or those that have fallen in wells and the like and one fears that it will die and be wasted, one may kill them in a different way such as by a sword or other means to fatally wound them.

2 *nāṣibī*: a religious slur used for those who display hatred towards the family of the Prophet, their descendants and followers.

3 *naḥr*: a form of slaughter used for camels and other large animals whose throats cannot be cut, whereby the blade is thrust into the throat at the juncture of the neck and chest.

الفصل الثاني: في الذباحة

ويشترط في الذابح الإسلام أو حكمه، ولو ذبح الذمي أو الناصب لم يحلّ الأكل، ويحلّ [من] المخالف.

وإنما يكون بالحديد مع القدرة، ويجوز مع الضرورة بما يفري الاوداج.

ويجب قطع المرى والودجين والحلقوم، ويكفي في المنحور طعنه في وهدة اللبة.

ويشترط في الذبيحة: استقبال القبلة، والتسمية، ولو أخل بأحدهما عمداً لم يحلّ، ولو كان ناسياً جاز.

ويشترط في الإبل النحر، وفي غيرها الذبح، وأن يتحرّك بعد التذكية حركة الأحياء، وأقله حركة الذنب أو تطرف العين، أو يخرج الدم المسفوح، ولو فقدا لم تحلّ.

ويستحب في الغنم ربط قوائمها عدى إحدى رجليه، وفي البقر إطلاق ذنبه، وربط أخفاف الإبل إلى الإبط، وإرسال الطير.

وما يباع في سوق المسلمين فهو ذكي حلال إذا لم يعلم حاله، ولو تعذّر الذبح أو النحر كالمتردي والمستعصي يجوز أخذه بالسيوف وغيرها مما يجرح إذا خشي التلف.

A fish is made permissible by taking it out of the water alive. If it dies in the water after it has been caught, it is not permissible to consume. The same applies to locusts, who must be caught alive. Being a Muslim or pronouncing God's name is not necessary. Crickets are forbidden to consume. If [locusts] are burned in a grass fire before they are caught, they are forbidden to consume.

The slaughter of the unborn offspring is carried by slaughtering its mother, and it is fit for consumption so long as the body has formed completely. If it is still alive after its mother has been slaughtered and her abdomen cut open, then it is only permissible to be consumed after being slaughtered in the proper manner.

وذكاة السمك إخراجه من الماء حياً ، ولو مات في الماء بعد أخذه لم يحّل . وكذا ذكاة الجراد أخذه حياً ، ولا يشترط فيهما الإسلام ولا التسمية .

والدبا حرام، ولو احترق في أجمة قبل أخذه فحرام .

وذكاة الجنين ذكاة أمه مع تمام الخلقة، ولو أخرج حياً لم يحلّ بدون الذكاة .

Section Three: Other Foods and Drinks

1. *Sea Animals:*

Only fish with scales are permissible for consumption. Those that float on the surface of the sea [dead fish] are forbidden, as are fish that feed on the droppings of other animals except after they are fed fresh clean grass for a day and a night, as well as eels, tortoises, frogs and lobsters. There is no problem with consuming mackerel, prawns, shrimp, wrasse and other types known as *ṭamir* and *iblāmī*.

Fish found inside the cavity of another fish may be consumed as long as it is of a permissible type, but not that which is found inside a snake, unless it was apprehended, and the fish is still intact [undigested].

Fish eggs follow the same ruling as their adult counterparts, and if in doubt, those that are coarse may be consumed.

2. *Land Animals:*

Domestic cattle are fit for consumption, as are wild buffaloes, mountain sheep, roes, gazelles and deer.

Eating horses, mules and donkeys is discouraged. From the permissible animals, those that have been feeding exclusively on human excrement are forbidden to eat, unless they have been made to abstain. In such cases, a camel must be fed fresh grass for forty days, a cow for twenty days and an ewe for ten days. Consuming the meat of an animal that has been suckled by a sow is disapproved, and if that has led to the growth of its flesh then both that animal as well as its offspring are impermissible for consumption.

Predators such as lions and foxes are forbidden to eat, as are rabbits, lizards, jerboa, insects, lice, beetles and fleas.

الفصل الثالث: في الأطعمة والأشربة

وفيه مباحث

الأول: في حيوان البحر

ولا يؤكل منه إلا سمك له فلس، ويحرم الطافي والجلّال منه حتى يطعَم علفاً طاهراً يوماً وليلةً، والجري والسلحفاة والضفادع والسرطان. ولا بأس بالكنعت، والربيثا، والطمر، والطبراني، والإبلامي، والاربيان. ويؤكل ما يوجد في جوف السمكة إذا كانت مباحة، لا ما تقذفه الحية إلا أن يضطرب ولم ينسلخ.

والبيض تابع، ومع الاشتباه يؤكل الخشن.

الثاني: البهائم

ويؤكل النعم الأهلية، وبقر الوحش، وكبش الجبل، والحمر، والغزلان، واليحامير.

ويكره الخيل، والبغال، والحمير.

ويحرم الجلّال من المباح، وهو ما يأكل عذرة الإنسان خاصة، إلا مع الاستبراء، وتطعَم الناقة علفاً طاهراً أربعين يوماً، والبقرة عشرين، والشاة عشرة. ولو شرب لبن خنزيرة كره، ولو اشتدّ لحمه كره هو ونسله.

ويحرم كل ذي ناب كالأسد والثعلب، ويحرم الأرنب، والضب، واليربوع، والحشرات، والقمل، والبق، والبراغيث.

3. *Birds:*

Birds of prey such as falcons and vultures are forbidden to consume, as are those who glide more than they flap their wings, along with those who have neither gizzard, crop nor spur. Bats and peacocks are also not allowed. Permissible birds that have fed exclusively on human excrement are forbidden until they have been made to abstain [and fed clean feed], so a duck and the like must be quarantined for five days and hens for three. [Consuming] Hornets, flies, eggs of prohibited birds, and those whose wings rub together are all dubious matters.

Consuming crows, swallows, hoopoes, shrikes, rooks, rollers, laughing doves and larks is disapproved.

4. *Inanimate Parts of Animals:*

Carrion and any parts thereof are forbidden for consumption, except for the wool, hair, fur and feathers of an animal that would be permissible to consume had it been alive. Similarly, its horns, bones, hooves, rennet and eggs, provided the hard shell has formed on top.

Of animals that have been properly slaughtered, the following is forbidden to consume: the phallus, testicles, spleen, digested food in the gut, blood, bladder, gall bladder, placenta, vulva, spinal cord, spinal marrow, glands, the webbing between the toes [of animals who have feet], the hard lump in the brain, and the pupils. Kidneys and the two auricles of the heart are discouraged to consume.

Essentially impure things such as faeces and any flesh that is severed from the animal whilst it is alive are forbidden to consume, as is soil except for a small pinch of the soil from the grave of al-Ḥusayn which may be consumed for its healing properties. Lethal poisons are also forbidden.

الثالث: الطيوس

ويحرم السبع كالبازي، والرخم، وما كان صفيفه أكثر من دفيفه، وما ليس له قانصة ولا حوصلة ولا صيصة، والخفاش، والطاووس، والجلّال من الحلال حتى يستبرأ، فالبطة وشبهها بخمسة أيام، والدجاجة بثلاثة. والزنابير، والذباب، وبيض المحرم، وما اتفق طرفاه في المشتبه.

ويكره: الغراب، والخطاف، والهدهد، والصرد، والصوام، والشقراق، والفاختة، والقبرة.

الرابع: الجامد

ويحرم الميتة وأجزاؤها، عدى صوف ما كان طاهراً في حياته، وشعره ووبره وريشه، وقرنه وعظمه وظلفه وبيضه إذا اكتسى الجلد الفوقاني، والأنفحة.

ويحرم من الذبيحة: القضيب، والأنثيان، والطحال، والفرث، والدم، والمثانة، والمرارة والمشيمة والفرج، والعلباء والنخاع والغدد وذات الاشاجع وخرزة الدماغ والحدق.

ويكره: الكلى، وأذنا القلب.

ويحرم الأعيان النجسة: كالعذرة، وما أبين من الحي، والطين، عدى اليسير من تربة الحسين عليه السلام للاستشفاء، والسموم القاتلة.

5. *Liquids:*

Every intoxicant is forbidden, be it wine or something else, as is juice that has fermented, beer, blood, a blood clot including that which is found in an egg as it is impure, and any other liquid that is impure or that has come into contact with an impure element. The parts of solidified liquids such as fat or honey that have directly come into contact with the impure thing are forbidden, but the rest is permissible.

Oil that become contaminated through an impurity may be used as fuel, for lighting purposes, but exclusively outdoors.

Urine of all animals is forbidden, except for camel urine if used for medicinal purposes. Similarly, drinking the milk of an animal that is forbidden to consume is also forbidden.

If the status of meat is unknown, it must be thrown in fire. If it contracts, then it is considered slaughtered meat and is permitted for consumption, but if not, then it is carrion. If there is a mixture of two different meats and it is not possible to tell apart [the permissible from the forbidden], they must both be avoided.

الخامس: المائع

ويحرم كل مسكر من خمر وغيره، والعصير إذا غلا، والفقاع، والدم، والعلقة وإن كانت في البيضة، وهي نجسة، وكل ما هو نجس من المائع وغيره وتلقى النجاسة وما يكتنفها من الجامد كالسمن والعسل ويحل الباقي.

والدهن النجس بملاقاة النجاسة يجوز الاستصباح به تحت السماء خاصة. ويحرم الأبوال كلها عدى أبوال الإبل للاستشفاء. وكذا يحرم لبن الحيوان المحرم.

ولو اشتبه اللحم ألقي في النار، فإن انقبض فذكي، وإلا فميتة، ولو امتزجا واشتبه اجتنبا.

Other Rulings

1) One is allowed to eat at the house of anyone mentioned in the Quranic verse[4] as long as he does not know of any reason for it to be discouraged.

2) When wine turns into vinegar it becomes pure [and consumable], be it for medicinal purposes or otherwise, as long as it is not mixed with an impurity.

3) Purées and pastes [of fruits] are not forbidden, even if one can smell the scent of fermentation therein.

4) If juice ferments by itself or through heat, it becomes forbidden until two-thirds of it evaporate or it turns into vinegar.

5) The one who is compelled [due to starvation] may consume just enough forbidden food to save his life, except for the rebel who revolts against the Imam and the vagabond, who is the highway robber.

6) It is recommended to wash the hands before eating food, to pronounce God's name, to eat with the right hand, to wash the hands thereafter, to praise God and lie down afterwards with the right leg resting on the left. It is prohibited to eat at a table where there is an intoxicant, and to over-indulge in foods that are harmful to one's health.

4 Quran 24:61

مسائل:

الأولى: يجوز للإنسان أن يأكل من بيت من تضمنته الآية خاصة مع عدم العلم بالكراهية.

الثانية: إذا انقلبت الخمر خلاً طهرت، بعلاج كان أو غيره، ما لم يمازجها نجاسة.

الثالثة: لا يحرم شيء من الربوبات وأن شم منها رائحة المسكر.

الرابعة: العصير إذا غلا من قبل نفسه أو بالنار حرم حتى يذهب ثلثاه، أو ينقلب خلاً.

الخامسة: يجوز للمضطرّ تناول المحرم بقدر ما يمسك رمقه، إلا الباغي، وهو: الخارج على الإمام عليه السلام، والعادي، وهو قاطع الطريق.

السادسة: يستحب غسل اليد قبل الطعام، والتسمية، والأكل باليمين وغسل اليد بعده، والحمد، والاستلقاء وجعل الرجل اليمنى على اليسرى. ويحرم الأكل على مائدة المسكر وإفراط الأكل المتضمن للضرر.

The Book of Inheritance

كتاب الميراث

This book contains several sections.

Section One: Grounds for Inheritance

There are two grounds upon which one person may inherit from another: biological and non-biological relations. Biological relations are further divided into three hierarchical groups:

The First Group: Parents and Children

If the deceased is survived by only his father, then he receives the entire inheritance; and if only by his mother, then she receives a third as her fixed share, and is given the remainder also [as there are no other heirs]. If they both survive him, then the father receives the residual amount [two-thirds].

If there is a spouse alongside them both, then he/she receives their fixed share[1], the mother receives one-third, and the residual goes to the father.

A single surviving son receives the whole inheritance, and similarly for two or more sons, who receive equal shares of it. A single surviving daughter receives half of the inheritance as her fixed share, and the remainder also [because there are no other heirs]. Two or more daughters would receive a fixed two-thirds in equal shares, and the residual in addition. If there are sons and daughters together, males receive twice the share of females.

If there are parents and sons, each parent receives one-sixth and the residual goes to the sons. If there are daughters as well, the residual is divided among the children where the male receives twice the share of the female.

If there is a single parent surviving along with a daughter, the parent receives one-quarter of the whole amount, and the daughter receives her fixed share and the residual. If there are two daughters or more, then the parent receives one-fifth.

If both parents and a daughter survive, the parents each receive one-fifth of the whole amount between them, and the residual goes to the daughter. If there are two daughters or more, then the parents receive one-third between them.

If there is also a spouse alongside them, then there is a reduction in the share of the daughter or daughters.

1 If the deceased is male, then the fixed share that is inherited by the wife is one quarter. If the deceased is female, then the fixed share that is inherited by the husband is a half.

وفيه فصول :

الفصل الأول: في أسبابه

وهي شيئان، نسب . وسبب . فالنسب مراتبه ثلاث :

الأولى: الأبوان والأولاد

فللأب المنفرد المال، وللأمّ وحدها الثُلث والباقي ردّ عليها ، ولو اجتمعا كان الباقي له .

ولو كان معهما زوج أو زوجة فله نصيبه وللأم الثُلث والباقي للأب .

وللابن المال، وكذا الاثنين فما زاد بالسوية . ولو انفردت البنت فلها النصف والباقي ردّ عليها ، وللاثنتين فما زاد الثُلثان والباقي ردّ عليهما ، فلو اجتمع الذكور والإناث من الأولاد فللذكر مثل حظّ الأنثيين .

ولكل واحد من الأبوين مع الذكور السُدس والباقي للأولاد ، ولو كان معهم إناث فالباقي بينهم للذكر مثل حظّ الأنثيين .

ولكل واحد من الأبوين منفرداً مع البنت الرُبع بالتسمية والردّ والباقي للبنت كذلك ، ومع البنتين فما زاد : الخُمس .

ولهما معاً مع البنت : الخُمسان تسمية ورداً والباقي لها ، ومع البنتين فما زاد : الثُلث .

ولو شاركهم زوج أو زوجة دخل النقص على البنت أو البنات .

Other Rulings:

1) If the deceased is survived by both his parents as well a brother, or two sisters, or four sisters, or two brothers, they prevent the mother from receiving any more than one-sixth, provided that they are: Muslims, neither murderers [of the deceased] nor slaves, already born and not *in utero*, children born of the same parents or of the father with the father still alive. If any of these conditions are unmet, then there is no obstacle [to her receiving more]. If, however, they are fulfilled, then even if the deceased has no children surviving alongside his parents, his mother still receives only one-sixth and the residual goes to the father. If there is a daughter alongside the parents, then each parent receives one-sixth, the daughter receives a half [two-sixths], and the residual sixth is further divided into quarters and inherited by the father and the daughter [in the ratio of 3:1].

2) The grandchildren take the children's place if the deceased's children are no longer alive, and each of them receives the share corresponding with whose children they are, so the daughter's children would receive one-third as opposed to the son's children [who receive two-thirds], since the male receives twice the share of the female. The closer relation excludes the further one from inheriting, in that those related through both parents have priority over those related solely through their shared father. The daughter's children receive the residual in the same proportion that she would have received, shared equally between them regardless of whether they are male or female.

3) The oldest male child receives the clothes that the deceased was wearing on his back, his ring, his sword and his copy of the Quran, provided that he is not impudent or corrupt in his beliefs, and only if the deceased has left more than just these things. He is also responsible for repaying any missed or fasts that were incumbent upon the deceased.

The Second Group: Siblings and Grandparents

If the deceased has no surviving children or grandchildren, nor either of his parents, then his estate is inherited by his siblings and grandparents. His full brother or brothers [i.e. of the same parents] would receive the entire inheritance. If the deceased is survived by a sister only [i.e. there are no male siblings], then she would receive half [as per her right in the inheritance], and the remainder would be given to her. Two or more sisters receive two-thirds shared between them, and the remainder is shared by both.

مسائل:

الأولى: إذا خلّف الميّت مع الأبوين أخاً وأختين أو أربع أخوات أو أخوين، حجبوا الأم عما زاد على السُدس، بشرط أن يكونوا مسلمين، غير قاتلين ولا مماليك، منفصلين غير حمل، ويكونوا من الأبوين، أو من الأب ويكون الأب موجوداً، فإن فقد أحد هذه فلا حجب، وإذا اجتمعت الشرائط فإن لم يكن معهما أولاد فللأم السُدس خاصة والباقي للأب. وإن كان معهما بنت فلكل من الأبوين السُدس وللبنت النصف، والباقي يردّ على الأب والبنت أرباعاً.

الثانية: أولاد الأولاد يقومون مقام الأولاد عند عدمهم، ويأخذ كل فريق منهم نصيب من يتقرّب به، فلأولاد البنت مع أولاد الابن الثُلث للذكر مثل حظّ الأنثيين، ولأولاد الابن الثُلثان كذلك، والأقرب يمنع الأبعد ويشاركون الأبوين كآبائهم، ويرد على أولاد البنت كما يرد عليها ذكوراً كانوا أو إناثاً.

الثالثة: يحبى الولد الذكر الأكبر: بثياب بدن الميت وخاتمه، وسيفه، ومصحفه، إذا لم يكن سفيهاً ولا فاسد الرأي، بشرط أن يخلف الميت غير ذلك، وعليه ما على الميت من صلاة وصيام.

المرتبة الثانية: الإخوة والأجداد

إذا لم يكن للميت ولد - وإن نزل - ولا أحد الأبوين، كان ميراثه للإخوة والأجداد، فللأخ من الأبوين فما زاد المال، وللأخت من قبلهما النصف والباقي ردّ عليهما، وللأختين منهما فما زاد الثُلثان والباقي ردّ عليهما.

If there are both brothers and sisters, then the male receives twice the share of the female. A single surviving uterine sibling [i.e. sharing only the same mother], male or female, receives one-sixth as well any residual. If there are two or more uterine siblings, they receive one-third, and the remainder is shared between them equally, regardless of gender.

Half-siblings who share only the same father [consanguine] are treated the same as full siblings without any differentiation between them, and the same rulings apply to them.

If the deceased is survived by full siblings as well as consanguine siblings together with uterine siblings, then a single uterine sibling receives one-sixth, or one-third if there are two or more, shared between them equally irrespective of gender. The residual is shared among the full siblings, the male receiving twice the share of the female. The consanguine siblings do not inherit in this case.[2]

If he is survived by only consanguine siblings together with uterine siblings [and no full siblings], then a single uterine sibling receives one-sixth, or two-thirds if there are more than two, divided equally between them. The residual then goes to his consanguine siblings, the male receiving twice the share of the female. If his consanguine siblings are all female, then any residual is divided between them and his consanguine siblings in a ratio of 3:1 or 4:1.

A surviving spouse receives the higher amount of whatever they are entitled to [as either a sole inheritor or an inheritor alongside other heirs], and the deduction would be incurred in the share of either the full siblings or the consanguine siblings.

The grandfather, if he is the only surviving relative, receives the entire inheritance. The same applies to the grandmother. If they are both paternal grandparents, then they divide it between them, the male receiving twice the share of the female. If they are both maternal grandparents, then they share the amount equally.

If both paternal and maternal grandparents survive him, then the maternal side receives one-third, whether it is one grandparent or both, and the residual goes to the paternal grandparents. If there is a surviving spouse additionally, then the deduction is incurred in the paternal grandparents' share. The closer relation prevents the further one from inheriting.

2 Due to the aforementioned ruling that the closer relation excludes the further one from inheriting, in that full siblings have priority over consanguine siblings.

ولو اجتمع الذكور والإناث فللذكر مثل حظّ الأنثيين، وللواحد من ولد الأم ذكراً أو أنثى السُدس والباقي رد عليه، وللاثنين فصاعداً الثُلث والباقي رد عليهم الذكر والأنثى سواء.

ويقوم من يتقرّب بالأب خاصة مقام من يتقرّب بالأبوين من غير مشاركة وحكمهم حكمهم.

ولو اجتمع الإخوة من الأبوين مع الإخوة من كل واحد منهما كان لمن يتقرّب بالأم السُدس إن كان واحداً والثُلث إن كانوا أكثر بينهم بالسوية وإن كانوا ذكوراً وإناثاً، ولمن تقرّب بالأبوين الباقي واحداً كان أو أكثر للذكر مثل حظّ الأنثيين وسقط الإخوة من الأب.

ولو اجتمع الإخوة من الأم مع الإخوة من الأب خاصة كان لمن تقرّب بالأم السُدس إن كان واحداً، والثُلث إن كان أكثر بالسوية، والباقي لمن تقرّب بالأب للذكر مثل حظّ الأنثيين.

ولو كان الإخوة من قبل الأب إناثا كان الرد بينهنّ وبين المتقرّب بالأم أرباعاً أو أخماساً وللزوج والزوجة نصيبهما الأعلى، ويدخل النقص على المتقرّب بالأبوين أو بالأب.

وللجد إذا انفرد المال، وكذا الجدة، ولو اجتمعا لأب فللذكر ضعف الأنثى، وإن كانا لأم فبالسوية. ولو اجتمع المختلفون فللمتقرّب بالأم الثُلث وإن كان واحداً والباقي للمتقرّب بالأب، ولو دخل الزوج أو الزوجة دخل النقص على المتقرّب بالأب والأقرب يمنع الأبعد.

If there are both grandparents and siblings present, then the grandfather is treated as the brother and the grandmother as the sister [in the division of shares]. The grandfather, or great-grandfather, would receive shares equal to the brothers.

The deceased's nephews and nieces take the place of the deceased's brothers and sisters if they are no longer alive and receive the same shares as the grandparents. They each receive shares corresponding with their relation to the deceased. If they are the children of his uterine siblings, then males and females receive equal shares, but if they are children of consanguine siblings, then the male receives twice the share of the female.

The Third Group: Paternal and Maternal Uncles and Aunts:

They only inherit in the absence of the previous two groups. A sole surviving paternal uncle would receive the entire inheritance, as would two or more uncles between them, or a paternal aunt, or two or more. If the paternal uncles and aunts are all from the same parents, then the male receives twice the share of the female. If they are from different parents, then a uterine paternal uncle or aunt receives one-sixth, and if there are two or more then they share one-third between themselves equally, and the consanguine uncles and aunts cease to inherit. If there are no full paternal uncles and aunts, then the consanguine uncles and aunts may take their place and receive their share.

A sole surviving maternal uncle would receive the entire inheritance, as would two or more uncles between them, or a maternal aunt, or two or more. If there are both full maternal uncles and aunts, they all receive an equal share [irrespective of gender]. If they are from different parents, then the uterine uncle or aunt alone receives one-sixth, or one-third divided equally if there are two or more of them. The residual goes to full uncles and aunts, divided equally between them. Consanguine maternal uncles do not inherit, unless there are no surviving full uncles and aunts, in which case the former can take their place.

ولو اجتمع الإخوة والأجداد كان الجد كالأخ والجدة كالأخت. والجد وإن علا يقاسم الإخوة.

وأولاد الإخوة والأخوات يقومون مقام آبائهم عند عدمهم في مقاسمة الأجداد، وكل واحد منهم يرث نصيب من يتقرّب به، ويقتسمون بالسوية إن كانوا لأم، وإن كانوا لأب فللذكر ضعف الأنثى.

المرتبة الثالثة: الأعمام والأخوال

وإنما يرثون مع فقد الأوّلين، فللعم وحده المال، وكذا العمان فما زاد، كذا العمة والعمتان والعمات، ولو اجتمعوا، فللذكر منهم مثل حظّ الأنثيين، ولو تفرّقوا فللواحد من الأم السُدس، وللزائد عليه الثُلث بالسوية والباقي لمن تقرّب بالأبوين واحداً أو أكثر للذكر ضعف الأنثى، وسقط المتقرّب بالأب، ولو فقد المتقرّب بهما قام المتقرّب بالأب مقامه وحكمه حكمه.

وللخال المنفرد المال، وكذا الخالان فما زاد، وكذا الخالة والخالتان والخالات، ولو اجتمعوا تساووا، ولو تفرّقوا فللمتقرّب بالأم السُدس وإن كان واحداً، والثُلث إن كان أكثر بالسوية، والباقي لمن يتقرّب بالأبوين واحداً كان أو أكثر بالسوية، وسقط المتقرّب بالأب، ولو فقد المتقرّب بهما قام المتقرب بالأب مقامه كهيئته.

If there are both paternal and maternal uncles surviving, then the maternal uncles receive one-third, regardless of number and gender. The residual goes to the paternal uncles. If the maternal uncles are from different parents, then a single surviving uterine maternal uncle receives one-sixth of three parts, and if there are two or more, then one-third of three parts, divided equally between them. The residual [of that part] goes to the full maternal uncles, and consanguine ones do not inherit. The residual of the total amount [two parts of the whole] goes to the paternal uncles. If they, in turn, are from different parents, then a single uterine paternal uncle receives one-sixth of it, and if there are two or more, then one-third divided equally between them. The residual goes to the full paternal uncles, and the consanguine paternal uncles do not inherit. A surviving spouse in addition to these would receive his/her share by priority, after which the uterine relations would receive a third of the original amount, and the residual would go to the full uncles and aunts, or in their absence, the consanguine ones.

Paternal and maternal cousins take the place of paternal and maternal uncles respectively in their absence, and they each receive a share corresponding to their relation to the deceased, whether there is one or more. The closer relation excludes the further one from inheriting, except in one particular instance, which is that the deceased's paternal cousin, whose father is his father's full brother [sharing both parents] has priority over the deceased's consanguine paternal uncle. In this case, only the full paternal cousin would inherit.

Paternal and maternal great-uncles and great-aunts would take the place of one's paternal and maternal uncles and aunts in their absence, whereby the closer relation excludes the further one from inheriting. Therefore, the presence of full paternal and maternal cousins excludes consanguine and uterine paternal and maternal uncles from inheriting.

If there are two grounds for which an heir may inherit, then he does so through both, such as being the deceased's consanguine paternal cousin as well as his uterine maternal cousin, or being the deceased's husband who is also her paternal or maternal cousin.

If one of these two relationships excludes the other one from being grounds for inheritance, then he may inherit through the former, such as being the deceased's consanguine paternal cousin as well as his mother's brother.

ولو اجتمع الأخوال والأعمام فللأخوال الثُلث وإن كان واحداً ذكراً أو أنثى، والباقي للأعمام، فإن تفرّق الأخوال فللمتقرّب بالأم سُدس الثُلث إن كان واحداً، وثُلثه إن كان أكثر بالسوية، والباقي لمن تقرّب بالأبوين، وسقط المتقرّب بالأب، وللأعمام الباقي، فإن تفرّقوا فللمتقرّب بالأم سُدسه إن كان واحداً، وإلا فالثُلث، والباقي للمتقرّب بهما، وسقط المتقرّب بالأب، وللزوج أو الزوجة نصيبه، وللمتقرّب بالأم ثلث الأصل، والباقي للمتقرّب بهما أو بالأب.

ويقوم أولاد العمومة والعمات والخؤولة والخالات مقام آبائهم مع عدمهم، ويأخذ كل منهم نصيب من يتقرّب به، واحداً كان أو أكثر، والأقرب يمنع الأبعد إلا في صورة واحدة، وهي: ابن عمّ من الأبوين مع العم من الأب، فإن المال لابن العم خاصة.

وعمومة الأب وخؤولته وعمومة الأم وخؤولتها يقومون مقام العمومة والعمات والخؤولة والخالات مع فقدهم، والأقرب يمنع الأبعد، وأولاد العمومة والخؤولة وإن نزلوا يمنعون عمومة الأب وخؤولته وعمومة الأم وخؤولتها.

ولو اجتمع للوارث سببان متشاركان ورث بهما، كابن عمّ لأب هو ابن خال لأمّ، أو زوج هو ابن عمّ، مع ابن عمّ أو ابن خال.

ولو منع أحدهما الآخر ورث من قبل المانع كابن عمّ لأب هو أخ لأمّ.

Section Two: Non-Biological Grounds for Inheritance

There are two such grounds: marriage and patronage (*walāʾ*).

The deceased's surviving husband, in the absence of a child, receives a half. If there is a child, or even a grandchild, then he receives one-quarter. The surviving wife would receive one-quarter with no children, and one-eighth with. If there are no other heirs, then the surviving husband gets the residual amount. In the case of the surviving wife, there are two opinions.[3] If there are two wives or more, then they share the one-quarter or one-eighth.

Spouses inherit from each other regardless of whether the marriage has been consummated or not, and during the process of a revocable divorce [in the waiting period].

The husband inherits the full inheritance [of his deceased wife], as does the wife if they had a child together. If they did not, then she only inherits from the value of his properties and land, so the buildings, contents, palms and trees must be valued, and she would receive a share of the value.

If a terminally ill person gets married, his surviving wife would inherit only if the marriage was consummated, but if not, then she neither receives the dowry nor the inheritance.

Patronage is Further Divided Into Three Types:

1) The patronage of the manumitted slave whereby the manumitter inherits from his ex-slave provided that he had manumitted him voluntarily and did not acquit himself of liability for any crimes committed by his freed slave, in the absence of any biological heirs. He would share the inheritance with the surviving spouse. If the ex-slave had numerous benefactors, they all have a share in the inheritance. If the manumitter is no longer alive, then the patronage passes on to his [i.e. the former master's] parents and male children by priority, and in their absence to his paternal relatives. If the patron was a woman, it passes on to her paternal relatives, but not her children.

Patronage is not inherited through maternal relations; nor can it be sold, gifted or set as a condition for a transaction.

3 That she either receives the remainder, or that it is remitted to the Imam.

الفصل الثاني: في الميراث بالنسب

وهو اثنان: الزوجية، والولاء.

فللزوج مع عدم الولد النصف، ومعه وإن نزل الرُبع، وللزوجة مع عدمه الرُبع ومع وجوده الثُمن، ولو فقد غيرهما ردّ على الزوج، وفي الزوجة قولان ويتشارك ما زاد على الواحدة في الثُمن أو الرُبع.

ويرث كل منهما من صاحبه مع الدخول وعدمه، ومع الطلاق الرجعي. ويرث الزوج من جميع التركة، وكذا المرأة إذا كان له ولد منها.

ولو فقد ورثت إلا من العقارات والأرضين، فيقوم الأبنية والآلات والنخيل والأشجار وترث من القيمة.

ولو تزوّج المريض ودخل ورثت، وإلا فلا مهر ولا ميراث.

وأما الولاء: فأقسامه ثلاثة:

الأول: ولاء العتق. ويرث المعتق عتيقه مع التبرّع وعدم التبري من الجريرة بعد فقد النسب. ويشارك الزوج والزوجة.

ولو كان المنعم متعدداً تشاركوا، ولو عدم فالأقرب انتقال الولاء إلى الأبوين والأولاد الذكور، فإن فقدوا فللعصبة.

ولو كان المنعم امرأة انتقل إلى عصبتها دون أولادها.

ولا يرث الولاء من يتقرّب بالأم.

ولا يصح بيعه ولا هبته ولا اشتراطه في بيع.

The transfer of patronage is valid, so if a female ex-slave bore a child with another slave, the child's patron would be her ex-master. If his father was then manumitted, the child's patronage would then transfer to his father's manumitter. In the latter's absence, patronage passes on to his parents and male children, and in their absence to his paternal relatives. In their absence, it would pass on to the manumitter's patron [if he has one], and if not then to the patron of his paternal relatives, and in his absence, to their guarantor or sponsor; and in his absence to the Imam. The patronage never returns to the mother's ex-master.

If the patron dies leaving two sons, followed by the death of one of his sons, followed by the ex-slave, the surviving son would share the inheritance with the heirs of the deceased.

2) Patronage on the basis of bailment for a crime, whereby the person's patron is one who stands as his guarantor, and patronage belongs to him and he would inherit in the absence of any biological relative and non-biological patron, but would share the inheritance with a surviving spouse, and he takes priority over the Imam [in inheriting] and cannot be bypassed. One can only be a bail guarantor for someone else who does not have a patron, like an ex-slave who was released from bondage out of obligation [and not voluntarily], or for someone whose only inheritor is the guarantor himself.

3) Patronage of the Imam: If the deceased has no surviving biological or non-biological heirs, his estate goes to the Imam who would use it as his discretion. ʿAlī would divide the estate among the poor people of his town and for his weak neighbours.

During the period of occultation [of the Imam], it is distributed among the poor.

وجرّ الولاء صحيح، فلو حملت المعتقة بعد العتق من مملوك حراً فولاؤه لمولاها ، فإذا أعتق الأب انجرّ الولاء إلى معتق أبيه، فإن فقد فلأبويه وأولاده الذكور ، فإن فقدوا فلعصبته : فإن فقد فلمولى مولى الأب ، فإن فقد فلمولى عصبة المولى ، فإن فقد فللضامن ، فإن فقد فللإمام.

ولا يرجع إلى مولى الأم.

ولو مات المنعم عن اثنين ثم مات المعتق بعد موت أحدهما شارك الحي ورثة الميت.

الثاني : ولاء تضمن الجريرة ومن توالى إنساناً يضمن حدثه، ويكون ولاؤه له، وورث مع فقد كل مناسب ومسابب ، ويشارك الزوجين ، وهو أولى من الإمام. ولا يتعدى الضامن.

ولا يضمن إلا سائبة كالمعتق واجباً ، أو من لا وارث سواه.

الثالث : ولاء الإمامة ، وإذا فقد كل مناسب ومسابب انتقل الميراث إلى الإمام يعمل به ما شاء وكان عليّ عليه السلام يضعه في فقراء بلده وضعفاء جيرانه.

ومع الغيبة يقسم في الفقراء.

Section Three: Things Which Prevent Inheritance

There are three such factors: being a disbeliever, murder and enslavement.

1) Being a disbeliever: The disbeliever cannot inherit the Muslim even if they are related, and the presence of a disbelieving relative does not exclude another from inheriting. So, if a [deceased] Muslim man was survived by a disbelieving son, who in turn had a Muslim child, the latter would inherit directly from his grandfather. If this Muslim man had no surviving Muslim heir, his estate would pass on to the Imam.

The Muslim can inherit from the disbeliever, and his presence excludes other disbelievers from having a share in the inheritance, so if a disbeliever is survived by a disbeliever son and a Muslim cousin, his estate would be inherited by his cousin. If the disbeliever embraces Islam before the inheritance is divided, he receives his share of it if he has equal right to it as someone else, and he may receive all of it if he has priority, regardless of whether the deceased was a Muslim or a disbeliever. If the deceased left only one [disbelieving] heir who embraced Islam afterwards, he would not inherit anything.

Muslims may inherit from each other, even if they differ in their beliefs, and disbelievers may inherit from each other, even if they differ in their religions.

The born-Muslim apostate (*murtadd fiṭrī*) must be killed immediately, and his wife must observe the waiting period of a widow from the time of his apostasy. His assets are divided and inherited, and repentance in this instance has no effect on these rulings.

The apostate who was not born a Muslim (*murtadd millī*) must be urged to repent. If he refuses, then he is killed. His wife must observe the waiting period of divorce upon his apostasy. His assets are only divided after he is killed. If he repeats the crime [after having repented], he is to be killed upon the fourth count.

If a woman apostatizes, she must be arrested and beaten at prayer times until she repents, or dies, even if she is a born-Muslim apostate.

The apostate's estate may be inherited by the Muslim, and if he has only disbeliever heirs, it passes to the Imam. The apostate cannot inherit a Muslim.

الفصل الثالث: في موانع الإرث

وهي ثلاثة: كفر، وقتل، ورقّ.

أما الكفر: فلا يرث الكافر المسلم وإن قرب، ولا يمنع من يتقرّب به، فلو كان للمسلم ولد كافر وله ابن مسلم ورث الجد ولو فقد المسلم كان الميراث للإمام. والمسلم يرث الكافر، ويمنع مشاركة الكفّار، فلو كان للكافر ولد كافر وابن عم مسلم فميراثه لابن العم، ولو أسلم الكافر قبل القسمة شاركه إن كان مساوياً وأخذ الجميع إن كان أولى، سواء كان الميت مسلماً أو كافراً، ولو كان الوارث واحداً وأسلم الكافر لم يرث.

والمسلمون يتوارثون وإن اختلفوا في الآراء، والكفّار يتوارثون وإن اختلفوا في الملل.

والمرتدّ عن فطرة يقتَل في الحال، وتعتدّ امرأته من حين الارتداد عدّة الوفاة، ويقسم ميراثه، ولا تسقط هذه الأحكام بالتوبة.

وعن غير فطرة يُستتاب، فإن تاب وإلا قتل، وتعتدّ زوجته عدّة الطلاق، ولا تقسم أمواله إلا بعد القتل، ولو تكرّر قُتل في الرابعة.

والمرأة إذا ارتدّت حُبست وضُربت أوقات الصلاة حتى تتوب أو تموت، وإن كانت عن فطرة.

وميراث المرتدّ للمسلم ولو لم يكن إلا كافراً انتقل إلى الإمام، والمرتدّ لا يرث المسلم.

2) Murder: If the murder was intentional and wrongful, the [murderous] heir is barred from inheriting from his victim. If it was accidental, then he is barred [only] from inheriting the blood-money based on one opinion. The murdered person's estate, therefore, is inherited by people other than his murderer [when the victim was murdered intentionally and without just cause by the murderer], even if they are distantly or closely related to the murderer. If there is no other heir, it passes on to the Imam.

The blood-money is inherited by paternal relatives, male and female, and spouses. There is a difference of opinion regarding maternal relatives inheriting it.

If the person who was murdered intentionally has no heir, the Imam cannot forgive the murderer, and must either demand he pay the blood-money or sentence the murderer to death. The blood-money can then be used to pay off the deceased's debts and bequests, even if they were wilful. The debtor has no right to block the execution of the punishment.

3) Enslavement: This bars inheritance from both sides.[4] If a free man and a slave both stand to inherit, the inheritance goes to the free man, even if he is distantly related. If a slave is manumitted before the division of inheritance, then he receives the same share as those in the same group as him, and if he has priority, then he receives the whole amount.

If the deceased has a sole heir who is then manumitted, he does not inherit, and if there is no other heir besides the slave, he may persuade his master to take the value of the assets inherited and use it to manumit him, and he may keep the rest. If the value of the estate falls short, then he is not freed.

The inheritance left by the slave belongs to his master, even though we mentioned previously that he may own certain things, so the *mudabbar* slave, the slave-mother (*umm al-walad*) and the *mukātab* slave – of both conditional and unrestricted manumission contracts, if he has not managed to buy any part of his freedom – are all treated as the fully-bonded slave.

4 That is, a slave can neither be inherited and nor can he inherit.

الثاني: القتل، وهو يمنع الوارث من الإرث إن كان عمداً ظلماً، ولو كان خطئاً منع من إرث الدية على قول، وميراث المقتول لغير القاتل وإن بعد أو تقرّب بالقاتل، ولو فقد فللإمام.

والدية يرثها من يتقرّب بالأب ذكوراً أو إناثاً والزوج والزوجة، وفي المتقرّب بالأم قولان.

ولو لم يكن للمقتول عمداً وارث لم يكن للإمام العفو بل أخذ الدية أو القتل، ويقضي من الدية الديون والوصايا، وإن كانت للعمد، وليس للديان المنع من القصاص.

الثالث: الرقّ، وهو مانع في الطرفين ولو اجتمع الحر مع المملوك فالمال للحر وإن بعد، ولو أعتق قبل القسمة شارك مع المساواة واختص مع الأولوية.

ولو كان الوارث واحداً وأعتق لم يرث، ولو لم يكن وارث إلا المملوك أجبر مولاه على أخذ القيمة من التركة وأعتق وأخذ الباقي، ولو قصرت التركة لم يفكّ.

وميراث المملوك لمولاه وإن قلنا أنه يملك، فالمدبّر وأم الولد والمكاتَب المشروط أو المطلَق إذا لم يتحرّر منه شيء كالقنّ.

Section Four: Ratios of Shares

The share of a half would be taken from two equal parts [of the estate]. The share of one-third or two-thirds from three parts of it; one-quarter from four parts, one-sixth from six parts and one-eighth from eight parts. If the fixed shares are one-quarter and one-sixth, then they are calculated from twelve parts [i.e. their lowest common denominator], and likewise one-eighth and one-sixth are calculated from twenty-four parts.

The base number of fixed shares may not be exactly divisible between the actual number of heirs, in which case the base number can be multiplied by the actual number of heirs if they do not have a common denominator, as in the case of both parents and five daughters surviving the deceased. Alternatively, the number of heirs is divided by a common divisor then multiplied by the base number of fixed shares, as in the case of both parents and six daughters, whereby three [the base number of six divided by the common divisor of two] is multiplied by the base number of fixed shares [i.e. by six, to get a total number of eighteen portions].

If the fixed shares fall short of the actual number of portions due to the inclusion of a spouse, then the daughter(s) and the full or consanguine sister(s) will receive less.

If the number of shares exceed the number of heirs with fixed shares, the remainder goes to residuary relatives besides either of the spouses or the mother in the presence of siblings.

Residuaries through both paths are more entitled to it than those by a single path.

If one of the deceased's heirs dies shortly after him, before the division of inheritance, the share of common heirs of both deceased persons is calculated by finding the common denominator of the second fixed share and the original fixed share. If there is no common denominator, then the two are multiplied together. This is the case whether the later deceased was survived by his own heirs in turn, or whether he was simply an heir to the former on equal footing with others.

الفصل الرابع: في مخارج السهام

النصف من اثنين، والثُلث والثُلثان من ثلاثة، والرُبع من أربعة، والسُدس من ستة، والثُمن من ثمانية. ولو كان في الفريضة رُبع وسُدس فمن اثني عشر، والثُمن والسُدس من أربعة وعشرين.

وقد تنكسر الفريضة فيضرب عدد من انكسر في أصل الفريضة - إن لم يكن بين نصيبهم وعددهم وفق - مثل: أبوين وخمس بنات وإلا ضربت الوفق من العدد كأبوين وست بنات تضرب ثلاثة وفق العدد مع النصيب في الفريضة.

ولو قصرت الفريضة بدخول الزوج أو الزوجة دخل النقص على البنت أو البنات والأخت أو الأخوات للأبوين أو للأب.

ولو زادت الفريضة ردت على غير الزوج والزوجة والأم مع الإخوة.

وذو السببين أولى بالرد من السبب الواحد.

ولو مات بعض الوراث قبل القسمة وتغاير الوارث أو الاستحقاق فاضرب الوفق من الفريضة الثانية في الفريضة الأولى، وإن لم يكن وفق فاضرب الفريضة الثانية في الأولى.

Section Five: The Inheritance of the Child of a Woman Separated By *Liʿān* Divorce,[5] the Person Born Out of Wedlock, the Unborn Foetus and the Assets of the Missing Person

The [wealth belonging to a] child of a woman accused of adultery and subsequently disowned by her former husband is inherited by his mother and anyone related through her, as well as his own children and his spouse, and he in turn inherits from them. There is no inheritance either way between him and the father or those related through him. If he is survived by both full siblings as well as uterine siblings, they all receive an equal share of the inheritance he leaves behind.

The [wealth of a] person born out of wedlock cannot be inherited by his adulterous mother or father, nor their relatives, and neither can he inherit from them. His wealth may only be inherited by his own children and his spouse, and he from them. If he is not survived by them, then his estate passes on to the Imam.

If the foetus is born alive, then he/she has a share of the inheritance, otherwise not. A share must be set aside for him/her equivalent to two males by way of precaution, and the other heirs in that category receive the lesser of the two shares. The blood-money paid for a foetus goes to its parents and full relatives, or in their absence consanguine relations.

The missing person's assets are only divided after the elapsing of sufficient time whereby someone like him would not be expected to still be alive.

5 Consequent to her being accused of adultery and the child's paternity by her husband denied.

الفصل الخامس: في ميراث ولد الملاعَنة والزنا والحمل والمفقود

ولد الملاعنة: ترثه أمه ومن يتقرّب بها وولده وزوجه أو زوجته، وهو يرثهم. فلا توارث بينه وبين الأب ومن يتقرّب به، ولو ترك إخوة من الأبوين مع إخوة من الأم تساووا في ميراثه.

وولد الزنا : لا يرثه الزاني ولا الزانية ولا من يتقرّب بهما ، ولا يرثهم، وإنما يرثه ولده وزوجه أو زوجته، وهو يرثهم، ومع عدمهم الإمام.

والحمل: إن سقط حياً ورث، وإلا فلا، ويوقف له قبل الولادة نصيب ذكرين احتياطاً، ويعطى أصحاب الفرض أقل النصيبين، ودية الجنين لأبويه ومن يتقرّب بهما أو بالأب.

والمفقود : يقسم أمواله بعد مضي مدّة لا يمكن أن يعيش مثله إليها غالباً.

Section Six: The Inheritance of the Hermaphrodite

This is someone who has both genitalia of the male and the female. Whichever part urine passes through first is the dominant gender in laws of inheritance. If it passes through both at the same time, then it must be investigated to see in which one the flow of urine stops last. If they both stop equally, then he/she receives half the share of a male and half the share of a female.

If the deceased is survived by two children: a son and a hermaphrodite, they are first considered as two males, and then a male and a female. Then their fixed shares are multiplied together to make allowances for both cases, making a total of twelve shares. The hermaphrodite receives five-twelfths and the boy seven-twelfths. If instead, it was a daughter and a hermaphrodite, the girl would receive five-twelfths and the hermaphrodite seven-twelfths. If both son and daughter survived with a hermaphrodite, then the total shares would be divided into forty parts [whereby the hermaphrodite receives thirteen parts, the boy eighteen and the girl nine]. The inheritance of a person without sexual organs is determined by a coin toss to decide his/her gender.

الفصل السادس: في ميراث الخنثى

وهو من له فرجان، فأيّهما سبق بالبول منه حكم له، ولو تساويا حكم للمتأخر في الانقطاع، فإن تساويا أعطي نصف سهم رجل ونصف سهم امرأة.

ولو خلّف ولدين ذكراً وخنثى فرضتهما ذكرين ثم ذكراً وأنثى، وضربت إحدى الفريضتين في الأخرى، ثم المجتمع في حالتيه فيكون اثني عشر، للخنثى خمسة وللذكر سبعة. ولو كان معه أنثى كان لها خمسة وللخنثى سبعة ولو اجتمعا معه فالفريضة من أربعين ولو فقد الفرجين ورث بالقرعة.

ومن له رأسان أو بدنان على حقو واحد، يصاح به فإن انتبها معاً فواحد وإلا فاثنان.

Section Seven: The Inheritance of People Who Drown or Die As a Result of Being Crushed

They may inherit from each other upon fulfilment of certain conditions: that both of two such people or one of them has assets and that they are one another's heirs, and that it is not possible to ascertain who died first. It is debatable whether the same ruling would apply to causes of death other than drowning or crushing accident.

Upon fulfilment of these conditions, each one inherits from the other's original assets and not from what he in turn inherited. Those in a weaker position, from the point of view of inheritance, are dealt with first. So, if both a father and his son drowned, we assume the death of the son first, and thus the father inherits his share [from the son] first [because the father's share is less than the son's had the father died and been survived by a son]. Then the son would inherit whatever he is due from his father's original assets, and each one's share of inheritance goes to the rest of his heirs. If only one of them has assets, they are automatically transferred to the heirs of the other. If there is no heir, then it passes to the Imam.

الفصل السابع: في ميراث الغرقى والمهدوم عليهم

وهؤلاء يتوارثون بشروط: أن يكون لهما أو لأحدهما مال وكانوا يتوارثون، ويشتبه المتقدّم. وفي ثبوت الحكم بغير الغرق والهدم إشكال. ومع الشرائط يرث كل منهم من صاحبه لا مما ورث منه. ويقدّم الاضعف في الإرث، فلو غرق أب وابن فرض موت الابن وأخذ الأب نصيبه ثم يرث الابن نصيبه من تركة الأب مما ورث وينتقل نصيب كل واحد منهما إلى وارثه، ولو كان لأحد الآخرين مال انتقل ماله إلى ورثة الآخر. ولو لم يكن وارث كان للإمام.

Section Eight: The Inheritance of Zoroastrians

Such people inherit from each other on biological as well as marital or non-biological grounds, both valid as well as immoral [i.e. incestuous for example] relationships, though there is a difference of opinion about this. If the deceased is survived by a mother, for example, who is also a wife, she would receive her share. If the existence of one of the relationships excludes inheriting from the other, then one inherits exclusively from the former, such as in the case of a daughter who is also a grand-daughter, in which case she would receive only the fixed share of a daughter.

الفصل الثامن: في ميراث المجوس

وهؤلاء يرثون بالنسب والسبب صحيحهما وفاسدهما - على خلاف، فلو ترك أمّاً هي زوجة فلها نصيبها، ولو كان أحدهما مانعاً ورث به خاصة: كبنت هي بنت بنت، فإنها ترث نصيب البنت خاصة.

The Book of Judiciary Laws, Witnesses and Punishments

كتاب القضاء
والشهادات والحدود

This book contains several sections.

Section One: The Qualities of a Judge

The judge must be a legal subject, a believer, just, knowledgeable, of legitimate birth, accurate and not merely availing himself with the legal edicts of scholars. He must also act with permission from the Imam, and in his occultation, the judgment of the jurist is binding if he encompasses all the above qualities.

It is recommended for his arrival to be announced, and for him to sit at court in the city centre with his back to the *qibla*, to enquire about the evidences, deposits in trust, and inmates along with the reason for their imprisonment. He should also separate witnesses in cases where there is a possibility that the witnesses agree to testify without knowledge; i.e. there is a possibility they may have colluded to testify against the defendant, and he should consult experts.

Judging cases when his mind is preoccupied with anger, hunger, thirst, worry, joy or anything else is disapproved. He should not, at the time of passing judgment, do so from behind closed doors.[1] It is also disapproved to appoint specific people to testify, and for the debtor to mediate with the judge to drop the case against him.

1 In other words, the case must be judged in an open court visible to the public should they desire to be present.

وفيه فصول:

الفصل الأول: في صفات القاضي

ولا بدَّ أن يكون: مكلفاً، مؤمناً، عدلاً، عالماً، طاهر المولد، ضابطاً. ولا يكفيه فتوى العلماء.

ولا بدَّ من إذن الإمام، وينعقد قضاء الفقيه مع الغيبة إذا جمع الصفات.

ويستحب الإعلان بوصوله، والجلوس وسط البلد مستدبر القبلة، والسؤال عن الحجج والودائع وأرباب السجن وموجبه. وأن يفرق الشهود مع التهمة، ومخاوضة العلماء.

ويكره القضاء مع شغل القلب بالغضب والجوع والعطش والهمّ والفرح وغيرها، واتخاذ حاجب وقت القضاء، وتعيين قوم للشهادة، والشفاعة إلى الغريم في إسقاطه حقه.

The Imam judges on the basis of his knowledge, and others must do so too when dealing with the rights of people [but not when dealing with the rights of God].[2] If he does not have knowledge about the case [i.e. he does not know whether the truth lies with the defendant or the claimant], then he must judge on the basis of testimony, provided he is convinced of the credibility and integrity of the witnesses [whose testimonies] must be heard unequivocally [i.e. if a witness is declared by others to be a credible witness, then their testimony would be heard without interrogating the claim that they are credible witnesses], but not unreliable witnesses [i.e. if a person declares a witness as a non-credible witness, then he will be interrogated for this claim and the testimony would not be heard until either the claimant provides evidence of the witness being non-credible, or no evidence – in which case his testimony is admitted in court], unless there is a conflict, in which case the witness's testimony will not be admitted [i.e. if someone says *x* is credible and another person says *x* is not credible, and hence there is a conflict, the claim of the discreditor will be given priority and the witness dropped from court.]

Accepting bribes is prohibited, and they must be returned even if the judgment was truly in the briber's favour. When the creditor requests for his adversary to be present, it must be granted, unless it is a woman who does not normally participate in external society or an ill person, in which case someone must be appointed to act on their behalf.

2 That is, a judge may rule in either the defendant's or claimant's favour if he knows the truth of the case, but he cannot judge in accordance with his knowledge in the absence of witnesses or clear evidence (*bayyina*) when it comes to fixed punishments such cutting off the hands, as this is the rights of God

ويقضي الإمام بعلمه. وغيره به في حقوق الناس، وإذا انتفى العلم حكم بالشهادة مع علمه بعدالة الشهود أو التزكية وتسمع مطلقة، بخلاف الجرح، ومع التعارض يقدم الجرح.

وتحرم الرشوة، ويجب إعادتها وإن حكم بالحق، وإذا التمس الغريم إحضار خصمه أجابه، إلا المرأة غير البرزة أو المريض فينفذ إليهما من يحكم بينهما.

Section Two: The Process of Administering Justice

The judge must treat both sides equally in his address to them, his greeting them, his treatment of them,[3] his looking at them, listening to them and in his administration of justice. The Muslim is allowed to be seated or be placed at a higher level than the non-Muslim who would then be standing or seated at a lower level.

A judge must not indicate to either of the disputing parties how not to be held accountable in court. If one side starts with a claim against his adversary, the claimant's statement is heard first, he must let him do so.[4] If they both start with different claims simultaneously, he must hear the one who is sitting to the right of his adversary.

If the defendant accepts the plaintiff's charge, the judge must hold him responsible for the consequences of his actions [payment of dues and the like], if he is convinced of his mental capacity and wilful acceptance of fault. If the defendant refuses to do so, he may imprison him at the request of the plaintiff. If the plaintiff requests for his claim to be documented, he must document it only after having confirmed his name and surname, or after his identity has been confirmed by two just witnesses, or by physical description.

If the defendant claims financial incapacity, which is then confirmed, he must be given respite to pay his dues. If his financial constraint cannot be corroborated and he is known to have assets, he must be made to pay whatever he owes after evidence, if the claim involves monetary payment of dues. If he does not [have assets], then his word must be accepted if he swears an oath of honesty.

If the defendant denies the claim, further evidence is requested from the plaintiff. If he provides it, the judges rules in his favour accordingly, but if he does not, then he must be asked to swear an oath for his claim. If the plaintiff insists that the denier should take an oath instead, then he must be made to do so.

3 That is, there should be no distinction or preference given to one party over the other in the manner in which they are seated or where they are located in court. For example, a judge should not tell the rich landowner to sit and the poor tenant or labourer to stand etc.

4 This is standard court procedure in Islam. If a claim is being made, then the claimant must give his statement first.

الفصل الثاني: في كيفية الحكم

وعليه أن يسوي بين الخصمين في الكلام والسلام والمكان والنظر والإنصات والعدل في الحكم، ويجوز أن يكون المسلم قاعداً أو أعلى منزلاً والكافر أخفض أو قائماً.

ولا يلقن الخصم، ولو بدر أحدهما بالدعوى قدمه فيها، ولو ادّعيا دفعة سمع من الذي على يمين خصمه.

فإن أقرّ خصمه ألزمه إن كان كاملاً مختاراً، فإن امتنع حبسه مع التماس خصمه، ولو طلب المدّعي إثبات حقه أثبته مع معرفته باسمه ونسبه أو بعد معرفة عدلين، أو بالحلية.

ولو ادّعى الإعسار وثبت أنظر، وإن لم يثبت ألزم بالبينة إذا عرف له مال، أو كان أصل الدعوى مالاً، وإلا قبل قوله مع اليمين.

وإن جحد طلبت البينة من المدّعي، فإن أحضرها حكم له، وإلا توجّهت له اليمين. فإن التمسها حلف المنكِر.

It is not permissible to make him take an oath until and unless the plaintiff requests it, so if the defendant volunteers to take an oath or the judge asks him to do so, it does not automatically free him from liability until and unless the plaintiff requests it. If he refuses to do so, it goes back to the plaintiff [to swear his honesty], and his claim is upheld if he swears to it. If he refuses, then his claim is void, and if the defendant, in turn, counter-suggests for the plaintiff to swear an oath but the latter refuses, his case is dismissed.

If the defendant agrees to swear an oath denying the charges, the plaintiff does not receive payment of his claimed dues, nor is any further evidence heard after that oath, unless the defendant were to go back on his word.

If the claim of a debt owed is against a person who has died, the plaintiff, in addition to providing evidence that the debt continues to exist, needs to swear an oath to that effect.

If the denier remains silent due to an impediment, the judge must arrive at a conclusion by ascertaining that he understands his charge to be able to accept or deny it. A single translator does not suffice. If the defendant is obstinate in his refusal to respond, he must be imprisoned until he does so.

ولا يجوز إحلافه حتى يلتمس المدّعي فإن تبرّع أو أحلفه الحاكم لم يعتدّ بها وأعيدت مع التماس المدّعي، فإن نكل ردّت على المدّعي وثبت حقه إن حلف، وإن نكل بطل وإن ردّ اليمين حلف المدّعي فإن نكل بطلت دعواه.

وإذا حلف المنكِر لم يكن للمدّعي المقاصة، ولا تسمع بينته بعد اليمين إلا أن يكذب نفسه.

ولو كان الدين على ميّت احتاج المدّعي مع البينة إلى يمين على البقاء استظهاراً.

ولو سكت المنكِر لآفة توصل إلى معرفة إقراره أو إنكاره، ولا يكفي المترجم الواحد، وإن كان عناداً حُبس حتى يجيب.

Section Three: Adjuring an Oath

Oaths can only be sworn upon Almighty God's Names. If the *dhimmī*'s oaths upon his own religion's sacredness are more binding on him, then it is allowed.

It is recommended for the judge to caution them [against taking false oaths], to alert them of the severe consequences of doing so and to emphasise the gravity of the matter, highlighting the scale of penalty due for the crime, such as cutting the hands of the thief for example, which will vary based on legal opinion, time and place. It suffices for the person to swear: "I swear by God that I do not owe him anything."

The mute person may swear an oath by gesturing. Oaths must be adjured in a public court of law if possible. The oath must be sworn in the affirmative, unless it is to deny someone else's actions, in which case it would be sworn to deny knowledge of it.

If the defendant counterclaims that he has indeed paid his dues or that the latter had written off his debt by gifting his loan to him, then he is now the plaintiff.

A fixed punishment cannot be administered on the basis of an oath, nor can one make an oath in the absence of knowledge, and nor to substantiate another's claim to property.

The testimony of a credible witness is accepted if the testimony begins before the oath has been sworn, in financial matters and debts, but not for moon-sighting, divorce and retributive punishments [because in all such cases two credible witnesses are required.]

If two credible witnesses testify to the judgment of a previous judge in front of second judge, then the second judge must enforce the ruling of the previous judge so long as it does not go against what has been permitted by the law.

الفصل الثالث: في الاستحلاف

ولا يجوز بغير أسماء الله تعالى، ولو كان أحلاف الذمي بدينه أردع جاز.

ويستحب الوعظ والتخويف والتغليظ، في نصاب القطع فما زاد بالقول والمكان والزمان ويكفي «والله ماله قبلي كذا».

ويمين الأخرس بالإشارة ولا يحلف إلا في مجلس القضاء مع المكنة.

واليمين على القطع، إلا في نفي فعل الغير فإنها على نفي العلم.

ولو ادّعى المنكر الإبراء أو الإقباض انقلب مدّعياً.

ولا يمين في حدّ، ولا مع عدم العلم، ولا يثبت مالاً لغيره.

وتقبل الشهادة مع اليمين إذا بدأ بالشهادة وعدل، في الأموال والديون، لا في الهلال والطلاق والقصاص.

وإذا شهد بالحكم عدلان عند آخر أنفذه الحاكم الثاني ما لم يناف المشروع.

Section Four: The Plaintiff

The plaintiff must be a legal subject, claiming on his own behalf or on behalf of someone over whom he has guardianship, regarding ownable property and is allowed to retrieve whatever belongs to him. He may likewise retrieve any debts he is owed despite the debtor's refusal, absence of evidence to the contrary, or by way of the debtor's willingness to repay the debt. If a plaintiff stakes a claim for an unclaimed asset [i.e. something not in the possession of any party or individual], the matter will be ruled in his favour when there is nobody to dispute his claim. In the case of a defendant who is absent, and whose adversary provides sufficient evidence, his assets are sold to cover his debt, but paid only to the plaintiff directly. If, however, he is making a claim on behalf of someone else [e.g. because he is that person's guardian or representative] then he will only be paid upon provision of a surety upfront.

If two people litigate a dispute over property that they jointly hold, then they have equal right to it, and each one must request the other to swear an oath. If the same property is held by only one of them, then it is ruled in favour of the one who possesses the property at the time of dispute with an oath of honesty. If it is property held by a third party, then it is ruled in favour of the one whose claim he substantiates, and the adversary would request him to swear an oath. If he substantiates both their claims to it, they have equal right to it, and each one must request his adversary to swear an oath. If he negates both their claims, then it remains in his possession.

If two spouses [divorcing] dispute over household belongings, it is said that those things fit for a man's use belong to the husband and those fit for a woman's use belong to the wife, and those that they both use are shared between them equally.

الفصل الرابع: في المدعي

ولا بدَّ أن يكون مكلّفاً مدعياً لنفسه أو لمن له الولاية عنه ما يصح تملّكه وله انتزاع العين، أما الدين فكذا مع الجحد وعدم البينة ومع عدم البذل. ولو ادّعى ما لا يد لأحد عليه قضي له به مع عدم المنازع. ويحكم على الغائب مع البينة ويباع ما له في الدين، ولا يدفع إلا بكفيل.

ولو تنازع اثنان ما في يدهما فلهما بالسوية، ولكل إحلاف صاحبه، ولو كان في يد أحدهما فللتشبث مع اليمين. ولو كانت في يد ثالث فهي لمن صدقه وللآخر إحلافه، فإن صدقهما تساويا ولكل إحلاف صاحبه، وإن كذبهما أقرت في يده.

ولو تداعى الزوجان متاع البيت قيل للرجل ما يصلح له وللمرأة ما يصلح لها، وما يصلح لهما بينهما

Shaykh al-Ṭūsī states in his *Mabsuṭ*, "If there is no evidence to prove the rightful owner of something and both sides claim it as theirs, it is ruled as belonging to both. If their proofs of evidence are contradictory, it is ruled in favour of the person living outside of the property unless the evidence of the one currently in possession of [i.e. residing in] the property is seen with justification [of his ownership] [i.e. the person residing in the house provides evidence that gives sufficient reasons for his ownership; e.g. property deeds, inheritance etc.]. If both their proofs are justified, then it is ruled in favour of the interfering party. If they both demonstrate convincing right to it, then it is ruled as belonging to them both, with each side receiving the part held by his adversary. If it was held by a third party, then it is ruled in favour of the most credible one, or the one who holds more of it. If they both hold equal parts, then the case is settled by drawing lots. The one who has drawn in his favour must swear an oath of honesty, after which it would be ruled in his favour. If he refuses, then his adversary must swear an oath [and the claim settled in his favour]. If they both refuse, then it is shared between them."

وقال في (المبسوط) : إذا لم تكن بينة ويدهما عليه فهو لهما ، ولو تعارضت البينتان قضى للخارج إلا أن تشهد بينة المتشبث بالسبب. ولو شهدتا بالسبب فللخارج، ولو تشبثا قضى لكل بما في يد صاحبه فيكون بينهما بالسوية ، ولو كان في يد ثالث قضي للأعدل فالأكثر عدداً ، وإن تساويا أقرع فيحلف من تخرجه القرعة ، فإن امتنع أحلف الآخر ، فإن امتنعا قسم بينهما .

Section Five: The Qualities of a Witness

There are six such qualities incumbent in a witness: maturity, mental capacity, faith, credibility, s/he should not have any conflicts of interest in the trial, and legitimate birth.

The testimony of children over ten years of age is accepted in cases of personal injury, provided their testimonies do not conflict and that there were not engaged in a prohibited act during the time when they witnessed the crime being perpetrated.

The testimony of *dhimmīs* is accepted for bequests in the absence of Muslim witnesses. The testimony of a person who behaves immorally is only acceptable after his repentance. Partners' testimonies for each other in matters related to their partnership is not admissible, nor of the executor of a will or the agent in matters over which he holds guardianship. The testimony of the enemy is equally not acceptable, and nor the son's testimony against his father though the reverse is allowed. Each of their testimonies in favour of the other, however, is acceptable, and the same goes for spouses.

The slave's testimony against his master is not accepted, but there is a difference of opinion regarding his testimony for other people. Once he is manumitted, it is accepted both for and against him [i.e. his ex-master].

If a person withholds their testimony due to an impeding factor, their testimony is accepted when that factor ceases.

The testimony of one who volunteers himself to be a witness is not accepted, nor the testimony of women in matters relating to moon-sighting, divorce and crimes with fixed penalties. Their testimonies are accepted accompanied by men's testimonies in personal and property law. Their testimonies are acceptable on their own in cases pertaining to ascertaining virginity and other private female matters.[5] The testimony given by a midwife in cases involving a quarter of the inheritance of a miscarried foetus born alive is accepted, as is the single testimony of a woman in cases involving a quarter of the bequest.

5 Such as childbirth, breastfeeding and menstruation.

الفصل الخامس: في صفات الشاهد

وهي ستة: البلوغ، وكمال العقل، والإيمان، والعدالة، وانتفاء التهمة وطهارة المولد.

وتقبل شهادة الصبيان في الجراح مع بلوغ العشر وعدم الاختلاف وعدم الاجتماع على المحرم.

وتقبل شهادة أهل الذمة في الوصية مع عدم المسلمين. ولا تقبل شهادة الفاسق إلا مع التوبة، ولا شهادة الشريك لشريكه فيما هو شريك فيه، ولا الوصي فيما له الولاية فيه، وكذا الوكيل، ولا العدو، ولا شهادة الولد على الوالد، ويجوز العكس، وتقبل شهادة كل منهما لصاحبه، وكذا الزوجان.

ولا تقبل شهادة المملوك على مولاه، وفي غيره قولان. ولو أعتق قبلت له وعليه.

ولو شهد من تحملها مع المانع بعد زواله قبلت.

ولا تقبل شهادة المتبرّع، ولا شهادة النساء في الهلال والطلاق والحدود، وتقبل مع الرجال في الحقوق والأموال، وتقبل شهادتهنّ بانفرادهنّ في العذرة وعيوب النساء الباطنة وشهادة القابلة في رُبع ميراث المستهل، وامرأة واحدة في رُبع الوصية.

Section Six: Miscellaneous Issues Pertaining to Testimonies

1) The witness is not permitted to give evidence without full knowledge. It is not sufficient to merely see written evidence without ascertaining its content, even if it has been submitted by others. In cases involving ownership of property, it suffices to have witnessed the person using it as theirs.

There are four matters that are substantiated through popularity: genealogy, independent ownership, *waqf* and marriage. If one hears someone make a confession, he may bear witness to it, but if he is told as hearsay, he may not testify to it.

2) A qualified witness is not allowed to withhold their testimony if they have knowledge about it and there is no risk of reasonable threat. If he is summoned to testify [in spite of other factors that may prevent his testimony from being accepted], then he must do so mandatorily by public duty (*wujūb kifāʾī*). He cannot testify against someone whom he does not know personally unless he is first identified by two credible witnesses. The witness is allowed to look at the face of a woman for the purpose of testifying.

3) A secondary testimony given to substantiate a primary testimony is accepted only in cases involving debts, property law and personal rights, but not crimes with fixed penalties.

There must be at least two primary just witnesses giving evidence, and it is acceptable for there to be two [secondary] witnesses to testify for each of the two primary witnesses, but a secondary witness is only accepted when the primary witness is unable to attend in person. If the primary witnesses reject the testimonies of the secondary witnesses, then the testimonies of the secondary witnesses are rejected, and the case is dismissed without judgment. A third testimony cannot be admissible at all.

الفصل السادس: في بقية مسائل الشهادات

الأولى: لا يحل للشاهد أن يشهد إلا مع العلم، ولا يكفي رؤية الخط مع عدم الذكر وإن أقام غيره، ويكفي في الشهادة بالملك مشاهدته متصرفاً فيه.

ويثبت بالسماع: النسب والملك الطلق والوقف والزوجية.

ولو سمع الإقرار شهد وإن قيل له لا تشهد.

الثانية: لا يجوز للشاهد كتمان الشهادة مع العلم وانتفاء الضرر غير المستحق ولو دعي للتحمل وجب على الكفاية، ولا يشهد على من لا يعرفه إلا بمعرفة عدلين، ويجوز النظر إلى وجه امرأة للشهادة.

الثالثة: تقبل الشهادة على الشهادة في الديون والأموال والحقوق لا الحدود.

ولا يكفي أقل من عدلين على أصل، ولو شهد اثنان على كل واحد من الأصلين قبلت، وإنما تقبل مع تعذر حضور شاهد الأصل. ولو أنكر الأصل ردّت الشهادة مع عدم الحكم، ولا تسمع الشهادة الثالثة في شيء أصلاً.

4) If the two witnesses retract their testimony before the case has been settled, it is dismissed. If they retract it after the ruling, it is not retracted but they are liable to pay a fine for their false testimony. If it turns out that they gave false evidence, the original ruling is amended accordingly. If there are any damages resulting or the ruling cannot be reversed, the witnesses must be held liable.

If the witnesses to a murder retract their testimony after the sentencing claiming that they were mistaken, they must be charged a fine. If, however, they intentionally gave false evidence, the heirs of the falsely convicted person have the right of retaliation against them all or against one of them. In the latter case, the surviving witness(es) would be obligated to pay compensation to the heirs of the false witness who was chosen to be killed. If there is any outstanding amount payable, their next of kin would be charged with paying it.

If only one of the witnesses retracted his testimony, if he intentionally lied, he must be killed in retaliation and the rest of them must pay compensation to the heirs of the wrongfully convicted person. If he mistakenly gave false evidence, he must be made to pay the blood-money compensation due to the heirs of the wrongfully convicted person.

If two witnesses gave testimony in the case of a theft whereby the hand of the convict was cut, then later retracted their testimony saying that they had been mistaken about his identity, they are fined the blood-money compensation due for the hand, and their testimony regarding the second suspect is not accepted.

5) Those who give false evidence should be publicly denounced as such and be given disciplinary punishments at the Imam's discretion.

الرابعة: إذا رجع الشاهدان قبل الحكم بطل، وإن كان بعده لم ينقض وغرمهما.

ولو ثبت تزويرهما استعيدت العين، فإن تلفت أو تعذّر الاستعادة ضمن الشهود.

ولو قال شهود القتل بعد القصاص «أخطأنا» غرموا، وإن قالوا «تعمّدنا» اقتصّ منهم أو من بعضهم ويرد البعض ما وجب عليهم، فإن فضل شيء أتمّه الولي، ولو قال بعضهم ذلك ردّ عليه الولي ما يفضل عن جنايته، واقتصّ منه إن كان عمداً وأخذ منه ما قابل فعله من الدية إن قال «أخطأت».

ولو شهدا بسرقة فقطعت يد المشهود عليه ثم قالا: «أوهمنا» والسارق غيره، غرما دية اليد، ولا يقبل قولهما على الثاني.

الخامسة: يجب شهرة شاهد الزور وتعزيره بما يراه الإمام رادعاً.

Section Seven: The Fixed Punishment for Fornication

Fornication is when the man's penis penetrates the woman's vulva to the point where the glans disappears, from the front or the rear, without any marital contract of any kind, nor concubinage. [For a man to be considered a fornicator], he must be legally mature and mentally sound, knowing of its prohibition and acting wilfully. If he intentionally contracts a marriage with a forbidden relative, knowing of its prohibition, then he is to be punished. If a strange woman fornicates with a man pretending to be his wife, she is punished but not him. If he claims to be married to her or claims that he was convinced that the woman he had sexual intercourse with was his wife but later on he discovered that she was not, he is not punished for it.

If he knowingly marries a woman in her waiting period, he must be punished if the marriage is consummated, as must the woman. If one of them claims probable ignorance, it is accepted.

The blind person is punished only after discounting the probability of mistaken identity whereby the strange woman is assumed to be one's wife, but not in spite of it.

Fornication is established either through the perpetrator's own confession four times, or through the testimony of four credible male witnesses, or three male witnesses and two females. If two male witnesses and four females give testimony, the convict may be sentenced to flogging but not stoning. A single male testimony alongside women's is not acceptable, even if there are many women. If there are less than four witnesses, they must be punished for false testimony.

The witnesses' testimonies must all match each other's in every aspect [i.e. time, place and session] and must be eye-witness accounts of the act of penetration itself.[6] If they testify based on having witnessed them lying together, embracing, kissing or fondling, then they are subject to a discretionary punishment.

6 Literally like "the wand going into the kohl pot".

الفصل السابع: في حدّ الزنا

وهو يثبت بإيلاج فرجه في فرج امرأة، حتى تغيب الحشفة، قبلاً أو دبراً، من غير عقد ولا شبه عقد ولا ملك، بشرط بلوغه وعقله وعلمه بالتحريم واختياره، ولو علم التحريم وعقد على المحرم ثبت الحدّ، ولو تشبّهت الأجنبية عليه حدّت دونه، ولو ادّعى الزوجية أو ما يصلح شبهة سقط الحدّ.

ولو تزوّج المعتدّة عالماً حدّ مع الدخول، وكذا المرأة، ولو ادّعى أحدهما الجهالة المحتملة قبل. ويحدّ الأعمى مع انتفاء الشبهة المحتملة لا معها.

ويثبت بالإقرار من أهله أربع مرات، أو بشهادة أربعة رجال عدول أو ثلاثة وامرأتين، ولو شهد رجلان وأربع نسوة ثبت الجلد دون الرجم، ولا يقبل رجل واحد مع النساء وإن كثرن. ولو شهد أقل من أربعة حدّوا للفدية.

ويشترط في الشهادة اتّفاقها من كل وجه والمشاهدة عياناً كالميل في المكحلة، ولو شهدوا بالمضاجعة والمعانقة والتقبيل والتفخيذ ثبت التعزير.

If the accused confesses to that for which he is sentenced to stoning [i.e. adultery], then retracts his confession, his sentence is lifted; but if he confesses to that which warrants capital punishment [e.g. an extramarital affair or any other crime such as drinking or murder], then his retraction makes no difference. If he confesses then repents, the Imam decides what to do with him. If he repents after the evidence has been submitted, the punishment is determined per the evidence, but if it is before, then he is not sentenced to the fixed punishment.

The man who fornicates with his mother, or one of his forbidden female relatives – related by blood or milk – or his father's wife, or a Muslim woman if he is a *dhimmī*, and any man who rapes a woman must be killed, regardless of whether he is married or not, slave or free, Muslim or non-Muslim.

As for the man who fornicates with other unrelated women, if he is married – meaning that he has access to sexual intercourse in either a permanent marriage or concubinage – and mentally sound, must be flogged one hundred lashes, then stoned [to death] if he fornicated with a female of mature age and sound mind. If it was with a minor or a mentally incapacitated woman, then he is only flogged.

Similarly, the married woman who fornicates must be stoned after being flogged, if she had physical access to lawful sexual intercourse as the man. If a man who has been divorced by his wife through *khulʿ* returns back to his wife and then has illicit sex with another woman, he is only flogged if he had not re-consummated the marriage with his wife, but if had re-consummated the marriage then he is flogged and stoned to death. The same rules apply to the manumitted slave, and the *mukātab* slave who has bought his full freedom.

If a married woman fornicates with a minor, she must be flogged, but if it is with a mentally incapacitated person, she is also stoned. If the fornicator is also a man who is either unmarried or married but has no access to his wife, he is flogged a hundred lashes, then his head is shaved, and he is banished from the town [for a year]. The woman and the slave's heads are not shaved, and they are not banished.

If a man fornicates again after having been punished, he must be punished again. If his crime is a repeat crime for one which he was not previously punished for, then one punishment suffices for both counts. If he fornicates a third time after the two previous counts, he must be killed. Others say he must be killed on the fourth count. The same applies to a woman who reoffends.

The slave who fornicates must be flogged fifty lashes, whether he is married or not, and the same applies to the female slave. For repeated counts of fornication, they are killed upon the eighth or ninth count after having been punished upon each count.

ولو أقرّ بما يوجب الرجم ثم أنكر سقط، ولو كان بحدّ لم يسقط. ولو أقرّ ثم تاب تخيّر الإمام، ولو تاب بعد البينة تحتمت الإقامة، ولو كان قبلها سقط الحدّ.

ويقتل الزاني بأمه أو بإحدى المحرمات نسباً أو رضاعاً أو بامرأة الأب، أو بالمسلمة إذا كان ذمياً، أو بمن أكرهها عليه، محصَّناً كان أو غير محصَّن عبداً أو حراً مسلماً أو كافراً.

أما الزاني بغير المحرمات نسباً أو رضاعاً، فإن كان محصّناً - وهو الذي له فرج مملوك بالعقد الدائم أو الملك يغدو إليه ويروح - ويكون عاقلاً، جلد مائة، ثم رجم إن زنى ببالغة عاقلة، وإن كان بصغيرة أو مجنونة جلد خاصة.

وكذا المرأة المحصّنة ترجم بعد الحدّ، وإحصانها كإحصان الرجل.

ولو راجع المخالع لم يرجم حتى يطأ، وكذا العبد إذا أعتق، والمكاتَب إذا تحرّر.

ولو زنت المحصّنة بصغير حدّت، ولو كان بمجنون رجمت، وإن كان غير محصّن جلد مائة سوط وحلق رأسه وغرب عن البلد. وليس على المرأة والمملوك جز ولا تغريب.

فإن زنى بعد الحدّ ثانية تكرّر الحدّ، وإن لم يحدّ كفى حدّ واحد، فإن زنى ثالثة بعد الحّدين قُتل، وقيل في الرابعة، وكذا المرأة.

أما المملوك فيجلد خمسين محصّناً كان أو غيره، وكذا المملوكة، ويقتل في الثامنة أو التاسعة مع تكرار الحدّ في كل مرّة.

Other Rulings:

1) The judge has the option of either sentencing *dhimmīs* [who commit illicit sexual acts] to fixed punishments or to return them to their own religious authorities to sentence them.

2) The pregnant woman cannot be punished until after she has given birth and the child is independent of her [i.e. fully weaned]. Similarly, the sick person and the woman bleeding irregularly must not be flogged [until after they have recovered]. However, [if they commit adultery] then they are stoned [immediately]. If the situation necessitates that the sick person be flogged [before they have recovered from their illness], then it is to be done with a stick having a hundred whips and he is lashed once.

The punishment must not be administered in extremely hot or cold weather, nor in enemy territories. The perpetrator who has taken refuge in the Sanctuary [of Mecca] must not be punished therein but instead denied food and drink until he is compelled to come out, whereupon it must be meted out. If he fornicated in the Sanctuary itself, then he is punished therein.

3) If a person is sentenced to both stoning and flogging, the flogging precedes. The man sentenced to stoning must be placed in the ground up to his hips, and the woman up to her chest. If either of them escapes, having been convicted through evidence against them, they must be brought back. If they were convicted by their own confession and they escape during the stoning, they are allowed to do so. The witnesses throw the first stones [if he/she was convicted through testimony], and the Imam if it was through confession.

4) The man is to be stripped, then flogged and must be lashed hard, avoiding his face. The woman must be flogged whilst seated and fully clothed, with her clothes bound tightly to her body.

5) A man who marries a slave woman in addition to being married to a free Muslim woman, and has intercourse with her without his wife's permission must be given an eighth of the penalty as the fixed punishment of a fornicator. If someone fornicates at a sacred time or in a sacred place, he must be flogged more than the stipulated fixed punishment.

مسائل:

الأولى: للحاكم إقامة الحدّ على أهل الذمة، ورفعه إلى أهل ملته ليقيموه عليه.

الثانية: لا يقام الحد على حامل حتى تضع، ويستغنى الولد، ولا المريض ولا المستحاضة وترجمان.

ولو اقتضت المصلحة تقديم حدّ المريض ضرب بضغث فيه مائة سوط دفعة.

ولا يقام في شدّة الحرّ ولا البرد، ولا في أرض العدو، ولا على الملتجئ إلى الحرم، ويضيق عليه في المطعم والمشرب حتى يخرج فيقام عليه الحدّ، ولو زنى في الحرم حدّ فيه.

الثالثة: لو اجتمع الجلد والرجم بدئ بالجلد، ويدفن المرجوم إلى حقويه والمرأة إلى صدرها، فإن فرّ أحدهما وقد ثبت بالبينة أعيد، وإن كان بالإقرار لم يعد مع إصابة الحجر. ويبدأ الشهود بالرجم، وفي الإقرار الإمام.

الرابعة: يجرّد للجلد، ويضرب أشد الضرب، ويتقي وجهه وتضرب المرأة جالسة وقد ربطت عليها ثيابها.

الخامسة: من تزوّج بأمة على حرة مسلمة فوطأها قبل الإذن كان عليه ثمن حدّ الزاني، ومن زنى في زمان شريف أو مكان شريف ضرب زيادة على الجلد.

Section Eight: The Punishment for Sodomy, Lesbian Sexual Acts and Pimping

The act of sodomy must be proven in the same way that fornication is proven, and if convicted of penetrative anal sex, the perpetrator is either killed, stoned to death, thrown from a towering height or burned. The Imam may also opt to burn him or sentence him to death by other means, even if it was with a minor or mentally incapacitated person.

If a mentally incapacitated person or minor commit the act with a sane adult, they must be disciplined but the latter is sentenced to death.

If a slave claims that his master raped him, his claim is accepted, but if found that he has not, he is killed. If a *dhimmī* commits the act with a Muslim, he must be killed even if the act was not penetrative.

The passive partner to the act is also killed if the act was penetrative, but if not, then he must be flogged one hundred lashes, whether he is a free man or a slave, and whether he was the active or passive partner. If the crime is repeated, he is killed on the fourth count.

Two unrelated men who are found naked under the same cover are to be sentenced to discretionary punishment of between thirty to ninety-nine lashes. If they repeat the offence, they are to be sentenced to death on the third count. A man who passionately kisses a young boy must also be sentenced to discretionary punishment.

The lesbian sexual act is proven the same way as fornication, and if convicted the active as well as the passive partner, free or slave, are sentenced to one hundred lashes. If she repeats the offence, she is killed on the fourth count.

If the person accused, of sodomy for example, repents before the case is taken to court, the punishment is not administered, but if he repents after it is proven in court, then it is not lifted.

Two women found naked under the same cover are to be sentenced to discretionary punishment, and if they repeat the offence twice, they are sentenced to the fixed punishment.

The fixed punishment for pimping is seventy-five lashes, after which his head must be shaved, he must be publicly defamed, then exiled, whether he is a free man or a slave, Muslim or non-Muslim. The female pimp's head is not shaved and nor is she exiled. The crime is convicted by testimony of two witnesses or a confession twice.

الفصل الثامن: في اللواط والسحق والقيادة

يثبت اللواط بما يثبت به الزنا إن أوقب قتل، أو رجم، أو أُلقي من شاهق، أو أحرق، وللإمام إحراقه أو قتله بغيره، وإن كان بصغير أو مجنون.

ولو لاط المجنون أو الصغير بعاقل أدبا، وقتل العاقل.

ولو ادّعى العبد إكراه مولاه قبل وإلا قُتل. ولو لاط الذمي بمسلم قُتل وإن لم يوقب.

ويقتل المفعول مع الإيقاب، ولو لم يوقب جلد مائة، حراً كان أو عبداً، فاعلاً أو مفعولاً. ولو تكرّر الحدّ قتل في الرابعة.

ويعزّر الأجنبيان المجتمعان في إزار واحد مجرّدين من ثلاثين إلى تسعة وتسعين، ولو تكرّر التعزير حدّ في الثالثة، ويعزّر من قبّل غلاماً بشهوة.

ويثبت السحق بما يثبت به الزنا، ويجب فيه جلد مائة على الفاعلة والمفعولة الحرة والأمة سواء، ولو تكرّر الحدّ قُتلت في الرابعة.

ويسقط الحدّ بالتوبة قبل البينة كاللواط ولا يسقط بعدها.

وتعزّر المجتمعتان تحت إزار واحد مجرّدتين، وتحدّان لو تكرّر التعزير مرتين.

ويحدّ القواد خمساً وسبعين جلدة، ويحلق رأسه، ويشهر وينفى، حراً كان أو عبداً مسلماً أو كافراً، ولا جز على المرأة ولا نفي. ويثبت بشاهدين، أو الإقرار مرتين.

Section Nine: The Punishment for False Accusation, Slander and Verbal Abuse

If a legal subject falsely accuses a free, mature, sane adult, married Muslim by saying, "You adulterer" or "You sodomiser" or "You who has been sodomised" or "You are an adulterer" or "sodomiser" in any language to that effect, knowing the full meaning of his words, he must be flogged eighty lashes, whether he is a free man or a slave.

If a person falsely accuses someone who he has previously acknowledged to be his son, saying to him, "You are not my son" or tells someone else "You are not your father's child" [thus implying that he is of illegitimate birth], he must be given a fixed punishment. If he falsely accuses another person saying, "You son of an adulterer and adulteress" or "You son of adulterous parents", the punishment of the false accuser is a right belonging to the parents if they were Muslims, even if their child is a disbeliever. If someone falsely accuses a Muslim, "You son of a disbelieving woman" or "Your mother is an adulteress", he must be given a discretionary punishment. If he falsely accuses someone, saying, "You husband of an adulteress" or "You brother of an adulteress" or "You father of an adulteress", the fixed punishment is the right of the individual who he is falsely accusing of illicit sexual practice, and not the person being addressed. If he were to falsely accuse someone by saying, "You committed adultery with *x* (woman)" or "So and so committed sodomy with you" or "you committed sodomy with him", he must be given a double fixed punishment [since he has accused two individuals of having engaged in illicit sexual practice and hence both have a right to demand his punishment].

Any verbal insult that humiliates someone [a Muslim] and refers to a cause of shame that is not known in public is subject to discretionary punishment, such as a man's saying to his wife, "I did not find you a virgin", or "I had a wet dream about your mother last night", or "You indecent person" or "You drunkard."

Similarly, anyone who slanders a child or a mentally incapacitated person, or calls someone of Muslim appearance a disbeliever, or falsely accuses someone of adultery, or a father who slanders his son must be given a disciplinary punishment.

In the case of someone making false accusations of a group of people, if they take it up against him collectively, he is punished once, but if they each make a claim against him separately, then he must be punished for each count.

الفصل التاسع: في حد القذف

من قال من المكلفين للبالغ العاقل الحر المسلم «يا زان» أو «يا لائط» أو «يا منكوحاً في دبر» أو «أنت زان» أو «لائط»، بأي لغة كانت، مع معرفة القائل بالفائدة حدّ ثمانين جلدة، حراً كان أو عبداً.

ولو قال لمن اعترف ببنوته «لست بولدي» أو قال لغيره «لست لأبيك»، وجب الحدّ. ولو قال: «يا بن الزاني أو الزانية» أو «يا بن الزانيين» فالحدّ للأبوين إذا كانا مسلمين ولو كان المواجه كافراً، ويعزّر لو قال للمسلم: «ابن الكافرة» أو «أمك زانية». ولو قال: «يا زوج الزانية» أو «يا أخ الزانية» أو «يا أب الزانية» فالحدّ للمنسوبة إلى الزنا دون المخاطَب. ولو قال: «زنيت بفلانة» أو «لاط بك فلان» أو «لطت به» وجب حدّان.

ويعزّر في كل قول موجب للاستخفاف، كقوله لامرأته «لم أجدك عذراء» أو «احتلمت بأمك البارحة» أو «يا فاسق» أو «يا شارب الخمر» إذا لم يكن المقول له متظاهراً.

وكذا يعزّر قاذف الصبي والمجنون والكافر والمملوك والمتظاهر بالزنا، والأب إذا قذف ولده.

ولو قذف جماعة، فإن جاؤوا به مجتمعين فعليه حدّ واحد، وإن جاؤوا متفرّقين فلكل واحد حدّ.

A person is charged with making a false accusation by their own confession twice, so long as they qualify as a legal subject, or by testimony of two just witnesses. The child and mentally incapacitated person are given discretionary punishment if they slander.

The right to demand a fixed punishment is inherited just as wealth is, but not between spouses. If one of the heirs chooses to forgo his right to demand punishment, the other heirs still retain their right to demand justice [from the just ruler].

If the offence is repeated three times, he must be sentenced to death on the fourth count.

If two people make false accusations of each other, they must both be given discretionary punishment.

Someone who insults the Prophet or one of the Imams must be killed. Anyone who hears him is allowed to kill him, ensuring his own safety from harm. Similarly, the false claimant to prophethood must also be killed, as is the one who says, “I do not know whether Muḥammad spoke the truth” and belies him whilst living as a Muslim, and the Muslim who practices sorcery. If the sorcerer is a disbeliever, he must be given a discretionary punishment.

ويثبت القذف بالإقرار مرتين من المكلّف، أو بشهادة عدلين. ويعزّر الصبي والمجنون إذا قذفا. والحد موروث كالمال، ولا ميراث للزوجين، ولو عفي أحد الوراث كان للباقي الاستيفاء على التمام. ولو تكرّر الحدّ ثلاثا قُتل في الرابعة. ولو تقاذف اثنان عزّرا.

ويُقتل من سبّ النبي ﷺ أو واحداً من الأئمة عليهم السلام. ويحل لكل سامع قتله مع أمن الضرر، وكذا يقتل مدّعي النبوة. ومن قال: لا أدري صدق محمد ﷺ، وكذّبه مع تظاهره بالإسلام أولاً، والساحر إذا كان مسلماً، ويعزّر الكافر.

Section Ten: The Fixed Punishment for Consuming Intoxicants

Any legally responsible person who willingly consumes an intoxicating spirit, beer or fermented grape juice that has not evaporated into vinegar [i.e. two-thirds of it], knowing of its prohibition, must be flogged eighty lashes on bare skin on his back and shoulders, avoiding the face and private parts, upon his sobriety, regardless of whether he is a free man or a slave or a disbeliever who drinks it publicly. If he repeats the offence thrice, he must be killed upon the fourth count.

If one consumes alcohol deeming it permissible, he is considered an apostate [and sentenced as such and therefore killed], and the one who deems other intoxicants besides alcohol permissible is sentenced to the fixed punishment [of eighty lashes].

If one sells alcohol deeming it permissible, he must be made to repent. If he refuses, he must be killed, and the seller of other intoxicants must be sentenced to discretionary punishment. If he repents before the case is taken to court, the punishment is not meted out, but if he does so after his conviction, it is upheld. If he confesses then repents, he is dealt with at the Imam's discretion.

The offence is convicted by testimony of two male witnesses or through one's own confession twice.

If someone consumes an intoxicant, inadvertent that it was such or ignorant of its prohibition, the fixed punishment is lifted.

Whoever deems permissible what Muslims unanimously know to be prohibited such as consuming carrion, for example, must be killed. If he consumes it, knowing that it is prohibited, he must be given a discretionary punishment.

There is no blood-money compensation owed for someone who has been given the death penalty or a discretionary punishment, and if it transpires that the witnesses gave false testimony [due to which he was wrongfully killed], the blood-money is paid [to his next of kin] from the public treasury.

الفصل العاشر: في حد المسكّر

من تناول مسكراً أو فقاعاً أو عصيراً قد غلا قبل ذهاب ثلثيه اختياراً مع العلم بالتحريم والتكليف حدّ ثمانين جلدة عارياً على ظهره وكتفه، ويتقي وجهه وفرجه بعد الإفاقة، حراً كان أو عبداً أو كافراً متظاهراً. ولو تكرّر الحدّ ثلاثاً قُتل في الرابعة.

ولو شرب الخمر مستحلاً فهو مرتدّ ويحدّ مستحلّ غيره.

ولو باع الخمر مستحلاً استتيب، فإن تاب وإلا قُتل، ويعزّر بائع غيره.

ولو تاب قبل قيام البينة سقط الحدّ، ولا يسقط بعدها. ولو أقر ثم تاب تخيّر الإمام.

ويثبت بشهادة عدلين، أو الإقرار مرتين من أهله.

ولو شرب المسكر جاهلاً به أو بالتحريم سقط الحدّ.

ومن استحلّ ما أجمع على تحريمه كالميتة قُتل، ولو تناوله محرّما عزّر.

ولا دية لمقتول الحد أو التعزير، ولو بان فسق الشهود فالدية في بيت المال.

Section Eleven: The Fixed Punishment for Theft

The conditions that must be fulfilled before amputating a thief's [hand] are: his status as a legal subject, the absence of any doubt [in the thief's mind that what he has taken does not belong to him], the breach of a well-secured property that is locked away, hidden behind closed doors or buried; the removal of the item whose value should exceed a quarter dinar of pure gold in the currency of the time; in person, illicitly.

Upon fulfilment of these conditions, the four fingers of the right hand are amputated. If he reoffends, his left foot is amputated from the ankle bone, leaving the heel. If he reoffends a third time, he must be sentenced to life imprisonment, and if he steals in prison then he is to be killed. If the re-offence of theft occurs with the previous count having gone unpunished, a single punishment suffices.

If a minor or mentally incapacitated person steals, they are sentenced to discretionary punishment. The slave is not amputated for stealing from his master, though the employee, the husband, wife, and guest are, if they steal secured property from their respective counterparts. The thief must be made to return what he has stolen.

The thief who steals from public places, such as public bath houses and mosques, may not be amputated, nor the one who picks pockets that are evidently visible. If they were hidden away or inside another person's clothing, then he must be amputated.

The stealer of shrouds must be amputated as well as the one who traffics [kidnapped] slaves and free people. Someone who digs up graves without taking anything must be sentenced to discretionary punishment, but if he reoffends, having escaped the authorities previously, he must be killed.

Theft is convicted by testimony of two just male witnesses or by confession of the perpetrator twice. A single confession and a single testimony accompanied by an oath of honesty only serve to prove the loss of the property [but not to convict for amputation]. If the accused repents before the charge is proven, the fixed punishment [of amputation] no longer applies, but if he repents afterwards, then it does. If he repents after his confession, his punishment is at the Imam's discretion.

الفصل الحادي عشر: في حدّ السرقة

ويشترط في قطع السارق: التكليف، وانتفاء الشبهة، وهتك الحرز - وهو المستور بقفل أو غلق أو دفن - وإخراج النصاب وهو ما قيمته رُبع دينار خالصاً مضروباً بسكة المعاملة بنفسه سراً.

ومع الشرائط تقطع أصابعه الأربع من يده اليمنى، فإن عاد قطعت رجله اليسرى من مفصل القدم ويترك له العقب، فإن عاد ثالثاً خلد السجن، فإن سرق فيه قتل. ولو تكرّرت السرقة من غير حد كفى حد واحد.

ولو سرق الطفل أو المجنون عزّرا، ولا يقطع العبد بسرقة مال السيد، ويقطع الأجير والزوج والزوجة والضيف مع الإحراز دونهم. ويستعاد المال من السارق.

ولا يقطع السارق من المواضع المنتابة كالحمامات والمساجد، ولا من الجيب والكم الظاهرين، ولو كانا باطنين قطع.

ويقطع سارق الكفن، وبايع المملوك والحر، ولو نبش ولم يأخذ عزّر، فإن تكرّر وفات السلطان قتله.

ويثبت بشهادة عدلين أو الإقرار مرتين من أهله، ويكفي في غرم المال المرة وشهادة الواحد مع اليمين. ولو تاب قبل البينة سقط الحدّ لا بعدها، ولو تاب بعد الإقرار تخيّر الإمام.

Other Rulings

1) If two people jointly steal something worth the minimum punishable value, the predominant opinion is that neither of them faces amputation, until and unless each of their share equals or exceeds the minimum value.

2) Punishment by amputation is dependent upon the owner pressing charges. If the victim of the theft does not take the case to court, the Imam cannot punish by amputation. If the victim chooses to let the thief keep the item as a gift or pardons him from [the punishment of his limbs] being amputated, he may do so only before the matter is taken to court and not afterwards.

3) Whether one steals property of minimum value all at once or over numerous instances, the predominant opinion is that he must be amputated.

4) If the father steals from his son's property, he is not amputated, but if the son steals from his father, he is.

5) The right hand must be cut, even if it is paralysed or both his hands are paralysed, or he does not have a left hand. If he does not have a right hand, then his left hand is cut, or according to another opinion, his left foot.

مسائل:

الأولى: لو سرق اثنان نصاباً فالأقوى سقوط الحدّ عنهما حتى يبلغ نصيب كل واحد النصاب.

الثانية: قطع السارق موقوف على المرافعة، فلو لم يرافعه المسروق منه لم يقطع الإمام. لو وهبه أو عفا عن القطع سقط إن كان قبل المرافعة وإلا فلا.

الثالثة: لو أخرج النصاب دفعة وجب القطع، وكذا لو أخرجه مراراً على الأقوى.

الرابعة: لو سرق الوالد من مال ولده لم يقطع، ولو سرق الولد قطع.

الخامسة: يقطع اليمين وإن كانت إحدى يديه أو هما شلاوين أو لم يكن له يسار ولو لم يكن له يمين قطعت يساره، وقيل رجله اليسرى.

Section Twelve: The Fixed Punishment of the *Muḥārib* (One Who Revolts Against the Islamic State) and His Like

Whoever takes up arms to terrorise people on land or sea, by night or day, it is at the Imam's discretion to either put him to death, crucify him, amputate his opposite limbs [i.e. right hand and left foot] or to banish him. If he repents before he is charged, he may be spared the fixed punishment against the state but is still liable for crimes against people. If he repents after the charge, the punishment is not lifted.

If he is banished, the authorities must write to all neighbouring towns that they must not transact, eat or socialise with him, until he repents.

The armed robber is treated as a *muḥārib*, and a person is allowed to defend himself against him if his own safety is not compromised.[7] If he kills the thief in self-defence, there is no blood-money compensation.

A woman or child who are the target of rape are allowed to defend themselves against their assailant, and if they kill him, there is no blood-money compensation.

If a thief enters a house and is prevented from entering by its inhabitants, but he refuses to desist, then they are not held liable for any harm caused to him or to any of his limbs.

Embezzlers, looters and imposters who give false testimony and the like to defraud people, as well as those who drug people to immobilise them, must all be sentenced to discretionary punishments and made to return whatever they stole.

7 If, however, there is a good chance that he could lose his life, then he should not confront the armed robber.

الفصل الثاني عشر: في حدّ المحارب

كل من جرّد السلاح للإخافة في برّ أو بحر ليلاً أو نهاراً، تخيّر الإمام بين: قتله، وصلبه، وقطعه مخالفاً، ونفيه. ولو تاب قبل القدرة عليه سقط الحدّ دون حقوق الناس، ولو تاب بعدها لم يسقط.

وإذا نفى كتب إلى كل بلد بالمنع من معاملته ومؤاكلته ومجالسته إلى أن يتوب.

واللص محارب يدفع مع غلبة السلامة، فإن قتل فهدر.

ومن كابر امرأة على فرجها أو غلاماً فلهما دفعه فإن قتلاه فهدر.

ومن دخل دار قوم فزجروه فلم ينزجر لم يضمنوا تلفه أو تلف بعض أعضائه.

ويعزّر المختلس والمستلب، والمحتال بشهادة الزور وغيرها، والمبنج، بما يرتدع غيره [به] ويستعاد منه ما أخذه.

Other Rulings:

1) If a sane person of legal responsibility has sexual intercourse with an animal, he must be sentenced to discretionary punishment. If it is an animal whose meat is consumable, its meat and the meat of its offspring is henceforth forbidden, and it must be slaughtered then burned, and its owner compensated for its value. If it is not possible to identify the animal out of the herd, the herd must be divided in half, then lots must be drawn until a single animal remains.

If its meat is not consumable, it must be taken out of the town and sold in another town, and its value paid to its owner if the owner was not the perpetrator. If the animal belongs to the perpetrator, then its value must be given away in charity. The perpetrator's act is proven by two reliable male witnesses or by the perpetrator confessing twice. If the offence is repeated, the perpetrator must be killed on the fourth count.

2) The crime of having sexual intercourse with a corpse is treated the same as fornicating with a living person in terms of the punishment and the consideration of whether the perpetrator is married or not, except that the crime is more harsh here and dealt with more seriously. If the corpse is the perpetrator's own wife, he is given a discretionary punishment. The crime is convicted by testimony of four male witnesses. The rulings for one who sodomises a corpse are the same as for the one who sodomises a living person, though his punishment is harsher.

3) The person who is caught masturbating with his hand must be sentenced to a discretionary punishment and is convicted by testimony of two male witnesses or him confessing once.

4) A person is allowed to defend himself, his dependants and his property by any means necessary. He must first resort to the safest option then move on to more aggressive options only if he cannot defend himself with it. If someone spies on private residences of a community whose residents request him to refrain from doing so, if that prohibition turns out to be futile and they subsequently retaliate by throwing stones or rocks at him, then they are not required to paid any compensation for any injury they cause.

مسائل:

الأولى: إذا وطأ البالغ العاقل بهيمةً عزّر، ثم إن كانت مأكولة اللحم حرم لحمها ولحم نسلها، وتذبح وتحرق ويغرم قيمتها لصاحبها، ولو اشتبهت قسم القطيع نصفين ثم أقرع ثم قسم الخارج بالقرعة إلى أن يقع إلى واحدة.

ولو كانت غير مأكولة أخرجت من البلد وبيعت في غيره، ويغرم قيمتها لصاحبها إن لم يكن له، ويتصدّق بالثمن على رأي.

ويثبت بشهادة عدلين، أو الإقرار مرتين. ولو تكرّر التعزير قُتل في الرابعة.

الثانية: من زنى بميتة فهو كمن زنى بحية في الحدّ واعتبار الإحصان، ويغلظ هاهنا العقوبة، ولو كانت الميتة زوجة عزّر، ويثبت بأربعة. وحكم اللائط بالميت حكم اللائط بالحي ويغلظ عقوبته.

الثالثة: من استمنى بيده عزّر، ويثبت بشهادة عدلين والإقرار مرة.

الرابعة: للإنسان الدفع عن نفسه وحريمه وماله ما استطاع. ويجب الأسهل فإن لم يندفع به انتقل إلى الأصعب، ومن أطلع على قوم فزجروه فلم ينزجر فرموه بحصاة أو عود فجني عليه فهدر.

The Book of Retributive Punishments and Blood-Money Compensation

كتاب القصاص والديات

This chapter contains several sections.

Section One: Homicide

This can be either:

a) intentional murder, which is to commit any act with the intention of killing another person, such as when a person intends to kill another human being with an action that is capable of leading to that person's death even if the act in question does not cause death most of the time; or if he intentionally carried out an action which on most occasions causes death, even if he did not intend to kill.

b) involuntary manslaughter, whereby an individual commits an action wittingly but not with the intention of causing death, such as beating to discipline someone, who then dies.

c) purely accidental, whereby the action and the death are both accidental, such as aiming at a bird but striking a person instead, and similarly for various categories of injuries.

Retributive punishment [i.e. death penalty] for the first type of homicide is convicted when the perpetrator is a sane, legally responsible person killing an innocent counterpart, regardless of whether the killing was done in person such as throat-slitting or strangling, or whether it was done by other means such as arrows, stones, repeated beating with a stick leading to death, or throwing someone to be preyed on by lions, and even fatally wounding someone to later cause their death. In the latter case, both retribution and blood-money compensation for the loss or injury of a limb are accounted for by the compensation payable for a life being taken or the retribution of murder. If he wounded him, then killed him, he is liable for both types of retribution if they were separate instances, but if not, then only for the loss of life.

If someone compels another person to kill, the killer is given the death penalty, as would be the case if he orders another person to kill, even if he is a slave being ordered to kill. The person who ordered the killing is sentenced to life imprisonment.

If one person holds him [i.e. the victim] down whilst a second kills him and a third watches, the killer is sentenced to death, the one who held him down to life imprisonment, and the bystander to having his eye gauged out.

وفيه فصول:

الفصل الأول: في القصاص

القتل: إما عمد. وهو أن يقصد بفعله إلى القتل، كمن يقصد قتل إنسان بفعل صالح له ولو نادراً، أو يقصد إلى فعل يقتل غالباً وإن لم يقصد القتل.

وإما شبيه عمد. وهو أن يكون عامداً في فعله مخطئاً في قصده، كمن يضرب تأديباً فيموت.

وإما خطأ محض. بأن يكون مخطئاً في الفعل والقصد معاً كمن يرمي طائراً فيصيب إنساناً، وكذا أقسام الجراح.

ويثبت القصاص بالأول مع صدوره من البالغ العاقل، في النفس المعصومة المتكافئة، سواء كان مباشرة كالذبح والخنق، أو تسبيباً كالرمي بالسهم والحجر والمتكرّر بالعصا بحيث لا يحتمله مثله، والإلقاء إلى الأسد فيفترسه، وكذا لو جرحه فسرت الجناية فمات، ويدخل قصاص الطرف وديته في قصاص النفس وديتها. ولو جرحه ثم قتله فإن فرق اقتصّ منهما وإلا فالنفس.

ولو أكره غيره على القتل اقتصّ من القاتل، وكذا لو أمر، ويخلد الآمر السجن به، وإن كان عبد الآمر.

ولو أمسكه واحد وقتله آخر ونظر ثالث قُتل القاتل وخلّد الممسك [السجن] وسملت عين الناظر.

Section Two: Pre-Conditions of the Death Penalty

There are five such conditions.

1) Freedom: If the killer is a free man, he is not sentenced to death for killing a slave, nor a *mukātab* slave, nor is a free woman sentenced to death for killing a slave mother (*umm al-walad*), nor a free man for killing a *mudabbar* slave. He must instead be made to pay their market value on the day that he killed them, and this must not exceed the amount of blood-money due for a free man.[1] Similarly, the market value of a slave woman must not exceed the blood-money due for a free woman,[2] nor must the blood-money due for a *dhimmī* slave exceed the blood-money due for his master,[3] and nor must the blood-money due for a slave woman exceed the blood-money due for a *dhimmī* woman.[4]

The free man is only sentenced to death for killing his own counterpart and for killing a free woman, upon which half of her blood-money may be redeemed [by his heirs]. The same applies to a free woman, whereby she faces the death penalty for killing her counterpart or a free man, whereupon her heirs cannot recover anything by way of blood-money.

The same ruling applies for retribution by talion for injured and mutilated limbs as long as their value does not exceed one-third of the blood-money due for a free man's life, whereby the blood-money for cases in which the amount exceeds 300 dinars is calculated on the basis of the total being halved. He would face retributive punishment for injuring her, half of which would then be recovered; and she would face retributive punishment for injuring him, without recovery of any excess.

The male slave is sentenced to death for killing another, whether male or female, and the female slave for killing her counterpart as well as killing a male slave.

If a slave kills a free man, the victim's next of kin may opt to either sentence him to death or to use him as a slave. His master has no say in the matter. If the slave wounds a free man, the latter may either seek retribution or use him as a slave if his market value covers the compensation due for the crime, but if not, then he may use him proportionately, or he can be sold and the compensation taken from his price.

1 The blood-money for a free man is 1,000 gold dinars.

2 The blood-money for a free woman is 500 gold dinars.

3 The blood-money of a free *dhimmī* is 800 dirhams.

4 The blood-money for a free *dhimmī* woman is 400 dirhams.

الفصل الثاني: في شرائط القصاص

وهي خمسة:

الأول: الحرية

إذا كان القاتل حراً، فلا يقتصّ من الحر للعبد، ولا للمكاتب، ولا لأم الولد، ولا المدبر، بل يلزمه قيمته يوم قُتل ولا يتجاوز دية الحر، ولا بقيمة الأمة دية الحرة، ولا بدية عبد الذمي دية مولاه، ولا بدية أمته دية الذمية.

ويقتل الحر بمثله، وبالحرة مع رد نصف الدية، والحرة بمثلها، وبالحر، ولا يؤخذ منها الفضل.

وكذا في قصاص الجراح والأطراف ما لم يبلغ ثُلث دية الحر فينتصف دية المرأة، ويقتصّ لها من الرجل مع رد الفضل، وله منها، ولا رد.

ويُقتل العبد بالأمة والأمة بمثلها وبالعبد.

ولو قتل العبد حراً كان ولي الدم مخيّراً بين قتله واسترقاقه، ولا خيار لمولاه، ولو جرح اقتصّ المجروح أو استرقّه إن استوعب الجناية قيمته وإلا فبالنسبة، أو يباع فيؤخذ من ثمنه الأرش.

His master can also bail him out by paying the compensation due. If the slave kills his master, the master's next of kin can choose to have him killed [or continue to remain a slave of theirs]. If he intentionally kills a fellow slave, he is sentenced to death, but if he kills him accidentally, his master may either ransom him for the value of the slave who was killed, or offer his slave to the master of the slave who was killed. He may keep the difference between the value of his slave and the murdered slave [if his slave has a greater market value] but he is not liable for the shortfall.

The *mukātab* slave – both the unrestricted type as well as the conditional – is treated as the fully bonded slave if he has not made any contributions towards his freedom. If he has made contributions towards it, then he is killed in retribution for a free man but not for a slave, in which case he must work in proportion to the share of his freedom, and is then sold; or he is used as a slave in proportion to the share of his bondage.

If he kills him [i.e. another slave] accidentally, the Imam must deal with him in proportion to the share of his freedom. His master has the option to either ransom him for the compensation, or to offer his services to the master whose slave he killed in proportion to his value.

If a free man kills two free people, he is sentenced to death for both.

If a slave kills two people, one after the other, the next of kin of the victims have equal right of retribution from him, as long as the first murder has not been compensated for. If it has, then only the next of kin of the second [have the right of retribution].

2) Islam: If the killer is a Muslim, he cannot be sentenced to death for killing a non-Muslim, even if he is a *dhimmī*. Instead, he must be sentenced to discretionary punishment and fined the blood-money for a *dhimmī*.

The *dhimmī* is sentenced to death for killing a fellow male *dhimmī* and for a female *dhimmī* after recovering half of the blood-money due for her. The same applies to the female *dhimmī* who kills a fellow female or male, except that there is no recovery of excess.

If a *dhimmī* kills a Muslim intentionally, his life and property are at the disposal of the victim's heirs whereby they may sentence him to death if they wish or use him as a slave if they wish. It has been said that his young children are also enslaved. If he embraces Islam after the killing, then he is treated as a Muslim.

If he killed accidentally, he must be made to pay the blood-money from his own assets, and if he has no assets, then the Imam acts as his clan (*ʿāqila*)[5]

5 *ʿāqila*: "clan": those of his heirs who would be beneficiaries to any compensation paid for his death.

ولو كانت الجناية خطأً فلمولاه أن يفديه بأرش الجناية، والأقوى: بأقل الأمرين من القيمة وأرش الجناية. ولو قتل مولاه قيد به إن اختار الولي، ولو قتل عبداً مثله عمداً قتل به، ولو قتل خطأً للمولى فكه بقيمته أو دفعه، وله فاضل قيمته عن قيمة المقتول، ولا يضمن النقص.

والمكاتَب المشروط أو المطلَق الذي لم يؤدّ شيئاً كالقنّ، وإن كان قد أدّى شيئاً قيد بالحر لا القنّ، بل يسعى في نصيب الحرية ويباع، أو يسترقّ في نصيب الرقية.

ولو قتل خطأً فعلى الإمام في نصيب الحرية، وللمولى الخيار بين فك الرقية بالأرش أو تسليم الرق للرقية.

ولو قتل الحر حرين قُتل بهما.

ولو كان القاتل عبداً، على التعاقب، اشتركا فيه ما لم يحكم به للأول فيكون للثاني.

الثاني: الإسلام

إذا كان القاتل مسلماً، فلا يُقتل مسلم بكافر وإن كان ذمياً، بل يعزّر ويغرم دية الذمي.

ويُقتل الذمي بمثله، وبالذمية بعد رد فاضل ديته، والذمية بمثلها، وبالذمي ولا رد.

ولو قتل الذمي مسلماً عمداً دفع هو وماله إلى أولياء المقتول إن شاؤوا قتلوه وإن شاؤوا استرقّوه، وقيل: يسترقّ أولاده الصغار أيضاً.

ولو أسلم بعد القتل فكالمسلم.

ولو قتل خطأً لزمته الدية في ماله، فإن لم يكن له مال فالعاقلة الإمام دون أهله.

responsible for paying it and not his family.

3) That the killer not be the victim's father: The father cannot be sentenced to death for killing his son but is instead charged with payment of blood-money, an atonement penalty and discretionary punishment. If the son kills his father, he is sentenced to death, as would be the mother if she were to kill her son.

4) Mental Capacity: If a mentally incapacitated person or a minor kill someone, they are not sentenced to death, but their clan must pay the blood-money, since their motive would be considered accidental. If a person of age of maturity kills a minor, he is sentenced to death, and if someone of sound mental faculty kills a mentally incapacitated person, he is charged to pay the blood-money, unless it was committed in self-defence, in which case he is not held liable.

The predominant opinion is that the blind person is treated the same as the person with sight.

5) That the person killed be innocent: Therefore, if someone were to kill an apostate or one who is legally allowed to be killed, he is not sentenced to death.

الثالث: أن لا يكون القاتل أباً

فلا يقتل الأب بالولد بل يؤخذ منه الدية، ويعزّر ويكفّر. ولو قتل الولد أباه قُتل به، وكذا الأم لو قتلت ولدها قُتلت به.

الرابع: العقل

فلو قتل المجنون أو الصبي لم يقتلا، بل أخذت الدية من العاقلة، لأن عمدهما خطأ. ولو قتل البالغ صبياً قُتل به، ولو قتل العاقل مجنوناً أخذ منه الدية. إلا أن يقصد دفعه فيكون هدراً، والأعمى كالمبصر على الأقوى.

الخامس: أن يكون المقتول معصوم الدم

فلو قتل مرتداً أو من أباح الشرع قتله لم يقتل به.

Section Three: Joint Killings

If a group of people jointly kill a free Muslim man, his next of kin can demand for all of them to be sentenced to death, but they must pay whatever remains of the blood-money due for killing each perpetrator. He also has the option to sentence only one [or some] of them to death, in which case the remaining accomplices would pay their share of blood-money to the heirs of the killed perpetrator. Any excess to be paid in compensation upon sentencing two of the perpetrators to death must be borne by the victim's next of kin, and any excess leftover after their heirs have been paid can be kept by him. The same rulings apply to retribution for mutilation of limbs.

If two women kill one man, they must both be sentenced to death, and no compensation is payable. If there are more than two women, they are all sentenced to death and the blood-money for the additional perpetrator is paid to their heirs. The victim's next of kin also has the option of sentencing only some of them to death, in which case the remaining accomplices are liable for paying their share of the blood-money [for the killed perpetrators to their respective heirs].

If a man and a woman jointly kill a man, his next of kin may sentence them both to death, and pay the man's heirs blood-money for his killing. He may alternatively sentence only the man to death and ask the woman to pay the blood-money for him, or he may kill the woman and ask the man to pay half of the blood-money due for her.

If a free man and a slave jointly kill a free man, the victim's next of kin can sentence them both to death and pay half the blood-money to the free man's heirs for his death, and pay the slave's master the difference between his market value and the blood-money due for his death. If he chooses to sentence only the free man to death, the slave's master would have to pay his heirs half the blood-money or give up his slave for their use. If the value of the slave exceeds the value of half the blood-money, the master may keep the excess. If the victim's next of kin chooses to sentence the slave to death, the free man must pay any additional cost beyond half of the blood-money for the slave to his master. Otherwise, the remainder goes to the victim's heirs.

الفصل الثالث: في الاشتراك

إذا اشترك جماعة في قتل حر مسلم كان للولي قتل الجميع بعد رد فاضل دية كل واحد عن جنايته عليه، وله قتل البعض ويرد الآخرون قدر جنايتهم على المقتصّ منه، ولو فضل للمقتولين فضل قام به الولي، وإن فضل منهم كان له، وكذا البحث في الأطراف.

ولو قتلت امرأتان رجلاً قُتلتا به ولا رد، ولو كنّ أكثر قُتلن به بعد رد الفاضل عليهنّ، وللولي قتل البعض، وترد الباقيات قدر جنايتهنّ.

ولو اشترك رجل وامرأة في قتل رجل فللولي قتلهما بعد رد الفاضل على الرجل، وله قتل الرجل، وترد المرأة ديتها عليه، وله قتل المرأة وأخذ نصف الدية من الرجل.

ولو اشترك عبد وحر في قتل حر فللولي قتلهما بعد رد نصف الدية على الحر وما يفضل من قيمة العبد عن جنايته على مولاه.

ولو قتل الحر رد السيد عليه نصف الدية، أو سلّم العبد إليه، ولو زادت قيمته على النصف كان الزيادة للمولى، ولو قتل العبد رد الحر على المولى ما فضل عن نصف الدية وإلا كان تمامها لأولياء المقتول.

If a free woman and a slave jointly kill a free man, his next of kin may sentence them both to death. If the market value of the slave exceeds the value of blood-money due for his crime, the victim's next of kin must return the excess to his master. He may alternatively sentence only the woman to death and use the slave in his own service if his market value is the same or less than the blood-money due. If not, then the excess goes to his master. If only the slave is sentenced to death, and his market value is the same as or less than the blood-money due for him, the victim's next of kin may take half the blood-money from the woman. If his market value exceeds, the woman pays the excess. Any excess left over after the blood-money due for the victim is covered, goes to his heirs.

ولو اشترك عبد وامرأة في قتل الحر فللولي قتلهما ، ولو فضلت قيمة العبد عن جنايته رد الولي على مولاه الفاضل ، وله قتل المرأة واسترقاق العبد إن كانت قيمته بقدر الجناية أو أقل ، وإلا كان الفاضل لمولاه.

ولو قتل العبد وقيمته بقدر الجناية أو أقل كان للولي أخذ نصف الدية من المرأة ، ولو كانت القيمة أكثر ردت المرأة عليه الفاضل ، فإن استوعبت دية الحر ، وإلا كان الفاضل لورثة المقتول.

Section Four: How Homicide is Convicted

This is done in three ways:

Firstly, by confession. A single confession from the perpetrator suffices. If someone confesses to having killed a person intentionally, after which someone else confesses to the same crime, and the former retracts his confession, neither one is given the death penalty and the blood-money is paid out of the public treasury. If someone confesses to having killed a person intentionally, and another person confesses to having killed the same person accidentally, the victim's next of kin may choose whichever of the two he wishes to believe, upon which he must not pursue charges against the other.

Secondly, through evidence provided by the testimony of two just male witnesses. To convict homicide punishable by payment of blood-money only – such as accidental death or mutilation of body parts – the testimony of one male witness and two females, or one male witness and an oath of honesty suffices.

Thirdly, by a collective oath (*qasāma*), which is used as supportive evidence to prove charges of probable cause, to support the truthfulness of the claimant, if he is the only witness, for example. In such a case, the next of kin of the victim would use the collective oath to prove the claim by swearing an oath alongside fifty people from the community.

If the claimant does not have a collective oath, he must be made to swear an oath of honesty repeatedly, and if he refuses to swear it, then the defendant must swear oaths of honesty alongside fifty members of the community. If he does not have anyone on his side, then he must be made to swear an oath fifty times. The claim against him is corroborated upon his refusal.

Grievous bodily harm to be retributed through blood-money is convicted in the same way as with the loss of life. If a body part is partially mutilated, the number of people in the collective oath are adjusted proportionately.

الفصل الرابع: فيما يثبت به القتل

وهو ثلاثة:

الأول: الإقرار. ويكفي المرة من أهله، ولو أقر بقتله عمداً فأقرّ آخر أنه الذي قتل ورجع الأول سقط القصاص وكانت الدية على بيت المال، ولو أقرّ واحد بقتله عمداً وأقرّ آخر أنه قتل خطأ كان للولي الأخذ بقول من شاء منهما ولا سبيل له على الآخر.

الثاني: البينة. وهي عدلان. ويثبت ما يوجب الدية - كالخطأ والهاشمة - بشاهد وامرأتين، أو بشاهد ويمين.

الثالث: القسامة. وهي تثبت مع اللوث وهو إمارة يغلب معها الظنّ بصدق المدّعي، كالشاهد الواحد. فللولي معه إثبات الدعوى بأن يحلف هو وقومه خمسين يميناً.

ولو لم يكن للمدّعي قسامة كررت عليه الأيمان، ولو لم يحلف حلف المنكِر خمسين يميناً هو وقومه، ولو لم يكن له أحد كررت الخمسون عليه، ولو نكل ألزم الدعوى.

والأعضاء الموجبة للدية كالنفس، ولو نقصت فبالحساب. ولا يثبت اللوث بالفاسق الواحد ولا الصبي ولا الكافر.

Probable cause cannot be proven through the claim of a single immoral witness, nor a minor, and nor a disbeliever. If a group of immoral people or women report on something whereby the possibility of their collusion in the matter can be precluded, then probable cause can be established, but if they are a group of disbelievers or minors, then probable cause cannot be proven unless their number reaches the extent of *tawātur*.[6]

If a murdered person is found in the home of a group of people, in their place of trade or their village, it is probable cause [to suspect them and they are held responsible until proven innocent]. If he is found between two villages, then probable cause lies with the one he is nearest to. If the body is equidistant from both, then they are both to be suspected of having probable cause. If he is found in a field, army barracks or a marketplace, and the killer is unknown, his blood-money is paid from the public treasury. If no probable cause can be precluded, then the claim [related to his death] is treated as all other claims.

6 *tawātur*: is a technical term which means roughly “broad authentication”. It is often used to indicate that something has been reported by such a large number of independent sources who could not possibly have colluded to fabricate it, and that its authenticity is thereby assumed to be guaranteed.

ولو أخبر جماعة الفساق أو النساء مع الظنّ بانتفاء المواطاة ثبت اللوث، ولو كانوا كفّاراً أو صبياناً لم يثبت اللوث إلا أن يبلغوا حدّ التواتر.

ولو وُجد قتيلاً في دار قوم أو محلّتهم أو قريتهم كان لوثاً، ولو وُجد بين قريتين وهو إلى أحدهما أقرب فهو لوث، ولو تساوت مسافتهما تساويا في اللوث، ولو وجد في فلاة وجهل قاتله، أو في عسكر أو سوق فديته على بيت المال، ومع انتفاء اللوث يكون الدعوى فيه كغيرها من الدعاوي.

Section Five: Administering Retributive Justice

Intentional homicide is punishable by retributive justice [i.e. the death penalty], and does not involve the payment of blood-money unless there is an agreement [between the killer and the victim's next of kin]. The same ruling applies to injuries [i.e. "an eye for an eye, a tooth for a tooth"]. The death penalty can only be administered with the sword and is restricted to decapitation. There is no indemnity for complications resulting from a retributive punishment in the absence of any transgression.

If the right of retribution rests with a group of people [i.e. the victim's heirs], it must be decided with their joint interests. If some of them opt to receive blood-money and the killer is willing to pay it, the rest of them may sentence him to death, upon which the share of blood-money accepted by the others is returned to the heirs of the perpetrator. The same would apply if some of them opted to forgive him.

If the killer dies before he is sentenced to death, the blood-money is taken from his estate. If the victim was missing his hands when he was killed, and he had previously been compensated for them or been justly retributed, his next of kin may sentence his killer to death but must return the value of his hands as compensation to his heirs. If the hands had previously been amputated without foul play and there was no compensation paid for them, then his next of kin need not remit anything.

Retribution for a mutilated limb applies to the same categories of people to whom retribution for loss of life applies, in the same way. A woman must be made to compensate for injuring a man, with no excess to be returned, and a man must be made to compensate for injuring a woman up to a third of the value of the full compensation; whatever exceeds that must be returned to him.

The health of the limb carries import in that a healthy limb may not be cut in retribution for a paralysed limb, though a paralysed limb may be cut in retribution for a healthy one if it can be severed completely. The span of a cut to the head in retribution for an open head injury must equal the latter in length and width but not in depth. The definition [i.e. type] of the injury should be used to determine the sentence.

الفصل الخامس: في كيفية القصاص

قتل العمد يوجب القصاص ولا يثبت الدية إلا صلحاً ، وكذا الجراح، ولا قصاص إلا بالسيف، ويقتصر على ضرب العنق، ولا يضمن سراية القصاص مع عدم التعدّي.

ولو كان القصاص لجماعة وقف على الاجتماع. ولو طلب البعض الدية ودفعها القاتل كان للباقي القصاص بعد ردّ نصيب الآخرين على القاتل، وكذا لو عفا البعض.

ولو مات القاتل قبل القصاص أُخذت الدية من تركته. ولو كان المقتول مقطوع اليد في القصاص أو أخذ ديتها كان للولي القصاص بعد ردّ دية اليد ، ولو قُطعت من غير جناية ولم يأخذ ديتها فلا رد.

ويثبت القصاص في الطرف لكل من يثبت له القصاص في النفس، ويقتصّ للرجل من المرأة ولا رد ، وللمرأة من الرجل مع الرد فيما زاد على الثُلث.

ويعتبر سلامة العضو ، فلا يقطع الصحيح بالأشلّ، ويقطع الاشلّ بالصحيح إذا كان مما ينحسم. وتساوي المساحة في الشجاج طولاً وعرضاً لا نزولاً بل يعتبر الاسم كالمُوَضِّحَة.

Retributive justice is meted out for crimes in which there are no discretionary punishments, and vice versa, there are no retributive punishments where discretionary punishments apply, such as in the case of a deep injury to the centre of the head, a deep wound to the body, and the breaking of bones. There is no retributive justice for the *dhimmī* from a Muslim, nor for a slave from a free man.

The [perpetrator's] nose must be cut off for causing someone else to lose his nose or sense of smell, or his whole ear for causing loss of hearing in his victim. Male genitals must not be cut in retaliation for causing impotence. The good eye of a one-eyed perpetrator must be gauged out for causing someone to lose their good eye, even if he becomes blinded by it. If a minor is made to lose a tooth, he must wait to see: if it grows back, the perpetrator is made to pay compensation, but if not, then he must be made to lose his.

The perpetrator who takes refuge in the Sanctuary of Mecca must be restricted in his access to food and drink until he comes out and can be punished. If he committed the crime within the Sanctuary, then he must be punished therein.

If he cut off one person's hand first and then another person's finger, he must be physically punished for the first crime and made to pay compensation for the second. If he cut the finger first, then its owner can demand for his finger to be cut in turn, and the owner of the cut hand may demand for his hand to be cut thereafter, upon which he must refund the blood-money for the finger.

ويثبت القصاص فيما لا تعزير فيه، ولا قصاص فيما فيه تعزير كالأمومة والجايفة وكسر العظام. ولا يقتصّ للذمي من المسلم، ولا للعبد من الحر.

ويقطع الأنف الشامّ بفاقده، والأذن الصحيحة بالصماء، ولا يقطع الذكر الصحيح بالعنين، وتقلع عين الأعور الصحيحة بعين السليم قصاصاً وإن عمي، وينتظر بسنّ الصبي سنة فإن عادت فالأرش وإلا فالقصاص.

والملتجئ إلى الحرم يضيق عليه في المطعم والمشرب ليخرج ويقتصّ منه، ولو جنى في الحرم اقتصّ منه فيه.

ولو قطع يد رجل وإصبع آخر اقتصّ للأول وكان للثاني الدية، ولو قطع الإصبع أولاً اقتصّ صاحبها ثم صاحب اليد ورجع بدية الإصبع.

Section Six: Blood-Money for the Loss of Life

The value of the blood-money payable for a free Muslim killed intentionally is: either one-hundred fully-grown camels, or two-hundred cows, or two-hundred high quality suits of clothing, which is four-hundred garments made from Yemeni fabric, or ten-thousand dirhams. This must be claimed in the same year from the perpetrator's assets and must be fixed by mutual agreement.

The value of the blood-money payable for a semi-intentional killing is thirty-three female camels of two or more years of age, thirty-three female camels of three years of age, and thirty-four female camels of four years of age and ready to conceive, or any of the other options mentioned above, all payable from the perpetrator's assets, and claimed over two years.

The value of blood-money payable for an accidental killing is: twenty female one-year old camels, twenty male two-year old camels, thirty female two-year old camels, and thirty female three-year old camels, or any of the other options mentioned above. His clan (*'āqila)* are responsible for paying it over three years.

The blood-money for a woman's life is half of the full compensations above.

The blood-money for a male *dhimmī* is eight-hundred dirhams, and for a female *dhimmī*, four-hundred dirhams.

The blood-money due for a slave is his market value, as long as it does not exceed the value of blood-money for a free man, in which case the excess is returned. The blood-money for a slave woman is her market value, as long as it does not exceed the value of blood-money for a free woman, in which case the excess would be returned.

The blood-money due for limbs is in proportion to their value, so a free man's limbs are valued as a fraction of the value of the blood-money for his life, and a slave's in proportion to his market value. The slave's master does not have the right to claim for them except after the slave has been sent to the perpetrator [to determine] the injuries that he was responsible for. All else is at the master's own cost. Grievous bodily harm that cannot be calculated [as a fraction of the blood-money for the whole person] is due indemnity.

The slave's perpetration of crimes is linked to his emancipation and not to his master, though he has the right to let him go through the compensation resulting from the crime.

الفصل السادس: في دية النفس

دية الحر المسلم في العمد : مائة من مسان الإبل ، أو مائتا بقرة ، أو مائتا حلة وهي أربعمائة ثوب من برود اليمن ، أو ألف شاة ، أو ألف دينار أو عشرة آلاف درهم وتستأدى في سنة واحدة من مال الجاني ، ولا يثبت إلا بالتراضي .

ودية شبيه العمد ، من الإبل : ثلاث وثلاثون بنت لبون ، وثلاث وثلاثون حقّة ، وأربع وثلاثون ثنية طروقة الفحل ، أو ما ذكرنا [ه] في مال الجاني . وتستأدى في سنتين .

ودية الخطأ ، من الإبل : عشرون بنت مخاض ، وعشرون ابن لبون ، وثلاثون بنت لبون ، وثلاثون حقّة . أو ما ذكرنا [ه] من باقي الاصناف . وتؤخذ من العاقلة في ثلاث سنين .

ودية المرأة : النصف من ذلك .

ودية الذمي ثمان مائة درهم . والذمية أربع مائة درهم .

ودية العبد : قيمته ما لم تتجاوز دية الحر فيرد إليها . ودية الأمة قيمتها فإن تجاوزت دية الحرة ردت إليها .

ودية الأعضاء بنسبة القيمة ، فكل ما في الحر كمال ديته ففي العبد كمال قيمته ، لكن ليس للمولى المطالبة بها إلا بعد دفع العبد إلى الجاني ، وما فيه دونه بحسابه ، وما لا تقدير فيه ففيه الأرش . وجناية العبد تتعلّق برقبته لا بالمولى ، لكن له فكّه بأرش الجناية .

Section Seven: Actions That Entail Liability for Blood-Money

There are two such liabilities:

1) Direct liability: which is where one may inflict harm unintentionally, such as a doctor treating a patient but causing him harm in the process; or a person turning over on top of another in their sleep and causing their death; or a person carrying something on their head which then falls on another person and breaks in the process, whereupon he is responsible for the goods. If someone falls from a height on top of another person, who then died, he would be responsible for his blood-money. If, however, he was pushed by someone else, then the latter is responsible for the blood-money.

If three people are breaking a wall down together, and it falls on one of them, killing him, then the remaining two are liable to pay two-thirds of his blood money. If someone takes another person out of his house at night, he is held responsible for him [if he dies], unless he can provide evidence for his natural death or that someone else killed him.

2) Indirect Liability: This is when someone digs a well in a place which does not belong to him, and someone falls into it; or he makes a dirt path or places obstacles on the road. If he did all that on his own property, then he would not be held liable. If someone enters a people's home with their permission, and is mauled by their dog, they are liable for any injury he sustains as a result, but if he entered without their permission, then they are not liable.

If someone rides a horse, he is liable for whatever damage it causes with its front legs, and similarly if he is leading it. If he is standing next to it, then he is liable for any injury it causes with its front and hind legs. If another person strikes it, then the striker is liable for damages. If two people ride a horse together, they are both held liable. If the owner of the horse accompanies it whilst it is being ridden by another, he is held liable and not the rider. If a horse throws its rider off, the owner is liable only if he was the one to frighten it. If not, then he is not.

If there is both direct and indirect liability together, the person directly liable is held responsible.

الفصل السابع: فيما يوجب ضمان الدية

وهو اثنان:

الأول: المباشرة. بأن يقع التلف من غير قصد، كالطبيب يعالج فيتلف المريض بعلاجه والنائم إذا انقلب على غيره فمات، ومن حمل على رأسه متاعاً فأصاب غيره، وكسر المتاع فإنه يضمنها، ولو وقع على غيره من علو فمات ضمن ديته، ولو أوقعه غيره فالدية على الدافع.

ولو اشترك ثلاثة في هدم حائط فوقع على أحدهم فمات كان على الباقين ثُلثا ديته. ولو أخرج غيره من منزله ليلاً ضمنه إلا أن تقوم البينة بموته أو بقتل غيره له.

الثاني: التسبيب. كمن حفر بئراً في غير ملكه فوقع فيها إنسان، أو نصب سكيناً أو طرح المعاثر في الطريق، ولو كان ذلك في ملكه لم يضمن. ولو دخل دار قوم بإذنهم فعقره كلبهم ضمنوا جنايته، ولو كان بغير إذن فلا ضمان.

ومن ركب دابة ضمن ما تجنيه بيديها، وكذا لو قادها، ولو وقف بها ضمن جنايتها بيديها ورجليها، وكذا لو ضربها غيره فالدية على الضارب، ولو ركبها اثنان تساويا في الضمان، ولو كان صاحبها معها ضمن دون الراكب، ولو ألقت الراكب ضمن المالك إن كان بتنفيره وإلا فلا.

ولو اجتمع المباشر والسبب كان الضمان على المباشر.

Section Eight: Blood-Money Compensation for Various Body Parts

For the hair on one's head: full compensation, and the same goes for the beard, if they do not grow back again. If they grow back, then damages must be paid. Full compensation must equally be paid for a woman's hair, and if it grows back then she must be paid equivalent to a bridal gift (*mahr*).

For the eyebrows: five-hundred dinars, and half of that amount for a single brow.

For eyelashes and any other hair, damages must be paid.

For each eye, half of the compensation for the eyes is payable, and a quarter for each eyelid. The full compensation is payable for a one-eyed man's good eye, whether his defect was from birth or resulting from an act of God. If a one-eyed woman is blinded thus, a third of the compensation is payable.

For [loss of] the nose, full compensation is due, as it is if it crushed or badly broken. If it can be fixed without any remaining defect, then one-hundred dinars are payable. If it is paralysed, then two-thirds of the compensation is payable, and if it is permanently blocked, then half the compensation. Half the compensation is payable for each nostril.

For every ear, half the compensation is payable, and for parts of the ear, a fraction of the compensation proportionately. For the earlobe, a third of its compensation is payable, and similarly for a perforated eardrum.

For every lip, half the compensation is due, and for part of it, proportionately. If the lips have been shrunk, then the full compensation is payable, and if they hang loosely, then two-thirds of it.

For the healthy tongue of an adult or child, full compensation is due. If part of it has been cut, he is made to pronounce the twenty-eight letters of the alphabet and compensated a proportionate fraction for each letter that he cannot pronounce. One-third of the compensation is due for the mute person's tongue, and a proportionate fraction for part of it. If someone whose tongue looks intact, but he claims that he cannot speak, he is made to provide a collective oath to corroborate his claim.

الفصل الثامن: في ديات الأعضاء

في شعر الرأس: الدية كاملة، وكذا في اللحية إذا لم ينبتا، ولو نبتا فالأرش وفي شعر المرأة ديتها، فإن نبت فمهرها.

وفي الحاجبين: خمس مائة دينار، وفي كل واحد النصف.

وفي الأهداب الأرش وكذا باقي الشعر.

وفي كل واحد من العينين نصف الدية، وفي كل جفن ربع الدية. أما عين الأعور الصحيحة ففيها الدية كاملة إن كان العور خلقة أو بشيء من قبل الله تعالى، وفي خسف العوراء الثُلث.

وفي الأنف الدية كاملة، وكذا في مارنه، أو كسر مفسد. ولو جبر على غير عيب فمائة دينار. وفي شلله ثُلثا ديته. وفي الروثة وهي الحاجز نصف الدية. وفي أحد المنخرين نصف الدية.

وفي كل أذن نصف الدية، وتقسط الدية على أجزائها. وفي الشحمة ثُلث ديتها وكذا في خرمها.

وفي كل شفة نصف الدية، وفي بعضها بحسابه، ولو تقلصت قال الشيخ: فيه ديتها، ولو استرختا فثُلثا الدية.

وفي لسان الصحيح أو الطفل الدية، ولو قطع بعضه اعتبر بحروف المعجم وهي ثمانية وعشرون حرفاً، فيقسط الدية عليها فما نقص أخذ قسطه. وفي لسان الأخرس ثُلث الدية، وفي بعضه بالحساب مساحة. ولو ادّعى الصحيح ذهاب نطقه صدق مع القسامة.

For the loss of teeth, full compensation is paid, and are counted as twenty-eight teeth. The twelve at the front are valued at fifty dinars each and the sixteen at the back are valued at twenty-five dinars each. If a single tooth is lost, he is entitled to one-third of the original compensation amount [for the teeth]. The loss of any additional teeth [to twenty-eight] does not entitle one to more compensation. Two-thirds of the compensation are due for the discolouration of the tooth, and for the damage of a tooth without its removal, two-thirds as well. The lost tooth of a minor who has not yet cut adult teeth is due payment of damages if a new tooth grows, but if not, then he is entitled to compensation for an adult tooth.

For the neck, if it is broken and the person is rendered paralysed, then the full compensation is payable, and similarly if he is injured in a way as to prevent him from swallowing. If this state ceases, then payment of damages applies.

For the jaws, full compensation is payable if they are broken independently of the teeth, such as in the case of a minor who has not yet cut teeth, and twice the full compensation if they are broken along with all teeth.

Each hand is worth half the compensation amount, up until the wrist. In the case of paralysis of a hand, two-thirds of its compensation are due, and for amputation of a previously paralysed hand, a third of the compensation for a good hand. The same ruling applies to the loss of an extra hand.

Each finger on either hand is worth one-tenth of the full compensation, divided into thirds for parts of the finger or halves for the thumb. A third of the original compensation [for a finger] is due for the loss of an extra finger or a paralysed one. Paralysis of the fingers entitles one to two-thirds of the original amount.

For the fingernails, ten dinars is payable for each one that does not grow back or grows back black. If it grows back normally, then five dinars is payable.

If the back is broken, full compensation is due, as is the case if the victim is rendered humpbacked and unable to stand straight, or unable to sit. If he recovers, then a third of the compensation is due. If he is rendered unable to walk or to have sexual intercourse, then twice the full compensation is payable. Damage to the spinal cord also results in full compensation.

For each of a woman's breasts, half the full compensation is paid, and similarly for each of her nipples. If her milk supply is cut-off [as a result of the injury] or her let-down reflex is affected, then damages must be paid. For a man's nipples, half the compensation is paid according to Shaykh [al-Ṭūsī], and an eighth according to Ibn Bābawayh [al-Qummī].

وفي الأسنان الدية، وهي ثمانية وعشرون، اثنا عشر مقاديم في كل واحد خمسون [ديناراً] وستة عشر مئاخير في كل واحدة خمسة وعشرون. وفي الزائد منفردة ثُلث دية الأصلية، ولا دية لها مع الانضمام.

وفي اسوداد السن ثُلثا ديتها، وفي انصداعها من غير سقوط ثُلثا ديتها.

وفي سنّ الصبي الذي لم يثغر الأرش إن نبت وإلا فدية المثغر.

وفي العنق إذا كسر وصار الإنسان أصور: الدية، وكذا لو جنى عليه بما يمنع الازدراد، ولو زال فالأرش.

وفي اللحيين: الدية لو انفردا عن الأسنان كالصبي وفاقد الأسنان، ومع الأسنان ديتان.

وفي كل يد: نصف الدية، وحدّها المعصم.

وفي شلل اليد ثُلثا ديتها، وفي الشلاء ثُلث الصحيحة، وكذا للزائدة.

وفي كل إصبع من اليدين عُشر الدية، ويقسط على ثلاث أنامل، وفي الإبهام على اثنين. وفي الزائدة ثُلث الأصلية، وكذا الشلاء، وفي الشلل الثُلثان.

وفي الظفر: عشرة دنانير إن لم ينبت أو نبت أسود، ولو نبت أبيض فخمسة.

وفي الظهر إذا كسر: الدية، وكذا لو أصيب فاحدودب أو صار بحيث لا يقدر على القعود، ولو صلح فثُلث الدية، ولو ذهب مشيه وجماعه فديتان.

وفي النخاع: الدية.

وفي كل واحد من ثديي المرأة نصف ديتها، وكذا في حلمتها، ولو انقطع لبنهما أو تعذر نزوله فالأرش. وفي حلمة الرجل نصف الدية عند (الشيخ) وثمنها عند (ابن بابويه).

For the loss of the penis, full compensation is payable, as it is for the glans. An impotent person is entitled to one-third of it. Full compensation is also payable for the loss of both testicles, or a half for each one. Four-hundred dinars are payable for a piercing injury to the scrotums of both, and if the injury is serious enough to impair his walking, then eight-hundred dinars.

For the loss of each of a woman's two labia half the compensation is payable. For injury resulting from raping a woman, full compensation is due. This does not apply to her husband after she reaches the age of puberty. If it is before that, then the husband is liable to pay the full compensation alongside the bridal gift and the cost of her upkeep throughout his or her life. If he was not her husband and he raped her, then he must pay full compensation along with the bridal gift, but if it was consensual then only the compensation. If the raped woman was a virgin, then she is also entitled to damages payable for her virginity.

For the loss of each of the buttocks, half the compensation is payable, and the same for each leg up until the ankle bone. The compensation for the toes is the same as per the fingers. Half the compensation is due for each thigh and each upper leg too.

For a broken rib, twenty-five dinars is payable for each rib surrounding the heart, and ten for each one adjoining to the upper arms.

For a broken tailbone, which results in faecal incontinence, full compensation is due, as is the case with a broken perineum leading to urinal or faecal incontinence.

For a broken collarbone, which is then repaired without disfigurement, four-hundred dinars.

If someone is punched in the stomach hard enough for him to lose control over his bowels, the perpetrator must be punched in the stomach in retaliation or made to pay a third of the compensation in damages.

He who injures a virgin by fingering her until he pierces her hymen whereby she suffers from incontinence must be made to pay the full compensation as well as the bridal gift of her contemporaries.

In the case of a fractured bone in a limb, one-fifth of the compensation due for that limb is payable. If it heals without lasting damage, then four-fifths of the compensation for the breakage. If it needs to be removed, then one-quarter of the compensation of the breakage is payable. For a contusion, one-third of the full compensation is due, but if it heals without lasting damage, then four-fifths of the compensation for a contusion. If the bone is dislocated whereby it no longer functions, two-thirds of the compensation for the whole limb are payable, and if it heals without lasting damage, then four-fifths of the compensation for the dislocation.

وفي الذكر: الدية، وكذا في الحشفة، وفي العنين ثُلث الدية.

وفي الخصيتين: الدية، وفي كل واحدة النصف، وفي أدرة الخصيتين أربع مائة دينار، فإن فحج فلم يقدر على المشي فثمانمائة.

وفي كل واحد من شفري المرأة نصف ديتها.

وفي إفضاء المرأة ديتها، ويسقط عن الزوج بعد بلوغها، ولو كان قبله ضمن الزوج مع المهر الدية والإنفاق عليها حتى يموت أحدهما، ولو لم يكن زوجاً وكان مكرهاً فالمهر والدية، ومع المطاوعة الدية، ولو كانت المكرهة بكراً فلها أرش البكارة أيضاً.

وفي كل واحدة من الأليتين: نصف الدية، وفي كل واحدة من الرجلين نصف الدية، وحدها مفصل الساق وإصبعهما كاليدين.

وفي كل واحد من الساقين والفخذين: نصف الدية.

وفي كسر الضلع: خمسة وعشرون ديناراً إن كان مما يخالط القلب، وإن كان مما يلي العضدين فعشرة.

وفي كسر البعصوص إذا لم يملك الغائط: الدية وكذا في العجان إذا لم يملك البول ولا الغائط.

وفي الترقوة إذا كسرت وجبرت على غير عيب: أربعون ديناراً.

ومن داس بطن إنسان حتى أحدث: ديس بطنه، أو يفتدي ذلك بثُلث الدية.

ومن افتضّ بكراً بأصبعه حتى خرق مثانتها فلم تملك بولها [فعليه] ديتها ومثل مهر نسائها.

وفي كسر عظم من عضو: خُمس دية ذلك العضو، فإن صلح على غير عيب: فأربعة أخماس دية كسره، وفي مُوضِّحَته رُبع دية كسره، وفي رضّه ثُلث ديته، فإن برئ على غير عيب: فأربعة أخماس دية رضّه، وفي فكّه من العضو بحيث تتعطّل: ثُلثا دية العضو، فإن صلح على غير عيب فأربعة أخماس دية فكّه.

Section Nine: Blood-Money Compensation Due for Loss of Faculties

For loss of mental faculty, full compensation is due, and for diminished mental faculty, payment of damages. If it returns subsequently, the compensation is not returned.

For the complete loss of hearing, full compensation is due, and half if the loss is only in one ear. For partial hearing loss in one ear, the hearing is tested and measured against the other, and the compensation is calculated according to the difference in hearing distance between the two. If there is partial hearing loss in both ears, then his hearing is measured against a counterpart of the same age.

For the complete loss of sight in each eye, half the compensation is due, and for partial loss of sight in one eye, it is calculated proportionately, and the same goes for partial loss in both eyes. They are tested and compared to a counterpart of the same age.

For the loss of the olfactory sense, full compensation is due, and if it is caused by cutting off the nose, then twice the compensation is payable. For the partial loss of it, damages are payable at the discretion of the judge.

For the loss of gustatory perception, full compensation is due, and damages for the partial loss thereof.

If someone is injured to such an extent whereby he is no longer able to ejaculate during intercourse, full compensation is due, and similarly, if an injury causes urinal incontinence.

For the complete loss of the voice, full compensation is due.

الفصل التاسع: في ديات المنافع

في العقل: الدية، وفي نقصه الأرش، ولو عاد لم يرتجع الدية.

وفي السمع: الدية، وفي سمع إحدى الأذنين: النصف، ولو نقص سمع إحداهما قيس إلى الأخرى ويؤخذ بحسب التفاوت بين المسافتين، ولو نقص سمعهما قيس إلى المساوي له في السن.

وفي ضوء كل عين: نصف الدية، وفي نقصان ضوء إحداهما بحسابه، وكذا في نقصان ضوئهما ويعتبر بالقياس إلى عيني مساويه في السن.

وفي الشمّ الدية ولو قطع الأنف فذهب الشمّ فديتان، وفي نقصانه الأرش بما يراه الحاكم.

وفي الذوق الدية، وفي نقصانه الأرش.

ولو أصيب فتعذر عليه الإنزال حالة الجماع: فالدية.

وفي سلس البول: الدية.

وفي الصوت: الدية.

Section Ten: Blood-Money Compensation Due for Various Wounds

There are eight types of head injuries:

al-ḥāriṣa: which is a wound to the surface of the skin, and for which a camel is due in compensation.

al-dāmiya: which is an open, bleeding wound, for which two camels are due.

al-mutalāḥima: which is a deep cut into the flesh, for which three camels are due.

al-samḥāq: which cuts into the flesh close to the bone, for which four camels are due.

al-muwaḍḍiḥa: which is a deep flesh wound that exposes the bone, for which five camels are due.

al-hāshima: which is a blow that fractures the bone, for which ten camels are due.

al-munaqqala: which is a wound that dislocates the bone, and for which fifteen camels are due.

al-ma'mūma: which is the deep head injury that reaches the brain, for which one-third of the full compensation is due. The same applies to the *jāyifa* category of injury, which is a bodily wound that pierces into internal organs.

A puncture to the nose entitles the victim to one-third of the compensation. If it then heals, then one-fifth of the compensation is due. If it spans from either nostril to the septum, then one-tenth of the compensation is payable.

If the lips are split such that the teeth can be seen, one-third of the compensation is due, and if they subsequently heal, then one-fifth; and half thereof for a single lip.

A puncture into either side of the leg, entitles one to one hundred dinars in compensation.

The bruising of the face resulting from grievous bodily harm is due one dinar and a half if the bruise is red, three dinars if it is green, and six if it is black. A bruise on the body is due half of that. Wounds to the head and the face are calculated equally, but wounds on the body are calculated based on the affected limb in relation to the head.

الفصل العاشر: في ديات الجراح

الشجاج ثمانية:

الحَارِصَة: وهي التي تقشر الجلد، وفيها بعير.

والدَامِية: وهي التي تأخذ يسيراً في اللحم، وفيها بعيران.

والمَتلاحِمة: وهي التي تأخذ في اللحم أكثر، وفيها ثلاثة أبعرة.

والسَمحَاق: وهي التي تنتهي إلى الجلدة المغشية للعظم، وفيها أربعة أبعرة.

والمُوَضِّحة: وهي التي توضِّح العظم، وفيها خمسة أبعرة.

والهاشِمة: وهي التي تهشم العظم، وفيها عشرة أبعرة.

والمُنقّلة: وهي التي تحوج إلى نقل العظم، وفيها خمسة عشر بعيراً.

والمأمومة: وهي التي تصل إلى أم الدماغ، وفيها ثُلث الدية.

وكذا الجايِفة: وهي التي تبلغ إلى الجوف.

ودية النافذة في الأنف: ثُلث الدية، فإن صلح فخُمس الدية.

وفي أحد المنخرين إلى الحاجز: عُشر الدية.

وفي شقّ الشفتين حتى تبدو الأسنان: ثُلث الدية، ولو برئت فالخُمس، وفي كل واحد نصف ذلك.

وفي النافذة في شيء من أطراف الرِجل: مائة دينار.

وفي احمرار الوجه بالجناية: دينار ونصف، وفي اخضراره: ثلاثة، وفي اسوداده: ستة، ولو كانت في البدن فعلى النصف. ويتساوى الشجاج في الرأس والوجه. أما البدن فبنسبة العضو الذي يتفق فيه من دية الرأس.

The female and male victim have equal right to compensation and retributive justice up to one-third of the full compensation. If the harm exceeds one-third, then the female receives half of the male's compensation.

Whatever [grievous bodily harm] to a man's body merits compensation, so it does to the woman's body, and similarly for the *dhimmī*. The slave's is calculated according to his market value, and whatever is calculated on the basis of a man's freedom, the corresponding ratios will apply to the compensations of a woman, a *dhimmī* and the market value of a slave.

The Imam acts as the next of kin or kin of whoever has no heirs, and can either take retributive justice or receive compensation, but cannot pardon the crime.

ويتساوى المرأة والرجل في الدية والقصاص فيما دون ثُلث الدية، فإذا بلغت الجناية ثُلث الدية صارت المرأة على النصف.

وكل ما فيه الدية من الرجل ففيه من المرأة ديتها، كذا من الذمي، ومن العبد قيمته. وما فيه مقدر من الحر فهو بنسبته من دية المرأة والذمي وقيمة العبد.

والإمام ولي من لا ولي له، يقتصّ أو يأخذ الدية، وليس له العفو.

Section Eleven: The Blood-Money Due for Killing an Unborn Child and Crimes Against the Deceased

For the killing of an implanted zygote, twenty dinars is payable in blood-money; for an embryo forty dinars; for a foetus sixty dinars, and after the development of the skeletal system, eighty dinars. Between the time that it is fully formed until it has an independent soul, one-hundred dinars is payable, and whatever falls between those stages is calculated proportionately.

The blood-money due for killing the unborn child of a *dhimmī* is one-tenth of the blood-money due for his father. And for the slave's, one-tenth of the market value of its slave mother, regardless of whether the child is male or female.

Once the soul has entered the body of the foetus, then full blood-money is payable for the male, and half for the female.

If a [pregnant] woman is killed, and her unborn child dies inside her as a result, then the blood-money for the woman is payable as well as the average of the two compensations [for a male and a female] if its gender is unknown.

If a woman aborts her child, either fully or partly, she must pay the blood-money for it to its next of kin, and she has no share in this compensation.

One who frightens a pregnant woman causing her to miscarry must pay ten dinars.

The blood-money for an unborn child is inherited by whoever would normally have inherited his estate, in ascending order of kinship.

The compensation due for any grievous bodily harm to it or loss of body parts is calculated in proportion to its full compensation.

Someone who beats a pregnant woman as a result of which she gives birth to a live foetus, who then dies shortly after the birth, must be killed if he did it intentionally. If not, then he must pay the blood-money for it.

For decapitating the corpse of a Muslim free man, one-hundred dinars is payable, and a proportion of the full compensation for amputating any of its limbs. The same applies to wounds and head injuries inflicted on the remains. This compensation is then spent on charitable causes on his behalf.

الفصل الحادي عشر: في دية الجنين والميت

في النطفة بعد استقرارها في الرحم: عشرون ديناراً، وفي العلقة: أربعون. وفي المضغة: ستون. وفي العظم: ثمانون. وإذا تمّت خلقته ولم تلجه الروح فمائة، وفيما بين ذلك بحسابه.

ودية جنين الذمي عشر دية أبيه.

والمملوك: عشر قيمة أمّه المملوكة سواء الذكر والأنثى.

ولو ولجته الروح فدية كاملة في الذكر ونصف في الأنثى.

ولو قُتلت المرأة ومات معها فدية للمرأة ونصف الديتين للجنين إن جهل حاله.

ولو ألقته المرأة مباشرة أو تسبيباً فعليها ديته لوارثه، ولا يسهم لها.

ومن أفزع مجامعاً فعزل فعليه عشرة دنانير.

ويرث دية الجنين من يرث المال، الأقرب فالأقرب.

ودية جراحاته وأعضائه بنسبة ديته.

ولوز ضُرب الحامل فألقت جنينًا حياً فمات بالإلقاء قُتل به إن كان عمداً، وإلا أُخذت الدية.

وفي قطع رأس الميت الحر المسلم مائة دينار، وفي قطع جوارحه بحساب ديته، وكذا في جراحه وشجاجه، وتصرّف هذه الدية في وجوه البر.

Section Twelve: Crimes Against Animals

Whoever kills an animal that is fit for consumption, through ritual slaughter, must pay its owner its value in damages. If he killed it by any other means, then he must pay him its market value on the day that it was killed. For cutting any of its limbs or breaking any part of its body, he must pay him damages due [i.e. the loss of its value].

If the animal is in a state whereby it cannot be consumed, even though it would normally have been consumable through ritual slaughter, if the killing was done through slaughter, then damages must be paid. The same applies if its limbs are cut and it remains alive. If it was not ritually slaughtered, then the market value of the animal must be paid.

If it is an animal whose meat is not fit for consumption, then its market value must be paid. The compensation for killing a hunting dog is forty dirhams; for a sheepdog and guard dog, twenty dirhams; for a farm dog, a dry measure (*qafīz*) of foodstuffs; and for the foetus of livestock, one-tenth of the value of an adult.

الفصل الثاني عشر: في الجناية على الحيوان

من أتلف حيواناً مأكولاً بالذكاة فعليه الأرش لمالكه، وإن كان بغيرها فعليه القيمة يوم الإتلاف، وفي قطع جوارحه أو كسر شيء من أعضائه الأرش.

وإن كان غير مأكول وهو مما يقع عليه الذكاة، فإن كان بالذكاة فالأرش، وكذا في قطع أعضائه مع استقرار الحياة، وإن كان بغيرها فالقيمة، وإن لم تقع عليه الذكاة فالقيمة، ففي كلب الصيد أربعون درهماً وفي كلب الغنم والحائط عشرون درهماً، وفي كلب الزرع قفيز من بر، وفي جنين البهيمة عُشر قيمتها.

Section Thirteen: About the Clan Responsible for Paying the Blood-Money

We explained previously that the perpetrator's clan is responsible for paying the blood-money for an accidental homicide. It is either: one's relatives, the manumitter, the guarantor of the offence or the Imam.

Relatives: These are the family members who are related to the deceased either through both parents or paternally. Fathers and sons are predominantly considered as being one's clan, and the perpetrator is not included among them. Women, minors and slaves are also excluded. The clan are not responsible for paying blood-money for an intentional homicide, nor for a slave, a *mudabbar* or a slave mother (*umm walad*); nor for injuries less than the *muwaḍḍiḥa*, nor for crimes that are convicted through confession, nor by compromise, nor for a person's crimes against his own self, nor for destruction caused by an animal and nor for damage to property.

The Imam acts as the clan of the *dhimmī* if he does not have assets [to pay the blood-money].

The blood-money payable is divided equally amongst relatives in ascending order of kinship, and evaluated by the Imam or whoever is appointed to judge in court. The clan do not resort to the perpetrator.

If the blood-money exceeds what the clan can afford to pay, then it is taken from the perpetrator's guarantors, and if it exceeds further, then from the guarantors' relatives, and then from their relatives and so on. If the blood-money payable exceeds their combined capacity to pay, then the excess is paid by the Imam. If the number of people in the clan exceeds the amount, then it is shared as a quota between them. If some of the clan are missing, then the ones present cannot be made to pay their share.

If a father intentionally kills his son, he must be made to pay the blood-money from his own assets to be given to another next of kin [of his son's]. If he has no other next of kin, then it goes to the Imam. If the killing was unintentional, then the clan is responsible for paying the blood-money.

الفصل الثالث عشر: في العاقلة

قد بيّنّا أن دية الخطأ على العاقلة، وهم: العصبة، والمعتِق، وضامن الجريرة، والإمام.

أما العصبة: فهم المتقرّبون إلى الميت بالأبوين أو بالأب، والأقرب دخول الآباء والأولاد في العقل، ولا يدخل القاتل فيه، ولا تعقل المرأة ولا الصبي ولا المجنون، ولا تعقل العاقلة عمداً ولا عبداً ولا مدبّراً ولا أم ولد ولا ما دون الموضّحة ولا ما يثبت بالإقرار ولا صلحاً ولا جناية الإنسان على نفسه ولا ما تجنيه البهيمة ولا إتلاف المال.

وعاقلة الذمي الإمام إن لم يكن له مال.

وتقسط الدية على الأقرب فالأقرب، وتقديره إلى الإمام، أو من ينصبه للحكومة، ولا ترجع العاقلة على الجاني.

ولو زادت الدية عن العصبة أخذت من الموالي، فإن اتسعت فمن عصبة الموالي، فإن اتسعت فمن موالي الموالي وهكذا، ولو زادت الدية عن العاقلة أجمع كان الزائد على الإمام، ولو زادت العاقلة وزع بالحصص، ولو غاب بعض العاقلة لم يخص بها الحاضر.

ولو قتل الأب ولده عمداً أخذت منه الدية لغيره من الوراث، وإن لم يكن وارث فالإمام، ولو كان خطأً فالدية على العاقلة.

This concludes what we have substantiated in this abridged work.

We ask Almighty God to designate it sincerely for His pleasure, for He is Ever-near, Responsive. And all Praise be to God, Lord of all created beings, and peace and blessings be upon our master, Muhammad, the chosen one, upon his vicegerent, ʿAlī, the selected one, and upon their good and pure families.

This work was completed on the eve of Tuesday, 25 Rabīʿ al-Thānī 759 AH [14 April 1358 CE] in the town of Ḥilla – May God protect it from calamities.

فهذا خلاصة ما أثبتناه في هذا المختصر.

ونسأل الله تعالى أن يجعل ذلك لوجهه خالصاً، إنه قريب مجيب.

والحمد لله رب العالمين، والصلاة السلام على سيّدنا محمد المصطفى، وعلى وصيه عليّ المرتضى، وآلهما الطيبين والطاهرين.

تمّ ذلك في ليلة الثلاثاء، خامس عشرين ربيع الثاني، لسنة تسع وخمسين وسبعمائة.

بمدينة حلّة حماها الله عن الآفات.

Bibliography

المصادر والمراجع

al-Amīn, Sayyid Muḥsin. *A'yān al-shī'a*. Edited by Ḥasan al-Amīn. 12 vols. Beirut: Dār al-Ta'āruf, 1403/1983.

Calder, Norman. "Doubt and Prerogative: The Emergence of an Imāmī Shī'ī Theory of Ijtihād." *Studia Islamica* 70 (1989): 58–78.

———. *Islamic Jurisprudence in the Classical Era*. Cambridge: CUP, 2010.

Gleave, Robert. *Scripturalist Islam: The History and Doctrines of the Akhbārī Shī'ī School*. Leiden: Brill, 2007.

al-Ḥillī, al-'Allāma al-Ḥasan b. Yūsuf Ibn al-Muṭahhar. *Tabṣirat al-Muta'allimīn fī aḥkām al-dīn*. Edited by Muḥammad Hādī al-Gharawī. Tehran : Intishārāt-i Faqīh, 1368Sh/[1989].

———. *The Way of Nobility: Knowledge of the Imam (Minhāj al-karāma fī ma'rifat al-imāma). A parallel English-Arabic text translated, introduced and annotated by Saiyad Nizamuddin Ahmad*. Birmingham: AMI Press, 2023.

———. *Khulāṣat al-aqwāl fī ma'rifat aḥwāl al-rijāl*. Edited by Jawād al-Qayyūmī. Qom: Mu'assasat Nashr al-Fiqāha, 1417/[1996].

———. *Taḥrīr al-aḥkām*. Edited by Ibrāhīm al-Bahādurī. 6 vols. Qom: I'timād, 1420/[1999].

Ibn Bābawayh. *Man lā yaḥḍuruhu al-faqīh*. Edited by 'Alī Akbar al-Ghaffārī. 4 vols. Qom: Mu'assasat al-Nashr al-Islāmī, 1404/[1983].

Al-Jamil, Tariq. "Ibn Taymiyya And Ibn Al-Mutahhar Al-Hilli: Shi'i Polemics And The Struggle For Religious Authority In Medieval Islam." In *Ibn Taymiyya and his times*, edited by Shahab Ahmed and Yossef Rapoport, 229–246. Oxford: OUP, 2010.

al-Kulaynī, Muḥammad b. Ya'qūb. *al-Kāfī*. Edited by 'Alī Akbar al-Ghaffar and Muḥammad Ākhūndī. 8 vols. Tehran: Dār al-Kutub al-Islāmiyya, 1367Sh/[1988].

al-Majlisī, Muḥammad Taqī. *Rawḍat al-muttaqīn*. Edited by Ḥusayn Mūsawī Kirmānī and 'Alī Panāh Ishtihārdī. 14 vols. Qom: Bunyād-e Farhang-e Islāmī, 1389/[1969].

Melchert, Christopher. *The Formation of the Sunni Schools of Law: 9th-10th centuries CE*. Leiden: Brill, 1997.

Modarressi, Hossein. "Rationalism and Traditionalism in Shî'î Jurisprudence: A Preliminary Survey." *Studia Islamica* 59 (1984): 141–158.

Pfeiffer, Judith. "Conversion Versions: Sultan Öljeytü's Conversion to Shi'ism (709/1309) in Muslim Narrative Sources." *Mongolian Studies* 22 (1999): 35–67.

Schmidtke, Sabine. *The Theology of al-Allama al-Hilli (d. 726/1325)*. Berlin: K. Schwarz, 1991.

al-Shīrāzī, Sayyid Ṣādiq. *Sharḥ Tabṣirat al-muta'allimīn fī aḥkām al-dīn*. 2 vols. Qom: Dār al-Tafsīr, 1430/[2008].

al-Ṭihrānī, Aghā Buzurg. *Al-Dharī'a ilā taṣanīf al-Shī'a*. 25 vols. Beirut: Dār al-Aḍwā', 1983.

al-Ṭūsī, Muḥammad b. al-Ḥasan. *Al-Tibyān fī tafsīr al-Qur'ān*. Edited by Aḥmad Shawqī al-Amīn and Aḥmad Ḥabīb Quṣayr al-'Āmilī. 10 vols. Najaf: Maktabat al-Amīn, 1964–69.

BIBLIOGRAPHY

www.ingramcontent.com/pod-product-compliance
Lightning Source LLC
LaVergne TN
LVHW051157210926
848486LV00021B/413/J